SECOND EDITION

BIOCHEMISTRY

A Functional Approach

R. W. McGILVERY, Ph.D.

Professor of Biochemistry,
University of Virginia School of Medicine

in collaboration with

GERALD GOLDSTEIN, M.D.

Professor of Internal Medicine,
University of Virginia School of Medicine

W. B. SAUNDERS COMPANY

Philadelphia / London / Toronto

W. B. Saunders Company: West Washington Square
Philadelphia, PA 19105

1 St. Anne's Road
Eastbourne, East Sussex BN21 3UN, England

1 Goldthorne Avenue
Toronto, Ontario M8Z 5T9, Canada

Library of Congress Cataloging in Publication Data

McGilvery, Robert W

Biochemistry, a functional approach.

Includes bibliographies and index.

1. Biological chemistry. I. Goldstein, Gerald, 1922– joint author. II. Title.

QP514.2.M3 1979 574.1'92 79–4729

ISBN 0–7216–5912–8

Biochemistry – A Functional Approach ISBN 0-7216-5912-8

Last digit is the print number: 9 8 7 6 5

TO ALICE

PREFACE TO THE SECOND EDITION

This book is intended for those who wish to understand living organisms, especially man. Biochemistry is essential for this purpose, but it would be almost impossible for a student to survey on his own the massive body of existing knowledge, constantly augmented by a remarkable torrent of brilliant discoveries. The purpose of the book, then, is to organize our knowledge into something that can be comprehended in a relatively short time and still convey a reasonably complete picture of the chemical structure and function of man. Readability without sacrifice of coverage has been a prime goal; I still agree with Ernest Hooton that you ought to treat science seriously, but you don't have to act as if you are in church. An important device in gaining that goal is to keep attention constantly focused on function, with repeated use of rationalization to show that the chemical facts are not isolated, but part of a whole.

Keeping the subject within bounds has involved some sacrifices. The subject is paramount, and the author's interests are secondary. We hope much of our present understanding of biochemistry will still be valid for students a century hence, and it is to that understanding that attention is directed, rather than to an examination of the experiments by which that understanding has been gained. These experiments have been exciting stuff, and it is painful to acknowledge that they rapidly acquire the musty stigma of historical detail to most students.

I hope the book serves its purpose well, but it clearly would have had much less chance of doing so without the intensive collaboration of Dr. Gerald Goldstein. Skilled internist by profession, perpetual student at heart, he has the most important qualification for a great teacher: the desire to share his joy upon learning new things. He contributed to the draft of every chapter, and every page has had the benefit of his repeated critical review. However, he is not to be blamed for any defects; I always had the last word.

I was also fortunate in inducing distinguished scientists to review parts of the manuscript. They include:

Robert Barker
Harris Busch
Thomas C. Bithell
E. Jack Davis
Edmond H. Fischer
Paul M. Gallop
John Gergely
Ching-hsien Huang
Richard T. Jones
Robert G. Kemp
Howard C. Kutchai

They eliminated errors and sharpened the presentation. However, they also are not responsible for the final draft. I especially absolve them of blame for those very few instances in which I thought it necessary to deviate from the usual nomenclature in order to aid comprehension by the novice. (These deviations are noted where they occur.)

Dr. Manuel Navia freely gave advice on stereo projections of immunoglobulin structure and generated new projections for this work. Drs. David E. Comings, Ching-hsien Huang, S. L. McKnight, O. L. Miller, Jr., and Lester J. Reed went out of their way to supply prints for reproduction. Drs. Joseph G. Cory and Lavell M. Henderson gave me the benefit of unpublished information and clarifying comments on certain points.

Many colleagues and students throughout the world kindly gave me the benefit of their comments on the first edition. I am especially indebted to Dr. H. D. Jackson for sending his extensively annotated copy of the book—may his kind increase.

Mrs. Roberta Kangilaski acted as editor of the book; her enthusiasm, skill, conscientious monitoring, and long hours of effort made her a true colleague. The book also had the benefit of the design skill of Lorraine Battista, the copy editing of Lloyd Black, and the skillful competence of the production department at the W. B. Saunders Company and York Graphic Services under the direction of Herbert Powell, Jr.

Finally, my biggest debt is to my wife Alice, diplomatic counselor and most patient of typists.

R. W. McGilvery
Charlottesville, Virginia

CONTENTS

1 INTRODUCTION

The essential contribution of biochemistry to an understanding of living beings will be obvious before we finish our limited survey. Organisms are marvelously complex assemblages of chemical compounds constantly involved in intertwining arrays of reactions, and one of the great delights in the study of biochemistry is the realization that it is possible to organize this complexity into comprehensible patterns. Much is to be learned, but one sees many large truths about all living things along with patches of finely detailed knowledge.

However, as with other branches of learning there are ill-defined boundaries to biochemistry beyond which its value as a device for description and explanation diminishes, even though its principles are determinative. We do not discuss the vagaries of the London gold market in terms of the biochemistry of the participants any more than we would discuss that biochemistry in terms of the physics of the constituent atoms upon which the chemistry rests. As we go from molecule to cell to tissue to organ to individual to society, we must introduce at each stage a more simplified system of statistics to handle the larger domain, statistics that we label as anatomy, physiology, psychology, economics, history, music, and so on. Even so, the basic biochemistry sometimes shows through in previously unsuspected ways. ''Chemistry'' has long been blamed for the human's chronic rut, but newer psychochemistry and its by-product drug culture have exposed the extent to which distinctively human activity can depend upon the concentration of single chemical compounds.

To illustrate the boundaries of application of biochemistry, consider the type behavior of the surgeon and the internist. The surgeon manipulates what he can see, and he thinks of his procedures in terms of anatomy and physiology, even in cases in which the ultimate effectiveness of his efforts is gauged in biochemical terms. On the other hand, the stock in trade of the internist largely consists of chemicals. He injects them or asks the patient to swallow them in order to alter the biochemistry of the patient or of some foreign organism, and biochemical language is often the natural medium for discussing what is happening.

Of course, we are making a very simplistic distinction between the specialties. The modern surgeon includes sophisticated biochemical thinking among his tools, and the internist always has probed and poked, and in modern times called in the radiologist, in order to visualize structure.

What kind of prospectus can we offer about biochemistry as a useful medium for describing the living? To begin, biochemistry describes the origin of form. The chemical constituents and the forces developed between them determine in describable ways the microscopic anatomy — the nature of cells and their constituent organelles. Biochemical principles are the tools of choice for that purpose. Biochemistry also describes the forces involved in the association of cells, its language merging into the language of anatomy as more complex levels of organization are considered.

Biochemistry describes heredity. The information on the nature and behavior of an organism that is passed from generation to generation and the mode of its transmission are molecular in nature. The complex summation of that information constitutes genetics, but the details are biochemistry.

Biochemistry describes much more. It is a social nicety to greet an acquaintance after several years of separation with the words, "You haven't changed a bit." We secretly note the coarsened skin, the deepened lines, the graying hair, but in a quantitative physical sense this conventional courtesy is frequently not far from the truth. Given a primitive environment during his absence, this one man, using the simplest of tools, could have diverted a stream, cut down hectares of forest, and in other ways changed his environment so drastically that even the most unobservant would know of his existence. Yet at the end of all this, we might well be hard put to measure more than a small change in the physical dimensions or chemical composition of the man himself.

All of this is well known, even trite, but many of the phenomena that biochemistry can address are contained in the small tale. Motion is a molecular phenomenon resulting from the cyclical formation and cleavage of chemical bonds. The energy for that motion is derived from reactions of compounds that enter the organism in a constant stream, with the products excreted into the environment. The entire process of energy generation and utilization is described by biochemistry.

The major point of the tale is the ability of the organism to maintain its character over long periods of time while acting as a chemical machine to change the environment, and this is also an important part of our story. We shall see how chemical reactions are used to constantly rebuild nearly all parts of the body; the molecular structure and the reactions proceeding within it are subject to continual review of need — a kind of zero-base budgeting, which enables adaptation to changing circumstances.

Finally, we see that the machinery does wear out. The species has built within it limiting devices that say, "Enough. Let the next generation take over." Most attempts at explaining aging are made in biochemical terms, but we know too little to say how effective this approach ultimately will be.

2 AMINO ACIDS AND PEPTIDES

We begin our study of the detailed chemistry of living organisms with an exploration of the proteins, including their nature, their formation, and the kinds of functions they carry out. We begin in this way because the proteins are the chemical compounds that define most of the properties we ascribe to life. They determine our metabolism, form our tissues, give us motion, transport compounds, and protect us from deleterious invasion. Even the heredity of an organism is nothing more than an expression of its ability to make various kinds of proteins at different rates.

What are proteins? In operational terms they are structures for placing reactive chemical groups in particular three-dimensional patterns and for controlling the access to these groups. Looking at proteins in this way, we have three tasks. We must consider how their structure is generated so as to create a particular three-dimensional pattern, or conformation, describe the constitution of the pattern, and then describe how that pattern performs a biological function. We shall make important beginnings on these tasks in this and the following chapter, but our understanding will mature throughout the book.

THE NATURE OF AMINO ACIDS

In chemical terms, a protein is a polymer of α-amino acids, that is, 2-amino carboxylic acids. (When we speak of amino acids, we shall mean these particular isomers.) Before we look at the way the amino acids are joined to form proteins, let us examine the nature of these building blocks as they occur in free form. This is not an academic exercise because tissues contain substantial amounts of each amino acid in solution, sometimes as much as several millimoles per kilogram. This pool of free amino acids is the source of material for constructing proteins, but it is also actively metabolized to many other products, and derangement of amino acid turnover frequently has severe consequences. Several amino acids have another important function: they are used as chemical messengers to transmit impulses between nerves.

Proteins are constructed from 20 different amino acids, which have this general structure:

$$\mathrm{H{-}\overset{\oplus}{N}H_2{-}CHR{-}C(\cdots O)_2^{\ominus}} \quad \text{or} \quad {}^{\oplus}\mathrm{H_3N{-}\underset{H}{\overset{R}{C}}{-}COO^{\ominus}}$$

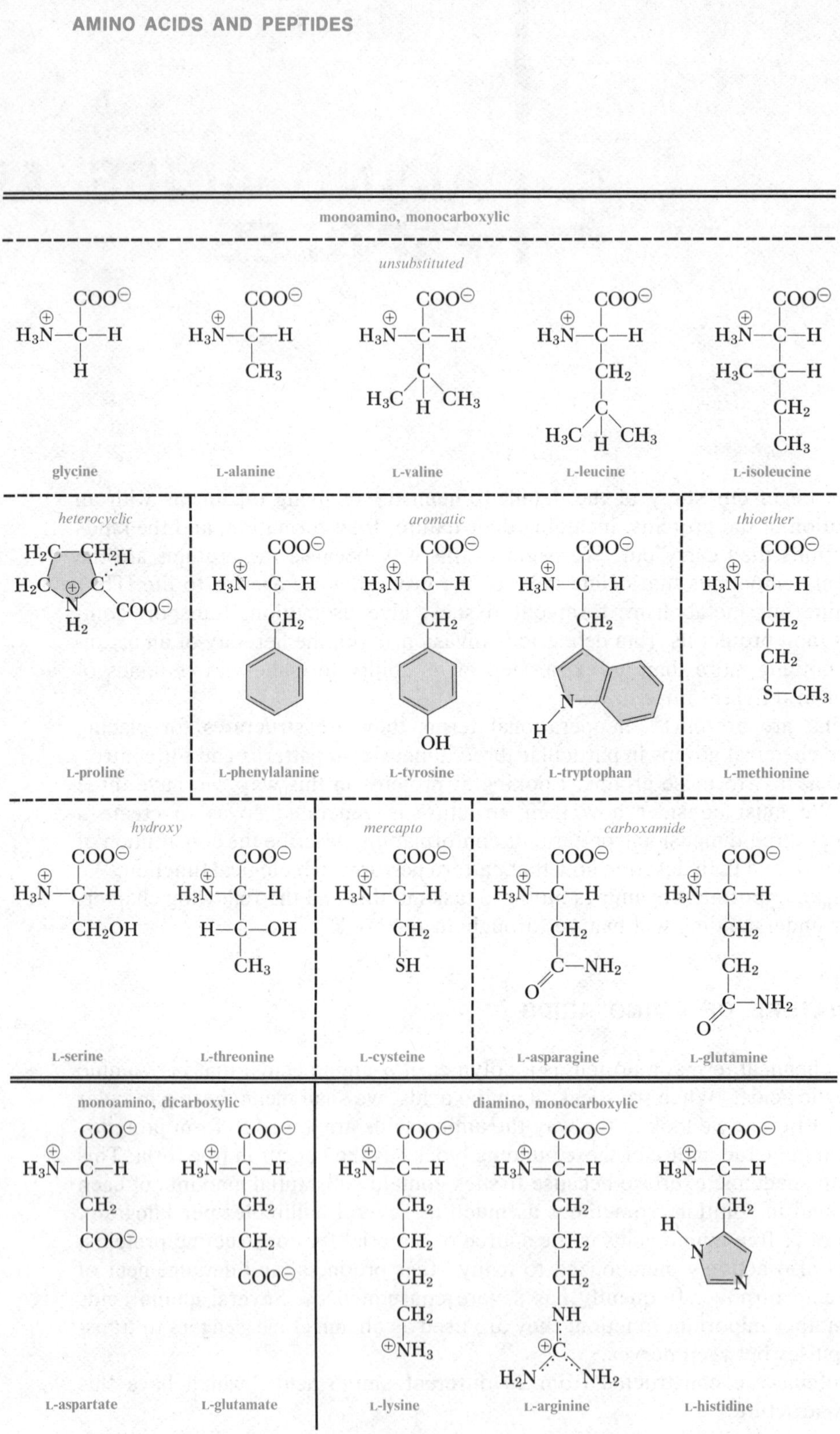

monoamino, monocarboxylic

unsubstituted

glycine **L-alanine** **L-valine** **L-leucine** **L-isoleucine**

heterocyclic

L-proline

aromatic

L-phenylalanine **L-tyrosine** **L-tryptophan**

thioether

L-methionine

hydroxy

L-serine **L-threonine**

mercapto

L-cysteine

carboxamide

L-asparagine **L-glutamine**

monoamino, dicarboxylic

L-aspartate **L-glutamate**

diamino, monocarboxylic

L-lysine **L-arginine** **L-histidine**

The molecule is shown in two ways, the first indicating the position of every bond, and the second using a common kind of shorthand notation in which substituent hydrogen atoms are lumped together without indicating bonds, and individual C=O bonds also are not drawn.

We see that amino acids draw their properties from a hydrogen atom and three substituent groups on C-2, the α-carbon atom. It is the nature of the R group that gives character to an individual amino acid, and we shall see that these groups, usually called the side chains, are all-important in determining the properties of proteins.

Properties Conveyed by Side Chains

The amino acids are commonly classified according to the character of their side chains, as is shown on the opposite page. Let us survey the nature and function of these side chains in general terms now and consider them in more detail as we encounter their specific effects in subsequent chapters. The chemical functions in the structures can be recognized without the necessity of rote memorization. First we have glycine, which has no side chain, and which therefore occupies the least space of all of the amino acids. This is an important property in itself.

Hydrophobic Bulk. Many of the amino acids are built to take up space without interacting with water. They are especially useful in shaping the interior of protein molecules. Side chains that serve in this way include alkyl hydrocarbon groups:

$^{\oplus}H_3N{-}C(H)(CH_3){-}COO^{\ominus}$
L-alanine

$^{\oplus}H_3N{-}C(H)(CH(CH_3)_2){-}COO^{\ominus}$
L-valine

$^{\oplus}H_3N{-}C(H)(CH_2{-}CH(CH_3)_2){-}COO^{\ominus}$
L-leucine

$^{\oplus}H_3N{-}C(H)(H{-}C(CH_3){-}CH_2{-}CH_3){-}COO^{\ominus}$
L-isoleucine

aromatic rings:

$^{\oplus}H_3N{-}C(H)(CH_2{-}C_6H_5){-}COO^{\ominus}$
L-phenylalanine

$^{\oplus}H_3N{-}C(H)(CH_2{-}\text{indole ring}){-}COO^{\ominus}$
indole ring
L-tryptophan

a heterocyclic ring, in which the side chain is also attached to the ammonium group on C-2:

$H_2C{-}C(H_2){-}CH_2$ ring; $^{\oplus}H_2N{-}C(H){-}COO^{\ominus}$
L-proline

and a thioether:

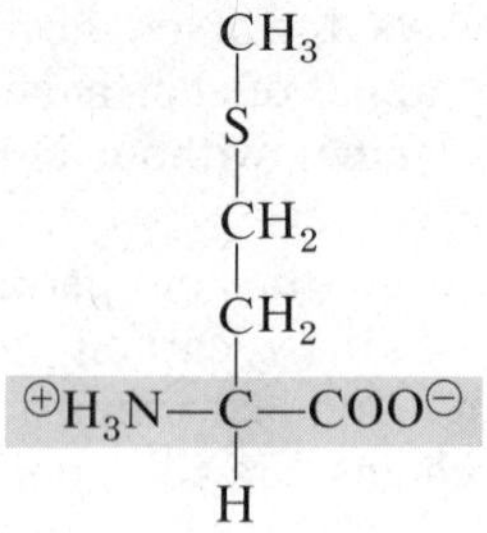

$$
\begin{array}{c}
CH_3 \\
| \\
S \\
| \\
CH_2 \\
| \\
CH_2 \\
| \\
{}^{\oplus}H_3N{-}C{-}COO^{\ominus} \\
| \\
H
\end{array}
$$

L-methionine

Π-Bond Interaction. When aromatic rings are stacked side by side, the π-electrons of the rings interact to form weak bonds. Some amino acids have aromatic rings that bond in this way with each other or with other flat resonant structures:

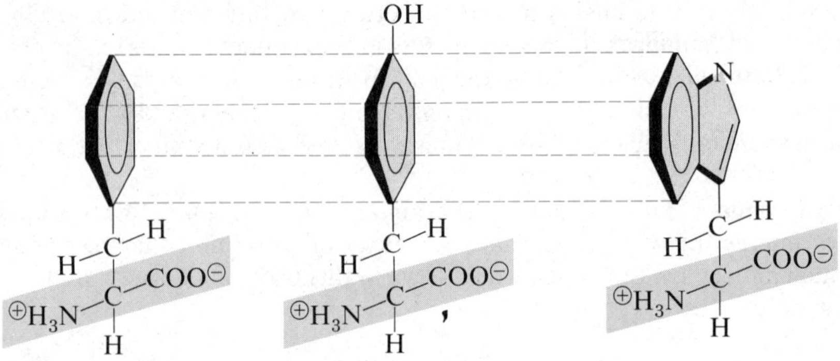

OH

N

H—C—H, ${}^{\oplus}H_3N$—C—$COO^{\ominus}$, H

L-phenylalanine L-tyrosine L-tryptophan

(The dashed lines are intended to convey the existence of interaction between the rings.)

Hydrogen Bonding. The hydrogen bond is one in which a proton is partially shared between two atoms containing unpaired electrons, such as O, N or S:

$$>N{-}H\cdots O<$$

The hydrogen bond is one of the most important in forming the structure of proteins, partially because many of the constituent amino acids have groups in their side chains containing such atoms, and all of these may form hydrogen bonds. Groups containing N and O make the side chains more polar, with a greater tendency to interact with water, but this tendency is counteracted by participation in hydrogen bonding. For example, burial of the alcoholic hydroxy groups of serine and threonine and the phenolic hydroxyl group of tyrosine in the interior of proteins is often facilitated by hydrogen bonding, sometimes to an adjacent group:

$$R{-}O{-}H\cdots O{=}C<$$

Strongly polar groups are frequently near the surface of protein molecules because of their ready association with water, but they also can form hydrogen bonds that sometimes enable them to be buried. They include, along with frankly ionic carboxylate and ammonium groups considered below, the carboxamide groups of asparagine and glutamine:

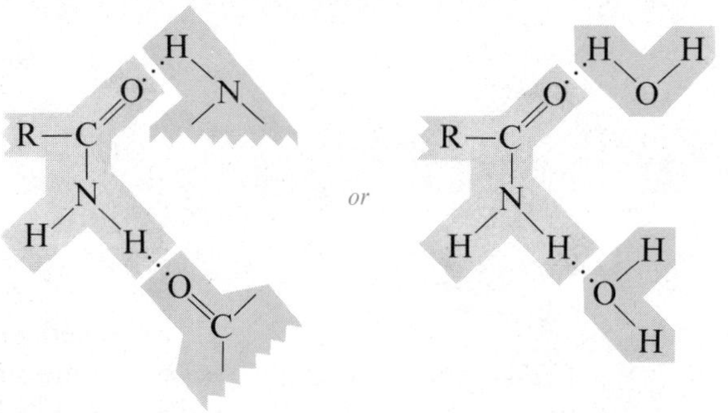

intramolecular bonding   or   bonding with water

Binding of Metallic Cations. Atoms with unshared electrons are sometimes used to bind metals, or groups containing metals. For example, the imidazole ring of a histidyl side chain binds the iron atom in hemoglobin, and the thiol ether group of methionine has a similar purpose in other proteins:

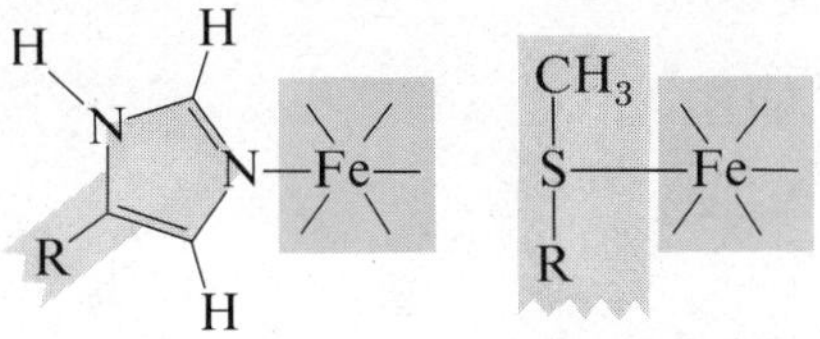

Carboxylate and amino groups also readily bond metallic ions, hence synthetic amino acids and related compounds were developed for this purpose. One that is widely used both as a reagent and as a drug is ethylenedinitrilotetraacetate or EDTA (also known as ethylenediaminetetraacetate):

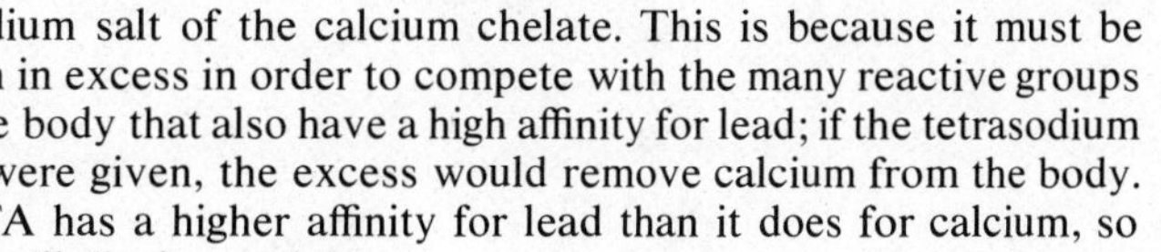

$$(^{\ominus}OOC{-}CH_2)_2N{-}CH_2{-}CH_2{-}N(CH_2{-}COO^{\ominus})_2$$

ethylenedinitrilotetraacetate (EDTA)

EDTA has a high affinity for metal cations with two or more negative charges. It is used to treat lead poisoning because the lead chelate is soluble and can be excreted. It is administered as the disodium salt of the calcium chelate. This is because it must be given in excess in order to compete with the many reactive groups in the body that also have a high affinity for lead; if the tetrasodium salt were given, the excess would remove calcium from the body. EDTA has a higher affinity for lead than it does for calcium, so lead will displace calcium:

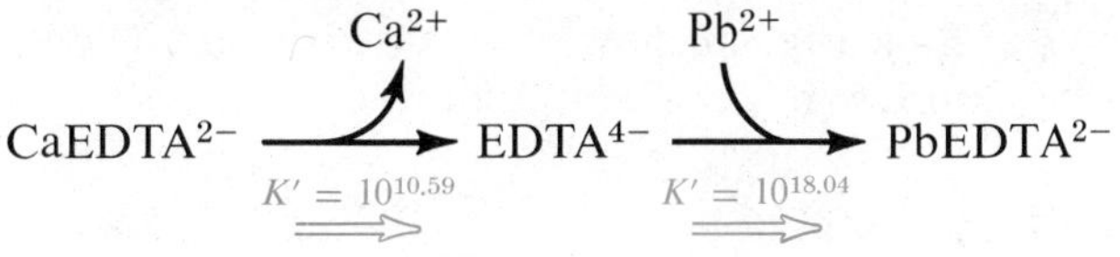

$$CaEDTA^{2-} \xrightarrow[K' = 10^{10.59}]{\nearrow Ca^{2+}} EDTA^{4-} \xrightarrow[K' = 10^{18.04}]{Pb^{2+} \searrow} PbEDTA^{2-}$$

Ionized Side Chains. Some amino acids have ionized groups in their side chains that cause strong affinity for water where they occur in proteins. These include amino acids with carboxylate groups:

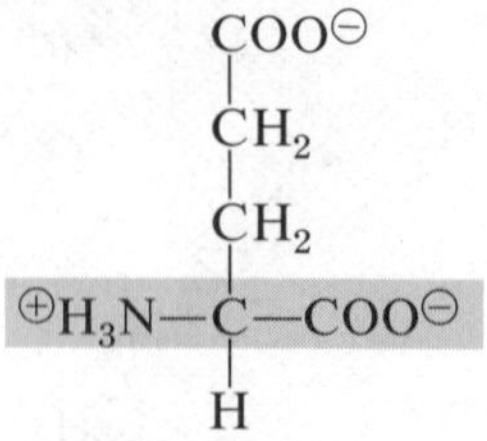

$$\begin{array}{c} COO^{\ominus} \\ | \\ CH_2 \\ | \\ CH_2 \\ | \\ {}^{\oplus}H_3N-C-COO^{\ominus} \\ | \\ H \end{array} \qquad \begin{array}{c} COO^{\ominus} \\ | \\ CH_2 \\ | \\ {}^{\oplus}H_3N-C-COO^{\ominus} \\ | \\ H \end{array}$$

L-glutamate 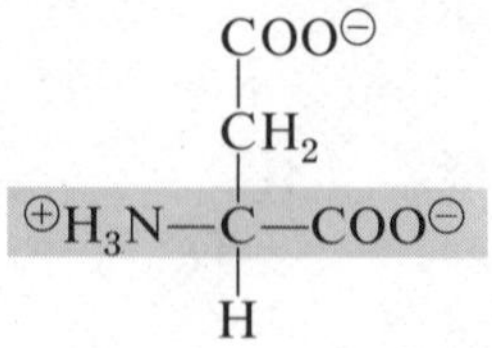 L-aspartate

and amino acids with positively charged groups containing nitrogen atoms:

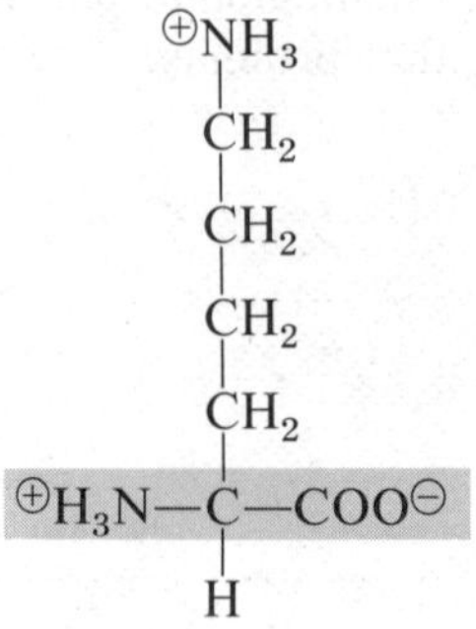

$$\begin{array}{c} {}^{\oplus}NH_3 \\ | \\ CH_2 \\ | \\ CH_2 \\ | \\ CH_2 \\ | \\ CH_2 \\ | \\ {}^{\oplus}H_3N-C-COO^{\ominus} \\ | \\ H \end{array} \qquad \begin{array}{c} H_2N \quad \oplus \quad NH_2 \\ C \\ \vdots \\ NH \\ | \\ CH_2 \\ | \\ CH_2 \\ | \\ CH_2 \\ | \\ {}^{\oplus}H_3N-C-COO^{\ominus} \\ | \\ H \end{array}$$

guanidine group

L-lysine 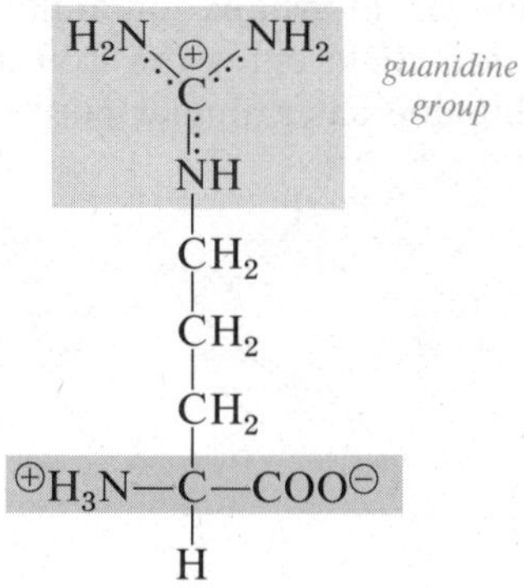 L-arginine

The positively and negatively charged side chains can form bonds through electrostatic interaction, and they can also form hydrogen bonds and bind metallic cations.

Amino Acids as Acids and Bases

Amino acids can both donate and accept protons; they are therefore said to be **amphoteric.** Every amino acid in neutral solution can behave as an acid because it contains at least one charged ammonium group from which a proton can dissociate:

$$R-NH_3^{\oplus} \rightleftharpoons R-NH_2 + H^{\oplus}$$

Similarly, it behaves as a base because it contains at least one charged carboxylate group that can accept a proton:

$$H^{\oplus} + R-COO^{\ominus} \rightleftharpoons R-COOH$$

Consider the simplest amino acid, glycine. It can equilibrate with H^+ in two ways:

$$^{\oplus}H_3N{-}CH_2{-}COOH \underset{K_1 = 10^{-2.35}}{\overset{H^{\oplus}}{\rightleftarrows}} {}^{\oplus}H_3N{-}CH_2{-}COO^{\ominus} \underset{K_2 = 10^{-9.78}}{\overset{H^{\oplus}}{\rightleftarrows}} H_2N{-}CH_2{-}COO^{\ominus}$$

A form, *Z* form, *B* form

The form shown in the middle is a **zwitterion,** meaning hermaphrodite ion, because it has equal numbers of positive ammonium groups and negative carboxylate groups, although its net charge is zero. It behaves as a base because the carboxylate groups will combine with increasing concentrations of H^+ to form uncharged COOH groups. The remaining ammonium group then gives the molecule a net positive charge (cationic form). On the other hand, the zwitterion can behave as an acid because the ammonium group will lose H^+ when the concentration of H^+ is lowered, leaving an uncharged amino group. The molecule then has a net negative charge from the remaining carboxylate group (anionic form).

What is the physiological form of glycine? The fraction of zwitterion present at a given H^+ concentration is shown in Figure 2–1. We see that glycine exists mostly as a zwitterion over a broad range centered near 10^{-6} M H^+ (pH 6). In general, amino acids with one ammonium group and one carboxylate group exist mainly as the zwitterion in physiological fluids. For example, here is the distribution of the various forms of glycine at pH 7.4 ($[H^+] = 10^{-7.4}$ M), which is the normal pH of blood plasma:

$^+H_3N{-}CH_2{-}COO^-$	zwitterion	99.58%
$H_2N{-}CH_2{-}COO^-$	anion	0.41%
$^+H_3N{-}CH_2{-}COOH$	cation	0.00089%
$H_2N{-}CH_2{-}COOH$	uncharged	0.0000037%

Other amino acids with one ammonium group and one carboxylate group behave as acids and bases in much the same way as does glycine with the zwitterion being the physiological form of each.

Why are the molecules called amino acids? Well, the original investigators thought that the uncharged form predominated. This form has an authentic amino group and a carboxyl group, which makes it both an amine and a carboxylic acid. We have known for many decades that it is almost non-existent, but many still draw amino acids that way.

Acidic and Basic Side Chains. The carboxylate and substituted ammonium groups on the side chains of some amino acids also behave as acids and bases. This is still true when the amino acids are combined to form proteins, so these are

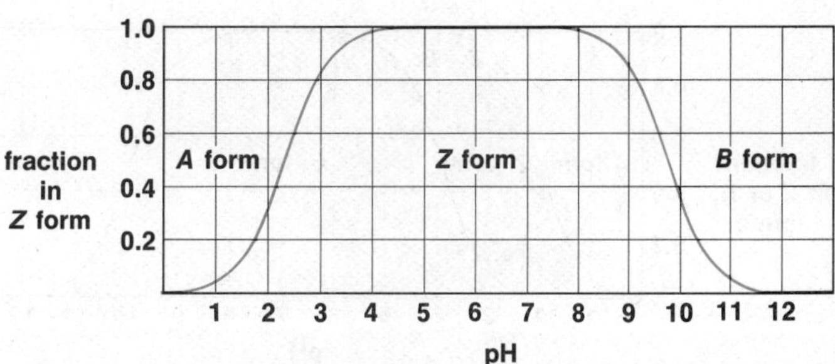

FIGURE 2–1

A plot of the fraction of glycine present as the zwitterion at various pH values. The fraction not present as a zwitterion exists as a cation in acidic solutions *(left)*, and as an anion in basic solutions *(right)*. Most monoamino, monocarboxylic acids behave in a similar manner upon titration.

the groups mainly responsible for giving amphoteric properties to proteins. The general principles can be grasped by examining the behavior of amino acids containing such side chains.

CARBOXYLATE SIDE CHAINS. Consider aspartic acid:

$$
\underset{A\ \text{form}}{\overset{+}{H_3N}\text{–}CH(COOH)\text{–}CH_2\text{–}COOH}
\underset{K_1 = 10^{-1.99}}{\overset{H^+}{\rightleftharpoons}}
\underset{Z\ \text{form}}{\overset{+}{H_3N}\text{–}CH(COO^{\ominus})\text{–}CH_2\text{–}COOH}
\underset{K_2 = 10^{-3.90}}{\overset{H^+}{\rightleftharpoons}}
\underset{B_1\ \text{form}}{\overset{+}{H_3N}\text{–}CH(COO^{\ominus})\text{–}CH_2\text{–}COO^{\ominus}}
\underset{K_3 = 10^{-9.90}}{\overset{H^+}{\rightleftharpoons}}
\underset{B_2\ \text{form}}{H_2N\text{–}CH(COO^{\ominus})\text{–}CH_2\text{–}COO^{\ominus}}
$$

Here we have three distinct equilibria. The ammonium group doesn't behave much differently than the ammonium group of glycine, but the side chain carboxylic group is a weaker acid than the group on glycine, and the 1-carboxylic group is a stronger acid. The shift in the proportion of the ionic forms with changes in H^+ concentration is shown in Figure 2–2. The zwitterions of aspartate and glutamate occur in acidic solution, whereas physiological fluids contain the fully ionized forms with one net negative charge. Monosodium glutamate, which is used as a food seasoning, gives a nearly neutral solution. (The use of monosodium glutamate also illustrates the importance of the biochemistry of the free amino acids. The "Chinese restaurant syndrome" afflicts some sensitive people with severe headaches, numbness, palpitation, and other symptoms of neurological disturbance owing to the effects of abnormally elevated glutamate concentrations. This was discussed with some levity until it was discovered that high glutamate can cause permanent damage to neurons of the embryonic hypothalamus in experimental animals. The use of monosodium glutamate as a flavor additive in baby foods is therefore being curtailed. Excess aspartate causes similar effects, but there has been no reason to add this amino acid in free form to foods.)

CATIONIC SIDE CHAINS. Three amino acids have side chains that may be positively charged under physiological conditions. **Lysine** has an ammonium group at the end of a hydrocarbon tail. (We shall see that twitching the tail to sweep the ammonium tuft over the surrounding surface is an important function of lysyl residues in some proteins.) This group is an even weaker acid than the ammonium group on C-2, which means that it will retain its positive charge at even lower H^+ concentrations (higher pH values). The result is that both of the ammonium groups of lysine, in addition to the carboxylate group, are charged in physiological fluids, and the physiological form is therefore a cation with one net

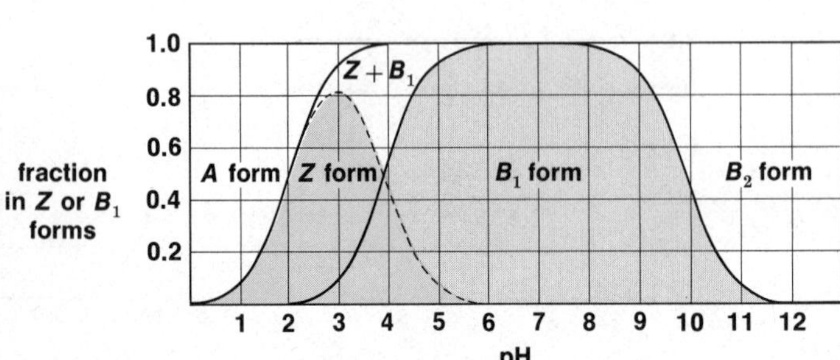

FIGURE 2–2

Monoamino, dicarboxylic acids such as aspartate can exist in four ionic forms. The fraction of the fully ionized form (B_1), with two negative carboxylate groups and one positive ammonium group, is shown by the central solid curve. The solid curve to the left plots the extent of conversion of the cationic form, with neither carboxylic acid group ionized, to more ionized species. The dashed bell-shaped curve is the fraction present as the zwitterion.

positive charge:

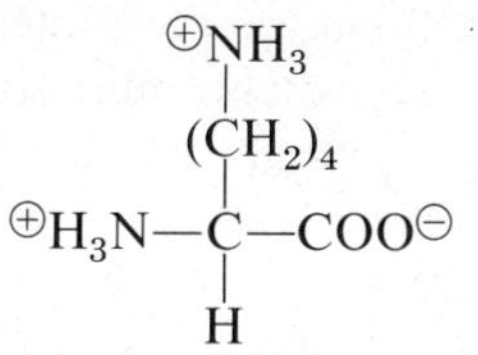

$$
\begin{array}{c}
\oplus NH_3 \\
| \\
(CH_2)_4 \\
| \\
{}^{\oplus}H_3N{-}C{-}COO^{\ominus} \\
| \\
H
\end{array}
$$

The guanidinium group on the side chain of **arginine** is an even weaker acid. Put another way, free guanidine groups are very strong bases, almost as strong as hydroxide ion itself, and they bind protons avidly. Therefore, the side chain of arginine retains its positive charge in all but strongly alkaline solutions, and the physiological form also is the cation with one net positive charge:

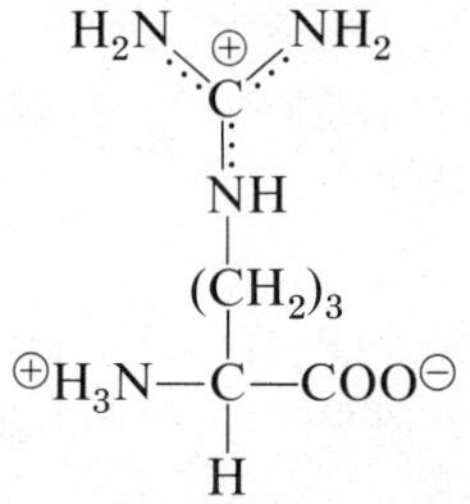

$$
\begin{array}{c}
H_2N \overset{\oplus}{\quad} NH_2 \\
C \\
\vdots \\
NH \\
| \\
(CH_2)_3 \\
| \\
{}^{\oplus}H_3N{-}C{-}COO^{\ominus} \\
| \\
H
\end{array}
$$

Histidine is Different. The imidazole group in its side chain is approximately half-ionized at pH 6.1 ($H^+ = 10^{-6.1}$ M) — sometimes as high as pH 7 when the amino acid is used to form proteins. This means that the physiological form of histidine is a mixture of the zwitterion and the cationic forms:

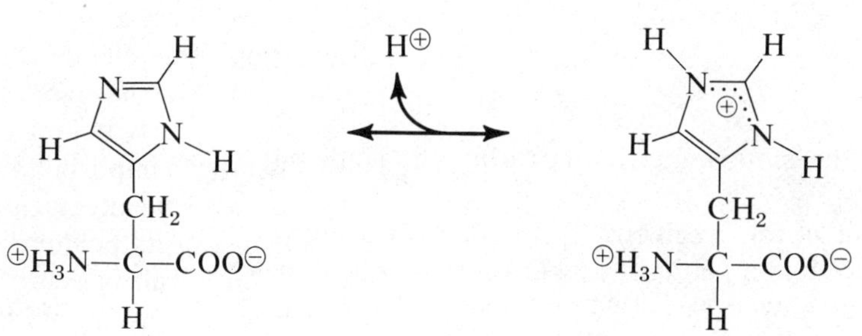

$$
\text{(imidazole, N–H)}\;\; {}^{\oplus}H_3N{-}\underset{H}{\overset{CH_2}{C}}{-}COO^{\ominus} \;\; \overset{H^{\oplus}}{\longleftrightarrow} \;\; \text{(imidazolium, } \oplus\text{, H–N, N–H)}\;\; {}^{\oplus}H_3N{-}\underset{H}{\overset{CH_2}{C}}{-}COO^{\ominus}
$$

The facility with which the side chain of histidine can switch from being an acid to being a base is an important feature for many biological functions, including the catalytic properties of enzymes.

Isoelectric Point. The charges on carboxylate ions can be repressed by increasing the concentration of H^+, while the ammonium groups (and similar cationic groups) can be made to lose their charge by decreasing the concentration of H^+. It follows that for each compound carrying both carboxylate and ammonium groups there is some value of the hydrogen ion concentration at which the number of negatively charged carboxylate groups will exactly equal the number of positively charged groups. This is true no matter how many of the respective groups there may be on a molecule. The H^+ concentration at which this occurs, usually expressed as a pH value, is known as the isoelectric point for the compound. It is the pH at which the molecule will fail to migrate in an electric field because it has no net charge. Some of the molecules may bear a net negative charge at a given moment, but they will be counterbalanced by an equal number of molecules bearing a net positive charge; the number of molecules that are zwitterions is greatest at the isoelectric point.

As the pH of a solution is raised above the isoelectric point of an amphoteric compound (decreasing acidity), an increasing number of molecules will bear a net negative charge, owing to the loss of H^+ from the counterbalancing cationic groups. The compound will then migrate to the positive pole in an electric field. When the pH is lowered below the isoelectric point (increasing acidity), an increasing number will bear a net positive charge, owing to the gain of H^+ by previously charged carboxylate groups to form the uncharged carboxylic acids. The compound will then migrate to the negative pole in an electric field.

Calculation of the Isoelectric Point. The ionizations of simple monoamino, monocarboxylic acids are described by two acidic dissociation constants:

$$K_1 = \frac{[\text{R—COO}^-][\text{H}^+]}{[\text{R—COOH}]} \qquad K_2 = \frac{[\text{R—NH}_2][\text{H}^+]}{[\text{R—NH}_3^+]}$$

The isoelectric point occurs when [R—NH_2]= [R—COOH], and a little algebraic manipulation shows that this happens when $[H^+] = \sqrt{K_1K_2}$. Put in logarithmic form,

$$\text{isoelectric pH} = \text{pI} = \tfrac{1}{2}(\text{p}K_1 + \text{p}K_2)$$

in which pK_1 and pK_2 are the negative logarithms of the respective dissociation constants. Consider leucine as an example:

$$K_1 = 10^{-2.36}; \quad K_2 = 10^{-9.60}$$
$$\text{pI} = \tfrac{1}{2}(2.36 + 9.60) = 5.98$$

Most of the monoamino, monocarboxylic acids have isoelectric points near pH 6.

What is the isoelectric point of a dicarboxylic, monoamino acid such as aspartate? It is a pH halfway between the pK values for the two carboxylic acid groups: pI = ½(1.99 + 3.90) = 2.95. At this acidic pH, sufficient protons add to the two carboxylate groups to leave only one remaining negative charge, while the ammonium group (pK = 9.90) is almost totally charged.

Similarly, the isoelectric point of a monocarboxylic, diamino acid such as lysine is a pH halfway between the pK values for the two ammonium groups: pI = ½(9.18 + 10.79) = 9.99. At this alkaline pH, only one H^+ remains attached to the two amino groups, while the lone carboxylate group (pK = 2.16) is totally ionized for all practical purposes.

Ionization and Solubility. The isoelectric points provide useful information for reasoning about the behavior of amino acids in solution because of the relationship they have to the various ionic forms, as is summarized in Figure 2–3. For example, the presence of charged groups on amino acids and proteins have important effects on their solubility.

Amino acids and proteins are least soluble at the isoelectric point, other things being equal. This is so because the zwitterion has no net charge, and it can

NATURE OF SOLUTION

	STRONGLY ACIDIC $> 10^{-2}M\ H^+$	*ACIDIC* $10^{-2} - 10^{-4}M\ H^+$	*WEAKLY ACIDIC* $10^{-4} - 10^{-7}M\ H^+$	*WEAKLY BASIC* $10^{-7} - 10^{-9}M\ H^+$	*BASIC* $10^{-9} - 10^{-11}M\ H^+$	*STRONGLY BASIC* $< 10^{-11}M\ H^+$
monoamino, monocarboxylic	$\oplus H_3N—R—COOH$ *cation*$^{+1}$	←	$\oplus H_3N—R—COO\ominus$ *zwitterion*	→	←	$H_2N—R—COO\ominus$ *anion*$^{-1}$ →
monoamino, dicarboxylic	$\oplus H_3N—R(COOH)—COOH$ *cation*$^{+1}$	$\oplus H_3N—R(COOH)—COO\ominus$ *zwitterion*	←	$\oplus H_3N—R(COO\ominus)—COO\ominus$ *anion*$^{-1}$ →	←	$H_2N—R(COO\ominus)—COO\ominus$ *anion*$^{-2}$ →
diamino, monocarboxylic (lysine)	$\oplus H_3N—R(\oplus NH_3)—COOH$ *cation*$^{+2}$	←	$\oplus H_3N—R(\oplus NH_3)—COO\ominus$ *cation*$^{+1}$	→	$H_2N—R(\oplus NH_3)—COO\ominus$ *zwitterion*	$H_2N—R(NH_2)—COO\ominus$ *anion*$^{-1}$
guanidino (arginine)	$H_2N\cdots\overset{\oplus}{C}\cdots NH_2$, NH, $\oplus H_3N—R—COOH$ *cation*$^{+2}$	←	$H_2N\cdots\overset{\oplus}{C}\cdots NH_2$, NH, $\oplus H_3N—R—COO\ominus$ *cation*$^{+1}$	→	← $H_2N\cdots\overset{\oplus}{C}\cdots NH_2$, NH, $H_2N—R—COO\ominus$ *zwitterion* →	$HN=C—NH_2$, NH, $H_2N—R—COO\ominus$ *anion*$^{-1}$
imidazolyl (histidine)	H–N(⊕)imidazole N–H, $\oplus H_3N—R—COOH$ *cation*$^{+2}$	H–N(⊕)imidazole N–H, $\oplus H_3N—R—COO\ominus$ *cation*$^{+1}$		N=imidazole N–H, $\oplus H_3N—R—COO\ominus$ *zwitterion*	N=imidazole N–H, $H_2N—R—COO\ominus$ *anion*$^{-1}$	

FIGURE 2-3 **Comparative behavior of various classes of amino acids with variation in H^+ concentration. The boundaries between the various ionic forms roughly indicate the $[H^+]$ at which the concentrations of two forms will be equal. The physiological forms are in the central column.**

therefore crystallize as such. The anionic or cationic forms can crystallize only as salts, such as sodium glycinate or glycine hydrochloride:

$$\underset{\text{sodium glycinate}}{H_2N{-}CH_2{-}COO^{\ominus}\ Na^{\oplus}} \qquad \underset{\text{glycine hydrochloride}}{{}^{\ominus}Cl\ {}^{\oplus}H_3N{-}CH_2{-}COOH}$$

Since these salts can freely dissociate in water, they are much more soluble.

Does this mean that amino acids with isoelectric points near the pH of physiological fluid are likely to crystallize in the tissues? No, it does not because the zwitterions, while having no net charge, do have an off-axis distribution of positive and negative charges that create strong dipoles in the molecule, thereby making nearly all quite soluble in water, even though they are less soluble than the ionic forms. This is generally true even when abnormally high concentrations of amino acids result from genetic defects (aminoacidopathies).

There is one conspicuous exception. Cystine is an amino acid that contains two ammonium groups and two carboxylate groups because it is formed by linking two molecules of another amino acid, cysteine, through side chain sulfur atoms (p. 142):

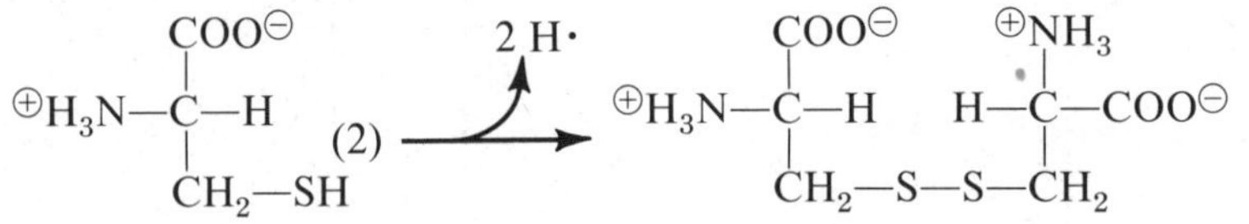

$$(2)\ \underset{\text{L-cysteine}}{{}^{\oplus}H_3N{-}CH(COO^{\ominus}){-}CH_2{-}SH} \xrightarrow{2\,H\cdot} \underset{\text{L-cystine}}{{}^{\oplus}H_3N{-}CH(COO^{\ominus}){-}CH_2{-}S{-}S{-}CH_2{-}CH({}^{\oplus}NH_3){-}COO^{\ominus}}$$

The crystal lattice of this molecule is so stable that the zwitterion is only soluble to the extent of 160 mg per liter of water at 37°C.

This low solubility causes trouble in people born with a genetic defect known as cystinuria that causes them to excrete high quantities of the amino acid in the urine. The H^+ concentration of urine frequently ranges near the isoelectric point for cystine (10^{-5} M, pH 5.0), at which the amino acid is least soluble. The presence of other compounds in the urine increases the solubility of cystine to approximately 300 mg per liter through "salting in" and the formation of complexes, but cystinuric patients frequently excrete even more than this, with the excess crystallizing into stones in the kidneys, ureters, and bladder. (The problem is not uncommon; over 1 per cent of the stones found in urinary tracts contain cystine as a major component.)

Could something be done to increase the solubility of cystine in the urine? One way of doing this would be to shift the pH from the isoelectric point. If we study Figure 2–1, we see that a shift of over 2 pH units (100-fold change in $[H^+]$) would be required in order to convert 10 per cent of the zwitterion to the anionic form. However, the solubility increases rapidly beyond that point, and it is indeed possible to keep substantially greater amounts of cystine in solution by prescribing repeated doses of sodium bicarbonate to the patient so as to raise his urinary pH over 7.0. (Unfortunately, the promise of this therapy proved somewhat illusory; the diminished acidity also increases the concentration of fully ionized phosphate in the urine, causing an increased precipitation of calcium phosphate in the stones. The result is the replacement of one rocky insult to the urinary tract by another.

Stereoisomerism of the Amino Acids

Let us briefly review what organic chemistry tells us about stereoisomers as it applies to the amino acids. All of the amino acids that are introduced into peptide chains by protein synthesis have one carbon (C-2) that is bonded to four different groups, except glycine, which has two H atoms. There are two possible arrangements of groups around such an asymmetric center:

$COO^{\ominus}$ / $H_3\overset{\oplus}{N}$—C—H / R

L-amino acid

↓ rotating plane of paper 180° ↓

R / H—C—$\overset{\oplus}{N}H_3$ / $COO^{\ominus}$

L-amino acid

$COO^{\ominus}$ / H—C—$\overset{\oplus}{N}H_3$ / R

D-amino acid

↓ spinning 180° on vertical axis ↓

$COO^{\ominus}$ / $H_3\overset{\oplus}{N}$···C···H / R

D-amino acid

Amino acids having one of the arrangements are said to have the L-configuration; those with the other have the D-configuration. The two configurations are mirror images of each other and like all mirror images of asymmetric objects cannot be superimposed no matter how they are turned. They are said to be enantiomorphic isomers, or **enantiomers.**

All of the asymmetric amino acids occurring in proteins belong to the L configurational family. This is true even though the D and L isomers have many identical chemical and physical properties. The important difference between them is that they cannot approach a fixed arrangement of groups, such as occur in another asymmetric compound, in the same way, and most biochemical reactions hinge upon mating arrangements between asymmetric groups. In the case of the amino acids, the distinction is so critical that many microorganisms deliberately use D-amino acids to create peptides that are highly toxic to other organisms; that is, they are antibiotics. The animal kidney has the ability to destroy D-amino acids, apparently to eliminate any possibility of forming toxic peptides.

Diastereoisomers. Some amino acids have more than one asymmetric carbon atom. Consider L-threonine, which has four possible stereoisomers:

$COO^{\ominus}$ / $H_3\overset{\oplus}{N}$—C—H / H—C—OH / CH_3

L-threonine

$COO^{\ominus}$ / H—C—$\overset{\oplus}{N}H_3$ / HO—C—H / CH_3

D-threonine

$COO^{\ominus}$ / $H_3\overset{\oplus}{N}$—C—H / HO—C—H / CH_3

L-allothreonine

$COO^{\ominus}$ / H—C—$\overset{\oplus}{N}H_3$ / H—C—OH / CH_3

D-allothreonine

Each of the four is a different compound, and only one, L-threonine, occurs naturally in proteins. There are two mirror-image pairs represented, and the two pairs are given different names. This is so because D- and L-threonine have similar behavior with symmetrical chemical reagents, and D- and L-allothreonine have similar behavior, but L-threonine and L-allothreonine behave differently and have different melting points, solubilities, and so on. They are said to be diastereoisomers.

It is common to designate a stereoisomer in structural formulas by a convention in which all vertical bonds are directed behind the plane of the paper and all horizontal bonds are directed in front of the plane of the paper. (We have followed the convention in this book, but a word of caution is necessary. Many people do this so long as they want to draw a carbon skeleton in a vertical position, but they simply turn the representation sideways if they want the carbon skeleton to be horizontal; of course, this now makes the vertical bonds in front of the paper and the horizontal bonds behind. The practice can cause great confusion with branched-chain molecules and has been avoided here.) There are 12 ways in which a molecule of L-alanine, for example, can be written with the vertical bonds behind the plane of the paper. Some of them are shown in the following:

$$
\begin{array}{c} COO^{\ominus} \\ | \\ {}^{\oplus}H_3N-C-H \\ | \\ CH_3 \end{array}
\qquad
\begin{array}{c} CH_3 \\ | \\ H-C-{}^{\oplus}NH_3 \\ | \\ COO^{\ominus} \end{array}
\qquad
\begin{array}{c} CH_3 \\ | \\ {}^{\oplus}H_3N-C-COO^{\ominus} \\ | \\ H \end{array}
\qquad
\begin{array}{c} {}^{\oplus}NH_3 \\ | \\ H-C-COO^{\ominus} \\ | \\ CH_3 \end{array}
\qquad
\begin{array}{c} CH_3 \\ | \\ {}^{\ominus}OOC-C-H \\ | \\ {}^{\oplus}NH_3 \end{array}
$$

A useful rule of thumb is that interchanging any two substituents on each asymmetric carbon in a conventional structural formula gives a representation of the other stereoisomer; making any two such interchanges gives another representation of the same stereoisomer.

DESIGNATING ROTATION. We designate all of the amino acids having the same arrangement of groups around C-2 found in L-alanine as L-amino acids regardless of the arrangement about other asymmetric centers. Some L-amino acids cause a rotation of a plane polarized light to the left, others to the right. The direction of rotation is shown when desired by lower case italic letters *(d)* or *(l)* in parentheses to designate dextro- or levorotatory, respectively. (It can also be shown by (+) for dextro- and (−) for levo-.) The older literature is confusing in this respect because there was a time of transition when the lower case letters were used for configurational family as well as for actual rotation.

R and S Nomenclature. Because of the difficulty sometimes created in designating configurational family with many types of compounds, a new nomenclature has been invented. Briefly, each of the four constituent groups about an asymmetric carbon is arranged in order of increasing atomic number of the nearest constituent atom or in order of increasing valence electron density. (N ranks higher than C, and O higher than N; ethylene carbons rank higher than saturated carbons, a —CH_2—COO^- group ranks higher than a —CH_2—CH_3 group, and so on.) One looks at the asymmetric center in such a way as to peer directly down on the substituent of lowest rank order, which is frequently —H with the amino

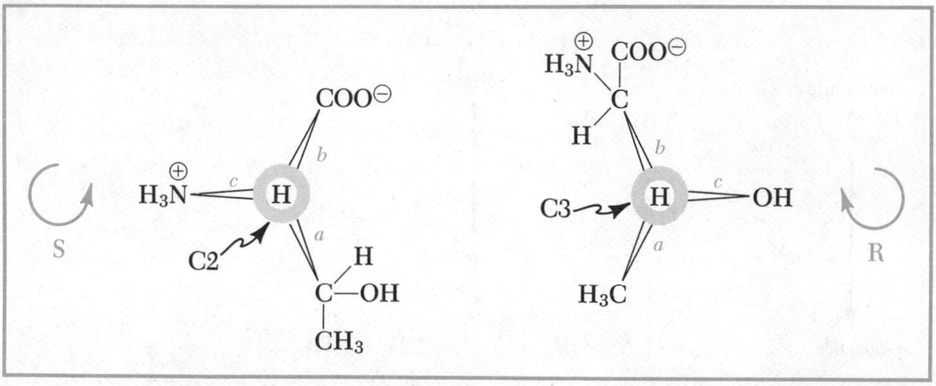

FIGURE 2–4 **Configuration in the (R) and (S) system can be determined by looking down on an asymmetric atom arranged so that the substituent of lowest atomic number is facing the viewer. In the example on the left, one is looking past the substituent H atom at C-2 of L-threonine. In order to go from the lowest to the highest substituent among the remaining three, one must go in a left-hand circle (COO outranks CH—OH(CH_3) because it has two O atoms attached to the carbon, and these higher atomic number atoms outrank the single O and C in the other group; the directly attached N of the amino group has a higher atomic number than C and therefore outranks both of the other groups). The left-hand circle is designated as (S) configuration.**

The example on the right shows the configuration about C-3 of L-threonine. One is again looking down on H as the substituent of lowest rank order in terms of atomic number. In this case, the methyl group is the lowest ranking of the remaining three substituents, and the hydroxyl group is the highest. Traveling in a right-hand circle carries one from lowest to highest ranking substituent, so this carbon has an (R) configuration.

acids. When this is done, the remaining three substituents will be arranged as spokes on a wheel, and one goes around the wheel from the lowest rank order to the highest.* If this is a clockwise direction, the configuration is **rectus** or **(R)**; if it is counterclockwise, the configuration is **sinistrus** or **(S).** The process is repeated for each asymmetric center. Designating the isomers does take some practice in visualization of the structures, but the nomenclature has the advantage of creating an unambiguous designation of the absolute configuration, no matter how many asymmetric centers there are. Under this system, L-threonine is (2S:3R)-threonine, or more systematically, (2S:3R)-2-amino-3-hydroxybutyrate (Fig. 2–4).

PEPTIDES

A peptide is a compound formed by linking amino acids with amide bonds, using the ammonium group of one molecule and the carboxylate group in another. The combination can be represented in a formal way:

$$^{\oplus}H_3N{-}R{-}COO^{\ominus} + {}^{\oplus}H_3N{-}R'{-}COO^{\ominus} \longrightarrow {}^{\oplus}H_3N{-}R{-}\overset{\displaystyle O}{\overset{\|}{C}}{-}\underset{\displaystyle H}{\underset{|}{N}}{-}R'{-}COO^{\ominus} + H_2O$$

*One can also arrive at the configurational designation by having the substituent of lowest rank away from the viewer, like the shaft of a steering wheel, and then going from highest to lowest in rank with the other substituents.

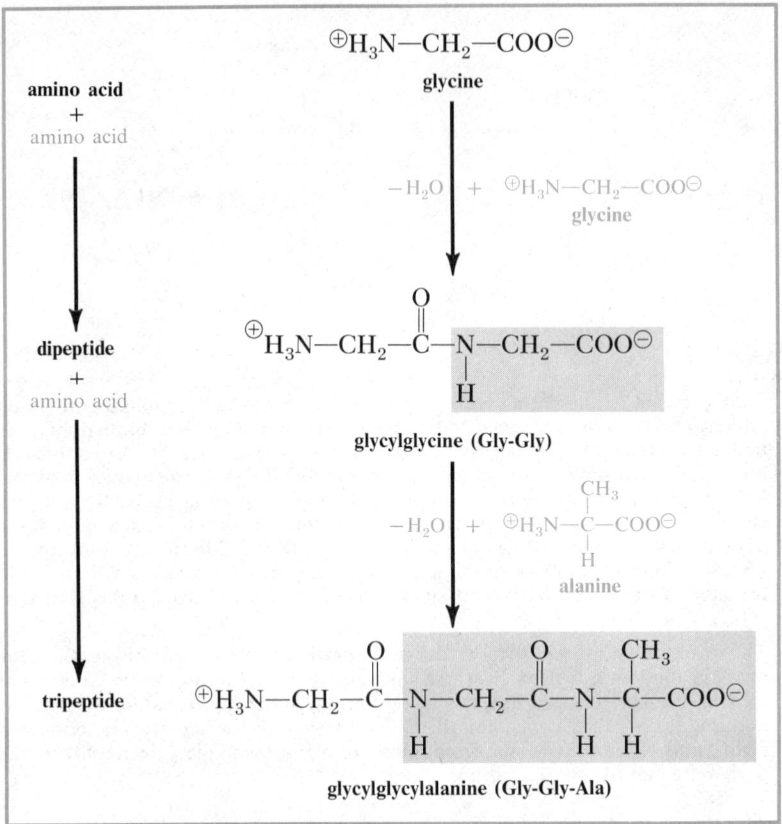

FIGURE 2–5 **Peptides are amides formed between the carboxylate group of one amino acid and the ammonium group of another. They are named from the ammonium end (the N-terminal) in terms of the constituent amino acid residues.**

and the relationship of peptides to the constituent residues of amino acids is also shown in Figure 2–5. (We shall see that the mechanism of biological synthesis is considerably more complicated.) A compound containing two amino acid residues is said to be a **dipeptide,** one containing three is a **tripeptide,** and so on. Relatively short chains of several amino acid residues are **oligopeptides** (**oligo** = few, little), whereas longer polymers are **polypeptides.** A polypeptide with more than 100 or so amino acid residues is said to be a **protein.** (The distinction is not sharp. Polypeptides with fewer amino acid residues are likely to be called proteins if they ordinarily have a well-defined conformation of the sort described in the next chapter.)

Nomenclature. Peptides are named from the terminal residue with a free ammonium group (N-terminal); the successive groups are designated by the name of the amino acid from which they are derived except for the replacement of the suffixes with -yl. (Glycyl in place of glycine, aspartyl in place of aspartate, but tryptophanyl, asparaginyl, cysteinyl, and glutaminyl in place of tryptophan, asparagine, cysteine, and glutamine.) Standard three-letter abbreviations are used to designate sequence in most instances, also beginning with the N-terminal. For example, methionylvalylaspartyllysyltyrosine, in which the remaining free ammonium group at one end is on a methionine residue and the free carboxylate group at the other end is on a tyrosine residue, is indicated as Met-Val-Asp-Lys-Tyr. All of the abbreviations are the first three letters of the name of the amino acid, except for

Asn — asparaginyl   Trp — tryptophanyl
Gln — glutaminyl   Ile — isoleucyl

(Note that the abbreviations signify a group, not the free amino acid, in the same way that Ac means acetyl, not acetic acid.)

Functions of Peptides. Most of the peptides with biological function are proteins, which typically have about 1000 amino acid residues. The average residue formula weight is near 110 in most proteins, so a molecular weight of the order of 100,000 is common, although the range is from 10,000 for small proteins to several million for large complexes made of several polypeptide chains. Most of the book will be about the functions of proteins.

However, the smaller peptides ought not be overlooked. They also can have important functions even though their total content in tissues is small compared to that of the proteins. Some are very potent substances. Most of the toxins in animal venoms and in plant sources are polypeptides, many of them too small to be considered proteins. Minute amounts of some oligopeptides with as few as three modified amino acid residues are effective as hormones. A derivative of a dipeptide, aspartylphenylalanyl methyl ester ("Aspartame") is 160 times as sweet as sucrose and is now being considered as a sugar replacement. (There is some discussion of possible deleterious effects from the extra aspartate and phenyl alanine released by breakdown of the compound.)

The Chemistry of Peptides

The character of a protein hinges upon the nature of its constituent chemical groups and their arrangement in space. How does this arrangement occur? One factor is the nature of the peptide backbone; there are favored orientations for a bare peptide skeleton. The other factor is the primary structure; varying sequences of amino acids create varying forces because of the different side chains. In other words, the nature of the constituent groups in itself helps determine their arrangement.

The Peptide Bond. The characteristic feature of peptides is the recurring amide bond linking C-2 atoms of successive amino acid residues, which is called a peptide bond. The backbone of a polypeptide chain is made of repeating C-2 and amide groups. The critical fact about this peptide bond is that it has some double bond character. We draw it as:

$$\begin{array}{cc} \mathrm{O} & \mathrm{H} \\ \| & | \\ -\mathrm{C}- & \mathrm{N}- \end{array}$$

when it would be more realistic to draw it as:

$$\begin{array}{cc} \mathrm{O} & \\ \vdots\!| & \\ -\mathrm{C}\overset{\cdots}{-} & \mathrm{N}- \\ & | \\ & \mathrm{H} \end{array}$$

The important consequence of this double bond character is that it freezes the entire amide group in a single plane. Slight deviations can occur, but for our purposes the planar configuration may be regarded as nearly fixed. The only possibilities for free rotation in the peptide backbone occur at the two bonds on the α-carbon atoms.

In other words, the arrangement in space of the backbone of a peptide chain is nearly completely defined by two angles of rotation on each α-carbon atom (Fig. 2–6), which are named Ramachandran* angles. These angles provide a handy way of defining the spatial orientation of amino acid residues in a peptide backbone. The actual numbers need not concern us, but we ought to note that they are not randomly scattered among all possible values. Some angles do not occur because steric hindrance prevents the particular orientation; adjacent carbonyl oxygen atoms collide before the angle can be reached. In addition, there are some orientations that occur much more frequently than others equally free of steric hindrance, and we shall see in the next chapter that the favored configurations permit the formation of additional forces between parts of the peptide chain.

We have noted that polypeptides are amphoteric because of the presence of amino acid residues with ionized side chains in addition to the terminal ammonium and carboxylate groups. That is, they titrate as acids and as bases, and they have an isoelectric point at which they are frequently least soluble and have the greatest tendency to aggregate.

Electrophoresis. Since different proteins frequently have different combinations of positively and negatively charged groups, they can often be separated by electrophoresis, that is, by causing them to move in an electric field (Fig. 2–7). This technique is frequently applied to the polypeptides comprising the proteins in blood plasma, and it is also used to diagnose the presence of abnormal hemoglobins.

BLOOD PLASMA PROTEINS. The proteins in the extracellular part of the blood, known as the plasma, are commonly measured in humans because they are readily accessible, have important functions, and sometimes reflect disturbances in other tissues. Most are soluble globular proteins; that is, they have a compact shape akin to a sphere, rather than a long fibrous shape. An important exception is fibrinogen, a protein used to form clots, which we shall discuss in Chapter 18.

*G. N. A. Ramachandran (1922–). Indian biophysicist.

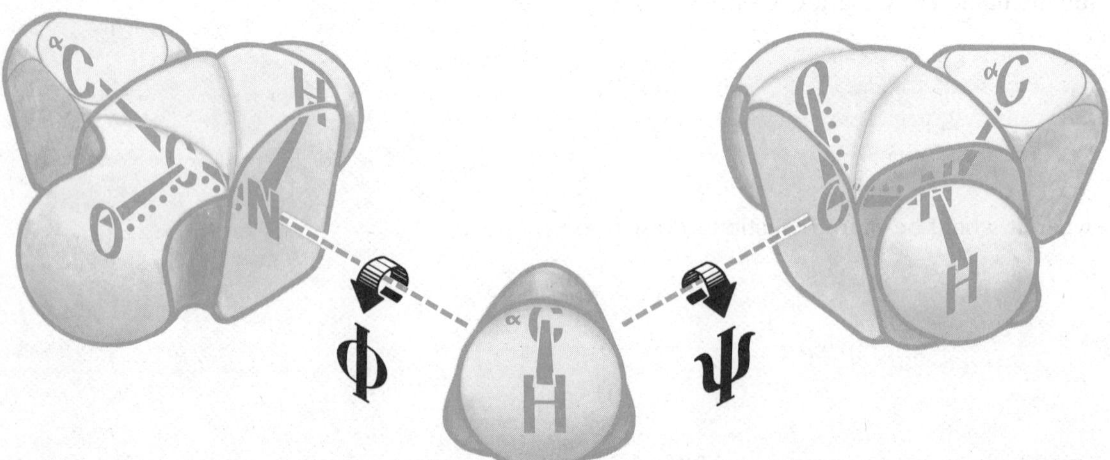

FIGURE 2–6 **Free rotation of a peptide chain occurs only around the bonds joining the nearly planar amide groups to the α-carbons. The angles of rotation, known as Ramachandran angles, therefore determine the spatial orientation of the peptide chain. By convention, the Φ and Ψ angles are assigned values of 180° in the illustrated conformation.**

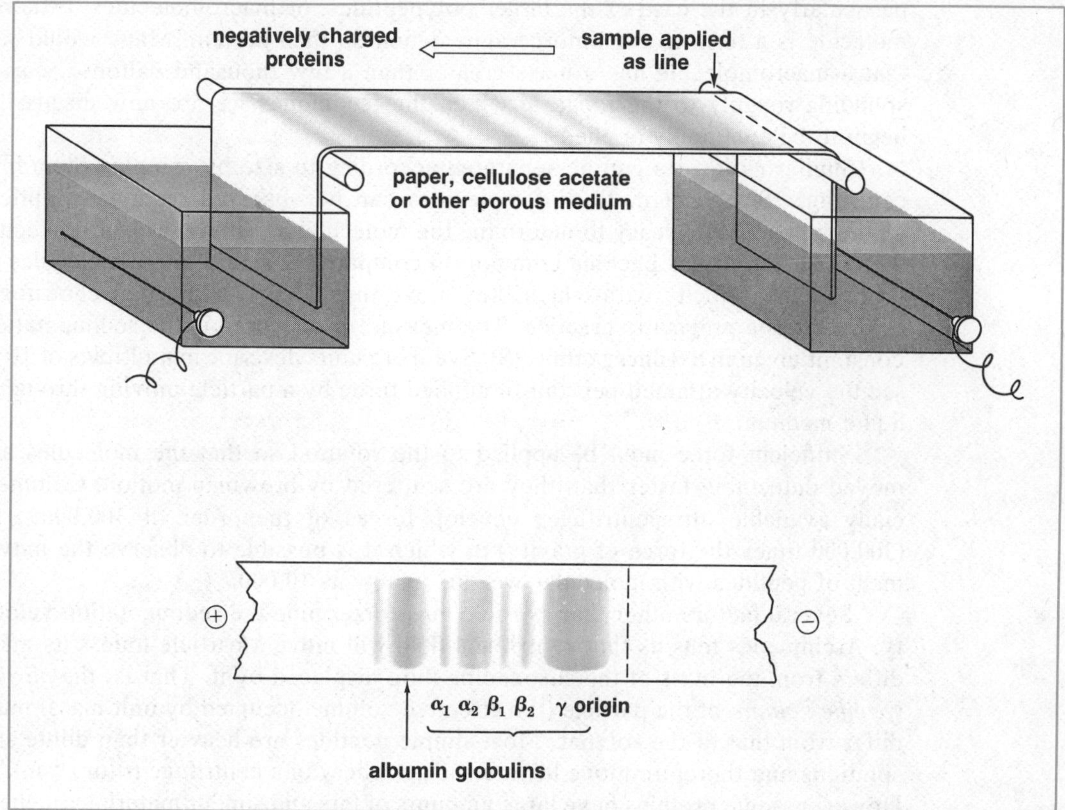

FIGURE 2–7 *Top.* Zone electrophoresis. A strip of porous material (e.g., paper or cellulose acetate) is saturated with a buffer solution and suspended between two containers of buffer that are fitted with inert electrodes. The sample containing a mixture of proteins is applied as a thin stripe; those proteins having different net charges separate when a voltage is applied. The buffer is usually made sufficiently alkaline to create a net negative charge on most proteins. Those with the most charges migrate most rapidly. *Bottom.* Results of electrophoretic separation of the proteins in human serum.

After a sample of blood clots, a clear serum is exuded that contains most of the globular proteins. The most abundant of these serum proteins are named as **albumins** or **globulins.** This is an old terminology stemming from early techniques for separating proteins; those soluble proteins that remained soluble in pure water were designated albumins, those that required the presence of dilute salt for solution were designated globulins.

Early application of electrophoresis in free solution to blood plasma revealed only one major kind of albumin, but several major groups of globulins with differing mobilities, which were termed α, β, and γ. Today, we know that there is indeed only one major kind of albumin in serum, but there are hundreds of different proteins among the globulins.

What do these proteins do? We shall say more about individual kinds as we go along. Suffice it now to note that many of the proteins are designed to bind specific small molecules for transport through the blood. Others bind foreign substances as a protective device. Still others catalyze particular reactions.

Characterization of Size by Centrifugation. Peptides range in size from glycylglycine, with a molecular weight of 132, to polypeptides containing as many as 1000 amino acid residues in a single chain. Describing the size of a given peptide, or sorting mixtures of peptides according to size, is a common and useful practice, particularly in the case of the larger polypeptides, or macromolecules. (Macromolecule is a term of even more vague definition than protein. Many would say that a macromolecule has a mass greater than a few thousand daltons*, corresponding roughly to the range at which the techniques we are now discussing begin to be commonly applied.)

Similar molecules can be separated according to size by spinning them in a centrifuge. Since net motion of molecules can be observed readily by optical methods, and it isn't easy to determine the molecular weight of a macromolecule with precision, it has become common to compare the size of large molecules in terms of the velocity with which they move through a solution in a centrifugal field. This is a pragmatic practice. The measure that is used is the sedimentation constant given in **Svedberg† units (S).** Svedberg units describe in multiples of 10^{-13} sec the velocity attained per unit of applied force by a particle moving through a liquid medium.

Sufficient force must be applied to the solution so that the molecules are moved detectably faster than they are scattered by brownian motion. Commercially available ultracentrifuges develop forces of the order of 300,000 × g (300,000 times the force of gravity) in which it is possible to observe the movement of peptides with molecular weights as low as 10,000.

Several factors other than particle mass determine the sedimentation velocity. Archimedes tells us that centrifugal field will move a particle unless its mass differs from the mass of the suspending fluid displaced by it. That is, the *partial specific volume* of the particle (the effective volume occupied by unit mass) must differ from that of the solvent. Most simple peptides are heavier than dilute salt solutions and therefore move toward the periphery of a centrifuge rotor ("sink"). However, some proteins have large amounts of fats and similar material attached, and these less dense materials cause such lipoproteins to move toward the axis of the rotor ("float") with a velocity expressed as an S_F.

Given the development of a force on a particle in a centrifugal field, it will accelerate, but there will be increasing frictional resistance to motion through the medium as the velocity increases. Acceleration will continue only until a velocity is reached at which the frictional counterforce just balances the applied centrifugal force. It is this limiting velocity that is described by the sedimentation constant. The sedimentation constant therefore depends upon size of particle and shape of particle as well as partial specific volume. We know from experience that a large chunk of material falls through water faster than an equal mass of fine powder, even though the total accelerating gravitational force is the same. The powder settles more slowly because its larger surface area is in frictional contact with the medium. Since the partial specific volume is nearly the same for the same material, the sedimentation constant is an index of size, provided that the particles have similar shapes. Shape is important because long, thin molecules have greater surface area to generate frictional force than do spherical molecules of the same mass. The asymmetric molecules of high axial ratio (ratio of long dimension to short dimension) therefore have smaller sedimentation constants for a given molecular weight.

In the succeeding chapters, we shall have occasion to refer to some cellular

---

*The dalton is a measure of the mass of individual molecules expressed in atomic units based on ^{12}C. That is, one atom of ^{12}C has a mass of 12 daltons. (But the atomic weight of ^{12}C is 12, not 12 daltons. Atomic and molecular weights are dimensionless ratios.)

†T. Svedberg (1884–1971). Swedish Nobel Laureate responsible for much of the theoretical and practical development of centrifugation as a tool for studying large molecules.

components by their S values for purposes of identification. To make sense out of the discussion, it is important to have in mind that S values, unlike molecular weights, are not additive when molecules combine. The union of two 5S particles does not create a 10S particle. (Two cannonballs glued together don't sink twice as fast as one alone.)

Although we can't directly compare sedimentation constants and molecular weights, it is helpful to have a rough idea of the corresponding ranges. A value of 2S will be obtained for nearly spherical proteins with molecular weights somewhat over 10,000; long, thin molecules having sedimentation constants of 2S may have molecular weights over 50,000. Similarly, 4S crudely corresponds to a molecular weight of 50,000 for spherical proteins, 8S to 160,000, and 16S to 400,000; the molecular weights of long, thin molecules with the same sedimentation constants are several-fold greater. (A more detailed description of the application of centrifugation and the necessary calculations is given by H. K. Schachman in Colowick, S. P., and N. O. Kaplan, eds.: (1957) *Methods in Enzymology,* Vol. 4, p. 32, Academic Press.)

FURTHER READING

McGilvery, R. W.: (1975) *Biochemical Concepts.* W. B. Saunders. Chapter 31 surveys the chemistry of peptides in more detail.

Readable Reviews

Dickerson, R. E., and I. Geis: (1969) *The Structure and Action of Proteins.* Harper and Row. Excellent introduction to the field.

Barker, R.: (1971) *Organic Chemistry of Biological Compounds.* Prentice-Hall. Useful survey of the properties.

Comprehensive Works for Reference

Greenstein, J. P., and M. Winitz: (1961) *Chemistry of the Amino Acids* Vol. 3. Wiley.

Putman, F., ed.: (1975–77) *The Plasma Proteins.* 2nd ed., 3 Vols. Academic Press. Contains introductory survey.

3 THE CONFORMATION OF POLYPEPTIDE CHAINS IN PROTEINS

PRIMARY STRUCTURE

Proteins are made from one or more polypeptide chains. The polypeptides in some proteins are conjugated with other kinds of molecules, such as carbohydrates and fats, but a substance isn't a protein unless it contains a polypeptide.

The fundamental principle of protein chemistry is that one protein differs from another because it has a particular arrangement of amino acid residues along its peptide chains. Each polypeptide in a protein consists of amino acids that are polymerized in a particular sequence, and this sequence is said to be its **primary structure.** Any change in the primary structure creates a different protein. Proteins that differ in only a single amino acid residue at some locations may have nearly identical properties, but there are other locations at which substitution of one amino acid for another can cause drastic changes in the nature of the protein.

Number of Different Polypeptides. Although there are only 20 different kinds of amino acids from which proteins are made, there is no danger of exhausting the possible primary structures for the creation of new proteins. A protein made by combining only 100 amino acid molecules is quite small as proteins go, and yet 20 different amino acids can be combined 100 at a time in 20^{100} different ways. This number is so large that every single protein molecule conceivable in our universe, existing now or at any time in the past, could be unique without using more than an infinitesimal portion of the possibilities.

In point of fact, however, every protein molecule is not unique. Informed guesses at the number of different kinds of proteins that a newborn infant makes range between 10,000 and 100,000. (These figures neglect differences caused by partial hydrolysis or other modifications of a polypeptide chain after it is formed.) Assuming there are 30,000, the infant will have on the average about 1×10^{17} identical molecules of each kind of protein. In addition, every one of the $\sim 1.5 \times 10^8$ infants born each year will contain identical copies of many of these kinds of protein molecules. Their common humanity is a result of this fact, and it is a marvelous thing that something like 20,000 atoms, to use a typical number, can repeatedly be assembled with so few errors in exactly the same way in each and every human conceived.

However, some protein molecules will be different in two infants taken at random (but not in identical twins). That is, there are many proteins of which one infant will have 10^{17} molecules and the other infant will have none. The second infant is likely to have molecules of a similar, but not identical, protein in its place. The individuality of the two infants comes from these differences in protein composition.

Similarly, the fly that buzzes around the infant will contain many copies of protein molecules that are identical to molecules found in another fly and some that differ in detail. However, it is almost certain that the human and the fly have no protein molecules in common; their functions are too different to be served in exactly the same way. Proteins are made to satisfy these diverse needs by varying the way in which the 20 kinds of amino acids are put together, that is, by altering the primary structure, and specification of the variations is a major part of the heredity of the organism.

A Semantic Point

It is common in biochemical literature to use the name of a protein as a class word when talking about properties shared by similar proteins from all biological sources. Thus, the statement, "hemoglobin is red," is automatically understood to apply to all related proteins from all species. Similarly, the term "human hemoglobin" without qualification usually means the collection of proteins found in most humans, even though there are two and sometimes more different compounds in each individual that would be named as hemoglobins. Proteins that differ in only a few amino acid residues are nearly always given the same general name.

SECONDARY STRUCTURE

The backbone of a polypeptide, like any long repeating polymer, will tend to form a coil. Why is this so? Because any repetitious displacement in space travels a spiral path. Imagine a polypeptide made of one kind of amino acid. Each residue in the middle of the chain is exposed to much the same environment as its neighbors. Therefore, they will tend to form the same Ramachandran angles relative to each other; the polypeptide backbone will be regularly bent in the same way. It so happens that particular spiral arrangements are quite common parts of protein molecules. These segments of regular geometry that appear in proteins are called secondary structures.

The kind of spiral formed will depend upon the forces determining the Ramachandran angles. Hydrogen bonds between the N—H and C═O groups of the peptide bonds are the most important forces arising only from the polypeptide backbone. A structure will be favored if it brings all of these groups into positions where they will form hydrogen bonds.

In addition, structures will be favored in which there is a minimum of unfilled space. Dispersion forces, also called London forces, stabilize structures in which atoms are tightly packed. Either the structure itself must be snugly nestled together, or it must have larger, open crevices that can be filled by solvent molecules.

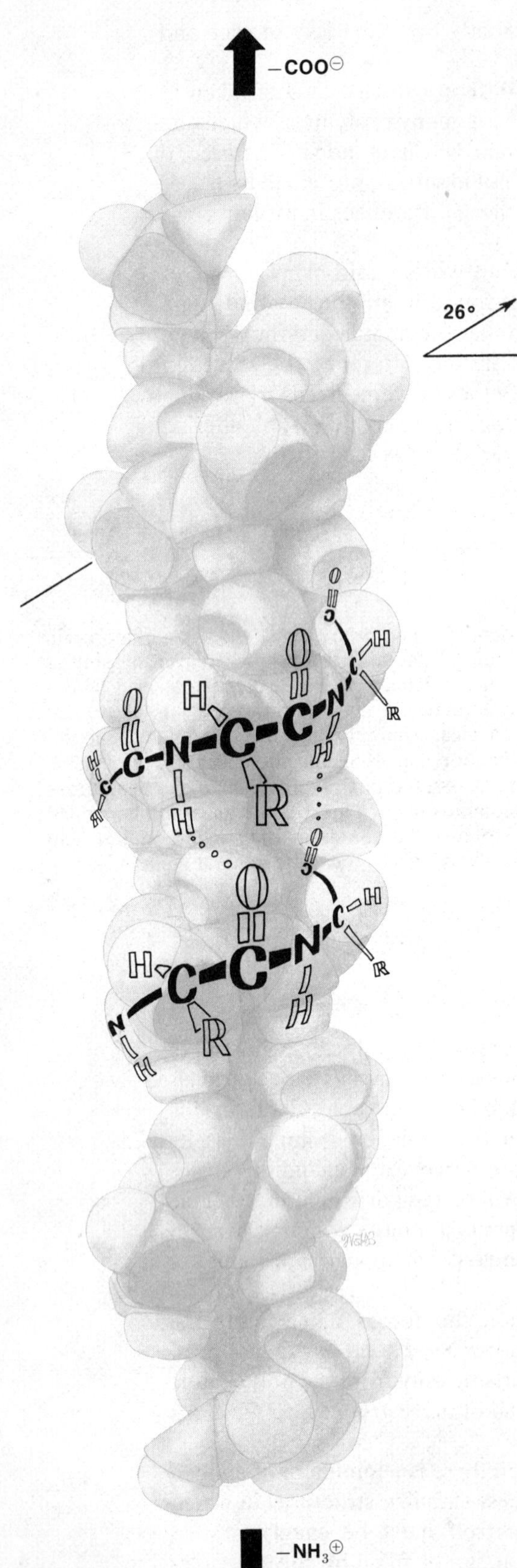

FIGURE 3–1

A model of an α-helix. The constituent atoms are labeled on two turns. The side chains of the amino acids have been omitted, but their positions are indicated by R. The helix has 3.6 residues per turn, bringing the fourth residue into a position where it can form a hydrogen bond to link the turns. Each —NH and C═O group is involved in forming such bonds in the middle of the helix.

The α-Helix. One of the most common ways in which a polypeptide chain forms all possible hydrogen bonds is by twisting into a right-handed screw with the —NH group of each amino acid residue hydrogen-bonded to the C=O groups of the fourth following residue, which is now on an adjacent turn of the helix (Fig. 3–1). This screw is the α-helix, and as the figure shows, it is a compact structure in which the peptide backbone forms a cylinder coated with amino acid side chains extending outward from its axis. The structure is obviously suitable for creating long fiber-like molecules, but we shall see that segments of α-helix are commonly a part of more globular protein molecules.

The structure of a polypeptide helix is sometimes designated by specifying the number of amino acid residues per turn of the helix, and the number of atoms in the ring created by formation of a hydrogen bond. According to this designation, an α-helix is a 3.6_{13} helix, because it has approximately 3.6 residues per turn, and a 13-member ring is created by the hydrogen bond (Fig. 3–2).

Other helices can be formed from single polypeptide strands, but they are not nearly as stable as the α-helix when formed from most amino acid residues. For example, the chain can be tightly twisted so as to have three residues per turn with hydrogen bonds formed with the third successive residue, creating 10-member rings (Fig. 3–2). This 3_{10} helix occurs as one or two turns in the chain, usually as a tightening of the end turns of an α-helix. (Sometimes the ends are tightened only enough to bring the —NH group into position for bond formation with the C=O groups on both the third and fourth residues, forming a turn of α_{II}-helix.)

Similarly, the α-helix may be loosened at the end to create a turn of π-helix, with hydrogen bonding to the fifth succeeding residue. These localized twists of 3_{10}, α_{II}, or π-helices have the same Ramachandran angles that would be found in a multi-turn helix. However, lengthy structures of these kinds are not common in real proteins.

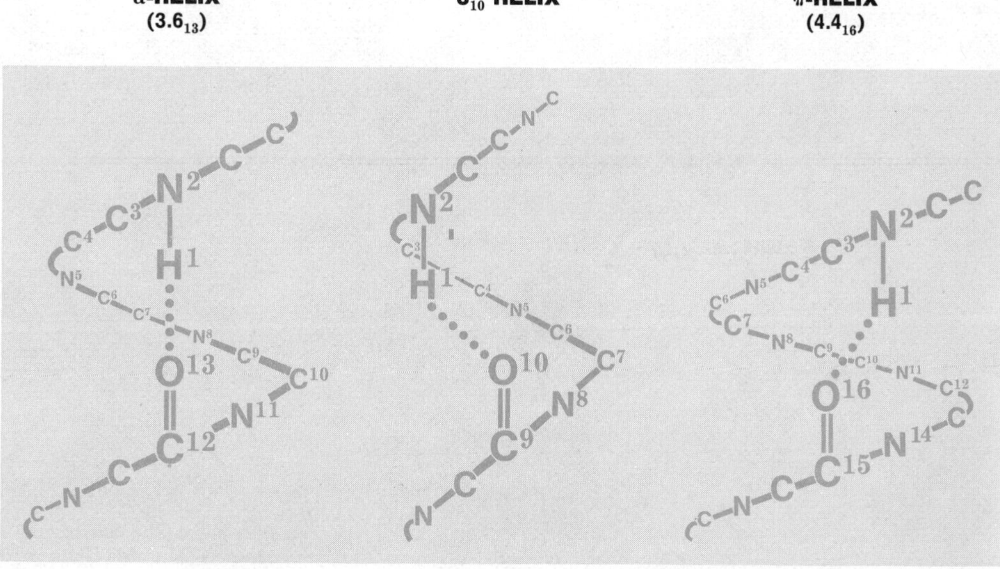

FIGURE 3–2 A schematic illustration of the position of the hydrogen bonds in three types of helices. Each helix may be designated by the number of residues per turn with a subscript indicating the number of atoms in each ring created by formation of a hydrogen bond. The α-helix is the most stable, but turns of the tighter 3_{10} helix or the looser π-helix are sometimes found at the ends of segments of α-helix.

ANTI-PARALLEL (β)

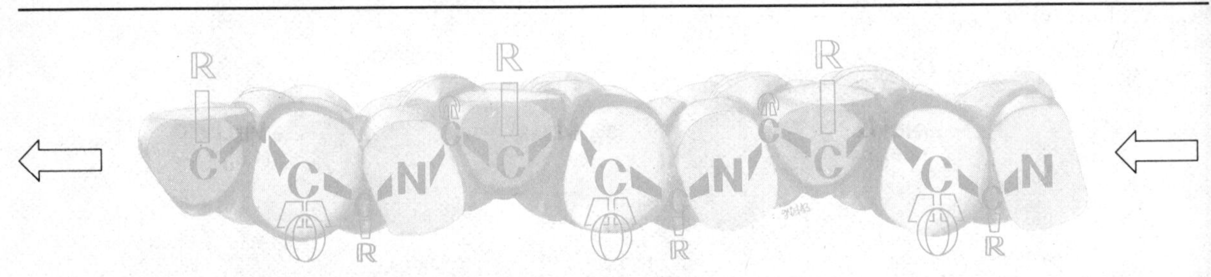

PARALLEL (β_p)

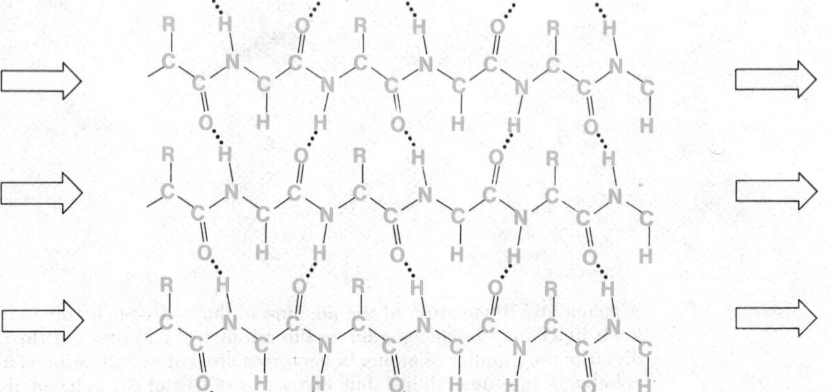

FIGURE 3–3 *See legend on opposite page.*

FIGURE 3–3 *Top.* **A view of one face of a model of an antiparallel pleated sheet (β-structure). Alternate side chains are directed toward the viewer, and the intervening side chains are out of sight on the other face of the sheet. Every NH and C═O group in the middle of a pleated sheet is involved in a hydrogen bond.** *Center.* **A nearly edge-on view of the sheet, illustrating the small pleats running the width of the sheet, and the alternating appearance of the side chains on the two sides. The hydrogen bonds are directed toward and away from the viewer.** *Bottom.* **Diagram of linkages in a parallel pleated sheet (β_P structure). Only the R and H substituents above the plane of the paper are indicated on the α-carbon atoms.**

β-Structure, or Pleated Sheet. All possible hydrogen bonds can also be formed between segments of peptide backbone that are stretched out to nearly maximum extension and then laid side by side (Fig. 3–3). The result is a β-structure, also known as the pleated sheet because of the zig-zag appearance when viewed on edge. The sheet may be **antiparallel,** with chains that run in the opposite direction as if the same chain were folded back and forth upon itself, or it may be made from segments of chain that are looped back to run in the same direction, creating the parallel pleated sheet, or β_P-structure.

The striking characteristic of the pleated sheet is that side chains of alternate amino acid residues appear on opposite sides of the sheet; the sheets, therefore, tend to form when every other side chain can interact in some effective way. The intervening side chains are segregated on the opposite side of the sheet.*

Reverse Turns. Important structural elements in proteins include bends that enable the chain to reverse direction and fold back on itself. Important forms of these bends involve hydrogen bonding from one residue to the third following residue. There are six ways in which this is done, one of which is shown in Figure 3–4. Two of the forms are equivalent to a turn of the 3_{10} helix, and the term

*We owe much of the credit for recognition of the helical and sheet structures to Linus Pauling (1901–), American physical chemist and Nobel Laureate. He earlier developed the concept of the hydrogen bond. His fame has been tarnished in his later years by his strong advocacy of vitamin C as a wonder drug, but this should not obscure the magnitude of his earlier scientific contributions.

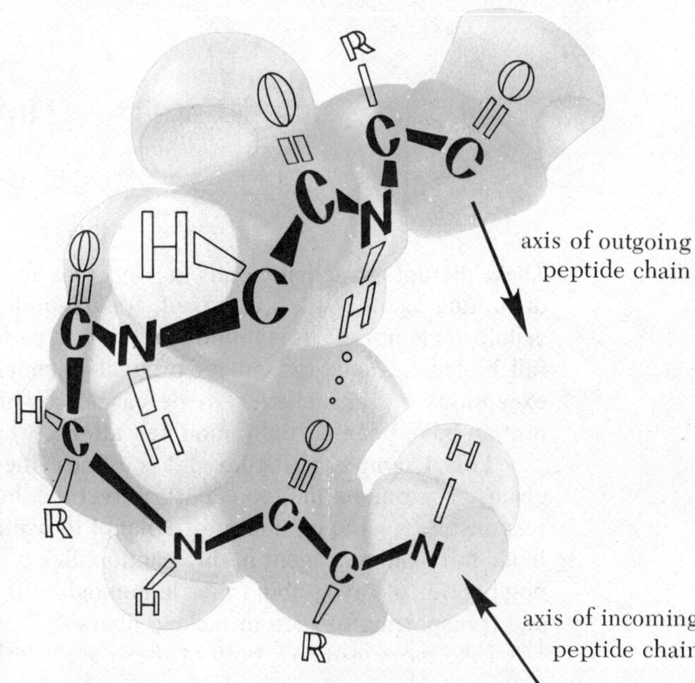

FIGURE 3–4

Some sharp bends in peptide chains are created by forming a hydrogen bond between the carbonyl oxygen of one residue and the amino hydrogen of the third following residue. There are six different ways in which the chain can twist to do this, of which one is illustrated here.

"3_{10}-bend" has therefore been applied to the group. Others have called some of the forms β-bends, but to avoid confusion, most now prefer more non-committal descriptions such as "reverse turn," "chain reversal," or "chain fold."

Non-ordered Arrangement. Part of a polypeptide chain may lack any recognizable geometrical order of the kinds we have been describing and is sometimes said to be in a random coil. This is a misleading term because the arrangement of a polypeptide chain is seldom random, even if it appears as tangled as a crushed coathanger. Lack of the repetitive elements of design that we call secondary structure simply reflects response to non-repetitive forces — non-repetitive because they arise from the side chains of the amino acid residues rather than from the peptide backbone.

TERTIARY STRUCTURE

The tertiary structure of a polypeptide chain is the complete form assumed by the chain — the combination of the various secondary structures and non-ordered segments. All the knowledge available to us at this time supports the notion that polypeptides assume the shape that will create the maximum forces — the shape that involves the greatest loss of energy in its formation. (There are some qualifications; it may be in some cases that the arrangement of the very lowest energy content appears so slowly that some other, more rapidly equilibrating conformation is the most stable practical form. The principle that the biologically functional molecule is the one that forms spontaneously is not violated by such circumstances.)

Evidence from Denaturation. The best evidence that the original native structure of natural polypeptides is the most stable configuration comes from experiments in which the polypeptides are disrupted, or denatured, without destroying any of the amino acid residues. This can be done by placing them in concentrated solutions (several molar) of urea or guanidinium chloride:

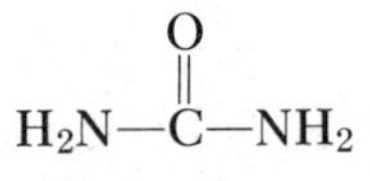

$$H_2N{-}\overset{\overset{\displaystyle O}{\|}}{C}{-}NH_2 \qquad\qquad H_2N{\cdots}\overset{\overset{\displaystyle NH_2}{\vdots}}{\underset{\oplus}{C}}{\cdots}NH_2$$

urea 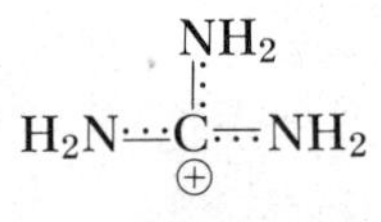 — guanidinium ion

These disrupt intrachain bonds in ways that are not completely understood. If the disrupting agent is then removed, for example, by dialyzing it away through a cellulose membrane, it is found that the polypeptide almost invariably recovers its full biological activity and its original chemical and physical properties. (The exceptions are all believed to be cases in which the amino acid residues of a protein have been partially modified after it was originally formed.)

Two Examples. Figure 3–5 outlines the backbone for two polypeptide chains. The one on the top is part of liver alcohol dehydrogenase, a protein that is responsible for the oxidation of ethanol to acetaldehyde. (The part that is shown binds the oxidizing agent in the reaction. See p. 480.) The one on the bottom is the polypeptide of myoglobin (muscle hemoglobin), a protein responsible for binding and transporting oxygen in muscle fibers.

Like most proteins, both of these are relatively globular in shape. (We shall discuss the more specialized fibrous proteins as we come to them.) However, the

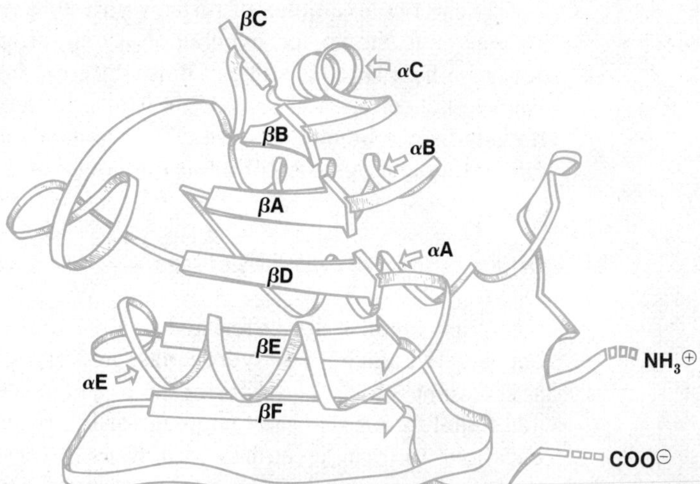

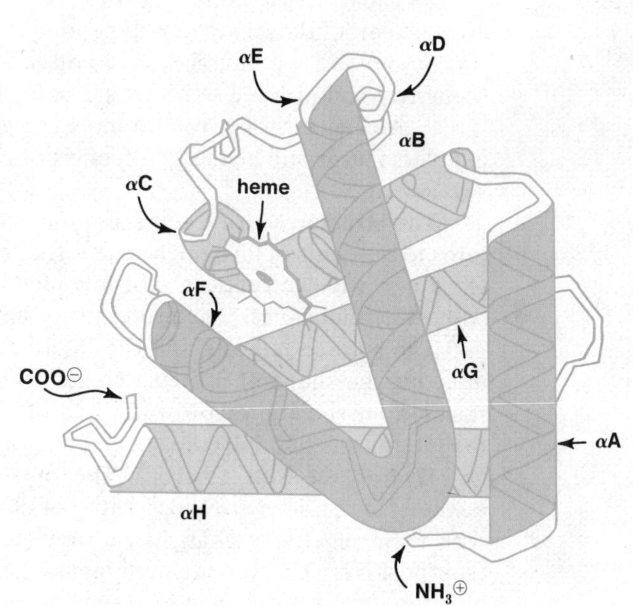

FIGURE 3–5

The tertiary structure of two polypeptide chains. *Top.* Part of the alcohol dehydrogenase found in the cytosol of liver. A central core of a twisted parallel pleated sheet is formed from six segments of polypeptide chain indicated by broad flat arrows. They are labeled βA, βB, βC, etc., according to the order in which the segments appear in the primary structure, with βA being made from a segment closest to the N-terminal. Around the central core are segments of α-helix, also labeled in sequence, and reverse turns, along with short non-ordered segments. *Bottom.* The complete molecule of myoglobin, which binds oxygen in muscle fibers. This chain is made entirely from eight segments of α-helix with short connecting pieces of reverse turns and non-ordered segments. The polypeptide is designed to fold around a molecule of heme so as to exclude water from it, as if the protein contained a pocket into which the heme slides. Hemoglobins in general contain polypeptide chains with a similar tertiary structure. Drawing of liver alcohol dehydrogenase by B. Furugren from C.-I. Brändén, et al., (1973) Proc. Natl. Acad. Sci. U.S.A., *70*: 2441. Myoglobin chain drawn from view given by R. E. Dickerson *in* H. Neurath, ed., (1964) *The Proteins*, Vol. 2. Academic Press, p. 634.

overall shape gives no clue to the quite different arrangements of the polypeptide chains from which the protein is created. The liver alcohol dehydrogenase contains a central core of a bent parallel pleated sheet made from six segments of chain around which are four recognizable segments of α-helix. (The segments are lettered according to the order in which they appear from the N-terminal of the chain. βA is the first segment used to form the pleated sheet, αA is the first segment appearing in an α-helix, and so on.) Between these segments are various twists and bends.

The myoglobin polypeptide, on the other hand, contains no pleated sheet and is made up almost entirely of segments of α-helix, with what appears to be a minimal amount of connecting chain between some segments.

These two examples of tertiary structure are not at all unusual. The general arrangement shown for liver alcohol dehydrogenase is similar to one found in several other proteins, some of quite different functions. The same arrangement of stacked helices seen in myoglobin is also found in other oxygen-binding proteins, ranging from a hemoglobin found in the nitrogen-fixing nodules of legumes to the circulating hemoglobins of adult humans.

Origin of the Tertiary Structures

What creates the difference in the conformation of lowest energy content between liver alcohol dehydrogenase and myoglobin? It is the difference in amino acid composition. The varying side chains create possibilities for developing additional forces that may augment or interfere with the tendency of the peptide backbone to form secondary structures of regular geometry. We are not able to analyze in detail how a particular primary structure gives rise to a particular geometry in globular proteins, but great progress has been made toward the development of the general principles.

Secondary Structures Are Preferred. There is a growing belief that the greatest part of the length of polypeptide chains forms secondary structures of one sort or another, with much less in non-ordered connecting segments. Most of the seemingly non-ordered portions are actually made from specific kinds of reverse turns. (An analysis of seven proteins showed one third each of the residues to be in reverse turns and helices, with one sixth each in pleated sheets and non-ordered segments.)

The Structures Are Compact. The same forces that cause secondary structures to be compact also operate in tertiary structures. Any openings are usually a few atoms wide and appear to be intended for access of solvent or for combination with other molecules. The protein may have bizarre forms, but it is not porous.

The surface is polar and the interior is non-polar. One of the major forces affecting the shape of proteins comes from the tendency of polar side chains to remain in association with water and of non-polar side chains to agglomerate in the interior of the molecule.

Hydrophilic Side Chains. The most hydrophilic, or water-loving, of the amino acid residues are those with net charges, such as aspartate and glutamate with their negatively charged carboxylate groups, and lysine and arginine with their positively charged ammonium and guanidinium groups. It is almost impossible for a polypeptide chain to assume a stable conformation with a charged group buried from access to water. (The only known exceptions occur when the charged character is dissipated through the formation of multiple hydrogen bonds.) Thus, a polypeptide will bend into a shape that keeps these groups on the surface. The carboxamide groups of asparagine and glutamine are also quite polar and stabilize contortions of the chain that enable them to remain in contact with water or to be in a position where they can form hydrogen bonds with other groups.

Hydrophobic Bonds. Non-polar side chains tend to cluster together in the interior of protein molecules so as to avoid exposure to water. The force causing this comes, not from any specific attraction of such groups for each other, but from the propensity of surrounding water to resist being forced into ordered structures by the presence of the non-polar groups. The so-called hydrophobic bond is really an association created by spurned water, and it is the same kind of force that causes insoluble oils to form compact spheres upon suspension in water. How does this happen?

Molecules of water have a strong tendency to form hydrogen bonds with each other as is shown by the release of heat when water freezes into the fixed lattice of ice. There is an opposing tendency of all molecules to stay dispersed, to be random and unordered. This tendency, measured as entropy, is increased by a rise in temperature. The two effects just balance at the melting point. At higher temperature, hydrogen bonds persist for shorter and shorter times; each molecule becomes more and more promiscuous.

A non-polar molecule introduced into liquid water represents a eunuch at an orgy. Since it cannot form hydrogen bonds or other bonds created by polarity, neighboring water molecules lose some of their possibilities for random union, and associate with each other longer than they would in the absence of the non-polar intruder. However, this longer association represents greater order, a step toward freezing into ice, and we have already said that creation of such order at temperatures above the melting point can be accomplished only by the introduction of energy. It follows that dispersing nonpolar molecules in water requires energy, and the non-polar molecules will aggregate when left to themselves so as to diminish their surface, thereby diminishing the number of water molecules with hindered random associations.

SOME NON-POLAR CHAINS ARE ON THE SURFACE. If the degree of polarity of the side chains were the only determinant of structure, a polypeptide would fold so that every group not interacting with water was buried in the interior. The tendency to make this separation is very strong and can aid in shaping complex surfaces, but the separation is not complete. There are usually more non-polar side chains than can be accommodated in the interior of a protein, and some are exposed on the surface of most proteins. This is an important part of the development of specific function, as well as shape. Having more reactive groups separated by the relatively unreactive non-polar side chains gives individual character to the surface geometry — the stars are bright because the sky is black.

Generation of Specific Structures

The general forces acting on a polypeptide chain seem clear, but what kind of amino acid sequence is required to generate particular shapes of protein molecules? What contribution is made by side-chain hydrogen bonding, electrostatic interactions, and so on? A good beginning has been made on understanding the origin of specific secondary structures, but specification of a complete tertiary structure from a knowledge of the primary structure is not yet in sight.

(The problem may not be quite as difficult as it once seemed, since we now know that similar structures are sometimes built from quite different amino acid sequences. There may be favored patterns that occur frequently, like those seen in our examples, but it is too early to be sure.)

A good test of what we know was recently made in this way: The three-dimensional structure of a protein, the enzyme AMP kinase, was calculated from the X-ray diffraction pattern. (The enzyme is of great importance in energy metabolism, and we shall discuss its action later.) Before the structure was announced, several groups of investigators who had devised methods for predicting secondary structures from the amino acid sequence were asked to apply their method to AMP kinase, using only the primary structure of the 194 residues in the polypeptide. Some of the methods relied mainly on empirical observations of the occurrence of particular residues or combination of residues in known structures in other proteins, while others used combinations of empiricism with quantitative estimations of bond energies. Three laboratories hazarded estimates of the position of reverse turns, four tried to locate pleated sheets, and eight were willing to assign residues to α-helices.

When the predictions were compared with the actual structure, it was found that out of 175 residues located in secondary structures, the participating laboratories predicted on the average 105 correctly, 55 incorrectly, and failed to include 70. Most did very well in locating the general position of the structures but had problems in defining exactly where helices and pleated sheets began and ended in the chain. In short, the general principles appear sound, but the application needs to be perfected. Some simple examples will show how some combinations of amino acid residues determine structure — limiting ourselves to purely local forces without any attempt to assess longer range interaction.

Generation of Reverse Turns. There is only one apparently infallible general rule for the prediction of structure. A residue of proline cannot occur in the middle of an α-helix:

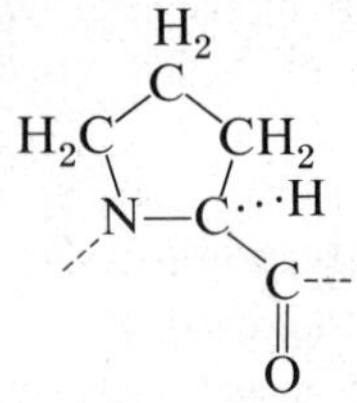

```
        H2
        C
   H2C     CH2
     N—C···H
          C---
          ‖
          O
```

prolyl group

The nitrogen atom of a prolyl group has no attached hydrogen atom with which to bond a preceding helical turn, and the fixed orientation of the bulky side chain is such that it gets in the way. The result is that the prolyl group can be in the first turn of an α-helix, but the preceding amino acid residues must occupy some other configuration. The prolyl group is therefore a "helix-breaker" that increases the probability of a reverse turn.

Reverse turns are also made more likely by the presence of residues with small side chains. Glycine, for example, makes very tight folding possible because it has no interfering side chain, and alanine only has a methyl group:

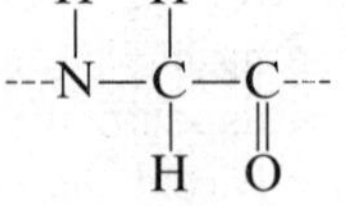

```
   H  H              H  CH3
   |  |              |  |
---N—C—C---      ---N—C—C---
      |  ‖              |  ‖
      H  O              H  O
```

glycyl group 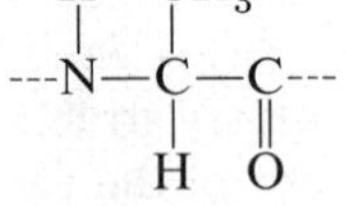   alanyl group

In addition, the side chains of serine, threonine, aspartate, and asparagine have hydrogen-bonding groups only a short distance removed from the peptide backbone, and these can form additional bonds that stabilize reverse turns:

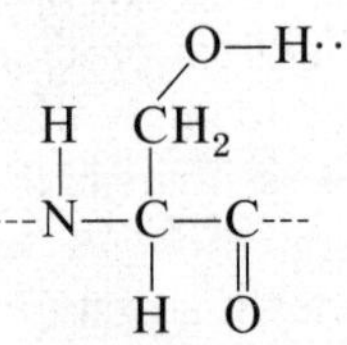

```
                                                                   H
                                          ···O   O···          ···O  N
                                              C                   C    H···
    O—H···         ···H—O   CH3               |                   |
 H  CH2              H  CH                H  CH2               H  CH2
 |  |                |  |                 |  |                 |  |
---N—C—C---       ---N—C—C---          ---N—C—C---          ---N—C—C---
     |  ‖                |  ‖                 |  ‖                 |  ‖
     H  O                H  O                 H  O                 H  O
```

seryl group 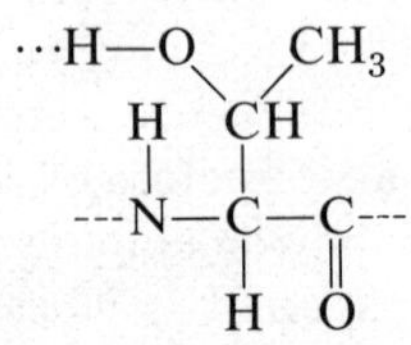   threonyl group 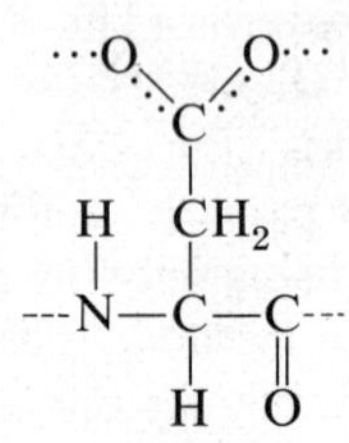   aspartyl group 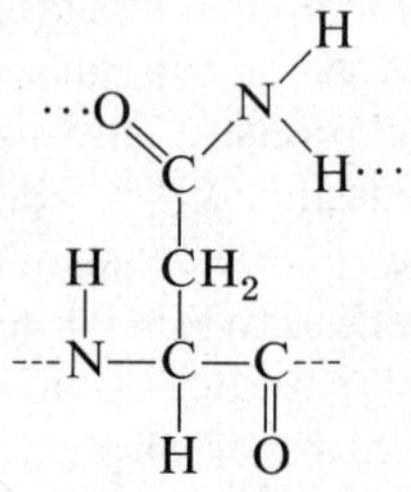   asparaginyl group

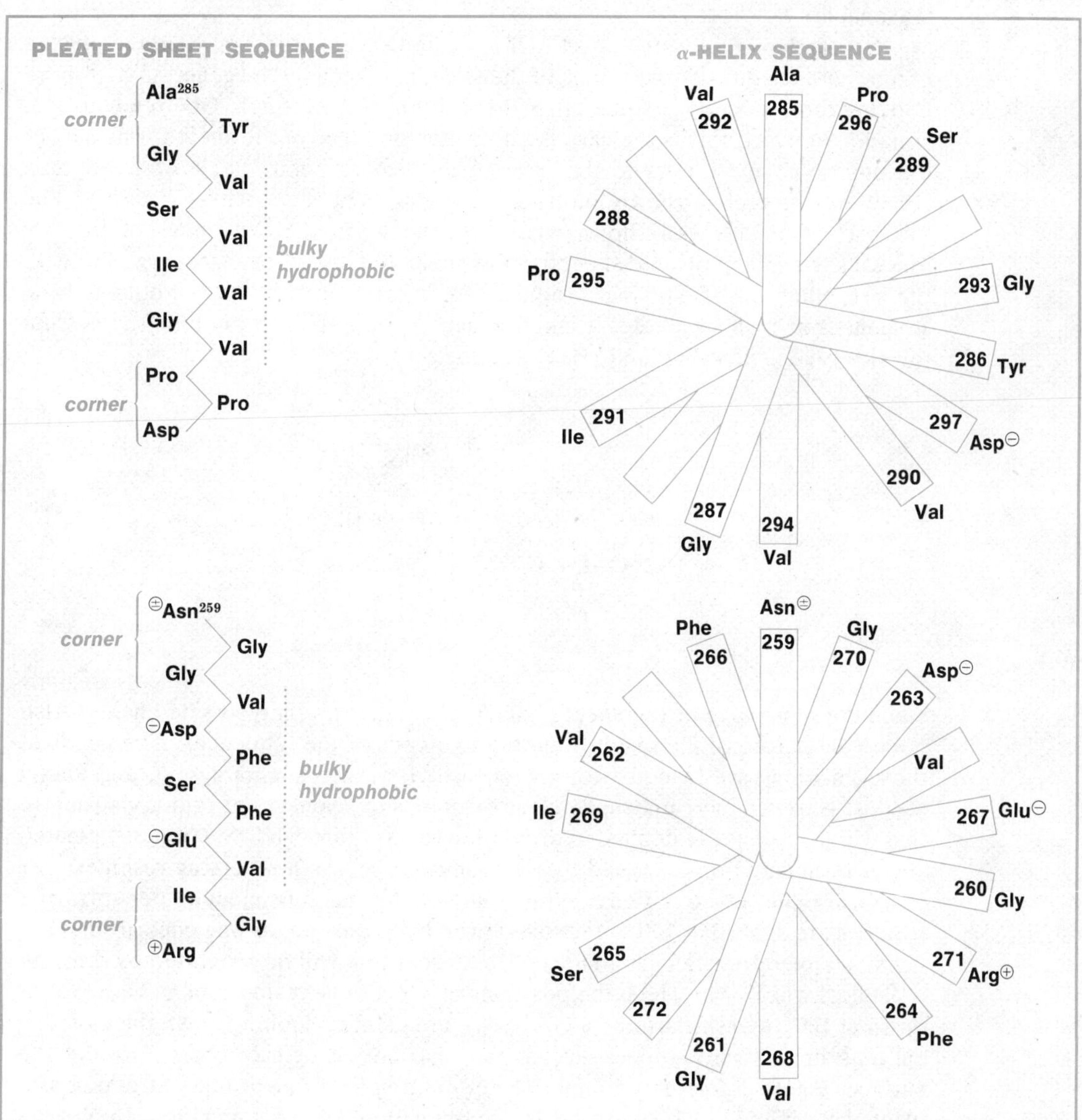

FIGURE 3–6 The forces creating segments of secondary structure can sometimes be visualized by plotting the amino acid sequence in the order it would assume in the structure. *Left.* Two segments of peptide chain from liver alcohol dehydrogenase are plotted as segments of pleated sheet. Such an arrangement creates a massive array of non-polar side chains on one side of the sheet *(to the right as shown),* which would tend to stay together in contact with another non-polar surface. The ends of the segments contain prolyl and glycyl groups that facilitate the formation of bends. *Right.* The same segments are plotted as if they were in α-helices. Successive tabs are spaced 100° apart, and numbered according to the position of the corresponding amino acid residue in the polypeptide chains. The distribution of side chains around the surface of the helix can then be seen as if one were looking down its axis from the N-terminal end. Compact clusters of hydrophobic side chains may indicate an interior contact that stabilizes the structure. No such clusters are evident here.

The presence of any one of these groups may make such a turn possible, but it does not guarantee it; however, when one sees three or so of these small residues close together in a primary structure, prediction of a sharp bend at that point is reasonable.

Generation of a Pleated Sheet. If a pleated sheet is to form, then the placing of alternate residues on two sides of the sheet must enable the generation of more bonding force than does some other disposition of the groups. One recognizable situation in which this is the case occurs when one face of the sheet forms part of the non-polar interior while the other face is more polar. Such situations can sometimes be recognized by plotting the amino acid sequence as a zigzag. The left part of Figure 3–6 shows what happens when the sequences of the two longest segments of pleated sheet from liver alcohol dehydrogenase, together with the preceding and following residues, are plotted in this way. Folding these sequences as a sheet creates a massive aggregation of hydrophobic groups from the side chains of valine and phenylalanine:

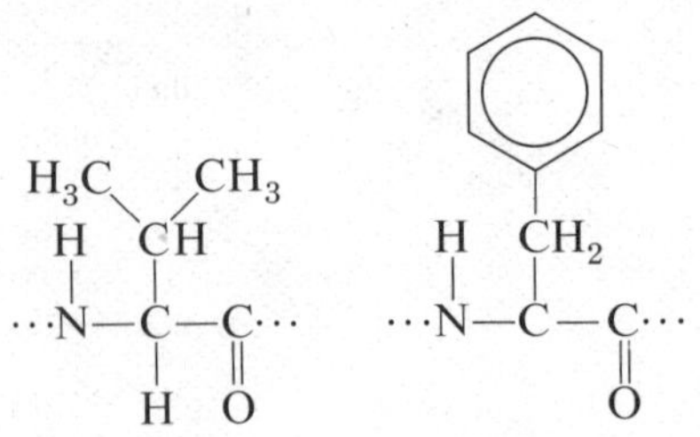

$$\cdots NH{-}CH[CH(CH_3)_2]{-}C({=}O)\cdots$$

valyl group

$$\cdots NH{-}CH(CH_2C_6H_5){-}C({=}O)\cdots$$

phenylalanyl group

while the other face of the sheet is made of smaller or charged side chains. Also notice the nature of the residues on the two ends of the sequences. The bends at those positions are loaded with glycyl, prolyl, aspartyl, asparaginyl, and alanyl groups. Here we have unusually clear examples of segments of primary structure that will make sharply defined hydrophobic regions through formation of a pleated sheet, terminated on each end by an abundance of chain-reversing residues.

Generation of an α-Helix. Since an α-helix has 3.6 residues per turn, the residues are spaced at 100° intervals viewed from the end, as one goes around the helix. We therefore can get an idea of the forces that will be generated by drawing a ''helical wheel,'' in which the positions of successive residues in a chain are set down at 100° intervals around a circle. We are in effect looking down the cylindrical axis of the helix and observing the distribution of side chains around the surface. Figure 3–6 gives a comparison of plotting the same primary structure as a pleated sheet and as a helical wheel. The thing to look for is some indication of the development of specific bonding forces along particular sides of the helix. Could, for example, one side of the helix be facing the hydrophobic interior and another side be facing the aqueous phase? If so, a helix becomes more likely.

When we look at the segments in those terms, they appear rather nondescript. The segments from 12 o'clock to 4 o'clock are relatively polar; the remaining portion is non-polar, but the groups are too widely scattered to be definitive. The first segment couldn't be a continuous α-helix in any case because of the prolyl residues that end it. The second segment possibly might be, but the apparent bonding forces are nothing like those seen with a pleated sheet configuration.

Now let us look at a sequence from myoglobin (Fig. 3–7). Here the pleated sheet configuration results in a rather random distribution of groups, while the wheel shows that an α-helix will create a well-defined region of bulky hydrophobic

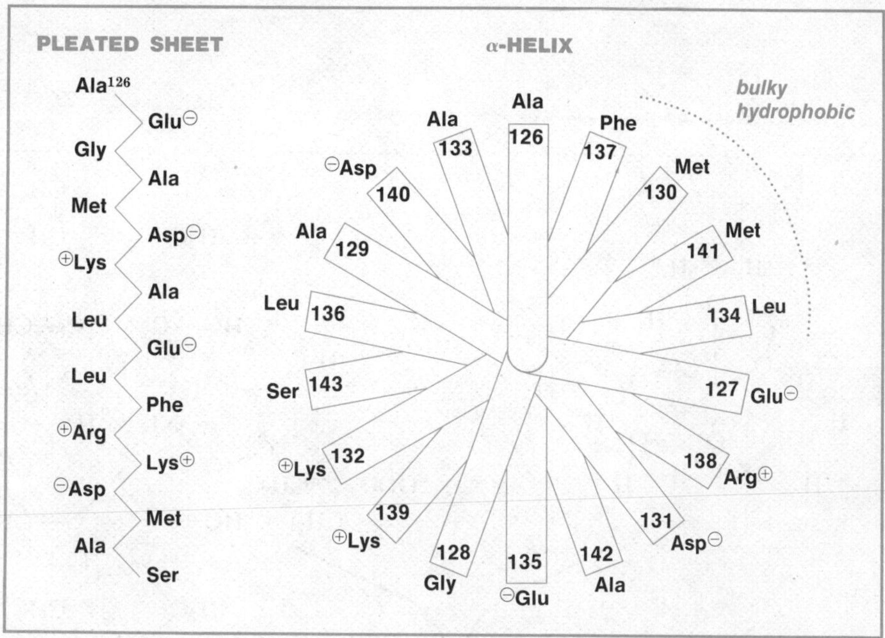

FIGURE 3–7 **Plots of an amino acid sequence from human myoglobin as a pleated sheet and an α-helix. There is no indication of separate hydrophobic and polar areas if folded as a pleated sheet, whereas folding as a helix produces a compact area of hydrophobic surface.**

groups. In fact, this segment of the chain is a portion of an α-helix, with the hydrophobic segment in contact with other parts of the chain in the interior, and the remainder of the surface being exposed to surrounding water. It is the potential for creating this distribution of groups that makes myoglobin fold in the way it does.

The Heme Pocket

There is another important aspect to the structure of myoglobin. It is a heme protein, meaning that it belongs to a class of conjugated proteins that contain heme groups in addition to polypeptide chains, and the primary structure must be arranged so as to hold the heme in position.

Heme is a combination of iron (II) and protoporphyrin IX (Fig. 3–8). We shall say a great deal more about the porphyrins later. The point of importance now is that the porphyrins contain a highly resonant planar ring system in which four nitrogen atoms are fixed in the center at a spacing that is ideal for bonding of their unshared electrons with a metal ion. The porphyrins, therefore, act as tetradentate (four-toothed) ligands, and protoporphyrin IX has an especial affinity for iron.

Iron(II) ions have a coordination number of six, which means that they can associate with six electron pairs per atom. When a ferrous ion forms a chelate with a porphyrin, two of its coordination positions are still unfilled, so free heme in aqueous solution will be hydrated as shown in the figure. One of the functions of the polypeptide portion of myoglobin (the "globin") is to provide a histidyl residue at a position where one of its nitrogen atoms will link to the fifth coordina-

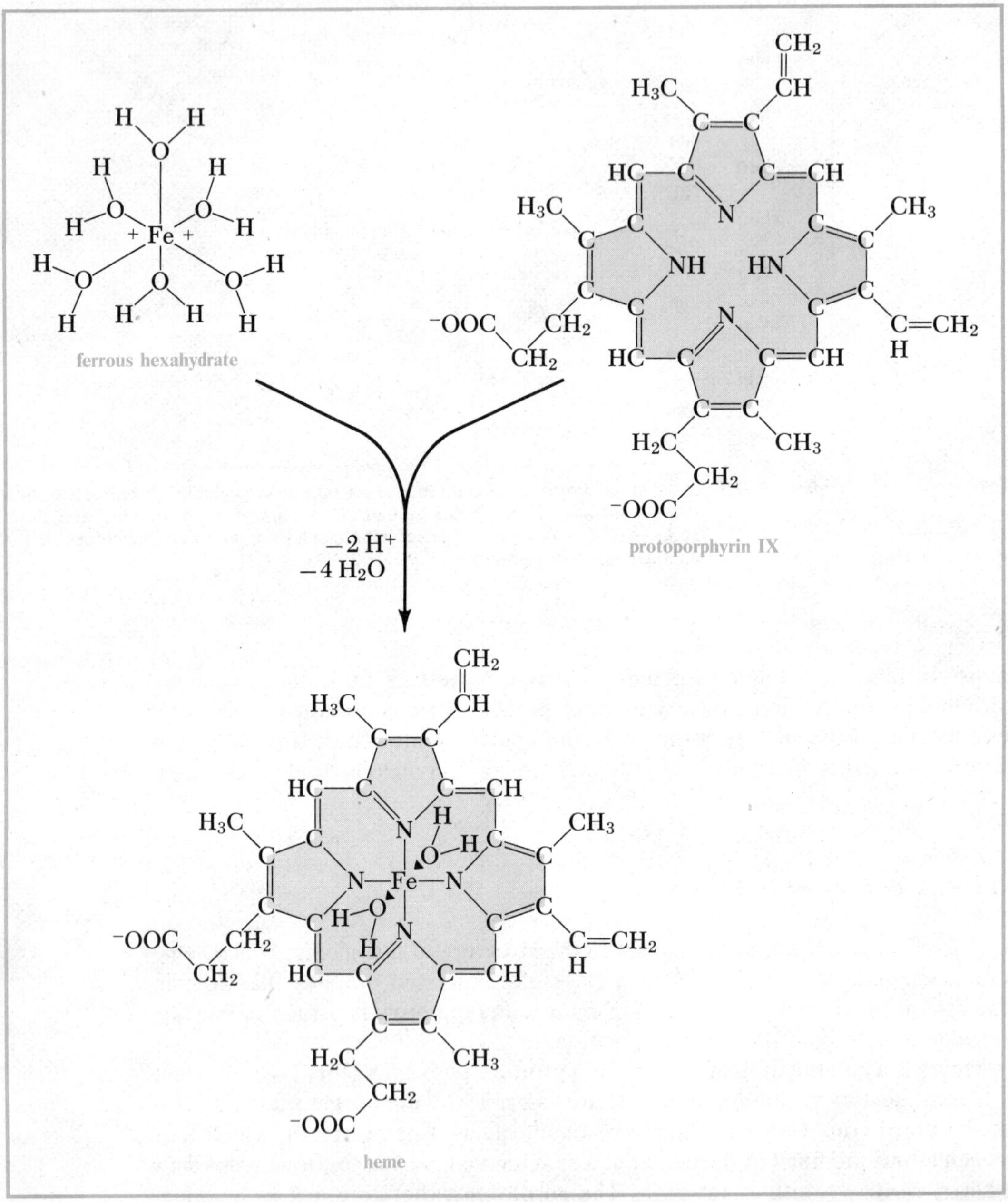

FIGURE 3–8 **The formal stoichiometry of the combination of iron (II) and protoporphyrin IX to form heme. The metallic ion has a coordination number of six, and any otherwise unfilled positions will be occupied by water or hydroxide ions.**

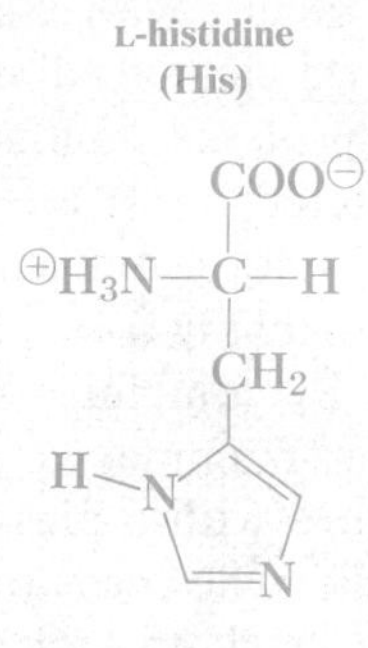

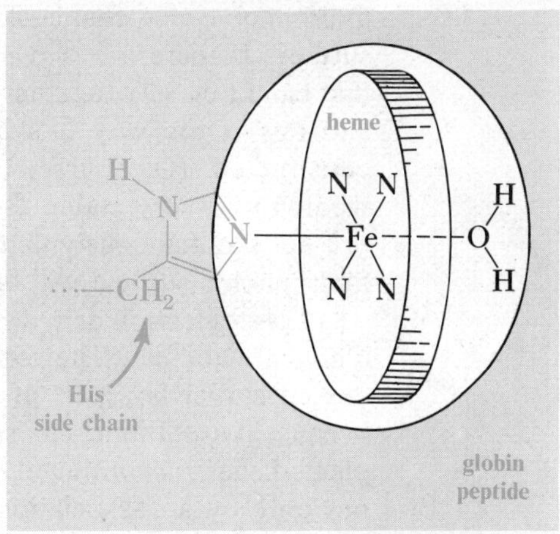

FIGURE 3–9 **Schematic illustration of heme in a pocket formed by the globin polypeptide. A histidyl side chain acts as an additional ligand for the iron atom in heme. The free amino acid, histidine, is shown at the left.**

tion position of the iron (Fig. 3–9). The sixth position of the iron is left open and now has the ability to bind oxygen.

The remainder of the peptide chain wraps around the porphyrin ring to hold it in position. The walls of the resultant heme pocket are created by hydrophobic side chains from several of the surrounding helical segments, closely fitting the hydrophobic porphyrin ring. In addition, the aromatic side chain of a phenylalanyl group lies parallel to the porphyrin for interaction of the π-electrons. The porphyrin has two charged side chains, but these protrude out of the heme pocket to contact the aqueous phase.

In sum, the polypeptide chain of myoglobin has an amino acid sequence that favors the formation of a particular set of α-helices that will snugly nestle around a heme group and hold it in four ways: chelation of the iron, hydrophobic exclusion of water, interaction of aromatic rings, and close packing of the atoms. Indeed, the polypeptide is built to fold into a particular tertiary structure in the presence of heme. (Otherwise, the hydrophobic heme pocket would be exposed to water.)

QUATERNARY STRUCTURE

Some protein molecules are complexes of more than one polypeptide chain; each chain with its tertiary structure then constitutes a subunit of a larger molecule. The geometry of the combination of subunits into the complete molecule is said to be its quaternary structure.

The forces available for the formation of quaternary structure are the same as those that create secondary and tertiary structure. One of the most important is the tendency for hydrophobic groups to combine so as to exclude water. Suppose that a polypeptide chain folds in such a way as to create a face with exposed hydrophobic side chains. That face would tend to stick to any other hydrophobic surface. If there are two polypeptide chains with areas of hydrophobic surface that can fit closely together, then the two chains will bond into a larger molecule, and this is the way in which many proteins containing multiple subunits are constructed. (Of course, tertiary structure with an exposed hydrophobic face wouldn't be very stable as an isolated entity, but when two or more chains are present simultaneously that can form matching faces, otherwise transient areas of hydrophobic surface will be stabilized by combining with each other.)

Liver alcohol dehydrogenase, for example, is a dimer that contains two identical subunits. The tertiary structure with a pleated sheet core that we discussed earlier is a part of each subunit. The formation of a dimer is facilitated because two subunits can stack back-to-back on each other in such a way that the pleated sheet is continued from one subunit to the other. The subunits are joined not only by an association of hydrophobic side chains, but also by hydrogen-bonding that now creates a continuous pleated sheet through both subunits. (The segment involved is the upper one in Figure 3–6.)

Hemoglobin

Perhaps the most thoroughly studied example of quaternary structure is that of hemoglobin. The molecules of hemoglobin that carry oxygen in human blood are tetramers made from a pair of each of two kinds of polypeptide chains (Fig. 3–10). These chains are designated α and β in the most abundant hemoglobin in adults, and this hemoglobin A_1 therefore has the chain formula $\alpha_2\beta_2$. Each of these chains has a tertiary structure much like that of myoglobin, which is made from only one polypeptide chain (Fig. 3–11). However, they differ from myoglobin in that they are built to associate in particular ways. That is, α chains tend to bind β chains, and vice versa. The result is a tetrahedral molecule with the four subunits in close contact, except for a hole in the center that is freely accessible to water.

The hemoglobin chains associate and the myoglobin chains do not because some of the amino acid residues on the surface are different. Of the residues in the hemoglobin that are in contact with each other, some 13 have hydrophobic side chains that are in positions comparable to those occupied by side chains bearing charges in myoglobin. Only three of the contacting side chains in hemoglobin have charges where hydrophobic side chains are present in myoglobin. In addition, some residues with polar side chains are introduced in hemoglobin at positions where they will form hydrogen bonds with the neighboring subunits.

The surface faces of the tetramer are designed, on the other hand, to minimize any tendency to associate. Red blood cells are stuffed with hemoglobin almost to the limit of its solubility, with adjacent molecules not much more distant that they are in a crystal. Even a moderate tendency to aggregate would not permit such a high concentration, and therefore would diminish the capacity to carry oxygen.

What is the point of building hemoglobin, or any other protein, out of multiple polypeptide chains when many respectable proteins have only one? Sometimes proteins are built this way simply to become large enough so that they won't leak through membranes, but we shall see when we turn to the biological functions of proteins that those composed of more than one subunit can respond to changes in environmental concentrations in a more complex way than can those made from only one polypeptide chain. But before we examine these functions we should

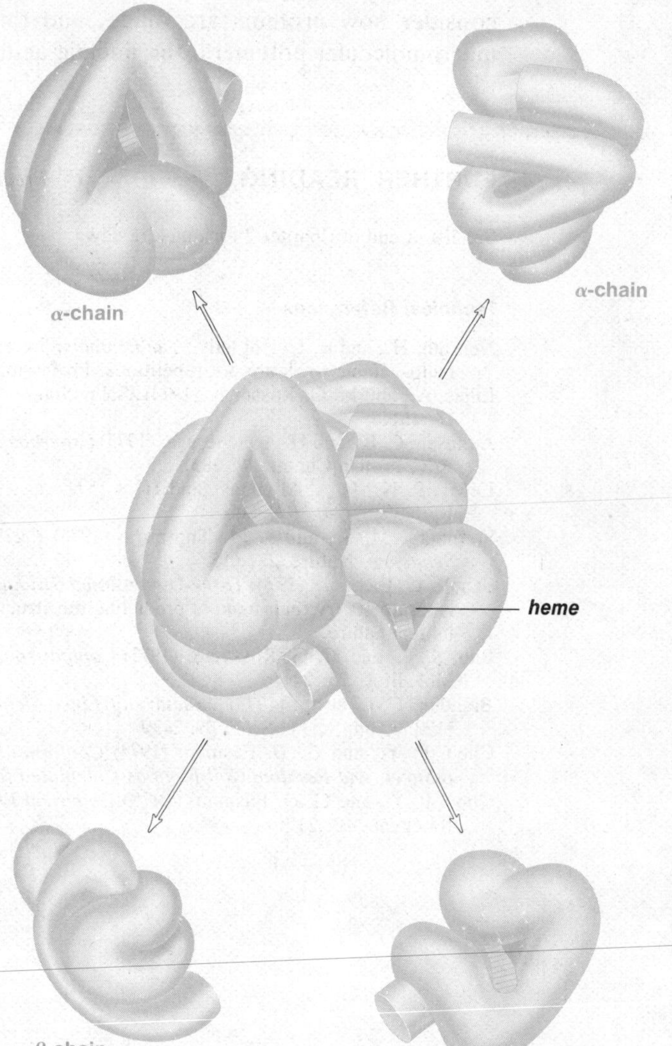

FIGURE 3–10

A schematic drawing of the most abundant adult hemoglobin. The molecule contains pairs of each of two kinds of peptide chains, termed α and β, which are also shown in exploded view at the approximate orientation they have in the complete molecule. Two of the four heme units are visible in pockets formed by folds of the peptide chains.

The drawing is of an idealized molecule – a sort of emaciated hemoglobin in which some of the external side chains have been plucked off to expose the underlying skeleton of the chains. In fact, it is difficult to distinguish chains at the surface of a complete model because the space between them is filled with protruding atoms.

The drawing is based on reports by Cullis, Muirhead, Perutz, and Rossman, Proc. Roy. Soc. (London), ser. A, *265*: 161 (1962); and by Perutz, J., Molec. Biol., *13*: 646 (1965); but the angle at which the molecule is viewed differs from those presented in the references.

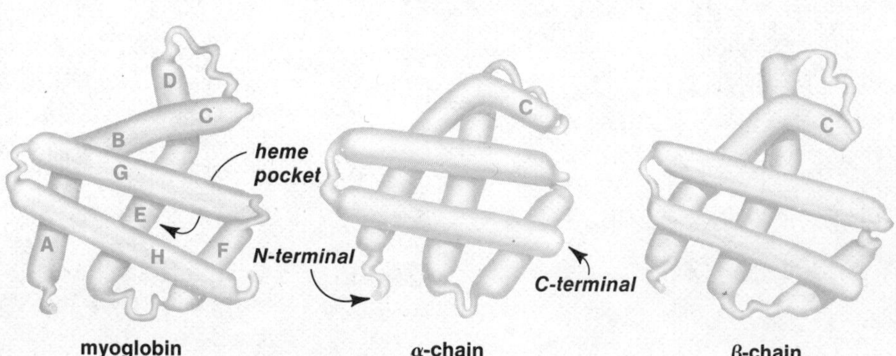

FIGURE 3–11 **The arrangement of the peptide backbone is quite similar in myoglobin and in the α and β subunits of hemoglobins. The α-helices are indicated as cylinders. These highly schematic renditions view the molecule from the opposite side of that given for myoglobin in Figure 3–5.**

consider how proteins are made, and this process involves another class of macromolecular polymers, the nucleic acids, discussed in the next chapters.

FURTHER READING

See list at end of Chapter 2 for quick reviews.

Technical References

Neurath, H., and R. L. Hill, eds.: *The Proteins,* 3rd ed. Academic Press. The various editions of this multi-volume work are not repetitious. Professional-level articles.

Liljas, A., and M. G. Rossman: (1974) *X-Ray Studies of Protein Interactions*. Annu. Rev. Biochem., *43*: 461.

Anfinsen, C. B., and H. A. Scheraga: (1975) *Experimental and Theoretical Aspects of Protein Folding*. Adv. Protein Chem., *29*: 205.

Lewis, P. N., F. A. Momany, and H. A. Scheraga: (1973) *Chain Reversals in Proteins*. Biochem. Biophys. Acta, *303*: 211.

Sternberg, M. S., and J. M. Thornton: (1978) *Prediction of Protein Structure from Amino Acid Sequence*. Nature, *271*: 15.

Schulz, G. E., et al.: (1974) *Three-Dimensional Structure of Adenyl Kinase*. Nature, *250*: 120. Various laboratories try their hand at predicting the structure of an enzyme in advance of information on its real nature.

Rao, S. T., and M. G. Rossman: (1973) *Comparison of Super-Secondary Structures in Proteins*. J. Mol. Biol., *76*: 241.

Brändén, C.-I., et al.: (1973) *Structure of Liver Alcohol Dehydrogenase at 2.9-Å Resolution*. Proc. Natl. Acad. Sci. U.S.A., *76*: 2439.

Chou, P. Y., and G. D. Fasman: (1974) *Conformational Parameters for Amino Acids in Helical, β-Sheet, and Random Coil Regions Calculated from Proteins*. Biochemistry, *13*: 211.

Chou, P. Y., and G. D. Fasman: (1978) *Empirical Prediction of Protein Conformation.* Annu. Rev. Biochem., *47*: 217.

4 DNA AND GENETIC INFORMATION

The potential of a fertilized ovum to grow into a human lies in the information carried within it; this single cell contains complete specifications for its own transformation into a human body with 10^{13} cells of many different kinds molded into specific tissues.* The information specifies not only the form, but also the patterns of behavior of tissues; cells in the adult respond to chemical and physical changes in their surroundings as the result of instructions in the primal cell. Only a mere six picograms — 6×10^{-12} g — of deoxyribonucleic acid (DNA) need be present to contain this information, both for structure and for function.

Nothing else in human experience is remotely comparable to the compact packaging of large quantities of information in DNA. Our most sophisticated integrated circuits have tens of thousands of electrical elements on silicon chips as small as a centimeter square and 0.2 mm thick, but the average volume occupied by two of these elements, microscopic though it is, is sufficient to contain the different molecules of DNA necessary to direct the growth of replacements for all of the medical students, house staff and medical faculty members presently in the United States — some 160,000 of them.

DNA Is a Polynucleotide. Nucleic acids in general are polymers of nucleotides, and nucleotides are made by combining nitrogenous heterocyclic rings, sugars, and phosphate groups (ionized derivatives of orthophosphoric acid, H_3PO_4). The heterocyclic constituents are referred to as the "**bases**" of nucleotides.

Nucleic acids are arranged in this way:

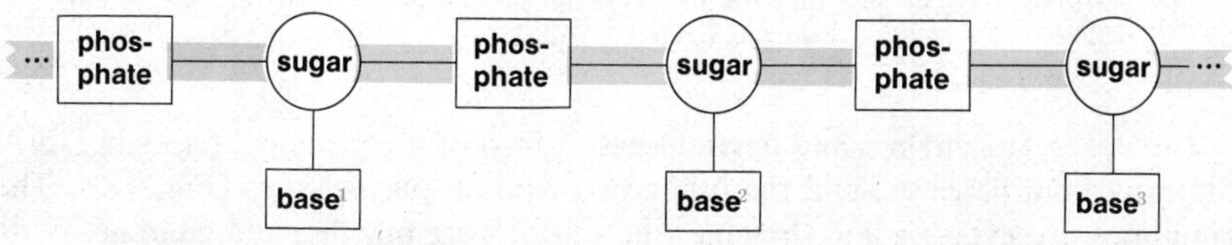

Each sugar residue in a repeating backbone of sugar-phosphate units has a heterocyclic base attached as a side chain.

*This approximation neglects the red blood cells. An adult human has roughly 2×10^{13} of these small modified cells, which have no nuclei or mitochondria, and therefore no DNA.

Two structural classes of nucleic acids occur in cells, differing in the nature of the sugar they contain. Some contain ribose, a five-carbon sugar*, and are known as ribonucleic acids (RNA), while others contain the 2-deoxy derivative of ribose and are known as deoxyribonucleic acids (DNA):

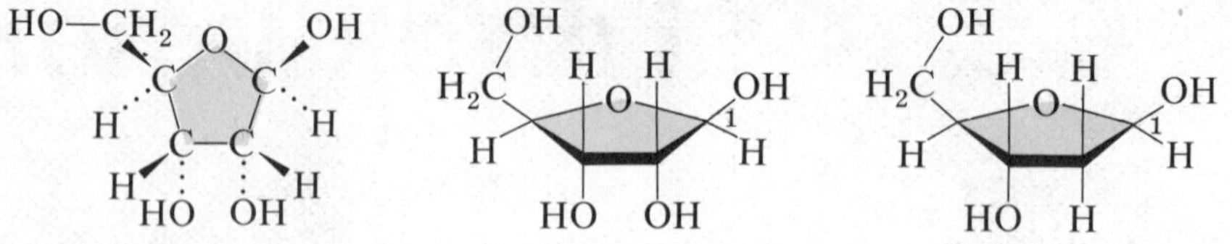

D-ribose
ring viewed face-on

D-ribose
ring viewed edge-on

2-deoxy-D-ribose
ring viewed edge-on

DNA Occurs as a Double Helix. The DNA of cells is made of two different, but related, polynucleotide chains, twisted together into a double helix (Fig. 4–1). In this stable structure, the heterocyclic bases from each chain are stacked in pairs, like plates of atoms, with the stack held together by two ropes of sugar-phosphate backbone running along the outside of the stack.

The information in DNA consists of the sequence in which different base-pairs occur in the center of the stack, and the length of a molecule in different organisms is roughly proportional to the amount of information it contains, although there are great variations. Nearly all of the DNA in a fertilized human ovum is in 46 large molecules, one for each chromosome, in the nucleus. (Much smaller molecules occur in the mitochondria and the centrioles.) If these large DNA molecules from one cell were stretched end to end, they would match the height of a tall man — nearly two meters, but they are folded into the small volume necessary to fit within the cell nucleus. They are only a thin two nanometers in diameter.

Since each base pair constitutes a plate only one atom thick, roughly 0.34 nanometers, the total length of two meters in a human cell must be created by stacking some 6×10^9 pairs (1 nanometer = 10^{-9} meter).

Base Sequence Determines Amino Acid Sequence. In polypeptides, it is the variation in sequence of amino acid side chains along the repeating polypeptide backbone that conveys individuality. Similarly, it is a variation in the sequence of four different bases along the repeating polynucleotide backbone that constitutes the information in DNA. Here we have the fundamental postulate of molecular biology:

> **Sequences of bases in DNA eventually determine the sequences of amino acids incorporated into polypeptide. Since the form and function of proteins depend upon the amino acid sequences of their constituent polypeptide chains and these functions include the construction and regulation of all other cellular constituents, DNA in this way specifies the character of the organism.**

The Bases Are Purines and Pyrimidines. Two of the principal bases in DNA contain pyrimidine rings, and the other two contain purine rings (Fig. 4–2). The pyrimidines are **cytosine** and **thymine**; the purines are **adenine** and **guanine.** Note them well because the shuffling of the sequences of these bases in the parent cells is the source of all of the differences between the smallest bacterium and the most sapient human. Other bases — derivatives of the major compounds — are also

*The structure of sugars is discussed in Chapter 10.

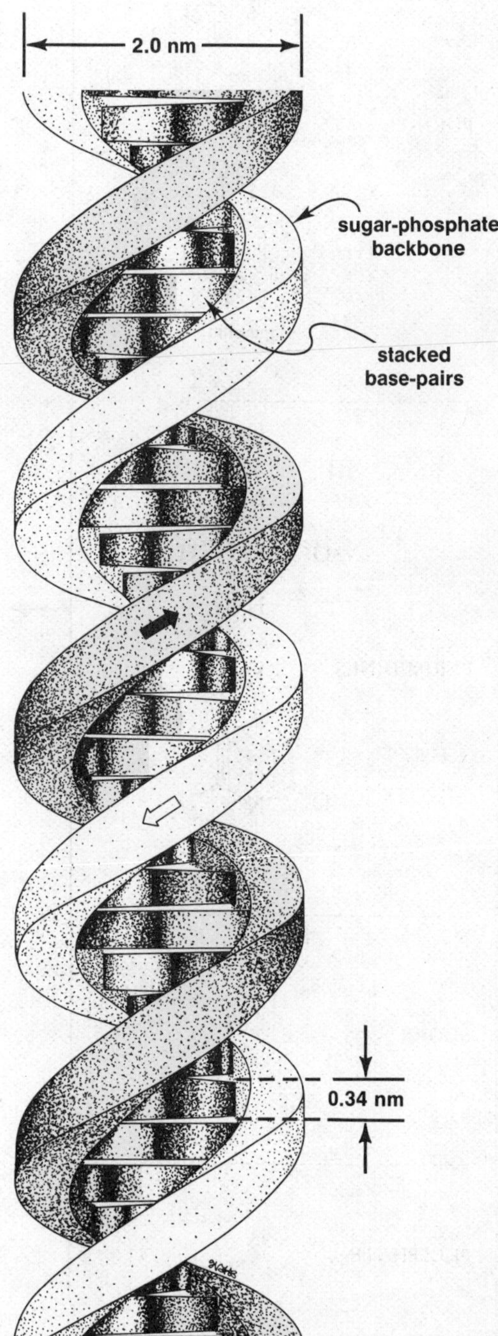

FIGURE 4–1

DNA is composed of two polynucleotide strands running in opposite directions that twist into a double helix so as to bring the bases into contact in the center. The sugar phosphate ester backbones of the strands are exposed on the outside of the helix. Ten pairs of nucleotide residues, with their associated paired bases in the center, form one complete turn of the helix.

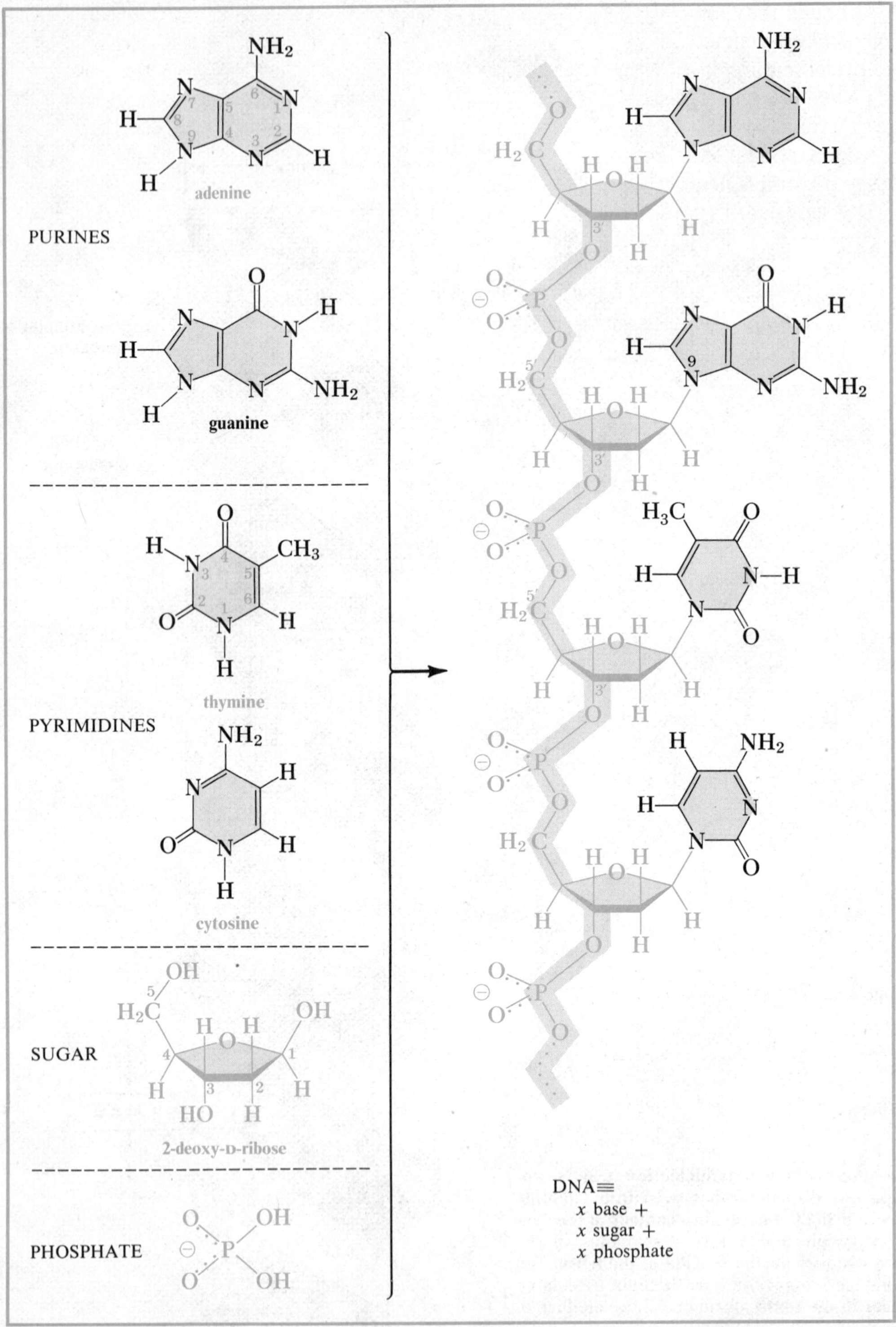

FIGURE 4–2 The formal stoichiometry for the combination of inorganic phosphate (P_i), 2-deoxy-D-ribose, and purines or pyrimidines into deoxyribonucleic acid (DNA). The conventional abbreviation for the polynucleotide fragment shown is d(A-G-T-C-). Adenine and guanine are substituted purines; thymine and cytosine are substituted pyrimidines.

The numbering of atoms in ribose, purines, and pyrimidines is shown. In the nucleotides, the atoms of the sugar are designated 1′, 2′, etc., to distinguish these positions from the numbered positions on the purine or pyrimidine rings.

present. Although they are a very small fraction of the total, they may represent important regulatory signals.

The critical feature of these heterocyclic compounds is that they form hydrogen bonds when fitted in a particular way (Fig. 4–3). In this configuration, adenine pairs with thymine by two hydrogen bonds, and guanine pairs with cytosine by three hydrogen bonds. Adenine and thymine, guanine and cytosine are said, therefore, to be **complementary bases**; the stacked bases in the center of the double helix in DNA consist of these pairs:

	adenine = thymine	
one	thymine = adenine	*other*
polynucleotide	guanine ≡ cytosine	*polynucleotide*
chain	cytosine ≡ guanine	*chain*

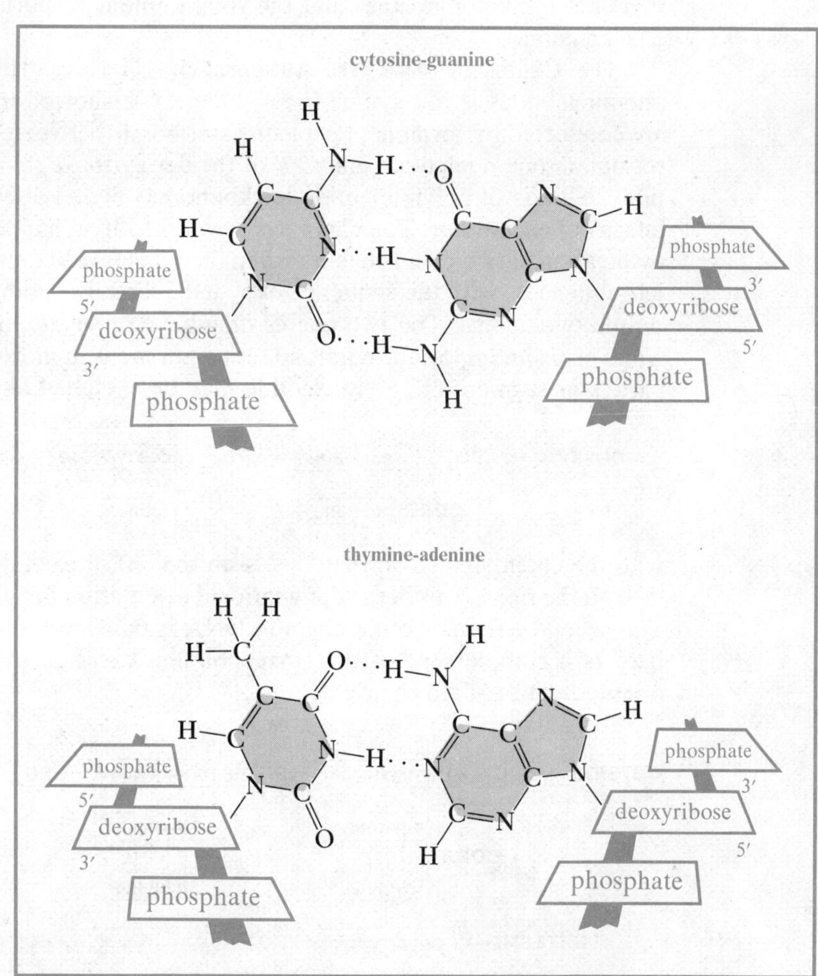

FIGURE 4–3 **Hydrogen bonding of complementary bases between adjacent nucleic acid strands. The symbols are spaced to scale the actual positions of the atoms according to Pauling and Corey, (1956) Arch. Biochem. Biophys., *65*: 164.**

DNA chains do not associate to form a double helix unless the base sequence in one chain is the complement of the sequence in the other chain, and this association of complements is the basis for the transfer of information.

DNA Contains Equal Amounts of Purines and Pyrimidines. Since double-stranded DNA contains a matching complement for each constituent base, it follows that the content of adenine equals the content of thymine, and the content of guanine equals the content of cytosine, with the total number of purine residues equal to the total number of pyrimidine residues. This is true for the total DNA in a cell, and for any segment of the double helix within it; it is an identifying characteristic of complementary nucleic acid chains. Suppose that a segment of 10,000 bases (5,000 base pairs) contains 3,000 adenine residues. It must also contain 3,000 thymine residues as complements. The remaining residues will be:

10,000 bases − 3,000 adenine − 3,000 thymine = 4,000 (guanine + cytosine)

Since the number of guanine residues equals the number of cytosine residues, there must be 2,000 of each. The total content of pyrimidines therefore is 3,000 thymine + 2,000 cytosine, and the total content of purines is 3,000 adenine + 2,000 guanine.

The Chains in DNA Are Antiparallel. The sugar-phosphate backbone of polynucleotides is not symmetrical; Figure 4–2 showed that the nucleotide units are connected by forming phosphate ester bonds between C-5′ of the deoxyribose residue in one nucleotide and C-3′ of the deoxyribose residue in the next nucleotide.* That is, a polynucleotide backbone has 5′ and 3′ ends that are not mirror images of each other. The chain has a polarity. It so happens that the most stable association of two chains occurs when their bases are complementary in antiparallel sequences, with the sequence of 5′ and 3′ linkages going in opposite directions on the two chains. The two chains then have opposite polarity.

This is an important point, so let us pursue it in more detail. The layout of a particular segment of polynucleotide may be sketched like this:

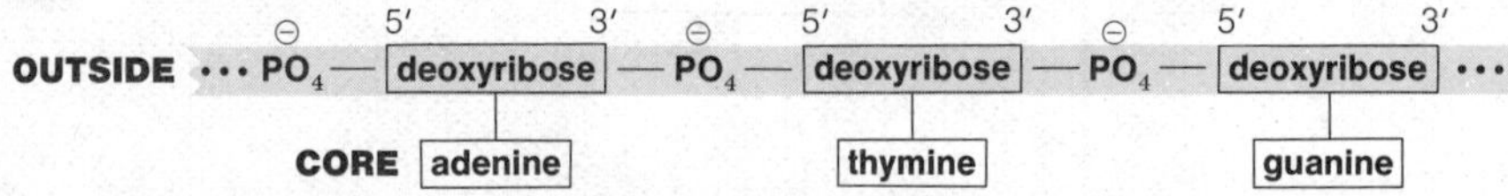

with the chain viewed so that C-5′ is on the left of each deoxyribosyl group, and C-3′ on the right. This is the conventional orientation for viewing polynucleotides. The second polynucleotide chain in DNA is constructed so that each successive base is a complement of the corresponding bases in the first chain when it is oriented with C-5′ to the right:

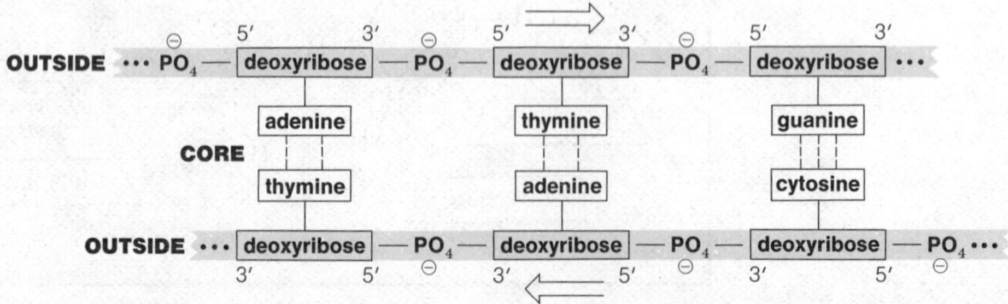

*The prime mark is used in numbering the sugar components of nucleotides to distinguish positions on them from positions having the same numbers on the base components.

These chains are complementary in antiparallel — the backbones run in opposite directions. The nature of DNA chains is such that only antiparallel chains can fold into a double helix with stable bonds between the complementary bases.

Nomenclature of Nucleotides

It is advisable to gain an easy fluency in the names and symbols of nucleotides before going further. Current nomenclature is based on the **nucleosides**; nucleosides are the combinations of heterocyclic bases and sugars. **Nucleotides** are nucleoside phosphates (base-sugar-phosphate). Each nucleoside constructed from ribose and the common bases of the nucleic acids has a trivial name and a corresponding letter that is used as a symbol:

Constituent Base	Ribonucleoside*	Symbol*	Corresponding Nucleotide*
adenine	adenosine	A	adenosine 5′-phosphate
guanine	guanosine	G	guanosine 5′-phosphate
cytosine	cytidine	C	cytidine 5′-phosphate
thymine	thymidine	T	thymidine 5′-phosphate

*Adding the prefix *deoxy-* to the names of the nucleosides or nucleotides designates the corresponding compounds containing deoxyribose. The symbol is then prefixed by d. For example, dA is 2′-deoxyadenosine.

One oddity of nomenclature persists. It was formerly believed that thymine only occurs as a base in DNA, and thymidine without a prefix was therefore used to designate the deoxynucleoside. We now know better, and it is advantageous to make the nomenclature internally consistent by using thymidine for ribosylthymine and deoxythymidine for the constituent of DNA, as is done in this book. However, many persist in using the older name while using dT as a symbol for the deoxynucleotide. Logic may ultimately prevail.

Another oddity of nomenclature must also be faced. The nucleotides have other trivial names, formerly in more common use, and the group names for residues in polynucleotides are derived from these older names:

Present Name	Older Name	Present Group Name
adenosine 5′-phosphate*	adenylic acid*†	adenylyl*
guanosine 5′-phosphate	guanylic acid	guanylyl
cytidine 5′-phosphate	cytidylic acid	cytidylyl
uridine 5′-phosphate	uridylic acid	uridylyl

*The prefix deoxy- is added when appropriate.

†Cognoscenti of structure modify these names so as to indicate the anionic form that actually exists under physiological conditions (adenylate, guanylate, etc.).

Unless otherwise specified, trivial names are understood to designate the commonly occurring nucleoside 5′-phosphates, although 2′ and 3′-phosphates are also known.

These group names can be used to construct names for polynucleotides, but the result is cumbersome and nearly everyone resorts to symbols when feasible.

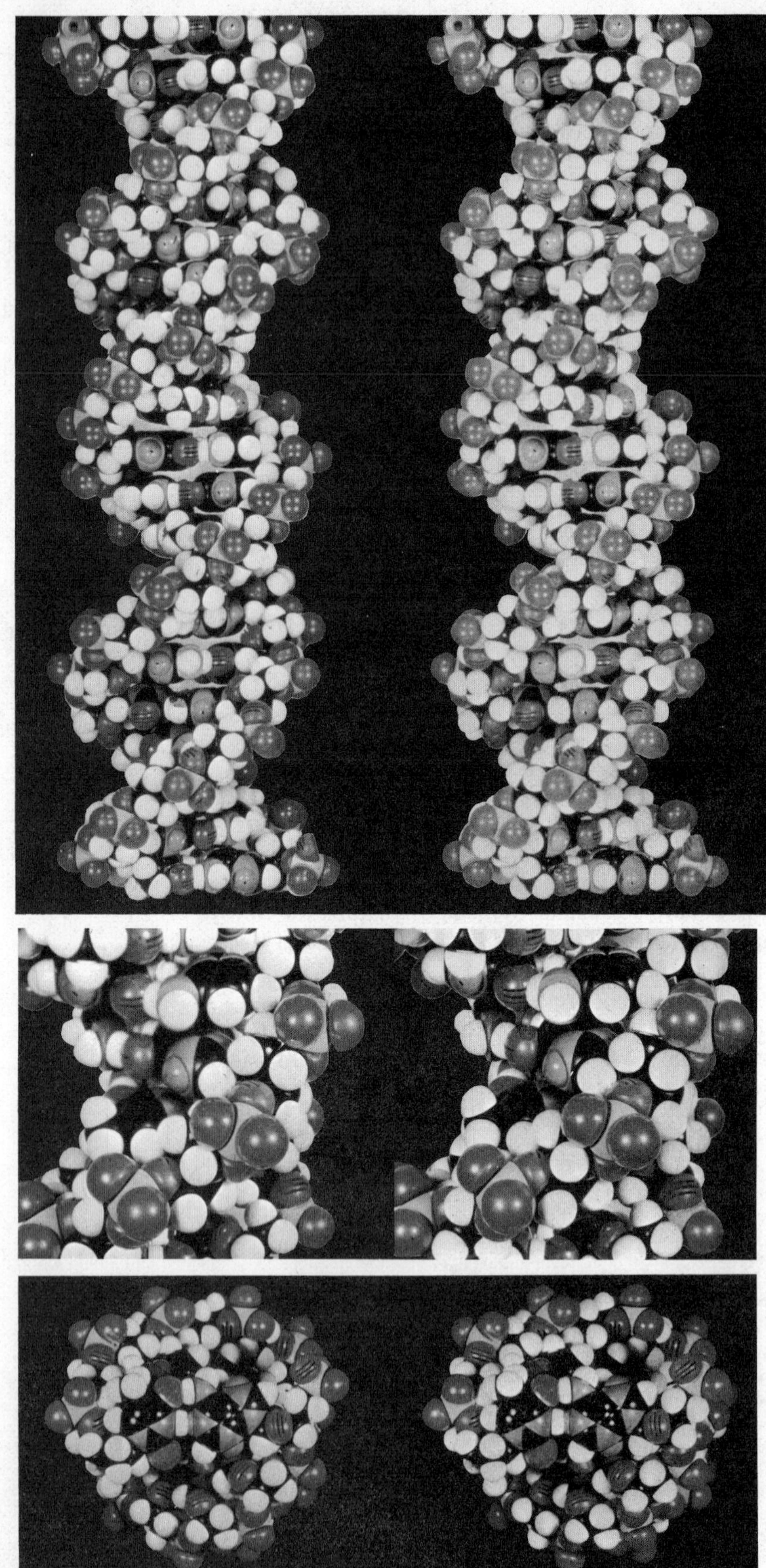

FIGURE 4–4 *See legend on facing page.*

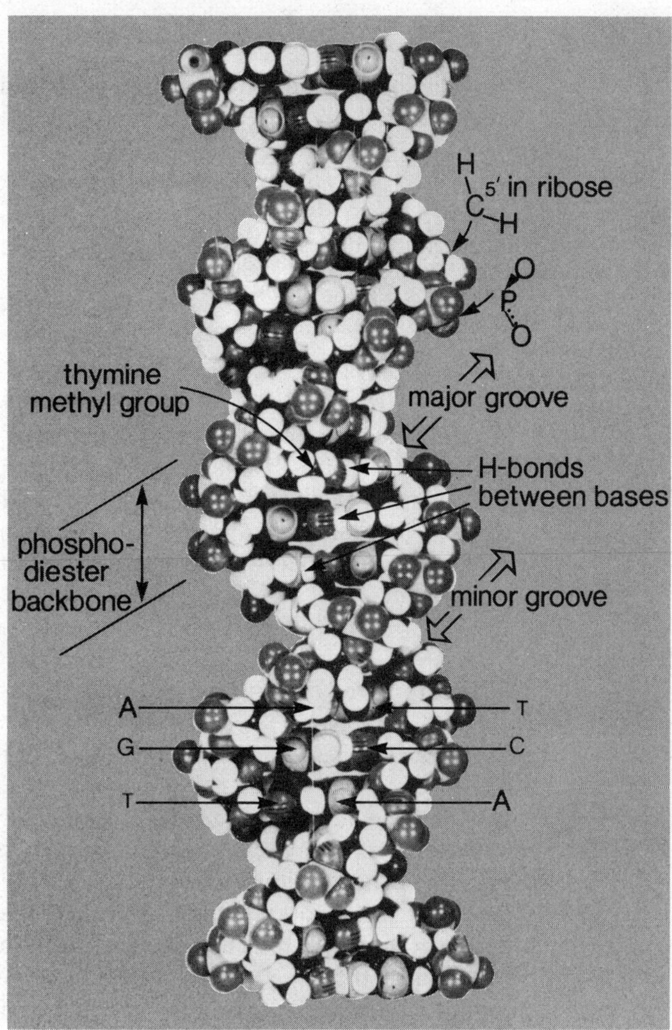

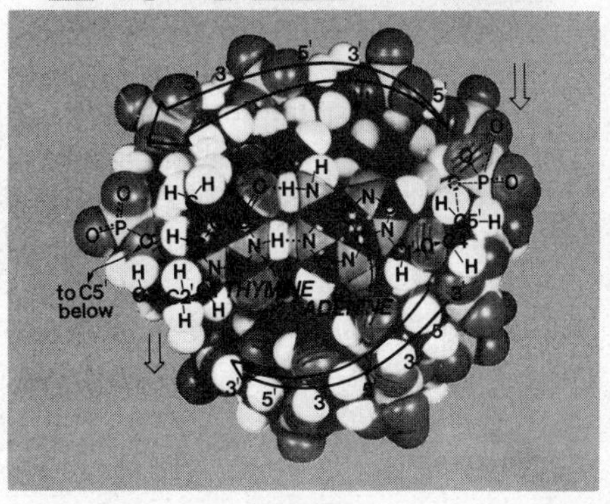

FIGURE 4–4

Stereo photographs of a space filling model of DNA. Some significant features are indicated in the keys (*right*). *A*. Two full turns of double helix. *B*. A closer view into the major groove showing details of the sugar-phosphate backbone and the stacked bases. *C*. End view, showing base-pairing. Author's model based on S. Arnott, (1970) Progr. Biophys. Mol. Biol., *21*:267.

***A note on viewing:* Stereo illustrations are widely used in modern biochemistry. It is helpful to learn to see them without optical aids—the illusion of depth is greater. One ought not expect to do this automatically; persistence is required, but once the skill is developed, it is easily applied. Use of cardboard tubes (~25 mm I.D. × 100 mm I.D.) helps some to begin. I find it helpful to bring the pair very close to the eyes, staring into the distance through the blurred images until they fuse, then withdrawing them slowly until focus is reached.**

For example, the segment we used as an example above is read from left to right as given (C-5′ on the left in each residue):

· · · deoxyadenylyldeoxythymidylydeoxyguanylyl · · ·

but it can be designated by:

· · · -dA-dT-dG · · ·

or

· · · d(-A-T-G) · · ·

in which the intermediate hyphens represent the phosphate groups connecting the deoxynucleosides. (It is sometimes advantageous to indicate if the ends of the chains are phosphorylated or have free hydroxyl groups by adding p or –OH. pdA · · · indicates a polynucleotide beginning with phosphate on C-5′ of a deoxyadenosyl group. HO–dA would indicate that the 5′-hydroxyl group is not phosphorylated.)

Characteristics of the Double Helix

Polar Exterior and Non-polar Core. Figure 4–4 shows different views of a model of DNA. All of the phosphate groups are on the outside. Diesters of phosphoric acid are very strong acids and DNA is fully ionized, so it has continuous skeins of negative charge running down each side that must be neutralized with associated cations. The bases, on the other hand, are compactly stacked in the middle so as to exclude water from contact with their hydrophobic faces. The hydrogen bonds connecting the base pairs are therefore in a non-polar environment where they can approach maximum strength. Isolated nucleosides will not form strong bonds with their complements in aqueous solution, because the faces of the rings are not shielded from disrupting contacts with water molecules; however, they do form tight pairs in non-aqueous solvents, and similar strong bonding is seen in the non-polar interior of the double helix, relative to the force developed between mismatched bases.

The stacking of the bases also develops additional forces through aromatic interactions and the close fit of adjacent rings that stabilize the particular helical configuration.

Wide and Narrow Grooves. A striking feature of the general outline of the double helix is the deep grooves formed between the phosphate-sugar backbones. One groove is substantially wider than the other in DNA, and they are frequently designated as the major and minor grooves. This is not to be taken as an indication of relative functional importance, and the minor groove is no small crack in the surface.

The grooves are the only "windows" through which the bases of a double helix can be "identified" by interaction with an external molecule. Each of the four bases has a particular orientation to the axis of the helix, so that the same atoms are always present in each groove. For example, the base of the wide groove is always paved with C-6, N-7, and C-8 of purine rings, and C-4, C-5, and C-6 of pyrimidine rings. The narrow groove is paved with C-2 and N-3 of the purine rings, and C-2 of the pyrimidine rings. It is at least possible that some proteins may be constructed to identify particular locations on the long DNA molecules through interaction with specific combinations of these exposed atoms even though the double helix remains intact, and in this way control the information to be transferred.

Physical Characteristics. The double strands of DNA may be separated (**denatured**) by increasing the temperature of a solution, and the temperature at which the transition is half-completed is the **melting point** of the DNA structure. The term also has some physiological connotations; for example, the transfer of information from DNA involves a separation of the two polynucleotide chains, and the factors necessary for this disruption of the double helix are said to lower the melting point below the temperature of the cell. Values of the melting point of DNA alone under typical experimental conditions range around 85°C; the actual value depends upon base composition since it requires more energy to break the three hydrogen bonds in G-C pairs than it does the two in A-T pairs.

Among the many properties of DNA, the one most frequently measured as an index of the loss or gain of double helical structure is the ultraviolet absorption spectrum. The nucleosides and nucleotides have characteristic ultraviolet spectra owing to the conjugated bonds in the purine and pyrimidine groups. The absorbancy is diminished by the hydrogen bonding occurring in the double helix, and this loss, or hypochromicity, can be used as a quantitative index of the fraction of the molecules existing in helical form.

If solutions of denatured, but complementary, DNA are held at temperatures just below the melting point, the double helix will again form. This process is referred to as **annealing** the DNA; the rate at which it occurs depends upon the fraction of the DNA molecules that are exact complementary matches — the higher the fraction, the greater the proportion of collisions between molecules that will bring complementary sequences near each other.

The buoyant density, ρ (rho), of DNA in concentrated salt solutions increases with the proportion of guanosine-cytidine pairs:

$$\rho = 1.660 + 0.098\ (\text{G} + \text{C})/(\text{total nucleosides})$$

The change is sufficient to permit the separation of DNA molecules into groups of similar base composition by centrifuging them through a gradient of increasingly

TABLE 4–1 SUMMARY OF THE NATURE OF DUPLEX DNA

Two antiparallel polynucleotide chains twisted into a double helix
Negatively charged phosphate-deoxyribose backbones on outside, complementary base pairs on inside
Adenine and thymine, guanine and cytosine paired together by hydrogen bonds that are stabilized in non-polar interior
Length of molecules: 14 to 73 mm in human cells, 1.6 mm in the bacterium *E. coli*; 0.34 nm per base pair; 3.4 nm per turn
Diameter: 2.0 nm
Base pairs per complete turn: 10 (36° turn per pair)
Two grooves, 2.2 nm and 1.2 nm across, measured parallel to axis of helix
Buoyant density: 1.660 + 0.098 (fraction of G + C)
Melting point: ranges around 85° C for native molecules, increases with greater G + C content
Mass: 618 daltons per pair of nucleotides (anion form)

concentrated salt solutions. (Cesium chloride, CsC1, is frequently used.) Those molecules with a greater fraction of G-C, and a correspondingly lower fraction of A-T, sediment farther before reaching a solution of equivalent density.

USE OF GENETIC INFORMATION

Given that the genetic information is the sequence of nucleotides in DNA, the question is: How is the nucleotide sequence converted into usable form for constructing and regulating the cells? The entire process hinges on the binding of complementary bases, and the principal features of the flow of information are described now as background for the detailed examination made in the next chapters.

Replication. When cells divide, each daughter cell must have faithful copies of the entire genetic instructions in the original cell. This is accomplished by replication of all of the DNA, in which new complementary polynucleotide chains are created for each of the original chains in the cell. In preparation for mitosis, the strands of each double helix are separated, and a new DNA strand is constructed on each previous strand. Each old strand then has a new complementary partner exactly like the other old strand, if replication proceeds without error.

Transcription. The use of the information in DNA to direct the life of a cell involves the synthesis of ribonucleic acid (RNA) molecules as working copies of segments of DNA.

Some of the RNA molecules are structural components of the apparatus of protein synthesis, for example, the **ribosomes** on which polypeptide chains are built. Other RNA molecules act as **messengers** from DNA in the nucleus, carrying instructions for assembly of cellular proteins; these instructions are used by enzymes catalyzing protein synthesis in the cytosol.

Translation. The translation of the nucleotide sequence on messenger RNA molecules into an amino acid sequence in a polypeptide chain involves attachment of each amino acid to an identifying small RNA molecule, **transfer RNA.** This molecule serves to transfer the amino acids from free solution to the point of assembly of polypeptide chains, but the transfer only occurs when directed by the instructions in messenger RNA molecules.

Arrangement of DNA Sequences

Some of the segments of DNA appear to act as spacers between the segments transcribed into complementary RNA strands. Other segments have no known function and may serve some structural role in organizing the DNA within a cell. The differences in function are reflected in the amount of repetition of identical, or similar, nucleotide sequences within the total DNA. Instructions for assembly of polypeptide chains are sometimes present in as few as two identical, or closely similar, copies, each being a sequence of perhaps 3,000 nucleotide pairs out of the total 6×10^9 pairs. In contrast, those portions of the DNA that appear to have only structural significance contain as many as 10^6 nearly identical sequences of base pairs.

The different degrees of repetition can be recognized in the following way: The total DNA of cells is sheared into relatively small lengths by rapid stirring of a solution. (Even gentle stirring can create sufficient moment on these long molecules to break some covalent bonds.) The fragmented DNA is then dissociated into single strands by raising the temperature above the melting point. The strands are then exposed to annealing conditions that favor recombination into double helixes. (Double helical segments can be identified by their failure to bind to adsorbents such as nitrocellulose filters or hydroxyapatite columns.) The chance that a fragment will encounter another fragment containing long sequences of complementary bases, thereby permitting stable recombination, is proportional to the number of these similar sequences in the original DNA. The more repetitive the sequence, the faster the recombination. Long annealing is required for recombination of nearly unique sequences.

DNA of eukaryotic cells has been classified in three categories according to the results of such experiments. The **non-repetitive** fraction includes sequences that exist in only a few copies; many of these sequences are instructions for building polypeptides. The **moderately repetitive** fraction includes sequences repeated many times in identical form, but it also includes sequences that are similar, but not completely identical. Spacer segments of the DNA that occur between non-repetitive segments apparently are isolated in the moderately-repetitive fraction; these segments may include regulatory sequences that control how often instructions for particular proteins are to be transcribed. Other moderately repetitive sequences include those from which ribosomal and transfer RNA molecules are transcribed; there are many copies of these genes.

The **highly repetitive** sequences, with short segments occurring in as many as 10^6 copies, include what is sometimes referred to as satellite DNA — DNA that has a density markedly different from the remainder owing to an unusual base composition, and therefore separates in a separate ''satellite'' band upon centrifugation. (The difference in composition is not apparent in all organisms.) The purpose of these sequences is unknown, but much of the DNA containing them appears to be associated with the nucleolus of the cell, and may have some structural role. Most of the repeated DNA in mammals exists in over 1,000 copies of each kind.

FURTHER READING

Freifelder, D.: (1978) *The DNA Molecule. Structure and Properties.* Freeman. Compendium of important past papers.

Arnott, S.: (1970) *The Geometry of Nucleic Acids.* Progr. Biophys. Mol. Biol., *21*: 267. Contains recent determinations of bond lengths and angles.

Langridge, R., et al.: (1960) *The Molecular Conformation of Deoxyribonucleic Acid.* II. J. Mol. Biol., *2*: 38.

Watson, J. D.: (1975) *Molecular Biology of the Gene,* 3rd ed. Benjamin.

Mahler, H. R., and E. H. Cordes: (1971) *Biological Chemistry,* 2nd ed. Harper-Row. Includes detailed discussion of nucleic acids.

Wu, R.: (1978) *DNA Sequence Analysis.* Annu. Rev. Biochem., *47*: 607. Explains how nucleotide sequences are determined.

5 RIBONUCLEIC ACIDS: FORMATION AND FUNCTION

THE NATURE OF RIBONUCLEIC ACIDS (RNA)

RNA and DNA, two kinds of polynucleotides distinguished mainly by the presence or absence of one oxygen atom on the sugar residues, serve different functions, which must be regulated separately. We have seen that molecules of DNA constitute the master file of genetic information carefully reproduced in successive generations of cells, whereas molecules of RNA are used as working copies and as tools for making proteins. These differences in function are reflected in the life-span of the molecules; the cell does not intentionally destroy DNA, whereas it constantly degrades and remakes RNA as a necessary part of its adjustment to changing circumstances.

The differing functions of the two classes of polynucleotides mean that they participate in different reactions; they have groups that form different bonds. Both classes have a sugar-phosphate backbone, with bases attached as side chains, but the sugar in RNA is ribose rather than deoxyribose, and the additional hydroxyl group is sufficient to make an RNA chain recognizably different from a DNA chain of identical base sequence. Also, RNA molecules are made with uracil rather than thymine, as the base complementary to adenine. Uracil differs from thymine only in the absence of the 5-methyl group, and it forms the same hydrogen bonds:

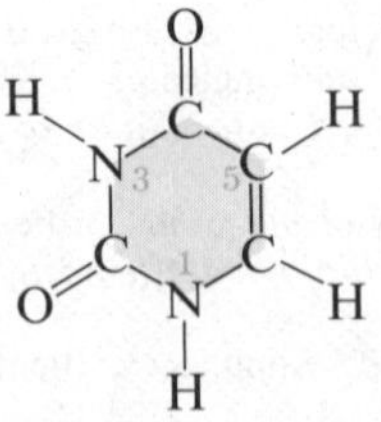

uracil

The other three bases laid down in RNA (adenine, guanine, and cytosine), are the same as those in DNA. (We shall see that these bases are frequently modified after the polynucleotide is formed.)

Molecules of RNA Are Smaller than Molecules of DNA. Each molecule of RNA is made as an antiparallel complement to only part of a DNA strand. The smallest molecule of DNA in the nucleus of a human cell contains 45,000,000 base pairs. The largest molecule of RNA discovered in cultures of human cells has approximately 50,000 nucleotide residues in a single chain — only a little more than 0.001 of the length of the shortest DNA chain. Still larger RNA molecules may be discovered, but there is no reason to believe they will have more than a very small fraction of the length of a DNA molecule, and some RNA molecules are transcribed with as few as 120 nucleotide residues.

Internal Folding of RNA

The behavior of polynucleotides is determined by base sequence in the same way that the behavior of polypeptides is determined by amino acid sequence. RNA chains are like DNA chains in that they can associate into a double helix, if a substantial number of the base side chains are in antiparallel complementary sequences. The additional hydroxyl groups in the RNA backbone do not permit stable formation of the 10-pair turn seen in DNA, but RNA can form an open-cored helix with 11 or 12 base pairs per turn.*

However, RNA molecules do not occur as double helices with two separate chains, except in certain viruses, because RNA is not made in complementary pairs like the chains in DNA. Separate complements do not exist for a given molecule of RNA in uninfected cells. Does this mean that molecules of RNA have no helical secondary structure? Not necessarily because a molecule of RNA frequently is partially complementary to itself.

It is possible for a single polynucleotide chain to bend back upon itself so as to bring different parts of the chain into antiparallel alignment. If these segments have complementary antiparallel sequences of bases, they will associate into an internal double helix (Fig. 5–1). Furthermore, additional hydrogen bonds can be

*The double helices of polynucleotides with 10, 11, and 12 pairs per turn are known respectively as B, A, and A′ helices. In addition to the longer pitch, the surface of the A helices seen with RNA is distinguished by the shallowness of the narrow groove, which is almost absent, and the conversion of the wide groove into a relatively deep cleft.

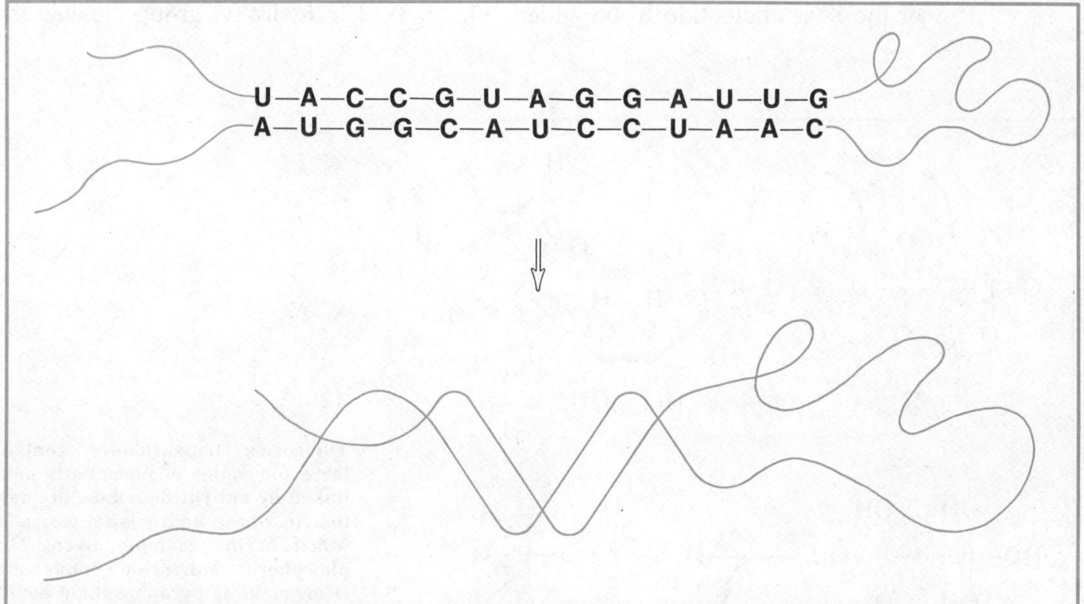

FIGURE 5–1 **Origin of secondary structures in ribonucleic acids. Antiparallel complementary sequences within a single chain can cause it to fold back upon itself to create a segment of double helix.**

formed between bases that are brought into proximity in different alignments than those seen in the double helix, so that a specific 3-dimensional conformation can be stabilized in RNA molecules by many of the same forces that stabilize conformations in proteins. The nucleic acid molecules may have regions of secondary structure (the internal double helix) joined by bends into a tertiary structure.

Here we have an important point: **Part of the base sequence in DNA molecules is present to create specific structures in RNA molecules by direct transcription.** Some of the short RNA molecules twist into compact shapes that are recognized as carriers for amino acids — these are the transfer RNA molecules. Some of the larger RNA molecules twist upon themselves so as to generate surfaces that associate with particular proteins and thereby create the ribosomes upon which proteins are made.

However, part of the nucleotide sequence in DNA contains the information for amino acid sequence in proteins, and the resultant sequence in messenger RNA molecules does not always have bases in the right places for creating compact secondary structures. Molecules of messenger RNA therefore are usually in a less ordered configuration, although they contain some structured segments.

FORMATION OF RNA BY TRANSCRIPTION OF DNA

The Chemical Reactions

RNA is constructed from nucleoside 5′-triphosphates, in which three residues of phosphoric acid are linked by anhydride bonds (Fig. 5–2). These linkages between phosphate groups are pyrophosphate bonds; they are related to inorganic pyrophosphates (PP_i).

Polynucleotides are built by addition of one nucleotide residue at a time, not in a concerted reaction. The polymerizing reaction involves the 3′-hydroxyl group at the end of the growing polynucleotide chain and the α, β-pyrophosphate bond of the next nucleotide to be added (Fig. 5–3). The hydroxyl group cleaves the

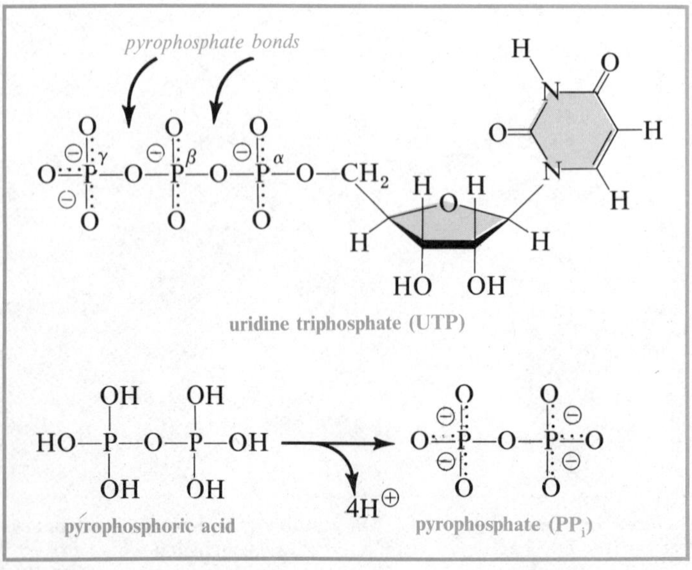

pyrophosphate bonds

uridine triphosphate (UTP)

pyrophosphoric acid

pyrophosphate (PP_i)

FIGURE 5–2

Nucleoside triphosphates contain three molecules of phosphoric acid linked by anhydride bonds, in addition to ribose and a base, which is uracil in the example given. The phosphoric anhydride bonds are referred to as pyrophosphate bonds because the same structure occurs in inorganic pyrophosphate, the ionized form of pyrophosphoric acid.

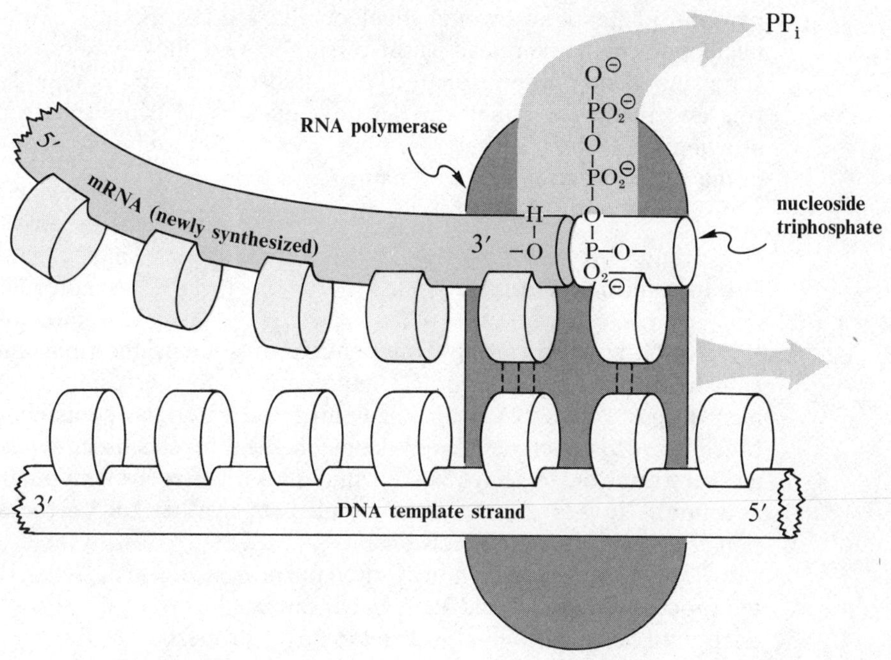

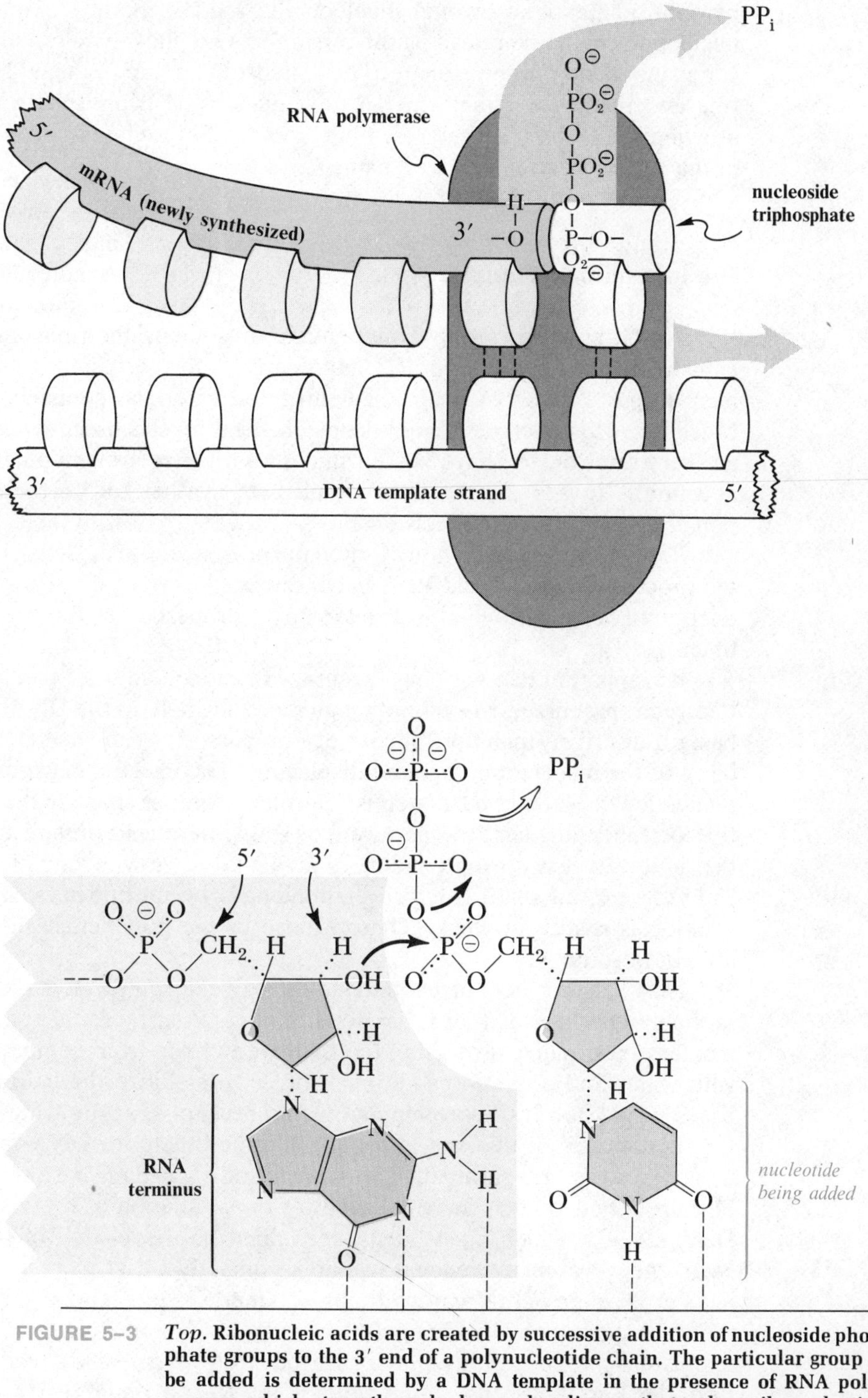

PP$_i$
5′
3′
RNA terminus
nucleotide being added

FIGURE 5–3 *Top.* Ribonucleic acids are created by successive addition of nucleoside phosphate groups to the 3′ end of a polynucleotide chain. The particular group to be added is determined by a DNA template in the presence of RNA polymerase, which strengthens hydrogen bonding so that only a ribonucleotide with a base that is the antiparallel complement of the corresponding base on DNA will be bound. The RNA is built from 5′ end to 3′ end, while the DNA template is read from 3′ end to 5′ end. *Bottom.* The precursors used to lengthen the RNA chain are ribonucleoside triphosphates, which lose a pyrophosphate group upon addition to the terminal 3′-hydroxyl group of the growing RNA chain.

pyrophosphate bond through nucleophilic attack, thereby extending the polynucleotide chain and releasing the remaining two phosphate groups in the precursor as inorganic pyrophosphate. The 3′-hydroxyl group of the added nucleotide is free and ready for attack on the next nucleoside triphosphate precursor. The growing RNA chain is therefore built from the 5′ terminal toward the 3′ terminal, in the same direction that it is named.

The Information Transfer

The straightforward polymerization of nucleoside triphosphates does not explain the critical feature of RNA synthesis — the perfect synthesis of a complement to part of a DNA chain, beginning and ending at precisely defined points. Much is yet to be learned about the mechanism for this process in which mistakes are very rare, but we have some sound ideas. There are essentially two separate questions: How is an error-free complement synthesized? How is the transcription begun and ended at precise points?

Perfect transcription hinges upon the presence of specific proteins to catalyze the process, **DNA-directed RNA polymerases.** We won't discuss catalysis by proteins until later, but we can develop some properties of the polymerase at this time.

It is apparent that very high accuracy must include some mechanism whereby "correct" precursors are bound much more strongly to the DNA template that is being transcribed than are "incorrect" precursors — so strongly that the probability of the incorrect compound displacing the correct is negligible. The nucleotide added to RNA at each step is "correct" when its base is the complement of the corresponding base in one strand of DNA. In other words, accurate transcription involves very strong bonding of nucleotides carrying complementary bases, and yet we know that the hydrogen bonding between free nucleotides in aqueous solution is relatively weak. The presence of the polymerase must increase the affinity in some way.

How can a protein promote bonding between complementary nucleotides? A rational hypothesis is that it has some non-polar surfaces that snugly fit complementary base pairs, protecting the hydrogen bonds from competing interactions with water and providing additional forces to stabilize the pairing. Mismatched bases would not fit the protein and would be kept apart by water.

Before any of this can happen, the double helix of DNA must be partially unwound so as to separate the two strands and expose the base sequence that is to be transcribed. This is accomplished by an association of protein with separated DNA strands, which must be stronger than the association of the strands with each other. Present evidence is that approximately 6 base pairs in the DNA double helix are broken in this way, with the polymerase advancing one base at a time as the RNA chain is built (Fig. 5–4). Since the binding of RNA precursor and DNA template is antiparallel, it follows that the RNA polymerase "reads" the template from the 3′ to the 5′ end, while building the RNA from 5′ to 3′ end.

SEGMENTAL TRANSCRIPTION. One aspect of RNA formation we do not understand. The information in DNA for a single polypeptide chain — one gene — may be divided by long nucleotide segments. It appears that the intervening segment is removed from the RNA after transcription, with the desired pieces joined by the action of an RNA ligase.

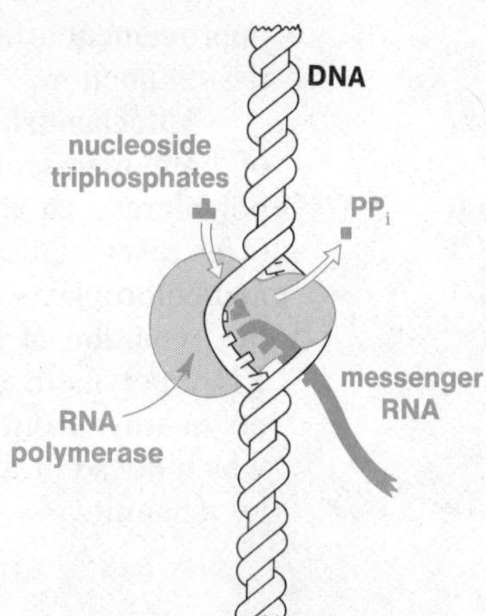

FIGURE 5–4

In order to expose the bases in DNA for use as a template in making RNA, it is necessary to uncoil the DNA double helix as RNA polymerase moves along it. Approximately six bases are exposed at a time, and only one of the DNA strands is used as a template.

Multiplicity of RNA Polymerases. Mammalian cells, like other eukaryotic cells, are known to have at least three distinct RNA polymerases in their nuclei. Type I RNA polymerase is associated with the nucleolus and generates large precursors of ribosomal RNA from specific segments of DNA. The types II and III RNA polymerases occur in the nucleoplasm and there is good evidence that they make precursors of messenger RNA and transfer RNA, respectively. (The type III enzyme also makes a small ribosomal RNA segment.)

The RNA polymerases can be differentiated experimentally by the action of α-amanitin, a compound found in the mushroom *Amanita phalloides:*

RNA Polymerases*	Transcribed Precursor	Effect of Amanitin
I or A	ribosomal RNA	small
II or B	messenger RNA	inhibited by 10^{-9} to 10^{-8} M
III or C	transfer RNA	inhibited by 10^{-5} to 10^{-4} M

*The enzymes are designated in various ways by different authors.

We see that amanitin has little effect on the RNA polymerase in the nucleolus, which synthesizes the large precursor of ribosomal RNA, but only small concentrations are necessary to stop the action of the polymerase that synthesizes precursors of messenger RNA. Intermediate concentrations block the formation of transfer RNA.

Amanitin is composed of eight amino acids joined in a ring. (Some of the amino acids are modified from the usual forms.) Its effect has practical consequences because the unwary who eat the Amanita mushroom are preventing the formation of messenger RNA, and thus the formation of proteins, in their cells. Unfortunately, they don't know all is not well until the depletion of the supply of proteins is great enough to cause acute gastrointestinal distress and other indications of damage; by that time the amanitin is thoroughly absorbed. A temporary improvement after the first day sometimes conceals irreversible and lethal loss of tissue function.

Mitochondrial RNA Polymerase. The RNA molecules found in mitochondria of both plants and animals, and in the chloroplasts of plants, and the RNA polymerase catalyzing their formation resemble similar molecules in bacteria. (This resemblance is one of the strongest reasons for believing that mitochondria and chloroplasts arose from symbiotic inclusion of prokaryotic organisms during the evolution of eukaryotes.) Both the organelles and the bacteria have a single RNA polymerase to catalyze the formation of all of the types of RNA. The polymerase from these sources is usually inhibited by an antibiotic, rifampicin, which does not affect the enzymes from eukaryotic nuclei, but it is not inhibited by amanitin.

Signals in DNA

How do the RNA polymerases transcribe sharply defined sequences of DNA, repeatedly making exact copies of particular RNA molecules? How do the different polymerases distinguish the specifications for different types of RNA? The only information in DNA is contained in the sequence of nucleotides, and there must be something recognizable in the sequence at particular points that causes specific types of polymerase molecules to be bound or released; that is, there must be recognizable initiation and termination points for transcription of each RNA molecule.

Specific recognition is also required in another way. DNA occurs as pairs of complementary strands, but RNA is transcribed from the sequence on only one strand. If it were transcribed from both, the product would be a double-stranded RNA helix, and one RNA strand would not be able to fold back upon itself to form the specific secondary and tertiary structures seen in ribosomal and transfer RNAs. Beyond that, messenger RNAs specify amino acid sequences that generate particular molecular architectures. If an antiparallel complement existed for a molecule of messenger RNA, it would specify a totally different amino acid sequence, and it is asking too much to expect this sequence also to define a functional protein. Specific signals must be present for binding of the RNA polymerase on one DNA strand, with the complement of the signals in the other strand not capable of binding the polymerase.

The Operon Concept. The nature of the identifying signals in mammalian cells is not known, but we have some clues from work with bacteria. The sequence of bases in bacterial DNA is continuous, but the bacterial RNA polymerase behaves as if the DNA being transcribed were divided into units, which are known as operons. Each operon has toward its 3′ end a sequence that identifies a precise binding site for RNA polymerase. The sequence that is to be transcribed is toward the 5′ end followed by a sequence identifying the termination of the transcribed portion. Some operons also include one or more sequences of bases at which other proteins may be bound to regulate the rate of initiation of new RNA chains (Fig. 5–5).

The sequence of nucleotides at the beginning of some bacterial operons is now known. There are no obvious peculiarities in some of these, but the early

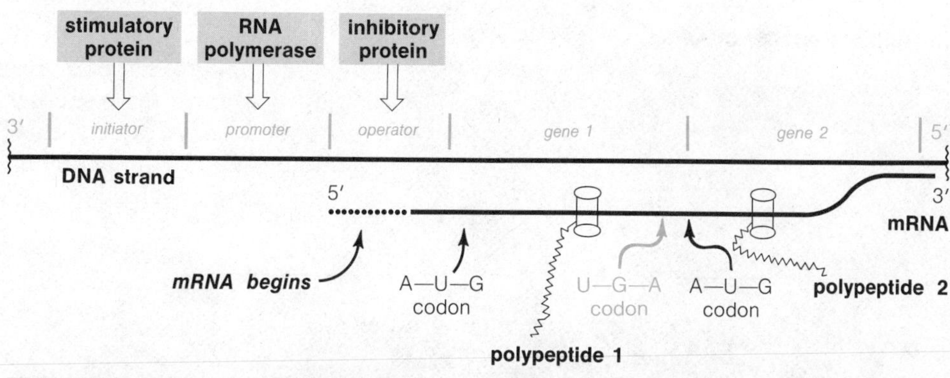

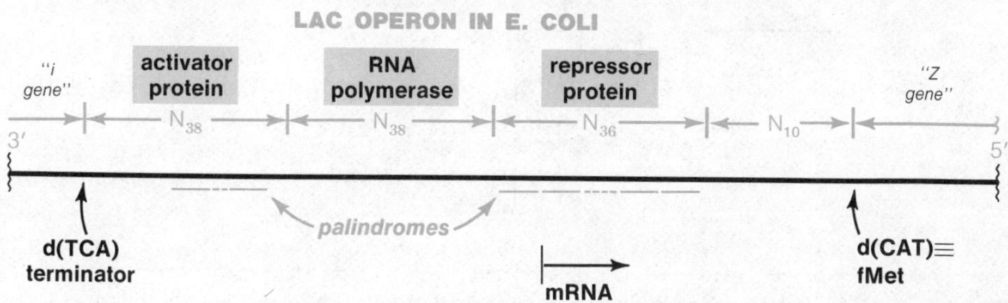

FIGURE 5–5 *Top.* Generalized structure of an operon in a bacterial DNA strand. All operons include a site for attachment of RNA polymerase (the promoter site) on the DNA strand followed toward the 5′ end by sequences that are transcribed into the form of RNA. In the example shown, these sequences include two successive genes. Both genes are transcribed as a single segment of messenger RNA, which has two initiation sites and two termination sites for making two different polypeptide chains, one for each gene.

Operons may also include initiator sites, usually toward the 5′ end from the promoter, at which signaling proteins are bound to stimulate the beginning of transcription by RNA polymerase. Operons frequently have operator sites, usually after the promoter, at which signalling proteins are bound to prevent transcription by RNA polymerase. These various binding sites sometimes overlap. Part of the operator or promoter sites may be transcribed in the initial segment of RNA that precedes polypeptide coding in messenger RNA or the structural sequences in transfer RNA and ribosomal RNA.

Bottom. An actual operon sequence that controls lactose metabolism by the bacterium *Escherichia coli.* (Lactose is the sugar of milk.) The terminator triplet of the preceding gene is followed by a sequence of 38 nucleotide residues at which an activator protein is bound. Another 38 residues included in the RNA polymerase binding site are followed by 36 residues in an operator at which a repressor protein is bound. The repressor protein is removed when the bacterium encounters lactose or related sugars, thereby permitting the polymerase to begin transcription of a messenger RNA that includes instructions for making enzymes necessary for lactose metabolism. Both the activator and repressor regions include palindromic sequences that are likely signals for binding of the specific proteins. Messenger RNA transcription begins well within one of these sets of sequences that precede the initiating A-U-G triplet.

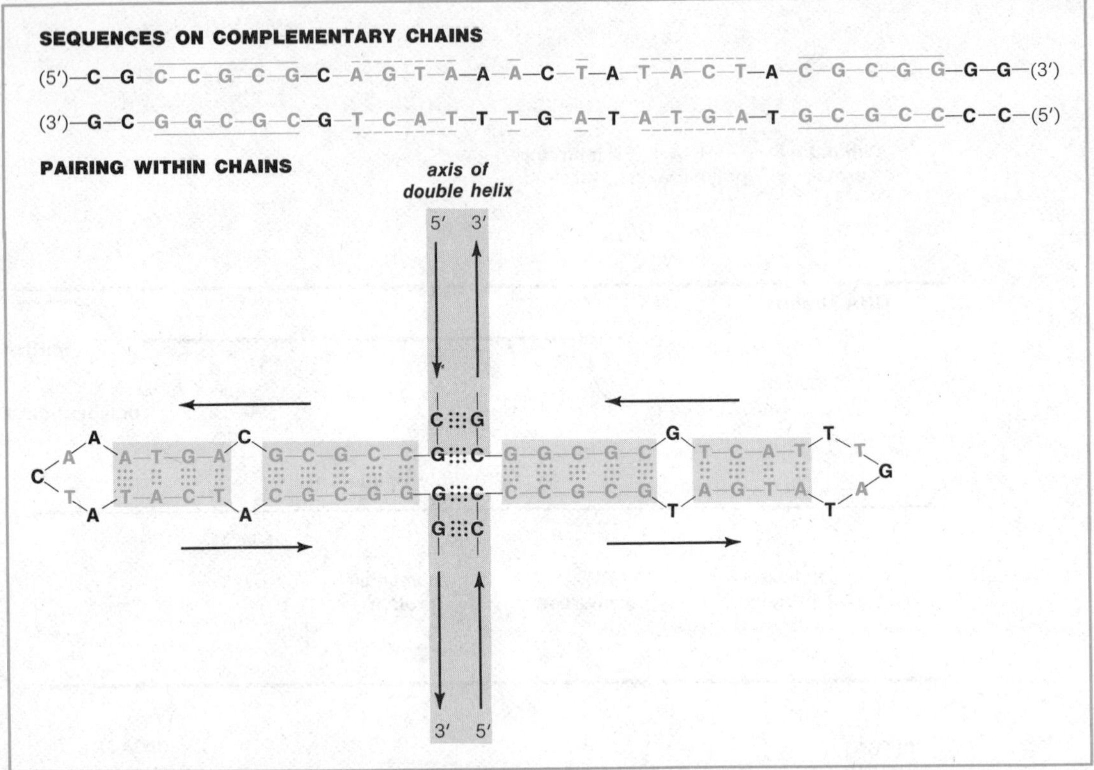

FIGURE 5–6 *Top.* **Palindromes in duplex nucleic acid chains are segments with the same sequence of nucleotides when read from opposite directions. Two sets of palindromes are illustrated, under dashed and solid lines, respectively. For example, the sequence C-C-G-C-G, read from the 5′ end, occurs to the left in the top chain, and the same sequence occurs to the right in the bottom chain, read from its 5′ end. In each case, the other chain has the antiparallel complement. Therefore, these two pieces of double helix appear identical from the corresponding 5′ ends.** ***Bottom.*** **Palindromes may theoretically form a double helix in two ways. Each strand may bind to the other strand as is usually seen in DNA, or the antiparallel complements within each strand may combine with themselves. Should this happen, the result would be two segments of double helix, each formed by one strand, that protrudes from the long two-strand double helix. Some matching bases would not be able to bond effectively owing to steric considerations, for example, the A-T pairs in the end loops.**

sequence in others has a recognizable symmetry that is not present in succeeding regions, creating what has been termed a palindrome.

The Nature of Palindromes. Palindromes are segments of double helix that have the same sequence when viewed in either direction.* Each strand of a palindromic sequence is complementary to itself, as well as to the other strand of the double helix, and it has been speculated that such sequences may form protruding whiskers on a monotonous double helix, making an obviously distinctive structure for RNA polymerase to recognize (Fig. 5–6). (When speaking of

*Palindrome is borrowed from the name for words or groups of words that read the same in either direction, for example, "level," or "Able was I ere I saw Elba." A spectacular, but less pure, example that neglects spaces and punctuation has been devised by Alastair Reid: "T. Eliot, top bard, notes putrid tang emanating, is sad. I'd assign it a name: gnat dirt upset on drab pot toilet." (Quoted by Brendan Gill in *Here at the New Yorker,* Random House (1975).)

molecules, "recognize" is to be interpreted as anthropomorphic shorthand for "bond in a specific way.") Others contend that the operons are in unbroken linear double helix, with the symmetrical segments perhaps binding correspondingly symmetrical arrangements of subunits on the RNA polymerase. Both of these notions are speculative, and the actual mechanism of recognition is as yet unknown.

Repeated Sequences. Several RNA molecules from bacteria infected with viruses, as well as uninfected bacteria, have been found to end with sequences such as · · · G-U-U-U-U-U-U-A–OH, or G-U_6-A, which suggests that an antiparallel complementary sequence in DNA such as · · · d(T-A-A-A-A-A-A-C) · · · may cause a termination of transcription as the RNA polymerase passes from right to left over the repeated dA groups toward the dT. As in the case of the palindromes, not enough information is available to know if such repeated sequences are widely used as signals, or even how they operate when they are used. We can say only that plausible examples are at hand to illustrate how a base sequence can be used to signal termination and initiation of transcription at a precise nucleotide in DNA, not one before nor one after.

THE GENETIC CODE

The mechanism by which genetic information is used for the synthesis of proteins is sketched in a general way in Figure 5–7. Specific instructions for the amino acid sequence of a polypeptide are transcribed from a segment of DNA — the gene for the polypeptide — and transferred from the nucleus to the cytoplasm as messenger RNA. Ribosomes, which carry the machinery for synthesis of peptide bonds, are bound to the messenger RNA. One molecule of messenger RNA frequently carries several ribosomes which appear as a cluster, or polysome, in electron micrographs.

Each kind of amino acid is carried to the site of peptide synthesis through covalent linkage to its specific kind of transfer RNA. The ribosomes provide a movable platform on which the molecules of transfer RNA associate with messenger RNA through complementary binding of bases. Each kind of transfer RNA has a particular base sequence that will only bind with a complementary sequence in messenger RNA, and this is the mechanism by which the amino acids carried on transfer RNAs are brought into position at the proper time for incorporation into a growing polypeptide chain.

Triplet Coding

Since it is the sequence of four different nucleotides that is used to convey information for the combination of twenty different amino acids into peptide chains, each amino acid must be represented by combinations of at least three nucleotides. This is true because there are only 16 different doublets of four nucleotides (4^2), but there are 64 triplets (4^3). In the polynucleotide chain of messenger RNA molecules directing the synthesis of polypeptides, and in the segments of DNA chains from which these molecules are transcribed, each successive group of three nucleotides, beginning at a precisely defined location, designates either a specific amino acid, or a chain-terminating punctuation mark. This is the molecular basis for the inheritance of protein composition.

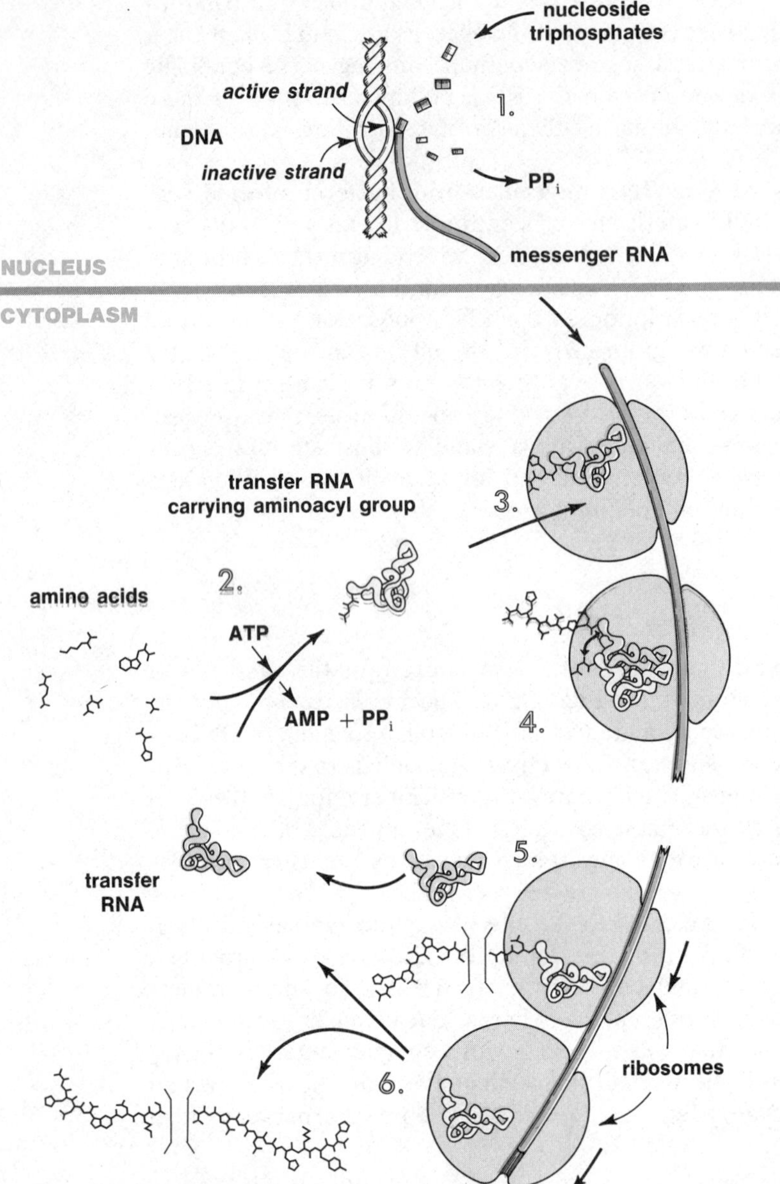

FIGURE 5–7 Schematic summary of protein synthesis. *Top, step 1.* A molecule of DNA in the nucleus is unfolded so that one of its strands can be used as a template to direct the formation of messenger RNA from nucleoside triphosphates, which lose inorganic pyrophosphate (PP$_i$) as they attach to the growing RNA chain. The completed mRNA moves to the cytoplasm *(bottom)*, where ribosomes combine with it and use it as a template for making polypeptide chains.

The following steps are shown on separate ribosomes for clarity, but in fact they are repeated in sequence on each ribosome. The successive ribosomes create longer and longer polypeptide chains as they move down the molecule of mRNA.

Step 2. Meanwhile, amino acids combine with specific molecules of transfer RNA (tRNA) in the cytoplasm by a reaction that also involves the cleavage of adenosine triphosphate (ATP) into adenosine monophosphate (AMP) and PP$_i$.

Step 3. The tRNA molecules, carrying the amino acids as aminoacyl groups, diffuse to the ribosomes, on which the growing polypeptide is attached to another molecule of tRNA. The incoming tRNA, which bears the next group required for

(Legend continued on opposite page)

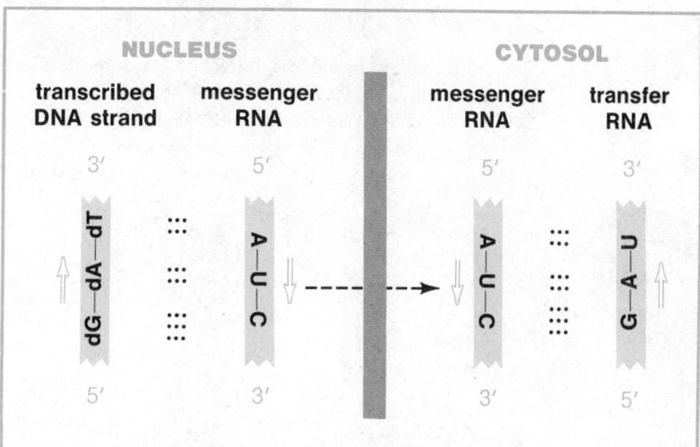

FIGURE 5–8

Antiparallel complements.

Table 5–1 lists the meaning of each of the 64 possible combinations of three successive bases in both messenger RNA and DNA. Remember that transcription involves antiparallel complementary bonding; the code is read in messenger RNA from the 5′ end toward the 3′ end — a happy coincidence of function with nomenclature — but the corresponding triplets in DNA must be read backwards to satisfy nomenclature. The triplets in transfer RNA also are bound to messenger RNA in antiparallel sequence, so they also must be read backwards with respect to messenger RNA. The sequence in messenger RNA is said to be a codon for the particular amino acid, and the matching complement in transfer RNA is said to be an anticodon. The facts are not as confusing as the telling, and the example in Figure 5–8 may help dispel any fog.

The Genetic Code Is Redundant. That is, all of the amino acids except methionine and tryptophan are represented by more than one triplet, and sometimes as many as six. The advantage of this is that every triplet means something, and if a mistake is made in transcribing or translating the nucleotide sequence, the growth of the polypeptide chain is not interrupted (unless the mistake is read as a terminator triplet).

The Genetic Code and Amino Acid Structure

Nucleotide triplets and amino acids are not randomly matched. Close study of the table of triplets will show that the code is constructed so that substitution of

---

FIGURE 5–7 *Continued*

the growing polypeptide (in this case a leucyl residue), has a sequence of nucleotides that causes it to complex with mRNA on the ribosome.

Step 4. **When the proper tRNA is in place, the polypeptide is transferred onto the amino group of the new residue brought in by tRNA, so that the chain is now one residue longer.**

Step 5. **When the transfer of the previous step is completed, the previously bound tRNA no longer carries a polypeptide and is free to dissociate from the ribosome, returning to the mixed pool of tRNA in the soluble cytoplasm, where it is available for transport of another molecule of its specific amino acid. The ribosome now moves along the mRNA molecule to the position where the placement of the next amino acid will be directed.**

Step 6. **Steps 3, 4, and 5 are repeated. As each amino acid residue adds to the polypeptide, the ribosome moves down the mRNA molecule. When a ribosome reaches a specific terminator sequence in mRNA, the now complete polypeptide is detached into the soluble cytoplasm. The ribosome itself can then move free of the mRNA and be available for attachment to the beginning of the message in yet another molecule of mRNA (not shown).**

TABLE 5–1 THE GENETIC CODE

The triplets in messenger RNA (codons) are listed in black ink. Beneath them in blue ink are the corresponding triplets in the DNA strand from which messenger RNA is transcribed. The messenger RNA and DNA triplets are antiparallel complements.

Listed by Triplet

A-A-A –Lys d(T-T-T)	C-A-A –Gln d(T-T-G)	G-A-A –Glu d(T-T-C)	U-A-A –Term* d(T-T-A)
A-A-G –Lys d(C-T-T)	C-A-G –Gln d(C-T-G)	G-A-G –Glu d(C-T-C)	U-A-G –Term* d(C-T-A)
A-A-C –Asn d(G-T-T)	C-A-C –His d(G-T-G)	G-A-C –Asp d(G-T-C)	U-A-C –Tyr d(G-T-A)
A-A-U –Asn d(A-T-T)	C-A-U –His d(A-T-G)	G-A-U –Asp d(A-T-C)	U-A-U –Tyr d(A-T-A)
A-C-A –Thr d(T-G-T)	C-C-A –Pro d(T-G-G)	G-C-A –Ala d(T-G-C)	U-C-A –Ser d(T-G-A)
A-C-G –Thr d(C-G-T)	C-C-G –Pro d(C-G-G)	G-C-G –Ala d(C-G-C)	U-C-G –Ser d(C-G-A)
A-C-C –Thr d(G-G-T)	C-C-C –Pro d(G-G-G)	G-C-C –Ala d(G-G-C)	U-C-C –Ser d(G-G-A)
A-C-U –Thr d(A-G-T)	C-C-U –Pro d(A-G-G)	G-C-U –Ala d(A-G-C)	U-C-U –Ser d(A-G-A)
A-G-A –Arg d(T-C-T)	C-G-A –Arg d(T-C-G)	G-G-A –Gly d(T-C-C)	U-G-A –Term* d(T-C-A)
A-G-G –Arg d(C-C-T)	C-G-G –Arg d(C-C-G)	G-G-G –Gly d(C-C-C)	U-G-G –Trp d(C-C-A)
A-G-C –Ser d(G-C-T)	C-G-C –Arg d(G-C-G)	G-G-C –Gly d(G-C-C)	U-G-C –Cys d(G-C-A)
A-G-U –Ser d(A-C-T)	C-G-U –Arg d(A-C-G)	G-G-U –Gly d(A-C-C)	U-G-U –Cys d(A-C-A)
A-U-A –Ile d(T-A-T)	C-U-A –Leu d(T-A-G)	G-U-A –Val d(T-A-C)	U-U-A –Leu d(T-A-A)
A-U-G –Met d(C-A-T)	C-U-G –Leu d(C-A-G)	G-U-G –Val d(C-A-C)	U-U-G –Leu d(C-A-A)
A-U-C –Ile d(G-A-T)	C-U-C –Leu d(G-A-G)	G-U-C –Val d(G-A-C)	U-U-C –Phe d(G-A-A)
A-U-U –Ile d(A-A-T)	C-U-U –Leu d(A-A-G)	G-U-U –Val d(A-A-C)	U-U-U –Phe d(A-A-A)

**Term* are terminator codons.

Note that amino acids within a block of codons, or in the corresponding position in the same row or column of blocks, are interchangeable by substitution of one nucleotide.

TABLE 5–1 THE GENETIC CODE *Continued*

Listed by Amino Acid							
Ala–	G-C-A d(T-G-C)	Gly–	G-G-A d(T-C-C)	Lys–	A-A-A d(T-T-T)	Thr–	A-C-A d(T-G-T)
	G-C-G d(C-G-C)		G-G-G d(C-C-C)		A-A-G d(C-T-T)		A-C-G d(C-G-T)
	G-C-C d(G-G-C)		G-G-C d(G-C-C)				A-C-C d(G-G-T)
	G-C-U d(A-G-C)		G-G-U d(A-C-C)				A-C-U d(A-G-T)
Arg–	A-G-A d(T-C-T)	Gln–	C-A-A d(T-T-G)			Trp–	U-G-G d(C-C-A)
	A-G-G d(C-C-T)		C-A-G d(C-T-G)				
	C-G-A d(T-C-G)			Met–	A-U-G d(C-A-T)	Tyr–	U-A-C d(G-T-A)
	C-G-G d(C-C-G)	Glu–	G-A-A d(T-T-C)				U-A-U d(A-T-A)
	C-G-C d(G-C-G)		G-A-G d(C-T-C)	Phe–	U-U-C d(G-A-A)		
	C-G-U d(A-C-G)				U-U-U d(A-A-A)	Val–	G-U-A d(T-A-C)
		His–	C-A-C d(G-T-G)				G-U-G d(C-A-C)
Asn–	A-A-C d(G-T-T)		C-A-U d(A-T-G)	Pro–	C-C-A d(T-G-G)		G-U-C d(G-A-C)
	A-A-U d(A-T-T)				C-C-G d(C-G-G)		G-U-U d(A-A-C)
		Ile–	A-U-C d(G-A-T)		C-C-C d(G-G-G)		
Asp–	G-A-C d(G-T-C)		A-U-U d(A-A-T)		C-C-U d(A-G-G)	Term*–	U-A-A d(T-T-A)
	G-A-U d(A-T-C)		A-U-A d(T-A-T)				U-A-G d(C-T-A)
				Ser–	A-G-C d(G-C-T)		U-G-A d(T-C-A)
Cys–	U-G-C d(G-C-A)	Leu–	C-U-A d(T-A-G)		A-G-U d(A-C-T)		
	U-G-U d(A-C-A)		C-U-G d(C-A-G)		U-C-A d(T-G-A)		
			C-U-C d(G-A-G)		U-C-G d(C-G-A)		
			C-U-U d(A-A-G)		U-C-C d(G-G-A)		
			U-U-A d(T-A-A)		U-C-U d(A-G-A)		
			U-U-G d(C-A-A)				

one base for another frequently creates a triplet for the same, or a structurally related, amino acid. This is esthetically pleasing, but it is also a coldly practical arrangement. Accidental mismatching or chemical alteration of a nucleic acid can cause substitutions of one base for another in the nucleic acids, as we shall discuss in detail in Chapter 7. In some cases, the substitution causes no change in amino acid sequence. For example, introduction of an alanyl group into a protein is coded by d(T-G-C) in DNA. The triplets made by changing the first nucleotide to dC, dG, or dA still codes for an alanyl group.

Even in those cases in which a base substitution creates a triplet designating another amino acid, the effects are often minimized by the similarity of the new amino acid. To continue our example, the table shows that a single base substitution in any Ala triplet will result in the incorporation of Ala itself, or Thr, Pro, Ser, Gly, Asp, Glu, or Val. Many of these groups are like the alanyl group in function; they are likely to occur in chain reversals, and they may match the function of an alanyl group closely enough to permit adequate use of the accidentally modified protein.

This conservation of function during accidental alteration of base sequence is especially clear in the case of the amino acids bearing large hydrophobic side chains, which are so important in developing interior geometries in a protein molecule. The triplets for Ile, Leu, Met, Phe, and Val differ by only one base, and at least two thirds of the accidental base substitutions that may occur in coding for these groups will result in incorporation of one of these amino acid residues at the same location.

Similar relationships can be seen for the polar amino acid residues. The pairs that can be interchanged by single base substitutions include Asn/Asp, Asp/Glu, Gln/Glu, Arg/Lys, Arg/His. In short, the genetic code has been evolved so as to diminish any adverse consequences of accidental base substitution.

STRUCTURE AND FUNCTION IN RNA

Ribosomes

The ribosomes on which protein synthesis occurs are particles composed of many different molecules of proteins and a few quite large molecules of RNA, which are built to associate in a particular way. The molecules of ribosomal RNA also contain sequences for binding transfer and messenger RNAs in appropriate configurations for synthesis of polypeptide chains.

Mammalian ribosomes have a diameter near 30 nm, and a sedimentation coefficient of 80S. (The ribosomes of prokaryotes are somewhat smaller.) They are made from two subunits, which associate and dissociate during protein synthesis. The subunits have sedimentation coefficients of 60S and 40S, and each is constructed from proteins and specific pieces of RNA:

Ribosomal Particle	40S	60S	80S
Mol. Wt. of Particle	1.5×10^6	3×10^6	4.5×10^6
Number of RNA Constituents	1 (18S)	3 (5S, 5.8S, 28S)	4
Fraction as RNA	47%	57%	54%
Number of Proteins (approximate)	30	40	70

Many of the proteins are enzymes that are required for the synthesis of peptide chains; other proteins have a structural function in forming the shape of the ribosomal particles.

Ribosomes Contain Pockets. The larger, 60S ribosomal particle appears in electron micrographs to have a dished top, with an asymmetrical protuberance holding up the 40S particle when the two are combined to form a complete ribosome (Fig. 5–9). The 40S particle evidently has a pinched waist, which leaves an opening where it is held away from the surface of the 60S particle. This opening is believed to be lined by the sites for binding messenger RNA and the transfer RNA molecules bearing the amino acids that will be combined during translation of the message. The complete ribosome is in effect a sleeve that passes over messenger RNA, transfer RNAs, and the growing peptide chain to effect their contact with the protein factors in the ribosome necessary for the translational process.

Ribosomal RNA (rRNA)

Formation of Precursors. In the formation of ribosomal RNA, we have the first example of a principle that applies to most, if not all, kinds of RNA. DNA is transcribed to create overly long precursors that must be cleaved and whittled by hydrolysis into the final lengths. The ribosomes contain four different molecules of RNA, but there are only two genes from which the RNA is transcribed. One of these is used to form a precursor for 5S RNA, and transcription of the other creates a much larger 45S precursor that is split into pieces from which the other three ribosomal RNA molecules are made.

The chromosomes contain many copies of these two genes, enabling the simultaneous synthesis of many identical rRNA precursors. This multiplicity of genes is necessary because ribosomes are the sites at which all proteins are synthesized; they must be present in abundance, with approximately 10^6 in a typical mammalian cell and 10^4 in a typical bacterium. The mammalian ribosomes have a half-life near 130 hours (rat liver), and in order to meet the demand, something over 10^4 molecules of each of the two rRNA precursors are synthesized per hour in a cell. To facilitate this rapid transcription, DNA in a liver cell contains an estimated 1660 copies of the gene for 5S rRNA and 330 copies of the gene for 45S rRNA.

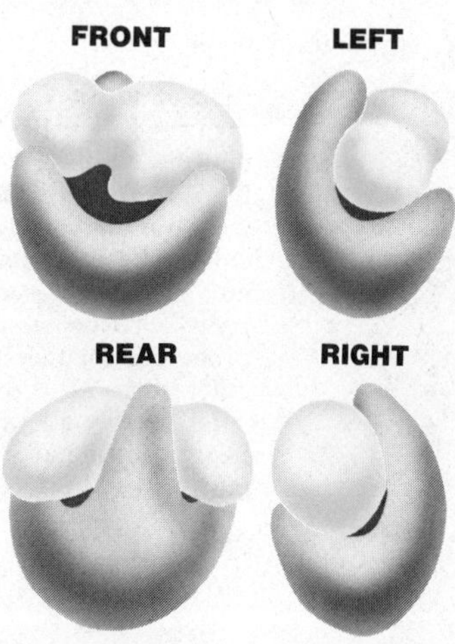

FIGURE 5–9

Ribosomal subunits in *E. coli.* (Eukaryotic ribosomes appear similar, but have not been studied in such detail.) The large subunit (gray) has a trough with a toothed margin on which the small subunit (blue) appears to rest. The small subunit has a constriction. Much study and a little imagination is necessary to generate detailed representations such as these from rather vague electron micrographs. There does appear to be an opening between the subunits that permits the ribosome to surround messenger RNA and its attached transfer RNA, while still being able to move over these structures. (Based on drawings by G. Stöffler and H. G. Wittman *in* H. Weissbach and S. Petska, eds., (1977) *Molecular Mechanisms of Protein Biosynthesis.* Academic Press, p. 117.)

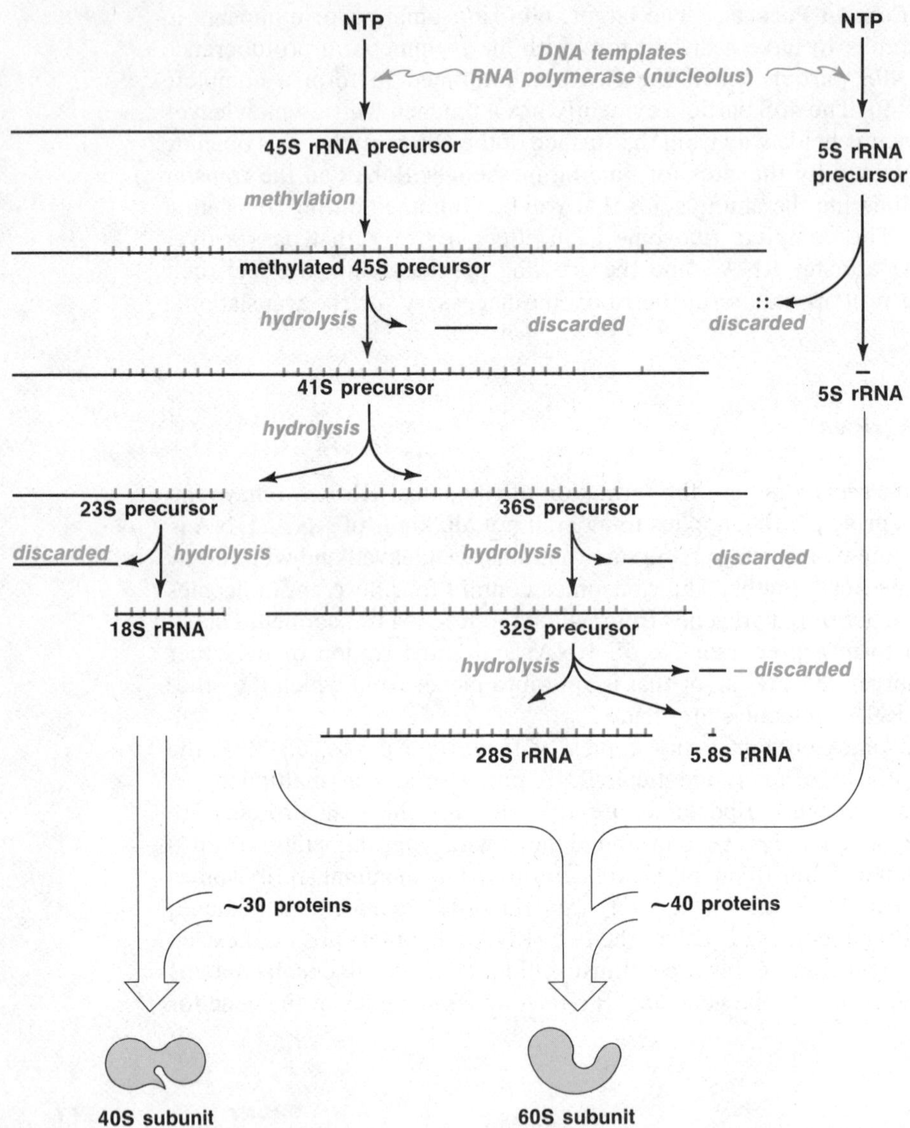

FIGURE 5–10 **Ribosomal RNA is transcribed in the nucleolus from two genes in DNA and then modified. *Right.* A small precursor is polymerized from nucleoside triphosphates (NTP), followed by partial hydrolysis of the ends to create 5S RNA. *Left and center.* A much larger precursor (45S) is separately polymerized; it includes the sequences of the 5.8S, 18S, and 28S ribosomal RNA molecules. These sequences are then methylated, while the spacer segments are not. The spacer segments are removed by hydrolysis, liberating the finished pieces of ribosomal RNA. The 18S RNA combines with approximately 30 protein molecules to form the light (40S) ribosomal subunit in the cytosol. The other three RNA molecules combine with approximately 40 protein molecules, nearly all different from those found in the light subunit, to form the heavy ribosomal subunit (60S). The combination with proteins actually begins before the 45S precursor is completely fragmented.**

Nature of the Precursors. Transcription to form 5S rRNA apparently begins with the first nucleotide of the finished product. How do we know? Because the 5S piece in completed ribosomes still has a 5′-triphosphate group. It is not certain how transcription ends. The 5S molecule in bacteria is made with extra nucleotides containing the U_6-Pu–OH transcript (Pu = purine nucleoside) of a termination signal, but such a signal has not been recognized in the human rRNA, or that of other eukaryotes. In any event, the extra nucleotides are hydrolyzed from the 3′ end until the proper total of 121 residues in the 5S piece has been reached.

The 45S precursor of the larger pieces of RNA in ribosomes is hydrolyzed at specific locations while the molecule is within the nucleolus (Fig. 5–10). Some of the fragments produced represent spacer sequences that are discarded; the others become the 28S and 5.8S components of the larger ribosomal particle, and the 18S component of the smaller particle, presumably in that order. The arrangement of spacers and retained sequences in the original precursor is not known for certain.

Of the approximately 14,000 nucleotides present in 45S precursor, somewhat over half appear in finished ribosomal RNA molecules, the remainder being discarded. There are approximately 5,000 nucleotides in the 28S piece, 2,000 in the 18S piece, and 160 in the 5.85A piece, as estimated in human HeLa cells.

HeLa cells were originally obtained by culture of a carcinoma of the cervix in a patient named Henrietta Lack; various strains of the progeny are now widely used for experimental purposes. The cells are aneuploid, meaning that their chromosome number is not an integer multiple of the normal haploid number, and they are also hyperploid, meaning they have more than the normal complement of chromosomes, but they are still valuable tools for many purposes.

Modification of the RNA. The 45S precursor of rRNA is modified through several covalent reactions before it is cleaved into pieces. The principal kind of modification is methylation; approximately 110 methyl groups are added, of which over 100 appear on the 2′-hydroxyl group of ribose moieties, with the balance going mainly on guanine and adenine rings (Fig. 5–11). All of the methyl groups appear to be placed on residues that are in the final ribosomal RNA; the intermediate spacer sequences are not modified. There is a further methylation of adenine groups in the 18S component after cleavage.

Another important modification involves a shift of the bond between the ribose and uracil moieties of uridine phosphate residues (Fig. 5–12) creating a pseudouridine group, which is abbreviated as ψ (psi).

Why do these modifications occur? We can get some clue by summarizing where they do not occur. All of the kinds of RNA contain some modified nucleotide residues, but they are not present in the center parts of messenger RNA molecules — the base sequences that code for amino acid sequence in polypeptides. Furthermore, we shall see that the modifications in transfer RNA molecules are more diversified and abundant than the modifications in ribosomal RNA. It seems likely that these differences between the classes of RNA are recognition signals by which the functions of the various RNAs are differentiated. Similarly, the unmarked spacers in the 45S precursor of ribosomal RNAs may indicate the positions at which the precursor is to be hydrolyzed. The test of these speculations is yet to come.

A fundamental point that must be kept in mind is that the modifications of RNA do not occur independently of the genetic information in DNA. The positions at which modification occurs are determined by the base sequences in RNA

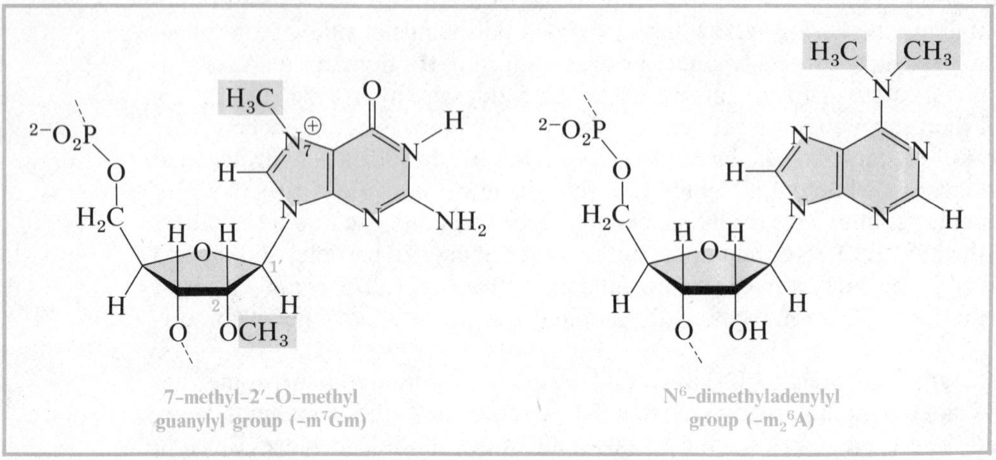

7-methyl-2′-O-methyl guanylyl group ($-m^7Gm$)

N^6-dimethyladenylyl group ($-m_2{}^6A$)

FIGURE 5–11 **Types of methyl derivatives found in ribosomal RNA. Methylation occurs mainly on the ribosyl moiety of nucleotide residues to form 2′-O-methyl derivatives, and to a much lesser extent on the bases, especially the purines. The left formula illustrates a guanine-containing nucleotide modified in both ways, while the right illustrates an adenine-containing nucleotide modified only on the purine by a double methylation. In standard abbreviations, an m before the nucleoside letter indicates a methylated base, with a superscript number to indicate position, and an m after the nucleoside letter indicates methylation of the 2′ oxygen on the ribosyl group. Thus, m_2^6A designates an adenosine moiety with two methyl groups on N-6 of the adenine; m^7Gm designates a guanosine moiety with a methyl group on position 7 of the guanine ring, and on the 2′ oxygen of its ribosyl group.**

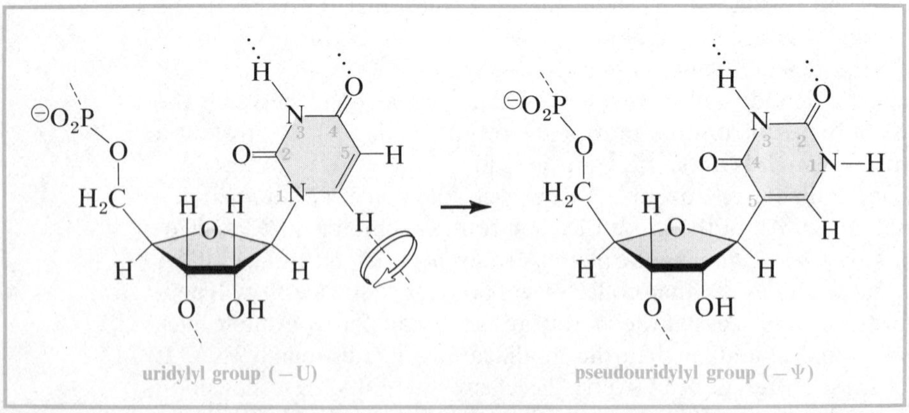

uridylyl group (–U)

pseudouridylyl group (–Ψ)

FIGURE 5–12 **Pseudouridylyl groups (Ψ) are created by flipping over the uracil ring of a uridylyl group in an intact nucleic acid. Uracil in pseudouridylyl groups still forms complementary bonds with adenine in adjacent antiparallel chains by using its other carbonyl oxygen atom.**

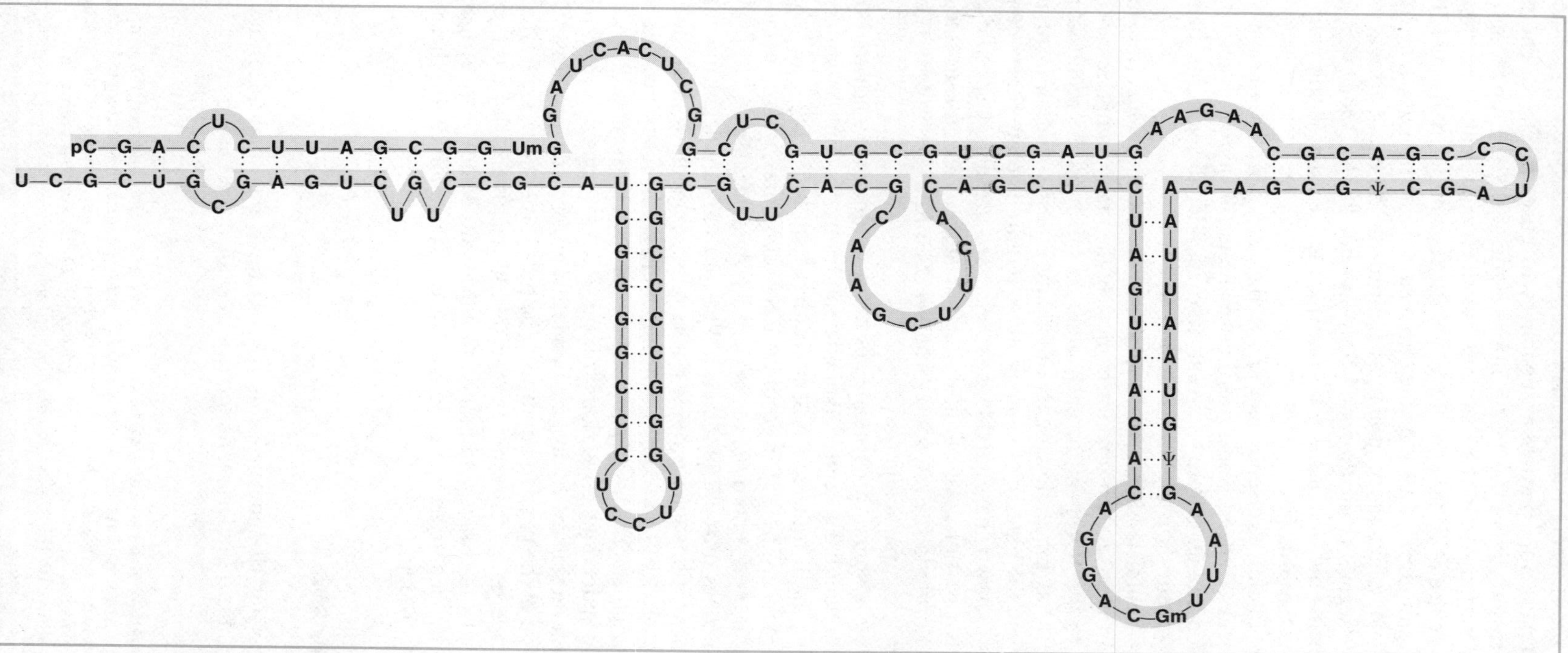

FIGURE 5–13 A proposed secondary structure for 5.8S rRNA. The drawing is modified from Nazar, Sitz, and Busch, (1975) J. Biol. Chem., *250*: 8591. The proposal is based on the primary structure of RNA from Novikoff ascites hepatoma, a malignancy of rat liver that is widely used for experimental purposes. Note that no break in the helical segments is indicated where U-G pairs occur. This is in accord with experimental findings with other kinds of RNA even though U and G are not strongly bonded.

and by the characteristics of the proteins that catalyze the modifications, both of which result from particular sequences of bases in DNA. The nature of DNA therefore determines the modifications made in RNA, even though DNA does not directly participate in the reactions by which modification occurs.

Structure of Ribosomal RNA. Each segment of ribosomal RNA folds back upon itself to form double helixes because it contains many lengths of nucleotides that are complementary in antiparallel order. The exact arrangement of complementary segments in ribosomal RNAs is not known, not even for the smallest components but one of the proposals for the 5.8S unit is shown in Figure 5–13 to illustrate the principle.

The folding into secondary and tertiary structures begins early in the maturation process. Indeed, some preribosomal and ribosomal proteins stick to the 45S precursor chains before cleavage into individual segments begins. (We shall discuss precursor proteins in Chapter 8.)

The Completed Ribosomal Particles. As the individual ribosomal RNA components appear in their final form, more of the many component protein molecules adhere to the RNA and to each other, completing the architecture of the ribosomes. It has been shown experimentally that association occurs in a particular order, with the binding of some proteins required before others can be added, and it is also known that individual proteins are present at specific locations in the completed particle. All of these findings support the reasonable deduction that each molecule in the ribosome, whether it be RNA or protein, is built to fit other molecules in a particular way.

The ribosomal RNA molecules are not completely covered by proteins; the 5S component has a sequence of nucleotides that is complementary to a sequence occurring in all transfer RNA molecules, thereby aiding attachment of tRNA to the ribosome. The 18S ribosomal RNA combines with a nucleotide sequence present at the 5′ end of all messenger RNA molecules.

Functions of the Ribosomes. What are the significant functional features of ribosomes? Ribosomes provide points of attachment for messenger RNA, and for the transfer RNA molecules bearing amino acids. They provide an environment that strengthens the hydrogen bonding between messenger RNA and the complementary anticodons on various transfer RNAs, probably a hydrophobic environment that stabilizes hydrogen bonds. They provide proteins that catalyze the formation of the peptide bonds and the provision of energy for the entire process. They undergo a structural change that moves them along the messenger RNA from triplet to triplet, and finally, they provide devices for recognizing the end as well as the beginning of the coding for a polypeptide chain, thereby releasing the finished product.

Transfer RNA

Transfer ribonucleic acids (tRNA) have the small size and high solubility required for transporting amino acid residues; the known examples have from 75

FIGURE 5–14 **The formation of an aminoacyl tRNA involves a reaction between the proper amino acid and adenosine triphosphate *(top)*, releasing PP_i and forming the aminoacyl adenosine monophosphate, which is a mixed anhydride of carboxylic and phosphoric acids. The aminoacyl group is then transferred to the terminal ribosyl moiety of the corresponding tRNA *(center right)*. The group is shown on the 3′ oxygen *(bottom center)*, but there is in fact an equilibration between the 2′ and 3′ positions. The AMP that is released *(center left)* is phosphorylated by the processes of oxidative metabolism, generating the original ATP.**

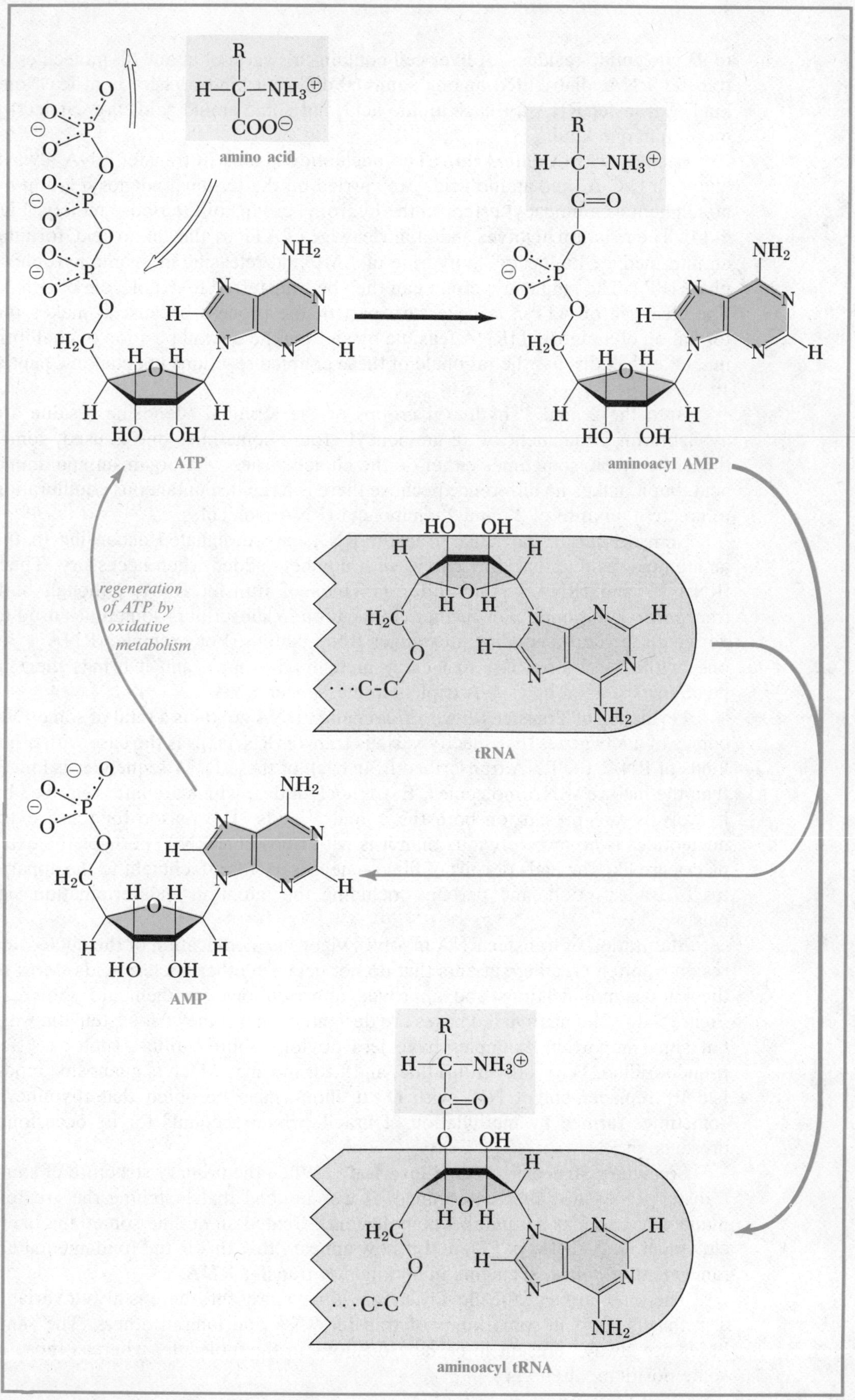

amino acid

ATP

aminoacyl AMP

regeneration of ATP by oxidative metabolism

tRNA

AMP

aminoacyl tRNA

to 93 nucleotide residues. A liver cell contains a swarm of about 10^9 molecules of transfer RNA, distributed among some 60 different kinds. There is at least one kind of transfer RNA for each amino acid, but some amino acids are carried by more than one kind.

Loading with Amino Acid. The nucleotide chains in transfer RNA always end with C-C-A, and amino acids are carried on the terminal adenosyl group by attachment as aminoacyl esters to the hydroxyl groups of the ribose moiety (Fig. 5–14). The reaction involves an initial cleavage of ATP by the amino acid, forming an intermediate aminoacyl anhydride of AMP and releasing inorganic pyrophosphate (PP_i). The aminoacyl group can then be transferred to a molecule of tRNA. The cleavage of ATP is a necessary part of the process because it makes the formation of aminoacyl tRNA feasible by shifting the overall position of equilibrium. (We shall discuss the rationale of these coupled reactions in detail in Chapter 19.)

Both the 2′ and 3′ hydroxyl groups of the terminal adenosine residue are available for attachment of an aminoacyl group; sometimes one is used, sometimes the other, sometimes either — the choice varies with organism and amino acid, but it makes no difference because there is a rapid spontaneous equilibration to create a mixture of 2′- and 3′-aminoacyl tRNA molecules.

Nomenclature. Particular transfer RNAs are designated according to the amino acid residue that they carry, with numbers added when necessary. Thus, $tRNA^{Val\text{-}1}$ and $tRNA^{Val\text{-}2}$ are different kinds of transfer RNA, although both transport valyl groups. For further clarification, a subscript is sometimes used to designate the corresponding messenger RNA codons. For example, $tRNA^{Gln-1}_{CAA}$ is one of the types of tRNA that carry glutaminyl groups, and it brings them to positions specified by C-A-A triplets in messenger RNA.

Formation of Transfer RNA. Mammalian DNA contains a total of some 1300 copies of many genes that specify various transfer RNAs; as is the case with other kinds of RNA, the RNA transcribed from each of these DNA sequences is longer than the mature tRNA molecules. Extra nucleotides, which are later removed by hydrolysis, are present on both the 3′ and 5′ ends. The reason for these extra nucleotides is no more obvious than it is with ribosomal RNA; perhaps the extra pieces are like the ends of rolls of film — necessary for attachment to the apparatus of transcription, and perhaps including the initiation and termination signals.

Maturation of transfer RNA involves vigorous modification of the nucleoside residues, often creating variants that do not occur in other nucleic acids. Most of the known modifications and approved abbreviations for them are shown in Figure 5–15. The methylated bases are designated in a somewhat systematic way, but some important examples have less obvious abbreviations, such as ψ for pseudouridine, D for dihydrouridine, and I for inosine, which is guanosine modified by replacement of NH_2 with O. It should also be noted that thymine is sometimes formed by methylation of uracil, which accounts for its occasional presence in RNA.

Secondary Structure — the Cloverleaf. When the primary structure of some transfer RNAs was first determined, it was noticed that matching the greatest number of bases as antiparallel complements created an outline something like a cloverleaf (Fig. 5–16, p. 82), and it now appears that this is the fundamental arrangement of helical segments in all kinds of transfer RNA.

The three arms of the clover leaf are always present; there is also a variable arm that is short in some kinds of transfer RNA and long in others. The same bases are always present in certain locations in the molecule, whereas those at some positions often vary.

Two parts of the clover leaf have especial functional significance. The 3′ end, with its constant -C-C-A – OH at which amino acids are attached by transport, is immediately preceded by a double helix, which is formed by pairing with the 5′ end of the chain. This structure is therefore known as the **acceptor stem** — it accepts amino acids.

The loop opposite to the acceptor stem in the cloverleaf contains the anticodon that is bound to messenger RNA. This loop, and its supporting stem, is therefore known as the **anticodon arm.**

1-methyladenosine (m^1A)

inosine (I)

1-methylguanosine (m^1G)

2-methyladenosine (m^2A)

1-methylinosine (m^1I)

N^2-methylguanosine (m^2G)

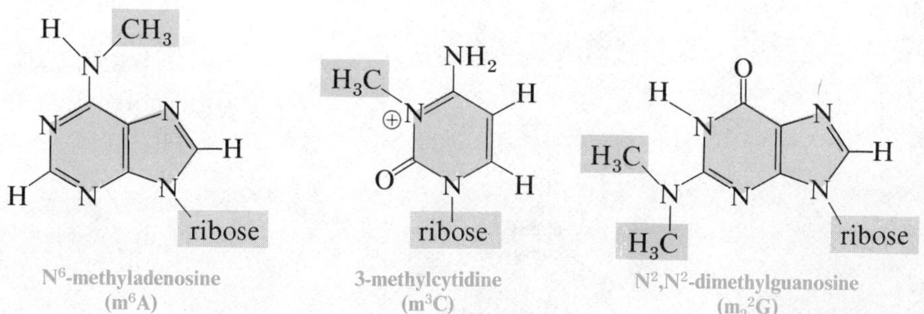

FIGURE 5–15 **Modified bases that are known to occur in various transfer RNAs and their accepted abbreviations. Pseudouridine (Ψ) was shown in Figure 5–12; 7-methyl guanosine (m^7G) and N^6,N^6-dimethyladenosine (m^6_2A) were shown in Figure 5–11. Inosine is a nucleoside containing hypoxanthine that is formed by hydrolysis of the 6-amino group in adenosine.**

Illustration continued on following page

2′-O-methylguanosine
(Gm)

2′-O-methylcytidine
(Cm)

5-methylcytidine
(m^5C)

N^4-acetylcytidine
(acC)

5-oxyacetyluridine
(acoU)

2-thiocytidine
(sC)

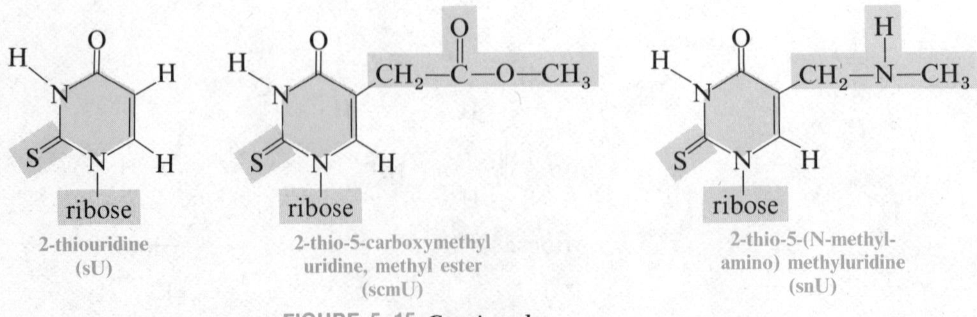

2-thiouridine
(sU)

2-thio-5-carboxymethyl
uridine, methyl ester
(scmU)

2-thio-5-(N-methyl-
amino) methyluridine
(snU)

FIGURE 5–15 *Continued*

N^6-isopentenyladenosine
(iA)

2-thio-N^6-isopentenyl
adenosine
(siA)

dihydrouridine
(hU or D)

2-methylthio-N^6-iso-
pentenyladenosine
(msiA)

N-(9-ribosyl-N-purin-
6-yl) carbamoyl threonine
(tA)

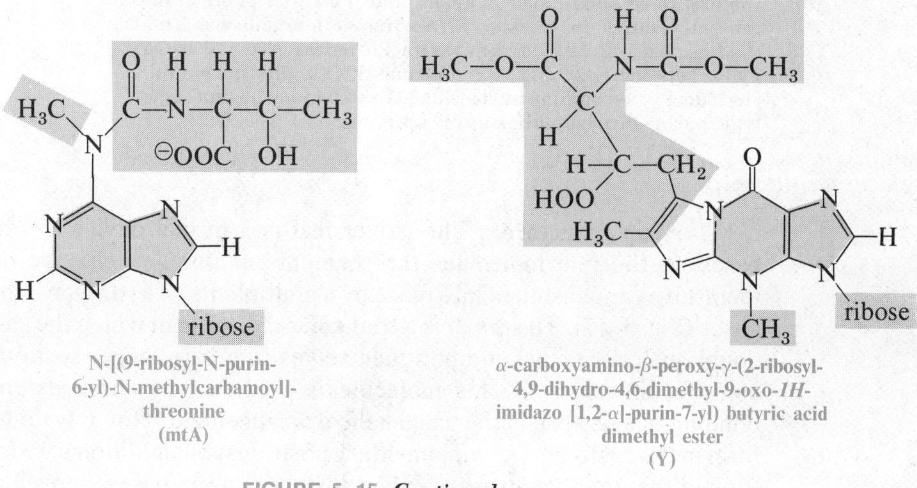

N-[(9-ribosyl-N-purin-
6-yl)-N-methylcarbamoyl]-
threonine
(mtA)

α-carboxyamino-β-peroxy-γ-(2-ribosyl-
4,9-dihydro-4,6-dimethyl-9-oxo-*1H*-
imidazo [1,2-α]-purin-7-yl) butyric acid
dimethyl ester
(Y)

FIGURE 5–15 *Continued*

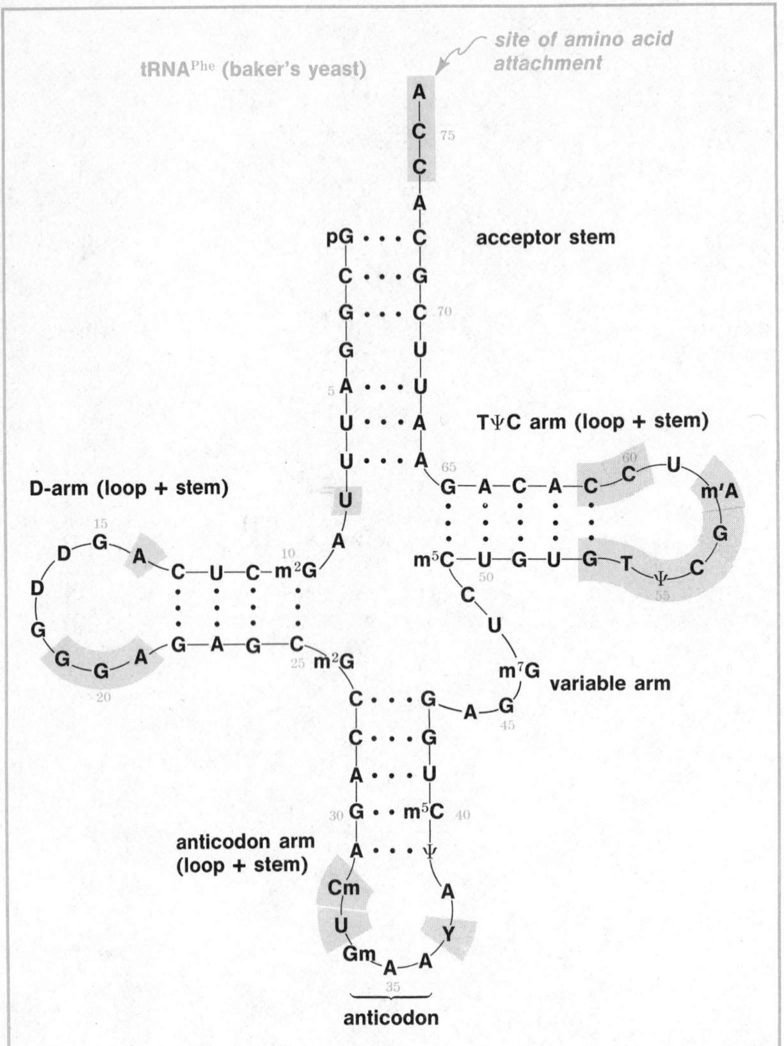

FIGURE 5–16 **The secondary structure of a transfer RNA. This example was the first to be determined in detail, and illustrates many of the general features in transfer RNAs from all organisms. *Blue shading*: base always present. Extra Pu before A-4 and extra Py-Pu between G-20 and A-21 in some tRNAs. (Pu, purine nucleoside; Py, pyrimidine nucleoside.) *Gray shading*: base of same type, purine or pyrimidine, always present.**

Tertiary Structure. The clover leaf is a formal device for depicting paired bases; in the real molecules the segments of double helix are bent back upon themselves and fastened into place by a multiplicity of hydrogen bonds between the bases (Fig. 5–17). The result is a triangular molecule in which the short segments of double helix are stacked upon themselves to create longer sections of helix. The compact structure of this molecule is achieved by more extensive hydrogen bonding than is seen in the usual helical arrangement. Not only do hydrogen bonds form in the customary complementary positions, but additional hydrogen bonds are formed that involve atoms at other positions in the bases, as well as atoms in the sugar and phosphate backbones. Every base in the molecule is involved in

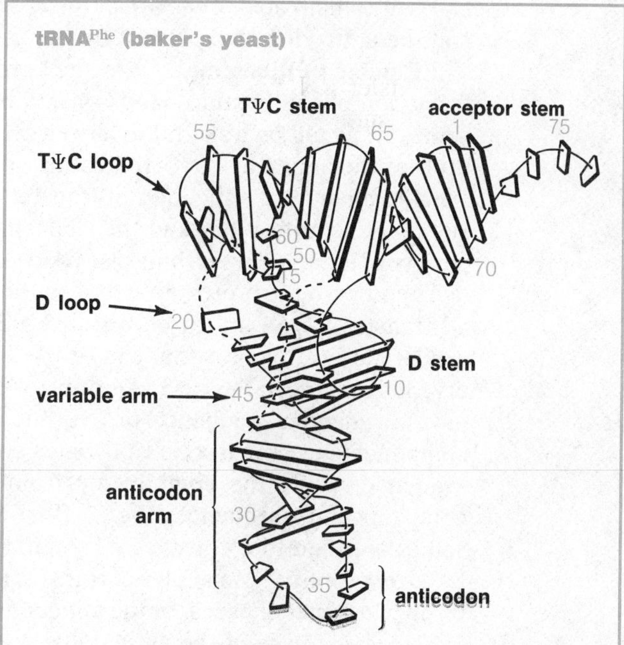

FIGURE 5–17

The tertiary structure of a transfer RNA. Transfer RNAs are bent upon themselves so as to stack the helical segments shown in Fig. 5–16 into two long helical segments at right angles to each other. This figure indicates pairs of bonded bases by bars, with the ribose-phosphate backbone threaded through them. Segments of double helix (secondary structure) are held in position by the formation of additional hydrogen bonds, involving the phosphate-sugar backbone as well as the bases. Modified slightly from an illustration by S. H. Kim, (1975) Nature, *256*: 680.

hydrogen bonding except for the C-C-A at the 3′ terminal and the three bases in the anticodon. Those that are not a part of antiparallel complements form other types of bonds. We have a preview in these small molecules of the complexity we ought to expect in the structures of the much larger ribosomal RNA molecules, which are yet to be determined in detail.

Recognition by Aminoacyl tRNA Synthetases. Molecules of transfer RNA have a complex form and contain a variety of unusual structures in order to perform their critical role in translating nucleotide sequences into amino acid sequences. They must form specific bonds with several components during translation, especially with the aminoacyl-tRNA synthetases, which are the proteins that catalyze the ATP-driven attachment of amino acids.

There is one synthetase for each amino acid, and it recognizes all of the kinds of tRNA that are built to carry that amino acid. When we think on it, we see that each of these enzymes must bind molecules of transfer RNA in two ways. It must recognize the C-C-A terminal to which amino acid groups are always attached; since all transfer RNAs are alike in this respect, it is probable that all of the synthetases form the same sort of bonds with the terminal region. On the other hand, the enzyme must recognize some specific regions in the transfer RNAs that are built to carry an individual amino acid; all of the $tRNA^{Val}$, no matter how many there may be, most have some specific surface features for recognition by the enzyme that attaches valine to the tRNA, but that will not be recognized by the other 19 synthetases. These specific features must arise from the variable regions of the molecule, that is, those base sequences that are not alike in all transfer RNAs, which are indicated in Figures 5–16 and 5–17.

Similarly, the aminoacyl synthetase proteins must also bind one kind of amino acid. We shall have more to say about the origin of specificity of enzymes in later chapters, noting now that it is crucial to the accuracy of translation for each synthetase to contain a region forming highly specific bonds with one amino acid and another region forming equally specific bonds with the transfer RNAs that carry the amino acid. Otherwise, the correct correspondence between amino acid and the anticodon also present on the transfer RNA will not exist.

Binding to Ribosomes. We shall see that the processes of translation — the formation of the peptide bonds — involves several protein components of ribosomes. It would be wasteful to have a separate machinery of peptide synthesis for each of the amino acids — the nature of the amide bond is the same in all. The specific binding of anticodon in transfer RNA with triplet on messenger RNA is sufficient identification, and the ribosome is so constructed that once this initial password is given, everything else proceeds in the same way. One amino acid looks like another to the proteins catalyzing the formation of peptide bonds. However, if all transfer RNAs are to be handled alike by ribosomes, they must have some similar surface features that can be used for attachment to the ribosomal machinery. This is indeed the case. All transfer RNAs except the initiator tRNA (below), have an identical sequence of five nucleotides in one arm; all known 5S RNA components in prokaryotic ribosomes contain the antiparallel complement of the sequence, while the components from eukaryotic ribosomes have a partially complementary sequence (Fig. 5–18). Transfer RNAs and the 5S RNA of ribosomes are therefore constructed to bind to each other.

Anticodons. Molecules of transfer RNA become attached to messenger RNA through a triplet of bases, or the anticodon, that is freely available at one apex (Fig. 5–17). The linkage of the antiparallel complements — anticodon and codon — is facilitated by the favorable environment provided in the ribosomes, but the positioning of the anticodon for easy consummation of the union is evidently aided in another way. The anticodons in tRNA, again save one, always occur in this sequence: -Py-U-N-N-N-Pu, in which Py and Pu are pyrimidine and purine nucleosides, respectively, and the Ns are the anticodon nucleosides. The flanking -Py-U- and -Pu- positions may well attach to specific positions on the ribosomes so as to hold the anticodon between them in an approachable posture.

The anticodons themselves have some common characteristics. The base in the initial nucleotide in the sequence — the one that binds to the 3′ end of the codon in messenger RNA — is often modified, and the second nucleotide sometimes contains pseudouridine instead of uridine, but no other alterations are known; the third nucleotide is never altered.

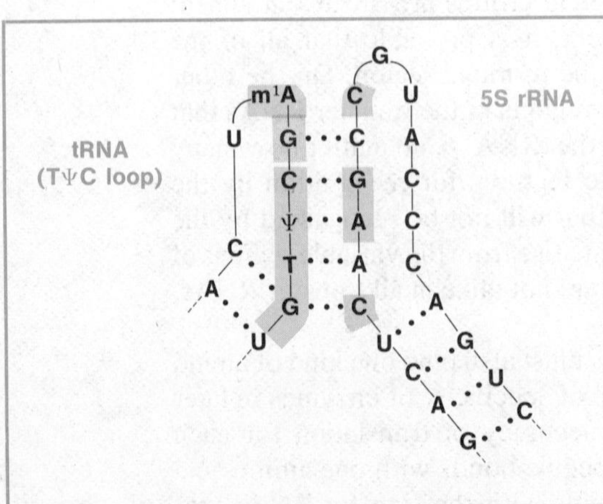

FIGURE 5–18

Those transfer RNAs bringing amino acids for elongation of a polypeptide chain are bonded to the 5S RNA in ribosomes through an antiparallel complementary sequence involving the TΨC loop. Prokaryotes appear to use nearly complete complementary binding as shown. Some of the bases, shaded with blue, are also present in eukaryotes, but not all. Initiator transfer RNAs do not have the TΨC complementary sequence and therefore do not bind to ribosomes at the same place. (*Blue shading:* base always present. *Gray shading:* similar base, Pu or Py, always present.)

The Wobble Hypothesis. The binding of anticodons and codons can occur when they are not exact antiparallel complements; the permitted exceptions involve the third base of the codon and the first base of the anticodon (Fig. 5–19). The observed facts are that initial nucleosides in the anticodon will bind with final nucleosides in the codons as follows:

Anticodon Position #1	*binds with*	Codon Position #3
C		G
G		C, or U
I		A, C, or U
U		A, or G

The explanation advanced for these apparent exceptions to the usual rules for antiparallel complements is that it is possible for the third base of the messenger RNA triplet to deviate ("wobble") from its customary orientation sufficiently to accommodate the formation of different hydrogen bonds.

Whatever its origin, the greater versatility of binding by the anticodons eliminates the necessity of constructing tRNA molecules that are antiparallel complements for each and every messenger RNA codon. For example, a $tRNA^{Ala}$ carrying the anticodon I-G-C can be bound to three different triplets, G-C-A, G-C-C, and G-C-U, designating alanyl groups on messenger RNA. Only one additional kind of $tRNA^{Ala}$ is required to match the remaining Ala codon (G-C-G).

The use of I-G-C illustrates a case in which the modification of the base in the anticodon (G changed to I) has extended its possibilities for binding. On the other hand, the freedom of binding of the first base can be limited by some modifications. Changing uridine to 2-thiouridine makes it incapable of binding any nucleoside

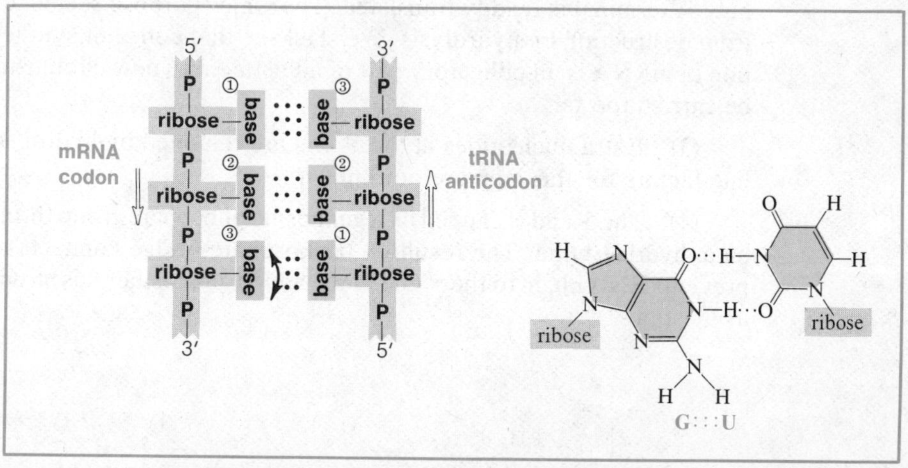

FIGURE 5–19 **The wobble hypothesis. The third base of the codon in messenger RNA need not be an exact complement of the first base in the transfer RNA anticodon in order to bond effectively. It is hypothesized that the third base in the codon can move (wobble) enough so as to permit effective bonding of tRNA in other cases. For example, guanine in an mRNA codon can bond with uracil in a tRNA anticodon as shown at the right. See the text for other examples and the references at the end of chapter for a recent alternate hypothesis.**

other than adenosine. Table 5–2, which lists the known anticodon sequences, has some other examples, such as sU-U-C, which will fit only the G-A-A codon for glutamyl groups.

Messenger RNA

The precursors of messenger RNA first appear in heterogeneous nuclear RNA (hnRNA), which is a mixture of many different molecules of RNA, with only a few of each kind. Since the genes for many different proteins are being transcribed at once, it is reasonable to expect the product of transcription to be a heterogeneous mixture. However, as little as 10 per cent of the mass of hnRNA survives to be transported as mRNA into the cytosol. The function of the discarded segments is not known; they probably include transcripts of many regulatory and spacer sequences between, and perhaps sometimes within, structural genes that are unnecessary for later translation of the intervening messages. The heterogeneous nuclear RNA may include some molecules that are not precursors of messenger RNA. In any event, those hnRNA molecules that are precursors are extensively modified before they leave the nucleus as mRNA.

The maturation of messenger RNA involves covalent modification of some nucleotides in addition to partial hydrolysis of overly long precursors; it also involves extension of the polynucleotide chain without direct instructions from DNA. The order of events is something like this:

(1) DNA is transcribed to form RNA in which the message for amino acid sequence is sandwiched between terminal sequences that include some kind of identification. The identification may include palindromes, repeated sequences, a particular tertiary structure, or some presently unknown signal.

(2) The RNA is recognized as a candidate mRNA by an enzyme that catalyzes the addition of adenosine phosphate groups to its 3′ end, as many as 200 of them. Not all candidate mRNAs are tagged in this way, and mRNA without a poly-A terminal is readily translated. The major purpose seems to be a protection from destruction by hydrolysis. We shall see that constant synthesis and destruction of mRNA is an obligatory part of adjustment to new circumstances, but it can be carried too far.

(3) Extra nucleotides at the 5′ end are removed by hydrolysis. The controlling factors for this step are not known.

(4) The 5′ end is capped by a guanosine phosphate group through formation of an anhydride bond. The resultant triphosphate bridge connects the 5′ end of the previous RNA chain to the 5′ end of the added guanosine; this structure is indicated as $G^{5'}$-ppp-. . . .

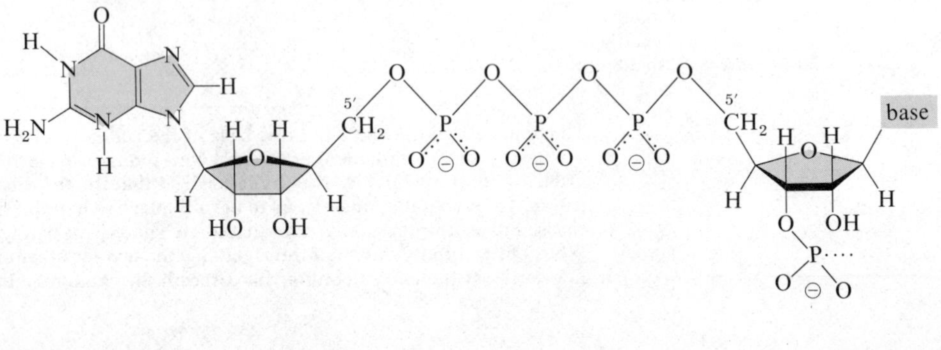

H N O N H H_2N N H N H H 5′ CH_2 O H H O H HO OH O P O ⊖ O O P O ⊖ O O P O ⊖ O O 5′ CH_2 H H O base H H O OH P.... O ⊖ O

TABLE 5–2 SOME KNOWN ANTICODON SEQUENCES

The anticodon is in boldface. Preceding and following residues also are given. The first nucleoside of the anticodon binds to the third nucleoside of the codon. When more than one codon will fit, the alternative third nucleosides are separated by diagonals.

Amino Acid Residue	Anticodon Sequence	Matching Codon	Organism
Ala	U-**I-G-C**-m^{1}I	G-C-A/C/U	baker's yeast
Asp	U-**G-U-C**-m^{1}G	G-A-U/C	brewer's yeast
Arg	U-**I-C-G**-m^{2}A	C-G-A/C/U	*E. coli*
Gln	U-**C-U-G**-m^{2}A	C-A-G	*E. coli*
	U-**sU-U-G**-m^{2}A	C-A-A	*E. coli*
Glu	U-**sU-U-C**-m^{2}A	G-A-A	*E. coli*
Gly	U-**C-C-C**-A	G-G-G	*E. coli*
	U-**U-C-C**-A	G-G-A/G	*E. coli*, phage T4
	U-**G-C-C**-A	G-G-U/C	*E. coli*
His	U-**xG-U-G**-m^{2}A	C-A-U/C	*E. coli*
Ile	U-**I-A-U**-tA	A-U-A/C/U	torula yeast
	U-**G-A-U**-xA	A-U-C/U	*E. coli*
Leu	U-**C-A-G**-xG	C-U-G	*E. coli*
	U-**xU-A-A**-siA	U-U-A/G	phage T4
	U-**m^{5}C-A-A**-m^{1}G	U-U-G	baker's yeast
Lys	U-**scU-U-U**-tA	A-A-A	baker's yeast
	U-**C-U-U**-xA	A-A-G	baker's yeast
Met$_f$	U-**C-A-U**-A	A-U-G	*E. coli*
	U-**C-A-U**-xA	A-U-G	baker's yeast
Met$_m$	U-**xC-A-U**-tA	A-U-G	*E. coli*
Phe	U-**Gm-A-A**-Y	U-U-C/U	baker's yeast, wheat
	U-**G-A-A**-msiA	U-U-C/U	*E. coli*
Ser	U-**I-G-A**-iA	U-C-A/C/U	baker's yeast, rat
	U-**aU-G-A**-msiA	U-C-A/G	*E. coli*
	U-**G-C-U**-tA	A-G-C/U	*E. coli*
Thr	U-**G-G-U**-tA	A-C-C/U	*E. coli*
Trp	U-**C-C-A**-mtA	U-G-G	*E. coli*
	U-**Cm-C-A**-A	U-G-G	brewer's yeast
Tyr	U-**G-Ψ-A**-iA	U-A-C/U	baker's yeast
	U-**xG-U-A**-msiA	U-A-C/U	*E. coli*
Val	U-**G-A-C**-tA	G-U-U/C	*E. coli*

Abbreviations:
aU = 5-oxyacetyluridine
I = inosine
Gm = 2′-O-methylguanosine
m^{2}A = 2-methyladenosine
m^{5}C = 5-methylcytidine
m^{1}G = 1-methylguanosine
m^{1}I = 1-methylinosine
msiA = 2-methylthio-6-isopentenyladenosine
mtA = 9-ribosyl-N-(purin-6-yl-N-methylcarbamoyl)threonine
scU = 2-thio-5-carboxymethyluridine, methyl ester
siA = 2-thio-6-isopentenyladenosine
snU = 2-thio-5-(N-methylaminomethyl)uridine
sU = 2-thiouridine
tA = 9-ribosyl-N-(purin-6-ylcarbamoyl)threonine
XA, XG, or XU = unknown derivative of A, G, or U
Y = fluorescent derivative of G
Ψ = pseudouridine

(5) Covalent modifications of the terminal segments, but not of the intermediate message, are made, especially methylations of bases and ribose units. Some of the modifications may occur earlier, but not all, because the initial capping groups are converted to m^7Gpppm^6Am. . . . This initial structure is important for later recognition as messenger RNA by ribosomes.

(6) The messenger RNA moves to the cytosol, with partial removal of some of the poly-A tail en route.

Initiation of Translation. Proper attachment of the messenger RNA to ribosomes hinges upon the methylguanosine cap at the 5′ end, but the most critical event for translation is the recognition of the nucleoside — the one and only nucleoside — at which the message begins. This nucleoside is always A at the beginning of the sequence . . . A-U-G · · ·, which is the codon for a methionyl group. The synthesis of all peptides in the cytosol of eukaryotic cells begins with the incorporation of methionine. Prokaryotic cells and the mitochondria and chloroplasts of eukaryotic cells have a somewhat different twist; they begin synthesis with the N-formyl derivative of methionine:

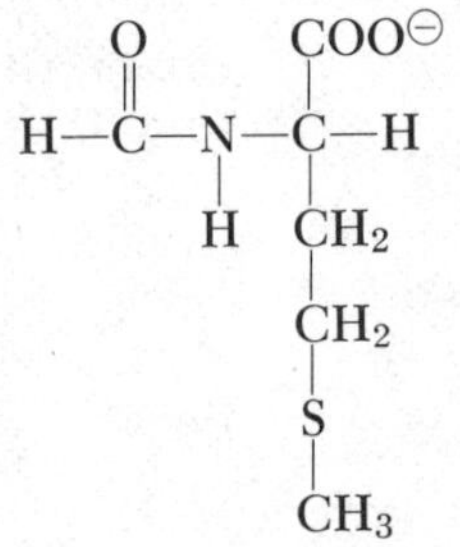

$$\begin{array}{c} \text{O} \qquad\quad \text{COO}^{\ominus} \\ \| \qquad\qquad | \\ \text{H—C—N—C—H} \\ | \qquad\quad | \\ \text{H} \quad \text{CH}_2 \\ | \\ \text{CH}_2 \\ | \\ \text{S} \\ | \\ \text{CH}_3 \end{array}$$

N-formyl-L-methionine

The methionyl group is often removed before the polypeptide chain is completed in eukaryotes so that the finished polypeptide chain begins with the amino acid coded after the A-U-G triplet; the N-formyl group is removed, but methionine remains as the initial amino acid in many proteins in prokaryotes.

A problem is obvious: methionyl groups occur at many points in most polypeptide chains, and all of these positions are designated by the single methionyl code, A-U-G, in messenger RNA. How is one of several identical triplets singled out as the correct starting point? Indeed, how is it distinguished from the same sequence that occurs in successive codons? G-A-U-G-C-C could be two successive codons meaning Asp-Ala, or it could be an initiating A-U-G sequence between other nucleotides. It is apparent that there is some previous signal in the messenger RNA that causes binding to ribosomes at one specific location.

THE TRANSLATION MECHANISM

General Description

Polypeptide chains are built one amino acid residue at a time on ribosomes bound to messenger RNA. The machinery of peptide synthesis contained in the ribosomes cranks through the same sequence of events, over and over again, with the ribosome advancing three nucleotide residues along the messenger RNA as each amino acid is added to the growing polypeptide. Molecules of transfer RNA carry amino acids to the site of peptide synthesis, where they are bound to messenger RNA, anticodon to codon. Each emptied molecule of transfer RNA is discharged from the ribosome before a loaded molecule corresponding to the next triplet in line can be bound.

TABLE 5-3 SUMMARY OF RIBONUCLEIC ACIDS

Relative sizes (shown as if compacted spheres)*:

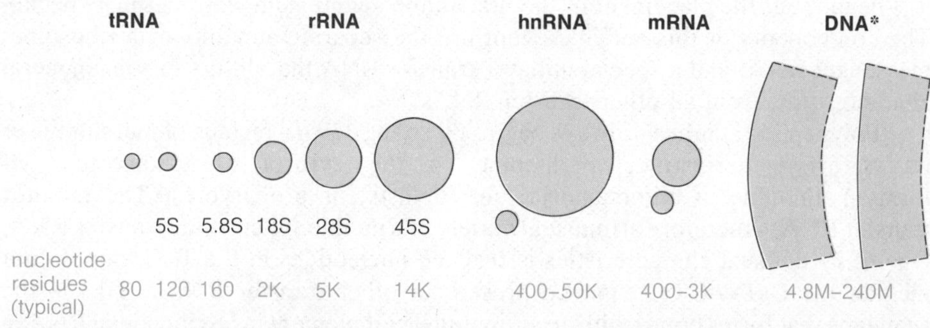

Ribosomal RNA (rRNA)

Formed as 45S component in nucleolus by amanitin-insensitive polymerase; 5S component formed separately
Functional sequences extensively methylated
Intervening spacers in 45S component removed by hydrolysis to create 5.8S, 18S, and 28S components
Internally complementary base sequences form A or A′ helices (11 or 12 base pairs per turn) within themselves
Ribosomes contain one each of the RNA components and many protein molecules.
Ribosomes contain two subunits—a 40S particle and a 60S particle (30S and 50S in bacteria), which combine to form 80S intact ribosomes in cytoplasm
40S ribosomal subunit contains 18S RNA with binding site for messenger RNA
60S ribosomal subunit contains 28S, 5.8S, and 5S RNAs; 5S component binds transfer RNAs
Ribosomes have a pocket through which mRNA and tRNA pass during synthesis of polypeptide chain

Transfer RNA (tRNA)

Precursors formed in nucleoplasm by polymerase moderately sensitive to amanitin
Bases and sugars extensively modified, and excess nucleotides on 5′ end removed by hydrolysis
Mature molecule contains 75 to 88 nucleotide residues
Cloverleaf pattern created by secondary structures when spread out
Molecule folds into compact tertiary structure resembling right triangle; anticodon and attachment site for amino acids at opposite ends of dished hypotenuse
Always has -C-C-A at 3′ end
Amino acids attached by ester bonds to terminal ribosyl group on 2′ or 3′ hydroxyl groups
Some areas of constant structure for attachment to ribosomes and maintenance of conformation
Some areas with variable composition to generate specific binding by one aminoacyl tRNA synthetase
One amino acid may be carried by different kinds of tRNA, all of which react with one aminoacyl tRNA synthetase

Messenger RNA (mRNA)

Precursors transcribed in nucleoplasm by polymerase highly sensitive to amanitin
Precursors included in heterogenous nuclear RNA (hnRNA)
Precursors shortened by hydrolysis and methylated without affecting central message for amino acid sequence
As many as 200 adenylyl groups added to 3′ end, apparently for stabilization after passage to cytosol; many of these later removed by hydrolysis
Terminal portions contain signals for initiation and termination of translation as polypeptide chain
Structure not known in detail; 5′ end may have ordered conformation, but central message believed to have less secondary structure than rRNA or tRNA in most cases
Size of molecule varies with length of polypeptide chain it represents; some with 100 nucleotide residues, some with 3,000, and more
Translation begins with incorporation of methionyl residue (formylmethionyl in prokaryotes and mitochondria)
Initiation occurs at triplet A-U-G that is located at specific point relative to uncharacterized initiation signals (palindromes?)

*Fragments of the largest and smallest human DNA molecules are depicted to show radii of curvature. The entire smaller sphere would be wider than the page at the scale used. None of the nucleic acids are present as compact spheres in real cells.

Initiation of Translation

The building of a polypeptide chain goes along more or less automatically once it is begun, but the placement of the first amino acid residue is something special. The components of this seminal event are the separate subunits of a ribosome, messenger RNA, and a special initiator transfer RNA that differs in some general characteristics from all other transfer RNAs.

Polypeptide synthesis always begins by laying down a residue of methionine or its N-formyl derivative; methionine in the cytosol of eukaryotes, N-formylmethionine in mitochondria, chloroplasts, or prokaryotes. The initiator transfer RNA is therefore an unusual variety of a methionyl-carrying transfer RNA. One of its unusual characteristics is that the nucleotides in the TψC loop do not include the G-T-ψ-C-G sequence present in other transfer RNAs. This is the sequence that forms bonds with an antiparallel complement in 5S ribosomal RNA, a component of the heavy ribosomal subunit (60S). Instead, the initiator transfer RNA is built to bind to the light ribosomal subunit (40S).

In order to distinguish the initiator transfer RNA from transfer RNA carrying methionyl groups for incorporation into the middle of peptide chains, the two types of molecules are designated as $tRNA_f^{Met}$ and $tRNA_m^{Met}$, respectively (f for formylmethionyl, even with the eukaryotes that use methionyl groups on the initiator). The latter, $tRNA_m^{Met}$, has the same G-T-ψ-C-G loop found in other transfer RNAs and does not bind to an isolated light ribosomal subunit.

Let us now examine the mechanism of translation in detail:

Step One

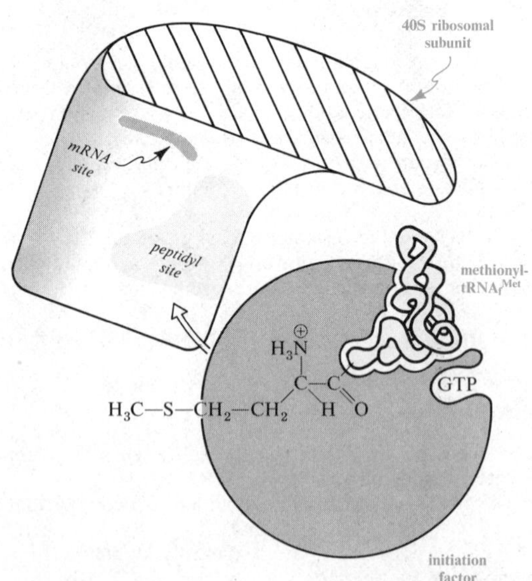

A molecule of the initiator transfer RNA carrying a methionyl group (Met-$tRNA_f^{Met}$) combines with an **initiation factor,** which is a protein that also binds a molecule of GTP. The complex that is formed fits a particular location on a light ribosomal subunit. The presence of the initiation factor prevents the light subunit from combining with a heavy subunit to form a complete ribosome. The initiator transfer RNA occupies an important part of a peptidyl-binding site to be discussed shortly.

Step Two

The initial segment of a molecule of messenger RNA binds in a specific way to the light ribosomal subunit through combination with an exposed complementary sequence in the 18S RNA of the ribosome. The messenger RNA is oriented so as to

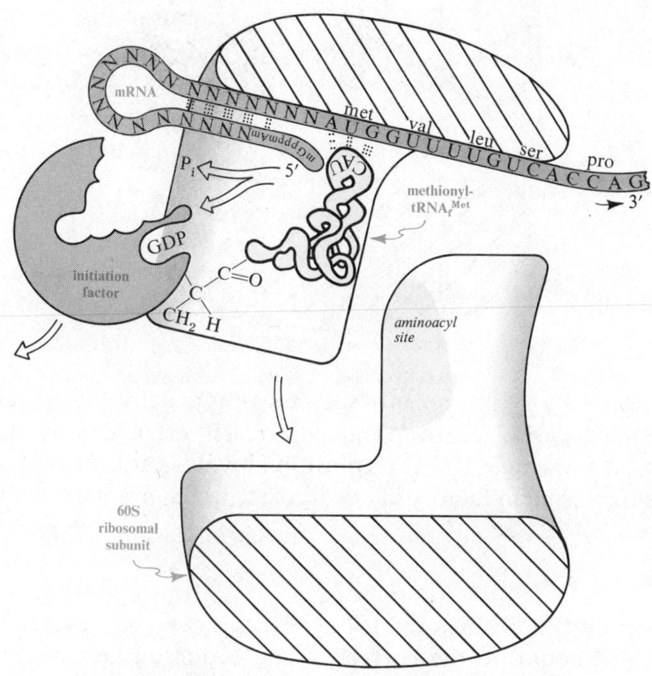

bring the initiator codon A-U-G in contact with the C-A-U anticodon of tRNA$_f^{Met}$. Other protein factors (not shown) combine to complete the initiation complex. (Messenger RNA is depicted with an unknown beginning sequence of nucleotides, – N, followed by coding suitable for creating the polypeptide component of hemoglobin.)

Completion of the complex brings GTP into contact with a protein in the light subunit that catalyzes its hydrolysis into GDP and inorganic phosphate (P_i). The inorganic phospate is released, but GDP remains bound to the initiation factor. However, the conversion of GTP to GDP causes an alteration in the conformation of the initiation factor so that it can no longer remain attached to transfer RNA and the light ribosomal subunit, and the protein dissociates from the subunit.

Release of the empty initiation factor has an important consequence; it exposes sites on the light ribosomal subunit that bind the heavy ribosomal subunit to form a complete ribosome. The heavy subunit brings to the complex the major part of an aminoacyl site used for binding another molecule of transfer RNA, as shown in the next step.

Step Three

When the two subunits of a ribosome combine around messenger RNA and the attached Met-tRNA$_f^{Met}$, they form the aminoacyl site that binds another molecule of transfer RNA — one with an anticodon corresponding to the next codon in the mRNA sequence. The second molecule of transfer RNA, and all succeeding molecules are brought to the ribosome in combination with an **elongation factor**, a protein that is similar to the initiation factor in that it also binds GTP. However, the

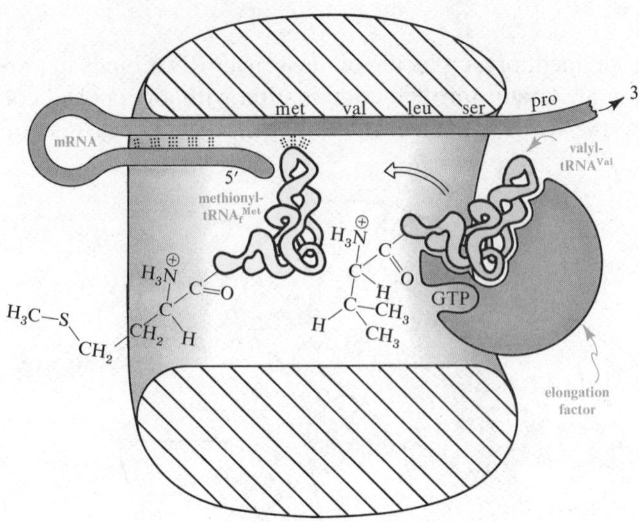

elongation factor binds to ribosomes in such a way as to deliver its attached transfer RNA to the aminoacyl site, where its anticodon can be brought into proximity to the second codon on messenger RNA. (Elongation involves at least one, and perhaps several, additional protein factors in eukaryotes, which are not shown in the diagram.)

Step Four

The correct transfer RNA, valyl-tRNA^{Val} in this case, becomes firmly attached because it is also bound to the corresponding codon on messenger RNA. The

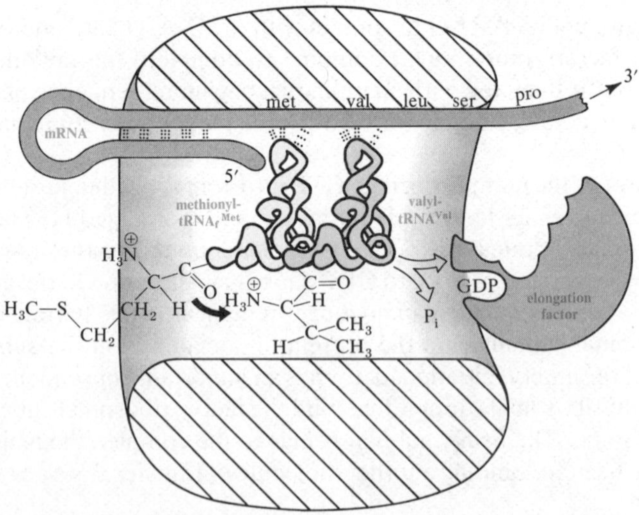

binding brings the associated GTP molecule near a protein catalyzing its hydrolysis into GDP and P_i, perhaps the same protein that catalyzes a comparable hydrolysis in the initiation step. The resultant GDP-elongation factor, like the GDP-initiation factor, has only a weak affinity for the ribosome, and it dissociates so as to leave behind the loaded transfer RNA it brought to the ribosome. The net effect is that the combination of transfer RNA molecules with ribosomes is driven by a hydrolysis of

GTP to GDP and P_i; the initiation and elongation factors are intermediaries to this end.

When the elongation factor leaves, the first two amino acid residues are in place on their respective tRNAs, ready for synthesis of the first peptide bond. Synthesis occurs by moving the methionyl group from its transfer RNA onto the ammonium group of the adjacent valine residue. (It is not known exactly how the transfer RNAs are aligned so as to bring the amino acid residues into proximity.) The reaction is catalyzed by peptide synthetase, another protein in the heavy ribosomal subunit.

Step Five

Transfer RNAs have high affinity for the peptidyl site only when they are loaded with amino acids. When the initiating $tRNA_f^{Met}$ loses its attached methionyl group, it dissociates from the ribosome into the cytosol. (Once released, it can be loaded again with a methionyl group for initiation of peptide synthetis on another ribosome.)

The $tRNA^{Val}$ still bound to the valyl codon now carries a methionylvalyl group, and the entire ribosome moves around it along the messenger RNA for a distance of three nucleotide residues. It is not shown, but this movement requires the presence of still another protein carrying GTP, a **translocation factor.** The hydrolysis of GTP and the resultant dissociation of the translocation factor causes the conformational shift by which the ribosome moves.

Translocation of the ribosome on messenger RNA brings methionylvalyl-$tRNA^{Val}$ to the peptidyl site of the ribosome, leaving the aminoacyl site vacant. The empty site is then filled by another loaded transfer RNA corresponding to the next codon in sequence, the codon for a leucyl group in this case. Here again, the transfer RNA is brought in by formation of a complex with an elongation factor and GTP.

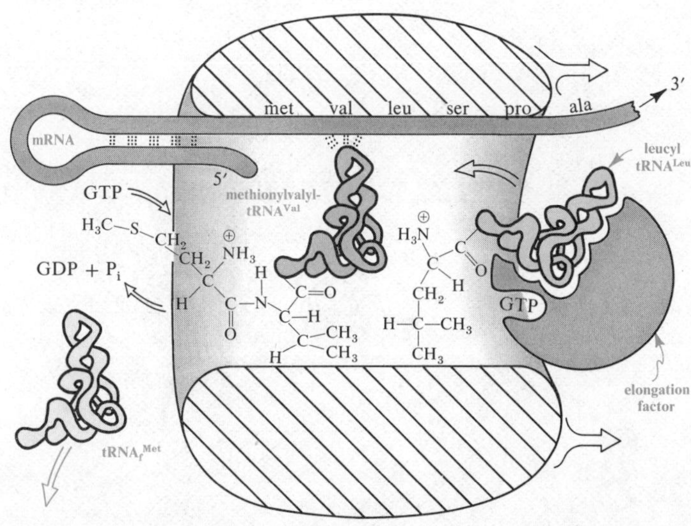

Step Six

The building of the polypeptide chain proceeds by a repetition of the steps outlined above. The elongation factor that brought in leucyl-$tRNA^{Leu}$ leaves when its associated GTP is hydrolyzed to GDP and P_i, but the tRNA remains on the ribosome with its anticodon bound to the Leu codon on messenger RNA. The peptidyl group (methionylvalyl) is then transferred from $tRNA^{Val}$ to the amino group on the leucine residue, creating the second peptide bond.

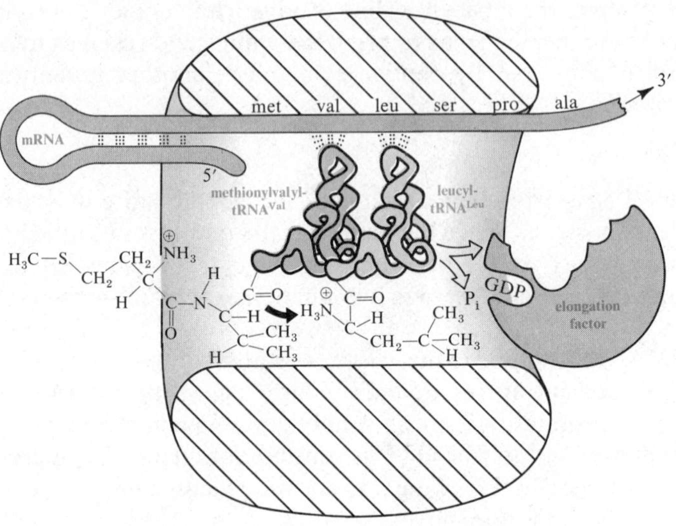

Step Seven

Once more, the emptied transfer RNA, $tRNA^{Val}$, dissociates from the ribosome, and the ribosome moves three nucleotide residues down the messenger RNA chain with a concomitant hydrolysis of GTP. This translocation brings the transfer

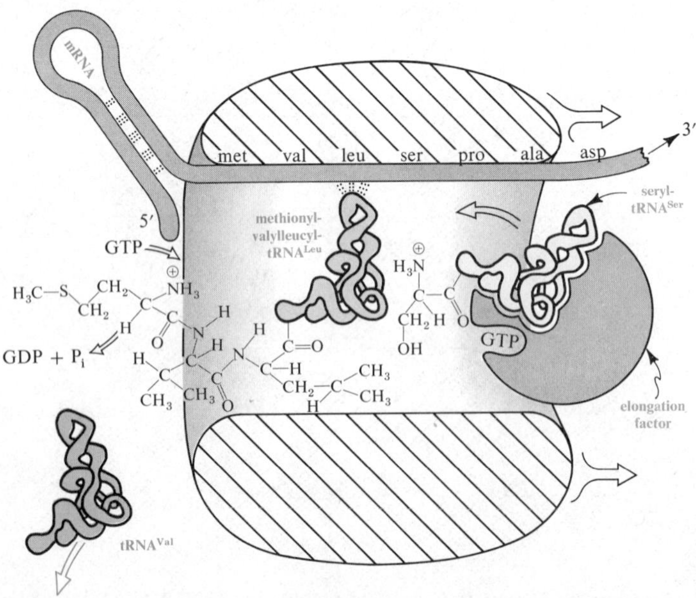

RNA with the newly formed peptidyl group (Met-Val-Leu-$tRNA^{Leu}$) onto the peptidyl site of the ribosome and exposes the aminoacyl site to entry by the next

loaded transfer RNA in complex with GTP and an elongation factor, the transfer RNA here carrying a seryl group.

Step Eight ff. (not shown)

Peptide synthesis, translocation, and charging the ribosome with the next aminoacyl tRNA are repeated until the polypeptide chain is completed. Growth of the chain ends when the ribosome moves over one of the termination codons. At this stage the completed polypeptide chain is attached to the transfer RNA that brought in the final amino acid residue, the C-terminal residue of the chain. The presence of a terminator triplet, perhaps reinforced by some additional signal in subsequent mRNA nucleotides, activates **termination factors.** These protein factors appear to be like the initiation, elongation, and translocation factors in that they appear to bind GTP in one conformation, and shift to another conformation upon hydrolysis of the GTP to GDP and P_i. Four separate steps, initiation, elongation, translocation, and termination, appear to proceed by related mechanisms, each driven by the hydrolysis of GTP. In the case of the termination factors, the conformational shift is associated with hydrolysis of the completed polypeptide chain from its ester linkage with transfer RNA, and with release of the free polypeptide and the emptied transfer RNA from the ribosome. The ribosome is then free to dissociate from messenger RNA by combination of the light subunit with initiation factors, ready to begin the entire sequence once more.

To repeat a point made earlier, the N-terminal portion of the polypeptide is modified during synthesis. In eukaryotes, such as man, the terminal methionyl group is usually hydrolyzed at some point, so that the released chain frequently begins with the amino acid specified by the second codon in messenger RNA, a valyl group in our example. Bacteria retain methionine as the first residue for many proteins, but frequently remove the masking formyl group, so that the N-terminus is a free ammonium group.

FURTHER READING

Reviews

Weissbach, H., and S. Pestka, eds.: (1977) *Molecular Mechanism of Protein Biosynthesis.* Academic Press.

Lewin, B.: (1974) *Gene Expression,* 2 vols. Wiley. The best detailed survey of nucleic acid function.

Maniatis, T., and M. Ptashne: (1976) *A DNA Operator-repressor System.* Sci. Am., *243*: 64.

Fiddes, J. C.: (1977) *The Nucleotide Sequence of a Viral DNA.* Sci. Am., *237*(6): 54. Disconcerting news of overlapping genes.

Chambon, P.: (1975) *Eukaryotic Nuclear RNA Polymerases.* Annu. Rev. Biochem., *44*: 613.

Nomura, M., A. Tissieres, and P. Lengyel, eds.: (1974) *Ribosomes.* Cold Spring Harbor Laboratory Press. Begins with general surveys. Contains versions of ribosome structure.

Kurland, C. G.: (1977) *Structure and Function of the Bacterial Ribosome.* Ann. Rev. Biochem., *46*: 173.

Sigler, P. B.: (1975) *An Analysis of the Structure of tRNA.* Annu. Rev. Biophys. Bioeng., *4*: 477. Excellent correlation of structure and function. Lack of subject index eliminates usefulness as reference.

Rich, A., and S. H. Kim: (1978) *The Three-dimensional Structure of Transfer RNA.* Sci. Am., *238*(1): 52.

Lewin, B.: (1975) *Units of Transcription and Translation: The Relationship between Heterogeneous Nuclear RNA and Messenger RNA.* Cell, *4*: 11.

Anderson, W. F., et al.: (1977) *International Symposium on Protein Synthesis.* FEBS Letters, *76*: 1. Valuable short summary of components involved in translation and their action as known in late 1976.

Brinacombe, R., G. Stöffler, and H. G. Wittman: (1978) *Ribosome Structure*. Annu. Rev. Biochem., *47*: 217.

Richards, E. G.: (1978) *Tilting at Wobble*. Nature, *273*: 488. Some of the responses of tRNA to multiple codons may be due to 2-base coding rather than wobble.

Leder, P.: (1978) *Discontinuous genes*. N. Engl. J. Med., *298*: 1079. Includes electron micrograph of globin mRNA and DNA hybrid, showing discontinuity of information.

Experimental Reports

Kim, S. H., et al.: (1974) *Three-dimensional Tertiary Structure of Yeast Phenylalanine Transfer RNA*. Science, *185*: 435.

Nazar, R. O., T. O. Sitz, and H. Busch: (1975) *Structural Analysis of Mammalian Ribosomal Ribonucleic Acid and its Precursors*. J. Biol. Chem., *250*: 8591. Structure of 5.8S RNA.

Fox, G. E., and C. R. Woese: (1975) *5S RNA Secondary Structure*. Nature, *256*: 205.

6 REPLICATION OF DNA AND THE CELL LIFE CYCLE

The fundamental event in the reproduction of cells is the formation of new molecules of deoxyribonucleic acid. Replication of DNA is the device for insuring continuity of genetic information from one generation to the next, and complete duplication of the information occurs only in preparation for cell division. However, partial replication also occurs in the interval between cell divisions in order to repair damage to DNA molecules.

THE REACTIONS OF REPLICATION

The primary mechanism for synthesizing DNA is much like the mechanism for synthesizing RNA: a strand of existing DNA is used as a template to guide the condensation of deoxyribonucleoside triphosphates onto the growing end of a new DNA strand, with an accompanying loss of inorganic pyrophosphate (Fig. 6–1). The protein that catalyzes the reaction, DNA polymerase, strengthens the bonding between the complementary bases (one in the template and one in the added deoxynucleotide residue) so that the reading of the template is extremely accurate; the estimated error of 1 mismatch in 10^9 pairings in humans corresponds to an energy of binding of approximately -50 kJ. (Energy and equilibrium are reviewed in Chapter 19). The strong bonding is probably obtained by containing the bases in a closely-fitting hydrophobic environment, as is also postulated for RNA polymerase.

However, the replication of DNA differs from the synthesis of RNA in that DNA occurs as a double helix, and new antiparallel complements are synthesized for both of the old strands at the same time.

Replication Is Semi-conservative. When new DNA is synthesized, the old DNA that serves as a pattern is unwound, and each of the old strands acts as a template for laying down a new antiparallel complement:

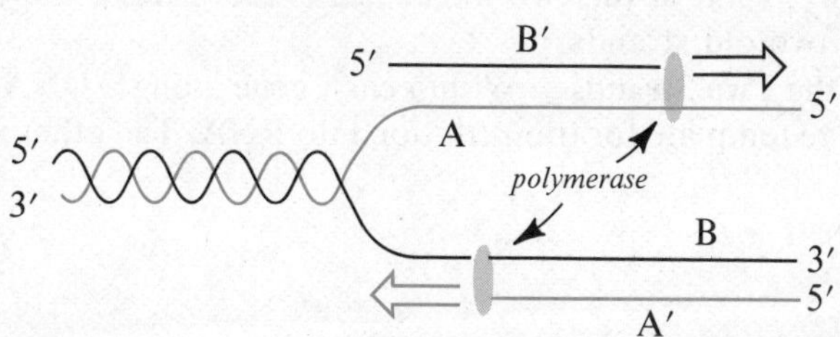

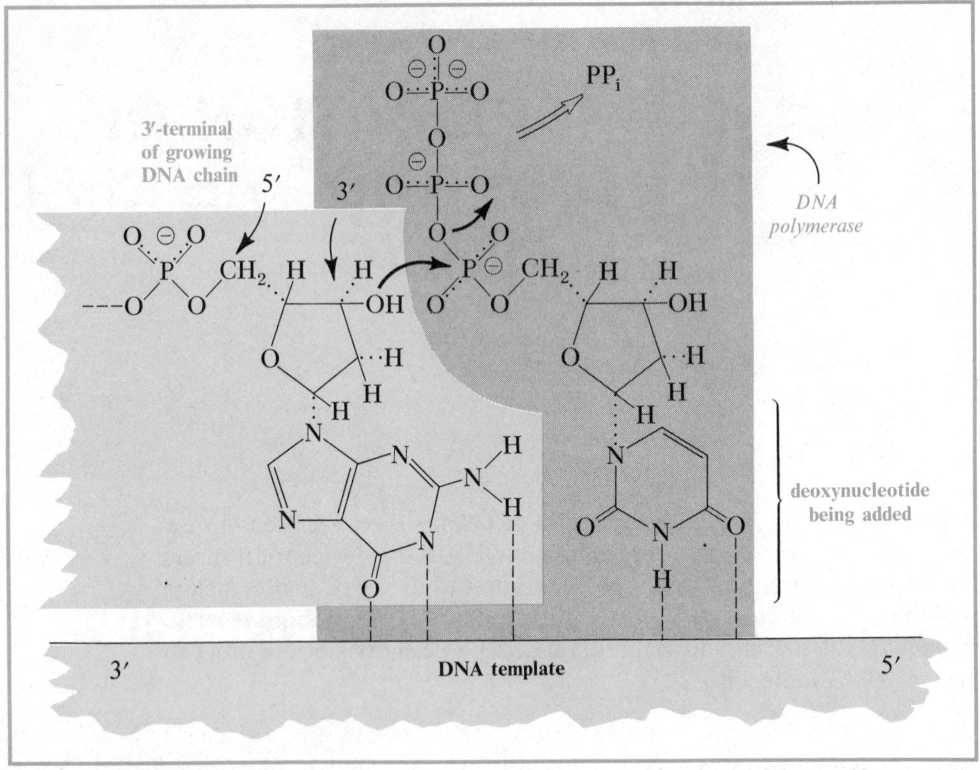

FIGURE 6–1 **DNA is replicated by the extension of a new strand in proximity to an existing strand, which is used as a template. Deoxynucleotide residues are brought to the site in the form of deoxyribonucleoside 5′-triphosphates (*right*), and are attached to the growing chain (*left*) at its free 3′-hydroxyl group. The attachment involves the loss of inorganic pyrophosphate (PP$_i$). The selection of the correct complementary nucleotide residue involves the formation of hydrogen bonds between it and the DNA template strand; the bonds are strengthened by the presence of DNA polymerase.**

The DNA being replicated contains complementary strands, here labeled **A** *(blue)* and **B** *(black)*, which have their 5′ terminals at opposite ends of the double helix. When these are unwound, molecules of DNA polymerase act independently on them to lay down new complements in opposite directions. A new complement, **B′**, is made for old strand **A**, and a new complement, **A′**, is made for old strand **B.** At the completion of replication, the full length of each old strand will be matched by a new complementary strand, and these pairs will twist into two molecules of double helix, each half old and half new (Fig. 6–2). Since replication usually proceeds perfectly, each of the two molecules is identical to the original double helix containing two old strands.

Function of the Two Strands. Within each gene along a DNA molecule, one strand is the active template for transcription into RNA. The other strand is not an

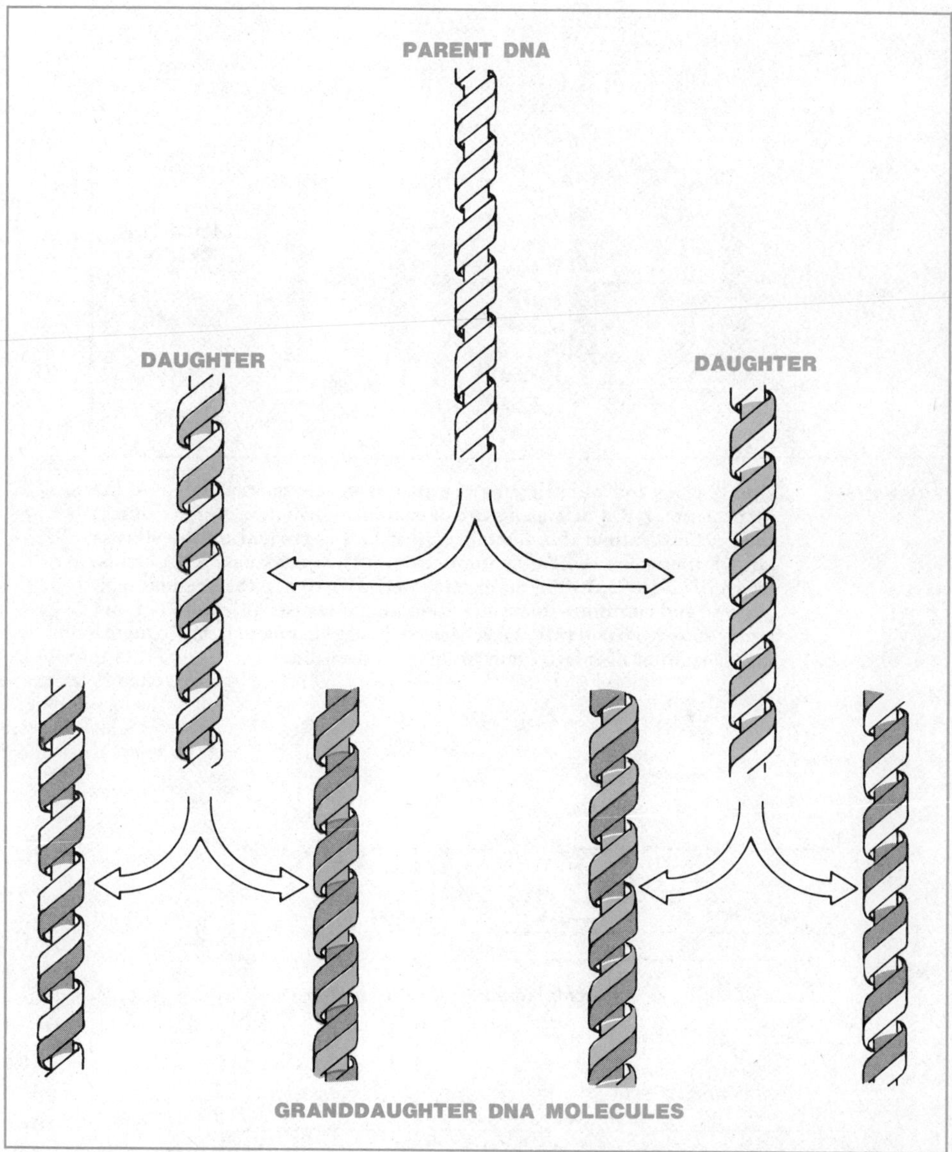

FIGURE 6–2 **Replication of DNA proceeds by a semi-conservative mechanism in which a new complementary strand is synthesized to match each parent strand. Each daughter molecule in the first generation contains one parent strand (*white*) and one newly synthesized strand (*blue*). In the next generation, these strands will be in turn separated and distributed among the granddaughter molecules, each of which will contain one newly-synthesized strand (*black*).**

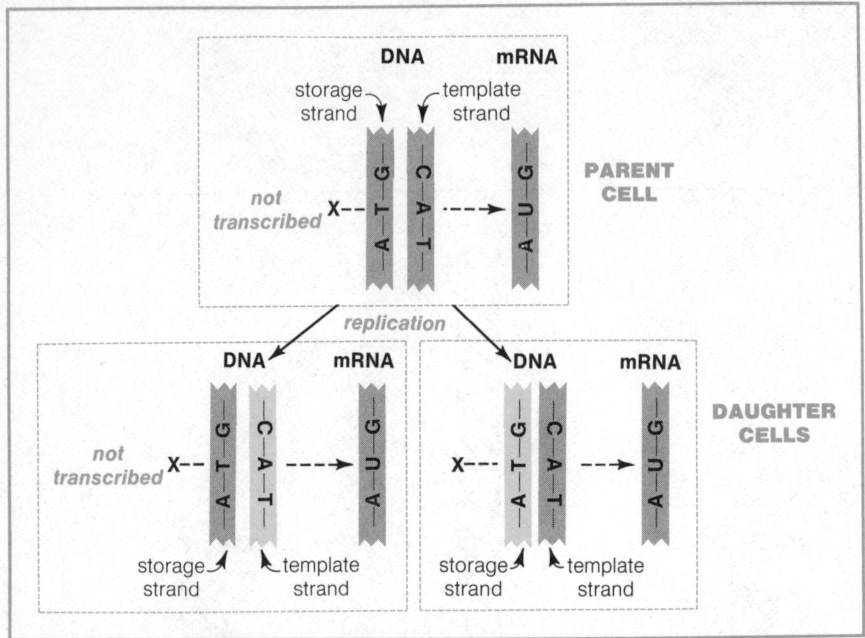

FIGURE 6–3 **One DNA strand of each gene is transcribed into messenger RNA; the other is not, but it acts as a store of genetic information. Upon replication, a new antiparallel complement (*blue*) is created on the storage strand, thereby making a template from which messenger RNA can be transcribed (*left*). In the other daughter cell (*right*), the original template strand continues to act as a template for transcription of RNA, but a new storage strand (*blue*) has been replicated to match it, and provide for transmittal of information in the next division.**

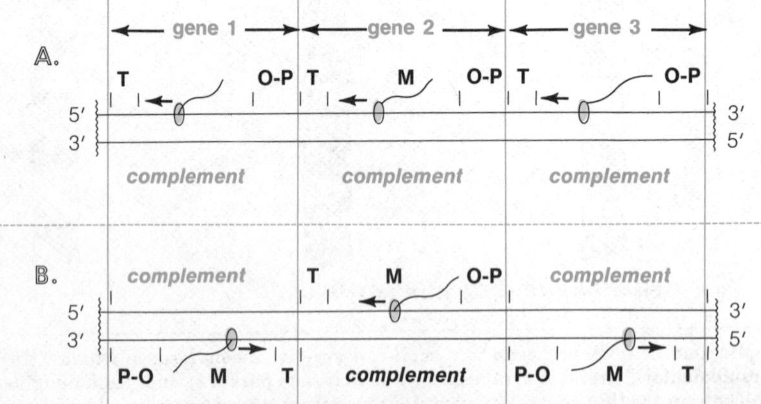

FIGURE 6–4 **Transcribed segments of DNA may be on both strands or on one. *A*. Genes may be read only from one strand of DNA, with the promoter-operator (P-O) of one gene adjoining the terminator (T) of the next in some cases. M is the transcribed message. The RNA polymerase reads each gene in the same direction. *B*. In bacteria, some genes are read from one strand, some from the other. (They do not necessarily alternate as shown here.: The RNA polymerase (*blue*) moves in opposite directions on the two antiparallel strands.**

inert support; it is the storage template for replication of a new active strand in preparation for cell division. What are to be the active messages in the DNA of daughter cells are created as complements of the inert segments. Figure 6–3 shows the relationship of deoxynucleotide sequence in the daughter and parent strands, using the C-A-T triplet for methionine as an example. (It is a useful exercise to compare sequence in both strands of DNA, in the messenger RNA transcribed from it, and in the transfer RNA anticodons that combine with messenger RNA. The actively transcribed DNA strand and the corresponding anticodons have comparable sequences; the storage DNA strand and messenger RNA have comparable sequences.)

This raises a question: Are all genes transcribed from one strand, with the other being used only for storage (Fig. 6–4*A*)? We know this is not so in *E. coli* and other bacteria in which the initiating signals and the succeeding message occur sometimes on one strand and sometimes on the other, being read in opposite directions because of the antiparallel arrangement (Fig. 6–4*B*). We also know it is not so in SV–40, a simian virus that can be incorporated into human DNA in vitro, but the question is not settled for mammalian cells.

The Steps in Replication

There is more to replication than the simple assembly of complementary polynucleotide chains. It is a complex process of many steps; let us consider them in sequence.

Multiple Initiation Sites — Replicons. A DNA molecule is too large to be made in one piece. Mammalian DNA polymerases create approximately one micrometer of DNA per minute. A polymerase molecule would therefore require about 55 days to make its way down the full length of the largest DNA molecule in human cells, whereas the observed time for total replication is frequently between four and eight hours, sometimes less. Replication is completed in a fraction of a day instead of many weeks by making new strands of DNA in short pieces, with polymerization proceeding in several places at once. Each molecule of DNA has several initiation points at which its replication begins. The segments that are replicated from one initiation point are termed replicons; there are perhaps 1000 replicons, each some 15 to 60 μm in length, per chromosome.

Mechanism of Initiation. We don't know how replication begins. Even the nature of the initiation points is unknown. They may sometimes coincide with sequences signaling the beginning of messages for transcription into RNA so that polymerizations forming DNA and polymerizations forming RNA start at the same site on template strands, but even this possibility is hypothetical.

One thing is certain: The duplex strand must unwind in order for replication of DNA to occur. Otherwise, the bases would be buried and not available for binding the complementary deoxyribonucleotides from which a new chain is constructed. Unwinding of the small circular chromosomes of bacteria has been demonstrated to require proteins with a high affinity for single-stranded DNA. These **unwinding proteins** hold the strands apart for replication; in physical terms, the melting point of duplex DNA is lowered below ambient temperature in the presence of these proteins, owing to the high stability of the complex between the proteins and dissociated DNA strands. Eukaryotic cells contain similar unwinding proteins. Unwinding also requires a nick in one strand—a hydrolysis of one phosphodiester bond by a **swivel enzyme** to relieve the torque in the long DNA molecule.

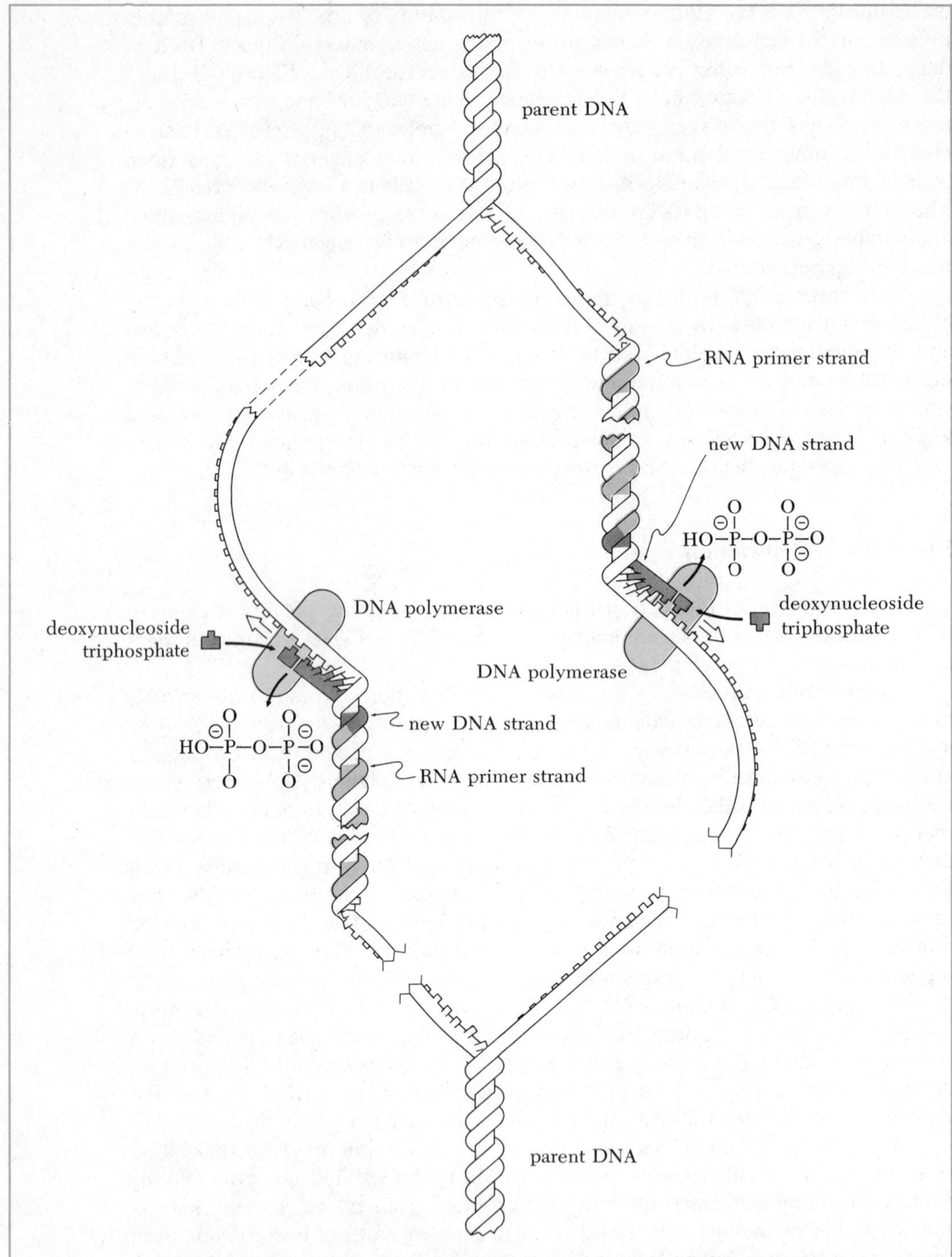

FIGURE 6–5 **DNA untwists for replication, creating two forks (*top and bottom*) that progressively separate. Replication proceeds simultaneously on both strands near each of the two forks, but the action is shown here on only one of the two—the top fork. Each of the two strands is read from the 3′ end toward the 5′ end, which is toward the fork from a distance away on one strand (*left side*), and away from the fork on the other strand (*right side*). Replication begins by first laying down a length of RNA as an antiparallel complement to the template DNA. This primer strand is then extended by DNA polymerase, using deoxyribonucleoside triphosphates as precursors, and liberating inorganic pyrophosphate (PP_i).**

Replication Forks. When the central portion of a duplex DNA fiber is unwound, two topologically identical forks are created at the points where the strands diverge (Fig. 6–5). The machinery of replication works on both strands at each of these forks. One strand will be replicated toward a fork, while its antiparallel complement is being replicated away from the same fork. Like events occur at the other fork, so replication is bidirectional, proceeding toward both ends of the chromosome from each initiation point. If the number of replicons per DNA duplex fiber is of the order of 1,000, replication could be proceeding simultaneously at 2,000 forks, with both strands being replicated at each.

RNA Primers and Okazaki Pieces. It now appears likely that the replication of a newly separated DNA duplex begins with its partial transcription into RNA! Cells that are preparing to divide apparently activate an RNA polymerase that forms complementary RNA with each strand of DNA as a template. Although details are not certain, the current assessment is that the enzyme functions until only some 10 to 12 ribonucleotides have been joined, and it is then superseded on the DNA template by a DNA polymerase that begins the addition of deoxyribonucleotide residues onto the end of the RNA chain. The DNA polymerase can't initiate the formation of DNA chains; it can create DNA only as an extension of RNA within cells, and it does so until approximately 120 deoxyribonucleotide residues have been polymerized. (Of course, each added residue contains the base complementary to the corresponding base in the existing DNA strand being used as a template.) These short fragments of DNA built on RNA primers are known as Okazaki pieces after the man who first discovered them and who recognized that DNA is synthesized in a discontinuous fashion.* (Okazaki pieces in bacteria appear to be ten times the size of those in mammals, with 50 to 100 residues in the RNA primer, followed by 1,000 residues of deoxyribonucleotide.)

There is a reason for making DNA in pieces. Antiparallel strands must be replicated in opposite directions because new chains are built by adding nucleotidyl groups onto 3′-hydroxyl groups (Fig. 6–5). As the template DNA is unwound, one strand could in theory be continuously elongated toward the fork, but the other must be replicated in pieces away from the fork as sufficient template becomes exposed:

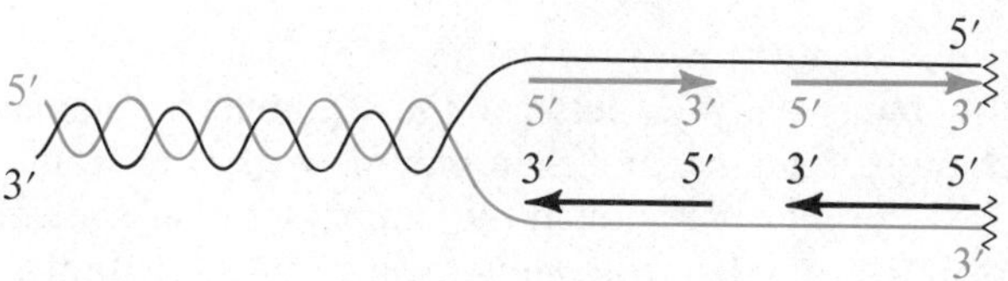

Since strands grow at their 3′ ends, the new segments shown at the bottom *(black)* might continue to grow as the duplex uncoils, but this is not true of the new segments at the top *(blue),* because they have their 5′ ends next to the duplex. In real life, the same mechanism appears to be used in both strands, with both being replicated in pieces.

Removal of RNA and Combination of Okazaki Pieces. The ultimate products of replication are continuous strands of DNA equal in length to the original templates. Although the precise events are not known, it is clear that the conversion of Okazaki pieces to continuous strands involves removal of the priming

*Dr. Reiji Okazaki at age 14 was in Hiroshima when the atomic bomb exploded. He later developed myelogenous leukemia and died August 1, 1975, at age 44.

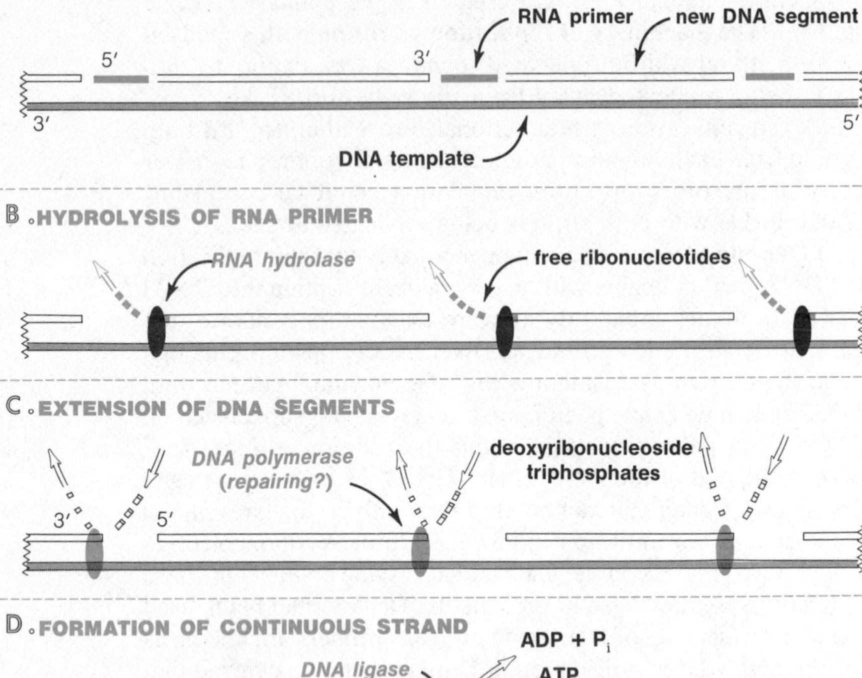

FIGURE 6–6 **A. DNA is replicated in discontinuous segments, known as Okazaki fragments, each of which is headed by a length of RNA primer. *B.* The RNA is removed by hydrolysis, catalyzed by a hydrolase specific for RNA-DNA hybrids. *C.* The gaps are then filled by increasing the length of the DNA segments through the action of another kind of DNA polymerase. *D.* Finally, the segments are joined through the action of a DNA ligase, which drives the reaction by simultaneously hydrolyzing a molecule of adenosine triphosphate (ATP).**

segments of RNA, filling the gaps between the segments by elongating them with more deoxynucleotide residues, and then joining them (Fig. 6–6).

The short RNA primer is removed by an enzyme that specifically catalyzes the hydrolysis of RNA attached to complementary DNA without affecting other RNA molecules in the cell. Once the priming RNA groups are removed, the space they occupied is filled by the action of another DNA polymerase, which extends the DNA chain to cover the now exposed template. After the gaps between DNA segments are filled, the abutting 3′ and 5′ ends are joined by still another enzyme, a DNA ligase, thereby creating a continuous strand. (A molecule of adenosine triphosphate is hydrolyzed to provide the energy for the ligase reaction; the utilization of ATP for such purposes is discussed in detail later.)

Modification of Bases and Restriction Enzymes. After replication of the base sequence is completed, a few of the bases in DNA are modified; methylation of some cytosine residues is especially common. The function of these changes in eukaryotes is not known; in prokaryotes they serve in part as a device for identifying the DNA of the organism. Cells produce restriction enzymes to catalyze the hydrolytic destruction of any DNA molecule that does not have the specific pattern of modified bases peculiar to that cell. If the DNA cannot be identified as a friend, it is taken to be a foe and is destroyed. This not only removes some viral invaders, but it also can remove parts of the cell's own DNA strands that have been damaged. (The segments destroyed by restriction enzymes are believed to be in or near palindromes.)

LOCALIZATION OF DNA: CHROMOSOMES

The two meters of DNA in human cells are folded into compact packages so as to fit in nuclei with typical diameters of four to six micrometers. These packages are chromosomes, created by condensing polynucleotide chains with proteins. The typical chromosome in a resting cell is monotene, that is, it contains one double-stranded fiber of DNA (duplex DNA), a fact that was not appreciated until recent years, owing to the difficulty of handling such very long molecules without breaking them by mechanical shear.

Most Human Somatic Cells Contain 46 Chromosomes. The largest has some 255 femtograms (255×10^{-15} g) of DNA in nearly 80 mm of duplex, while the smallest has 46 fg in 15 mm of duplex. The somatic cells are diploid, meaning that they carry a double set of chromosomes, one derived from each parent, with most polypeptides being independently specified by DNA from each parent.

As an example of the degree of compaction in chromosomes, human Group D chromosomes (numbers 13 to 15 in standard nomenclature) contain an estimated 31,000 μm of double helix, whereas the chromosomes are 4 to 6 μm in length.

The Chromatin Fiber

The name chromosome was first applied to the densely stained structures visible with the light microscope in dividing eukaryotic cells. It was later shown that these compact mitotic chromosomes are composed of fibers looped into bundles (Fig. 6–7). The constituent fibers are complexes known as chromatin.

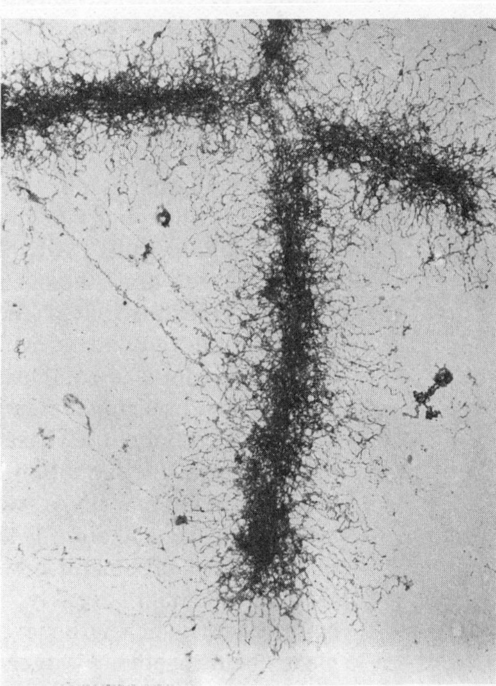

FIGURE 6–7

An electron micrograph of a human chromosome during mitosis (10,000×). This preparation was gently spread out on water to show the individual chromatin fibers that constitute the highly condensed chromosomes. The junction of the two chromatids at the centromere is clearly shown. (From D. E. Comings and T. E. Okada, (1970) *Cytogenetics*, 9:440. Reproduced by permission.)

TABLE 6–1 TYPICAL COMPOSITION OF CLASSES OF HISTONES

Class	H1	H2A	H2B	H3	H4
Total amino acid residues	210	129	125	135	102
% Lys groups	28	11	16	10	10
% Arg groups	2	9	6	13	14
% Ala groups	25	13	10	13	7
% Gly groups	7	11	6	5	17
Number Cys groups	0	0	0	2	0

Histones. The major component of chromatin is protein. Much of the shortening of the DNA fiber is caused by association with an equal mass of a particular class of proteins, the histones. Histones are relatively small proteins (M.W. 11,000 to 21,000) that contain a high proportion of amino acids with cationic side chains; histones are therefore isoelectric in basic solutions. (The other proteins of chromosomes, including many enzymes and regulatory proteins, are usually isoelectric in more neutral or acidic solutions.)

NATURE OF HISTONES. Every nucleated cell that has been examined — plant, animal, or protist – contains at least five general classes of histones that may be distinguished by size and amino acid composition (Table 6–1). (The nomenclature of these classes has gone through several metamorphoses, and earlier versions are also presented to aid in reading the literature.)

It appears safe to infer that the rather odd amino acid sequence of some of the histones is of fundamental importance to the life of nucleated cells, because it is similar in all kinds of organisms. Indeed, the arginine-rich histones, H3 and H4, show the least change of all known proteins over the course of evolution; H3 of the common garden pea *(Pisum sativum)* is identical in all except 4 of its 135 residues to H3 of the common cow *(Bos taurus)*. There is evidently little room in these small proteins for deviations from the long-tested functional arrangement.

Histones Stabilize DNA in a Compact Form. This protects the genetic information from damage while at the same time providing a device for regulating access to the DNA for replication or transcription. Histones do this because they are built with most of the cationic (''basic'') side chains in one or both ends of the polypeptide chain, leaving the middle of the chain relatively neutral. The concentration of like charges in each end prevents it from forming a stable secondary structure, while the middle is free to generate a segment of α-helix. The freely available regions of positive charge can combine with the many negatively charged phosphate groups in the backbone of DNA. The resultant neutralization of net charge permits the central helical regions of the histones to associate with each other and draw the chromosome together.

The Repeating Unit of Chromosomes. The chromatin fiber is made by coiling DNA strands around or within a complex containing four kinds of histones. Successive globular complexes are linked through somewhat weaker bonds by another kind of histone, creating a repetitious structure in which each unit differs from the next only in the base sequence of its constituent DNA. (It is important to recognize that the repetition of fiber structure does not represent a repetition of information in the DNA; structural organization may bear no relationship to information organization.) This arrangement accounts for the shortening and thickening of chromatin relative to naked DNA. Mild treatment of chromatin removes the more weakly associated constituents and creates a shish-kebab appearance, which is believed to represent the remaining stable histone complexes held together by intermediate lengths of naked DNA (Fig. 6–8).

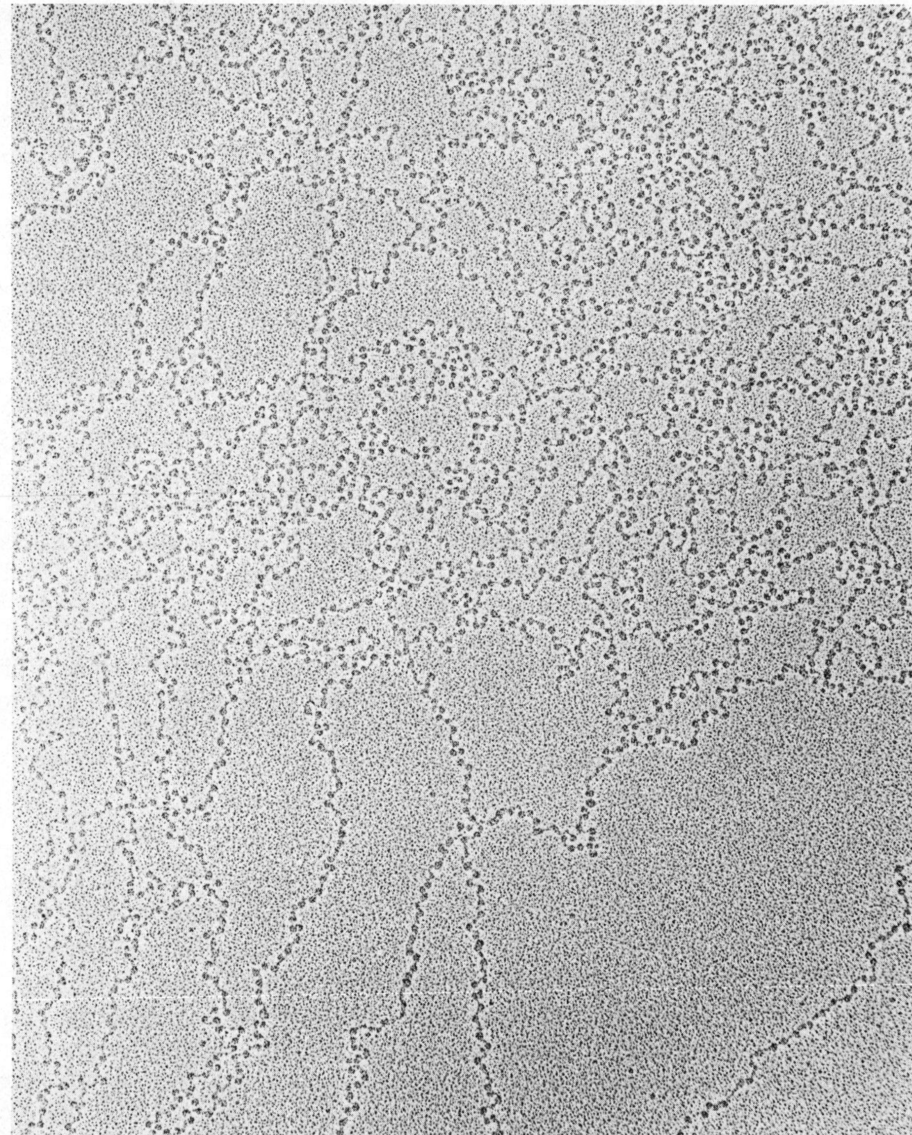

FIGURE 6–8 **Dispersed chromatin showing bead-like nucleosomes on a thread of DNA (73,200×). Prepared from growing embryo (blastoderm stage) of *Drosophila melanogaster*. Unpublished print courtesy of S. L. McKnight. Details of the methodology and other superb electron micrographs showing replication forks and nascent ribonucleo-proteins growing along the DNA fibers may be found in S. L. McKnight and O. L. Miller, Jr., (1977) *Electron Microscopic Analysis of Chromatin Replication*, Cell, *12*: 795.**

The stable histone complex has a tetrameric core containing two molecules each of H3 and H4 histones (those that are highly conserved during evolution), with two pairs of H2A and H2B spaced around the core. One or two molecules of H1 is believed to be a link between the complexes, causing further condensation.

The known dimensions are such as to require 70 nm of double helix (205 nucleotide pairs) to be folded within ~ 10 nm of chromatin fiber. It is not known how this is done, although it is theoretically possible for DNA to be kinked so as to change direction without disrupting any base pairs.

Euchromatin and Heterochromatin

Most of the chromatin fibers constitute the euchromatin and disperse in the nucleus between cell division, but some fibers remain in a condensed form that can be intensely stained by basic dyes and are known as heterochromatin. Some of the condensed chromatin localized to the inner layer of the nuclear envelope in female cells, known as the Barr body, is the remnant of one X chromosome. (The other X chromosome is dispersed, and its DNA is actively used for transcription of RNA.) We are more concerned with the constitutive heterochromatin that occurs in both males and females. Much of the DNA in constitutive heterochomatin is highly repetitive, containing as many as 10^6 identical sequences of nucleotides.

The function of heterochromatin is still not clear. Most is evidently not transcribed as RNA. It is closely associated with the centromeres of the mitotic apparatus (see below), and it also forms bands in the condensed chromosomes appearing at mitosis; it is therefore likely to have a structural function of some kind.

We already noted (p. 55) that highly repetitive DNA is likely to include satellite DNA, differing in gross base composition, and therefore in buoyant density, from much of the other DNA.

Other Chromosomes

Bacteria contain single chromosomes, in which the double helix is **supercoiled,** with its ends then joined to form a duplex circle. The total length of DNA in *Escherichia coli* is only 0.1 that in the smallest human chromosome, but it still must be tightly folded to fit within the bacterial cell.

Eukaryotic cells also contain **circular chromosomes,** which are found in those cytoplasmic organelles that have their own apparatus for protein synthesis, such as the **mitochondria** in plants and animals, and the **chloroplasts** in plants. These organelles have a separate heredity carried by their circular DNA; they may be descendants of symbiotic prokaryotic organisms that once resided in primitive eukaryotic cells. Their ribosomes, as well as their DNA, are more like corresponding constituents of bacteria than those found elsewhere in the same cell.

Even though mitochondrial DNA has a molecular weight of only 10^7, which can code for a few dozen proteins at most, it represents a significant heredity that can be passed only through the maternal line. (Spermatozoa do not contribute significant numbers of mitochondria to the fertilized ovum.) The importance of this cytoplasmic inheritance, at least in the plant kingdom was forcefully emphasized when a major fraction of the U. S. corn crop was lost to a blight, owing to altered mitochondrial heredity in some widely used hybrids. Comparable defects in human cytoplasmic genes are yet to be demonstrated.

THE CELL LIFE CYCLE

A cell activated to undergo cell division performs a series of orderly, integrated biochemical events culminating in the formation of two daughter cells. It has been customary to view these events as occurring in discrete phases. The microscopically observable **mitotic phase (M)** and the biochemically detectible phase of **DNA synthesis (S)** were separated by two phases called **gaps.** The phase between mitosis and DNA replication was called $\mathbf{G_1}$, and the phase between DNA replication and mitosis was labeled $\mathbf{G_2}$.

It is almost self-evident that rigorous controls must exist to limit the size of organs and tissues. Hence, a normal organ or tissue only permits its cells to proceed in the cell cycle as required for its functions at the time. A major purpose of the physical examination of a patient is to determine if the individual organs have in fact exceeded their normally limited size. Enlargement may represent excessive growth, as well as inflammatory swelling.

Organs that rapidly sustain cell losses such as the hematopoietic system and the gastrointestinal tract have a large fraction of their cells in almost continuous cycles of cell replication. These organs appear to have the program leading to cell division continually active, and are considered to enter a G_1 period after mitosis (Fig. 6–9). Other organs such as kidney and liver have a very small fraction of cells in division while the majority of cells have long quiescent periods between cell divisions. Such cells appear to inactivate the program for cell division at the completion of mitosis. A separate phase, G_0, has been proposed for such cells.

Variations in the proportion of cycling cells occur in many organs. The number of cells forming leukocytes increases in response to bacterial infection. Partial hepatectomy rapidly signals the onset of cell division in the remaining liver. Changes in the proliferative activity of the uterus regularly accompany changes in the estrogen and progesterone levels in the body, and are the basis for the menstrual cycle.

Requirements for Macromolecular Synthesis. Interference with ribonucleic acid synthesis or protein synthesis at any time in the cycle up to the onset of mitosis will either destroy the cell or delay the progression through the cycle. The program of cell cycle is arranged to provide for the transcription of various RNAs and the translation into proteins at specific times so that the appropriate components are available when required. In contrast, deoxyribonucleic acid synthesis is

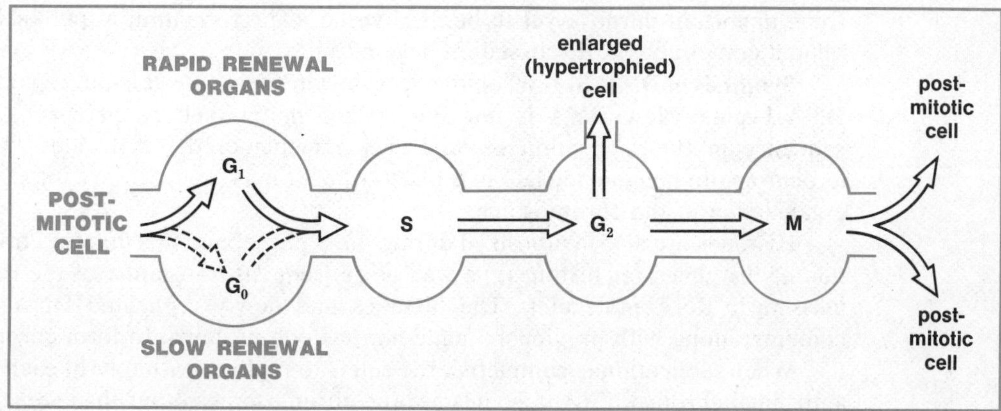

FIGURE 6–9 **Outline of the cell cycle. Post-mitotic cells in some organs enter a brief gap phase (G_1) as a preliminary to renewed division. In other organs, the cells carry out their function over a long period (G_0) without division. Pre-mitotic events involve DNA synthesis (S phase) and a second gap phase (G_2) from which enlarged cells are sometimes diverted. Otherwise, mitosis (M) proceeds, with the two daughter cells ready to begin the cycle again.**

scheduled to occur only in the S phase. During mitosis the synthesis of all macromolecules decreases markedly.

Duration of Phases. There is considerable variation in the time spent in G_0, G_1, S, and G_2. Neurons and muscle cells may spend a lifetime in G_0. G_1 may be as short as two hours or last for days. S in human cells is known to vary from 11 to 34 hours; the G_2 phase occupies from 3 to at least 16 hours. Mitosis continues for about 30 minutes to 1 hour.

More information about the duration of the cell cycle phases of individual cancer cells is mandatory if exploitation of these differences in the treatment of malignant diseases is to be possible.

G_0–G_1 Phase. The fulfillment of mature cells' metabolic and structural functions do not require mitosis, and a majority of the cells in the body are in G_0 or G_1 phase. The decisive early events in the transition from these phases have been difficult to determine. They include activation of genes leading to synthesis of regulatory non-histone proteins in the cytoplasm. These proteins move to the nucleus, are modified (see Chapter 8), and activate new areas of the genome resulting in the transcription of new RNA. By the sequential unfolding of this pattern, the cell has increased the synthesis of ribosomal RNA as well as new messenger RNA. The enzymes for DNA and histone synthesis appear, and the cell is ready for the S phase.

Putrescine and spermidine, which are aliphatic polyamines, are also produced in increased quantities:

$$\overset{\oplus}{H_3N}\text{—}CH_2\text{—}CH_2\text{—}CH_2\text{—}CH_2\text{—}\overset{\oplus}{NH_3}$$

putrescine

$$\overset{\oplus}{H_3N}\text{—}CH_2\text{—}CH_2\text{—}CH_2\text{—}\overset{\oplus}{NH_2}\text{—}CH_2\text{—}CH_2\text{—}CH_2\text{—}CH_2\text{—}\overset{\oplus}{NH_3}$$

spermidine

Since these compounds are multivalent cations at physiological pH, they are presumed to react with the negatively charged backbone of DNA or RNA for some important purpose yet to be discovered. (The formation of putrescine and related compounds is discussed on page 627.)

Synthesis of DNA, S. When the necessary proteins are at hand, replication of DNA begins. New DNA is not synthesized helter-skelter; there is a defined sequence in the replication of different segments of the molecules, with heterochromatin being made late in S phase when it may be necessary for structural organization of the chromosomes.

Histones are also synthesized during the S phase, and this involves transcription of the genes for histones, as well as subsequent translation of the resultant messenger RNA molecules. The histones and newly replicated DNA rapidly combine, along with previously made non-histone proteins, to form chromatin.

When replication is completed, the cell is tetraploid, with four of each kind of autosomal chromatid. (Autosomes are the chromosomes other than sex chromosomes; a chromatid is the fraction of a chromosome carrying one duplex DNA molecule.) The number of sex chromatids also has been doubled.

Second Gap, G_2. Complete replication of DNA is followed by another period in which the necessary RNA and proteins are synthesized. Less is known about the character of these proteins, which may be required to accelerate the condensation of DNA into compact chromosomes. In addition, there is a decrease in the synthesis of the H2A, H2B, H3, and H4 histones.

Some cells may develop differently in G_2, growing larger without division. This may be a common event in muscle, where there are many tetraploid cells — cells with twice the normal content of DNA. In the adult heart, 95 per cent of the cells are tetraploid; only 15 per cent are tetraploid in children under three years of age. (An enlargement of tissues without increase in cell number is called hypertrophy; an increase in cell number is a hyperplasia.)

Mitosis, M. The terminal event of the cell cycle is the segregation of the constituents of one cell into two daughter cells. It is this phase that is most dramatic under the light microscope, but it is usually complete within one-half hour, whereas some 8 to 12 hours are required to prepare for mitosis.

It is not necessary for us to review in detail the complicated sequence of events by which the constituents of one cell are segregated into two cells — the functionally important event from the standpoint of biochemical genetics is the division of the chromosomes. Replicated chromatin condenses into thick diploid chromosomes, in which the chromatin fibers are looped into cylinders some 400 to 800 nm in diameter. The chromosomes are assembled into a planar array and then are split into haploid pairs, with one of each pair drawn into opposite sides of the cell. This assembly and separation is accomplished by the spindle apparatus, made from microtubules, and including the proteins actin and myosin, which contribute motility (see Chapter 20).

STRUCTURE OF MICROTUBULES. Microtubules are polymers of a protein, tubulin, which is a dimer of two different polypeptide chains. (The synthesis of tubulin and the centriole proteins increases during the S phase.) The dimers associate into long filaments that aggregate side-by-side so as to form a 13-filament cylinder (Fig. 6–10). (The association is driven by a hydrolysis of guanosine triphosphate, GTP.) The filaments are staggered relative to each other so as to create the surface appearance of a three-fiber helix, but the microtubule is not made as a stack of helical ropes — it is a hollow bundle of straight rods.

Genetic recombination. Similar DNA molecules are in close proximity during mitosis until chromatid separation, and are subject to exchange of segments between molecules. This may happen, for example, by breaking polynucleotide chains at precisely comparable places on the molecules, and resealing the breaks between the molecules rather than in a single molecule:

This crossover is usually of minor importance in dividing somatic cells, because perfect replication creates identical copies, and exchanging portions of these copies will still result in identical molecules, if there is no error in the process. However, genetic recombination is the all-important device for reshuffling the properties of parents among their children, and the molecular mechanism hinges upon the peculiar life history of germ cells.

Germ cells go through a different sequence of divisions by which the tetraploid chromatin created in the S phase is parceled out among four haploid daughter cells, each of which contains only one copy of one gene. Included in this sequence is a meiosis — a division in which replicated sister chromatids from both parents are fused into a single chromosome. The related, but not identical, DNA molecules derived from the parents are then attached to each other, so that

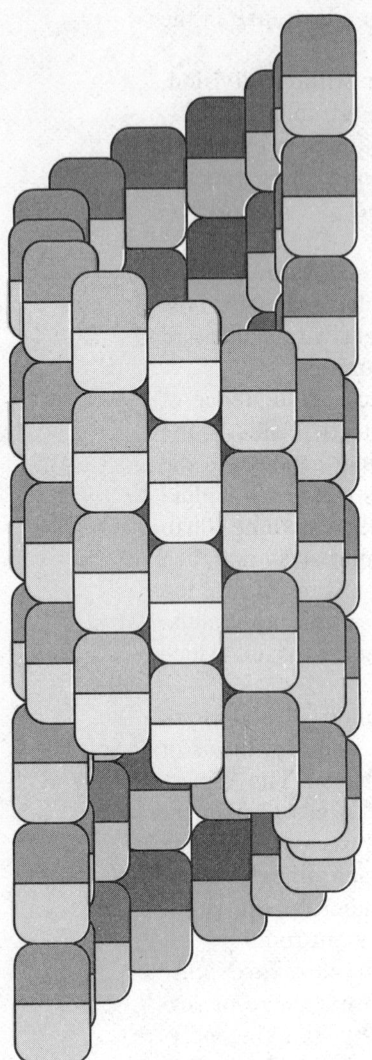

FIGURE 6–10

Schematic arrangement of microtubules. A microtubule is a hollow bundle of 13 filaments, each made from stacked molecules of tubulin. Tubulin is a protein containing two different subunits (*gray and blue shading*). One subunit lies slightly behind the other in the real molecule (not shown here). Adjacent filaments are displaced from each other so as to create the effect of a cylinder made from three stacked helical fibers, but the fundamental unit is a straight filament, not a helix.

exchange of whole gene segments by crossover creates new combinations of genes (Fig. 6–11). We infer that our survival is evidence for specific crossover regions between genes — our proteins are not a random hodge-podge of pieces.

To recapitulate, each eukaryotic cell contains two similar, but not identical, molecules of DNA specifying the same genes. Upon replication for mitosis, each of these molecules is converted to two identical copies. The cell then contains two different and separate pairs of DNA molecules; each pair is made of identical molecules that are attached to each other until separation of the haploids during mitosis. During their attachment, molecules may exchange segments of DNA within a pair, but this does not alter the identity within a pair or the differences between the pairs — the new cells will still be identical to the old, barring accidents.

During meiosis, two identical copies of each of the two different DNA molecules are also made, but the four resultant molecules are now fused into one grand chromosome, in which exchange of segments occurs among all four. After exchange, the four molecules are all different, with each having a unique combination of genes from the two parent molecules. When these four molecules are

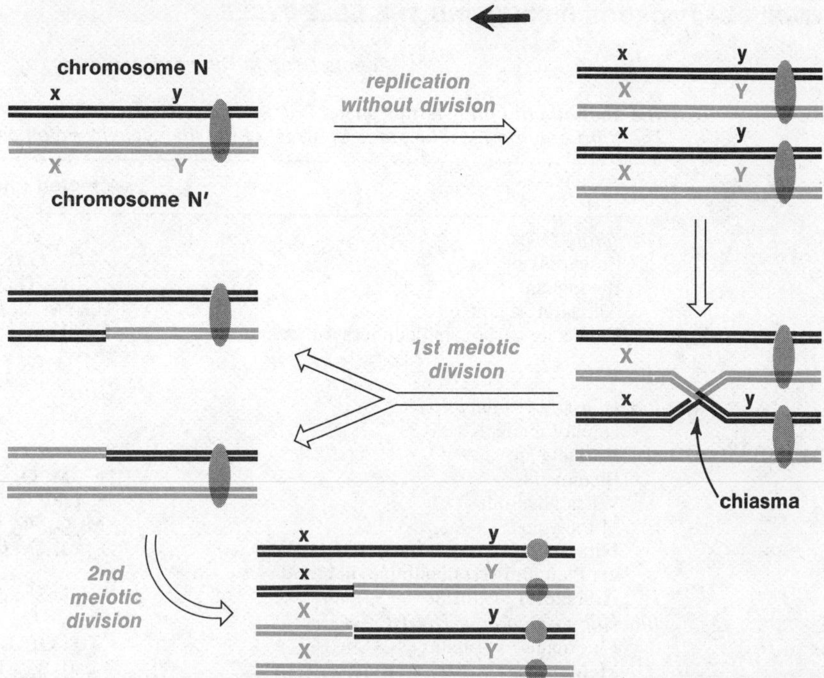

FIGURE 6–11 **The crossovers occurring during meiosis recombine segments of DNA so that one molecule contains some genes derived from one parent, and some from the other. When a germ cell precursor replicates without division (*top*), the resultant bivalent contains two identical pairs of duplex DNA molecules. After crossover and the first meiotic division (*center*), the pairs of DNA molecules are no longer identical. Separation of these pairs to create haploid germ cells (*bottom*) results in four different chromatids. Two of these are like the parents on the segment shown, but other segments on these molecules would probably be unlike the parents in their combination of genes.**

distributed among four germ cells, each will contain a different heredity. Since the exchange process itself may occur in many different ways, the odds are high that no two germ cells will be exactly alike.

Disruption of the Eukaryotic Cell Life Cycle

Compounds that kill eukaryotic cells are powerful drugs commonly used in the practice of medicine, and are also invaluable tools for the analysis of the complex biochemical events in the history of cells. Many of the most valuable are compounds affecting the cell life cycles. (Such drugs are referred to as chemotherapeutic agents, although those synthesized by microorganisms are more commonly called antibiotics.) The clinical situations for which such potentially toxic compounds are used can be differentiated into two categories. In one category are those diseases where abnormal behavior of cells results in serious illness. Malignant growths — cancers — represent the largest group of such diseases. The second group of diseases are those in which normal cell activities have undesirable consequences. When an individual receives an organ transplant the natural response includes the proliferation of lymphocytes capable of destroying the transplanted foreign organ. Chemotherapeutic drugs are used to preserve the foreign tissue by interfering with the proliferation. Such drugs are sometimes used

TABLE 6–2 DRUGS DISRUPTING THE CELL CYCLE

*Agents Used in Cancer Chemotherapy**

The second column lists the phase in the life cycle at which the cells are sensitive to the drug. The third column lists the stage of arrest of the life cycle in which cells tend to accumulate.

	Affected Phase	Accumulating Cells
Disrupting DNA		
Procarbazine	G_0, G_1/S	
Bleomycin	M. G_2	G_2
Alkylating agents	G_0, M, G_1/S	
Imidazole-carboxamidedimethyl triazeno	None	G_2 (in vivo) G_1 (in vitro)
Blocking DNA replication		
Daunorubicin	S	
Adriamycin	S	
Bleomycin	M, G_2	G_2
Vinca alkaloids	S, Late G_1, M	M
Mitomycin	M, G_1, G_2 (?)	S/G_2
Nitrosoureas	G_0, M, G_1, G_1/S	
cis Platinum (II) diammine dichloride		
Arabinosyl cystosine	S	G_1/S, S/G_2
Blocking formation of DNA Precursors		
Fluorinated pyrimidines	S, G, G_2	G_1/S
Methotrexate	S, early G_1, G_1/S	G_1/S
6-Mercaptopurine		
6-Thioguanine		
Azathioprine		
Arabinosyl cytosine	S	G_1/S, S/G_2
Hydroxyurea	S	G_1S
Inhibiting synthesis of 45S ribosomal RNA		
Actinomycin D	G_0	G_1/S, S/G_2, G_2
Daunorubicin	S	
Adriamycin	S	
Mithramycin		
Nitrogen mustard	G_0, M, G_1/S	G_2
Chloro-ethyl cyclohexylnitrosourea (CCNU)	M, G_1	
Inhibiting processing of 45S RNA		
CCNU	M, G_1	
bis-Chloro-nitrosourea (BCNU)	G_1/S	
5-Fluorouracil deoxyriboside (5 FUdR)	S, G_1, G_2	G_1/S
Inhibiting formation of hnRNA		
Actinomycin D		G_1/S, S/G_2, G_2
Nitrogen mustard	M, G_1/S	G_2

*For completeness, a few drugs not used in patients are included because they affect specific events in the cycle.

in treating autoimmune diseases, a heterogeneous group in which the end result of immune reactions causes damage to apparently normal tissue.

Chemotherapeutic drugs and antibiotics are also widely used in the treatment of bacterial, fungal, and parasitic infections. Drugs for these purposes are designed to exploit the biochemical differences between prokaryotic and eukaryotic cells.

Selectivity and Specificity. The Jericho of malignancy has not crumbled to repeated blaring of new chemotherapeutic drugs, but significant cracks in the walls have appeared, leading in some instances to cures. What is the problem that prevents more general success? Despite the differences between malignant and

DRUGS DISRUPTING THE CELL CYCLE *Continued* **TABLE 6–2**

*Agents Used in Cancer Chemotherapy**

The second column lists the phase in the life cycle at which the cells are sensitive to the drug. The third column lists the stage of arrest of the life cycle in which cells tend to accumulate.

	Affected Phase	Accumulating Cells
Inhibiting mRNA polyadenylation		
3′deoxyadenosine (experimental)		
Inhibiting transport of mRNA or cleavage of mRNA		
BCNU		
Inhibiting tRNA synthesis		
Formycin* (not used in patients)		
Inhibiting mitochondrial RNA synthesis		
Ethidium bromide* (not used in patients)		
Undefined locus in RNA synthesis		
Vinca alkaloids	S, late G_1, M	M
Mitomycin	M, G_1, G_2 (?)	S/G_2
Disaggregating polyribosomes		
BCNU	G_1/S	
Preventing formation of polyribosomes		
Actinomycin D		G_1/S, S/G_2, G_2
Inhibiting peptide chain initiation and elongation		
Cycloheximide* (not used in patients)	G_1/S	G_2
Inhibiting elongation of peptide chains		
Emetine* (not used in patients)		
Terminating peptide chain synthesis prematurely		
Puromycin* (not used in patients)	G_1/S	G_2/M
Interfering with supply of amino acids		
Vinca alkaloids	S, late G_1, M	M
L-Asparaginase		
Methotrexate (?)	G_1/S	S, early G_1, G_1/S

Agents Used in Infections
(bind bacterial ribosomes)

Chloramphenicol	blocks peptidyl transferase
Streptomycin	affects initiation complex, prevents normal triplet recognition
Tetracycline	blocks binding of aminoacyl tRNAs

*For completeness, a few drugs not used in patients are included because they affect specific events in the cycle.

normal cells, the biochemical characteristics of malignancy are not yet known. Consequently, any chemotherapeutic attack on malignant cells also affects normal cells; the toxicity of these drugs is therefore a major limitation to their use. Here is a case in which new insights into biochemical events has been of critical importance in the design and use of new drugs.

Such insights include an understanding of the specificity of the actions of the drugs. Table 6–2 lists some of the many available drugs in an order approximating the sequence of events in the cell cycle. Some drugs are most active against cells in a particular phase of the cell life cycle. Some drugs (non-cycle active) will interfere with quiescent cells as well as cells in cycle. A drug may be most effective if a cell is exposed to it during a particular phase, the phase of greatest lethal sensitivity, but the effect may not become manifest until another phase occurs. Frequently the drug acts near the indistinct boundaries we ascribe to the cell cycle, indicated in the table by listing both phases separated by a diagonal. Many of these drugs act during several phases.

Many drugs are listed several times. No attempt is made to hazard a statement as to which event is decisive in its cytotoxic action.

Many drugs acting during the S phase interfere with the formation of nucleic acids, particularly DNA. **Methotrexate** is one of the more widely used; it is effective in slowing the development of cancers derived from white blood cells (leukemias) and of solid tumors. (Its structure and mechanism of action are discussed in detail on p. 664.) Methotrexate is also used in the treatment of psoriasis, but only when the condition is serious enough to warrant exposure to the dangerous side effects of the drug.

Bleomycin, a drug acting during the G_2 phase, illustrates the potential of more sophisticated strategies. This drug can be used to stop the cycle in all of the cells as they reach G_2. If the drug is then withdrawn, the cells will continue the cycle in synchrony. Most, if not all, will enter the S phase at nearly the same time, so it could be that if one waited a proper interval after cessation of bleomycin and then

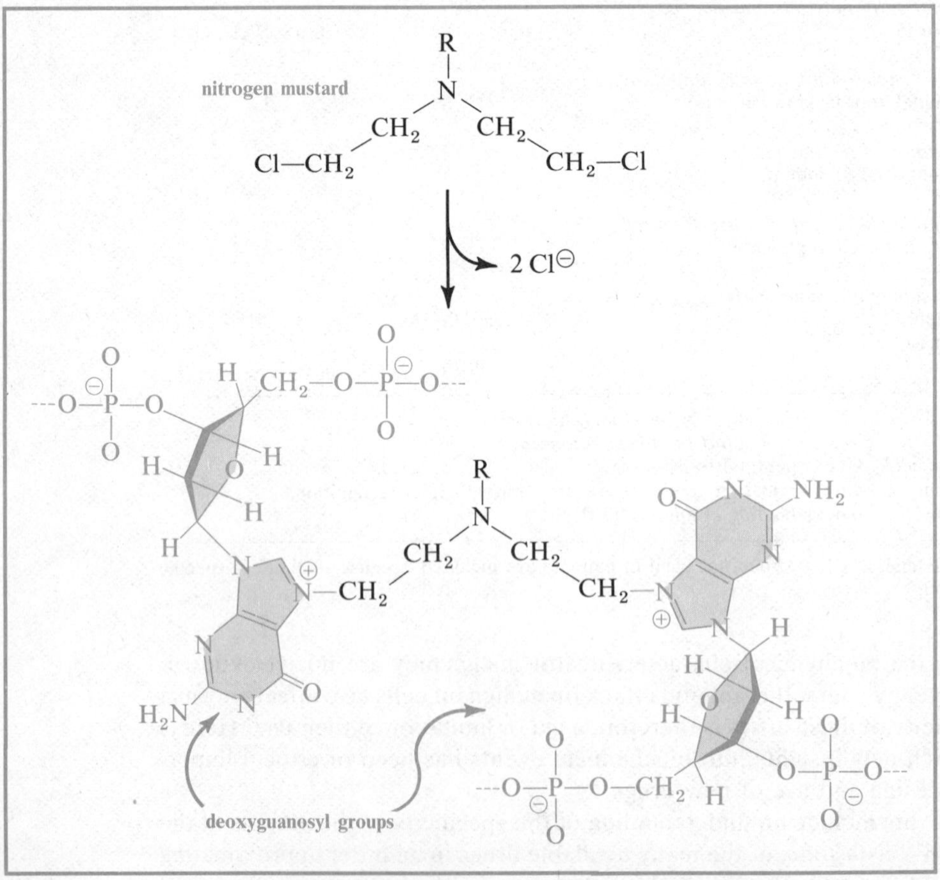

FIGURE 6–12 **Nitrogen mustards are alkylating agents that link two deoxyguanosyl residues in DNA, thereby preventing the untwisting necessary for replication or transcription. A principal mode of action is shown here.**

administered a single dose of a drug acting during the S phase, it might be effective against nearly all of the malignant cells.

Drugs acting during the M phase include compounds that form complexes with tubulin, and therefore dissociate the mitotic spindle, such as **vinblastine** and **vincristine,** which are alkaloids derived from a periwinkle plant, *Vinca rosea*.

Drugs that prevent the transcription of RNA also prevent tumor growth, and act throughout the cell cycle. Some modify DNA; for example, alkylating agents form covalent derivatives. Nitrogen mustards, which are analogues of the sulfur-containing mustard gas used during World War I, react with guanine residues (Fig. 6–12), and they are especially effective because they are bivalent and therefore form cross-links between adjacent segments of DNA. The covalent modification prevents both replication and transcription.

Transcription can also be prevented by compounds containing large planar ring systems that will slip between base pairs in DNA (Fig. 6–13). **Intercalation** of DNA with these rings extends the double helix into a more stiff rod, and prevents transcription by RNA polymerase. One of the effective intercalating agents, actinomycin D, is known to fit best between successive G-C pairs so that both of its two peptide chains can form hydrogen bonds with amino groups on the guanosyl residues. Actinomycin is one of several antibiotics produced by species of Streptomyces, which are soil bacteria resembling fungi in their growth habit.

Chemotherapy for Infections. Many agents are available to block the cell cycle of bacteria without substantial effect on the cell cycle of the human host. Most of these agents act by binding one or more components of the bacterial ribosome. Superficially, this appears to account for the selectivity of the agents, since the bacterial and eucaryotic ribosomes differ in composition and structure. **Chloramphenicol** and **streptomycin,** for example, have little effect on ribosomes from the cytosol of mammalian cells at concentrations that totally prevent protein synthesis by bacterial ribosomes. However, this is not enough to account for the differences in toxicity because the mitochondria of the mammalian cells also have ribosomes that resemble the bacterial ribosome, and isolated mitochondrial ribosomes are susceptible to inhibition by many of the same agents, such as chloramphenicol, streptomycin, and tetracycline. Antibiotics such as these are safe to use only because they enter mammalian cells less readily than they enter bacterial cells, or because they do not cross mitochondrial membranes. Even so, some of the toxic side effects of these antibiotics may be due to their entrance into host mitochondria.

VIRUSES AND THEIR REPLICATION

Viruses are simple organisms composed of one or more molecules of nucleic acid, usually associated with at least a few protein molecules. The proteins may have only a structural role, such as the coat of proteins that polymerize as a capsid around the nucleic acids in some viruses, or they may have a catalytic function during the life cycle of the virus. The viral nucleic acids specify the formation of these proteins, and sometimes others that are necessary for the replication of the virus, but are not included in it. The difference between viruses and other organisms is that the virus must invade other cells and use part of the machinery of the host in order to reproduce itself. Any cell — plant, animal, or protist — is subject to attack by viruses.

Even these simple organisms are like other forms of life in that some controls are exerted over the expression of their genetic information so that all of the

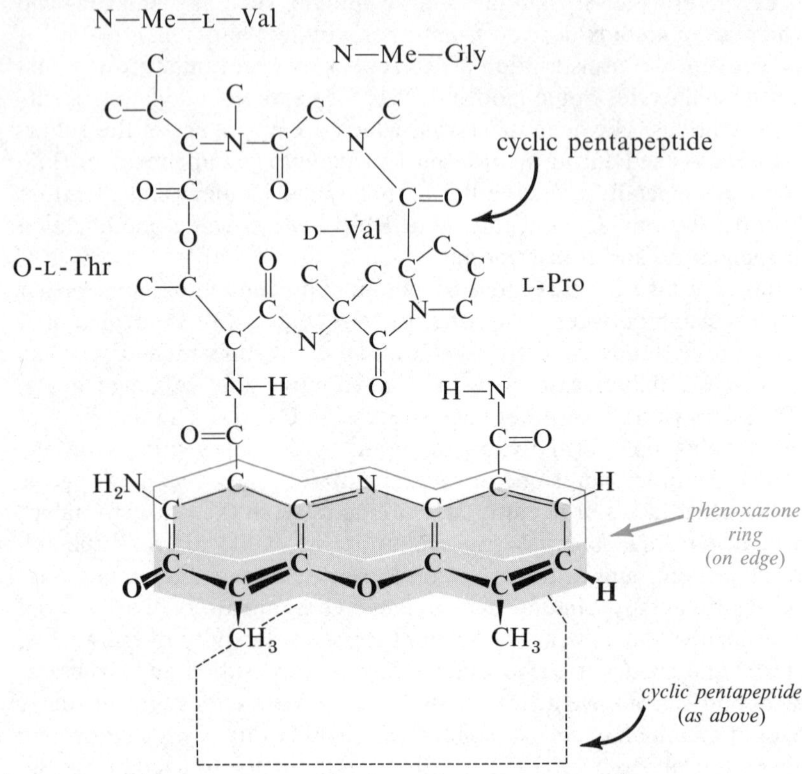

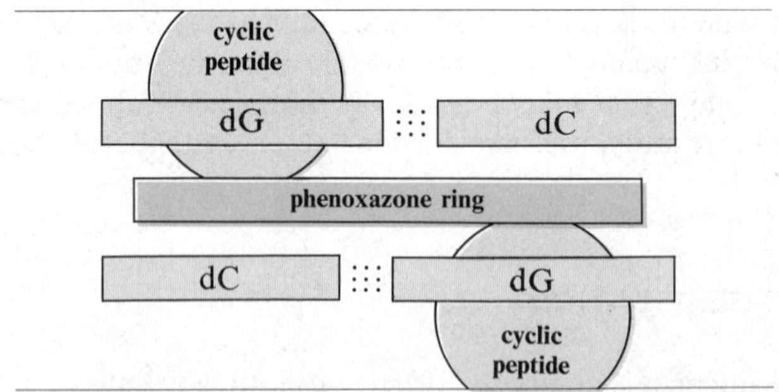

FIGURE 6–13 **Actinomycin D is an example of an intercalating agent, containing a planar ring system that will slip between base pairs in DNA, causing it to partially untwist and become rigid. The top formula shows the phenoxazone ring of actinomycin D on edge. Attached to the ring are two identical cyclic pentapeptide units containing modified amino acid residues. (Substituent H atoms are omitted for clarity.) These pentapeptide groups, only one of which is shown in detail, are roughly at right angles to the intercalating ring, and are so disposed that they fit in the grooves of the DNA double helix.**

The structure of actinomycin D is such that it forms stable bonds when inserted between two dG-dC pairs, as shown at the bottom.

nucleic acids are not transcribed at one time. Their life cycle may be divided into early and late appearances of functions.

Classes of Viruses

Viruses are also unusual in that they may carry their genetic information as RNA or DNA, and either nucleic acid may be single-stranded or double-stranded; viruses are classified on this basis:

(1) **double-stranded DNA** of 3 to 160 megadaltons occurs in adenoviruses, herpes viruses, and the pox viruses; some of these viruses are oncogenic, that is, they sometimes cause invaded cells to become malignant;

(2) **single-stranded DNA** of 1.5 to 2.4 megadaltons is best known in viruses that attack bacteria, such as some coliphages (many coliphages contain double-stranded DNA);

(3) **double-stranded RNA** of the order of 15 megadaltons is known in reoviruses that cause respiratory and enteric infections;

(4) **single-stranded RNA** of 0.4 to 13 megadaltons occurs in polio and coxsackie viruses, in rabies virus, in myxoviruses such as those that cause influenza, in measles and mumps viruses, and in leukoviruses that may cause leukemias (a malignant proliferation of white blood cells). Some bacteriophages also contain single-stranded RNA, and the sequence of 1,355 nucleotides has been determined for one of them, a remarkable accomplishment!*

Mechanism of Replication (Fig. 6–14)

Viruses containing DNA reproduce much in the same manner as other cells. When viral DNA is injected into a host cell, it acts as a template both for transcribing messenger RNA and for replicating additional DNA. Viral DNA replication involves the fundamental use of a DNA polymerase directed by a DNA template strand, but the mechanism of strand separation and combination is not always like the mechanism we described earlier in the chapter.

It is the RNA viruses that use a novel mechanism in which an RNA chain is used as a template upon which to construct an antiparallel complement. Essentially, they fall into two major groups. One contains information for creating an **RNA-directed RNA polymerase** that directly replicates the viral RNA. This polymerase has the same function as the RNA polymerase of cellular organisms, except that it uses RNA rather than DNA as a template. The other group also uses a different polymerase, but it is an **RNA-directed DNA polymerase** ("reverse transcriptase"), which makes DNA, utilizing RNA as a template. This opens a pathway by which the information from infectious RNA can be integrated with the host DNA.

It is the last group that has aroused much attention recently because it includes viruses that appear to cause malignancy in animals. The genetic information of these viruses is introduced into the host cell as RNA. The virus itself includes the polymerase that makes DNA complementary to the viral RNA; the viral RNA strand is then destroyed and a second strand of DNA is formed on the new DNA strand. (This sequence resembles the destruction of the RNA primer and its replacement by a DNA strand during the formation of Okazaki fragments.) The result is the appearance of the genetic information of the virus within the host

*W. Fiers, et al., (1975) Nature, *256*: 273.

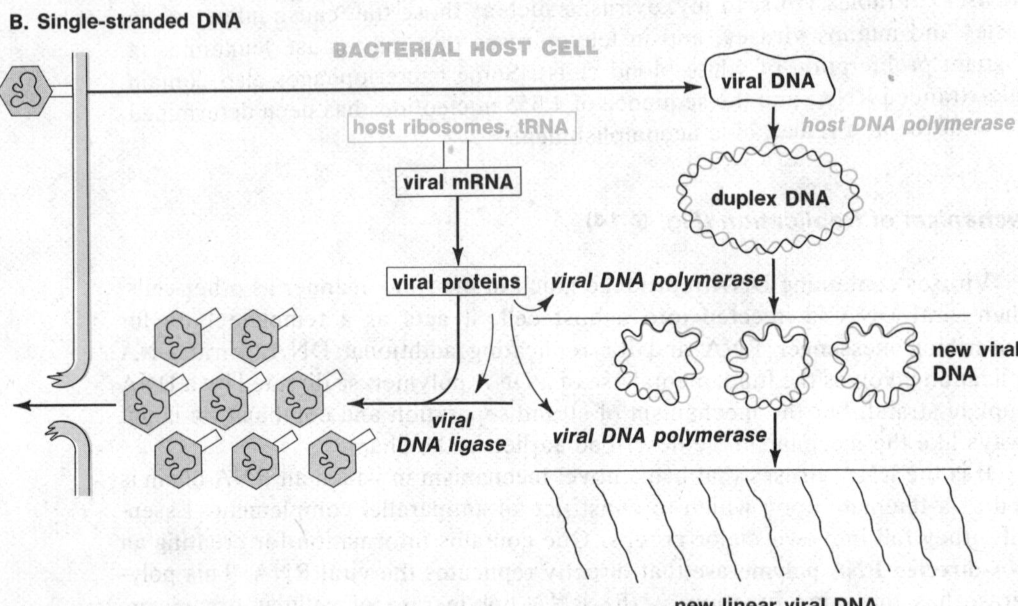

FIGURE 6–14 **The replication of viruses. Not all mechanisms are shown.**

A. **Double-stranded DNA from a single virus particle (virion) enters the nucleus of a host cell. It is transcribed first by the host cell's RNA polymerase to create messenger RNA corresponding to viral genes. These RNA molecules are translated in the cytosol of the host cell to create viral proteins, which re-enter the nucleus. Some of the viral proteins are enzymes, catalyzing both translation and replication of the viral DNA. The newly replicated DNA combines with other viral proteins to create new virions, which are released upon dissolution of the host cell.**

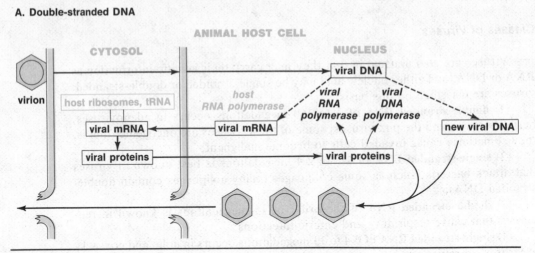

B. **Single-stranded viral DNA infecting bacteria is replicated by the host cell's DNA polymerase to create a double-stranded molecule, which is transcribed by the host cell's RNA polymerase to create messenger RNA and then viral proteins. The viral proteins include a DNA polymerase that replicates the duplex viral DNA, and another polymerase that transcribes these duplex circular molecules into new single-stranded linear DNA molecules. These linear molecules are converted to circular molecules by a DNA ligase, which then combine with viral proteins to create new virions.**

Illustration continued on opposite page

2 NEGS

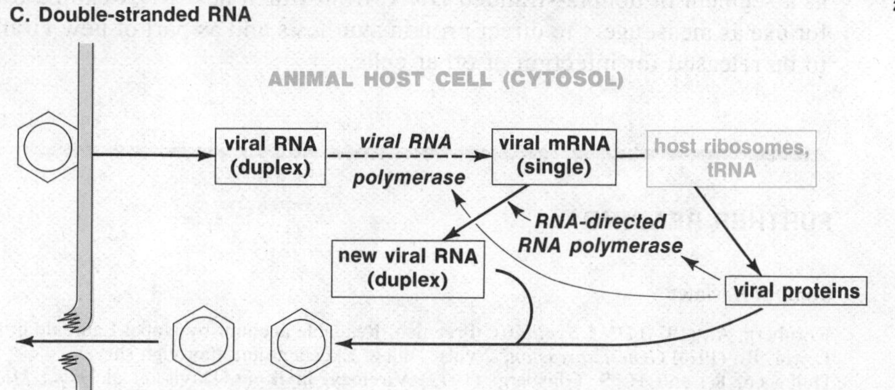

D. Single-stranded RNA (DNA intermediate)

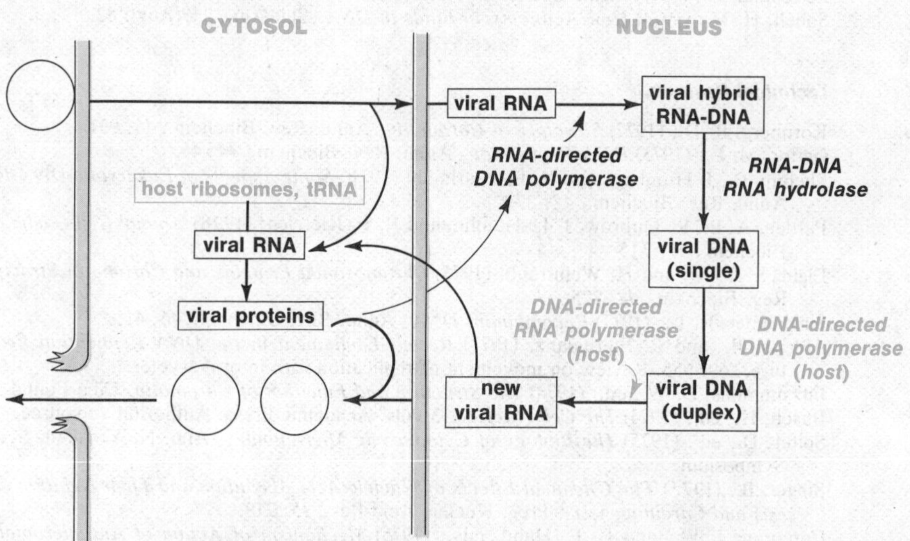

FIGURE 6–14 *Continued* **C. Double-stranded RNA passes into the cytosol of host cells from the virion, where it is transcribed by an RNA-directed RNA polymerase to create a single-stranded messenger RNA. This mRNA is translated to form viral proteins, but it is also replicated by an RNA-directed RNA polymerase to create new duplex viral RNA molecules, which combine with viral proteins to create new virions.**

***D.* Some single-stranded RNA molecules from viruses are replicated through double-stranded RNA intermediates, much as is the messenger RNA in *C*, above. Molecules from other viruses, including some that create cancers in the host, are replicated through DNA intermediates, as shown here. The viral RNA is replicated by an RNA-directed DNA polymerase. (This enzyme is often called reverse transcriptase, but names of this sort are not consistent with normal enzyme nomenclature.) The result is a duplex hybrid molecule, in which one strand is DNA and the other RNA. The polymerase has associated hydrolase activity, which causes the destruction of the RNA component of the hybrid, leaving a single DNA strand – the complement of the original virus. This strand is then replicated by the host cell's DNA polymerase to create a duplex DNA, which is then transcribed by the host's RNA polymerase to create more molecules of viral RNA. Some of these are used as messenger RNA to direct the formation of viral proteins, which then combine with the viral RNA to create new virions.**

The intermediate duplex DNA is sometimes incorporated into the host cell's chromosomes (not shown); this is sometimes associated with its transformation into a malignant cell.

as a segment of double-stranded DNA, from which new RNA can be transcribed for use as messengers to direct protein synthesis and as part of new viral particles to be released for infection of other cells.

FURTHER READING

General Reviews

Kornberg, A.: (1974) *DNA Synthesis*. Freeman. Readable account by Nobel Laureate in field.

Lewin, B.: (1974) *Gene Expression*, 2 vols. Wiley. Excellent and thorough survey.

Dulbecco, R., and H. S. Ginsberg: (1973) Virology, in B. D. Davis, et al., eds.: *Microbiology*. Harper-Row, p. 1007. Authoritative textbook.

Stahl, F. W.: (1969) *The Mechanics of Inheritance*, 2nd ed. Prentice-Hall. Excellent elementary introduction to mitosis, meiosis, recombination.

Lewin, B.: (1975) *The Nucleosome: Subunit of Mammalian Chromatin*. Nature, *254*: 651.

Felsenfeld, G.: (1975) *String of Pearls*. Nature, *257*: 177. More on chromatin.

Sobell, H. M.: (1974) *How Actinomycin Binds to DNA*. Sci. Am., *230*(Aug): 82.

Technical Reviews

Kornberg, R. D.: (1977) *Structure of Chromatin*. Annu. Rev. Biochem., *46*: 931.

Gefter, M. L.: (1975) *DNA Replication*. Annu. Rev. Biochem., *44*: 45.

Sheinin, R., J. Humbert, and R. E. Pearlman: (1978) *Some Aspects of Eukaryotic DNA Replication*. Annu. Rev. Biochem., *47*: 277.

Pardee, A. B., R. Dubrow, J. L. Hamlin, and R. F. Kletzien: (1978) *Animal Cell Cycle*. Annu. Rev. Biochem., *47*: 715.

Elgin, S. C. R., and H. Weintraub: (1975) *Chromosomal Proteins and Chromatin Structure*. Annu. Rev. Biochem., *44*: 725.

Sinsheimer, R. L.: (1977) *Recombinant DNA*. Annu. Rev. Biochem., *46*: 415.

Alberts, B., and R. Sternglanz: (1977) *Recent Excitement in the DNA Replication Problem*. Nature, *269*: 655. Review on movement of replication fork in prokaryotes.

Fitzsimmons, D. W., ed.: (1974) *The Structure and Function of Chromatin*. Ciba Found. Symp. 28.

Busch, H., ed.: (1974) *The Cell Nucleus*, 3 vols. Academic Press. Authoritative source.

Soifer, D., ed.: (1975) *The Biology of Cytoplasmic Microtubules*. Ann. N. Y. Acad. Sci., vol. 253. Symposium.

Singer, B.: (1975) *The Chemical Effects of Nucleic Acid Alkylation and Their Relation to Mutagenesis and Carcinogenesis*. Prog. Nucleic Acid Res., *15*: 219.

Corcoran, J. W., and F. E. Hahn, eds.: (1975) *Mechanism of Action of Antimicrobial and Antitumor Agents*. Springer-Verlag.

Cozzarelli, N. R.: (1977) *The Mechanism of Action of Inhibitors of DNA Synthesis*. Annu. Rev. Biochem., *46*: 641.

7 ALTERATIONS OF DNA

The character of cells and the organisms constructed from them depend upon the DNA that they contain; alterations in the arrangement of chromatin or in the sequence of bases will frequently cause obvious changes in the nature of the creature, be it one of the blessed or a mere bacterium.

The DNA of an organism contains much more information than is used at a given moment. Signals in the form of changing concentrations of particular chemical compounds govern the choice of particular segments of DNA to be transcribed, the number of molecules of the resultant messenger RNA that will be made, and the rate of translation of each mRNA molecule. The character of an organism, therefore, hinges not only on the peptides that it is able to make, but also on the way in which it responds to changes in its environmental circumstances. Environment controls the expression of heredity, but the response to environment is in itself a hereditary characteristic that is subject to alteration.

CHANGES IN THE GENETIC MESSAGE

Damage to DNA

It is easy to visualize something going wrong with the giant DNA molecules. Chemical reactions may change some of the 6×10^9 base pairs present in each somatic cell. Some of the long strands may rupture. An occasional mistake in replication may cause the wrong base to be inserted in a new strand. The wrong strands may be connected during recombination. Such errors do occur. Many of them, maybe most, result in the death of the individual cell, and there are so many cells that the loss is trivial. Some disrupt the genetic program of the cell so as to promote unrestrained division of the cell into a malignant growth and the causes of such errors are therefore said to be oncogenic (onchos = tumor). Others result in the development of more benign anomalies that are perpetuated as the cell divides.

Errors in the DNA of germ cells are more serious because they may be replicated in all of the cells of the offspring and continued in the descendants. Such errors are said to be **mutations,** and these constitute a major social problem. In a survey of hospitalized children in Baltimore, 6.4 per cent were in the hospital because of mutations in a single gene, 0.7 per cent were there because of a more gross aberration in chromosome structure, and 31.5 per cent were judged to have some gene-influenced condition. Only 53.5 per cent were believed to have a non-genetic condition or disease, with the remaining 8.2 per cent not assigned to any category.

Data on the low frequency of some genetic conditions can mislead the unwary on the prevalence of the altered genes that cause them. Consider phenylketonuria, to be discussed in detail later (p. 581), which results from one of several alterations in a single protein, and produces mental retardation if treatment is not begun in early infancy. From 1,000,000 live births, only 54 infants will have phenylketonuria. Each of the 54 is unable to make the normal protein. Because the condition is recessive, both chromosomes carrying specifications for the protein must be defective, and both parents must have contributed defective versions of the gene. In such cases the probability that a child will have two defective genes is the product of the probabilities for the presence of one defective gene in each chromosome. Since the gene is autosomal, that is, it does not occur on a sex-determining chromosome, the probabilities are the same for males as they are for females in large populations. We can therefore estimate the gene frequency for phenylketonuria to be $(54 \times 10^{-6})^{1/2}$, which is 7.3×10^{-3}. Approximately one out of every 68 humans has a gene for phenylketonuria in one of his paired chromosomes, an impressive incidence, and there are many deleterious conditions of similar prevalence.

Origin of Errors

Accidents in Replication. Something occasionally goes wrong with the polymerization of new DNA strands. A nucleotide may be omitted, or an extra one inserted, but the most common error is the incorporation of a nucleotide that is not the complement of the corresponding base on the template strand. When one purine is substituted for the other, or one pyrimidine for the other, the error is known as a **transition**. The substitution of a purine for a pyrimidine, or vice versa, is known as a **transversion**.

Since the accuracy of replication hinges upon the strength of the bonding between complementary nucleotides in a ternary complex with the DNA polymerase and not between themselves alone, the accuracy therefore also depends upon the nature of the polymerase, and some polymerases are more likely to make mistakes than others. The best estimate is that the DNA polymerases principally responsible for replication in animals make something less than one error in 10^9 nucleotide pairings, or in 10^6 replications of a 1000-nucleotide gene. The RNA-directed DNA polymerase associated with a virus causing myeloblastosis (a form of leukemia) in birds is an example of an enzyme that makes many more errors. In experiments using synthetic templates, this enzyme made one error in as few as 600 nucleotide pairings. Some argue that a certain rate of error is necessary for adequate evolutionary response to new situations — random modifications of proteins permit the selection of changes enabling the organism to compete better, and rapid changes may be especially appropriate for viruses that must generate new forms in order to survive as their hosts become immune to the original versions. Even if so, flexibility may be gained at a price, for most of the errors in coding are believed to be deleterious rather than advantageous.

Some errors result from the existence of the bases in more than one form. For example, we draw the ring of adenine in a fully aromatic form but draw the other bases as keto tautomers because these are the most abundant forms. However, a small amount of each exists as the other tautomer at a given moment (Fig. 7–1), and the tautomers may form hydrogen bonds with the "wrong" complementary base. Similarly, if the imide nitrogen of the rings of either guanine or thymine is ionized, these bases may bond with each other.

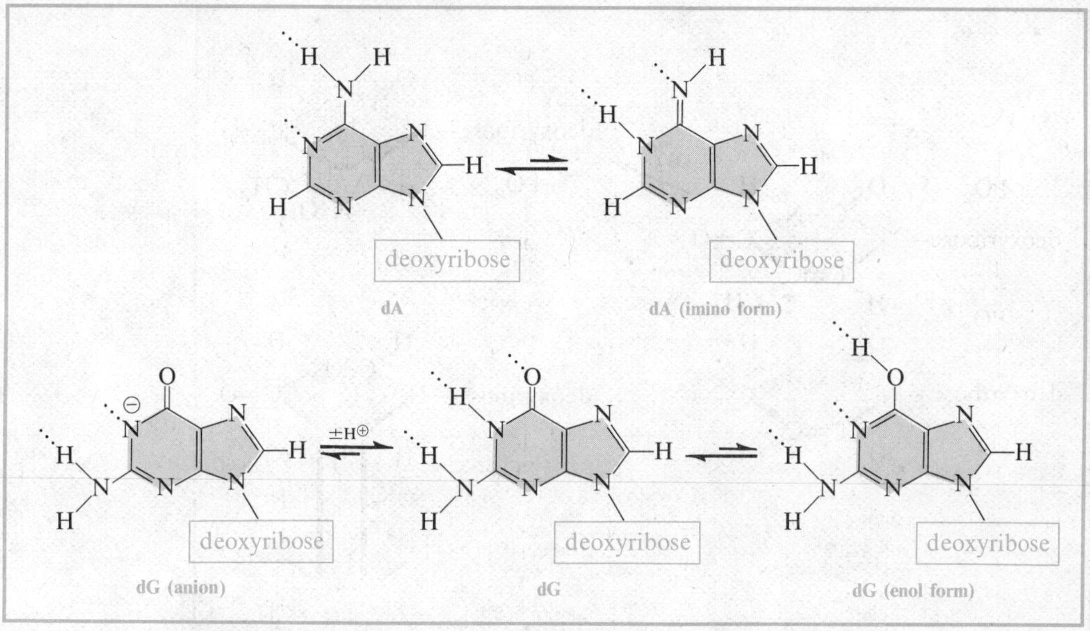

dA

dA (imino form)

dG (anion)

dG

dG (enol form)

FIGURE 7–1 The bases in nucleic acids have a slight tendency to exist in tautomeric forms, with a corresponding shift of donor and acceptor atoms for making hydrogen bonds. (The paired bases are not shown.) For example, the tautomer of adenine (*above right*) has atoms in position to form two hydrogen bonds with cytosine, whereas the tautomer of guanine (*lower right*) has atoms in position to form three hydrogen bonds involving both carbonyl oxygens and the imide hydrogen of thymine. The bases can also ionize to a small extent with the same result; the anionic form of guanine (*lower left*) can form hydrogen bonds with thymine.

Chemical Modification. The extent to which bases in DNA are modified through spontaneous chemical reactions is not known. For example, hydrolytic deamination of adenine produces hypoxanthine, which behaves as if it were guanine in the formation of complements (Fig. 7–2). (Deamination can be produced experimentally by a different mechanism through treatment with nitrous acid.)

Radiation Damage. DNA is also damaged by absorption of photons of ultraviolet, or shorter, wavelengths. In the ultraviolet range, absorption mainly affects the pyrimidines, especially thymine, by activating the ethylene bond in the ring. If the activated pyrimidine is next to another pyrimidine, a bond may form between

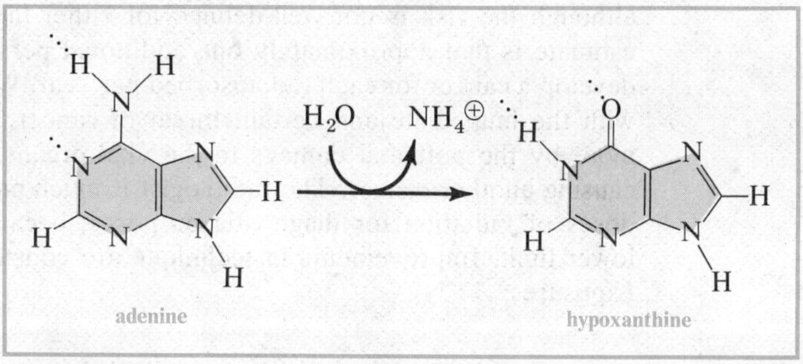

H_2O $NH_4^{\oplus}$

adenine

hypoxanthine

FIGURE 7–2

Hydrolytic deamination of adenine forms hypoxanthine, which can form two hydrogen bonds with cytosine (not shown).

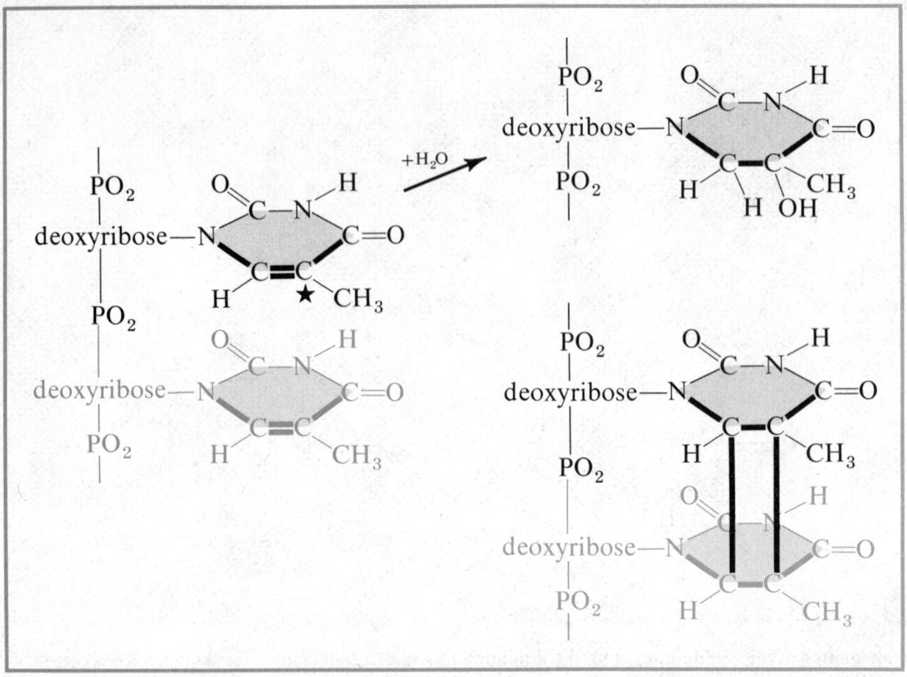

FIGURE 7–3 **A residue of thymine (*black*) that is activated by absorption of a quantum of ultraviolet light may react with water (*top*) or with a neighboring second residue of thymine (*blue*). The hydration reaction is relatively innocuous, but the formation of a thymine dimer clamps DNA into a biologically inert configuration.**

them, creating a dimer; if not, water may add to the molecule (Fig. 7–3). The formation of a dimer is the much more disruptive change; it may occur within a strand or between strands.

More energetic radiation, such as X-rays, gamma rays, or ionized particles, forms intermediates by electron expulsion. The resultant reactions include breakage of the polynucleotide chain through disruption of phosphodiester bonds, and opening of the rings in the bases. When oxygen is present, it is also activated by radiation, and damage is significantly increased through the formation of additional oxidized products.

The destruction of DNA is the basis for the use of X-rays and other radiation against cancers. Approximately one hit out of 300 causes lethal damage to mammalian cells. The absorption of radiation also increases the risk of malignancy; although the risk is not well-defined for either high or low doses, a reasonable estimate is that approximately one additional person in each million people will develop a cancer for each rad absorbed per year. When a radiotherapist is dealing with the immediate and certain threat of cancer, the dosages given are limited more by the potential damage to the vital organs than by the prospect of later causing another cancer. The radiologist is much more cautious in using even low doses of radiation for diagnostic purposes, because there is no definitely safe lower limit. Improvements in technique are constantly diminishing the required exposures.

Repair of DNA

Almost all living cells contain apparatus for recognizing distortions in the double helix of DNA and correcting the errors that cause them. These distortions may be caused by substitution of the wrong base during replication, by covalent linkage of bases, or by intercalation of a foreign compound. (Since many errors are corrected, it follows that the rate of damage to DNA is substantially greater than the rate projected from the remaining uncorrected lethal or mutagenic events.)

Light-Activated Repair. The formation of pyrimidine dimers by ultraviolet radiation is sometimes reversed by using a thief to catch a thief. At least some cells contain a protein that associates with the dimers and also absorbs ultraviolet light. Activation of the protein by the light causes it to break the bond in the dimer and release the constituent pyrimidines in their original form. Since the cells in which the dimers are formed must be exposed to light, it follows that light is available in the same cells for a reversal of the process. This mechanism had been shown in all phyla except the placental mammals, and it has also been demonstrated recently in human leukocytes, and may be present in other human tissues.

Excision Repair. A general mechanism for repairing damaged DNA involves the removal of a segment of the damaged strand, with the gap then being filled by the action of a DNA polymerase, followed by the action of DNA ligase (Fig. 7–4). At least two versions of this mechanism exist — long-patch or short-patch repair.

The critical step is recognition of the error, and enzymes are present that combine with distorted, but unbroken double helixes, and then catalyze a hydrolytic cleavage of the strand containing the error. These enzymes are **endonucleases,** attacking the center of a continuous polynucleotide chain near a site of distortion. After endonucleases act in long-patch repair, the nicked strand is subject to attack by an **exonuclease** that successively removes more nucleotide residues from the broken end so as to enlarge the gap. The gap may be as many as 100 residues long before it is filled by a DNA polymerase. (The polymerase that is involved may be the same enzyme that fills the gaps between Okazaki pieces during replication; see p. 104.)

When polynucleotide chains are broken by ionizing radiation, only as few as one or two nucleotide residues are hydrolyzed by an exonuclease before the gap is filled and the break eliminated (short-patch repair). Paradoxically, these seemingly small defects are the least likely to be repaired, although the job is sometimes completed in less than an hour once it is undertaken.

Long-patch repair as described here is more reliable than short-patch repair, even though it requires as much as 20 hours for completion. Both long and short-patch repair are used to repair accidental changes in one of the DNA strands. Breaks in both strands are likely to be lethal unless they happen to occur between cistrons in a way permitting continued transcription.

Defective Repair. The best evidence for the importance of the DNA repair mechanisms to humans is provided by a condition known as **xeroderma pigmentosum.** This rare disease is characterized by sensitivity to ultraviolet light (290 to 320 nm), with exposed skin developing a spectrum of disturbances ranging from excessive freckling through horny growths (keratoses), dilated capillary networks (telangiectasia), and ulceration to the appearance of carcinomas in the first decade of life.

Patients with xeroderma pigmentosum have one of at least five different biochemical defects in the repair mechanism. This can be demonstrated by com-

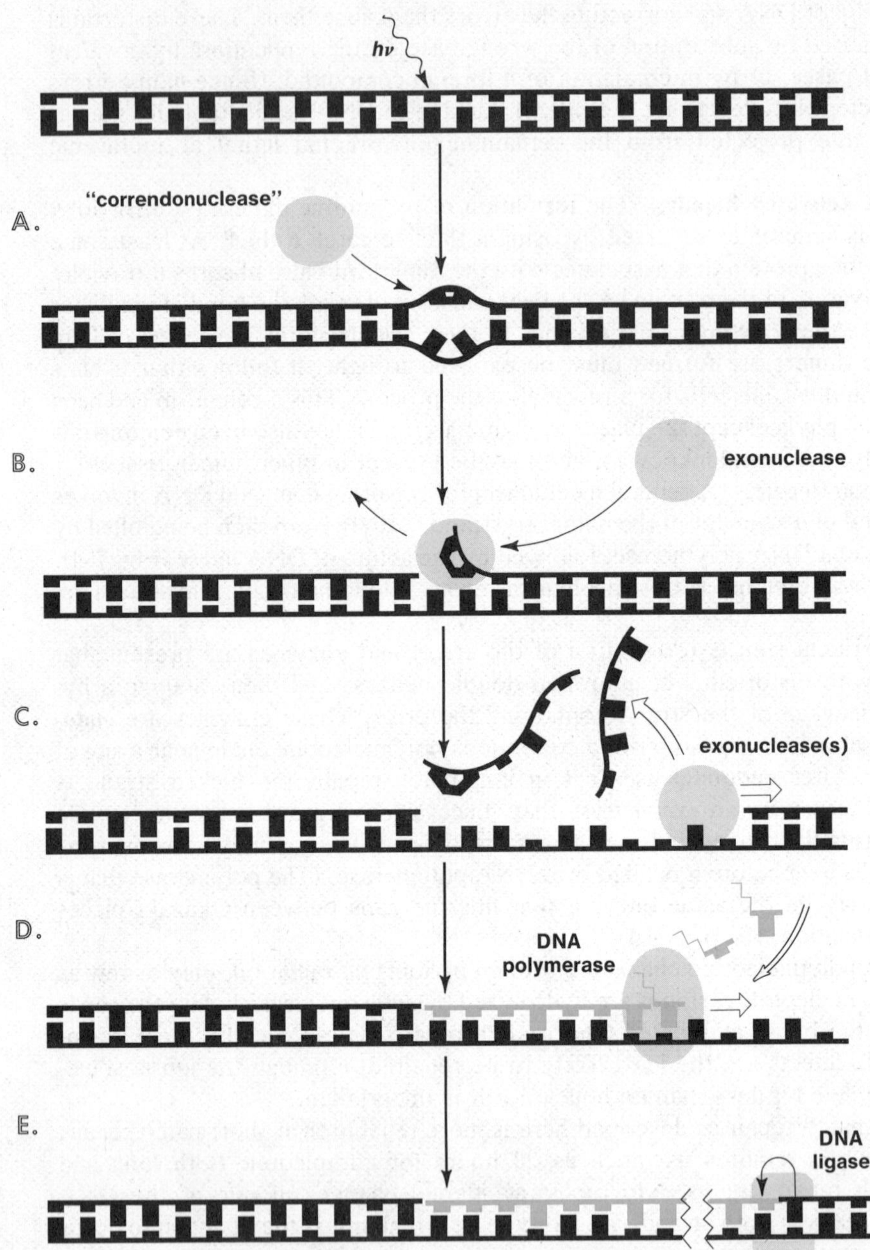

FIGURE 7–4 **Thymine dimers, and similar errors, can be removed from DNA by the combined action of four enzymes. *A.* The critical first step involves the binding of an endonuclease, "correndonuclease," at a distorted segment of double helix, followed by hydrolysis of one phosphodiester bond in the damaged strand. *B.* Following incision of the strand, the endonuclease is replaced by an exonuclease. *C.* The exonuclease breaks phosphodiester bonds five to eight residues removed from the site of distortion, liberating oligonucleotide fragments (oligonucleotides = polynucleotides containing a few residues). *D.* DNA polymerase proceeds to fill the gap with the correct complementary residues (*blue*). *E.* The final gap is closed by the action of DNA ligase.**

bining cells from various patients with recovery of normal repair in the combination. This supports the concept of repair processes consisting of several steps, each requiring its own protein, because a defect in the synthesis of any one of the necessary proteins would disrupt repair, but not in the same way.

> *Xeroderma Pigmentosum*
>
> Incidence of 1 in 250,000 people
>
> Autosomal recessive inheritance
>
> Widespread separate involvement includes malignancy in skin, with from 2 to 100 separate cancers appearing in a patient; freckles; damage to eye
>
> Prevention of damage in afflicted patients by minimizing ultraviolet exposure through screening lotions, clothing, and special glasses and by frequent skin examination
>
> Occasionally linked to central nervous system disease with small head, mental deficiency, hearing loss, and faulty gait

MUTATIONS

Any change in the DNA from which additional cells are constructed represents a mutation. If the term is not qualified, it refers to alterations in the germ cells that affect subsequent generations, but there are also somatic mutations that result in changes in the tissues of one individual. Let us examine the molecular basis for these changes, and some of the results that are manifested by differences in one kind of human protein, hemoglobin.

Errors in Crossover

The process of crossing-over during meiosis introduces the possibility for several types of misadventure, including the loss of segments of DNA from a strand, unequal exchanges from one strand to the other, and exchanges involving segments that are not homologous.

Hemoglobins Lepore. Some humans have hemoglobins with peptide chains containing unusual sequences that were created by crossover. Normal human adults contain two principal kinds of hemoglobin, HbA_1 and HbA_2, in their erythrocytes. Each has the same pairs of α chains, but they differ in the second pair, which is designated β in HbA_1 and δ (delta) in HbA_2. The chain formulas are therefore $\alpha_2\beta_2$ for HbA_1 and $\alpha_2\delta_2$ for HbA_2. The β and δ chains differ in only 10 out of the 146 amino acid residues, but these differences are sufficient to alter function.

The subtleties of chain composition will be discussed in Chapter 13; suffice it now to note that the genes for α chains are on one chromosome, and the genes for β and δ chains are adjacent to each other on another chromosome, where it is possible for errors in alignment and crossover to occur between them (Fig. 7–5). Such errors sometimes create chains in which one end has the sequence of a β chain and the other end has the sequence of a δ chain. Hemoglobins containing

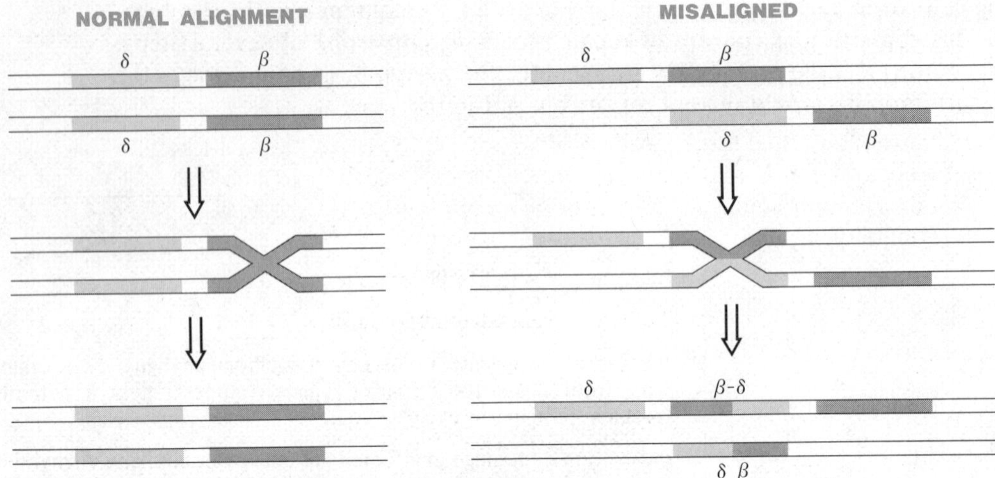

FIGURE 7–5 **Genes for δ chains in hemoglobin are adjacent to the genes for β chains. *Left.* A crossover within one of these genes between sister chromosomes during meiosis ordinarily has no effect on the nature of hemoglobin in the progeny. *Right.* Misalignment of chromosomes before crossover occurs creates one chromosome that contains both β and δ genes as well as a gene for a hybrid hemoglobin chain that begins with the β sequence and ends with the δ sequence. The other chromosome contains only a gene for the other type of hybrid hemoglobin, known as Hb Lepore, which begins with the δ sequence and ends with the β sequence.**

these abnormal chains are known as Lepore hemoglobins. (Variant hemoglobins are named after the city in which the first example is studied. Strictly, hemoglobin Lepore is a δ-β hybrid, with the δ sequence toward the N-terminal; β-δ Lepore variants, with the β sequence toward the N-terminal are hemoglobin Miyada, also known as anti-Lepore.)

Frame-Shift Mutations

An error in replication or crossover may cause the insertion or deletion of nucleotide residues in DNA. If the error occurs in a segment from which messenger RNA is transcribed, the result will depend upon the number of residues affected. Suppose that three residues, or some multiple of three, are added or subtracted. The remaining residues will still be in the correct triplet sequence for coding the intended amino acids, so the result will be the formation of a peptide chain that has some residues missing, or some additional residues inserted, in an otherwise correct polypeptide (Fig. 7–6).

On the other hand, error in replication that changes the number of nucleotide residues by one or two, or some multiple thereof not divisible by three, causes all of the remaining length of messenger RNA to enter a ribosome at the wrong place in the triplet sequence, so that the wrong amino acid will be incorporated into the resultant polypeptide chain at nearly every position. These errors are said to be frame-shift mutations; the product of such mutations will have little of the character of the intended protein unless the alteration occurs near the end of the amino acid sequence, leaving the bulk of the structure intact.

Hemoglobins created by both types of nucleotide deletion are known. Some variants lack one or more amino acids with the remainder of the sequence correct, indicating that one or more multiples of three nucleotides have been deleted from the DNA directing hemoglobin formation. Other variants have extensive changes in amino acid composition caused by frame-shift mutations. A well-characterized example is Hb Wayne, which contains five additional residues at the end of an α chain because the normal terminator codon lost its meaning. Here are the nucleotide sequences in DNA that are transcribed for the end of α chains, and the corresponding amino acids, in normal hemoglobin and in Hb Wayne, with the normal terminator in blue:

Normal DNA

5′ d(· · · X-G-C-T-A-C-C-G-A-G-G-C-T-C-C-A-G-C-T-T-A-A-C-G-G-T-A-$\frac{C}{T}$-T-T · · ·) 3′

transcribed but not translated	Term	Arg	Tyr	Lys
		141	140	139

Altered DNA

5′ d(· · · X-G-C-T-A-C-C-G-A-G-G-C-T-C-C-A-G-C-T-T-A-A-C-G-G-T-A-T-T · · ·) 3′

Term	Arg	Pro	Glu	Leu	Lys	Val	Thr	Asn · · ·
	146	145	144	143	142	141	140	139

Remember that the active strand of DNA is read from the 3′ end, that is, from right to left as written above, so the normal C-terminal of the α chain is Lys-Tyr-Arg-COO^- as specified here, with the arginyl group being the last of the 141 amino acid residues composing the chain. (The X at the left end and the $\frac{C}{T}$ in the Lys triplet at the right end signify that the exact triplet is unknown at these positions.)

The deletion of one nucleotide from the Lys triplet in DNA creates a messenger RNA in which all subsequent triplets will be read one residue late. The sequence U-A-A normally read as a terminator codon in RNA will now be read as the end of one triplet and the beginning of another, each of which designates amino acids rather than termination. The reading continues on the abnormal

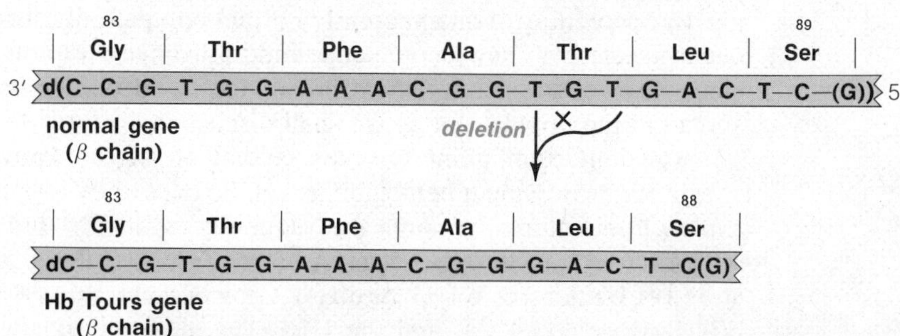

FIGURE 7–6 **When meiosis creates a gene that differs in nucleotide number from the parent by some multiple of three, the gene product is a polypeptide that may differ from the normal only in the absence or presence of some amino acid residues, or only one residue, if the change coincides with the triplet coding for amino acids. In the example shown, the deletion of the three nucleotide residues from which the codon for threonyl groups is ordinarily transcribed into the messenger RNA for the β chain of hemoglobin causes the formation of a shortened chain with an otherwise normal sequence. Combination of this chain with a normal α chain creates Hb Tours.**

mRNA, extending the polypeptide chain until a chance combination of nucleotides falls into a terminator sequence. This chance combination occurs in what is a post-terminator spacer sequence in normal messenger RNA.

Point Mutations: Base Changes

A point mutation is the substitution of one base for another in DNA. Since there are many genes coding for each transfer and ribosomal RNA, it is likely that a point mutation in one of them will pass unnoticed. Many of the genes coding for messenger RNA are unique, but some point mutations in these sequences also are not detectable. This is so because of the redundancy of the genetic code; many base substitutions create triplets coding for the same amino acid. For example, changing d(A-A-A) to d(G-A-A) in DNA merely converts one triplet for phenylalanine to another, and the composition of the resultant polypeptide will not be altered.

The point mutations that do cause alterations in amino acid composition have variable effects; some may never be seen because they so drastically affect the function of critical proteins that they cause early death of the germ cells in which they appear. Others are known only in the heterozygous state, in which the production of an abnormal protein can be tolerated because the corresponding unaltered chromosome continues to direct the formation of messenger RNA from which the normal protein is translated. There is sufficient margin of safety for many functions to permit survival under most circumstances with half the normal activity of a protein.

Still other mutations have little or no discernible effects because the substituted amino acid residue does not alter the function of the protein in any substantial way, and we discussed in Chapter 5 (p. 70) how the code is constructed so as to minimize damage from point mutations.

Point mutations in hemoglobin exemplify the range of effects that are seen. Dozens of different hemoglobins are known in humans that differ from the more common hemoglobins in only one amino acid residue. Some have lost so much function that they are seen only in heterozygous individuals, who still have one normal gene to produce a functional peptide chain. Others are so nearly like the common hemoglobins that their existence is only detected by accident or by massive screening of an apparently normal population. Most variant hemoglobins occur in relatively few people and tend to disappear rapidly from the population; some have conveyed sufficient advantage in special environments for them to spread in the population, as we shall discuss in Chapter 13.

The important point for now is that all of the known single amino acid substitutions in human hemoglobins can be caused by a single change in a coding triplet. For example, variants are frequently created by replacing a histidyl group with a tyrosyl group (His→Tyr). A histidyl group is designated in the transcribed strand of DNA by d(A-T-G) or d(G-T-G), whereas a tyrosyl group is designated by d(A-T-A) or d(G-T-A), and the His→Tyr mutation would result from an error causing the replacement of dG by dA.

On the other hand, we never see a phenylalanyl group inserted in place of a histidyl group, because the phenylalanyl group is designated by d(A-A-A) or d(G-A-A), and His→Phe would therefore require both a substitution of dA for dT and of dA for dG. Such double mutations are very improbable. Suppose that one in every 100,000 infants is born with a new single base substitution at some position in the cistrons for α and β chains in hemoglobin. (This is a high incidence, requiring a 50-fold faster mutation rate than is usually estimated.) In order for a

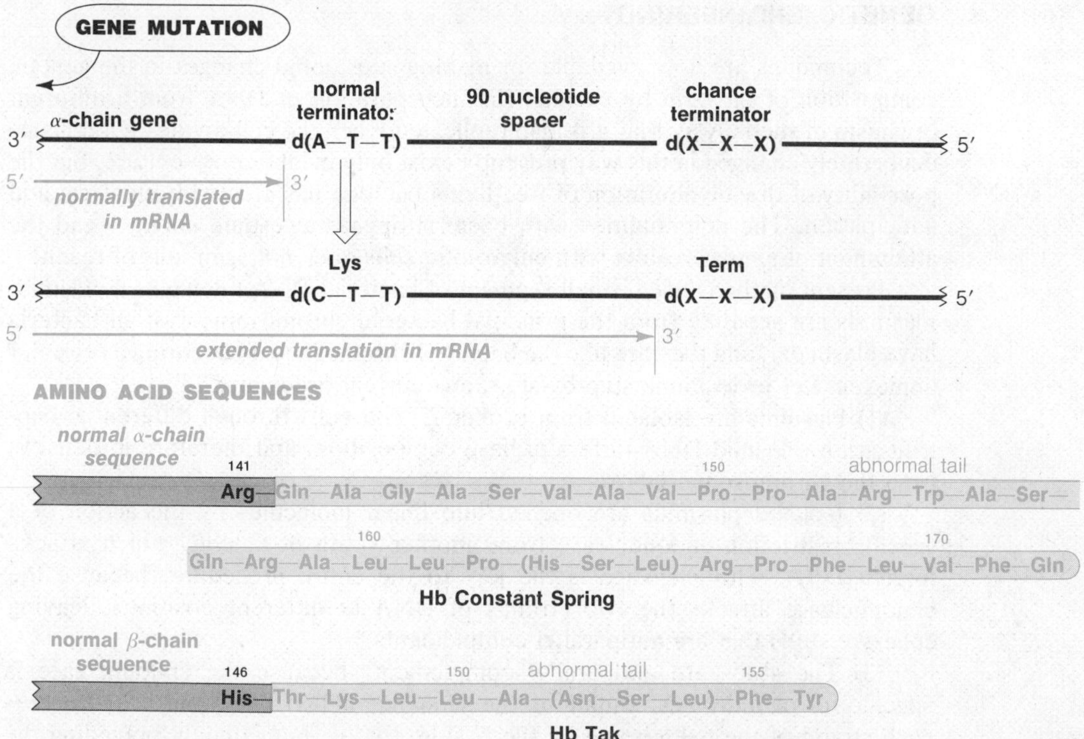

FIGURE 7–7 **Point mutations may change the template for a terminator codon to one for a codon designating an amino acid residue. When this happens, the resultant messenger RNA will continue to be translated as a polypeptide until a chance sequence of nucleotides gives a termination signal. In the gene for the α chain of hemoglobin, 90 nucleotides occur in the spacer segment between the normal terminator and the next chance occurrence of a terminator. Consequently, mutation of the normal terminator to the sequence designating a lysyl group results in the incorporation of an additional 31 amino acid residues beyond the normal end of the α chain. (One for the modified terminator, and 30 for the intervening 90 nucleotides.) The lengthened chain, occurring in Hb Constant Spring, is shown with the normal terminus in black, and the added residues in blue. A similar mutation, Hb Tak, is known for the β chain. See Baralle, F. E.: (1977) *Complete Nucleotide Sequence of the 5′ Non-coding Region of Human α- and β-Globin mRNA.* Cell, *12*: 1085.**

double mutation to occur, one of these infants also must have a second base substitution in the same triplet altered by the first substitution. This would occur only in fewer than one out of 10^{12} infants, even with the high mutation rate assumed; the odds are that this infant is still to be born.

Point mutations in the terminator triplet frequently result in the continued incorporation of amino acids beyond the normal end of a polypeptide chain. The extension of the chain continues until the chance occurrence of a terminator codon in a region ordinarily not transcribed. The effect is much like that caused by a frame-shift mutation, except that the proper amino acids appear in the polypeptide before the terminator position, and the number and kind of extra amino acids incorporated is the same for all point mutations in the terminator. (Why? Because the point mutation does not alter the reading of subsequent triplets, and the next terminator will still occupy the same position.) Examples of such mutations are known in the genes for both the α and β chains of human hemoglobin (Fig. 7–7).

GENETIC ENGINEERING

Techniques are now available for making substantial changes in the genetic composition of bacteria by causing the incorporation of DNA from a different organism in such a way that it is also replicated when the cell divides. Organisms deliberately changed in this way presently exist only in laboratory culture, but the possibility of drastic alteration of free-living bacteria has created both alarm and anticipation. The potentialities with bacteria appear awesome enough, and the attainment of similar results with eukaryotic cells does not seem out of reach.

Present methods use a small segment of bacterial DNA known as a plasmid. Plasmids are separate from the principal bacterial chromosome (not all bacteria have plasmids), and they are like the bacterial chromosome in occuring as circular duplexes. Let us examine step-by-step, one current procedure.

(1) Plasmids are isolated from broken *E. coli* cells through differential centrifugation. Plasmid DNA differs in base composition, and therefore in density, from the chromosomal DNA.

(2) Isolated plasmids are opened into linear molecules by the action of a specific restriction endonuclease from another strain of *E. coli,* which attacks foreign DNA. This cleavage is the key to the entire procedure, because the endonuclease attacks the two strands of DNA at different positions, leaving cohesive stubs that are antiparallel complements.*

(3) The stubs are antiparallel complements because the endonuclease is specific for palindromes containing the sequence d(T-T-A-A), and it hydrolyzes each strand of the palindrome on the 5′ side, that is, immediately preceding the first dT. The endonuclease attacks any kind of DNA at positions containing this sequence, but the plasmid usually used for genetic engineering has only one such sequence and is therefore opened into a single linear molecule.

(4) Donor DNA that contains the desired genes is isolated from any source, and it is also subjected to the action of the specific endonuclease. The donor DNA is thereby broken into segments, with each end containing d(T-T-A-A).

(5) The segments of donor DNA are mixed with the opened plasmid DNA. (The two kinds of DNA may be mixed before treatment with endonuclease.) Since the plasmid DNA and the donor segments end with the same sequence, they will sometimes associate. (The melting temperature for this four-nucleotide duplex is approximately 2° C, but a fraction will cohere at higher temperatures.)

(6) The mixture is combined with DNA ligase, which seals the strands together when they cohere.

(7) The result is a mixture of circular DNA molecules, many of which are recombinant DNA containing both the plasmid DNA and one or more donor segments. The mixture is added to recipient cells of *E. coli* that have been treated with $CaCl_2$, which makes them more permeable to plasmid DNA.

(8) The treated cells are spread on a solid medium favorable for growth of those containing the new gene. Those that grow can then be selected for propagation in larger cultures. Initial experiments demonstrated the technique by transferring genes that conveyed resistance to antibiotics into susceptible strains of *E. coli,* followed by growth of the bacteria on a medium containing the antibiotic. Only those cells in which the resistance gene was introduced could grow on the medium.†

---

*Many restriction endonucleases of differing specificities have now been isolated; they are proving to be valuable tools for determining the actual base sequence in DNA.

†Nearly all educated people must be aware that questions have been raised about the safety of the recombinant techniques. As is usual in public debates, unattractive human behavior has occasionally obscured the issues. Labels of Chicken Little Activism and Nietzschean Big Science have been applied by opponents. Drawing a line for acceptable risk is fortunately not our task. No matter the outcome, development of the techniques is a magnificent accomplishment, and its potential appears enormous.

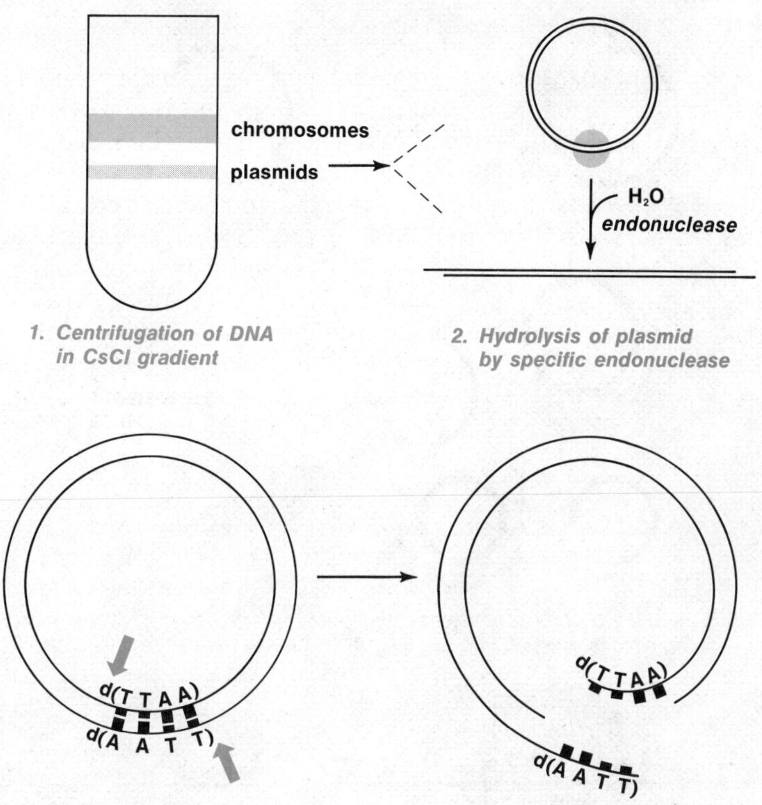

1. Centrifugation of DNA in CsCl gradient

2. Hydrolysis of plasmid by specific endonuclease

3. The endonuclease attacks opposite ends of a specific palindrome, creating complementary stubs on the linear product.

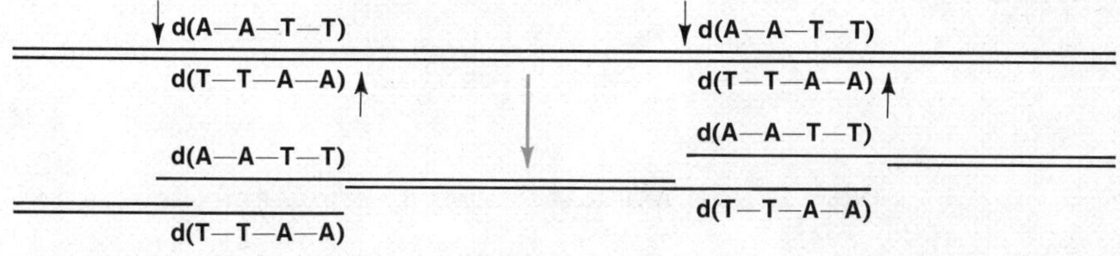

4. DNA from a selected donor is cleaved by the endonuclease at all sites containing the d(A—A—T—T) sequence.

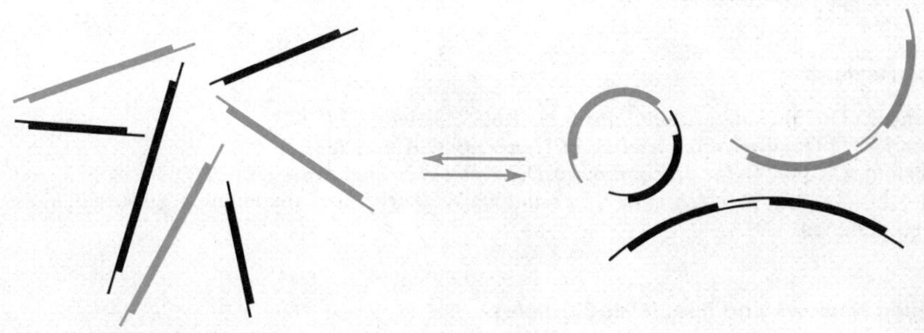

5. The cleaved segments of plasmid and donor DNA partially associate at the complementary stubs present on each segment. Various combinations form and re-form.

Illustration continued on following page

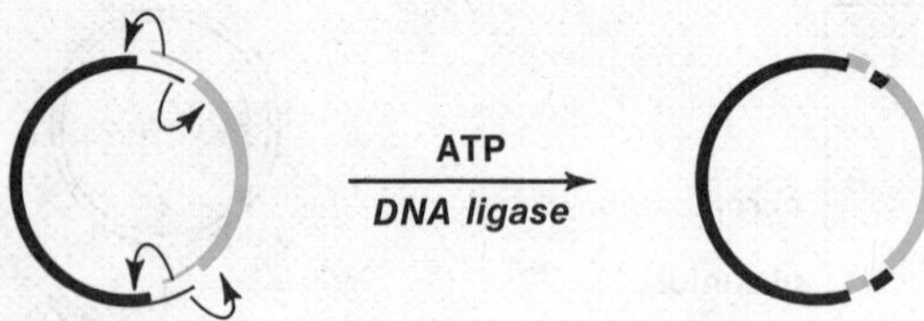

6. DNA ligase combines with associated DNA segments and closes the gaps in the joints.

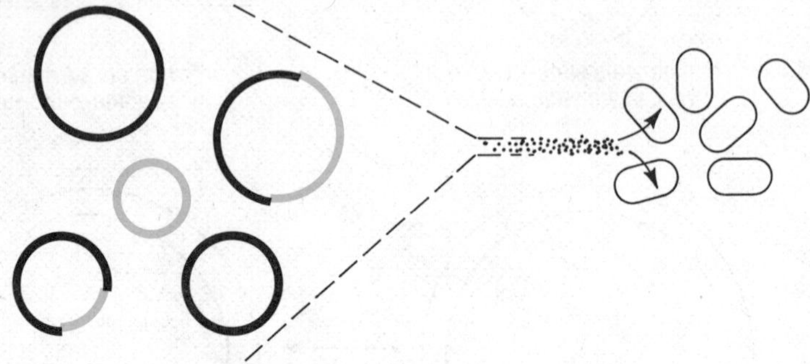

7. Reconstituted plasmids, some containing donor DNA fragments, are mixed with E. coli that have been treated with $CaCl_2$ to make them permeable to the plasmids.

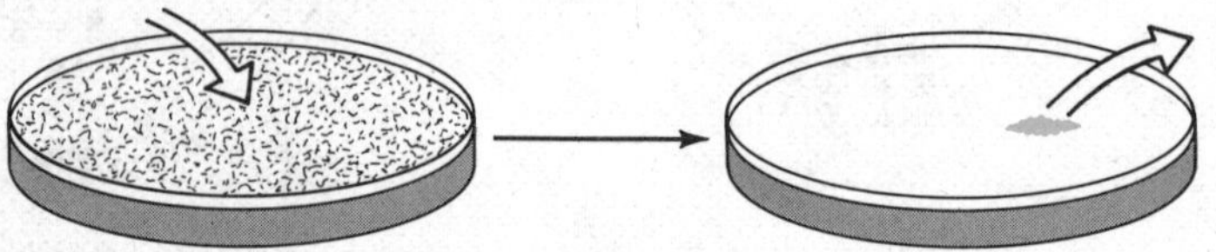

8. Treated E. coli are spread on the surface of an agar medium containing an antibiotic. Only those bacteria that contain a plasmid combined with a donor gene for antibiotic resistance can reproduce. Cells can be taken from the visible colony for culture on a larger scale.

FURTHER READING

Quick Reviews

Cohen, S.: (1976) *Gene Manipulation*. N. Engl. J. Med., *294*: 883.
Cohen, S.: (1975) *The Manipulation of Genes*. Sci. Am., *223*: 25.
Grobstein, C.: (1977) *The Recombinant DNA Debate*. Sci. Am., *237*: 22.
Wade, N.: (1977) *The Ultimate Experiment*. Walker, Short distillation of journalistic articles on recombinant DNA.

Detailed Reviews and Specialized Articles

Hanawalt, P. C., and R. B. Setlow, eds.: (1975) *Molecular Mechanism for Repair of DNA*. Plenum.
Grossman, L., et al.: (1975) *Enzymatic Repair of DNA*. Annu. Rev. Biochem., *44*: 19.
Roth, J. R.: (1974) *Frameshift Mutations*. Annu. Rev. Genet., *8*: 319.
Wood, J. F.: (1975) *Molecular Mechanisms of Radiation-induced Damage to Nucleic Acids*. Adv. Radiat. Biol., *5*: 182.

8 REFINEMENT OF PROTEINS

A major theme in our discussion to this point has been the expression of genetic information through creation of polypeptide chains with specific arrangements of 20 amino acid residues and through the regulation of the rate at which these chains are constructed. However, many polypeptides undergo further covalent reactions not involving the nucleic acids before they become finished mature proteins. Some of the reactions are hydrolyses of portions of the polypeptide chain; others modify the nature of the side chains so that there are more than 20 different kinds of residues. Is this an extra-genetic determination of structure? No, it is not. The modifications are comparable to those seen in the nucleic acids; the modifying agents themselves are created according to genetic instructions, and the sites at which they act on the polypeptide chains are fixed by the original amino acid sequences. Environment can influence the rate and the time at which these changes occur but not their nature.

In this chapter, we shall examine some typical examples to limn the principles without making a detailed catalog; other examples will be mentioned when we later treat the functions involved.

MODIFICATION BY PARTIAL HYDROLYSIS

Many newly synthesized polypeptide chains are promptly cleaved by hydrolysis, with one or more fragments being discarded. Proteins that are destined for secretion from the cell, and perhaps some that are included in lysosomes or other specialized cellular structures, are particularly likely to be constructed as oversize polymers to be refined later by hydrolysis into active products. Examples are many of the blood plasma proteins, which are synthesized in the liver; several polypeptides that act as hormones; and enzymes that are secreted by the pancreas or the liver. The purposes of this seemingly wasteful procedure are not the same for all such proteins.

Identification for Compartmentation. Some proteins are packaged for export in secretory granules; the proteins cross the endoplasmic reticulum membrane into the cisternal spaces and are transferred to the Golgi apparatus for enclosure. Insulin, the parathyroid hormone, and the digestive enzymes of the pancreas are handled in this way, along with other proteins. Those examples are specifically cited because they are known to undergo two stages of hydrolysis in transit (Fig. 8–1). The first hydrolysis occurs immediately after formation of the peptide chain; it may even begin before translation is completed. The discarded segment consists mostly of residues carrying bulky hydrophobic groups; in the pancreatic proteins

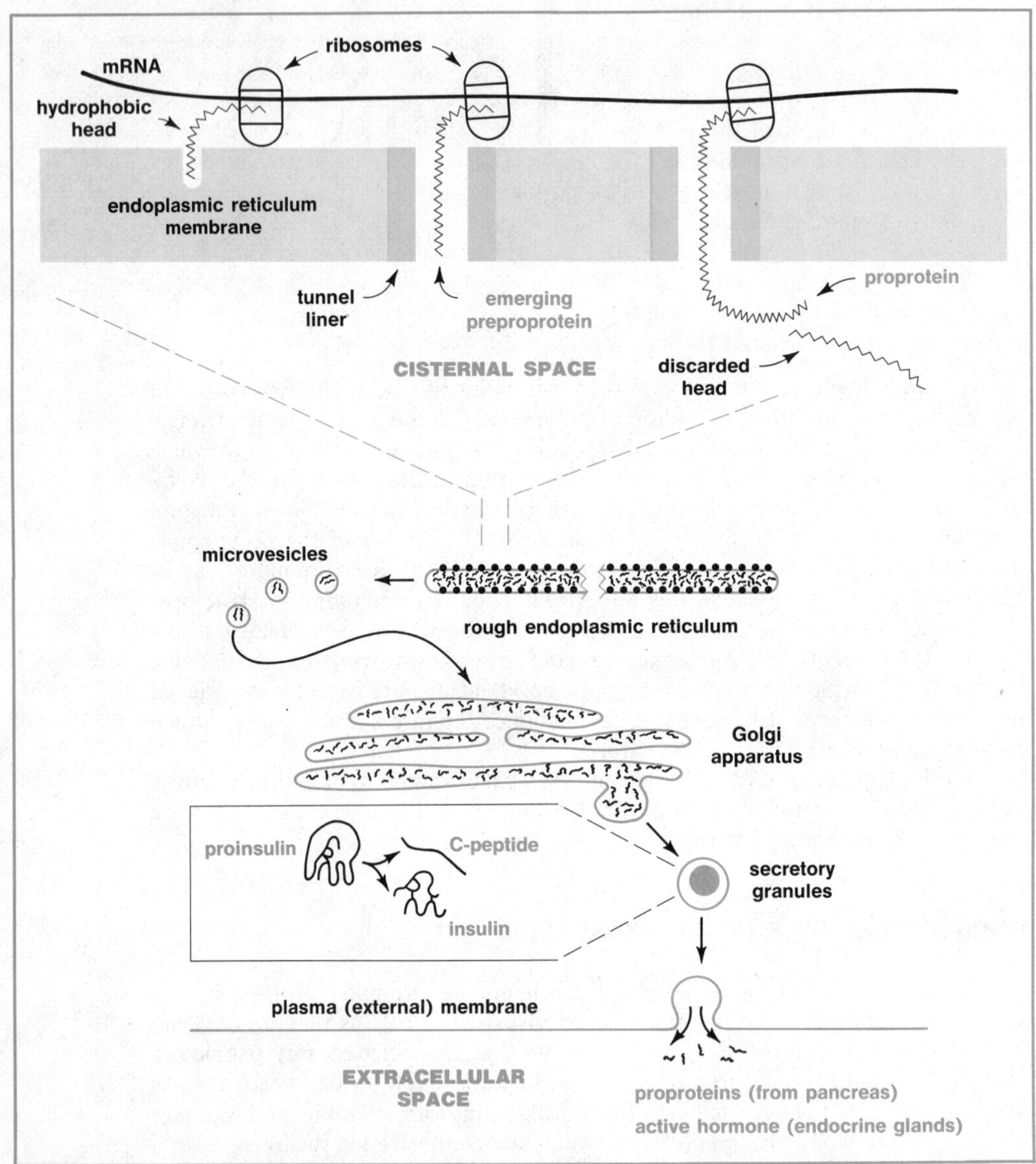

FIGURE 8–1 The synthesis of proteins for export from the cell. *Top.* The protein first appears in precursor form, in which the initially translated sequence is rich in hydrophobic side chains, which facilitate penetration of the membrane of the endoplasmic reticulum, thereby binding the attached ribosome and messenger RNA. The entry of the precursor chain (preproprotein) may promote the gathering of proteins to line a tunnel through the membrane. In any event, the preproprotein passes into the cisternal space of the endoplasmic reticulum, and the hydrophobic head of the polypeptide is discarded, leaving a proprotein. *Center, smaller scale.* The proprotein is packaged in microvesicles for transport to the Golgi apparatus. *Bottom.* Secretory granules containing the proprotein are released from the Golgi apparatus and fuse with the plasma membrane for discharge of the contents from the cell. In the case of proinsulin and proparathyroid hormone, the proprotein is partially hydrolyzed to create the active form within the secretory granules and the Golgi apparatus, as illustrated in a magnified view for insulin. The hydrolytic enzymes from the pancreas are not activated until they have been discharged into the lumen of the intestine.

there is at most one charged residue in this segment out of a total of a dozen or so.

Present belief is that the non-polar sequence at the beginning of the chain provides a structure to be buried in the endoplasmic reticulum as soon as it appears, thereby attaching the ribosome and the growing peptide chain to a particular site where a tunnel through the membrane can develop. The growing chain is then fed through the tunnel, and the now worthless initial segment is removed. This picture provides an explanation for the synthesis of many secreted proteins by ribosomes attached to the endoplasmic reticulum. (The rough endoplasmic reticulum is studded with ribosomes, the smooth is not.) Fixation of the initial segment of the peptide chain would hold the ribosome, the messenger RNA attached to the ribosome, and all of the other ribosomes on the messenger RNA to the surface of the endoplasmic reticulum. There is good evidence for the secondary attachment of ribosomes through messenger RNA.

It is possible that the mechanism described for transfer through the endoplasmic reticulum to the Golgi apparatus is a general one for handling all exported proteins. It does not necessarily follow that removal of the initial segment of polypeptide occurs with all secreted proteins; the segment may have some additional function in some.

Masking of Biological Activity. Some proteins are threats to the tissue in which they are made, and enzymes that hydrolyze proteins, such as those involved in digestion or blood clotting (see Chapters 17 and 18), are particularly dangerous. Such enzymes are made with an extra segment of polypeptide that prevents them from attacking other proteins. This inactive form of the enzyme is said to be a **proenzyme,** which is converted to the active form by itself being partially hydrolyzed to remove the inhibiting segment of polypeptide chain.

The first stage of hydrolysis of these enzymes converts a **preproprotein** — the form initially translated and transported into the endoplasmic reticulum — into a **proprotein,** which is secreted from the cell. The second stage of hydrolysis, the conversion of proprotein to the active form, occurs outside the cell under circumstances making it desirable for the protein to carry out its function. For example, digestive enzymes are activated after entry into the intestines; tissue blood clotting enzymes are activated when tissues are damaged.

Fixation of Tertiary Structure. Most proteins attain their physiological form by spontaneously folding into a conformation of minimal energy content. However, some desirable configurations evidently cannot be attained in this way; the active form of some proteins is not the one with minimal energy content, and it exists only because it is cemented in place with additional covalent bonds between segments of polypeptide chain. Insulin (next section) and collagen (next chapter) are examples.

The problem is to hold the desired conformation long enough for it to be pinned together by some further reaction; it is solved by synthesizing the protein as a proprotein, in which the initial segment contributes forces necessary for proper folding of the entire molecule. The to-be-active segments of the chain are then fixed by forming covalent bridges. After this is done, the initial segments have served their purpose and are discarded by hydrolysis, thereby revealing the activity of the remaining molecule.

Pyroglutamate Residues. Some polypeptide chains begin with a residue of the heterocyclic compound variously known as pyroglutamate, oxoproline, or pyrrolidone carboxylate (Fig. 8–2). The presence of this residue is an indicator that the polypeptide is made from a longer precursor because the pyroglutamate residue is formed when a peptide bond preceding a glutamine residue is hy-

glutamine residue

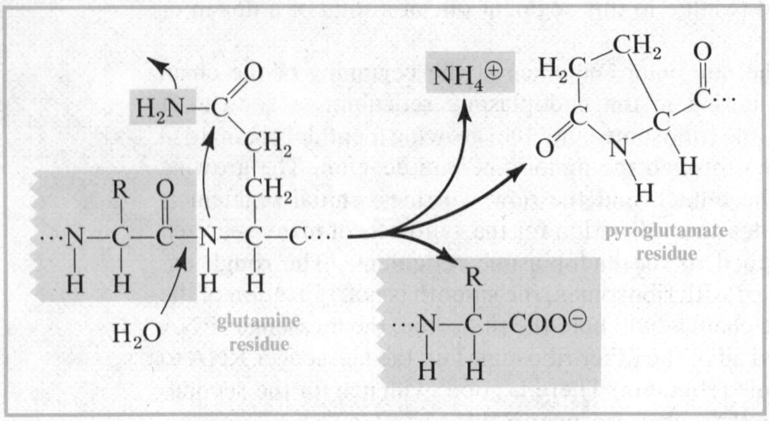

pyroglutamate residue

FIGURE 8-2 **Hydrolysis of polypeptide chains at the peptide bond preceding a glutamine residue causes the conversion of the residue to the uncharged pyroglutamate residue. The effect is an exchange of the peptide nitrogen (which would appear as an ammonium group in hydrolysis of other peptide bonds) for the side chain amide nitrogen of the glutamine. The amide nitrogen is released as a free ammonium ion. The other product of the hydrolysis is the initial segment of the chain, with a free carboxylate group.**

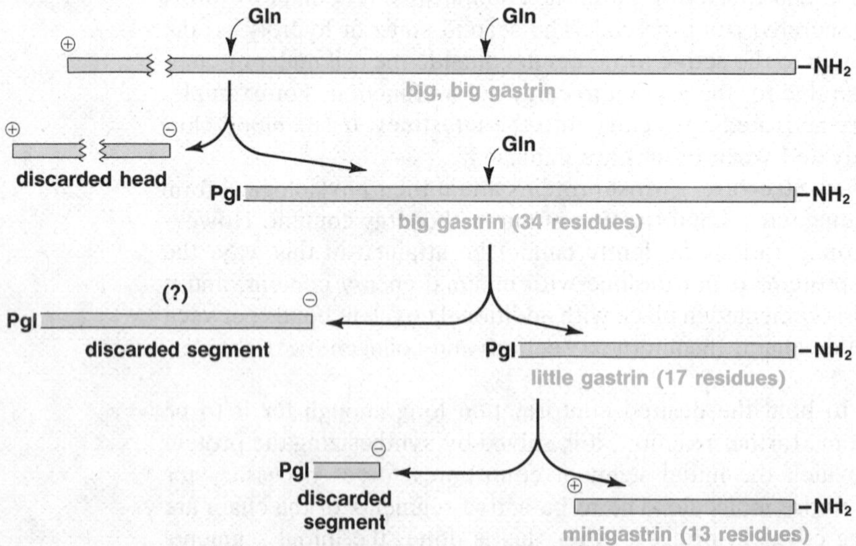

FIGURE 8-3 **Various species of the hormone gastrin are created by successive hydrolyses. The first two hydrolyses occur before glutaminyl groups, creating products that begin with pyroglutamate (Pgl) residues. The final hydrolysis occurs at a leucine residue, with the product containing the charged ammonium group typical of most polypeptides. The terminal carboxyl group is present as an amide (see p. 145).**

drolyzed. Hydrolyses at other locations do not cause distinctive changes in the terminal groups.

For example, pyroglutamate occurs in some **gastrins,** which are polypeptide hormones secreted by cells from the distal part of the stomach (antrum) and the initial part of the small bowel (duodenum). Gastrins provoke increased secretion of acid by the stomach. The activity of the hormones resides in the C-terminal end (Fig. 8–3), so hydrolytic removal of the initial portions of the chain makes little

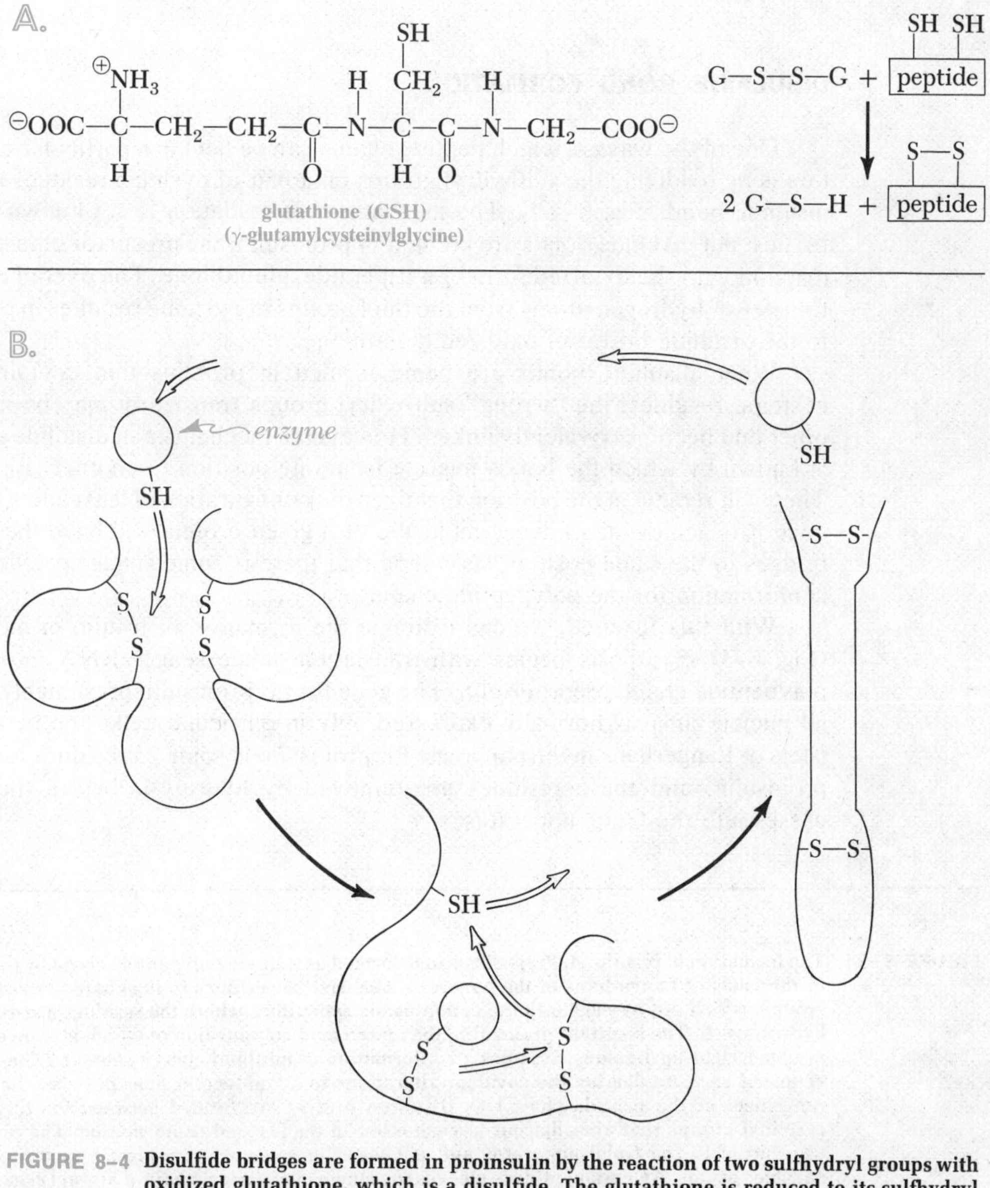

A.

glutathione (GSH)
(γ-glutamylcysteinylglycine)

$$G{-}S{-}S{-}G + \text{peptide(SH SH)} \longrightarrow 2\,G{-}S{-}H + \text{peptide(S{-}S)}$$

B.

enzyme

FIGURE 8–4 **Disulfide bridges are formed in proinsulin by the reaction of two sulfhydryl groups with oxidized glutathione, which is a disulfide. The glutathione is reduced to its sulfhydryl form by oxidizing the cysteinyl groups to their disulfide form. *B.* Disulfide bridges are rearranged by reaction with a sulfhydryl-containing enzyme that splits the various disulfide bonds and reforms them. Successive recombinations occur until the arrangement of the peptide chain with the lowest energy content (greatest internal bonding) is achieved.**

difference. The first gastrins isolated are now known as little gastrin and minigastrin, with minigastrin formed by hydrolysis of little gastrin. It is obvious that little gastrin is itself the product of hydrolysis of a still larger polypeptide, because the N-terminal residue is pyroglutamate. Such a polypeptide was indeed discovered (big gastrin), with a glutaminyl group in the right place to correspond with the pyroglutamyl group of the little gastrin, but it also began with a pyroglutamate residue, indicating the existence of still another precursor, which has also been demonstrated (big big gastrin), although its amino acid sequence has not been determined.

DISULFIDE BOND FORMATION

One of the ways in which peptide chains can be held in a particular configuration is by oxidizing the sulfhydryl groups on a pair of cysteine residues to form a disulfide bond (Fig. 8–4*A*). The mechanism of oxidation is not known for most tissues, but disulfide bonds are created in proinsulin, the precursor of insulin, by a reaction with the oxidized form of a tripeptide, glutathione. The overall effect is a transfer of hydrogen atoms from the thiol groups of cysteine residues in proinsulin to the disulfide bridge of oxidized glutathione.

When disulfide bonds are being formed in proteins that contain several cysteine residues, the "wrong" sulfhydryl groups transiently may be near each other and become covalently linked. However, a mechanism of disulfide exchange is known by which the bonds migrate from one position to another (Fig. 8–4*B*). They will remain in the position that fixes the configuration of least energy content once it is achieved, so every molecule of a given protein will have the disulfide bridges in the same position, assuming that there is some single specific favored conformation for the polypeptide chain.

With this in mind, we can examine the formation of insulin in more detail (Fig. 8–5). Synthesis begins with translation of messenger RNA into a single polypeptide chain, preproinsulin. The gene for preproinsulin presumably exists in all nuclei, but it is normally expressed only in particular cells, the beta cells of islets of Langerhans in the pancreas. Preproinsulin is some 23 residues longer than proinsulin, and these residues are removed by hydrolysis before the peptide passes into the Golgi apparatus.

FIGURE 8–5 **The formation of insulin. *A.* Preproinsulin is formed as a single polypeptide chain by the beta cells in the islets of Langerhans of the pancreas. The first 23 residues in the chain are rich in leucyl groups, which aid passage into the endoplasmic reticulum, where the residues are discarded by hydrolysis. *B.* The resultant proinsulin, the amino acid composition of which is shown in the box at upper right for humans, assumes a conformation of minimal energy content. (The drawing is arranged so as to display the covalent alterations to advantage; it does not give the actual arrangement of the peptide chain.) *C.* Disulfide bridges are formed between the three pairs of cysteinyl groups that are adjacent to each other in the favored conformation. The proinsulin is transferred to the Golgi apparatus and packaged in secretory granules. Attack by hydrolytic enzymes begins. An endopeptidase cleaves the single chain near a pair of arginyl groups into two pieces, which are held together by the disulfide bonds. *D.* Another enzyme, an exopeptidase, removes the two arginine residues from the end exposed by the endopeptidase. Meanwhile, the chain is attacked by the endopeptidase at another site, thereby exposing adjoining lysyl and arginyl groups to attack by the exopeptidase. *E.* The result is the removal from proinsulin of a large central segment of polypeptide chain, the C-peptide, along with four basic amino acid residues. The remaining ends of the chain are held together as a single insulin molecule by the previously formed disulfide bonds, and it is this molecule that appears in the blood upon discharge of the secretory granules.**

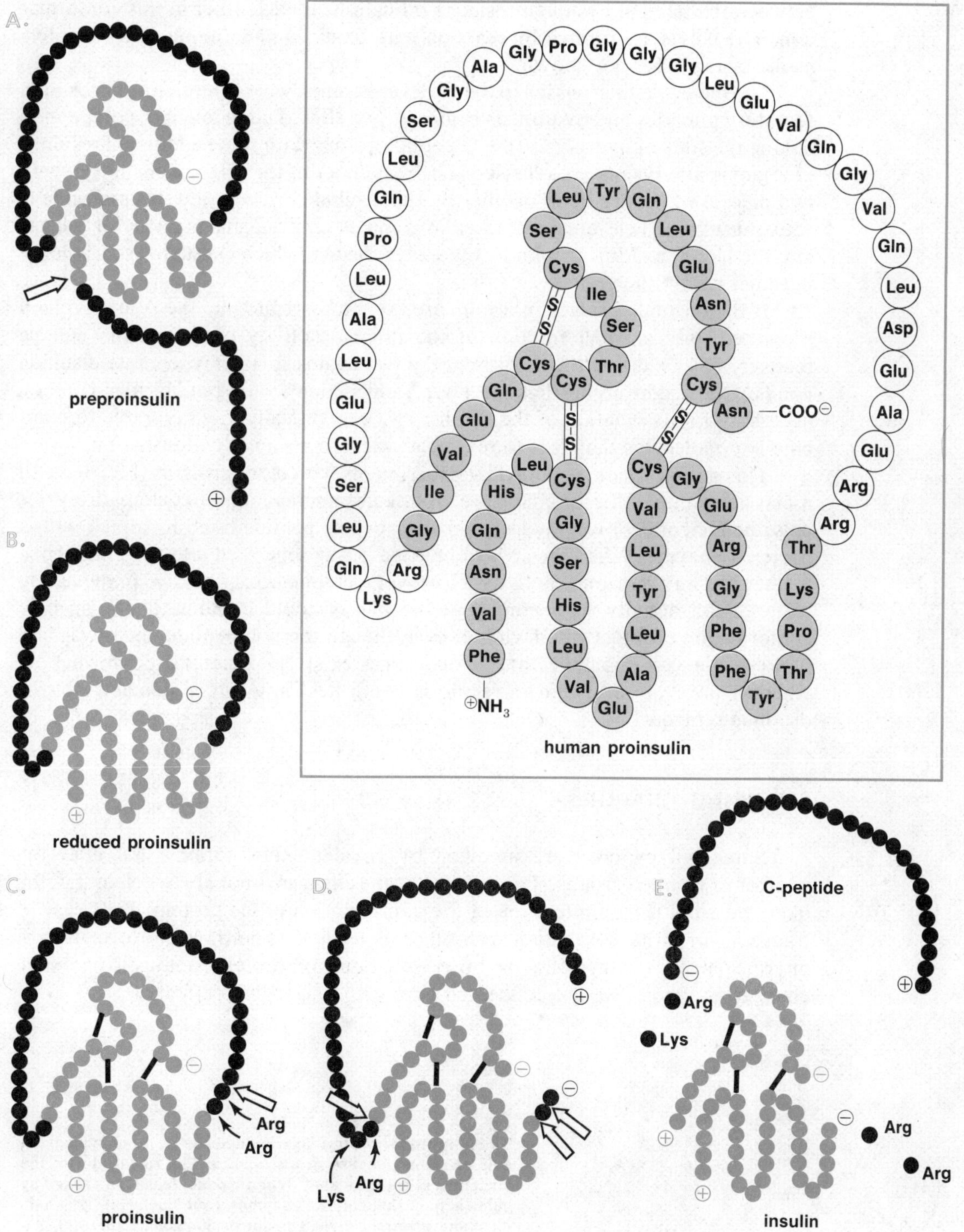

FIGURE 8–5 *See legend on opposite page.*

At some stage, probably before transit of the membrane is completed but after hydrolysis of the initial segment, the proinsulin polypeptide folds into its conformation of minimal energy content, and three disulfide bonds are formed between the pairs of cysteinyl residues brought near each other in this conformation. The bonds are formed by reaction with oxidized glutathione, and the intermediate stages are not known.

The molecule then passes to the Golgi apparatus, where the chain is packed in secretory granules and hydrolysis begins at two sites. The segment of polypeptide linking the sites of hydrolysis (the C-peptide), along with some additional residues of arginine and lysine, are released. The remainder of the original chain is now in two pieces, which are held together by the disulfide bridges, and this molecule is active insulin, which remains packed in granules until a signal is given for release into the blood (p. 720). Insulin forms a zinc chelate, which crystallizes as a dense hexamer within the granules.

If the disulfide bridges in insulin are broken by reduction, the A and B chain separate. Only a small fraction of the original activity of the insulin can be recovered by oxidizing the mixture of the two chains so as to regenerate disulfide bonds. One might argue that the poor yield is due to the bimolecular reaction necessary for association of the chains; it is energetically less favorable to combine two molecules than it is to rearrange one into a similar conformation.

This notion is not sustained by experiments with other proteins (Fig. 8–6). In a classic example, the ribonuclease from calf pancreas (an enzyme hydrolyzing polyribonucleotides) was hydrolyzed at a specific peptide bond by attack with a bacterial enzyme. When the split ribonuclease was separated into two fragments, neither had any remaining activity. However, combination of the two immediately restored full activity, indicating that the pieces could combine in the specific conformation of intact ribonuclease, even though the polypeptide chain was still broken. We can conclude that ribonuclease exists in a natural conformation, whereas the mature insulin molecule is being held in a forced geometry by its disulfide bridges.

CHANGING CHARGES

Some polypeptides are modified by covalent substitutions that alter the number of charged groups. The purpose of the changes is not always clear, but the likely possibilities are alteration of the conformation of the protein, or its ease of transport, or of its association with other molecules. (The transient formation of intermediates in enzymatic or other reactions of proteins may also involve changes in charge; we are concerned here with longer-lived effects.)

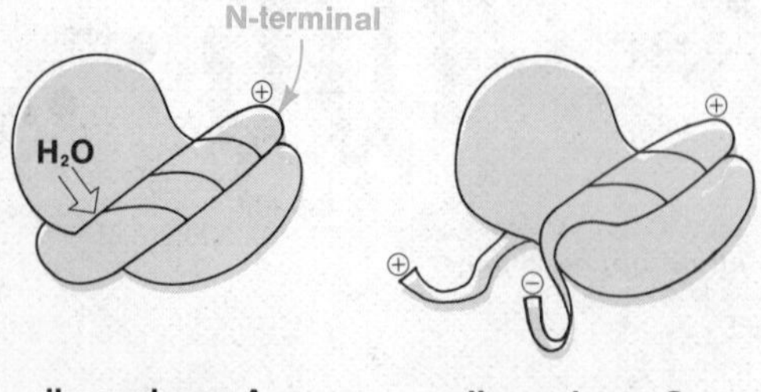

FIGURE 8–6

The N-terminal portion of ribonuclease A, an enzyme from calf pancreas, occurs in an α-helix. The remainder of the structure is not indicated. When ribonuclease is attacked by subtilisin (a hydrolytic enzyme from bacteria), a single cleavage occurs between residues 20 and 21 (*arrow*). Although the cut ends swing away, the small N-terminal segment remains attached, and its helical structure is retained because the combined structure is still in the conformation of lowest energy content; the cleaved molecule (S for subtilisin-modified) retains its biological activity.

Removing Charge

We have already seen one way of diminishing charge on a polypeptide chain—hydrolysis to create the uncharged pyroglutamyl group. Charges can be eliminated in a more general way by converting ammonium or carboxylate groups to neutral amides. For example, the terminal ammonium groups or the lysyl side chains may be acetylated:

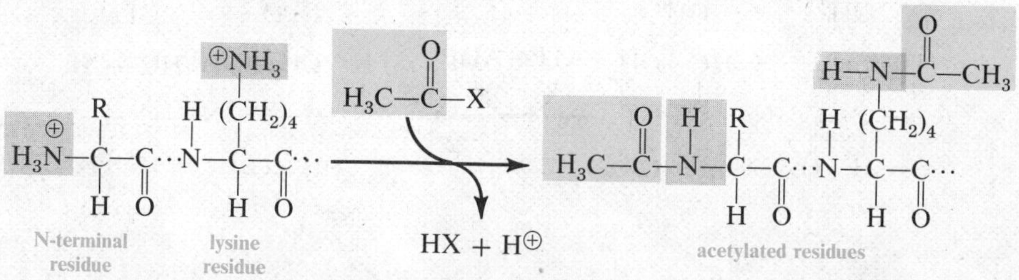

N-terminal residue lysine residue $HX + H^{\oplus}$ acetylated residues

Similarly, a terminal carboxylate group can be converted to a simple amide:

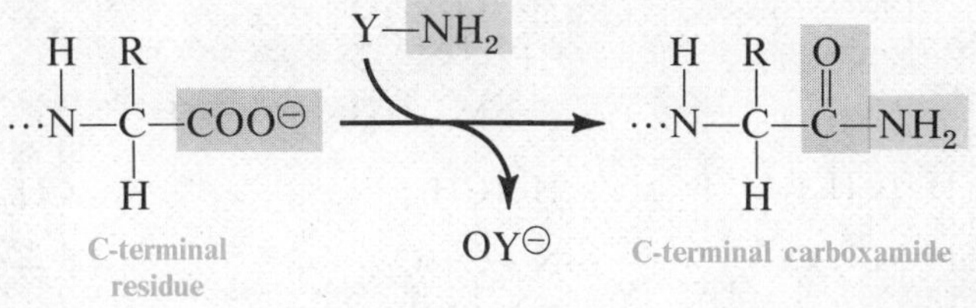

C-terminal residue $OY^{\ominus}$ C-terminal carboxamide

(Aspartate and glutamate residues modified in this way would be indistinguishable from directly incorporated residues of asparagine or glutamine, in which the amide group is already present.) Gastrins are examples of peptides modified in this way.

These changes, and those discussed in the following, are not exceedingly rare. Several proteins are known to occur as acetyl derivatives. Terminal amidation occurs in several small peptide hormones. An extreme example of loss of charge is seen in **thyrotropin releasing factor,** which contains only three amino acid residues. The first is a pyroglutamyl group derived from a glutamine residue by hydrolysis of a precursor, and the last is a proline residue with the carboxylate group amidated:

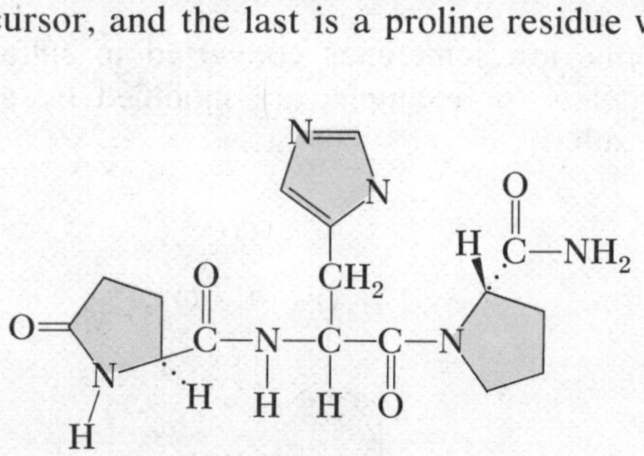

thyrotropin releasing factor
(pyroglutamyl histidyl proline amide)

(The releasing factor causes the anterior pituitary gland to release thyrotropin, a hormone that stimulates the thyroid gland to release still another hormone, thyroxine (p. 715.)

Adding Charges

Some proteins are modified by adding anionic groups, such as phosphate or sulfate groups, in the form of amides or esters. For example, the hydroxyl groups of serine or threonine residues are common acceptors for phosphate, creating phosphate esters:

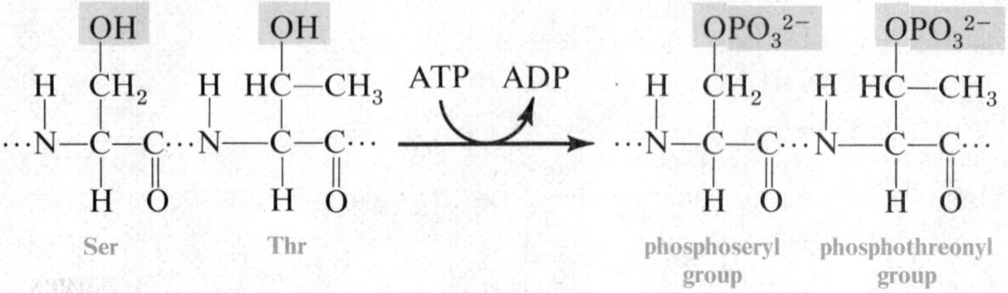

Ser Thr ATP ADP phosphoseryl group phosphothreonyl group

Less common, but not rare are formation of amides by phosphorylation of lysyl, arginyl or histidyl groups:

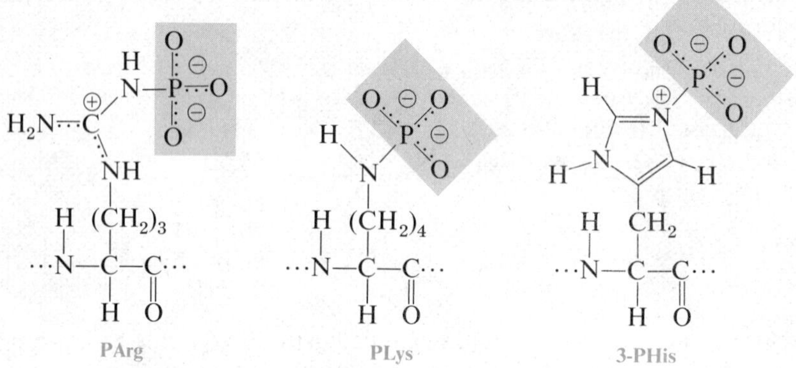

PArg PLys 3-PHis

There are ambiguities in the numbering of the imidazole ring. Biochemists tend to number the nitrogen atom next to the substituent alanyl chain as #1 and the other nitrogen as #3. Rules of organic nomenclature specify the reverse numbering so as to minimize the number of side chain substitution (4 instead of 5).

Similarly, proteins are sometimes converted to sulfate esters or amides. Some gastrin molecules, for example, are modified by esterification of their tyrosyl groups with sulfate:

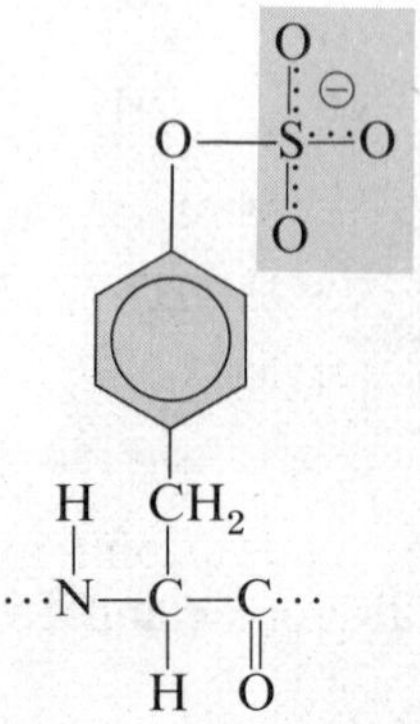

Functions of Phosphoproteins

Nutrition for the Young. Some proteins are heavily phosphorylated as a device for carrying phosphorus and associated calcium from the mother to the offspring. Animals that reproduce via a closed egg include heavily phosphorylated proteins in the yolk for the nourishment of the embryo, and calcium ion is also provided by chelation with these negative groups. For example, **phosvitin** is a protein with 100 to 120 phosphate groups per molecule of 35,500 daltons. (Ten per cent of the mass of phosvitin is contributed by phosphorus atoms.) The polypeptide chain contains many clumps of seryl residues, which are the sites of phosphorylation. Chicken eggs are important sources of both phosphorus and calcium in the human diet owing to the presence of phosvitin and similar proteins.

Mammalian infants have another source of phosphorus and calcium in milk, which contains **casein.** Casein is not a single entity; it is several different polypeptides occurring in combination with calcium and some small anions, a micelle. The composition varies with species, and the polypeptides from cows are the most thoroughly described. The bovine caseins occur in a micelle that has a core of two classes of phosphorylated polypeptides (α and β), containing from 0.5 to 1.0 per cent P, and they are associated with Ca^{2+}. The core is covered with a disulfide-linked coating of another class of polypeptide (κ) containing much less phosphorus. The coating prevents micelles from aggregating. Human casein is believed to have a similar structure, although the components differ from those in bovine milk. (There is also considerable polymorphism of casein within species, indicating that many genes are involved in its formation.)

When casein is delivered to the infant stomach, exposure to the increased $[H^+]$, together with partial enzymatic hydrolysis of the coating peptides, causes a disruption of the micelles. The constituent polypeptides have little secondary structure, and the length of the polypeptide chain is readily exposed for digestion. The accompanying phosphate and calcium ions are then available for absorption.

Regulatory Functions. The change in charge that accompanies phosphorylation is a means of altering biological function. We shall explore individual examples in more detail; suffice it for now that we shall see cases in which phosphorylation causes constituent groups of the polypeptide to become oriented in an active array; we shall see other cases in which the opposite occurs, and phosphorylation causes either a disruption or masking of the active site. The extent of phosphorylation necessary to cause these changes is frequently small, with only a few critically located seryl or threonyl residues being affected.

METHYLATIONS

The amino acid residues in polypeptides, like the nucleotide residues in polynucleotides, are subject to modification by methylation. The commonly affected groups are the cationic side chains. Lysyl residues may have from one to three methyl groups added (Fig. 8–7). Histidyl residues are methylated, especially on the 3-position, and arginyl residues are methylated on any one of the three nitrogen atoms in the side chain.

The purpose of these methylations is not known; they affect acidic ionization as well as spatial arrangement. A few residues are affected in some proteins; well known examples are the contractile proteins of muscle (p. 356).

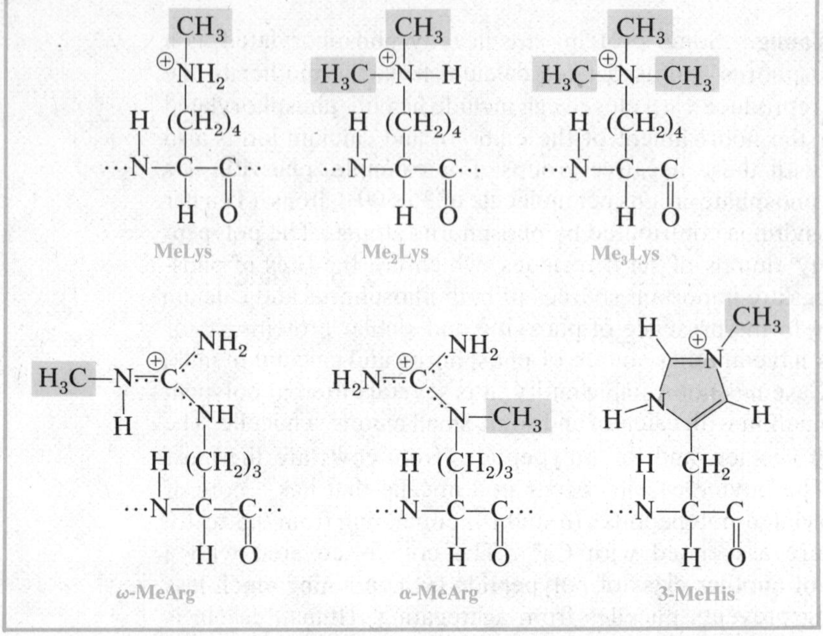

MeLys Me$_2$Lys Me$_3$Lys

ω-MeArg α-MeArg 3-MeHis

FIGURE 8–7 **Nitrogenous side chains in polypeptides are sometimes methylated. The top line illustrates the mono-, di-, and tri-methyl derivatives of lysine residues. The bottom line illustrates methylation of the ω and α nitrogen atoms, respectively, of arginine residues, and methylation of the 3-N position in a histidine residue.**

MODIFICATION OF HISTONES

The histones are especially subject to modification. They contain methylated amino acid residues and undergo charge modification through acetylation and phosphorylation.

The kind of change and the type of histone that is affected has a definite association with particular stages in the cell cycle. It is not possible to draw a good picture of the exact function of these changes, so there is little value in detailing the isolated facts that are known, but it is well to have in mind the probable generalizations.

The association between the positively charged histones and the negatively charged phosphodiester backbone of DNA is mainly electrostatic. Modification of the histones by acetylation or phosphorylation of ammonium groups diminishes the net positive charge and therefore weakens the association with DNA. This is consistent with other indications that the charge-diminishing changes accompany increased transcription or replication of DNA. The histones must dissociate from DNA if it is to be transcribed or replicated.

The modifications cannot in themselves control the transcription of specific genes because the same histones occur repeatedly, and it is not known to what extent the occurrence of more specific proteins affects the sites of histone modification. The histones that are modified during the S phase, when replication occurs, are not all the same as those modified during other phases, when transcription occurs. Specific changes coincide with the initiation of mitosis. Cause and effect relationships for these various alterations are yet to be worked out.

FURTHER READING

Quick Reviews

Steiner, D. F.: (1976) *Errors in Insulin Biosynthesis*. N. Engl. J. Med., *294*: 952.
Walsh, J. H., and M. I. Grossman: (1975) *Gastrin*. N. Engl. J. Med., *292*: 1324.
Anfinsen, C. B.: (1973) *Principles that Govern the Folding of Peptide Chains*. Science, *181*: 223. Nobel lecture.

Detailed Reviews

Taborsky, G.: (1974) *Phosphoproteins*. Adv. Prot. Chem. *28*: 1.
Rubin, C. R., and O. M. Rosen: (1975) *Protein Phosphorylation*. Annu. Rev. Biochem., *44*: 831.
Elgin, S. C. R., and H. Weintraub: (1975) *Chromosomal Proteins and Chromatin Structure*. Annu. Rev. Biochem., *44*: 725. Modification of histones is briefly discussed on p. 747 ff.
Steiner, D. F., et al.: (1974) *Proteolytic Processing in the Biosynthesis of Insulin and Other Proteins*. Fed. Proc., *33*: 2104.
Palade, G.: (1975) *Intracellular Aspects of the Process of Protein Synthesis*. Science, *189*: 347. Nobel lecture.

Specific Papers

Williams, R. E.: (1976) *Phosphorylated Sites in Substrates of Intracellular Protein Kinases: A Common Feature in Amino Acid Sequences*. Science, *192*: 473.
Habener, J. J., et al.: (1975) *Pre-proparathyroid Hormone Identified by Cell-free Translation of Messenger RNA from Hyperplastic Human Parathyroid Tissue*. J. Clin. Invest., *56*: 1328.
Friedman, M., ed.: (1977) *Protein Crosslinking*. Adv. Exp. Biol. Med., vol. 86A.
Devillers-Thiery, A., et al.: (1975) *Homology in Amino Terminal Sequence of Precursors to Pancreatic Secretory Proteins*. Proc. Natl. Acad. Sci. U.S.A., *72*: 5016.
Russell, J. H., and D. M. Geller: (1975) *The Structure of Rat Proalbumin*. J. Biol. Chem., *250*: 3409.
Sharp, G. W., et al.: (1975) *Studies on the Mechanism of Insulin Release*. Fed. Proc., *34*: 1537.
Shields, D., and G. Blobel: (1978) *Efficient Cleavage and Segregation of Nascent Presecretory Proteins*. . . . J. Biol. Chem., *253*: 3753.

9 FIBROUS STRUCTURAL PROTEINS

In our initial discussion of proteins, we emphasized those of globular structure. The ready solubility in water of most of these proteins is not always a desirable property. Some proteins are used to provide structural elements, and maintenance of strucutral integrity requires low solubility. This is as true for an organelle within a microorganism as it is for the integument of an elephant. Many structural elements also must have mechanical strength. It is desirable for the organism to resist the slings and arrows of outrageous fortune, and it is imperative that it not thaw and resolve itself into a dew.

Molecules meeting structural requirements often have a high ratio of length to diameter and are constructed to form fibers by associating side-by-side. Vertebrate skin contains striking examples of such molecules. Much of the skin comprises two populations of cells, separated sharply by a thin layer made of a network of fibers — the basement membrane. Between the membrane and the surface the cells produce an insoluble protein aggregate, **keratin.** These cells and their products constitute the epidermis.

Beneath the membrane is the dermis, the deep layer of the skin. Here the cells are producing fibers made of the protein **collagen**, and the fibers are crisscrossed in a felt-like mat.

The division between dermis and epidermis is as sharply marked by the separate location of collagen and keratin as it is by the visible differences between the cells of the two layers. Both of these proteins are very insoluble. Both have high tensile strength. Both form fibers. But here the resemblance ends, and it is interesting to see how their structures vary to satisfy the separate requirements of the tissues in which they occur.

Fibers and Matrix

Both the dermis and the epidermis illustrate supporting structures made by embedding fibers in an amorphous matrix, thereby gaining advantageous properties. The fibers may range from flexible shafts to rigid rods, and the matrix may be a hard casting or a resilient bed. Such structures appeared early in the evolution of multicellular organisms. (Man has consciously used the principle for many millennia; we think of fiberglass and reinforced concrete and tend to forget why the followers of Moses were indignant when required to make bricks without straw.)

TABLE 9–1

	Fiber	Matrix	Typical Locations
keratin	protein (α-keratin)	protein (keratohyalin)	epidermis, nails, hair, horn, hoof
collagenous tissue	protein (collagen)	polysaccharide (chondroitin sulfate)	intercellular space, cartilage, bone, basement membranes
wood	polysaccharide (cellulose)	polymerized polyhenols (lignin)	higher plants
chitinous shell or cuticle	polysaccharide (chitin)	tanned protein (sclerotin)	exoskeleton of crabs, insects, spiders

The physical properties of biological combinations of fiber and matrix vary with the materials in two phases. These materials are commonly polymers of amino acids, the proteins, or of carbohydrates, the polysaccharides. The principle examples in humans are the keratins of skin, in which both the fibers and the matrix are made from polypeptide chains, and connective tissues in which fibers of collagen, a protein, are buried in a matrix of complex polysarccharides. We shall examine keratin and collagen in this chapter and discuss the carbohydrate matrix in the next chapter.

Other organisms have different combinations. For example, the connective tissue of woody plants contains fibers of cellulose, a carbohydrate polymer, buried in a matrix of condensed polyphenols. The exoskeleton (shell or cuticle) of arthropods contains fibers of chitin, a polysaccharide, buried in a matrix of tanned protein. Fiber and matrix combinations are listed in Table 9–1.

KERATINS

Keratins are hard and tough proteins in the epidermis and related structures (nail, hair, horn, and hoof). The entire fiber-matrix complex is frequently called keratin, but present usage leans toward designating the fibers as **α-keratins** and the matrix as **keratohyalins.**

The general strategy for creating the keratin complex is first to make the fibers and then to polymerize the matrix around them. For example, young cells in the epidermis form tonofilaments of keratin within the cells — fibers that run from one cell interface to another. These filaments grow more dense as the cell matures and moves toward the surface of the skin, and finally the filaments become surrounded by a relatively hard matrix, which is in turn enclosed by a quite hard and impermeable modified cell membrane.

Keratin Fibrils

The keratin fibers are composed of bundles of fibrils, each fibril made from three polypeptide chains. As many as seven different precursor chains may be present in a single tissue, so the fibrils are a mixture rather than a single entity, but all appear to have similar properties.

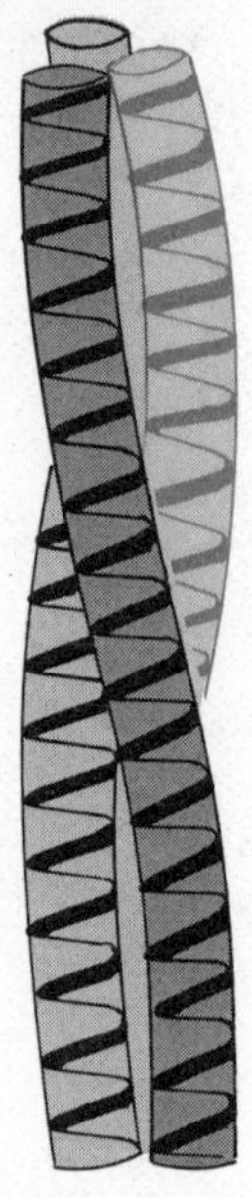

FIGURE 9–1

α-Keratin is made from coiled coils. Each polypeptide chain is in a right-handed α-helix, and a bundle of three α-helices is given a left-handed twist to create a strong rope-like structure. The three chains are usually different.

The precursor chains contain regions in which the constituent amino acid side chains favor the formation of α-helix, separated by other regions rich in residues that prevent α-helix formation — proline, glycine, and serine (the N-terminal residue is always N-acetylserine for unknown reasons). The composition of the helical regions is such that three chains associate to form a coiled coil (Fig. 9–1), with the right-handed coils of the individual chains counter-twisted into left-handed triple helices, making a strong, rope-like bundle.

The non-helical regions presumably favor association of the fibrils with each other and with the matrix. In any event, approximately 10 to 12 fibrils do aggregate into bundles, which usually have an ill-defined core and a surrounding ring.

Keratohyalin Matrix

The matrix in which keratin fibers are buried is made from cysteine-rich polypeptide chains, which appear as the cell is pushed toward the skin surface. The sulfhydryl groups are oxidized to form disulfide bridges that connect polypeptide chains. The result is an interlocking network of protein that surrounds the keratin fibers in a dead cell that has lost most of its structure; even the cell membrane is converted to a layer of disulfide-linked protein, a layer more tightly cross-linked than keratohyalin.

Variations in Keratins

The properties of keratin complexes differ between tissues owing to variations in the cross-links that hold the structure together. Simply increasing the proportions of disulfide-rich keratohyalin will make a harder and hornier structure.

Further stabilization is sometimes achieved by forming covalent cross-links within the α-keratin fibrils or between the fibrils and the surrounding keratohyalin. Such links are less common in the epidermis in which only secondary forces are necessary to link polypeptide chains in the coiled coils and to bind the resulting fibril to keratohyalin, but stronger bonds are more abundant in hair and other modified structures.

The cross-links include a few disulfide bridges formed between cysteinyl groups in the fibril and the matrix; they also include a type of covalent linkage we have not previously encountered, involving a lysyl group on one polypeptide chain and a glutaminyl group of another (Fig. 9–2). An enzyme catalyzes the substitution of the amine group in a lysine residue for the amide nitrogen in glutamine. The result is the formation of an amide bond, sometimes called an isopeptide bond, between the two residues. Such cross-links are relatively rare in skin but are more common in hair, especially in the cells constituting the core

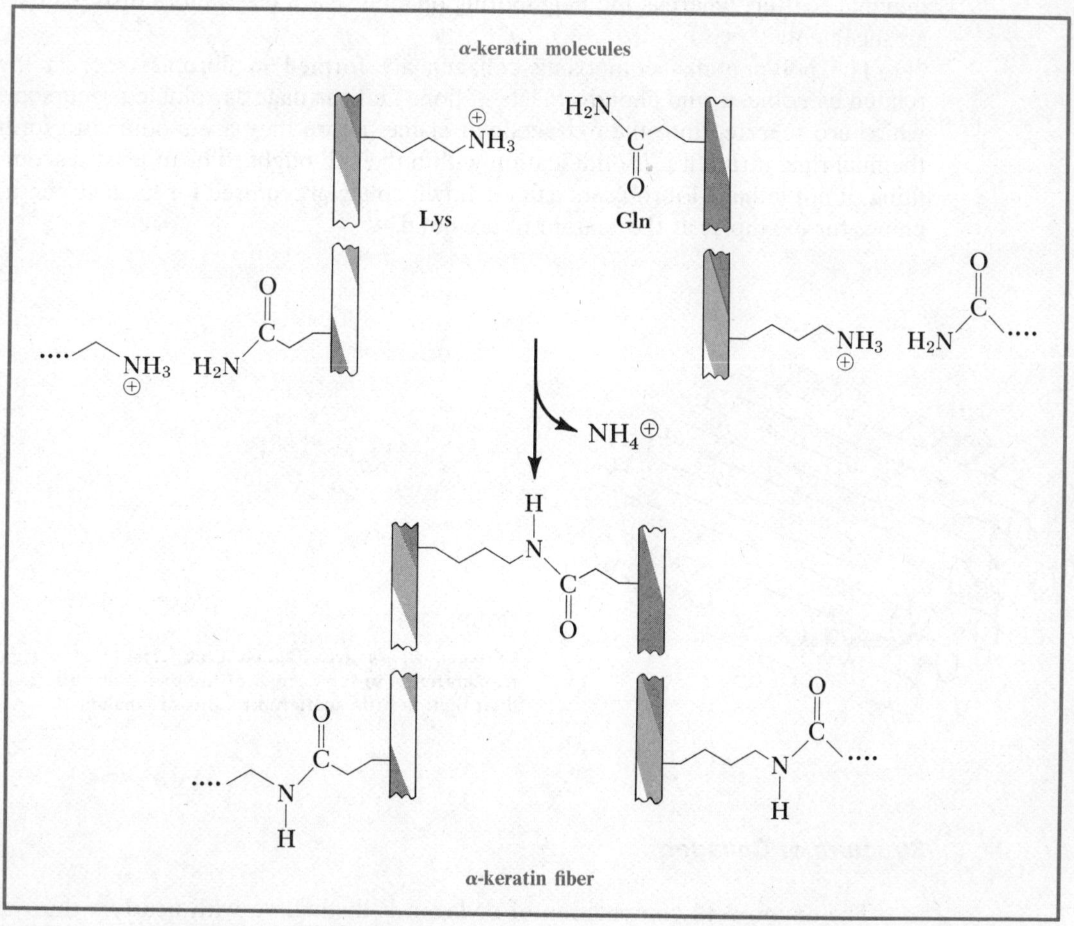

FIGURE 9–2 **The side by side association of α-keratin molecules into fibrils is frequently strengthened by the formation of covalent bonds between them. These bonds are created by transferring the glutamyl group of a glutamine residue from its amide nitrogen to the nitrogen atom of a lysine residue. The amide nitrogen of the original glutamine is released as a free ammonium ion.**

(medulla). The cross-links, disulfide or amide, have the effect of preventing slippage of the fibers through the matrix but are not so abundant as to destroy their fibrous character. (We shall see (p. 325) that similar cross-links are used to stabilize blood clots.)

COLLAGENS

The collagens are proteins that occur as insoluble rigid rods in all organs. They may be dispersed in a matrix when stiffening of a gel is all that is required, as in the ground substance around cells, or in the vitreous humor of the eye, or they may be neatly bundled in tight parallel arrays when great strength is required, as in tendons.

Indeed, collagens are the predominant proteins in the human body. They are an important constituent of bone in which the fibers are arranged at an angle to each other so as to resist mechanical shear from any direction. About half the dry weight of cartilage is collagen, from which the cartilage acquires its toughness. Even the cornea of the eye contains a high proportion of collagen arranged in neatly stacked arrays (Fig. 9–3) so as to transmit directly impinging light with minimal scatter, whereas the neighboring opaque sclera has a more disorganized arrangement.

The polypeptides composing collagen are formed in fibroblasts or in the related osteoblasts and chondroblasts of bone and cartilage as soluble precursors, which are secreted into the extracellular space before they are modified to form the final rigid structure. (Solidification within the cell ought to be at least discomfiting, if not lethal.) Fibroblasts can lay down collagen required for local strengthening, for example, at the margin of a wound.

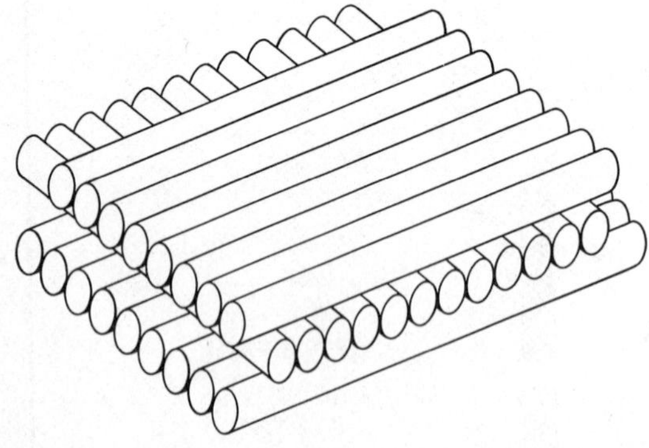

FIGURE 9–3

Collagen fibers are stacked crosswise to give rigid transparency to the cornea of the eye. Normally incident light is little scattered by this arrangement.

Structure of Collagen

The amino acid composition of collagen is distinctive with nearly a third of the residues being glycine. For example, one collagen polypeptide chain has a total of 1,052 residues in which the entire sequence from residue 17 through residue 1,027 can be represented as $(\text{Gly-X-Y})_{337}$.

Furthermore, over one in five of the total residues is proline, and nearly half of these are modified by hydroxylation at the 4 position and to a lesser extent at the 3 position of the ring:

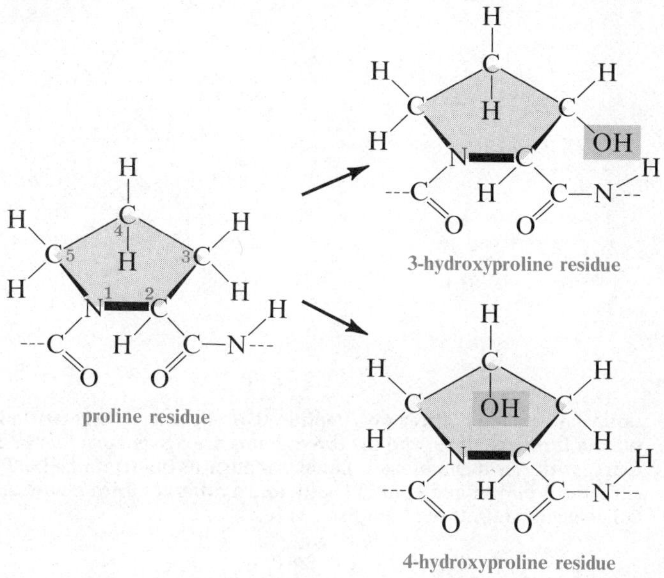

proline residue

3-hydroxyproline residue

4-hydroxyproline residue

These hydroxyprolines, abbreviated Hyp, occur mainly in collagen. A few are found in another connective tissue protein, **elastin,** and still fewer in the C1q component of the complement system of blood (p. 336), but in general the content of hydroxyproline in a tissue is an index of its content of collagen. To summarize, of the total amino acid residues in the known kinds of collagen,

> 33 to 35 per cent are **glycine**
> 20 to 24 per cent are **proline + hydroxyproline,** distributed as
> 6 to 13 per cent proline, and
> 9 to 17 per cent hydroxyproline

The 18 other amino acids occur in fewer than half of the residues.

The polypeptides of collagen form a triple helix. Collagen molecules are made from three polypeptide chains twisted into a rope-like structure (Fig. 9–4). The major factor in creating this structure is the occurrence of glycine as every third residue, because it enables twisting each chain one turn for each three residues, thereby bringing all of the glycyl residues to the inside of the rope. Since glycine has only a hydrogen atom for a side chain, polypeptides can make a close fit. The twist within chains is left-handed, but the bundle of three chains is given one right-handed twist for each ten turns of a single chain, thereby strengthening the bundle.

The prolyl and hydroxyprolyl groups can be accommodated in a triple helix without distortion, whereas we earlier emphasized that they will not fit in the less-tightly twisted α-helix of a single polypeptide chain. In addition, full stability of the triple helix depends upon the formation of hydrogen bonds with the hydroxyl group on residues of hydroxyproline. (Much of the stabilizing force comes from the formation of hydrogen bonds between the peptide backbones; there is one direct bond per turn of a chain and an additional bond involving an inserted water molecule, but these are not quite sufficient in themselves to give a full stability.)

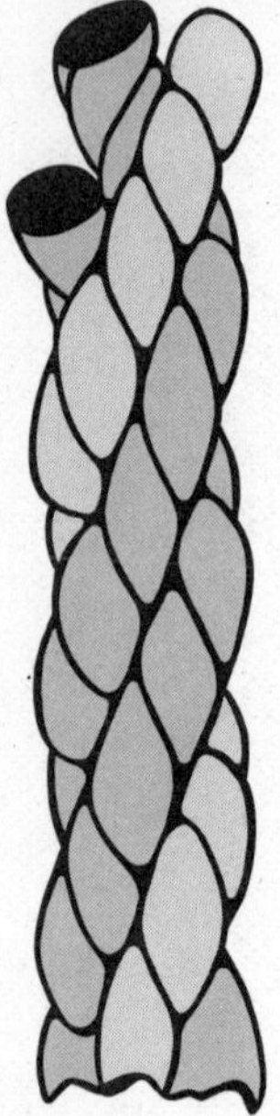

FIGURE 9–4 **Collagen is made of three polypeptide chains. Each chain is twisted to the left one turn in three residues, and the three chains are twisted together in a right-handed helix, with ten turns of each chain per turn of the triple helix. Type I collagen contains a pair of one kind of chain, and a different third chain, as shown; other collagens contain three identical chains.**

Hydroxylysine and Sugars. Hydroxyl groups are also formed on some of the lysine residues in collagen:

$$\begin{array}{c} COO^{\ominus} \\ | \\ {}^{\oplus}H_3N{-}C{-}H \\ | \\ CH_2 \\ | \\ CH_2 \\ | \\ HO{-}C{-}H \\ | \\ CH_2 \\ | \\ {}^{\oplus}NH_3 \end{array}$$

5-hydroxy-L-lysine

They also occur in the C1q component of complement but not in elastin, so they are even more distinctive than the hydroxyproline residues. These groups are not as abundant as those on proline residues, and the hydroxylysine side chains apparently have two main functions: to participate in the formation of cross-links (next section) and to act as sites for the attachment of sugar groups. We shall consider glycoproteins — proteins containing attached residues of sugar — at some length in the next chapter. Collagen contains two such residues, galactose and glucose, linked with the hydroxyl group of hydroxylysine residues. The number of added sugar molecules varies from one type of collagen to another.

Heterogeneity. A given molecule of collagen is not always hydroxylated at all potential sites on proline or lysine residues, and all of the possible carbohydrate residues are not always attached. The result is some variation of composition within one type of collagen, as well as between types. This variation probably has a functional role. For example, the collagens in a rat tail are less fully hydroxylated than the same kind of molecule in the remainder of the body. The tail is more exposed, and the fewer hydroxyl groups enable the cartilage to remain flexible at the resultant lower temperature.

Different Kinds of Collagen. There are at least five different genetic types of polypeptide chains in collagens, and the collagens of different tissues vary according to the types of chains being made. An unfortunate and confusing nomenclature has developed employing Greek letters in an entirely different way than they are used with other proteins. To avoid further conflict, let us simply designate these chains as a, b, c, d, and e:

Our designation	a	b	c	d	e
Literature designation*	$\alpha1$ (I)	$\alpha2$	$\alpha1$ (II)	$\alpha1$ (III)	$\alpha1$ (IV)

*The literature uses β for a dimer of triple helical collagen molecules.

One of the collagens that occurs in skin, bone, and tendon contains two a chains and one b chain, that is, it is a_2b. The remaining collagens are made from three identical chains — c_3, d_3, or e_3. We cannot at this time describe the functional significance of the differences in composition of these collagens, but their importance is evident from their varying occurrence:

Type I collagen, a_2b, occurs in bone, tendon, soft tissue, and scars: it contains the least carbohydrate.

Type II collagen, c_3, occurs in cartilage and is made by chondrocytes.

Type III collagen, d_3, predominates in fetal skin, but a_2b is plentiful at time of birth; d_3 contains disulfide bridges between chains, whereas a_2b and c_3 do not. Type III collagens are present in scars and all soft tissues examined but not in bone or tendon (except in scars).

Type IV collagen, e_3, is the collagen of the basement membranes. It also contains disulfide bridges. It has the highest carbohydrate content.

Fiber Formation and Cross-Linking

Collagen molecules are stacked together into rigid, strong fibers after secretion from the cell. The molecules are built to be stacked in a very regular parallel way (Fig. 9–5), with the ends of each molecule contacting definite regions on the sides of adjacent molecules. The exact arrangement in the third dimension is not known, but it may be a spiral, as if a sheet of rods had been rolled up like a carpet.

FIGURE 9–5 **Individual molecules in a collagen fibril are staggered by ~23 per cent of their length. Molecules in line are spaced by gaps equal to ~14 per cent of their length. This diagram exaggerates the cross-fibril dimension by ~75 per cent, compared to the length, but representing the molecules by lines does not show that each is a triple helix, as shown in Fig. 9–4.**

The fibers are strengthened through the formation of covalent bonds between adjacent molecules.

These cross-links are formed by oxidatively deaminating the side chains of selected lysine or hydroxylysine residues (Fig. 9–6), creating aldehyde groups. (The modified residues are referred to as allysine or hydroxyallysine. The older literature used lysinal, but this term is now reserved for the synthetic analogue of lysine in which the α-carboxyl group is reduced to an aldehyde.) The aldehydes react in a variety of ways with groups on other side chains of adjacent molecules,

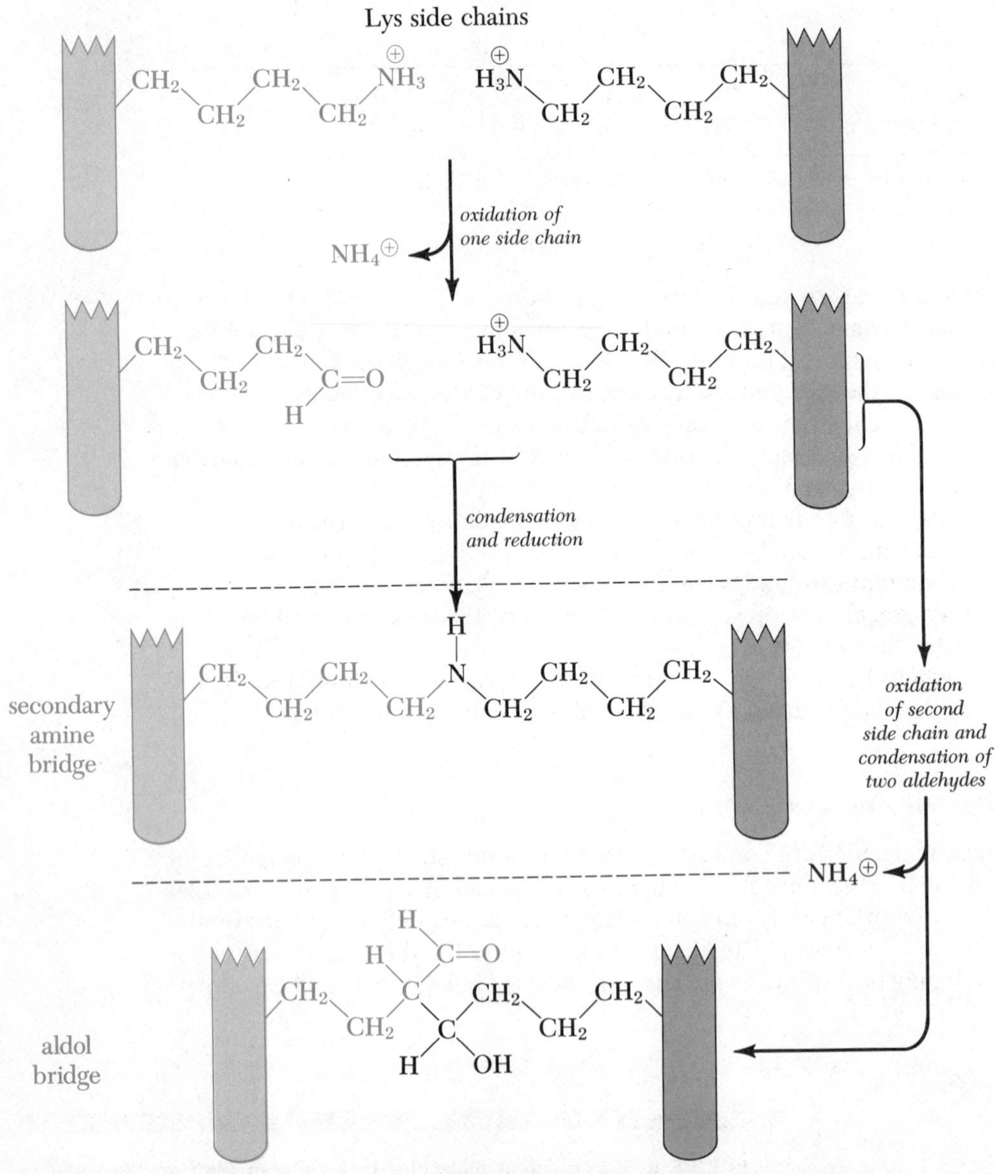

FIGURE 9–6 **Cross-links are formed between collagen molecules by oxidizing lysyl side chains to aldehydes. The aldehyde groups may condense with additional unmodified lysyl side chains (*middle*), or with each other (*bottom*), or with histidyl side chains (not shown). The condensation products are sometimes reduced to saturated structures, as also shown in the middle. Several variants are known.**

including other aldehyde groups, amino groups in lysine or hydroxylysine residues, and the imidazole groups of histidine residues. The reactive aldehydes are created mainly near the ends of the polypeptide chains, outside of the triple helical region, and combine with groups that appear to be specifically located for this purpose in the more central portions of adjacent molecules.

Synthesis of Collagen

The synthesis of collagen is almost a catalog of various kinds of post-translational modification of polypeptide chains. Some of the modifications begin before the synthesis of the polypeptides is completed; other do not occur until the molecules are secreted from the cells. A summary of the temporal sequence of the changes is given in Figure 9–7. Let us examine some of them in more detail.

The Procollagen Polypeptides. The polypeptides of collagen are synthesized as precursors, including long segments at both the N- and C- terminals that are later discarded by hydrolysis (Fig. 9–8). (Collagen monomers begin with pyroglutamate residues as a result of cleavage of procollagen before a glutamine residue.) These extra segments are not rich in glycine or proline, and do not occur as a triple helix. The function of the initial N-terminal segment is not clear, but the final C-terminal portion contains cysteinyl groups, which form disulfide bridges, and it is these bridges that initially hold three polypeptide chains together and facilitate the formation of a triple helix to create a procollagen molecule.

Hydroxylation. The hydroxylation of proline and lysine residues begins before synthesis of the polypeptide chains is completed. The hydroxylation of proline residues is necessary for initiation of triple helix formation in the central portion of the procollagen chains. The hydroxylation reactions (which are discussed further on p. 581) are complex processes involving the simultaneous oxidation of another compound, α-ketoglutarate, to succinate.

The Hydroxylations Do Not Occur At Random. The hydroxylations are under genetic control in that the enzymes catalyzing them affect groups occurring only in specific amino acid sequences. Prolyl groups are converted to **4-hydroxyprolyl** groups when they occur immediately before glycyl groups, (-Gly-X-Pro-Gly-, but not -Gly-Pro-X-Gly-). A few, but not all, prolyl groups are converted to **3-hydroxyprolyl** groups, particularly in type IV collagens, when they occur between glycyl and 4-hydroxyprolyl groups, as in -Gly-Pro-4Hyp- → -Gly-3Hyp-4Hyp-. The determining factors are not known, but there is only one 3-hydroxyproline in an a chain, whereas there as many as 16 in an e chain. Similarly, lysyl groups are hydroxylated only at specific sites, which vary in different types of collagen.

Secretion and Partial Hydrolysis. The secretion of procollagen involves transport through the Golgi apparatus where sugar residues are added to selected hydroxylysyl groups. Procollagen, like other secreted proteins, is packaged in vesicles, and partial hydrolysis begins before secretion.

In any event, specific enzymes are secreted along with most procollagens to catalyze the hydrolysis of the N-terminal and C-terminal segments. The loss of the C-terminal portion removes the disulfide bridges between the three polypeptide chains in most collagens, but stabilization is still provided by the prolyl groups that have already been hydroxylated.

The type IV procollagen of the basement membranes and perhaps some of the type III procollagens in soft tissues apparently retain their C-terminal segments because the finished protein still has disulfide bridges.

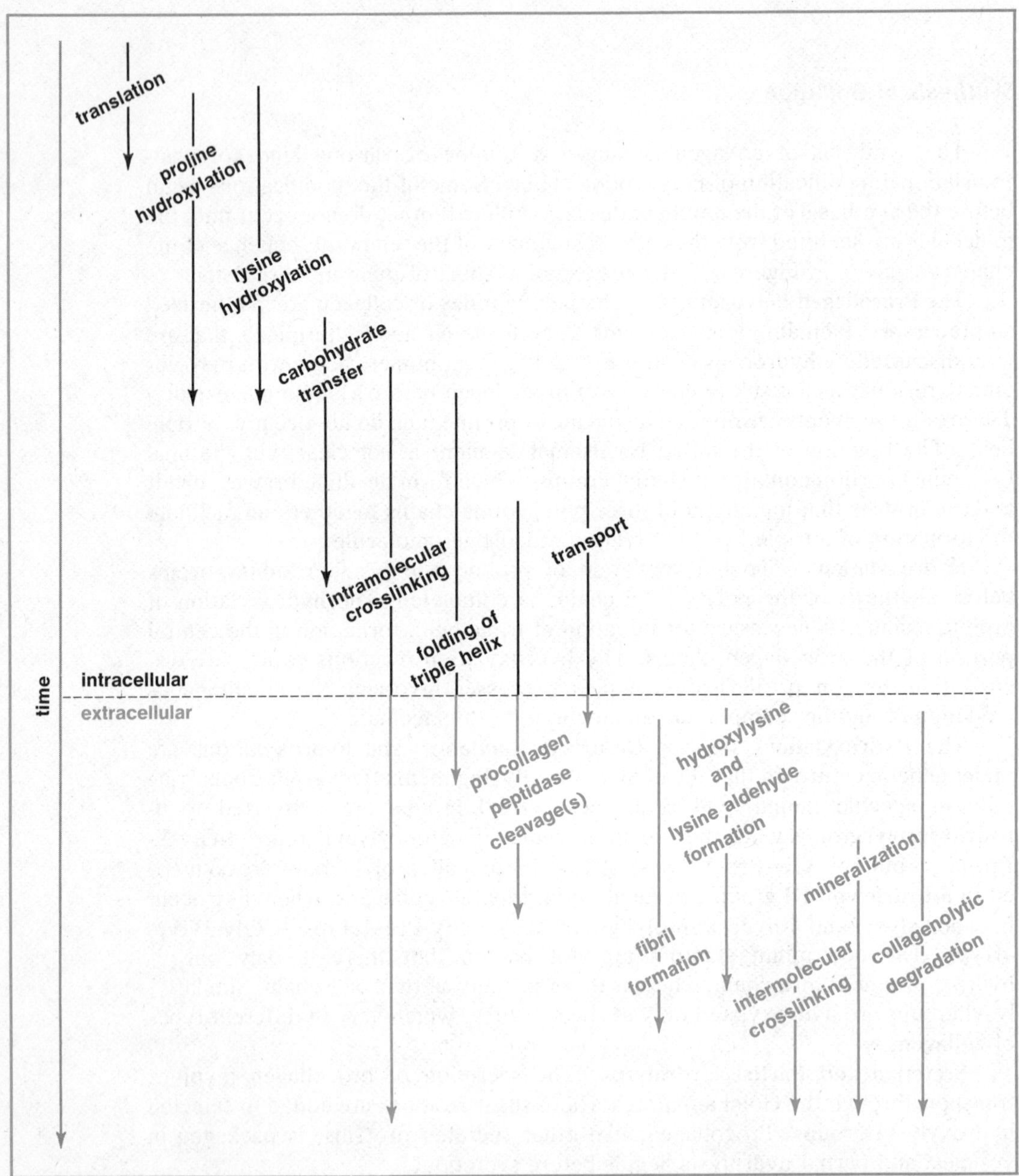

FIGURE 9–7 **The time course of events in collagen synthesis. Redrawn from P. M. Gallop and M. A. Paz, (1975) Physiol. Rev., *55*:473. © American Physiological Society. Used by permission.**

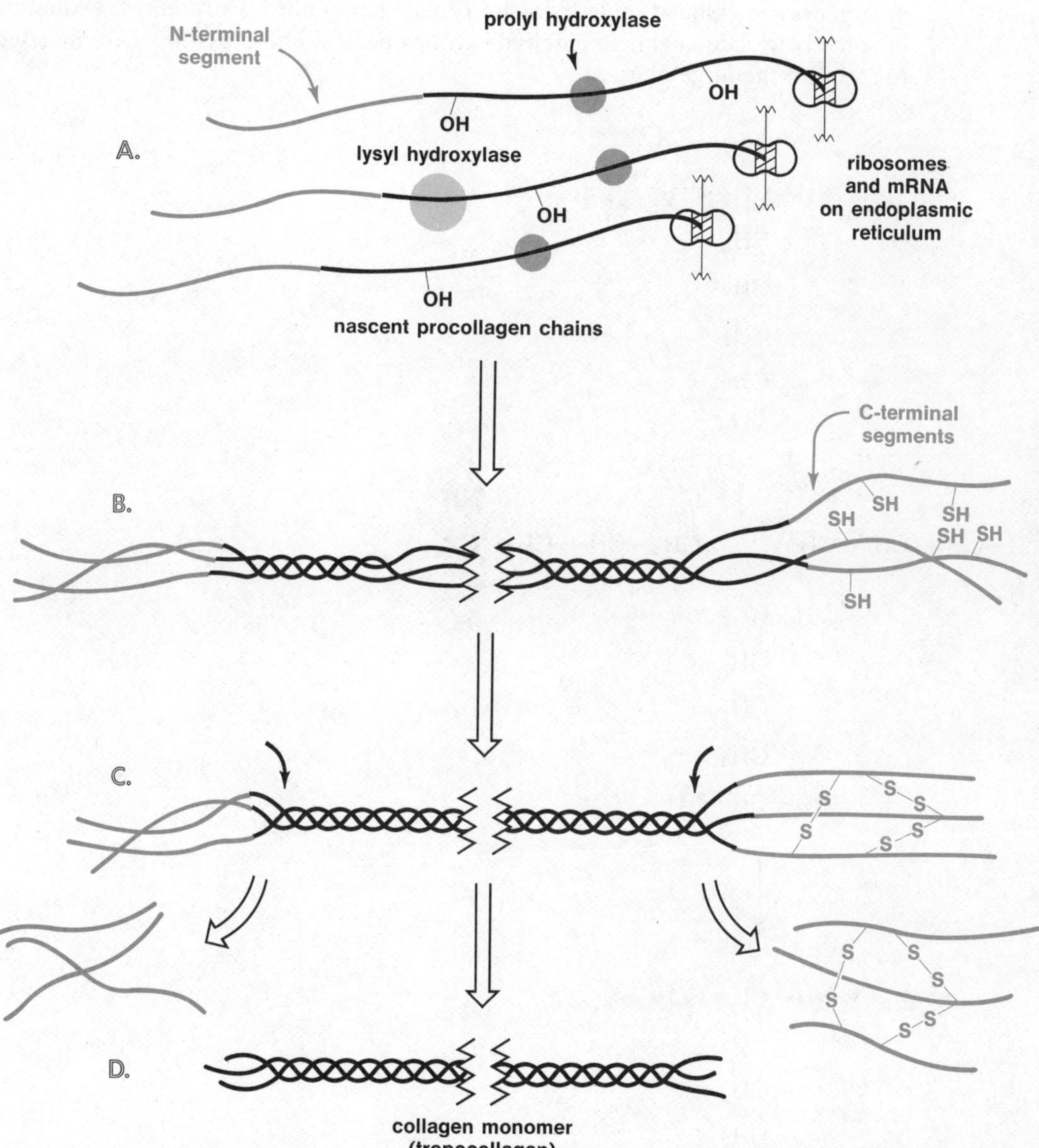

FIGURE 9–8 The collagen molecules are first synthesized as long procollagen polypeptide chains. *A*. Hydroxylation of prolyl and lysyl residues begins before the completion of translation, and aids and promotes (*B*) the combination of chains into a triple helix. *C*. The triple helix is stabilized during structural refinement by covalent disulfide bridges between the chains. *D*. During and after the secretion of the molecule, specific enzymes (procollagen peptidases) catalyze the cleavage of the now superfluous N- and C-terminal polypeptide segments, leaving the mostly helical central portion of the monomers available for association into fibers (not shown).

The other procollagens are converted by hydrolysis to a collagen monomer, also known as **tropocollagen,** in which all but a small portion of the residues at the two ends of the chains are in a triple helix.

Fibril Formation and Cross-Linking. Once the ends are lost, the collagen monomers can associate side-by-side to create larger fibrous structures. Oxidation of some lysyl side chains to aldehyde groups begins, and cross-links are inserted to stabilize the large fibers.

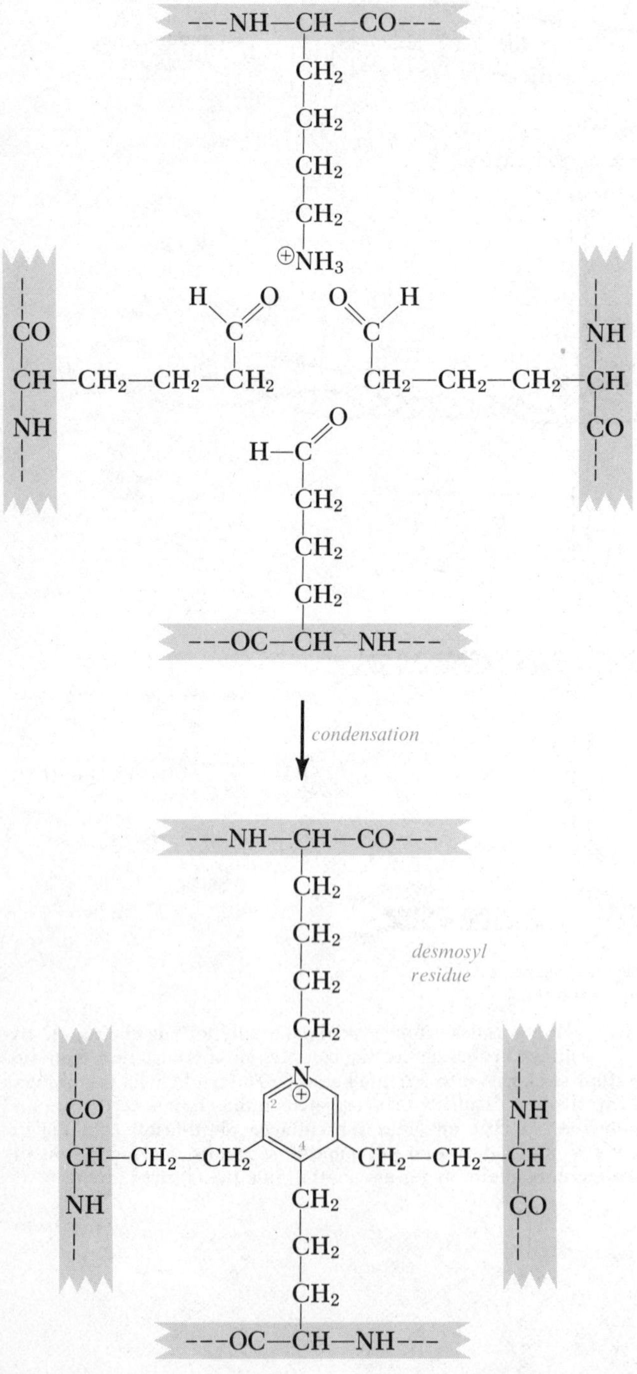

---NH—CH—CO---

CH_2
CH_2
CH_2
CH_2
$^{\oplus}NH_3$

CO ... CH—CH_2—CH_2—CH_2—C(H)=O ... O=C(H)—CH_2—CH_2—CH_2—CH ... NH

H—C=O
CH_2
CH_2
CH_2

---OC—CH—NH---

condensation

---NH—CH—CO---

CH_2
CH_2
CH_2
CH_2

desmosyl residue

$N^{\oplus}$

CO ... CH—CH_2—CH_2— ... —CH_2—CH_2—CH ... NH

CH_2
CH_2
CH_2

---OC—CH—NH---

FIGURE 9–9

The formation of desmosine or isodesmosine residues in elastin involves the combination of three allysine residues (the aldehyde derivative of lysine) with one unmodified lysine residue. The result could be covalent linkage of four different polypeptide chains, but it is currently believed that two polypeptide chains are joined by using a pair of lysyl groups from each.

ELASTIN

The lungs, walls of arteries, and some ligaments contain elastic fibers that provide an ability to stretch without tearing. These fibers are made from a protein, elastin, which like collagen has **glycine** in one third of its residues. However, glycine does not occur in regular sequence in elastin, so the protein is not in a triple helix. Elastin also has a high proportion of **alanine, proline,** and **valine** that together with glycine account for over 80 per cent of the amino acid residues. The alanine residues are concentrated around the cross-links, with as many as eight preceding those derived from modified lysine residues. Glycine frequently alternates with valine or proline, with sequences such as Pro-Gly-Val-Gly-Val being repeated as many as six times. A few of the prolyl groups are hydroxylated.

The formation of elastin involves the formation of cross-links from allysine residues (the aldehyde form of lysine) and unmodified lysine residues, but these differ from cross-links in collagen in that many of them link four segments of peptide chain through a heterocyclic ring (Fig. 9–9).

Although the structure of elastin in not known in detail, its elastic character is believed by some to result from its occurrence as a random three-dimensional network that can be stretched in any direction; others think that the strong preponderance of hydrophobic residues causes it to form a lattice of interconnected balls, with elasticity coming from a deformation of the lattice. (Deformation would cause increased exposure of hydrophobic groups to water, thereby creating the restorative force.)

DEFECTS IN STRUCTURAL PROTEINS

We have seen that the collagens and elastin are major components of many tissues (Fig. 9–10), and it follows that any disruption in the formation of these proteins may have a variety of consequences, depending upon the relative degree of impairment of particular structures. Specialized disturbances in collagen and elastin synthesis, as opposed to general impairment of protein synthesis, may occur at any stage, from transcription of the genes through the hydroxylation reactions, partial hydrolysis of the polypeptide chains, and oxidation of lysyl side chains to create cross-links. Many clinical entities involving defects in synthesis

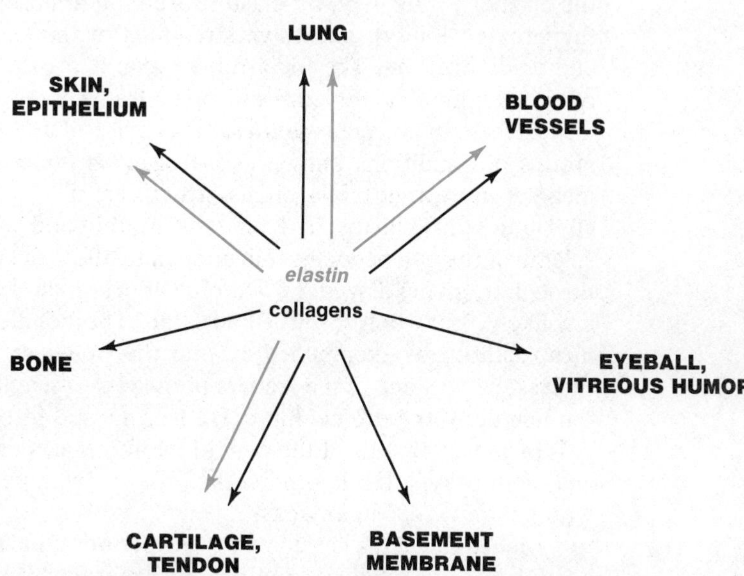

FIGURE 9–10

The key role of collagen as a structural element makes a variety of supportive tissues vulnerable to any defect in the synthesis of the protein. Elastin is less widely used, but it has a vital function in the arteries, in which it is a major fraction of the total protein.

of structural proteins have been recognized, but the biochemical basis for many of the defects is just beginning to be understood.

Disturbances in Collagen Synthesis

Scurvy is the result of a deficiency of **ascorbate (vitamin C)** in the diet; it is the price we sometimes pay for abandoning the free-living fruit-picking life of our primate ancestors. Most of the effects are those of defective collagen synthesis. Infants are afflicted with painful tenderness in their limbs so that they draw up their legs and lie quietly. Bone formation is defective, and the blood vessels are weakened, with hemorrhages common. Adults also tend to bleed readily, especially from the gums, around hair follicles in the skin, and into the joints. Wounds remain open. The teeth loosen.

Scurvy has been known for millennia as an affliction of organized traveling groups of men: soldiers on the march, sailors at sea, and exploring parties who have depended upon easily shipped supplies of non-perishable foods. Western civilization didn't make the connection between the disease and diet until the eighteenth century, when a series of studies by the British Navy brought scurvy under control on its vessels.

The exact mechanism of ascorbate action is not known. Ascorbate is a ready donor of electrons; that is, it is a good reducing agent, and this property is probably the basis for its physiological function, but definitive proof is lacking. It in some way facilitates the smooth hydroxylation of prolyl and lysyl groups in collagen and elastin. Normal connective tissue cannot be laid down in its absence.

Ehler-Danlos syndrome is a group of rare hereditary disorders in which collagen formation is impaired. They are distinct conditions of different origin, presumably affecting different steps in collagen synthesis, although the clinical results are similar. For example, one kind of patient appears to lack the ability to hydrolyze procollagen, so the normal association into fibers cannot occur; another is unable to hydroxylate lysyl groups, thereby preventing normal attachment of carbohydrate and formation of cross-links.

The lack in these people of the rigidity conveyed by collagen without loss of the elasticity conveyed by elastin results in abnormally free motion of the joints (hyperextensibility), excessive stretching of the skin, coupled with easy tearing and ready bruising. The India-rubber man and Etta Lake, the Elastic Lady, both of circus fame, had Ehlers-Danlos syndrome.

Defects in collagen synthesis may arise in a less obvious way. People with a groups of conditions known as **osteogenesis imperfecta** have improper development of more rigid collagenous structures. The total incidence is about one in 40,000 to 60,000 births. The tendons are thin and subject to rupture. The opaque sclerae of the eye becomes thin enough to allow the choroidal pigments to be seen through it, giving a Wedgewood blue appearance. Multiple fractures of the bones are likely, both before birth and later. The condition is not necessarily totally incapacitating as exemplified by Ivar the Boneless, who could not walk. Nevertheless, he was one of the leaders of the great assault of the Vikings on England in the middle of the 9th century. (He was carried into battle on a shield.)

In some versions of the condition, fibroblasts produce less of type I collagen, and more of type III; in others, the production appears normal, and what is going wrong is not clear in any case.

Diabetes mellitus is a common and serious condition we shall consider in more depth later; it is a failure to utilize the sugar glucose normally (p. 721). One of the dangerous consequences of diabetes is a thickening of the basement membrane in arterioles generally and in the glomerulus of the kidney specifically. The membrane serves as a filtration barrier and as a scaffold separating parenchymal cells and connective tissue in many organs. Seamless repair of injuries by replacement of damaged cells requires a normal basement membrane; scars form when it is defective. The alteration of this collagenous tissue is the source of some of the irreversible consequences of diabetes that may result despite good management.

Disturbances in Elastin Synthesis

The oxidation of lysyl side chains to create cross-links in both elastin and collagen involves catalysis by a protein that contains **copper.** Any condition affecting the availability of copper to the cells will interfere with both elastin and collagen synthesis. The effect on collagen formation may include a disturbance of ascorbate availability, as well as an impairment of lysinal formation.

Dietary deficiencies of copper are rare, but failure to absorb copper can be caused by intestinal disease or by genetic defects such as the **steely hair syndrome**, also known as **Menke's disease.** Impaired elastin formation results in defective arterial intima, the internal layer of the arterial wall. Impaired collagen formation is manifested by bone defects similar to those seen in scurvy.

FURTHER READING

General Reviews

Bradbury, J. H.: (1973) *The Structure and Chemistry of Keratin Fibers.* Adv. Protein Chem., *27*: 211.

Piez, K. A.: (1972) *The Chemistry and Biology of Collagen. In* C. B. Anfinsen, R. F. Golberger, and A. N. Schechter, eds.: *Current Topics in Biochemistry.* Academic Press, p. 101.

Tanzer, M. L.: (1973) *Cross-linking of Collagen.* Science, *180*: 561.

Gallop, P. M., and M. A. Paz: (1975) *Post-translational Protein Modifications, with Special Attention to Collagen and Elastin.* Physiol. Rev., *55*: 418.

Bornstein, P.: (1974) *Disorders of Connective Tissues. In* P. K. Bondy, and C. E. Rosenberg, eds.: *Duncan's Diseases of Metabolism,* 7th ed. Saunders, p. 881.

Specialized Reviews and Articles

Steinert, P. M., and W. W. Idler: (1975) *The Polypeptide Composition of Bovine Epidermal α-Keratins.* Biochem. J., *151*: 603.

Matolsky, A. G.: (1975) *Desmosomes, Filaments, and Keratohyaline Granules.* J. Invest. Dermatol., *65*: 127.

Ramachandran, G. N., and A. H. Reddi, eds.: (1976) *Biochemistry of Collagen.* Plenum.

Friedman, M., ed.: (1977) *Protein Cross-linking.* Adv. Exp. Biol. Med., Vol. 86B. Symposium articles.

Folk, J. E., and J. S. Finlayson: (1977) *The ϵ-(γ-Glutamyl) Lysine Cross-link and the Catalytic Role of Transglutaminases.* Adv. Protein Chem., *31*: 2.

Bornstein, P.: (1974) *The Biosynthesis of Collagen.* Annu. Rev. Biochem., *43*: 567.

Vitto, J.: (1977) *Biosynthesis of Type II Collagen.* Biochemistry, *16*: 3421.

Barnes, M. J., and E. Kodicek: (1972) *Biological Hydroxylations and Ascorbic Acid with Special Regard to Collagen Metabolism.* Vitam. Horm., *30*: 1.

Spiro, R. G.: (1973) *Biochemistry of the Renal Glomerular Basement Membrane and its Alterations in Diabetes Mellitus.* N. Engl. J. Med., *288*: 1337.

Martin, G. R., P. H. Byers, and K. A. Rez: (1975) *Procollagen.* Adv. Enzymol., *42*: 167.

Cardinale, G. J., and S. Undenfriend: (1974) *Prolyl Hydroxylase.* Adv. Enzymol., *41*: 245.

Sykes, B., M. J. O. Francis, and R. Smith: (1977) *Altered Relation of Two Collagen Types in Osteogenesis Imperfecta.* N. Engl. J. Med., *296*: 1200.

Gray, W. R., L. B. Sandberg, and J. A. Foster: (1973) *Molecular Model for Elastin Structure and Function.* Nature, *246*: 641.

Fessler, J. H., and L. I. Fessler: (1978) *Biosynthesis of Procollagen.* Annu. Rev. Bioch., *47*: 129.

10 CARBOHYDRATES AS STRUCTURAL ELEMENTS

Carbohydrates include the simple sugars, modified derivatives, and polymers of one or more of these compounds. One of the sugars, D-glucose, is the premier fuel for most organisms, and it is also the most important single precursor of other body constituents. These sweeter functions of glucose will occupy much of our later attention, but we are now concerned with a more sinewy role, the use of carbohydrates as structural elements.

THE NATURE OF SUGARS

The sugars are formally defined as polyhydric aldehydes or ketones — compounds with a hydroxyl group and a carbonyl function on separate carbon atoms. We shall see that even the simple sugars, the **monosaccharides**, exist as a mixture of tautomeric forms in equilibrum with each other, and the free hydroxy aldehyde or ketone is frequently only a minor component. Even so, it is handy to classify sugars in terms of this form. For example, many of the structural carbohydrates are derivatives of three sugars that are considered to be **aldohexoses** — aldehyde sugars with six carbon atoms:

D-galactose	D-glucose	D-mannose
H—C=O	H—C=O	H—C=O
H—C—OH	H—C—OH	HO—C—H
HO—C—H	HO—C—H	HO—C—H
HO—C—H	H—C—OH	H—C—OH
H—C—OH	H—C—OH	H—C—OH
CH_2—OH	CH_2—OH	CH_2—OH

These sugars are identical in empirical formula and in the kind of substituents on each carbon atom, which means that they are by definition **stereoisomers**, compounds that differ only in the spatial arrangement of the substituent groups. However, they are not mirror images, which is the special class of stereoisomers known as **enantiomers**. For example, the mirror image of D-glucose is L-glucose, which is very rare in nature:

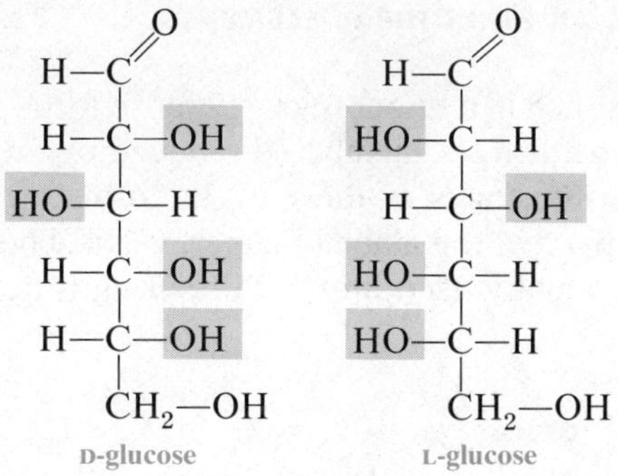

D-glucose

L-glucose

Despite their similar structures, glucose, galactose, and mannose are quite different compounds as is shown by the following physical properties.*

	D-Galactose	D-Glucose	D-Mannose
solubility in water (*g/100 ml*)	10.3 (0°)	32.3 (0°)	
		83 (17.5°)	248 (17°)
melting point (°C)	167°	146°	132°

D-glucose and D-galactose and D-glucose and D-mannose are related to each other as epimers—stereoisomers differing in configuration on only one carbon atom. (D-mannose and D-galactose are not epimers.)

Furanose and pyranose. Little of the three aldohexoses exists in the free aldehyde form drawn above because a six-carbon chain readily folds upon itself so as to bring the aldehyde group into proximity with a hydroxyl group on either C-4 or C-5, and the two groups readily react to form a **hemiacetal** ring:

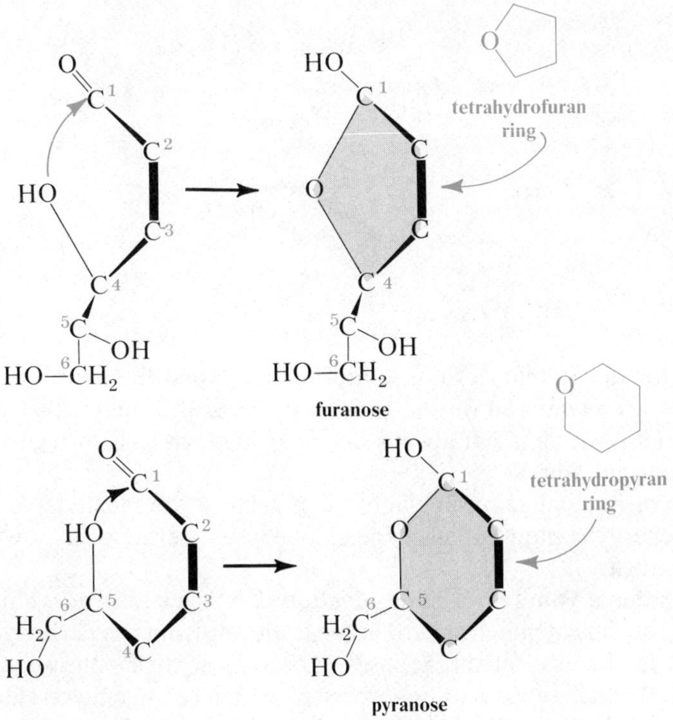

tetrahydrofuran ring

furanose

tetrahydropyran ring

pyranose

*We are cheating a little here because the data given apply to the tautomeric form in which these compounds readily crystallize and not to the free aldehyde forms. These tautomers are discussed in the next pages.

The sugar is then in a furanose or pyranose form, with either a five-membered reduced furan ring or a six-membered reduced pyran ring. It is difficult to represent the steric arrangement of these cyclic forms by a Fischer convention (horizontal bonds in front of the plane of paper, vertical bonds in back), and a **Haworth convention** is frequently used, in which the ring is depicted nearly on edge:

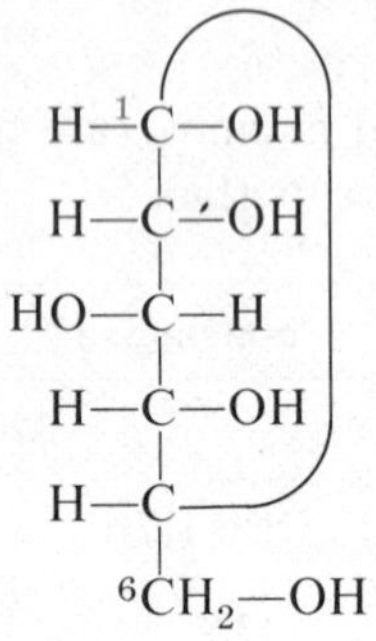

Fischer convention 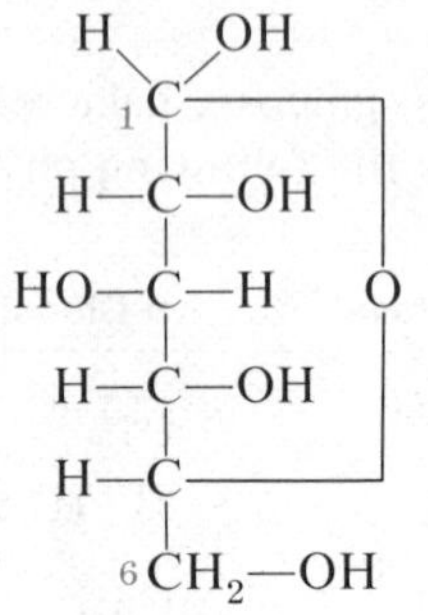 | modified Fischer convention 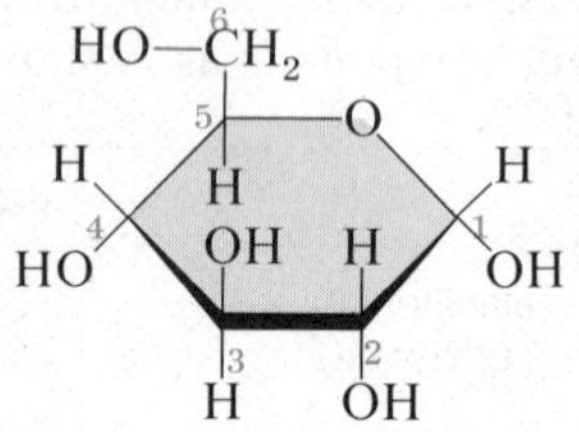 | Haworth formula

The Haworth convention is useful for many purposes, but even it does not accurately represent the true conformation of the molecule because the rings are not planar. It is becoming more common to indicate the pucker of the ring by a conformational formula; D-glucose, for example, is predominantly in a chair form:

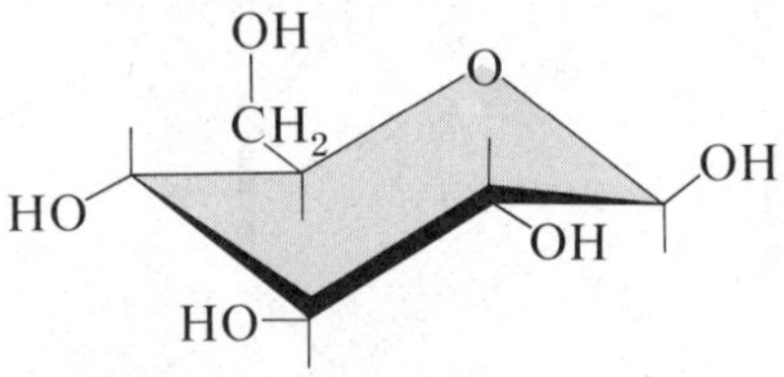

One reason for the stability of this conformation is that the hydroxyl groups on C-2 3, and 4 are **equatorial** (in the general plane of the ring) rather than **axial.** (The equatorial position is not always the most stable at C-1, owing to proximity to the ring oxygen atom.)

Because of its configuration glucose can achieve the most stable conformation (lowest energy content) of all of the aldohexoses; perhaps this is why it is the predominant sugar.

Configurational Family. The designation of a sugar as D or L hinges on the configuration of the asymmetric carbon most distant from the carbonyl function, which is C-5 in the case of the hexoses. The reason is that the sugars may be regarded in a formal sense as derivatives of either D- or L-**glyceraldehyde,** the simplest aldoses. For example, adding one formaldehyde unit would create the tetroses:

```
      O                         O
H—C//                     H—C//
   |                         |
   H                         H
   +                         +
      O                         O
H—C//                     H—C//
   |                         |
HO—C—H                    H—C—OH
   |                         |
   CH2OH                     CH2OH
L-glyceraldehyde          D-glyceraldehyde
```

```
      O              O              O              O
H—C//          H—C//          H—C//          H—C//
   |              |              |              |
HO—C—H         H—C—OH       HO—C—H         H—C—OH
   |              |              |              |
HO—C—H        HO—C—H         H—C—OH         H—C—OH
   |              |              |              |
   CH2OH          CH2OH          CH2OH          CH2OH
L-erythrose    L-threose      D-threose      D-erythrose
```

Further additions would create the pentoses, then the hexoses, and so on. Each addition doubles the number of possible aldose isomers, so there are 16 possible aldohexoses in eight DL pairs.

The notion of glyceraldehyde as the parent compound turned out to be a happy concept; most of the sugars in nature are indeed derived from D-glyceraldehyde by reactions that retain the D configuration on the next to last carbon atom.

Anomers. When the ring closes to form a furanose or pyranose, the resultant hydroxyl group on C-1 may be above or below the plane of the ring, yielding isomers that are designated as α or β (Fig. 10–1). (When drawn in the Fischer convention, the α-isomer has the hydroxyl group on the same side as the C-5 oxygen atom.)

The crystalline glucose of commerce is α-D-glucose, but when it is dissolved in water, the ring is free to open into the aldehyde form. The aldehyde form in turn is free to condense into the furanose or the pyranose forms, and either of these may be the α or β isomers, as shown in Figure 10–1. What we regard as a single compound in solution is in fact a mixture of at least five different compounds. In the case of glucose, circumstances are somewhat simplified because less than 1 per cent of the total is in the furanose form and less than 0.1 per cent is in the aldehyde form. A solution of glucose may be regarded as a mixture of α-D-glucose, and β-D-glucose, constantly equilibrating through the formation of trace amounts of the aldehyde. This equilibration can be observed by measuring the optical activity of the solution (its ability to rotate the plane of polarization of light); the rotation decreases as a fresh solution of α-D-glucose comes to equilibrium, and the equilibration of the sugar is called mutarotation.

The furanose isomers form more rapidly than do the pyranose isomers, but they are such a small fraction of the total in the case of glucose that the actual concentration is not known. This is not true of all sugars; 5 per cent of galactose is present as furanoses at equilibrium.

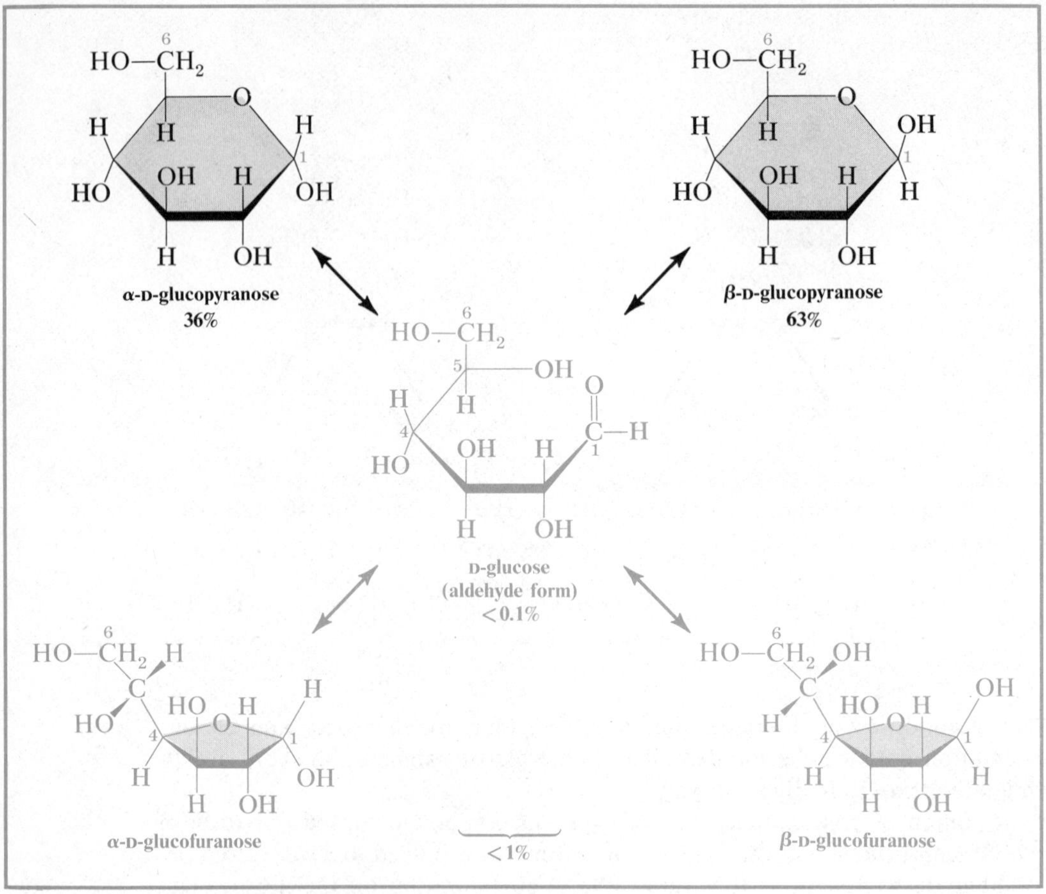

FIGURE 10–1 **The aldehyde form of glucose (*center*) reacts to form a pyranose ring (*top*), or a furanose ring (*bottom*). Closure of the ring creates a mixture of two configurations on C-1, designated α and β. All of these forms equilibrate in solution, but the two pyranose forms (*black*) predominate by far.**

MODIFIED SUGARS

Sugars are modified for structural purposes to provide other types of reactive groups. The principal modified compounds are **hexosamines,** in which the oxygen atom on C-2 is replaced by nitrogen; the **hexuronates**, in which C-6 is oxidized to a carboxylate group; and the **deoxy sugars**, in which one or more carbon atoms, usually C-6, has an H in place of the usual OH group.

Hexosamines

The common hexosamines of animal cells are D-glucosamine and D-galactosamine:

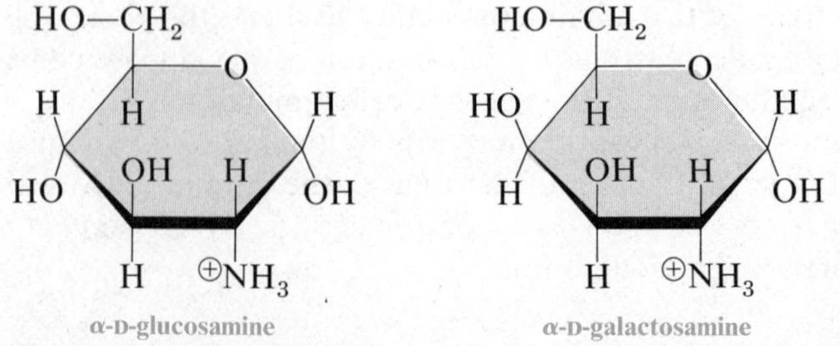

They sometimes occur as such in structural carbohydrate, but more often as the N-acetyl derivatives:

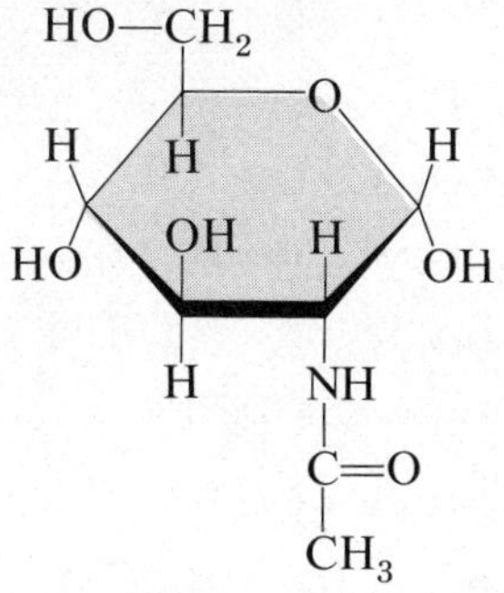

N-acetyl-D-glucosamine 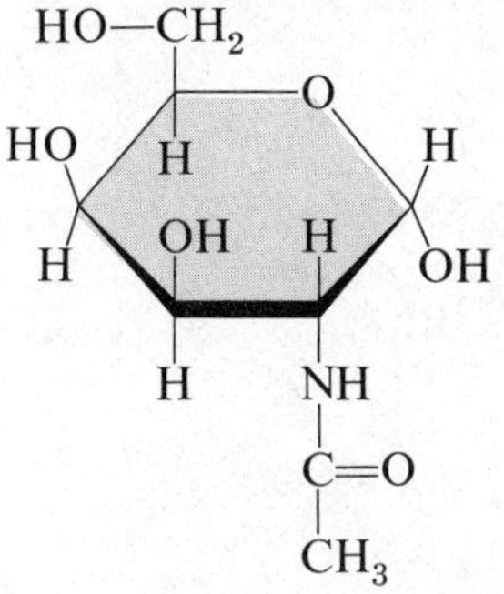 N-acetyl-D-galactosamine

Only one isomer is shown in the preceding illustrations, but these compounds undergo the same equilibration of free aldehyde, pyranose, and furanose with α and β forms as do the simple sugars; that is, they can mutarotate.

Hexuronates

Two hexuronates, in which C-6 is oxidized to the anionic form of a carboxylic acid, are of importance in animals. These are the anions of D-glucuronic acid, derived from D-glucose, and its 5-epimer, L-iduronic acid:

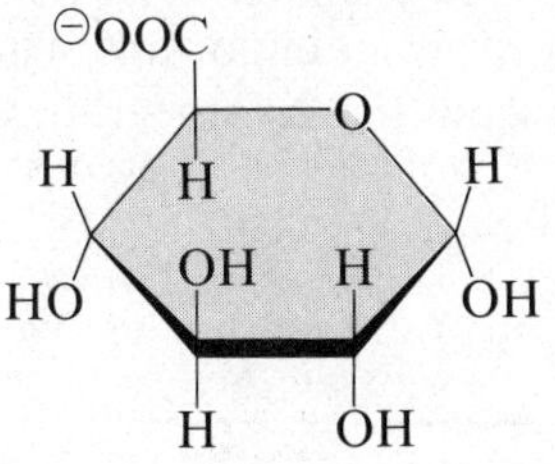

α-D-glucuronate 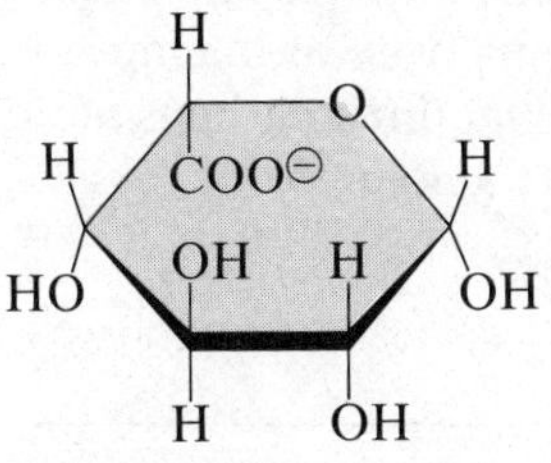 β-L-iduronate

L-Iduronic acid is named as a derivative of the hexose, L-idose, but in fact the L-iduronate residues in structural carbohydrates are formed by epimerization of residues of D-glucuronate (p. 680). The uronates also occur as anomers and can mutarotate.

Deoxy Sugars

Two deoxy hexoses of importance are L-fucose and L-rhamnose:

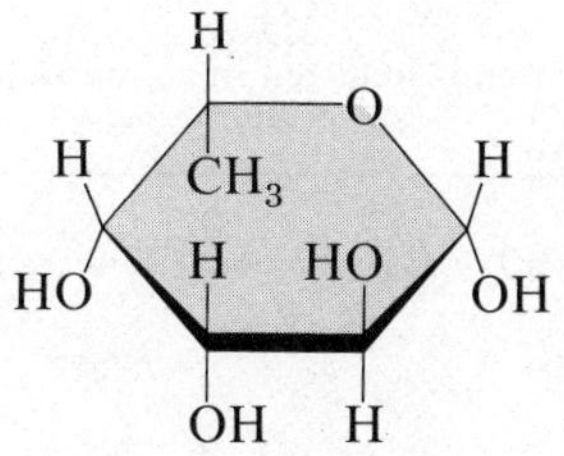

β-L-fucose (6-deoxy-β-L-galactose) β-L-rhamnose (6-deoxy-β-L-mannose)

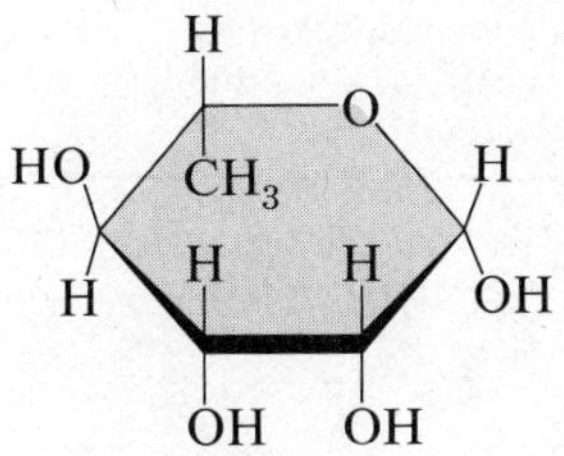

Although these compounds have a formal relationship to L-galactose and L-mannose, they are actually formed from D-hexoses (p. 682).

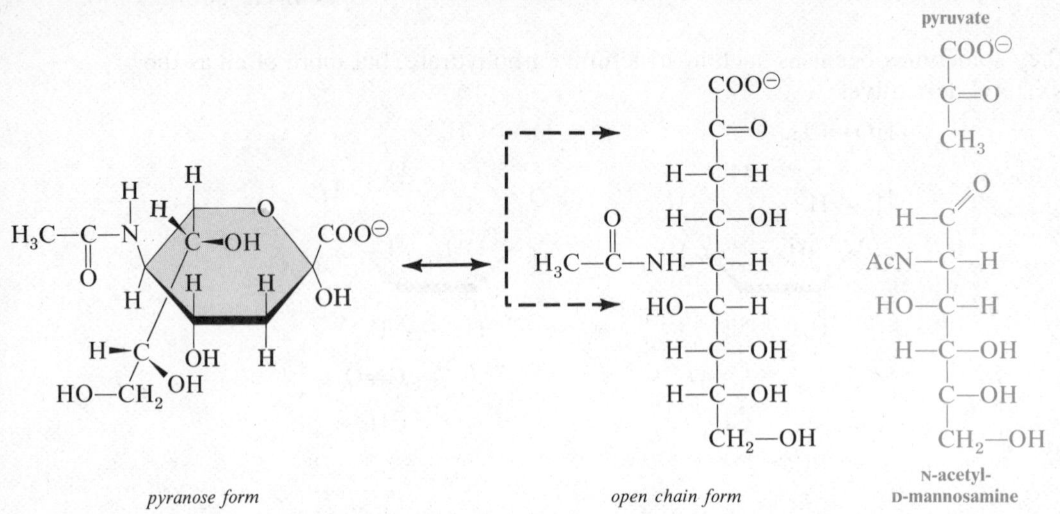

pyruvate

pyranose form

open chain form

N-acetyl-D-mannosamine

FIGURE 10–2 **N-Acetylneuraminate, an important constituent of glycoproteins, exists in a pyranose form. The nine-carbon neuraminate chain is made by condensing derivatives of pyruvate and mannosamine (*right*).**

Sialic Acids (Neuraminates)

The final component in our list of major structural carbohydrates, N-acetylneuraminate, or sialate, is the most complex (Fig. 10–2). The terminal six carbons in the nine-carbon chain are derived from N-acetyl-D-mannosamine. (This amino analogue of mannose is not in itself an important structural component.) The initial three carbons are derived from the 2-ketocarboxylic acid, pyruvic acid, in its anionic form.

SUGARS

An **aldose** has an aldehyde group; a **ketose** has a ketone group.

A **hemiacetal** (or **hemiketal**) is the condensation product of a carbonyl group with one alcohol group.

A **furanose** is the hemiacetal (or hemiketal) form of a sugar with a five-membered furan ring (four carbon atoms and one oxygen atom in ring).

A **pyranose** is the hemiacetal (or hemiketal) form of a sugar with a six-membered pyran ring (five carbon atoms and one oxygen atom in ring).

An **epimer** has a different steric arrangement on one asymmetric carbon atom.

Enantiomers are mirror images; they have a different steric arrangement on all asymmetric carbon atoms.

An **anomer** of a furanose or pyranose is the stereoisomer with a different steric arrangement on the potential carbonyl carbon (C-1 with aldoses, C-2 with ketoses).

Mutarotation is the equilibration of one anomer of a sugar with the other possible forms through the intermediate formation of the free carbonyl form.

Hexosamines are sugars in which the oxygen atom on C-2 is replaced by a nitrogen atom; that is, they are 2-amino-2-deoxy-aldohexoses.

Hexuronates are sugars in which the terminal carbon atom is oxidized to an ionized carboxylic acid group.

POLYMERS OF SUGARS

Glycosides

Sugars polymerize by forming acetals. We know from organic chemistry that the hemiacetals of aldehydes can react with another molecule of alcohol to yield an acetal:

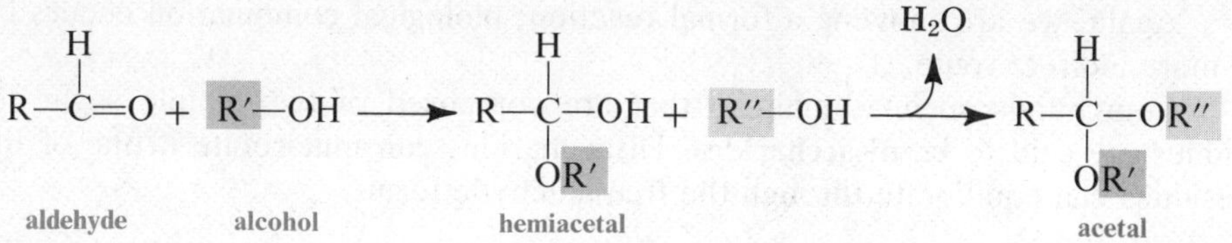

aldehyde alcohol hemiacetal acetal

The furanose or pyranose forms of sugars are hemiacetals, and they can condense this way in the test tube to form acetals, which are termed **glycosides**:

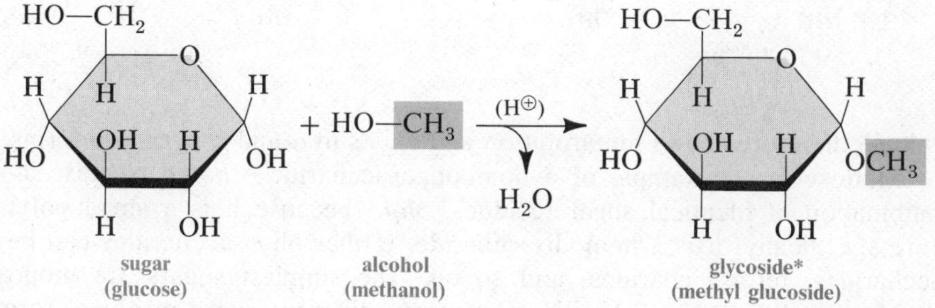

sugar (glucose) alcohol (methanol) glycoside* (methyl glucoside)

Many glycosides with this type of structure occur in nature, although they are formed in a less direct way (Chapter 27). The plant kingdom includes an especially varied assortment, with a great variety of compounds bearing hydroxyl groups combined with many kinds of sugars. Some of these plant glycosides have profound physiological effects at low concentrations in animals, such as the cardiac glycosides derived from *Digitalis*, the common foxglove.

When sugars combine, one molecule contributes the hemiacetal and the other molecule contributes the hydroxyl group. For example, two molecules of glucose can combine to form maltose, a glucose glucoside:

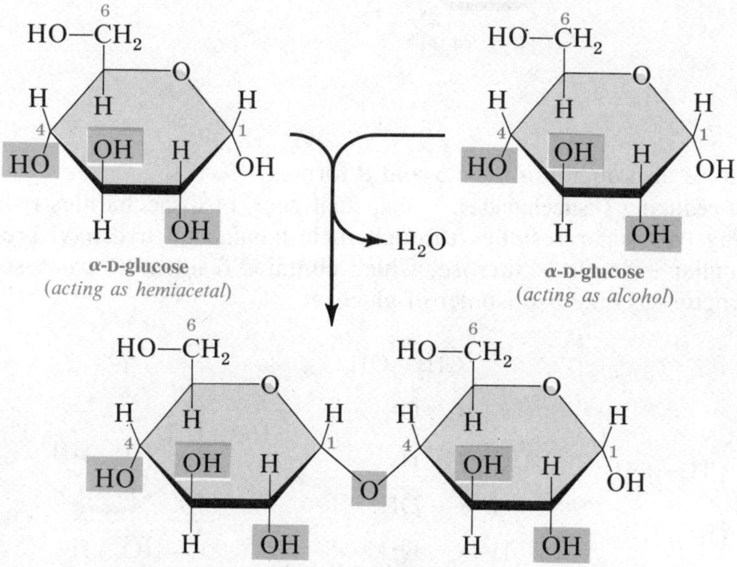

α-D-glucose (*acting as hemiacetal*) α-D-glucose (*acting as alcohol*)

α-maltose
α-D-glucosyl-(1→4)-α-D-glucose
(O-α-D-glucopyranosyl-(1→4)-α-D-glucopyranose)

Again, we are showing a formal reaction; biological combination occurs by a more indirect route.

Compounds such as maltose that are composed of two simple sugar residues are said to be **disaccharides**. Disaccharides can mutarotate if one of the residues can equilibrate through the free aldehyde form:

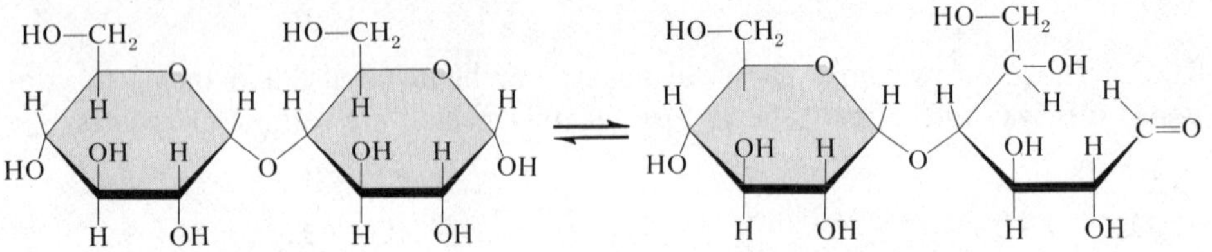

maltose
(*α-pyranose form*)

maltose
(*open-chain form*)

Maltose therefore shows mutarotation and exists in α and β pyranose forms.

Maltose is an example of a homooligosaccharide — *homo* because it is a combination of identical sugar residues, *oligo* because it is a small polymer. More specifically, it is a homodisaccharide. (Other oligosaccharides can be trisaccharides, tetrasaccharides, and so on. The simplest sugars are monosaccharides.) Polymers can also be made from different sugar residues, forming heterooligosaccharides or longer heteropolysaccharides. An example of a heterodisaccharide is lactose, the sugar in milk:

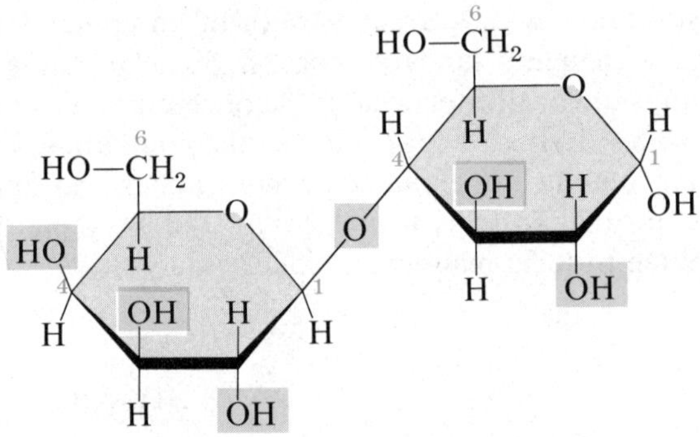

α-lactose
β-D-galactosyl-(1→4)-α-D-glucose

Lactose also equilibrates in α and β forms.

Non-reducing Disaccharides. A special class of disaccharides is formed by combining two sugar residues through their hemiacetal hydroxyl groups. The most familiar example is sucrose, which contains residues of fructose and glucose. Fructose is a ketose isomer of glucose:

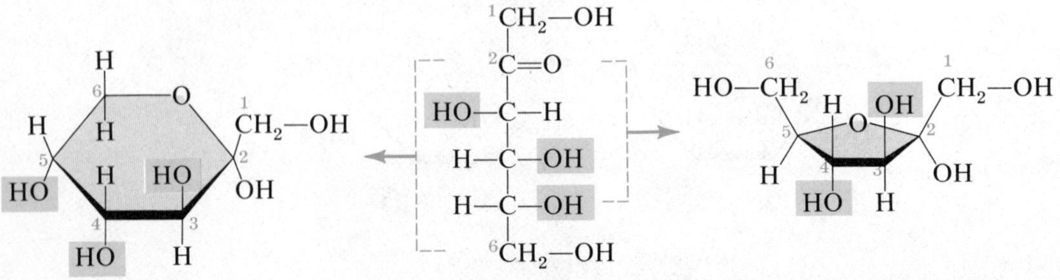

α-D-fructopyranose

D-fructose
(*open-chain form*)

α-D-fructofuranose

Forming a linkage between C-1 of the pyranose form of glucose and C-2 of the furanose form of fructose creates sucrose:

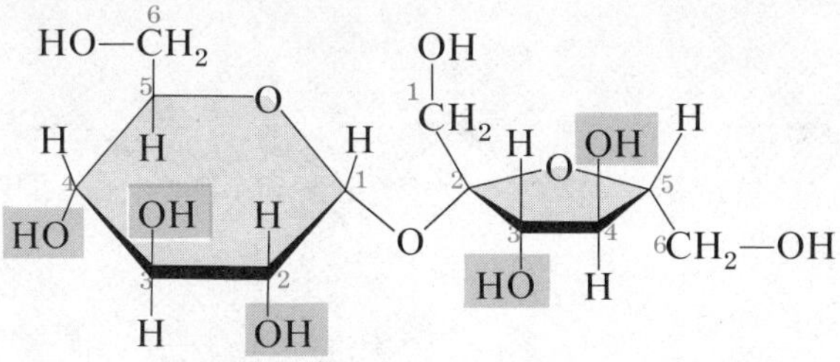

sucrose
(β-D-fructofuranosyl-α-D-glucopyranoside)

Sucrose is synthesized by plants as a transportable form of carbohydrate; it is a device for moving fuel from the leaves to other parts of the plant, analogous to lactose as a device for moving fuel from mother to infant.

Sucrose has become an important dietary fuel for humans, accounting for over 15 per cent of the total energy consumption of Americans.

Chemically, sucrose is distinctive because both of the potential carbonyl groups are locked in acetal linkage. The rings cannot open, and sucrose does not mutarotate or exist as anomers. Sucrose, therefore, does not exist transiently as a free aldehyde or ketone. The free carbonyl group in other sugars has a reactivity that is accentuated by the neighboring hydroxyl groups; they are readily oxidized by relatively mild reagents, such as alkaline cupric ion solutions. Hence, they are termed **reducing sugars** because they reduce cupric ion to cuprous ion, whereas sugars such as sucrose are said to be non-reducing sugars. (Boiling samples of urine with blue cupric reagents to detect glucose used to be a familiar part of laboratory examinations as a screening test for diabetes. Automated quantitative analysis for blood glucose concentration and some qualitative tests still hinge on the reducing power of glucose.)

Another non-reducing disaccharide of general interest is **trehalose**, which is α-glucose-(1→1)-α-glucoside. Trehalose is the storage and transport fuel in insects. The general use of disaccharides for transport arises from the need to separate this function from the more general functions of the monosaccharides.

DISACCHARIDES

reducing	maltose = O-α-D-glucopyranosyl-(1→4)-D-glucopyranose
	lactose = O-β-D-galactopyranosyl-(1→4)-D-glucopyranose
non-reducing	sucrose = β-D-fructofuranosyl-α-D-glucopyranoside*
	trehalose = α-D-glucopyranosyl-α-D-glucopyranoside*

*The formal names end in pyranoside, rather than pyranose, to indicate that the combination involves C-1; that is, the compound is a glycoside of both residues.

Higher Polymers

The combination of sugar residues through acetal formation can be extended almost indefinitely. For example, glucose residues can be combined, first to form homooligosaccharides, and then longer homopolysaccharides:

HAWORTH FORMULA

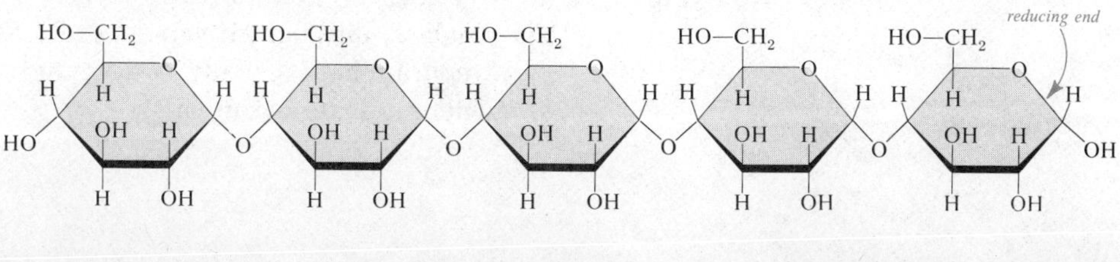

reducing end

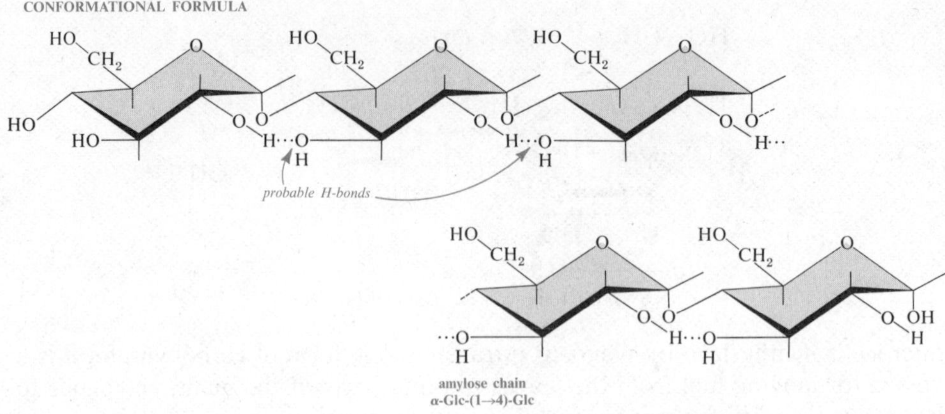

This particular polymer is known as amylose; similar combinations of glucose residues by $\alpha(1 \to 4)$ linkages are important structural features in the stored fuels, glycogen in animals and starch in plants. Only one residue at the end of these polymeric chains is able to mutarotate and behave as a reducing sugar. Various kinds of carbohydrate residues may be combined to form hetero-oligosaccharides and heteropolysaccharides, and it is in these compounds that one finds important structural components.

Abbreviations. Description of the carbohydrate polymers is greatly aided through the use of abbreviations for the carbohydrate residues:

Aldohexoses	Gal =	D-galactose
	Glc =	D-glucose
	Man =	D-mannose
Ketohexoses	Fru =	D-fructose
Osamines	GalN =	D-galactosamine
	GalNAc =	N-acetyl-D-galactosamine
	GlcN =	D-glucosamine
	GlcNAc =	N-acetyl-D-glucosamine
Uronates	GlcUA =	D-glucuronate
	IdoUA =	L-iduronate
Deoxysugars	Fuc =	L-fucose
	Rha =	L-rhamnose
Neuraminates	Neu =	D-neuraminate
	NeuNAc or NAN =	N-acetyl-D-neuraminate
	Sia =	sialate; unspecified substituted neuraminate (usually N-acetyl-neuraminate in humans)

The linkage between the groups is then specified in the ordinary way. For example, amylose, the glucose polymer illustrated immediately above, is:

$$\alpha\text{-Glc-}(1\rightarrow4)\text{-}\alpha\text{-(Glc)}_n\text{-}(1\rightarrow4)\text{-Glc}$$

Lactose is β-(Gal)-(1→4)-Glc; sucrose is β-(Fru)-(2→1)-α-Glc, and so on.

GLUCANS: HOMOPOLYMERS OF GLUCOSE

There are many theoretical polymers of glucose; condensation is possible with the hydroxyl group on any of the carbon atoms and may involve either the α or the β anomers. Only a few of the possibilities are known to occur in nature, and let us examine why.

Permissible Configurations

Theoretical analysis of polysaccharide structure is not as advanced as it is for polypeptides, but it is now possible to perceive the structural features of importance and relate them to function. Most of the carbohydrate polymers have a hemiacetal bridge between two pyranose rings, much as we saw in maltose. Since the motion of atoms within the rings is restricted, major alterations in conformation of such polymers are limited to rotations around the hemiacetal ether oxygen. That is, the configuration is mainly fixed by the relative positions of successive rings in the polymer. These positions may be described by a pair of dihedral angles, analogous to the Ramachandran angles used to describe polypeptide conformation. The preferred conformations of lowest energy content cluster near certain angles.

The favored angles in polysaccharides vary with the kind of sugar, the anomer that is involved, and the carbon atoms that are connected because the hydroxyl groups must be oriented so as to favor formation of hydrogen bonds. It now appears that there are four possible secondary structures in polymers of aldopyranoses:

Type A polymers are extended ribbons, such as are seen in cellulose. Cellulose is a polymer of glucose with β-(1→4) linkage, and the type A ribbon is the conformation of minimum energy content for such polymers. It is the ideal conformation for making fibers, because parallel polysaccharide chains can readily be linked by hydrogen bonds (Fig. 10–3). The β-configuration causes adjacent rings in a chain to be rotated 180° relative to each other.

Chitin, a polymer of N-acetylglucosamine residues, has a structure similar to that of cellulose. Chitin is the fibrous component of the exoskeleton of arthropods: crabs, spiders, insects.

Type B polymers occur as a helix of variable dimensions, usually left-handed and with an open core (Fig. 10–4). This is the preferred conformation of amylose and of some heteropolysaccharide chains occurring in connective tissues. In each of these cases, there are too many easily interconvertible forms, some involving extensive bonding with water, available to these polymers for them to be useful in making rigid structures, but they are suitable for use as fuels and as a matrix in connective tissues. It is interesting to note the drastic differences in properties between an α(1→4) glucan such as amylose and a β(1→4) glucan such as cellulose.

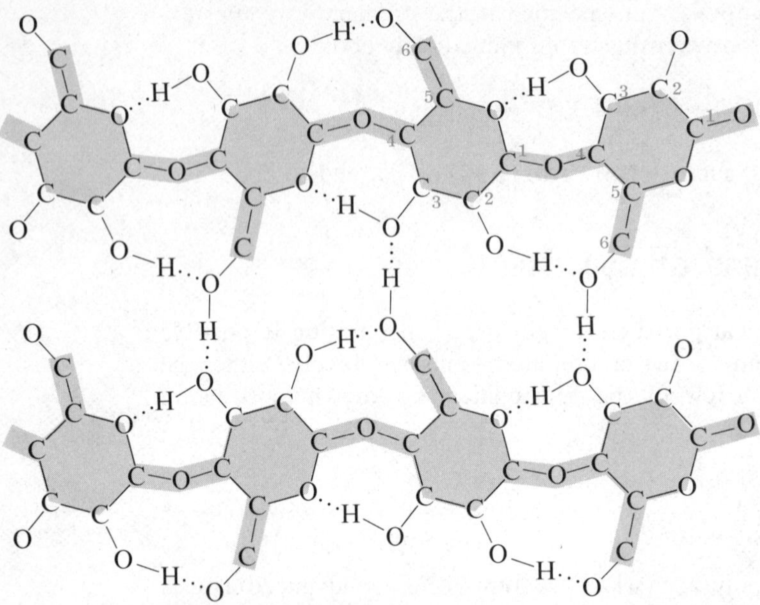

FIGURE 10–3 **Cellulose is made from parallel ribbons of glucose residues linked by β-(1→4) bonds. The arrangement is stabilized by hydrogen bonds involving all of the oxygen atoms except those in the ether bridges between rings. The rings are being viewed face-on in this drawing; it is *not* a Haworth projection. It is based on the arrangement shown in K. H. Gardner and J. Blackwell, (1974) *The Structure of Native Cellulose.* Biopolymers, *13*: 1975.**

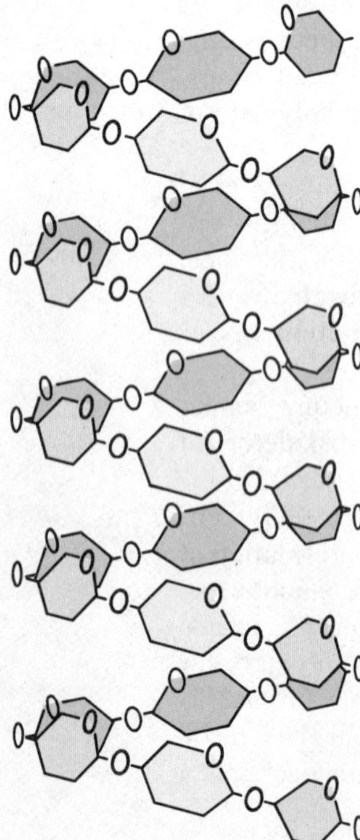

FIGURE 10–4

Amylose chains form open-cored helices, probably left-handed, of varying sizes. One possible arrangement is shown here.

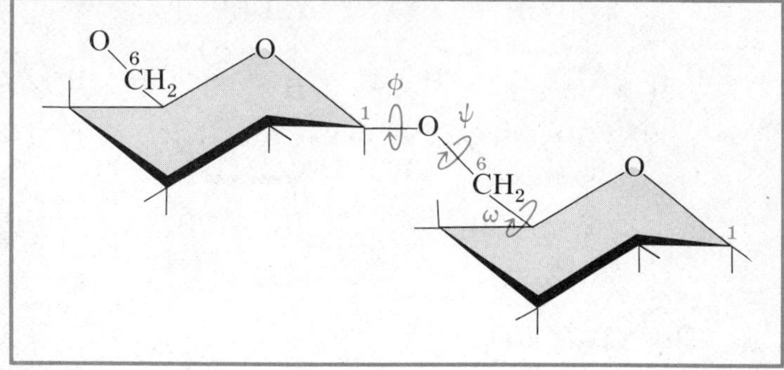

FIGURE 10–5

Glucose residues that are linked by 1→6 bonds can rotate relative to each other around the C-6 to C-5 bond, and three angles must be specified in order to describe their relative position. The greater freedom of motion permits 1→6 polymers to assume open configurations not available to polymers linked at other positions.

Type C polymers have a crumpled structure, somewhat like a folded bellows. There are no known natural examples.

Type D polymers have an open structure created by linking sugars through 1→6 bonds. Since an additional degree of freedom is created by rotation around these bonds (Fig. 10–5), such polymers would probably exist in an open extended form in crystals and be in a random form in solution. No pure examples are known, but 1→6 chains are a part of the structure of dextrans, which are polysaccharides made by some bacteria.

The Natural Glucans

Glycogen is a polymer that acts as a reservoir of glucose residues for use as a fuel and a precursor of other compounds. The fundamental structure of glycogen is the amylose chain, with glucose residues joined in a α-(1→4) linkage, but it differs from amylose in having 1→6 branches at every fourth glucose residue in the interior of the molecule (Fig. 10–6). (Longer branched sequences occur at the periphery.) The branches prevent any formation of a helix, and the result is a tree-like structure, sketched in cross section in Figure 10–6*C*. (The role of glycogen as a fuel is discussed in Chapter 27.)

Starch is the storage carbohydrate of plants. It consists of two discrete kinds of molecules, with a minor fraction composed of pure amylose chains and the bulk made of amylopectin, which has a structure similar to that of glycogen but less highly branched. Starch is the major fuel in the diet of most humans.

Cellulose occurs in a few invertebrates but mainly in plants as a structural component. Mammalian cells lack the ability to break the β-(1→4) bonds of cellulose to yield glucose; only ruminants and similar animals that harbor a special bacterial flora capable of attacking cellulose are able to utilize it as a dietary fuel. However, the very indigestibility of cellulose makes it important in providing bulk to the intestinal contents of humans.

The *dextrans* are gelatinous polymers that coat some bacteria. They have an α-(1→6) backbone with varying numbers of 1→3 and 1→4 cross-links, creating a three-dimensional network that is freely accessible to water. They are widely employed as molecular sieves in the separation of proteins and other large molecules by column chromatography because the degree of cross-linking determines the size of the molecule that can penetrate the gels. They are also used to restore blood plasma volume on those infrequent occasions when blood cannot be used. Some species of bacteria that populate the teeth have the abil-

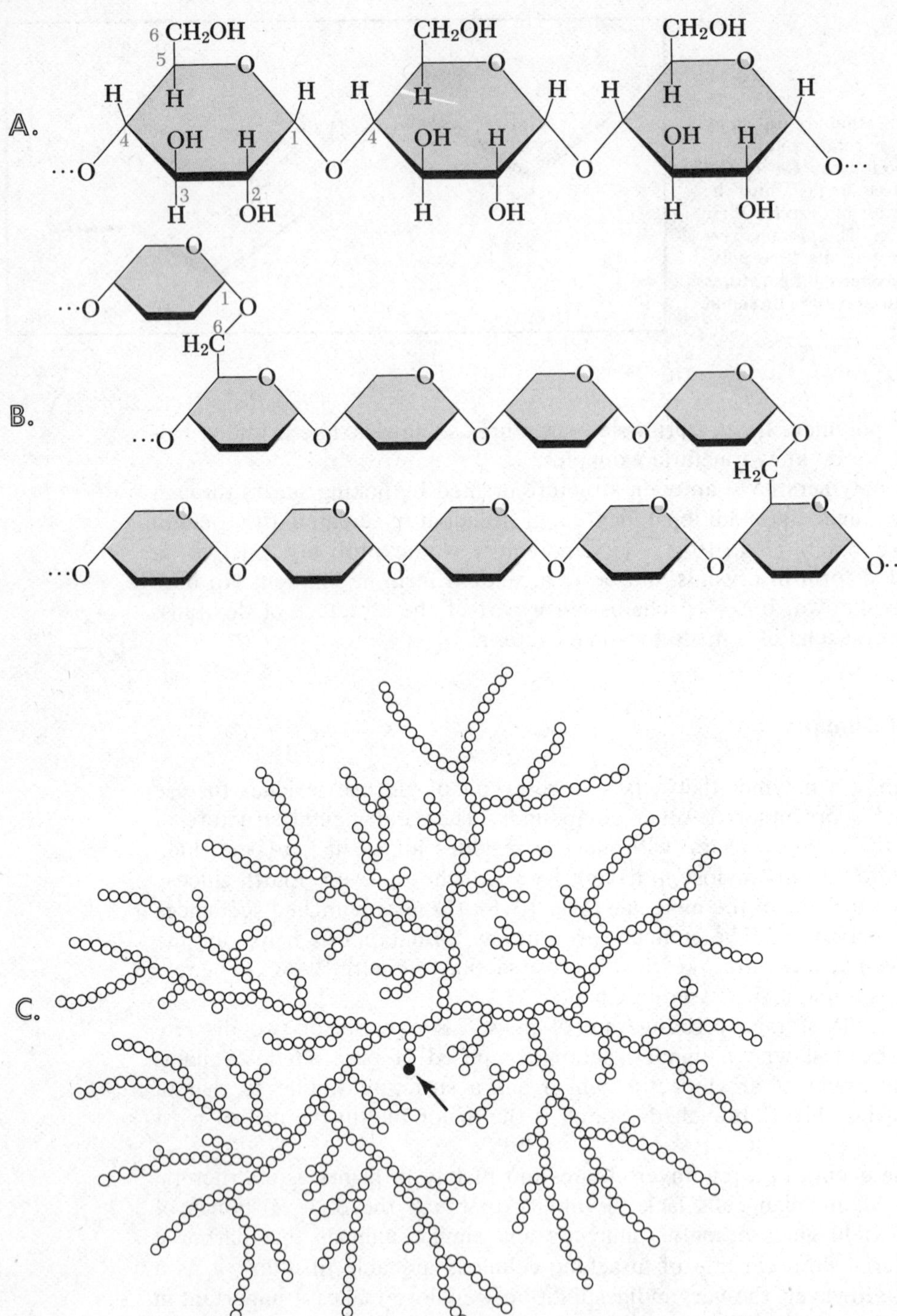

FIGURE 10–6 *A.* Glycogen contains linear amylose chains, made by linking carbons 1 and 4 of glucose through oxygen. *B.* Branches occur in the amylose chains where carbon 6 of a residue is also linked through oxygen to the C-1 terminal of another chain segment. *C.* A cross section through glycogen showing the tree-like structure created by branched amylose chains. The short inner segments are part of branches extending above and below the cross section. The circles represent glucose residues; only one residue (*arrow*) may assume an open-chain form and mutarotate because it is the only residue that does not have C-1 attached to another residue through an ether linkage.

ity to convert sucrose, but not glucose, to dextrans, forming plaques. The sticky dextrans aid in fixing the bacteria to the teeth, where they promote decay. The child with the sweet tooth is indeed more likely to find it rotting.

PROTEOGLYCANS

Many compounds are made by covalent combination of proteins and polysaccharides, with a spectrum of composition ranging from mostly protein to mostly polysaccharides. At one end of the spectrum are the **proteoglycans**, also called **proteinpolysaccharides**, in which the protein components are such a minor part of the total mass that their existence was overlooked for many years. These proteoglycans have long heteropolysaccharide chains covalently attached to a protein core, much like bristles on a brush.

Mucous secretions owe their viscous lubricating properties to proteoglycans and glycoproteins, and the proteoglycans formerly were called **mucopolysaccharides**. The major occurrence of proteoglycans is in the matrix in which collagen, elastin, and bone minerals are embedded. The character of connective tissue depends upon the relative proportions of fiber and matrix, as well as upon the nature of the two kinds of components; the intercellular cement is mostly ground substance with a few fibers, whereas a tendon is mostly fibers with minimal ground substance filling the spaces between them.

Chondroitin Sulfates

The most abundant proteoglycans in cartilages and arterial walls contain a heteropolysaccharide, chondroitin, esterified with sulfate groups. Let us examine the important features of these chondroitin sulfates as typical examples of proteoglycans.

The Carbohydrate Chains. Chondroitin is mainly composed of alternating residues of N-acetyl-D-galactosamine and D-glucuronate and therefore may be written as a repeating disaccharide unit:

$$(\text{GalNAc-GlcUA})_n$$

Other proteoglycans also consist of repeating disaccharide units containing an osamine, with the other residue often being a uronate. (We shall later see that the carbohydrates are built one residue at a time, not by combining preformed disaccharide units.)

Chondroitin occurs as a straight chain, unlike the branched structure of glycogen. All of the residues are joined in the β-configuration at C-1, and the glycosidic bonds are alternately 1→3 and 1→4:

$$[(1\rightarrow3)\text{-}\beta\text{-GalNAc-}(1\rightarrow4)\text{-}\beta\text{-GlcUA}]_n$$

The result is a structure in which all of the linkages between residues are equatorial (Fig. 10-7). Chondroitin chains, like those of the other proteoglycans, tend to assume a helical type B configuration in the solid state and are presumably even more open when in free contact with water.

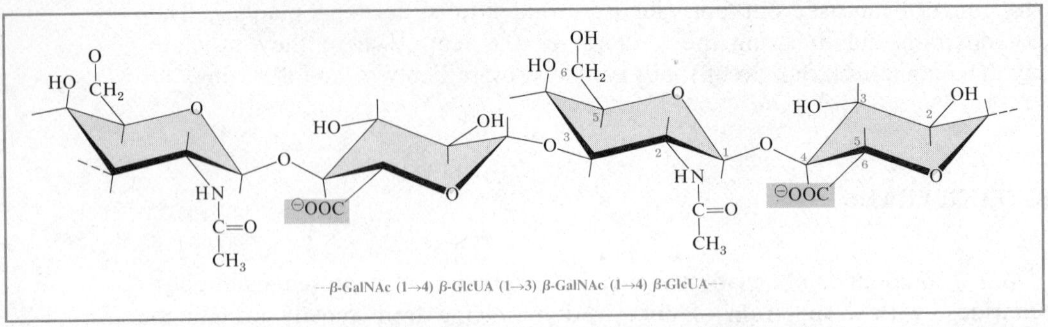

---β-GalNAc (1→4) β-GlcUA (1→3) β-GalNAc (1→4) β-GlcUA---

FIGURE 10–7 **The main chain of the chondroitin sulfates contains alternating residues of N-acetylgalactosamine and glucuronate. Notice that the β-linkages between residues cause alternate residues to be twisted by half-turns. By linking the glucuronate residues through C-4 and the N-acetylgalactosamine residues through C-3, an all-equatorial conformation is achieved, which is similar to the conformation of an amylose chain. Like amylose, the chondroitin chain tends to form helices.**

The linkage between the chondroitin chain and the core protein is made by a special sequence of carbohydrate residues containing D-galactose and the pentose **D-xylose** attached to seryl side chains in the polypeptides (Fig. 10–8).

Sulfate groups are present on C-4 of the galactosamine residues in one kind of chondroitin and on C-6 in another (Fig. 10–9). They add negative charge to the chains in much the same way that phosphate groups add negative charge to polypeptide chains. The proportions of chondroitin 4-sulfate and chondroitin 6-sulfate vary from one tissue to another, but we don't know why as yet.

Composition. Typical protein-chondroitin sulfate molecules will have compositions in this range:

20 to 60 polysaccharide chains per molecule; with
20 to 50 disaccharide residues in each chain, attached to
a core protein with a molecular weight of 110,000 to 140,000, giving
a total molecular weight of 2×10^5 to 2×10^6.

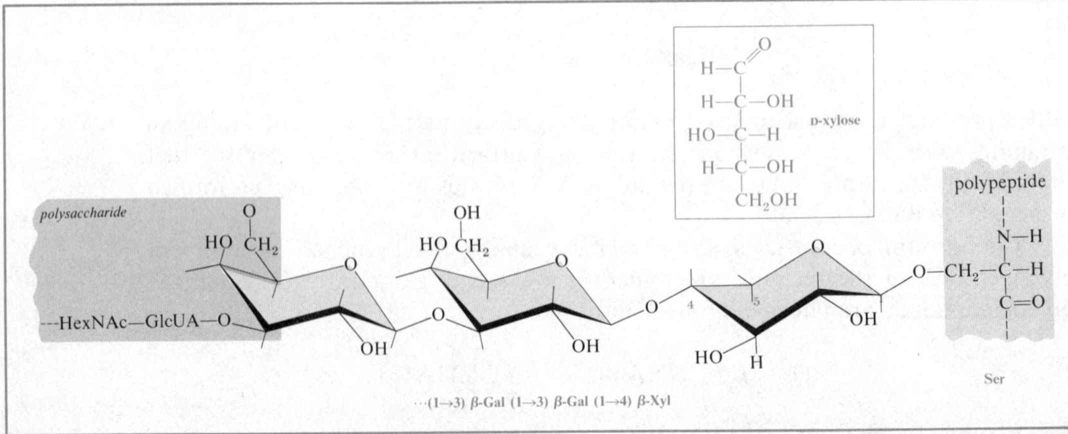

FIGURE 10–8 **In the chondroitin sulfate proteoglycans, the polysaccharide chain is joined to the polypeptide chain through a terminal sequence containing residues of galactose and xylose. (The open-chain formula of xylose is shown in blue.) The xylose in turn is linked to the hydroxyl group of a serine residue in the polypeptide.**

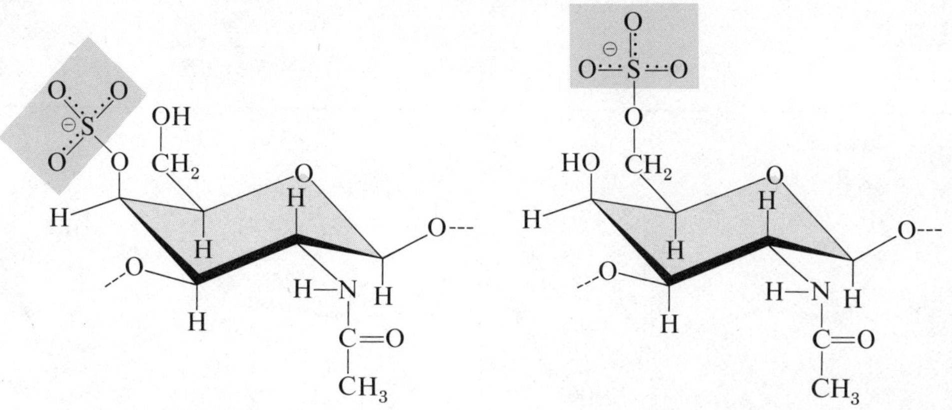

FIGURE 10–9 **Sulfate groups are linked to C-4 of N-acetylgalactosamine residues in some chondroitin sulfates, and to C-6 in others. The result in either case is the anionic form of an ester of sulfuric acid, contributing extra negative charges to the polysaccharide chain.**

Properties. The chondroitin sulfates, like other proteoglycans, have an abundance of negative charges on already hydrophilic carbohydrate chains. Charge repulsion will tend to keep the chains extended from the protein core and separated from each other, but they are free to sweep through the surrounding solution so that the volume occupied by a molecule in solution is much greater than the partial specific volume of the dehydrated solid. Schubert and Hamerman express it beautifully*:

> Statistically such a domain may have a fairly well defined size and shape. It can be visualized somewhat like the definitely shaped head of a tree, a black oak or a lombardy poplar, with branches extending throughout many cubic yards of space though the wood of its branches may occupy only a few percent of the volume of the head. Small birds can easily fly through the head of the tree, but not through the wood of its branches. Small molecules can easily swim through the domain of a proteinpolysaccharide molecule, larger ones may encounter frequent obstructions, and very large ones could not even enter.

Here we have substances with the ability to exclude large molecules and let small ones through. They have a large negative charge that will attract cations. In short, they may act as molecular sieves and as cation exchangers. It is attractive to assume that the proteoglycans have important functions as conduits for the selective transport of materials, but the degree to which this is true is still conjectural.

One clear function of the chondroitin sulfates is to act as a water-bed around cells and fibers, creating a gel with the resilience to disperse shocks without permanent deformation. It seems likely that they interact in specific ways with the collagen fibers from which the structures gain rigidity, but the study of these complex molecules has not progressed to the point where more positive statements can be made.

---

*Schubert, M., and D. Hamerman: (1968) A Primer on Connective Tissue Biochemistry. Lea and Febiger. Quoted by permission.

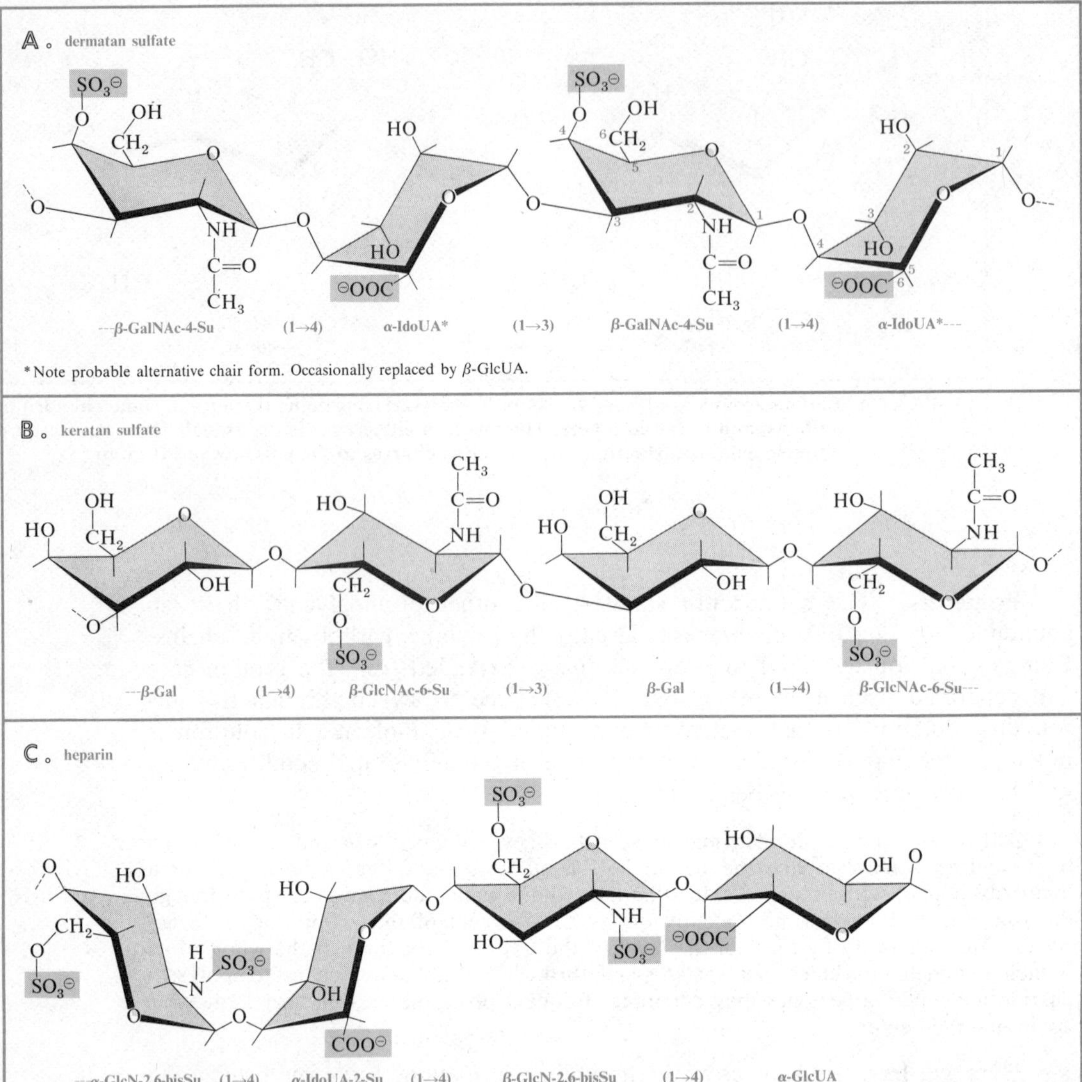

FIGURE 10–10 **A. Dermatan sulfate closely resembles chondroitin 6-sulfate in general structure, except that configuration is reversed on C-5 of the uronate residues, so that they become residues of iduronate.**

B. Keratan sulfate is made by polymerizing galactose with N-acetylglucosamine residues, followed by esterification with sulfate. The molecule is drawn here with a configuration resembling that of chondroitin sulfates, but this is conjectural.

C. Heparan sulfate in connective tissues is believed to resemble the heparin of mast cells in composition and configuration. Heparin contains both iduronate and glucuronate residues, along with N-acetylglucosamine residues. A speculative interpretation of a six-residue repeating unit is given here, with one residue of iduronate not shown at either end. Heparin, and presumably heparan sulfate, unlike other heteropolysaccharides shown here, contains axial links between residues as well as the more common equatorial bonds. These create a more looping conformation.

Other Sulfated Proteoglycans

Connective tissues contain several other sulfated heteropolysaccharides, in addition to the chondroitin sulfates, which differ in the nature of the constituent residues (Fig. 10–10). The proportion of the different compounds varies from tissue to tissue, and we make the obvious inference that appropriate properties are gained by these differences, but we are not yet able to make specific statements on what these properties are or the contribution made by specific heteropolysaccharides. The composition of the proteoglycan fraction can be changed by altering the proportions of two or more proteoglycans, each carrying only one kind of heteropolysaccharide, or it may be changed by adding more than one kind of heteropolysaccharide chain to the same protein core. Both mechanisms appear to be used.

Dermatan sulfates resemble chondroitin sulfates, except that α-L-iduronate residues occur in place of many, but not all, of the β-D-glucuronate residues. This does not represent a physical replacement of one kind of residue with another, but occurs by an epimerization of the residues after formation of the polysaccharide. The H and COO^- are interchanged on C-5 of many of the glucuronate residues, so that dermatan sulfate contains both of the hexuronates. Owing to the peculiarities of nomenclature, the interchange also changes the designation of the anomer from β-D to α-L, even though the configuration on C-1 has not been altered. The carbohydrate chains are linked to the protein by the same Gal-Gal-Xyl-Ser sequence seen in chondroitin sulfate.

The dermatan sulfates are especially common in skin, hence the name, but they also occur in other tissues.

Keratan sulfates occur as minor constituents along with chondroitin sulfates and are perhaps bound to the same protein cores. They are known to be present in cartilage, the cornea of the eye, and the pulpy nucleus of intervertebral discs. Keratans contain D-galactose, rather than a uronate, in combination with N-acetyl-D-glucosamine:

$$[(1\rightarrow3)\text{-}\beta\text{-Gal-}(1\rightarrow4)\text{-}\beta\text{-GlcNAc}]_n$$

A sulfate group is present on C-6 of the osamine, so the keratan sulfates are also polyanions, but with approximately half of the charge of chondroitin or dermatan sulfates, which also have uronate groups.

Heparan sulfates are more complicated heteropolysaccharides that contain both glucuronate and iduronate residues, along with alternating glucosamine residues that are only partially acetylated. They appear to be related to a more widely studied compound, **heparin**, which has a similar structure with more sulfate groups attached and no acetyl groups on the glucosamine residues.

Heparin is not a constituent of connective tissues; it occurs within the mast cells that line arteries, especially in the liver, lungs, and skin. A repeating unit of six carbohydrate residues has been proposed:

$$[\underset{a\ \ a}{(1\rightarrow4)}\text{-}\alpha\text{-IdoUA-}\underset{e\ \ e}{(1\rightarrow4)}\text{-}\alpha\text{-GlcN-}\underset{a\ \ a}{(1\rightarrow4)}\text{-}\alpha\text{-IdoUA-}\underset{e\ \ e}{(1\rightarrow4)}\text{-}\beta\text{-GlcN-}$$

$$\underset{e\ \ e}{(1\rightarrow4)}\text{-}\alpha\text{-GlcUA-}\underset{a\ \ e}{(1\rightarrow4)}\text{-}\alpha\text{-GlcN-}]_n$$

a = *axial* *e* = *equatorial*

in which each glucosamine residue contains sulfate groups on both C-6 (ester linkage) and the nitrogen atom (amide linkage), and each of the iduronate residues contains sulfate in ester linkage on C-2, making a total of eight sulfate groups per six carbohydrate residues. The chains are relatively short with only about 50 carbohydrate residues.

An important property of heparin, and presumably of heparan sulfates, is that some of the residues are linked by α-axial bonds, rather than the β-equatorial bonds common in other proteoglycans. The result is a more looping chain (Fig. 10–10*C*).

Heparin inhibits clotting of blood (p. 330), and commercial preparations are widely used for this purpose. It is sometimes used in the management of patients with myocardial infarctions and strokes to aid in prevention of further clot formation. It is also used during surgery on the heart or blood vessels. Lower doses of heparin are being increasingly used as a prophylactic measure prior to surgery in order to prevent pulmonary embolism (dissemination of clots to the lung). Heparin is also used for the less dramatic purpose of preventing obstruction of an indwelling catheter.

Hyaluronate

Hyaluronic acid, occurring as a polyanion, probably qualifies as a proteoglycan by containing a small amount of protein, but the bulk of the molecule is made of a long heteropolysaccharide chain containing some 5,000 carbohydrate residues, with a molecular weight of 1 to 3 million. The chain has no sulfate groups and contains alternating residues of D-glucuronate and N-acetyl-D-glucosamine:

$$[(1\rightarrow3)\text{-}\beta\text{-GlcNAc-}(1\rightarrow4)\text{-}\beta\text{-GlcUA}]_n$$

The long chain of hyaluronate gives it even greater conformational mobility than is seen in the chondroitin and other sulfates. With negative charges only on alternating residues, there is somewhat less charge repulsion. Hyaluronate molecules therefore sweep through very large domains, with approximately 1,000 times the volume of the anhydrous molecule. That is, one gram of hyaluronate in excess water will exclude other large molecules from about one liter of space.

Hyaluronate is an effective lubricant, as well as being a resilient buffer against mechanical damage. It is therefore an important constituent of the synovial fluid in joints, and it also is a major component of the vitreous humor of the eye, of arterial walls, of umbilical cords, and a variety of other connective tissues.

THE GLYCOPROTEINS

The glycoproteins are proteins that contain attached carbohydrates, but they differ from the proteoglycans in that the carbohydrates are not polymers of repeating units. They are in shorter chains, often highly branched. Glycoproteins may contain only a few carbohydrate chains or so many that they amount to more than half the mass of the molecule.

The attachment of carbohydrate residues to polypeptide chains does not convey a particular type of function. Glycoproteins include enzymes, hormones, antibodies, structural proteins, and so on. Despite the diversity of function, most of the glycoproteins are found in a few kinds of locations: in the extracellular fluids, the lysosomes within cells, and in the plasma membrane surrounding cells.

The purpose of the carbohydrate is not always clear. It gives a lubricating property to the carbohydrate-rich proteins of mucous secretions and protects some circulating proteins from removal by the liver. Distinctive carbohydrates on the surface of cells act as receptors for binding specific compounds. Beyond this, we must lean mainly on conjecture. Since many secreted proteins have attached carbohydrate, it has seemed reasonable to suppose that the carbohydrate in some way aids the secretory process, although many proteins are secreted very well without attached carbohydrates. In some cases, the presence of the carbohydrate groups has been shown to be necessary for biological function; they maintain proper conformation, or bind the protein at appropriate sites.

We shall discuss the diverse functions later. Our purpose now is to describe some common elements in the structure of the attached carbohydrates.

Types of Carbohydrate Linkage

The carbohydrate groups in mammalian glycoproteins are linked to the hydroxyl groups of serine and threonine residues, or to the amide nitrogen of asparagine residues. These two types of linkages are associated with quite different kinds of carbohydrates.

Serine/threonine-linked carbohydrates are frequently simple. We have already noted that type I collagens contain a few sugar residues; typically, there may be one or two β-galactosyl groups and one or two α-glucosyl-(1→2)-β-galactosyl groups attached to hydroxylysyl groups in each 1,000 amino acid residues:

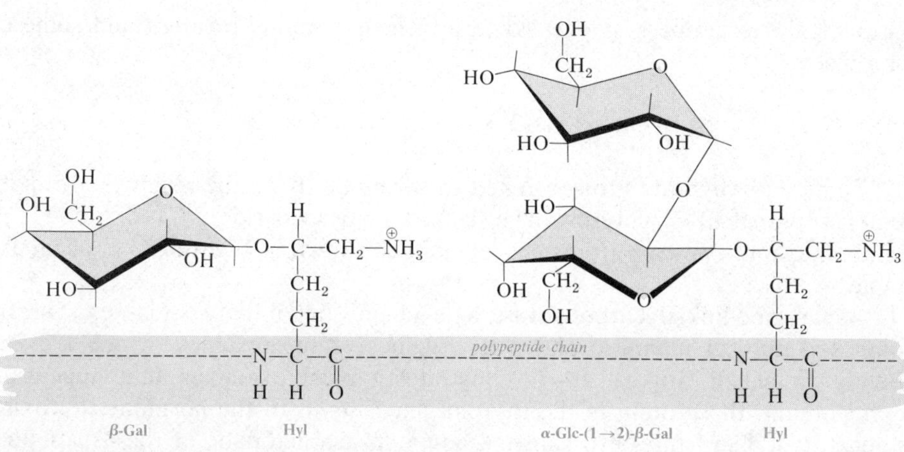

polypeptide chain

β-Gal Hyl α-Glc-(1→2)-β-Gal Hyl

At the other extreme, the submaxillary glycoprotein that has been characterized from sheep contains over 800 disaccharide units — one every six amino acid residues, on the average — that amount to ~40 per cent of its mass. However, nearly all of these units are made of N-acetylneuraminate and

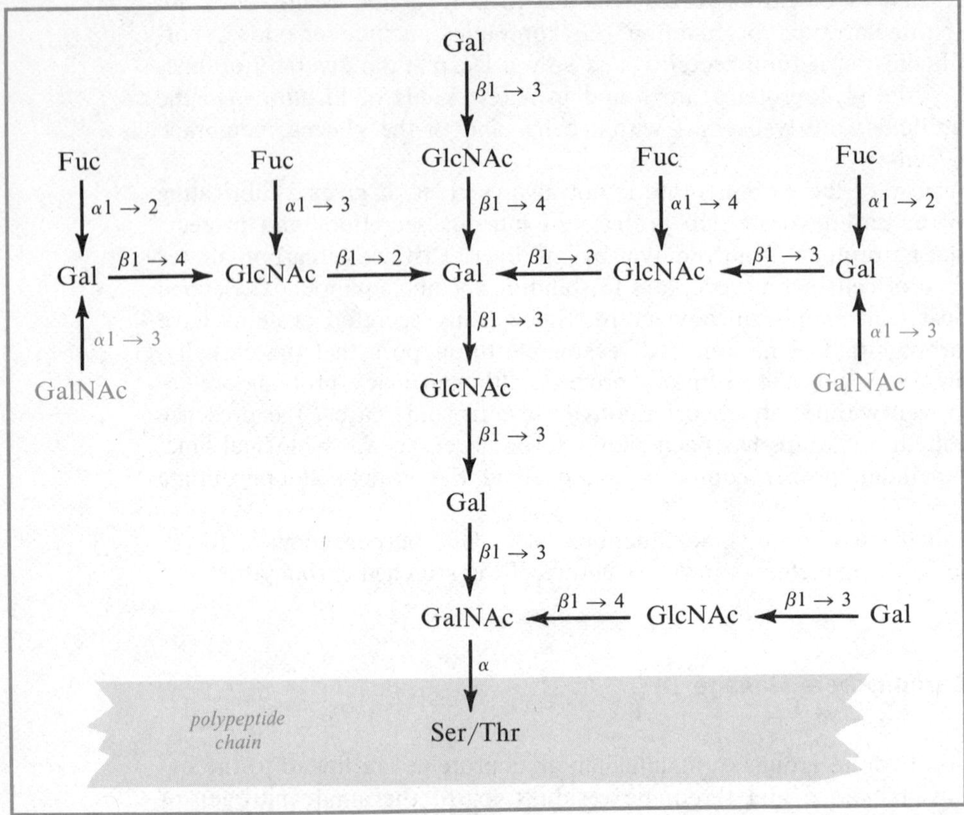

FIGURE 10–11 **Red blood cells contain oligosaccharides that are responsible for blood-group specificity. The carbohydrate responsible for type A specificity, shown here, can be linked to seryl or threonyl groups on a polypeptide chain. Individuals with type B blood differ in having galactose residues in place of two N-acetylgalactosamine residues (*blue*). The structure is shown here with the maximum number of side-chain fucose residues, all of which do not occur in every individual.**

N-acetylgalactosamine residues, some attached to serine residues and some to threonine:

$$\alpha\text{-NeuNAc-}(2{\rightarrow}6)\text{-}\alpha\text{-GalNAc-Ser/Thr}$$

Some carbohydrate groups linked to serine or threonine residues are quite complex branched structures. The hetero-oligosaccharide in red blood cell membranes that conveys group A specificity is shown in Figure 10–11 as an example.

Asparagine-linked carbohydrate side chains fall into two classes. One is composed only of mannose and N-acetylglucosamine residues, which are frequently branched. Figure 10–12 illustrates typical examples that appear in thyroglobulin, the protein in the thyroid gland in which the hormone thyroxine is generated. The figure also shows a common characteristic of the asparagine-linked glycoproteins — they frequently contain sheared versions of the characteristic side chains that lack one or more carbohydrate residues.

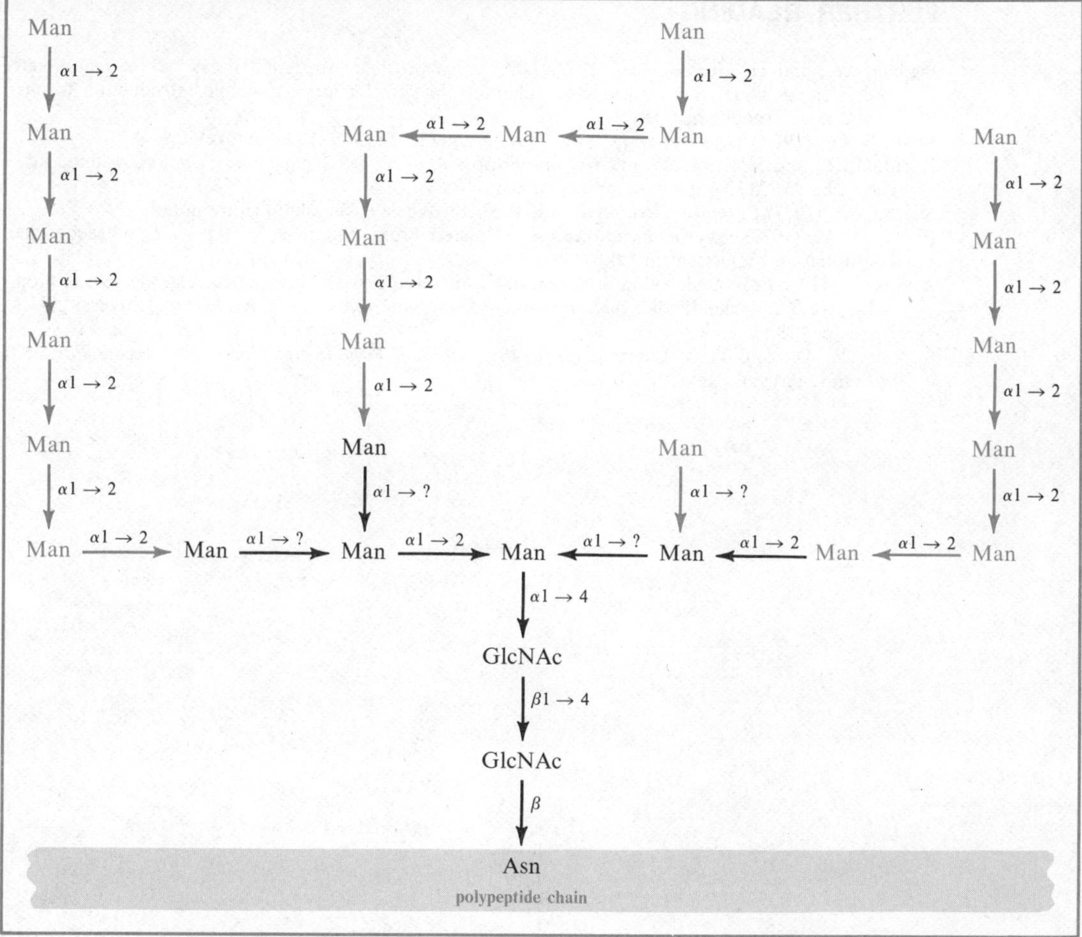

FIGURE 10-12 **Some of the carbohydrate groups found in thyroglobulin, the protein of the thyroid gland from which thyroid hormones are released, are made of branched mannose polymers linked to an asparagine residue of the polypeptide chain through two N-acetylglucosamine residues. Mannose is commonly present at branch points in the carbohydrates of glycoproteins. The residues shown in black always occur; varying numbers of those shown in blue are missing in some chains.**

The second class of asparagine-linked polysaccharides contains a wider variety of carbohydrate residues in branched structures. Many of the complex side chains are capped by residues of either N-acetyl-D-neuraminate or L-fucose. The function of fucose is not known, but the presence of N-acetylneuraminate apparently prevents recognition by sites on the membrane of liver parenchymal cells. If the sialate residue is removed, which sometimes happens as a protein ages in the circulation, the liver takes it up for destruction.

Some glycoproteins contain more than one of these types of carbohydrate branches. Human thyroglobulin contains all three: serine-linked, asparagine-linked and mannose-rich, asparagine-linked and complex composition.

FURTHER READING

Pigman, W., and D. Horton, eds.:(1972) *The Carbohydrates.* Academic Press. A yet incomplete multi-volume work. The introductory chapters in vol. 1A are a readable introduction to carbohydrate stereochemistry.

Spiro, R. G.: (1973) *Glycoproteins.* Adv. Protein Chem., 27: 349. Excellent review.

Kornfeld, R., and S. Kornfield: (1976) *Comparative Aspects of Glycoprotein Structure*. Annu. Rev. Biochem, *45*: 217. A good snappy summary.

Sharon, N.: (1975) *Complex Carbohydrates.* Addison-Wesley. Detailed lecture notes.

Rees., D. A.: (1977) *Polysaccharide Shapes*. Halsted Press. Summary of types of polysaccharide conformation by pioneer in field.

Atkins, E.: (1975) Molecular Conformation of Connective Tissue Mucopolysaccharides. *In* Holton, J. B., and J. T. Ireland, eds.: *Inborn Errors of Skin, Hair, and Connective Tissue.* University Park Press, p. 119.

Comper, W. D., and T. C. Laurent: (1978) *Physiological Function of Connective Tissue Polysaccharides*. Physiol. Rev., *57*: 313.

11 LIPIDS AND MEMBRANES

The lipids are the waxy, greasy, and oily compounds of the body. They repel water, and this hydrophobic nature is used as a tool for a variety of purposes. Some lipids — the fats — are important fuels, and their coalescence into nearly anhydrous droplets creates a reserve of potential energy that is a much lighter burden to carry than an equal reserve of waterlogged carbohydrate. Still other lipids are major structural components, and their ability to associate so as to exclude water and other polar compounds makes complex organisms possible through the formation of membranes.

Membranes separate cells within tissues and organelles within cells, creating compartments with separate chemistries so as to permit distinct organization and regulation; each compartment becomes an individual part of a more complex whole. Biochemical study of such membranes involved only a few people until recent years, but the fundamental principle has been known for many decades: Plasma membranes and intracellular membranes are made with a hydrophobic core and polar surfaces.*

NATURE OF LIPIDS

Fatty Acids

The fundamental building blocks of stored fats and many structural lipids are straight chain aliphatic carboxylic acids, which may be saturated or may contain one or more double bonds:

$$H_3C{-}(CH_2)_n{-}\overset{\overset{\displaystyle O}{\|}}{C}{-}OH \qquad H_3C{-}(CH_2)_m{-}\left(CH_2{-}\overset{\overset{\displaystyle H}{|}}{C}{=}\overset{\overset{\displaystyle H}{|}}{C}\right)_x{-}(CH_2)_n{-}\overset{\overset{\displaystyle O}{\|}}{C}{-}OH$$

saturated fatty acid — *cis*-unsaturated fatty acid

When double bonds are present, they are nearly always in the *cis* configuration, and if there is more than one, they are spaced at three-carbon intervals.

Nomenclature. Both the systematic and trivial names of fatty acids of metabolic or structural importance are listed in Table 11–1. The predominant compo-

*In discussing membranes, we are concerned with those bounding cells and organelles and not with structural sheets, such as basement membranes, or more complex serous membranes, such as the peritoneum.

TABLE 11–1 ALKYL CARBOXYLIC ACIDS

Components	Trivial Name	Systematic Name
1	formic	
2:0	acetic	
3:0	propionic	
4:0	butyric	
5:0	valeric	pentanoic
6:0	caproic	hexanoic
8:0	caprylic	octanoic
10:0	capric	decanoic
12:0	lauric	dodecanoic
14:0	myristic	tetradecanoic
16:0	**palmitic**	hexadecanoic
16:1(9)	**palmitoleic**	*cis*-9-hexadecenoic
18:0	**stearic**	octadecanoic
18:1(9)	**oleic**	*cis*-9-octadecenoic
18:1(11)	**vaccenic**	*cis*-11-octadecenoic
18:2(9,12)	**linoleic**	*all cis*-9,12-octadecadienoic
18:3(9,12,15)	**linolenic**	*all cis*-9,12,15-octadecatrienoic
20:4(5,8,11,14)	**arachidonic**	*all cis*-5,18,11,14-eicosatetraenoic
24:0	**lignoceric**	tetracosanoic

nents of structural lipids contain 16 or more carbon atoms and are shown in bold-face type. Trivial names are frequently used for these and for the shortest-chain compounds.

Since lipids usually contain an assortment of different fatty acids, shorthand designations have been devised for indicating composition in terms of the number of carbon atoms and the number of double bonds. Thus 16:0 is **palmitic acid,** with 16 carbon atoms and no double bonds; it is the most abundant saturated fatty acid. 18:1 designates the most abundant unsaturated fatty acids (**oleic** and **vaccenic** acids), with 18 carbon atoms and one double bond. The position of the double bonds is indicated in parenthesis. The important unsaturated fatty acids are shown in Figure 11–1.

Fatty Acids Are Amphipathic. Fatty acids have both hydrophobic and hydrophilic regions. This duality of response to water is the key to the function of biological lipids in general. The hydrocarbon tails try to agglomerate so as to expose a minimum surface, while the polar carboxyl groups attempt to maintain contact with the watery world around them. The length of the hydrocarbon chains determines the dominant behavior; hydrophobic interactions are so strong in palmitic acid (16:0) that it is only soluble to the extent of eight parts per million in an acidic solution at 30° C.

Micelles. If palmitic or another long-chain fatty acid is exposed to a neutral solution, it ionizes to form anions, or soaps, and the resultant negatively charged carboxylate group has a much stronger tendency to associate with water. The sodium or potassium salts of long-chain fatty acids are many-fold more soluble than the undissociated acids, especially at elevated temperatures. (A true solution of sodium palmitate can reach one millimolar at 60° C, whereas a solution of palmitic acid is saturated at less than 5 micromolar.)

The tendency of the charged head groups to remain in contact with water is so great that the soaps do not precipitate when their solubility is exceeded; instead they form small clusters, or micelles. The simplest form of micelle is a sphere in which the hydrocarbon tails are grouped in the center, and the polar head groups are associated with water and counterions on the surface (Fig. 11–2). Other shapes

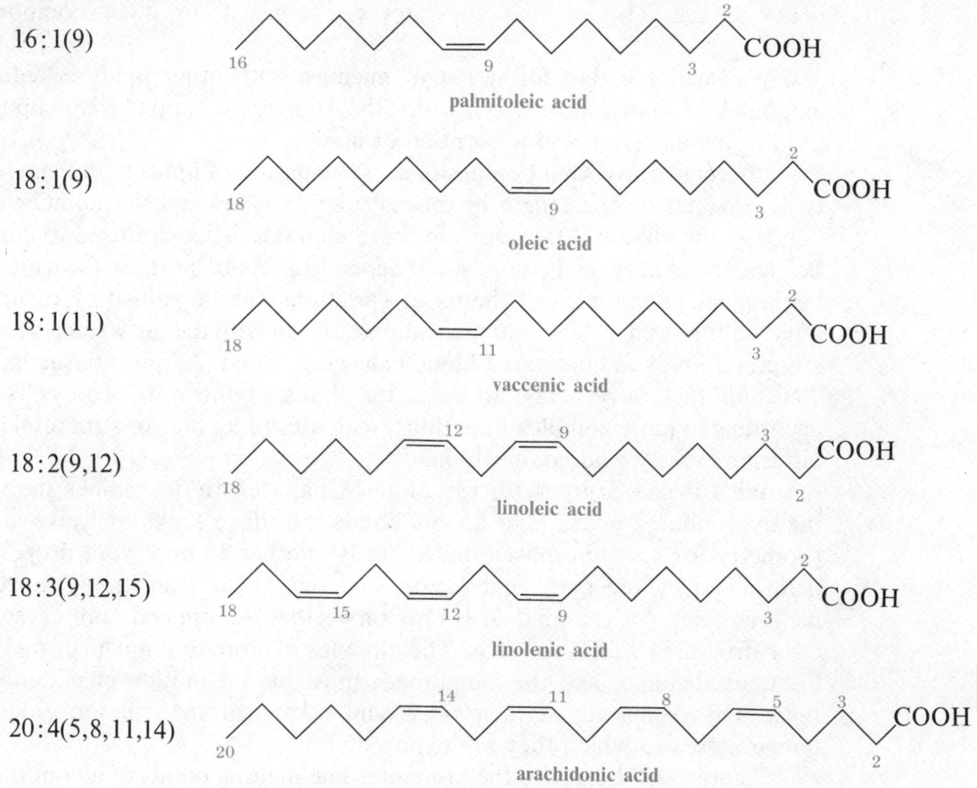

16:1(9) — palmitoleic acid (16, 9, 3, 2, COOH)

18:1(9) — oleic acid (18, 9, 3, 2, COOH)

18:1(11) — vaccenic acid (18, 11, 3, 2, COOH)

18:2(9,12) — linoleic acid (18, 12, 9, 3, 2, COOH)

18:3(9,12,15) — linolenic acid (18, 15, 12, 9, 3, 2, COOH)

20:4(5,8,11,14) — arachidonic acid (20, 14, 11, 8, 5, 3, 2, COOH)

FIGURE 11–1 **Some common unsaturated fatty acids. The configuration is *cis* around each double bond, and multiple double bonds are spaced at three-carbon intervals.**

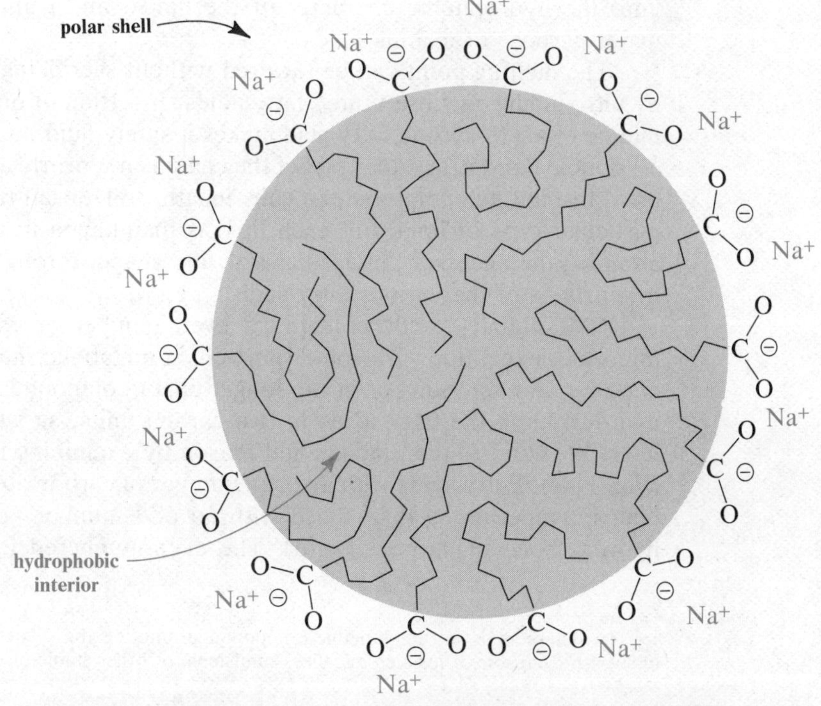

FIGURE 11–2 **The anions of long-chain acids, the soaps, form micelles in water.**

occur as the ratio of soap to water is increased, or other components are added.

We shall see that formation of micelles with other lipids is aided by the amphipathic character of fatty anions; their detergent action* is an important tool used in the digestion and absorption of lipids.

Effects of Fatty Acid Composition. The nature of lipids built from fatty acids is determined by the length of the fatty acid chains and the number of double bonds in the chains. Alterations in these characteristics are used to control lipid behavior, and fatty acid composition depends upon the purpose for which a lipid is being made. Some idea of the usual variations can be gained by comparing the fatty acids stored as fats in human adipose tissue with the fatty acids used to make structural lipids in human red blood cells (Fig. 11–3). Adipose tissue mainly uses 16:0 and 18:1 fatty acids, whereas the lipids of the red blood cells are built according to more complex directions, with different kinds of structural lipids also differing in fatty acid composition.

What is gained by these variations? Chain length determines the volume of the hydrophobic phase, and double bonds introduce kinks that give a favorable geometry for some purposes, but there is another all-important property that is determined by the fatty acid recipe, and this is the transition temperature, or melting point, for the lipid. It is imperative that the lipids do not crystallize into rigid structures within the cells. The globules of stored fat must be freely accessible upon demand, and the membranes must permit motion of material through them. Organisms adjust their fatty acid composition to fit the environmental temperatures to which they are exposed.

Figure 11–4 illustrates the principle. The melting points of all but the shortest fatty acids increase in a regular way with increasing chain length. (The shortest have anomalous behavior because their crystal structures are heavily dependent on hydrogen bond formation.) One way to maintain a liquid state would therefore be to use relatively short-chain fatty acids in building lipids. This, however, would limit the hydrophobic character of the lipids, and gaining hydrophobicity is a major purpose of making lipids.

The melting point can be lowered without sacrificing hydrophobic character by introducing *cis*-unsaturated fatty acids. Insertion of one double bond near the middle of an 18-carbon fatty acid makes it safely fluid. (A *trans* double bond or a *cis* double bond near either end of the chain is not nearly as effective.) The effects on transition temperature of chain length and unsaturation depend upon the particular type of lipid, but each lipid is maintained in a fluid state, while also attaining the necessary molecular size and shape, through insertion of particular proportions of the various fatty acids.

Natural fatty acids contain an even number of carbon atoms. The only important exceptions are some short-chain metabolic intermediates. The preference for an even number in the longer-chain compounds is satisfied by forming and degrading the fatty acids in two-carbon units, as we shall discuss in detail later. We can see the fundamental reason by examining the melting points again (Fig. 11–4). Fatty acids with an even number of carbon atoms consistently melt at higher temperatures than those with an odd number. This indicates that their hydrocarbon chains pack better. The even-numbered fatty acids are therefore

*A detergent is an amphipathic compound promoting the dispersion in water of substances otherwise incapable of forming micelles, emulsions, or other stable mixtures.

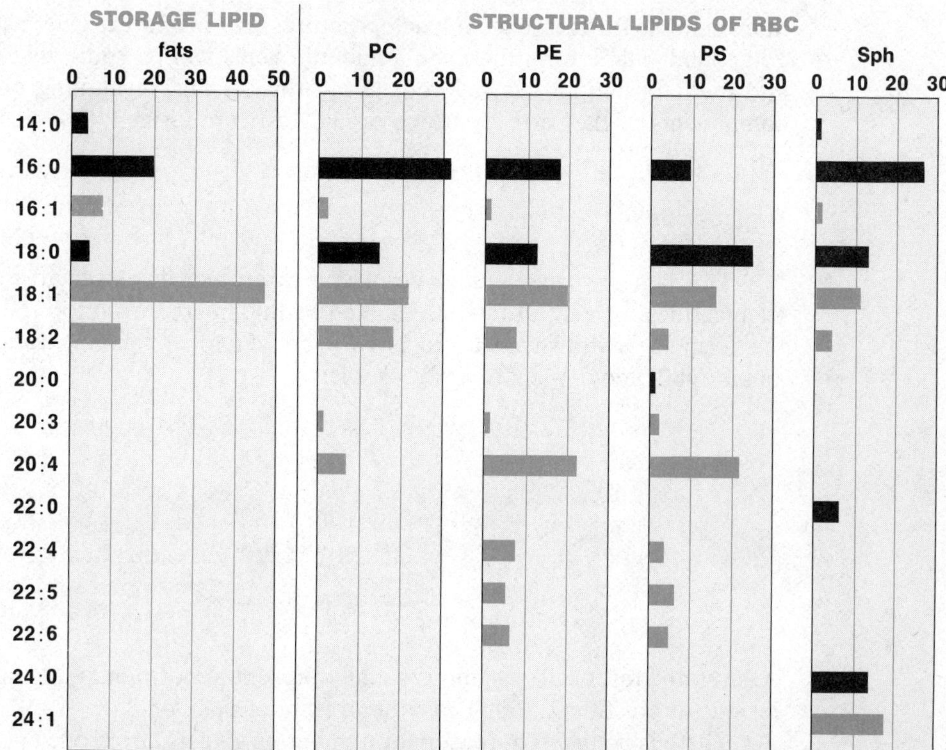

FIGURE 11–3 **Profiles of fatty acid composition of lipids. Saturated fatty acids are shown in black, unsaturated in blue. PC, PE, PS are phosphatidylcholine, phosphatidylethanolamine, and phosphatidylserine, respectively. Sph is sphingomyelin.**

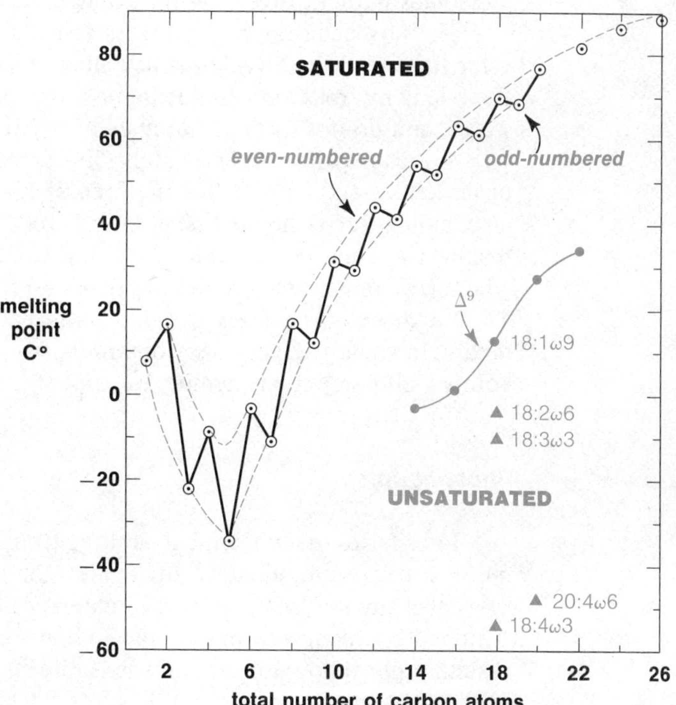

Figure 11–4

The melting points of fatty acids.

more suitable for creating compact structures. Put in another way, associations of compounds made from the even-numbered acids will be somewhat more stable, and their formation was favored during the original reshuffling of the organic components of the earth by which organisms were formed.

Triglycerides

The fats are esters of glycerol and three molecules of fatty acids. These triglycerides, or neutral fats, are named and numbered from top to bottom when the formula is drawn so as to have a conventional L-configuration. Thus, 1-palmitoyl-2-oleoyl-3-stearoyl glycerol is:

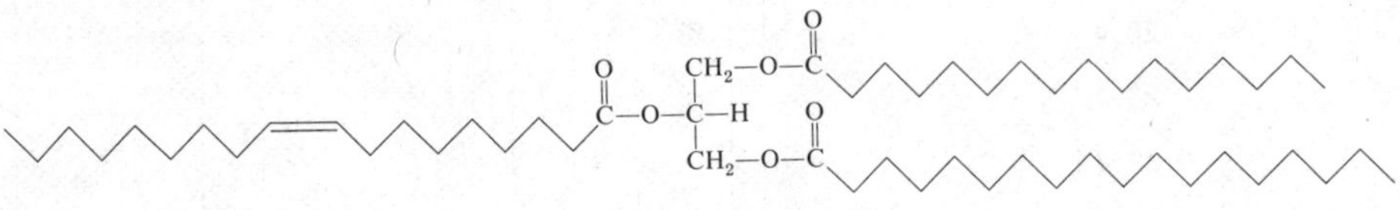

1-palmitoyl-2-oleoyl-3-stearoyl glycerol

A saturated fatty acid residue is usually present at position 1, and an unsaturated residue at position 2; either may be present at position 3.

There is no inherent reason for numbering glycerol from one end or the other, according to systematic organic nomenclature. The rule described above is a convention, and its application is sometimes indicated by the letters sn, for stereospecific numbering, as in sn-1-palmitoyl-sn-2-oleoyl, etc., but this is really not necessary if the basis for the convention is understood. Describing the isomer as L removes all doubt. If you think there is no structural difference between the two ends of glycerol, mentally try to superimpose opposite ends of two identical molecules without reversing the configuration of the central carbon atom.

The fats occur as oily droplets in cells. The polar contribution of the three ester linkages is relatively small compared to the hydrophobic character of the three long hydrocarbon chains in most fats, so the fats are only slightly soluble in water, and do not form stable micelles by themselves.

A reason why approximately 50 per cent of the residues in most fats are unsaturated, and why 16:0 is the prevalent saturated acid, may be deduced from an examination of the melting points (Table 11–2). A triglyceride made of two 16:0 residues and one 18:1 residue melts at 35 or 36° C, barely below body temperature, whereas a triglyceride containing two 18:0 residues and one 18:1 residue melts at 42° C and would be solid at body temperature. A safety factor is provided by mixing in some triglycerides containing two 18:1 fatty acids, which have melting points well below body temperature.

Phospholipids

In order to make useful structures from lipids, these structures must include more strongly amphipathic molecules than the triglycerides so that they can present a greater surface to the aqueous phase. Increased affinity for water is obtained by incorporating charged phosphate ester groups into the lipids. The resultant phospholipids are indispensable components of membranes and are also used as detergents to coat fat droplets for transport within the body.

TABLE 11–2 The melting points of triglycerides with various fatty acids esterified with the three carbons of glycerol. These values are for synthetic compounds, presumably DL mixtures of the asymmetric forms. Blue shading indicates unsaturated residues.

Fatty Acid Residue			
	Position		
1	*2*	*3*	**Melting Point**
6:0	6:0	6:0	−25
8:0	8:0	8:0	10
10:0	10:0	10:0	31.5
12:0	12:0	12:0	44
14:0	14:0	14:0	55
16:0	16:0	16:0	66
16:0	18:1(9)	16:0	36
18:1(9)	16:0	16:0	35
16:0	18:1(9)	18:1(9)	18
18:1(9)	16:0	18:1(9)	18
18:0	18:1(9)	18:0	42
18:0	18:1(9)	18:1(9)	23.5
	*(18:2(9,12))(16:0)$_2$		27
	*(18:3(9,12,15))$_2$(16:0)		−3

*Position not specified.

Phosphatidyl Compounds. Many of the phospholipids are derivatives of phosphatidic acid, which contains glycerol that is esterified with fatty acids on two carbon atoms and with phosphoric acid on the third:

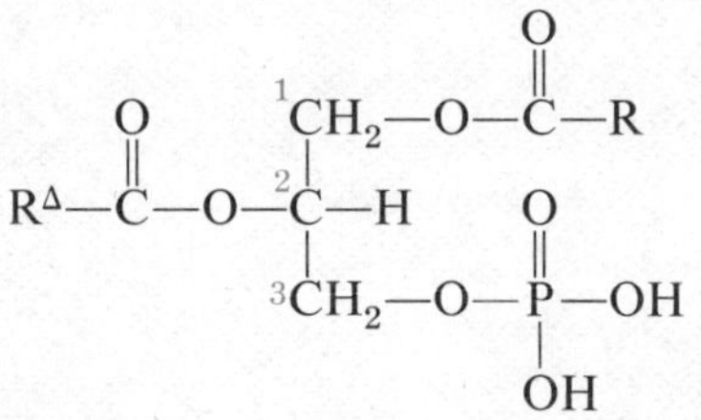

```
                                O
                                ‖
            O        ¹CH₂—O—C—R
            ‖          ²|
   Rᐞ—C—O—C—H        O
                        |          ‖
                       ³CH₂—O—P—OH
                                  |
                                 OH
```

phosphatidic acid
(R$^{\Delta}$ = unsaturated chain)

One example is shown here, but the fatty acid composition varies. However, most phosphatidyl compounds have a saturated fatty acid linked to C-1 and an unsaturated fatty acid linked to C-2, as shown.

The phosphatidic acids, like other phospholipids, ionize at physiological H^+ concentrations; they carry between one and two positive charges in neutral solution, and we shall refer to them as the phosphatidates.

Phosphatidylcholine, phosphatidylethanolamine, and **phosphatidylserine** are structurally related compounds (Fig. 11–5) that carry both positive and negative charges. We speak of them as if they were single compounds, but there are many phosphatidylcholines and related compounds that vary in fatty acid composition. Table 11–3 shows the fatty acid residues found at each position of the phosphatidylcholines in human erythrocytes.

These structures with highly charged heads and hydrophobic tails have a strong detergent action. The phosphatidylcholines, also known as lecithins,* are frequently incorporated into processed foods to keep the lipids dispersed.

*Phosphatidylethanolamines were formerly named cephalins.

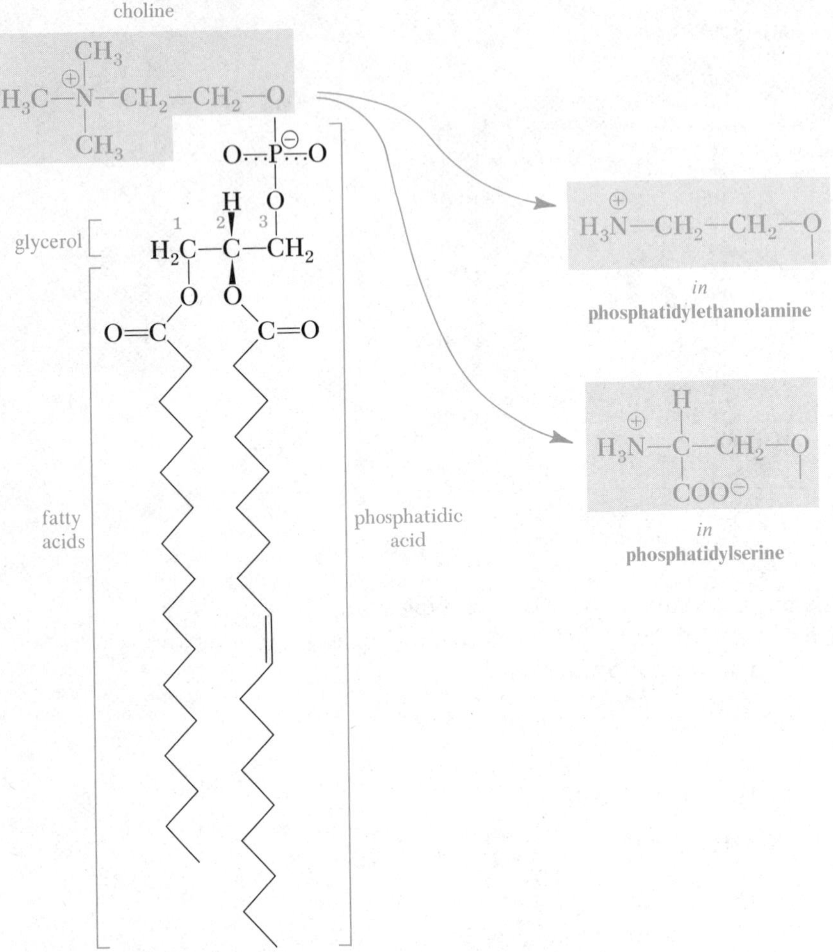

FIGURE 11–5 **Major phospholipids in animals include phosphatidyl derivatives of choline, ethanolamine, or serine (*blue shading*). These phospholipids contain both positive and negative charges, and therefore have strongly hydrophilic heads.**

TABLE 11–3 FATTY ACID COMPOSITION OF PHOSPHATIDYLCHOLINES IN HUMAN RBC*
(listed as mole percent)

Component	16:0	16:1	18:0	18:1	18:2	18:3	20:3	20:4	20:5	22:6
Position 1	61		24	10	0.6					
Position 2	9	1.8		26	35	1.0	4	12	0.5	1.6

*Means of values from several laboratories.

Phosphatidyl inositols are esters of a cyclic derivative of glucose, with a high proportion of 1-stearoyl-2-arachidonyl species (18:0) (20:4):

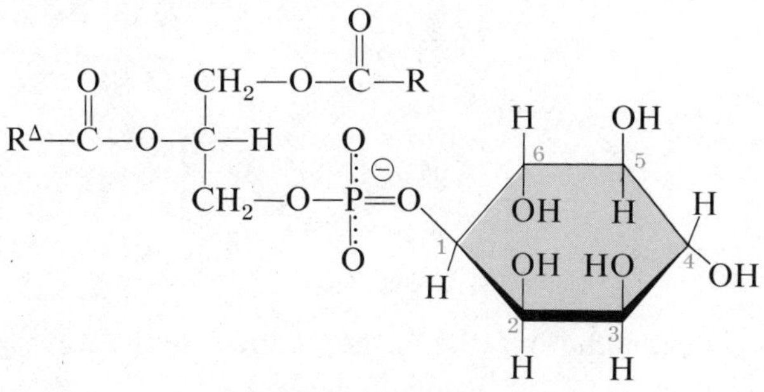

phosphatidylinositol

Phosphatidylglycerol and diphosphatidylglycerols (cardiolipins) are esters of phosphatidates with an additional molecule of glycerol:

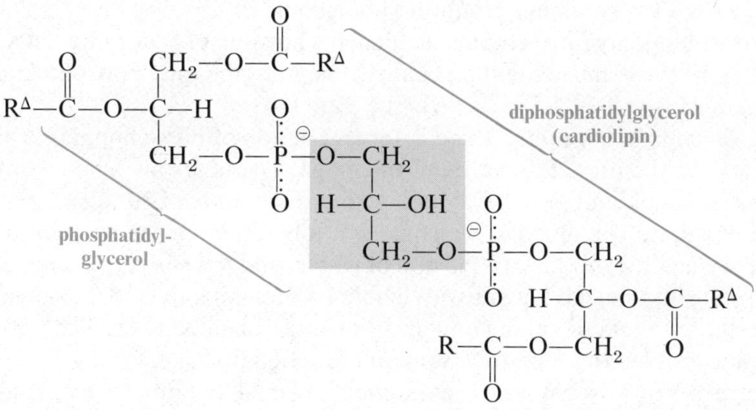

phosphatidyl-glycerol

diphosphatidylglycerol (cardiolipin)

These compounds are rich in polyunsaturated residues; linoleic acid (18:2(9,12)) accounts for 70 to 80 per cent of the fatty acid residues in visceral diphosphatidyl glycerol.

Lysophosphatidate derivatives are compounds lacking a fatty acid residue on C-2 of glycerol:

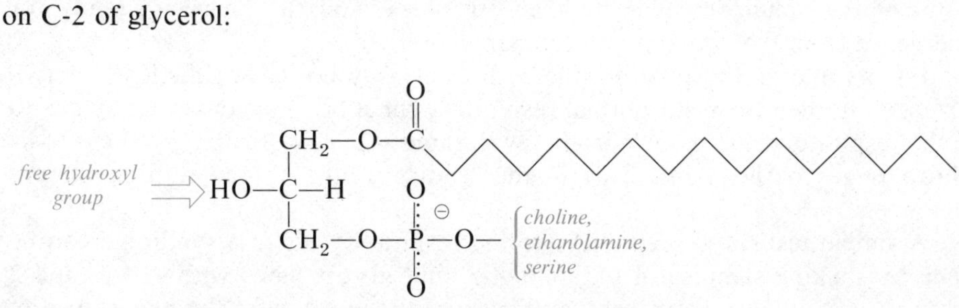

free hydroxyl group

choline, ethanolamine, serine

lysophosphatidyl compound

Thus, there are lysophosphatides, lysophosphatidylcholines, and so on. (The name was coined because the lysophosphatidylcholines, or lysolecithins, are especially effective detergents that cause the lysis of red blood cells.) Phosphatidal compounds (phosphatidalcholine, etc.), also known as plasmalogens, are relatives of the phosphatidyl compounds in which the fatty acid residue on C-1 is replaced by a long-chain aldehyde:

```
                                   H  H
                                   |  |
                 O      CH2—O—C=C/\/\/\/\--/\/\/\
                 ||     |
 /\/\/--/\/\/\C—O—C—H        O
                        |           :     { choline,
                        CH2—O—P—O— { ethanolamine,
                                    :     { serine
                                    O
```

phosphatidal compound (plasmalogen)

The phosphatidal compounds may be further reduced to the corresponding alkoxy compounds.

Sphingomyelins are closely similar in shape to phosphatidylcholines (Fig. 11–6) but are of quite different chemical structure, with the polar backbone and one of the hydrocarbon tails contributed by a dihydroxy amine, sphingosine. Sphingosine, therefore, replaces glycerol and one of the fatty acids seen in the phosphatidyl lipids. The other hydrocarbon tail comes from a fatty acid bound by amide linkage to the amino group of sphingosine.

The sphingomyelins contain a high proportion of saturated fatty acid residues, some with unusually long chains. They are the only phospholipids known to contain 24:0 and 24:1 fatty acids (see Fig. 11–3).

Surfactants in the Lung. The detergent action of phospholipids is absolutely necessary for the normal function of the lungs. The lungs are a collection of huge numbers of small bubbles, the alveoli, generating a total of 80 to 100 m^2 of surface area in an adult. The alveolar walls are not inherently strong enough to maintain their shape against the surface tension of water, and the surface tension is reduced by secreting an unusual phosphatidylcholine containing two 16:0 chains, together with lesser amounts of sphingomyelins, into the lung chamber. The phospholipid surfactant enables the lung to assume its functional shape.

Since a fetus is bathed in amniotic fluid and obtains oxygen through the maternal circulation, it has no need for the surfactant until exposure to air at delivery. It forms little of the dipalmitoyl phosphatidylcholine before the 30th week of gestation, and analysis of the amniotic fluid shows a higher concentration of sphingomyelin than of the phosphatidylcholine. Synthesis of the phosphatidylcholine increases rapidly thereafter so that its concentration is twice that (or more) of the sphingomyelins by the 35th week, and the lungs are ready for emergence from the aqueous environment.

Infants that are born before the 35th week may not have sufficiently active surfactant in their lungs for normal respiratory function, and consequently develop the respiratory distress syndrome, with rapid, shallow breathing and cyanosis (bluish tinge to the skin). This is the leading cause of death in premature babies.

A simple test for susceptibility to the respiratory distress syndrome can be made by shaking samples of the amniotic fluid in test tubes with 1, 1.3, and 2 volumes of ethanol, respectively, and examining the tubes for the persistence of bubbles after several minutes. Dipalmitoyl phosphatidylcholine makes a very stable foam under these conditions, whereas those containing unsaturated fatty acids and the sphingomyelins do not. Fluid bathing mature lungs will generate persistent bubbles even with the highest concentration of alcohol; the presence of bubbles at only the lowest concentration of alcohol indicates a high risk that the infant will develop respiratory distress syndrome.

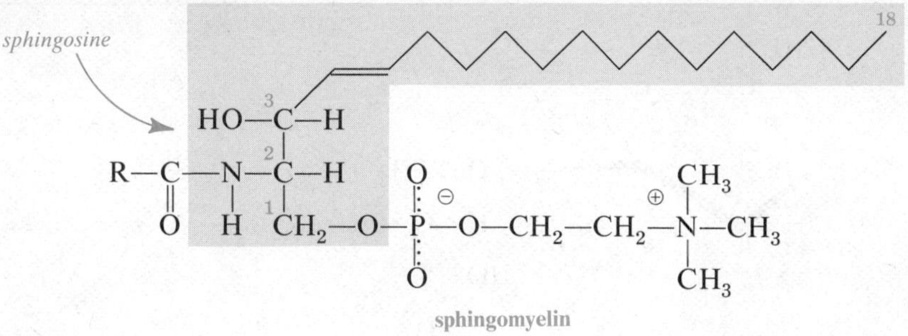

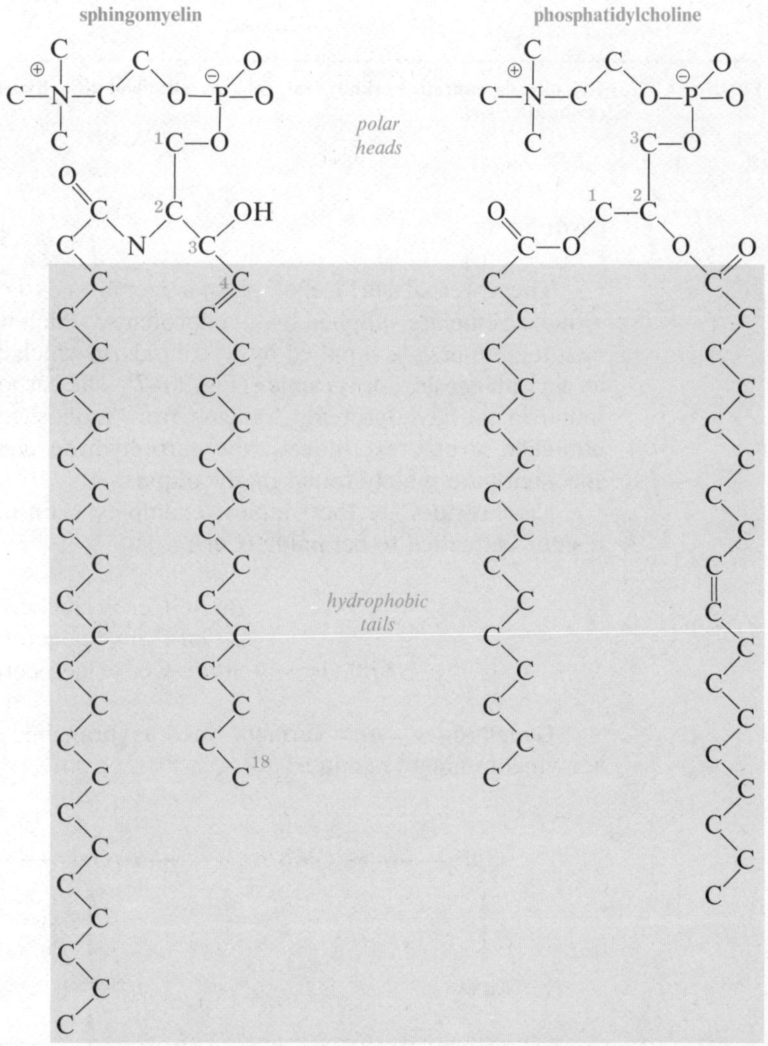

FIGURE 11–6 **Sphingomyelins are phospholipids containing sphingosine, a long-chain amino alcohol (*top, blue shading*), with a fatty acid attached to the amino group in amide linkage. The 1-hydroxyl group is attached to phosphate and choline by ester linkage. The conventional formula does not accurately convey the actual geometry of either sphingomyelin or of phosphatidylcholine, which closely resemble each other. A projection of the location of some of the atoms in atomic models is indicated in the bottom drawing. Sphingomyelin frequently contains a much longer (24-carbon) fatty acid than is common in the phosphatidyl compounds.**

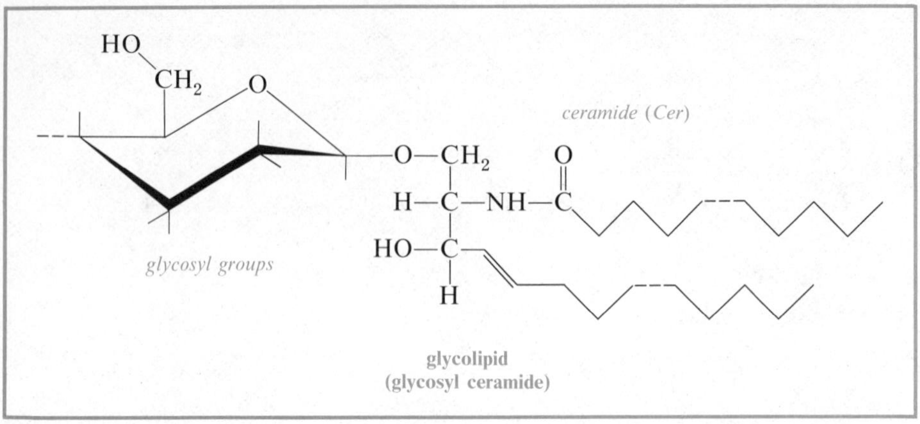

FIGURE 11–7 **Glycolipids contain carbohydrate chains attached to a hydroxyl group of a ceramide (Cer).**

Glycolipids

The external surface of plasma membranes is studded with carbohydrate groups. Some are supplied by glycoproteins, which we discussed in the preceding chapter. Others are supplied by glycolipids, in which carbohydrates are attached to an **acylsphingosine,** or **ceramide** (Fig. 11–7). The carbohydrates are similar to those found in the glycoproteins, ranging from monosaccharide residues to complex branched structures. Indeed, the carbohydrate chains conveying blood group specificity are mainly found in glycolipids.

Cerebrosides are the simplest examples, with one or more monosaccharide residues attached to ceramide (Cer):

$$\beta\text{-Glc-Cer} = \text{glucocerebroside},$$
$$\beta\text{-Gal-Cer} = \text{galactocerebroside},$$
$$\beta\text{-Gal-}(1 \rightarrow 4)\text{-}\beta\text{-Glc-Cer} = \text{lactocerebroside}.$$

Gangliosides are variants of a branched structure containing N-acetylneuraminate residues:

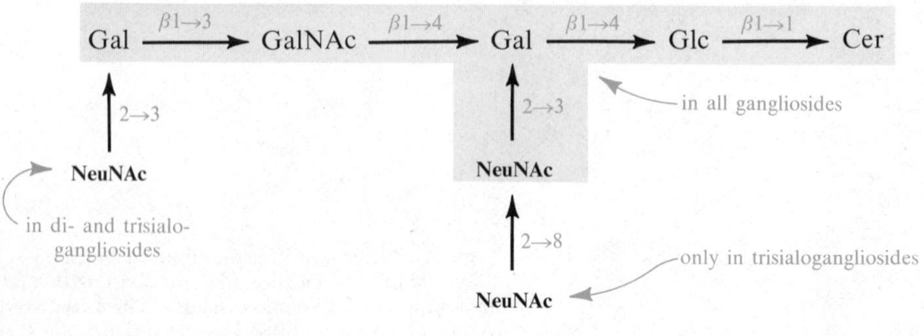

Cholesterol

Animal membranes contain a steroid alcohol, cholesterol (Fig. 11–8). We shall have much more to say about the steroids because they include many

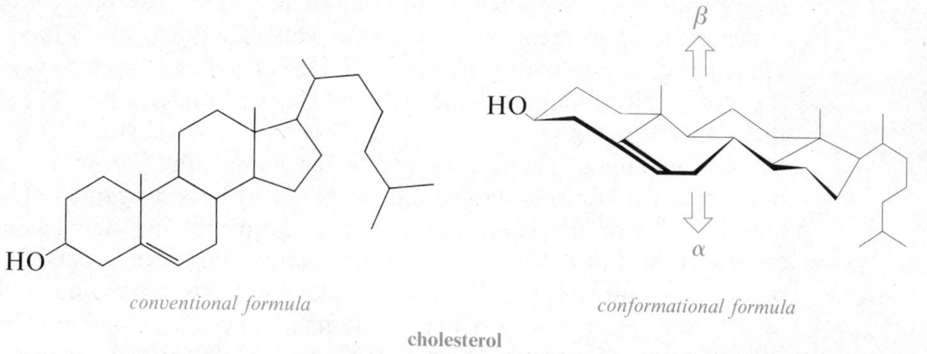

FIGURE 11–8 **Cholesterol is a steroid containing one hydroxyl group and a branched aliphatic side chain. The puckers of the nearly planar polycyclic steroid ring system are better appreciated in a conformational formula (*right*). Groups oriented toward the top as drawn are said to be in the β-configuration; those toward the bottom are in the α-configuration.**

hormones, but cholesterol is the progenitor of them all in the human body, as well as serving as an important structural component. The critical feature of cholesterol from a structural standpoint is the relatively rigid steroid ring, a nearly planar structure with the polar hydroxyl group at one end, two methyl groups on one side, and a freely movable hydrocarbon tail at the other end.

MEMBRANES

The essential ingredients of membranes are lipids (primarily phospholipids) to generate a hydrophobic core with polar faces, and proteins to carry out biological functions associated with the membrane, such as catalysis of reactions that may be facilitated by the hydrophobic phase, or selective transport from one compartment to another. **Membranes are asymmetric;** the two sides are different. Plasma membranes, which are the cell boundaries, have the carbohydrate chains of glycoproteins and glycolipids exposed on the exterior surface. Some proteins are firmly embedded in one surface; others pass through so as to be exposed on both surfaces; still others are less tightly bound to the surfaces. The membrane may also be attracted to intracellular structural proteins such as contractile filaments and microtubules.

The nature of membranes differs widely from organelle to organelle within a cell, from tissue to tissue within an organism, and from organism to organism. All membranes have the hydrophobic core and polar surface with embedded proteins, but the nature of the core, the proteins, and the surface is not the same.

The Lipid Core

The central structure of membranes is a double layer of phospholipid molecules, each molecule oriented with its hydrocarbon tails toward the central plane and its polar head toward the surface (Fig. 11–9). Large areas of such double layers cannot survive even small motions in water; they break up into patches, which close upon themselves to create water-filled balloons within the solution. Homogenization of phospholipids in dilute salt solutions creates vesicles, most of them of a uniform microscopic size that can be isolated as a clear suspension — these are the prototypes of membrane-bound biological structures.*

Composition. There is no pat recipe for making the lipid core of a membrane. The core must have a certain degree of structural rigidity but be sufficiently fluid so as to permit lateral movement of both the constituent lipids and those embedded proteins that are not fixed to some components outside of the membrane. (By contrast, exchange of lipid components from one side of the membrane to the other is very slow.) Fluidity is maintained in animal membranes by the incorporation of phospholipids containing 18:1 fatty acid residues, along with

*Such dispersed preparations were first created by A. D. Bangham in England. His vesicles, known in some circles as bangosomes, are walled by several bilayers of lipid like a hollowed-out onion. Ching-hsien Huang, now our colleague at Virginia, discovered how to make vesicles with only one enclosing bilayer, and these huangosomes are widely used in model studies of membrane phenomena. The generic term for such experimental particles is liposome.

TABLE 11–4 THE STRUCTURAL LIPIDS

Phospholipids	
phosphatidates	glycerol; saturated fatty acid on C-1, unsaturated fatty acid on C-2, phosphate on C-3
phosphatidylcholine, phosphatidylethanolamine, phosphatidylserine, phosphatidylinositol, di- or monophosphatidyl glycerol	choline, ethanolamine, serine, inositol, or glycerol attached in ester linkage to phosphate of phosphatidate
lysophosphatidyl compounds	same as above, but lacking unsaturated fatty acid group on C-2
plasmalogens, or phosphatidal compounds	same as above, except fatty aldehyde bound to C-1 of glycerol by vinyl ether linkage
alkyl phospholipids	same as above except fatty alcohol bound to C-1 of glycerol by ether linkage
sphingomyelin	sphingosine; saturated fatty acid on N atom; phosphate-choline on C-1
Glycolipids	
ceramide	sphingosine; saturated fatty acid on N atom; same as sphingomyelin without phosphate-choline
cerebrosides	mono- or oligosaccharide; ceramide
gangliosides	N-acetylneuraminate branch on oligosaccharide; ceramide
fucose-containing	resemble blood-group glycoprotein oligosaccharide; ceramide
Cholesterol	steroid ring with one hydroxyl group, two methyl group branches, and hydrocarbon tail

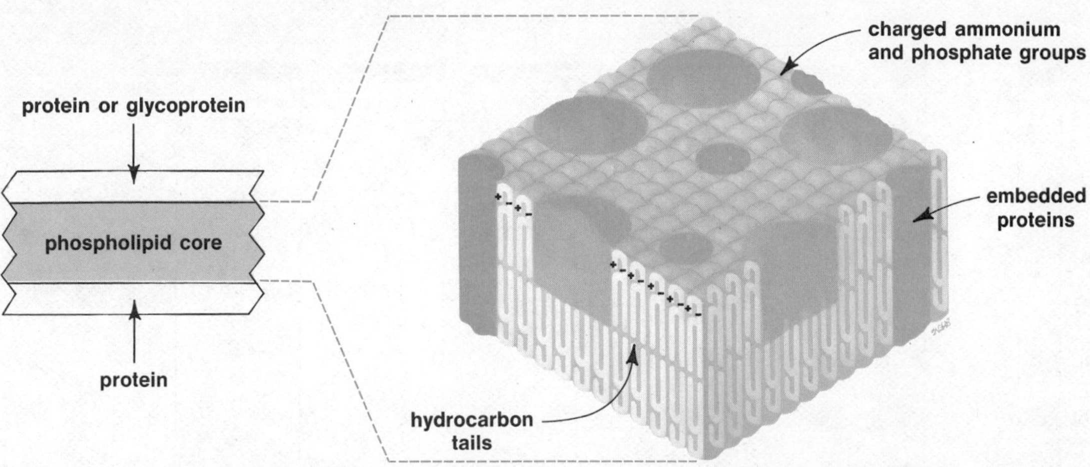

FIGURE 11–9 Generalized structure of membranes, with a phospholipid core and polar faces. Some of the embedded protein molecules are exposed on both sides of the core, others only on one side. The drawing is highly schematic; the actual shapes of the proteins are not known.

some more unsaturated residues. The lipid core need not be homogeneous, and it behaves as if it had localized fluid patches ("liquid crystal") embedded in a more stable gel-like phase.

The core must also conform to, and perhaps help determine, the curvature of the membrane surface, which may vary from the relatively broad sweep of a large cell envelope to the tight reversals of direction found in the inner membrane of mitochondria. The particular requirements may be met by alterations in the relative proportions of different lipid components (Fig. 11–10) or by alterations in the fatty acid composition of particular kinds of lipids.

Some trends can be noted in the composition of membranes in mammalian cells. Glycolipids are found only on the plasma membrane. The content of sphingomyelin tends to decrease as one goes from the plasma membrane through the interior of the cell to the nucleus, as does the content of cholesterol. The content of diphosphatidylglycerol (cardiolipin) tends to increase in internal membranes: the mitochondrial inner membrane is especially rich in the compounds, which may reflect its postulated origin from symbiotic bacteria (p. 108) because the bacteria also have high contents of these phospholipids in their membranes. However, the mitochondrial membrane also contains phosphatidylcholine, which is absent from bacterial membranes.

Functions of the Lipids. We have some clues as to the functions of some major components in mammalian plasma membranes, especially the red blood cell membrane, but we have to confess total ignorance of the roles of the tantalizing minor ingredients — the plasmalogens, lysophosphatidates, and so on, which are not likely to be present because of mere whim.

The relative proportions of phosphatidyl amines and sphingomyelin appear to depend upon the radius of curvature of the membrane. Phosphatidylcholine and sphingomyelin have larger head groups than do phosphatidylethanolamine and phosphatidylserine and therefore tend to occur on the outer surface of the lipid core where there is more area per residue (Fig. 11–11).

Cholesterol appears to fit between the hydrocarbon chains of the phosphatidyl lipids. The saturated chains in the phospholipid can occur as a straight

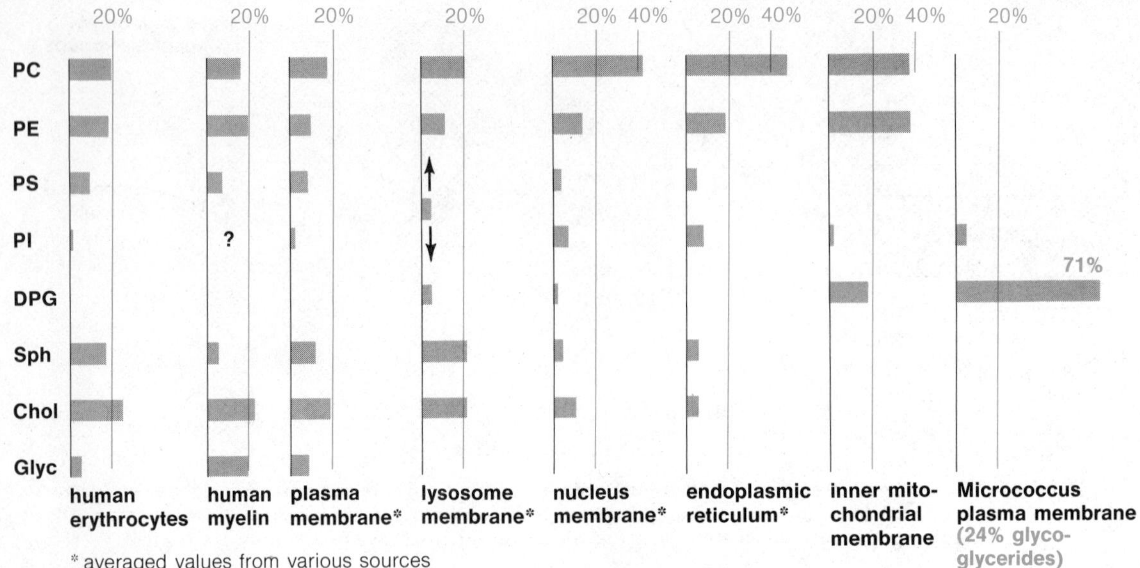

* averaged values from various sources
PC = phosphatidylcholine
PS = phosphatidylserine
DPG = diphosphatidylglycerol + phosphatidate
Chol = cholesterol + cholesteryl esters
PE = phosphatidylethanolamine
PI = phosphatidylinositol
Sph = sphingomyelin
Glyc = glycolipids

FIGURE 11–10 **The proportions of different phospholipids differ from one membrane to another. The inner structures of a cell tend to have membranes with a higher content of diphosphatidylglycerol and a lower content of sphingomyelin than does the plasma membrane.**

extended form, but the unsaturated chain must have a bend or a kink (Fig. 11–12), which makes room for cholesterol to slip in between molecules, with its relatively flat side against the saturated chain and the protruding methyl groups on its other side filling the gap created by the kink. The polar hydroxyl group is then in a position to form hydrogen bonds with other polar groups toward the surface of the lipids. Insertion as described here would explain why cholesterol stabilizes bilayer structures, and why the hydrocarbon chains appear locked in position toward the surface of the bilayer, while being more fluid in the center.

The lipids also interact with the hydrophobic surfaces of proteins in the membranes. Some kinds of lipids associate preferentially with the proteins, creating a region of different lipid composition in their neighborhood. This region may be less fluid than the bulk lipid phase, aiding in the sealing of the membrane around the protein. Indeed, there is experimental evidence that association with particular lipids is necessary for the biological activity of some membrane-bound proteins.

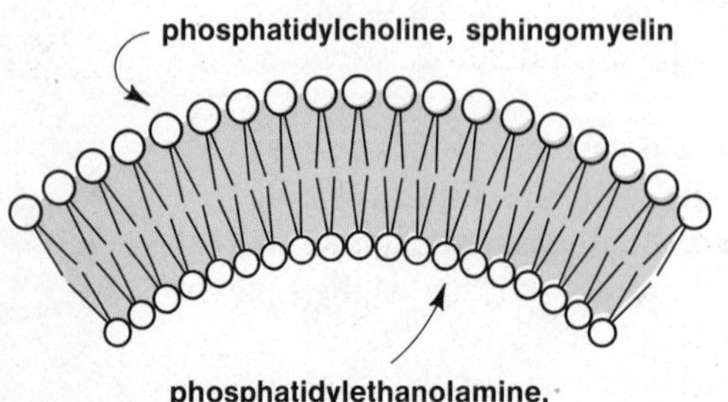

FIGURE 11–11

Since there is more area in the outer surface of a curved membrane than in the inner surface, phospholipid molecules with large head groups will tend to appear preferentially in the outer surface; these include phosphatidylcholine and sphingomyelin.

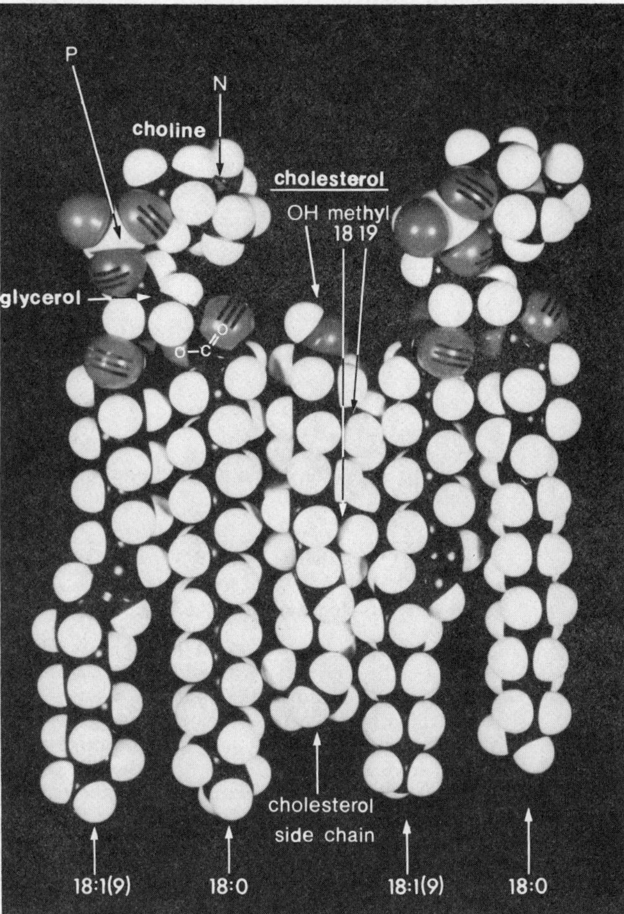

FIGURE 11–12

A molecular model showing cholesterol lying between two molecules of phosphatidylcholine. The top half of the cholesterol model is the ring system viewed edge-on; the bottom half is the hydrocarbon side chain. The polar hydroxyl group of cholesterol is at the nonpolar–polar interface, where it can form a relatively strong hydrogen bond with a carbonyl oxygen atom on the adjacent fatty acid residue. The two methyl groups on the right side of the steroid ring are accommodated in a snug fit by the kink in the unsaturated side chain of phosphatidylcholine; the saturated fatty acid side chain lies against the relatively flat face of the steroid to the left, which has no substituent groups. Photograph supplied by Prof. Ching-hsien Huang.

Anesthetics and the Lipid Core. Loss of consciousness or of local sensation can result from any kind of disturbance of neural transmission; if the disturbance disappears when the cause is removed and there is no permanent damage to the tissues, the causative agent may find application as an anesthetic, either general or local. Some anesthetics — the barbiturates are examples — appear to act by interfering with metabolic processes that provide energy. Others are believed to function through effects on membranes.

The common characteristic of a wide variety of anesthetic agents is a preferential solubility in lipids. Indeed, the solubility in olive oil is a direct indicator of the anesthetic potency for many compounds acting as general anesthetics, including several in clinical use (Fig. 11–13).* It seems almost certain that the anesthetic action is a result of appearance of the compounds in the lipid phase of membranes, but the exact molecular mechanism of the effects is not known. It may result from a simple expansion of volume of the lipid core, effects on local viscosity, a

*An important factor in selecting anesthetics is inflammability. To avoid blowing up all hands in the operating room, anesthesiologists welcomed the introduction of halogenated hydrocarbons, but these also have their perils in the form of potential liver damage upon repeated exposure.

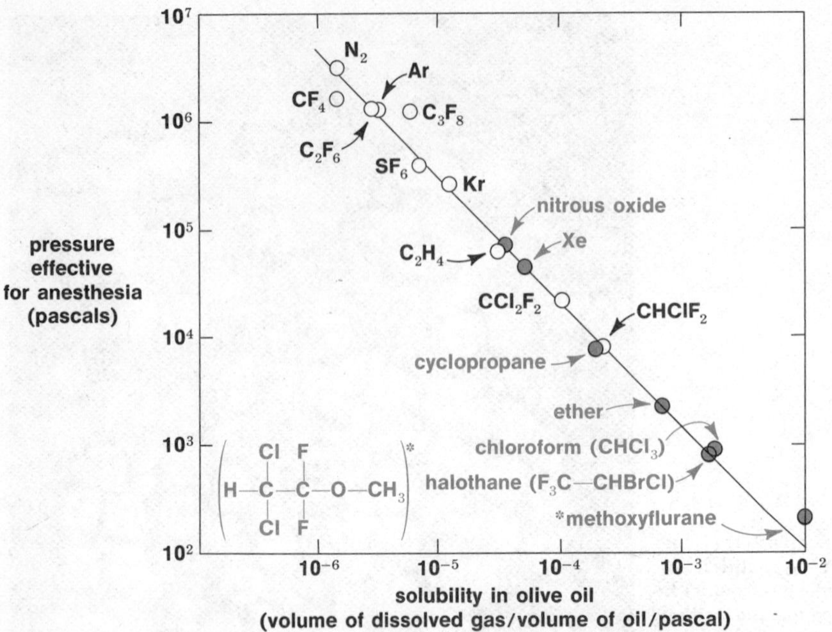

FIGURE 11–13 **The effectiveness of general anesthetics administered in the vapor phase is linearly proportional to their solubility in olive oil, which is used as an analogue of the lipid core of membranes. The effectiveness is indicated here as the pressure of gas necessary to achieve a certain level of anesthesia in 50 per cent of the mice used to test the agent. Modified from E. B. Smith *in* M. J. Halsey, R. A. Millar, and J. A. Sutton, (1974) *Molecular Mechanisms in General Anesthesia.* Churchill Livingston.**

disturbance of the interaction with protein components, or other changes. Most attention is concentrated on the plasma membrane, since the conduction of nerve impulses involves the generation of a potential across the membrane and its discharge through the flow of ions, all of which could be influenced by changes in the lipid core.

Other anesthetics may act by selective interactions at the polar-hydrophobic interface in membranes. These include compounds more widely used as local anesthetics, such as compounds related to benzocaine:

$$H_2N-C_6H_4-\overset{O}{\overset{\|}{C}}-O-CH_2-CH_3$$

ethyl *p*-aminobenzoate (Benzocaine)

Other compounds that may act in this way include some steroids that recently have been introduced as general anesthetics for intravenous administration.

Membrane Proteins

Each type of membrane has specific biological functions due to the presence of particular proteins; membranes differ in function because they contain different proteins. Despite these differences, there seems to be a general kind of structure

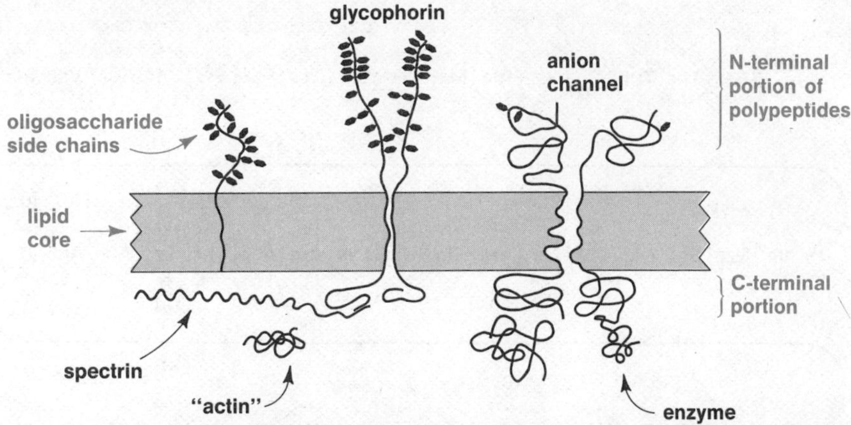

FIGURE 11–14 **The distribution of proteins in membranes as exemplified by the human red blood cell membrane. The drawing is based upon reasonable inferences from present information. Three major glycoproteins are known, all with their carbohydrate groups on the outside of the cell. (The carbohydrates are more complicated branched oligosaccharides than are indicated here.) The N-terminal portion of each polypeptide chain is outside, with the C-terminal portion directed into, and in some cases, through the phospholipid layer.**

One of the proteins, glycophorin, protrudes through the inner surface, where it appears to be connected with a structural protein named spectrin, as a part of a contractile apparatus.

Another protruding protein appears to contain two polypeptide chains and to be responsible for the transport of anions through the membrane.

In addition to spectrin other proteins that are bound to the inner surface of the membrane include an enzyme (glyceraldehyde-3-phosphate dehydrogenase) involved in carbohydrate metabolism, and actin, another component of the contractile apparatus.

Modified from a drawing in V. T. Marchesi, H. Furthmayr, and M. Tomita, (1976) *The Red Cell Membrane*, Annu. Rev. Biochem., *45*: 667. Reproduced by permission.

in that some proteins are embedded in one or the other face of the lipid core, while others pass through the core and are exposed on both sides of the membrane to aqueous phases (Fig. 11–14). Still other proteins may be more loosely bound to one face, or both, depending upon the membrane.

Little is known about the structure of most membrane proteins, but reasonable assumptions about their nature are confirmed by some examples at hand. It is likely that the proteins are associated with the lipid core because they have hydrophobic surfaces that cannot otherwise be shielded from water. A collection of hydrophobic groups at one end of a molecule would cause that end to become buried in the lipid core, whereas a hydrophobic middle section of sufficient length would cause a molecule to span the core. This is a simple extension of the principles that govern conformation of proteins in aqueous solution; the exclusion of water from hydrophobic groups is achieved in some proteins by placing the groups in a bulk lipid phase, instead of wrapping them into the center of the protein.

The meager information that is at hand on the nature of membrane proteins supports the predictions on structure. Two proteins involved in electron transfers in the endoplasmic reticulum — cytochrome b_5 and cytochrome b_5 reductase — have polypeptide chains in which one end takes care of the business of the protein and the other end is loaded with hydrophobic amino acid residues to be buried in

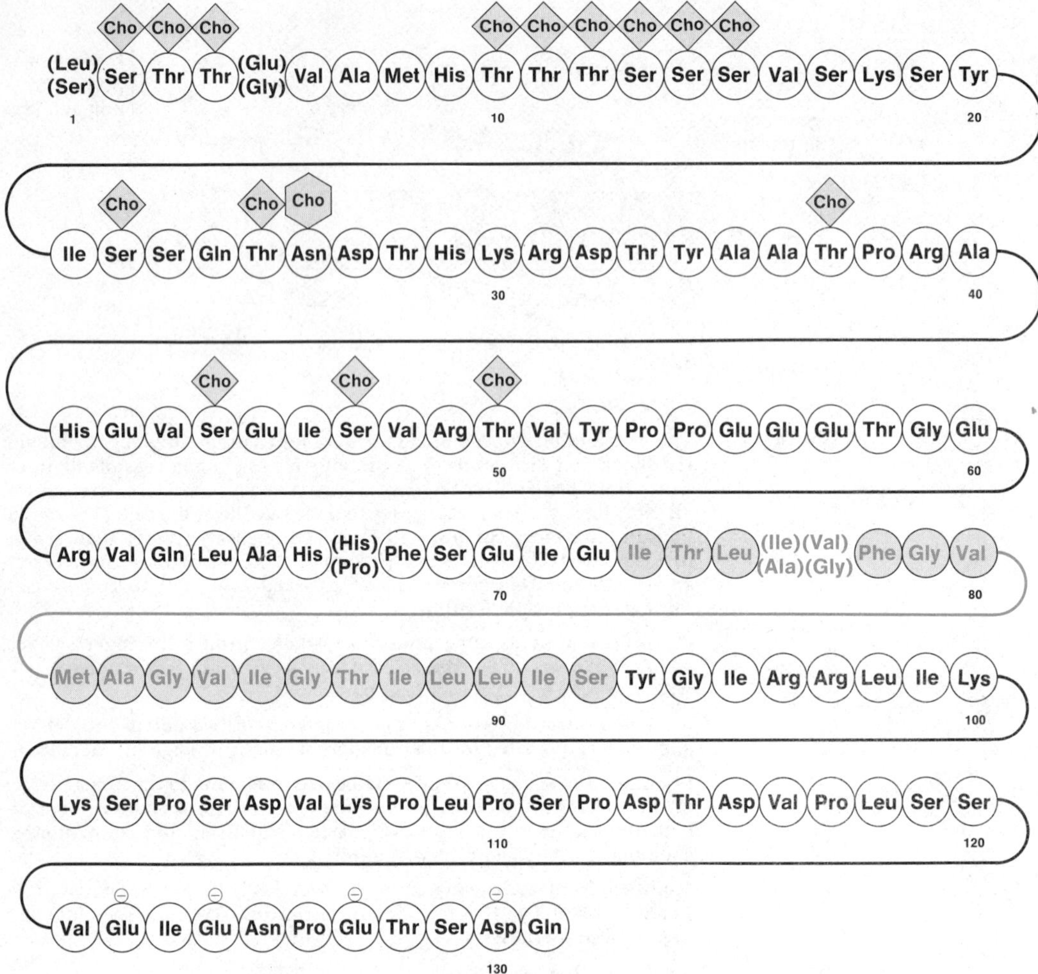

FIGURE 11–15 **The primary structure of glycophorin A, a protein that crosses the red blood cell membrane, includes a sequence of hydrophobic amino acid residues that is long enough to span the lipid core. The N-terminal segment appearing on the outside of the cell has many attached hydrophilic carbohydrate groups, while the C-terminal segment on the inside is rich in negatively-charged residues that can bind calcium ions within the cell.**

Modified from a drawing in V. T. Marchesi, H. Furthmayr, and M. Tomita, (1976) ***The Red Cell Membrane*****, Annu. Rev. Biochem., *45*: 667. Reproduced by permission.**

the lipid core. The proteins are catalytically active if the hydrophobic tail is experimentally removed, but they cannot attach to membranes. The tail serves as a binder for the active end.

The amino acid sequence is known for the protein glycophorin that spans the red blood cell membrane (Fig. 11–15). Here the hydrophobic residues are concentrated in the center of the peptide chain; the N-terminal portion has many attached carbohydrate groups, and the C-terminal portion is rich in negatively charged groups.

The contributions of proteins to the structural character of membranes is yet to be determined. Natural membranes with the proteins in place are stronger and can have larger radii of curvature than single phospholipid bilayers. The contribution of the proteins to function will be touched upon repeatedly in our later discussions.

FURTHER READING

General Reviews

Parsons, D. S., ed.: (1975) *Biological Membranes.* Oxford University Press. Short textbook.
Harrison, R., and G. B. Lunt: (1975) *Biological Membranes.* Wiley. Another short textbook.
Jackson, R. L., and A. M. Gotto, Jr.: (1974) *Phospholipids in Biology and Medicine.* N. Engl. J. Med., *290*: 24, 87.
Bretscher, M. S., and M. C. Raff: (1975) *Mammalian Plasma Membrane.* Nature, *258*: 43.

Specialized Reviews

Wallach, D. F. H.: (1957) *Membrane Molecular Biology of Neoplastic Cells.* Elsevier. Includes detailed reviews of general membrane functions.
Chapman, D., ed.: (1968) *Biological Membranes.* Academic Press.
Nelson, G.: (1972) *Blood Lipids and Lipoproteins.* Wiley-Interscience.
Ansell, G. B., J. N. Hawthorne, and R. M. C. Dawson: (1973) *Form and Function of Phospholipids.* Elsevier.
Prince, L. M., and D. F. Sears: (1973) *Biological Horizons in Surface Science.* Academic Press.
Villee, C., D. B. Villee, and J. Zuckerman, eds.: (1973) *Respiratory Distress Syndrome.* Academic Press.
Moscona, A. A., ed.: (1974) *The Cell Surface in Development.* Wiley.
Cooper, R. A.: (1977) *Abnormalities of Cell-membrane Fluidity in the Pathogenesis of Disease.* N. Engl. J. Med., *297*: 371.

Journal Articles

Huang, C.: (1976) *Roles of Carbonyl Oxygens at the Bilayer Interface in Phospholipid-sterol Interaction.* Nature, *259*: 242. Describes cholesterol-phospholipid geometry.

12 ANTIBODIES: DEFENSIVE PROTEINS

The world swarms with alien species, many of predatory bent. The vertebrates have markedly improved their chances for survival through the development of defenses against actual invasion of the flesh. Before a microorganism can gain entrance, it must pass sturdy mechanical barriers, such as the skin, the filters in the nose, and the constantly sweeping cilia, and survive exposure to destructive chemical agents, like H^+ in the stomach, fatty acids in the skin, and the lysozyme in tears. (Lysozyme is an enzyme that catalyzes hydrolysis of bacterial cell walls.)

An invader that circumvents the primary barriers mobilizes a second line of defense in the immune system, attracting specialized cells and substances to the site of the breach. The usual result is quiet and efficient disposal of the alien organisms, unless the invasion is massive. In addition to its linebacker role for the primary defensive screens, the immune system also may eliminate any aberrant cells that develop within the body.*

The Immune Reaction

The initial pattern of the immune response is largely stereotyped and non-specific; such diverse materials as splinters, foreign serum proteins, and bacterial toxins, in addition to intact microorganisms, will cause the same general effects. A visible inflammation frequently marks the site of substantial encounters between the obtruding substance and phagocytic cells, lymphoid cells, and the specialized proteins known as antibodies with which this chapter is mainly concerned.

Antigens and Antibodies. The non-specific natural immune system is backed up by a remarkably specific response in which resistance to a particular foreign substance develops after an initial exposure to it. A substance causing this kind of response is an antigen.

Two kinds of specific responses develop upon exposure to an antigen. In one, specific protein reagents, the antibodies, are synthesized and released from lym-

*The temptation to draw analogies between the functions of the immune system and the Department of Defense, the FBI, and the CIA is increased by knowing that the system sometimes exceeds its genetically prescribed duty — to discriminate between self and non-self — and attacks its own host's cells, thereby diminishing, rather than improving, the chances of survival.

phocytes and plasma cells. Antibodies are proteins constructed to combine with antigens; their structure and properties can be determined in detail by the same methods used with other proteins.

The second kind of specific immune response involves a direct mediation of immune function by cells, predominantly lymphocytes. This system of cellular immunity is less accessible to detailed analysis, and is beyond the scope of our discussion from a chemical point of view, but it ought to be remembered that immunity results from the harmonious interplay of all of the immune systems, specific and non-specific.

Individual Variations. There is a spectrum of effectiveness of the immune response in humans. A small, but significant number of individuals have defects in one or more areas of the immune systems and are consequently subject to repeated bouts of infection. At the other end of the spectrum are those who develop immune reactions to their own cells and therefore are prone to autoimmune diseases.

Most humans have more appropriate responses; serious infectious diseases can be prevented in them by bringing the specific immune system to a primed state of readiness through exposure to vaccines that contain some of the antigenic characteristics of a pathogenic organism or lethal toxin but lack the deleterious biological activity.

ANTIGENS

Proteins, polysaccharides, nucleic acids, lipids, synthetic polymers — all include examples of antigens. There is no common chemical feature other than a certain minimum size to define an antigen, and even here antigens may be as small as 1,000 daltons, or as large as millions of daltons. The structural components of one individual, which normally do not act as antigens within him, will provoke an immune response upon transplantation to another individual, except an identical twin.

Antigenic Determinants

Although a molecule must have a certain size to act as an antigen and cause the formation of antibodies, each antibody can bind only to a relatively small exposed portion of the surface of the antigen; this reactive portion is said to be the antigenic determinant for the antibody. That is, an antibody is constructed to fit a particular exposed arrangement of chemical groups in the antigen, and it will not fit other regions of the antigen's surface. One antigen molecule often contains several antigenic determinants because several areas are sufficiently exposed and reactive to cause the formation of antibodies. The result is that there may be several kinds of antibodies reacting with one kind of antigen, each binding in a different way. A toxin or organism may be modified sufficiently to become harmless and still retain enough of its original conformation to induce antibodies with a high affinity for some determinants on the native material. This is the basis for using killed organisms as vaccines.

Determinants in Myoglobin. The nature of the determinants has been studied in myoglobin from sperm whales. (It is abundant in whale muscles, hence its isolation from this source.) The structure of myoglobin and the sequence of the 153 amino acid residues in its single polypeptide chain are known.

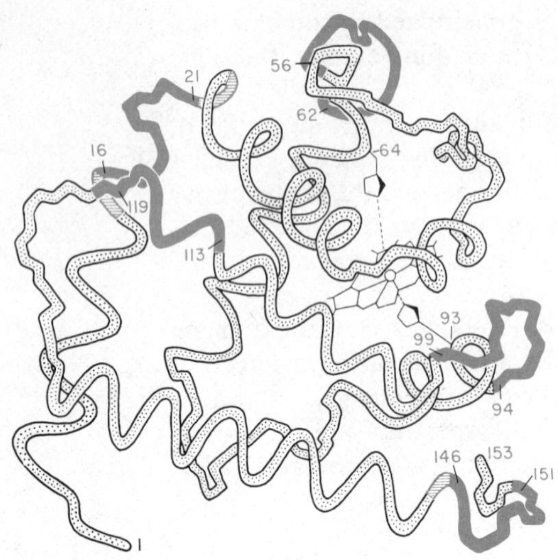

FIGURE 12–1

Antigenic determinants in myoglobin. Those portions of the polypeptide backbone bearing determinant side chains are shown in blue; those in striped blue are sometimes also involved. Modified from M. Z. Atassi, (1975) Immunochemistry, *12*: 423. Copyright by Pergamon Press. Reproduced by permission.

Rabbits and goats can be immunized to sperm whale myoglobin, meaning that the protein can be injected in a form such that it will act as an antigen in either species and provoke the formation of antibodies. Study of the interaction of the resultant antibodies and myoglobin showed that either species produced a mixture of antibodies interacting with the same five distinct regions on the myoglobin molecule, which are shown in Figure 12–1. Each of these regions consisted of no more than six or seven amino acid residues. However, each of the regions in itself behaved as several antigenic determinants. Some antibodies interacted only with four particular residues in the region, others with more, but in no case exceeding the six or seven residue limit. Estimates of the maximum determinant size in various antigens range from $0.6 \times 0.6 \times 1.5$ to $0.7 \times 1.2 \times 3.4$ nanometers.

A picture develops: Given an exposed area with several amino acid side chains, one antibody may be built to fit a certain combination of these groups, while another antibody is built to fit a different combination. It is the combination of groups that is the antigenic determinant: one group on the antigen may participate in several determinant combinations, each best fitting a different antibody. Put another way, different antibodies may fit different areas on an antigen, and some of the areas overlap.

A substance must have regions of stable geometry if it is to behave as an antigen. Only a few groups are needed to act as a determinant, but the groups must remain in nearly the same relative positions if they are to provoke the formation of a closely-fitting antibody. Small molecules are not effective antigens. However, once an antibody has been generated by a molecule sufficiently large to be an antigen, a small molecule bearing the same groups occurring in the determinant region may also interact with the antibody; it need assume the right configuration only transiently in order to be bound. Such small molecules are said to be **haptens** (haptein = to fasten) and have been useful reagents for studying antigen-antibody interactions.

For example, if a protein substituted with dinitrophenyl groups is used as an antigen, some of the resultant antibodies will bind to determinants including the dinitrophenyl groups. These antibodies may also bind free dinitrophenol, even though the free phenol will not act as an antigen by itself. Dinitrophenyl compounds are therefore haptens for these antibodies, and the nature of the hapten-antibody interaction can be studied without the distracting influence of the remainder of the original large antigen molecule.

Specificity of Antibodies

It is estimated that a mature individual can synthesize as many as 10^6 kinds of antibodies, each differing in the antigenic determinant with which it combines best. The discrimination of antibodies is often exquisite. Proteins of like function in closely related species usually differ in only a few amino acid residues, but these differences may cause sufficient change in the surface conformation so that antibodies reacting with one have markedly different affinity for the other.

Use of Antibodies for Assay. The discriminatory power of antibodies makes them valuable analytical reagents, used to detect small molecules as well as specific proteins. For example, it is sometimes desirable to determine the blood concentrations of digoxin or digitoxin (Fig. 12–2), drugs commonly used in the treatment of congestive heart failure. The toxic effects of these drugs mimic the symptoms and signs of heart failure so that it is difficult to know if the patient has too much or too little of the drug. However, the drug concentrations can be measured with specific antibodies, which are made by injecting rabbits with a protein to which digoxin or digitoxin has been covalently coupled so that the free drugs then act as haptens. The antibodies are so specific that they can discriminate between digoxin and digitoxin, which differ only in the presence of a 12-hydroxyl group on the steroid ring. Antibodies prepared with one of the drugs will bind it at only one-hundredth the concentration required for equal binding to the other drug.

The actual analysis involves **radioimmunoassay**; the central feature is competition for antibody binding between a radioactively-labeled antigen and its unlabeled counterpart in the unknown sample. A complex of antibody with a radioactive drug is added to the unknown sample. The unlabeled drug in the sample displaces part of the radioactive drug from the complex; the measured amount of displaced radioactivity at equilibrium is an index of unlabeled drug concentration. Results from the assay are available in short enough time to make it a useful clinical procedure.

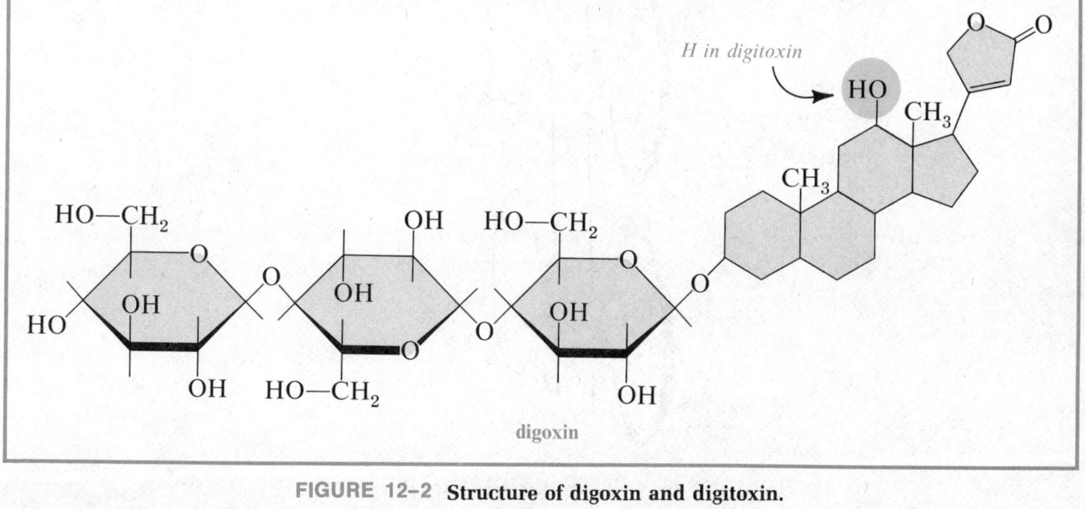

FIGURE 12–2 **Structure of digoxin and digitoxin.**

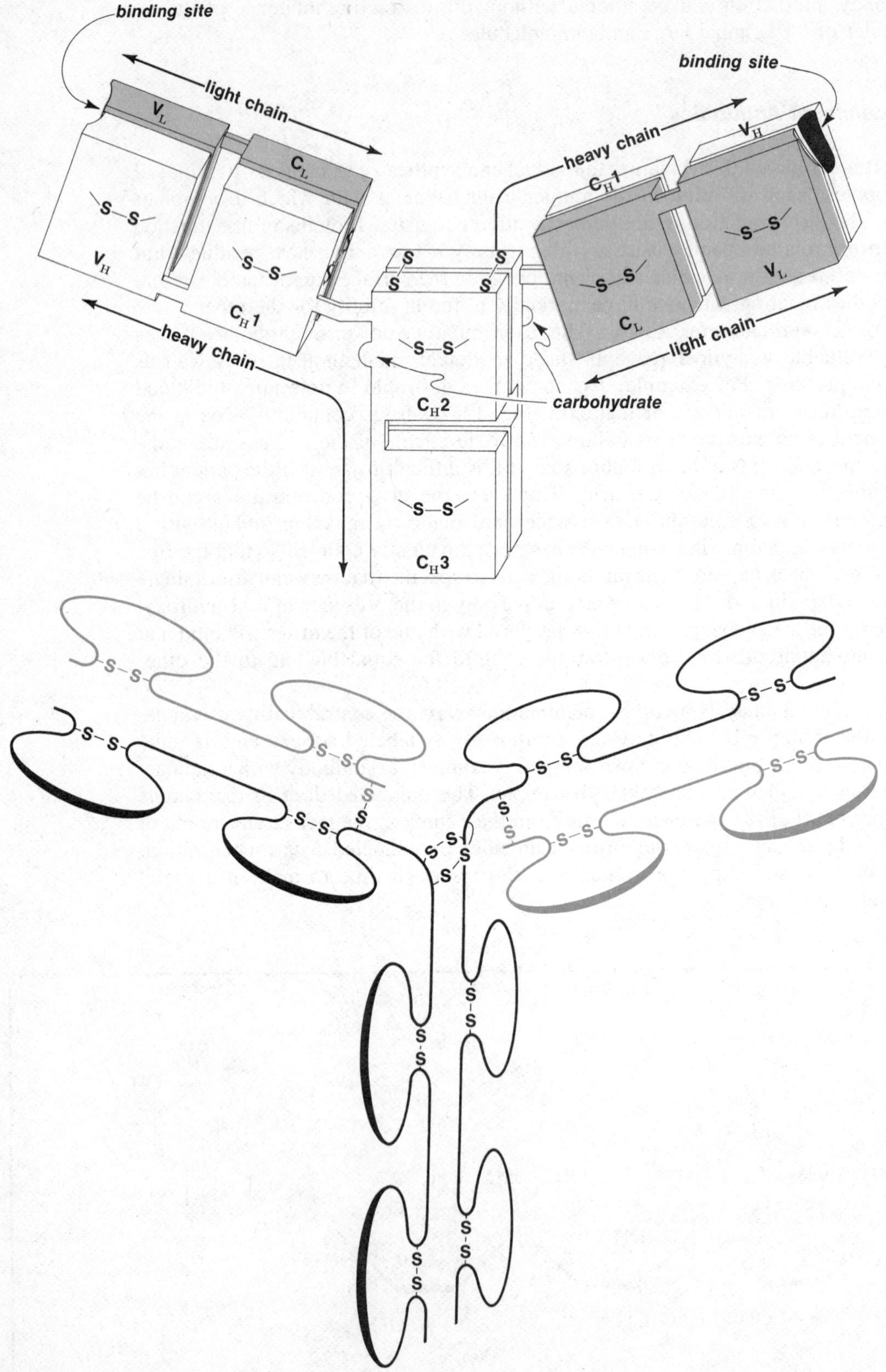

FIGURE 12–3 *Legend on opposite page.*

STRUCTURE OF ANTIBODIES

Antibodies react with specific antigens because they are built with combining sites to fit the antigen. The interactions between antibody and antigen depend upon the same forces involved in other biological functions of proteins and knowledge of structure in antibodies clarifies many general features of protein function.

Antibodies are referred to as **immunoglobulins (Ig)**, or as **gamma globulins,** a term derived from the behavior of the most abundant class, IgG, upon electrophoresis (p. 21). The gamma globulins of blood serum are mostly antibodies; antibodies are also found in the beta-globulin region, and to a considerably lesser extent in the alpha-globulins.

Antibodies Are at Least Divalent. The fundamental unit of an antibody is a Y- or T-shaped molecule with an identical site at the end of each arm for binding an antigen (Fig. 12–3); the antibody can therefore bind two antigen molecules simultaneously. The antibody molecule has an axis of symmetry; rotating one half of the molecule 180° superimposes it on the other half.

Heavy and Light Chains. An antibody contains equal amounts of two kinds of polypeptide chains, heavy and light, with the heavy at least twice the size of the light. One pair of identical heavy chains and one pair of identical light chains combine in this way: Each arm of the molecule contains the entire length of a light chain bound to a nearly equal-sized portion of the amino terminal sequence of a heavy chain. The remaining carboxyl terminal portions of the heavy chains combine to form the stem. The union of the chains is usually stabilized by disulfide bonds.

Domains

Each arm of an antibody is composed of two globular masses of folded polypeptide chain connected by more extended segments. The stem is made from at least two, and sometimes more, globular masses of similar size, also connected to the arms by extended segments and to each other. Here we have an illustration of an important feature of many proteins — the antibodies are composed of domains that are distinct but connected. A domain is a folded compact structure that behaves as an architectural unit. Figure 12–3 showed the domains in schematic form; they can be seen in more detail for one kind of antibody in Figure 12–4.

The Immunoglobulin Fold. The globular domains of antibodies are quite similar in size because each is constructed according to the same architectural plan. Each is made by folding parts of two polypeptide chains: one heavy and one light chain in the arms and two heavy chains in the stem. The folds in each chain are constructed in a strikingly similar pattern, even though their functions differ. (Some refer to the characteristic folds in individual chains as the domains; in this

---

FIGURE 12–3 **Schematic outline of antibody structure. Two light chains (*blue*) and two heavy chains (*black*) generate a T- or Y-shaped structure with an axis of symmetry. The chains are organized into domains (*top*). The domains are partially created by disulfide bonds, and disulfide bonds are also used as links between chains, but not always at the positions shown here (*bottom*).**

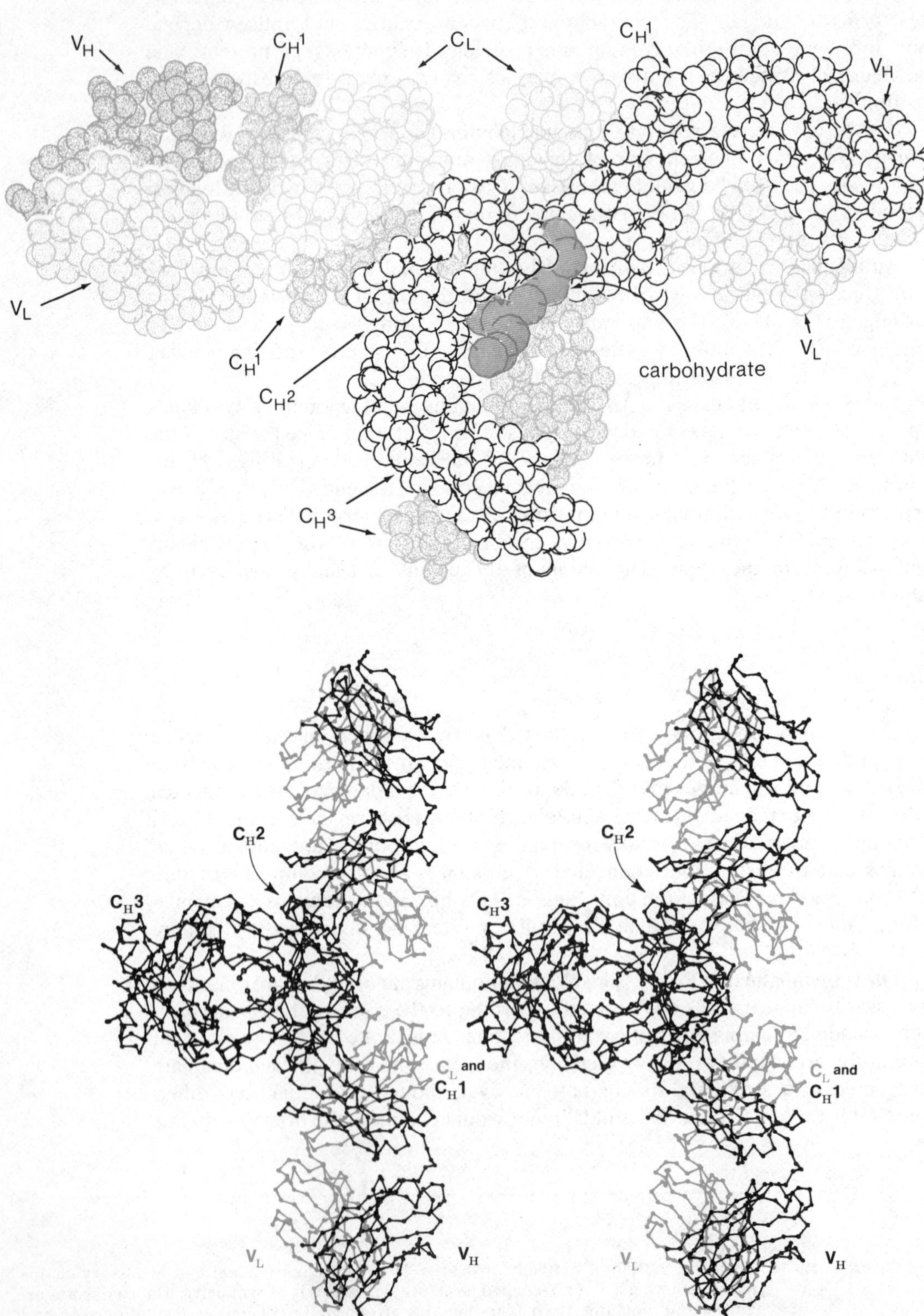

FIGURE 12–4 *Legend on opposite page.*

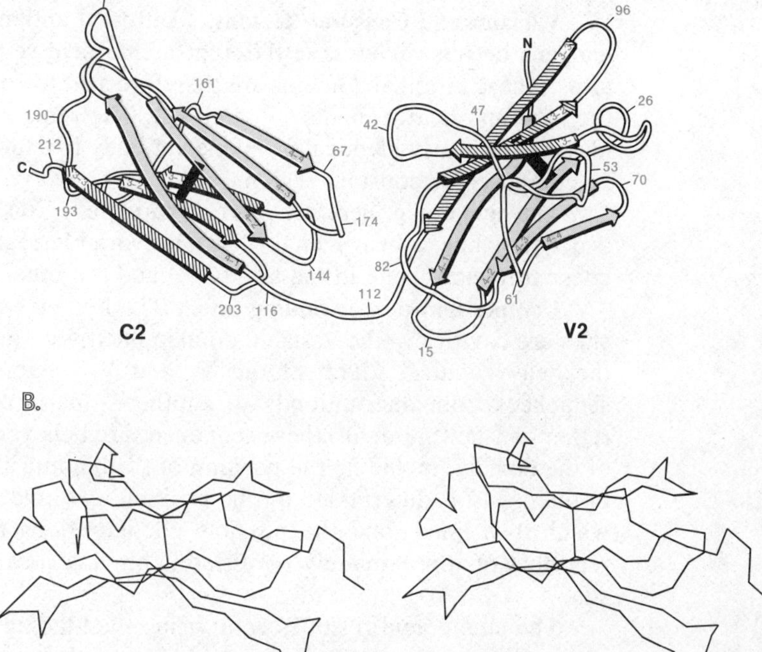

FIGURE 12–5 *A.* Outline of chain folding in a Bence-Jones protein containing the variable and constant domains of a light chain. Segments of pleated sheet are indicated by *arrows.* Each domain contains two layers of pleated sheet, one shaded *blue* and the other *striped black.* Note the similarities of the folding in the two domains. Reprinted with permission from A. B. Edmundson, et al., *Rotational Allomerism and Divergent Evolution of Domains in Immunoglobulin Light Chains.* (1975) Biochemistry, *14*: 3593. Copyright by the American Chemical Society.

B. Stereo drawing of the detailed course of the polypeptide chain in a constant domain. Lines connect α-carbon atoms. Modified from D. R. Davies, E. A. Padlan, and D. M. Segal. Reproduced with permission from the Annual Review of Biochemistry, Vol. 44, p. 654. Copyright 1975 by Annual Reviews, Inc.

usage, the globular regions we call domains would be regarded as pairs of domains.) Each involves a sequence of approximately 110 amino acid residues, of which 50 to 60 per cent is in two layers of antiparallel pleated sheet arranged somewhat like a sandwich, with the two slices of bread linked by a stabilizing intrachain disulfide bond (Fig. 12–5). Hydrophobic amino acids are invariably present around the S–S bond to further stabilize the structure.

FIGURE 12–4 Detailed structure of an antibody molecule. *Top.* The location of individual residues is indicated by circles; light chains are in *blue*, and heavy chains are in *black*, with one of the chains shaded. Carbohydrate residues are shown by *heavy black shading. Bottom.* Stereo view of the course of the polypeptide chains in the same molecule. Circles locate α-carbon atoms. The top view is modified from E. W. Silverton, M. A. Navia, and D. R. Davies, (1977) Proc. Natl. Acad. Sci. U.S.A., *74*: 5140. Reproduced by permission. The bottom stereo view was kindly generated by Dr. Navia for this work.

The Binding Site

Variable and Constant Regions. Different antibodies combine with different antigens because they have different amino acid sequences at the ends of their arms. These terminal domains are therefore said to contain the variable regions of the light and heavy chains (V_L and V_H). The remainder of the domains may be identical or nearly identical from one antibody to another and are therefore said to be composed of constant regions.

The light chain therefore has one variable region and one constant region (V_L and C_L), while a heavy chain has one variable region (V_H) and at least three constant regions, one in the arm (C_H1) and the others in the stem (C_H2, C_H3).

Composition of the Binding Site. The known examples of antibody binding sites are cavities in the variable domain lined with portions of both the light and the heavy chains. Much of the V_L and V_H regions have similar amino acid sequences from one antibody to another — parts of the variable domains are rather constant, after all; these sequences are believed to create the general shape of the domain, including the position of the binding site cavity. The specific form of the cavity is determined by the position and kind of amino acid residues with which it is lined, and the possible configurations are sufficient to generate a repertory of approximately 10^6 different binding sites in the antibodies of one person.

The amino acid residues comprising the binding site are scattered along the polypeptide chain. Their locations in the chains were first recognized because these were the positions at which amino acid composition differed in all antibodies and were designated as **hypervariable sequences.**

Hypervariable regions were found at positions 26–32, 48–55, and 90–95 of some light chains and positions 31–37, 51–68, 84–91, and 101–110 of the corresponding heavy chains; the remainder of the chains in the variable domain is folded so as to bring these hypervariable sequences near each other at the end of the antibody arms, creating a binding site.

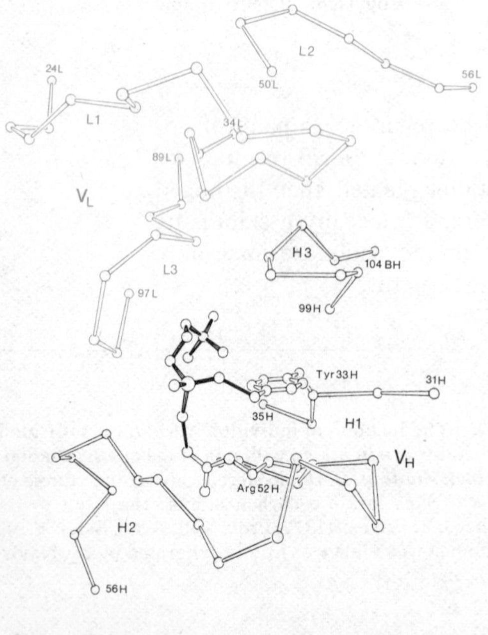

FIGURE 12-6

Outline of the path of the polypeptide chains around an antibody binding site. This site fixes phosphocholine (*heavy black*) as a hapten, but the natural antigen is not known. Modified from D. R. Davies, E. A. Padlan, and D. M. Segal. Reproduced with permission from the Annual Review of Biochemistry, Vol. 44, p. 663. Copyright 1975 by Annual Reviews, Inc.

An example of a portion of a binding site is shown in Figure 12–6. The site fits O-phosphocholine as a hapten, although the natural antigen for this antibody is not known. The site appears to be $0.6 \times 0.6 \times 0.7$ nm in dimensions; two sides of the entrance are formed by the first and third hypervariable sequences of the light chain, and the other two sides are formed by the second and third hypervariable sequences of the heavy chain.

TYPES OF IMMUNOGLOBULINS

Differences in the biological behavior of antibodies were recognized long before anything was known about their molecular architecture and were used as a basis for classification; the terminology gradually is being translated into molecular descriptions.

Classes of Immunoglobulins. Antibodies exist in general structural types that appear to be related to different functional roles in the body. Five such classes are known in humans: A, D, E, G, and M. IgG is largely confined to the intravascular and extracellular space; IgA is the predominant class secreted at mucosal surfaces; IgE binds to the plasma membrane of mast cells and basophils; and IgM is mainly in the intravascular space. The primary structural difference in the classes is in the kind of heavy chain; these chains designated by the corresponding Greek letters, α, δ, ϵ, γ, and μ. The light chains in any of the five classes are one of two kinds, κ or λ. For example, the chain composition of an IgG molecule may be $\kappa_2\gamma_2$ or $\lambda_2\gamma_2$.

The types of heavy chains differ in the nature of their constant domains, and this leads to further differences between them, which are summarized in Table 12–1. IgA associates into dimers and IgM into pentamers, in which a further single small polypeptide of about 15,000 daltons, the J piece, is incorporated. The IgA dimers include still another polypeptide chain, a secretory component of 70,000 daltons that appears to facilitate secretion through mucous membranes.

The classes also differ in their content of carbohydrate, which makes up from 2.5 to 10 per cent of the weight and is commonly linked from N-acetylglucosamine to asparagine in the constant domains of the heavy chains.

Subclasses. It is possible to distinguish subclasses with finer structural differences among immunoglobulins of one general class, which may be created by the existence of heavy chains that differ in amino acid residues present at certain

PROPERTIES OF HUMAN IMMUNOGLOBULIN CLASSES TABLE 12–1

Class	IgA	IgD	IgE	IgG	IgM
Heavy chain (H)	α	δ	ϵ	γ	μ
Estimated M. W.					
complete	55K	62K	70K	50K	70K
without carbohydrate	48K	56K	58K	49K	58K
Oligosaccharide chains/heavy chain	2–3		5	1	5
Chain composition	L_2H_2 or L_4H_4	L_2H_2	L_2H_2	L_2H_2	$L_{10}H_{10}$
Extra components	SC, J				J
Approximate total M.W.	162K, 390K	178K	188K	146K	880K
Normal adult serum concentration					
mg/100 ml	200	3	0.05	1,000	120
Range	90–450			800–1,800	60–275

positions in their constant domains. Four such subclasses of IgG, two of IgM, and two of IgA have been recognized. The serum of a normal individual contains all classes and subclasses of immunoglobulins.

Allotypes. Allotypes of immunoglobulins, like those of other proteins, differ in amino acid composition at a few positions in the constant domains of either light or heavy chains. In the human, they are structural variations in the constant regions; unlike subclasses, each allotype is not found in all individuals. For example, one person may inherit kappa light chains with allotypes designated as Inv that differ in amino acid composition at two positions:

Allotype	Residue	
	153	*191*
Inv 1	Val	Leu
Inv 1,2	Ala	Leu
Inv 3	Ala	Val

Like other allotypes, they are gained from both the maternal and paternal inheritance, that is, they are part of the respective haplotypes (the inheritance from haploid parent germ cells). A person may have immunoglobulins from two of the Inv systems, but not from three.

BINDING OF ANTIBODIES

The function of an antibody begins with its binding of an antigen. The specificity is not absolute; the binding site of an antibody will often "recognize" determinants of closely related structure with at least a partial fit into the site, but in general the tightest combination of antibody and antigen is achieved only when the fit is very good and complete. Antigen and antibody must be close friends, not mere acquaintances, for the best union.

Insoluble Antigen-Antibody Complexes

Agglutination. Antibodies can cause cells to agglutinate into large, readily sedimentable clumps. The cells may be microorganisms, erythrocytes, or nucleated cells because the only requirements for agglutination are a particulate antigen with multiple determinant sites and antibodies with at least two binding sites, requirements that may be met by any normal cell and antibody to it. The multivalent antibodies physically cross-link the polyvalent antigens (Fig. 12–7).

Agglutination of antigen *in vivo* is probably extremely rare, but the use of agglutination reactions *in vitro* to type blood is a critical key to the success of transfusions. Their use is extremely important in the diagnosis of some infectious diseases, and in the general identification of many bacteria. The antigenic determinants causing agglutination frequently involve the carbohydrate side chains of glycoproteins in the outer cell membrane. At least nine major blood group systems have been identified in the human erythrocyte on the basis of variations in glycoprotein structure (p. 188). The recipient of transfused red blood cells must have the same determinants found in the donor.

Some invertebrates and plants contain high molecular weight substances, usually proteins, that mimic antibodies in their ability to agglutinate red blood

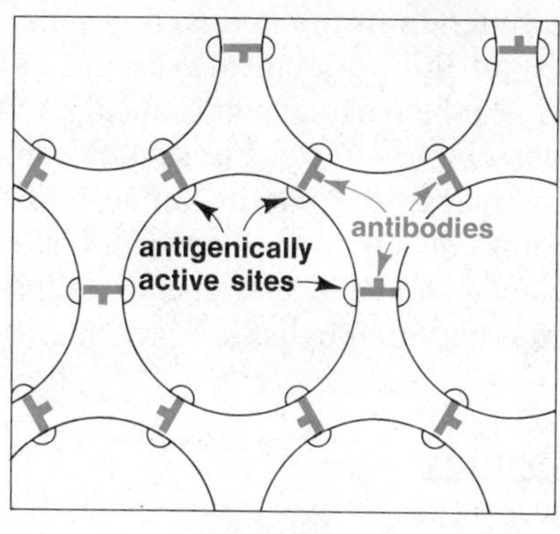

FIGURE 12–7

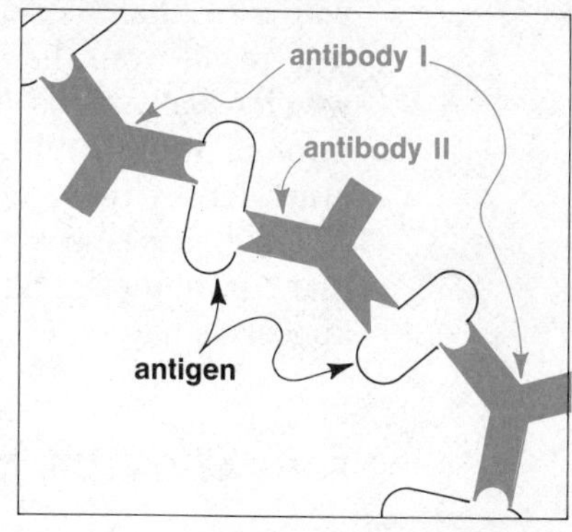

FIGURE 12–8

FIGURE 12–7 **Cells and other large antigens usually contain many identical antigenic sites. The divalent antibodies may therefore link the antigens into a three-dimensional lattice.**

FIGURE 12–8 **A precipitin reaction may result from the formation of a lattice with soluble antigens that contain different antigenic determinants through combination with different antibodies.**

cells. These substances, known as **lectins,** react with specific carbohydrates in some instances and can be used as an analytical tool to detect the presence of the carbohydrates; some are so specific that they can identify specific blood group antigens. Another property of some lectins is their ability to induce DNA synthesis in cells of the lymphocytic type. **Concanavalin A,** from jack beans, is a lectin widely used for experimental purposes.

Precipitin Reaction. Antibodies also may cross-link soluble multivalent antigens to form precipitates (Fig. 12–8). This reaction is used as a precise quantitative tool to determine amounts of antibody and antigen and is also useful in the analysis of the structural requirements for specificity of an antibody.

Biological Effects

If there is any protective value to the individual in the agglutination of precipitin reactions, it is yet to be demonstrated. Indeed, adverse consequences follow when mismatched blood is transfused. However, the appearance of antigen-antibody complexes has many powerful consequences that are for the most part purposeful and lead to the inactivation and removal of foreign materials. The signal for initiation of these events appears to be alterations in the conformation of the stem of the antibody, caused by binding of antigen to the arms.

Activation of Complement. The complement system is a group of at least 11 serum proteins that are activated in sequence following the union of antigens with some antibodies, and we shall say more about this cascade mechanism in Chapter 18. The end result is lysis of a foreign cell. Activation is especially efficient and rapid upon combination of IgM with erythrocytes or certain bacteria; one molecule of bound IgM can initiate the lysis of one complete cell.

Earlier stages in the activation of complement act as signals for other biological events, such as migration of macrophages to the locale of the antigen, increased phagocytosis of the foreign material, and even visible inflammation.

Deleterious Effects. The effects of antigen-antibody combination are not always good; inflammation may cause extensive damage to basement membranes, with a resultant severe malfunction in the involved organs. Illness, mild to fatal, may result from the activity of IgE antibodies. All normal individuals have IgE, which usually binds strongly to the plasma membranes of mast cells and basophils through its stem without detectable effects. Subsequent binding of antigen to the arms causes these cells to release the potent amines, histamine and serotonin (Chapter 32), and a compound of unknown nature, the slow reactive substance, that contributes to anaphylactic shock. The effects in unduly sensitive individuals appear as hay fever, asthma, and anaphylactic reactions.

FORMATION OF ANTIBODIES

There is no reason to believe antibodies are not made in the same way as other proteins; the problem that has provoked speculation is how different antibodies to seemingly innumerable antigens can be formed.

Selective vs Instructive Theories. Paul Ehrlich* postulated that cell surfaces had receptors complementary to antigens and that upon combination with antigen, the cell was stimulated to synthesize and release additional receptors. In effect, the antigen thus selected those cells capable of forming antibodies to it. These receptors were conceived to be chemical side chains protruding from some structural core.

The side chain theory survived until many laboratory-synthesized derivatives, not known to resemble any natural compound, were also found to trigger specific antibody synthesis. It seemed improbable that the body would contain natural receptors to unnatural antigens, and the notion arose that the antigen somehow instructed an antibody to assume the proper configuration for binding or created an antibody with that configuration.

This instructional theory reigned supreme until it became apparent that the conformation of antibodies, like that of other proteins, is determined by amino acid sequence, which is in turn coded in DNA. The problem then became one of explaining how the coding for perhaps a million proteins could be incorporated in a genome believed to contain fewer than 100,000 different genes, perhaps only 40,000. Various ways in which an antigen might cause a proliferation of genes were explored, but a theory based on selection of cells has resurfaced, and it is now accepted that antigens produce antibodies by selecting cells that have receptors for the antigen.

One Cell — One Immunoglobulin. Antibodies are made and released by plasma cells and lymphocytes. At a given stage in its normal development, one of these cells makes a homogeneous single kind of antibody; every antibody molecule it produces is identical. In other words, each normal cell produces one specific sequence of light chains and one specific sequence of heavy chains at a time.

Two Genes — One Polypeptide Chain. The light and heavy chains are single polypeptides, but it does not seem likely that the information for amino acid sequence of the entire chain is contained in a single sequence of nucleotides representing one gene. The reason is that the amino terminus of each chain is variable and must be coded by distinct genes, whereas the carboxyl terminus is relatively constant and could be coded by only a few genes — genes corresponding to the various subclasses of heavy chains and genes for the kappa and lambda

*Paul Ehrlich (1854–1915), Nobel Laureate, had many ideas in advance of the technics necessary for their exploitation. He was a pioneer in chemotherapy and the development of biological stains, as well as in immunology.

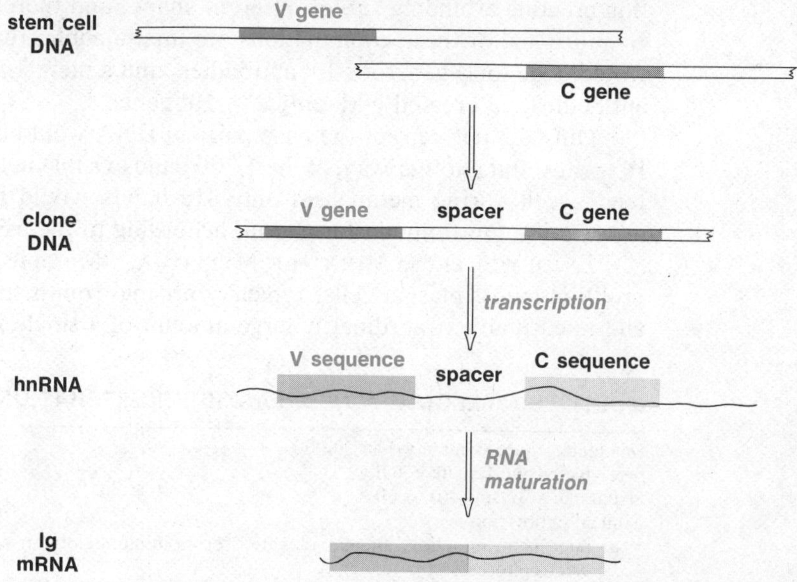

FIGURE 12–9 **Genes for variable and constant regions of antibodies occur at different locations in chromosomes of stem cells. Upon maturation into an individual lymphocyte, DNA is spliced so as to bring one V-gene and one C-gene in'o proximity, although separated by a spacer segment. Subsequent divisions of the lymphocyte generate a clone of cells. Antibody formation by the clone involves transcription of the entire sequence into hnRNA, which is then matured by eliminating the spacer transcript, and ligating the V and C sequences into a translatable mRNA. Spacer segments have also been found within the V-gene. (See J. C. Marx: (1978) *Antibodies: New Information about Gene Structure.* Science, *202*: 298.)**

light chains. If single genes coded for the entire length, the genetic information for the constant portions would have to be repeated along with the information for each of the many variable regions. Evidence is accumulating that this is not the case; the variable region and the complete constant regions are separately coded, and the information from two parts of the genome is combined to create single polypeptides (Fig. 12–9).

Genetic analysis now indicates three loci for constant regions in the human chromosomes, one each for the two kinds of light chains, kappa and lambda, and one for the heavy chains. The linkage of the genes in the heavy chain locus has been deduced from some abnormal immunoglobulins created as a result of abnormal chromosome recombinations:

$$C\mu_1\text{-}C\mu_2\text{-}C\gamma_4\text{-}C\gamma_2\text{-}C\gamma_3\text{-}C\gamma_1\text{-}C\alpha_1\text{-}C\alpha_2\text{-}C\delta\text{-}C\epsilon.$$

(The subscript numbers refer to subclasses, not domains.)

With this arrangement, nearly all of the additional genetic information for immunoglobulins is available for coding many variable regions, with their unique sequences. Furthermore, production of one million different antibodies does not require one million separate genes. Different binding sites can be made from different combinations of light and heavy chains. That is, 10^4 variable light chains and 10^4 variable heavy chains could combine in 10^8 permutations, each combination creating a binding site of different shape, and therefore different specificity. Even if most of these combinations are unattainable, the use of only 10 per cent would code for a generous 10^7 antibodies, and a mere 1 per cent would create 10^6 antibodies, all created with only 2×10^4 genes.

Only 0.2 per cent of the base pairs in DNA would be required to create 2×10^4 genes. Put another way, of the 1,200 bands visible in the chromatin of a human female cell during metaphase, only 1½ bands would be required to carry the information for antibody formation, according to the assumptions used above.

EVIDENCE FROM MULTIPLE MYELOMA. Multiple myeloma is a malignant proliferation of plasma cells, typically arising from a single cell, that synthesize and release an extraordinarily large amount of a single immunoglobulin.

MULTIPLE MYELOMA: A NEOPLASTIC PROLIFERATION OF PLASMA CELLS

Incidence—approximately 1 in 20,000
Sex—males and female equally
Usual age at onset—40 to 60
Clinical expression—
- (a) bone pain, infection, anemia, elevated serum immunoglobulin with homogeneous peak on electrophoresis;
- (b) characteristic punched-out lesions in bone on X-ray;
- (c) plasmacytosis (bone marrow contains increased numbers of plasma or related cells);
- (d) urine may contain large amounts of light chains referred to as Bence-Jones protein;

Treatment—usually alkylating agents and prednisone (a synthetic hormone similar in action to cortisol).

However, approximately 1 per cent of the patients have a biclonal multiple myeloma, that is, they have two populations of abnormally proliferating plasma cells that make excessive amounts of two antibodies. In some cases, the two accumulating antibodies belong to different classes (A and G, for example). Part of the structure of the variable regions in the heavy and light chains has been determined in some, and the important finding was that the same variable region sometimes occurred in both antibodies. That is, two kinds of light chains or two kinds of heavy chains would have the same variable regions.

This finding is easy to rationalize in terms of two genes being used to direct the formation of one polypeptide chain. In the usual monoclonal myeloma cells, a given variable region is being spliced to a given constant region to form one kind of antibody. If one of these cells now mutates so as to bring a different constant region gene into proximity with the variable region gene, a second line of neoplastic cells will arise and form another kind of antibody with the same variable region.

FURTHER READING

McBride, G.: (1976) *Antibodies Yield Their Secrets and Display Therapeutic Versatility.* J.A.M.A., *235*: 583. Useful introduction.

Poljak, R. J.: (1975) *Three-dimensional Structure, Function, and Genetic Control of Immunoglobulins.* Nature, *256*: 373.

Silverton, E. W., M. A. Navia, and D. R. Davies: (1977) *Three-dimensional Structure of an Intact Human Immunoglobulin.* Proc. Natl. Acad. Sci. U.S., *74*: 5140.

Davies, D. R., E. A. Padlan, and D. M. Segal: (1975) *Three-dimensional Structure of Immunoglobulins.* Annu. Rev. Biochem., *44*: 639.

Kabat, E. A., E. A. Padlan, and D. R. Davies: (1975) *Evolutionary and Structural Influences on Light Chain Constant (C_L) Region of Human and Mouse Immunoglobulins.* Proc. Natl. Acad. Sci. U.S., *72*: 2785.

Goodman, M.: (1975) *Analogies between Hemoglobin and Immunoglobulin Evolution*. Immunochem., *12*: 495.
Edmundson, A. B., et al.: (1975) *Rotational Allomerism and Divergent Evolution of Domains in Immunoglobulin Light Chains*. Biochemistry, *14*: 3953.
Gilmore-Hebert, M., and R. Wall: (1978) *Immunoglobulin Light Chain mRNA Is Processed from Large Nuclear RNA*. Proc. Natl. Acad. Sci. U.S., *75*: 342.
Gilbert, W.: (1978) *Why Genes in Pieces?* Nature, *271*: 501.
Brack, C., and S. Tonegawa: (1977) *Variable and Constant Parts of the Immunoglobulin Light Chain Gene of a Mouse Myeloma Cell Are 1250 Nontranslated Bases Apart*. Proc. Natl. Acad. Sci. U.S., *74*: 5262.
Anon.: (1977) *Immunoglobulin Genes and the Immune Response*. Nature, *269*: 648. Report of meeting.
Atassi, M. A.: (1975) *Antigenic Structure of Myoglobin*. Immunochem., *12*: 423. Review.
Williamson, A. R.: (1976) *The Biological Origin of Antibody Diversity*. Annu. Rev. Biochem., *45*: 467.

13 BINDING FOR TRANSPORT: HEMOGLOBIN AND OXYGEN

Substances frequently are moved from one place to another in the body by binding to a protein carrier. The transport of oxygen by hemoglobin and the nature of this carrier will illustrate many general principles.

Let us first ask why a carrier is necessary. Why isn't simple diffusion of oxygen from the atmosphere to the cells sufficient? Rapid diffusion requires a large difference in concentration over a short distance. The distance from the lungs to most cells is so great that this steep gradient can be achieved for only a small part of the route, the short distance across the plasma membrane. Not only is the distance small, but oxygen has a much higher solubility in the lipid core than it does in the surrounding aqueous phases, making the absolute magnitude of the concentration difference greater (Fig. 13–1).

The delivery of oxygen from the atmosphere to the intercellular fluid is another matter. The distance from cell surface to atmosphere is sufficiently short for effective diffusion in only the simplest organisms, and oxygen is delivered in several stages in the larger organisms (Fig. 13–2). Animals as low in the evolutionary scale as the annelids (for example, earthworms) became possible only by the development of hemoglobin and related proteins as carriers for oxygen, together with a pumped circulation to move the carrier rapidly between surfaces exposed to the atmosphere and the more deeply buried cells.

Individual cells with very high demands for oxygen, such as those in some striated muscles and the heart, also acquired an internal hemoglobin, myoglobin, to facilitate rapid transport from the inside surface of the plasma membrane to the mitochondria, the organelles consuming most of the oxygen. Short as the distance is, diffusion is not sufficient to sustain the necessary rapid flux of oxygen within those cells.

Magnitude of the Task

How fast must oxygen be transported? Under basal conditions (rest, overnight fast), a typical rate of oxygen consumption by 22-year-old Americans of median size is:

12 millimoles (260 ml) per minute by males (177 cm, 73 kg)
9 millimoles (200 ml) per minute by females (163 cm, 58 kg).

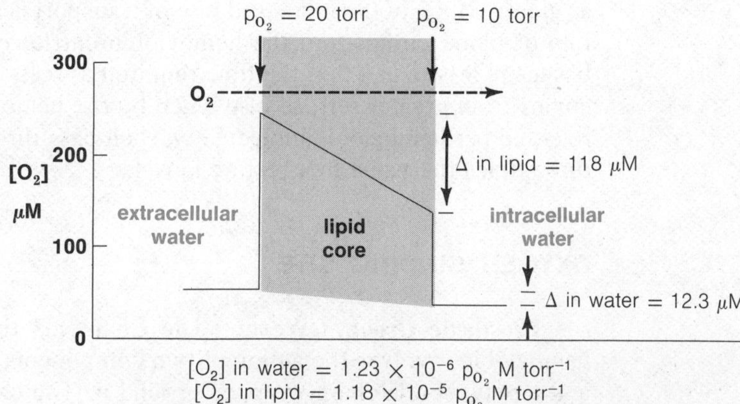

FIGURE 13–1 **The higher solubility of O_2 in the lipid core (*blue*) of a membrane creates a larger concentration gradient, which accelerates diffusion. The concentrations of dissolved O_2 are plotted for the extracellular water and its adjacent lipid face at a partial pressure of 20 torr (mm Hg), and for the intracellular water and its adjacent face at a p_{O_2} of 10 torr (*right*). The difference in concentration across the lipid core at these tensions is 9.6 times the difference between the two aqueous phases (118/12.3), and diffusion would rapidly diminish the difference to a small value across a real membrane. An exaggerated Δp_{O_2} is used here to illustrate the point.**

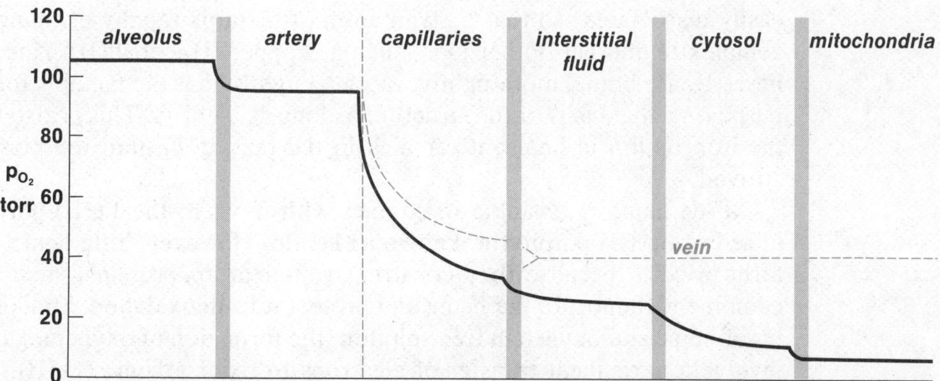

FIGURE 13–2 **The transport of oxygen in higher organisms requires the presence of concentration gradients at several sites, with the oxygen tension progressively falling from the lung alveoli to the mitochondria in the peripheral tissues where oxygen is consumed. The p_{O_2} in the venous return can be sampled as an index of tissue consumption, but it is usually higher than it is in many capillaries, because some circulatory channels pass through areas of lower consumption, and also because there is some direct diffusion of oxygen from the arterial to the venous side.**

Consumption is increased many-fold over the basal rate by exercise, reaching a maximum of over 200 mmoles min^{-1} in the male of median size who is a trained athlete. (The maximum consumption of an untrained male of that size is about 2/3 as great.) Much of the increased rate of transport is accounted for by an increased rate of blood circulation; the hemoglobin carrier is shuttled between lungs and tissues in less than 0.2 of the time required at rest. The remainder of the increase comes from greater release of oxygen by the hemoglobin; the amount of oxygen released per hemoglobin molecule on each pass through a working muscle can be three times the value in a resting muscle.

OXYGEN BINDING SITE

Prosthetic Group. We noted in Chapter 3 that the binding of oxygen by hemoglobin involves the action of two components — heme (iron(II)-porphyrin), and a polypeptide chain wrapped around it. The oxygen molecule is carried in a crevice between these two. The use of heme illustrates a general principle. The performance of some biological functions requires chemical groups that are not present in the 20 amino acids from which proteins are made. Additional needs for chemical reactivity are met by building the necessary groups into some low molecular weight compound, and then constructing a polypeptide chain to associate strongly with this compound. The result is a protein composed in part of polypeptide, known as the **apoprotein,** and in part of another compound, the **prosthetic group** (from *prosthesis,* an additional part). We shall encounter many kinds of apoproteins and their associated prosthetic groups.

The word hemoglobin is a contraction meaning blood globulin, and when the protein was found to contain a porphyrin bound to iron in addition to polymerized amino acids, it was a logical extension to call the iron porphyrin heme and the apoprotein **globin.**

The iron atom in hemoglobin is like many metallic ions that participate in biological processes by forming complexes with proteins or with other components of the cell. Some complexes, such as those of magnesium and calcium, are easily dissociated, so that a given atom of metal is rapidly exchanged from one complex to another. Other ions, such as copper(II), cobalt(II), zinc, and manganese(II), are bound more tightly, often so tightly that the metal is for all practical purposes a fixed part of the structure as long as it exists. This is also the case with the iron(II) ion in heme; it remains in the porphyrin until the porphyrin is destroyed.

Free heme is unstable in contact with oxygen; the Fe(II)-porphyrin is oxidized to Fe(III)-porphyrin, known as hemin. However, little heme exists in free form in cells, because they construct sufficient apoproteins, such as globin, to combine with most of the heme and protect it from oxidation. Unlike the encounter of heme and oxygen in free solution, the formation of oxyhemoglobin does not involve a permanent transfer of electrons to oxidize heme (Fe(II)-porphyrin) to hemin (Fe(III)-porphyrin); the iron in hemoglobin remains as Fe(II) during oxygen transport.

MECHANISM OF TRANSPORT

Binding for transport differs from binding by antibodies in an important way: Transport implies not only a picking up of the substance at the point of supply, but also a letting go at the point of demand, and it is the release that antibodies are not built to do. We saw in the antibodies how binding is achieved through specific arrangement of reactive groups; now let us consider what must be done in order to release a bound molecule.

Some transported molecules are acquired by cells in a rather drastic way; the entire carrier complex is engulfed by pinocytosis and then degraded within the cell so as to detach the load. Such a mechanism is prohibitively expensive for transport of something in as high a demand as is oxygen. If hemoglobin had to be destroyed in order to give up its oxygen, our hard-working median man would consume nearly 50 grams of this protein per second.

We noted before (p. 40) that the internal hemoglobin in muscle, myoglobin, is a single polypeptide chain wrapped around a heme molecule, while circulating hemoglobin contains four subunits, two α chains and two β chains, each similar in size to myoglobin. The transport behavior of these two kinds of hemoglobin is quite different; the difference deserves close examination not only because of its inherent importance, but also because similar differences in the biological function of proteins will be seen in catalysis by enzymes.

Transport by Myoglobin

The Myoglobin Dissociation Curve. One possibility for transport down a concentration gradient is through simple equilibration of the carrier with the transported molecule. Myoglobin is designed to carry oxygen in this way down a relatively steep concentration gradient within cells:

$$Mb + O_2 \rightleftharpoons MbO_2$$

Where the concentration of oxygen is high, the carrier binds more; where the concentration is low, the carrier binds less and will release the oxygen that it brought to the region of low concentration (Fig. 13–3). The carrier is **facilitating diffusion** in a simple way.

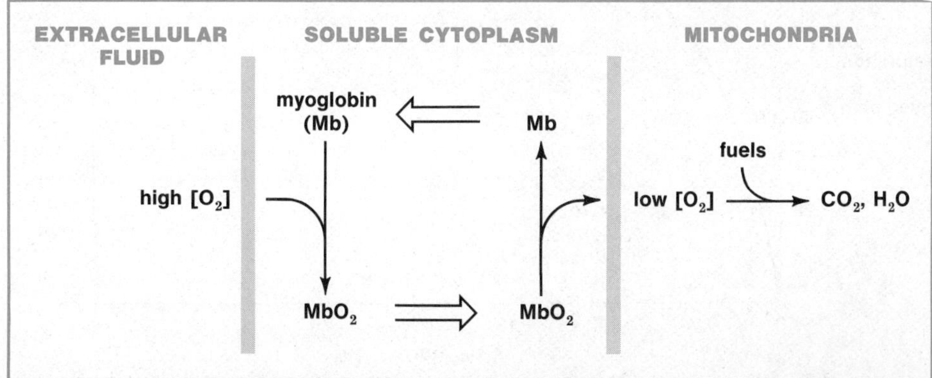

FIGURE 13–3 **The presence of myoglobin can facilitate diffusion within a cell. If the concentration gradient of O_2 is in a range over which the concentration of oxygen greatly affects the degree of its binding to myoglobin, much more of the myoglobin will be oxygenated on the high side of the gradient. Effective concentration gradients will therefore be established for oxymyoglobin (MbO_2) in one direction and for deoxymyoglobin (Mb) in the opposite direction, so that molecules of the carrier will go to and fro across the concentration gradient, picking up O_2 on the high side, and releasing it on the low.**

The behavior of myoglobin toward oxygen can be described by the ordinary equilibrium expression:

$$\frac{[MbO_2]}{[Mb]\,[O_2]} = K_{eq}$$

Given a fixed supply of myoglobin within a cell, we can calculate the fraction of the total myoglobin that will be oxygenated at varying oxygen concentrations from the above equation and obtain a curve with the hyperbolic shape shown in Figure 13–4.

> *A note on units*
>
> Concentration of oxygen in solution, the **oxygen tension,** is frequently given as the partial pressure, p_{O_2}, of a gas phase in equilibrium with the solution. The customary unit has been the **torr** (mm of mercury). Some now express pressures in **millibars** (dynes cm^{-2}), or **pascals** (newtons m^{-2}). The pascal is the unit of the International System. One torr = 1.333 millibars = 133.3 pascals. One millibar = 100 pascals = 0.750 torr.

The problem with myoglobin, or with any carrier functioning by simple mass-action equilibrium, is that it can be efficient only when it operates over a large concentration gradient; the ratio of oxygen concentrations at the points of uptake and release must be high. We can see this by comparing the approximate

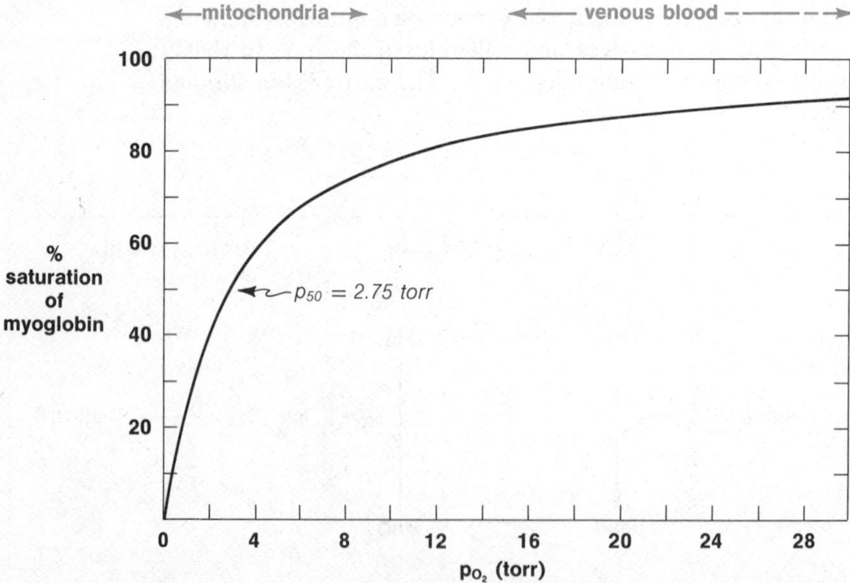

FIGURE 13–4 **The effect of oxygen concentration upon the oxygenation of myoglobin. This association curve is a plot of the percentage of myoglobin molecules binding oxygen (per cent saturation) as a hyperbolic function of the $[O_2]$, given as the partial pressure of O_2 gas in equilibrium with the solution. Half of the total myoglobin is oxygenated, that is, $[Mb] = [MbO_2]$, when the p_{O_2} is 2.75 torr. (This is termed the p_{50} for 50 per cent saturation, or the $p_{0.5}$ for half-saturation.) Typical ranges for the measured p_{O_2} of venous blood, and the estimated range in mitochondria are shown at the top. (p_{50} from E. Antonini, (1965) Physiol. Rev., *45*: 123.)**

oxygen concentrations required for saturating myoglobin to varying extents:

Saturation*	$[O_2]$
0.90	24.75 torr
0.75	8.25
0.50	2.75
0.25	0.92
0.10	0.31

*Saturation is the fraction of the myoglobin molecules containing bound oxygen.

The oxygen concentration must fall to one ninth of its original value, or below, for two molecules of myoglobin to release one molecule of oxygen. (Compare values at 0.75 saturation and 0.25 saturation, for example.)

This is not an impossible requirement; such conditions may occur within working muscle cells, in which the gradient of oxygen tension may range from 10 torr at the cell surface to 1 torr at the mitochondria, and myoglobin is indeed used within these cells to facilitate the diffusion of oxygen. (An oxygen molecule can move much faster than the bulky myoglobin molecule, but the solubility of oxygen is so low that there are far fewer molecules of it in solution in the muscle.)

Transport by Hemoglobin

Hemoglobin discharges oxygen at a partial pressure that is high relative to the pressure at which it takes on oxygen, sufficiently high to provide the concentration gradient necessary for transport from the blood vessels into the recipient tissues. The construction of hemoglobin from multiple subunits is a critical device for generating this behavior.

Cooperative Interaction. The four subunits in hemoglobin, each with its heme, behave much like myoglobin when separated in the laboratory, but they do not act as independent entities in the complete molecule in red blood cells. Hemoglobin behaves differently because the subunits interact in a cooperative way; although the tetramer has a lower affinity for oxygen than do its separate monomers or myoglobin, the binding of oxygen by one subunit in the tetramer enhances further binding of oxygen by the others. Similarly, the loss of one molecule of oxygen from the saturated carrier, $Hb(O)_4$, makes it easier to lose additional molecules, and the final oxygen molecule begins to depart at what is still a relatively high oxygen tension. A plot of oxygen binding as a function of tension therefore has a complex sigmoid shape (Fig. 13–5). Hemoglobin is said to exhibit **positive cooperativity** because a change in binding by one subunit makes it easier for a similar change to occur in the other subunits.

Events in oxygen transport can be deduced from Figure 13–5. Exposure of hemoglobin molecules to oxygen tensions near 100 torr in the lungs causes them to become approximately 97 per cent saturated, with an average of $0.97 \times 4 = 3.88$ molecules of oxygen per hemoglobin. Little oxygen is released until the erythrocytes encounter lower oxygen tensions in the capillaries of peripheral tissues. This tension will be 30 to 40 torr in tissues of moderate oxygen consumption, and only one quarter of the transported oxygen will be released, roughly one molecule per molecule of hemoglobin. However, in tissues with high rates of consumption, such as the heart or rapidly working skeletal muscles, the further drop in oxygen

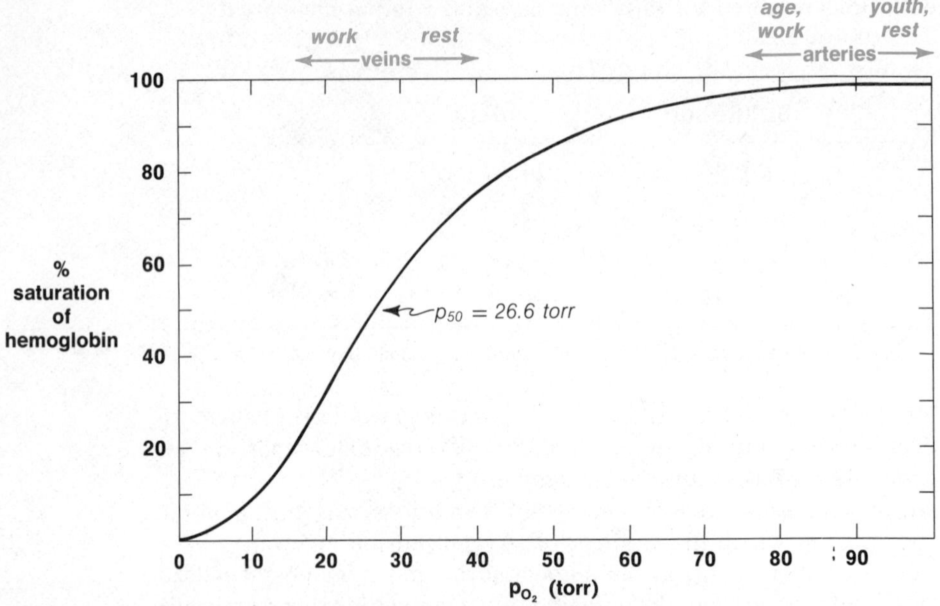

FIGURE 13–5 **The oxygen association curve for hemoglobin in red blood cells. The sigmoidal shape is characteristic for multiple binding sites that interact cooperatively. The p_{50} differs by ±2 torr in individuals; 26.6 torr is a "standard" value for intact red blood cells. The pO_2 in the venous circulation at given fluxes of oxygen is determined by this value. The value in arterial blood also depends upon the transport capacity of the lungs, which declines with age, and is severely impaired by many pulmonary diseases. ("Standard" association curve in J. W. Severinghaus, (1966) J. Appl. Physiol., *21*: 1108.)**

tension toward 15 torr will cause hemoglobin to release 80 per cent of its oxygen, or an average of 3.1 molecules. The release is greater in some muscle capillaries and in the heavily working heart; a fall to 10 torr causes a 90 per cent release. Hemoglobin is therefore constructed so as to provide a large reserve of oxygen-carrying capacity for use during periods of high demand, without a drastic fall in oxygen tension. (Contrast the concentration ratio of 10 required for 90 per cent delivery by hemoglobin with the concentration ratio of 100 required for 90 per cent delivery by myoglobin.) Upon return of the blood to the lungs through the venous circulation, the whole process begins over again.

The Hill Coefficient. If hemoglobin bound oxygen one molecule at a time without cooperative interaction, the combination could be expressed by the same simple association equation developed for myoglobin: $[HbO_2]/[Hb][O_2] = K$, in which Hb represents one subunit. However, if the binding goes through n successive stages, each with a different affinity between hemoglobin and oxygen, the dissociation equation will approach this form:

$$\frac{[Hb(O_2)_n]}{[Hb]\ [O_2]^n} = K$$

in which all of the bound oxygen is assumed to be present as $Hb(O_2)_n$. This equation can be recast logarithmically:

$$n \log[O_2] = \log \frac{[Hb(O_2)_n]}{[Hb]} - \log K$$

According to this equation, a plot of the logarithm of oxygen tension against the logarithm of the ratio of oxyhemoglobin to deoxyhemoglobin will give a straight line with a slope of n, the Hill* coefficient. The Hill coefficient is useful in describing the action of proteins with cooperative interactions, because it is an indication of the minimum number of sites that influence each other. The value for hemoglobin is 2.8, showing at least three contacts between subunits that change upon oxygen binding. However, the Hill equation is not a theoretically rigorous quantitative description.

MODIFICATION OF TRANSPORT

Allosteric Effects

A red blood cell is not a packet of pure hemoglobin solution; it contains other compounds, some of which are of critical importance in adjusting the amount of oxygen that is released at given tensions. The components of interest in humans are H^+, CO_2, and the phosphate ester, **2,3-bisphosphoglycerate:**

$$\begin{array}{c} COO^{\ominus} \\ | \\ H—C—O—PO_3^{2-} \\ | \\ CH_2—O—PO_3^{2-} \end{array}$$

2,3-bisphosphoglycerate

(2,3-Bisphosphoglycerate is sometimes abbreviated DPG for its older name, 2,3-diphosphoglycerate.) Each of the three components diminishes the affinity of hemoglobin for oxygen, much as if it competes for a place on the carrier (Fig. 13–6). The major physiological function of these effects is to alter the amount of oxygen liberated in the peripheral circulation. Increasing the concentration of any one lowers the amount of oxygen that can be retained by hemoglobin.

Here we have the basis for the regulation of oxygen transport within an individual and also between species. The oxygen tension at the tissues, given a certain demand, is determined by the affinity of hemoglobin for oxygen. A certain base line is established for a species, or for developmental stages of an individual, through the genetically-determined construction of a particular hemoglobin. In humans, and in many other animals, the hemoglobin is made with inherently too strong an oxygen affinity for proper function. Adjustments to properly weaken the

*Hill, Archibald Vivian (1886 —), English muscle physiologist and Nobel Laureate.

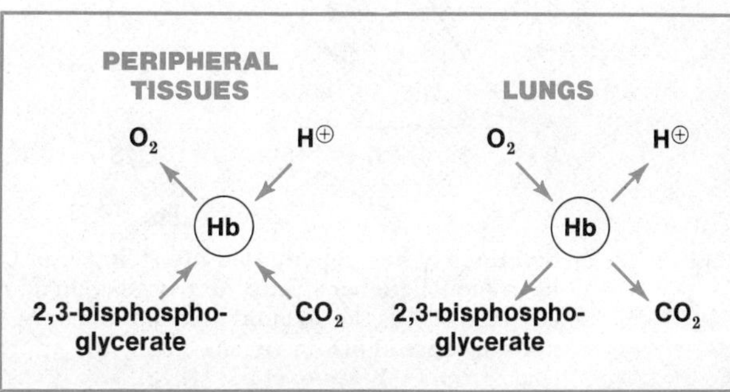

FIGURE 13–6

Oxygen is effectively pushed off hemoglobin by H^+, CO_2, or 2,3-biphosphoglycerate (*left*). Oxygen, in turn, diminishes the binding of all three when blood is returned to the lungs, where its higher concentration forces it onto hemoglobin (*right*).

binding are made through changes in the environment of the hemoglobin molecule — alterations in the concentrations of H^+, CO_2, and 2,3-bisphosphoglycerate.

Definitions. A change in the biological behavior of a protein caused by compounds not directly involved in its function is an allosteric effect (*allo* = other).* Thus H^+, CO_2, and 2,3-bisphosphoglycerate are allosteric effectors, which modulate the binding of oxygen by hemoglobin. The term has been modified by usage, as we shall see when we discuss the regulation of enzymes; it is especially pertinent to regulatory responses caused by interactions between multiple binding sites, including the response of hemoglobin to oxygen, itself. Allosteric effects can frequently be recognized by a sigmoidal plot of response to changing reactant concentrations.

The magnitude of allosteric effects in human hemoglobin is large. Hemoglobin that has been stripped of CO_2 and phosphate compounds in the laboratory retains a sigmoid response to oxygen tension, but its affinity for oxygen becomes much closer to that of myoglobin (Fig. 13–7). However, addition of physiological concentrations of CO_2 and 2,3-bisphosphoglycerate to a hemoglobin solution at constant H^+ concentration makes the oxygen saturation curve nearly identical to that of intact red blood cells, showing that most of the allosteric effects in the complete system arise from these compounds (at constant pH). The cooperative effects change the association so that hemoglobin will release oxygen at a higher tension in the capillaries. Let us now summarize the action of the individual effectors.

Effect of H^+ – the Bohr† Effect. An acidification of blood causes a discharge of oxygen from the red blood cells. The greater the amount of H^+ being

*The term was introduced by Jacques Monod and Francois Jacob, French Nobel Laureates.

†Christian Bohr. Danish pioneer respiratory physiologist. His son Niels had some success as a physicist.

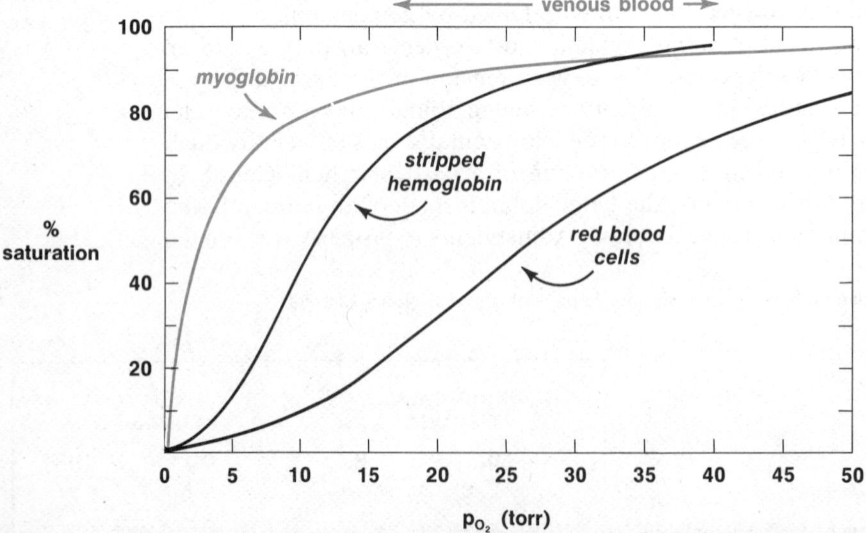

FIGURE 13–7 **Solutions of hemoglobin that are stripped of organic phosphates and CO_2 have a much higher affinity for oxygen than do red blood cells. The stripped carrier still has the sigmoid association curve, but the p_{50} is well below the normal range of pO_2 in venous blood, and approaches the p_{50} of myoglobin. (Data from O. Brenna, et al., (1972) Adv. Exp. Biol. Med., 28: 19.)**

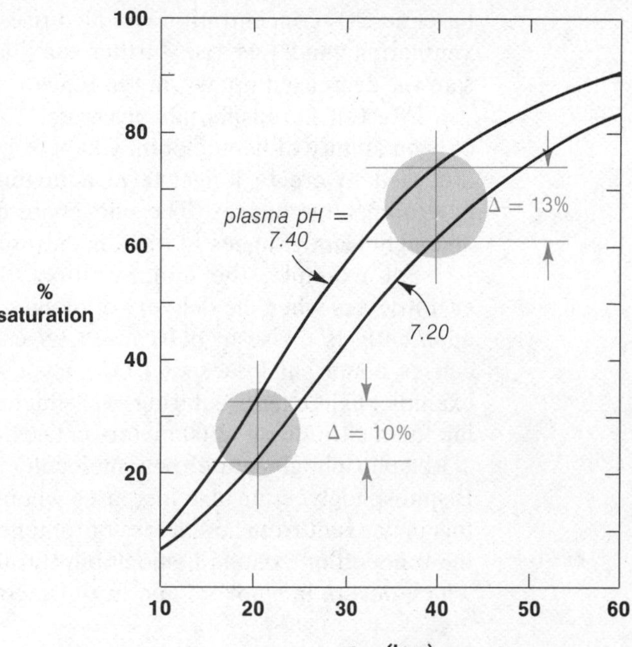

FIGURE 13–8

The Bohr effect is the decrease in affinity of hemoglobin for O_2 at higher H^+ concentrations (lower pH). A fall in the measured plasma pH from 7.4 to 7.2, such as seen in blood draining from hard-working muscles, causes corresponding changes within the erythrocytes, which result in a further 10 per cent discharge of oxygen while holding P_{O_2} at 20 torr—a typical value during heavy exercise. The discharge is even greater at 40 torr (13 per cent), showing that acidosis—a general increase in $[H^+]$ throughout body fluids—would tend to maintain a higher P_{O_2} in the tissues.

poured into the blood from the tissues, the greater the release of oxygen in the capillaries. We shall see (Chapter 39) that there is some acidification of venous blood from CO_2 produced by the tissues, but the most important effects occur when there is an unusually high demand for oxygen, as in strenuously exercising muscles, or when the oxygen supply is impaired, as in circulatory defects. Under such circumstances, tissues produce large amounts of H^+, primarily in association with lactate formation. For example, the plasma pH of blood passing through normal hard-working muscles falls from 7.4 to 7.2 or lower. (The pH within erythrocytes is approximately 0.2 unit lower, but the extracellular pH, which is readily accessible to measurement, is used as a guide to the intracellular pH.) The consequences of such a change are shown in Figure 13–8; when H^+ concentration increases (lower pH), hemoglobin can retain less oxygen at a given oxygen tension. Put another way, the increased acidity enables delivery of more oxygen without a fall in oxygen tension; the additional delivery is equivalent to 10 per cent saturation (0.4 molecule of O_2 per molecule of hemoglobin) at the 20 torr partial pressure common in the venous drainage of trained working muscles. (The tension is lower in muscles of the flabby who are made to work hard.)

If H^+ decreases the affinity for oxygen, why doesn't it impair the uptake of oxygen by hemoglobin in the lungs? It does, but the saturation curve is so nearly flat at the higher oxygen tensions that the effect is small; hemoglobin is 1 to 2 per cent less saturated at pH 7.2 than it is at pH 7.4 in the lungs. In sum, the Bohr effect provides a built-in delivery of more oxygen in response to demand for oxygen, as manifested by increased H^+ formation.

Effect of CO_2. An important part of the decreased affinity for oxygen of hemoglobin in blood is due to the high concentration of CO_2. The total CO_2, which includes not only the dissolved gas but also carbonic acid and the bicarbonate ion in equilibrium with it, is 25 millimolar in adult arterial blood. The tissues discharge between 0.7 and 1.0 mole of CO_2 for each mole of O_2 consumed, causing a maximum increase of approximately 25 per cent in the total CO_2. Similarly, the baseline CO_2 concentration will also rise if there is any impairment in pulmonary ventilation, and this rise further facilitates the release of oxygen to compensate for decreased uptake in the lungs.

Effect of 2,3-Bisphosphoglycerate. This compound, like H^+, diminishes the oxygen affinity of hemoglobin. Changes in 2,3-bisphosphoglycerate concentration are used to create longer-term adjustments of oxygen affinity in response to environmental changes. The phosphate ester is retained within the erythrocyte, and significant changes in its concentration require hours.

For example, the concentration of 2,3-bisphosphoglycerate increases in erythrocytes when the delivery of oxygen is impaired (hypoxia). (The actual signal apparently is a rise in pH caused by excessive ventilation in the lungs, which causes abnormal losses of CO_2.) Exposure to high altitude is one cause. For example, light exercise during experimental exposure to low pressure corresponding to an altitude of 4,500 meters causes an increase from 0.8 to 1.1 molecules of 2,3-bisphosphoglycerate per molecule of hemoglobin within 24 hours. 2,3-Bisphosphoglycerate also increases when there is a loss of functional hemoglobin; this may result from actual loss of total hemoglobin (anemia), or by impairment of the function of existing hemoglobin through combination with carbon monoxide, which occurs in smokers and in those constantly exposed in other ways.

Structural Basis

What happens to a hemoglobin molecule when it binds oxygen? We have direct evidence from X-ray crystallography for the following:

The Iron Atom Shifts. The iron atom in deoxyhemoglobin, lacking a ligand for the sixth coordination position, has a high spin state, and lies out of the plane of the porphyrin ring. Binding of an oxygen molecule stabilizes a low spin state, with the iron atom in the plane of the ring. There is a corresponding movement of the histidine residue attached to the iron (see p. 39), and a conformational change occurs in the globin polypeptide chain. (The lower net spin occurs because the strong ligand field of oxygen forces more electrons into paired orbitals in the iron atom.)

The subunits of hemoglobin shift position relative to each other. The forces creating the roughly tetrahedral shape of hemoglobin are developed by contacts between the subunits, and some contacts change upon oxygenation. We can describe the changes most readily by labeling the subunits, α^1, α^2, β^1, and β^2, even though there is no difference between the two α chains or the two β chains:

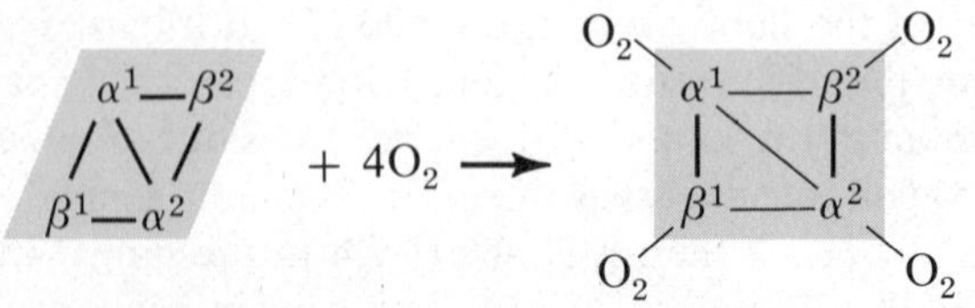

$\alpha^1\beta^2\beta^1\alpha^2 + 4O_2 \rightarrow \alpha^1\beta^2\beta^1\alpha^2(O_2)_4$

deoxyhemoglobin oxyhemoglobin

Each α subunit is bonded to one β subunit by contacts that do not change appreciably, so the $\alpha^1\beta^1$ and $\alpha^2\beta^2$ interfaces remain nearly constant. There are no contacts between the β subunits. The interfaces that do shift when oxygen is added are $\alpha^1\beta^2$, $\alpha^2\beta^1$, and $\alpha^1\alpha^2$, and the shift weakens the contacts.

The overall result is the rotation of one $\alpha\beta$ pair relative to the other, along with a small change in shape of each subunit. It is possible to make reasonable interpretations of the allosteric behavior of hemoglobin in terms of these conformational shifts.

Tense and Relaxed Conformations. In the original Monod-Jacob formulation of allosteric effects, deoxyhemoglobin was said to have a tense conformation that impeded combination with oxygen, while fully oxygenated hemoglobin has a relaxed conformation with no impediments to oxygen binding. Some structural basis for the notion has since been developed; access to the heme in β subunits of deoxyhemoglobin is blocked by a protruding valine residue, which shifts out of the way as oxygen is added, but the α subunits also have a lower oxygen affinity in deoxyhemoglobin. The intermediate steps in the transition between the two forms have not been worked out. It is worth a moment to examine two possibilities, because the same considerations will apply to other allosteric proteins.

Oxygen molecules may approach hemoglobin like the proverbial camel in the tent. The first one gets into a heme cleft with difficulty, so that combination becomes probable only at relatively high oxygen tensions, but after it is bound to one hemoglobin subunit, the ensuing conformational changes shift bonds to other subunits, making access to them easier. Each addition of oxygen breaks more bonds between subunits, making further addition easier and accounting for the sigmoid rise in the association curve (Fig. 13–9*A*).

On the other hand, there may be an inherent equilibration of relaxed and tense forms occurring in both the presence and the absence of bound oxygen (Fig. 13–9*B*). According to this view, oxygen combines more readily with the relaxed form; the combination shifts the equilibrium to make more relaxed form, thereby facilitating the addition of more oxygen molecules.

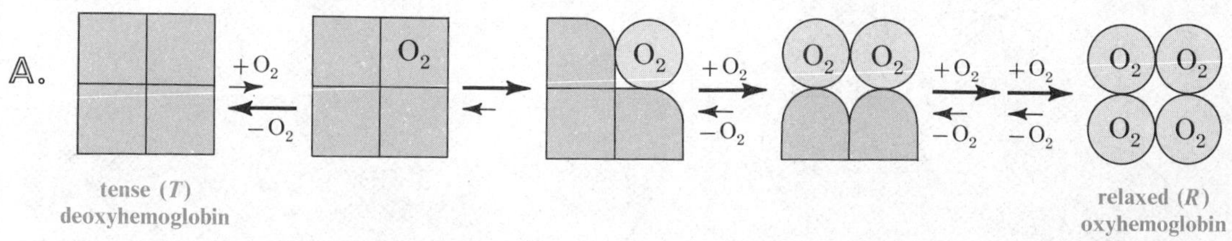

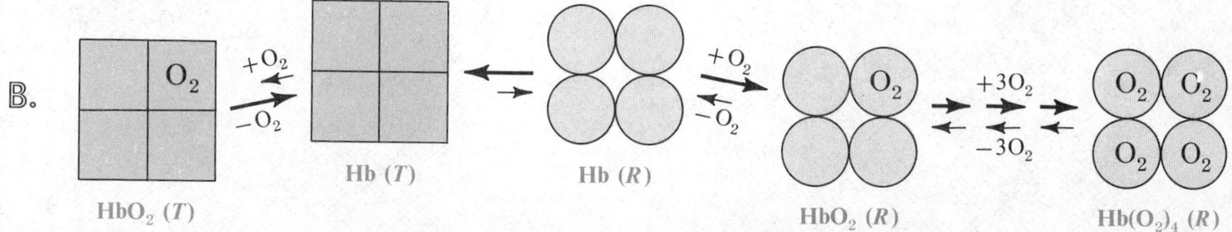

FIGURE 13–9 Allosteric effects can arise in at least two general ways. *A*. The initial binding of O_2 to deoxyhemoglobin occurs with the tense conformation, but the equilibrium is unfavorable. However, when one is bound (*far left*), it causes a shift in conformation of the subunit to a relaxed state (*gray circle*), which modifies the bonds of the adjacent subunits, so that they can take on additional O_2 more readily to form the fully oxygenated molecule (*far right*). *B*. Another possibility is that the tense (*T*) and relaxed (*R*) conformations constantly equilibrate, even in the absence of oxygen (*second and third compounds from left*), but the equilibrium favors the *T* form. However, the *R* form has the higher affinity for O_2; therefore, raising the pO_2 shifts the equilibrium in favor of the *R* conformation, which is then readily available for binding more O_2.

Either mechanism may involve intermediate steps with distinct intermediates, or all of the oxygen molecules may be bound in a rush, once one is put in place.

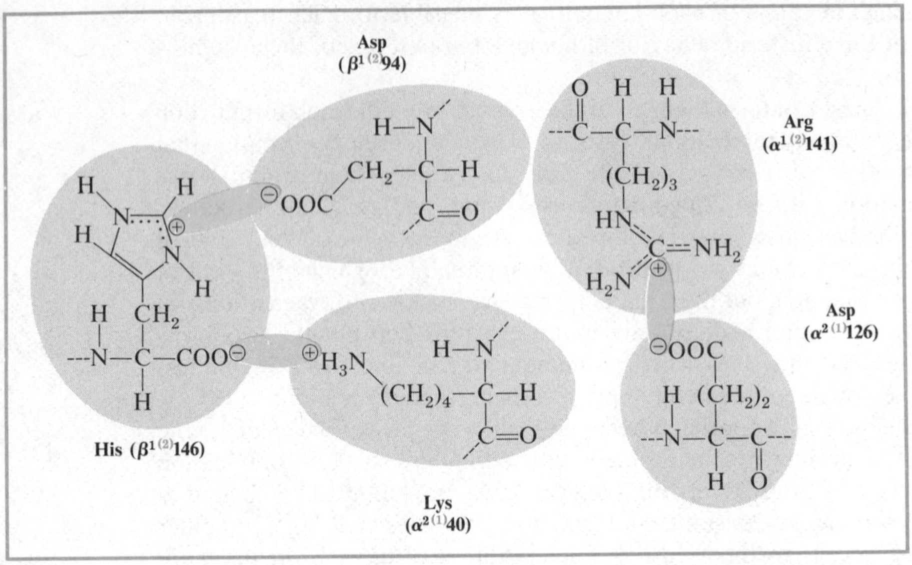

Asp ($\beta^{1(2)}94$)

Arg ($\alpha^{1(2)}141$)

Asp ($\alpha^{2(1)}126$)

His ($\beta^{1(2)}146$)

Lys ($\alpha^{2(1)}40$)

FIGURE 13–10 **Formation of three pairs of salt bridges is favored by the loss of oxygen from human oxyhemoglobin. Only one example of each of the three kinds of bridges is shown. These kinds are (*1*) associations between imidazolium ions and carboxylate ions within the same β chains (*top left*); (2) associations between terminal carboxylate groups of the same β-His residues, and side chain ammonium groups (on Lys residues) of the α chains (*bottom left*); (3) associations between guanidinium groups of arginine side chains on one α chain and carboxylate groups of aspartate side chains on the other (*right*). (The $\alpha^{1(2)}$, and similar designations, indicate that the α^1 and α^2 chains form the same bonds.)**

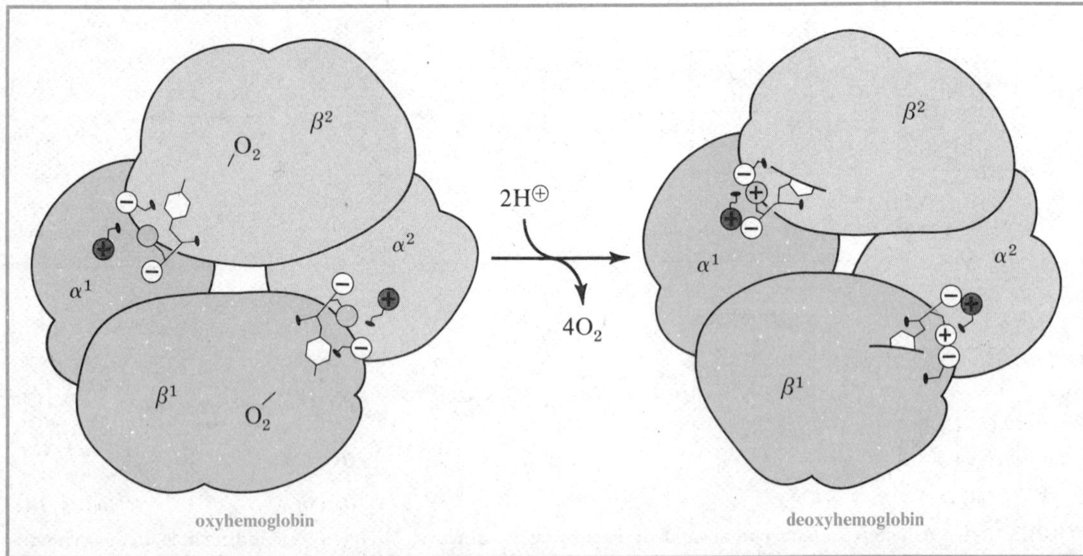

FIGURE 13–11 **The effect of oxygenation on $\alpha^1\beta^2$ bonds. The conformation of oxyhemoglobin (*left*) leaves a β-tyrosyl group exposed on the surface; its location is such that the adjacent carboxyl-terminal histidine residue is prevented from participating in salt bridges. When oxygen is lost, the resultant conformational change buries the tyrosyl side chain (*right*), and pulls the adjacent histidine (*dark blue*) to a location where its COO^- group forms a salt bridge with an α Lys^+, and its imidazole group is close to a β Asp^-. The imidazole group therefore retains a proton more readily to become positively charged and establish another salt bridge. These bridges lock the molecule in the tense conformation. (Drawing modified from R. E. Dickerson, (1972) Annu. Rev. Biochem., *41*: 815.)**

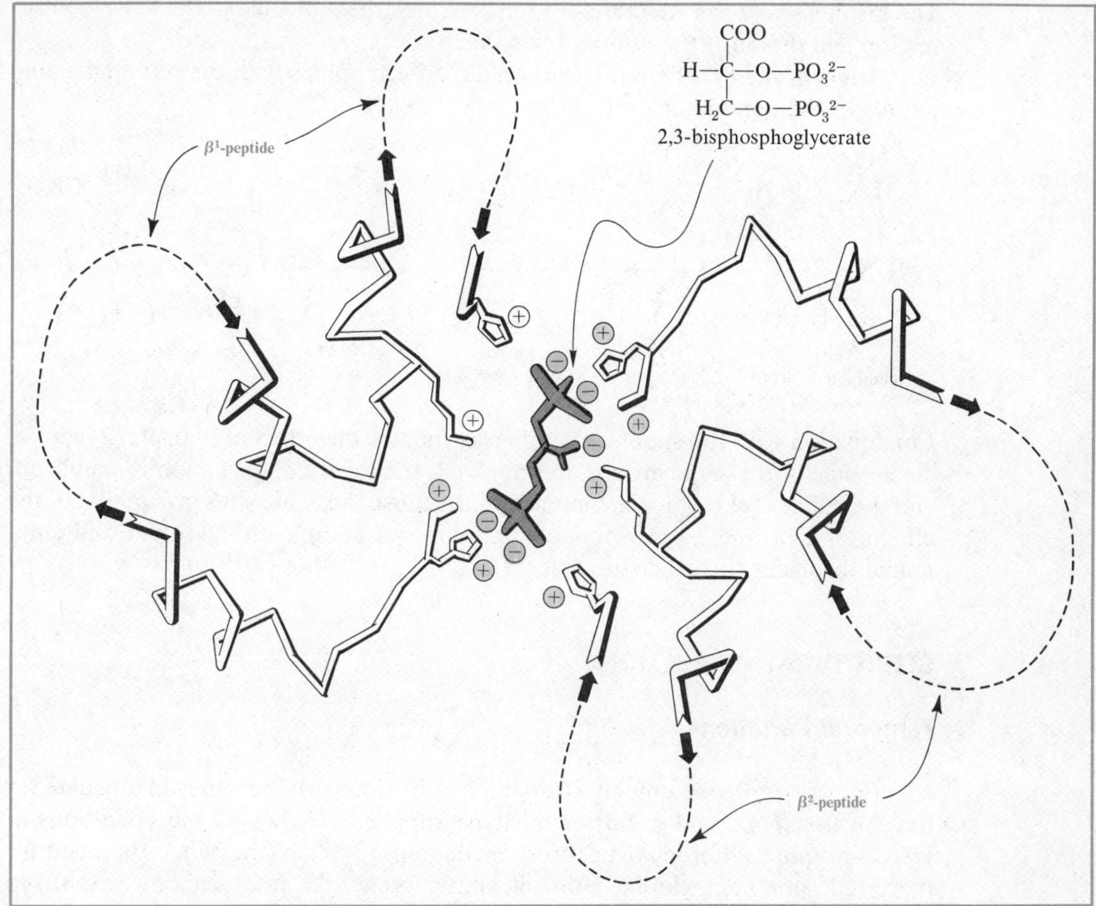

FIGURE 13–12 **2,3-Bisphosphoglycerate, with its five negative charges, combines with positively charged side chains in the cleft between β-subunits of hemoglobin, keeping them separated in the tense, deoxygenated conformation. The compound therefore diminishes the affinity of hemoglobin for oxygen. (Adapted from A. Arrone, (1972) Nature, *237*: 146.)**

Origin of the Bohr Effect. The displacement of oxygen by protons and vice versa is accomplished by making and breaking salt bridges between the subunits. The molecule is constructed so that some bridges present in deoxyhemoglobin (tense conformation) are absent in oxyhemoglobin (relaxed conformation). The bridges involve positively charged imidazolium groups (from histidine residues) and ammonium groups (from terminal valine residues and side chains of lysine residues) combined with negatively charged carboxylate groups (from terminal histidine or arginine residues and side chains of aspartate residues) (Fig. 13–10).

Increasing the concentration of H^+ increases the tendency for the imidazole and amino groups to gain a positive charge and form the salt bridges. Therefore, H^+ promotes a shift from the relaxed to the tense conformation, thereby decreasing the affinity for oxygen (Fig. 13–11).

Origin of the Bisphosphoglycerate Effect. Eight positively charged groups line the cleft between β subunits. The groups are so positioned that they will form strong electrostatic bonds with the negative charges on 2,3-bisphosphoglycerate when the molecule is in the tense conformation (Fig. 13–12). The bonds are weakened when the molecule is in the relaxed conformation with oxygen bound.

The bisphosphoglycerate therefore tends to lock the molecule in the tense conformation and diminish the affinity for oxygen.

Origin of the CO_2 Effect. Carbon dioxide combines with the terminal amino groups of the peptide chain to form carbamates:

Val-1 (ammonium form) $\rightleftharpoons$ Val-1 (amino form) + $H^{\oplus}$; Val-1 (amino form) + CO_2 $\rightleftharpoons$ N-carboxyl-Val-1 + $H^{\oplus}$

Val-1
(ammonium form)

$H^{\oplus}$

Val-1
(amino form)

$H^{\oplus}$

N-carboxyl-Val-1

Combination with the subunits in this way mimics the effect of phosphoglycerate, since some of the same groups are involved; the tense conformation is stabilized, and the affinity for oxygen is diminished. Because the same sites are involved, the effects of CO_2 and bisphosphoglycerate are not completely additive; one compound displaces the other to some extent.

STRUCTURAL VARIATIONS

Temporal Variations

Embryos grow so quickly that simple diffusion early becomes inadequate for oxygen transport, and a carrier must be supplied. However, the conditions *in utero* are quite different than those encountered after birth; hence the need for more than one hemoglobin. Humans contain genes for producing at least seven different hemoglobin subunits, which are expressed at different times during development. The seven are designated α, β, ${}^{A}\gamma$, ${}^{G}\gamma$, δ, ϵ, and ζ. The ζ chains apparently act in place of α chains; the others are all substitutes for β chains. A general chain formula for the human hemoglobins will therefore be:

$$\begin{array}{r|l} & \beta_2 \\ \alpha_2 & \text{or } {}^{A}\gamma_2 \\ \text{or } \zeta_2 & \text{or } {}^{G}\gamma_2 \\ & \text{or } \delta_2 \\ & \text{or } \epsilon_2 \end{array}$$

Embryonic hemoglobins are characterized by the presence of either ζ or ϵ chains, in place of the α and β chains, respectively, found in adult hemoglobins. The complete amino acid sequence and functions of these hemoglobins is yet to be determined.

Fetal hemoglobins, Hb F, contain α chains the same as those found in adults, paired with either ${}^{A}\gamma$ or ${}^{G}\gamma$ chains. These γ chains differ from each other only in the presence of alanyl or glycyl groups at position 136.

Adult hemoglobin is mostly **Hb A_1**, which is the $\alpha_2\beta_2$ tetramer we have discussed at length. This is the chain composition for 97 to 98.5 per cent of the hemoglobin in a sample of adults (mean 97.5 per cent). The remaining hemoglobin is mostly **Hb A_2**, which has δ chains in place of β chains, and is therefore $\alpha_2\delta_2$. Hb A_2 is only 1.5 to 3.2 per cent of the total hemoglobin in adults (mean 2.48 per cent), and the reason for the existence of this minor component is unknown.

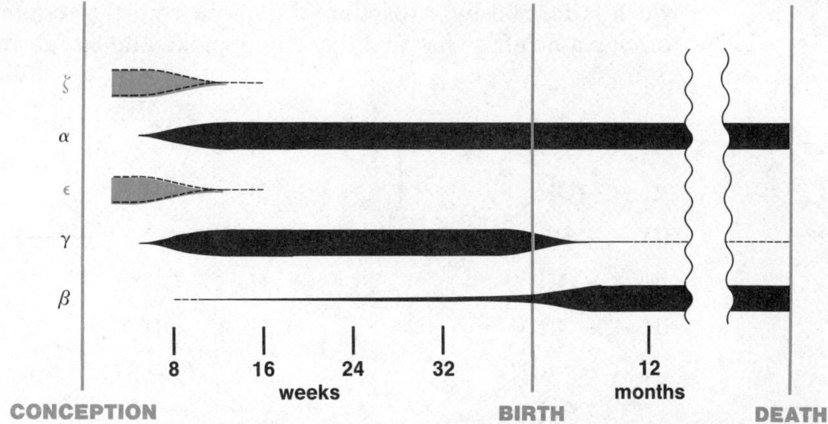

FIGURE 13–13 **The sequence of development of hemoglobin. The earliest events following conception are conjectural. Perhaps there is a $\zeta_2\epsilon_2$ in the smallest embryos, but the first demonstrable appearance of these chains has been as $\zeta_2\gamma_2$ and $\alpha_2\epsilon_2$, showing that their production probably overlaps that of Hb F, $\alpha_2\gamma_2$, which is the principal hemoglobin throughout the remainder of fetal life. The β chains of Hb A first appear at 8 weeks after conception, but are not a significant part of the total until term and shortly thereafter, when the production of Hb F falls rapidly. The formation of δ chains is not shown–it would make a very thin line, at best–but it presumably parallels that of β chains.**

The early maturation sequence is not worked out; the ζ chains have been identified only in Hb-Portland, $\zeta_2\gamma_2$, a hemoglobin found in fetuses unable to make α chains, but a likely succession is shown in Figure 13–13. The content of Hb F rises until it is half of the total hemoglobin by the time (2 months) the embryo has grown to 30 mm length (crown to rump), and 90 per cent of the total at 50 mm length (2.5 months). The formation of adult hemoglobin, Hb A, begins about the 8th week of gestation but does not exceed 10 per cent of the total until the 30th week, and is 10 to 30 per cent at term. Thereafter, the content of Hb A_1 and Hb A_2 rise, and the content of Hb F falls until it is less than 10 per cent of the total at 6 months after birth and less than 2 per cent at 12 months. For some reason the γ chain genes are still turned on in precursors of some erythrocytes, but most adult cells contain no detectible Hb F. The Hb F is 0.03 to 0.7 per cent of the total hemoglobin in adults.

The necessity for a different hemoglobin before birth is clear; the oxygen tension available to the fetal blood in the placenta is lower than the oxygen tension available to the maternal blood in the lungs. Therefore, fetal erythrocytes must have a higher affinity for oxygen than adult cells. The oxygen association curve of Hb F is much like that of Hb A; the important distinction is that Hb F binds 2,3-bisphosphoglycerate less tightly and, therefore, binds oxygen more tightly.

Post-translational Modification. Approximately 10 per cent of the molecules of Hb F are altered by acetylating the terminal amino group of α chains. Blocking the amino groups destroys the reaction with CO_2 and weakens the binding of 2,3-bisphosphoglycerate so that the affinity for oxygen is stronger. (Acetylation is also seen in adult felines; it is strange that the oxygen transport in cats differs from that in all other adult mammals examined.)

Some modifications appear to occur accidentally. Approximately 5 to 8 per cent of the hemoglobin in adult human cells is composed of several minor components, A_{1a}, A_{1b}, A_{1c}, etc. At least two of these arise from spontaneous combinations with other cellular components. A component of special interest is A_{1c}, which is formed by a reaction of glucose with the terminal amino groups, first forming a Schiff's base and then an aminoketone by rearrangement:

glucose		Schiff's base (aldimine)		amino ketone
				H—N—Hb
H—C=O	H_2N—(Val)Hb	H—C=N—Hb		H—C—H
H—C—OH		H—C—OH	*Amadori rearrangement*	C=O
HO—C—H	⟶	HO—C—H	⟶	HO—C—H
H—C—OH		H—C—OH		H—C—OH
H—C—OH		H—C—OH		H—C—OH
CH_2—OH		CH_2—OH		CH_2—OH

HUMAN HEMOGLOBINS*

Component	Chain Composition	M.W. (Anhydrous)	Comments
A_1	$\alpha_2\beta_2$	64,450	also minor A_{1a}, A_{1b}, A_{1c}; A_{1c} has hexose attached, and is increased in diabetes mellitus
A_2	$\alpha_2\delta_2$	64,564	unknown function
^{A}F	$\alpha_2{}^{A}\gamma_2$	64,734	differ only in Ala or Gly at #136
^{G}F	$\alpha_2{}^{G}\gamma_2$	64,706	part of Hb F's is acetylated
embryonic	$\alpha_2\epsilon_2$	?	known only in embryos
Portland-1	$\zeta_2\gamma_2$	?	known only in embryos

Occurrence: Adult blood: 97.5–98.5% Hb A_1 (including modified forms)
1.5 – 3.0% Hb A_2
0.03–0.7% Hb F
Fetal blood: >90% Hb F from ~10 weeks to 30 weeks gestation
$^{G}\gamma/^{A}\gamma = 0.7$ in newborn
p_{50} ~ 26.6 torr in adult, 22.7 in newborn (whole blood)
Concentration: 2.0–2.5 mM (13–16 g/100 ml) in adult blood (8–10 mM to heme)
lower in females than in males
$4.2–5.0 \times 10^{-16}$ moles/erythrocyte ($27–32 \times 10^{-12}$ g)

Subunits	Amino Acid Residues	M.W. (Including heme)
α	141	15,742
β	146	16,483
$^{A}\gamma$	146	16,625
$^{G}\gamma$	146	16,611
δ	146	16,540
myoglobin	153	17,669

Effectors (all decrease affinity for O_2):

H^+:
- 0.44 H^+ taken up per O_2 released (Haldane coefficient)
- $\Delta \log p_{50}/\Delta pH = -0.48$ (Bohr coefficient)
- 80% taken up on α1Val, β146His

2,3-bisphosphoglycerate:
- 0.9 molecule/molecule of hemoglobin
- bound in cleft between β subunits
- 40% bound in arterial blood; 90% would be bound if all O_2 discharged
- content rises in conditions causing O_2 deprivation

CO_2:
- 1.3 molecules/molecule of hemoglobin in venous blood
- bound as carbamate to terminal amino groups

*Note: Quantitative values are typical for adult blood and vary between individuals.

The concentration of this component rises as a red cell ages, showing the relatively slow rate of combination, and the velocity of the reaction is dependent upon the concentration of glucose. Hb A_{1c} is formed much faster in patients with diabetes mellitus, owing to the higher concentrations of glucose that occasionally occur even with the best of control, and may rise to 12 per cent or more of the total hemoglobin. The level of this component has been suggested as a means for following the effectiveness of treatment of diabetics. (We shall say much more about diabetes in Chapter 38.) HbA_{1b} appears to be a mixture of compounds, similar to HbA_{1c} except that they are phosphorylated derivatives, mainly derived from glucose phosphate, but also including triose phosphates. It is possible that HbA_{1c} is formed by hydrolysis of the phosphate group from HbA_{1b}.

Individual Variations

Human blood is probably more readily accessible for study than any other experimental tissue, and samples are drawn each year from millions of individuals. Human hemoglobin therefore represents an excellent opportunity for studying genetic variation of a single protein in a single species, each variation representing an alteration in those DNA molecules affecting the synthesis of the protein.

Most mutations are deleterious, and will gradually be eliminated, the rate depending upon the severity of the resulting reproductive disadvantage. Most hemoglobin variants, therefore, occur in only a few individuals, but some occur in many, even though they are life-threatening. We must look in those cases for a compensatory advantage conveyed by the mutation.

Thalassemias. Impairments of the formation of a single kind of hemoglobin chain result in thalassemias, so-called because the conditions are common in Mediterranean countries (*Thalassa* = sea), and are designated α- or β- thalassemias to show the specific deficiency. They are also prevalent in Southeastern Asia, with an occurrence of some 1 in 100 in Thailand!

Thalassemias may result from mutations affecting any function of DNA, including the regulator genes that determine the rate of synthesis of particular kinds of messenger RNA.

α-CHAIN THALASSEMIAS. It is easier to understand the thalassemias if the gene distribution is kept in mind. Humans may have one or two α-chain genes on a chromosome. Therefore, an individual may have a total of two, three, or four genes, depending upon the inheritance of one or two genes from each parent. Current evidence is that usually Caucasians and Orientals have four genes, Negroes three, and Melanesians two. All of these genes are identical in most humans. (Lower animals commonly make two different kinds of α-chains.)

The effects of inheritance of defective genes or their deletion depends upon what proportion of the total is affected. Total loss of α-chains prevents the formation of Hb F, as well as Hb A_1 and Hb A_2. Fetuses survive for a time by making increased amounts of embryonic hemoglobin (Hb-Portland, $\zeta_2\gamma_2$), but they commonly die before term, or shortly after delivery. (The condition is named hydrops fetalis.)

The genes for making the β-like chains — β, ${}^{A}\gamma$, ${}^{G}\gamma$, δ, and perhaps ϵ — are linked on a different chromosome than that carrying the α-chain genes and are not affected by deletion or impairment of the latter. When the β-like genes are turned on in the normal way during development and there is insufficient α-chains to combine with their products, the accumulated excess forms homotetramers — β_4,

COMMON HUMAN HEMOGLOBIN MUTANTS

Thalassemias are impairments of gene transcription.

α-thalassemias: one or more defective α-chain genes out of apparent total of 3 genes in Negroes, 4 in Caucasians and Orientals

Impair formation of Hb A_1, A_2, F, embryonic.

All genes affected—hydrops fetalis, with death *in utero* or shortly after birth. Accumulate $\zeta_2\gamma_2$(Portland), γ_4 (Bart's).

Three genes affected—Hb H disease, with 5–20 per cent β_4 and some γ_4 in adults. Inclusion bodies, abnormal red cells, mild anemia.

Two genes affected—thalassemia trait, with 5 per cent Hb H in cord blood at birth, some mild anemia, and altered red cell morphology in adults.

One gene affected—detectable γ_4 in cord blood at birth, no discernible effects in adults.

Common in S.E. Asia.

β-thalassemias: one or two defective β-chain genes, with partial or total impairment of affected gene.

Impaired formation of Hb A_1.

One gene affected—thalassemia minor, if gene totally defective. Increased formation of Hb A_2.

Both genes affected—Cooley's anemia, thalassemia major, if both genes totally defective. 20–80 per cent Hb F in adults, with total hemoglobin falling to 0.3–0.5 mM (2–3 g/100 ml).

Common in many parts of the malaria belt.

Incidence among U.S. Blacks: trait 15 (5–20) per 1,000 births; thalassemia major, 0.06 (0.006–0.1) per 1,000 births.

Hemoglobin variants with amino acid substitutions caused by point mutations. (All incidence figures are for U.S. blacks.*)

S–S. Sickle cell anemia. β6Glu→Val. Deoxy form less soluble than is A_1. Precipitates in venous blood, causing sickling and early cell destruction. Painful crises and short life. 1.6 (0.9–4.9) per 1,000 births; 0.5 (0.3–1.6) per 1,000 all ages.

S–A. Sickle cell trait. No effect except when impairment of oxygen supply increases [deoxy Hb S] beyond solubility. 80 (60–140) per 1,000 births.

C–C. (β6Glu→Lys); D–D (β121Glu→Gln); E–E (β26Glu→Lys). Hb C, D, or E disease. Mild anemia in most instances with little impairment. 0.2 (0.02–0.5) of C–C per 1,000 births.

C–A. No detectable effects. 30 (10–46) per 1,000 births.

S–C. Causes sickling (S–C disease). 1.2 (0.3–3.2) per 1,000 births; 0.8 (0.2–2.1) per 1,000 all ages.

D–A, E–A. No detectable effects.

S–β-thalassemia. Causes sickling. 0.6 (0.15–1.4) per 1,000 births; 0.3 (0.08–0.7) per 1,000 all ages.

C–β-thalassemia. 0.2 (0.02–0.5) per 1,000 births.

Persistent Hb F trait. Failure of Hb F production to fall to normal adult levels. 1 per 1,000 births.

Persistent Hb F homozygote. 0.00025 per 1,000 births.

S-persistent Hb F. Causes sickling. 0.04 (0.03–0.07) per 1,000 births.

C-persistent Hb F. 0.015 (0.005–0.02).

β-thalassemia–persistent Hb F. 0.008 (0.0025–0.01) per 1,000 births.

*Incidence figures adapted from A. O. Motulsky: (1973) *Frequence of Sickling Disorders in U.S. Blacks.* N. Engl. J. Med., *288*: 31.

γ_4, or δ_4. Two have been given names: γ_4 is Hb Bart's*, and β_4 is Hb H. There are several lessons to be learned from these tetramers. Only traces of Hb Bart's occur in normal fetuses, and none of the homotetramers in adults. Therefore, the rates of production of α-like and β-like chains must be closely matched in normal individuals. We can also see that the affinity of unlike chains for each other is greater than the affinity of like chains. In addition, all of the homotetramers have the functional characteristics of myoglobin, with high oxygen affinity and none of the cooperative interactions, showing that the allosteric effects of normal hemoglobin come not from mere subunit association, but from specific interactions of the two different kinds of chains. The homotetramers are unstable and tend to form precipitates (inclusion bodies) within red blood cells; not only is there less normal hemoglobin produced, but the life of the cells is also shortened.

The effects of partial impairment of α-chain genes range from no discernible symptoms with only one out of four genes affected (the silent carrier state, detectable by the presence of 1 to 2 per cent Hb Bart's in umbilical cord blood)

*Bart's is the very English way of designating St. Bartholomew's Hospital in London.

through α-thalassemia trait with two out of four or one out of three genes affected and only mild symptoms, to hemoglobin H disease, in which only one of four genes is normal and one quarter of the hemoglobin in fetal cord blood is Bart's (or 4 to 30 per cent in adult blood is H), and there is a moderately severe inclusion body anemia.

Since the homozygous condition is fatal and heterozygotes may be substantially impaired, one would think the defective genes would disappear. Instead, they are quite common in those parts of the world where malignant malaria is common. Indeed, an estimated 25 per cent of the Thai have one or more defective α-chain genes, and homozygous α-chain thalassemia (hydrops fetalis) is the leading cause of stillbirth among them. Why should this be? The heterozygous thalassemias, like the better-known sickle cell trait discussed below, give partial protection against *Plasmodium falciparum,* the most dangerous malarial parasite, which spends part of its life cycle in red blood cells. Here is a classic case of balanced polymorphism, with the incidence of the deleterious gene increasing toward a level at which the enhanced death rate from its presence will just balance the decreased death rate from malaria.

β-THALASSEMIAS. These are common in the malarial regions from the Mediterranean across northern India into Indonesia and in tropical West Africa, but not in central Africa. The incidence is 10 per cent or higher in some Italian villages. There are at least three types of defects, one causing partial loss of β-chain production, another total loss, and a less common one causing partial loss of function of both β- and δ-chain genes, which are closely linked on the same chromosome. Since there are only two genes in most individuals for β-chains, there is only one heterozygous state possible for each type, and the effects are usually minor. The condition may be discovered only upon examination of the cells. Homozygotes develop frank anemias, with precipitation of excess α chains. In either case, the genes for γ chains are not affected, and those for δ chains may not be, so that even heterozygous adults are likely to have increased amounts of Hb F and A_2 in their blood.

Coding Variants. Now let us consider mutations that change the amino acid sequence in one or more hemoglobin subunits. We discussed Hb Lepores and Miyada, in which parts of the genes for β and δ chains are combined, and Hb Wayne, created by a frame-shift mutation, in Chapter 7. The more common mutations alter a single base in DNA so as to substitute an amino acid at one position. The effect depends upon the amino acid and its location. Many such variants have substantially different properties (oxygen affinity, cooperativity, stability, solubility), but even variants with apparently normal properties may be synthesized at a slower rate, or destroyed faster, so that a heterozygote will have detectably lower concentrations of the mutant hemoglobin and an increased amount of Hb F. Homozygotes are then likely to be frankly anemic.

Humans with a long ancestry in the malarial belt commonly have one of four variant hemoglobins — **Hb C, D Los Angeles** (or Punjab), **E,** or **S.** All are so common that it is difficult to regard them as abnormal, although their effects are not innocuous. As many as 30 per cent of the inhabitants of some areas contain a gene for one or more of these variants. All are believed to ameliorate malaria in heterozygotes. All are created by substitution of one residue in each β chain of HbA_1.

For example, the first abnormal hemoglobin discovered, Hb S, has a valine residue at position 6 in place of the glutamate residue found in HbA_1. It may informatively be designated as $\alpha_2\beta_2{}^{6Glu \rightarrow Val}$ (read as Glu becoming Val), or more simply as $\alpha_2\beta_2{}^{6Val}$, or $\alpha^A{}_2\beta^S{}_2$.

The occurrence of these hemoglobins is shown in Figure 13–14.

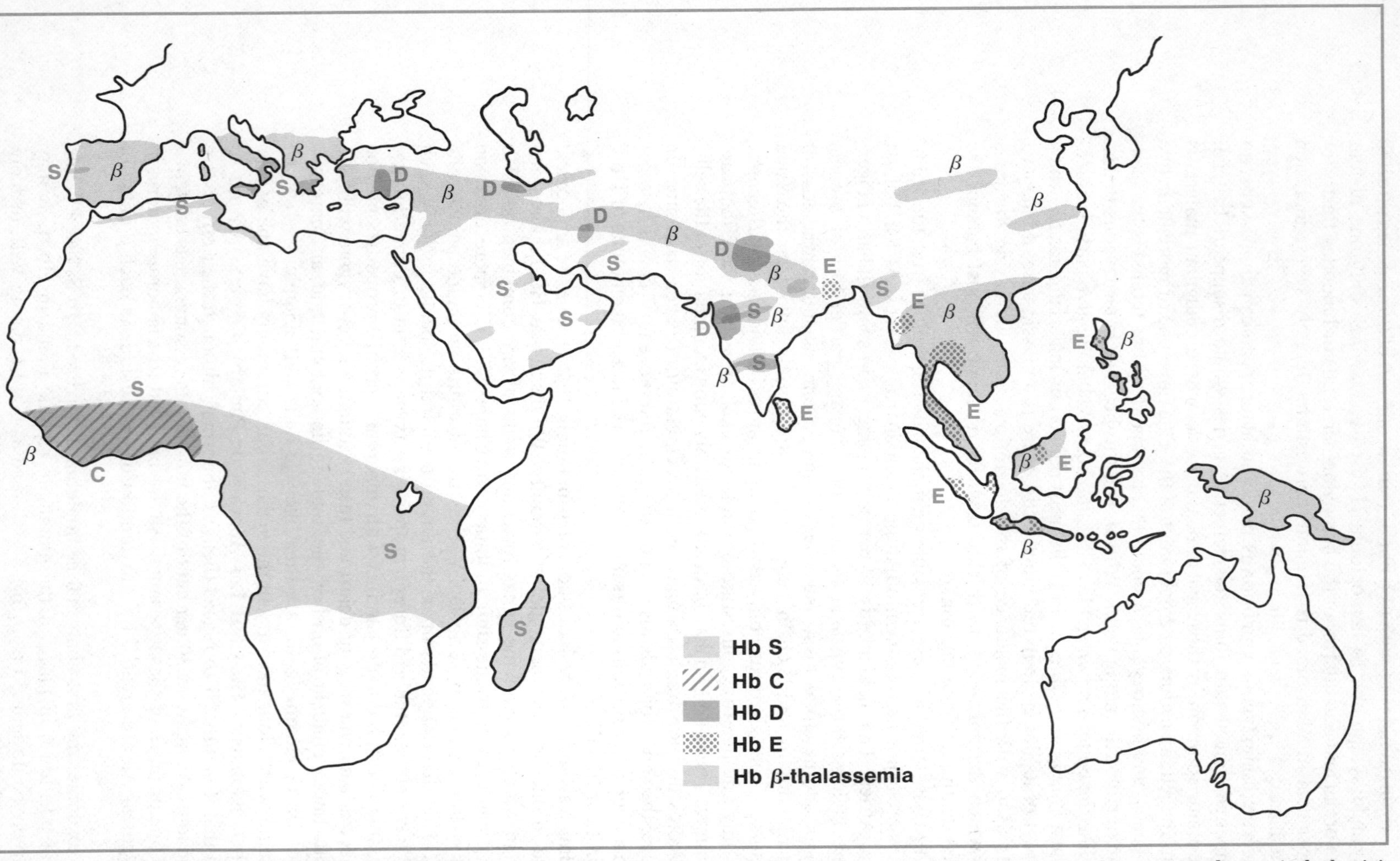

FIGURE 13–14 The geographical distribution of the common human hemoglobin mutants. The incidence of α-thalassemia has not been surveyed extensively, but it is known to be common in parts of Africa and southeastern Asia. The combined location of these mutants roughly outlines the regions of risk for falciparum (malignant) malaria. (This general sketch is based on information from H. Lehmann and R. G. Huntsman, (1974) *Man's Hemoglobins*, Lippincott.)

SICKLE CELL ANEMIA AND TRAIT. The disease appears in a homozygous individual with Hb S, but no Hb A_1. Hb F is increased. A heterozygote has the trait with a diminished amount of Hb A_1 and less than an equal amount of Hb S. The mutation has little effect on inherent oxygen affinity, but it sharply diminishes the solubility of deoxyhemoglobin, causing it to form long tubules. The resultant diminution of the molar volume causes the usual discoid erythrocyte to pucker into a sickle cell when the oxygen tension is diminished. Circumstances causing this to occur are critical for the individual because some of the sickled cells are so severely distorted that they cannot recover and are destroyed, especially in the spleen. The average lifetime of the cell is shortened in homozygotes to 17 days from the 120 days found in other people. A frequent consequence is repeated infarction of the spleen.

The Hb S at higher concentration in the homozygotes begins to precipitate at oxygen tensions near 50 torr, and most cells sickle at 40 torr. Sickling is slow, however, and cells usually rush back to the lungs fast enough to be reoxygenated before they are damaged. However, sickling can begin in any sluggish spots in the circulation, with the sickled cells obstructing the flow and aggravating the condition until the oxygen supply for the surrounding tissues is lost.

The anemic person can maintain reasonable oxygenation even with low hemoglobin concentrations owing to a shift in the oxygen association curve caused by the condition. Hb S and Hb A have much the same inherent affinity for oxygen, but the precipitation of deoxyhemoglobin S changes the equilibrium to favor discharge of oxygen.

There is a catch. The oxygen supply to the lungs must be maintained. Living at high altitudes, flight in unpressurized aircraft, anesthesia, any condition causing fluid to accumulate in the lungs — all can be life-threatening.

The concentration of Hb S in heterozygotes — those with sickle cell trait —is usually too low to cause sickling under physiological conditions at sea level but is still great enough to be a potential source of trouble under any conditions restricting oxygen supply. Four black soldiers died after vigorous exercise in El Paso, elevation 1,200 meters. (On the other hand, trained blacks with the trait had no problems at 3,400 meters in the Olympics at Mexico City.)

RARE POINT MUTATIONS. Over 270 point mutations in human hemoglobins have been described, affecting all of the known subunits except ϵ and ζ. Most are rare and known only in the heterozygous condition, but they have been useful in understanding the structural basis for hemoglobin function.

Mutations that change the lining of the heme pocket, particularly those causing the introduction of polar residues, often make the heme susceptible to oxidation, creating methemoglobins in which at least two of the groups are oxidized to hemin (Fe III). Such hemoglobins are designated **Hb M**: for example, M Osaka ($\alpha_2^{58\text{His}\rightarrow\text{Tyr}}\beta_2$) or M Milwaukee ($\alpha_2\beta_2{}^{67\text{Val}\rightarrow\text{Glu}}$). (The histidyl group altered in M Osaka is near the site at which oxygen is bound; the valyl group altered in M Milwaukee is in contact with the porphyrin ring.) These hemoglobins behave like myoglobin, and it would not be possible for a homozygote to survive.

Other instructive mutations have aided in pinpointing the location of bonds in cooperative interactions and the sources of the protons liberated in the Bohr effect.

Variations between Species

The number of different hemoglobins and their transport properties differ among organisms; it is frequently possible to rationalize the alterations in terms of

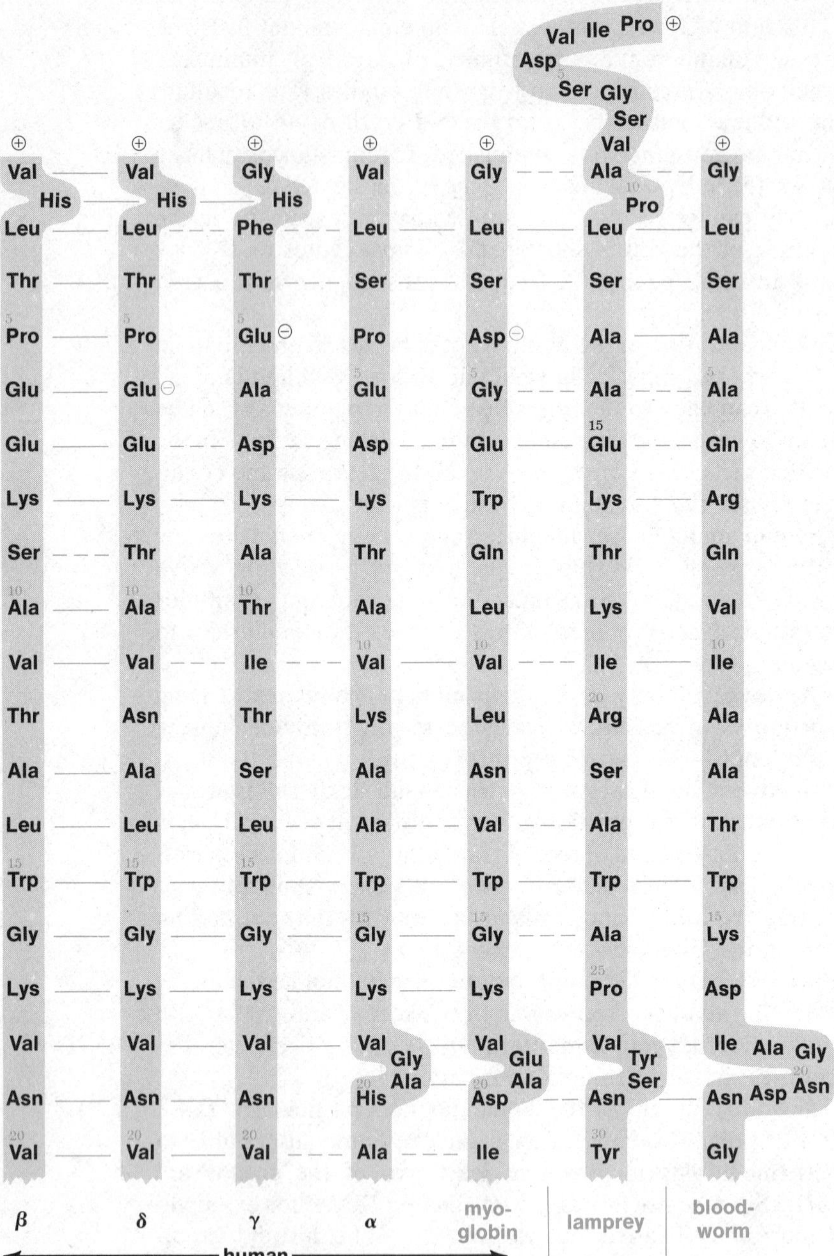

FIGURE 13–15 The amino acid sequences of the amino-terminal portions of different hemoglobin chains are arranged here so as to align residues of comparable function ("maximizing the homology"). Identical residues in neighboring chains are connected by *solid blue lines;* residues of similar character (hydrophobic, H-bonding, etc.) are connected by *dashed blue lines.* Analysis of the relationships allows some deductions to be made about the evolution of the chains. For example, the few differences between β and δ chains indicates a relatively recent divergence, but the common occurrence of 2-His in these and the γ chains indicates they are at least kissing cousins. The additional residues present at α-18,19, and at the corresponding positions in myoglobin, and in the lamprey and bloodworm hemoglobins, are not in the β, δ, or γ chains, indicating that the latter three and the α chain had separate ancestors.

the physiological requirements of the particular species. In general, the larger the animal, the smaller the flux of oxygen required by a given amount of tissue and, therefore, the lower the oxygen tension required to maintain that flux. Large mammals have hemoglobins with a higher affinity for oxygen so that the tension drops to lower levels before oxygen is released; their hemoglobins also have a smaller Bohr effect.

Some primitive organisms have monomeric hemoglobin units that generate cooperative interactions through reversible associations rather than through the permanent interactions within tetramers seen in mammalian hemoglobins.

Evolution of Hemoglobins. Amino acid sequences are known for hemoglobins from a wide variety of organisms, and it is possible to trace the evolution of the different chains by comparison of the sequences. The shorter the time since the divergence of any two species from a common ancestor, the fewer the changes that will have occurred in the amino acid composition of a given protein.

For example, the number of differences at comparable positions between human subunits and those in other species are:

Subunit	Chimpanzee	Gorilla	Rhesus monkey	Pig	Cow
α-chain	0	1	4	18	17
β-chain	0	1	8	23	25

The evidence from hemoglobin composition (and from protein composition in general) is that the chimpanzee is a close relative of man, the rhesus monkey satisfyingly more distant, and the cow can be eaten with a clear conscience.

Maximizing the Homology. Mutations may cause deletion or insertion of amino acids into a polypeptide, in addition to substitutions. In order to compare two chains, they must be matched in such a way as to allow for the differences in chain length. This is done by aligning the chains with the maximum number of similar amino acids in comparable positions. It is relatively easy to do this with the hemoglobin chains. Figure 13–15 shows how the known sequences from the various human chains can be matched with those from a primitive vertebrate and an even more primitive invertebrate. The lines on the figure show where identical or functionally similar amino acids have been aligned in neighboring chains. Study will show some alignments of more distant chains that are not indicated.

Information like that in Figure 13–15 is also a guide to the evolution of the different hemoglobin subunits. The β and δ chains are more like each other than

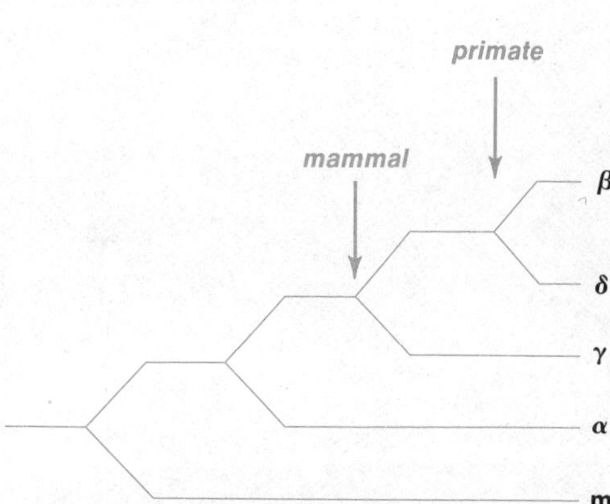

FIGURE 13–16

The order of evolution of various hemoglobin subunits. No indication of relative times is intended. The positions of ε and ζ chains on this family tree are not known.

like the γ chain, but the three of them have more resemblances to each other —the insertion of 2-His, for example — than they have to α chains or to myoglobin. Both the α chains and myoglobin have two additional residues (18 and 19) similar to the primitive lamprey hemoglobin, and the bloodworm has four additional residues in a comparable location. Study of the complete sequences along these lines indicates that the hemoglobins have evolved as shown in Figure 13–16.

FURTHER READING

Bunn, H. F., B. G. Forget, and H. N. Ranney: (1977) *Human Hemoglobins*. Saunders. Excellent treatment of oxygen transport.

Balwin, J. M.: (1975) *Structure and Function of Hemoglobin*. Prog. Biophys. Mol. Biol., *29*: 225.

Fermi, G.: (1975) *Three-dimensional Fourier Synthesis of Human Deoxyhemoglobin at 2.5 Å Resolution*. J. Mol. Biol., *97*: 237. Contains a description of structural features.

Edelstein, S. J.: (1975) *Cooperative Interactions of Hemoglobin*. Annu. Rev. Biochem., *44*: 209.

Riggs, A.: (1976) *Factors in the Evolution of Hemoglobin Function*. Fed. Proc., *35*: 2115.

Kilmartin, J. V., and L. Rossi-Bernardi: (1973) *Interaction of Hemoglobin with Hydrogen Ions, Carbon Dioxide and Organic Phosphates*. Physiol. Rev., *53*: 836.

Benesch, R. E., and R. Benesch: (1974) *The Mechanism of Interaction of Red Cell Organic Phosphates with Hemoglobin*. Adv. Protein Chem., *28*: 211.

Rørth, M., and P. Astrup, eds.: (1972) *Oxygen Affinity of Hemoglobin and Red Cell Acid Base Status*. Academic Press.

Brewer, G. J., ed.: (1972) *Hemoglobin and Red Cell Structure and Function*. Adv. Exp. Biol. Med., *28*. Esp. Brenna, O., et al.: *The Interaction Between Hemoglobin and its "Oxygen-linked" Ligands*, p. 19.

Siggard-Andersen, O., and N. Salling: (1971) *Oxygen-linked Hydrogen Ion Binding of Human Hemoglobin*. I, II, III. Scand. J. Lab. Clin. Invest., *27*: 351, 361.

Siggard-Andersen, O., et al.: (1972) *Oxygen-linked Hydrogen Ion Binding of Human Hemoglobin*. Scand. J. Lab. Clin. Invest., *29*: 185.

Paulsen, E. P., and M. Koury: (1976) *Hemoglobin AI_c Levels in Insulin-dependent and -independent Diabetes Mellitus*. J. Am. Diab. Assoc., *25*(Suppl. 2): 890. Also notes formation of tight hemoglobin-membrane complex in diabetes.

Koenig, R. J., et al.: (1976) *Correlation of Glucose Regulation and Hemoglobin A_{1c} in Diabetes Mellitus*. N. Engl. J. Med., *295*:417.

Stevens, V. J., et al.: (1977) *Nonenzymatic Glycosylation of Hemoglobin*. J. Biol. Chem., *252*:2998.

Kitchen, H., and S. Boyer, eds.: (1974) *Hemoglobins: Comparative Molecular Biology Models for the Study of Disease*. Ann. N.Y. Acad. Sci., 241.

Orkin, S. H., and D. G. Nathan: (1976) *The Thalassemias*. N. Engl. J. Med., *295*: 710.

Lehmann, H., and R. G. Huntsman: (1974) *Man's Hemoglobins*. Lippincott.

Dayhoff, M.: (1972, 1976) *Atlas of Protein Sequence and Structure*. Vol. 5 and suppl. 2. Best source for amino acid sequences and mutants.

Lessin, L. S., and W. N. Jensen, eds.: (1974) *Sickle Cell Symposium*. Arch. Int. Med., *133*, no. 4. Many informative articles on molecular biology, physiological effects, and the clinical conditions.

Kan, Y. W., et al.: (1977) *Nondeletion Effect in α-Thalassemia*. N. Engl. J. Med., *297*: 1081. Demonstration that not all impairments of α-chain synthesis can be ascribed to structural gene deletion.

14 THE ENZYMES

A living organism is a magnificent assembly of chemical reactions, an entity rather than chaos because most of its components make an orderly contribution to existence of the whole. Many of these reactions occur only at significant rates because specific proteins — the enzmes — are present to catalyze them. The enzymes are constructed to control both the kinds of reaction and the rates at which they occur, and it is the existence of these proteins that creates a complex harmony of chemical function.

Enzymes contain the same amino acids found in other proteins, and they have a three-dimensional structure that represents the conformation of least energy content, as do other proteins. All of our present information supports the view that enzymatic catalysis involves the same kinds of bonds that appear during ordinary organic reactions, using structures supplied by amino acid residues and by prosthetic groups. The binding of prosthetic groups is itself determined by conformation of amino acid residues, so the study of the nature of enzymes is in many ways only an extension of the principles of protein structure that we have already developed.

Why then do some proteins act as very effective catalysts — more effective than any other known compounds in making some reactions go faster — whereas other proteins containing the same amino acid side chains do not? The obvious thing the enzymes can do that other proteins cannot is to bring several chemical groups to bear in a particular geometric conformation. This is the basis for enzyme function; the spatial arrangement of amino, carboxyl, and other bonding groups increases catalytic effectiveness by orders of magnitude.

Specificity. Enzymes, like antibodies, are frequently highly specific in binding particular compounds, because the amino acid residues in the binding site closely fit only a few compounds, perhaps one. To use the technical nomenclature, an enzyme may have a high specificity for its substrates.

Substrate is an old term, firmly embedded in the biochemical literature, for a compound whose reaction is catalyzed by an enzyme:

$$\text{substrate} \xrightarrow{enzyme} \text{product}$$

Rephrasing the concept of specificity, one enzyme may catalyze the reaction of only a few different substrates. Put still another way, only a few compounds may be capable of acting as substrates for an enzyme.

The specificity of individual enzymes is an important feature of biological control because changes in the activity of particular enzymes will affect only a few compounds. However, there is another aspect of specificity that concerns catalytic ability itself, and this is how it comes about. A selectivity toward a few compounds–high specificity–implies a close molding of the binding site to a particular geometry, with participation of several side chains from the enzyme to create a much stronger force with "good" substrates than with other compounds. The existence of this stronger force is an important factor in easy deformation of the substrate to break old bonds and create new ones.

Catalytic Groups. Another clue to the catalytic mechanism comes from the presence in proteins of groups known to be effective catalysts for ordinary organic reactions. For example, many amino acid side chains are capable of donating or accepting protons so as to act as general acid or general base catalysts (Fig. 14–1). Other groups, such as hydroxyl or amino groups, may act as nucleophiles, and surrounding structures in an enzyme are often designed to reinforce their catalytic

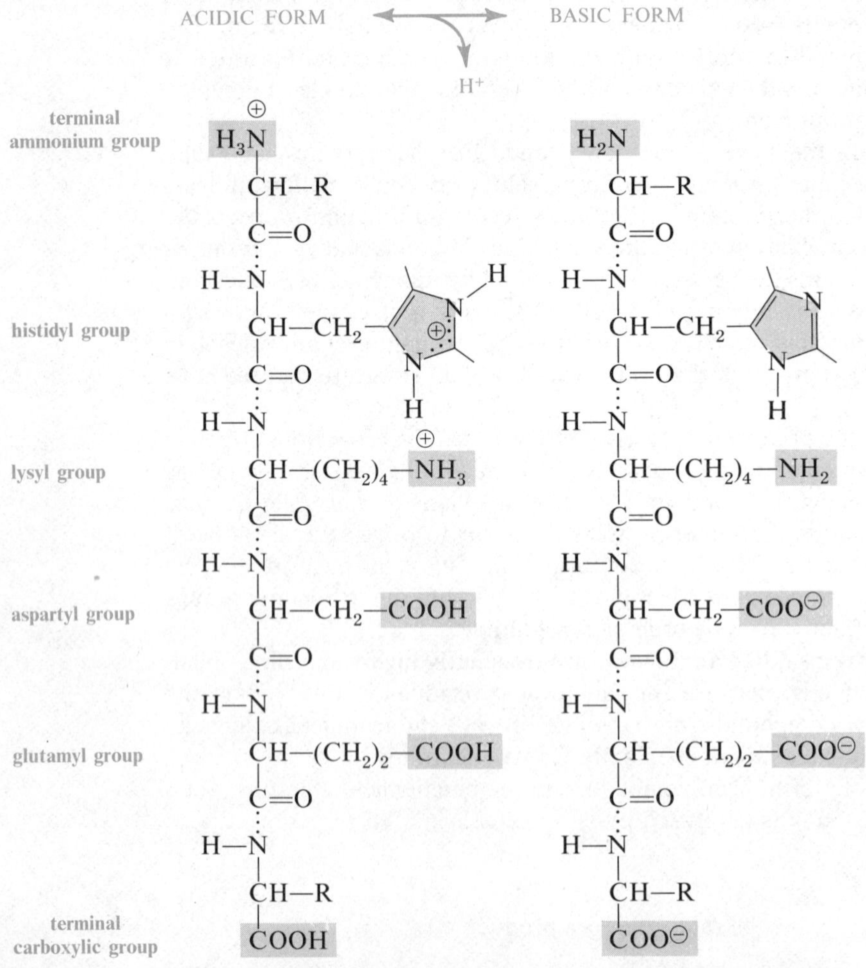

FIGURE 14–1 **Groups available for use as general acids or general bases in enzymatic reactions include side chain substituents and the terminal ammonium or carboxylate groups. Each may in some instances act as an acid and in others as a base; the two forms are shown.**

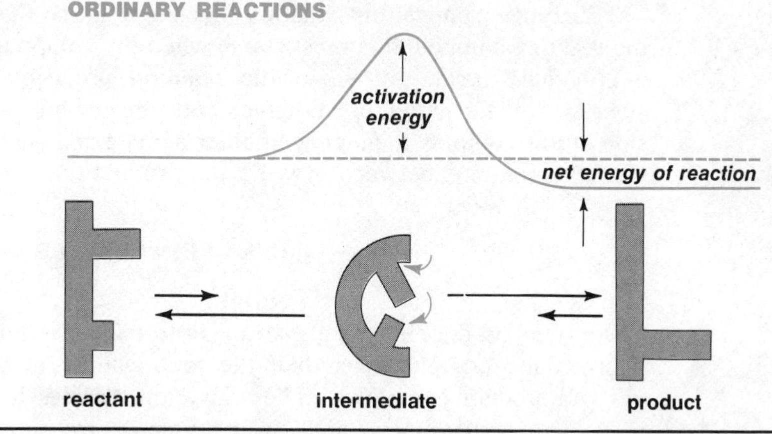

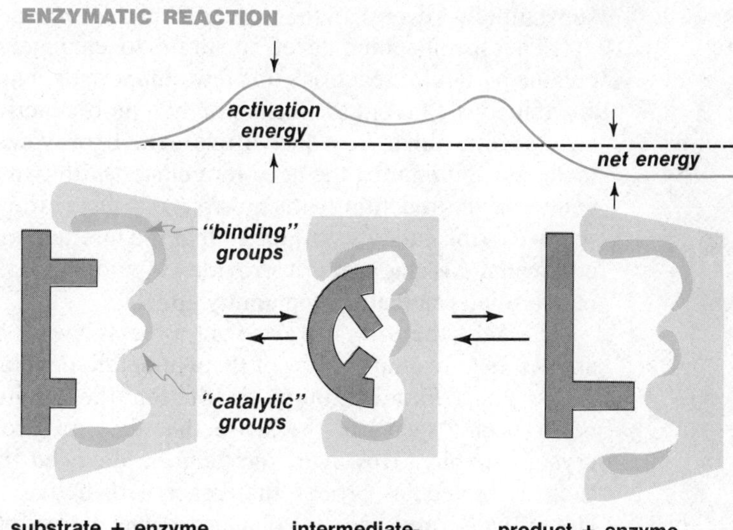

FIGURE 14–2 *Top.* **In ordinary chemical reactions, a molecule must assume some unstable intermediate form before it changes into a product. The probability of occurrence of the intermediate is low as shown by the arrows, which is equivalent to saying that it represents a state of high energy content. A molecule must cross this activation energy barrier before reaction occurs, even if the overall process has a substantial net loss of energy.**

Bottom. **In enzymatic reactions, the binding and catalytic groups of the enzyme combine with the substrate in such a way as to make the intermediate-enzyme complex have a lower energy content than the complexes of substrate or product with the enzyme. The activation energy for formation of the intermediate-enzyme complex is therefore much lower than the activation energy for the reaction in the absence of enzyme.**

properties. (Nucleophiles in effect donate electrons to bond other groups lacking filled electron orbitals.)

General Catalytic Mechanism. Enzymes owe their peculiar catalytic effectiveness to a combination of specific binding and the presence of catalytic groups. In an ordinary chemical reaction, a compound must contort into an unfavorable configuration before it changes into something else (Fig. 14–2); the formation of the less probable intermediate state represents an activation energy barrier that must be overcome before reaction can occur.

Enzymes change this situation by being built so that their combination with the reacting compound is most stable when the compound is in a ''high-energy'' intermediate form. An enzymatic reaction proceeds through at least three stages — the formation of a complex between enzyme and substrate, the conversion of this complex to an enzyme-intermediate complex, and the further conversion to a complex between enzyme and product that can dissociate:

$$\mathbf{E + S \rightleftarrows ES \rightleftarrows EI \rightleftarrows EP \rightleftarrows E + P}$$

The groups on the enzyme are arranged to make the total energy of the enzyme intermediate complex less than the total energy in the enzyme-substrate or enzyme-product complexes. The activation barriers in converting enzyme-substrate to enzyme-intermediate, or enzyme-intermediate to enzyme-product are substantially lower than the barrier without enzyme.

The specific binding of substrate to enzyme therefore serves not only to confine particular reactions to a few compounds, but also to create a complex that spontaneously favors the formation of a more reactive configuration. The enzyme does several things to achieve this end. It provides catalytic groups in a single molecule, eliminating the need for collison with two or more molecules during the course of the reaction (**entropy effect**); it aligns the substrate at favorable angles relative to the catalytic groups so that the interaction with these groups is facilitated (**orbital steering**), and it provides binding groups at positions that stabilize a reactive intermediate (**propinquity effect**).

Perhaps the easiest way to get a feel for what enzymes do and how they go about it is to examine a few of them in detail, developing the general principles as we go along. First, we ought to note that the details of enzyme mechanism have never been ''seen'' in the sense that they are frozen for inspection by X-ray crystallography. However, they can be deduced through the use of techniques such as molecular probes that react with active sites, destruction of selected amino acids, and spectral changes. Fine details of mechanism are not always certain, but the principal features are clear.

AN ENZYME SAMPLER

Pancreatic Ribonuclease A

The pancreas of cattle and other ruminants makes an enzyme that catalyzes the hydrolysis of RNA. Ribonuclease is an example of a **hydrolase**—an enzyme catalyzing a hydrolysis — and, more specifically, it is a **phosphodiesterase**, hydrolyzing one of two ester linkages on a single phosphate group. (Hydrolytic enzymes are commonly named by adding the suffix **-ase** to the name of the substrate they attack.)

The enzyme is believed to be useful to cattle because bacteria in the rumen convert part of the host's nitrogen supply to RNA, which must be degraded to recover the nutrient. It is a relatively small protein, containing only 124 amino acid residues in a known sequence, and is easily isolated in crystalline form from a readily available and inexpensive source. Its geometry is known from X-ray crystallography. We have already noted (p. 144) its value in deducing the forces determining protein structure, and it has been equally useful for study of the mechanism of enzyme action.

The Overall Reaction. Ribonuclease A catalyzes the hydrolysis of any 5′-phosphate ester bond in RNA in which the phosphate group is also attached to the 3′-carbon of a pyrimidine nucleoside, such as cytidine or uridine:

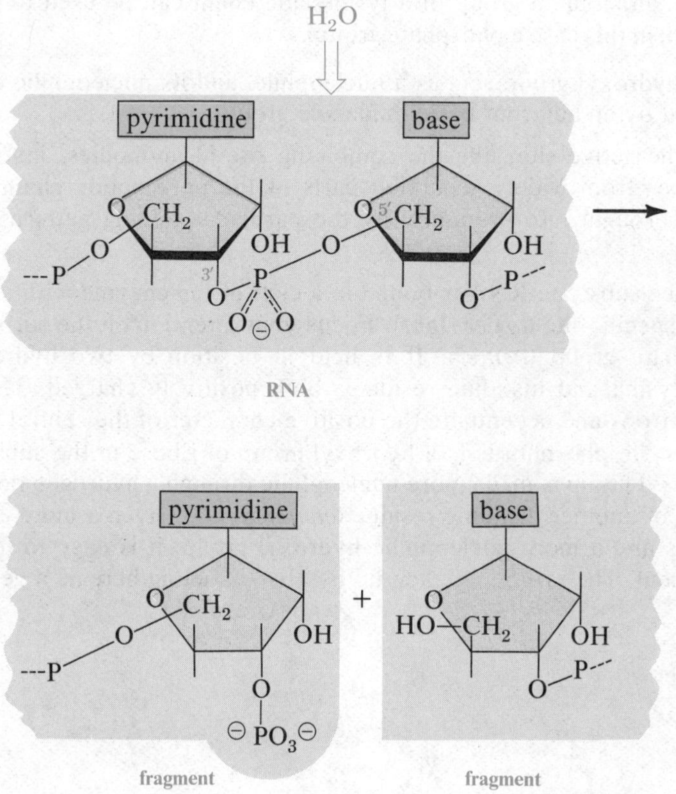

H₂O
pyrimidine
base
O CH₂
O 5′ CH₂
OH
OH
---P
O
3′
O
P
O
P---
O
O⊖ O
RNA

pyrimidine
base
O CH₂
+
O
OH
HO—CH₂
OH
---P
O
O
O
P---
⊖ PO₃⊖
fragment
fragment

Continued action of ribonuclease A hydrolyzes a polynucleotide into fragments, all except the terminal pieces bearing a free hydroxyl group at the 5′ end, and an ester phosphate group at the 3′ end, which must be on a pyrimidine nucleotide residue. (The 5′ residue will be a purine nucleotide if hydrolysis is complete.) Thus the hydrolysis of a portion of an RNA chain might proceed like this:

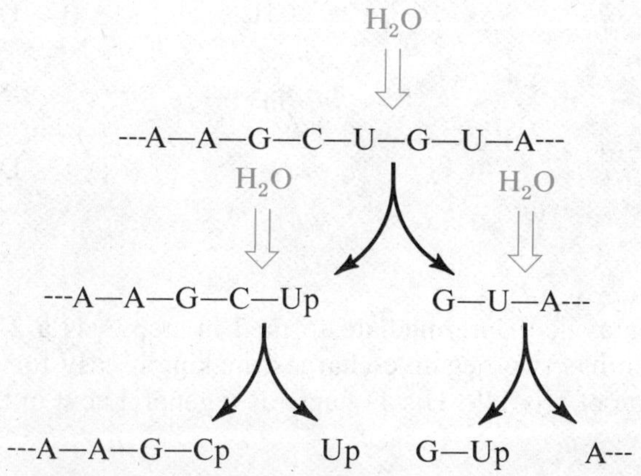

H_2O

---A—A—G—C—U—G—U—A---

H_2O H_2O

---A—A—G—C—Up G—U—A---

---A—A—G—Cp Up G—Up A---

The Catalytic Mechanism. These are the important features illustrated on the next pages:

(1) Imidazole groups in histidine side chains act as general acid and as general base catalysts;

(2) the ammonium group of a lysine side chain can be used to bind a substrate anion, in this case a phosphate group;

(3) a hydroxyl group acts as a nucleophile, and its nucleophilic character is accentuated by an adjacent basic imidazole group.

(4) The active site, like the combining site in antibodies, involves amino acid residues from widely separated parts of the polypeptide chain. These residues are brought into proximity by the particular folding pattern of the protein.

A. The substrate RNA is bound in a cleft of the enzyme with a pyrimidine group in a specific site (*lower right*). Focus your attention on the surroundings of the phosphate group (*center*). It is held in position by two hydrogen bonds involving lysine and histidine residues, both positively charged. These groups attract electrons and accentuate the positive character of the central phosphorus atom. Near the phosphorus is a hydroxyl group of ribose in the substrate itself. This hydroxyl group is made more nucleophilic through a hydrogen bond with the basic form of another histidine residue (*center right*). Given a more electrophilic phosphorus and a more nucleophilic hydroxyl group, it is easy to form a bond between them. The proton is taken up by His 12, acting here as a general base.

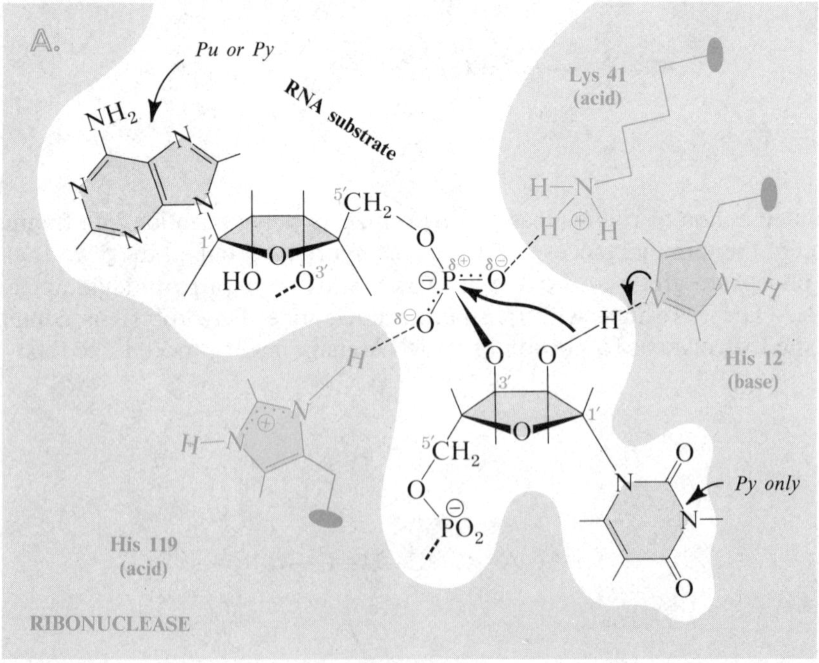

B. The pentavalent intermediate formed in step A is a 2′,3′-cyclic phosphodiester, which has two negative charges, making it easy for an O–P bond to shift to the proton of His 119. His 119 acts as a general acid in this step.

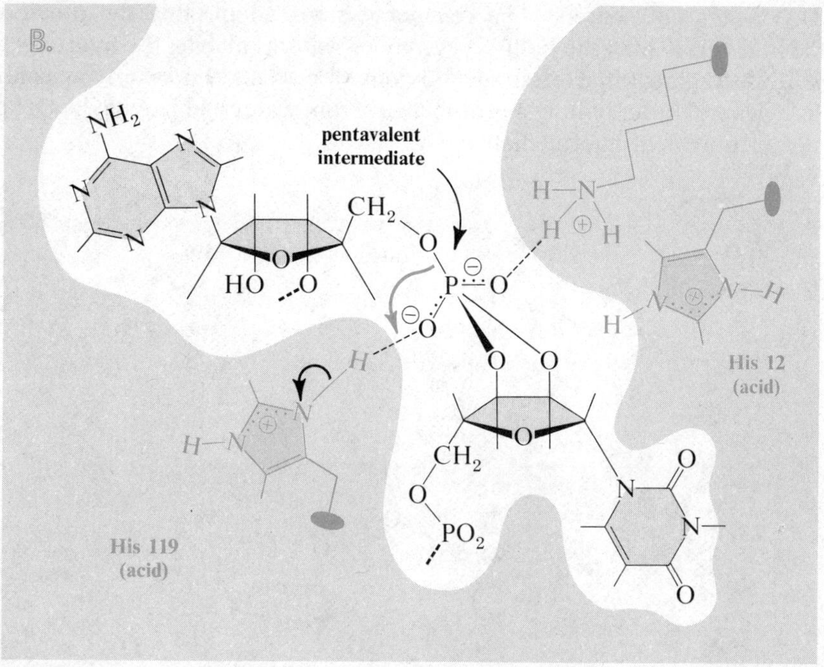

C. One end of the substrate, which bears the now free 5′ hydroxyl group, can leave the enzyme. The remainder of the substrate is still present as a 2′, 3′-cyclic phosphodiester, held by its negative charge to the adjacent lysine residue. At this point, half of the reaction has been completed, with two energy barriers crossed in forming and rearranging the pentavalent unstable intermediate of step B.

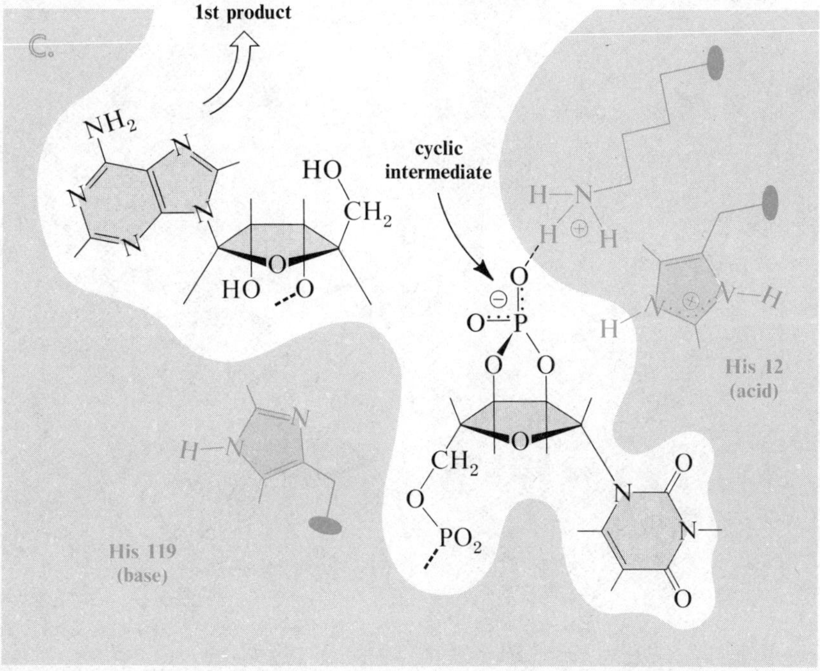

D. Water now enters. The remaining steps go through the previous sequence in reverse, with the hydroxyl group of water replacing the hydroxyl group on the RNA fragment that left in the previous step. His 119, formerly an acid, now acts as a general base, pulling a proton away from water and leaving its OH group in a new pentavalent intermediate.

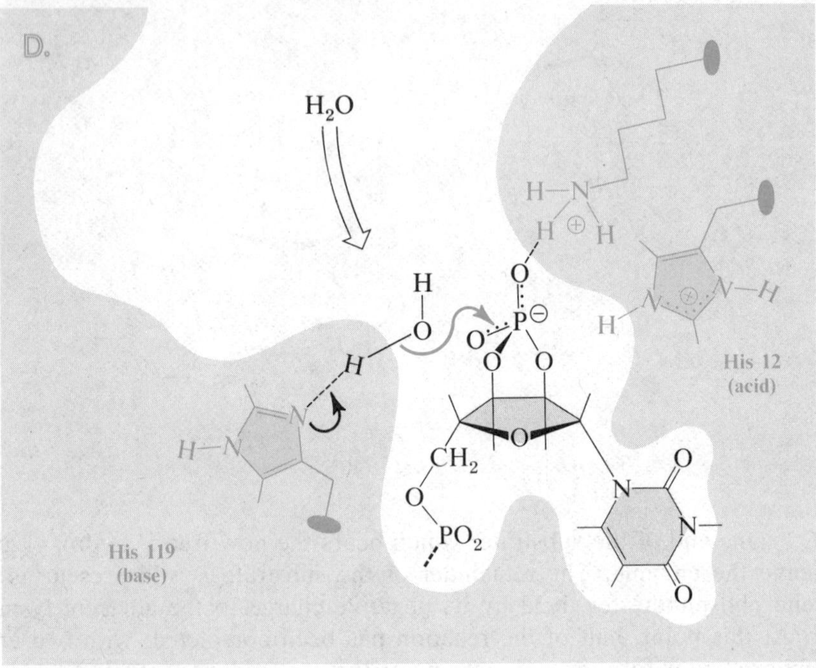

E. His 12, formerly a base, now acts as a general acid, donating a proton to disrupt one of the O–P bonds in the pentavalent intermediate.

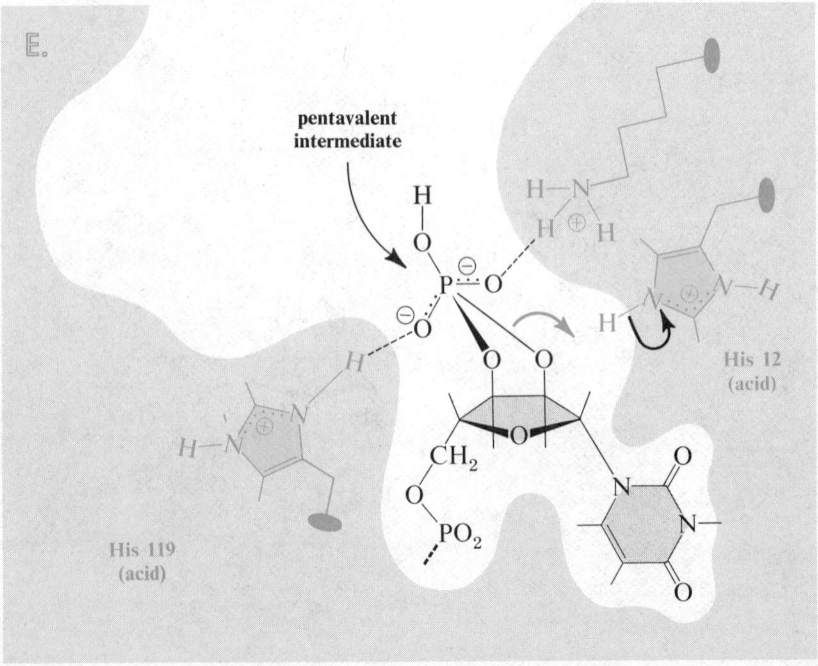

F. The result is the conversion of the remainder of the RNA to a stable form that can be released and the restoration of the enzyme to its original form, ready to react with another molecule of RNA.

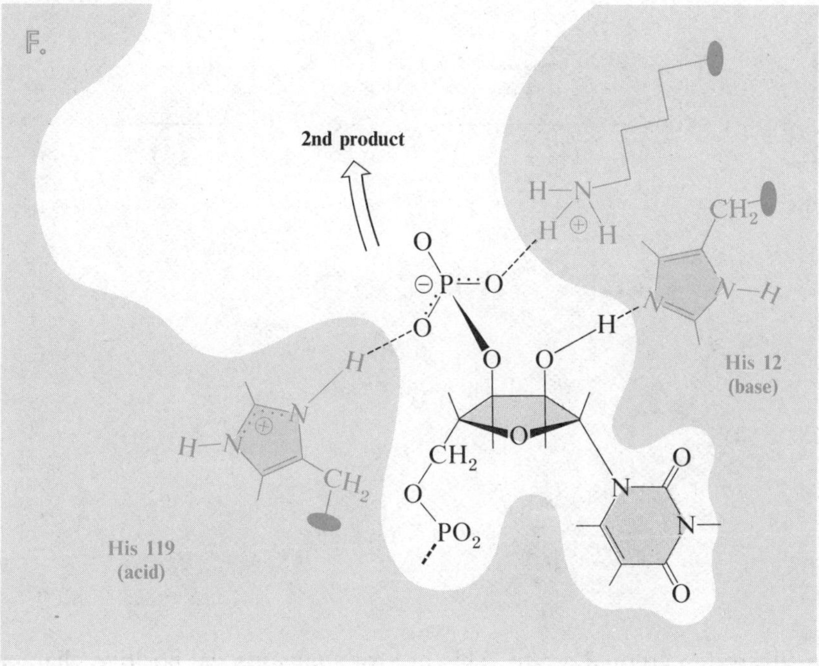

F.
2nd product
His 12 (base)
His 119 (acid)

Carboxypeptidase

The carboxypeptidases are pancreatic enzymes that hydrolyze the peptide bond closest to the carboxyl terminal of a polypeptide chain and liberate the terminal residue as a free amino acid (*top, page 262*). These hydrolases therefore are peptidases; peptidases attacking terminal peptide bonds are called exopeptidases. Their function in digestion is covered in the next chapter.

The points of interest are:

(1) a glutamate residue is used as a nucleophilic group to accept an acyl residue;

(2) the enzyme contains zinc and is an example of a metalloprotein;

(3) an arginine residue is used to fix the carboxylate terminal of the substrate in place, and arginine commonly occurs in enzymes to hold anionic substrates in position (especially phosphate monoesters);

(4) a tyrosine residue is placed to form hydrogen bonds with the substrate.

The mechanism apparently involves the transfer of the peptide acyl group to a glutamate residue in the enzyme, creating a transient acid anhydride, and liberating the terminal amino acid (*center, page 262*). The anhydride is then cleaved by hydrolysis to liberate the shortened polypeptide chain.

The Reaction Mechanism (*bottom, page 262*). The zinc atom is in the enzyme to act as an electrophile; it is chelated with neighboring histidine and glutamate residues in the enzyme, which promote its attraction for the negative oxygen atom in the substrate carbonyl group. This in turn augments the positive character of the substrate carbonyl carbon atom, facilitating its bonding to a nucleophilic glutamate residue. Cleavage of the peptide bond is also aided by hydrogen bond formation between a tyrosine hydroxyl group and the amide nitrogen, accentuating its good leaving group properties.

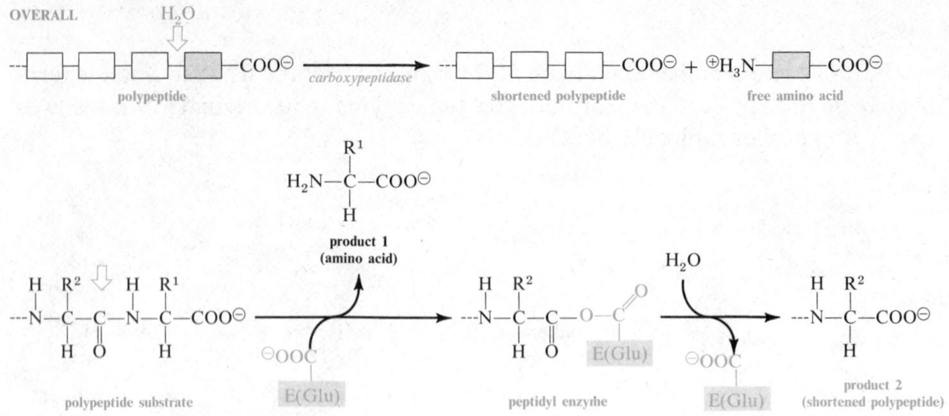

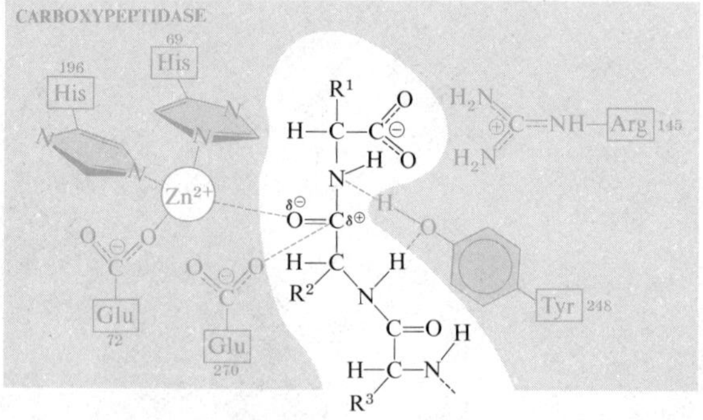

Serine Proteases

Endopeptidases are enzymes catalyzing hydrolysis of interior bonds in polypeptide chains (*top opposite*). Many have identical mechanisms involving transfer of the peptide acyl group to a serine residue, forming an intermediate ester with the enzyme, which is then cleaved by water (*center opposite*). These functions in digestion are discussed in Chapter 17.

The particular feature of interest here is an augmentation of the nucleophilic character of the serine hydroxyl group by a chain of hydrogen bonds, linking it through a histidine side chain to an aspartate side chain. This triad of Asp-His-Ser occurs in many endopeptidases from a variety of organisms, ranging from bacteria to humans. In ribonuclease, a substrate hydroxyl group acted as a nucleophile; in the serine proteases, an enzyme hydroxyl group serves that function.

A. In effect, the negative aspartate side chain pushes electrons through the intervening histidine side chain by means of hydrogen bonds to make the serine hydroxyl group a more effective nucleophile, which then readily bonds with the electrophilic carbon in a peptide bond of the substrate.

B. The product of the electron shift in step A is a tetrahedral intermediate. It has a negative charge, while the aspartate residue has lost its charge. Electrons now move from the peptide group toward the aspartate, breaking the amide bond.

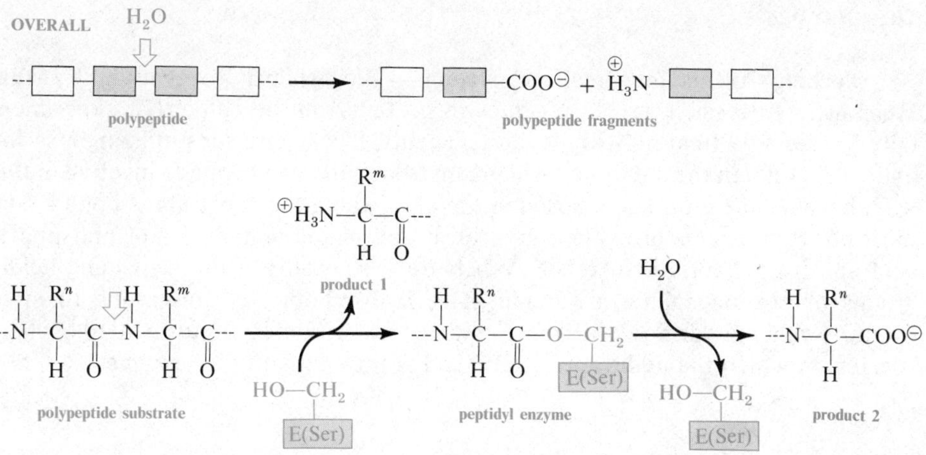

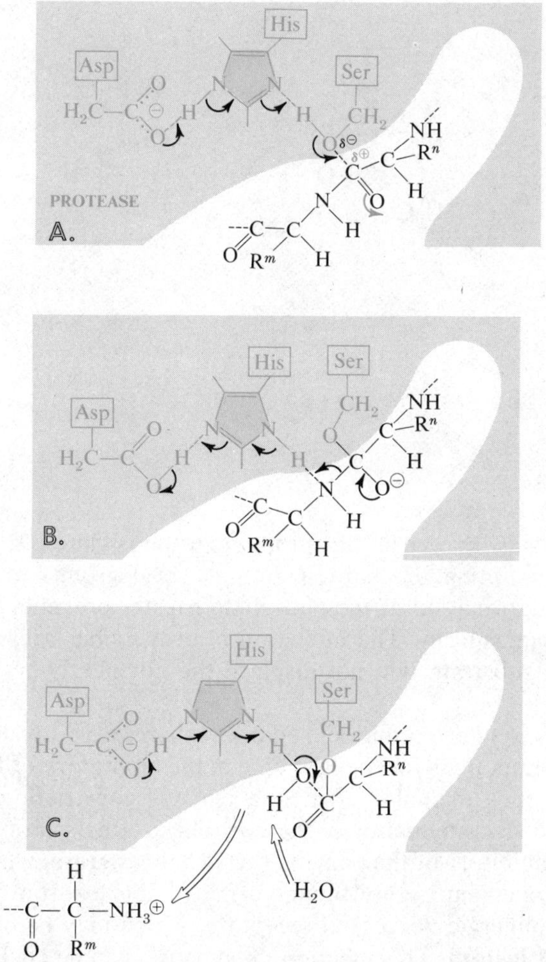

OVERALL H_2O

polypeptide → polypeptide fragments

polypeptide substrate → peptidyl enzyme + product 1 → (H₂O) product 2

PROTEASE

A.

B.

C.

C. A free amino group has appeared, and the fragment bearing it easily leaves and is replaced by water. The acyl fragment of the original substrate peptide remains as an ester of the serine hydroxyl in the enzyme. The electron movements of the previous steps now reverse, and the result, not shown, is formation of a new tetrahedral intermediate, followed by release of the remaining polypeptide chain with a free carboxylate group.

Blocking Agents for the Seryl Group — Biochemical Reagents and Lethal Weapons. Accessible serine residues in proteins can be converted experimentally to esters by treating with acylating agents. Those enzymes utilizing a serine hydroxyl group in the catalytic mechanism frequently can be made inactive in this way, because the group is exposed in the cleft where the substrate is bound. The most effective agents are acid anhydrides, and mixed anhydrides of phosphoric acid and hydrofluoric acid react even better, especially if the remaining acidic groups on the phosphate are made more hydrophobic by converting them to esters. Such anhydrides will react with serine residues to form a phosphate triester, in which the additional ester bond is formed with the enzyme:

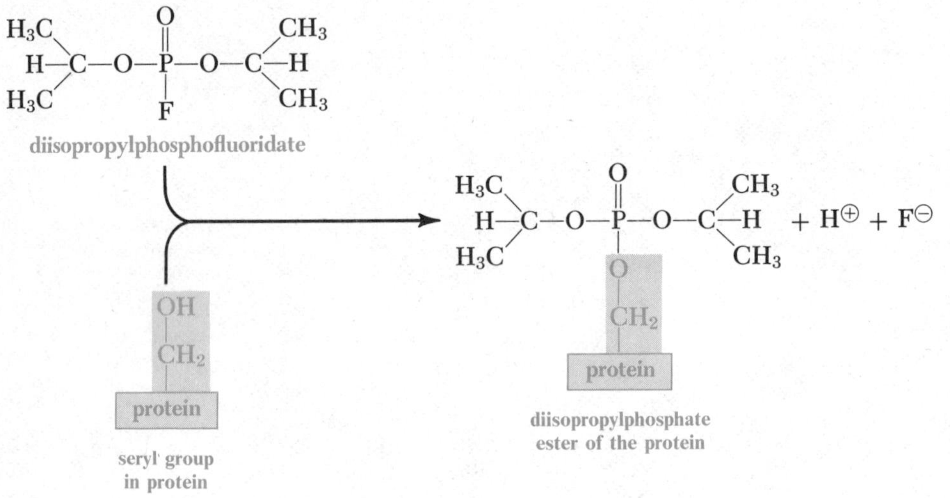

$$(H_3C)_2CH{-}O{-}\overset{O}{\overset{\|}{P}}(F){-}O{-}CH(CH_3)_2 + HO{-}CH_2{-}\text{protein} \longrightarrow (H_3C)_2CH{-}O{-}\overset{O}{\overset{\|}{P}}(O{-}CH_2{-}\text{protein}){-}O{-}CH(CH_3)_2 + H^{\oplus} + F^{\ominus}$$

diisopropylphosphofluoridate

seryl group in protein

diisopropylphosphate ester of the protein

The particular agent shown, usually known as diisopropylfluorophosphate (DFP), is an effective tool for demonstrating the participation of seryl groups in an enzymatic reaction because the formation of the phosphate triester prevents the completion of the hydrolytic mechanism. The inhibition is irreversible because even high concentrations of a substrate will not displace the covalently-bound inhibiting group.

The effectiveness of esters of phosphofluoridic acid was discovered by accident. Chemists preparing the compounds for the first time in the laboratory of Dr. Willie Lang in Germany developed mental confusion, a sense of constriction of the larynx, and a painful loss of accommodation of the eye. These effects are now known to be characteristic of inhibitors of the enzyme **acetylcholinesterase**, which utilizes a seryl group in its mechanism. Acetylcholine carries impulses from the terminus of certain nerves (cholinergic nerves) to receptors, which may be on a striated muscle fiber or another neuron. This junction, or synapse, can be cocked to fire again only if the acetylcholine is hydrolyzed by action of the acetylcholinesterase:

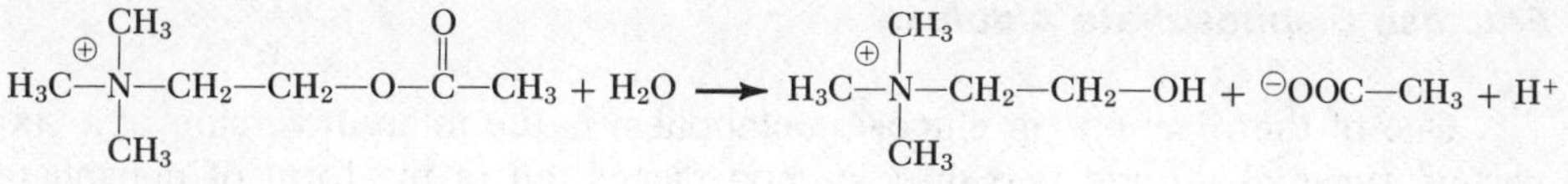

$$H_3C{-}\overset{\oplus}{\underset{CH_3}{\overset{CH_3}{N}}}{-}CH_2{-}CH_2{-}O{-}\overset{O}{\overset{\|}{C}}{-}CH_3 + H_2O \longrightarrow H_3C{-}\overset{\oplus}{\underset{CH_3}{\overset{CH_3}{N}}}{-}CH_2{-}CH_2{-}OH + {}^{\ominus}OOC{-}CH_3 + H^+$$

acetylcholine choline acetate

Acetylcholinesterase inhibitors therefore cause paralysis of striated muscles, among other effects.

The original workers concluded on empirical grounds that the compounds might have value as insecticides. This suggestion was ignored, but with the approach of the Second World War, it was realized that they could provide very potent war gases. A variety of volatile derivatives, many still not discussed in public, were prepared in several of the belligerent countries during the war and after it.

The original suggestion that the compounds ought to be good insecticides was resurrected in the postwar period and proved to be correct. The problem with insecticides is one of lowering the volatility and yet retaining the small aliphatic groups that permit access into an enzyme's cleft, thereby diminishing hazards to humans without sparing the bugs. The better compounds are still dangerous, but malathion, a sulfur analogue that is converted to the oxy compound in tissues, is thought to be safe enough to permit general sale. Its formula is:

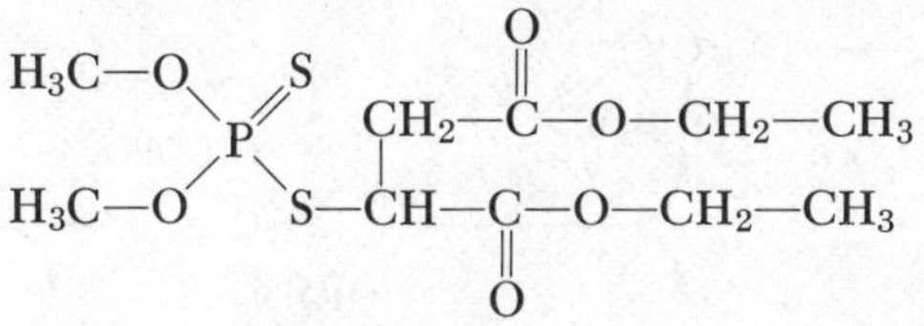

$$(H_3C{-}O)_2P({=}S){-}S{-}\underset{CH_2{-}\overset{O}{\overset{\|}{C}}{-}O{-}CH_2{-}CH_3}{CH}{-}\underset{\underset{O}{\|}}{C}{-}O{-}CH_2{-}CH_3$$

malathion

Since the most effective phosphate derivatives are dangerous for routine use in the laboratory, other anhydrides such as phenylmethylsulfonyl fluoride:

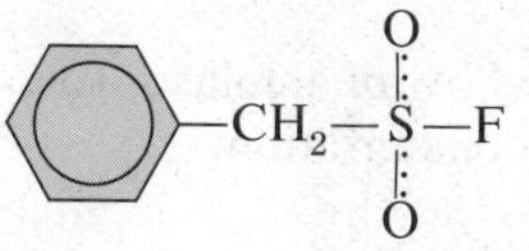

$$C_6H_5{-}CH_2{-}\overset{O}{\underset{O}{S}}{-}F$$

phenylmethylsulfonyl fluoride

have been developed to inhibit serine proteases. They are commonly employed during the isolation of proteins that do not themselves contain active seryl groups, in order to inhibit the proteases that are also present in the source tissues. These proteases frequently damage the desired proteins if they are left unchecked.

Fructose Bisphosphate Aldolase

One of the reactions in glucose metabolism is the interconversion of a six-carbon sugar chain and two three-carbon pieces, all in the form of phosphate esters (*below*). Condensation occurs during glucose synthesis, and the reverse cleavage is required for use of glucose as a fuel; we shall consider these metabolic processes in detail later.

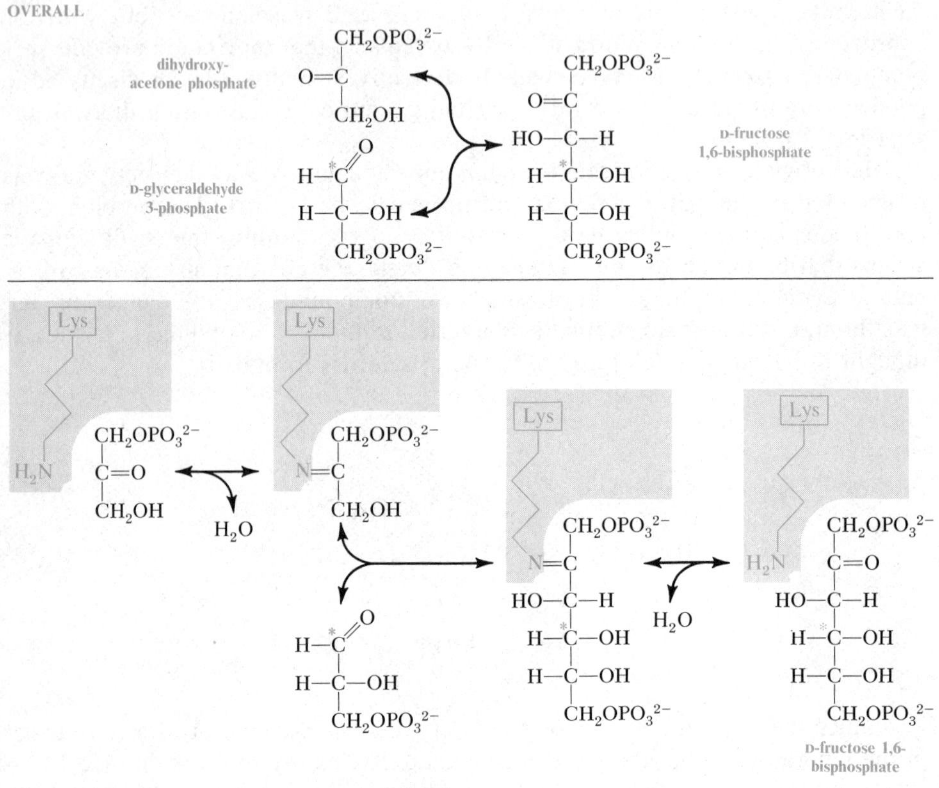

OVERALL

dihydroxy-acetone phosphate

D-glyceraldehyde 3-phosphate

D-fructose 1,6-bisphosphate

Lys

H_2O

D-fructose 1,6-bisphosphate

*The asterisk identifies a particular C atom for orientation only.

The reaction is catalyzed by an aldolase, so-called because the overall reaction is a variant of an aldol condensation:

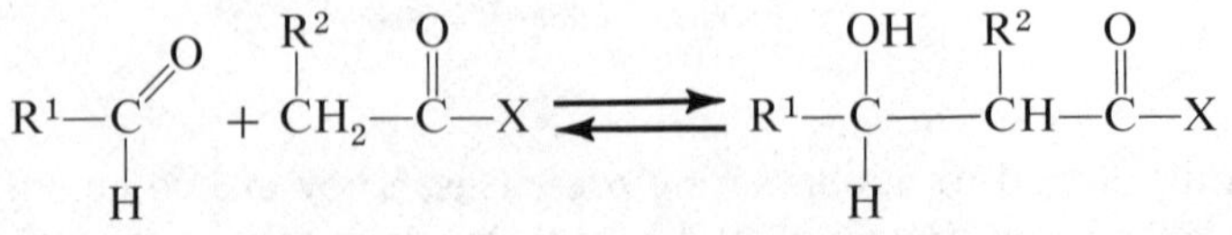

$$R^1{-}CHO + R^2{-}CH_2{-}C(=O){-}X \rightleftharpoons R^1{-}CH(OH){-}CH(R^2){-}C(=O){-}X$$

The point of interest in this reaction is that the enzyme uses a lysine residue to react with a ketose substrate, which is the phosphate ester of either dihydroxyacetone or fructose. The lysine residue supplies an amino group on a tail of carbon atoms waving out some distance from the remainder of the peptide, and in this case it acts as a nucleophile to attack the carbonyl carbon atom of the ketoses. The substrate becomes covalently bound as a **ketimine** (**Schiff's base**) to the enzyme, labilizing the adjacent carbon atom for condensation or cleavage. (A histidine residue, not shown, acts as a proton carrier to catalyze these covalent reactions, much as it does in the serine proteases.)

PROVISION OF REACTIVE GROUPS BY COENZYMES

Many enzymes differ from those we have discussed previously in that effective catalysis requires the presence of chemical groups that do not occur in the side chains of the common amino acids. These chemical groups must be added in the form of an additional compound, in other words, as a **prosthetic group** or **coenzyme**. In such cases the enzyme consists of the coenzyme and the associated polypeptide chains, which are called the **apoenzyme.** Apoenzyme + coenzyme = enzyme.

The polypeptide chains in such enzymes have a third function. In addition to binding the substrate and providing groups required in the catalytic mechanism, they also must bind the coenzyme. Let us first discuss an example of an important group of enzymes, the **aminotransferases,** to illustrate these functions.

Aspartate Aminotransferase

The Reaction and the Coenzyme. A critical reaction in all organisms is the transfer of an amino group from one carbon skeleton to another by the process of transamination. Frequently, the most active exchange is between glutamate and aspartate:

$$\underset{\alpha\text{-ketoglutarate}}{\begin{array}{c} COO^{\ominus} \\ | \\ C{=}O \\ | \\ CH_2 \\ | \\ CH_2 \\ | \\ COO^{\ominus} \end{array}} + \underset{\text{L-aspartate}}{\begin{array}{c} COO^{\ominus} \\ | \\ H_3\overset{\oplus}{N}{-}C{-}H \\ | \\ CH_2 \\ | \\ COO^{\ominus} \end{array}} \longleftrightarrow \underset{\text{L-glutamate}}{\begin{array}{c} COO^{\ominus} \\ | \\ H_3\overset{\oplus}{N}{-}C{-}H \\ | \\ CH_2 \\ | \\ CH_2 \\ | \\ COO^{\ominus} \end{array}} + \underset{\text{oxaloacetate}}{\begin{array}{c} COO^{\ominus} \\ | \\ C{=}O \\ | \\ CH_2 \\ | \\ COO^{\ominus} \end{array}}$$

(α-Ketoglutarate is named as 2-oxoglutarate in approved nomenclature, in which the English have a dominant voice. The English use of oxo for keto is spreading in this country, but is not yet common.) This reaction is catalyzed by **aspartate aminotransferase** (also known as glutamate-oxaloacetate transaminase, glutamate-aspartate transaminase, and similar permutations).

The aminotransferases contain **pyridoxal phosphate,** a coenzyme frequently required by enzymes attacking free amino acids. Pyridoxal phosphate is fixed by polypeptide chains of the aminotransferases in such a way that its aldehyde group is brought into proximity with a lysine side chain, with which it reacts to form an aldimine, the reactive structure in the enzyme (Fig. 14–3).

The various groups in pyridoxal phosphate have not been added for decoration. The ionized phenolic group and the positively charged nitrogen of the pyridine ring enhance the reactivity of the aldimine, while the phosphate ester and methyl substituents serve to fix the coenzyme at a specific site. Pyridoxal phosphate illustrates a general property of coenzymes, which is to have a small part of

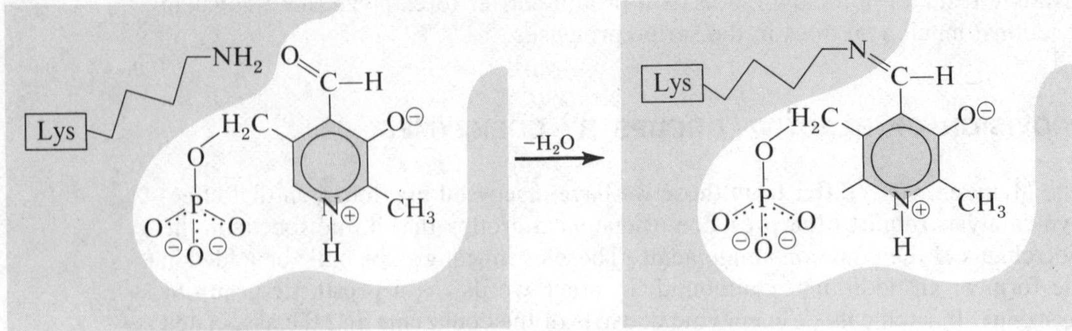

FIGURE 14–3 **The combination of the apoenzyme of an aminotransferase with its coenzyme, pyridoxal phosphate. A lysine residue in the protein reacts with the aldehyde group of the coenzyme to form an aldimine.**

the structure, perhaps a single group, actively participating in the reaction, with the remainder of the molecule serving to accentuate that reactivity and to provide a framework for specific binding to the apoenzyme.

The Steps in Transamination. We shall detail the individual covalent changes in the transamination reaction to show how the coenzyme contributes to catalysis but without giving the actual mechanism, which probably involves acid-base catalysis by histidine side chains.

A. A molecule of L-aspartate collides with the enzyme and is bound, probably by its carboxylate groups, so that its amino group is near the aldimine structure on the enzyme.

B. The double bond shifts from the lysine N atom to the aspartate N atom, releasing the lysine amino group, and forming the aldimine of aspartate. It is much easier to form an aldimine by exchange of groups with a preexisting aldimine than it is by *de novo* combination of amine and aldehyde.

C. The double bond of the aldimine shifts, forming a ketimine. This tautomerization is the rate-limiting step of the sequence.

D. The ketimine is hydrolyzed to form oxaloacetate and **pyridoxamine phosphate.** The oxaloacetate is free to dissociate, but the pyridoxamine phosphate is still held to the enzyme by other groups on the coenzyme. However, the loss of the covalent binding makes it easier to experimentally dissociate coenzyme and apoenzyme at this stage, and this aided the discovery of the mechanism of the reaction.

E. A molecule of α-ketoglutarate (2-oxoglutarate) combines with the protein, also by attachment of carboxylate groups.

F. A ketimine is formed between the α-ketoglutarate and the pyridoxamine phosphate.

G. The double bond again shifts, now forming the aldimine of L-glutamate and pyridoxal phosphate.

H. The aldimine double bond shifts from the nitrogen of L-glutamate to the lysine residue on the protein, leaving glutamate free to dissociate and restoring the original condition of the enzyme so that it can begin the whole process over again with a new molecule of aspartate.

The mechanism can operate in the exact reverse of the steps given so that one can start with glutamate + oxaloacetate and arrive at aspartate + α-ketoglutarate.

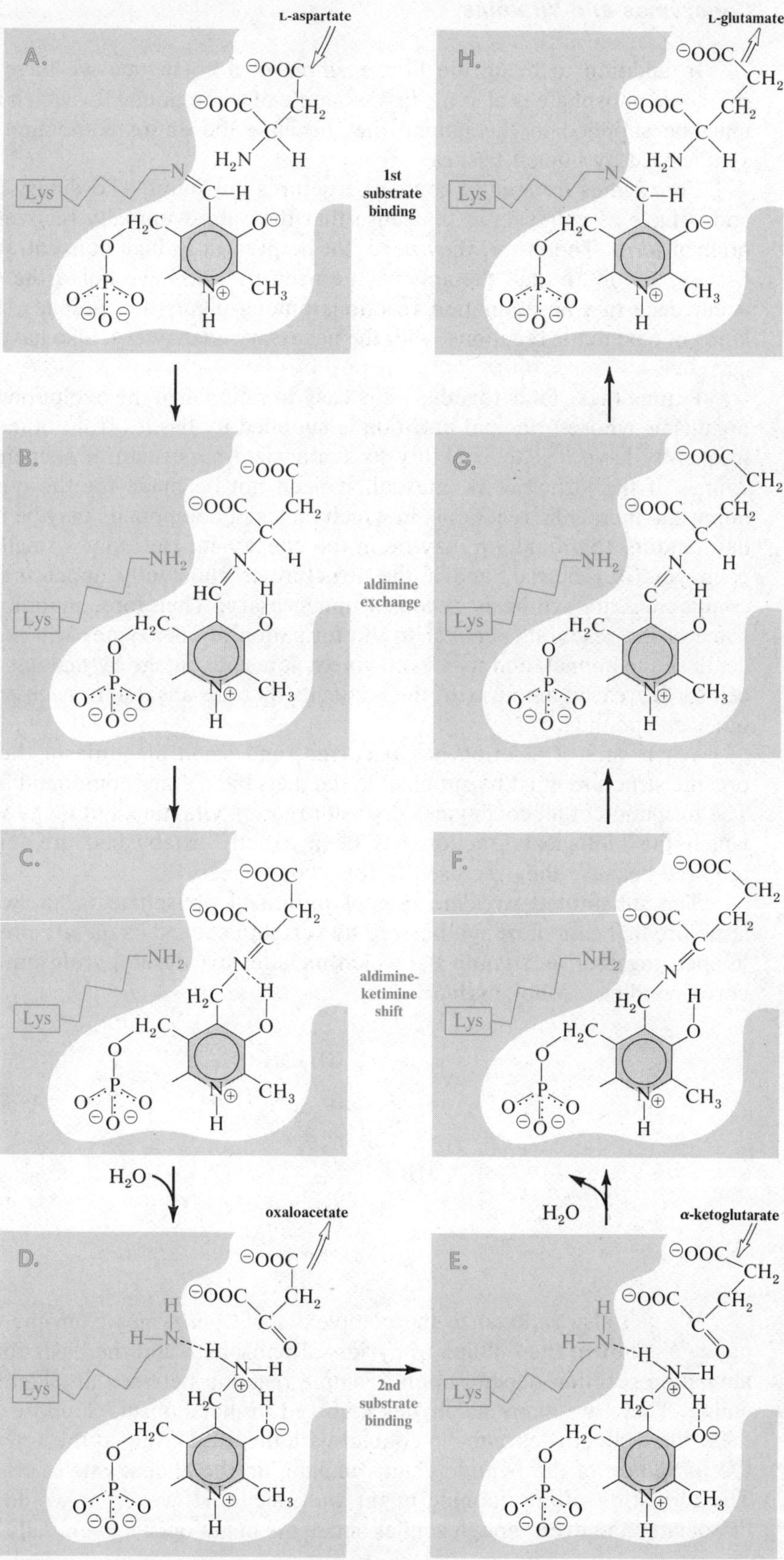
L-aspartate
A.
1st
substrate
binding
B.
aldimine
exchange
C.
aldimine-
ketimine
shift
H₂O
oxaloacetate
D.
2nd
substrate
binding
E.
α-ketoglutarate
H₂O
F.
G.
H.
L-glutamate
Lys

Coenzymes and Vitamins

In addition to being the first example of a coenzyme we have discussed, pyridoxal phosphate is also the first example of a compound for which a precursor must be supplied in the human diet because the entire compound cannot be synthesized by human tissues.

Coenzymes frequently contain structures not found in ordinary metabolites and behave as catalysts in the sense that they are eventually recovered in their original form. Therefore, they need not be present in high concentrations in the tissues, any more than the apoenzyme need be. We have noted the widespread occurrence of transamination reactions among plants and animals. Most other kinds of enzymatic reactions, with the necessary coenzymes, also have a wide occurrence.

Putting these facts together, it is easy to rationalize the evolutionary truth — organisms whose principal nutrition is supplied by tissues from other organisms frequently have lost their ability to synthesize the structural elements of coenzymes. If the structure is unusual, it need not be made for the quantitatively important metabolic reactions, in which moles of compounds may be handled per day, and its sole function may be in the coenzyme. But only a small amount of coenzyme is required, and if the structure is constantly appearing in dietary compounds, its synthesis becomes unnecessary. Therefore, mutations deleting some of the reactions peculiar to the formation of coenzymes will not cause the death of the animal, and may even give it some advantage by making room in the cell for increased amounts of the enzymes that are absolutely necessary to form other compounds.

When such a deletion has occurred, and small amounts of the necessary organic structure must be supplied in the diet, the dietary compound is a *vitamin*. The formation of all coenzymes does not require vitamins, but many vitamins for which the biological function has been exactly established are known to be required because they are used to form coenzymes.

The substituted pyridine ring of pyridoxal phosphate is an example of a structure that cannot be synthesized by vertebrates, and its dietary precursors are lumped together as **vitamin B_6**, which includes **pyridoxal, pyridoxamine,** and the corresponding alcohol, **pyridoxine:**

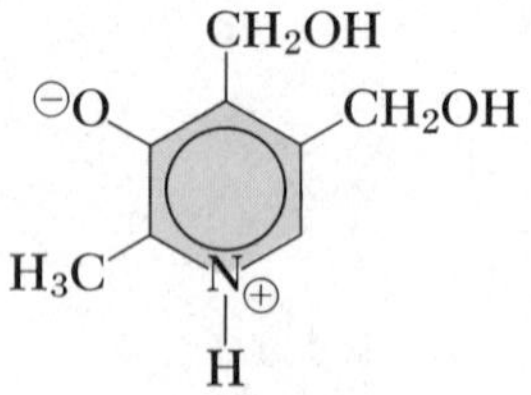

CH_2OH
$^{\ominus}O$ CH_2OH
H_3C $N^{\oplus}$
H

pyridoxine

We can also rationalize the complexity of coenzymes from the example at hand. As shown, the binding of pyridoxal phosphate and the basis for the transamination reaction depend upon a simple reaction between an aldehyde and an amine. There are many aldehydes involved in metabolism. Suppose that one of these many aldehydes came in contact with the lysyl group at the active site after the formation of the peptide chains making up the apoenzyme of transaminase. The formation of an aldimine might well occur. However, it would eventually dissociate, and even though similar accidents might occur, eventually a molecule of pyridoxal phosphate would collide with it. Once this happened, dissociation would become infrequent, because the other binding groups on the molecule fit the particular peptide configuration and hold the coenzyme in place. Two things have been gained. The combination of coenzyme and apoenzyme has been stabilized by the use of multiple binding groups. Just as importantly, a variety of aldehydes can be used in the general metabolism without any significant disruption of the transamination reaction, because only the one very specific aldehyde structure will fit the apoenzyme.

The same kind of an advantage applies to the lysine residue. We shall see that these residues are employed in other enzymes to bind completely different coenzymes, participating in reactions bearing little resemblance to transamination. Each of these kinds of coenzymes has a unique configuration, not capable of being confused with any of the others, and each apoenzyme can't bind the wrong coenzyme to its lysyl groups because it is built to conform to the complex structure of the proper coenzyme.

This represents a beginning on an important concept in metabolism: *The kinds of structures used in biological compounds and the types of reactions they undergo are relatively few.* The close regulation of the complex assembly of reactions that we lump together as metabolism doesn't depend upon each compound's having a unique kind of structure. It depends upon the compound's having a particular *combination* of structures, and the matching of the combination by a configuration built into only a few, perhaps one, of the hundreds of enzymes made by the cell.

THE NATURE OF OXIDATIONS AND REDUCTIONS

Let us briefly review oxidation-reduction reactions before considering a biological example. Three kinds of oxidations are important for our purposes:

(1) Addition of Oxygen. Combination with elemental oxygen is an oxidation and this was the original meaning of the word. A compound is said to be oxidized when its relative content of oxygen is increased, but oxidations are not to be confused with hydrations, in which water is added.

This is an oxidation of an organic compound:

$$\begin{array}{c} \mathrm{H}\;\;\mathrm{H} \\ |\;\;\;\;| \\ -\mathrm{C}-\mathrm{C}- \\ |\;\;\;\;| \\ \mathrm{H}\;\;\mathrm{H} \end{array} + \frac{1}{2}\,O_2 \longrightarrow \begin{array}{c} \mathrm{H}\;\;\;\mathrm{H} \\ |\;\;\;\;\;| \\ -\mathrm{C}-\mathrm{C}- \\ |\;\;\;\;\;| \\ \mathrm{HO}\;\;\mathrm{H} \end{array}$$

This is a hydration and not an oxidation:

$$\begin{array}{c} \mathrm{H}\;\;\mathrm{H} \\ |\;\;\;\;| \\ -\mathrm{C}{=}\mathrm{C}- \end{array} + H_2O \longrightarrow \begin{array}{c} \mathrm{H}\;\;\;\mathrm{H} \\ |\;\;\;\;\;| \\ -\mathrm{C}-\mathrm{C}- \\ |\;\;\;\;\;| \\ \mathrm{HO}\;\;\mathrm{H} \end{array}$$

(2) Dehydrogenations. A compound may also be oxidized by the removal of hydrogen, including both the proton and its associated electron. These dehydrogenations are not to be confused with simple ionizations of acids.

This is an oxidation:

$$\begin{array}{c} \text{H}\;\text{H} \\ |\;\;| \\ -\text{C}-\text{C}- \\ |\;\;| \\ \text{H}\;\text{H} \end{array} \longrightarrow \begin{array}{c} \text{H}\;\text{H} \\ |\;\;| \\ -\text{C}=\text{C}- \end{array} + H_2$$

This is an oxidation:

$$\begin{array}{c} \text{H}\;\text{H} \\ |\;\;| \\ -\text{C}-\text{C}- \\ |\;\;| \\ \text{H}\;\text{H} \end{array} + \frac{1}{2}O_2 \longrightarrow \begin{array}{c} \text{H}\;\text{H} \\ |\;\;| \\ -\text{C}=\text{C}- \end{array} + H_2O$$

(Notice that this reaction creates the starting material for the hydration given above—the compound with the double bond is already as oxidized as is the corresponding alcohol.)

This is also an oxidation:

$$\text{H}-\overset{\overset{\text{O}}{\|}}{\text{C}}-\text{OH} + \frac{1}{2}O_2 \longrightarrow CO_2 + H_2O$$

This is an ionization, and not an oxidation:

$$\text{H}-\overset{\overset{\text{O}}{\|}}{\text{C}}-\text{OH} \longrightarrow \text{H}-\text{C}\begin{array}{l}\diagup\!\!\!\!\cdot\,\text{O} \\ \ominus \\ \diagdown\!\!\!\!\cdot\,\text{O}\end{array} + H^{\oplus}$$

(Only a proton is liberated here, and the corresponding electron is retained in the organic anion.)

(3) Electron Transfer. A compound may be oxidized by a simple removal of electrons with a concomitant increase in positive charge. In this case, there must be another compound present to accept electrons, and this compound is said to be **reduced** by the gain of electrons. (The name came from the practice of reducing metallic oxide ores to the free metals.)

This is an oxidation-reduction:

$$Fe^{3+} + Cu^{+} \longrightarrow Fe^{2+} + Cu^{2+}$$

In this case, the cuprous ion is being oxidized by the ferric ion. The ferric ion is being reduced by the cuprous ion. We can say that the ferric ion is an oxidizing agent, which becomes reduced when it oxidizes other compounds. The cuprous ion is a reducing agent, which becomes oxidized when it reduces other materials.

The idea that every oxidation must be accompanied by a reduction, and vice versa, has been extended from the simple example we have here to other cases in which it is not obvious. Thus, we say that oxygen is reduced when it adds to organic compounds in the way illustrated in example (1) above, or hydrogen is being reduced in the first reaction given under example (2), while the carbon skeleton is being oxidized. The terminology is somewhat of a formalism in these cases, but the value of the practice will become more apparent as we become familiar with the actual sequences of biological oxidations, in which similar final results are sometimes achieved by the use of the differing types of oxidations.

Lactate Dehydrogenase — An Oxidoreductase

Enzymes that catalyze oxidation-reduction reactions always require a coenzyme, prosthetic group, or bound metallic ion for activity. For example, one of the most common types of reaction is the equilibration of alcohols and carbonyl compounds:

$$\mathrm{H{-}\overset{|}{\underset{|}{C}}{-}OH} \longleftrightarrow \mathrm{\overset{|}{\underset{|}{C}}{=}O} + \mathrm{H} + 2\mathrm{H}^{\oplus} + 2e^{\ominus}$$

In order for this reaction to proceed, there must be some acceptor (or donor, in the reverse direction) of electrons, and the electron carrier for equilibrating alcohols and ketones (or aldehydes) is usually **nicotinamide adenine dinucleotide (NAD),** or a close relative. The name implies its structure in which nucleotides containing nicotinamide and adenine are linked through a pyrophosphate bond:

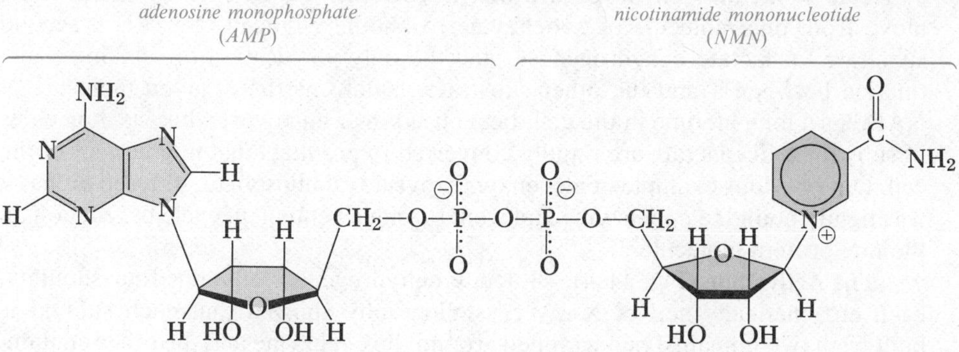

adenosine monophosphate (AMP) — *nicotinamide mononucleotide (NMN)*

nicotinamide adenine dinucleotide (NAD)

The pyridine ring of NAD is the part of the molecule that changes during oxidation and reduction. The reduction of NAD requires the addition of a **hydride ion** — a proton with two electrons — and the nucleotide loses a positive charge in the process:

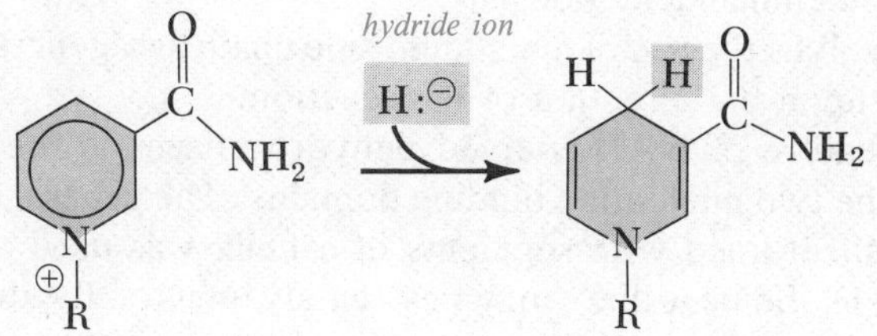

hydride ion $\mathrm{H}{:}^{\ominus}$

$\mathrm{NAD}^{\oplus}$ (nicotinamide adenine dinucleotide) → **NADH (reduced nicotinamide adenine dinucleotide)**

(The reaction probably goes through free radical intermediates that are not shown.)

A typical example of the use of NAD is in the equilibration of **lactate** and **pyruvate**, which are important products of the degradation of glucose:

$$\underset{\text{L-lactate}}{\text{HO}-\overset{\overset{\displaystyle COO^{\ominus}}{|}}{\underset{\underset{\displaystyle CH_3}{|}}{C}}-\text{H}} + \text{NAD}^{\oplus} \xleftrightarrow{\textit{lactate dehydrogenase}} \underset{\text{pyruvate}}{\overset{\overset{\displaystyle COO^{-}}{|}}{\underset{\underset{\displaystyle CH_3}{|}}{C}}=\text{O}} + \text{NADH} + \text{H}^{+}$$

The abbreviations, NAD^+ and NADH, are used in reactions to show only the change in charge, not the total charges on the molecules. (When used in text, the charge is omitted from the abbreviation.) The reaction is catalyzed by a lactate dehydrogenase, more particularly by a lactate:NAD oxidoreductase. The name is misleading because the enzyme may catalyze the oxidation of lactate or the reduction of pyruvate, depending upon the relative concentrations of reactants.

Although NAD is consumed during the oxidation of lactate, it is said to be a coenzyme for lactate dehydrogenase because the same molecules of NAD can be used repeatedly to oxidize many molecules of lactate, provided that there is some additional reaction in which NADH is converted back to the oxidized NAD. The semantics here are a little cloudy. A molecule of NAD is not so tightly bound to a molecule of lactate dehydrogenase that it "belongs" to it. It may, and it does, move from one molecule of apoenzyme to another. Indeed, NAD is a second substrate for lactate dehydrogenase, and the only justification for making a distinction between it and the other substrate, lactate, is that a given molecule of NAD has a long lifetime in the cell, being used over and over, whereas molecules of substrates like lactate are rapidly converted to products that migrate out of the cell. Our previous example of a coenzyme, pyridoxal phosphate, differed in that it was tightly bound to apoenzyme and went through a complete cycle of reaction on the one protein molecule.

The Active Site (Fig. 14–4). Lactate dehydrogenase contains four subunits, each enzymatically active. X-ray crystallography showed that each subunit is built from two domains, one wrapped around the coenzyme and the other containing the substrate binding site. Each of these domains contains an array of side chains to hold the coenzyme or substrate in position. These include hydrogen-bonding carboxylate groups of aspartate and glutamate, ammonium groups of lysine, and a hydroxyl group from tyrosine. The anionic groups of the substrate and of the coenzyme are held in position by arginine side chains from the polypeptide. The coenzyme binding domain itself is a combination of two domains, one wrapped around the nicotinamide nucleotide part of the coenzyme and the other around the adenine nucleotide part.

Catalysis involves the use of a histidine side chain as a general acid or general base, depending upon the direction of the reaction.

A striking feature of NAD-coupled dehydrogenases is the close structural resemblance of the two nucleotide binding domains. The general plan, involving a twisted pleated sheet lined with segments of α-helix was used as an example in Chapter 3 (p. 31). Some other enzymes, catalyzing totally different types of reactions, are also built according to this plan. It appears that some exceedingly useful primordial peptide appeared early in evolution, and many variants on the structure arose, perhaps in part by gene duplication followed by separate mutations.

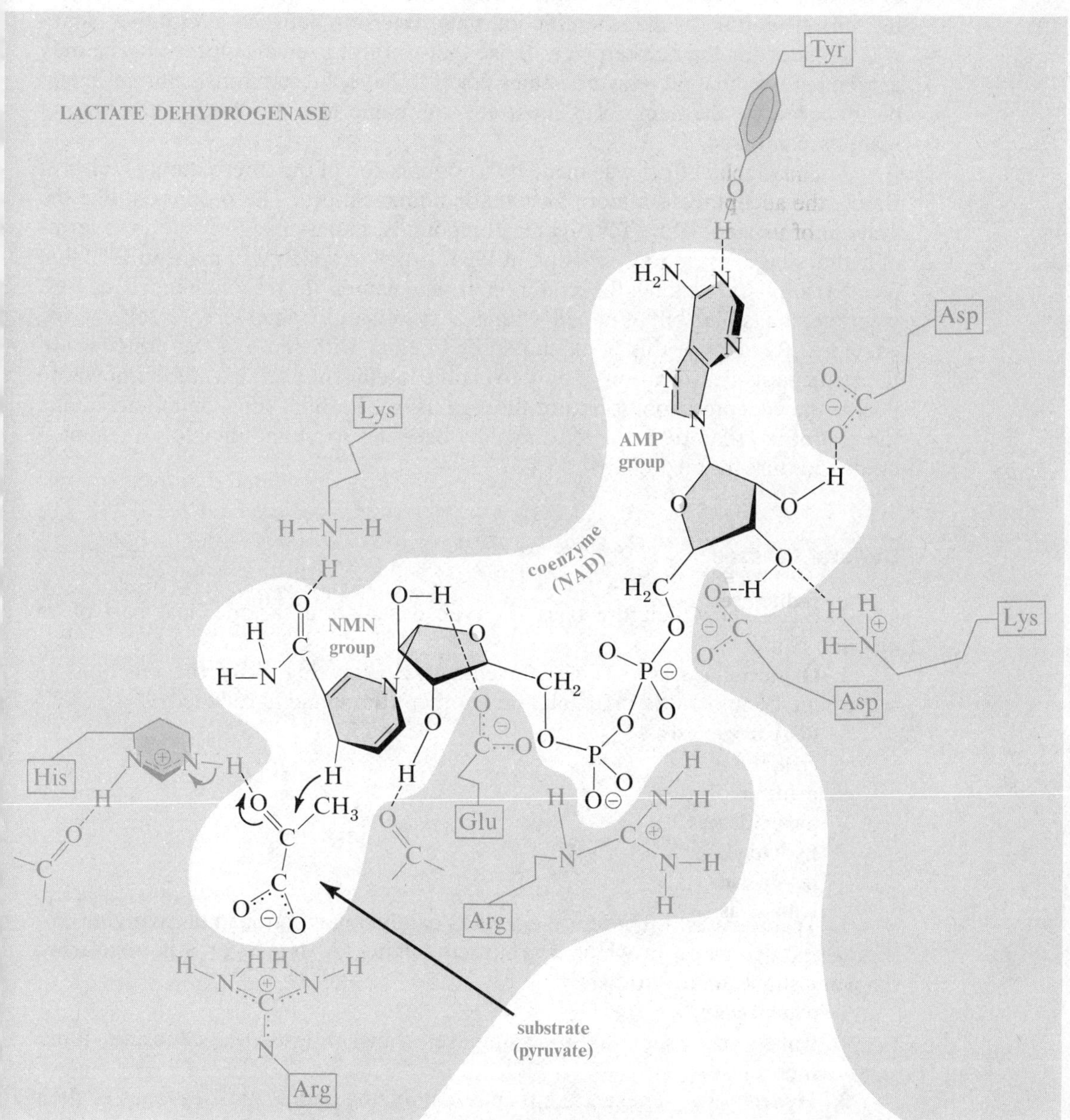

FIGURE 14–4 **The binding sites of lactate dehydrogenase for the coenzyme, NADH, and the substrate, pyruvate. (The particular structure is for the muscle enzyme, which usually catalyzes the reduction of pyruvate to lactate, rather than the reverse reaction.) The binding site is contributed by three domains in the dehydrogenase, one wrapped around the adenosine portion of the coenzyme (AMP group), one around the nicotinamide nucleotide portion (NMN group), and the third around the substrate (*lower left*). (Based on figure by J. J. Holbrook, A. Liljas, S. J. Steindel, and M. G. Rossman, (1975) Lactate dehydrogenase, 11:240, in P. D. Boyer, ed., *The Enzymes*, 3rd ed. © 1975 by Academic Press. Reproduced by permission.)**

ENZYME NOMENCLATURE

Until this decade, the creation of names for enzymes was mainly the responsibility of the discoverer, subject to modification by later investigators and to the taste of the various editors of the scientific journals. Decision between alternative names was a matter for the marketplace; those that weren't used disappeared. The only agreement was that all enzyme names ought to have the suffix *-ase,* but this might be preceded by the name of a substrate, the name of a product, or the type of reaction catalyzed.

After 1956, an effort was made by a commission of the International Union of Biochemistry to create a more systematic nomenclature. The proposals, like the systematic nomenclature for organic compounds, had a mixed reception in terms of actual usage, even after revision in 1964. A new revision was made in 1972 that faced reality by making the common trivial names the recommended names, wherever possible, but grouped within a systematic framework for classifying enzymes. Readers of this book should be familiar with some of the principles of the systematic classification, if only to enable intelligent exercise of their power to decide the accepted nomenclature through usage. Perhaps the Commissions, like the rest of us, tend to oscillate too wildly between strict and dictatorial regimenation and unrestrained chaos.

General Classes

The IUB nomenclature uses six type designations as the final part of an enzyme name:

1. Oxidoreductases. These are enzymes catalyzing all of the reactions in which one compound is oxidized and another compound is reduced:

dehydrogenases
reductases
oxidases
peroxidases
hydroxylases
oxygenases

2. Transferases. These are enzymes catalyzing reactions not involving oxidation and reduction in which a group containing C, N, P, or S is transferred from one substrate to another:

transferases
trans – ases (such as transaminase, transketolase, transaldolase, transmethylase)

3. Hydrolases. These are enzymes catalyzing hydrolytic cleavages or their reversal. (They do not include enzymes catalyzing the addition or removal of water from one compound, such as fumarase or enolase.) This includes all of the enzymes named as:

esterases
amidases
peptidases
phosphatases
glycosidases

4. Lyases. These are enzymes catalyzing the cleavage of C — C, C — O, C — N bonds, etc., without a hydrolysis or oxidation-reduction:

decarboxylases

aldolases
synthases (but not properly named synthetases)
cleavage enzymes (such as citrate or 3-hydroxy-3-methylglutaryl CoA cleavage enzymes)
hydrases or hydratases or dehydratases
deaminases
nucleotide cyclases

5. Isomerases. These are enzymes catalyzing intramolecular rearrangements not involving a net change in the concentration of compounds other than the substrate:

isomerases
racemases
epimerases
mutases

6. Ligases. These are enzymes catalyzing all of the reactions involving the formation of bonds between two substrate molecules that are coupled to the cleavage of a pyrophosphate bond in ATP or another energy donor. They include all of the enzymes now named as synthetases, except in cases where this name has been misapplied. The name synthetase was originally defined to include those enzymes now being defined as ligases, but it was later applied to enzymes catalyzing the formation of compounds by other mechanisms. Thus, one sometimes sees glycogen synthetase or δ-aminolevulinate synthetase even though the reactions catalyzed by these enzymes do not involve the cleavage of a high-energy phosphate bond. The term *synthase* was later coined to cover these enzymes, which are lyases in the IUB nomenclature.

Individual Enzymes

Within each type of enzyme, further subdivisions are made and given group numbers. The groups are divided into numbered sub-groups, and the individual enzymes within a sub-group are given another number. Thus, each enzyme is designated by four numbers. Some examples will make it clear, and also show advantages and disadvantages of the systematic nomenclature. The recommended trivial name is shown in italics.

1. Oxidoreductases
   1. Acting on the CH — OH group of donors
      1. With NAD or NADP as acceptor
         27. L-lactate:NAD oxidoreductase
             EC 1.1.1.27 *(lactate dehydrogenase)*

2. Transferases
   4. Glycosyltransferases
      1. Hexosyltransferases
         18. α-1, 4-Glucan:α-1, 4-glucan 6-glycosyltransferase
             EC 2.4.1.18 (1,4-*α-glucan branching enzyme*)

   7. Transferring phosphorus-containing groups
      1. Phosphotransferases with an alcohol group as acceptor
         2. ATP:D-glucose 6-phosphotransferase
            EC 2.7.1.2. *(glucokinase)*

3. Hydrolases
   4. Acting on peptide bonds
      21. Serine proteases
         4. No systematic name given
            EC 3.4.21.4 *(trypsin)*

4. Lyases
   2. Carbon-oxygen lyases
      1. Hydro-lyases
         13. L-Serine hydro-lyase (deaminating)
            EC 4.2.1.13 *(serine dehydratase)*

5. Isomerases
   1. Racemases and epimerases
      3. Acting on carbohydrates and derivatives
         1. D-Ribulose 5-phosphate 3-epimerase
            EC 5.1.3.1 *(ribulosephosphate 3-epimerase)*

6. Ligases
   2. Forming C – S bonds
      1. Acid-thiol ligases
         3. Acid: CoA ligase (AMP-forming)
            EC 6.2.1.3 *(acyl CoA synthetase)*

FURTHER READING

Bender, M.L., and L.J. Brubacher: (1973) *Catalysis and Enzyme Action.* McGraw-Hill. Succinct and clear review of general principles of catalysis.

Citri, N.: (1973) *Conformational Adaptability in Enzymes.* Adv. Enzymol., *37*: 397.

Jencks, W.P.: (1975) *Binding Energy, Specificity, and Enzymatic Catalysis: The Circe Effect.* Adv. Enzymol., *43*: 219. Lengthy review of relation between binding and catalysis.

Sigman, D.S., and G. Mouser: (1975) *Chemical Studies of Enzyme Active Sites.* Annu. Rev. Biochem., *44*: 889.

Roberts, G.C.K., et al.: (1969) *The Mechanism of Action of Ribonuclease.* Proc. Natl. Acad. Sci. U.S.A., *62*: 1151.

Richards, F.M., and H.W. Wychoff: (1971) *Bovine Pancreatic Ribonuclease.* Vol. 4, p. 647. *In* P.D. Boyer, ed.: *The Enzymes,* 3rd ed. Academic Press.

Harsuck, J.A., and W.N. Lipscomb: (1971) *Carboxypeptidase A.* Vol. 3, p. 1. *In* P.D. Boyer, ed.: *The Enzymes,* 3rd ed. Academic Press.

Koeppe, R.E., II, and R.M. Stroud: (1976) *Mechanism of Hydrolysis By Serine Proteases.* Biochemistry, *15*: 3450.

Hunkapiller, M.W., M.D. Forgac, and R.H. Richards: (1976) *Mechanism of Action of Serine Proteases.* Biochemistry, *15*: 5581. This and the preceding reference are interesting illustrations of different approaches to the fine points of mechanism.

Kraut, J.: (1977) *Serine Proteases: Structure and Mechanism of Catalysis.* Annu. Rev. Biochem., *46*: 331. Presents an alternative view of mechanism, de-emphasizing the hydrogen bond between His and Ser.

Beckett, A.H. and A.A. Al-Badr: *A Modified Structure for the Acetylcholine-Esterase Receptor.* J. Pharm. Pharmacol., *27*: 855.

Rose, I.A.: (1975) *Mechanism of the Aldose-ketose Isomerase Reaction.* Adv. Enzymol., *43*: 491.

Lai, C.Y. N. Nakai, and D. Chang: (1974) *Amino Acid Sequence of Rabbit Muscle Aldolase and the Structure of the Active Center.* Science, *183*: 1204.

Riordan, J.F., K.D. McElvany, and C.L. Borders, Jr.: (1977) *Arginyl Residues: Anion Recognition Sites in Enzymes.* Science, *195*:884.

Rossman, M.G., A. Liljas, C.-I. Bränden, and L.J. Banaszak: (1975) *Evolutionary and Structural Relationships among Dehydrogenases.* Vol. 11, p. 61. *In* P.D. Boyer, ed.: *The Enzymes,* 3rd ed. Academic Press.

Enzyme Nomenclature. (1973) Elsevier. This report of the 1972 recommendations of the IUB is valuable, even though Anglicized, reference work for more general purposes than nomenclature.

15 RATES OF ENZYMATIC REACTIONS

We dissected enzymatic reactions into individual steps and made qualitative analyses of the mechanisms in the previous chapter because it is important to understand how enzymes work. Now we want to understand what is accomplished by the action of enzymes. One aspect is the nature of the reactions that are catalyzed, but it is a mistake to assume that memorization of sequences of reactions is enough for understanding the metabolic economy. To be sure, the metabolic economy is composed of chemical reactions, many of which we must learn, but an inseparable part of the economy is the maintenance of quantitative balance. In short, we cannot begin to understand the physiological function of enzymes until we have some idea of the factors influencing reaction rates, and this idea requires numbers as well as qualitative facts. (Some make the equally serious error of presuming that facility in symbolic manipulations can totally substitute for knowledge of factual detail, but we shall have no trouble in avoiding this mistake in our later discussions.)

The Kinetic Parameters

We saw in the previous chapter that even an enzymatic reaction involving a single substrate and a single product will proceed in at least four steps: The enzyme and substrate combine to form a complex, **ES; ES** is transformed by internal catalysis to an intermediate complex, **EI; EI** changes into an enzyme-product complex, **EP,** which decomposes into the enzyme and product. Summarizing these reactions, and the reverse reactions that can also occur, we have:

$$\mathbf{E + S \rightleftarrows ES \rightleftarrows EI \rightleftarrows EP \rightleftarrows E + P.}$$

There are eight different reactions in this example, which is much simpler than many enzymatic reactions actually are, and each of these reactions will involve a separate mathematical description of its rate.

V_{max} and K_M. The kinetic analysis of enzymatic reactions has been simplified by combining the constants for individual steps into two parameters that describe the overall process in a useful way and which can be estimated from routine laboratory determinations. One of these parameters is the maximum velocity, V_{max}, which is the theoretical limit for the rate of reaction under defined conditions when the substrate concentration is so high that the active site is constantly occupied by substrate. V_{max} is an inherent property of a given enzyme, and it only depends upon the amount of the enzyme that is present in a given solution. It is a theoretical limit because it can be reached only when the substrate is present at infinite concentration, and no product is present to occupy the active site. It is the velocity of the reaction when the enzyme is **saturated** with substrate.

The second useful parameter is the Michaelis* constant, K_M, which is defined as the substrate concentration at which the actual velocity is ½ of the maximum velocity with no product present:

$$K_M = [S], \text{ when } v = \tfrac{1}{2} V_{max}, \text{ and } [P] = 0.$$

With these two parameters in mind, we are in a position to describe much of the physiologically important behavior of enzymes.

The Michaelis-Menten Equation

The major complication in analyzing the rate of any enzymatic reaction is the reverse reaction. The mathematics can be greatly simplified if the reverse reaction is negligible compared to the forward reaction. We shall discuss how this ideal is approached under laboratory conditions, but for the moment let us assume that we are dealing with a steady state condition, in which substrate is constantly supplied to maintain its concentration at a fixed level, and the product concentration is kept near zero by whisking it away as fast as it formed. In this steady state, the rate of formation of the intermediate complexes, **ES, EI,** and **EP,** will just balance the rate of their removal so that their concentrations remain fixed.

Under these arbitrary conditions, the velocity of the reaction of a single substrate is described by the Michaelis-Menten equation:

$$v = \frac{V_{max}S}{K_M + S}$$

in which v is the velocity, V_{max} is the maximum velocity with the enzyme concentration and the conditions used, and S is the concentration of substrate. The mathematical form of the relationship between velocity and substrate concentration was recognized long before much was known about the mechanisms of enzymatic reactions, and the equation is valid even though it was originally derived in 1913 using erroneous assumptions.

According to the equation, when $S = K_M$,

$$v = \frac{V_{max}K_M}{K_M + K_M} = \frac{V_{max}}{2}$$

which is as it must be, since we defined the Michaelis constant as the substrate concentration at which the velocity is half of the maximal.

*Leonor Michaelis (1875–1949), German (later American) biochemist.

Relation of Substrate Concentration and Rate. We can calculate the effect of substrate concentration on rate:

When $\frac{S}{K_M}$ =	0.1	0.2	0.4	1.0	2.0	4.0	10.0	20.0	50.0	100.0
$\frac{v}{V_{max}}$ =	0.09	0.17	0.29	0.50	0.67	0.80	0.91	0.95	0.98	0.99

An important point emerges. More efficient use of the enzyme without sacrificing control of rate is obtained if the range of substrate concentrations is centered roughly equal to K_M. Within that range, a substantial fraction of the enzyme's catalytic capability is being used, and it is still capable of sensitive response in rate to changes in substrate concentration. Outside of that range either velocity of reaction or its control will suffer. **Most enzymes are built with affinities for their substrates such that their K_M values will lie within an order of magnitude of the physiological concentrations of their substrates.**

Effect of the Reverse Reaction

The simple kinetic analysis we have given depends upon the assumption that the reverse reaction can be neglected, but the concentration of the product is rarely negligible in real biological systems. Since enzymes catalyze attainment of equilibrium from either direction, what we call "product" can also act as a substrate for the reverse reaction. It follows that there will be kinetics of a similar mathematical form when "product" reacts to yield "substrate," complete with its own Michaelis constant and maximum velocity, but operating in the reverse direction.

This gives an additional dimension to the evolutionary design of an enzyme. In effect, there is a competition between what we call substrate and product for the active site on an enzyme. How much time the site is occupied by one or the other will depend upon the relative affinity of the enzyme for the two compounds and their concentrations. The greater the affinity (the lower the K_M) and the higher the concentration, the greater the fraction of V_{max} that can be realized in one direction.

However, the V_{max} and K_M values for forward and reverse reactions are not independent entities. They are related to the equilibrium constant for the reaction according to the Haldane* equation:

$$K_{eq} = \frac{V_F}{K_F} \div \frac{V_B}{K_B} = \frac{V_F K_B}{V_B K_F}$$

in which: V_F and V_B are the maximum velocities (V_{max}) for the forward and back reactions, respectively;

K_F and K_B are the Michaelis constants (K_M) for the forward and back reactions, respectively.

Let us use the Haldane equation to make two points. Firstly, an enzyme will use the largest fraction of its catalytic ability if the relative affinities for substrate and product are adjusted so that V_{max} is the same in both directions. Secondly, it is

*J. B. S. Haldane (1892–1964). English biochemist and geneticist; also known as far-left politicizer turned disillusioned exile. Not to be confused with his father, J. S. Haldane (1860–1936), respiratory physiologist and source of the Haldane coefficient for H^+ liberation from hemoglobin.

not always desirable to catalyze a reversible reaction equally well in both directions; therefore, the enzymes may be built to favor one direction or the other. Now, it is important to remember that the direction in which a reaction can go is determined by the concentrations of reactants relative to the position of equilibrium inherent in the reaction. **A catalyst can't change the equilibrium constant; it can only make the reaction go faster in whatever direction it is capable of going.** However, an enzyme is a catalyst that can be constructed to drag its feet when concentrations favor reaction in one direction, while happily going to work when concentrations shift so as to favor reaction in the other direction. This property is gained only at some sacrifice of the inherent catalytic potential.

We can develop the principle from a simple example. Imagine a reversible reaction:

$$\mathbf{F} \rightleftarrows \mathbf{B}$$

for which $K_{eq} = 1$.

Let us imagine three different enzymes catalyzing this reaction. The first enzyme has equal values of V_{max} and of K_M for the forward and the back reactions. The second enzyme has values of V_{max} and K_M for the forward reaction that are 10 times those for the back reaction. The third enzyme is the converse case, with values of V_{max} and K_M for the back reaction that are 10 times those for the forward reaction. Let us assume that the largest value of V_{max} is identical for each enzyme, and identify this value as **V**. To summarize:

$$\text{enzyme 1: } K_F = K_B;\ V_F = V_B = \mathbf{V}$$

$$\text{enzyme 2: } K_F = 10\ K_B;\ V_F = 10\ V_B = \mathbf{V}$$

$$\text{enzyme 3: } K_B = 10\ K_F;\ V_B = 10\ V_F = \mathbf{V}$$

Now, let us test these three enzymes in two hypothetical circumstances. In one case, the concentration of **F** is 10 times the concentration of **B,** so the reaction must go forward. In the second case, the concentration of **B** is 10 times the concentration of **F,** so the reaction must go backward. Let us further assume that the concentration of the substrate in each case is equal to the K_M for the enzyme in the direction the reaction is going so that each enzyme in each circumstance is put at equal advantage in this respect. What will happen? (The equation used is given below.)

The first enzyme, with equal kinetic parameters, catalyzes the reaction at 43 per cent of V_{max} in the forward direction and at 43 per cent of V_{max} in the reverse direction.

This enzyme is well designed to catalyze the reaction in either direction. The second enzyme, with larger kinetic parameters in the forward direction, catalyzes the forward reaction at 30 per cent of **V,** but the back reaction goes at only 4.5 per cent of **V,** even though it has a higher affinity (lower K_M) for **B.** This enzyme is designed to catalyze the forward reaction when concentrations favor it, but not the back reaction when concentrations shift to favor it.

Similarly, the third enzyme, with larger kinetic parameters in the backward direction, catalyzes the back reaction at 30 per cent of **V,** but catalyzes the forward reaction at only 4.5 per cent of **V.** It is built to do its job when concentrations favor the back reaction. Designing the molecule to have different kinetic parameters has altered the capability of the latter two enzymes to catalyze the same reaction in one direction as opposed to the other.

We can also see from the values that making the enzyme favor catalysis in one direction or the other has been done at the expense of catalytic efficiency. The latter two enzymes only catalyze the reaction at 30 per cent of **V** in the favored direction, whereas the first enzyme catalyzes the reaction at 43 per cent of **V** in either direction.

The velocities in the above examples were calculated from an equation for reversible reactions obeying Michaelis-Menten kinetics:

$$v = \frac{V_F \dfrac{S}{K_F} - V_B \dfrac{P}{K_B}}{1 + \dfrac{S}{K_F} + \dfrac{P}{K_B}}$$

Isozymes

An organism need not be limited to one particular enzyme for catalyzing a given reaction. Indeed, the same reaction may be catalyzed by completely different proteins that differ in mechanism as well as in structure, a circumstance frequently true when the same reaction occurs in mitochondria and in the soluble cytoplasm. However, in some cases, cells may contain structurally related enzymes utilizing the same mechanism but with different kinetic parameters to fit particular requirements. Such enzymes are said to be isozymes, or isoenzymes.

The lactate dehydrogenases of mammals are good examples. These enzymes are tetramers made from polypeptide chains occurring in two principal forms, designated H for heart and M for muscle. (A third kind is in spermatozoa, but it is of minor importance in the total fuel economy, whatever its other consequences may be.) These polypeptide chains are so similar that they are interchangeable in the tetramer, giving five theoretical possibilities for the major forms of lactate dehydrogenase: H_4, H_3M, H_2M_2, HM_3, and M_4. All occur in tissues. H_4 is predominant in tissues of high oxidative capacity, and it has kinetic parameters that favor catalysis of the oxidation of lactate to pyruvate over the reduction of pyruvate to lactate, and this is appropriate for utilization of lactate as a fuel. Other tissues, such as striated muscle fibers built for quick twitches (white fibers) contain mostly the M_4 isozyme, which has no particular preference for the direction of catalysis. We shall see that the concentrations of reactants usually favor the reduction of pyruvate to lactate in those tissues.

Reactions with Two or More Substrates

Many enzymatic reactions involve more than one substrate:

$$\mathbf{A} + \mathbf{B} \xrightarrow{E} \mathbf{P}, \text{ or } \mathbf{A} + \mathbf{B} \xrightarrow{E} \mathbf{P} + \mathbf{Q},$$

and so on. The velocity of these reactions depends upon the concentrations of all of the components, and the kinetic equations are more complex. Furthermore, the form of the kinetic equations depends upon the mechanism of the reaction.

For example, two substrates may react in a **sequential** fashion (Fig. 15–1), with both becoming attached to the enzyme before any products are formed. Lactate dehydrogenase and other NAD-coupled dehydrogenases function in this way. The mechanism is said to be **ordered** if the substrates add and the products

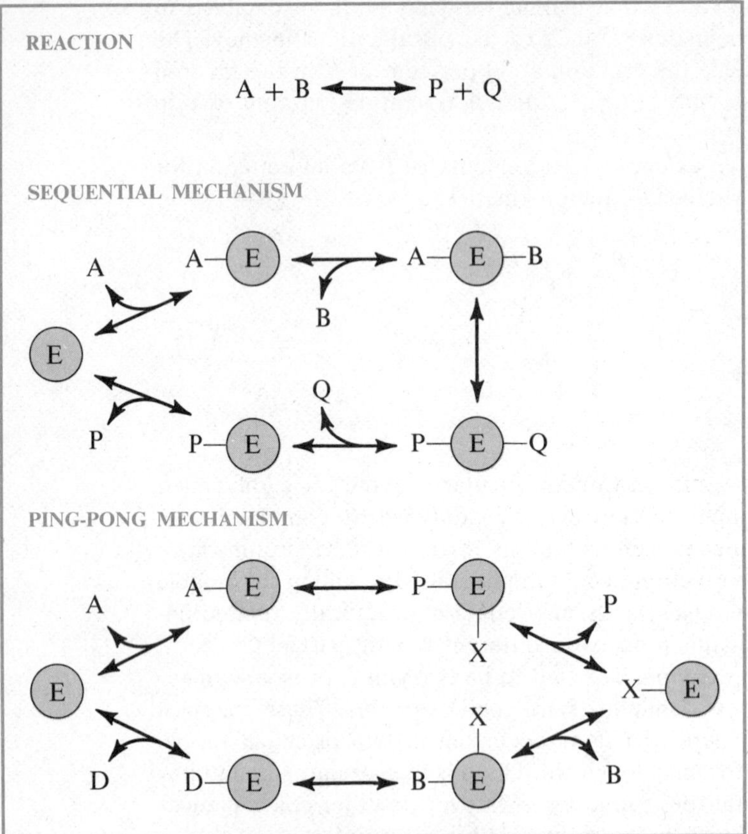

FIGURE 15–1 **Reactions involving two substrates may proceed by a sequential mechanism (*top*) in which the substrates successively add to the enzyme before a reaction forming the enzyme-products complex occurs. The products then dissociate, one at a time, from the enzyme. Alternatively, the reactions may proceed by a ping-pong mechanism (*bottom*), in which one substrate reacts on the enzyme to transfer a functional group, X, to it. The first product, P, then leaves before the second substrate, B, is added. The functional group X then is transferred from the enzyme to B, forming the second product, Q.**

leave in a defined way (**A** adds before **B, P** leaves before **Q,** for example) and **random** if it makes no difference which substrate is bound first by the enzyme.

In other cases, substrates react by a **ping-pong** mechanism in which one substrate reacts with the enzyme so as to transfer a functional group onto the enzyme, leaving one product. The second substrate then reacts to pick up the functional group, forming a second product. We saw such a mechanism with the aminotransferases, in which the amino group passes from an amino acid substrate to the pyridoxal phosphate coenzyme, followed by its transfer to a keto acid substrate.

MEASUREMENT OF KINETICS

Measurements of the rates of enzymatic reactions in the laboratory are done for two purposes. One is to determine the kinetic parameters for both theoretical and practical use; the other is to assay the concentration of the enzyme in unknown samples.

Determination of Kinetic Parameters

The Michaelis constant and maximum velocity may be determined for an enzyme most easily by making measurements under circumstances in which the reverse reaction is negligible. One wants to have a minimum formation of product and a minimum fractional change in substrate concentration during the course of the reaction. This is facilitated when a precise and sensitive method is available for measuring the concentration of the product of the reaction. The product obtained per unit time will then be an estimate of the initial velocity of the reaction. With luck, this velocity will be the rate of the enzymatic reaction after steady state is reached but before the overall reaction is slowed down by changes in the product and substrate concentration. (The time required for reaching steady state is usually less than the time required for mixing substrate and enzyme in routine analyses.)

Suprisingly, this approach frequently gives useful information. If a series of determinations of initial velocity are made at varying substrate concentrations, a graph of the sort shown in Figure 15–2 can be constructed from the results. This is the typical hyperbolic curve obtained when Michaelis-Menten kinetics are obeyed.

The data can be analyzed in two ways. One is through algebraic analysis, and computer programs have the advantage that they can take into consideration the changing product concentrations as well as the changing substrate concentrations.

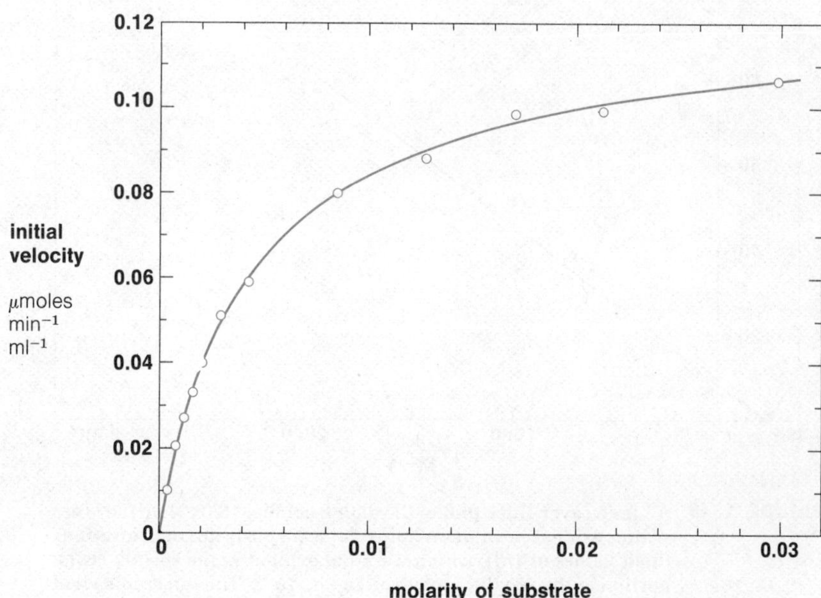

FIGURE 15–2 **Measured values of the initial velocity of an enzymatic reaction with varying concentrations of substrate. The scatter apparent at the higher substrate concentrations amounts to 2 per cent of the highest velocity, which is quite good for measurements of most enzymes with standard techniques.**

The other method, almost universally used in the past, is through graphical analysis that neglects the product concentration.

One such analysis is based on inversion and rearrangement of the Michaelis-Menten equation:

$$\frac{1}{v} = \frac{K_M + S}{V_{max}S} = \frac{K_M}{V_{max}} \times \frac{1}{S} + \frac{1}{V_{max}}$$

If we now take $1/v$ and $1/S$ as our variables, we have a simple linear equation of the type $y = ax + b$ which can be solved either by a linear regression analysis or by graphical analysis.

The reciprocals of the initial velocities are plotted as functions of the reciprocals of substrate concentrations. If Michaelis-Menten kinetics describe the reaction, the result will be a straight line, such as is shown in Figure 15–3, which was derived from the same data shown in Figure 15–2. These reciprocal plots are known as Lineweaver-Burk* plots, after the men best known for developing this type of analyses.

As is indicated on the figure, the intercept on the ordinate is the reciprocal of V_{max} and the intercept on the abscissa is the negative reciprocal of K_M, so both of these constants can be estimated from the plot.

When a reaction involves two or more substrates, the respective Michaelis constants can be estimated with the same sort of procedure by adding a large excess of one, effectively saturating the enzyme, and varying the concentration of the other across its K_M value.

*Hans Lineweaver and Dean Burk, American biochemists, now retired.

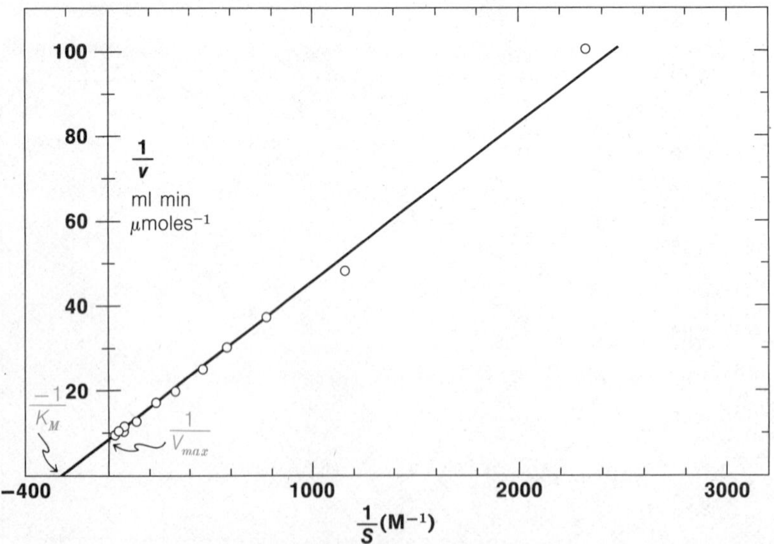

FIGURE 15–3 **A Lineweaver-Burk plot of the data from Fig. 15–2. Note the gross scatter evident in values obtained at low substrate concentrations (high values of $1/S$), which were not evident in the steeply rising portion of the regular plot given in Fig. 15–2. The intercepts yield the following values:**

$$\text{abscissa} = -\frac{1}{K_M} = -230\ \text{M}^{-1};\ K_M = 0.0043\ \text{M}$$

$$\text{ordinate} = \frac{1}{V_{max}} = 8.3\ \text{ml min}\ \mu\text{moles}^{-1};\ V_{max} = 0.120\ \mu\text{moles min}^{-1}\ \text{ml}^{-1}$$

Other rearrangements of the Michaelis-Menten equation can be used to generate linear plots, but a discussion of their advantages is not necessary for our purpose. (See Further Reading for sources.)

Assay of Enzymes

The assay of enzyme concentrations by kinetic measurements is a common procedure in many kinds of laboratories — clinical, industrial, and experimental. The most common clinical tests involve assay of enzymes in blood, because conditions causing damage to specific tissues usually result in the release of abnormally large amounts of characteristic enzymes, but the concentration of particular enzymes is also determined in tissue samples or in cultured cells, especially to diagnose genetic abnormalities.

It is not yet practical to make a full mathematical analysis of kinetics for routine assays, which frequently involve single incubations of one sample under fixed conditions, so the problem in devising useful procedures is one of minimizing the effect of changing substrate and product concentrations while the enzyme is acting. This can be done by using a high substrate concentration. For example, if one starts with a substrate concentration of 10 K_M, and allows no more than 0.1 of this amount to be converted to product, the rate will drop by only 1 per cent during the assay if the product has no effect (S drops from 10 K_M to 9 K_M, and v from 0.91 V_{max} to 0.90 V_{max}). Half of the amount of enzyme in a sample will therefore come within 1 per cent of forming half the amount of product in a fixed time under these conditions, so the amount of product is nearly a linear function of the amount of enzyme in the sample. This is the ideal circumstance that the enzyme assayist strives for, but it is not always easy to obtain.

Figure 15–4 shows the time course of reactions that have identical kinetic parameters for the forward reaction but that differ in equilibrium constant, and therefore in kinetic parameters for the back reaction. *Curve A* illustrates the type of reaction beloved by those who assay enzymes, with a high equilibrium constant and low affinity for the product creating nearly linear kinetics for a large part of the reaction course. The amount of product obtained in a single determination is directly proportional to the amount of enzyme present over a wide range with reactions of this type.

Curve D shows a kind of reaction that would be almost impossible to tackle in the ordinary way. The affinity of the enzyme is so much higher for the product than it is for the substrate that the formation of only a little product causes a large drop in reaction rate. (Note that this is true even though the maximum velocity in the forward direction is 1,000 times greater than that in the reverse direction.) Fortunately, enzymes with kinetics like this are very rare, if they ever occur. They wouldn't be very useful catalysts in an organism. Curve B illustrates the action of an equally inefficient enzyme, but the position of equilibrium is so far in favor of product, that the catalysis still may be useful.

Curve C shows a more typical time course for a readily reversible reaction. One must either have a sensitive analytical procedure for the product, or take steps to remove the product as fast as it is formed. The latter procedure is preferable, and it may be accomplished by adding an excess of another enzyme to catalyze an additional reaction of the product, or by adding some reagent that selectively reacts with the product and converts it to another compound. Either of these methods also makes it possible to use enzymes for analyzing the concentrations of compounds in samples, even when the equilibrium is not favorable.

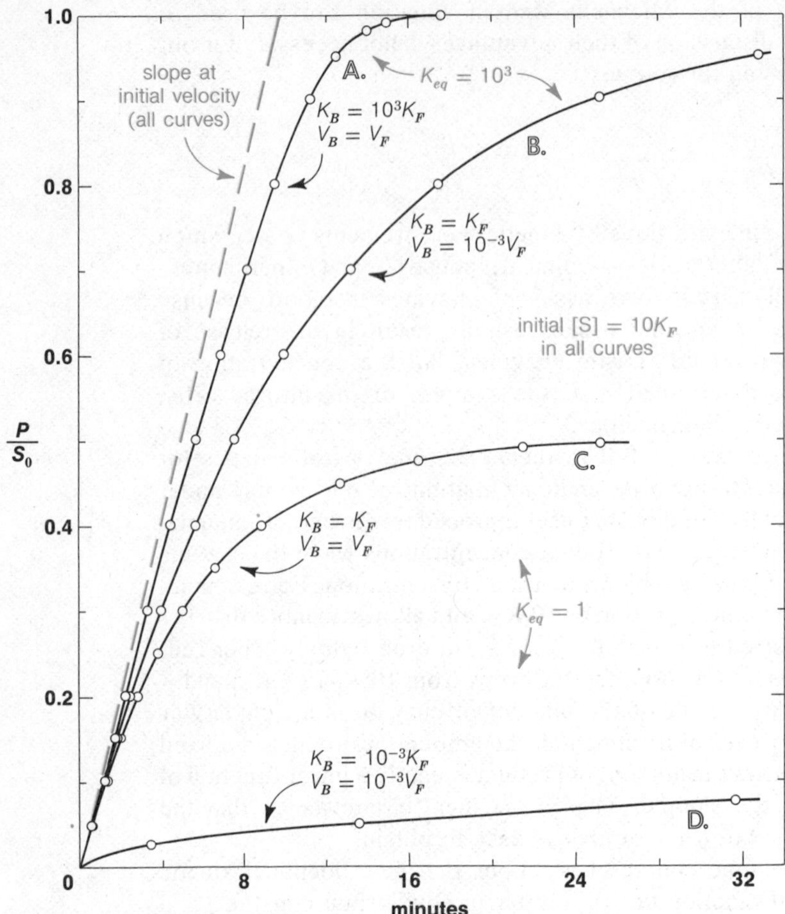

FIGURE 15–4 **The effect of the back reaction and the equilibrium constant on the time course of enzymatic reactions. The fraction of the initial substrate that has been converted to product (P/S_0) is plotted as a function of time of reaction with four different enzymes. The relative kinetic parameters for forward and back reactions are given by each curve. All enzymes are assumed to have the same V_{max} in the forward direction; note that all of the reactions would follow the dashed line if substrate was continually added to maintain its concentration, and if the product was removed as fast as it formed.**

For example, the equilibrium of lactate dehydrogenase favors the reduction of pyruvate to lactate:

$$\text{pyruvate} + \text{NADH} + \text{H}^+ \rightarrow \text{lactate} + \text{NAD}^+$$

NADH has a peak absorption of light at 340nm, and NAD does not, so the course of the reaction can be followed by noting the decreased absorption of light at this wave length. The changes in absorbance may be followed over time intervals for assaying lactate dehydrogenase, or excess enzyme may be added to an unknown sample, and the reaction allowed to go to completion to measure the amount of pyruvate in the sample.

The reaction cannot be used as such to determine lactate because the equilibrium is unfavorable for the formation of pyruvate, but this determination becomes possible if hydrazine is included in the reaction mixture to remove pyruvate, thereby making the overall equilibrium favor lactate oxidation:

$$\underset{\text{L-lactate}}{\mathrm{HO{-}C(COO^{\ominus})(H){-}CH_3}} \xrightarrow[K'_{eq} = 3.4 \times 10^{11}]{\mathrm{NAD^{\oplus}} \;\rightarrow\; \mathrm{H^{\oplus} + NADH}} \underset{\text{pyruvate}}{\mathrm{COO^{\ominus}{-}C{=}O{-}CH_3}} \xrightarrow[K'_{eq} \cong 2.4 \times 10^{14}]{\underset{\text{hydrazine}}{\mathrm{H_2N{-}NH_2}} \;\rightarrow\; \mathrm{H_2O}} \underset{\text{pyruvate hydrazone}}{\mathrm{COO^{\ominus}{-}C{=}N{-}NH_2{-}CH_3}}$$

$$K'_{eq} \cong 7 \times 10^{2}$$

Units of Enzyme Activity. The common enzyme unit (U) for expressing activity of enzymes has the dimensions of micromole of substrate consumed per minute. A new unit, the katal (kat) is moles per second (kat $= 6 \times 10^7$ U). These values can be translated into estimates of actual enzyme concentration if its turnover number is known under the same conditions. The turnover number is the number of reaction cycles one molecule of enzyme catalyzes per unit time (min or sec).

Effect of Temperature

Reactions catalyzed by enzymes are like other reactions in that the rate is increased by increasing temperatures — up to a point. Beyond a critical range, the transition temperature, the activity of an enzyme declines sharply (Fig. 15–5). Not only enzymes, but nearly all proteins lose conformation readily above certain temperatures. When thermal energy becomes great enough to cause the rupture of a few bonds, the neighboring bonds are weakened, and the whole molecule becomes unzippered. This gross loss of conformation is known as denaturation of the protein.

It happens that most of the proteins in an organism become denatured at temperatures only 10 to 15° C above the usual cellular temperatures. It may be that this is related to a certain flexibility of motion that is necessary in order to

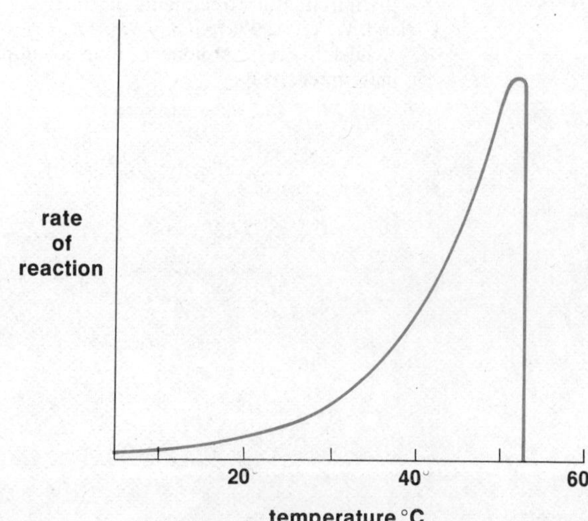

FIGURE 15–5

Enzymatic reactions are like most reactions in that their rates accelerate as the temperature rises above the freezing point. However, unlike most reactions, the rate abruptly falls to low values when the temperature is reached that causes disruption of the conformation of the enzyme.

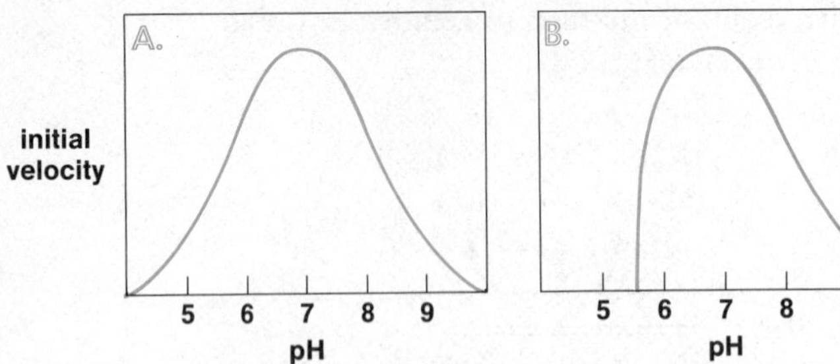

FIGURE 15–6 **The effect of pH on the rate of enzymatic reactions. *A.* The enzyme is stable over the pH range tested, and the mechanism of catalysis requires both acidic and basic groups. *B.* The mechanism is similar to that of the above enzyme, but the transition temperature for disruption of the enzyme's structure falls at pH 5.7 below the temperature used for assay.**

carry out enzymatic functions, or others depending upon conformational shifts. The tiger may have to be tickled to a rippling alertness that is not far removed from a destructive frenzy.

Effect of pH

A typical enzyme catalyzes its reaction most effectively at a specific concentration of H^+ (Fig. 15–6), not only because the mechanism depends upon acidic and basic groups being present in optimum proportions, but also because the conformation of the protein is also dependent upon the state of charge of bonding groups in general. Severe changes in pH cause substantial lowering of the transition temperature for disruption of conformation.

FURTHER READING

Segal, H. L.: (1959) *The Development of Enzyme Kinetics.* Vol. 1, p. 1; and Alberty, R. A.: (1959) *The Rate Equation for an Enzymic Reaction.* Vol. 1, p. 143. *In* P. D. Boyer, H. A. Lardy, and K. Myrback, eds.: *The Enzymes,* 2nd ed. Academic Press. These articles remain among the best of the introductory treatments of kinetics.

Cleland, W. W.: (1970) *Steady State Kinetics.* Vol. 2, p. 1. *In* P. D. Boyer, ed.: *The Enzymes,* 3rd ed. Academic Press. A pioneer in the development of computer analyses of kinetics treats the subject more intensively.

16 REGULATION OF ENZYMES

REGULATION BY SUBSTRATE CONCENTRATION

The rates of many enzymatic reactions are changed only as a result of changing substrate concentrations. When enzymes are constructed and metabolite concentrations are maintained so that [S] is approximately the same as K_M, the rates of reactions adjust so as to minimize changes in concentration of reactant and product. The adjustment may be gross; the direction of easily reversible reactions may shift because concentrations on first one side, and then the other, exceed the equilibrium values. The effect is to remove the predominant compound and alleviate any deficit of the other.

Even with reactions that always go in the same direction, simple Michaelis-Menten kinetics minimize concentration changes. A model is shown in Figure 16–1, in which an enzyme catalyzes a readily reversible reaction ($K_{eq} = 1$) that is part of a sequence. In case 1 its substrate, **F,** is constantly being formed and its product, **B,** is constantly being removed so that there is a steady flux through the reaction equal to 0.1 of the maximum velocity of the enzyme. (The velocity is only 0.1 V_{max} because the product concentration is being maintained at a relatively high level. This is typical for easily reversible reactions.)

What happens if the enzyme is suddenly presented with its substrate at twice the former rate (case 2)? The rising concentrations of substrate will accelerate the

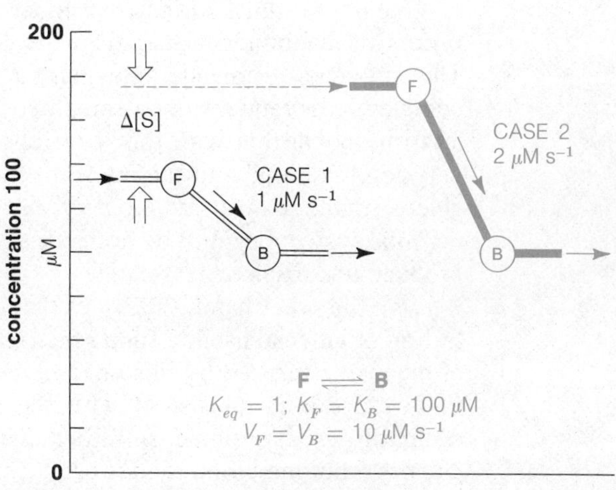

FIGURE 16–1

The kinetic behavior of enzymes dampens fluctuations in metabolite concentrations. In the example shown here, a doubling of the rate of flow of a compound through an enzymatic reaction (case 1 to case 2), only causes a 30 per cent increase in concentration of the substrate, *F*. (For clarity, the concentration of the product, *B*, is considered to be held constant by subsequent reaction. The kinetic parameters used in the example are not critical; there would be even less rise in substrate concentration if its initial concentration were somewhat below the K_F value.)

reaction until the rate of removal of the substrate just balances the rate at which it is being formed from preceding reactions, and no further rise in [S] will occur. The concentration of the substrate, F, increases only 31 per cent before the new flux through the system is doubled, presuming [B] is held constant.

Liver Glucose Utilization. A prime example of effective physiological use of substrate concentration as a controlling mechanism is found in the enzymes catalyzing the transfer of phosphate from ATP to glucose:

$$\textbf{ATP + glucose} \rightarrow \textbf{ADP + glucose 6-phosphate}$$

This is the first reaction in the utilization of glucose by cells. The reaction is catalyzed by enzymes called **hexokinases** because they are not very specific for the kind of hexose. The suffix, *-kinase,* was invented to suggest that the reaction results in the activation of glucose for metabolism. As is implied by the single arrow, the reaction is essentially irreversible under the conditions prevalent in tissues.

Four hexokinases occur in tissues. The types predominant in brain and liver parenchymal cells have been most intensively studied; the measured Michaelis constants are as follows:

	K_M (ATP)	K_M (glucose)
brain hexokinase (type I)	4×10^{-4} M	5×10^{-5} M
liver hexokinase (type IV)	1×10^{-4} M	2×10^{-2} M

The concentration of ATP in these tissues is held near 10^{-3} M or greater. This is substantially greater than K_M for either enzyme, and ATP is therefore not limiting in either case.

Now, the concentration of glucose in the blood supply to the brain and other peripheral tissues usually ranges around 5×10^{-3} M, although it may increase to around 9×10^{-3} M after consumption of a large amount of carbohydrate and may drop to 3×10^{-3} M during fasting or heavy muscular work. The liver operates under different conditions. It receives blood from the intestine through the portal circulation as well as from the arterial circulation, and the portal blood has substantially higher concentrations of glucose during absorption of a carbohydrate-containing meal.

All of the values for glucose concentration we have mentioned are nearly two orders of magnitude higher than the K_M for glucose with the brain hexokinase. This provides a considerable margin of safety for the brain, which is almost completely dependent upon the glucose supply for its major fuel. The observed facts are consistent with this if we allow for a gradient of glucose concentration between the blood and the interior of the neurons, because the concentration of glucose in blood must drop to nearly 2×10^{-3} M before disturbances of the central nervous system begin to be noticed, and even lower concentrations are needed to produce unconsciousness.

On the other hand, the K_M of the liver hexokinase is greater than the concentration of glucose usually found in arterial blood. Therefore, the phosphorylation of glucose catalyzed by this enzyme will be very sensitive to glucose concentration, and this is consistent with the physiological role of the liver. The liver ordinarily produces glucose rather than consuming it, and it only removes more glucose from the blood than it produces during temporary overloads after eating.

The balance between production and utilization occurs with blood concentrations near 0.006 M.

A rise in glucose concentration will accelerate the liver hexokinase reaction because the enzyme is operating far below its saturating substrate concentration, whereas the brain hexokinase is always nearly saturated and is relatively insensitive to fluctuations in glucose concentration.

REGULATION OF CATALYTIC ABILITY

A broader range of control over enzymatic reactions can be obtained by changing the V_{max}. This may be done by changing the catalytic ability of individual enzyme molecules or by changing the number of enzyme molecules. Let us now consider the former.

Concept of the Rate-Limiting Step

The idea that an enzyme catalyzing one reaction in a sequence may be the principal control point at which the rate of the entire sequence is governed is an important principle in understanding metabolic regulation. The key reaction is referred to as the rate-limiting step.

Since the rates of each reaction in a sequence must be identical if there is not to be a gross gain or loss of some intermediate compound, it may seem strange to speak of a step as being rate-limiting. The rate-limiting step is not one that goes slower than all of the others but is the step at which changes in enzymatic activity – in V_{max} – will cause the largest changes in the actual overall velocity of the sequence. Water flows through a pipe at the same rate at all points throughout its length, but it is the size of a constriction that determines what the rate is. Let us consider an example.

The metabolism of glucose includes the following sequence of reactions:

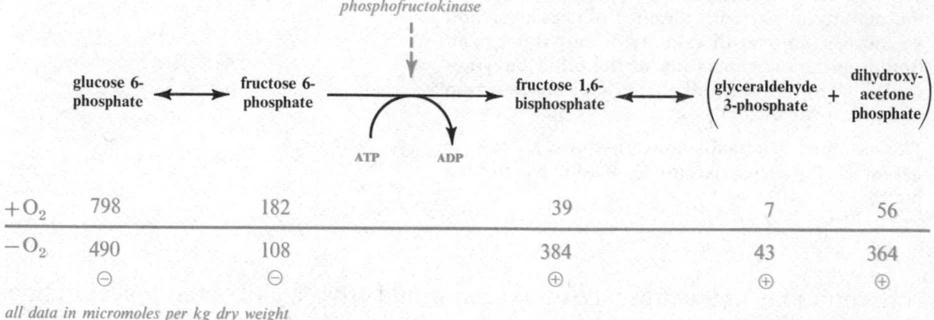

	glucose 6-phosphate	fructose 6-phosphate	fructose 1,6-bisphosphate	glyceraldehyde 3-phosphate	dihydroxyacetone phosphate
$+O_2$	798	182	39	7	56
$-O_2$	490	108	384	43	364
	⊖	⊖	⊕	⊕	⊕

all data in micromoles per kg dry weight

Listed below the intermediate compounds are their concentrations measured in rat hearts perfused with and without a supply of oxygen*. Deprivation of oxygen caused a drop in the concentration of the first two compounds in the sequence, and a rise in the concentration of subsequent compounds. A plot of successive

---

*Data from J. R. Williamson (1965) in B. Chance, ed.: *Control of Energy Metabolism.* Academic Press, p. 335. Data such as these are often more difficult to interpret than is implied here because a large fraction of compounds present in low amounts is bound to enzymes, and the true concentrations in solution are not measured. This is inherent in the materials and in no way implies lack of skill by the analyst.

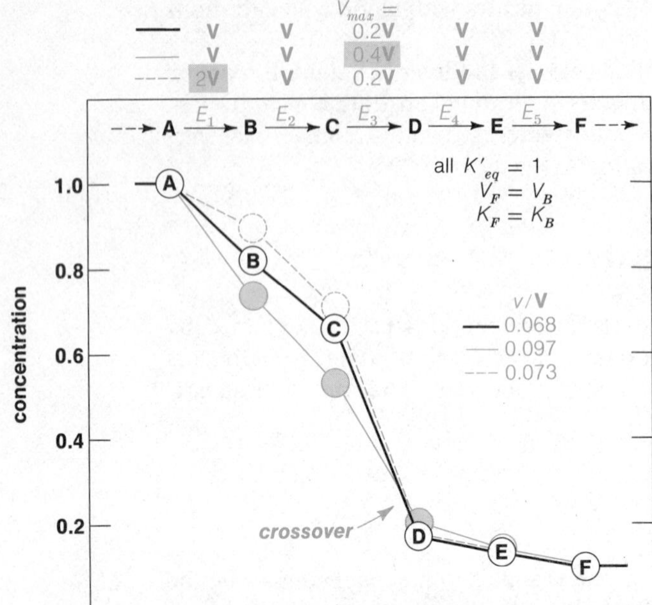

FIGURE 16–2 **Demonstration of a rate-limiting step. This example postulates a sequence of five consecutive reactions, each with an equilibrium constant of one. All of the enzymes have equal catalytic potential (equal V_{max}) except the third; which in the basal condition (*black line*) has only 0.2 the V_{max} of the others. The concentrations of the initial substrate, A, and the final product, F, are assumed to be constant.**

The *black line* shows the steady state concentrations of the various reactants under these conditions. The *solid blue line* shows the effect of doubling the activity of E_3 and the resultant "crossover" in concentrations. The overall velocity increases by 43 per cent.

The *dashed blue line* shows the effect of doubling the activity of E_1, with absence of crossover and an increase in overall velocity of only 6 per cent. Doubling the concentration of the other enzymes would have similar small effects, making detection of a crossover difficult.

The assumed Michaelis constants are: $K_F = K_B$ = 1.0 for E_1, 0.8 for E_2, 0.6 for E_3, 0.3 for E_4, and 0.2 for E_5.

concentrations in the presence of oxygen would cross a plot of the concentrations in the absence of oxygen between fructose 6-phosphate and fructose 1,6-bisphosphate.

Crossover points are taken as evidence of the rate-limiting steps in a sequence. The easiest explanation for the changes shown above is that phosphofructokinase, the enzyme catalyzing the reaction at the crossover, is rate-limiting and that the deprivation of oxygen in some way caused the activity of the enzyme to increase. (We shall discuss the mechanism shortly.)

For those with interest, the principle can be developed by mathematical models. Let us define a sequence of reactions in which the central reactions (E_3) has a maximum velocity only 0.2 that of all the others:

$$\mathbf{A} \xrightarrow{E_1} \mathbf{B} \xrightarrow{E_2} \mathbf{C} \xrightarrow{E_3} \mathbf{D} \xrightarrow{E_4} \mathbf{E} \xrightarrow{E_5} \mathbf{F}$$

To make it simple, let us assume that $K_{eq} = 1$ and that each enzyme has forward and back reactions with identical kinetic parameters. Let us further assume that the initial substrate, **A,** is being fed in so as to maintain a constant concentration, and the final product, **F,** is being removed so as to maintain its concentrations at 0.1 that of **A**. The model used here assumes that the Michaelis constants roughly fit the declining concentrations of intermediate compounds, but none of these assumptions are imperative to make the important points. The intermediate concentrations and the kinetic parameters are summarized in Fig. 16–2. (Note that the steadily declining concentrations of intermediates in the model do not occur in real sequences, such as our example above, in which equilibrium constants for successive reactions are not equal.)

The important observation is that doubling the activity of the intermediate enzyme, E_3, causes a much larger increase in the overall velocity of the system than does doubling the activity of any of the other enzymes. In the example, doubling V_{max} for E_3 increases the flux by 43 per cent (0.068 to 0.097 of the highest V_{max}), whereas doubling the activity of E_1 only increases the flux by 6 per cent (0.068 to 0.073). The change in E_3 caused a seven times larger change in velocity than did the change in E_1. The rate-controlling nature of E_3 is shown by the crossover in concentrations when its activity is altered.

Irreversible Reactions

Many rate-limiting steps involve reactions with positions of equilibrium so far in favor of the product that a net back reaction is not feasible in any real system. (Having a net formation of one compound does not mean that the back reaction is not occurring; it means only that the forward reaction is faster.)

The advantage of an effectively irreversible reaction is that a greater range of control is possible without approaching equilibrium. Figure 16–3 shows what happens in a sequence identical to that we just considered (Fig. 16–2), except that the rate-limiting central step is now assigned a $K_{eq} = 10{,}000$. The basal condition is hardly distinguishable from the circumstances seen in which all reactions are reversible, except that the overall velocity is somewhat faster. Doubling the velocity of this rate-limiting enzyme causes a modest 56 per cent increase in

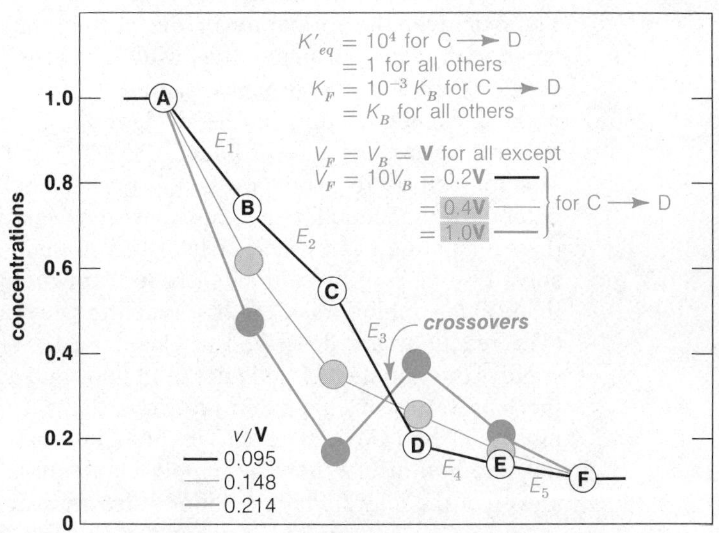

FIGURE 16–3

Control of a sequence containing an effectively irreversible reaction. In this postulated sequence, $K'_{eq} = 1$ in all reactions but the third, for which it is 10,000. The basal condition (*black line*) is much like that of Fig. 16–2, with V_{max} for the third reaction 0.2 that of the others. Increasing V_{max} to 0.4 (*thin blue*) and 1.0 (*heavy blue*) that of the other causes large changes in the overall rate in this case, compared to the changes shown in Fig. 16–2.

The assumed values of K_F are 1.0 for E_1, 0.8 for E_2, 0.6 for E_3, and 0.2 for E_4 and E_5.

overall rate, but the important point is that further increases in the activity of this enzyme still cause increases in the overall rate, even though the concentration of its product, (**D**), is exceeding the concentration of its substrate, (**C**). The exaggerated crossovers clearly show the rate-limiting character of E_3.

In this model sequence, the presence of the irreversible reaction can easily stretch the capacity of the neighboring reversible enzymes toward their maximum velocity so that they become the limiting enzymes. In real systems, this situation is commonly avoided by having those enzymes catalyzing readily reversible reactions in very high concentrations, compared to those catalyzing irreversible reactions. These enzymes can then maintain a resonable flux even though they function at only a small fraction of their V_{max} when substrate and product concentrations are not far from equilibrium.

Our model calculation invoked only a few-fold change in activity of the regulated enzyme; changes *in vivo* may be 100-fold or more. Regulation of phosphofructokinase, cited above, is a real example of such changes in enzymes catalyzing effectively irreversible reactions.

REGULATION BY EFFECTORS

Fine control of the activities of existing enzyme molecules frequently involves combination of the enzyme with some other compound, which may block the active site or alter the conformation of the enzyme so as to change its catalytic activity. Compounds altering rate in this way are said to be **effectors.** or modulators, of the activity; they may be **activators** that increase the activity or **inhibitors** that decrease the activity. Rate-limiting enzymes at key control points are frequently constructed so that they have distinctive binding sites for the effectors, even though the effector does not participate in the catalytic reaction. This is analogous to the presence of a binding site for 2,3-bisphosphoglycerate in hemoglobin.

In general, activations are devices used to mobilize stored compounds; fuels are made available for combustion in response to effectors that act as signals of the need for increased energy production. Inhibitions are frequently used to stop the formation of an unnecessary excess of some body constituent.

Negative Feedback. The synthesis of components of the organism frequently involves a succession of intermediates that are not used for any other purpose. For example, the formation of the purines incorporated into nucleotides does begin with common metabolites used for many purposes, but the synthesis proceeds by adding one group at a time to form intermediates that are good for nothing except making purine nucleotide. Each of the steps in the formation requires an enzyme. The problem is how to regulate the rate of a long sequence of reactions so as to produce enough purine nucleotides, but not too much, and without ever accumulating the otherwise useless intermediate compounds. Rather than regulating each reaction involved in such long sequences, the problem is solved by making the ultimate products of the sequence, the purine nucleotides themselves, inhibitors of the first reaction peculiar to purine synthesis (Fig. 16–4). (The reactions are described in Chapter 34.) When ATP and GTP are in short supply, the initial enzyme is not inhibited, and a flow of substrate begins through the long series of enzymatic reactions that ultimately result in the formation of more ATP and GTP. (The activity of all intermediate enzymes is high, so they are not rate-limiting.) When the supply is adequate, the initial enzyme is inhibited, and all of the intermediate reactions also slow down because the enzymes run out

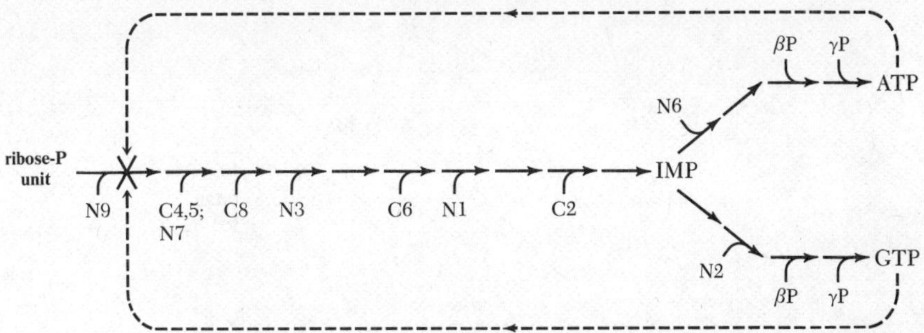

FIGURE 16–4 **The purine component of purine nucleotides is assembled one piece at a time on a ribose phosphate unit. The numbers of the successively added atoms indicate their positions in the completed molecule. This long sequence is regulated through feedback inhibition of the first reaction by the final products, thereby preventing wasteful accumulation of the many intermediate compounds.**

of substrates. The general process by which a compound inhibits its own formation is called "negative feedback" after a term borrowed from electronics.

Competitive Inhibitors. Some effectors compete with the substrate for an enzyme, either by occupying the catalytic site, or by favoring an enzyme conformation that is unfavorable for substrate binding. The inhibitor and substrate compete for binding, and raising the concentration of one diminishes the binding of the other. Indeed, this is the hallmark of competitive inhibition. Given a fixed concentration of inhibitor, increasing the concentration of substrate will make the velocity approach the same V_{max} seen in the absence of the inhibitor.

Recognition of competitive inhibition by synthetic compounds has been valuable in deducing the nature of active sites in enzymes, since they frequently are bound by the same groups as a substrate. If an enzyme with Michaelis-Menten kinetics combines with a competitive inhibitor, I, the initial velocity of the reaction is described by:

$$v = \frac{V_{max}S}{S + K_M(1 + I/K_I)}$$

in which $K_I = \frac{[\mathrm{E}][\mathrm{I}]}{[\mathrm{EI}]}$, and $I = [\mathrm{I}]$.

The equation can be rearranged in terms of reciprocals of initial velocity and substrate concentration, as we did with the simple Michaelis-Menten equation, but the essential features are apparent as written. Mathematically, the presence of inhibitor has the effect of increasing K_M in the equation by the factor $\left(1 + \frac{I}{K_I}\right)$; *but it does not change the value of* V_{max}. This is what would be expected, because an infinite concentration of substrate should effectively displace all of a competitive inhibitor.

The equation also shows us that K_I is equal to the concentration of inhibitor that doubles the *apparent* value of K_M. Now, if the inhibitor doesn't change the maximum velocity but does alter the apparent value of the Michaelis constant, differing concentrations of inhibitor should give a series of straight lines intersecting at $1/V_{max}$ on the ordinate, when Lineweaver-Burk plots are made of the effect of substrate concentration on rate of reaction.

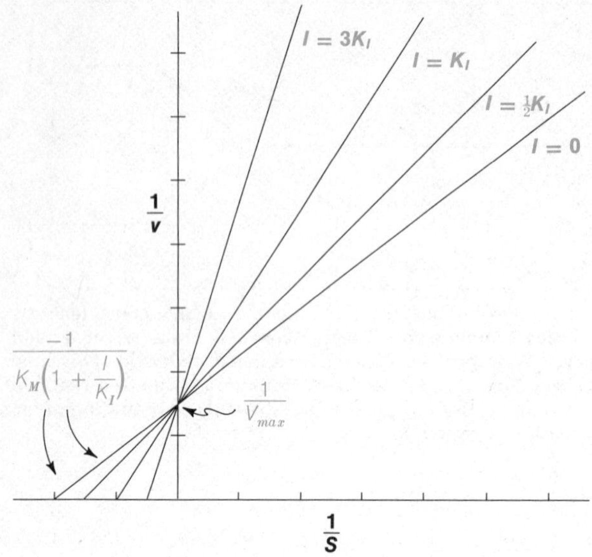

FIGURE 16–5

Lineweaver-Burk plots obtained with varying concentrations, *I*, of a purely competitive inhibitor. There is a single intercept on the ordinate (vertical axis) because V_{max} is the same in all cases, but the intercept on the abscissa (horizontal axis) changes.

The value of K_I can be estimated from the value of the apparent K_M in the presence of inhibitor, and this value is calculated from the negative reciprocal of the intercept on the abscissa. Figure 16–5 shows an example.

Non-competitive Inhibitors

Compounds may inhibit an enzymatic reaction by combining with an enzyme in such a way that they are not displaced by the substrate but prevent the enzymatic reaction from occurring. For example, the reaction of the phosphofluoridate esters with acetylcholinesterase is essentially irreversible, and the esterified enzyme will not catalyze the hydrolysis of acetylcholine. The effect of

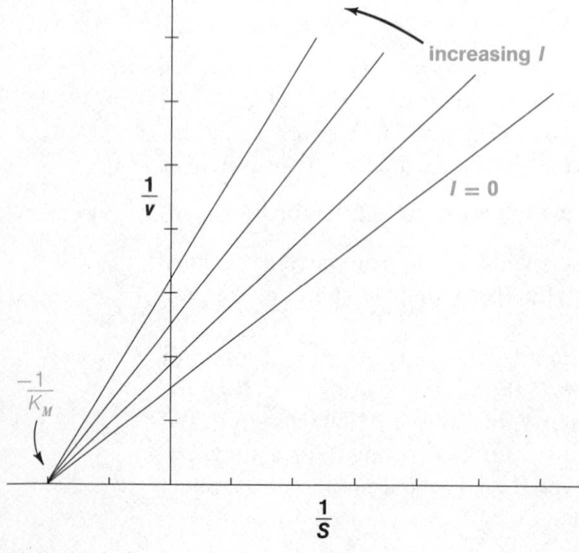

FIGURE 16–6

Lineweaver-Burk plots obtained with varying concentrations of a purely non-competitive inhibitor. K_M is constant in all cases, so there is a single intercept on the abscissa. The inhibitor has the same effect as removal of part of the enzyme, so V_{max} and the intercepts on the ordinate change.

these inhibitors is to remove enzyme from the solution, thereby lowering V_{max}, but the remaining enzyme displays the same kinetics as does a more dilute solution of the enzyme with no inhibitor added. In other words, K_M will not be affected, and Lineweaver-Burk plots at varying inhibitor concentrations will give a series of lines that intersect on the abscissa (Fig. 16–6).

We now have two straightforward cases that can be recognized from Lineweaver-Burk plots of the kinetics at different inhibitor concentrations: straight lines intersecting on the ordinate indicate classic competitive inhibition, and straight lines intersecting on the abscissa indicate purely non-competitive inhibition.

However, these cases do not by any means exhaust the possibilities for inhibition. An inhibitor may react more readily with the intermediate enzyme-substrate complexes than it does with the free enzyme. The inhibitor-enzyme complex may be catalytically active, but with altered values of K_M or V_{max}. The result of the various possibilities may be that varying inhibitor concentrations will produce Lineweaver-Burk plots that are straight lines, but which do not all intersect at one point, so that the inhibition is "mixed" rather than being competitive or non-competitive. In other cases, Lineweaver-Burk plots produce curved, rather than straight, lines. Permutations of these various kinds of effects can be analyzed in as much detail as patience permits, but such extensive analysis is not required for our purposes. More complete treatment can be found in the references cited at the end of the chapter.

Regulation with Interacting Subunits

Homotropic Kinetics. Some enzymes are built like hemoglobin, with multiple subunits that interact cooperatively so as to create more complex responses to changes in substrate concentration. When the interacting subunits are identical, each with its catalytic site, the resultant kinetics are said to be homotropic. If the binding of one or more molecules of substrate stabilizes a new conformation in which the affinity for the substrate is increased at the remaining binding sites, the cooperativity is said to be positive. With **positive cooperativity,** the kinetic response to changing substrate concentration is *sigmoidal,* much like the oxygen association curve of hemoglobin (Fig. 16–7). (Some enzymes with multiple subunits do not show cooperative interactions and have Michaelis-Menten kinetics. Lactate dehydrogenase is an example.)

Cooperativity can also be **negative,** with the binding of one substrate molecule hindering the binding of additional molecules. The presence of cooperative interactions and their identification as positive or negative can be recognized by characteristic kinetic plots (Fig. 16–8).

Physiological Implications. An enzyme with positive homotropic interactions cannot effectively catalyze its reaction until the substrate concentration has built up to a level approaching the steep portion of the velocity-concentration curve. This is a desirable device for preventing the depletion of a substrate when it undergoes several reactions catalyzed by different enzymes, or when it is irreversibly converted to a product. With a homotropic enzyme the rate of reaction falls rapidly toward zero at substrate concentrations well above zero.

Heterotropic Kinetics. Some enzymes that contain cooperatively interacting subunits can also be regulated by binding effectors at sites other than the catalytic sites. Conformational shifts are then caused by changes in concentration of the controlling effector, as well as the substrate.

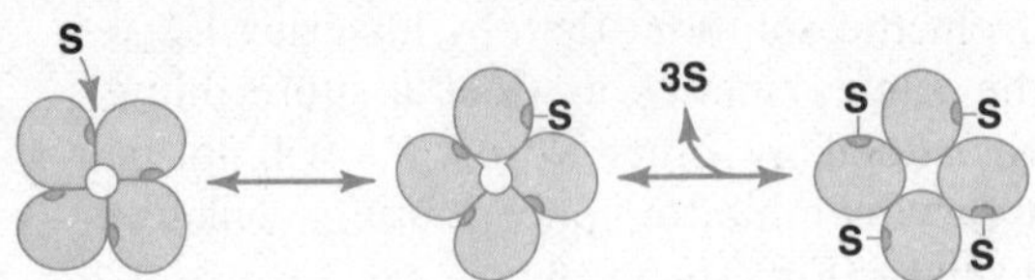

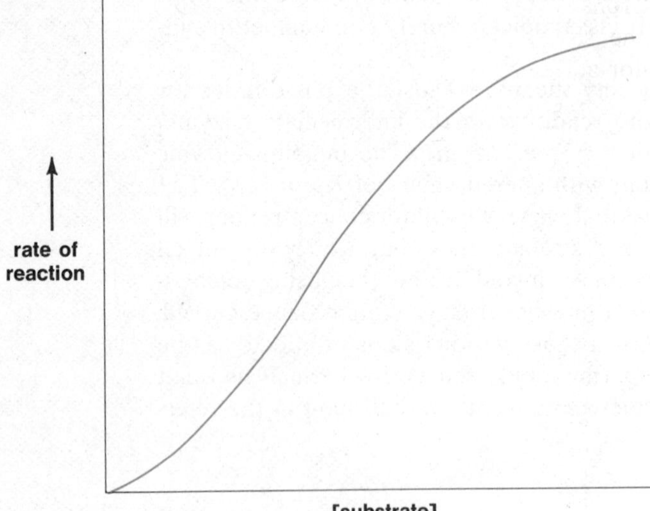

FIGURE 16–7

Enzymes that are constructed from multiple subunits may exhibit complex kinetics, such as the positive cooperativity shown here. In this example, the free tetramer has a conformation in which access to the catalytic sites is impeded in some way (*upper left*). When the substrate concentration is elevated enough to stabilize a conformation with one of the subunits in a more open form (*top center*), the remaining subunits can more readily shift to the fully active conformation that combines with substrate (*upper right*). The result is a sigmoidal response of rate of reaction to substrate concentration.

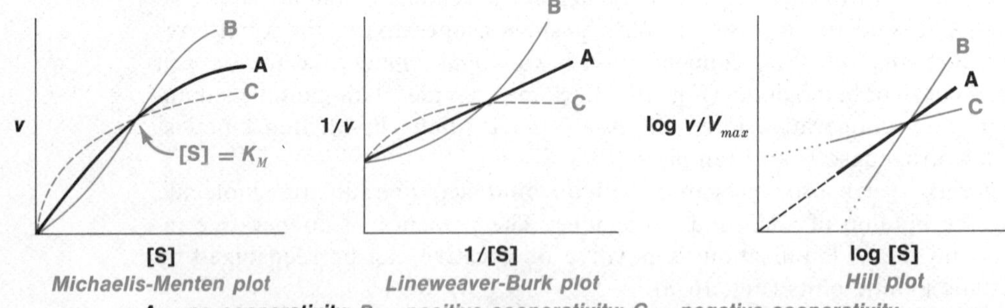

FIGURE 16–8 The presence or absence of cooperativity gives characteristic kinetic plots. In the ordinary Michaelis-Menten plot of initial velocity against substrate concentration (*left*), positive cooperativity is easily distinguished by its sigmoid character (*curve B*), but it is difficult to distinguish Michaelis-Menten kinetics (*curve A*) from kinetics of negative cooperativity (*curve C*) by inspection. The differences become more obvious in a Lineweaver-Burk plot (*center*) in which positive cooperativity gives a curve that is concave upward, while negative cooperativity gives a curve that is concave downward, in contrast to the straight line of Michaelis-Menten kinetics. A logarithmic Hill plot (*right*) also distinguishes the mechanisms by the position of the projected intercept, which is on the abscissa for positive cooperativity and on the ordinate for negative cooperativity, but the greatest value of the Hill plot is that the slope of the central part of the curve gives the Hill coefficient, *n*, which is the minimum number of interacting sites. (The ends of the Hill plot, not shown, cannot be used because they always curve to approach a slope of 1.)

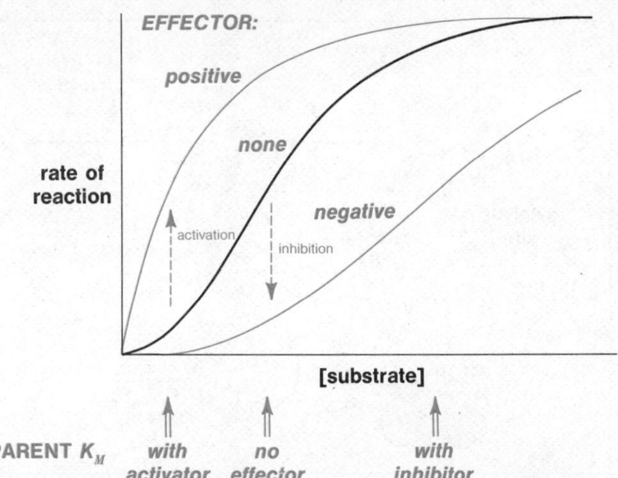

FIGURE 16–9

Kinetics with heterotropic effectors. The plotted examples alter the apparent K_M of the enzyme without substantial effect on the V_{max}. The positive effector in this example eliminates homotropic interactions so that the enzyme has simple Michaelis-Menten kinetics. The negative effector exaggerates the interactions.

A heterotropic effector changes the sigmoidal response of velocity to substrate concentration. A frequent effect is to alter the apparent K_M for a substrate without substantial change in V_{max} (Fig. 16–9), but in some cases, the converse is true. When the effector activates an enzyme, it may do so by diminishing internal constraints on catalysis, so that each subunit of the enzyme acts as an isolated molecule with simple Michaelis-Menten kinetics. If the effector inhibits the enzyme, it may do so by stabilizing the interactions between subunits, so that the kinetics become more sigmoidal, and higher concentrations of substrate are required to overcome the interactions.

The binding site for the effector need not even be on the same subunit as the active catalytic site. Many examples are known of enzymes that are aggregates of separate catalytic and regulating subunits, all interacting (Fig. 16–10).

Let us now consider two examples of heterotropic regulation.

PHOSPHOFRUCTOKINASE AND GLUCOSE COMBUSTION. The combustion of glucose residues, derived either from glucose itself or from stored glycogen, involves several key control points. We have already noted one, phosphofructokinase, which catalyzes the transfer of a phosphate group from ATP to fructose 6-phosphate:

$$\text{fructose 6-phosphate} + \text{ATP} \rightarrow \text{fructose 1,6-bisphosphate} + \text{ADP}$$

The purpose of the combustion is to generate usable energy, and we shall later see that the state of the energy balance is signaled by the relative concentration of AMP, which is not a substrate or product of the reaction. When the energy supply is low, the concentration of AMP increases. (We shall look at this important signal in detail in Chapter 25.) An increase in the concentration of AMP therefore indicates a need for more fuel, and phosphofructokinase catalyzes an irreversible reaction on the route by which fuel is supplied. This enzyme is constructed so that AMP is an activating effector, which changes the sigmoidal kinetics of the enzyme to a simple Michaelis-Menten form through loss of cooperative interaction (Fig. 16–11). The V_{max} of the enzyme is also increased to some extent, but the important result is that the affinity of enzyme for its substrate is greatly increased; the K_M is shifted into the normal physiological range of concentration for fructose 6-phosphate.

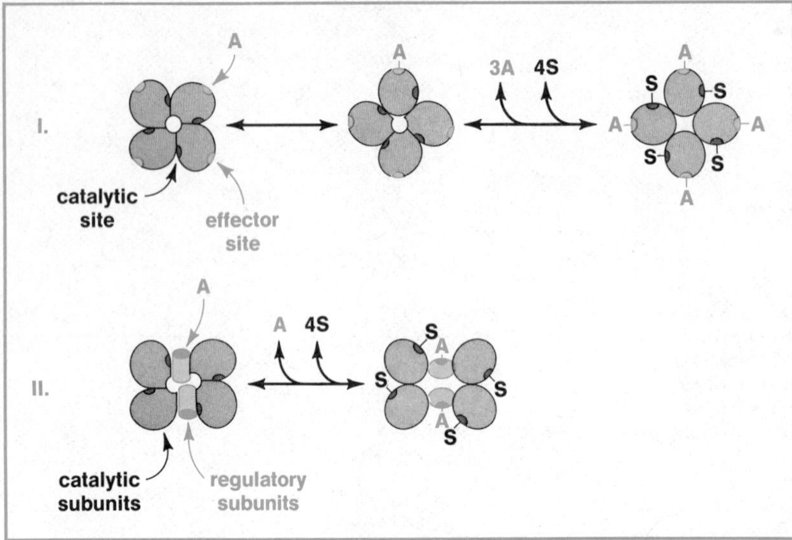

FIGURE 16–10 **I. Effector sites may be on the same polypeptide subunits containing the catalytic sites with binding of the effector altering the conformation of the subunits by stabilizing a more reactive form.**

II. Effector sites also may be present on separate regulatory polypeptide subunits which interact with the catalytic subunits so as to fix them in a particular conformation. Combination of an activating effector with the regulatory subunits then modifies the interaction so as to permit the catalytic subunits to assume an active conformation. As is indicated here, the number of regulatory sites need not be equal to the number of catalytic sites when they occur on separate polypeptide chains.

AMIDOPHOSPHORIBOSYL TRANSFERASE AND PURINE SYNTHESIS. The initial step in purine synthesis at which control by negative feedback occurs (Fig. 16–4) involves the displacement of an attached pyrophosphate ester group from a ribose residue by transfer of an amide nitrogen from glutamine:

glutamine + 5-phosphoribosyl pyrophosphate $\xrightarrow{\textit{amidophosphoribosyl transferase}}$
glutamate + 5-phosphoribosylamine + pyrophosphate

We shall discuss these compounds in detail in Chapter 34, but the point of interest now is the nature of the control. The enzyme has cooperative interaction only in the presence of one of the purine nucleotides. Therefore, an increase in the concentration of any of the adenosine or guanosine phosphates will cause a shift from Michaelis-Menten kinetics to sigmoidal homotropic kinetics and increase the K_M for the ribosyl substrate to a value much above the physiological concentration range. (Fig. 16–12 illustrates the action of AMP as an example; ATP and GTP are probably the important effectors in cells.) The action of purine nucleotide on this enzyme is therefore the opposite of the action of AMP on phosphofructokinase, which makes another important point. The same compound may act as an effector on several enzymes; it may activate some and inhibit others.

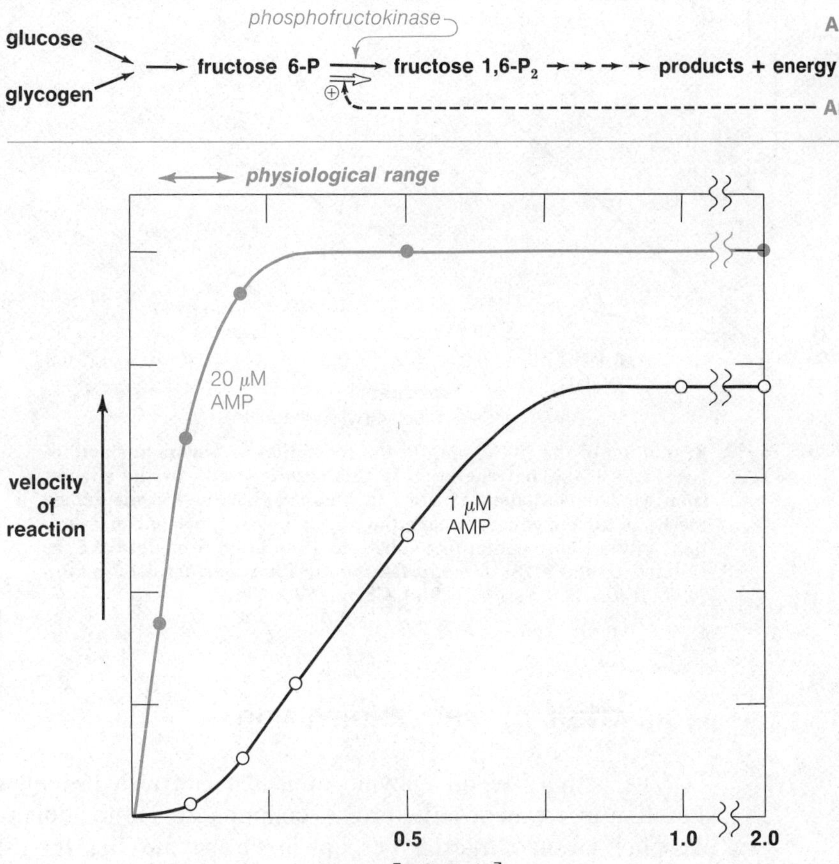

FIGURE 16–11 Regulation of phosphofructokinase by AMP. (Other effectors are not shown.) The activation of phosphofructokinase by AMP is indicated by the broad arrow and plus sign in the reaction sequence at the top. AMP has its effect by changing the homotropic sigmoidal kinetics of the enzyme to simple Michaelis-Menten kinetics, in which the K_M is now in the physiological range of substrate concentration, making the enzyme many-fold more active. (Replotted from data of K. Tornheim and J. M. Lowenstein, (1976) *Control of Phosphofructokinase from Rat Skeletal Muscle.* J. Biol. Chem., *251*: 7322.)

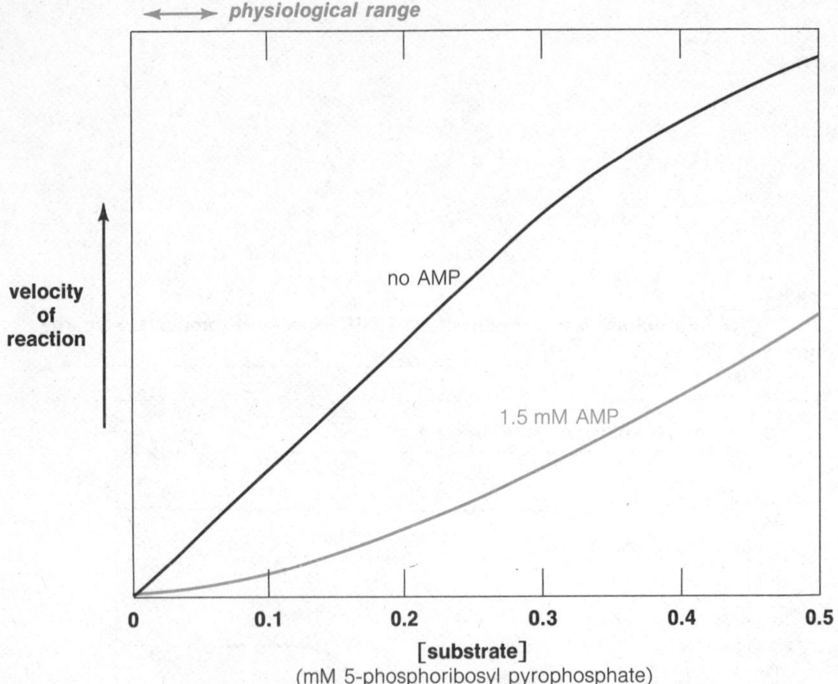

FIGURE 16–12 **Regulation of the initial step in the formation of purine nucleotides (see Fig. 16–4). The regulation in this case is a shift by the effector from Michaelis-Menten kinetics to homotropic kinetics, displacing the K_M of the enzyme far above the physiological range of concentrations of the ribose-containing substrate. (Replotted from data of E. H. Holmes, et al. (1973), *Human Glutamine Phosphoribosyl Pyrophosphate Amidotransferase*. J. Biol. Chem., *248*: 144.)**

REGULATION BY PHOSPHORYLATION

The activity of an enzyme molecule can also be changed through some alteration in covalent structure. A common device for doing this is through the phosphorylation of hydroxyl groups in the enzyme (Fig. 16–13). These groups are in side chains of residues of serine, threonine, or tyrosine in the polypeptide chain. The transfer of phosphate groups to the regulated enzyme from ATP is in itself catalyzed by another enzyme, a protein kinase. The phosphate groups are removed to restore the original state by still another enzyme, a **phosphoprotein phosphatase.**

The affected enzymes resemble other enzymes with cooperative interaction by existing in active and inactive conformations. The difference is that the shift

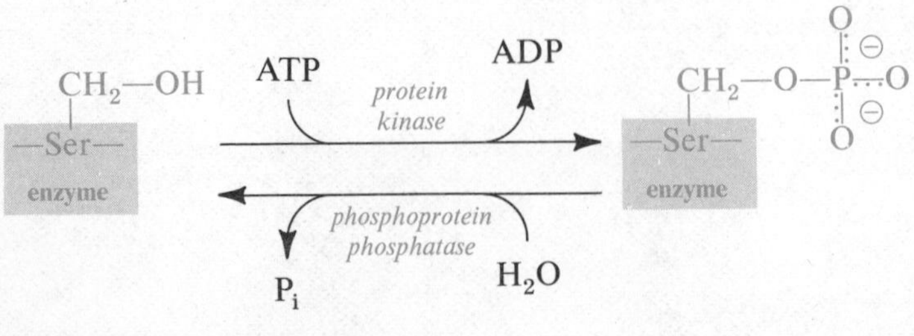

CH_2—OH, —Ser—, enzyme; ATP → ADP (*protein kinase*); CH_2—O—PO_3^{2-}, —Ser—, enzyme; H_2O → P_i (*phosphoprotein phosphatase*)

FIGURE 16–13

Regulation of enzymes through phosphorylation. A protein kinase catalyzes addition of phosphate groups, and a hydrolase catalyzes their removal.

from one conformation to another is stabilized by the presence or absence of the phosphate ester group on the enzyme (Fig. 16–14). The effect of phosphorylation may be an activation or an inhibition of the activity; we shall encounter examples of each.

The device of regulation through addition and removal of phosphate groups adds another dimension to control. Whether these groups will be present or absent on the regulated enzyme depends upon the relative activities of the protein kinase and the phosphoprotein phosphatase that catalyze the competing reactions. Here we have enzymes regulated by other enzymes, which are now in turn subject to regulation by effectors. The important difference between this and other effector-mediated mechanisms we have been discussing is that **changes in effector concentration are amplified** by the intervening enzymes.

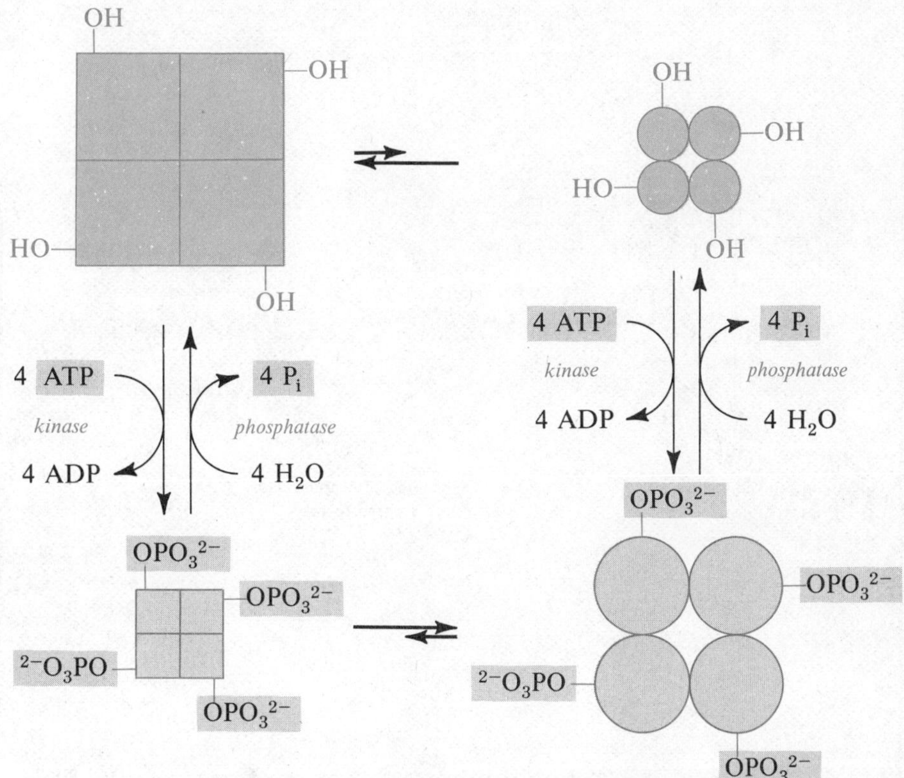

FIGURE 16–14 **Enzymes are probably regulated by phosphorylation through the stabilization of an otherwise unfavorable conformation. In the mechanism illustrated, a tetrameric enzyme exists predominantly in one conformation (*upper left*) when it is not phosphorylated. (Concentration of a conformation is indicated by relative size of the drawing.) Either the favored or the unfavored conformation may be phosphorylated by the transfer of phosphate groups from ATP, the transfer being catalyzed by a kinase. The presence of the phosphate groups shifts the equilibrium between the two conformations, so that the other form (*lower right*) is now favored. When the phosphate groups are removed by hydrolysis (catalyzed by a phosphatase), the conformational shift is reversed. Although the diagram shows the predominant free form in the conventional squares indicating the less reactive species (*upper left*), and the predominant phosphorylated form in the circles indicating the more reactive molecule, this is not true of all enzymes regulated by phosphorylation. Some enzymes are less active when phosphorylated.**

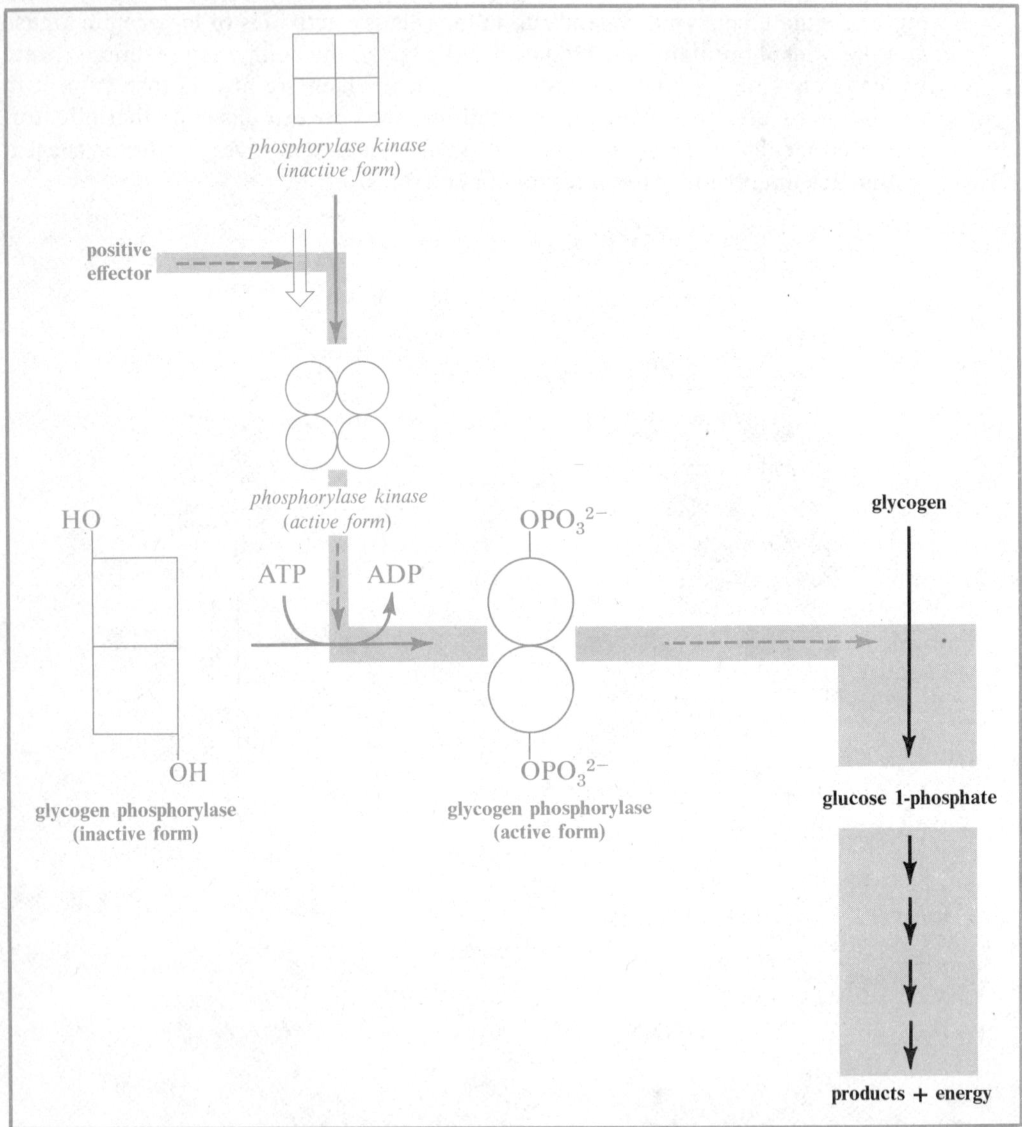

FIGURE 16–15 **The action of effectors is magnified when they influence the phosphorylation of an enzyme. In the example shown here, a positive effector increases the activity of phosphorylase kinase (*top center*), which then accelerates the phosphorylation of glycogen phosphorylase (*center left*). Each molecule of the active kinase can create many active molecules of glycogen phosphorylase. Each of the active phosphorylase molecules in turn catalyzes the cleavage of many glucose residues from glycogen (as glucose 1-phosphate). The magnification of the effector action is indicated by blue shading.**

For example, consider the activity of **glycogen phosphorylase,** which is the enzyme controlling the breakdown of glycogen for use of its glucose residues as fuel. We shall later see that it is under complex control. The critical feature is a stabilization of the active conformation of the enzyme when the enzyme is phosphorylated (Fig. 16–15). The transfer of phosphate groups to glycogen phosphorylase is catalyzed by another enzyme, a kinase. Activation of the kinase by an effector enables the phosphorylation of many molecules of the glycogen phosphorylase. Each of these many molecules of active enzyme can then catalyze the breakdown of many glucose residues. Without the intervening kinase, a molecule of effector would have to be provided for each of the molecules of glycogen phosphorylase to be activated, but the kinase has multiplied the action of the effector. We shall later see that this cascade multiplication of effector action is the basis for the potency of several hormones.

INDUCTION OR REPRESSION OF ENZYME SYNTHESIS

The ultimate method of regulating the rates of enzymatic reactions is to control the number of enzyme molecules present. Occasionally this may be done by regulating the rate at which particular enzymes are destroyed, but the more common method of control is to regulate the rate of synthesis. We have already discussed in Chapter 5 how this kind of regulation is achieved in prokaryotic organisms through operon control of gene transcription, but the mechanism for comparable control in eukaryotes is not known. However, the results are quite real. We shall frequently have occasion to refer to metabolic adaptations in which the amount of enzyme is adjusted to fit new environmental circumstances. Many hormones are active because of their effects on synthesis of particularly enzymes, and more will be said about their mechanism and effects in Chapter 38.

FURTHER READING

Boyer, P. D., ed.: (1970) *The Enzymes,* vol. 1. *Structure and Control.* Academic Press. Includes detailed discussion of regulatory mechanisms.

Segel, I. H.: (1975) *Enzyme Kinetics.* Wiley. Sophisticated analysis of inhibitions and allosteric regulation.

17 HYDROLASES: DIGESTION

Enzymes that catalyze hydrolysis are ubiquitous. They break bonds with a liberation of energy, and though these bonds could have been formed only by a correspondingly greater input of energy, the destruction is a necessary part of the biological scheme of things. Many of the hydrolases discussed here are extracellular enzymes, secreted to hydrolyze large molecules from the diet into smaller components that can be absorbed. Intracellular hydrolases, discussed later, are used in a constant reshaping of cellular composition that is a necessary part of adaptation to changing circumstances. Many of these enzymes are packaged in special organelles, the lysosomes.

Our concern now is with the hydrolases that attack the major classes of foodstuffs, the carbohydrates, fats, and proteins. The enzymes of interest, therefore, break the hemiacetal bonds of glycosides, the ester bonds of triglycerides, or the amide bonds of polypeptides; they are representatives of the general classes of glycosidases, esterases, and amidases:

$$H_2O + \overset{\text{(ring-O)}}{R^1}\text{—O—}R^2 \xrightarrow{\textit{glycosidase}} \overset{\text{(ring-O)}}{R^1}\text{—OH} + HO\text{—}R^2$$

$$H_2O + R^1\text{—}\overset{\overset{\displaystyle O}{\|}}{C}\text{—O—}R^2 \xrightarrow{\textit{esterase}} R^1\text{—COO}^{\ominus} + H^{\oplus} + HO\text{—}R^2$$

$$H_2O + R^1\text{—}\overset{\overset{\displaystyle O}{\|}}{C}\text{—}\overset{\overset{\displaystyle H}{|}}{N}\text{—}R^2 \xrightarrow{\textit{amidase}} R^1\text{—COO}^{\ominus} + H_3\overset{\oplus}{N}\text{—}R^2$$

The gastrointestinal tract is organized so as to make the food accessible to attack by hydrolases, to secrete them at the proper time and place, to absorb the products of hydrolysis, to recover components of the digestive secretions, and to dispose of the indigestible remains as feces (schematically illustrated in Fig. 17–1).

Most of the contribution of the mouth is mechanical; food is dispersed and mixed with the saliva, rich in lubricating glycoproteins. The saliva also contains

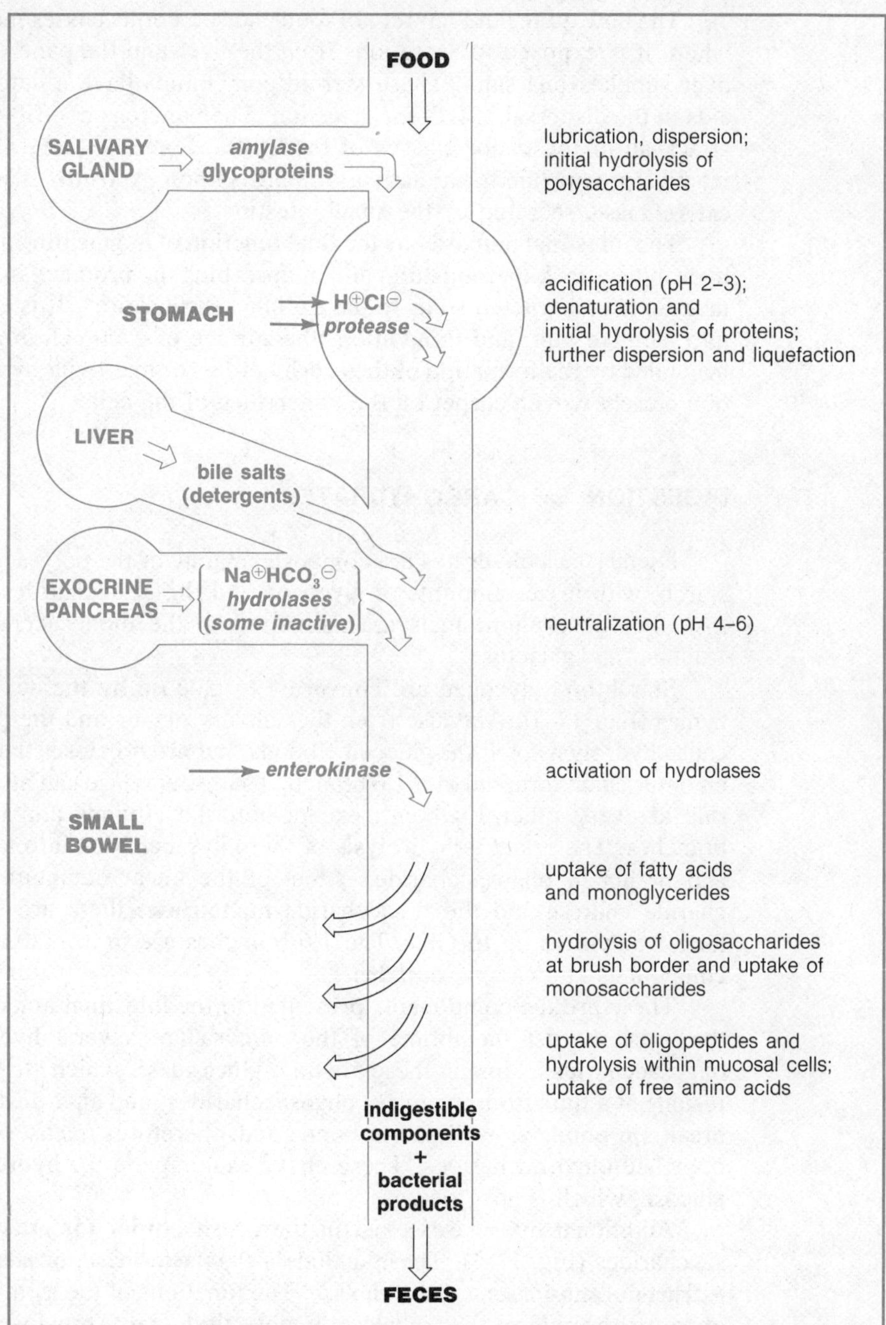

FIGURE 17–1 **Schematic diagram of sequence of digestion.**

amylase, a hydrolase attacking glucans, but the action of this enzyme is interrupted by the high acidity of the stomach when the food is swallowed.

The high concentration of HCl secreted by the gastric mucosa, to which food is thoroughly exposed by mechanical agitation, destroys many of the hydrogen bonds in proteins. The proteins unfold, that is, they are denatured and become more accessible to attack by hydrolases. The stomach also secretes proteases to begin this attack. (The high acidity kills many of the microorganisms in the food.)

The now quite fluid mixture of food and secretions passes into the duodenum, where it is exposed to secretions from the liver and the pancreas. Bile from the liver supplies bile salts. These steroid compounds have a detergent action that aids in the dispersal of fats for digestion. The pancreas contributes hydrolases to attack all of the major classes of foodstuffs. Some of these are secreted in an inactive proenzyme form; activation hinges upon exposure to another hydrolase, **enterokinase,** secreted by the small intestine.

The intestinal mucosa has the dual function of augmenting and continuing the hydrolytic attack on foodstuffs and of absorbing the products in usable form. The mucosa is constructed so as to have a huge surface area. It is extensively folded into flap-like villi, and in addition, the surface of each cell in a villus is further magnified by the formation of the brush border, or microvilli, which is like the pile of a closely woven carpet on the lumen face of the cell.

DIGESTION OF CARBOHYDRATES

Dietary carbohydrates are composed mainly of the polysaccharide known as starch, with lesser amounts of glycogen and the disaccharides sucrose and lactose. For absorption, fuels are converted to the monosaccharides — glucose, fructose, and galactose.

Starch and glycogen are converted to glucose by the action of several enzymes (Fig. 17–2). Amylase from the salivary glands and the pancreas can only cause hydrolysis of 1,4-α-glucoside bonds that are no closer than two residues to an outer chain terminal or a 1,6 branch. That is, amylase can attack a polysaccharide at every other 1,4 bond, except outermost bonds and those next to the branches. The result is hydrolysis of the polysaccharide into a mixture of linear and branched oligosaccharides. Most of the linear compounds are the disaccharide **maltose** and the trisaccharide **maltotriose**; there are small amounts of longer polymers up to Glc_9. The 1,6 branches are in a mixture of Glc_5 to Glc_9 compounds known as α-**dextrins.**

These are the components presented to the intestinal mucosa. Embedded in the brush border membrane of the mucosa are several hydrolases attacking oligosaccharides. Among these are an α-**glucosidase,** which strips off one glucose residue at a time from the linear oligosaccharides, and an α-**dextrinase,** which can break 1,6 bonds as well as 1,4 bonds and, therefore, releases glucose from the branched oligosaccharides. These enzymes complete the hydrolysis of starch to glucose, which is absorbed.

Additional hydrolases exist in the brush border for attacking other oligosaccharides (Fig. 17–3). These include a β-**galactosidase,** or **lactase;** a **sucrase,** or β-**fructofuranosidase;** and a **trehalase.** The functions of the lactase and sucrase are obvious; the role of the trehalase is more titillating. Trehalose occurs widely in insects as a storage and transport sugar, and insects may have been of more importance in the primate diet than one might assume from current efforts against cockroaches and the like. More fastidious epicures will encounter trehalose in mushrooms.

Sucrase occurs in a complex with α-dextrinase in the brush border. Why the two polypeptide chains are built to associate with each other is not clear.

Indigestible Carbohydrates. Many polysaccharides cannot be attacked in the primate small bowel. These include cellulose, inulin (a fructose polymer from the Jerusalem artichoke), and agar (a heteropolysaccharide from seaweeds), and various other heteropolysaccharides of plant origin, including some pentose poly-

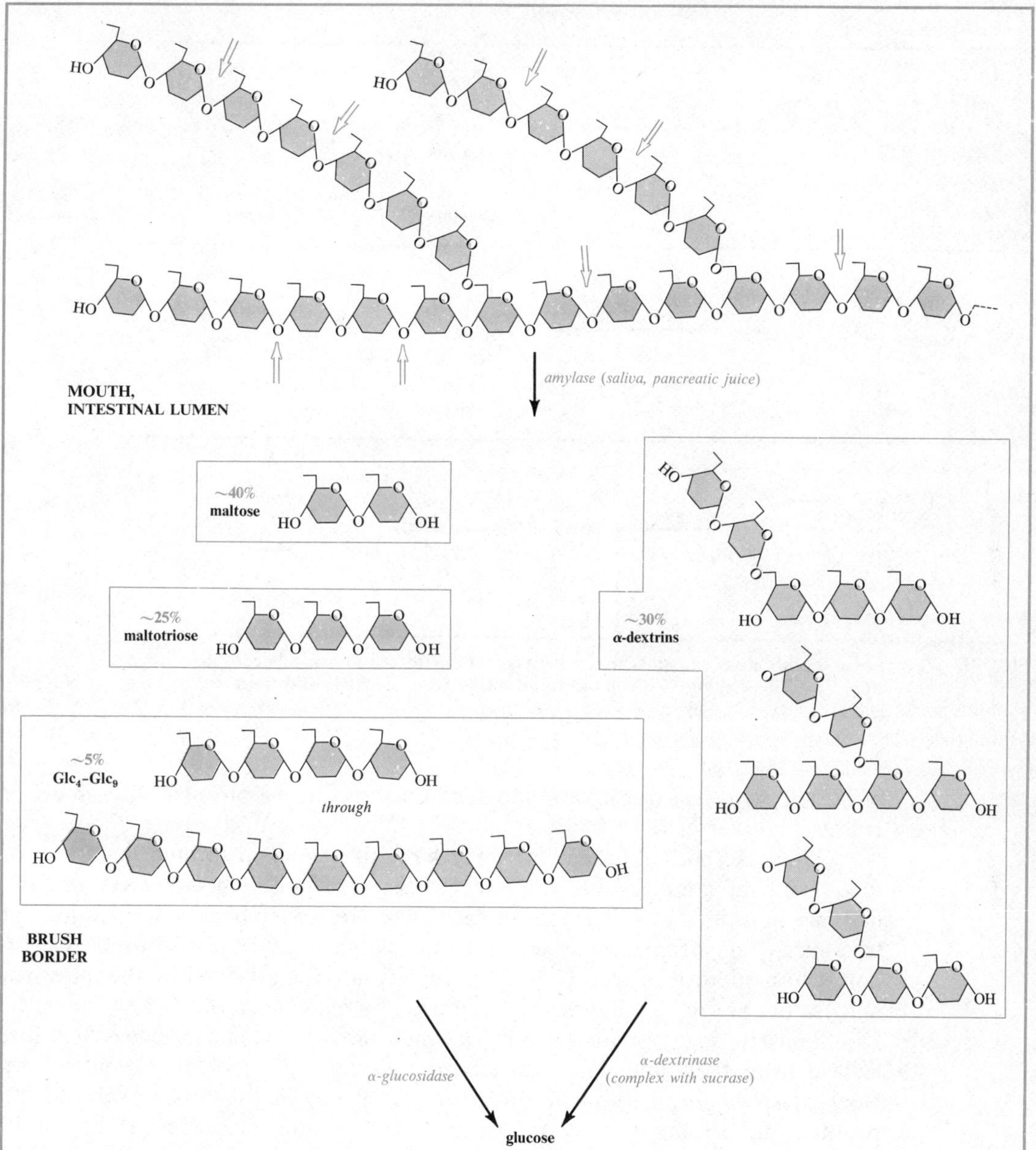

FIGURE 17–2 **Digestion of the branched polysaccharides, starch and glycogen.**

mer from *Psyllium* seeds that are widely sold to add fecal bulk. In most diets, cellulose is the most important component of indigestible fiber, and considerable emphasis is placed today on promoting easy passage of feces through the large bowel by increasing the cellulose intake.

However, some carbohydrates that are denied to the host as a fuel provide a banquet for the bacterial flora of his gut, and here lies a problem. Some bacteria that are given a plentiful supply of useable carbohydrate wear out their welcome through the production of large volumes of gas and lactic acid, along with other products that irritate the intestines. Audible rumblings (given the impressive name

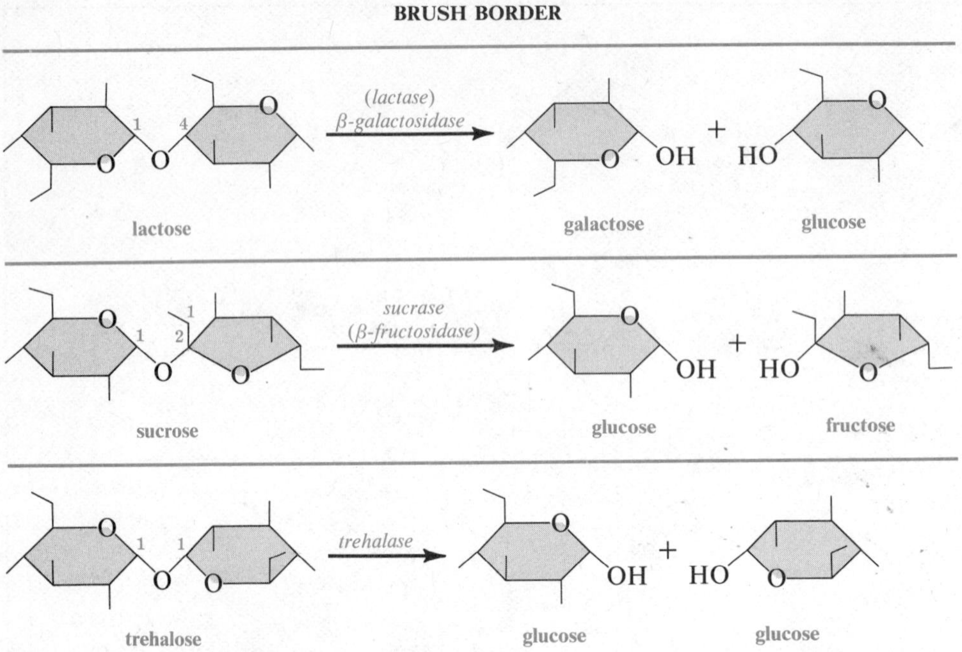

BRUSH BORDER

FIGURE 17–3 **Some dietary disaccharides are hydrolyzed at the intestinal brush border for absorption. Specific enzymes are present to attack lactose, sucrose, and trehalose.**

of borborygmi) and frank farts add social distress to the physical discomfort of cramps and diarrhea.

Otherwise normal individuals may have this problem through ingestion of foods containing indigestible oligosaccharides. The pulses (beans, peas, and the like) are notorious offenders; these seeds have relatively high concentrations of tri-, tetra-, and pentasaccharides containing α-galactosyl units attached by 1,6 bonds to a sucrose residue (Fig. 17–4), which are not attacked by the intestinal mucosa but are readily hydrolyzed and used by many bacteria.

Similarly, many adults become ill when they drink milk because they lost lactase from their intestinal brush border as they grew older. Orientals and Negroes are more likely to be forced to put aside childish foods owing to the resultant lactose intolerance. In Americans of various ancestries, 19 out of 20 Orientals, 14 out of 20 Negroes, and only 2 out of 20 Caucasians developed symptoms after eating 50 grams of lactose, the amount in a quart of cow's milk.

LACTOSE INTOLERANCE*

Occurrence

Common; usually seen in adults

Prevalence (=incidence) varies from 65 percent in blacks, Mexican-Americans, Orientals, Ashkenazic Jews, and Eskimos to 5 to 15 percent in Northern Europeans

Rarely congenital

Symptoms

Onset 1/2 to 4 hours after as little as one glass of milk

Abdominal distention, cramps, pain, diarrhea, increased flatulence

*T. M. Bayless, et al.: (1975) *Lactose and Milk Intolerance: Clinical Implications.* N. Engl. J. Med., *292*: 1156. Also see N. Engl. J. Med., *294*: 1057 (1976).

α-Gal-(1→6)-α-Gal-(1→6)-α-Gal-(1→6)-α-Glc-(1→2)-β-Fru	verbascose
α-Gal-(1→6)-α-Gal-(1→6)-α-Glc-(1→2)-β-Fru	stachyose
α-Gal-(1→6)-α-Glc-(1→2)-β-Fru	raffinose

FIGURE 17–4 **Some pulses, such as beans, peas, and soybeans, contain oligosaccharides with (1,6)-linked galactose residues that cannot be hydrolyzed for absorption. They may be regarded as sucrose with one, two, or three galactose residues attached.**

Later studies confirmed that intolerance is widespread among people of varied ancestries. In short, milk is not a very good food for many adults, and it has been pointed out that shipping dried milk to impoverished lands may cause more griping guts than gratitude. The potential magnitude of the problem can be appreciated from one patient, who, upon drinking two liters of milk on each of two days, had 141 passages of flatus per day, including 70 in four hours. His output of 346 ml of gas per hr, mostly H_2 and CO_2, exceeded the previous record of 168 ml per hr achieved by subjects ingesting 50 per cent of the caloric intake as baked beans. (Recognition has been sought from the Guinness Book of World Records.)

There is a more serious side to the problem. We shall see later that the pulses are among the better sources of protein readily available in the world, but the pesky carbohydrates that come along with the protein inhibit their more widespread use. Attempts are being made to selectively breed out the undesirable characteristic.

DIGESTION OF FATS

Fats often constitute even more of the potential energy in human diets than do carbohydrates. Fatty acids can be liberated by simple hydrolysis of the ester bonds in triglycerides, but the insolubility of the triglycerides presents a problem; digestion must occur at a phase boundary.

The first step is dispersion of the dietary fat into small particles with sufficient exposed surface area for rapid attack by a hydrolase. Detergent action and mechanical mixing do the job, with the detergent effect being supplied by several components, both in the diet and in the digestive juices, but especially by partially digested fats (fatty acid soaps and monoglycerides) and by bile salts.

Bile salts. Bile acids, which are steroid derivatives with a carboxyl-containing side chain, are converted to powerful detergents, the bile salts, by forming amide linkage with either glycine or taurine (Fig. 17–5). The principal bile acids are **cholic** and **chenodeoxycholic** acids, either of which may be conjugated with glycine or taurine. For example, cholic acid may form glycocholic or taurocholic acids, occurring as the anions. In addition, some of the compounds occur as the 3-sulfate esters, and are dibasic anions.

The detergent character of the bile salts comes from the presence of hydroxyl or sulfate ester groups on only one side of the molecule. The steroid ring is a puckered plane; the polar groups are on one side, giving a hydrophilic face; the other face is hydrophobic. The hydrocarbon side chain, culminating in a C-24 carboxyl group, adds to the mixture of hydrophobic and hydrophilic character.

Action of Lipase. The pancreas secretes a powerful esterase catalyzing the hydrolysis of triglyceride bonds. This pancreatic lipase by itself cannot interact

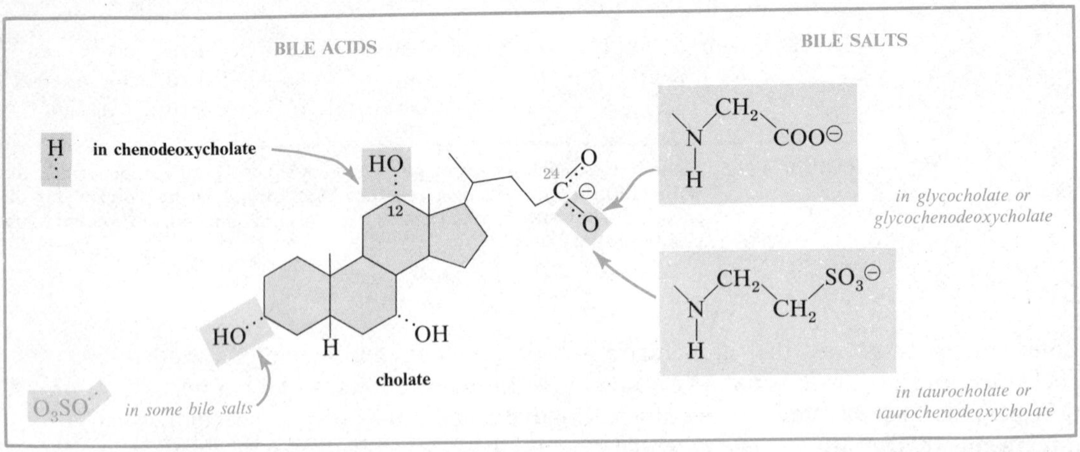

BILE ACIDS

BILE SALTS

H in chenodeoxycholate

cholate

O_3SO in some bile salts

in glycocholate or glycochenodeoxycholate

in taurocholate or taurochenodeoxycholate

FIGURE 17–5 Bile salts are detergents made by linking bile acids (*black*) with glycine or taurine (*blue*) in amide linkage. The predominant bile acids are cholic acid or chenodeoxycholic acid, which differ in the presence or absence of a 12-hydroxyl group; either may be conjugated with glycine or taurine in bile salts. Some bile salts are further modified by forming sulfate esters with the 3-hydroxyl group on the steroid ring.

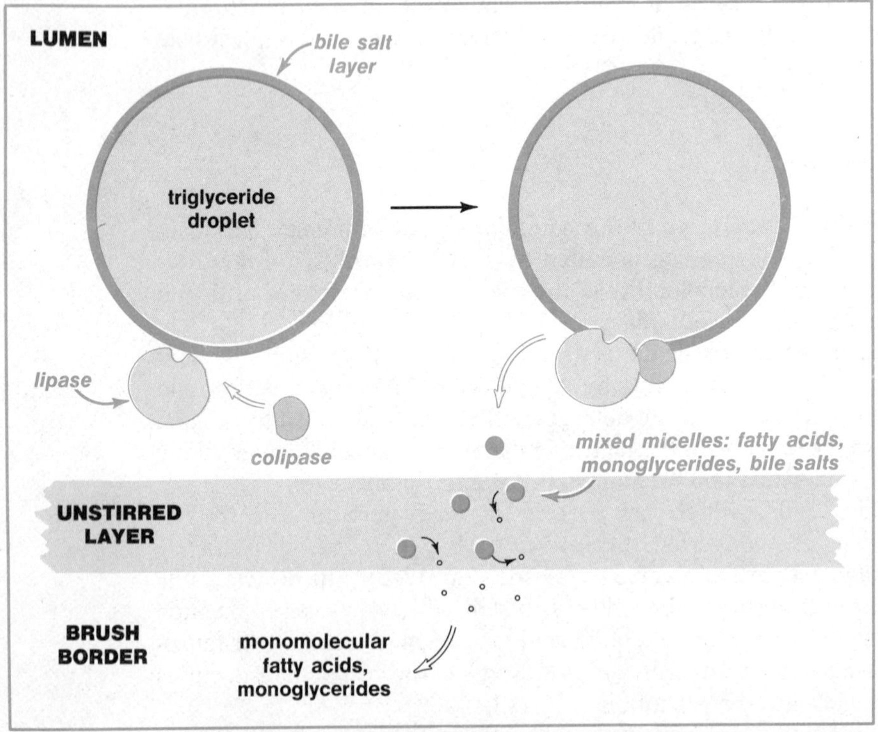

FIGURE 17–6 Triglyceride droplets become coated with a layer of bile salts, which prevents pancreatic lipase from making contact. (The lipase also complexes with bile salts.) An additional small protein, colipase, combines with lipase, bile salts, and the triglyceride droplet in such a way as to bring the lipase active site near the triglycerides, permitting hydrolysis to begin, and maintaining the lipase in its native conformation. The products of hydrolysis leave as mixed micelles, which diffuse readily through the unstirred water layer on the surface of the brush border.

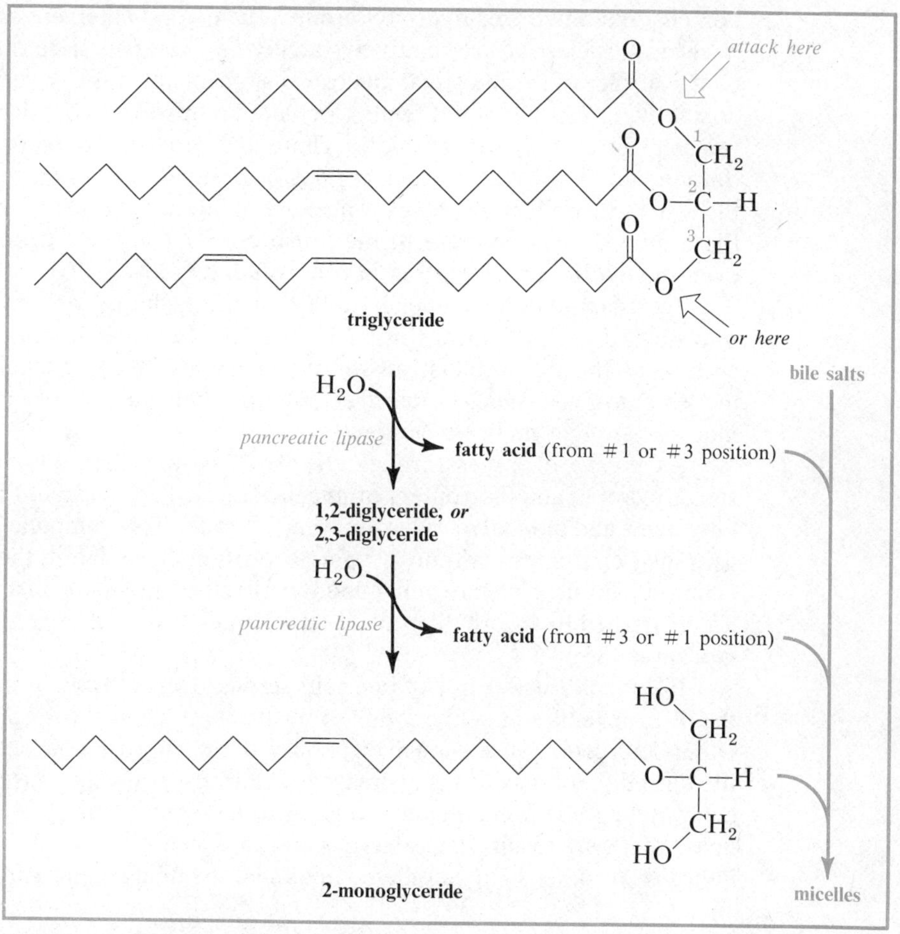

FIGURE 17–7 **Digestion of triglycerides begins with an attack on the C-1 or C-3 ester bond by pancreatic lipase. The resultant 1,2- or 2,3-diglycerides are further attacked by the hydrolase to leave the 2-acyl-glycerol, or monoglyceride. The liberated free fatty acids and monoglycerides form micelles with bile salts, which readily diffuse into proximity with the intestinal mucosa.**

with triglyceride droplets even in the presence of bile salts, which complex both the enzyme and the substrate (Fig. 17–6). However, the pancreas secretes an additional small protein, a colipase, that penetrates the surface of the fat droplet more readily and fixes the lipase so as to make triglycerides accessible to its active site. (Either colipase or bile salts also can act to stabilize the lipase at the interface.) Lipase, like other proteins, tends to denature at oil-water boundaries where a force exists to separate its hydrophobic and polar side chains.

Once brought into contact with the triglycerides, pancreatic lipase causes hydrolysis of either the 1 or the 3 ester bonds, but not the bond in the central 2 position (Fig. 17–7). Complete digestion by this enzyme therefore causes a conversion of a triglyceride to two fatty acid residues and a monoacyl glycerol (monoglyceride). In the laboratory, a slow spontaneous shift of the fatty acid from C-2 to either C-1 or C-3 can be shown, permitting removal by lipase of the final fatty acid residue, but this is probably not of great physiological importance.

Absorption. Bile salts form micelles with the fatty acids and monoglycerides liberated at the surface of a fat droplet. (Monoglycerides, with only one hydrocarbon chain and two free hydroxyl groups, have good detergent action.) The small micellar particles diffuse relatively rapidly and can penetrate the unstirred layer on the surface of the intestinal mucosa at a reasonable rate. (Every solid surface in contact with moving liquid tends to create an unstirred boundary layer in which transfer occurs only by diffusion.) There is a constant equilibration of the soaps (fatty acid anions) and monoglycerides in true solution as single molecules with those in the micellar complexes. The compounds are liberated from the micelles to the unstirred layer so close to the brush border that the diffusion time is short, even though the concentration in true solution is low.

To recapitulate in sequence, a fatty acid residue in dietary fats is dispersed into small droplets, appears either in free fatty acids or in monoglycerides at the surface of the oil droplets, moves into micelles in association with bile salts, moves out of the micelles into free solution, and then passes into the intestinal mucosa through the brush border.

As the food passes through the small bowel, the content of fat droplets steadily drops, and the content of micelles rises. These micelles are at first rich in fatty acids and monoglycerides, but they liberate these components to the mucosa as transit continues with most rapid absorption occurring in the jejunum. In the ileum, the fat droplets are gone, and the micelles are mostly bile salts. (The ileum is constructed to absorb bile salts, and we shall discuss this circulation of steroids in Chapter 35.)

If there is a deficiency of bile salts, either because of failure in production by the liver or failure in delivery due to obstruction, normal transport cannot occur. Other detergents cause enough dispersion of fats to permit the action of lipase, but the formation of micelles is decreased. Oil droplets mainly consisting of free fatty acids still persist in the ileum and go into the feces. Deficiency of bile salts or of lipase therefore results in **steatorrhea** — foul-smelling fatty stools, with a concomitant loss of other lipid soluble compounds, including some vitamins.

DIGESTION OF PROTEINS

Most of the proteins ingested by an animal differ from those in its cells. The objective of protein digestion is to convert these foreign polypeptide chains into their constituent amino acids, which can be used to construct the proper tissue components. Hydrolysis of the amide bond between amino acid residues seems straightforward, but getting it done is not so simple.

The variety of combinations of amino acids creates one problem. With 20 amino acids, there can be 400 different combinations of side chains in the two residues immediately adjacent to the peptide bond that is to be hydrolyzed. It is possible to construct a hydrolase that will ignore the adjacent structures and concentrate only on the amide linkage. Such enzymes will hydrolyze almost any peptide; some bacteria make proteolytic enzymes like this, and their lack of specificity makes them very valuable laboratory tools for degrading proteins; subtilisin, from *B. subtilis,* is a commercially available example. (Proteolytic enzyme is a synonym for protease, an endopeptidase.) However, the weak bonding by these enzymes results in relatively poor catalysis with low affinity for the substrate compared to other enzymes. Such broad specificity therefore would require large amounts of enzyme to hydrolyze food proteins in a given length of time. On the other hand, very narrow specificity, with a particular enzyme attacking only one, or a few, combinations of amino acids, would require producing dozens, perhaps hundreds, of different enzymes.

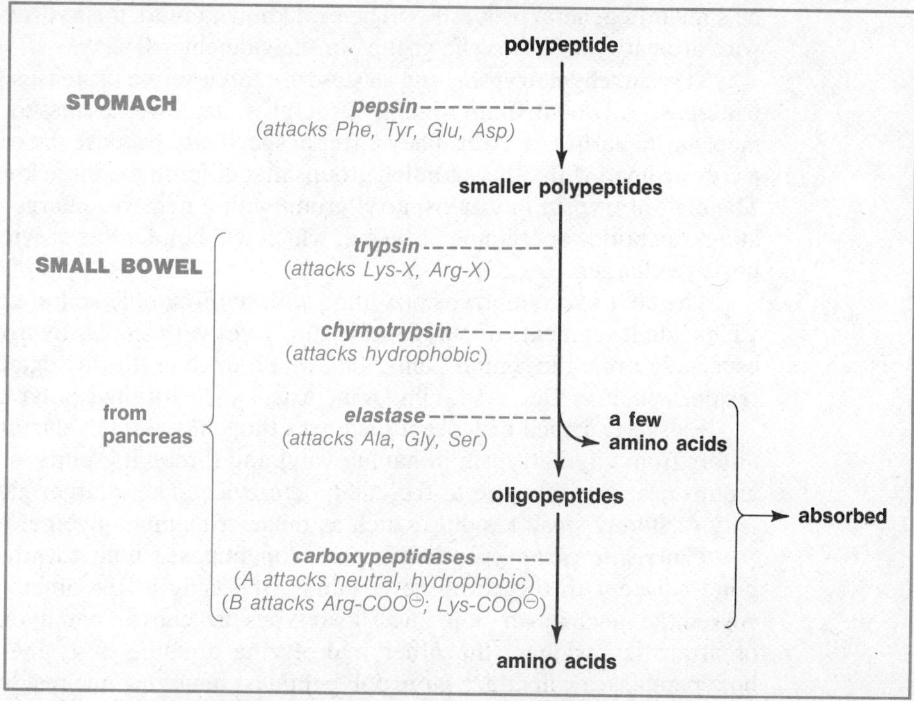

FIGURE 17–8 **Protein digestion involves attack on polypeptide chains by endopeptidases with differing specificities. Pepsin, trypsin, chymotrypsin, and elastase cause hydrolysis at enough different sites along the chains so as to convert dietary proteins (along with proteins from the digestive juices) to a mixture of oligopeptides. The oligopeptides are further shortened by the action of carboxypeptidases, which release the carboxyl terminal residues. The ultimate result is a mixture, mostly of di- and tripeptides along with some free amino acids, all of which are absorbed into the brush border.**

Whatever the specificity, enzymes poured into the gastrointestinal tract must be produced in considerable excess. Because the enzymes are also proteins and are subject to hydrolytic digestion, they destroy each other along with the dietary proteins.

The general plan of protein digestion is a compromise (Fig. 17–8). Four powerful types of endopeptidases and two of exopeptidases are supplied by the digestive juices. These enzymes have specificities broad enough to convert most proteins to a mixture of small oligopeptides and some free amino acids, but the bonding is specific enough to have a reasonably high substrate affinity (low K_M) and catalytic effectiveness (high turnover number). The products are small enough to be absorbed readily, and once absorbed, the oligopeptides can be attacked by a wide battery of intracellular exopeptidases, which need not be produced in the large quantities that would be necessary to maintain activity within the small bowel under attack by the powerful endopeptidases.

Let us now examine some of the components of the general plan. The readily available enzymes from the cow and pig have been studied in detail, but enzymes from other sources, including the human, appear to be quite similar.

Pepsins are secreted by the gastric mucosa. These endopeptidases are constructed to catalyze hydrolysis under acidic conditions, using side-chain carboxyl groups in the mechanism. More than one enzyme is secreted, but they appear to be similar in general properties. The best known favors the hydrolysis of peptides with aromatic or carboxylic groups in the side chains.

Trypsin, chymotrypsin, and **elastase** are three serine proteases secreted by the pancreas. All have quite similar structures, and we discussed their catalytic mechanism earlier (p. 262). They differ in specificity because the catalytic site is in a crevice lined with other bonding groups that differ in the three kinds of enzymes. The cleft in trypsin has an aspartyl group with a negative charge and, therefore, attacks peptides of arginine or lysine, which will bond in the crevice through their positive charges.

The cleft in chymotrypsin is lined with hydrophobic amino acid side chains, which admit segments of polypeptide substrates with similar hydrophobic groups, especially aromatic amino acids, but which exclude those segments containing residues such as Glu, Asp, Gln, Asn, Arg, Lys, with their polar side chains.

Elastase, named because it once was thought to attack elastin preferentially, differs from chymotrypsin in having valyl and threonyl groups in place of glycyl groups near the entrance to the catalytic crevice. These larger groups will admit only relatively small residues, such as those of alanine, glycine, or serine.

Pancreatic carboxypeptidases are exopeptidases built to attack the peptide bond adjacent to the carboxyl terminus, releasing a free amino acid. (We discussed the mechanism on p. 261.) Two types are known, one hydrolyzing neutral or aromatic residues, the other hydrolyzing arginine or lysine residues. Carboxypeptidases will attack large polypeptides, removing one residue at a time, but they are much more effective as digestion proceeds because the endopeptidases split the original chains into small pieces, each with its carboxyl terminus. In other words, the endopeptidases effectively increase the substrate concentration for the carboxypeptidases.

The exopeptidases of the intestinal mucosa attack oligopeptides after absorption and convert them to free amino acids for transfer into the blood. Some of these enzymes occur in the brush border, but most are in the interior cytoplasm. Not all have been described. Those that are known are mainly **aminopeptidases** — enzymes that remove the amino terminal residue by hydrolysis. (For example, the most thoroughly studied is a leucine aminopeptidase, an enzyme attacking amino terminal leucyl groups, or similar residues.) Some of the exopeptidases are dipeptidases attacking oligopeptides with only two residues; others are tripeptidases attacking those with three residues. There is also a **prolidase,** a carboxypeptidase specific for carboxyl terminal proline residues. In any event, the result is that nearly all of the dietary protein is converted to free amino acids before reaching the blood.

Zymogen Activation

Proteases are a menace to any cell. In order to protect the cells that make them, they are synthesized as **proenzymes,** or **zymogens,** which are secreted from the cell before conversion to the active form. The stomach is in less danger, because pepsin is not active at the more neutral intracellular pH, but the pancreas is a loaded bomb, rich in potentially active enzymes.*

A dramatic illustration of the potency of proteases is seen in some people with acute inflammation of the pancreas. The cause is usually gallbladder disease or alcoholism, which appear to obstruct pancreatic secretion. A particularly

*It is not known if Ralph Nader is aware of this common menace.

virulent form of this disease with a high mortality rate occurs when there is massive activation of the proteases. These people present with severe pain, vomiting, abdominal rigidity, fever, shock, and coma. In the fatal cases the pancreas may be completely liquefied; blood vessels are commonly destroyed (hence the name **acute hemorrhagic pancreatitis** that is applied).

Fortunately, the exocrine pancreas functions placidly in most people until they die of other causes, and the secret lies in effective control of zymogen activation. These proenzymes, like other proproteins, are synthesized with extra-long polypeptide chains. The additional residues prevent effective catalysis, either by obstructing access to the active site or by altering the conformation of the protein. Activation involves hydrolysis of the proenzyme, itself, to remove the inhibiting segment of polypeptide.

Pepsinogen has some 41 more amino acid residues than does pepsin. It has some proteolytic activity as it is made, sufficient for one molecule of pepsinogen to attack another and convert it to pepsin by hydrolyzing the extra pieces from the molecule. However, there is an important qualification. Pepsinogen is active only below pH 5. Furthermore, although the redundant segment of the molecule is split into pieces, one of the fragments, some 29 residues in length, still remains bound to pepsin and inhibits its activity until the pH falls toward 2. Molecules of active pepsin created by accident within the mucosa do nothing more until exposed to the acidity of the gastric contents, not only because they are built to be most active exclusively in acid solution but also because the inhibitory peptide clings to them.

Once the proteins are secreted, a few molecules of pepsin can rapidly activate many molecules of pepsinogen through proteolytic attack. The activation is autocatalytic; each molecule of pepsinogen that is converted to pepsin will then attack more pepsinogen, and so on.

The pancreatic peptidases are also secreted as proenzymes. The triggering event in their activation is exposure to secretions from the intestinal mucosa after passage from the pancreatic duct. These secretions include a highly specific protease, known as **enterokinase,** constructed to remove a particular sequence of residues in **trypsinogen,** the proenzyme form of trypsin (Fig. 17–9). Trypsinogen begins with one to three non-polar residues, depending upon the species, followed by the distinctive negatively charged Asp-Asp-Asp-Asp-Lys, with the next residue Ile or Val. Enterokinase hydrolyzes the bond between Lys and Ile (or Val) in this sequence. The negative charges in the sequence hold the molecule in an inactive conformation; once enterokinase breaks the bond, the remainder of the chain can swing into a new position, creating the active trypsin conformation.

When active trypsin appears, it attacks the other zymogens — **proelastase, chymotrypsinogen,** and **procarboxypeptidases,** as well as trypsinogen itself, — to convert them to their active forms. For example, a single cleavage of chymotrypsinogen by trypsin permits it to assume a fully active conformation (Fig. 17–10) even though disulfide bonds hold the molecule together. (Further cleavages by chymotrypsin molecules of each other can cause the release of small dipeptide fragments without important modification of activity; still further cleavages destroy the enzyme.)

The key in all of this is the activation of trypsinogen by enterokinase. Once a little active trypsin appears, autocatalytic activation of trypsinogen will create more trypsin, and full activation of the other zymogens rapidly follows. The thing that saves the pancreas is the absence of active trypsin until the juice has safely entered the small bowel. Even so, it is a chancy thing. Trypsinogen, like pepsinogen, has some propensity to assume an active conformation, and exposure to

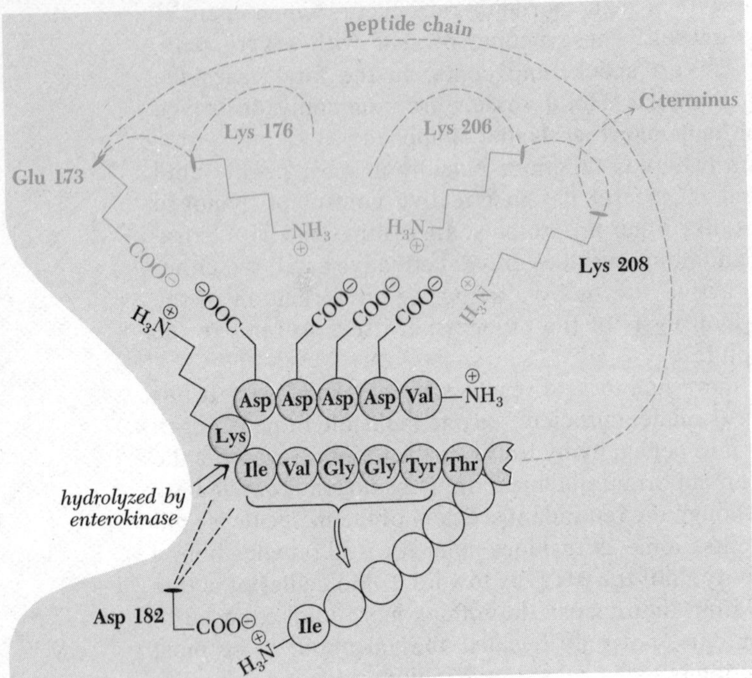

FIGURE 17–9

The activation of trypsinogen by enterokinase involves the hydrolysis of a specific lysyl peptide bond. The segment preceding that bond contains four aspartyl groups in succession, held by their negatively charged chains to positively charged lysyl groups in the remainder of the trypsinogen molecule. These electrostatic bonds (along with another involving the lysyl group being attacked) hold trypsinogen in an inactive strained conformation. When enterokinase breaks the attachment of this sequence to the remainder of the molecule, the new N-terminal segment shifts position to create a more relaxed and active conformation. Trypsin will also attack the same bond in trypsinogen molecules, once activation begins.

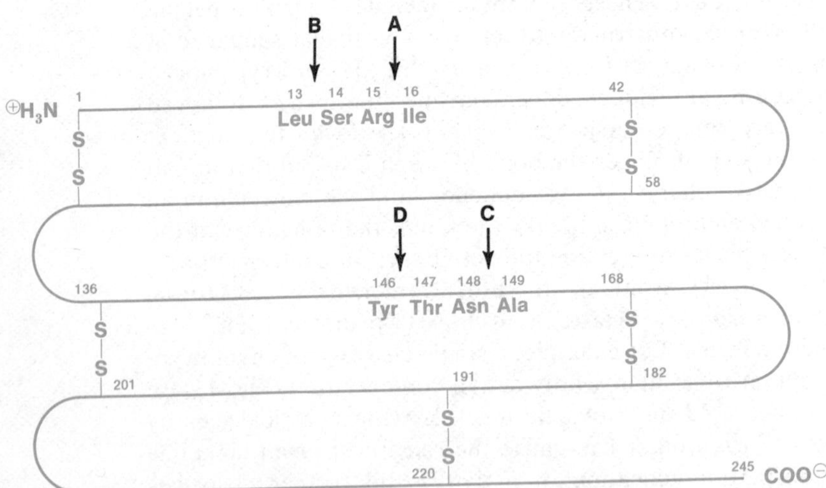

FIGURE 17–10 **The topology of chymotrypsinogen activation. The molecule contains five disulfide bridges. Initial activation by trypsin at point A releases the molecule from its strained inactive conformation, but the peptide segments are still held together by the disulfide bonds. Active chymotrypsin will cause further cleavages in chymotrypsinogen molecules, as indicated by B, C, and D.**

active proteases from other sources is conceivable. A critical safety factor is provided by the synthesis within the pancreas of another small protein that binds active trypsin tightly but cannot be hydrolyzed. This protein is therefore a **trypsin inhibitor** that prevents any accidentally activated molecules from creating more activated molecules. It is only when acute irritation of the pancreas results in the formation of an excess of active trypsin over the available supply of inhibitor that disaster occurs.

FURTHER READING

Gray, G. M.: (1975) *Carbohydrate Digestion and Absorption*. N. Engl. J. Med., *292*: 1225. Clear review.

Conklin, K. A., K. Y. Yamashiro, and G. M. Gray: (1975) *Human Intestinal Sucrase-isomaltase*. J. Biol. Chem., *250*: 5735.

Gray, G. M., K. A. Conklin, and R. R. W. Townley: (1976) *Sucrase-isomaltase Deficiency*. N. Engl. J. Med., *294*: 750. Account of digestive disturbances caused by recessive gene, not rare in heterozygous form.

Sipple, H. L., and K. W. McNutt, eds.: (1974) *Sugars in Nutrition*. Academic Press. Includes discussions of gas-forming oligosaccharides in foods.

Levitt, M. D., et al.: (1976) *Studies of a Flatulent Patient*. N. Engl. J. Med., *295*: 260. A case of lactose intolerance.

Spiro, H. M.: (1975) *The Rough and the Smooth*. N. Engl. J. Med., *293*: 83. A suavely witty analysis of the food fiber fad.

Rommel, K., and R. Bohmer, eds.: (1976) *Lipid Absorption: Biochemical and Clinical Aspects*. University Park Press. Useful review articles.

Nair, P. B., and D. Kritchevsky, eds.: (1971) *The Bile Acids*. Plenum. Three volume treatise on structure and function.

Momsen, W. E., and H. L. Brockman: (1976) *Effects of Colipase and Taurodeoxycholate on Pancreatic Lipase*. J. Biol. Chem., *251*: 378, 384.

Verger, R., J. Rietsch, and P. Desnuelle: (1977) *Effects of Colipase on Hydrolysis of Monomolecular Films by Lipase*. J. Biol. Chem., *252*: 4139.

Mathews, D. M., and J. W. Payne, eds.: (1975) *Peptide Transport in Protein Nutrition*. North-Holland.

Mathews, D. M.: (1975) *Intestinal Absorption of Peptides*. Physiol. Rev., *55*: 537.

Boyer, P. D., ed.: *The Enzymes*, 3rd ed. Academic Press. Articles on properties of specific enzymes: B. S. Hartley and D. M. Shotton, *Pancreatic Elastase*, Vol. 3, p. 349 is especially valuable.

Kassell, B., and J. Kay: (1973) *Zymogens of Proteolytic Enzymes*. Science, *180*: 1022.

Alpers, D. H., and B. Seetharam: (1977) *Pathophysiology of Diseases Involving Intestinal Brush-border Proteins*. N. Engl. J. Med., *296*: 1047.

18 HYDROLASES: BLOOD CLOTTING AND COMPLEMENT

The life of mammalian cells is absolutely dependent upon proximity to flowing blood as a means of fetching and removing compounds. This traffic is so vital that it is protected and regulated through a number of interdependent devices built into the blood itself and the vessels containing it. Some of these devices depend upon responses by cells and by organs that are beyond the scope of our inquiry, but among the devices is a powerful biochemical tool — the enzymatic cascade. When enzymes act in cascade, one enzyme attacks another, which in turn attacks a third, and so on; each step magnifies the effect of the preceding.

We are concerned here with four specific, but connected, phenomena that involve cascades: the clotting of blood, the lysis of clots, the formation of agents (kinins) that dilate vessels and increase permeability, and the lysis of cells upon attachment of antibodies. The mechanism of these events have these common features (Fig. 18–1):

(1) They involve limited proteolysis by specific hydrolases in the blood, most of which are serine endopeptidases attacking Arg-X bonds;

(2) the active hydrolases are created from proenzymes by attack with still other endopeptidases of similar character;

(3) the initial signals for the events trigger a cascade of proenzyme activation in which one enzyme partially hydrolyzes the proenzyme of another, making an active enzyme that in turn activates still another proenzyme, and so on;

(4) inhibitors are present to combine with accidentally activated enzymes and to prevent premature triggering of the cascade;

(5) the status quo is restored through constant removal of the activated enzymes, mainly by the liver;

(6) most of the components in these mechanisms are glycoproteins.

We can see in all of this considerable extension of the processes by which pancreatic hydrolases are activated while limiting the risk of premature appearance of functional enzymes.

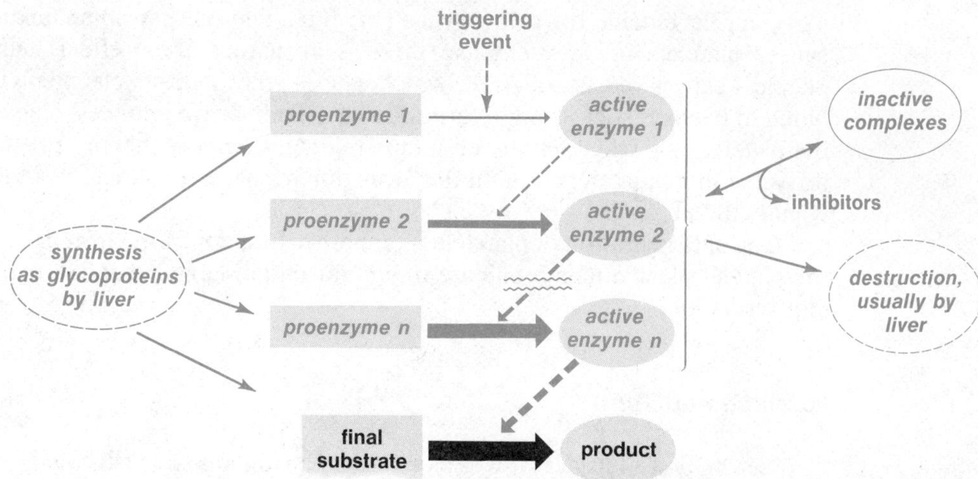

FIGURE 18–1 **General outline of cascade activation of endopeptidases in the blood.**

BLOOD CLOTTING

The formation of clots to plug defects in blood vessels is one of the more important defense mechanisms of the body. We readily perceive the loss of blood when it pours from gaping wounds, but the potential loss through constant minor injuries to the small vessels is only appreciated when we see the multiple little hemorrhages (petechiae) in people deficient in platelets or the large localized hematomas that occur in people with a defective clotting mechanism. These injuries can result from the simple stresses of motion and the accompanying contacts with physical objects, from ordinary chewing, from the normal movement and loss of cells in the gastrointestinal tract, and so on.

Minimizing blood loss is accomplished by three events. One is a clumping of platelets in the blood at the site of injury so as to plug the opening temporarily. This may be sufficient to seal small breaks. Another is a vasoconstriction of the injured vessel immediately to reduce the flow through the break. The third event is aggregation of a protein, fibrin, into a clot — a stable three-dimensional lattice that is strong enough to seal the damaged vessel while repairs are being made.

The Platelet Plug

The blood platelets, or thrombocytes, are disk-shaped cells about 1 × 3 microns in size that lack nuclei but contain mitochondria, other cytoplasmic organelles, and distinctive granules, or dense bodies. The blood normally contains about one platelet for each 20 erythrocytes. They arise from pieces of megakaryocytes in the bone marrow. The platelets have the property of adhering tightly to collagen fibers. So long as the endothelial lining of blood vessels is intact, platelets and collagen don't meet, but if the lining is disrupted by overt injury or pathological change, platelets stick to the exposed collagen fibers.

Platelets adhering to collagen undergo a remarkable morphological change. They become spiny spheres. The dense bodies disappear from the cytoplasm, and their contents, notably adenosine diphosphate (ADP) and the amine, serotonin (p.

624), appear outside the cell (release reaction). The ADP in some unknown way causes platelets to be sticky, so there is an autocatalytic effect; adhesion of platelets causes release of ADP, which causes still more platelets arriving in the blood to become sticky and in turn to release their dense granules. The growth of the platelet plug will continue until blood flow is stopped, thereby preventing the arrival of more platelets or until the formation of the fibrin lattice, described next, engulfs the plug.

The agglomeration of platelets has another function. Lipoproteins exposed in the platelet plasma membrane are important initiating and accelerating agents in clot formation.

Formation of Fibrin

Let us first consider how a clot forms and then go back through the earlier events that caused its formation. Clotting occurs because a soluble blood plasma protein, **fibrinogen,** is partially hydrolyzed to form fibrin. Fibrin can associate into a three-dimensional lattice, but fibrinogen cannot, even though it has a large surface (45 nm length, 9 nm maximum diameter, M.W. = 340,000). The extra polypeptide segments in fibrinogen carry negative charges so that electrostatic repulsion aids in keeping the molecules apart.

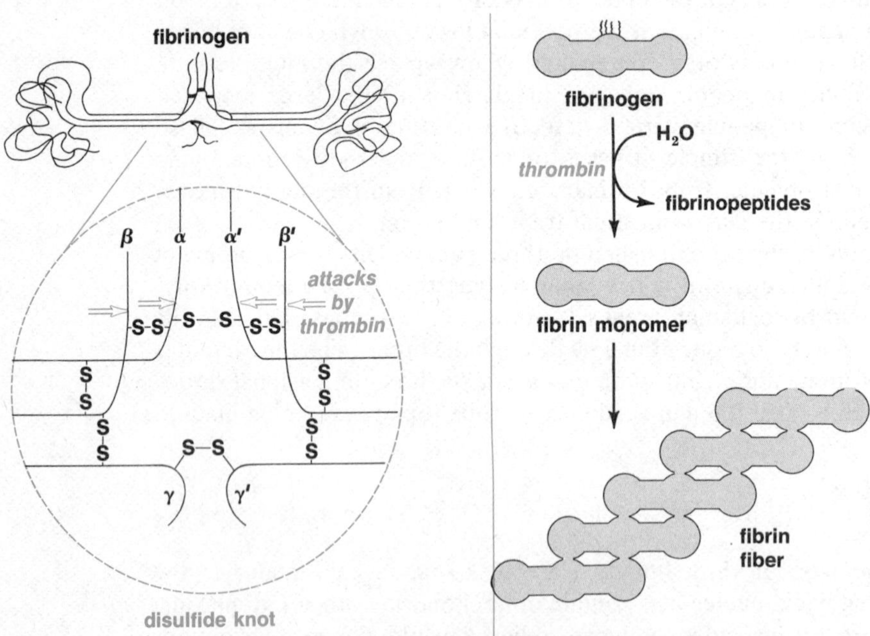

FIGURE 18-2 **Blood clotting involves a partial hydrolysis of fibrinogen (*right*) to release approximately 3 per cent of the polypeptide chains as fibrinopeptides. The remaining fibrin monomer is then free to associate into fibrous structures, apparently in a staggered side by side arrangement. The fibrinogen molecule is constructed of three pairs of unlike polypeptide chains (*left*) forming three globular regions connected by more fibrous segments. The two identical sets of three chains are joined in the center region by a "disulfide knot," shown in enlarged schematic view. Conversion of fibrinogen to fibrin involves hydrolysis of the N-terminal segments (*blue arrows*) of four of the six chains, two α and two β chains. Relative dimensions are not shown in these highly schematic drawings.**

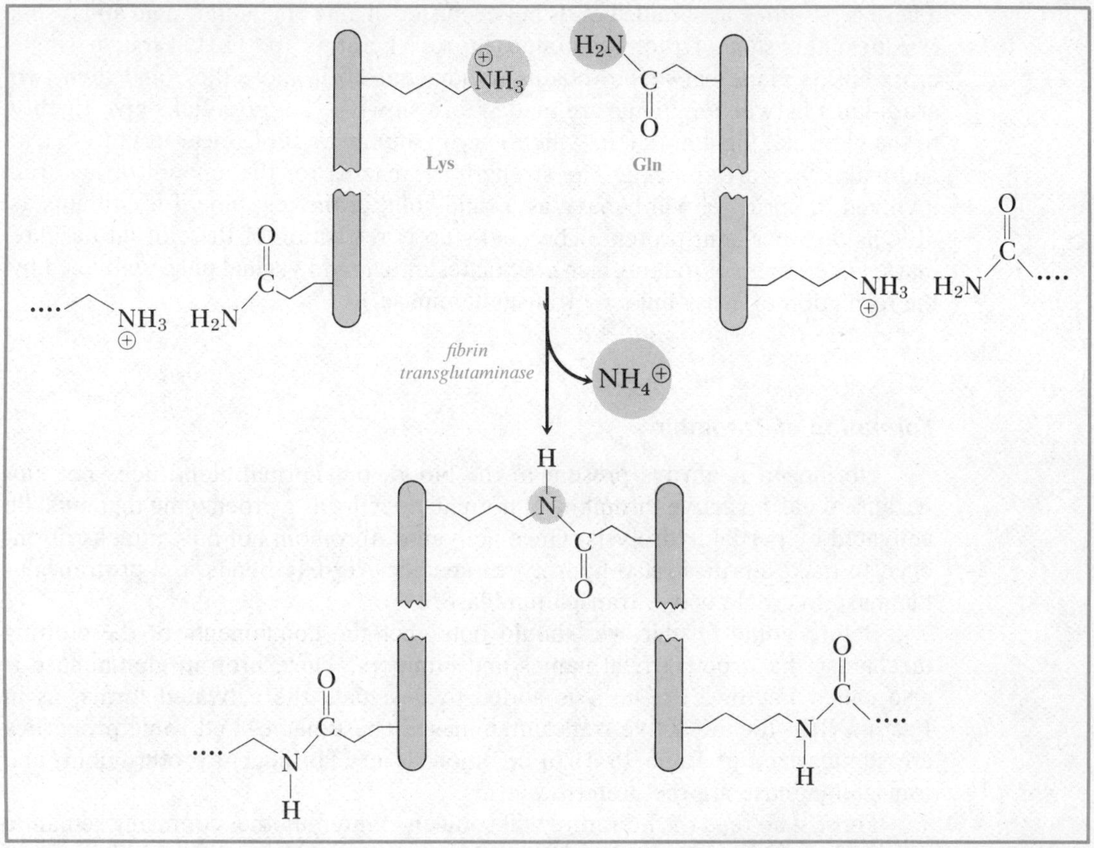

FIGURE 18–3 **Fibrin monomers are linked by covalent bonds in a fibrin clot through the action of a fibrin transglutaminase. This enzyme in effect transfers the glutamyl portion of a glutamine side chain to the nitrogen atom of a lysine side chain on an adjacent polypeptide, leaving the amide nitrogen of the original glutamine free as ammonium ion.**

The chain formula of fibrinogen is $\alpha_2\beta_2\gamma_2$, and the two sets of three unlike chains are linked head to head and side by side by disulfide bonds (the disulfide knot). The tails of the chains are believed to form globular knobs, creating a dumbbell with a swollen center (Fig. 18–2), which includes the N-terminal segments removed from the α and β chains to create fibrin. Once these segments are gone, fibrin monomers can associate side by side to form fibers, and the fibers can make an open lattice.

The hydrolysis of fibrinogen to form fibrin is catalyzed by **thrombin,** which is a serine endopeptidase very similar to the digestive endopeptidases in topology. Thrombin specifically attacks Arg-Gly bonds in each of the two α and β chains of fibrinogen, releasing approximately 3 per cent of the original molecule in the form of four polypeptide fragments. Ca^{2+} must be present for polymerization of fibrin monomers, and clotting can be prevented in blood samples by the addition of chelating agents such as citrate, oxalate, or ethylene diamine tetraacetate (EDTA).

The aggregation of fibrin monomers is too weak to serve as a semipermanent plug until final repairs are made. The hydrogen bonds, hydrophobic interactions, and similar forces between monomers are strengthened by the formation of covalent bonds (Fig. 18–3). They are formed by the action of a **transglutaminase,** which transfers the glutamyl portion of a glutamine residue to a lysine side chain.

These cross-links are amide bonds between side chains of glutamic acid and lysine residues, the same structures found in hair keratins (p. 153). First, a single cross-link is made between γ-chains in adjacent fibrin molecules, and then two cross-links between α-chains are made more slowly. The cross-links give rigidity to the clot, making a lattice in which varying numbers of platelets, erythrocytes, and leukocytes are trapped. The strength is amazing for the amount of material involved; a fibrin gel will behave as a rigid solid in the test tube if it contains as little as 0.05 per cent protein. The final step is retraction of the clot into a hard mass. (The semen of rodents also coagulates into a rigid vaginal plug, stabilized by the formation of cross-links by transglutaminase.)

Formation of Thrombin

Fibrinogen is always present in the blood, but normal blood does not clot because it has no active thrombin. It normally occurs as a proenzyme that must be activated by partial hydrolysis. Once activated, thrombin not only attacks fibrinogen to make fibrin, it also hydrolyzes specific Arg-Gly bonds in a protransglutaminase to create active transglutaminase.

Before going further, we should note that the components of the clotting mechanism have both trivial names and numbers. Thus, protransglutaminase is also called Factor XIII; an *a* is added to designate the activated forms, as in Factor XIII*a* for the active transglutaminase. These names and some properties are summarized in Table 18–1. In common usage, fibrinogen, prothrombin, and transglutaminase are the preferred terms.

Those who read the literature will soon encounter another confusing semantic point. The endopeptidases involved in blood clotting, like many peptidases, will also hydrolyze amino acid esters. Synthetic esters are frequently more convenient substrates for laboratory study of these enzymes. Hence, it is common to talk about the esterolytic activity when referring to clotting factors, sometimes in a context implying that this is a physiological function of these proteins. It is not; the only known purpose of these enzymes is to hydrolyze peptide bonds.

TABLE 18–1 FACTORS IN BLOOD CLOTTING

Factor* Designation	Trivial Name	M. W. $\times 10^{-3}$	Concentration g/liter	Function of Active Form
I	**fibrinogen**	340	1.5–3.5	precursor of fibrin lattice
II	**prothrombin**	69	?	hydrolyzes fibrinogen
III	tissue factor	220	–	hydrolyzes Factor VII
IV	**Ca^{2+}**	–	0.085–0.105	cationic cofactor
V	proaccelerin	290	?	protein cofactor for Factor Xa
VI	accelerin – redundant designation no longer used			
VII	proconvertin	63	0.001	hydrolyzes Factor X
VIII	anti-hemophiliac factor	1100	0.005–0.010	protein cofactor for Factor IXa
IX	Christmas factor	55.4†	?	hydrolyzes Factor X
X	Stuart factor	55†	0.010	hydrolyzes prothrombin
XI	plasma thromboplastin antecedent	160	?	hydrolyzes Factor IX
XII	Hageman factor	90	?	hydrolyzes Factor XI
XIII	**protransglutaminase**	320	?	forms cross-links in fibrin lattice

*preferred name in bold face
†bovine; data otherwise for human

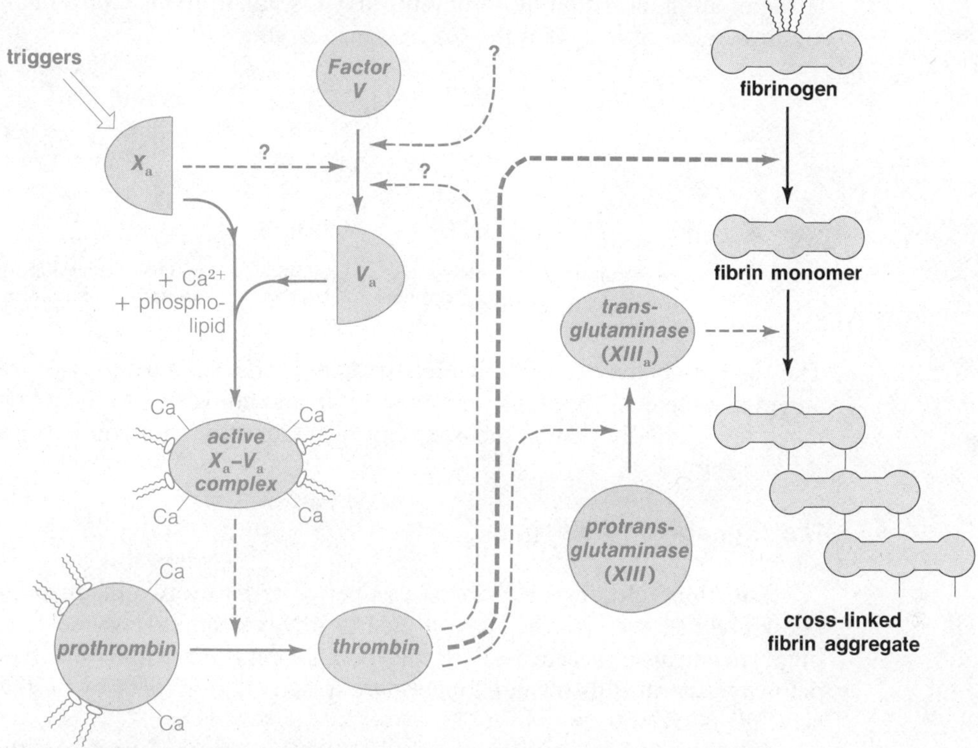

FIGURE 18–4 **The terminal events in blood clotting involve partial hydrolysis of several proteins.** *Solid lines* **indicate the reactions involved;** *dashed lines* **indicate the responsible enzyme.**

Prothrombin is converted to thrombin by two hydrolyses, both catalyzed by **Factor Xa,** another serine endopeptidase. The first hydrolysis removes an N-terminal segment, and the second cleaves an Arg-Ile bond within a disulfide loop, creating a two-chain thrombin molecule held together by the disulfide bond. Prothrombin and Factor Xa interact to cause hydrolysis when they are present in a complex with Ca^{2+} and phospholipid micelles, along with still another protein, **Factor Va.** Factor Va has no enzymatic properties; it creates only a favorable environment for hydrolytic activation of prothrombin by Factor Xa. The phospholipid micelles are available in activated platelets; combination with these micelles in effect gives a high local concentration of proteins ordinarily present in very low concentrations in blood plasma.

Let us summarize events at this point (Fig. 18–4). Clotting Factor V is converted to Va by limited proteolytic attack. Factor Va then combines with Ca^{2+}, phospholipids, and Factor Xa to make an active enzyme complex, which converts prothrombin to thrombin by release of a peptide fragment. (This fragment contains the oligosaccharide and the calcium binding parts of prothrombin.) Thrombin then converts fibrinogen to fibrin and protransglutaminase to transglutaminase, which forms covalent bonds between polymerized fibrin monomeric units.

The activation of transglutaminase illustrates a more general regulatory mechanism. The enzyme in the blood is composed of two pairs of chains, $\alpha_2\beta_2$. The potential catalytic site is in the α chains; the β chains act only as regulatory peptides to hinder activation by thrombin. However, once the α chains in the tetramer are hydrolyzed by thrombin, they lose their affinity for β chains, which are now released to reveal the fully active enzyme:

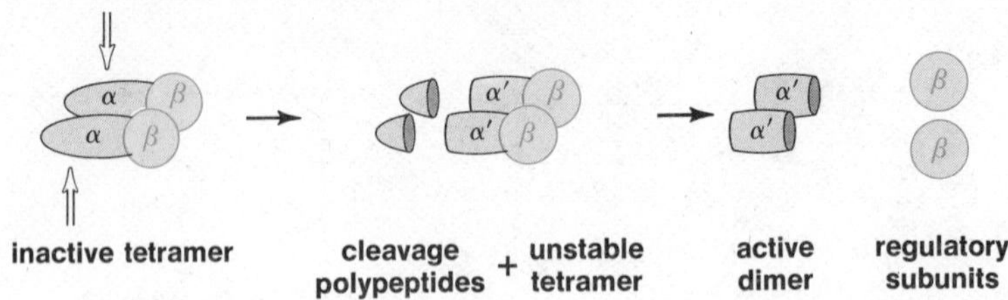

The same α-chains occur in platelets without β chains. When α_2 is released from adhering platelets, it is attacked by thrombin and converted to the active enzyme much more rapidly than is the $\alpha_2\beta_2$ tetramer found in the blood plasma.

The Triggers for Clotting

The blood clotting mechanism can be triggered by two independent routes, one known as extrinsic because it clearly involves an extravascular factor, the other as intrinsic because all of the components are within the blood. Each pathway generates its own endopeptidase to activate Factor X by partial hydrolysis (Fig. 18–5).

Extrinsic activation is the lesser characterized route. It involves the interaction of a lipoprotein tissue factor, which is exposed when the endothelium and other tissues are damaged, with **Factor VII,** a protein in blood plasma. Factor VIIa is an endopeptidase that activates Factor X. The tissue factor probably is also an endopeptidase that activates Factor VII, but this has not been established.

Intrinsic activation involves a cascade of three other endopeptidases, the last of which activates Factor X. These three are Factors XII, XI, and IX. An auxiliary protein Factor VIII, calcium ions, and platelet phospholipids are also required. The initial event in this sequence is the exposure of Factor XII to collagen or activated platelets, upon which it becomes an active endopeptidase, Factor XIIa, which converts Factor XI to XIa. Factor XIa in turn activates Factor IX in the presence of Ca^{2+}.

Finally, Factor IXa, in the presence of an auxiliary protein (perhaps modified to be VIIIa), platelet phospholipid, and Ca^{2+}, will activate Factor X.

People are known who have a hereditary absence of Factor XII, the first component of the sequence, but they usually have little, if any, trouble from bleeding. This is very puzzling, since there is ample evidence for the ability of Factor XII to trigger clotting in the test tube* and for some related functions that we shall mention below. Does this mean the intrinsic pathway is not something with physiological significance? Not necessarily. It could easily be that there is a protective redundancy of activation. There is evidence for a bypass on the intrinsic pathway by which platelets directly activate Factor IX, and this route may prevent routine problems with excessive bleeding.

The importance of the intrinsic pathway is best demonstrated by a congenital absence of normal Factor VIII, causing classic **hemophilia.** The affected gene is on the X chromosome and is defective in roughly one in 10,000 infants. Hemophil-

---

*Indeed, Factor XII was discovered because the blood of an individual failed to clot when drawn into a glass test tube; glass activates Factor XII.

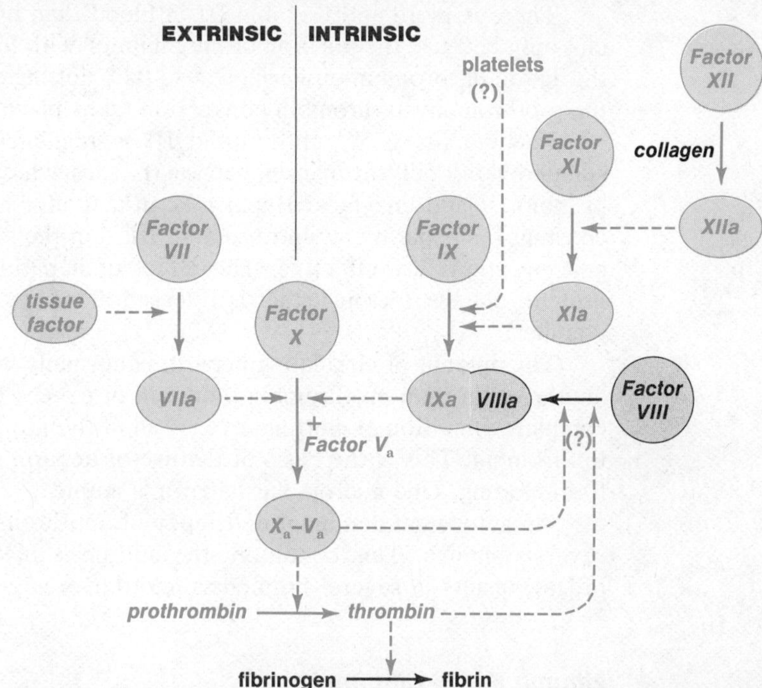

FIGURE 18–5 **The formation of Factor Xa may be caused by either of two pathways. The extrinsic pathway involves uncharacterized proteases from tissues. The intrinsic pathway begins upon exposure of Factor XII in the blood to collagen fibers.**

ia, therefore, appears at this rate in males and is very rare in females (calculated incidence of one in 100,000,000).

"The first record of hemophilia appeared around A.D. 500 in the writings of the Talmudic rabbis. They stated that a third male child would be excused from the ritual of circumcision if two previous male siblings had died from bleeding after their circumcision. By the 12th century, Maimonides stated if a woman had two sons who had bled from circumcision, she could postpone circumcision of future sons whether they be from the same husband or from another husband; thus, it was recognized that the disease was transmitted by the asymptomatic mother."* Excessive bleeding results from relatively trivial causes in these patients. Not only is there a loss of blood, there is also damage to tissues in which bleeding occurs. Hereditary deficiencies of other components in the intrinsic pathway cause similar problems.

Inhibitors of Clotting

Over 10 per cent of the proteins in blood plasma are inhibitors of the enzymes involved in blood clotting and of other proteases. Some of the clotting inhibitors react with Factors XIIa and XIa to block the intrinsic pathway. The best understood is **antithrombin III,** a protein that reacts with thrombin, Factor Xa, or Factor VIIa to form inert enzyme-substrate intermediates.

---

*Quotation from Harvey R. Gralnick, (1977) *Factor VIII,* Ann. Int. Med., *86*: 598.

There is more antithrombin III in blood than prothrombin; blood is able to clot only because the reaction of the inhibitor with thrombin is much slower than the action of thrombin on fibrinogen so that clotting can occur at locations where the prothrombin to thrombin conversion takes place rapidly.

The reactivity of antithrombin III is regulated by combination with the sulfated heteropolysaccharide, heparin (p. 185), which causes a rapid combination of inhibitor and enzymes. Heparin is an activator of the inhibitor. The binding constant for heparin and antithrombin III is in the range of 10^7 to 10^8 M, so low concentrations are effective. The effect of heparin can be counteracted by a protein, platelet factor 4, that is released along with the contents of the dense granules.

The amount of circulating heparin is normally very low; since there is more antithrombin than prothrombin, addition of excess heparin can cause rapid and complete inhibition of clotting, even if all of the prothrombin becomes converted to thrombin. This is the basis of the use of heparin in preventing blood samples from clotting. One micromolar heparin is ample.

An autosomal dominant deficiency of antithrombin III has been reported in several families. This condition, the antithesis of hemophilia and similar disorders, results in severe thrombosis (clotting *in vivo*).

Vitamin K and Clotting

Prothrombin, and Factors VII and IX become activated only in the presence of Ca^{2+}. Calcium can bond each of these proenzymes to its activator because the N-terminal segments of the proenzymes have glutamyl residues that have been modified by carboxylation (Fig. 18–6). These N-terminal segments, which are quite similar and probably have a common evolutionary origin, are blithely discarded by the activating hydrolysis — their usefulness has ended. Factor X also has these modified residues to bind Ca^{2+}, but they are not discarded during the activation because phospholipid is attached to Factor Xa through Ca^{2+}.

The insertion of carboxyl groups in the precursors of these clotting factors by the liver is known to require derivatives of vitamin K, although the mechanism of

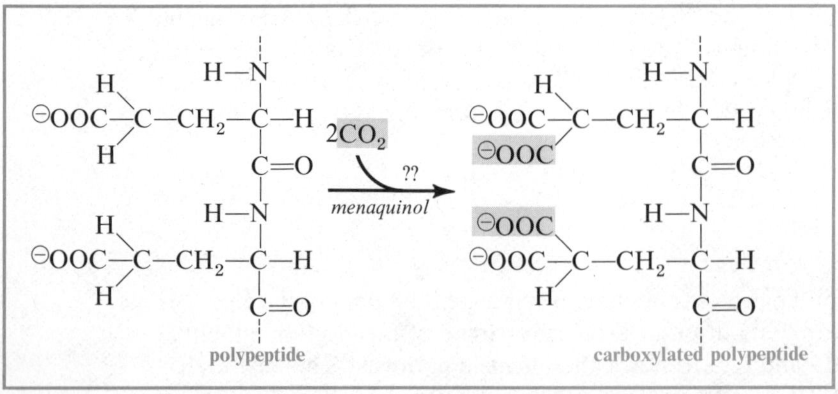

$2CO_2$?? *menaquinol*

polypeptide → carboxylated polypeptide

FIGURE 18–6 Several clotting factors are carboxylated on C-4 of specific pairs of glutamate residues in the liver before being released into the blood. The carboxylation reaction cannot occur unless menaquinol, or similar reduced derivatives of vitamin K, are available in the liver.

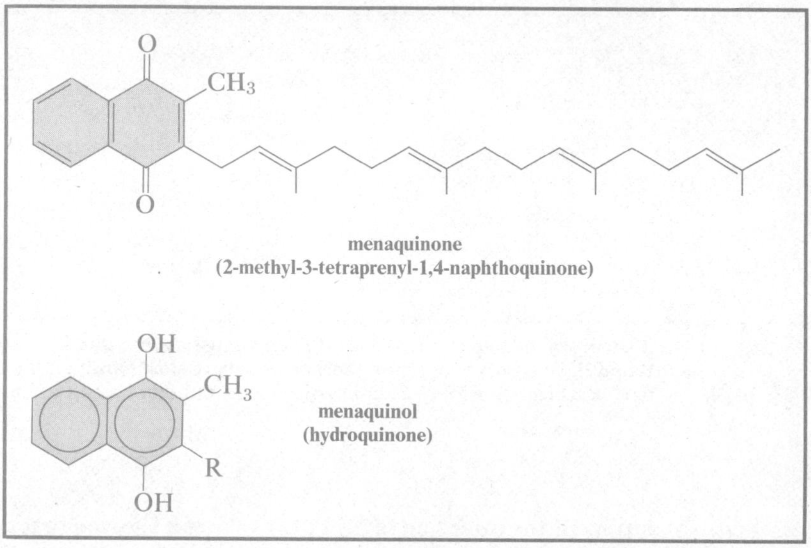

O
CH3
O

menaquinone
(2-methyl-3-tetraprenyl-1,4-naphthoquinone)

OH
CH3
R
OH

menaquinol
(hydroquinone)

FIGURE 18–7 **Menaquinone is the common form of vitamin K occurring in mammalian liver. Its biological activity involves reduction to the corresponding hydroquinone, menaquinol.**

carboxylation is not known. By definition, a compound has vitamin K activity when it enables the synthesis of clotting factors. Such compounds are substituted naphthoquinones, of which a typical example is menaquinone (Fig. 18–7). They are hydrophobic compounds that are absorbed from the mixed micelles formed during fat digestion. Any defect in fat absorption will also cause a failure in absorption of vitamin K, as with other hydrophobic vitamins, with resultant prolongation of blood clotting time. Patients with obstruction of the bile duct are routinely given vitamin K before surgery, for this reason. The nutritional aspects of the vitamin are discussed in Chapter 42.

Pathological Clotting

Inappropriate formation of thrombi is a common cause of disability and death. Stroke and heart attack are in the common language. A red thrombus containing red blood cells and other cells trapped in a lattice of fibrin may form and grow in some people's veins without any known provocation. Approximately 50,000 people die each year in the United States, and another 300,000 are hospitalized as a result of venous thromboembolisms. The formation of clots on a foreign surface is also a major unsolved problem in the use of artificial organs, even with such relatively simple prostheses as artificial heart valves.

White thrombi grow in more compact form along arterial walls, with fewer included erythrocytes, and are major contributors to **myocardial infarction** (loss of blood supply to parts of the heart muscle), **stroke** (loss of blood supply to parts of the brain), and **renal damage.**

We have already discussed the use of heparin in management of clotting (p. 186). Its great value is that its effects are immediate. Slower, and longer-lasting inhibitors of clotting are available in antagonists of vitamin K, and herein lies a tale. Cattle eating spoiled sweet clover develop a hemorrhagic disease, which

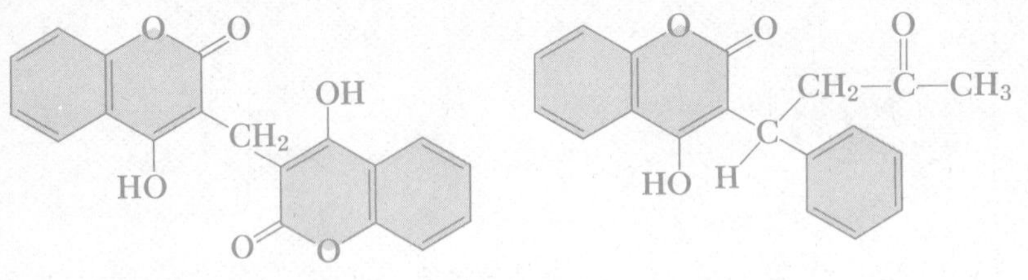

bishydroxycoumarin (Dicoumarol)

Warfarin

FIGURE 18–8 **Bishydroxycoumarin (Dicoumarol) is an antagonist of vitamin K function. It and its derivatives are used therapeutically to slow clotting. One derivative, Warfarin, is widely used to poison rats and other small animals.**

attracted the attention of K. P. Link.* Link showed that the disease was due to the formation of *bis-hydroxycoumarin* during fermentation of the clover (Fig. 18–8). This compound was shown to compete with vitamin K and prevent the formation of prothrombin and other vitamin K-dependent blood clotting factors. It and related compounds are now used to prevent prothrombin formation in patients prone to abnormal clot formation. Since response depends upon loss of already existing prothrombin, it takes two or more days for a useful effect to appear, and heparin is used initially for prompt results.

One of the coumarin derivatives, *Warfarin,* remains an effective rat poison even after continued use because it is difficult for the rats to associate the delayed death of their companions from internal bleeding with the original cause. (However, there are disturbing reports of the appearance of rats with genetic resistance to Warfarin.)

Ordinary aspirin also can prevent normal clotting by interfering with platelet aggregation. It isn't used intentionally for this purpose, but microscopic to gross loss of blood from the gastrointestinal tract frequently accompanies its continued consumption.

LYSIS OF CLOTS

Clots must be removed to restore circulation after repair of an injured vessel. This is accomplished in part by hydrolysis of fibrin into soluble pieces. **Plasmin,** the responsible endopeptidase, occurs as plasminogen in the blood, and plasminogen is converted to plasmin by hydrolysis of some arginyl bonds, much in the same way as are most blood proenzymes. The nature of the activating enzymes is uncertain. Apparently, there are proactivators in blood plasma, which are made active by some process involving Factor XIIa. According to one intrepretation, an autocatalytic chain of events begins with the formation of some active plasmin. This further hydrolyzes Factor XIIa to a modified form that lacks blood clotting activity, but which can attack the proactivators of plasminogen so that they become active and cause the formation of more plasmin (Fig. 18–9).

Blood contains several plasmin inhibitors, but the way in which inhibition is overcome to initiate clot dissolution is not known.

*K. P. Link, American biochemist, and striking feature of the University of Wisconsin. Now retired.

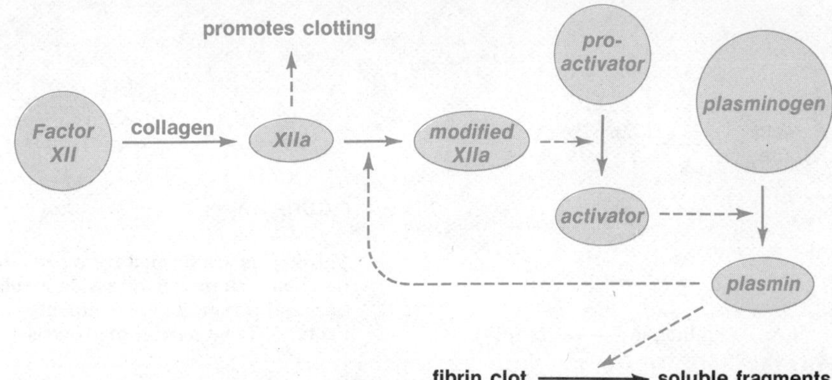

FIGURE 18–9 **The activation of plasmin, which converts clots to soluble fragments through partial hydrolysis.**

BRADYKININ AND KALLIKREINS

Bradykinin is a small peptide with only nine amino acid residues that is produced in the blood to dilate blood vessels and increase vessel permeability. It is a very potent physiological effector; doses as low as 200 nanograms per kg body weight can cause a measurable fall in blood pressure in small animals. It is also a smooth muscle constrictor and has the added effect of causing intense peripheral and visceral pain. Indeed, some suggest that production of bradykinin is the mechanism for stimulation of pain receptors upon injury of tissues. It is formed in conjunction with a variety of inflammatory conditions.

Bradykinin is synthesized by hydrolyzing a particular sequence out of the middle of relatively large precursors, the kininogens (Fig. 18–10). (Two sizes of kininogens are known, with 50,000 and 250,000 M.W.) The enzymes catalyzing these specific hydrolyses are kallikreins. All kallikreins split kininogens at the carboxyl end of the bradykinin sequence, but the position of the amino terminal cleavage varies with the kallikrein. The enzyme from plasma releases bradykinin; glandular tissues cleave kininogen one residue earlier, releasing Lys-bradykinin. A still longer Met-Lys-bradykinin also appears, perhaps due to cleavages by

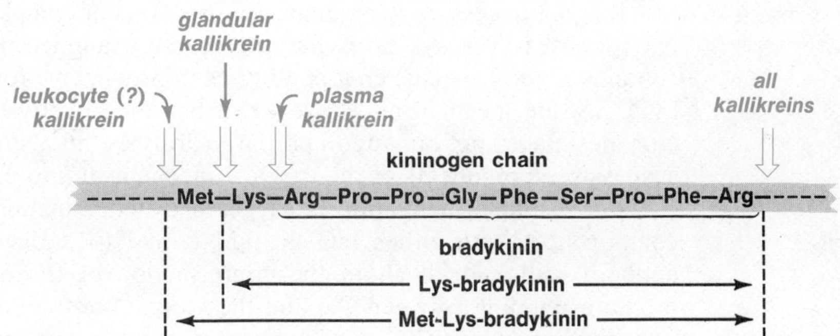

FIGURE 18–10 **The long polypeptide chains of kininogens in the blood and tissues contain a specific sequence of amino acids that is released by partial hydrolysis to form bradykinin. The partial hydrolyses are catalyzed by kallikreins.**

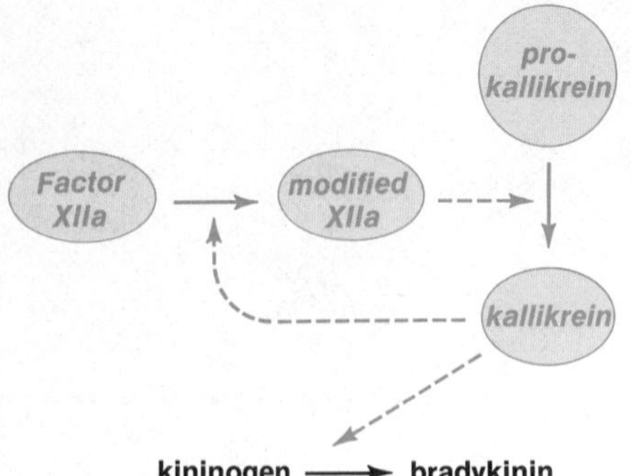

FIGURE 18-11

Kallikreins are formed from prokallikreins by the same modified blood clotting Factor XIIa involved in plasminogen activation. Kallikrein, like plasmin, may catalyze the modification of Factor XIIa by partial hydrolysis.

leukocytes. The extra N-terminal residues in these derivatives are removed by aminopeptidases in the liver and lungs, forming bradykinin itself.

The activation of kallikreins is triggered by the same events that activate lysis of fibrin clots, and it is argued that restoration of circulation is the common purpose, served by dilation of the vessels and concomitant removal of an obstructing clot. Prokallikrein is hydrolyzed to kallikrein by the same modified Factor XIIa that triggers plasminogen activation (Fig. 18–11). There is a positive feedback effect; kallikrein catalyzes the modification of Factor XIIa.

The action of kallikrein is blocked by at least three inhibitors in blood plasma, but the factors controlling the balance of inhibition and activation are not known. Bradykinin must be formed continuously to maintain its effect because 10 per cent is removed by one pass of the blood through the lungs, which contain a **peptidase** that removes the C-terminal Phe-Arg groups to destroy bradykinin activity.

COMPLEMENT

The complement system is a collection of some 18 proteins in which interactions occur as a result of partial hydrolyses. The system may be triggered by contact with antibody-antigen complexes or by clotting Factor XIIa, causing the formation of a protein complex that ruptures cell membranes. Complement is a system designed to destroy foreign cells and to otherwise aid in the removal of foreign materials. Its complexity may stem in part from the necessity of preventing false stimulation that would destroy the host's own cells.

It is not necessary to examine the functions of complement in great detail for our purposes. The system is like the others examined in this chapter in that it contains several specific endopeptidases in proenzyme form, which are activated by a cascade mechanism. However, it has the novelty that the action of these enzymes hinges not only upon partial hydrolysis but also upon association of the successive components of the cascade mechanism into complexes.

For example, activation is triggered by combination of the Clq subunit in component 1 with either IgG or IgM bound to antigen. The combination of antibody and antigen alters the conformation of the antibody in a way that promotes binding between Clq and the second constant domain of the antibody. The binding causes Clq to become an active endopeptidase that now partially hydrolyzes another subunit, Clr. Clr gains endopeptidase activity with which to attack still another subunit, Cls, which in turn becomes an active endopeptidase (Fig. 18–12).

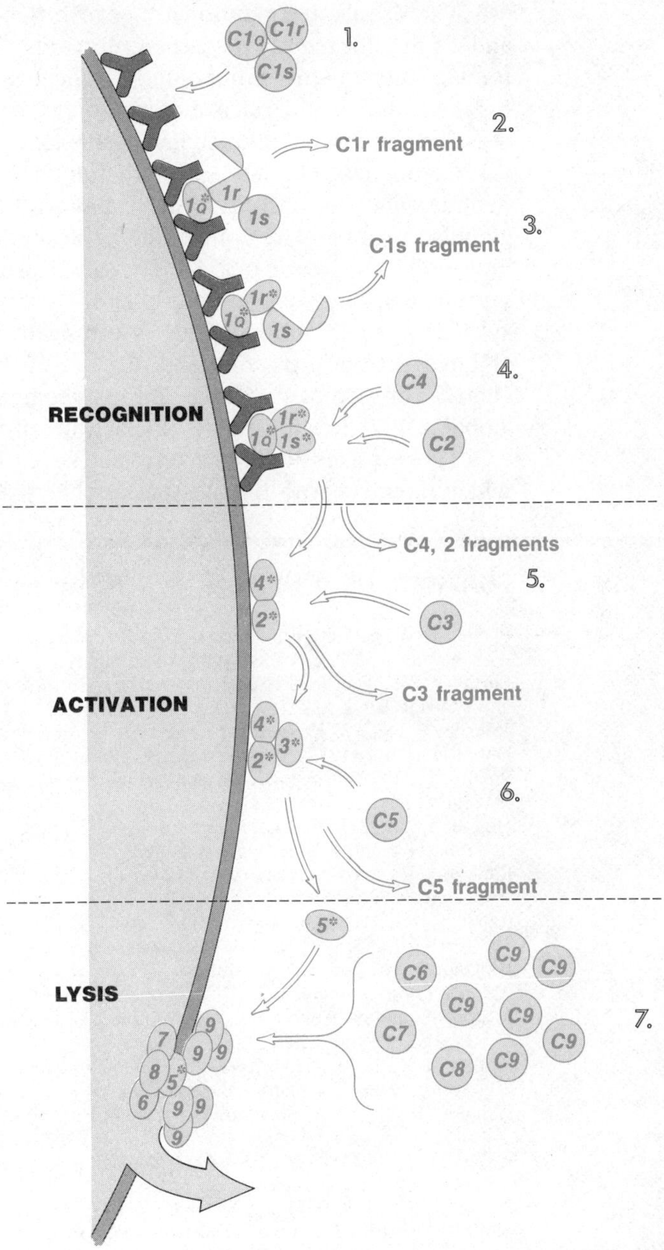

FIGURE 18–12

(*1*) The first component of complement binds to antibodies that are attached to antigens; the C1q subunit of the first component recognizes the altered conformation in the antibody stem caused by reaction with antigen.

(*2*) The C1q subunit becomes an active endopeptidase upon attachment, and partially hydrolyzes the C1r subunit.

(*3*) The modified C1r subunit is an active endopeptidase that attacks the C1s subunit, partially hydrolyzing it.

(*4*) The modified C1s subunit is also an active endopeptidase. It attacks the C4 and C2 components of complement.

(*5*) The partially hydrolyzed C4 and C2 components combine to make still another active endopeptidase, which attacks C3, releasing a fragment from it that acts as a signal to attract macrophages and to cause the release of histamine from neighboring mast cells.

(*6*) The larger portion of the hydrolyzed C3 combines with the C2-C4 complex to create a new endopeptidase activity that will attack C5. The smaller fragment released from C5 also is a signal for macrophages and for histamine release.

(*7*) The larger modified C5 fragment combines with one molecule each of components C6, C7, and C8, and with six molecules of C9 to form a complex that combines with the plasma membrane of the foreign cell, weakening it and causing discharge of its contents.

The Cls subunit of the active complex then attacks two other components, C2 and C4, which associate with each other. The C42 complex (read as C4 + C2) is in turn an endopeptidase that attacks a C3 component, releasing a small fragment (C3b). The larger remainder of C3 combines with active C42 and in turn gains endopeptidase activity to attack C5. C5 is also split to release a smaller fragment. The larger component combines with the remaining components, C6, C7, C8, and six molecules of C9 to make the final membrane-damaging complex. The common practice of designating the products of activating hydrolyses as component 3a or component 5b, for example, has been avoided here because it is completely inconsistent with usage in similar systems, such as clotting factors.

The details are mentioned because the peptide fragments released from C3 and C5 are not merely discarded remnants. They are the compounds responsible for the chemotactic summoning of phagocytic cells to the location of a foreign organism and for the release of histamine, with consequent contraction of smooth muscles and increased vascular permeability.

Component Clq has an interesting architecture. It is composed of six polypeptide subunits, each of which has its N- and C-terminal segments folded into globular regions. The connecting central segments are fibrous segments resembling collagen in composition. The entire protein has 5 hydroxyproline, 2 hydroxylysine, and 18 glycine residues per 100 and contains 9.8 per cent glucosylgalactosyl side chains. The molecule is arranged in a star-like configuration, with the globular regions from one end of each of the six subunits coalescing to form a central core from which the fibrous segments radiate outward to six separated globular units, each capable of reacting with a bound antibody molecule.

There is a less well-characterized second pathway for activating complement, which interacts with the classical pathway at C3; it is not discussed here.

FURTHER READING

Davie, E. W., and K. Fujikawa: (1975) *Basic Mechanisms in Blood Coagulation.* Annu. Rev. Biochem., *44*: 799. Excellent detailed summary.

Suttie, J. W., and C. M. Jackson: (1977) *Prothrombin Structure, Activation, and Biosynthesis.* Physiol. Rev., *57*: 1.

Jaffe, E. A.: (1977) *Endothelial Cells and the Biology of Factor VIII.* N. Engl. J. Med., *296:* 377.

Doolittle, R. F.: (1977) *Structure and Function of Fibrinogen.* Horiz. Biochem. Biophys., *3*: 164.

Stenflo, J.: (1976) *Vitamin K-Dependent Carboxylation of Blood Coagulation Proteins.* Trends Biochem. Sci., *1*: 256.

Nossel, H. M.: (1976) *Radioimmunossay of Fibrinopeptides in Relation to Intravascular Coagulation.* N. Engl. J. Med., *295*: 428.

Radcliffe, R., and Y. Nemerson: (1976) *Mechanism of Activation of Bovine Factor VII.* J. Biol. Chem., *251*: 4797.

Reich, E., D. B. Rifkin, and E. Shaw, eds.: (1975) *Proteases and Biological Control.* Cold Spring Harbor Laboratory. Continues discussion of all of the systems mentioned in this chapter.

Violand, B. N., and F. J. Castellano: (1976) *Mechanism of the Urokinases-catalyzed Activation of Human Plasminogen.* J. Biol. Chem., *251*: 3906.

Moroi, M., and N. Aoki: (1976) *Isolation and Characterization of α_2-Plasmin Inhibitor from Human Plasma.* J. Biol. Chem., *251*: 5956.

Stathakis, N. E.: (1977) *Familial Thrombosis due to an Antithrombin III Deficiency in a Greek Family.* Acta Haematol. (Basel), *57*: 47.

Colman, R. W.: (1974) *Formation of Human Plasma Kinin.* N. Engl. J. Med., *291*: 509. Clear review of kallikrein-bradykinin system.

Weiss, H. J.: (1975) *Platelets: Physiology and Abnormalities of Function.* N. Engl. J. Med., *293*: 531, 580. Clear review.

Simon, E. R., ed.: (1977) *Molecular Basis of Heparin Action.* Fed. Proc., *36*: 9ff. Useful symposium.

Müller-Eberhard, H. J.: (1975) *Complement.* Annu. Rev. Biochem., *44*: 697. More detailed discussion.

Porter, R. R.: (1977) *Structure and Activation of the Early Components of Complement.* Fed. Proc., *36*: 2191.

Mibashan, R. S., ed.: (1977) *Hemostasis and Thrombosis.* Semin. Hematol., *14*: 263–440. Reviews.

19 BIOCHEMICAL ENERGETICS

FREE ENERGY

In our discussion of metabolism, we shall constantly be concerned with defining the status of a reaction, or a set of reactions, in relation to the position of equilibrium. No matter how effective an enzyme may be, it cannot make a reaction go in a direction that will displace concentrations away from the equilibrium values.

As reactions go toward equilibrium they can do work. The farther from equilibrium that they start, the greater the amount of work that can be obtained from each mole of reacting compound. A convenient quantitative expression is the free energy change that occurs during the reaction, since this is by definition the capacity to do work under the conditions of constant temperature and pressure occurring in organisms.

Exergonic and Endergonic. Real reactions always liberate free energy and are said to be exergonic. A reaction that takes up free energy is an endergonic reaction, but reactions that do this don't exist as independent entities. Why do we waste time defining something that exists only in theory? We do so because it turns out to be a handy device for reasoning about some complex reactions that can be imagined to be made up of at least two components, one exergonic and the other endergonic. We treat many real biological reactions as the sum of a highly exergonic reaction, producing lots of free energy, and an endergonic reaction, which consumes part, but not all, of that free energy. Enough free energy must be liberated so that the overall result is at least somewhat exergonic, and it therefore can occur as a real reaction.

This brings us onto more familiar ground, because we are taught from an early age that we burn fuels not only for the energy with which to move but also for the energy necessary for the building of tissues and the like. This isn't too far from the truth; the major defect of the notion is its tendency to equate the concept of energy with a concrete physical entity that can be transferred in a bundle from one reaction to another. The concept is a very useful abstraction, but it is still an abstraction, not a substance.

Biological Examples. What are these exergonic and endergonic reactions, or rather the exergonic and endergonic components of biological reactions? Here are some conspicuous examples:

Exergonic (energy-producing)	Endergonic (energy-consuming)
Oxidation of fuels (carbohydrates, fats, and proteins)	Mechanical movement
	Synthesis of cellular constituents
Photosynthesis	Creation of concentration gradients
Fermentations	Storage of fuels

Anabolism and Catabolism. The exergonic components characteristically have large equilibrium constants and would cause a ''wasteful'' liberation of free energy if they occurred by themselves. Their sum is sometimes referred to as catabolism—the disruptive component of metabolism. The endergonic components characteristically cannot occur by themselves and have equilibrium constants less than one. The sum of these synthetic processes is sometimes referred to as anabolism.

Driving Endergonic Reactions with Exergonic Reactions

Impractically of Isolated Exergonic Reactions. Since the free energy liberated during a reaction depends only upon the distance from equilibrium at which it occurs, in principle any reaction could be made mildly exergonic by adjusting the concentrations of reactants and products so as to be slightly displaced from the equilibrium condition. In fact, this cannot be done for many reactions. Those reactions listed as exergonic have such high equilibrium constants that only unmeasureable amounts of starting material exist at equilibrium. They represent a sort of chemical overkill in which the reaction can be unnecessarily complete for biological purposes.

For example, the oxidation of the sugar, glucose, is a very common reaction in many organisms. In the test tube, occurring by itself, it has a very high equilibrium constant:

$$\text{glucose } (C_6H_{10}O_6) + 6\ O_2 \longrightarrow 6\ CO_2 + 6\ H_2O$$

$$K_{eq} = \frac{[CO_2]^6[H_2O]^6}{[\text{glucose}][O_2]^6} = 10^{495}.$$

It isn't possible to bring this system to equilibrium in real terms. At the highest concentration of water and CO_2 we can achieve, the calculated equilibrium concentration of glucose and oxygen is such that only one molecule of each could exist in a sphere with a diameter some 50 times that of the solar system.

Impracticality of Isolated Endergonic Reactions. On the other hand, the reactions listed as endergonic have very low equilibrium constants in water, and it would require impossibly large concentrations of starting materials to make them proceed to even a small extent. Consider an equilibrium involving a peptide chain that is 100 identical amino acid residues long, its constituent amino acids, and water:

$$100 \text{ amino acids} \longrightarrow \text{peptide} + 99 \text{ water}$$

$$K^{\circ}_{eq} = \frac{[\text{peptide}][H_2O]^{99}}{[\text{amino acid}]^{100}} = 10^{140}.$$

The large equilibrium constant makes it appear that the reaction will go readily, but wait, let us see what happens when the actual concentration of water, 55.6 M, is inserted in the equation:

$$K^{\circ}_{eq} = \frac{[\text{peptide}][55.6]^{99}}{[\text{amino acid}]^{100}} = 10^{140}.$$

From this, we can define a new "equilibrium constant":

$$K_{eq} = \frac{[\text{peptide}]}{[\text{amino acid}]^{100}} = \frac{K^\circ_{eq}}{[55.6]^{99}} = 10^{-33}.$$

Since the concentration of water is always near the same value in dilute solutions, it is common practice to calculate "equilibrium constants" that are adjusted for water concentration, as we have just done, and these are the values generally quoted in tables. Having made the calculation, we find that the reaction looks much less feasible. We have already seen in Chapter 5 that proteins are not made by this simple reaction, but let us consider why. A concentration of 0.01 M is a high value to expect for most intracellular constituents, including amino acids, and even if we take this generous estimate, we can calculate from the equilibrium equation that not even one molecule of peptide would be formed in an entire universe of solution by merely reversing hydrolysis. Indeed, we noted in discussing proteolytic enzymes that equilibration with water proceeds toward the hydrolysis of peptides, not toward their creation. Something else must be done if the formation of peptides from amino acids is to be feasible.

Coupling Endergonic Reactions to Exergonic Reactions. The biological problem is one of somehow combining wildly exergonic reactions, such as the oxidation of glucose, with impossibly endergonic reactions, such as the combination of amino acids into a peptide. It isn't a simple matter of having the two sets of reactions going in the same solution. We could make a solution of glucose, oxygen, and amino acids, along with appropriate catalysts, and find carbon dioxide being vigorously evolved from the oxidation of the sugar while the amino acids sit there essentially unchanged. Once more, energy is not a concrete thing that is automatically transferred from compounds of high chemical potential to those of low potential.

However, there is a way in which the oxidation of glucose can be made to drive endergonic reactions, and that is to modify both of the reactions so as to involve an additional compound that is produced by one and consumed by the other. We have already seen one of the most common of the modifications in connection with protein synthesis. The formation of polypeptides involves not only a combination of amino acids but also a sequence of reactions in which the nucleotides, ATP and GTP, are cleaved:

$$\text{amino acid} + \text{ATP} + \text{tRNA} \longrightarrow \text{aminoacyl-tRNA} + \text{AMP} + \text{PP}_i$$

$$n(\text{aminoacyl-tRNA} + \text{GTP}) \xrightarrow[mRNA]{ribosomes} \text{polypeptide} + n(\text{GDP} + \text{P}_i).$$

It is the exergonic cleavage of the nucleotides that converts what would be an endergonic formation of peptides into an exergonic combination. If the ATP and GTP being consumed for peptide synthesis are now remade in conjunction with the oxidation of glucose, the synthesis of peptides will be effectively driven by the oxidation of glucose, and this is exactly what happens.

We can make a general picture that applies to all metabolic processes:

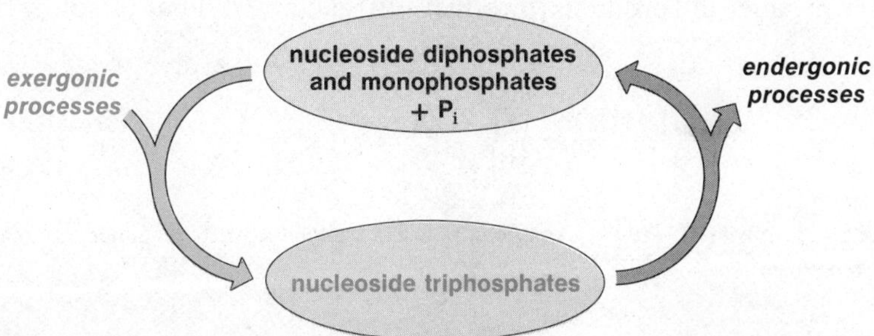

What would be highly exergonic processes, such as the oxidation of glucose, are modified so that nucleoside diphosphates and monophosphates are phosphorylated to create nucleoside triphosphates. This diminishes the liberation of free energy that ordinarily accompanies the oxidation of glucose.

The nucleoside triphosphates are then used in otherwise endergonic reactions, and they are cleaved in these reactions to the corresponding diphosphates or monophosphates so that the reaction becomes exergonic.

The overall effect is as if there were a direct transfer of energy, but the real mechanism is the modification of the exergonic reactions so as to create products of higher chemical potential and to use these products as reactants in the formerly endergonic reactions. The exergonic and endergonic processes became coupled through the involvement of the same nucleotides in both processes, which therefore have the function of energy carriers.

Quantitative Values

Numerical values of free energy change (ΔG, for Gibbs* free energy) are used in several ways in biochemistry. Students of macromolecular structure use classical chemical thermodynamics to explore probable conformation in a way that is outside our scope. Students of metabolism use the values in two somewhat simple ways. One is an index of the displacement from equilibrium. Given two reactions in which the ratio of reactants to products is equally displaced from equilibrium, the reactions will have equal free energy changes per mole reacting, no matter what the equilibrium constant may be. Expressed mathematically, if for

$$\text{A} \rightarrow \text{B}, [\text{B}]/[\text{A}] = Q_a, \text{ with } K_{eq} = K_a, \text{ and}$$

$$\text{X} \rightarrow \text{Y}, [\text{Y}]/[\text{X}] = Q_x, \text{ with } K_{eq} = K_x, \text{ then}$$

$\Delta G_{A \rightarrow B} = \Delta G_{X \rightarrow Y}$ when $K_a/Q_a = K_x/Q_x$.

Indeed, the numerical value of the free energy change can be calculated from this ratio of the equilibrium constant to the actual concentration ratios; and this is one of our most important equations in reasoning about energetics:

$$(1) \qquad \Delta G = -RT \ln \frac{K_{eq}}{Q} = -5{,}706 \log \frac{K_{eq}}{Q} \text{ joules mole}^{-1} \text{ at } 25^\circ\text{C}.$$

in which:

ΔG = Gibbs free energy change, which is negative for spontaneous reactions in which energy is liberated,

R = gas constant = 8.314 joules per mole per degree, or 1.987 calories per mole per degree,

T = absolute temperature,

Q = product of [products]/product of [reactants]. That is, for

$$\text{A} + \text{B} + \text{C} \cdots + \text{n} \longrightarrow \text{P} + \text{Q} + \text{R} \cdots + \text{z}$$

$$Q = \frac{[\text{P}][\text{Q}][\text{R}] \cdots [\text{z}]}{[\text{A}][\text{B}][\text{C}] \cdots [\text{n}]}.$$

*J. Willard Gibbs (1839–1903): American physical chemist who single-handedly created a large portion of chemical thermodynamics.

Standard Free Energy Changes. Another use for quantitative values is the comparison of the potential of different reactions for generating energy under the same conditions. This comparison is made in terms of a standard free energy change, ΔG^0, which is:

$$\Delta G^0 = -RT \ln K_{eq}.$$

This may be regarded as the free energy change under the hypothetical conditions when all reactants and products are at unit activity (thermodynamic concentration).

This can be seen by rearranging our equation for free energy changes:

$$\Delta G = -RT \ln \frac{K_{eq}}{Q} = -RT(\ln K_{eq} - \ln Q) = -RT \ln K_{eq} + RT \ln Q$$

when $Q = 1$, $\Delta G = \Delta G^0 = -RT \ln K_{eq}$.

Given two reactions, the one with the more negative standard free energy change has the higher equilibrium constant.

Free Energy Changes Are Additive. The principal value of free energy changes in our discussion of metabolic reactions comes from their logarithmic character, which makes the overall change in free energy for any series of reactions the sum of the free energy changes in the individual reactions. Free energy is therefore an especially useful concept for reasoning about the complex series of reactions that constitute metabolism.

The additive nature of free energy changes can be demonstrated by assuming three consecutive reactions, A → B → C → D:

(1) A ⟶ B $\quad K_{eq_1} = \frac{[B_{eq}]}{[A_{eq}]} \quad -\Delta G_1^0 = RT \ln K_{eq_1}$

(2) B ⟶ C $\quad K_{eq_2} = \frac{[C_{eq}]}{[B_{eq}]} \quad -\Delta G_2^0 = RT \ln K_{eq_2}$

(3) C ⟶ D $\quad K_{eq_3} = \frac{[D_{eq}]}{[C_{eq}]} \quad -\Delta G_3^0 = RT \ln K_{eq_3}$

(4) (SUM) A ⟶ D $\quad K_{eq_\omega} = \frac{[D_{eq}]}{[A_{eq}]} \quad -\Delta G_\omega^0 = RT \ln K_{eq_\omega}$

If we solve the first equilibrium expression for $[B_{eq}]$, substitute this value in the equation for the second reaction, solve this for $[C_{eq}]$, and substitute it in the third, we arrive at this expression:

$$K_{eq_1} K_{eq_2} K_{eq_3} = \frac{[D_{eq}]}{[A_{eq}]}.$$

This ratio of concentrations also gives the equilibrium constant for the overall reaction. Therefore, the product of the individual equilibrium constants is the overall equilibrium constant:

$$K_{eq_1} K_{eq_2} K_{eq_3} = K_{eq_\omega}$$

Taking the logarithm of the equation, and multiplying by $-RT$, we have:

$$-RT \ln (K_{eq_1} K_{eq_2} K_{eq_3}) = -RT \ln K_{eq_\omega}.$$

Since the logarithm of a product is the sum of the individual logarithms, this is equivalent to saying:

$$\Delta G_1^0 + \Delta G_2^0 + \Delta G_3^0 = \Delta G_\omega^0$$

and this is what we set out to show.

Not only is the additive nature of free energy changes useful in analyzing metabolic processes, but it can also be used to calculate the free energy change for a reaction by simply adding the values for two completely independent reactions whose theoretical sum happens to be the reaction of interest.

For example, it is important to have some idea of the free energy change for the reaction:

$$(1) \qquad \text{ATP} + \text{H}_2\text{O} \longrightarrow \text{ADP} + \text{P}_\text{i}.$$

The equilibrium for this reaction lies so far to the right that it is technically difficult to measure the remaining concentrations of reactants with sufficient accuracy for reasonable estimates of the equilibrium constant. However, values can be obtained from two other enzymatic reactions having lower equilibrium constants, thereby enabling reasonable measurements of equilibrium concentrations and calculation of their standard free energies:

(2) ATP + glucose $\longrightarrow$ ADP + glucose 6-phosphate

$$\Delta G^{0\prime} = -23{,}000 \text{ J mole}^{-1}$$

(3) glucose 6-phosphate + H_2O $\longrightarrow$ glucose + P_i

$$\Delta G^{0\prime} = -13{,}800 \text{ J mole}^{-1}$$

If we add reactions (2) and (3), the overall result is reaction (1), and we can find its free energy change:

$$\Delta G_2^{0\prime} + \Delta G_3^{0\prime} = \Delta G_1^{0\prime} = -36{,}800 \text{ J mole}^{-1} \text{ (at pH 7 and 25°C)}.$$

The prime symbols are used to designate free energy changes at defined conditions, for example, constant pH in this case. When H^+ is a reactant or a product, changes in its concentration would obviously change the free energy and the position of equilibrium for a reaction. Its fixed concentration is already taken into consideration when the prime symbol is a part of the standard free energy designation. This is a convenience for biochemical purposes, where the pH of tissues remains relatively constant.

Use of Numerical Values

It is handy to have in mind a rough idea of the relationship between numerical values for standard free energy changes and equilibrium constants. Figure 19–1 plots the relationship at 0°, 25°, and 38° C (273°, 298°, and 311° K). Looking at the figure we see that a reaction with $\Delta G^0 = -1$ joule per mole is likely to be freely reversible, whereas a value of $\Delta G^0 = -40{,}000$ joules per mole means that the

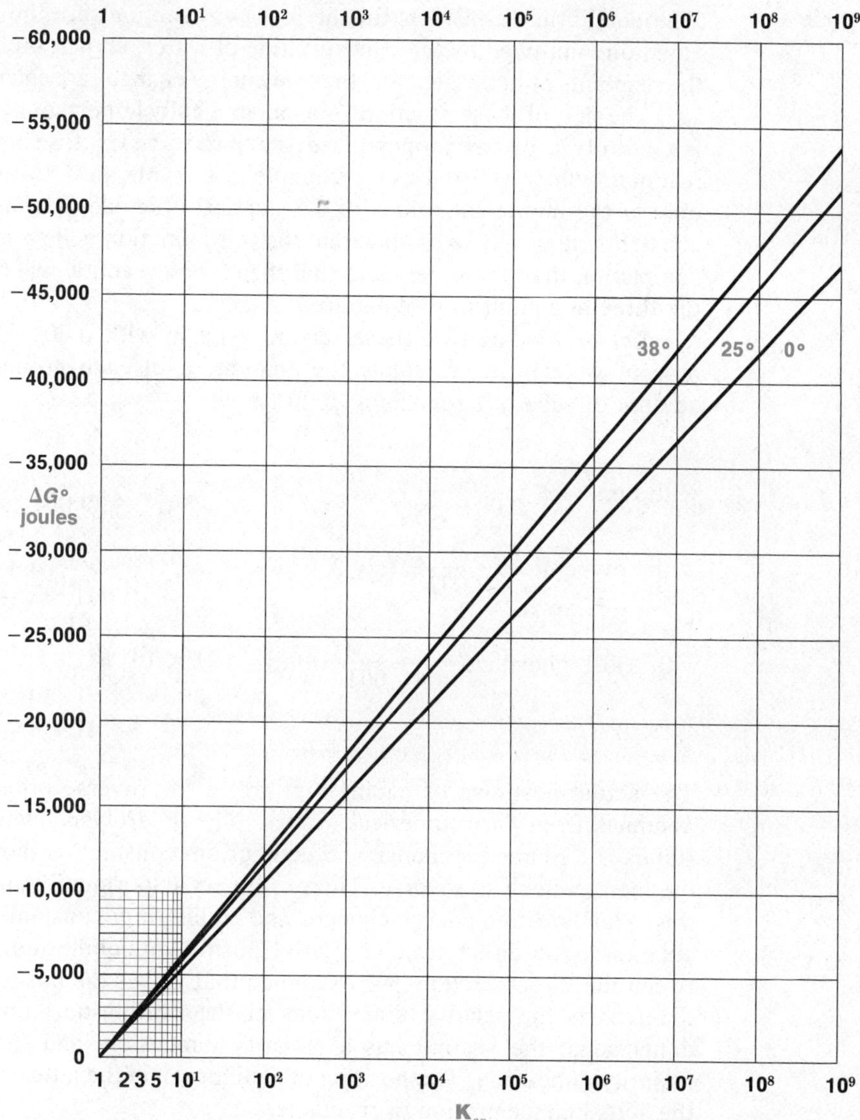

FIGURE 19–1 **The relation of standard free energy change and equilibrium constant at three temperatures. The ΔG^0 scale is linear; the K_{eq} scale is logarithmic. The same plot relates positive free energy changes and the reciprocal of the corresponding equilibrium constant.**

equilibrium lies so far in one direction that it would be difficult to measure the reverse reaction. But what about intermediate values? Here we have to be careful because the relative position of equilibrium depends upon the number of reactants and products. Whether one uses ΔG^0 or K_{eq}, there is frequently no substitute for actual calculation of the equilibrium position in making judgments about the feasibility of a reaction or the equilibrium concentrations of its components. Let us compare three important physiological reactions:

	$\Delta G^{0'}$	K'_{eq}
(1) fumarate + H_2O $\longrightarrow$ malate	−3,670	4.4
(2) polypeptide + H_2O $\longrightarrow$ 2 peptide fragments	−1,880	2.1 M
(3) 2 triose phosphates $\longrightarrow$ fructose 1,6-biphosphate	−22,850	1.01×10^4 M^{-1}

(The equilibrium constants for the first two reactions are those for the complete equation multiplied by the concentration of water; all pertain to the ionic forms of the reactants at neutrality, and the free energy changes are calculated accordingly.)

The first of these reactions is a metabolically important hydration; the second is a hydrolysis by an endopeptidase; the third is the fructose bisphosphate aldolase reaction, which we used as an example of enzyme mechanism (p. 266). Looking only at the numerical values of the standard free energy changes or equilibrium constants, it is easy to assume that the third reaction will go much farther toward completion than the others and that the second reaction will be least complete of the three at equilibrium. What are the facts?

Let us assume that these reactions begin with 0.001 M of each substrate (except water), and calculate the amount, x, of each product formed, and the amount of substrate remaining, $0.001 - x$:

with fumarate, $\frac{x}{0.001 - x} = K_{eq} = 4.4$

$$x = 0.81 \times 10^{-3}\text{ M}$$
$$0.001 - x = 0.19 \times 10^{-3}\text{ M}$$

with polypeptide, $\frac{x^2}{0.001 - x} = 2.1\text{ M}$

$$x = 0.99952 \times 10^{-3}\text{ M}$$
$$0.001 - x = 0.00048 \times 10^{-3}\text{ M}$$

with triose phosphates, $\frac{x}{(0.001 - x)^2} = 1.01 \times 10^4\text{ M}^{-1}$

$$x = 0.73 \times 10^{-3}\text{ M}$$
$$0.001 - x = 0.27 \times 10^{-3}\text{ M}$$

The actual positions of equilibrium are in the reverse order of what might be assumed from the numerical values. The hydrolysis, which has the lowest standard free energy change and equilibrium constant, is the most complete and the most difficult to reverse. There is a very important lesson to be learned from this: Standard free energy changes and equilibrium constants are a basis for calculation, not a direct scale of relative position of equilibrium. The distinction between the three reactions we examined that makes the position of equilibrium so different is the relative dimensions of the equilibrium constants. One has no dimensions, the second has a molarity dimension, and the third a reciprocal molarity dimension. The position of equilibrium of the latter two will depend upon the actual concentration of reactants.

There is one rule of thumb that is useful. Most hydrolyses are effectively irreversible under physiological conditions. The low concentrations of reactants found in real tissues favor cleaving one compound into two, even though the equilibrium constant and standard free energy change are low.

There is an additional complication. Even if the overall free energy change for a sequence would be negative, it is not always possible for it to occur. That is, a sequence may be impractical even if the overall reaction is going toward equilibrium. Consider $A \rightarrow B \rightarrow C$, with ΔG^0 for $A \rightarrow B$ of +50,000 J mole^{-1}, and for $B \rightarrow C$ of −50,000 J mole^{-1}. The overall standard free energy change is 0, and $K_{eq} = 1$. If the concentration of A is substantially greater than the concentration of C, one might think the reaction would go. It can't with real enzymes, because the K_{eq} for $A \rightarrow B$ is 1.7×10^{-9} (Fig. 19–1). Given a concentration of A at a typical value of 0.001 M, the concentration of B can never exceed 1.7×10^{-12} M. It is not possible to construct an enzyme to catalyze an effective reaction of B at that concentration, and the second reaction of the sequence therefore cannot happen at a significant rate, even though it has a very negative standard free energy change.

ENERGY-RICH PHOSPHATES

The use of nucleoside triphosphates and similar compounds as a sort of currency for energy exchange hinges upon the fact that relatively large amounts of free energy are liberated when these compounds are hydrolyzed. If we think on it, we can see that a high equilibrium constant for the cleavage of these compounds is necessary if cleavage is to be used for converting an otherwise endergonic process into a feasible exergonic reaction.

Suppose that $A + B \rightarrow C + H_2O$ is an endergonic reaction, with the equilibrium highly favoring hydrolysis of C rather than its synthesis. Perhaps the reaction can be made to go by coupling it with the cleavage of adenosine triphosphate (ATP) to adenosine diphosphate (ADP) and inorganic phosphate (P_i):

$$A + B + ATP \longrightarrow C + ADP + P_i.$$

If this is to be feasible, the hydrolysis of ATP:

$$ATP + H_2O \longrightarrow ADP + P_i$$

must be even more exergonic than the synthesis of C is endergonic. The sum of the free energy changes must be negative.

Let us examine the characteristics of those phosphate bonds that have a relatively high free energy of hydrolysis.

Phosphate Anhydrides and Esters. ATP can be hydrolyzed through several routes involving the cleavage of P—O bonds, some of which are summarized in Figure 19–2. The corresponding standard free energy changes and equilibrium constants for these reactions at 25°C are as follows:

Reaction	$\Delta G^{0\prime}$ (joules/mole)	K'_{eq}
(A) $ATP + H_2O \rightarrow ADP + P_i$	−36,800	2.8×10^6 M
(B) $ADP + H_2O \rightarrow AMP + P_i$	−36,000	2.0×10^6 M
(C) $ATP + H_2O \rightarrow AMP + PP_i$	−40,600	1.3×10^7 M
(D) $PP_i + H_2O \rightarrow 2P_i$	−31,800	3.7×10^5 M
(E) $AMP + H_2O \rightarrow A + P_i$	−12,600	1.6×10^2 M

$\mathbf{P_i}$ and $\mathbf{PP_i}$ are abbreviations for inorganic orthophosphate and pyrophosphate, respectively. The corresponding ionic forms that are most abundant at pH 7.4 are HPO_4^{2-} and $HP_2O_7^{3-}$.

The prime marks on the symbols for the standard free energy change and the equilibrium constant signify that these values apply to some empirical conditions; in this case, they were determined at a concentration of H^+ of $10^{-7.4}$ M (pH 7.4), and a concentration of Mg^{2+} of 10^{-4} M. The concentration of magnesium ion is important because it forms complexes with the phosphates, thereby altering the apparent equilibrium. The empirical equilibrium constants are calculated with the total nucleotide concentrations, including all ionic forms and the magnesium complexes.

The reactions are given here, and in most of the book, in terms of the predominant free ionic form present at pH 7.4, even though 0.9 of the ATP exists as the Mg complex. This is done in order to simplify recognition of reactions involving H^+. It is rather difficult to see what is going on when all of the ionic species are specified; for example, here is a more complete approximation of the hydrolysis of ATP at pH 7.4 and 0.5 mM Mg^{2+}:

$$0.88\ MgATP^{2-} + 0.014\ HATP^{3-} + 0.106\ ATP^{4-} + H_2O \longrightarrow 0.40\ MgADP^- + 0.06\ HADP^{2-} + 0.54\ ADP^{3-} + 0.04\ MgHPO_4 + 0.10\ H_2PO_4^- + 0.86\ HPO_4^{2-} + 0.44\ Mg^{2+} + 0.85\ H^+.$$

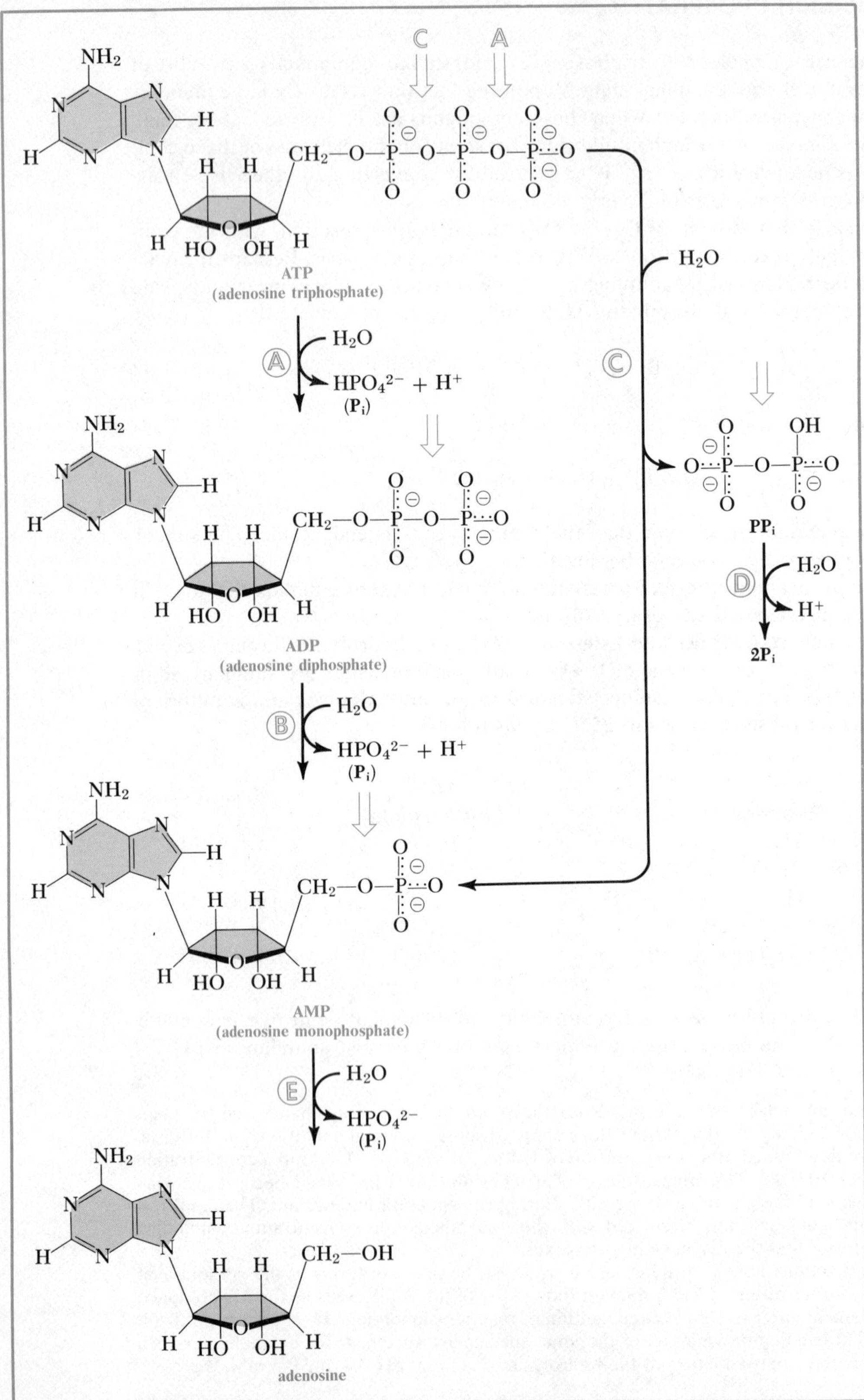

FIGURE 19–2 The pathways for hydrolysing phosphate bonds in ATP. The reactions are listed in the text by the same letters.

We see right away that there is something different about the hydrolysis of the phosphate of AMP, which forms free adenosine. If we look at the structures shown in Figure 19–2, we see that it alone represents the hydrolysis of a phosphate ester, a compound formally made from an alcohol (adenosine) and an acid (phosphoric acid). The other hydrolyses all represent the cleavage of an oxygen bridge between two phosphorus atoms, in other words, the cleavage of a **pyrophosphate** bond.

Formally, pyrophosphoric acid is an acid anhydride; it represents the combination of two molecules of phosphoric acid with the loss of water. More molecules of phosphoric acid may be added with the further loss of water to make extended chain polyphosphoric acids, which are polyanhydrides:

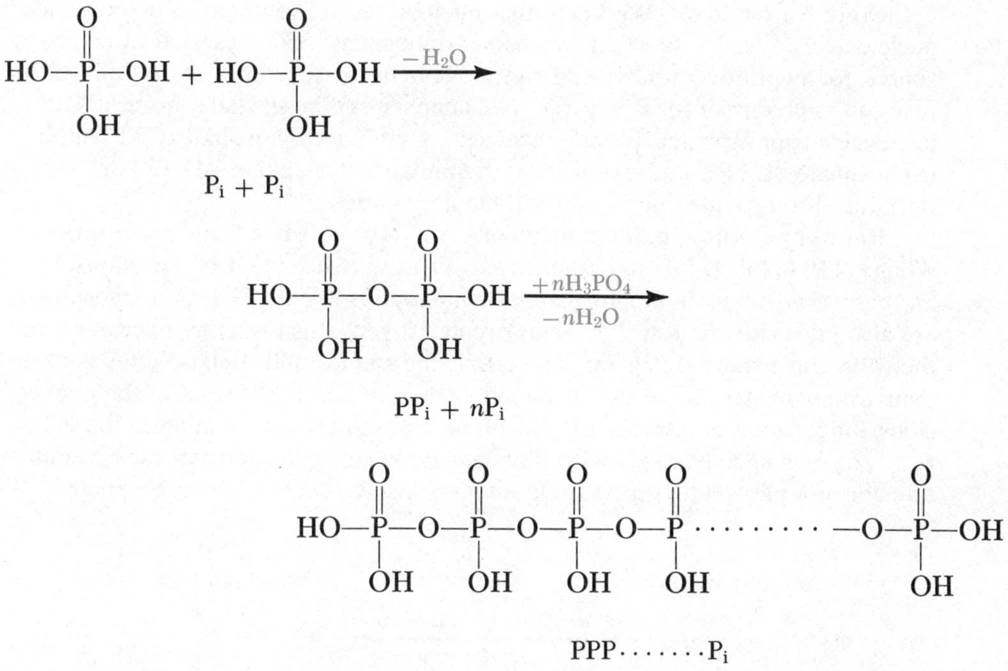

$$\mathrm{HO{-}\overset{\overset{\displaystyle O}{\|}}{\underset{\underset{\displaystyle OH}{|}}{P}}{-}OH + HO{-}\overset{\overset{\displaystyle O}{\|}}{\underset{\underset{\displaystyle OH}{|}}{P}}{-}OH} \xrightarrow{-H_2O}$$

$P_i + P_i$

$$\mathrm{HO{-}\overset{\overset{\displaystyle O}{\|}}{\underset{\underset{\displaystyle OH}{|}}{P}}{-}O{-}\overset{\overset{\displaystyle O}{\|}}{\underset{\underset{\displaystyle OH}{|}}{P}}{-}OH} \xrightarrow[-nH_2O]{+nH_3PO_4}$$

$PP_i + nP_i$

$$\mathrm{HO{-}\overset{\overset{\displaystyle O}{\|}}{\underset{\underset{\displaystyle OH}{|}}{P}}{-}O{-}\overset{\overset{\displaystyle O}{\|}}{\underset{\underset{\displaystyle OH}{|}}{P}}{-}O{-}\overset{\overset{\displaystyle O}{\|}}{\underset{\underset{\displaystyle OH}{|}}{P}}{-}O{-}\overset{\overset{\displaystyle O}{\|}}{\underset{\underset{\displaystyle OH}{|}}{P}}} \cdots\cdots\cdots {-}O{-}\overset{\overset{\displaystyle O}{\|}}{\underset{\underset{\displaystyle OH}{|}}{P}}{-}OH$$

$PPP \cdots\cdots P_i$

High-Energy, or Energy-Rich, Phosphates. We see by comparing the table and Figure 19–2 that the standard free energy is quite highly negative in each case in which a pyrophosphate bond is hydrolyzed. For this reason, such bonds and the compounds that contain them are said to be energy-rich or high-energy. (Some physical chemists become quite upset by this usage, believing the terms imply that the liberated free energy is concentrated in the bond or that the compounds are activated to unusual energy states, and neither of these conditions occurs. Most people who use the terms understand that they simply designate bonds or compounds with large standard free energies of hydrolysis, and this free energy change is the result of the changes in chemical potential of all components of the system, including water and the hydrogen ion.)

Bonds are sometimes designated with a squiggle symbol if their cleavage by hydrolysis has a large negative standard free energy change. Thus, ATP might be designated as A—P ~ P ~ P, because the final two bonds are of the energy-rich anhydride type, while the bond between adenosine and phosphorus is of an ordinary ester type. (We shall later see that there are other kinds of groups in addition to phosphate groups that are attached by energy-rich bonds in the sense that their hydrolysis has a large equilibrium constant.)

The Phosphate Pool

The conclusion we have reached is that pyrophosphates, especially the nucleoside pyrophosphates, act as a sort of medium of exchange for energy among reactions, being formed by what would otherwise be highly exergonic reactions and being cleaved to drive what would otherwise be endergonic processes.

Adenine Nucleotides. The nucleotides that are most commonly used to couple endergonic and exergonic processes are the adenosine phosphates. Thus, we shall see that the oxidation of fuels, photosynthesis, and related exergonic processes are coupled to the formation of ATP from ADP, whereas the formation of many compounds involves the simultaneous conversion of ATP to ADP or AMP, along with P_i or PP_i.

Other Nucleotides. We know that nucleotides are required as precursors of nucleic acids. We also saw that guanosine triphosphate (GTP) is used as an energy source for peptide synthesis and the movement of ribosomes in addition to its function as precursor for RNA. We shall later see syntheses that require still other nucleoside triphosphates as energy carriers. Cytidine triphosphate (CTP), uridine triphosphate (UTP), and even deoxythymidine triphosphate (dTTP) are sometimes used to provide energy for particular syntheses.

REGENERATION OF TRIPHOSPHATE BY NUCLEOSIDE DIPHOSPHOKINASE. When CTP, GTP, UTP, and other triphosphates are used to drive reactions, they are converted to their diphosphates or monophosphates. The monophosphates are also released when nucleic acids are destroyed. It is therefore necessary that there be some mechanism for converting the mono- and diphosphates back to their triphosphates for re-use, if metabolism is to continue. Most of the conversions hinge upon a transfer of phosphate from ATP. For example, there is a relatively non-specific nucleoside diphosphate kinase that catalyzes the reversible transfer of a phosphate group from ATP to various nucleoside diphosphates:

Nu—P∼P
nucleoside diphosphate

A—P∼P∼P
adenosine triphosphate

⇌ *nucleoside diphosphate kinase*, Mg^{2+}

Nu—P∼P∼P
nucleoside triphosphate

A—P∼P
adenosine diphosphate

The effect of the action of this enzyme is to make the ratio of concentrations of the triphosphates and diphosphates nearly the same for all the nucleotides.

NUCLEOSIDE MONOPHOSPHATE KINASES. Similarly, nucleoside monophosphate kinases catalyze the transfer of a phosphate group from ATP to various nucleoside monophosphates:

Nu—P
nucleoside monophosphate

A—P∼P∼P
adenosine triphosphate

⇌ *nucleoside monophosphate kinases*, Mg^{2+}

Nu—P∼P
nucleoside diphosphate

A—P∼P
adenosine diphosphate

Each monophosphate reacts with a specific enzyme (an AMP kinase, a GMP kinase, a CMP kinase, and so on). The AMP kinase is of particular importance because of the involvement of the adenosine phosphates in so many processes. It is the most active of the monophosphate kinases and catalyzes the reaction:

$$\text{ATP} + \text{AMP} \longleftrightarrow 2\ \text{ADP}.$$

We see that it would require two phosphorylations to convert the resultant ADP completely to ATP. Therefore, we can make a mental reminder that any reaction forming AMP from ATP is effectively spending 2 moles of high-energy phosphate per mole of AMP formed, because it will take an additional mole of ATP to convert the resultant AMP back to ADP before it can be further phosphorylated to regenerate the original ATP.

Complete Interrelationships. The interplay between these various high-energy phosphates is summarized in Figure 19–3. The involvement of ATP in the re-phosphorylation of the other kinds of nucleotides effectively means that the expenditure of any of the high-energy phosphates for endergonic processes, as represented on the right of the figure, is equivalent to the conversion of ATP to ADP, even though the rates of these processes may be independently regulated. The ultimate result in each case is the replenishment of the high-energy phosphate through the action of highly exergonic processes that are coupled to the phosphorylation of ADP, as shown on the left.

Phosphagens. The concentrations of the nucleoside di- and triphosphates represent only a small reserve for the energy requirements of cells. These compounds are intermediates, not stored fuels. Typical concentrations in cells range around several millimoles of ATP, tenths of a millimole of ADP, and 0.01 (or less) of a millimole of AMP per kilogram of active tissue (excluding gross structural components such as cartilage in animals, extracellular structures in protists, and woody tissue in plants). The concentrations of the guanine, cytosine, and uracil

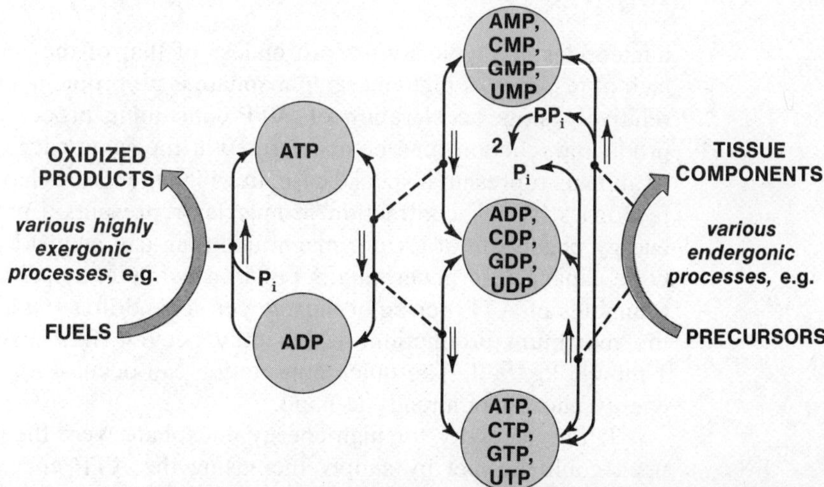

FIGURE 19–3 **The nucleoside phosphate pool. Exergonic processes, such as the oxidation of fuels, drive the phosphorylation of ADP to form ATP (*left*). The ATP is used in turn to phosphorylate nucleoside monophosphates to form nucleoside diphosphates, and to phosphorylate nucleoside diphosphates to form nucleoside triphosphates. The nucleoside triphosphates are used to drive various endergonic processes, such as the formation of tissue components from precursors. In doing so, either a phosphate or a pyrophosphate group is split from the nucleoside triphosphate.(Note that double-headed arrows are used to indicate potentially reversible reactions. The large arrows alongside indicate the direction of physiological flow.)**

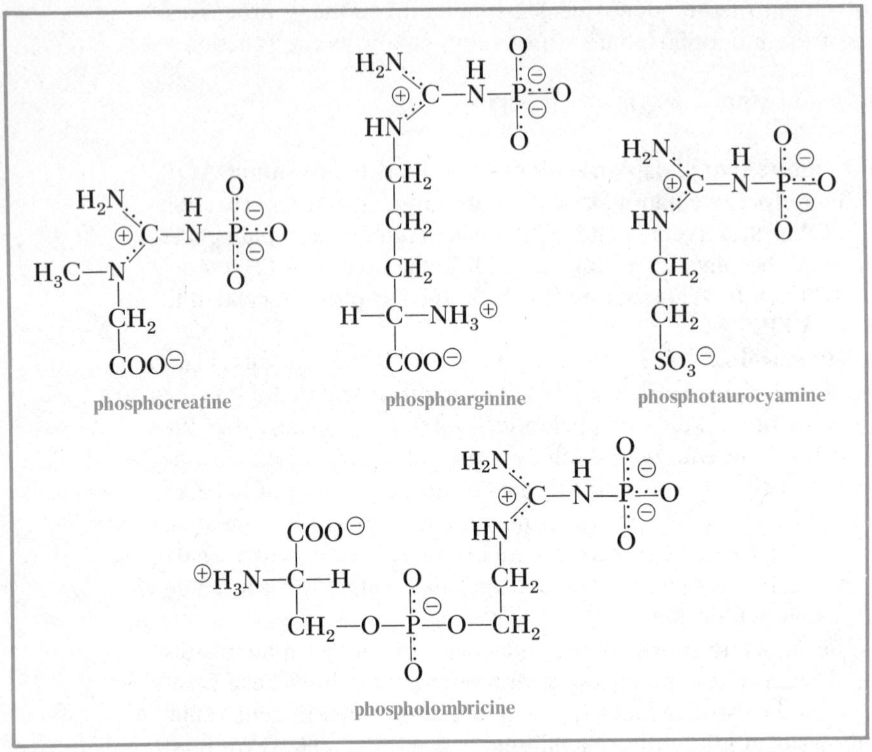

phosphocreatine

phosphoarginine

phosphotaurocyamine

phospholombricine

FIGURE 19–4 **The phosphagens are phosphorylated derivatives of guanidinio compounds, used to store high-energy phosphate, especially in muscles. The compounds used as phosphagens differ in various phyla.**

nucleotides are typically a tenth or less of that of the adenine nucleotides. The lack of reserves of high-energy phosphate is no problem for cells that require only relatively slow acceleration of ATP-consuming processes, because the ATP-producing reactions can be accelerated at the same pace. However, the muscles of animals represent a special case, in which the tissues have evolved to give rapid responses. Quick contraction of muscles represents a prompt demand on high-energy phosphate of greater magnitude than can be met by any reasonable metabolic capacity for generating ATP. For example, a man can expend as much as 6 mmoles of ATP per kg of muscle per second in a sudden burst of activity, but the maximum production, which only occurs after several seconds' delay, is 1 mmole $kg^{-1}s^{-1}$. The quick contraction can occur only at the expense of high-energy phosphate already at hand.

If the necessity for high-energy phosphate were the only consideration, the need could be met by simply increasing the ATP concentration in the tissue. However, this would affect the rate of all of the processes in which ATP is involved, and cause an intolerable disruption of the general metabolism. Hence, different compounds have been evolved to serve as a high-energy phosphate store. These compounds are known collectively as the phosphagens, and they are substituted phosphoguanidinium compounds (Fig. 19–4). (These compounds are named both as the phosphoguanidines and as the guanidine phosphates. For example, phosphocreatine and creatine phosphate mean the same thing.)

Phosphocreatine is the phosphagen found in most vertebrates and in some invertebrates. Phosphoarginine is found in many other invertebrates. The other compounds listed occur in fewer phyla. The phosphagens typically occur in tens of millimoles per kilogram of muscle. They also occur in brain, presumably to protect this vital tissue from an accidental transitory lack of ATP. Their function as energy stores hinges upon enzymes that catalyze the equilibration with ADP, such as the creatine kinase found in mammalian muscles and brain:

$$\underset{\text{creatine}}{H_2N{\cdots}\overset{\oplus}{C}(NH_2){\cdots}N(CH_3)-CH_2-COO^{\ominus}} \xrightleftharpoons[\text{creatine kinase}]{\text{ATP} \quad Mg^{2+} \quad \text{ADP}} \underset{\text{phosphocreatine}}{H_2N{\cdots}\overset{\oplus}{C}(-NH-PO_3^{2\ominus}){\cdots}N(CH_3)-CH_2-COO^{\ominus}}$$

This reaction effectively expands the supply of high-energy phosphate in a readily available way, and the supply can be rebuilt at rest for use in the next rainy day.

Role of Mg^{2+}. The kinases mentioned here, like most enzymes catalyzing the transfer of phosphate groups, require Mg^{2+} for activity. (This ion is required for some other enzymes, as well.)

FURTHER READING

Boyer, P. D., H. A. Lardy, and K. Myrbäck, eds.: (1960) *The Enzymes,* 2nd edition, Academic Press. The following articles are of interest: M. J. Johnson, *Enzyme Equilibria and Thermodynamics,* Vol. 3, p. 407. Written by one of the pioneers in applying clear thinking to the subject. Directed toward the beginning professional student. R. M. Bock, *Adenine Nucleotides and Properties of Pyrophosphate Compounds,* Vol. 2, p. 3. Detailed summary.

Lipmann, F.: (1941) *Metabolic Generation and Utilization of Phosphate Bond Energy.* Adv. Enzymol., *1*: 99. The classic introduction of the concept.

Interunion Commission on Biothermodynamics: (1976) *Recommendations for Measurement and Presentation of Biochemical Equilibrium Data.* Quart. Rev. Biophys., *9*: 439.

20 MUSCULAR CONTRACTION

High-energy phosphate is used for three general purposes: mechanical work, the generation of concentration gradients, and chemical syntheses. All are important, but of the three, mechanical work represents by far the greatest peaked load in most animals. A human working with a power output that can be continued for several hours may be cleaving pyrophosphate bonds in his skeletal muscles at six or more times the rate at which they are being cleaved in all other tissues; a sudden burst of activity, as in a jump, may push the rate of cleavage in muscles toward 100 times that in other tissues. This chapter is concerned with the mechanism by which the conversion of ATP to ADP and P_i is transduced into rapid application of force over a distance.

STRUCTURE OF MUSCLES

Muscles contract because they contain sets of filaments that are forced to slide over each other by a cleavage of ATP into ADP and P_i. This is the basic arrangement in all kinds of muscles, but there are variations on this theme. Smooth muscle cells, which contract relatively slowly and maintain tension over long periods of time without much additional expenditure of energy, are packed with filaments. The structural details of smooth muscle cells are now beginning to be described. Striated muscle cells, themselves organized into fibers, are subdivided into **myofibrils.** The myofibrils are made by lengthwise repetition of a fundamental arrangement, the **sarcomere.** A sarcomere is the longitudinal segment of a myofibril between crosswise divisions known as **Z-bands,** which are rigid, thin circular plates of protein (Fig. 20–1). The Z-band is an anchor between sarcomeres; a set of **thin filaments** bristles out lengthwise from each side; these filaments are polymers of a protein known as **actin.** In effect, two cylindrical bundles of fibers are cemented head-to-head by the Z-band. Successive paired bundles, with their central Z-bands, are stacked in the myofibril at a spacing such that there is a gap between their free ends. This gap is filled with a bundle of another kind of filament, the **thick filament,** which intermeshes with the ends of the thin filament bundles on each side. Thick filaments are composed of a protein known as **myosin;** they are held in a bundle by central protein cross-links known as the M-line.

The striated muscle fibril therefore can be pictured as alternating bundles of thick filaments and double-headed bundles of thin filaments, pushed together so that the ends of thick and thin filaments overlap. The separate filaments and the

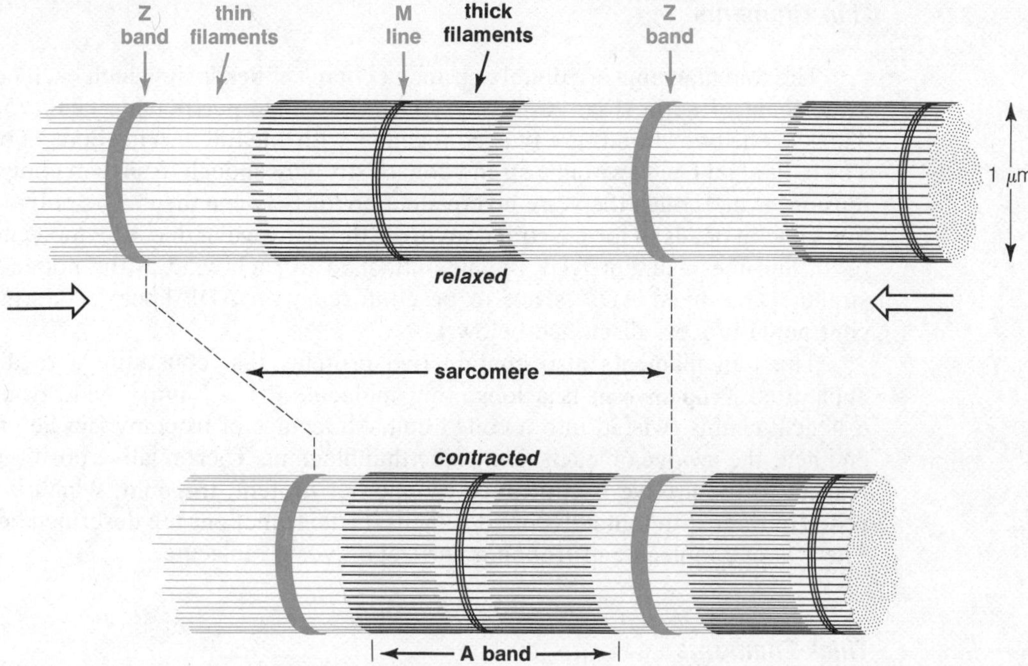

FIGURE 20–1 **Structure of striated muscles.**

overlap create microscopically visible striations in the myofibrils of striated muscle.

In sum, a striated muscle fiber is a bundle of a few thousand myofibrils that run its length. Each myofibril is in turn an end-to-end stack of a few thousand sarcomeres. Each sarcomere is a bundle of a few thousand filaments, which are the actual contractile elements of the muscle.

Types of Striated Muscles. Although we shall emphasize their common features, not all striated muscles are alike. The fibers of the heart (cardiac fibers) differ from those in skeletal muscles. Skeletal muscles are composed of at least three types of fibers in many animals. Two of the types are red in color, compared to a relatively white third type. The **red fibers** have a high capacity for using oxygen and readily burn fats as well as carbohydrates. The **white fibers** utilize glycogen or glucose almost exclusively. The red color comes from higher concentrations of myoglobin and other hemoproteins used for oxygen transport and consumption. One of the red fiber types is found mainly in muscles that contract slowly and steadily. The other red fibers and the white fibers exhibit faster twitches. (We shall make much of these distinctions in later chapters.) Migratory birds such as ducks have mostly red fibers in their flight muscles, whereas ground-living birds, like chickens, have mostly white fibers. (However, chickens run, and have mostly red fibers in their thigh muscles.) The hind-leg muscles of hares are rich in white fibers. However, if muscles rich in white fibers are experimentally stimulated at a steady low frequency over extended periods, the fibers become predominantly the red type. Other fibers are intermediate in color and composition; they predominate in muscles with relatively slow and sustained contractions.

Human muscles have a mixture of red and white fibers that resemble those in other animal muscles, but the distinctions are not as extreme.

Thin Filaments

The thin filaments are double-stranded chains of beads, in which each bead is a molecule of actin (Fig. 20–2*A*). Actin has a single polypeptide chain (M. W. 42,000). The two strands are twisted together with one full turn in about 13 beads. The individual beads within a strand cohere strongly enough to sustain contractile tension, even though there are no covalent bonds between them. The polymerization of actin beads to form a strand involves the cleavage of one ATP molecule per bead, and the resultant ADP remains attached to each bead in the polymerized strand. (This fixed ADP is not to be confused with ADP liberated during the contractile process discussed below.)

The thin filaments also contain two proteins that constitute a regulatory apparatus. **Tropomyosin** is a long, thin molecule (41 x 2 nm), made from two α-helical chains twisted into a coiled coil. Molecules of tropomyosin lie end-to-end near the groove on each side of the thin filament. Their relative position with respect to the groove is controlled by another protein, **troponin,** which is composed of three different polypeptide chains. (Their functions are described below.) There is one molecule of troponin per tropomyosin molecule.

Thick Filaments

The thick filaments are built from stacked molecules of myosin (Fig. 20–2*A* and *B*). Myosin is a large protein (470,000 M.W.) with twin globular heads attached to a single long shaft. The shafts are stacked into bundles from which the heads protrude (Fig. 20–2*B*); adjacent molecules overlap in such a way that

---

Myosin
470,000 M.W.
Combined to form thick filaments, 1,600 × 14 nm
Long shaft with two globular heads at one end
Heavy chains (two, identical):
- C-terminal α-helix paired in coiled coil to form 130 × 2 nm shaft
- N-terminal portion in separate globular heads, 20 × 5 nm
- Some lysine and histidine side chains are methylated

Light chains (four, two each of two kinds; different myosins have different kinds):
- One of each kind in each globular head
- One chain phosphorylated

Catalyzes hydrolysis of ATP to ADP and P_i in isolated form (ATPase activity)

Actin
42,000 M.W. globular protein, 4–5 nm diameter
Polymerize into double chains, like twisted strands of beads, to make thin filaments
Polymerization involves cleavage of ATP, and ADP remains on each monomeric unit

Tropomyosin
35,000 M.W. coiled coil of two α-helical chains
41 × 2 nm length covers seven actin molecules
Draped lengthwise on each side of thin filaments
Lateral position determines if contraction can occur

Troponin
Regulates position of tropomyosin
78,000 M.W. globular molecule
One molecule per tropomyosin molecule
Three subunits
- T subunit binds tropomyosin
- C subunit binds Ca^{2+}
- I subunit also required to regulate contact of actin and myosin by shifting attached tropomyosin

---

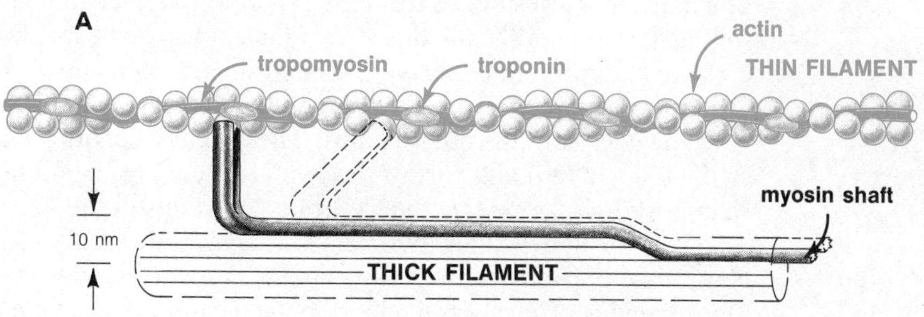

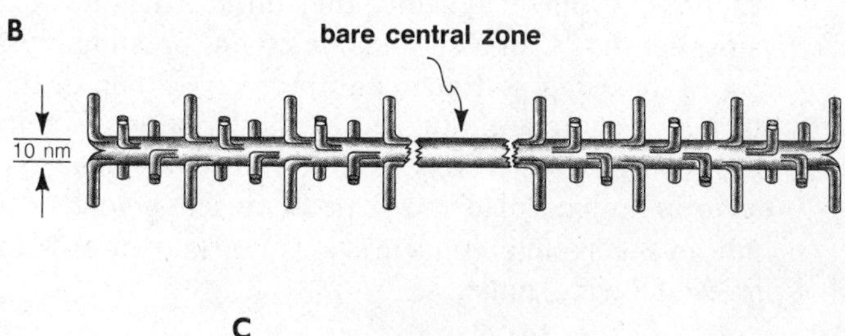

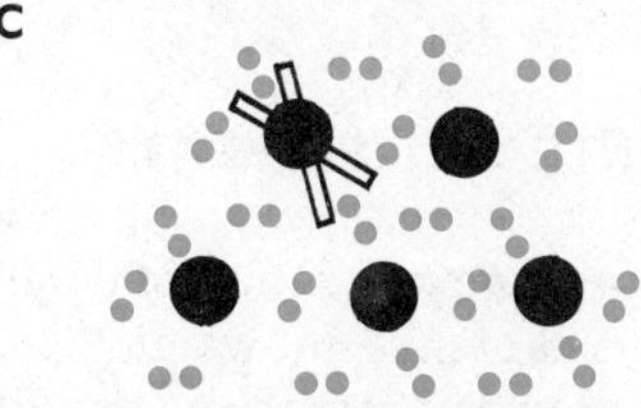

FIGURE 20–2 **A. The thin filaments (*blue*) are polymers of globular actin molecules in two twisted strands like chains of beads. Long molecules of tropomyosin lie near the groove of the strands on each side. Each molecule of tropomyosin, which covers seven actin monomers, is bound with a molecule of troponin, a more globular protein. Troponin and tropomyosin constitute the regulatory apparatus controlling contraction. The thick filaments (*black*) are bundles of myosin molecules. Myosin is a protein with a long shaft, made of coiled α-helixes, and a pair of identical globular heads. The heads protrude at right angles in a relaxed muscle. When a muscle contracts, the heads make contact with an actin monomer and flex toward the shaft (dashed lines), the end of which lifts from the thick filament. The flexure moves the thin filament toward the center of the thick filament.**

B. Myosin molecules are spaced within the thick filaments so that the heads protrude in a spiral arrangement. The heads are oriented toward the ends of the filaments with the shafts toward the center, so there is a zone in the middle devoid of protruding heads.

C. Each thick filament in a myofibril is surrounded by a hexagonal array of thin filaments, as sketched here in cross section. The two actin strands in a thin filament are oriented so that one faces each myosin head. (The head is shown to clasp the actin monomer, but the actual arrangement is not known.) Since the thin filaments are twisted in a spiral and the myosin heads also occur in a spiral arrangement, a cross section through the next head beyond the one shown would have all of the bead pairs and the contacting head rotated from the position shown.

successive heads appear at 60° angles around the shaft and farther toward the ends. The positions of the heads therefore make spirals around the bundled shafts. The molecules are stacked from the center of the thick filaments toward their ends so that a portion of the middle of the filament has no heads, but the two ends do. A cross-linking protein connects the central bare segments in the M-line.

Myosin is made from six polypeptide chains. Two of these are identical and much heavier than the other chains. These heavy chains are used to construct the shaft (130 × 2 nm) and part of the two identical heads of a myosin molecule. The shaft portion is formed by the C-terminal portions of the heavy chains folding into α-helixes, which further twist into a long coiled coil. A shorter part of the N-terminal portions of the heavy chains is incorporated into the globular heads. The reminder of each identical globular head is constructed from two different light chains. The myosin molecules in different muscle fiber types have the same heavy chains.

The light chains in each myosin head evidently determine the contractile response in some way, since they differ from one fiber type to another. One kind is phosphorylated on a serine side chain, presumably for regulatory purposes.

Myosins can differ in another way. Some of the lysyl and histidyl groups in the heavy chain are modified by methylation of their side chains, and the number of added methyl groups is determined by fiber type. Trimethyllysine residues occur in both cardiac and skeletal muscle myosin, but monomethyllysine is found only in the skeletal muscles. It is not known how these changes affect the properties of the molecule.

THE CONTRACTILE EVENT

The fundamental action of contraction is a binding of the actin chains by the globular head of myosin, which then bends on its base pulling the actin chain into a new position closer to the center of the sarcomere. In order to accommodate this movement, the shaft of myosin also can bend at another location indicated by a dashed outline in Figure 20–2*A*. Since the bundles of thin filaments overlap opposite ends of one bundle of thick filaments, the effect of the movement is to pull both sets of thin filaments toward each other over the thick filaments, which do not move. The sarcomere is shortened without altering the length of either the thin or the thick filaments. The displacement is accompanied by a release of ADP and P_i.

Calcium Ions Trigger Contraction. In the absence of stimulation, muscles are prevented from contracting by the positions of troponin and tropomyosin. One of the three subunits in troponin has a high affinity for tropomyosin, and a second subunit acts as an inhibiting component. This second subunit must be present if the troponin complex is to react with the actin heads in such a way as to drag tropomyosin into a position where it blocks the interactions between actin and myosin that are necessary for contraction to occur. One tropomyosin covers seven actin heads.

The third subunit of troponin can bind Ca^{2+}. When it does, the troponin conformation changes so as to move the attached tropomyosin within the thin filament groove. Effective contact between actin and myosin can then occur.

In a resting fiber, the 0.1 μM concentration of Ca^{2+} is well below the half-saturation level of troponin (5 μM), and troponin holds tropomyosin out of the groove. When a neuron stimulates the fiber through the motor endplate, Ca^{2+} is discharged from the sarcoplasmic reticulum surrounding the myofibrils, raising the concentration to 10 μM, or more, and causing troponin to move tropomyosin out of the way.

Sequence of Events. Although the details of mechanism are not known, the major events during muscle contraction appear to be these:

Contracted myosin heads are in a low-energy conformation, which is converted to the high-energy relaxed conformation by reacting with ATP (as the Mg complex) (Fig. 20–3*A*). The reaction probably involves phosphorylation of the myosin head, but the MgADP that is produced also remains bound to the myosin, along with phosphate, upon creation of the high-energy conformation (Fig. 20–3*B*). (When isolated myosin reacts with ATP, the intermediate complex reacts

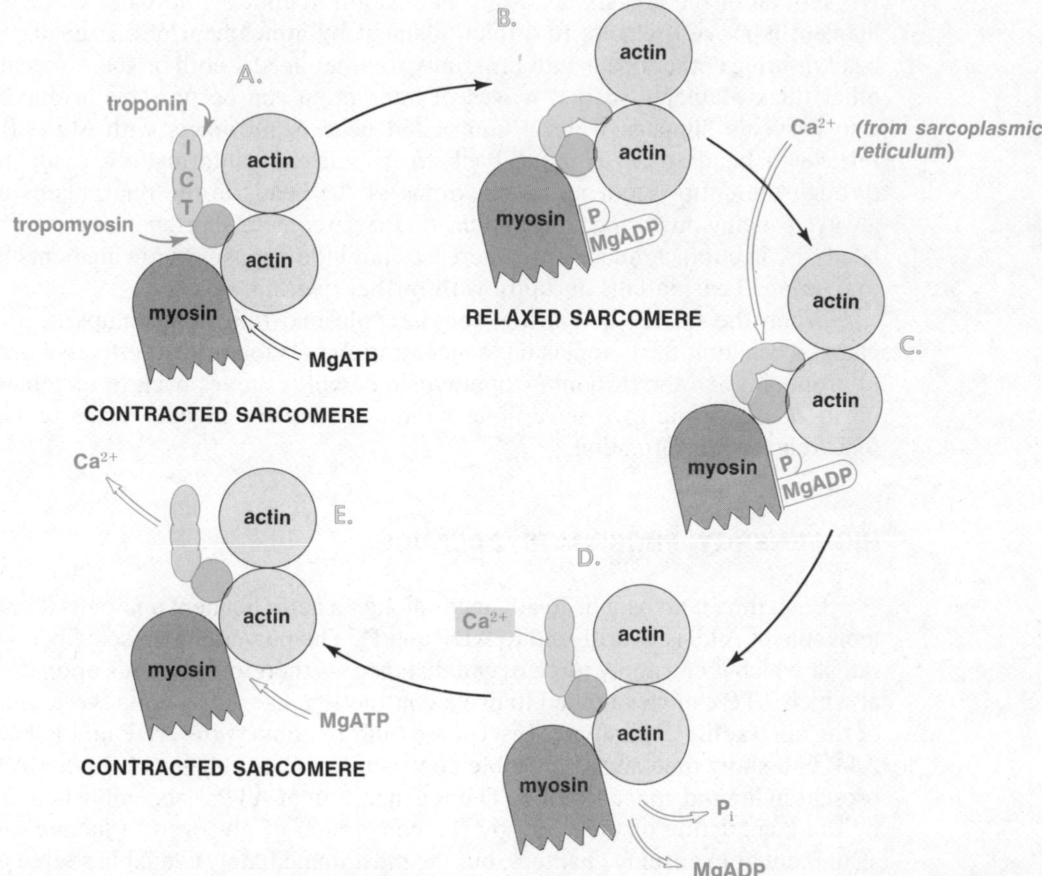

FIGURE 20–3 Events in muscular contraction. *A.* In a contracted muscle, myosin heads are in contact with actin monomers. The tropomyosin molecule, seen in cross section, does not interfere with the contact. Relaxation begins with a binding of ATP as the magnesium chelate.

B. ATP is cleaved on myosin to create a high-energy relaxed conformation in which the protein is phosphorylated and MgADP is still bound. Troponin binds to actin as a result, dragging the attached tropomyosin between actin and myosin so as to prevent their interaction.

C. Contraction is stimulated by a binding of Ca^{2+}, released from the sarcoplasmic reticulum, to the C component of troponin.

D. The binding of Ca^{2+} to troponin changes its conformation so that it no longer holds tropomyosin in place between actin and myosin. When actin and myosin interact, myosin reverts to its low-energy contracted conformation, with loss of the phosphate group and MgADP.

E. When Ca^{2+} is released from troponin and MgATP is bound, the cycle can begin again.

with water to release both ADP and P_i. Free myosin therefore acts as an ATP hydrolase, or ATPase.)

Ca^{2+} is released from the sarcoplasmic reticulum upon excitation and is bound by the troponin-tropomyosin complex, which moves so that actin can interact with myosin (Fig. 20–3*C*). The combination of the myosin heads with actin displaces the phosphate group as P_i and the bound ADP (Fig. 20–3*D*). The loss of phosphate and ADP causes the myosin heads to shift to their bent, or low-energy, configuration, pulling the attached actin along and contracting the muscle. The twin heads of a myosin molecule presumably act in some concerted mechanism with one actin bead, but details are not known. After contraction, Ca^{2+} leaves, and MgATP is once more bound to initiate a new cycle of relaxation and contraction (Fig. 20–3*E*), which will repeat so long as the $[Ca^{2+}]$ remains high.

Not all of the myosin heads are in position to contract at once. When a thin filament is moved relative to a thick filament by attachment of one head to one bead, it brings other beads into proximity to other heads, both on the same and on other thick filaments so that waves of interaction can occur. This arrangement also prevents slippage when a contracted head again reacts with MgATP and releases a bead so as to move back to its relaxed conformation, ready to go through the entire sequence again. Some of the heads in the thick filament are always hanging on during contraction. As the sarcomere shortens, more and more heads are in contact, and tension increases until the opposing actin filaments begin to overlap. Tension falls abruptly with further overlap.

When the nerves quit firing, the sarcoplasmic reticulum pumps in its discharged calcium; the cytoplasmic concentration falls too low for effective binding by troponin, and the troponin-tropomyosin complex moves back to its inhibitory position. Actin and myosin no longer interact, and the myosin heads remain in their relaxed conformation.

HIGH-ENERGY PHOSPHATE BALANCE

Each time a myosin head advances along an actin filament and pulls it back, a molecule of ATP is hydrolyzed to ADP and P_i. The power of a muscle fiber — the rate at which it can apply force over a distance — therefore depends upon the rate at which ATP can be supplied to drive contraction, as well as upon the properties of the contractile apparatus. Muscles are built to convert the ADP and P_i back to ATP in a short time so as to enable contraction to continue even though ATP is present in limited concentration. The regeneration of ATP is accomplished in part by the combustion of fuels and by the conversion of glycogen to lactate, as we shall recount in ensuing chapters, but the most immediately available source is the store of phosphocreatine (or other phosphagens in lower animals) contained within the muscle cells. Phosphocreatine is not used for any purpose other than the phosphorylation of ADP, and it has a relatively low affinity for Mg^{2+} or Ca^{2+}, so its concentration can be varied over wide ranges without disturbing other metabolic processes or grossly changing the divalent cation concentration.

Creatine Kinase Equilibrium

Phosphocreatine is a store of high-energy phosphate that can be used to create more ATP when ATP is being hydrolyzed by muscular contraction and

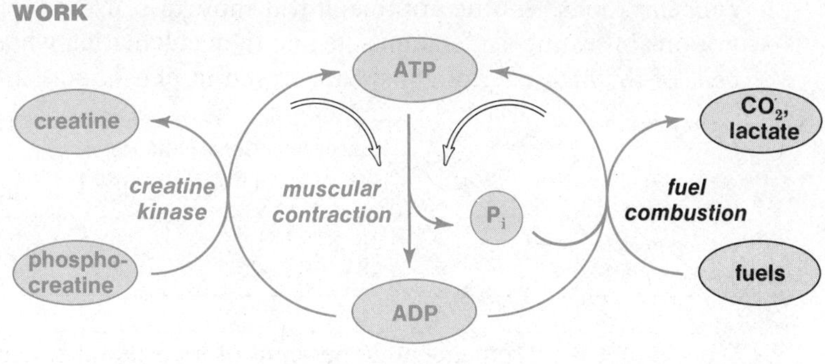

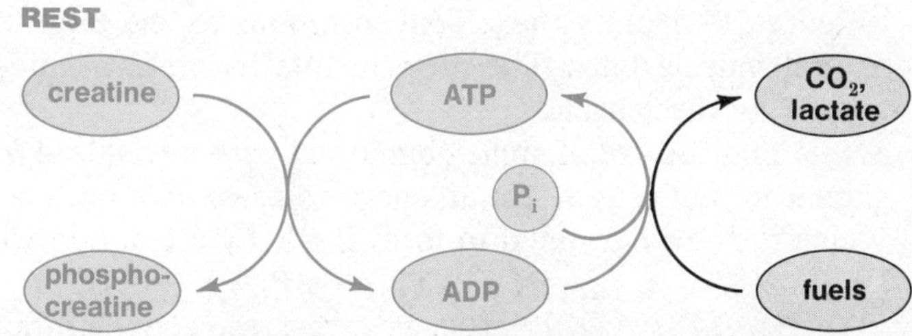

FIGURE 20–4 **The controlling event in a working muscle is the hydrolysis of ATP to ADP and P_i by muscular contraction (*center top*). The resultant increase in concentration of ADP will cause phosphate to be transferred from phosphocreatine by creatine kinase (*left top*), and will also cause increased fuel utilization to convert the ADP and P_i back to ATP (*right top*). Both processes serve to provide more ATP for continued muscular contraction.**

At rest, the utilization of fuels (*bottom right*) can create a high concentration of ATP relative to ADP, since ATP is not being hydrolyzed rapidly by muscular contraction. The resultant shift in the creatine kinase equilibrium causes phosphate to be transferred from ATP to creatine, thereby rebuilding the store of phosphocreatine.

regenerated when the muscle is at rest (Fig. 20–4). There is sufficient creatine kinase in striated muscle to keep its reaction near equilibrium, except perhaps in the most violent bursts of activity.

There is more to the use of phosphocreatine than a simple equilibration with ATP; it is a compound designed to maintain the concentration of ATP relatively constant until a muscle nears exhaustion. This property hinges upon the equilibrium of the creatine kinase reaction favoring the formation of ATP:

$$\text{ADP} + \text{phosphocreatine} \rightarrow \text{ATP} + \text{creatine} \qquad K'_{eq} = 54.$$

(The equilibrium constant hinges upon the H^+ and Mg^{2+} concentration; the quoted value is typical for intracellular conditions, with pH = 6.9 and $[Mg^{2+}]$ = 0.5 μM.)

At first glance, the high equilibrium constant may seem to make phosphocreatine a less effective energy store. In resting muscles, only two thirds of the total creatine is present as phosphocreatine; one third still remains as free creatine even though the ratio [ATP]/[ADP] is 100 or more. It is precisely this disparity of concentration ratios that enables the maintenance of nearly constant ATP concentrations. A little arithmetic will show this, using typical values for concentrations in resting skeletal muscle and then calculating what happens when 90 per cent of the high-energy phosphate stored in phosphocreatine is discharged.

	Millimolar Concentrations				
	[PCr]	[Cr]	[ATP]	[ADP]	[ATP][creatine] / [ADP][phosphocreatine]
resting muscle	20	10	5.94	0.055	54
working muscle	2	28	4.52	1.17	54

We see that ATP remains at 76 per cent of its original concentration, even though the phosphocreatine concentration has fallen to 10 per cent of its original value. (The sum of the ATP and ADP concentrations is slightly less in working muscle because of the AMP kinase reaction discussed below.)

Figure 20–5 plots these concentrations as the total high-energy phosphate store in a muscle falls. The drop in total store is a guide to the power being developed by the muscle.

Note that the use of some compound with a standard free energy of hydrolysis similar to that of ATP for an energy store would not do. Assume a compound X, which is phosphorylated to form P–X by a reaction with $K'_{eq} = 1$:

$$X + ATP \longleftrightarrow P - X + ADP$$

When [ATP]/[ADP] = 100, as at rest, the concentration of P–X will also be 100 times that of X. However, in order to use half of the stored P–X, the concentration of ATP must also fall to half of its original value. This explains why phosphagens, with their seemingly unfavorable equilibrium, have evolved.

AMP Kinase Equilibrium

Striated muscles contain as much AMP kinase, catalyzing the reaction:

$$ATP + AMP \xrightarrow{Mg^{2+}} 2\ ADP \qquad K'_{eq} = 1.0$$

as there is creatine kinase, so the three adenosine phosphates are kept nearly at equilibrium.

The AMP kinase provides a means of using the second high-energy phosphate bond in ATP for muscular contraction, like this:

$$2\ ATP \xrightarrow{\text{muscular contraction}} 2\ ADP + 2\ P_i$$

$$2\ ADP \xrightarrow{\text{AMP kinase}} ATP + AMP$$

$$\text{SUM:} \quad ATP \longrightarrow AMP + 2\ P_i$$

In order for this reaction to contribute a significant amount of high-energy phosphate, the ADP concentration must be substantially elevated. We saw in the preceding section that this can occur only when the phosphocreatine supply is nearly exhausted. Therefore, AMP only becomes a major fraction of the total adenosine phosphates when a muscle is working near its limit, and this is also shown in Figure 20–5.

AMP Deaminase. The amount of high-energy phosphate that can be drained for extreme exertion is increased in another way. An enzyme is present in muscles

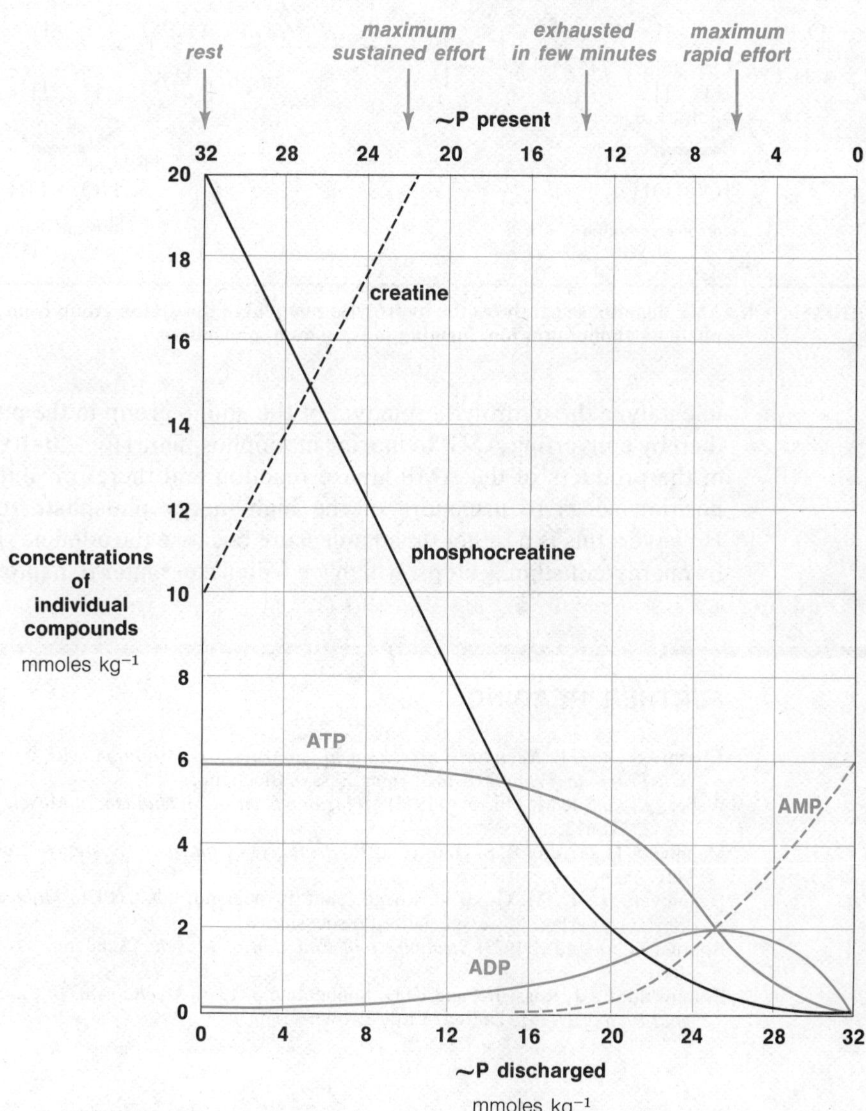

FIGURE 20–5 **Changes in the high-energy phosphate pool of working skeletal muscles, as increasing amounts of high-energy phosphate are depleted. The concentration of total high-energy phosphate represents a steady-state balance between the rate of depletion by the muscle power output and the rate of production by utilization of fuels. The greater the work load, the lower the steady-state concentration. Estimates of the concentrations corresponding to various loads are indicated by blue arrows at the top.**

Most of the depletion of high-energy phosphate is due to a fall in phosphocreatine concentration even at quite heavy work loads. Marked drops in ATP concentration, with concomitant rises in the concentration of ADP and AMP, occur only when the muscles are working at rates causing exhaustion in minutes, or less.

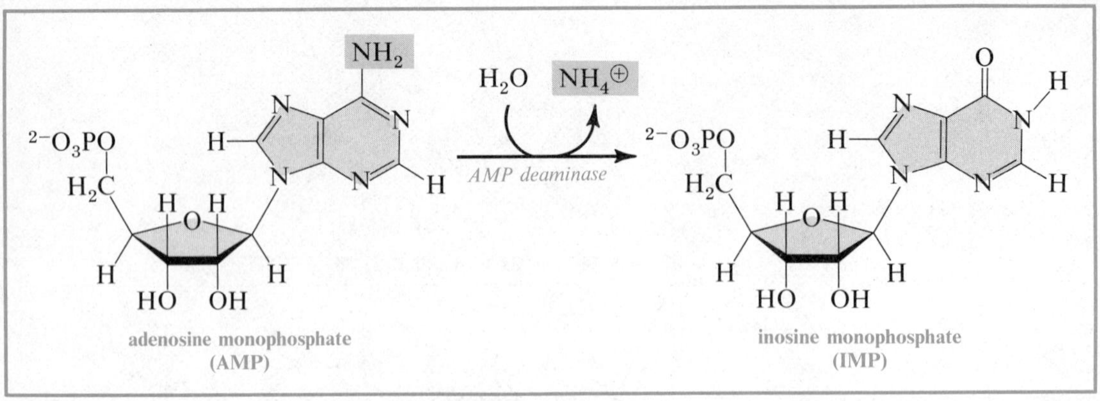

FIGURE 20–6 **AMP deaminase catalyzes the hydrolytic removal of the amino group from adenosine monophosphate as ammonium ion, forming inosine monophosphate.**

to catalyze the hydrolytic removal of the amino group in the purine ring of AMP, thereby converting AMP to inosine monophosphate (Fig. 20–6). This removes one of the products of the AMP kinase reaction and therefore shifts the equilibrium position so as to use more of the high-energy phosphate remaining in ADP. However, this is a rather drastic measure because the adenine ring must be rebuilt by energy-consuming steps, which we will discuss later (Chapter 34).

FURTHER READING

Ebashi, S.: (1974) *Regulatory Mechanism of Muscle Contraction with Special Reference to the Ca-Troponin-Tropomyosin System.* Essays Biochem., *10*: 1.

Weber, A., and J. M. Murray: (1973) *Molecular Control Mechanism in Muscle Contraction.* Physiol. Rev., *53*: 612.

Mannherz, H. G., and R. S. Goody: (1976) *Proteins of Contractile Systems.* Annu. Rev. Biochem., *45*: 427.

Heilmeyer, Jr., L. M. G., J. C. Ruegg., and T. Wieland, eds.: (1976) *Molecular Basis of Motility.* Springer-Verlag. More specialized discussion.

Bourne, G. H., ed.: (1973) *Structure and Function of Muscle.* Academic Press. Multi-volume treatise.

Braunwald, E., J. Ross, Jr., and E. H. Sonnenblick: (1976) *Mechanisms of Contraction of the Normal and Failing Heart.* 2nd ed. Little-Brown.

21 ION TRANSPORT ACROSS MEMBRANES

(Na^+ + K^+)-ADENOSINETRIPHOSPHATASE (ATPase)

We will discuss in this chapter how the cleavage of ATP to ADP and P_i can generate ionic concentration gradients across cell membranes and some of the uses to which these gradients are put. A prime example of such gradients is the difference in concentration of Na^+ and K^+ across the plasma membrane (Fig. 21–1). The extracellular environment is rich in Na^+, and the interior of the cell is rich in K^+. This is true for every cell in the body.

If the plasma membrane were freely permeable to these ions, K^+ would move out of the cell, and Na^+ in, until the ratios of their concentrations inside and outside of the cell became identical. The concentration gradients represent a displacement from equilibrium — a store of free energy that can be used by discharging the gradients. It follows that free energy must have been expended from some process in order to create the gradients, and this process is the hydrolysis of ATP to ADP and P_i. A substantial part of the fuel consumption at rest is used to generate high-energy phosphate for this purpose.*

*Some estimate that more than half of the energy output of the body at rest is used to maintain the cation concentration gradients. Specialists tend to appropriate most of the ATP for their pet processes, and all such claims add up to much more than the total ATP available, so skepticism is warranted. However, the demand for ATP to maintain ion concentration gradients is clearly not trivial.

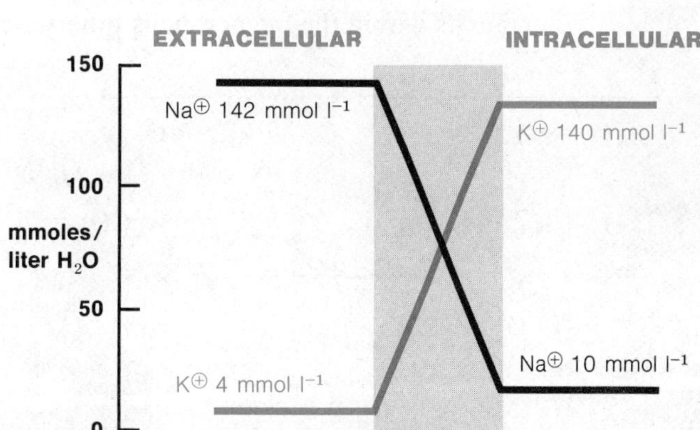

FIGURE 21–1

Large gradients in Na^+ and K^+ concentrations exist across the plasma membranes of cells. The sodium ion concentration is high outside of the cells, and the potassium ion concentration is high inside. The cited values are typical for mammalian cells.

Properties of the Pump

Sodium and potassium ions are pumped in opposite directions with a simultaneous hydrolysis of ATP by a complex of proteins that spans the plasma membrane. This complex is named (Na^+ + K^+)-adenosine-triphosphatase, sometimes shortened to Na,K-ATPase, because isolated preparations hydrolyze ATP, but only in the presence of Na^+ and K^+. As with many membrane proteins, the action of the complex has been difficult to study in detail since it must be associated with phospholipid to maintain its activity. Furthermore, X-ray diffraction patterns from which the structure could be deduced are not attainable because the lipid-protein complexes do not form rigidly ordered arrays. Much of the surface of the protein components is shielded from combination with reagents that are useful as probes in deducing the structure of more hydrophilic proteins. These difficulties are dwelt upon because they create similar problems in understanding the action of other membrane-bound systems that we shall encounter. However, in this case and in the others, the important results are clear, even though the detailed mechanism is not.

The overall reaction involves the transport of more Na^+ out than K^+ in, with the ratio approaching 3:2 for each ATP hydrolyzed:

$$3\ Na^{\oplus}_{(in)} + 2\,K^{\oplus}_{(out)} + ATP + H_2O \longrightarrow 3\ Na^{\oplus}_{(out)} + 2\ K^{\oplus}_{(in)} + ADP + P_i.$$

There is an important consequence of the unequal distribution of cations in the two directions: The concentration gradient is used to make a typical cell interior become 50 to 90 millivolts more negative than the external environment. The energy expended by the cation pump is, therefore, used to create a chemical potential, which is then used to generate an electrical potential across the plasma membrane.

The operation of the pump involves a phosphorylation of one of its protein constituents; phosphate is transferred from ATP to an aspartate side chain carboxylate group on the protein, creating a high-energy phosphate-carboxylate anhydride bond (Fig. 21–2).

An important feature of the ATPase complex is that it spans the plasma membrane. The ATP binding sites are within the cell where ATP is available from fuel metabolism. There appear to be two such sites in the complex on identical protein subunits. The complex contains a third subunit, a glycoprotein exposed on the outside of the cell. This subunit is the site of combination with the ATPase of digoxin and similar cardiac glycosides (p. 215), some of the effects of which may be due to an inhibition of Na^+ and K^+ exchange. (One of the glycosides, ouabain, is frequently used as an experimental tool.) The arrangement of these subunits within the membrane is otherwise unknown. It is possible that they are

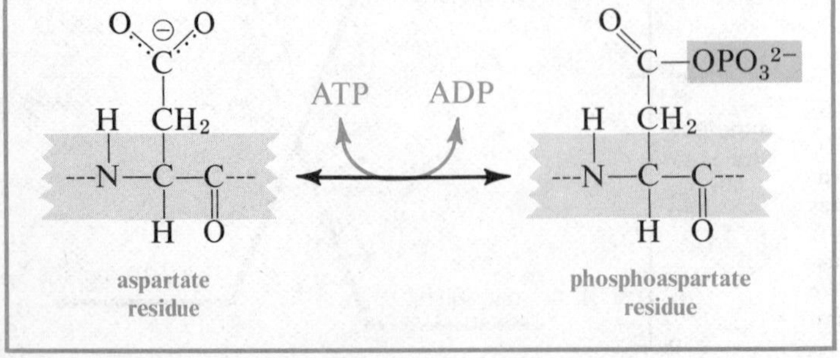

FIGURE 21–2

The action of (Na^+ + K^+)-adenosine-triphosphatase (ATPase) involves phosphorylation of an aspartate side chain in the enzyme by ATP. The phosphoaspartate residue formed is an anhydride of a carboxylic acid and phosphoric acid; its free energy of hydrolysis is comparable to that of ATP.

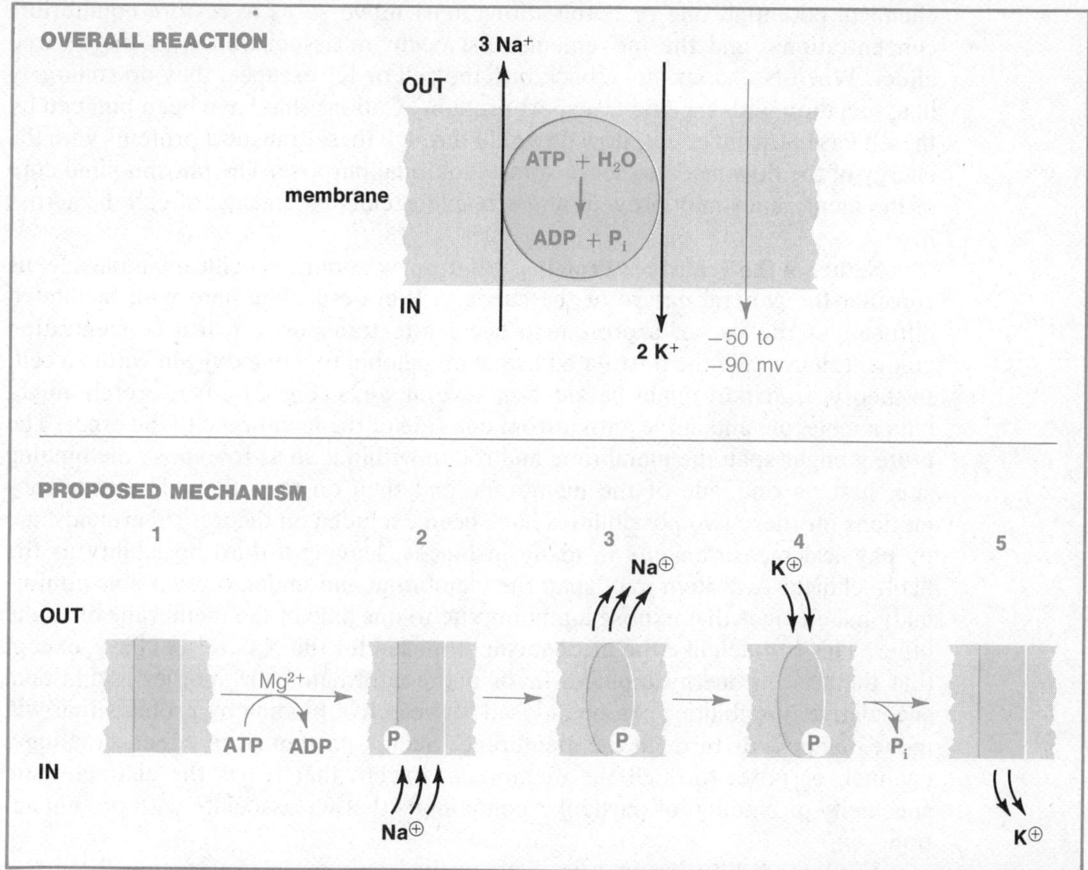

FIGURE 21–3 **A postulated mechanism for the Na^+, K^+-ATPase. The overall reaction is shown at the top, and the resultant chemical potential is used to produce an internal cell potential of −50 to −90 millivolts relative to the outside. Upon phosphorylation by ATP, the ATPase proteins (*1, bottom*), are converted to a form (*2*) with a high affinity for Na^+ within the cell. This form shifts conformation (*3*) in such a way as to make the binding sites have access to the outside of the cell, while diminishing the affinity for Na^+ and increasing the affinity for K^+ (*4*). Binding of K^+ causes a release of the phosphate group, and a reversion to the original conformation (*5*). In this conformation, the cation binding site is exposed to the interior of the cell, and has a low affinity for the cation, causing release of K^+ within the cell.**

built to provide an interior channel, more hydrophilic in character than the surrounding membrane core.

The phosphorylation is associated with a binding of intracellular Na^+ for export. Cleavage of the phosphate group on the protein to form P_i is associated with the final release of K^+ within the cell. In any event, some conformational change must occur upon addition and removal of the phosphate group that changes the affinity of the proteins for the cations, thereby forcing them to move against the electrochemical gradient. Many suggestions have been advanced for the nature of the intervening steps; one is shown in Figure 21–3.

UTILIZATION OF THE Na^+, K^+ GRADIENT

Either the chemical or the electrical potential differences, or both, created by the cation gradient can be used to drive other processes. In order to use the chemical potential, one or both cations must move so as to restore equilibrium concentrations, and the movement must occur in association with the desired effect. When Na^+ does move back into the cell or K^+ escapes, they do so largely in association with specific transport proteins. Cations that have been pumped by the ATPase now, in effect, flow downhill through these transport proteins with the energy of the flow made to serve some additional purpose. The phospholipid core of the membranes may prevent any sizeable accidental leakage of cations across it.

Nature of the Transport Proteins. Before examining specific examples, let us consider the general nature of the carriers. We are dealing here with facilitated diffusion — the use of a protein to accelerate transport down a concentration gradient, much like the postulated use of myoglobin to move oxygen within a cell. In theory, transport might be aided in several ways (Fig. 21–4). A protein might bind a molecule and move with it from one side of the membrane to the other. The protein might span the membrane and rotate within it so as to expose the binding site, first on one side of the membrane and then on the other. The necessary motions for these two possibilities have been excluded on theoretical grounds and by physical measurements in many instances, leaving a third possibility as the likely choice: A protein may span the membrane and undergo reversible conformational changes that expose a binding site to one side of the membrane or to the other. This is much like the mechanism proposed for the Na^+,K^+-ATPase, except that there is no energy input to favor one conformation or another. Additional speculative possibilities can be devised for selective binding by proteins that will move compounds through the membrane. Such a protein is, in effect, creating a channel, or pore, through the membrane, except that it has the characteristic specificity of binding for particular compounds that we associate with protein action.

Facilitated diffusion can be distinguished from passive diffusion through a truly open hole in the membrane by its specificity and on kinetic grounds. Like

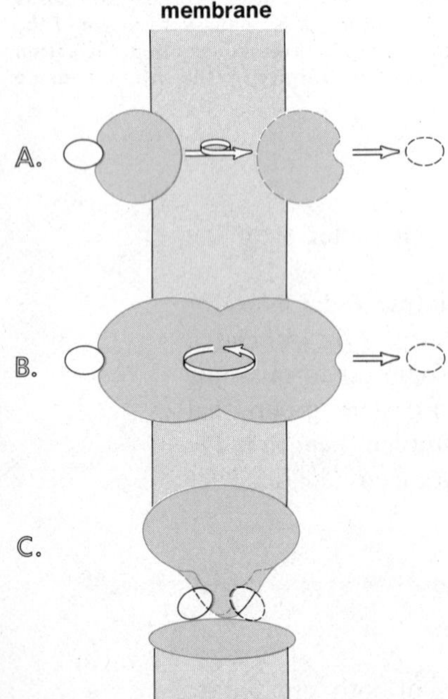

FIGURE 21–4
A transport protein (*blue*) might move a bound molecule through a membrane in any one of several ways. *A*. The protein may move through the lipid core while rotating to expose the binding site on the opposite face. *B*. The protein may extend through the membrane so that simple rotation moves the binding site from one side to the other. *C*. The protein may represent a closed pore or channel that exists in two conformations. A shift from one conformation to the other exposes the binding site to the opposite side of the membrane. Most transport proteins are now believed to function in this or similar ways involving conformational shifts without gross physical movement of the entire protein complex. (The sketch is diagrammatic.)

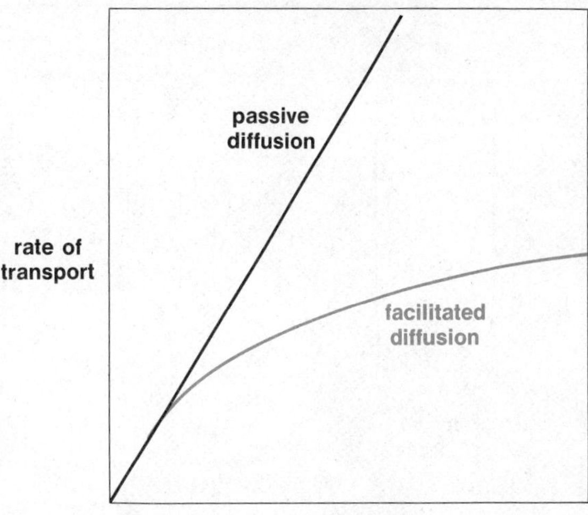

FIGURE 21–5

The use of a transport protein to facilitate diffusion can be distinguished from passive diffusion through an open pore by kinetic measurements. A protein can be saturated, so an increased concentration will cause the rate of transport to approach a maximum velocity, like the rate of an enzymatic reaction. Simple diffusion through an open pore will show a linear increase of velocity with an increasing concentration.

enzymes, transport proteins approach saturation with increasing substrate concentrations until further increases cause little change in the rate of transport, whereas the rate of passive diffusion will increase linearly (Fig. 21–5). Active transport, such as that catalyzed by the Na^+,K^+-ATPase, can often be distinguished from facilitated diffusion by its ability to create concentration gradients, that is, by changing concentrations or electrical potentials across a membrane away from the equilibrium values.

Excitable Tissues

One important use of the Na^+,K^+ gradient is to support excitability in some cells, for example, nerves and muscle fibers. These cells have specific channels through which the ions can move to discharge the gradients, but these channels are **gated.** That is, their permeability to the ions is altered by a signal. The signal can result from an increased concentration of a specific neurotransmitter (Chapter 32) or from a change in electrical potential across the membrane near the channel. It is the response to the changing potential that makes it possible for a signal to travel down a nerve.

Transmission depends upon the diminution of the electrical potential across a nerve membrane — the membrane is said to be **depolarized** — by allowing Na^+ to flow through specific gated Na^+ channels. In order to understand depolarization, it is necessary to understand the origin of the potential. The existence of different concentrations of ions across a membrane is not enough in itself for the generation of a potential; the ions must also move through the barrier. We are dealing here with **diffusion potentials** — electrical gradients created by the movement of ions down concentration gradients. When an ion moves from one side of the membrane to the other, it is carrying charge and creating a potential, but in order for this to happen, the membrane must be permeable to the ion (Fig. 21–6). The membrane of neurons is constructed so that channels for K^+ are slightly open in the resting state, but the channels for Na^+ are nearly completely closed. The result is that K^+ tends to move out of the neuron until it has created a potential that just balances

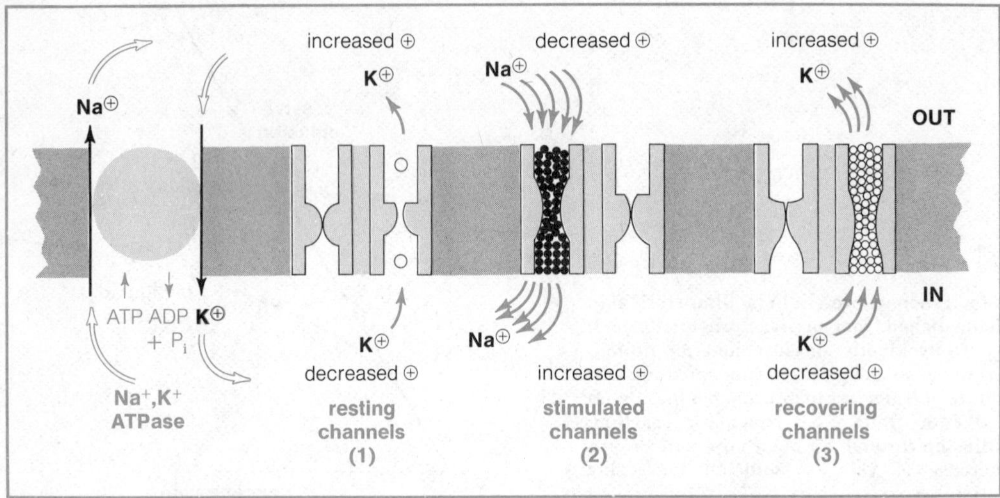

FIGURE 21–6 **Transmission along nerves involves movements of ions across the nerve membrane. Energy for transmission is obtained from the concentration gradients for Na^+ and K^+ created by the Na^+, K^+-ATPase (*left*). The nerve contains gated channels through which these gradients can be discharged. (*1*) In a resting nerve, the channel for K^+ is partially open, creating a potential across the membrane that is negative on the inside and positive on the outside. (2) When a stimulus is received, the channel for Na^+ is thrown wide open for a brief period. The resultant flux of positive charges from outside to inside creates a sharp potential pulse of opposite polarity—positive inside, and negative outside. (3) This is immediately followed by an opening of the K^+ channel and a closing of the Na^+ channel. The rate of flow of K^+ is not as great as the rate of the preceding flow of Na^+, but it is still sufficient to restore the original electrical potential (positive outside, and negative inside) in a relatively short time.**

the concentration gradient for K^+, thereby making the interior of the nerve approximately 85 millivolts negative.

When the nerve is stimulated, the gate of the Na^+ channel adjacent to the site of stimulation opens wide, and the rush of these cations toward the interior of the cell now creates a local potential that is some 30 millivolts negative on the outside of the cell, completely overcoming the potential created by the small current of K^+ in the opposite direction. The shift in potential, the action potential, in some way causes the gates of neighboring Na^+ channels to open. In the ordinary course of propagation, preceding gates will already be open, so the wave of increased conductance goes in one direction. (Propagation may involve displacement of Ca^{2+} near the gates to act in a manner resembling its effect on the contractile apparatus of muscles.)

After the action potential peaks, the Na^+ channel for some reason closes, and the gate of a K^+ channel opens wide. K^+ now rushes through to restore the original membrane potential. The maximum flux of K^+ is still much less than the maximum flux of Na^+, so recovery is slower than the initial activation: this is an important feature for unidirectional propagation. Preceding channels cannot refire until some time after the passage of the wave of stimulus.

The total displacement of ions during the cycle of stimulation and recovery is only a small fraction of the concentration gradients. If the ATP supply is stopped so that a nerve cannot replenish the Na^+,K^+ gradient by action of the ATPase, it is still capable of firing a large number of times before the existing gradient is exhausted.

Active Transport

It is often desirable for cells to accumulate other compounds in excess of the concentrations found in the external environment. For example, the intestinal mucosa absorbs glucose and some neutral amino acids from the lumen even when the concentration of these compounds is well below the concentration in the cell. This permits the recovery of the greatest possible amount of these important metabolites before the intestinal contents are attacked by the large bacterial populations in the lower bowel or are discharged in the feces.

Creating these concentration gradients requires energy, and the energy is supplied by a simultaneous discharge of the Na^+ concentration gradient. This gradient is created by the ATPase pump on the serosal side, away from the lumen, of the mucosal cells. These two processes are coupled by constructing transport proteins that interact both with Na^+ and with the desired compound (glucose or amino acid) in such a way that one crosses the cell membrane only slowly without the other (Fig. 21–7). The equilibrium position for transport by this mechanism is achieved only when the potential gradient (chemical + electrical) for glucose (or amino acid) is equal and opposite to the potential gradient for Na^+.

We see here an example of a **symport** — a simultaneous transport of two compounds in one direction. We will see in later chapters examples of **antiports,** in which the movement of a compound in one direction drives the movement of a second compound in the opposite direction (Fig. 21–8), but here, too, one concentration gradient is being used to create the other. (Transport devices for a single compound are sometimes called **uniports.**)

The association between intestinal absorption of glucose and sodium ions led to the development of a technique for the management of **cholera.** This infection

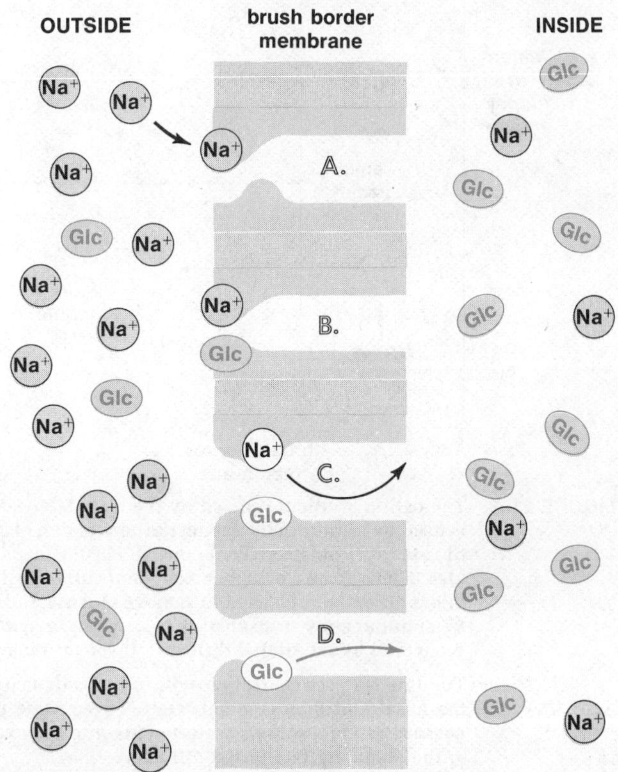

FIGURE 21–7

A possible mechanism whereby the concentration gradient for Na^+ across the brush border membrane in the intestinal mucosa is used to pump glucose into the cell from the intestinal lumen. A transport protein is used that binds both glucose and Na^+. *A.* Binding of Na^+ is not sufficient for transport of the cation. *B.* The binding of Na^+ increases the affinity of the protein for glucose, which is bound to a second site. *C.* The binding of both Na^+ and glucose facilitates equilibration with a conformation in which the binding sites are exposed to the other side of the membrane, where the (Na^+) is low. *D.* Loss of Na^+ weakens the affinity for glucose, which also leaves the protein, even though its concentration is relatively high in the neighboring solution.

The necessary ionic concentration gradient is generated by the Na^+, K^+-ATPase in the basal-lateral membrane of the mucosal cells (not shown). Glucose can cross that membrane by facilitated diffusion, since its concentration inside the cell has been made higher than the concentration in blood.

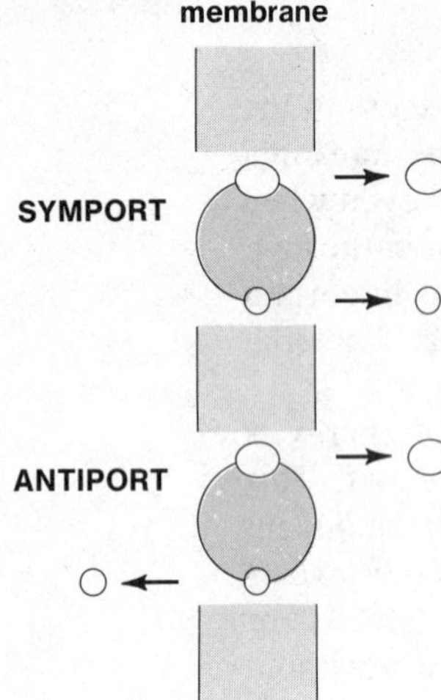

FIGURE 21–8

A concentration gradient for one compound can be made to create a concentration gradient for another compound in two ways. Symports will transport both compounds in the same direction through a membrane, with the flow of one down its pre-existing gradient carrying the other up a gradient. The Na^+-coupled transport of glucose into the intestinal mucosa is an example of a symport. Antiports couple transport of one compound to transport of the other in the opposite direction; the pre-existing gradient generates another gradient in the same direction, so that the concentrations of both compounds become high on the same side of the membrane.

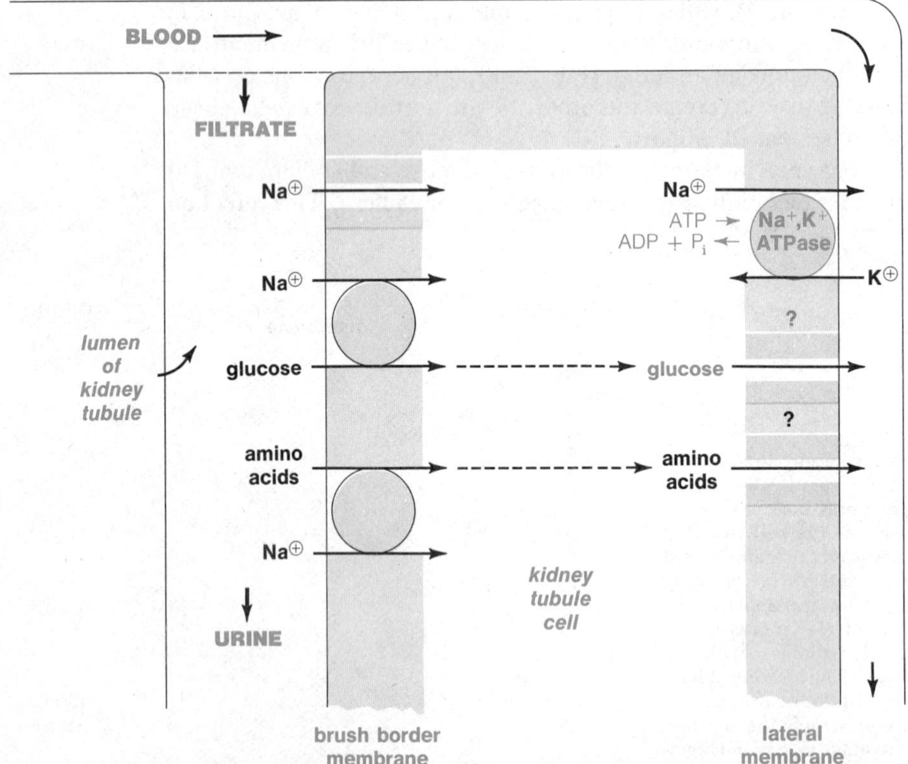

FIGURE 21–9 The cation gradient created by the Na^+, K^+-ATPase in the plasma membrane is used by kidney cells during the conversion of a blood filtrate into urine. The filtrate (*left*) passes over the brush border membranes of cells in kidney tubules. Since it has a higher Na^+ concentration than the interior of the cells, this gradient can be used to remove glucose and amino acids from the filtrate by simultaneous transport of Na^+, using a symport. Additional Na^+ can be recovered by facilitated diffusion through uncharacterized channels.

The low Na^+ concentration within the cells is created by Na^+, K^+-ATPase in the basal and lateral membranes. These same membranes have carriers for passage of glucose and amino acids into the extracellular space, and thence to the blood, by facilitated diffusion.

causes diarrhea so severe that the fecal stream issues from the anus like water from a faucet; the depletion of water and salts may be lethal. However, a constant infusion of glucose into the gut by means of a tube through the mouth causes a reabsorption of salt and associated water with a dramatic reduction in the amount of intravenous fluid required to maintain water and salt balance.

The kidney utilizes the same kind of mechanism for recovery of glucose and some neutral amino acids from filtered blood plasma as it passes through tubules in the organ. Some specialized cells have a brush border in the lumen; the other sides of these cells next to the blood vessels have a highly active Na^+,K^+-ATPase creating a gradient for Na^+ from the cells to the blood stream, and from the lumen of the tubules into the cells (Fig. 21–9) in the same way that similar cells do in the intestine. Recovery of Na^+ from the tubules can, therefore, occur by facilitated diffusion, utilizing a Na^+ carrier. (We shall discuss regulation of salt and water balance in Chapter 39). Specific symport carriers use the Na^+ gradient for recovery of glucose and amino acids from the filtrate.

Associated Anion Gradients

We have concentrated on the movement of cations, but each side of a membrane must have negative charges almost equal in number to the positive charges. The electrical potentials represent only a minute quantitative imbalance, and it follows that any net movement of cations must be associated with an equivalent movement of anions. This movement is accomplished through specific channels for Cl^- and HCO_3^-. The membranes of the cell are nearly impermeable to many other anions, such as phosphate esters, polycarboxylate compounds, and the like. Such exchanges of these compounds as do occur involve relatively slow symport and antiport mechanisms.

$(Ca^{2+} + Mg^{2+})$-ADENOSINETRIPHOSPHATASE

An essential feature of muscular contraction is the storage of Ca^{2+} in the sarcoplasmic reticulum at concentrations much greater than those in the cytoplasm around the contractile filaments. Stimulation of contraction involves opening channels so that these calcium ions can pour out rapidly. The recovery phase of contraction involves pumping the Ca^{2+} back into the sarcoplasmic reticulum. This is accomplished by an ATP-driven mechanism that in many respects resembles the mechanism for pumping Na^+ and K^+. Ca^{2+} is pumped inside the reticulum and Mg^{2+} is pumped out, and the apparent stoichiometry is:

$$2\ Ca^{2+}{}_{(out)} + Mg^{2+}{}_{(in)} + 2\ K^{+}{}_{(in)} + ATP \longrightarrow 2\ Ca^{2+}{}_{(in)} + Mg^{2+}{}_{(out)} + 2\ K^{+}{}_{(out)} + ADP + P_i.$$

The pump consists of two components; one reacts with ATP outside the reticulum, and the other is a protein-lipid component that may serve as a channel through the lipid core of the membrane to the inside of the reticulum. The mechanism of the transfer involves phosphorylation of an aspartyl residue, as in the Na^+,K^+-ATPase.

The storage of Ca^{2+} within the sarcoplasmic reticulum is made easier by the presence of **calsequestrin,** a calcium-binding protein on the interior surface of the membrane. This protein is very rich in carboxylate side chains (Glu and Asp), which chelate Ca^{2+}, thereby diminishing its effective concentration while keeping it readily available for use through dissociation.

FURTHER READING

Guyton, A. C.: (1971) *Textbook of Medical Physiology,* 4th ed. W. B. Saunders. pp. 52–57.

The following are more comprehensive discussions:

Whittam, R., and A. R. Chipperfield: (1975) *The Reaction Mechanism of the Sodium Pump.* Biochem. Biophys. Acta, *415*: 149. A review.

Skou, J. C.: (1974) *The (Na^+ + K^+-Activated Enzyme and its Relationship to Transport of Sodium and Potassium.* Quart. Rev. Biophys., *7*: 401.

Glynn, I. M., and S. J. D. Karlish: (1975) *The Sodium Pump.* Annu. Rev. Physiol., *37*: 13.

Schwartz, A., ed.: (1977) *Newer Aspects of Cardiac Glycoside Action.* Fed. Proc., *36*: 2207. A symposium on the 200th anniversary of the introduction of digitalis in therapeutics.

Anderson, D. K.: (1976) *Cell Potential and the Sodium-potassium Pump in Vascular Smooth Muscle.* Fed. Proc., *35*: 1294.

Fass, S. J., M. R. Hammerman, and B. Sacktor: (1977) *Transport of Amino Acids in Renal Brush Border Membrane Vesicles.* J. Biol. Chem., *252*: 583.

Hopfer, U.: (1976) *Sugar and Amino Acid Transport in Animal Cells.* Horiz. Biochem. Biophys., *2*: 106.

Shamoo, A. E., ed.: (1975) *Carriers and Channels in Biological Systems.* Ann. N. Y. Acad. Sci. vol. 264. Symposium.

MacLennan, D. H., and P. C. Holland: (1975) *Calcium Transport in Sarcoplasmic Reticulum.* Annu. Rev. Biophys., *4*: 377. Usefulness seriously compromised by lack of subject index.

Kretsinger, R. H.: (1976) *Calcium-binding Proteins.* Annu. Rev. Biochem., *45*: 239.

Rothstein, A., Z. I. Cabantchik, and P. Knauf: (1976) *Mechanism of Anion Transport in Red Blood Cells. Role of Membrane Proteins.* Fed. Proc., *35*: 3.

Wilson, D. B.: (1978) *Cellular Transport Mechanisms.* Annu. Rev. Biochem., *47*:933.

22 OXIDATIVE PHOSPHORYLATION

We have developed at this point the idea that living organisms can go about their business of growing, reproducing, and moving because the chemical reactions involved in these events include the cleavage of pyrophosphate bonds in ATP or in other nucleoside triphosphates that are equilibrated with ATP. The other side of life that isn't so obvious includes the means of generating the pyrophosphate bonds that are being expended to look alive.

Over a day's time, more than 85 per cent of the pyrophosphate bonds in the entire body are created by the reaction of fuels with molecular oxygen. The formation of ATP by phosphorylation of ADP is a concomitant of the combustion of fats, of carbohydrates, and even of proteins. The total oxidations of a typical fat and of glucose go like this in the laboratory:

$$\text{(fat)}\quad C_{55}H_{102}O_6 + 77\tfrac{1}{2}\, O_2 \longrightarrow 55\, CO_2 + 51\, H_2O$$

$$\text{(glucose)}\quad C_6H_{12}O_6 + 6\, O_2 \longrightarrow 6\, CO_2 + 6\, H_2O.$$

Cells modify these reactions so as to cause the phosphorylation of a large number of molecules of ATP per molecule of fuel consumed; the overall biological process can be approximated by these equations:

$$\text{(fat)}\quad C_{55}H_{102}O_6 + 77\tfrac{1}{2}\, O_2 + 437\,(ADP + P_i) \longrightarrow 55\, CO_2 + 437\, ATP + 488\, H_2O$$

$$\text{(glucose)}\quad C_6H_{12}O_6 + 6\, O_2 + 36\,(ADP + P_i) \longrightarrow 6\, CO_2 + 36\, ATP + 42\, H_2O.$$

How is this accomplished? Each of these processes must represent the summation of many individual reactions, because hundreds of molecules simply don't collide simultaneously in one grand reaction. There is a clue buried in the equations. Suppose that we calculate the number of molecules of high-energy phosphate produced per atom of oxygen gas consumed; this ratio, abbreviated as the **P:O ratio,** is commonly used as an index of the efficiency of energy conversion in oxidative metabolism. We find that it is 2.82 when fats are oxidized and 3.00 when glucose is oxidized. These values are nearly the same, and the reason is that the same mechanism of generating high-energy phosphate is used over and over again regardless of the nature of the fuel being burned. Most of the phosphorylations are associated with the consumption of oxygen.

The combination of the utilization of oxygen and the phosphorylation of ADP is known as oxidative phosphorylation, and it constitutes the principal source of high-energy phosphate for nearly all animals, and also for plants during the dark hours.

General Nature of Oxidative Phosphorylation

A fundamental feature of oxidative phosphorylation is this: Only a few coenzymes are used as oxidizing agents for a wide variety of organic compounds. The coenzymes become reduced when they are involved in any one of these many reactions, and the reduced forms of the coenzymes are mainly re-oxidized by transferring electrons from them to molecular oxygen. These transfers of electrons are made through particular intermediates, and they are the reactions used to drive the phosphorylation of ADP (Fig. 22–1).

Essentially there is one set of enzymes catalyzing a sequence of reactions through which electrons removed from a variety of organic substrates are fed, one pair at a time, to reduce molecular oxygen. Since the substrates are oxidized in small steps, with only two electrons or two hydrogen atoms being removed at a time, a long series of reactions is required before the last carbon and hydrogen atoms of the original fuel appear as CO_2 and H_2O. These more complicated sequences will concern us in the following chapters; we are here concerned with one of the major purposes for which they exist — the transfer of electrons to oxygen so as to cause the formation of ATP.

The Role of Membranes. The process of oxidative phosphorylation is localized in membranes in all kinds of cells in which it occurs. Oxidative phosphorylation occurs in the inner membranes of mitochondria in animals and in other eukaryotes, including fungi and the higher plants. In bacteria, it occurs in the cytoplasmic membrane, the innermost of the envelopes enclosing the cell.

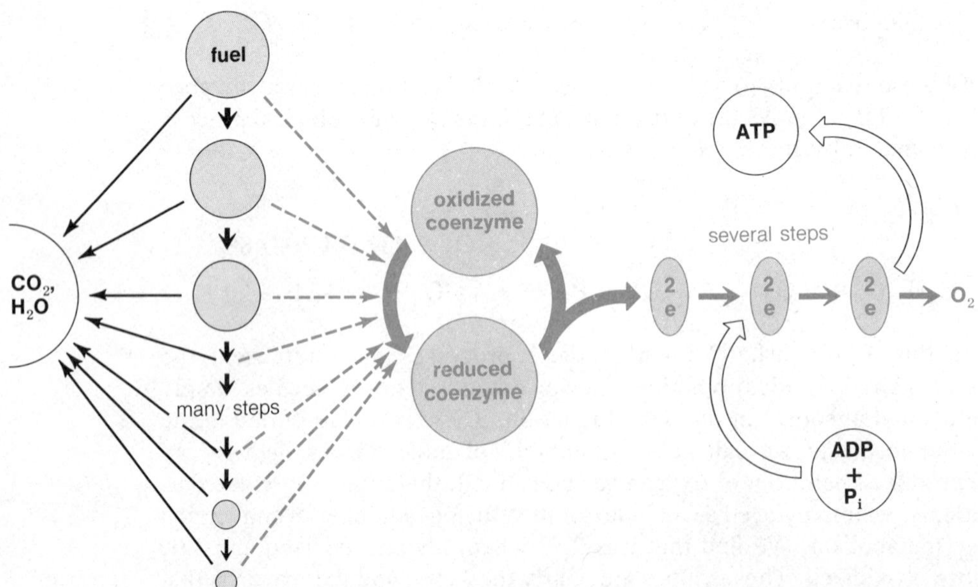

FIGURE 22–1 **Many different fuels can be attacked by the same oxidized coenzymes. The fuels are chipped away, two electrons at a time, with their substance gradually appearing as CO_2 and H_2O. The coenzymes become reduced, and their oxidized form is regenerated by the transfer of electrons to molecular oxygen through several intermediate carriers. These transfers are highly exergonic by themselves, and are used to drive the phosphorylation of ADP to form ATP.**

MITOCHONDRIAL MEMBRANES. A mitochondrion consists of two sacs, one inside the other. The outer membrane is relatively unwrinkled, but the inner membrane has bulges into the center of the mitochondrion, which greatly extend its entire surface. The number and shape of these bulges, which are known as cristae, vary from one kind of a cell to another, even in differing cells of the same organism (Fig. 22–2), but in all cases the membrane is continuous. The material enclosed by the inner membrane is the matrix of the mitochondrion.

The outer membrane is freely permeable to molecules up to about 400 M.W. and has a relatively small number of associated enzymes. A few other enzymes are located in the intermembrane space, but most of the catalytic activity of mitochondria is found within the inner membrane or its enclosed matrix. The inner membrane contains the enzymes responsible for transferring electrons to oxygen and for the associated phosphorylation of ADP to create ATP. The matrix is rich in other enzymes, many of which are necessary for the catabolism of fuels.

THE ELECTRON TRANSFERS

The reaction of typical substrate with molecular oxygen involves enough change in chemical potential to sustain the formation of two or three molecules of ATP for each atom of oxygen consumed (equivalent to four or six molecules of ATP per molecule of O_2). However, it is practical to make only one molecule of ATP at a time, and it is therefore necessary for the transfer of electrons from substrate to oxygen to be broken up into a stepwise sequence, if the greatest possible amount of ADP is to be converted to ATP. We can summarize this sequence as follows (Fig. 22–3): Electrons are transferred from substrates to **nicotinamide adenine dinucleotide (NAD,** p. 273), and then in succession to **ubi-**

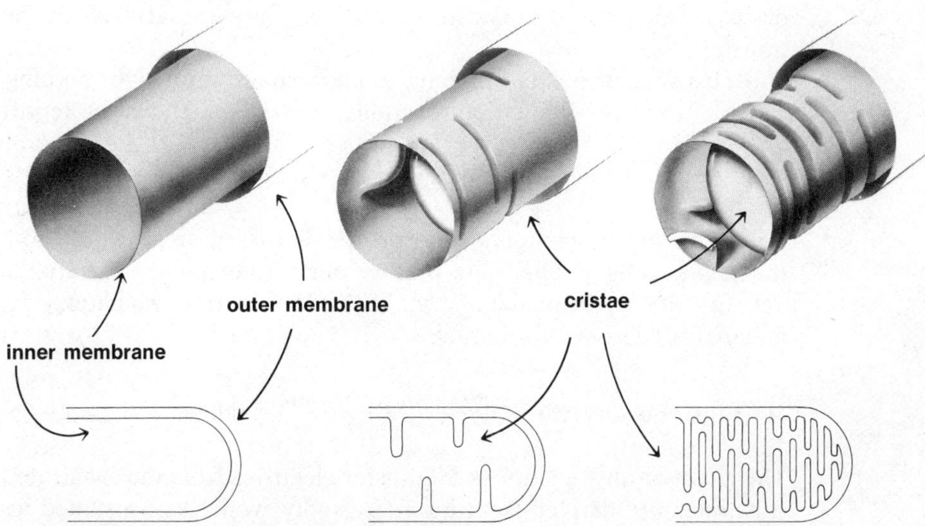

FIGURE 22–2 **Idealized gross structure of mitochondria, shown in cutaway view and in longitudinal section. The structure is made from two concentric bags, and the membrane comprising the inner bag may be indented into ridges, tubes, or pouches, called cristae. As shown, there may be few cristae or there may be so many that they nearly fill the interior, with cross sections appearing to have nearly even laminations.**

Illustration and legend continued on following page.

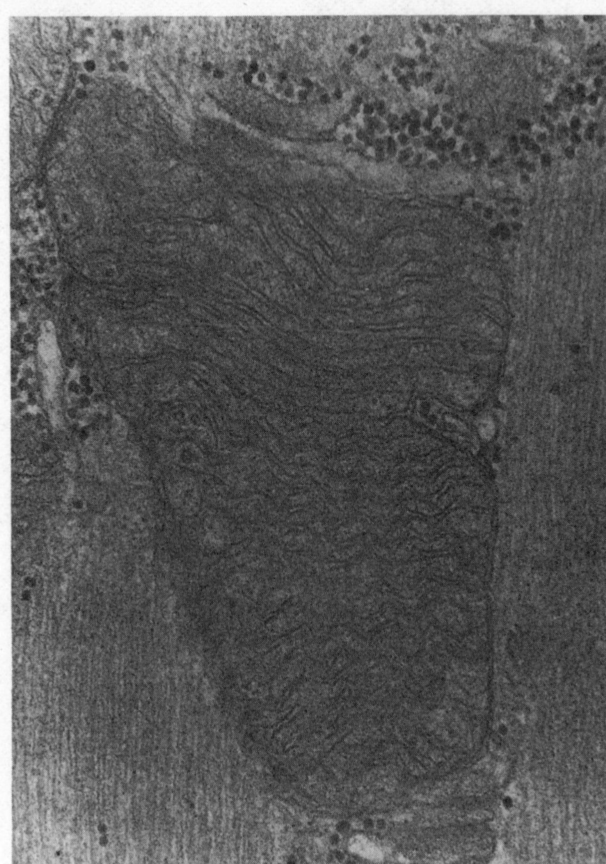

Figure 22–2
Continued

A mitochondrion in rat thigh muscle. The mitochondrion is enclosed within the outer membrane, and is filled with cristae of the inner membrane. The mitochondrion is intimately associated with muscle fibers to either side. More mitochondria lie in a band above and below the one seen here. The large surface of the inner membrane is characteristic of mitochondria in tissues having a high oxygen consumption, such as muscles built for extended periods of work. Compare the mitochondria in liver (p. 483). Magnification 60,000×. (Courtesy of Dr. Carlo Bruni.)

quinone, cytochrome c, and **oxygen.** (The nature of ubiquinone and cytochrome c is discussed below.) A molecule of ATP is generated at each of the latter three transfers.

This sequence is not difficult to understand; it involves nothing more than a series of oxidation-reduction reactions, each of which can be represented as:

$$\text{reduced A} + \text{oxidized B} \rightarrow \text{oxidized A} + \text{reduced B}.$$

This is a typical reaction of the sort we discussed on pages 271 to 275. Even so, the sequence is so important that we ought to examine it carefully, and we will. The first step is an oxidation by NAD of a substrate, which may be one of many metabolites derived from a fuel:

$$\text{(1) substrate} + \text{NAD}^{\oplus} \xrightarrow{\textit{substrate dehydrogenase}} \text{oxidized substrate} + \text{NADH} + \text{H}^{\oplus}.$$

The effect of this reaction is to transfer electrons from the substrate to NAD. The substrates are derived from the fuel supply, which is consumed in the process. There is only a limited supply of the NAD that is also required, and it is regenerated from NADH by the next reaction in the sequence:

$$\text{(2) NADH} + \text{H}^{\oplus} + \text{ubiquinone (Q)} \longrightarrow$$
$$\text{NAD}^{\oplus} + \text{ubiquinol (QH}_2) + {\sim}\text{P (high energy phosphate)}.$$

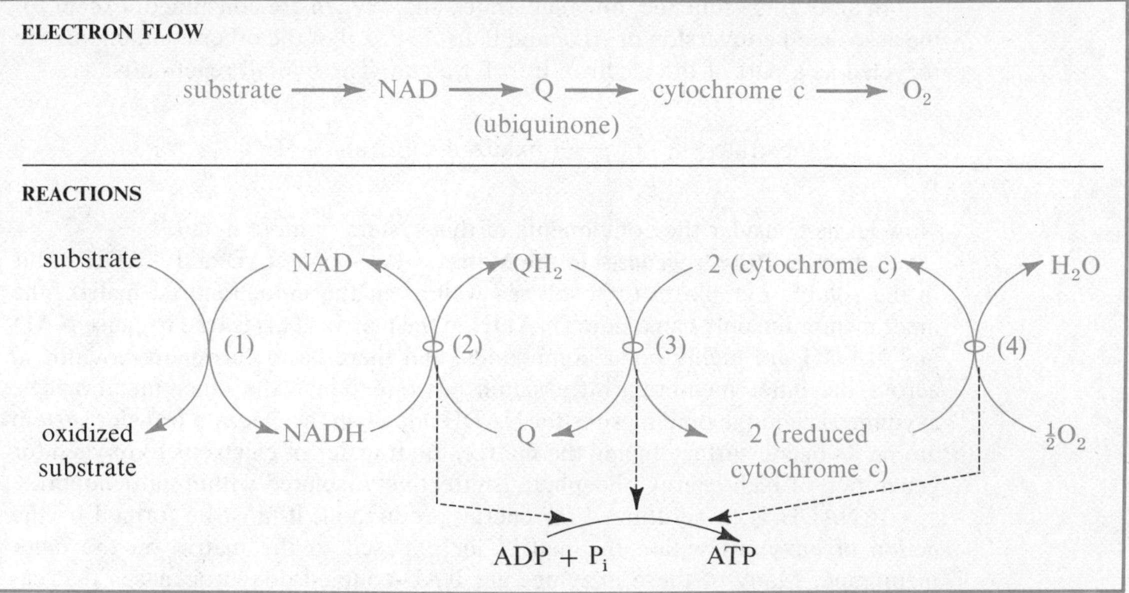

FIGURE 22-3 **General outline of electron flow in mitochondria. Many substrates are oxidized by NAD, which becomes reduced to NADH. NADH is oxidized by molecular oxygen in three major stages, involving passage of electrons through ubiquinone (Q) and cytochrome c. The scheme at the top shows the direction of electron flow, and the reactions are given in more detail at the bottom. Each electron carrier is successively reduced and re-oxidized:**

(1) substrate + NAD → oxidized substrate + NADH

(2) NADH + ubiquinone (Q) → NAD + ubiquinol (QH_2)

(3) QH_2 + 2 cytochrome c → Q + 2 reduced cytochrome c

(4) 2 reduced cytochrome c + $\frac{1}{2}O_2$ → 2 cytochrome c + H_2O

Each of the electron transfers after NADH involves the phosphorylation of ADP to create ATP.

NADH in the matrix is oxidized by ubiquinone in the inner membrane, producing a molecule of high-energy phosphate. As a result, ubiquinone (Q) is converted to its reduced form, ubiquinol (QH_2) (also called dihydroubiquinone). Since there is a limited supply of ubiquinone, it also must be regenerated from its reduced product:

(3) QH_2 + 2 (cytochrome c) → Q + 2 H^+ + 2 (reduced cytochrome c) + ~ P.

This reaction is also arranged so as to produce high-energy phosphate. Each molecule of cytochrome c can carry only one electron, whereas NAD and ubiquinone can carry two, hence the requirement for two molecules of cytochrome c.

As before, the limited supply of oxidized cytochrome c is regenerated from its reduced form with the production of high-energy phosphate. The oxidizing agent in this case is molecular oxygen, and the reaction completes the sequence:

(4) 2 (reduced cytochrome c) + 2 H^+ + ½ O_2 ⟶ 2 (cytochrome c) + H_2O + ~ P.

In all of this sequence, only substrates and oxygen are consumed, except for the associated conversion of ADP and P_i to ATP. All of the other components are recycled as a part of the electron transfer chain. The overall reaction is:

$$\text{substrate} + \tfrac{1}{2}\,O_2 \longrightarrow \text{oxidized substrate} + H_2O + 3 \sim P.$$

Now let us consider the components of this system in more detail.

Substrate Dehydrogenases in the Matrix. Although NAD and NADH occur in the soluble cytoplasm (cytosol), as well as in the mitochondrial matrix, the inner membrane only can oxidize NADH in the matrix. This is true because NAD and NADH are highly polar compounds, and there is no carrier to move them across the inner membrane at a significant rate. Since the inner membrane is asymmetric and the binding sites for NADH and O_2 in the electron transfer system are on its inside surface facing the matrix, the transfer of electrons to oxygen for generation of high-energy phosphate is effectively isolated within mitochondria.

If NADH is to be utilized for energy production, it must be formed by the action of enzymes within the matrix, or exposed to the matrix on the inner membrane. Many of these enzymes are NAD-coupled dehydrogenases that catalyze equilibrations of alcohols and ketones, very similar to the equilibration of pyruvate and lactate (p. 274) catalyzed by lactate dehydrogenase in the cytosol.

For example, a malate dehydrogenase in the matrix catalyzes this reaction, which is necessary for a large number of metabolic processes:

$$\begin{array}{c} COO^{\ominus} \\ | \\ HO{-}C{-}H \\ | \\ CH_2 \\ | \\ COO^{\ominus} \\ \text{L-malate} \end{array} \; \underset{\text{malate dehydrogenase}}{\overset{NAD^{\oplus} \quad H^{\oplus} + NADH}{\longleftrightarrow}} \; \begin{array}{c} COO^{\ominus} \\ | \\ C{=}O \\ | \\ CH_2 \\ | \\ COO^{\ominus} \\ \text{oxaloacetate} \end{array}$$

Reactions like this, and other types of NAD-coupled oxidations, constantly supply NADH to the mitochondrial inner membrane for use in oxidative phosphorylation.

NADH Dehydrogenase

The oxidation of NADH by ubiquinone is catalyzed by a highly organized collection of proteins known as NADH dehydrogenase, which is partially buried in the lipid core of the inner membrane. Here we encounter an enzyme complex in which individual components catalyze parts of the overall reaction. The advantage of a complex is that it enables coordinated action of several kinds of enzymes; the constituent enzymes are kept near each other in a specific pattern; the reaction intermediates can be passed directly from one to another, promptly and without loss.

The Flavoproteins. The component of NADH dehydrogenase that is on the inner surface, where it is accessible to NADH in the matrix, is a flavoprotein. That is, it has **riboflavin-5′-phosphate** bound to it (Fig. 22–4). Riboflavin is an example of a vitamin that serves as a constituent of coenzymes used in oxidation-reduction reactions. The oxidized form has an intense yellow color (flavus means yellow),

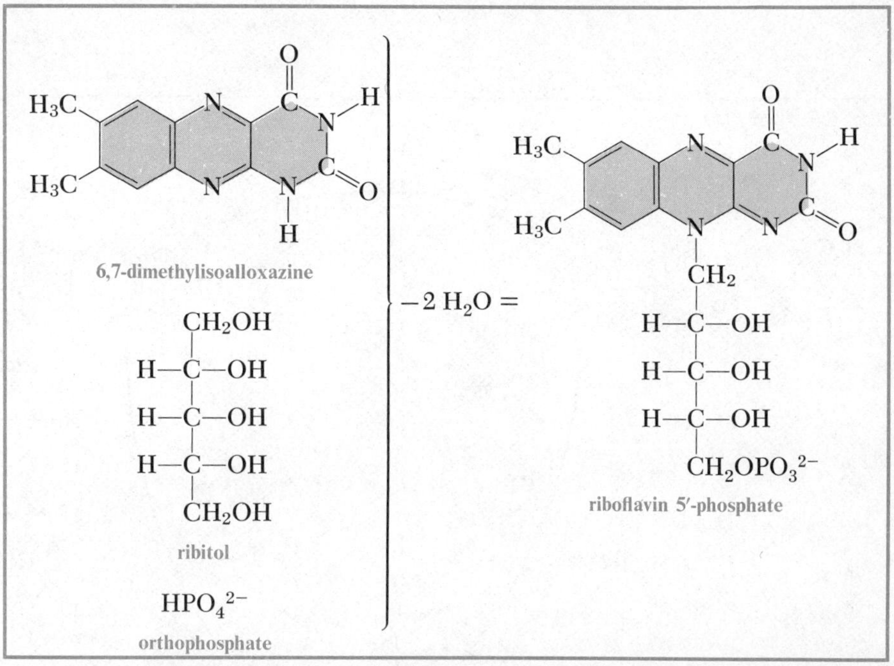

6,7-dimethylisoalloxazine

ribitol

HPO_4^{2-}
orthophosphate

$-2\ H_2O =$

riboflavin 5′-phosphate

FIGURE 22–4 **Riboflavin 5′-phosphate can be formally regarded as a combination of a dimethylisoalloxazine, ribitol (the alcohol corresponding to ribose), and inorganic phosphate with the loss of 2 H_2O. (The actual biological synthesis is more complex.)**

which is characteristic of flavoproteins; the reduced form is colorless. The monophosphate is often referred to as **flavin mononucleotide (FMN),** even though this is not strictly correct. (Riboflavin contains a polyhydric alcohol, not a sugar.) In general, the flavin coenzymes are tightly bound to their apoproteins so that a cycle of reduction and oxidation must be completed by each enzyme molecule before it can begin to repeat its catalytic mechanism.

The flavoproteins are a versatile group of enzymes. They can form quite stable semiquinones, which are free radicals with the flavin reduced by only one electron, so there are three distinct mechanisms by which they might transfer electrons (Fig. 22–5): They may never form the free radical and transfer two electrons to change the fully oxidized to the fully reduced form. They may transfer only one electron to change the fully oxidized form to the free radical form, or they may transfer one electron to change the free radical form to the fully reduced form. The mechanism of NADH dehydrogenases is not fully defined, although it appears to involve the two-electron transfer without an intermediate free-radical.

The Iron-sulfur Proteins. NADH dehydrogenase contains at least four different proteins with attached lattices of iron and sulfur atoms that are used to transfer single electrons. The exact structures of these lattices in NADH dehydrogenase are not known. However, the structures of related proteins are known. Among these are the ferredoxins from bacteria and plants, which differ in the number of iron atoms involved (Fig. 22–6). In all of these, the iron atoms are chelated with sulfur atoms, which are in part supplied by cysteinyl groups in the associated protein and in part as inorganic sulfide ions.

The entire lattice carries only one electron. Iron-sulfur proteins were referred to in the older literature as **non-heme iron.** The important feature of the iron-sulfur

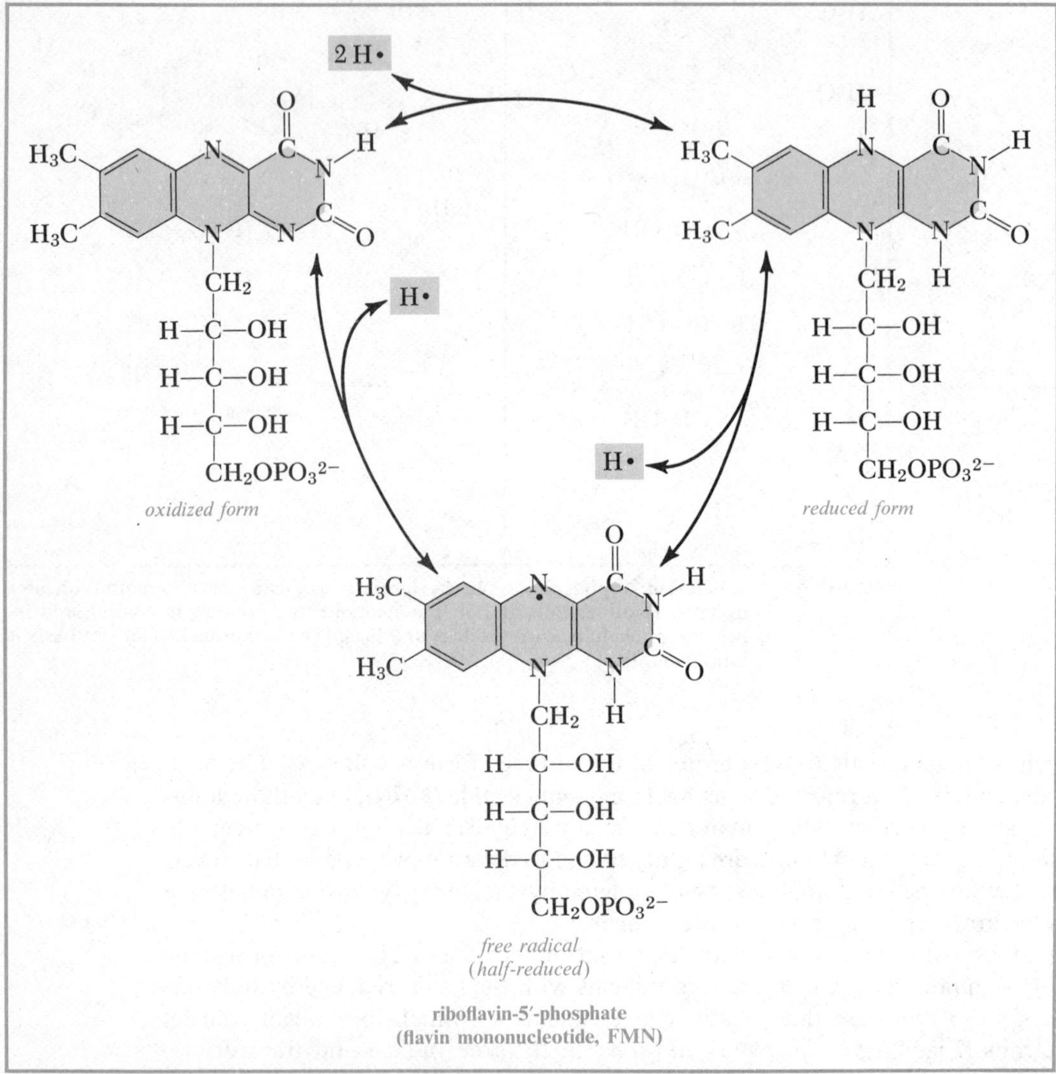

FIGURE 22–5 **The flavin coenzymes may transfer either two electrons (*top*), or one electron. The transfer of one electron involves an intermediate free radical, or semiquinone form (*bottom*), and it may occur in either of two ways (*right* or *left*).**

FIGURE 22–6 **The iron-sulfide complexes in animal tissues are not fully characterized. *A*. Some are believed to involve two iron atoms and two sulfide ions, resembling compounds known as ferredoxins that are found in plant chloroplasts. *B*. Others are believed to involve four each of iron atoms and sulfide ions, resembling ferredoxins found in certain bacteria. In either case, each iron atom is bound to four sulfur atoms, some contributed by cysteine residues.**

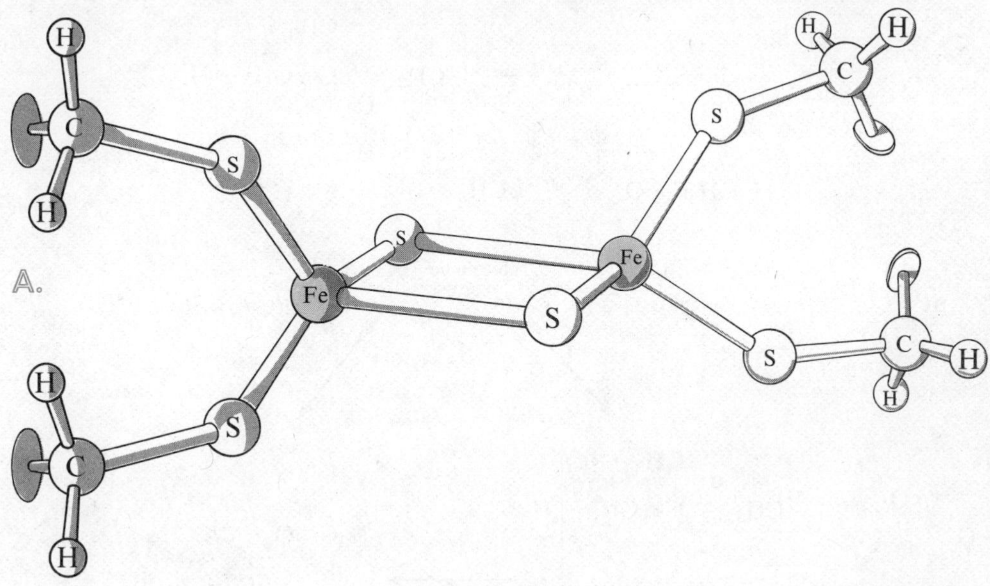

$Fe_2S_2Cys_4$ complex

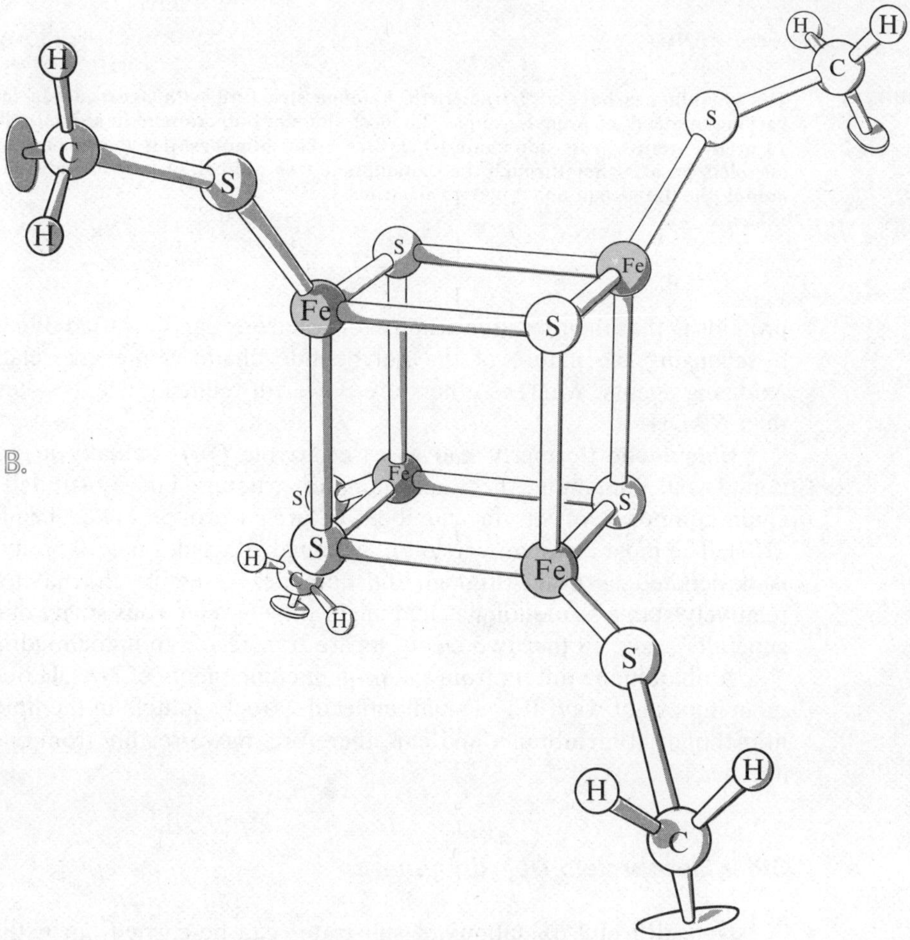

B.

$Fe_4S_4Cys_4$ complex

See legend on opposite page.

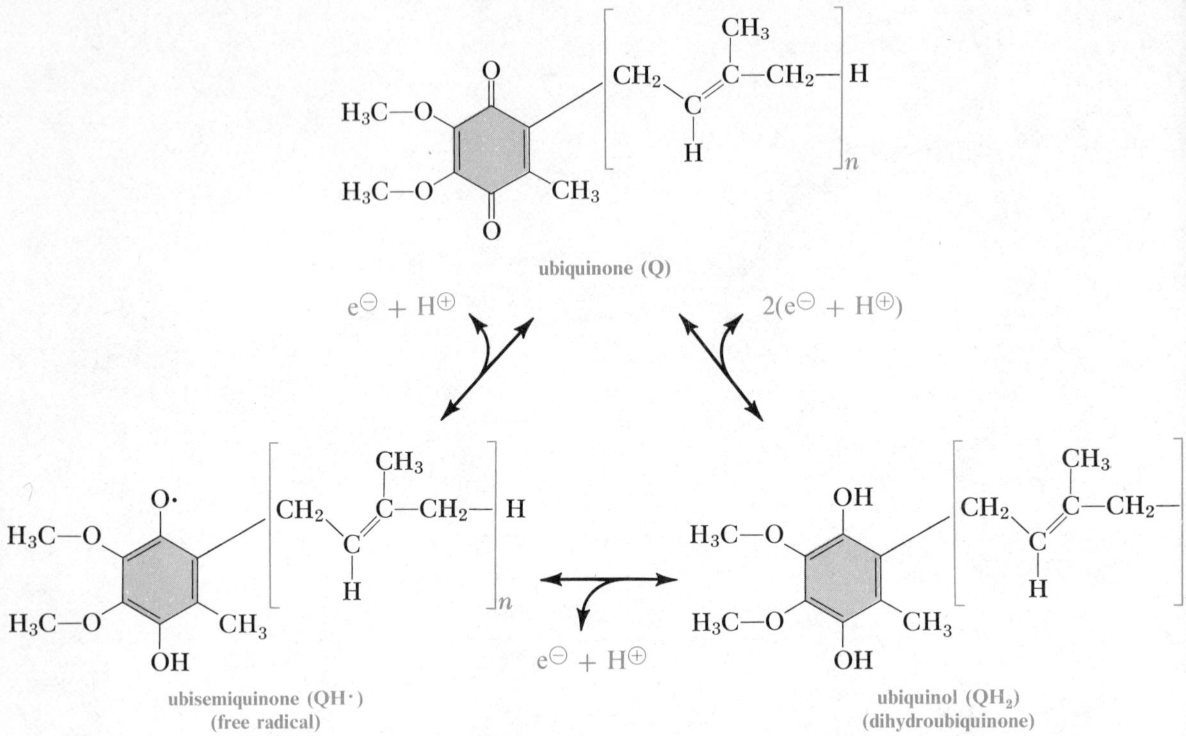

ubiquinone (Q)

$e^{\ominus} + H^{\oplus}$ $\quad$ $2(e^{\ominus} + H^{\oplus})$

$e^{\ominus} + H^{\oplus}$

ubisemiquinone (QH·)
(free radical)

ubiquinol (QH_2)
(dihydroubiquinone)

FIGURE 22–7 **The ubiquinones have a characteristic quinone structure with hydrocarbon tails made from varying numbers of prenyl groups. The most abundant ubiquinone in animal mitochondria has 10 prenyl groups in its side chain (Q_{10}). Like other quinones, the ubiquinones may be reduced one electron at a time through the semiquinone free radical, or may be reduced directly to the quinol (the dihydroquinone) by two electrons.**

proteins is that their relative affinity for electrons can be varied over a wide range by changing the nature of the polypeptide chain. Some are relatively strong oxidizing agents, whereas others are powerful reducing agents — even stronger than NADH.

Ubiquinone (formerly known as **coenzyme Q**) is actually a group of compounds, all containing the same quinone structure but substituted with a side chain composed of varying numbers of **prenyl** groups linked head-to-tail (Fig. 22–7). The most common version in mammalian tissues has 10 prenyl groups and is designated as Q_{10}. Although the quinones resemble the flavins in forming relatively stable semiquinones and might transfer electrons singly or in pairs, it is generally assumed that two electrons are transferred in mitochondria.

A ubiquinone differs from the protein components of NADH dehydrogenase in an important way. It is a small molecule, freely soluble in the lipid core of the mitochondrial membrane, and can, therefore, move readily from one location to another.

Other Flavoprotein Dehydrogenases

Not all useful oxidations of substrates can be carried out with NAD as an electron carrier. Some substrates must be attacked by a stronger oxidizing agent that can be regenerated without forming high-energy phosphate. Such oxidations are frequently catalyzed by separate flavoprotein dehydrogenases on the matrix surface of the inner mitochondrial membrane, and these agents directly remove hydrogen atoms, with the associated electrons, from fuel metabolites and transfer them to ubiquinone. Typical substrates are saturated acyl compounds, which are oxidized to the corresponding unsaturated derivatives; an important example involves the oxidation of succinate to fumarate:

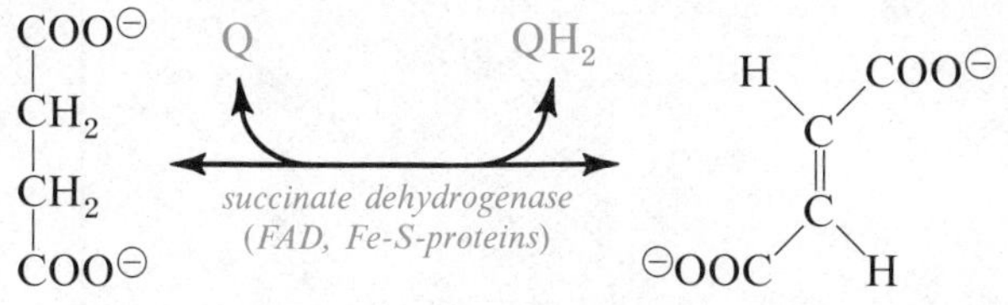

$$^{\ominus}OOC{-}CH_2{-}CH_2{-}COO^{\ominus} \xrightarrow[\text{succinate dehydrogenase (FAD, Fe-S-proteins)}]{Q \rightarrow QH_2} {}^{\ominus}OOC{-}CH{=}CH{-}COO^{\ominus}$$

succinate fumarate

Succinate dehydrogenase is a complex of proteins in the inner mitochondrial membrane that contains a flavoprotein in which the prosthetic group is a **flavin adenine dinucleotide (FAD)** rather than the riboflavin 5′-phosphate found in NADH dehydrogenase:

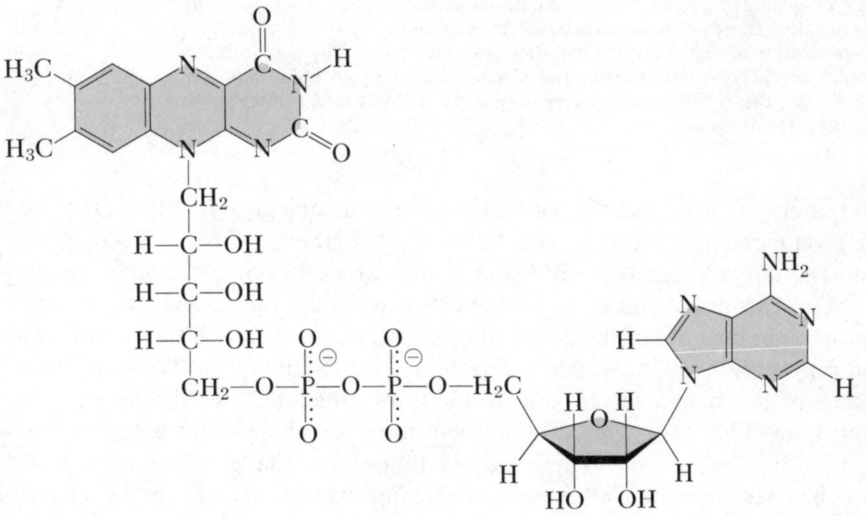

flavin adenine dinucleotide (FAD)

(It also differs in that the FAD is covalently bound to the apoprotein.) At least two iron sulfide proteins are also found in succinate dehydrogenase.

Ubiquinone therefore may be reduced by two routes (Fig. 22–8). Electrons from some substrates go to NAD and through the NAD dehydrogenase complex with a concomitant generation of high-energy phosphate. Electrons from other substrates, which are less powerful reducing agents, go through specific flavoprotein dehydrogenase complexes without passing through NAD, and without the coupled generation of high-energy phosphate.

THE CYTOCHROMES

Many of the carriers of electrons between ubiquinone and oxygen are cytochromes — hemoproteins differing from hemoglobin in that the iron atom is

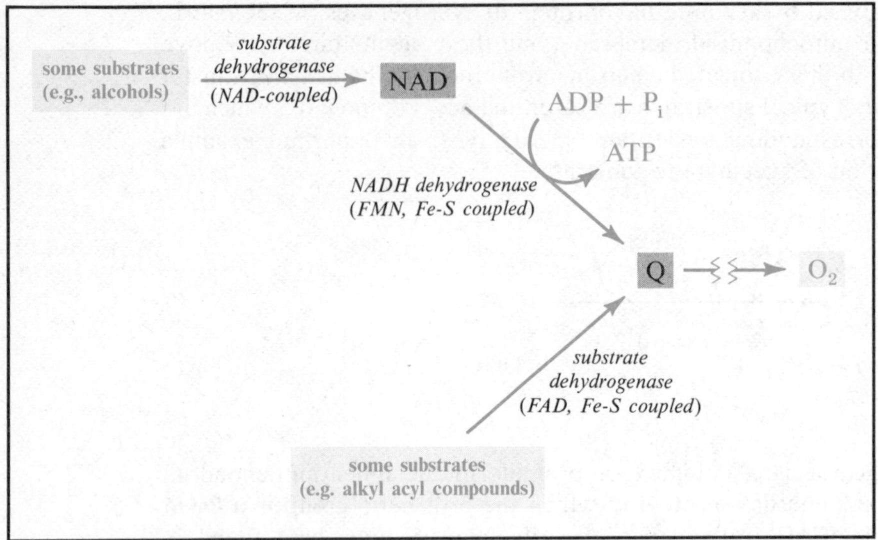

FIGURE 22–8 **Electrons may be transferred from substrates to ubiquinone by one of two routes. Substrates with more negative standard potentials (stronger reducing agents) are oxidized by NAD, and the electrons are passed through the ATP-generating NADH dehydrogenase complex. Other substrates with less negative standard potentials (weaker reducing agents) cannot support the generation of high-energy phosphate, and are attacked by metalloflavoprotein complexes that transfer the electrons directly to ubiquinone. In either case, the resultant ubiquinol is reoxidized by transfer of electrons to molecular oxygen.**

alternately oxidized and reduced during their physiological function. The prosthetic group changes from an Fe(III)-porphyrin (hemin) to an Fe(II)-porphyrin (heme) and back again. Each hemin group can carry one electron.

Cytochromes differ in their strength as oxidizing agents. The most important factor, as in the case of the iron-sulfide proteins, is the interaction between hemin and polypeptide chain, which is determined by the nature of the apoprotein. The nature of the hemin group and its mode of attachment to the apoprotein also contribute (Fig. 22–9, *top*). The cytochromes are given letter designations — a, a_3, b, c, c_1, and so on. These designations were made in the days when the cytochromes were mainly known as changing absorption bands in the spectrum of

FIGURE 22–9 **(*Top*) Many cytochromes contain one of three kinds of hemin group. Cytochromes b contain hemin B, which is the oxidized form of the heme also found in hemoglobin. The hemin C of cytochromes c is identical to hemin B, except that the cysteinyl groups from the surrounding polypeptide chains are covalently added to the vinyl groups of the hemin. The hemin A of cytochromes a differs substantially; one methyl group is oxidized to a formyl group, and a triprenyl hydrocarbon chain is condensed with a hydrated vinyl group.**

(*Bottom*). Idealized structure of cytochrome c. The polypeptide chain folds around a hemin molecule so as to fill the two remaining coordinating positions on the iron of the hemin with the sulfur atom of a methionine residue and a nitrogen atom in a histidine side chain. The hemin is also covalently bound to the polypeptide through the formation of a thioether bridge formed by condensation of cysteine sulfhydryl groups with the vinyl side chains of the hemin. Only one of these bridges is shown; the other lies in front of the histidine side chain.

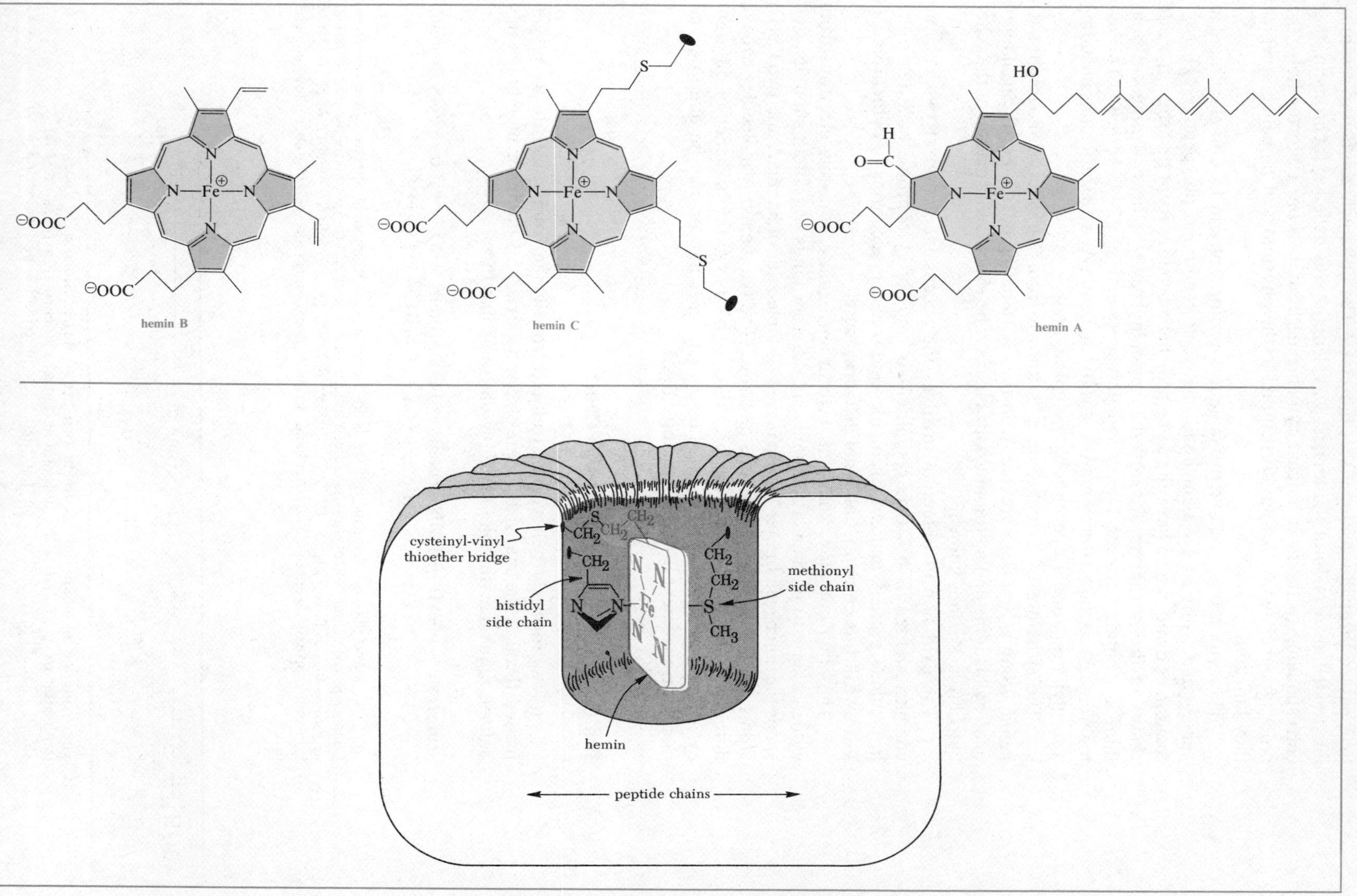

FIGURE 22–9 ***See legend on opposite page.***

light passed through working insect muscles.* The unknown compound responsible for the absorption of the longest wavelengths was called cytochrome a, that corresponding to the next longest cytochrome b, and so on. Unfortunately, the order of wavelength does not correspond to the physiological sequence in which they function.

The cytochromes b have the same iron-protoporphyrin IX complex found in hemoglobin, with no covalent bonds between hemin and polypeptide. The cytochromes c and c_1 also have the same prosthetic group, except that the vinyl side chains of the porphyrin are added to cysteinyl side chains, creating covalent thioether bonds between hemin and polypeptide (Fig. 22–9, *bottom*). The cytochromes a contain a different iron-porphyrin, hemin A, with side chains including a formyl group and a hydrophobic polyprenyl tail.

The structure of cytochrome c is known from X-ray crystallography; its hemin group is in a deep crevice, with the coordinating position of the iron atom filled by histidine and methionine side chains. The exact conformation of the other cytochromes is not known.

The oxidation of ubiquinol involves the successive action of two enzyme complexes (Fig. 22–10). One contains two kinds of cytochrome b, at least one iron-sulfide protein, and cytochrome c_1 and transfers electrons to cytochrome c. Let us call this complex **ubiquinol dehydrogenase.**

The other complex, known as **cytochrome oxidase,** transfers electrons from cytochrome c to oxygen. It contains six different kinds of protein subunits, and spans the inner membrane. Within it are two molecules of hemin A and two Cu(II) atoms, each of which accepts one electron. (The two hemin A molecules behave differently during electron transfer; this may be due to the presence of different cytochromes, commonly indicated as cytochromes aa_3, or may be due to asymmetric arrangement of identical molecules.)

The Complete Electron Transfer Chain

It was formerly believed that all electrons going from substrate to oxygen passed through the many different electron carriers of the inner membrane in a defined sequence, and one saw flow diagrams like this:

substrate → NAD → FMN-protein → Fe-S protein → Q → cyt b → Fe-S protein
cyt c_1 → cyt c → cyt a → Cu(II) → cyt a_3 → O_2.

*Our knowledge of electron transfer is in many ways a triumph of spectroscopic detection. Changes in visible light absorption have been used to discover cytochromes and follow their behavior as well as that of flavoproteins. Ultraviolet spectroscopy is used to follow reduction of NAD. Many iron-sulfide proteins were discovered as bands in electron paramagnetic spectrograms at low temperatures.

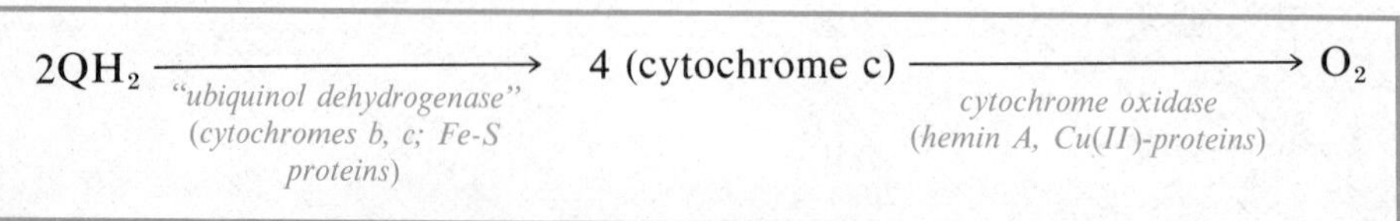

FIGURE 22–10 **The oxidation of ubiquinol occurs in two major stages. Since it is dioxygen (molecular oxygen) that is the ultimate acceptor, cytochrome oxidase must be able to channel four electrons per complete cycle of oxidation and reduction.**

The exact route is still in doubt, but emphasis is now shifting toward a view that the enzyme complexes act as units, and the definitive stages of transfer are those we discussed initially:

$$\textbf{substrate} \rightarrow \textbf{NAD} \rightarrow \textbf{Q} \rightarrow \textbf{cyt c} \rightarrow \textbf{O}_2.$$

We can better understand the reasoning in terms of the free energy changes that are involved. These changes in oxidation-reduction reactions are frequently expressed in terms of the electrical potential generated by the reaction. Let us review the basis for these numbers.

Reduction Potentials

An electrical cell consists of two electrodes, with an oxidation occurring at one and a reduction at the other. It is convenient to separate the reaction occurring at each electrode and express its potential individually, although this cannot be done in an absolute way. For purposes of comparison, the potential of an electrode at which hydrogen gas (1 atmosphere) and protons (1 M) are at equilibrium is taken as zero. Therefore, when one sees a value of −0.315 volts for the reaction:

$$NAD^{\oplus} + H^{\oplus} + 2\ e^{\ominus} \longrightarrow NADH$$

it is really the potential for a cell in which the other electrode reaction is:

$$H_2 \longrightarrow 2\ H^{\oplus} + 2\ e^{\ominus}$$

and the overall reaction becomes:

$$NAD^{\oplus} + H_2 \longrightarrow NADH + H^{\oplus}.$$

Unless otherwise stated, reduction potentials are given in which the oxidized form of the specified compound gains electons from its electrode, and electrons are obtained from H_2 at the other electrode of the cell. If the reaction runs in the opposite direction, an equal oxidation potential of opposite sign is created.

Reduction potentials are like free energies in that they depend upon the concentrations of the cell components; they are usually cited as standard potentials, $E^{0\prime}$, analogous to standard free energies. $\Delta G^{0\prime}$, in which the concentration of H^+ is fixed at a specified value around the electrode in question, and all other reactants, including the H^+ around the hydrogen electrode, are taken to be at 1 molar concentration. The standard reduction potential, $E^{0\prime}$, is related to the change in standard free energy by the equation:

$$\Delta G^{0\prime} = -nFE^{0\prime}$$

in which n = number of electrons transferred in the reaction, and F = Faraday constant = 96,487 joules per mole per volt.

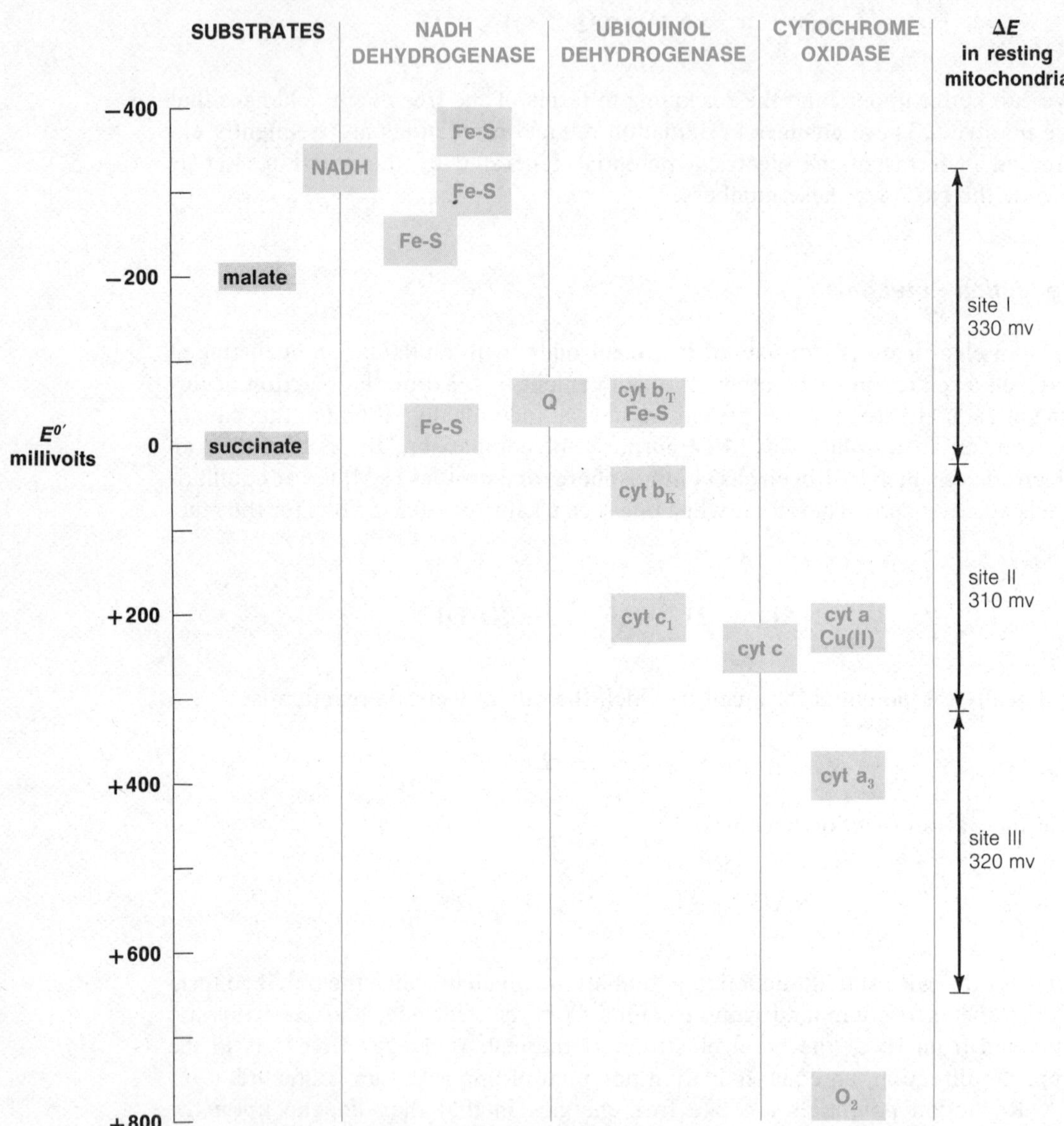

FIGURE 22–11 **A plot of the components of the electron transfer chain on a standard reduction potential scale. Beginning with the reduced form of the substrates (malate or succinate), each successive complex of electron-transferring proteins contains at least one component matching in potential the electron donor from which it will gain electrons, and one component matching the electron acceptor to which it will transfer electrons. The gap in potential between these components provides the energy to generate high-energy phosphate.**

The right column (*black*) shows estimates of the actual potential drop (not standard potential) across the three major stages in mitochondria of resting muscles when the high-energy phosphate concentration is near maximum.

Cell potentials can be added, so the standard potential for the reaction between NADH and oxygen can be obtained from the individual electrode standard potentials at 25° C, pH 7.0:

$$
\begin{array}{lll}
 & NADH + H^{\oplus} \longrightarrow NAD^{\oplus} + H_2 & E^{0\prime} = 0.315 \text{ volts} \\
 & \tfrac{1}{2}O_2 + H_2 \longrightarrow H_2O & E^{0\prime} = 0.815 \text{ volts} \\
\hline
\text{SUM:} & NADH + \tfrac{1}{2}O_2 + H^{\oplus} \longrightarrow NAD^{\oplus} + H_2O & E^{0\prime} = 1.130 \text{ volts}
\end{array}
$$

The standard free energy change for the oxidation of NADH by O_2 at pH 7 and 25° C can be calculated from that value:

$$-2 \times 96{,}487 \times 1.130 = -218{,}060 \text{ J mole}^{-1}$$

Carrier Potentials. A pattern emerges when the potentials of the electron carriers are plotted (Fig. 22–11). Many of the carriers within an enzyme complex have potentials clustered near the same value, but others differ more markedly. The range of potentials spanned within one complex overlap the range spanned by the next.

The current interpretation is that some carriers in each complex are "tuned" to match the carriers donating electrons to and accepting electrons from the complex. For example, many of the iron-sulfide carriers in NADH dehydrogenase have potentials near that of NAD, whereas at least one has a potential near that of ubiquinone. This arrangement is believed to facilitate efficient transport of electrons between the complexes. On the other hand, some carriers within a complex differ sharply in potential, which is the same as saying that there could be a large liberation of free energy upon transfer of electrons between them. These large jumps in potential are believed to provide the energy for the generation of high-energy phosphate.

According to this view, an ordered pathway for electrons among all of the electron-carrying components may not exist. There are many molecules of iron-sulfide proteins for each molecule of flavoprotein, for example, and therefore there are many possible channels within NADH dehydrogenase by which an electron may be brought to the brink of its energy-generating fall.

At the side of Figure 22–11 are indicated the potential spans believed to be associated with each of the three energy-conserving sites in the electron transport chain. Each of these relatively equal values is sufficient for the generation of high-energy phosphate.

COUPLED PHOSPHORYLATION

The fundamental fact about oxidative phosphorylation is that the generation of ATP from ADP and P_i occurs in association with three stages of electron transfer in the inner membranes of mitochondria. If each stage results in the formation of one molecule of ATP, then passage of a pair of electrons through all three stages can cause the generation of three molecules of ATP, and this is the expected result when electrons from substrates are brought to the inner membrane in the form of NADH. In other words, the P:O ratio is typically near 3.0 for mitochondrial oxidations in which NAD is involved, such as the oxidation of hydroxyl compounds, and the overall reaction becomes:

$$\text{CH—OH group} + \tfrac{1}{2}\,O_2 + 3\,(ADP + P_i) \rightarrow \text{C=O group} + 4\,H_2O + 3\text{ ATP.}$$

However, the P:O ratio for the mitochondrial oxidation of such things as the ethylene group in succinate is only 2.0 because electrons are removed from these groups by flavoprotein dehydrogenases that directly transfer the electrons to ubiquinone in the inner membrane without passage through the first phosphorylation site:

$$CH_2{-}CH_2 \text{ group} + \tfrac{1}{2}O_2 + 2(ADP + P_i) \rightarrow CH{=}CH \text{ group} + 3H_2O + 2ATP.$$

To reiterate this important fact, there are three discrete sites in the inner membrane for conservation of energy from electron transfers, one of which is involved only in the oxidation of NADH, with the other two involved in the transfer of electrons from all substrates, including NADH. The energy that is conserved may be used to generate ATP in each case.

Mechanism of Phosphorylation

Despite decades of work by many gifted individuals and much contentious, often rancorous, discussion, the mechanism by which electron transfers in membranes are coupled to phosphorylations still escapes us. The chemical potential dissipated by the electron transfers must somehow be captured if it is to force the synthesis of ATP. The transfers evidently create some sort of a high-energy state within the inner membrane prior to the generation of ATP.

Several phenomena can most readily be explained by assuming that this undefined high-energy state is in equilibrium with the oxidized and reduced forms

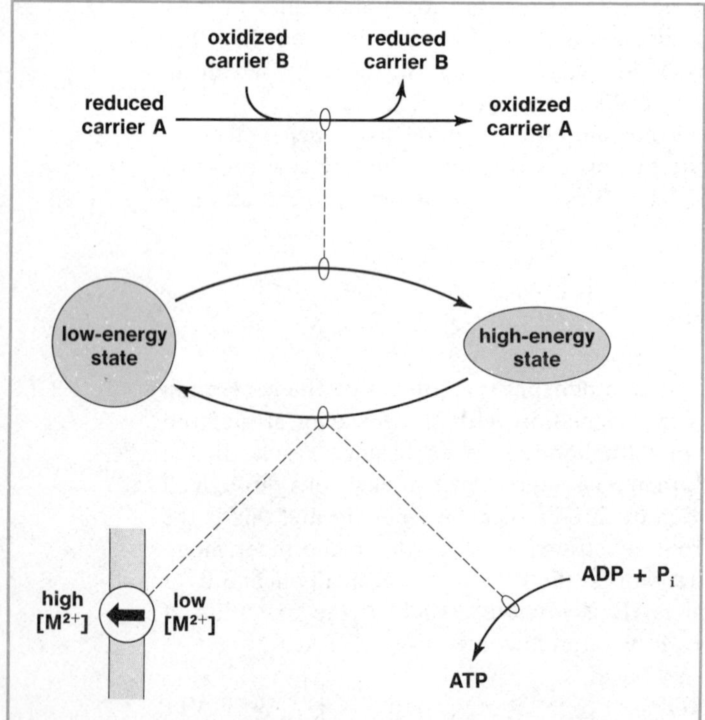

FIGURE 22–12

The transfer of electrons within the inner membrane of mitochondria creates some sort of a high-energy state within the membrane. This high-energy state may be discharged by transporting ions against a concentration gradient (*lower left*), or by phosphorylating ADP to create ATP (*lower right*).

of the electron carriers, with ionic concentration gradients across the membrane, and with ATP and ADP + P_i (Fig. 22–12). These phenomena are:

(1) Electron transfers cause the expulsion of H^+ from the matrix into the cytosol, so that the cytosol is more acidic by as much as 1.5 pH units.

(2) Calcium ions can be actively transported from cytosol to matrix, creating a higher concentration in the matrix.

(3) Both of these ion pumps can be driven by the electron transfers in the absence of ATP and with phosphorylation experimentally blocked.

(4) When ionic concentration gradients are artificially established within mitochondria in the absence of electron transfers, ADP and P_i are converted to ATP.

(5) Contrariwise, added ATP is hydrolyzed to ADP and P_i to pump ions when their concentration gradients are experimentally dissipated.

(6) Under still other experimental conditions, added ATP can be shown to drive the normal flow of electrons backward. For example, succinate will reduce NAD at the expense of high-energy phosphate (Fig. 22–13).

To summarize, the intermediate high-energy state can be reversibly generated or dissipated by electron transfers, ion pumps, and phosphorylation reactions; under physiological conditions it is generated by electron transfer and dissipated by the other routes.

This picture conveys most of the information we will need to understand metabolism; the electron transfers serve to maintain the adenine nucleotides in the form of ATP, but part of the resultant free energy may be diverted to maintain concentration gradients and the like. The diversion may occur directly from the high-energy state without the formation of ATP, or it may occur at the expense of ATP, as with the (Na^+ + K^+)-ATPase in plasma membranes (p. 363). In any event, the bulk of the chemical potential of the reduced electron carriers is used for the generation of ATP.

The High-energy State. Even if we do have most of the information necessary for further reasoning, it is unsatisfactory to have blind faith in a mystical high-energy state. Biochemistry is supposed to be on a sounder basis than parapsychology.

For some time after the discovery of oxidative phosphorylation, it was postulated that oxidation-reduction reactions were coupled to phosphorylation through the formation of covalent intermediates. No such intermediates could be found, despite intensive search and some premature claims. Two other explanations are now popular. One is the chemiosmotic hypothesis, and the other is a

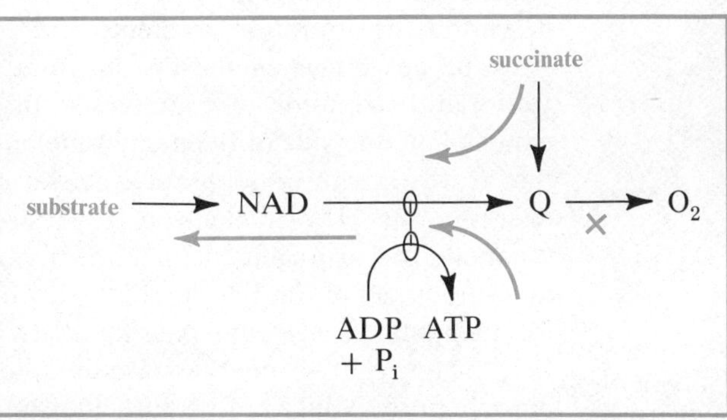

FIGURE 22–13

The near-equilibrium condition of the electron transfers and the phosphorylating mechanism in mitochondria can be demonstrated under conditions in which the flow of electrons from ubiquinone to oxygen is impeded. Addition of succinate and a high concentration of ATP in the absence of added ADP will then cause the reduction of NAD and of substrates ordinarily oxidized by NAD. Hydrolysis of ATP drives the backward flow of electrons.

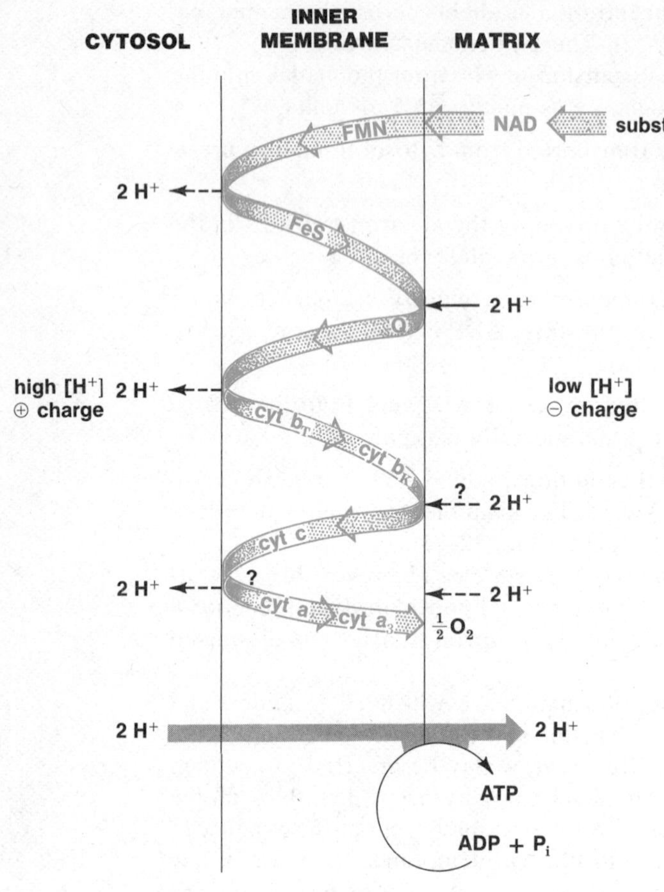

FIGURE 22–14

The chemiosmotic hypothesis of oxidative phosphorylation envisions electron flow from substrates to oxygen occurring through three loops that cross the inner membrane. Each loop is constructed so as to discharge H^+ outside the inner membrane (the cytosol side), and to take up H^+ on the matrix side. (The easily permeable outer membrane on the cytosol side is not shown in this or in subsequent illustrations of transport across the inner membrane.) The result is a creation of a concentration gradient for H^+ across the membrane, and an associated electrical potential. In this scheme, ATP is generated by what may be imagined as the reversal of an ATP-linked proton pump. That is, the combined force of the concentration gradient and potential gradient is sufficient so that propulsion of protons back into the matrix by this force can in some way cause ADP and P_i to combine as ATP.

conformational coupling hypothesis. (Neither of these completely excludes the possibility of some high-energy intermediate.)

The Chemiosmotic, or Electrochemical, Hypothesis (Fig. 22–14). This hypothesis supposes that the proton concentration gradient across the inner membrane is created directly by the electron transfers:

$$2\text{–}4\ H^{\oplus}_{inside} + \text{reduced A} + \text{oxidized B} \longrightarrow 2\text{–}4\ H^{\oplus}_{outside} + \text{oxidized A} + \text{reduced B}.$$

In this view, the high-energy state is the proton gradient, which represents a difference not only in concentration but also in electrical potential across the membrane. ATP is generated when the protons flow back into the mitochondria down the concentration gradient.

The advantage of the chemiosmotic hypothesis is that it accounts for the vectorial distribution of components of the electron transport chain, with some exposed on one side of the membrane and some on the other. The disadvantage is that it invokes three loops of electron carriers across the membrane, each discharging one H^+ per electron traversing the loop. It has not been possible to reconcile the arrangement of known carriers with this requirement, and recent measurements of the H^+ stoichiometry indicate that four, not two, H^+ are moved for each pair of electrons passing a site of energy conservation. Indeed, calculations show that discharge of only two H^+ across the measured gradients in mitochondria would not be sufficient to generate ATP.

The Conformation Hypothesis. This might be called a reverse contraction hypothesis with a Bohr effect. That is, it supposes that the high-energy state is a conformational change in some membrane proteins caused by the electron transfers, analogous to the high-energy state represented by relaxed muscle fibers. As a part of this conformational change, protons are discharged from the proteins on the outside (cytosol side) of the inner membrane, analogous to the way in which they are discharged from hemoglobin upon oxygenation. In this conception, the proton concentration gradient and resultant electrical potential are a part of the high-energy because they tend to push the protein back toward its low-energy conformation.

One can imagine a binding of ADP and P_i by a protein, followed by a shift to a high-energy conformation caused by electron transfer with an associated liberation of protons. When this occurs, the ADP and P_i become freely interconvertible with bound ATP. Release of the bound ATP then causes the protein conformation to shift to its low-energy state and take up the protons previously released.

ATP Synthase. Whatever may be the nature of the high-energy state, it is translated into formation of ATP by a well-characterized complex of enzymes exposed on the inner surface of the inner membrane. This complex comprises a globular headpiece on a stalk attached to a base piece buried in the membrane (Fig. 22–15). When the headpiece is experimentally detached from the membrane, it behaves as an ATPase, catalyzing the decomposition of the bound ATP with water, but the ATP synthesizing activity can be reconstituted by incorporating the complex, along with associated proteins, into phospholipid vesicles.

Stoichiometry of Phosphorylation

Since the exact mechanism of energy-transfer is not known, we cannot state with certainty that there is the production of exactly one high-energy conformation, or high-energy bond, per electron pair passing through each energy-conserving stage of electron transfer. Furthermore, even if a one-to-one stoichiometry is as presumed, the high-energy state can be discharged in other ways than through the production of ATP. Therefore, it may be incorrect to say that exactly three molecules of ATP are produced by the oxidation of one molecule of NADH

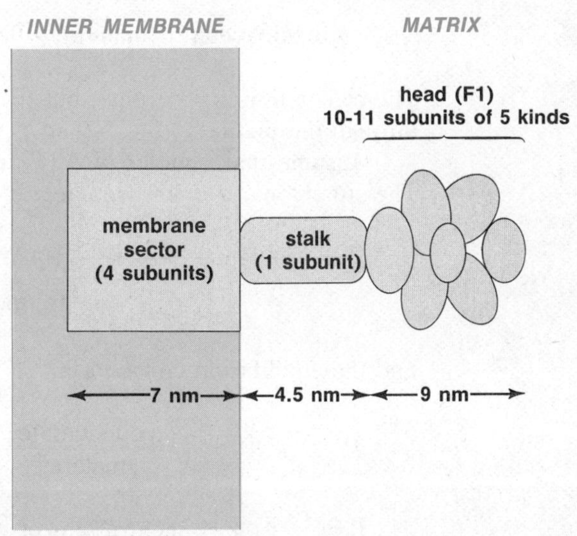

FIGURE 22–15

The high-energy phosphate is generated during oxidative phosphorylation by an elaborate complex of proteins partially embedded in the matrix surface of the inner membrane. This complex has been separated into three major parts, a head that contains the phosphorylating enzymes, a stalk that somehow links the phosphorylation to the high-energy state of the membrane, and a basepiece, or membrane sector, of undelineated function. The diagram is schematic; actual shapes are unknown. The membrane sector, in particular, is not a solid mass of the relative size shown.

with molecular oxygen, or that two molecules of ATP are produced per molecule of succinate oxidized. Even so, we shall use these values as a first approximation in all of our later discussion. Measured values of the P:O ratio approach these empirical integers. We can get some reassurance by a theoretical assessment of the possible yield.

Phosphate Potential. The proportion of the energy of electron transfer used to combine ADP and P_i to make ATP can be increased in two ways—the number of molecules of ATP formed can be increased or the ratio of ATP to ADP and P_i can be increased. Remember that the actual free energy change during a reaction depends upon the relative concentrations of reactants and products. In the case at hand, the actual free energy change during formation of one mole of ATP will be defined by $\Delta G^{0\prime} - RT \ln [\text{ADP}][\text{P}_i]/[\text{ATP}]$, in which $\Delta G^{0\prime}$ is +36,800 J mole^{-1} (the standard free energy change for reversing the hydrolysis of ATP). This actual free energy change is referred to by some as the phosphate potential, although strictly speaking, phosphate potential is the equivalent electrode potential. Some use it in this sense and still others apply it to the concentration ratio itself. The intent in any case is a quantitative measure of the force overcome by electron transfers. It follows that a fixed amount of free energy generated by electron transfers can be used to make more moles of ATP when the ratio $[\text{ADP}][\text{P}_i]/[\text{ATP}]$ is high than when it is low.

Let us test the two possibilities. We can phrase the first question this way: When the ATP and ADP concentrations are equal, what is the theoretical yield of high-energy phosphate from the oxidation of malate:

$$\text{malate} + \tfrac{1}{2}\,O_2 \longrightarrow \text{oxaloacetate} + H_2O \qquad \Delta G^{0\prime} = -192{,}000 \text{ J mole}^{-1}?$$

Since the standard free energy of formation of ATP is only +36,800 J mole^{-1}, one might think it should be possible to make 192/36.8 = 5.2 moles of ATP per mole of malate oxidized. However, we have already learned that standard free energies cannot be used for this kind of reasoning when there are molarity dimensions in the equilibrium constants, and the actual equilibrium conditions must be calculated.

Let us see what happens if we assume that all reactant concentrations, except those of ATP and ADP, are fixed at values similar to those found in living cells:

$$[\text{oxaloacetate}]/[\text{malate}] = 0.01;\ [\text{P}_i] = 0.01\ \text{M};\ [O_2] = 0.001\ \text{atm}.$$

The calculation is laborious, but the result is important, and let us rapidly skim through the method.

Assume that x moles of ATP are generated per atom of oxygen consumed. The stoichiometry then becomes:

$$\text{malate} + x(\text{ADP} + \text{P}_i) + \tfrac{1}{2}\,O_2 \longrightarrow \text{oxaloacetate} + x\text{ATP}$$

$$\Delta G^{0\prime} = 36{,}000x - 192{,}000 \text{ J mole}^{-1}$$

and the equilibrium constant is:

$$K'_{eq} = e^{\frac{-\Delta G^{0\prime}}{RT}} = \left(\frac{[\text{oxaloacetate}]}{[\text{malate}]}\right) \times \left(\frac{[\text{ATP}]}{[\text{ADP}]}\right)^x \times \left(\frac{1}{[\text{P}_i]}\right)^x \times \left(\frac{1}{[O_2]}\right)^{1/2}$$

If we set the concentrations of ATP and ADP to be equal and solve for x, we find x = 4.04 moles of ATP per atom of oxygen consumed. This means that as many as four molecules of ATP could be generated per molecule of malate oxidized if the cell will tolerate the concentration of ATP never being greater than the concentration of ADP.

Let us now ask the second question. What would be the theoretical ratio of ATP and ADP concentrations that could be generated by the oxidation of malate if the yield of ATP is held at 3.0 moles per atom of oxygen consumed (P:O = 3.0)? Substituting $x = 3.0$ in the above equation, and solving for the [ATP]/[ADP] ratio, we find that equilibrium is not reached until [ATP] / [ADP] = 858.

Evolution has made a choice between these extremes of high molar yield and high concentration ratios. In our discussion of high-energy phosphate stores in muscle, we noted that [ATP]/[ADP] had to be over 100 in order to charge the phosphocreatine store to the levels actually present. If the ratio is 150, then only 3.2 molecules of ATP can be made per molecule of malate oxidized (P:O = 3.2). It is apparent that the choice has been made for a high ratio of [ATP]/[ADP], and a P:O ratio of 3.0 means that phosphorylation is proceeding at nearly its maximum efficiency under existing steady state concentrations. This choice appears to be an absolute one, inherent in the mechanism used; that is, the P:O ratio is not increased when the ATP/ADP ratio is experimentally maintained at a low level.

ATP-ADP Antiport

For the most part, ATP is generated within mitochondria and used in the cytosol. ADP and P_i are constantly entering the mitochondrial matrix, and ATP is leaving (Fig. 22–16). Transport of these polar compounds across the inner mem-

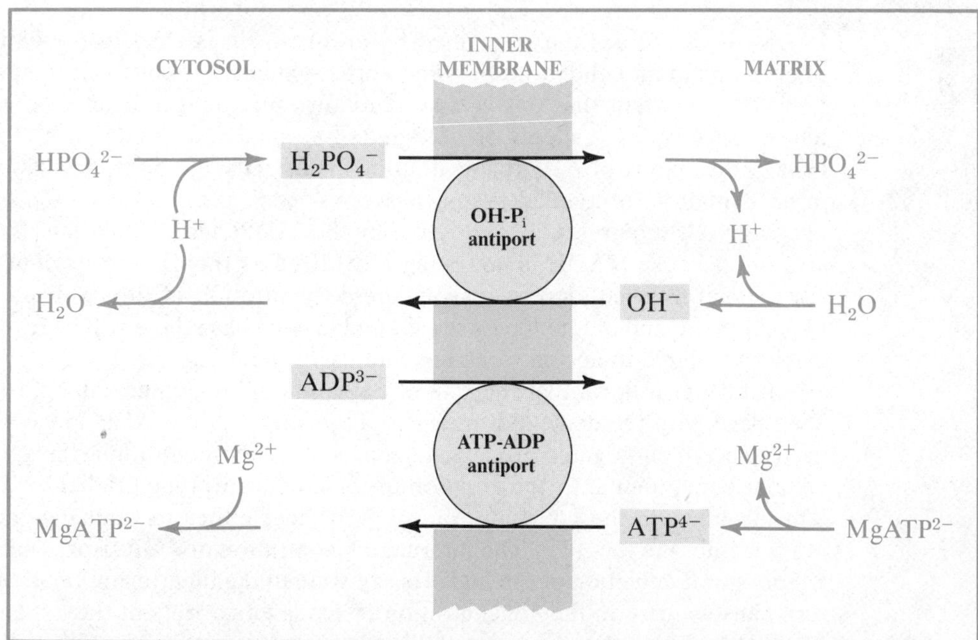

FIGURE 22–16 **Transport of substrates and products of the phosphorylation process across the inner mitochondrial membrane by antiport carriers. One exchanges P_i for OH^- and is driven by the OH^- concentration gradient accompanying the H^+ concentration gradient. The other exchanges ADP^{3-} for ATP^{4-} and is driven by the electrical potential across the membrane.**

brane involves carriers that function by an antiport mechanism. The entrance of P_i is coupled to the exit of OH^-, and the entrance of ADP is coupled to the exit of ATP. (There is another carrier that couples the entrance of P_i to the exit of dicarboxylate anions, but this is less important in the supply of P_i.) Both of these antiports create concentration gradients. The driving force for the exchange of P_i and OH^- is evidently the proton concentration gradient created by electron transfers. It appears that $H_2PO_4^-$ is the ion actually moved by the carrier so that one monovalent anion is being exchanged for another. The carrier is therefore electroneutral — the charge differential across the membrane does not make it go one way or the other. It is the differences in concentration of H^+ and OH^- on the two sides of the membrane that provide the energy.

ATP-ADP exchange is usually said to be electrogenic. That is, the driving force is ascribed to the difference in charge of the two compounds and the potential across the membrane, as is shown in the figure, but the mechanism is not certain. (Owing to the presence of Mg chelates, the total charge on all ionic forms of ATP and ADP can nearly be equal in some circumstances, and we do not have enough detailed quantitative information to make an exact analysis.)

In any event, there is an active transport by which the ratio of concentrations of ATP and ADP is made much larger in the cytosol (as much as 150) than it is in the mitochondrial matrix, where it is near 4. This means that utilization of the energy of electron transfers for phosphorylation is partitioned. Part is used to generate ATP within the matrix, and another part is used to raise its concentration outside the mitochondria. The end result, according to our approximation, is the appearance of 3 ATP outside of the mitochondria per NADH oxidized inside.

The Regulation of Oxygen Consumption

Most of the oxygen consumed by an organism is used for oxidative phosphorylation in mitochondria. In other words, oxidation is coupled to phosphorylation. To the extent that this is true, it follows that oxygen cannot be consumed unless there is also a supply of ADP and of P_i to be combined into ATP, and it is usually the supply of ADP that is limiting in our cells. No phosphorylation results in no oxidation in coupled systems. More specifically, the oxidation of NADH back to NAD requires ADP and P_i. When the [ADP] and [P_i] fall, the [NAD^+] will also fall because NADH is not being reoxidized as rapidly and its concentration rises. For the same kind of reason, the concentration of the oxidized forms of cytochrome c and ubiquinone will fall, and in each case there will be less oxidized carrier available to accept electrons and carry them toward O_2.

It further follows that the rate of oxidation of substrates will adjust so as to keep pace with the demand for ATP. ATP is cleaved into ADP and P_i when it is used to drive endergonic processes such as muscular contration, the synthesis of cellular constituents, or the creation of ionic concentration gradients. The resultant elevation in the concentration of ADP accelerates its exchange across the inner membrane for ATP. The internal concentration of ADP rises, which causes a more rapid depletion of the high-energy state in the inner membrane, and this in turn causes a rise in the concentration of oxidized carriers at the expense of the reduced forms, with a more rapid transfer of electrons toward oxygen. Therefore, the rate of oxidation of fuels is controlled by the requirements for driving endergonic processes, so long as oxidation is tightly coupled to phosphorylation. The coupling is indeed tight under most circumstances, and we might expect this, since a more efficient utilization of fuel ought to convey a great evolutionary advantage. The only exception we might expect is the utilization of oxygen for synthetic reactions, which is usually small.

The electron transfer reactions are apparently always near equilibrium with the phosphorylation reaction within the mitochondria, and deviations from equilibrium by small changes in the ratio of internal ATP and ADP concentrations can have large effects on the rate of electron transfer. This is so because phosphorylation occurs at three successive stages of electron transfer, and changes in the rate of electron transfer will therefore be nearly proportional to the third power of changes in the ADP concentration.

Extramitochondrial change in ATP/ADP ratio causes changes in the same direction within the mitochondria. Opinions differ on whether the ATP-ADP antiport can maintain full equilibration, or whether it is the rate-limiting step, but in either case rises or falls in external [ADP] result in corresponding alterations of internal [ADP].

Adenylate Energy Charge. The critical point is that combustion of fuels is regulated by changes in the ratio $[ATP]/[ADP][P_i]$, which can be expressed quantitatively as changes in the phosphate potential (p. 394). Another measure of this effect, the adenylate energy charge, has been used, but it has flaws. Essentially, adenylate energy charge is an indication of the fraction of the possible high-energy phosphate bonds that actually are present in the adenine nucleotides. This fraction will be 1.0 when ATP is the only adenine nucleotide present, and zero when AMP is the only one present. Since ADP has only one high-energy bond and a full charge is two high-energy bonds per adenine nucleotide,

$$\text{Adenylate energy charge} = \frac{2[\text{ATP}] + [\text{ADP}]}{2([\text{ATP}] + [\text{ADP}] + [\text{AMP}])}.$$

Unfortunately, the numerical values are not a useful guide to the metabolic state. When the [ATP]/[ADP] ratio is very high, as in resting muscle, very large changes in phosphate potential, and therefore in the rate of oxygen consumption, can occur with relatively small absolute changes in the ADP concentration, but the calculated adenylate energy charge changes only little. (We already noted that large changes in the ATP concentration do not occur in muscle until it approaches exhaustion.)

INHIBITORS OF OXIDATIVE PHOSPHORYLATION

Any compound that interferes with oxidative phosphorylation is surely lethal if the interference is severe, since ATP is required for so many vital functions, and oxidative phosphorylation supplies most of the ATP. We might expect such compounds to have their major application outside of the laboratory as pesticides for killing animals. We also might expect them to be toxic to all animals if they can enter the tissues. This is indeed the case. However, some of the compounds have important uses inside the laboratory as experimental tools, and some are even employed therapeutically in sublethal doses. Let us consider some examples.

Inhibitors of Electron Transfer

Cyanides are among the poisons better known by the general public. They are not extraordinarily potent (the minimum lethal dose for humans is estimated to be of the order of 1 to 3 millimoles), but the small HCN molecule rapidly enters tissues, so that a sufficient quantity may be lethal within a few minutes. It is this quick effect, forestalling effective countermeasures in the absence of advance preparation, that has gained the cyanides so much respect, and has led to their use as pesticides against such diverse enemies as rats and insects in ships, moles in lawns, and murderers in some states.

His lawyer probably doesn't realize it, but the felon executed in a gas chamber is having his mitochondrial electron transport blocked at its terminal reaction. The cyanide ion, CN^-, combines tightly with cytochrome oxidase ($K_i \sim$ 8–9 × 10^{-8} M). (Cyanide also combines with other iron porphyrins, especially in the ferric state, but with much less affinity than it has for cytochrome oxidase.) The consequence is the cessation of transfer of electrons to oxygen. The previous electron carriers in the chain accumulate in their reduced state, and the generation of high-energy phosphate ceases.

The effect of cyanide is as fundamental as deprivation of oxygen, and like the latter, causes rapid damage to the brain. Indeed, there have been clinical trials of sub-lethal doses of cyanide as a means of causing corrective disturbances in the brains of psychotics, much as with electric or insulin shocks used for the same purpose. No advantage appeared to counterbalance the disadvantage of the dangerously small margin of safety in dosage.

Cyanide is also encountered in other circumstances. Cassava root, an important foodstuff for many people, contains glycosides that liberate dangerous quantities. Cyanide also is liberated from a glycoside in apricot pits, presently notorious as laetrile. It also may be a hazard in the therapeutic use of sodium ferrinitrosocyanide, $Na_2Fe\ NO(CN)_5$, (sodium nitroprusside) for hypertensive crisis.

Hydrogen sulfide will also combine with cytochrome oxidase, and few realize that it is as toxic as HCN. Its bad odor gives more warning, but there have been fatalities from only a few inspirations at high concentrations of the gas. (In the test tube, 0.1 mM sulfide inhibits cytochrome oxidase more than does 0.3 mM cyanide — 96 per cent against 90 per cent.)

The antimycins are antibiotics produced by species of streptomyces. They strongly associate with mitochondria and block the passage of electrons from cytochrome b to cytochrome c_1; 0.07 micromole of antimycin A_1 per gram of mitochondrial protein is effective. As laboratory tools, they permit experimental distinction between events in the earlier and later parts of the electron transfer chain. Since they kill the host as well as invaders, they are not therapeutic agents. However, they have been employed to kill fish in small lakes for restocking. Only one microgram per liter is required and the compound disappears within a day.

Piericidin A, another antibiotic produced by streptomyces, has an action similar to that of rotenone. **Rotenone** is a compound extracted from the roots of tropical plants (*Derris elliptica, Lonchoncarpus nicou*) that complexes avidly with NADH dehydrogenase; only 30 nanomoles per gram of mitochondrial protein are effective. It is a valuable tool for distinguishing routes of electron flow beginning with NAD-coupled dehydrogenases from those beginning with flavoproteins since it does not affect the latter. It acts between the iron-sulfide proteins and ubiquinone. The results of its effect on oxidative metabolism have long been exploited by primitive people. Rotenone is relatively non-toxic to mammals because it is absorbed poorly, although exposure of the lungs to the dust is a little more dangerous. However, the compound readily passes into the gills of fish and the breathing tubes of insects, and is intensely toxic to these animals. Fish-eaters apply preparations of the appropriate plant roots to ponds, collect the floating fish whose mitochondrial NAD remains reduced, and eat them with impunity. Rotenone was in favor as a relatively safe insecticide, with low toxicity to land vertebrates and a short lifetime, but it is now frowned upon by monitors of public safety using more rigorous standards.

Barbiturates also block NADH dehydrogenase, but much higher concentrations are required. The sedative actions of these compounds appear to depend on other actions on neural membranes, but inhibition of respiration may augment the effect.

Inhibitors of Oxidative Phosphorylation

Oligomycins. These antibiotics from various streptomyces inhibit the transfer of high-energy phosphate to ADP. Therefore, they also inhibit electron transfers coupled to phosphorylation, but they have no effect on oxidation-reduction reactions that are not coupled. They are widely used as experimental tools for discriminating between the two kinds of reactions. They combine with a protein that seems to be in the stalk of the phosphorylating particles, and block any conversion of ADP to ATP, as well as the use of ATP for creating ionic concentration gradients, but they have no effect on the direct formation of ionic concentration gradients by electron transfer.

Atractyloside. This compound from *Atractylis gummifera,* a plant native to Italy, attracted renewed attention when several children were poisoned by eating rhizomes of the plant, resulting in three deaths. Atractyloside blocks oxidative phosphorylation by competing with ATP and ADP for a site on the ADP-ATP antiport of the inner membrane. It, therefore, prevents renewal of the ATP supply in the cytosol. In the laboratory, it is used as a device for separating intra- and extramitochondrial changes in the ATP balance.

Bongkrekate is a toxin formed by a bacterial species (Pseudomonas spp.) in a Javan coconut preparation (''bongkrek''). It also blocks the ATP-ADP antiport and has caused human fatalities. Two micromoles per gram of mitochondrial protein suffice.

Uncouplers of Oxidative Phosphorylation

If electron transfers in mitochondria can somehow be dissociated from phosphorylation, the supply of ATP will be impaired as effectively as if phosphorylation were inhibited. This uncoupling can be achieved, in effect, by causing a constant discharge of the high-energy state that is ordinarily used to generate ATP. Uncoupling can be distinguished from inhibition in this way: Uncoupling causes an increased oxygen consumption in the absence of increased utilization of ATP. Inhibition of phosphorylation, or inhibition of the ATP-ADP antiport, diminishes oxygen consumption in normal coupled mitochondria.

2,4-Dinitrophenol acts as an uncoupler at a concentration of 10 micromolar. It apparently does so by its ability to dissolve in the lipid core of the inner membrane where it can diffuse from one face to the other. Since the phenol can ionize, the effect will be to equilibrate H^+ concentrations on the two sides of the membrane, thereby continuously discharging the high-energy state and causing more consumption of oxygen.

Valinomycin is an example of an ionophore, a compound that binds ions for transport. This particular example, produced by streptomyces, is an antibiotic with a ring structure that closely fits K^+, but with a sufficiently hydrophobic exterior to permit passage through the inner mitochondrial membrane. It combines with K^+ from the cytosol side of the membrane, and the charged complex is driven by the electrical potential to the inside surface where K^+ can be released. This, in turn, dissipates part of the potential, permitting more H^+ to be pumped out of the mitochondria. The result is a constant exchange of K^+ for H^+ across the membrane, which also discharges the high-energy state, thereby causing increased consumption of oxygen.

FURTHER READING

Sanadi, D. R., ed.: (1976) *Chemical Mechanisms in Bioenergetics*. Am. Chem. Soc.

Tedeschi, H.: (1976) *Mitochondria: Structure, Biogenesis, and Transducing Function*. Springer Verlag.

Hatefi, Y., and L. Djavadi-Ohaniance, eds.: (1967) *The Structural Basis of Membrane Function*. Academic Press.

Hatefi, Y., and D. L. Stiggall: (1976) *Metal-containing Flavoprotein Dehydrogenases*. Vol. 13, p. 175, *in* P. D. Boyer, ed.: *The Enzymes*, 3rd. ed. Academic Press.

Caughey, W. S., et al.: (1975) *Heme A of Cytochrome C Oxidase*. J. Biol. Chem., *250*: 7602.

Yasunobu, K. T., H. F. Mower, and O. Hayaishi: (1976) *Iron and Copper Proteins*. Adv. Exp. Biol. Med., *5*: 74.

Caughey, W. S.: (1976) *Cytochrome C Oxidase*. Vol. 13, p. 299, *in* P. D. Boyer, ed.: *The Enzymes*, 3rd ed. Academic Press.

Packer, L., and A. Gómez-Puyou, eds.: (1976) *Mitochondria – Bioenergetics, Biogenesis, and Membrane Structure*. Academic Press.

Boyer, P. D., ed.: (1975) *Biological Energy Transductions*. Fed. Proc., *34*: 1699.

Pressman, B. C.: (1976) *Biological Applications of Ionophores*. Annu. Rev. Biochem., *45*: 501.

Dutton, P. L., and D. F. Wilson: (1974) *Redox Potentiometry in Mitochondrial and Photosynthetic Bioenergetics*. Biochim. Biophys. Acta, *346*: 165. Review.

Wilson, D. F., M. Erecinska, and P. L. Dutton: (1974) *Thermodynamic Relationships in Mitochondrial Oxidative Phosphorylation*. Annu. Rev. Biophys. Bioeng., *3*: 203.

Ciba Foundation: (1975) *Energy Transformations in Biological Systems*. Elsevier.

Senior, A. E.: (1973) *The Structure of Mitochondrial ATPase*. Biochim. Biophys. Acta, *301*: 249. Review.

Robin, E. E.: (1977) *Dysoxia*. Arch. Intern. Med., *137*: 905. Contains discussion of cyanide poisoning.

Palmer, R. F., and K. C. Lasseter: (1975) *Sodium Nitroprusside*. N. Engl. J. Med., *292*: 294. Discusses liberation of cyanide.

Reynafarje, B., M. D. Brand, and A. C. Lehninger: (1976) *Evaluation of the H^+/Site Ratio of Mitochondrial Electron Transport from Rate Measurements*. J. Biol. Chem., *251*: 7442.

Boyer, P. D., et al.: (1977) *Oxidative Phosphorylation and Photophosphorylation*. Annu. Rev. Biochem., *46*: 955, ff. A collection of small reviews by several authors.

Davis, E. J., L. Lumeng, and D. Bottoms: (1974) *On the Relationships between the Stoichiometry of Oxidative Phosphorylation and the Phosphorylation Potential of Rat Liver Mitochondria as Functions of Respiratory State*. FEBS, Lett., *39*: 9.

23 THE CITRIC ACID CYCLE

GENERAL DESCRIPTION

The combustion of fuels can be sketched very simply if we use only bold strokes (Fig. 23–1). Each kind of major fuel is converted to **acetyl** groups, which are handled by attachment to a particular coenzyme known as **coenzyme A.** The acetyl groups, regardless of source, can then be oxidized to CO_2 and H_2O.

Simple as this sketch is, it describes the major part of metabolic processes in tissues such as muscles and nerves, in which metabolism is mainly geared to the production of ATP for physical movement or for transmission of neural impulses. We shall see later that the oxidation of acetyl groups accounts for roughly two thirds of the total oxygen consumption and ATP production in most animals.

The sequence of reactions by which acetyl groups are oxidized is known as the citric acid cycle. The reactions function as a unit to remove electrons from acetyl groups within mitochondria and to feed the electrons into the oxidative phosphorylation pathway, with the ultimate transfer to molecular oxygen. The prime purpose of the complete citric acid cycle is the complete oxidation of acetyl groups in a way that will cause the formation of ATP.

The citric acid cycle gets its name because it begins with the formation of **citrate,** the anionic form of citric acid. (The cycle is also known as the Krebs cycle, after Sir Hans A. Krebs, its discoverer, and as the tricarboxylic acid cycle.) Citrate has six carbon atoms, and it is formed by the condensation of the two-carbon acetyl group removed from acetyl coenzyme A with a four-carbon compound, **oxaloacetate** (Fig. 23–2). The sequence proceeds with an oxidation that causes the loss of one carbon atom as CO_2, leaving the five-carbon compound, **α-ketoglutarate (2-oxoglutarate).** A second carbon atom is then lost as CO_2 in

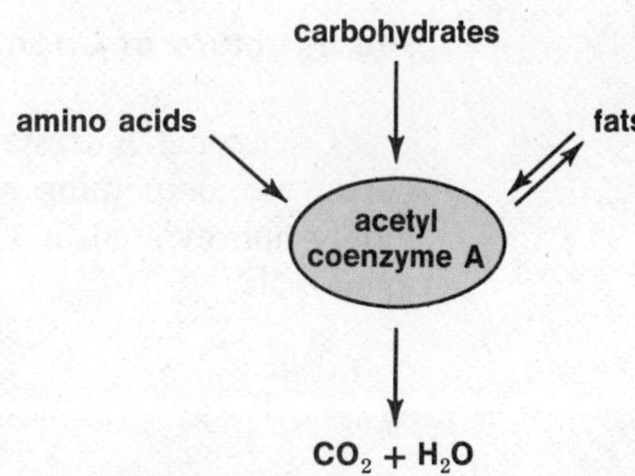

FIGURE 23–1

Acetyl coenzyme A is formed from carbohydrates, fats, and amino acids. Whatever the source, a large fraction of it is oxidized to CO_2 and H_2O, although any excess may be used to form fats for storage.

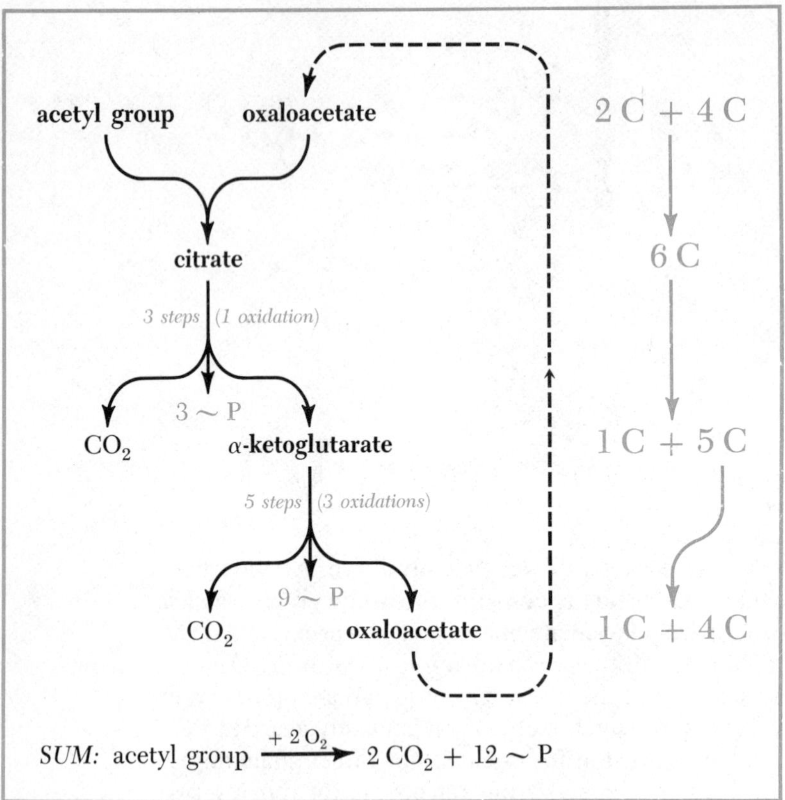

FIGURE 23–2 **Outline of the citric acid cycle.**

another oxidation, and the four-carbon remainder undergoes two more oxidations and is finally transformed into oxaloacetate, which replaces the oxaloacetate originally consumed and completes the cycle.

The dehydrogenases responsible for the four oxidations catalyze the transfer of hydrogen atoms, including the associated electrons, to NAD or directly to ubiquinone. In either case, the electrons are used for oxidative phosphorylation with an ultimate consumption of oxygen. Coupling with oxidative phosphorylation is possible because the enzymes involved in the cycle are found on the inner membrane of mitochondria or in the matrix. The overall result of all of this, as we shall see, is the complete combustion of the acetyl group.

ACETYL COENZYME A: THE COMPOUND

The Structure of Coenzyme A

Coenzyme A contains one sulfhydryl group that is the reactive part of the coenzyme, combining with acyl groups to form thiol esters. It is therefore commonly abbreviated as CoA-SH. Using this shorthand description, acetyl coenzyme A is:

$$CH_3{-}\overset{\overset{\Large O}{\|}}{C}{-}S{-}CoA.$$

This handy way of abbreviating the structure properly emphasizes the acyl group being transported and its linkage as a thiol ester, but coenzyme A is actually quite complex. The reactive sulfhydryl group is at one end of a molecule comprising an adenine nucleotide and the three other compounds named in Figure 23–3. We have already noted that coenzymes frequently contain unusual structures — unusual in the sense that animals have lost the ability to form them, and they must be supplied in the diet. Coenzyme A has such a structure, contained in the vitamin, **D-pantothenic acid.**

Surprisingly, the vitamin does not in this case contain the reactive group of the coenzyme. The unusual branched chain structure of pantothenate may act as a marker primarily for those enzymes fabricating the complete coenzyme from its components rather than for the enzymes with which it acts as a coenzyme. Animals lacking panthothenate die, but a natural deprivation of panthothenate is very rare because coenzyme A is present in all living cells and therefore in the natural foods of all animals.

Energetics of Thiol Esters

Thiol esters such as acetyl coenzyme A are high-energy compounds, with the standard free energy for hydrolysis being near −35,700 joules per mole at pH 7

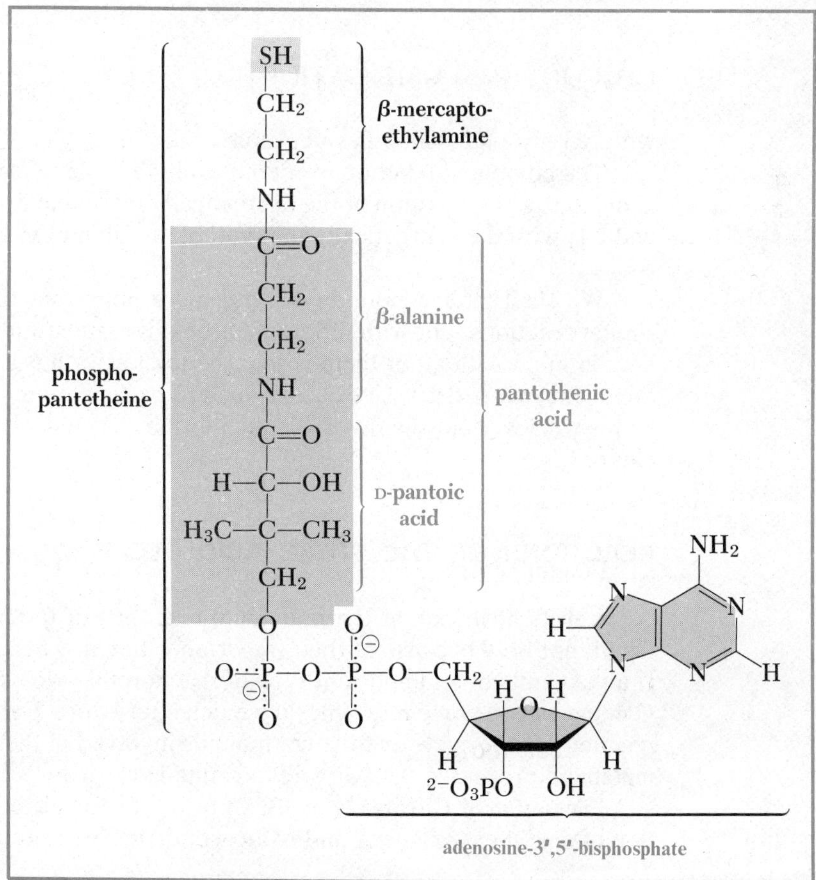

FIGURE 23–3 **The structure of coenzyme A and the names of its constituents.**

and 38°C. This corresponds to a K'_{eq} of 9.9×10^5, so acetyl coenzyme A won't form simply because acetate and coenzyme A are present in the same solution.

In other words, acetyl coenzyme A must be made by the same sort of general process used for the formation of peptides — coupling the formation to some reaction that would by itself have a highly negative standard free energy, thereby counterbalancing the highly positive standard free energy for the combination of acetate and coenzyme A. An example will show the principle.

Acetic acid occurs in the diet; vinegar is a familiar source. Ingested acetic acid is brought into the mainstream of intracellular metabolism by converting it to acetyl coenzyme A. The conversion is accomplished by an enzyme that catalyzes the simultaneous conversion of ATP to AMP and PP_i:

$$\text{Acetate} + \text{CoA-SH} + \text{ATP} \rightarrow \text{acetyl CoA} + \text{AMP} + PP_i.$$

The enzyme is called **acetyl CoA synthetase.**

The inorganic pyrophosphate produced by the reaction is in turn hydrolyzed by the ubiquitous inorganic pyrophosphatase with an additional liberation of free energy, so the formation of acetyl coenzyme A is another process in which there is an effective cleavage of two high-energy phosphate bonds during the synthesis of a compound because of the cooperative action of the primary enzyme catalyzing the synthesis and the pyrophosphatase acting on PP_i. (The pyrophosphatase increases the loss of free energy by diminishing the concentration of PP_i.) The overall reaction is:

$$CH_3COO^{\ominus} + \text{CoA-SH} + \text{ATP} + H_2O \longrightarrow CH_3{-}\overset{\overset{\displaystyle O}{\|}}{C}{-}S{-}\text{CoA} + \text{AMP} + 2P_i$$

with $\Delta G^{0\prime} = -22{,}000$ joules per mole.

The combined reaction is energetically feasible, and it is convenient to think about it as a combination of the energetically favorable hydrolysis of ATP to AMP and 2 P_i with the energetically unfavorable combination of acetate and coenzyme A.

We shall later encounter several more important thiokinases that catalyze similar reactions, but with different carboxylate substrates. An important thing to bear in mind with all of them is that the thiol ester has an energy-rich bond, and the coenzyme A derivatives are formed to participate in reactions that would not be possible without the favorable equilibrium created when the thiol ester is later cleaved.

REACTIONS OF THE CITRIC ACID CYCLE

Let us now look at the individual reactions of the citric acid cycle in some detail, not only because of their importance but also because each and every one is an example of an important type of reaction that we will repeatedly encounter. Gaining a good grasp now will save much later effort. Furthermore, the individual reactions of the cycle and the compounds involved in them are also parts of other metabolic processes that we will examine later.

Formation of Citrate (2C + 4C → 6C). The cycle begins with the condensation of acetyl coenzyme A and oxaloacetate to form citrate (Fig. 23–4), catalyzed

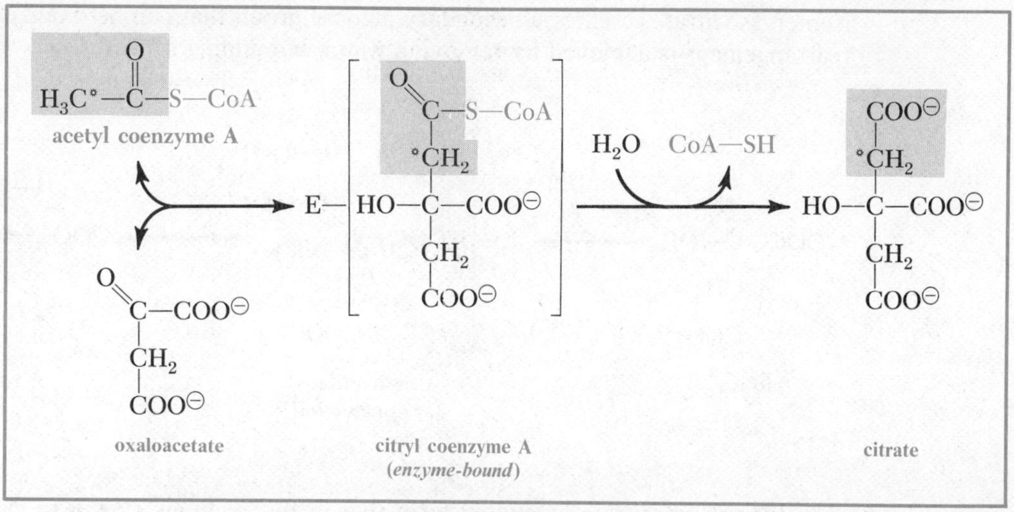

FIGURE 23–4 **Citrate synthase catalyzes a condensation of acetyl coenzyme A and oxaloacetate to form citryl coenzyme A as an enzyme-bound intermediate. The asterisk indicates the position of a particular carbon atom during the reaction.**

by the enzyme, **citrate synthase.** As the figure shows, the reaction proceeds in two parts on the enzyme surface; the initial condensation results in the formation of citryl coenzyme A (the coenzyme A thiol ester of citric acid). The citryl coenzyme A is then hydrolyzed to release free citrate and coenzyme A into solution.

The formation of a carbon to carbon bond by the condensation of a carbonyl group with an activated methylene carbon is a common type of reaction in metabolism. These are familiar reactions to the organic chemist, and are variants of the classical aldol condensation, which we examined in connection with the fructose bisphosphate aldolase reaction (p. 266).

It so happens that free acetate can be made to condense with oxaloacetate, so what is the point of evolving a reaction that utilizes the coenzyme A ester since it is destroyed before the reaction is over? The answer lies in the position of equilibrium of the reaction. The condensation of acetate and oxaloacetate by themselves has an equilibrium position such that only relatively high concentrations of the starting materials will result in significant formation of citrate. However, by making acetyl coenzyme A the starting material, we add the highly exergonic hydrolysis of the coenzyme A ester to the reaction, and the sum of these two processes makes the formation of citrate an almost irreversible reaction, so much so that it can readily proceed even when the intramitochondrial concentration of oxaloacetate falls to an estimated 4 micromolar or lower. The $K'_{eq} = 2.24 \times 10^6$.

Formation of α-Ketoglutarate (6C + NAD → 5C + CO_2 + NADH)

Citrate → Isocitrate. The metabolism of citrate begins by converting it into an alcohol that can be oxidized. Citrate has a tertiary alcohol group, which could not be attacked without breaking a carbon bond, but rearrangement into the isomer, isocitrate, creates a secondary alcohol group that can be oxidized. This rearrangement is achieved by removing water and adding it back:

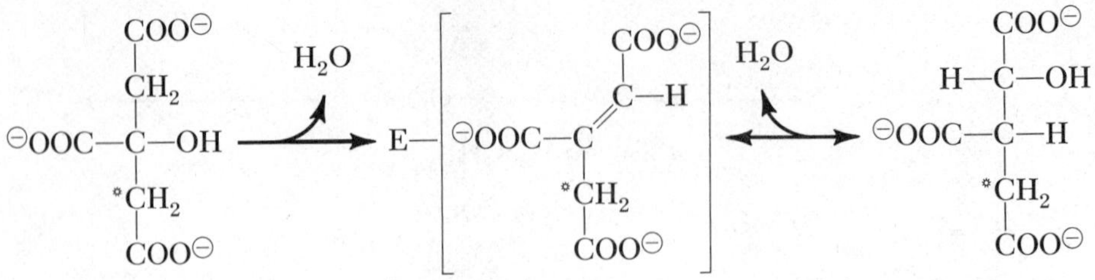

$$\begin{matrix} COO^{\ominus} \\ | \\ CH_2 \\ | \\ {}^{\ominus}OOC{-}C{-}OH \\ | \\ {}^{\circ}CH_2 \\ | \\ COO^{\ominus} \end{matrix} \xrightarrow{H_2O} E{-}\left[\begin{matrix} COO^{\ominus} \\ | \\ C{-}H \\ \| \\ {}^{\ominus}OOC{-}C \\ | \\ {}^{\circ}CH_2 \\ | \\ COO^{\ominus} \end{matrix}\right] \xleftrightarrow{H_2O} \begin{matrix} COO^{\ominus} \\ | \\ H{-}C{-}OH \\ | \\ {}^{\ominus}OOC{-}C{-}H \\ | \\ {}^{\circ}CH_2 \\ | \\ COO^{\ominus} \end{matrix}$$

citrate — *cis*-aconitate (*enzyme-bound*) — isocitrate

(The formula of citrate is rotated from that in the previous sketch to show the reacting end at the top.) The dehydration and hydration involved in the rearrangement are catalyzed by the same enzyme, aconitase. When aconitase catalyzes the addition of water to the double bond of *cis*-aconitate, the hydroxyl group may be added to either carbon; in one case citrate is formed and in the other, isocitrate. The enzyme will also catalyze the removal of water from either citrate or isocitrate to form *cis*-aconitate as an intermediate bound on the enzyme. Aconitase is an example of a hydratase, an enzyme catalyzing the reversible hydration of double bonds. (It is unlike many enzymes of this type, which usually require no cofactors, in that it contains Fe^{2+} for binding the substrates.) The interconversion of alcohols and unsaturated compounds by addition or removal of water is a common type of biochemical reaction, and the standard free energy change is usually small enough so that the reaction can go in either direction to a biologically significant extent. In this case the equilibrium mixture contains about 91 per cent citrate, 3 per cent *cis*-aconitate, and 6 per cent isocitrate. Since the intramitochondrial concentration of citrate is greater than 1 millimolar, the equilibrium concentration of isocitrate would be of the order of 0.1 mM, a reasonable level for metabolic intermediates.

Under ordinary circumstances, the reactions go through the sequence as given: citrate → (aconitate) → isocitrate. There is nothing peculiar about aconitase to cause this: it only catalyzes the attainment of equilibrium. The net flow of metabolites goes in one direction because citrate is constantly being formed by the irreversible citrate synthase reaction, and isocitrate is constantly being removed by the next reaction, in which isocitrate is oxidized.

Oxidative Decarboxylation of Isocitrate. The alcohol group of isocitrate is oxidized by NAD in a reaction catalyzed by an isocitrate dehydrogenase:

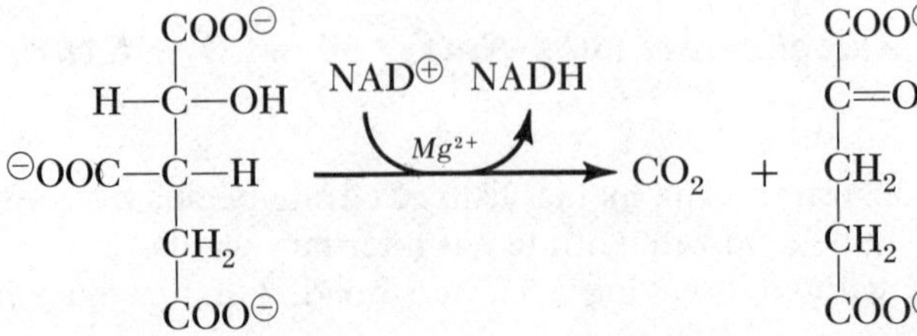

$$\begin{matrix} COO^{\ominus} \\ | \\ H{-}C{-}OH \\ | \\ {}^{\ominus}OOC{-}C{-}H \\ | \\ CH_2 \\ | \\ COO^{\ominus} \end{matrix} \xrightarrow[Mg^{2+}]{NAD^{\oplus} \quad NADH} CO_2 + \begin{matrix} COO^{\ominus} \\ | \\ C{=}O \\ | \\ CH_2 \\ | \\ CH_2 \\ | \\ COO^{\ominus} \end{matrix}$$

isocitrate — α-ketoglutarate

This reaction results not only in the conversion of the alcohol to the corresponding ketone, but also in the loss of one of the carboxyl groups as CO_2 to form α-ketoglutarate. The reaction is therefore an example of an **oxidative decarboxylation.** It is quite common, although not obligatory, for the enzymatic oxidation of the hydroxyl group in a 3-hydroxy carboxylic acid to cause the simultaneous loss of CO_2. In the example at hand, this not only occurs, but it does so without even a transitory appearance of the intermediate β-keto compound.

The enzyme we are discussing apparently catalyzes both the oxidation and the decarboxylation in a concerted way, but there are related enzymes that catalyze the two parts separately. When this happens, the initial oxidation of isocitrate by NAD causes the temporary formation of the corresponding ketone known as **oxalosuccinate** on the enzyme:

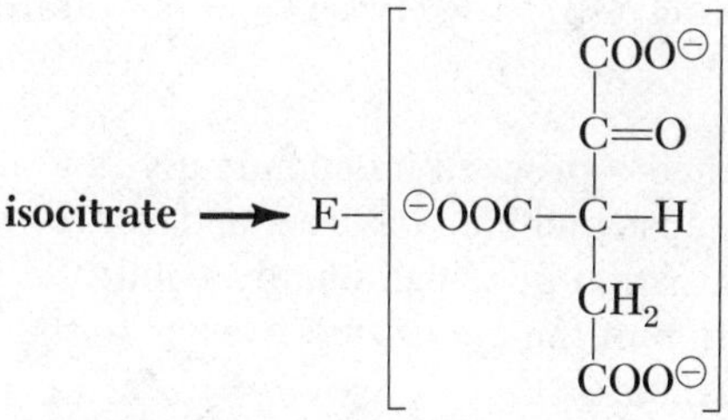

$$\textbf{isocitrate} \longrightarrow E\text{—}\left[\begin{array}{r} COO^{\ominus} \\ | \\ C{=}O \\ | \\ {}^{\ominus}OOC\text{—}C\text{—}H \\ | \\ CH_2 \\ | \\ COO^{\ominus} \end{array}\right]$$

This is a straightforward alcohol dehydrogenase type of reaction, but notice that the carbonyl group of the intermediate oxalosuccinate is in the β-position relative to the middle carboxylate group. (Of course, it is also in the α-position relative to the top carboxylate group, but that is not relevant to our point.) Now, we know from organic chemistry that β-ketoacids are readily decarboxylated:

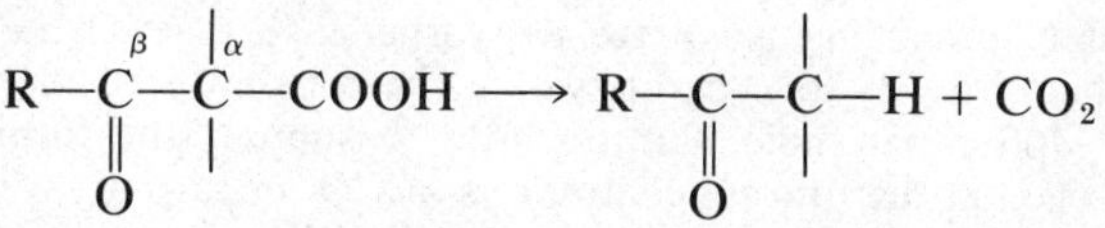

$$R\text{—}\overset{\beta}{\underset{\underset{O}{\|}}{C}}\text{—}\overset{|\,\alpha}{\underset{|}{C}}\text{—}COOH \longrightarrow R\text{—}\underset{\underset{O}{\|}}{C}\text{—}\overset{|}{\underset{|}{C}}\text{—}H + CO_2$$

Therefore, we might expect the intermediate oxalosuccinate to readily lose CO_2, and it does.

Role of Mg^{2+}. Isocitrate dehydrogenase requires Mg^{2+} for activity like many enzymes that catalyze β-decarboxylations. Enzymes catalyzing simple dehydrogenations without decarboxylation do not require metallic ions for activity.

Oxidative Decarboxylation of α-Ketoglutarate ($5\ C + NAD \rightarrow 4C + CO_2 + NADH$)

The next reaction of the citric acid cycle is an oxidation of α-ketoglutarate by NAD in such a way that the first carboxyl group appears as CO_2. The resultant four-carbon succinyl group is transferred onto coenzyme A, thereby capturing part of the energy of the oxidation in the energy-rich thiol ester, succinyl coenzyme A:

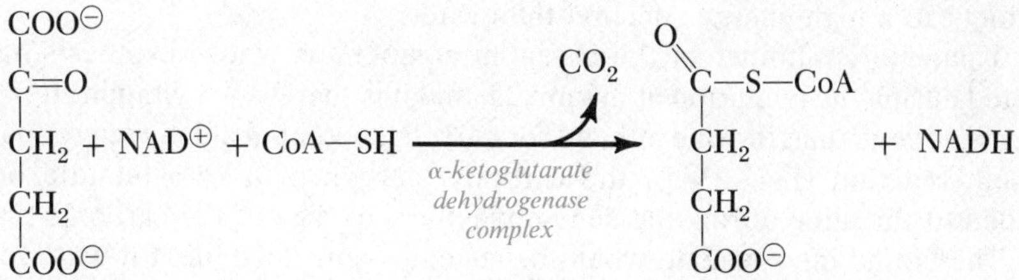

$$\begin{array}{l} COO^{\ominus} \\ | \\ C{=}O \\ | \\ CH_2 \\ | \\ CH_2 \\ | \\ COO^{\ominus} \end{array} + NAD^{\oplus} + CoA\text{—}SH \xrightarrow[\substack{\alpha\text{-ketoglutarate} \\ \text{dehydrogenase} \\ \text{complex}}]{CO_2} \begin{array}{l} O \\ \diagdown \\ C\text{—}S\text{—}CoA \\ | \\ CH_2 \\ | \\ CH_2 \\ | \\ COO^{\ominus} \end{array} + NADH$$

α-ketoglutarate **succinyl coenzyme A**

α-Ketoglutarate is an example of a 2-ketocarboxylate, and it is a convenient formalism to regard the oxidation of these compounds as the sum of two organic reactions. One is a simple decarboxylation to produce CO_2 and an aldehyde:

$$\text{R—}\overset{\overset{\displaystyle\text{O}}{\|}}{\text{C}}\text{—COOH} \longrightarrow \text{R—}\overset{\overset{\displaystyle\text{O}}{\|}}{\text{C}}\text{—H} + \text{CO}_2 \quad \Delta G^{0\prime}\ -23{,}000\ \text{J mole}^{-1}$$

and the other is the oxidation of the aldehyde:

$$\text{R—}\overset{\overset{\displaystyle\text{O}}{\|}}{\text{C}}\text{—H} + \text{NAD}^{\oplus} + \text{H}_2\text{O} \longrightarrow \text{R—COO}^{\ominus} + \text{NADH} + 2\text{H}^{\oplus} \quad \Delta G^{0\prime} \sim -50{,}000\ \text{J mole}^{-1}.$$

These steps could in theory proceed independently, and do so in some microorganisms, but the two steps combined have a standard free negative energy change that is ample for the creation of a high-energy bond; the reaction is modified to provide for the formation of the coenzyme A ester, rather than the free anion of the carboyxlic acid product.

The formation of the high-energy thiol ester bond is quite independent of the concomitant formation of high-energy phosphate by mitochondrial electron transport; it serves to tap the extra free energy available from the oxidative decarboxylation of the 2-ketocarboxylate, which is substantially greater than the free energy released during the oxidation of alcohols.

The immediately preceding isocitrate dehydrogenase reaction is also an oxidative decarboxylation, but of a 3-hydroxycarboxylate. The standard free energy change for that reaction is only –400 J per mole, far too little to support the formation of an extra high-energy bond even at the low concentrations of CO_2 occurring in tissue.

α-Ketoglutarate Dehydrogenase Complex. We can see that there are at least three things going on during the conversion of α-ketoglutarate to succinyl coenzyme A: decarboxylation, oxidation, and formation of the thiol ester, and all of this is too much to be accomplished by a single enzymatic reaction. Instead, cells construct a marvelous example of an integrated enzyme complex in which a sequence of discrete reactions is carried out in a coordinated way. This complex is a distinctive arrangement of different protein molecules. Each kind of protein catalyzes a separate part of the reaction sequence, but the association into a single complex makes it possible to link the reactions together without release of intermediates into solution.

The decarboxylating dehydrogenase is the component of the complex that initially reacts with α-ketoglutarate; it contains a coenzyme, **thiamine pyrophosphate,** we have not previously discussed. This protein catalyzes two successive reactions. The first reaction is a decarboxylation releasing the 1-carboxyl group as CO_2. The second reaction is the oxidation of the remaining bound four-carbon fragment to a high-energy succinyl thiol ester.

Thiamine is another of the essential compounds whose synthesis has been deleted during the evolution of mammals, and it is therefore a vitamin. It is used in other reactions than the one at hand for carrying potential aldehyde groups. In the present reaction (Fig. 23–5), the carbonyl carbon of α-ketoglutarate becomes bonded to thiamine during the accompanying cleavage of the carboxyl group.

The bound intermediate would be succinic semialdehyde if it were released. Instead, the potential succinaldehydate group is oxidized by transfer to a bound coenzyme, **lipoic acid,** on another protein in the complex, **transsuccinylase** (Fig.

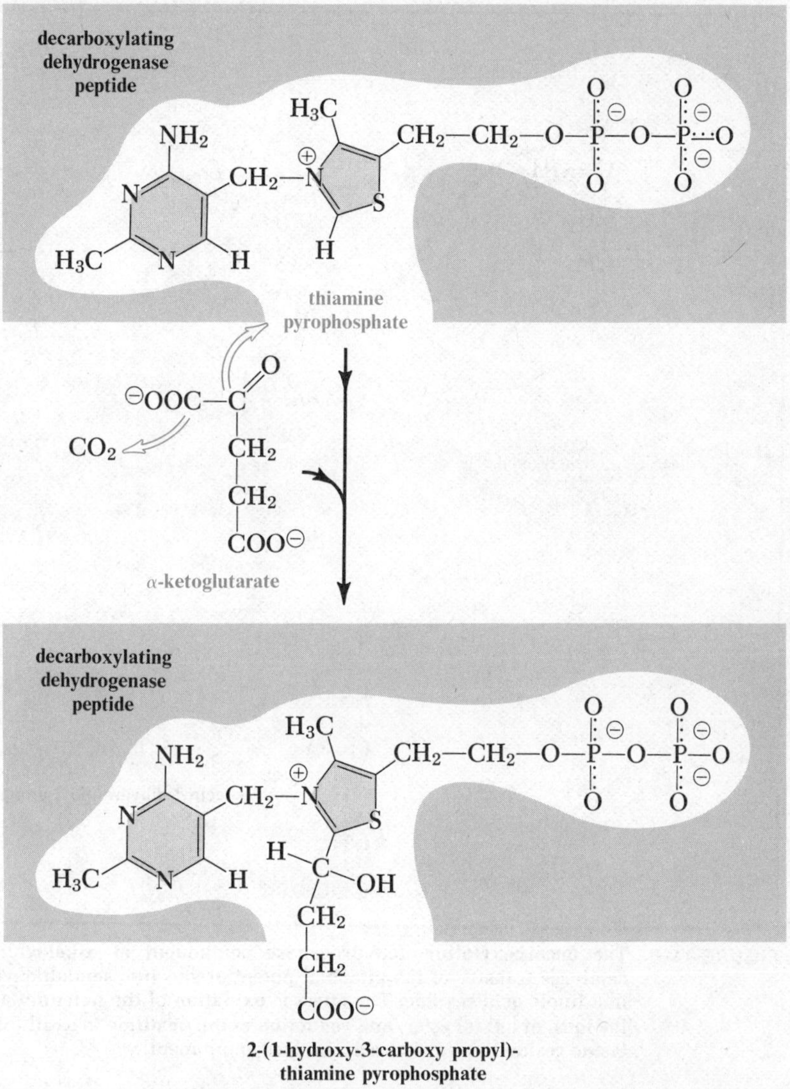

FIGURE 23–5 Thiamine pyrophosphate, attached to the decarboxylating dehydrogenase component of α-ketoglutarate dehydrogenase complex, accepts a potential succinic semialdehyde group from α-ketoglutarate, with a simultaneous loss of CO_2.

23–6). The reactive part of lipoic acid is a disulfide group, which acts as both an oxidizing agent for the substrate "aldehyde" and as a receptor for the resultant succinyl group. Two things are happening in this single reaction: The lipoyl group is being reduced to a dihydrolipoyl group, and the resultant succinyl group is attached as a thiol ester.

TRANSFER OF SUCCINYL GROUP TO COENZYME A. The transsuccinylase component carries lipoyl groups as oxidizing agents, but the protein is also an enzyme, and its function is to convert the high-energy succinyl thiol ester to a usable form by transferring the succinyl group to an incoming molecule of coenzyme A, which then leaves as succinyl coenzyme A (Fig. 23–7).

DIHYDROLIPOAMIDE DEHYDROGENASE. The action of transsuccinylase completes the oxidative decarboxylation of α-ketogluterate and the capture of

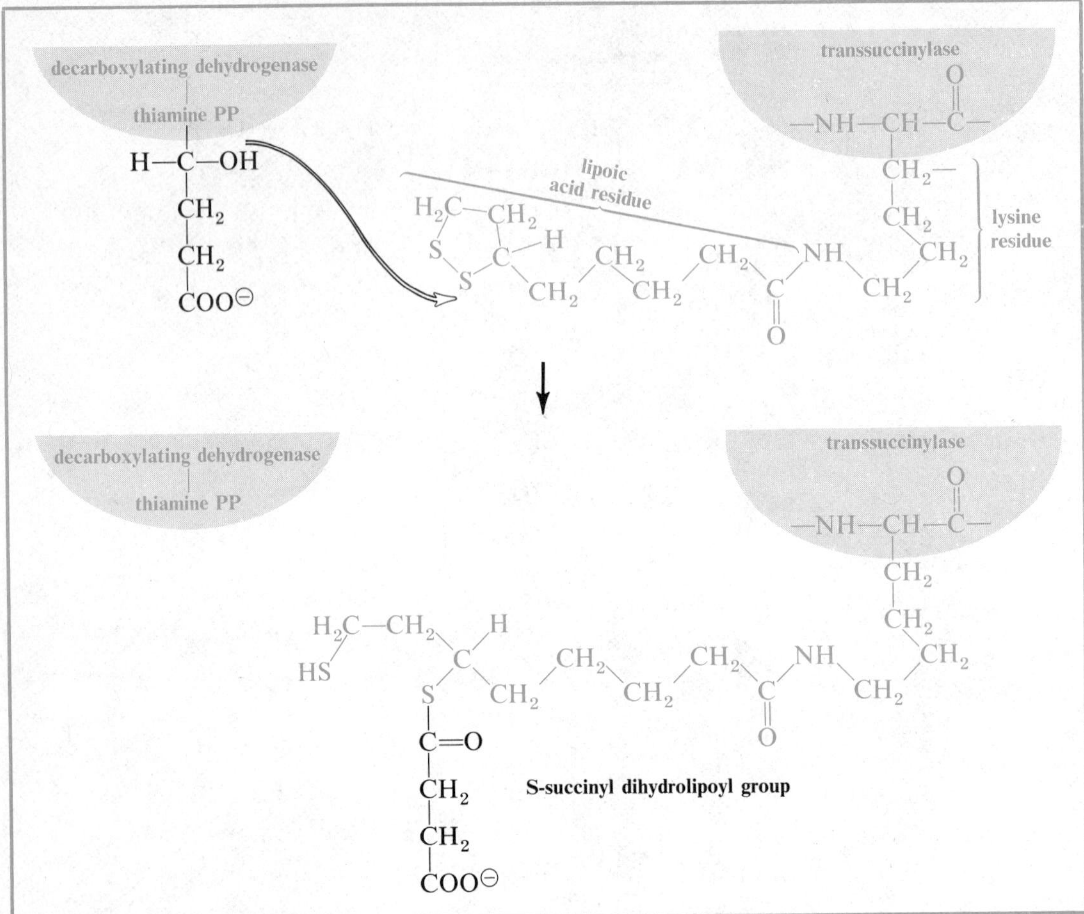

FIGURE 23–6 **The decarboxylating dehydrogenase component of α-ketoglutarate dehydrogenase complex catalyzes transfer of the attached potential succinic semialdehyde group to the disulfide group in a lipoic acid residue. The effect is oxidation of the potential aldehyde to a carboxylic acid in the form of a thiol ester, and reduction of the disulfide to a dithiol. The lipoic acid is attached to a lysine residue in the transsuccinylase component.**

part of the energy as a thiol ester. However, the electrons removed from the substrate are now present in the two thiol groups of a dihydrolipoic acid residue. The disposal of these electrons is handled by the third kind of protein in the enzyme complex, which is a dihydrolipoamide dehydrogenase. This dehydrogenase catalyzes the oxidation of dihydrolipoyl groups by NAD (Fig. 23–8), completing the overall reaction of the complex. The two results are the final restoration of the transuccinylase to its original form, and the formation of NADH, which can be used for oxidative phosphorylation in the inner membrane of the mitochondria.

Dihydrolipoamide dehydrogenase is a flavoprotein. It seems odd for a flavoprotein to be used in generating NADH, since most flavoproteins are stronger oxidizing agents than NAD, but the transfer of electrons from two thiol groups to NAD is apparently a special case, because other reactions of this sort also are catalyzed by similar flavoproteins. The reaction involves the cooperative action of the FAD coenzyme and a residue of cystine in the enzyme; the mechanism is not clear, but it may involve a covalent combination between an SH group and the flavin so that a pair of electrons is, in effect, shared between them.

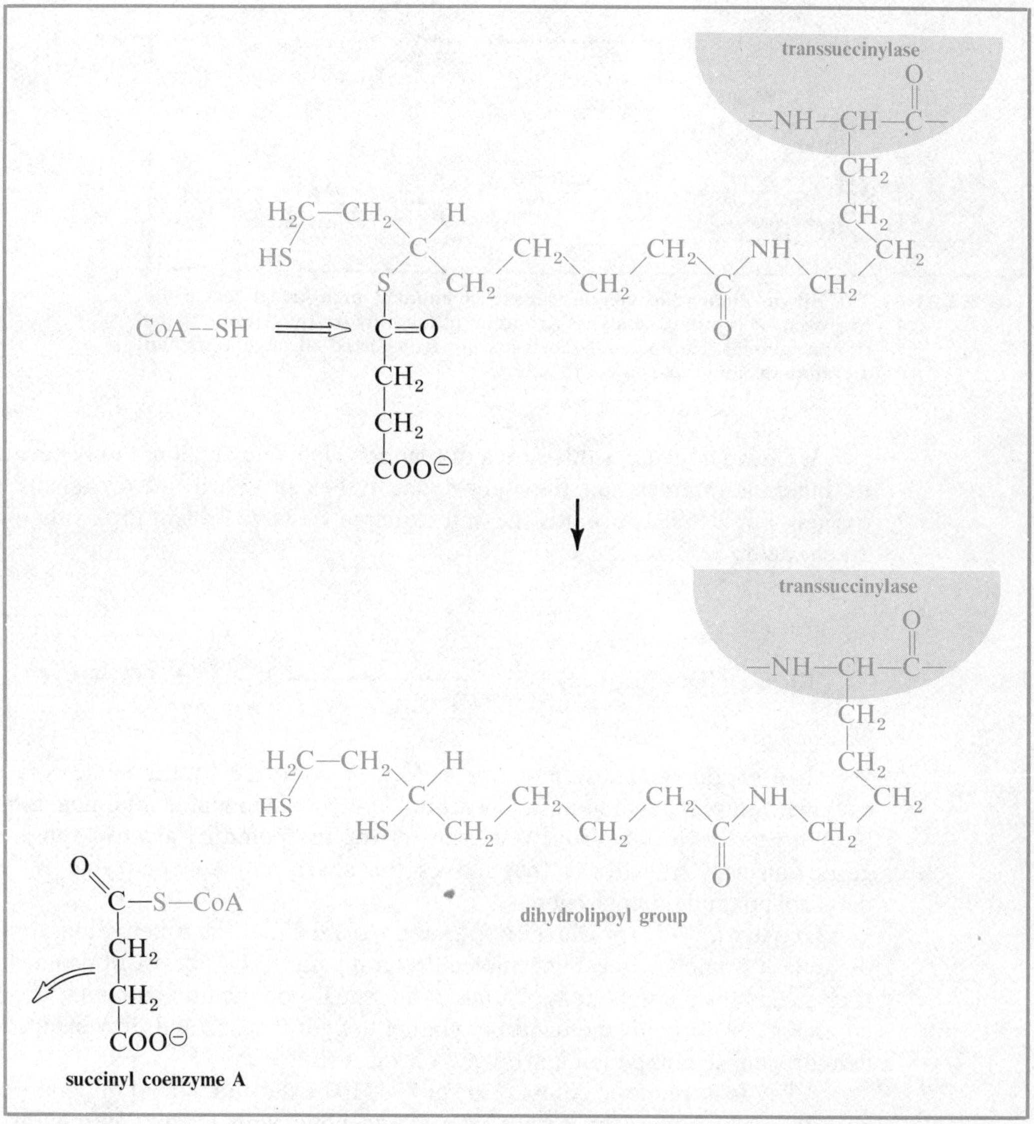

FIGURE 23–7 **The transsuccinylase component of α-ketoglutarate dehydrogenase complex catalyzes transfer of the attached succinyl group to coenzyme A, releasing succinyl coenzyme A as a product.**

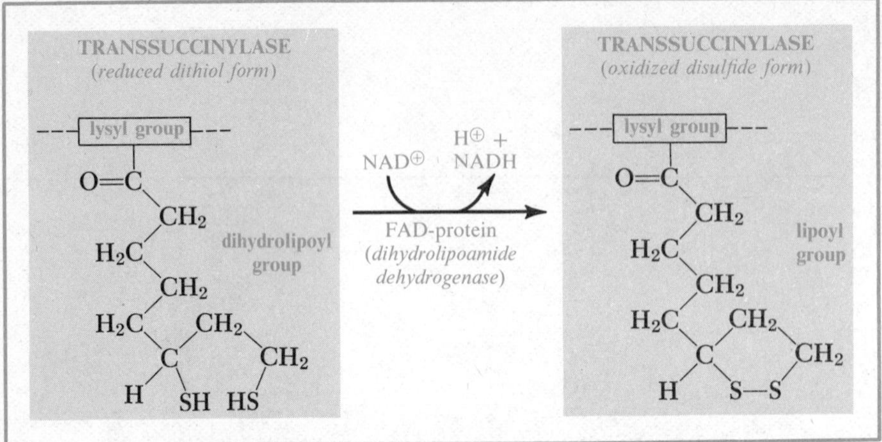

FIGURE 23–8 **The dihydrolipoamide dehydrogenase component of α-ketoglutarate dehydrogenase complex catalyzes oxidation of the dihydrolipoyl group in the transsuccinylase component. Electrons are transferred through FAD and a cystine residue (not shown) to NAD$^+$.**

We have gone over this series of reactions in some detail not only because of its inherent interest and the importance it has in nearly all organisms using oxygen, but also because it is the first example we have seen of the oxidation of a 2-keto acid:

$$\mathrm{R{-}\overset{\overset{\displaystyle O}{\|}}{C}{-}COO^{\ominus}} + \mathrm{CoA{-}SH} \xrightarrow{\mathrm{NAD^{\oplus}} \quad \mathrm{NADH}} \mathrm{R{-}\overset{\overset{\displaystyle O}{\|}}{C}{-}S{-}CoA} + \mathrm{CO_2}$$

We shall encounter many more, and each is catalyzed by the same sort of an enzyme complex, including a decarboxylating component containing thiamine pyrophosphate, a dehydrogenase transferring the potential aldehyde to a lipoyl group, an acyl transferase that moves the acyl group to coenzyme A, and a dihydrolipoamide dehydrogenase.

GEOMETRY OF THE DEHYDROGENASE COMPLEX. The α-ketoglutarate dehydrogenase complex is a huge molecule, comparable in size to ribosomes. The transsuccinylase components, some 24 identical polypeptide subunits, are in its core (Fig. 23–9), with the decarboxylating dehydrogenase and dihydrolipoamide dehydrogenase components arranged on the periphery.

A key feature of the complex is believed to be the attachment of lipoate to the core transsuccinylase by forming an amide bond with lysine side chains. The result is the formation of a long flexible chain with the reactive disulfide group removed some 12 atoms from the polypeptide backbone, where it can be swung so as to contact different proteins in the complex (Fig. 23–10).

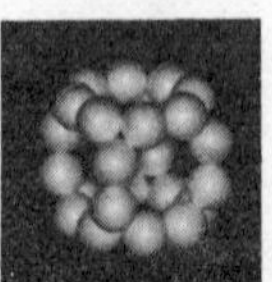

FIGURE 23–9

The α-ketoglutarate dehydrogenase complex is built around a transsuccinylase core. (*Left*) An electron micrograph of one molecule of complete complex as isolated from bovine kidney. From such seemingly indistinct images of the complex and its components, it is possible to deduce a model for the core (*right*). (Both photographs courtesy of Prof. Lester J. Reed.)

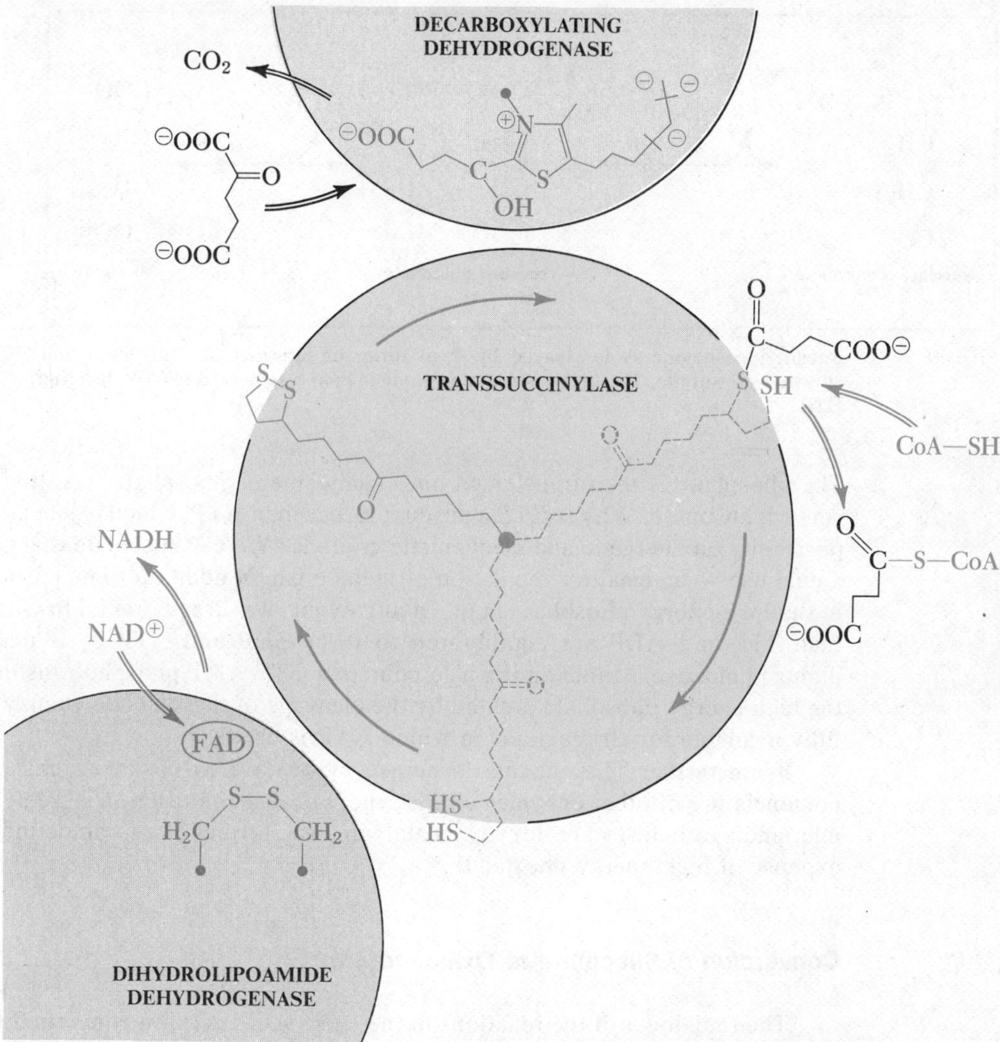

FIGURE 23–10 **Integrated action of the α-ketoglutarate dehydrogenase complex is believed to depend upon the wide area that can be covered by the long lipoyl-lysyl chains (*center*) on the transsuccinylase polypeptides. The substrate, α-ketoglutarate, is shown entering at the upper left with the loss of CO_2 and attachment of the resultant succinic semialdehyde to thiamine pyrophosphate on the decarboxylating dehydrogenase component of the complex. The subsequent transfer to a lipoyl group and then to coenzyme A is shown at *center right*. The reoxidation of a dihydrolipoyl group by a separate dehydrogenase component is shown at *lower left*, with transfer of the electrons to NAD, forming NADH.**

Recovery of Energy from Succinyl Coenzyme A

The next step in the citric acid cycle is the recovery of the potential energy of thiol ester bond as a high-energy phosphate (Fig. 23–11) by the succinyl CoA synthetase reaction. Here we have an example of a **substrate-level phosphorylation,** in which a high-energy phosphate is created in a metabolite, rather than being created by electron flow on the mitochondrial membrane. The reaction involves the cleavage of succinyl coenzyme A by P_i on the enzyme surface, perhaps with the formation of an intermediate succinyl phosphate, followed by the transfer of the phosphate onto the imidazole side chain of a histidyl residue on the enzyme.

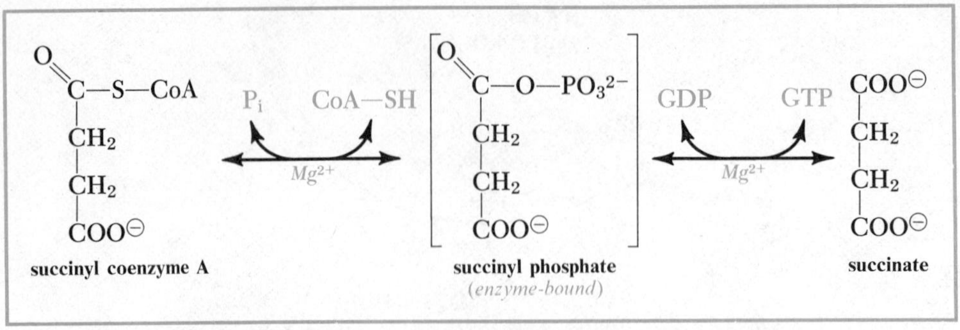

$$\begin{array}{c} O \\ \| \\ C{-}S{-}CoA \\ | \\ CH_2 \\ | \\ CH_2 \\ | \\ COO^{\ominus} \end{array} \xrightleftharpoons[Mg^{2+}]{P_i \quad CoA{-}SH} \left[\begin{array}{c} O \\ \| \\ C{-}O{-}PO_3^{2-} \\ | \\ CH_2 \\ | \\ CH_2 \\ | \\ COO^{\ominus} \end{array}\right] \xrightleftharpoons[Mg^{2+}]{GDP \quad GTP} \begin{array}{c} COO^{\ominus} \\ | \\ CH_2 \\ | \\ CH_2 \\ | \\ COO^{\ominus} \end{array}$$

FIGURE 23–11 **Succinyl coenzyme A is cleaved by P_i to form an intermediate enzyme-bound succinyl phosphate. The high-energy phosphate is then transferred to GDP, forming GTP.**

The phosphate is then transferred onto guanosine disphosphate, creating guanosine triphosphate. Why is GTP a product rather than ATP, which is just as likely a possibility on energetic and mechanistic grounds? We can speculate that GTP has a dual use — a signal for regulation of metabolism in addition to its participation in the high-energy phosphate pool. In any event, we already noted in Chapter 19 that GTP and ADP are equilibrated to form GDP and ATP by a nucleoside diphosphokinase. Mitochondria also contain a GTP:AMP phosphotransferase, so the high-energy phosphate created by the cleavage of the succinyl coenzyme A is fully available for all reactions in which ATP is used.

Synthetases. The enzyme is named synthetase to be consistent with the nomenclature of other enzymes of this type, because the reaction is freely reversible, and synthetases are enzymes catalyzing the formation of compounds at the expense of high-energy phosphate.

Conversion of Succinate to Oxaloacetate

The remainder of the reactions in the citric acid cycle are concerned with the regeneration of oxaloacetate from succinate, with a concomitant use of the electrons for oxidative phosphorylation.

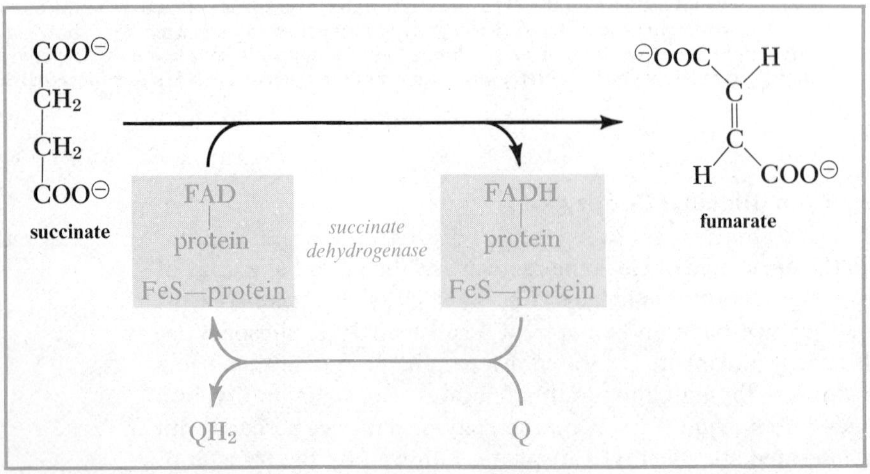

FIGURE 23–12 **The oxidation of succinate to fumarate.**

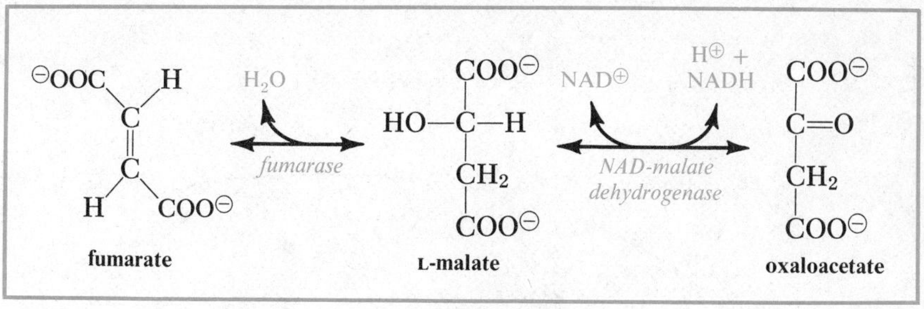

$$\underset{\text{fumarate}}{{}^{\ominus}OOC{-}CH{=}CH{-}COO^{\ominus}} \xrightleftharpoons[\text{fumarase}]{H_2O} \underset{\text{L-malate}}{{}^{\ominus}OOC{-}CH(OH){-}CH_2{-}COO^{\ominus}} \xrightleftharpoons[\text{NAD-malate dehydrogenase}]{NAD^{\oplus} \;\to\; H^{\oplus} + NADH} \underset{\text{oxaloacetate}}{{}^{\ominus}OOC{-}C({=}O){-}CH_2{-}COO^{\ominus}}$$

FIGURE 23–13 **Fumarate is equilibrated with oxaloacetate by a hydration to form malate, followed by an oxidation with NAD.**

Oxidation of Succinate. Succinate is oxidized to fumarate by succinate dehydrogenase (Fig. 23–12). We have already noted (p. 383) that this dehydrogenase is a flavoprotein and iron-sulfide protein complex on the inner mitochondrial membrane. (Flavin adenine dinucleotide is the coenzyme.) The dehydrogenase transfers electrons directly to ubiquinone, bypassing the first phosphorylation site in the electron transfer scheme. It follows that the overall transfer of electrons from succinate to oxygen will result in the generation of only two high-energy phosphates.

Regeneration of Oxaloacetate. Water is added across the double bond of fumarate to form L-malate, which then undergoes the final oxidation of the citric acid cycle to become oxaloacetate (Fig. 23–13). We have here typical hydratase and alcohol dehydrogenase reactions. The NADH produced by the dehydrogenase is available for oxidative phosphorylation, and the oxaloacetate replaces the molecule originally used to create citrate.

THE COMPLETE CYCLE

Total Stoichiometry

Now that we have gone over the individual steps, let us survey the complete cycle as it is portrayed in Figure 23–14. If the cycle is really oxidizing the acetyl group of acetyl coenzyme A to CO_2 and H_2O, the sum of all the equations for the individual reactions must add up to the formal stoichiometry:

$$H_3C{-}\overset{\overset{\displaystyle O}{\|}}{C}{-}S{-}CoA + 2\,O_2 \longrightarrow 2\,CO_2 + CoA{-}SH + H_2O$$

after any high-energy phosphate that is produced has been utilized and converted back to the starting materials. If we add all of the reactions in the cycle, we arrive at the sum:

$$H_3C{-}\overset{\overset{\displaystyle O}{\|}}{C}{-}S{-}CoA + 3\,NAD^{\oplus} + Q + GDP + P_i + 2\,H_2O \longrightarrow 2\,CO_2 + CoA{-}SH + 3\,NADH + QH_2 + GTP.$$

We see here the production of two molecules of CO_2 and the four transfers of pairs

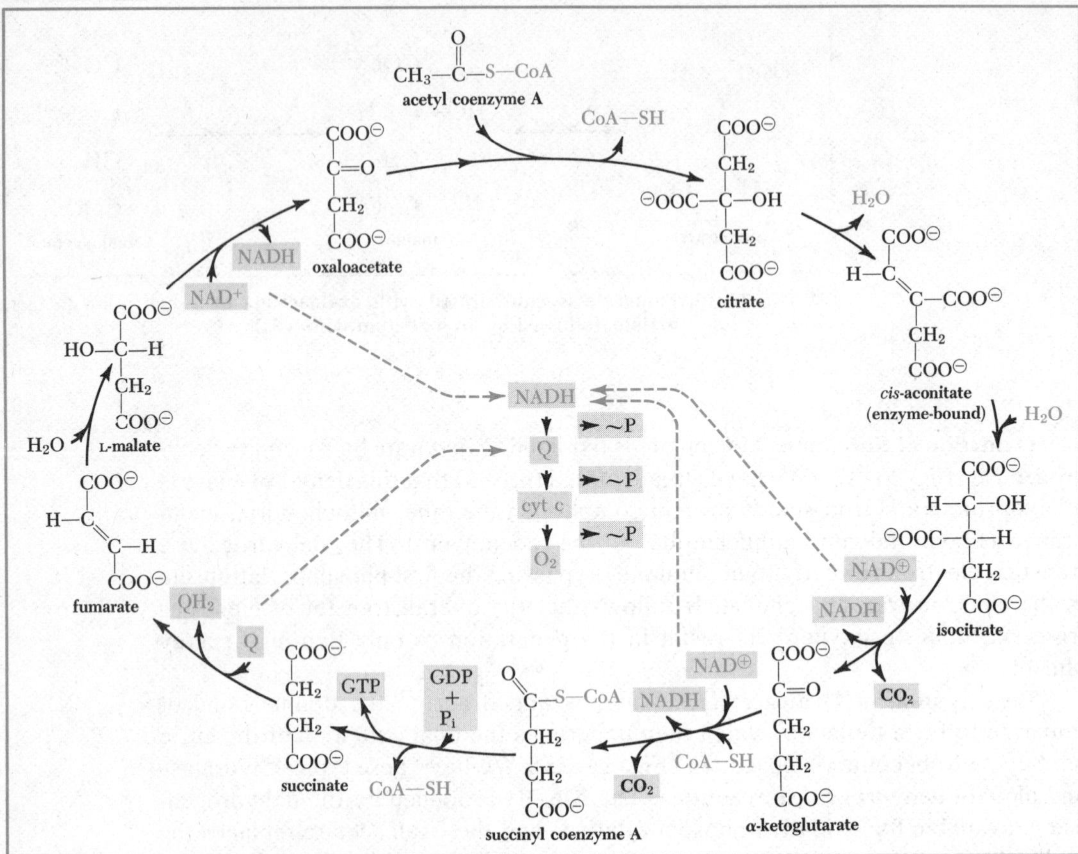

FIGURE 23–14 **The citric acid cycle. Citrate is oriented so as to correspond atom-for-atom with *cis*-aconitate and isocitrate; when pictured this way, the atoms in citrate derived from acetyl coenzyme A are at the bottom of the formula. The coenzymes involved in electron transfers are shown in blue shaded boxes. Components of the high-energy phosphate pool, and also the CO_2 produced by the cycle, are shown in gray shaded boxes.**

of electrons necessary for the oxidation. Neglecting ATP production, electrons will be transferred eventually to oxygen according to the equation:

$$3\ NADH + QH_2 + 2\ O_2 \rightarrow 3\ NAD^{\oplus} + Q + 4\ H_2O.$$

When GTP is utilized in other reactions, the effect will be:

$$GTP + H_2O \rightarrow GDP + P_i.$$

If we add these two equations to the sum for the cycle we indeed have the formal stoichiometry for the oxidation of acetyl groups.

In addition to showing the oxidation of the acetyl group, there is another important lesson in the stoichiometry. **The citric acid cycle does not involve the net production or consumption of oxaloacetate or of any other constituent of the cycle itself.** The reactions of the citric acid cycle do not provide a route for making additional oxaloacetate from acetyl groups. Failure to appreciate this point has led intelligent men into serious error in the past, and it must be kept firmly in mind to avoid making the same mistakes in the future.

High-Energy Phosphate Yield. A grand total of twelve molecules of ATP can be formed during complete oxidation of the acetyl group if there is perfect coupling between oxidation and phosphorylation. Three molecules of NADH are formed. The oxidation of each of these by O_2 results in the formation of three molecules of ATP from ADP and P_i, for a total of nine molecules of ATP. In addition, one molecule of ubiquinol is directly formed and the oxidation of this by O_2 results in the formation of two more molecules of ATP. Finally, the molecule of GTP formed in the succinyl CoA synthetase reaction is equivalent to an additional molecule of ATP:

$$\text{GTP} + \text{ADP} \rightarrow \text{GDP} + \text{ATP}.$$

Since 2 O_2 are consumed in the citric acid cycle, the P:O ratio, the ratio of high-energy phosphates produced to atoms of oxygen consumed, is 12:4, or 3.00. We later shall have occasion to use this number in assessing the relative energy production obtained from various fuels.

The Main-Line Sequence

Much of the citric acid cycle is made from a sequence of reactions that occurs again and again in biochemistry, sometimes complete and sometimes as fragments. This sequence begins with the anion or ester of a saturated carboxylic acid, which is successively modified by oxidation to the unsaturated compound, followed by hydration of the double bond to form an alcohol group, and subsequent oxidation of this group to form a keto compound (Fig. 23–15). The figure shows how this sequence occurs in the citric acid cycle, with the formulas of the compounds arranged to match the general scheme.

Regulation of the Citric Acid Cycle (Fig. 23–16)

Since the citric acid cycle is in the major routes of fuel combustion in many cells, there must be some control of the rate at which it proceeds. It wouldn't do to have the oxidative machinery going full tilt like a runaway boiler at times of rest, nor would it do to have it only sluggishly responsive when there is an immediate demand for ATP.

Stoichiometric Regulation by ADP. Since a major consequence of the action of the citric acid cycle is the conversion of ADP and P_i to ATP through the mechanism of oxidative phosphorylation, it follows that the cycle won't function unless there is a supply of ADP and P_i. It is the concentration of ADP that is usually limiting in animal tissues. Electrons can't be transferred to O_2 unless the cells are doing something that requires ATP (such as moving, synthesizing proteins, or transporting ions), thereby causing the production of ADP. Oxygen consumption diminishes in resting cells, and the electron carriers accumulate in their reduced form. To be specific, a lack of ADP at rest prevents NADH from being reoxidized back to NAD. The resultant lack of NAD will slow the oxidations of isocitrate, α-ketoglutarate, and L-malate. Ubiquinol will also accumulate because it cannot be re-oxidized to ubiquinone. Not only will the lack of ubiquinone slow the oxidation of NADH, but it will also slow the oxidation of succinate.

The rate of exchange of ATP and ADP across the inner membrane also is increased by a rise in external ADP concentration. Since changes in the rate of

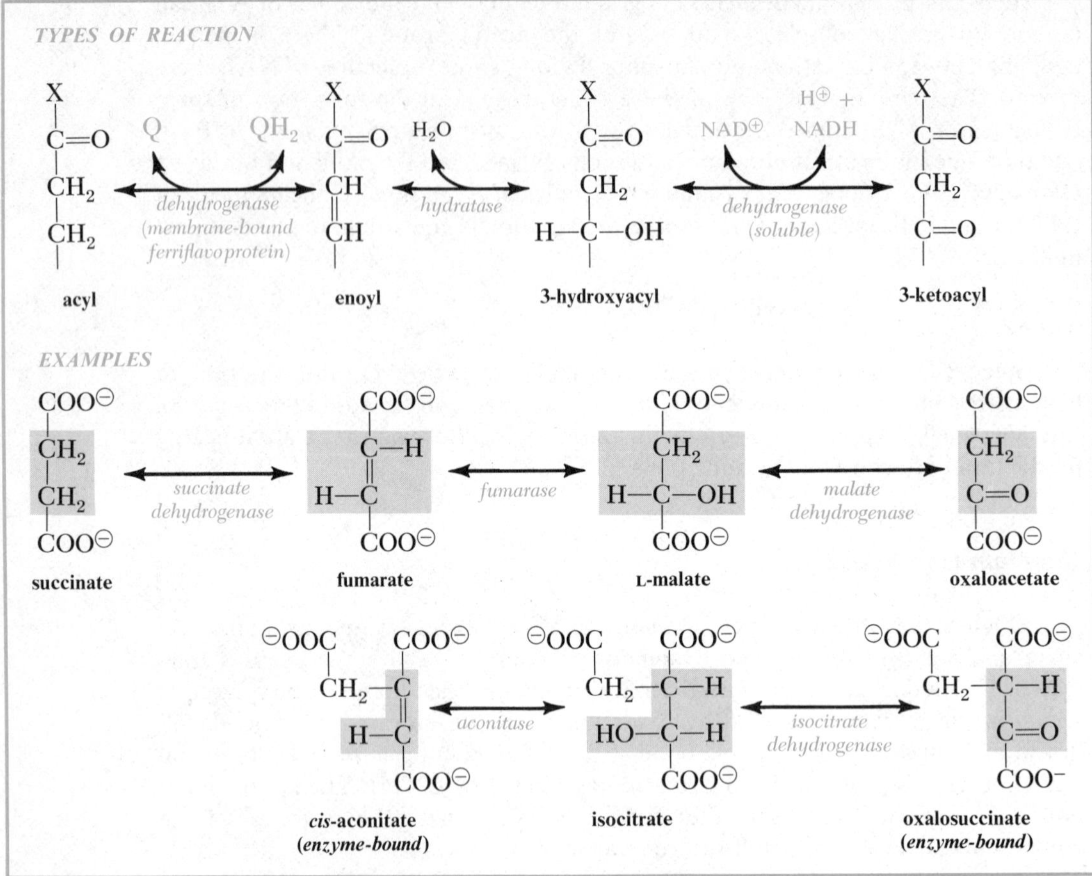

FIGURE 23–15 **The mainline sequence of metabolic reactions. This sequence, in which a compound containing a segment of saturated hydrocarbon chain (acyl group) is successively converted to the unsaturated analogue (enol group), an alcohol (hydroxy acyl group), and then to a carbonyl compound (ketoacyl group) is common in metabolism. The citric acid cycle is mainly composed of reactions of this sort, as shown in the examples.**

oxidative phosphorylation tend to be proportional to the third power of changes in ADP concentration (p. 397), only small shifts in the internal ATP-ADP balance of the mitochondrion suffice to control the rate of the oxidative metabolism.

We have emphasized ADP concentration as the normal regulating factor, but it may be that the muscular weakness or impaired cardiac performance sometimes seen in patients with a low P_i concentration (hypophosphatemia) is due to P_i becoming limiting. (The concentration may be lowered by prolonged use of antacids, with resultant passage of insoluble phosphates through the bowel, for example.)

Regulation by Effectors. Some of the reactions in the citric acid cycle require individual regulation because they are essentially irreversible under physiological conditions. There is always the danger that such reactions will continue until they have consumed the available supply of substrate or coenzyme.

Consider, for example, the **citrate synthase** reaction. Its position of equilibrium is so far toward the formation of citrate that all available acetyl coenzyme A or oxaloacetate could be converted to citrate, even when the remainder of the citric acid cycle is ticking along at the resting rate. This is prevented by making the rate

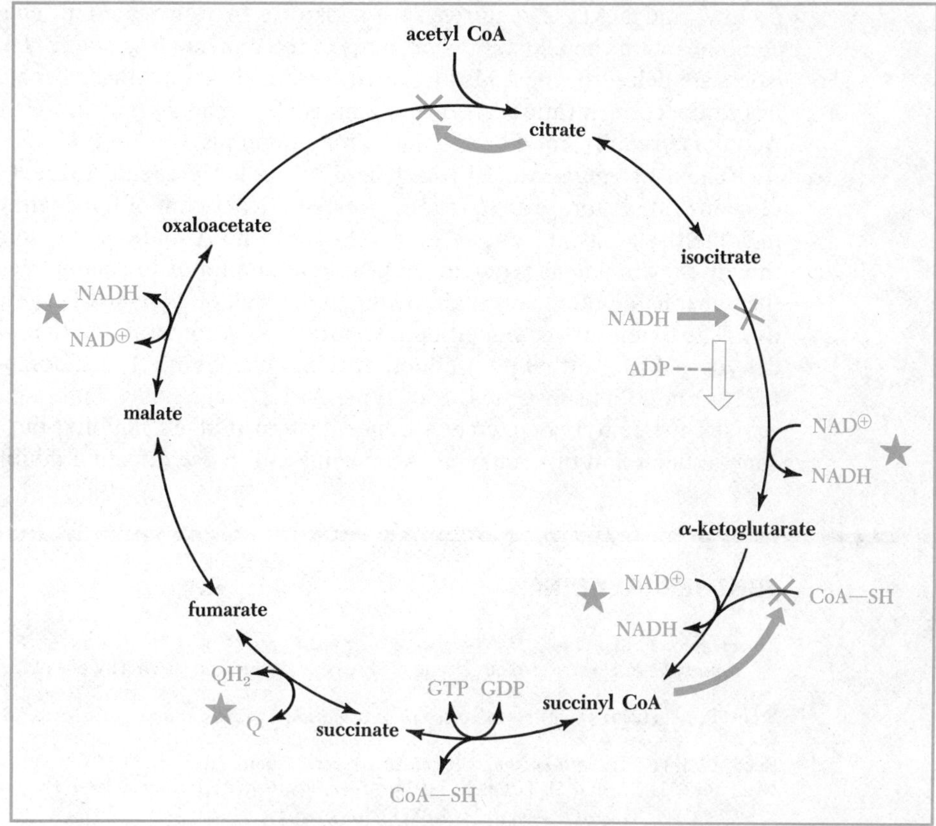

FIGURE 23–16 **Regulation of the citric acid cycle. The starred reactions require oxidized coenzymes, and the ratio of oxidized to reduced coenzyme is governed by the availability of ADP and P_i for oxidative phosphorylation. Other specific controls prevent irreversible steps in the cycle from depleting the supply of cofactors and cycle intermediates.**

of the reaction critically dependent upon the oxaloacetate concentration. The K_M for oxaloacetate is in the physiological concentration range; furthermore, citrate is a competitive inhibitor for oxaloacetate on the enzyme. The effect is double-barreled. An accumulation of citrate raises its concentration as an inhibitor, but it also lowers the concentration of oxaloacetate as a substrate. Why? It is because the complete cycle must function at the same rate to restore the oxaloacetate consumed in the first step. Any accumulation of intermediates in the cycle represents a depletion of oxaloacetate.

Now consider **isocitrate dehydrogenase.** It is especially important because it is the first committed reaction in the citric acid cycle — it has no other function than to participate in the cycle, and it is irreversible at intracellular concentrations of its reactions. (Citrate synthesis is used in fatty acid synthesis (Chapter 28), as well as in the cycle.) Two important allosteric effects are used here. ADP is a specific activator, which lowers the K_M for isocitrate. A rise in ADP concentration is a signal of a need for more high-energy phosphate, and the response to this signal includes an accelerated injection of substrate into the citric acid cycle.

Isocitrate dehydrogenase is also inhibited by NADH at an allosteric site. Any slowing of oxidative phosphorylation, which leads to an accumulation of NADH, therefore slows the enzyme in two ways. If the [NADH] is high, the [NAD] must be low, and NAD is required as a substrate for the reaction. The NADH also combines with the allosteric site to make the enzyme less active. (Any control of isocitrate dehydrogenase also tends to control citrate synthase because changes in isocitrate concentration are accompanied by changes in citrate concentration. Aconitase rapidly equilibrates the two compounds.)

The α-**ketoglutarate dehydrogenase** reaction represents a threat to the supply of coenzyme A for other reactions. Indeed, 70 per cent of the coenzyme A supply in some tissues is present as succinyl coenzyme A under some conditions, even though the enzyme is regulated. The compound tends to accumulate at rest when the phosphate potential is high, owing to the lack of GDP and P_i for its subsequent use. The extent of accumulation is controlled by constructing the α-ketoglutarate dehydrogenase so that its product, succinyl coenzyme A, is a competitive inhibitor for one of its substrates, coenzyme A. Here again is a double-barreled effect. A rise in succinyl coenzyme A concentration in itself inhibits, but it also represents a depletion of coenzyme A, causing still more effective inhibition.

FURTHER READING

Lowenstein, J. M.: (1967) *The Tricarboxylic Acid Cycle*. Vol. 1, p. 147, *in* D. E. Greenberg, ed.: *Metabolic Pathways*. 3rd ed. Academic Press. A detailed discussion by one of the more lucid writers.

Srere, P. A.: (1975) *The Enzymology of the Formation and Breakdown of Citrate*. Adv. Enzymol., *43*: 57.

Reed, L. J. (1974) *Multi-enzyme Complexes*. Accts. Chem. Res., *7*: 40.

Nishimura, J. S., and F. Grinnell: (1973) *Mechanism of Action and Other Properties of Succinyl Coenzyme A Synthetase*. Adv. Enzymol., *36*: 183.

Williams, Jr., C. H.: (1976) *Flavin-containing Dehydrogenase*, Vol. 13, p. 89, *in* P. D. Boyer, ed.: *The Enzymes*, 3rd ed. Academic Press.

Harvey, R. A., J. I. Heron, and G. W. E. Plaut: (1972) *Regulation of Diphosphopyridine Nucleotide-linked Isocitrate Dehydrogenase from Bovine Heart*. J. Biol. Chem., *247*: 1801 (Diphosphopyridine nucleotide is an old name for NAD, now camp.)

Singer, T. P., E. B. Kearney, and W. C. Keaney: (1973) *Succinate Dehydrogenase*. Adv. Enzymol., *37*: 189.

O'Connor, L. R., W. S. Wheeler, and J. E. Bethune: (1977) *Effect of Hypophosphatemia on Myocardial Performance in Man*. N. Engl. J. Med., *297*: 901.

24 THE OXIDATION OF FATTY ACIDS

Fatty acids are major fuels for animals, both as a primary supply from the diet and as a secondary supply created from other dietary components. They are stored within cells as fats (triglycerides) and later released through the blood stream to meet the demands of many tissues, especially muscles. The importance of the oxidation of fatty acids is not limited to the obese or to devotees of greasy foods; it is a critical part of the metabolic economy in the lean as well as the lardy.

The complete combustion of fatty acids to CO_2 and H_2O occurs in the mitochondria, where the transfer of electrons from the fatty acids to oxygen can be used to generate ATP. The combustion occurs in two stages; the fatty acid, in the form of a coenzyme A thiol ester, is first oxidized so as to convert all of its carbons to acetyl coenzyme A. The acetyl coenzyme A is then oxidized by the reactions of the citric acid cycle. Both stages generate ATP by oxidative phosphorylation.

INTRACELLULAR TRANSPORT OF FATTY ACIDS

Free fatty acids appear within a cell by absorption from the extracellular fluid, or by intracellular hydrolysis of triglycerides. Those used as fuels are mostly long-chain compounds with 16 or 18 carbon atoms and from zero to two double bonds. (The routes we shall describe also handle those with more double bonds and with either longer or shorter chains.) Whatever the disposition of a fatty acid is to be — formation of a structural component, storage as fat, or combustion as a fuel — it it is first combined with coenzyme A to form the highly polar thiol ester, which is easily soluble in the aqueous phases of the cell.

Formation of Acyl Coenzyme A. The acyl coenzyme A compounds with which we are now concerned are identical to acetyl coenzyme A in structure, except that they are made from long-chain fatty acids instead of the two-carbon acetic acid. They are like acetyl coenzyme A in being high-energy thiol esters, with a standard free energy of hydrolysis comparable to that of ATP, and they are made from the free fatty acids at the expense of high-energy phosphate:

$$\underset{\text{fatty acid}}{H_3C{-}(CH_2)_n{-}COO^{\ominus}} + \text{CoA—SH} + \text{ATP} \xrightarrow[Mg^{2+}]{\text{acyl CoA synthetase}} \underset{\text{acyl coenzyme A}}{H_3C{-}(CH_2)_n{-}\overset{\overset{\displaystyle O}{\|}}{C}{-}S{-}CoA} + \text{AMP} + PP_i$$

$$PP_i + H_2O \xrightarrow{\text{inorganic pyrophosphatase}} 2\,P_i$$

The conversion of the long-chain fatty acids to acyl coenzyme A compounds is like the conversion of acetate to acetyl coenzyme A; the reaction can go nearly to completion because the concentration of the resultant inorganic pyrophosphate is kept low by the action of inorganic pyrophosphatase.

Fatty acids in the cytosol are converted to acyl coenzyme A at two locations: the endoplasmic reticulum and the outer membrane of mitochondria. The distribution of function is not defined exactly, but it may be that most of acyl coenzyme A created on the endoplasmic reticulum is used to make triglycerides (Chapter 28), whereas most of the fatty acids converted to acyl coenzyme A on the outer mitochondrial membrane are used as fuel within the mitochondria (Fig. 24–1). (A small fatty acid binding protein within the cytosol may be a critical factor, with a concentration of fatty acids in excess of binding capacity being taken up by the mitochondria, but this is speculative at present.)

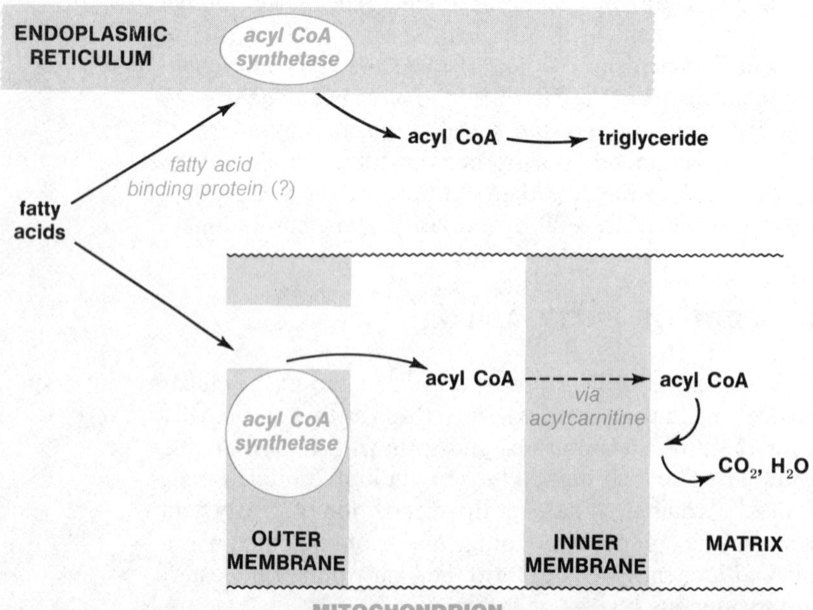

FIGURE 24–1 **Acyl CoA synthetase is present in both the endoplasmic reticulum and the outer membrane of mitochondria, but acyl coenzyme A formed from fatty acids in the endoplasmic reticulum is believed to be used mainly for making fats (triglycerides), while that made on mitochondria is used as a fuel.**

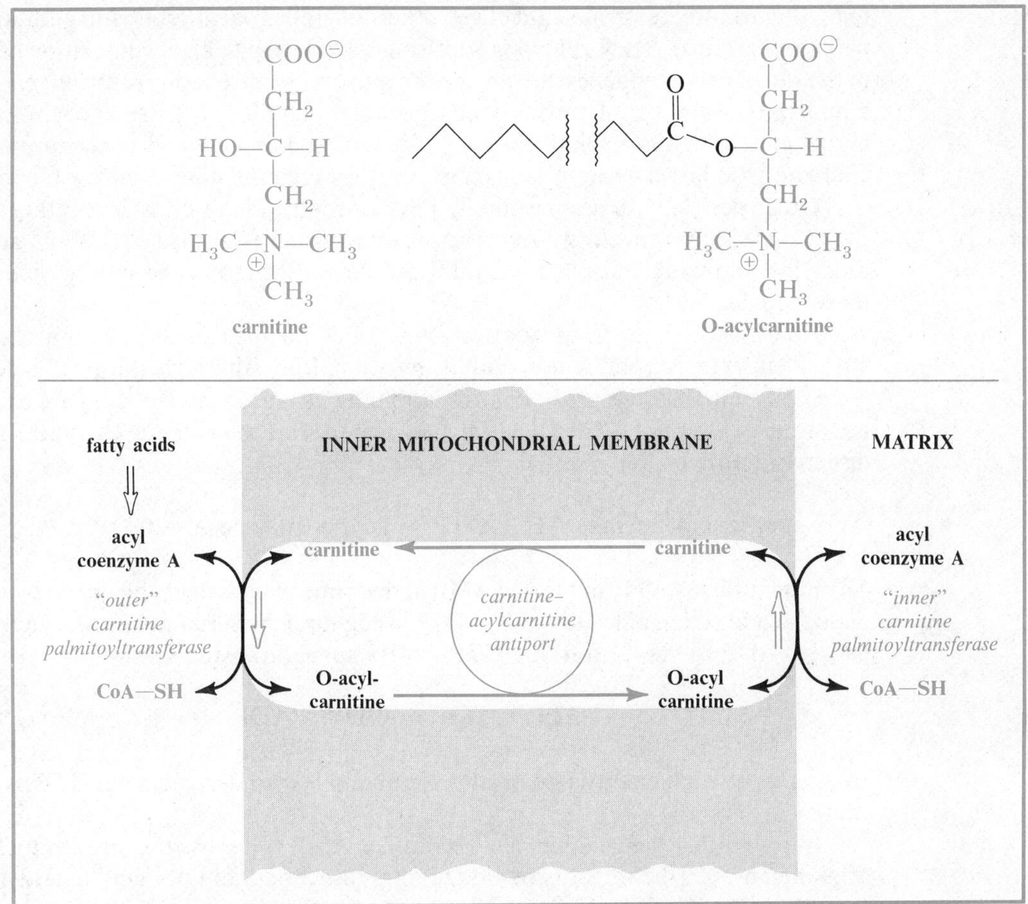

FIGURE 24–2 **Acyl groups are transferred across the inner mitochondrial membrane by combination with carnitine (*upper left*) to form O-acylcarnitine (*upper right*). Acyl groups are transferred from coenzyme A to carnitine on the outer surface of the membrane (*lower left*), and the O-acylcarnitine formed is moved to the inner surface by exchange with free carnitine using an antiport mechanism (*lower center*). The acyl group is then transferred from carnitine to coenzyme A within the mitochondrion, creating acyl coenzyme A that can be attacked by the acyl coenzyme A dehydrogenases exposed on the inner surface.**

Transport as Acyl Carnitine. The mitochondrial inner membrane is nearly impermeable to coenzyme A and its derivatives, permitting separate regulation of the concentrations of these compounds within and without mitochondria. Acyl coenzyme A therefore cannot move as such from the cytosol to the mitochondrial matrix. Instead, the acyl groups are moved across the inner membrane by transferring them from coenzyme A to another compound **carnitine,** at the outer surface of the inner membrane (Fig. 24–2). An antiport within the inner membrane then exchanges acyl carnitine on one side of the membrane for carnitine on the other. The acyl carnitine that appears on the inner surface of the inner membrane is once more equilibrated with acyl coenzyme A, this time utilizing coenzyme A from the mitochondrial matrix. The fatty acids therefore appear as acyl coenzyme A within the matrix without compromising the independent controls within the separate cellular compartments.

Some people are unable to use long-chain fatty acids in their skeletal muscles because of a congenital absence of **carnitine palmitoyl transferase** from their mitochondria (the enzyme exchanges other long-chain acyl groups in addition to palmitoyl groups). They get along suprisingly well despite an accumulation of fats in the blood and a tendency to damage the muscles by strenuous exertion, causing cramps. Destruction of muscle cells becomes manifest by the appearance of myoglobin, which is a relatively small protein, and an elevated concentration of characteristic larger protein molecules, such as creatine kinase, in the blood.

Other people with a congenitally low concentration of carnitine, rather than the enzyme, have developed more serious impairment of muscle function, sometimes in childhood. Fat-filled vacuoles are demonstrated in the muscle fibers of these people.

FORMATION WITHIN MITOCHONDRIA. Fatty acids can also be converted to acyl coenzyme A compounds within mitochondria. Although the physiological purpose is not clear the processes are intriguing because the syntheses are carried out at the expense of GTP. One of the acyl coenzyme A synthetases utilizes GTP directly in a reversible reaction:

$$\text{fatty acid} + \text{CoA—SH} + \text{GTP} \leftrightarrow \text{acyl coenzyme A} + \text{GDP} + \text{P}_i.$$

The other utilizes ATP and forms AMP in the same way as does the enzyme on the endoplasmic reticulum. However, AMP in the mitochondrial matrix is converted back to ADP by the action of a GTP:AMP phosphotransferase:

$$\text{GTP} + \text{AMP} \leftrightarrow \text{GDP} + \text{ADP}$$

so one of the high-energy phosphates expended is also derived from GTP by this route.

Since neither the guanine nucleotides nor AMP will cross the inner mitochondrial membrane, the effect is to isolate intramitochondrial fatty acid uptake from the extramitochondrial process. As we noted in the discussion of succinyl coenzyme A synthetase in the citric acid cycle, the guanine nucleotides probably are involved in some regulatory process that we presently do not understand.

OXIDATION OF SATURATED CHAINS

When the molecules of acyl coenzyme A reach the inner surface of the inner mitochondrial membrane, oxidation can begin. The overall scheme is a beautiful example of economy of design. The reaction sequence, illustrated for palmitoyl coenzyme A in Figure 24–3, is essentially the main-line sequence we discussed in the last chapter, followed by a thiolysis. A saturated acid derivative (acyl coenzyme A) is oxidized to the unsaturated derivative (enoyl coenzyme A) by a flavoprotein, followed by hydration to form an alcohol group (3-hydroxyacyl coenzyme A), which is then oxidized by NAD to a ketone (3-ketoacyl coenzyme A). The 3-ketoacyl coenzyme A is finally cleaved with another molecule of coenzyme A, liberating acetyl coenzyme A and a new acyl coenzyme A compound that is two carbon atoms shorter than the original substrate.

The marvelous thing is that the shorter acyl coenzyme A can then undergo the same sequence of reactions, clipping off another two-carbon unit as acetyl coenzyme A, with the whole process repeated until the entire long-chain acyl group has been chopped into two-carbon acetyl groups, which can be oxidized by the citric acid cycle.

Let us now examine some of the details of this elegant and simple process.

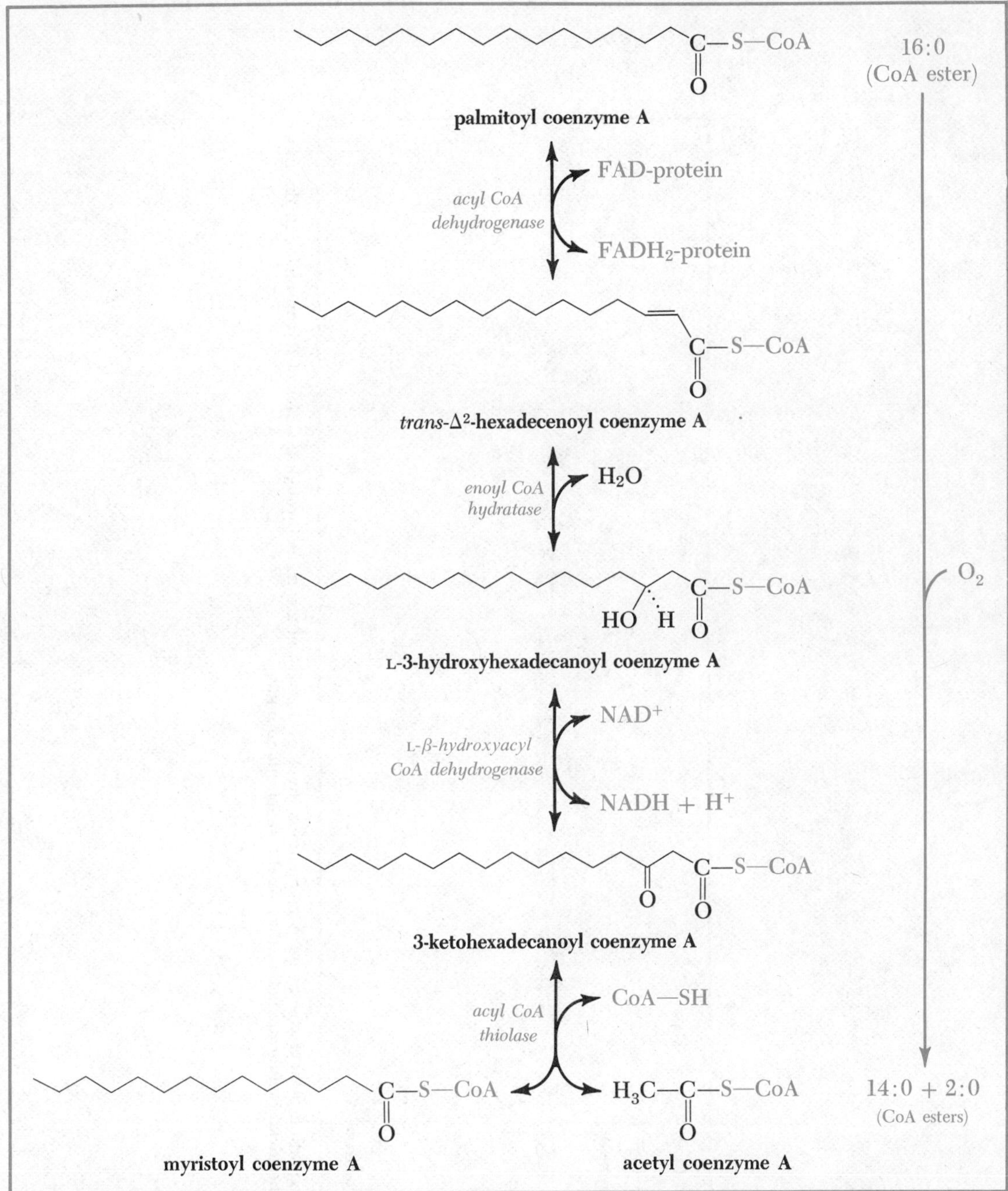

FIGURE 24–3 **The reactions of fatty acid oxidation shown with palmitoyl coenzyme A as substrate.**

Acyl coenzyme A dehydrogenases, found in the inner mitochondrial membrane, are flavoproteins because the oxidation of an acyl coenzyme A, like the oxidation of succinate in the citric acid cycle, does not release enough free energy to support the formation of 3 ATP. Therefore, it is necessary to bypass the first phosphorylation site by transferring electrons directly from the substrate to ubiquinone. The dehydrogenases remove two hydrogen atoms from acyl groups to form *trans* isomers of enoyl groups; three are known, with different specificities for substrate chain length. One works best on short chain compounds (C_4 to C_8),

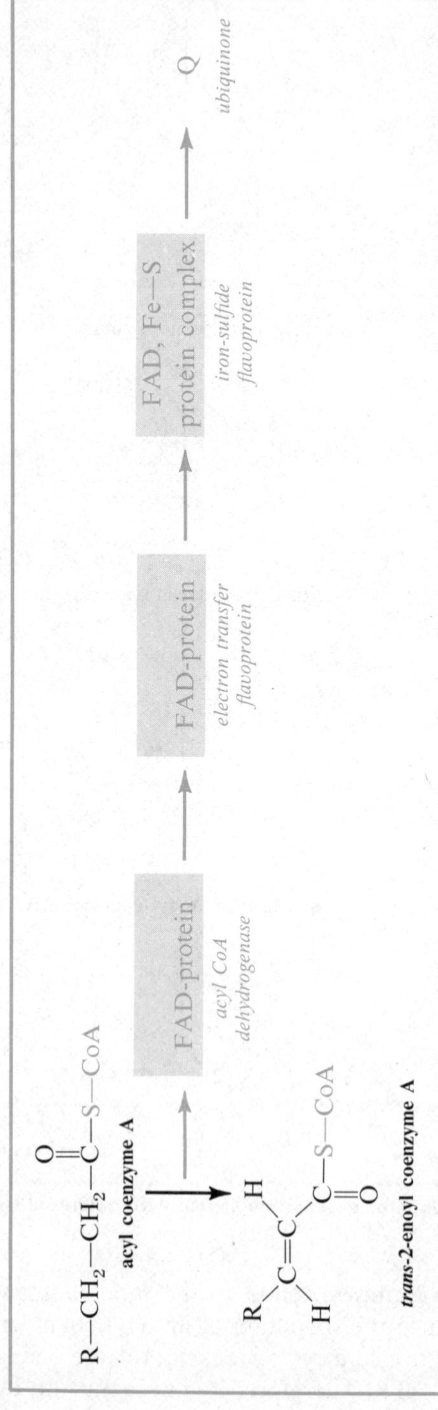

FIGURE 24–4 **The oxidation of acyl coenzyme A compounds involves passage of electrons through a series of flavoproteins in the inner mitochondrial membrane before they finally reach ubiquinone.**

another on intermediate size compounds (C_6 to C_{16}), and another on long-chain compounds (C_6 to C_{18} with a peak at C_{14}).

Acyl coenzyme A dehydrogenases contain no iron and transfer electrons to ubiquinone by a route that first involves another flavoprotein, the **electron-transfer flavoprotein (ETF),** and then still another flavoprotein and iron-sulfide complex (Fig. 24–4). Why these extra flavoproteins are used for fatty acid oxidation is not clear, since the end result is similar to that of succinate oxidation: Electrons are introduced into the mitochondrial electron transport chain at the level of ubiquinone with two moles of high-energy phosphate generated per mole of acyl coenzyme A oxidized. The distinctive pathway may enable fatty acids to be oxidized in preference to other fuels at some times, but no direct experimental test of this speculation has been devised.

Hydration and Dehydrogenation. The next two reactions occur in the mitochondrial matrix. The hydration of unsaturated acyl coenzyme A compounds to form the corresponding L-3-hydroxyacyl derivatives is straightforward and quite similar to the hydration of fumarate in the citric acid cycle, but it is catalyzed by a different enzyme, **enoyl CoA hydratase.**

The oxidation of the alcohol group by NAD also has no unusual features. There is no possibility of an accompanying decarboxylation such as we saw in the oxidation of isocitrate because there is no free carboxylate group — it is fixed as a coenzyme A ester. NAD for this reaction comes from the same pool supplying all NAD-coupled dehydrogenases in the matrix, and the NADH that is produced is oxidized with the generation of 3 ATP.

Cleavage and Coenzyme A. Esters of 3-keto acids are subject to C–C scission through attack by a nucleophilic reagent, both in the test tube and in biological systems. (The familiar synthesis of ethyl acetoacetate by a Claisen condensation is the reverse of such reaction.) The biological nucleophilic reagent frequently is coenzyme A, and the cleavage therefore is a **thiolysis** — a splitting by a thiol aided by enzymatic catalysis.

This reaction releases acetyl coenzyme A in a reaction that is highly favored at intracellular concentrations, while generating a new fatty acyl coenzyme A ester from the remaining carbon atoms.

The Complete Sequence (Fig. 24–5). This sequence therefore involves a repetition of oxidation, hydration, oxidation, and cleavage until the final four-carbon remnant (butyryl coenzyme A) has been oxidized and cleaved, at which point all of the carbon atoms in the original fatty acids will have been converted to acetyl groups.

ODD-CHAIN FATTY ACIDS. What happens if a fatty acid with an odd number of carbon atoms is introduced into mitochondria? There are such fatty acids in nature, although they are much less common than those with an even number of carbon atoms. The sequence of reactions proceeds just as before (We expect this because it is asking too much to expect an enzyme to be specific for substrates with chains of, say, 14, 16, or 18 carbon atoms without the enzyme also binding those of 15 or 17 carbon atoms.) Acetyl coenzyme A units are successively cleaved from these odd-numbered chains until the very end of the sequence, when the three-carbon remainder is in the form of **propionyl coenzyme A**:

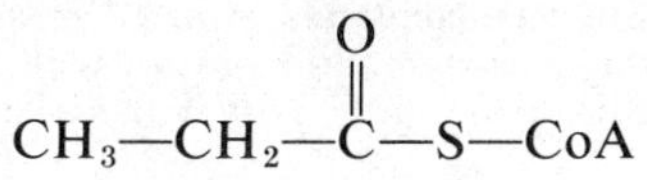

$$CH_3{-}CH_2{-}\overset{\overset{\displaystyle O}{\|}}{C}{-}S{-}CoA$$

propionyl coenzyme A

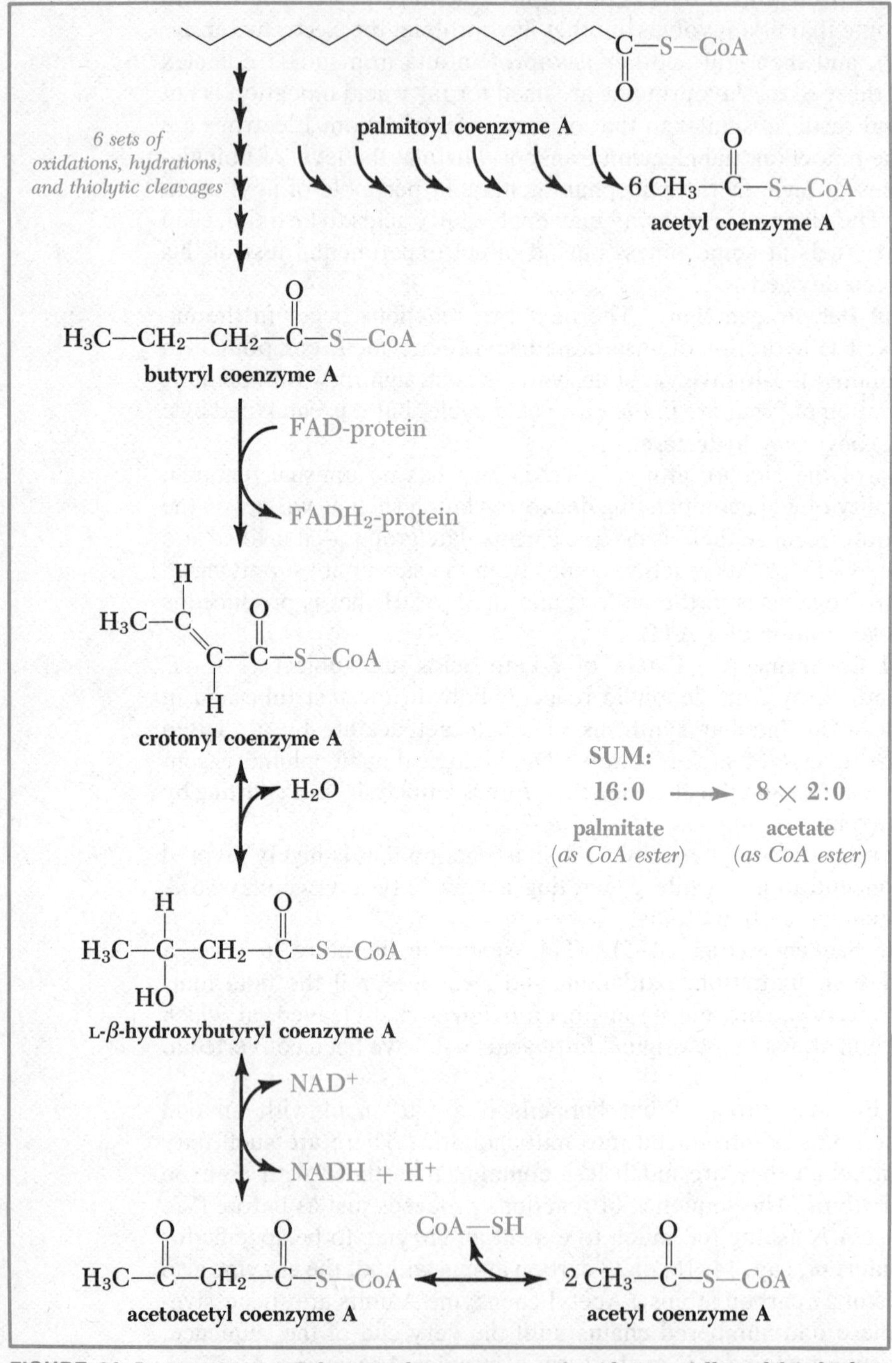

FIGURE 24–5 **Repetitions of the main-line sequence of oxidations followed by thiolytic cleavage finally converts the last four carbons of a fatty acid to butyryl coenzyme A, with the others having been converted to acetyl groups in acetyl coenzyme A. A repetition of the sequence and cleavage with butyryl coenzyme A completes the total conversion of the fatty acid carbons to acetyl groups.**

This compound is not oxidized directly in the citric acid cycle, but it has a metabolism that is important in the breakdown of amino acids and will be considered in connection with them (p. 570). The odd-chain fatty acids are only a small fraction of the total, and only the terminal three carbons appear as propionyl coenzyme A. The metabolism of propionyl coenzyme A is, therefore, not of quantitative significance in fatty acid oxidation.

OXIDATION OF UNSATURATED FATTY ACIDS

Since over half of the fatty acid residues in body lipids are unsaturated, a large part of the high-energy phosphate generated in many tissues must come from the oxidation of this type of compound. Oleic, 18:1(9), and vaccenic, 18:1(11), acids are the most abundant of the unsaturated fatty acids, accounting for approximately half of the residues in a typical triglyceride, but others with more double bonds are not trivial. The general route includes all of the reactions involved in the oxidation of saturated acids, with the addition of two more reactions that convert the unsaturated structures into intermediates along the regular pathway. Double bonds at odd-numbers of carbons are in the wrong position for the oxidative pathway, and even those in the correct position on even-numbered carbons are in the *cis* configuration.

As an example, let us consider the metabolism of the linoleic acid anion — *all cis*-$\Delta^{9,12}$ octadecadienoate. Its metabolism illustrates all of the reactions involved. The double bonds are well removed from the carboxylate end, so it will begin to be metabolized as if it were an ordinary saturated fatty acid, beginning with the formation of the coenzyme A ester, which is then carried three times through the regular sequence of oxidation, hydration, oxidation and cleavage to liberate acetyl coenzyme A:

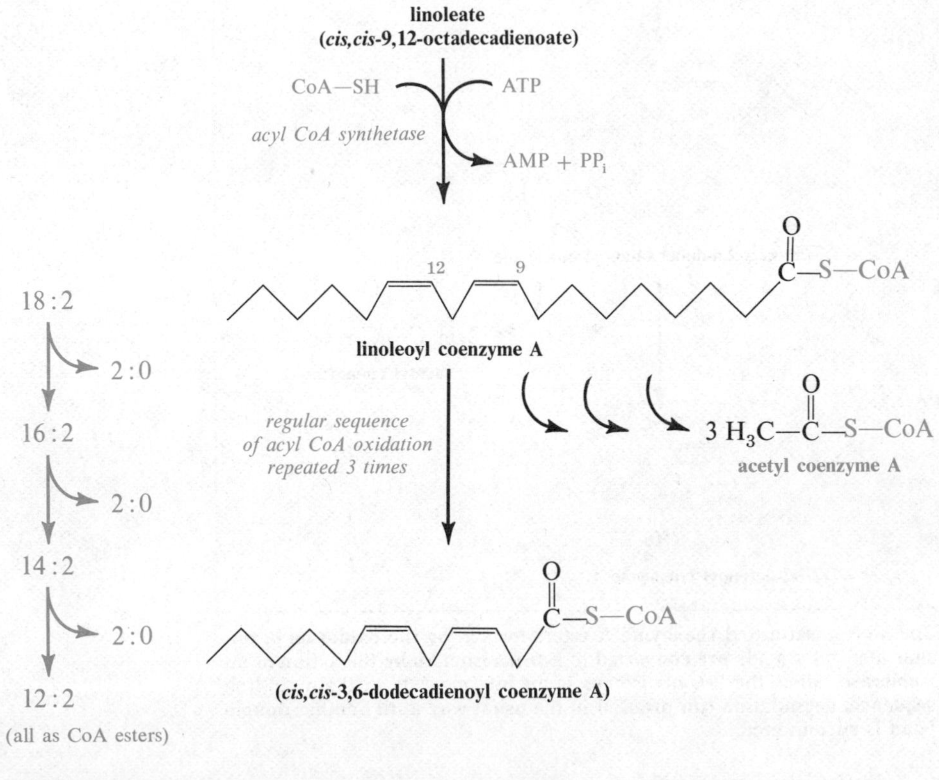

linoleate
(*cis,cis*-9,12-octadecadienoate)

CoA—SH, ATP → *acyl CoA synthetase* → AMP + PP_i

18:2 — linoleoyl coenzyme A (double bonds at 12 and 9; C(=O)—S—CoA)

→ 2:0

regular sequence of acyl CoA oxidation repeated 3 times → $3\ H_3C—C(=O)—S—CoA$ **acetyl coenzyme A**

16:2 → 2:0

14:2 → 2:0

12:2 — **(*cis,cis*-3,6-dodecadienoyl coenzyme A)** (C(=O)—S—CoA)

(all as CoA esters)

The sequence can proceed until three molecules of acetyl coenzyme A have been cleaved, but no farther, because the chain is then shortened so that the double bond that was originally on the ninth carbon of linoleic acid is now on the third carbon of the remaining chain, thereby preventing oxidation by the flavoprotein of acyl coenzyme A dehydrogenase, which attacks carbons 2 and 3.

Isomerization of Double Bond. One of the additional enzymes then comes into play. This enzyme catalyzes a migration of the double bond from the third to the second carbon atom, and at the same time changes the configuration from *cis* to *trans* (Fig. 24–6). Equilibrium for this reaction is reached with about 7/8 in the *trans* form, so no energy donors or receptors are involved. After migration, the double bond is in the same position and has the same configuration as the regular intermediates of fatty acid oxidation, so the compound will once more enter the general pathway, beginning with a hydration.

Hydration of *cis*-Enoyl CoA. After two more molecules of acetyl coenzyme A are cleaved from the chain, the second of the original double bonds in linoleic acid is now between C2 and C3, but it is still in its *cis* configuration. However, the enoyl hydratase that catalyzes the addition of water during the regular sequence is not specific for configuration, and it will also cause the addition to the *cis* form.

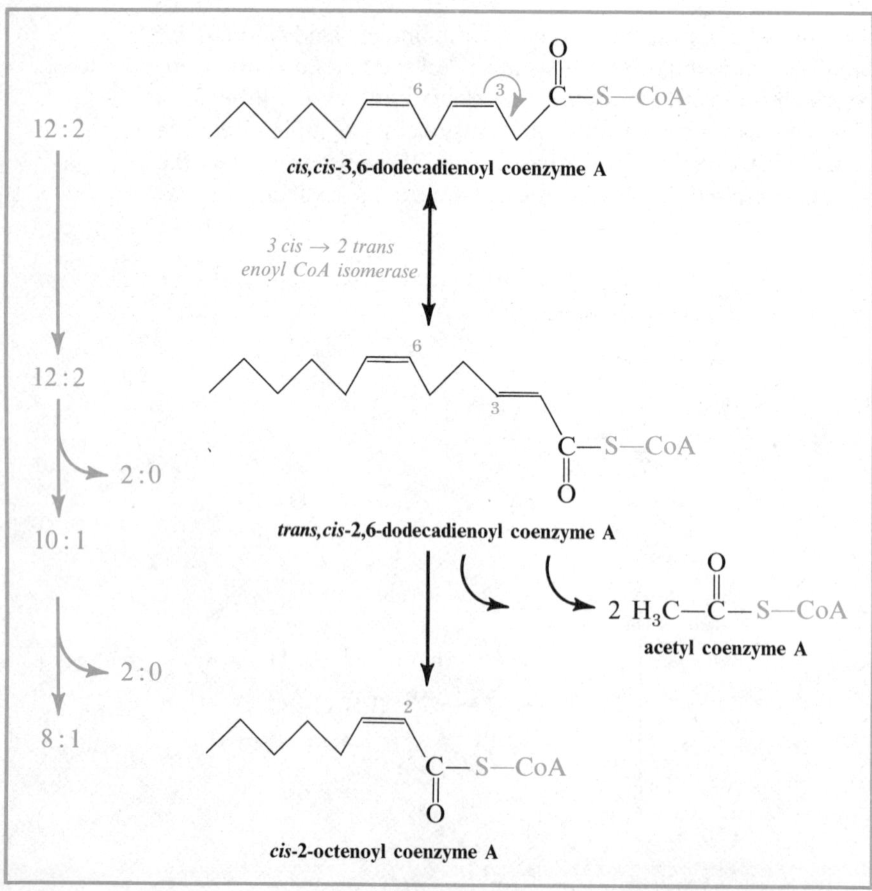

FIGURE 24–6 The 3-*cis* unsaturated coenzyme A esters formed by the oxidation of unsaturated fatty acids are converted to 2-*trans* isomers by the action of an isomerase. Since the 2-*trans* isomer is an intermediate in the main-line sequence, degradation can proceed in the usual way until another double bond is encountered.

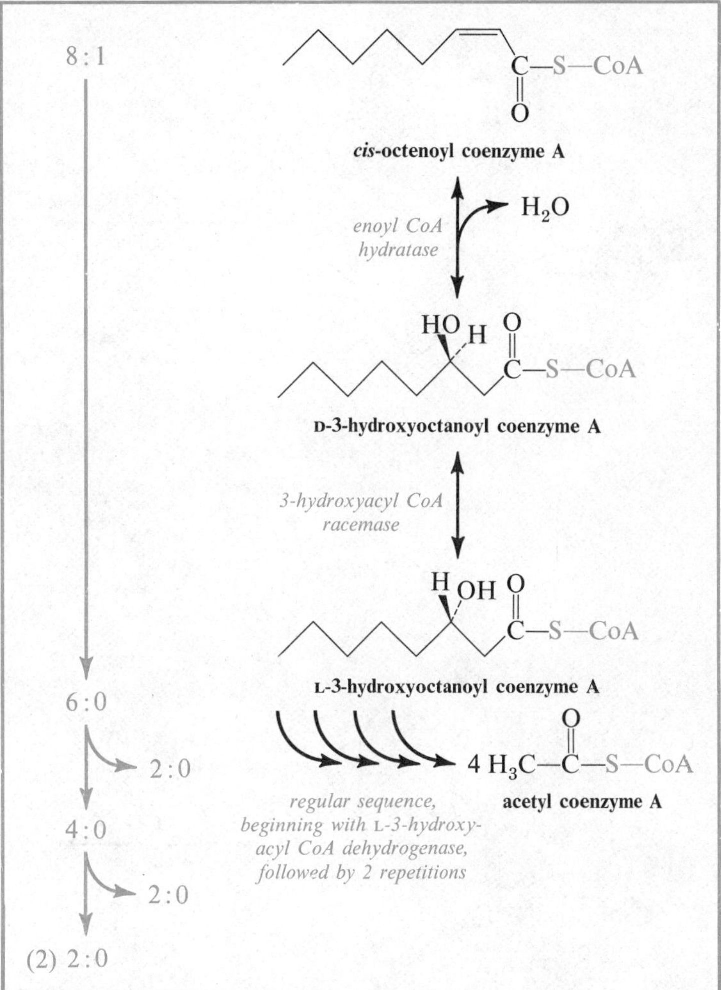

FIGURE 24–7

The action of a 3-hydroxyacyl CoA racemase.

The rub is that hydration of a *cis* Δ^2-enoyl coenzyme A by this enzyme results in the formation of a D-3-hydroxyacyl CoA, rather than the L-isomer formed in the regular sequence, and D-3-hydroxyacyl coenzyme A derivatives are not substrates for the L-3-hydroxyacyl CoA dehydrogenases used in the regular route. However, at this point the second of the additional enzymes, a racemase, brings the D-3-hydroxy compounds into the regular route by catalyzing the interconversion of the D- and L-isomers. The resultant L-isomer is constantly removed by the action of the hydroxyacyl CoA dehydrogenase, and the racemase will continue to act to convert more of the D-isomer to the L, so the effect is to divert all of the D-compound into the regular pathway of fatty acid oxidation (Fig. 24–7).

Here is another example of the beautiful economy of organization of metabolism. The introduction of two additional types of enzymes, an enoyl coenzyme A isomerase and a 3-hydroxyacyl coenzyme A racemase, makes it possible to handle any combination of double bonds found in an unsaturated chain through the same route used for saturated fatty acids. (It is an interesting exercise in mental gymnastics to test this on a random assortment of double bonds in a long chain, except that the structure, —C=C=C—, is forbidden. Allenes are rare among natural compounds, being known only as products of the metabolism of a few kinds of microorganisms.)

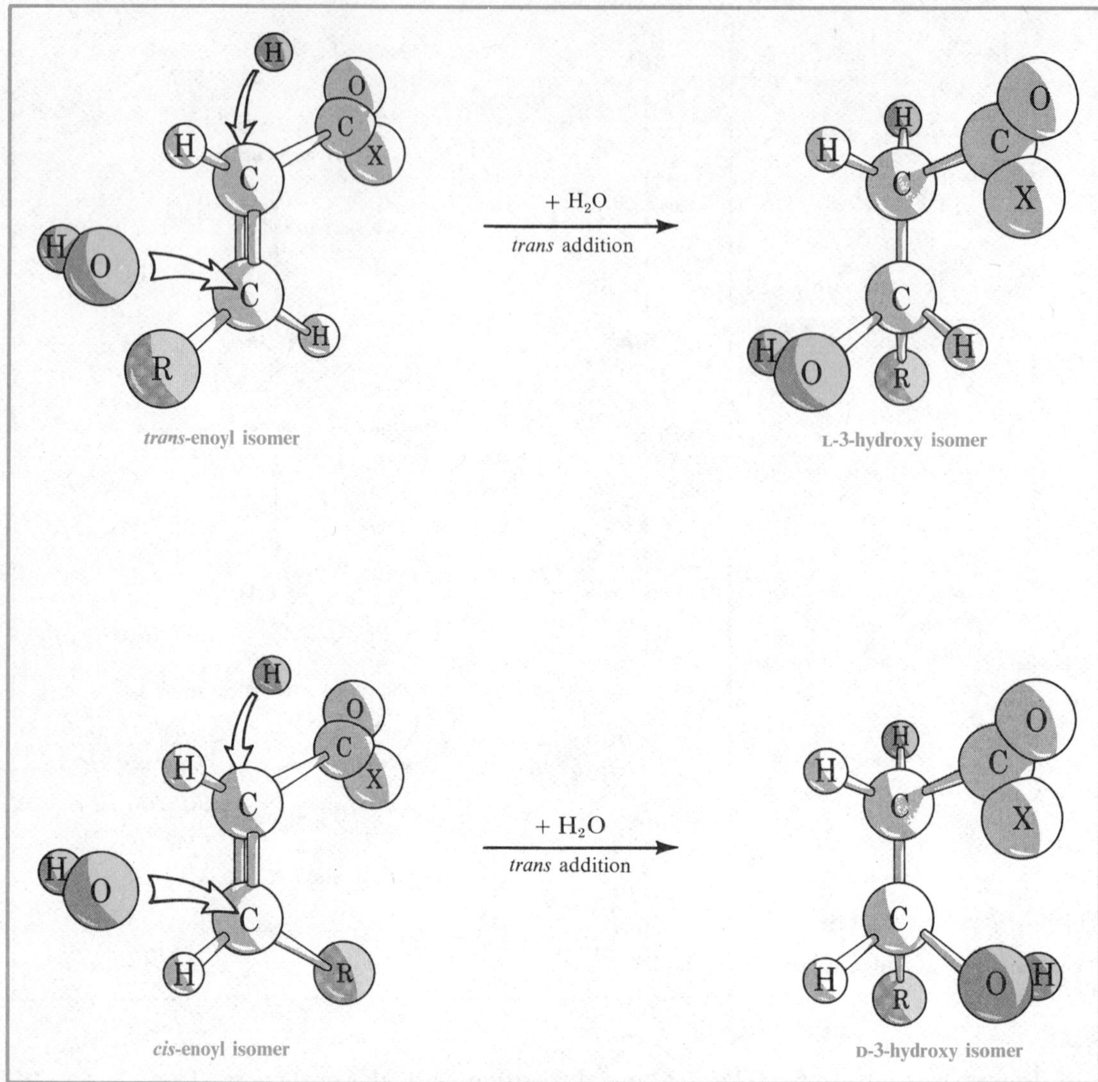

FIGURE 24–8 **If an enzyme catalyzes *trans* addition of the elements of water to a double bond, that is, addition of the H behind and OH in front of the bond as shown, in such a way that a *trans*-unsaturated compound is converted to an L-hydroxy compound, then the enzyme must catalyze the conversion of the *cis*-unsaturated isomer to a D-hydroxy compound. (Assuming specificity permits the enzyme to react with both isomers.)**

Mechanism of Hydration. Why do the *cis* and *trans* isomers yield different stereoisomers upon hydration? Hydratases are like other enzymes in that they are constructed to bind substrates in a particular way. If the addition of the elements of water to a *trans* bond create an L-isomer of the resultant hydroxy compound, then addition of water in a similar way to a *cis* bond must create a D-isomer. Why this is so is shown in Figure 24–8.

We noted in the previous chapter that hydration of *cis*-aconitate produces D-isocitrate; and hydration of fumarate, which is a *trans* isomer, produces L-malate. The hydration of *cis* to form D and of *trans* to form L is a handy mnemonic in these cases, but it is more of an accident than a general mechanism, and we shall see an important exception to this ''rule'' when we consider fatty acid synthesis.

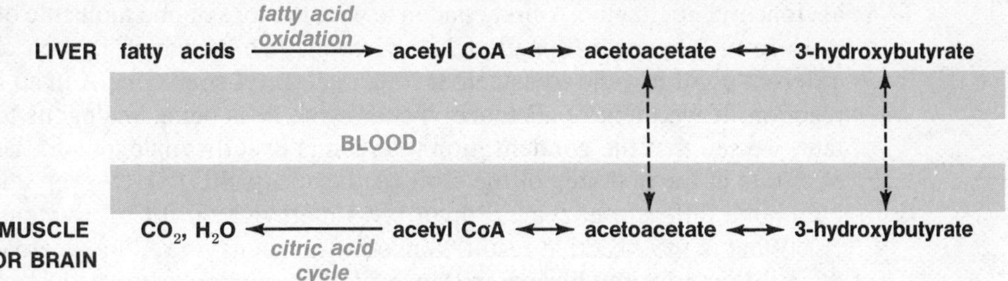

FIGURE 24–9 **Fatty acids can be oxidized to CO_2 and H_2O by the cooperative action of the liver and the muscles or brain, involving intermediate formation of a mixture of acetoacetate and D-3-hydroxybutyrate.**

OXIDATION OF FATTY ACIDS VIA 3-OXYBUTYRATES

We have outlined the major route for using fatty acids as fuels, in which the fatty acids are oxidized and cleaved to acetyl coenzyme A, followed by combustion of the acetyl groups through the citric acid cycle in the same cell. In other words, this route involves the total combustion of a fatty acid by single cells. It is especially important in muscles, including the heart.

However, there is another way in which fatty acids can be oxidized that involves the generation of acetyl groups in the liver or kidneys and their combustion in other tissues (Fig. 24–9). The acetyl groups are transported through the blood in the form of acetoacetate (3-oxybutyrate in systematic nomenclature) or D-3-hydroxybutyrate. Let us speak of these two compounds collectively as the 3-oxybutyrates.

Offhand, it might appear that this circuitous route offers no advantages to the organism over the direct oxidation of the fatty acids in the receptor tissues. The important difference is that the 3-oxybutyrates, unlike the fatty acids, readily enter neurons and can be utilized as a fuel by nervous tissue in place of the glucose that ordinarily is the major fuel for this tissue. Conditions in which the supply of glucose is impaired, such as starvation, cause an increased production of acetoacetate and D-3-hydroxybutyrate.

Formation of Acetoacetate. Acetoacetate is produced in the liver and kidneys by a simple two-step process (Fig. 24–10), in which acetyl coenzyme A and

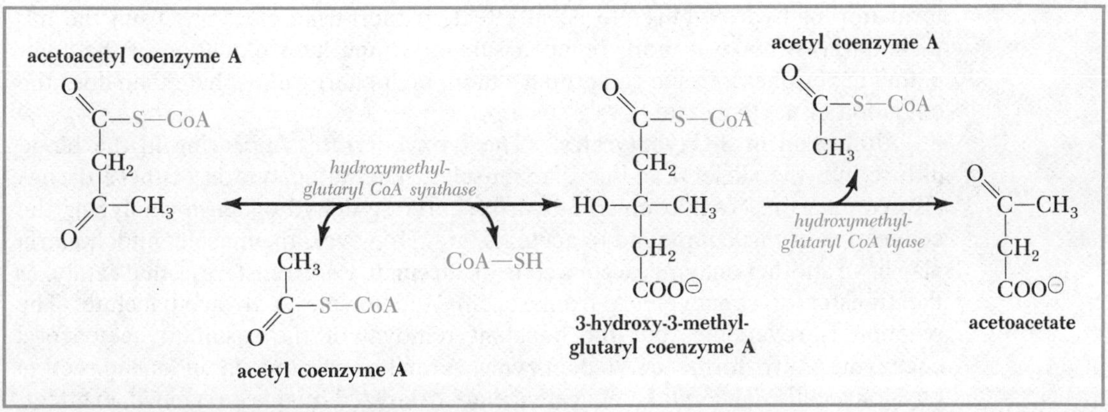

acetoacetyl coenzyme A
hydroxymethyl-glutaryl CoA synthase
acetyl coenzyme A
CoA—SH
3-hydroxy-3-methyl-glutaryl coenzyme A
acetyl coenzyme A
hydroxymethyl-glutaryl CoA lyase
acetoacetate

FIGURE 24–10 **The formation of acetoacetate.**

acetoacetyl coenzyme A first condense with the loss of one molecule of coenzyme A to form 3-hydroxy-3-methylglutaryl coenzyme A, which is then cleaved at a different point to yield free acetoacetate and acetyl coenzyme A in an irreversible reaction. If we think of acetoacetyl coenzyme A as being analogous to oxaloacetate, we see that the condensation reaction is exactly analogous to the formation of citrate in the first step of the citric acid cycle (p. 405). (However, it is catalyzed by a quite different enzyme, 3-hydroxy-3-methylglutaryl CoA synthase.)

What is the effective result? Since acetyl coenzyme A and acetoacetyl coenzyme A are in equilibrium because of the reaction catalyzed by acetoacetyl coenzyme A thiolase (p. 428), all of the carbons of 3-hydroxy-3-methylglutaryl coenzyme A can be formed from acetyl coenzyme A:

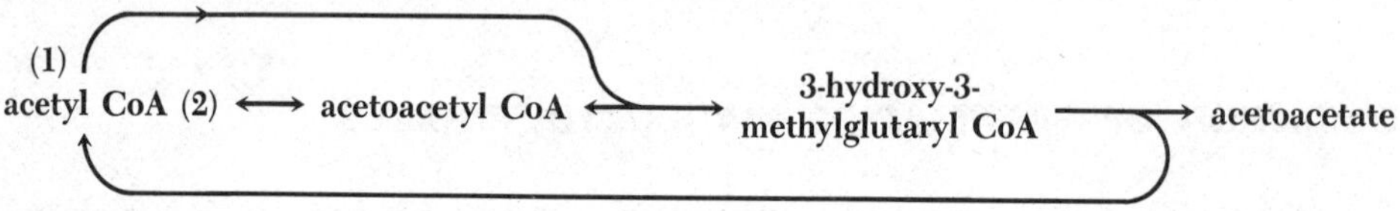

(1)
acetyl CoA (2) ⟷ acetoacetyl CoA ⟷ 3-hydroxy-3-methylglutaryl CoA ⟶ acetoacetate

Formation of D-3-Hydroxybutyrate. Part of the acetoacetate formed in the liver is converted to D-3-hydroxybutyrate because the mitochondrial cristae contain an NAD-coupled D-3-hydroxybutyrate dehydrogenase, which is a typical alcohol dehydrogenase catalyzing the reversible reaction:

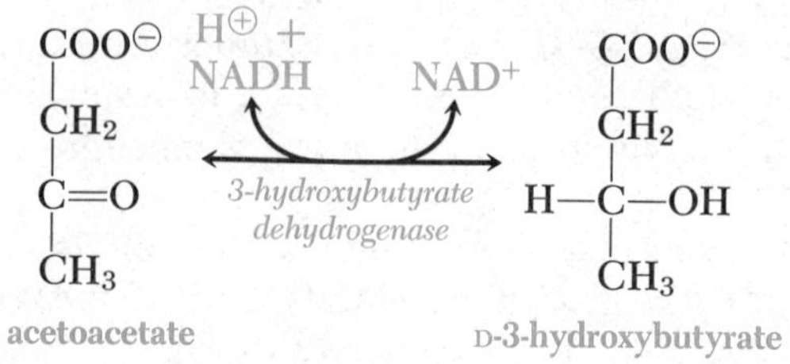

$$\underset{\text{acetoacetate}}{\begin{array}{l}COO^{\ominus}\\ |\\ CH_2\\ |\\ C{=}O\\ |\\ CH_3\end{array}} \quad \xrightleftharpoons[\text{3-hydroxybutyrate dehydrogenase}]{H^{\oplus} + NADH \quad NAD^{+}} \quad \underset{\text{D-3-hydroxybutyrate}}{\begin{array}{c}COO^{\ominus}\\ |\\ CH_2\\ |\\ H{-}C{-}OH\\ |\\ CH_3\end{array}}$$

(Why is the dehydrogenase built to form the D-3-hydroxy isomer? Presumably so as to differentiate the 3-oxybutyrate route from the usual route of fatty acid oxidation, which involves the L-3-hydroxy derivatives. No good rationalization has been given for the enzyme's occurrence in tight association with phospholipids in the inner membrane.) The amount of conversion of acetoacetate to D-3-hydroxybutyrate is determined by the ratio of NADH to NAD in the mitochondria. Indeed, the relative amounts of the two oxybutyrates produced by the liver is used as an index of the state of reduction of NAD in mitochondria. The formation of hydroxybutyrate is, in effect, withdrawing electrons from the mitochondria to make a more reduced substrate, and later oxidation of the compound in peripheral tissue can produce more high-energy phosphate than does the oxidation of acetoacetate.

Utilization of 3-Oxybutyrates. The 3-oxybutyrates appearing in the blood diffuse into the skeletal and cardiac muscles. The mitochondria of these tissues also contain the NAD-coupled 3-hydroxybutyrate dehydrogenase catalyzing the conversion of this compound to acetoacetate. However, the muscle mitochondria also have another enzyme, **acetoacetate – succinate CoA transferase,** that catalyzes the transfer of coenzyme A from succinyl coenzyme A to acetoacetate. The reaction is reversible, but the constant removal of the resultant acetoacetyl coenzyme A to form acetyl coenzyme A makes it proceed in one direction physiologically. The uptake of acetoacetate catalyzed by this enzyme is, in effect, at the expense of one mole of high-energy phosphate. This is so because it

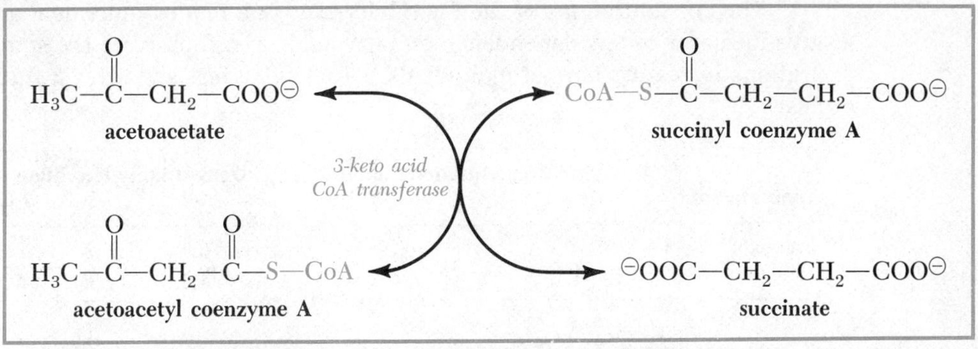

$H_3C-\overset{O}{\overset{\|}{C}}-CH_2-COO^{\ominus}$ (acetoacetate) + $CoA-S-\overset{O}{\overset{\|}{C}}-CH_2-CH_2-COO^{\ominus}$ (succinyl coenzyme A)

3-keto acid CoA transferase

$H_3C-\overset{O}{\overset{\|}{C}}-CH_2-\overset{O}{\overset{\|}{C}}-S-CoA$ (acetoacetyl coenzyme A) + $^{\ominus}OOC-CH_2-CH_2-COO^{\ominus}$ (succinate)

FIGURE 24–11 **Muscles and brain convert acetoacetate to acetoacetyl coenzyme A for use as a fuel by transferring coenzyme A from succinyl coenzyme A.**

involves the conversion of succinyl coenzyme A to succinate, and GTP would otherwise be obtained by this conversion through the succinyl CoA synthetase reaction in the citric acid cycle (p. 414).

In sum, the muscles have a mechanism for converting D-hydroxybutyrate to acetoacetate, and the carbons of acetoacetate obtained in this way and by direct diffusion from the blood are injected into the citric acid cycle in the form of acetyl coenzyme A. Two moles of acetyl coenzyme A are obtained from one mole of acetoacetate (Fig. 24–12).

Ketonemia and Ketosis. Acetoacetate is constantly undergoing spontaneous decarboxylation to form **acetone.** The reaction is slow, but if the concentration of acetoacetate becomes high, enough acetone may be formed to make its odor detectable in the breath. This is part of the reason that acetoacetate, D-3-hydroxybutyrate, and acetone were collectively called "the ketone bodies" by early investigators even though acetone is a minor part of the total. The term now seems quaint, but it is still in use, and an increase in blood concentrations of the compounds is called a ketonemia.

If the formation of 3-oxybutyrates is so rapid that large concentrations of the oxybutyrates begin to appear in the urine, an individual is said to have ketonuria, and to be in a state of ketosis. (Since H+ is produced along with the oxybutyrates, ketosis is frequently accompanied by **acidosis.**)

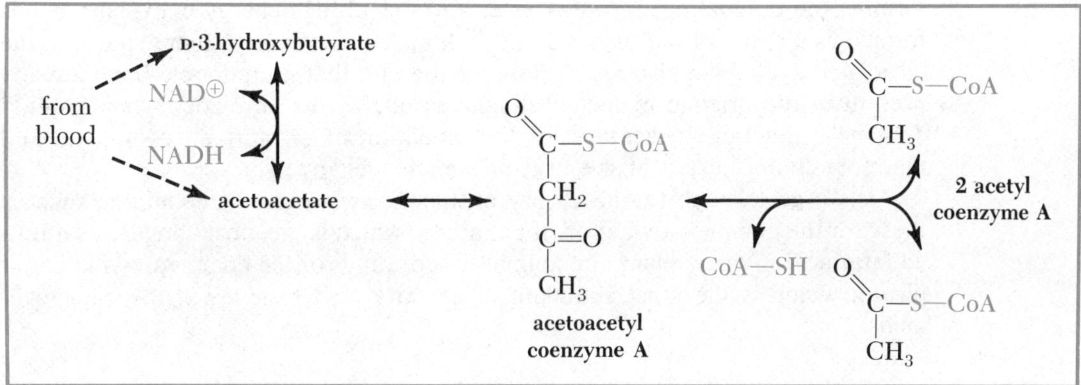

FIGURE 24–12 **Muscles and brain also can equilibrate D-3-hydroxybutyrate with acetoacetate by an NAD-coupled dehydrogenase, and can therefore use both acetoacetate and D-3-hydroxybutyrate obtained from the blood, with the formation of two molecules of acetyl coenzyme A from either compound.**

The concentrations of the 3-oxybutyrates in a resting individual are a sensitive indicator of this dependence on fatty acids as a fuel. Here are some concentrations in resting normal individuals before breakfast and after a further 7-day fast:

Time of Fast	Blood Concentrations (mM)		Daily Urinary Excretion (mmoles)	
	AcAcO⁻	*3-OH-Bu*	*AcAcO⁻*	*3-OH-Bu*
overnight	0.013	0.016	0.049	0.027
180 hours	1.1	4	10.9	77.1

(The preceding is based on data from G. F. Cahill, et al.: (1966) *Hormone-fuel Interrelationships during Fasting*. J. Clin. Invest., *45*: 1751.)

Typical values in the blood of a diabetic entering an emergency ward with acidosis would be 3 mM acetoacetate and 10 mM D-3-hydroxybutyrate. Before the introduction of insulin, higher values were commonly seen in patients with diabetic ketoacidosis.

YIELD OF HIGH-ENERGY PHOSPHATE FROM FATTY ACID OXIDATION

Yield from Direct Combustion. As we saw, each cleavage of acetyl coenzyme A from a saturated acyl chain is preceded by a pair of oxidations, one catalyzed by a flavoprotein and the other by an NAD-coupled alcohol dehydrogenase. Electrons from the flavoprotein are inserted into the mitochondrial electron transport system at the ubiquinone level and therefore yield two high-energy phosphate bonds per molecule of substrate oxidized. Electrons from NADH go through the complete electron transport mechanism and yield three high-energy phosphate bonds per molecule of substrate oxidized. The sum of the two reactions is therefore five high-energy phosphate bonds generated per acetyl coenzyme A unit cleaved. How many such cleavages occur with a given fatty acid? There is one less than the number of pairs of carbon atoms in the chain, because the cleavage of two carbons from the final four-carbon unit in the chain also leaves the terminal two carbons in the form of acetyl coenzyme A. (It only takes $n - 1$ cuts to divide a string into n pieces.)

We must take into account the two high-energy phosphate bonds expended in forming the original coenzyme A ester and the 12 high-energy phosphate bonds formed as a result of the oxidation of each molecule of acetyl coenzyme A in the citric acid cycle. We also must allow for the fact that each double bond already present in the original molecule eliminates one of the acyl coenzyme A dehydrogenase reactions necessary for the oxidation of a saturated compound and therefore diminishes the high energy phosphate yield by two.

If we go through the necessary arithmetic, we arrive at equations such as these for the complete oxidation of palmitate, which is the most abundant saturated fatty acid in many plant and animal tissues, and for the complete oxidation of oleate, which is the most abundant of all fatty acid residues in the higher animals:

Palmitate:

$$C_{16}H_{31}O_2 + 23\ O_2 + 129\ (ADP + P_i) \rightarrow 16\ CO_2 + 129\ ATP$$

Oleate:

$$C_{18}H_{33}O_2 + 25\tfrac{1}{2}\ O_2 + 144\ (ADP + P_i) \rightarrow 18\ CO_2 + 144\ ATP$$

We see that the P:O ratios are 129/46 and 144/51, or 2.80 and 2.82, respectively.

Yield from Combustion *via* 3-Oxybutyrates. The equations just given are those that apply to the complete oxidation of a fatty acid within the mitochondria of one cell. How is the picture changed if the fatty acids are oxidized to acetoacetate or to D-3-hydroxybutyrate in the liver, and these compounds are transported to the peripheral tissues for oxidation? The total balance is changed very little, but the oxidations, and therefore the production of high-energy phosphate, are distributed between the liver and the peripheral tissues. When one mole of palmitate is oxidized *via* acetoacetate, 0.30 of the required total oxygen consumption and 0.26 of the total ATP production occur in the liver; when it is oxidized *via* 3-hydroxybutyrate, 0.22 of the oxygen consumption and 0.17 of the ATP production hydroxybutyrate, 0.22 of the oxygen consumption and 0.17 of the ATP production occur in the liver. In either case the P:O ratio for oxidation of the oxybutyrate in the peripheral tissues is 2.89, so the peripheral tissues have higher yield of ATP per molecule of O_2 consumed in oxidizing the 3-oxybutyrates than they do when they completely oxidize palmitate. This may be another reason why the brain is adapted to use 3-oxybutyrates, but not fatty acids, thereby conserving its oxygen supply.

REGULATION OF FATTY ACID OXIDATION

The oxidation of the fatty acids to acetyl coenzyme A, like the subsequent oxidation of acetyl coenzyme A via the citric acid cycle, requires ADP to be available for coupled oxidative phosphorylation. If there is no demand for high-energy phosphate, then there is no production of ADP, no electron transport, and no fatty acid oxidation. Very simple.

However, given a demand for high-energy phosphate, the primary regulation in animals appears to hinge on the amount of substrates available. The 3-oxybutyrates are preferentially used by peripheral tissues when they are available, and high concentrations of the free fatty acids promote the formation of the 3-oxybutyrates in the liver. We shall have more to say about the relative utilization of fatty acids after we examine the other kinds of fuels available.

FURTHER READING

Wakil, S. J., ed.: (1970) *Lipid Metabolism*. Academic Press. Detailed summaries of several aspects.

Garland, P. B., et al.: (1969) *Interactions between Fatty Acid Oxidation and the Tricarboxylic Acid Cycle*. Pages 163–212 *in* J. M. Lowenstein, ed.: *Citric Acid Cycle*. Academic Press.

Hall, C. L., and H. Kamin: (1975) *The Purification and Some Properties of Electron Transfer Flavoprotein and General Fatty Acyl Coenzyme A Dehydrogenase from Pig Liver Mitochondria*. J. Biol. Chem., *250*:3470.

Pandi, S. V., and R. Parvin: (1976) *Characterization of Carnitine Acylcarnitine Translocase System of Heart Mitochondria*. J. Biol. Chem., *251*:6683.

Bank, W. J., et al.: (1975) *A Disorder of Muscle Lipid Metabolism and Myoglobinuria*. N. Engl. J. Med., *292*:443. Cases of deficiency of mitochondrial carnitine palmitoyl transferase.

Ruzicka, F. J., and H. Beinert: (1975) *A New Iron-sulfur Flavoprotein of the Mitochondrial Electron Transfer System*. Biochem. Biophys. Res. Comm., *66*:622.

25 THE OXIDATION OF GLUCOSE

Glucose is a major fuel for most tissues and is especially important for the brain. Much of the glucose is derived from starch, which accounts for over half of the fuel in the diets of most humans, although less than this in the United States and in other highly developed countries. Glucose is also produced from other dietary components, especially amino acids, by the liver and, to a lesser extent, by the kidneys.

The combustion of glucose is considerably more complex than the combustion of fatty acids, and it is convenient to consider it in parts (Fig. 25–1): (1) the transport of glucose into cells, followed by the formation of glucose 6-phosphate; (2) the transformation of glucose 6-phosphate into a form that can be split to yield two triose phosphates (phosphate esters of 3-carbon sugars); (3) the conversion of the triose phosphates into the 2-keto acid anion, pyruvate; and (4) the oxidation in mitochondria of pyruvate to acetyl coenzyme A, which can be further oxidized by the citric acid cycle.

The extramitochondrial part of the sequence, the conversion of glucose to pyruvate, is frequently called the **Embden-Meyerhof*** **pathway,** and it involves enzymes that are in solution in the cytoplasm, or loosely attached to membrane surfaces. High-energy phosphate is generated by substrate level phosphorylation

*Gustave Embden (1874–1933): German biochemist; one of the great pioneers in the study of metabolism. Otto Meyerhof (1884–1951): Another outstanding German biochemist, who received the Nobel prize in 1922, and who sought refuge in the United States in 1938.

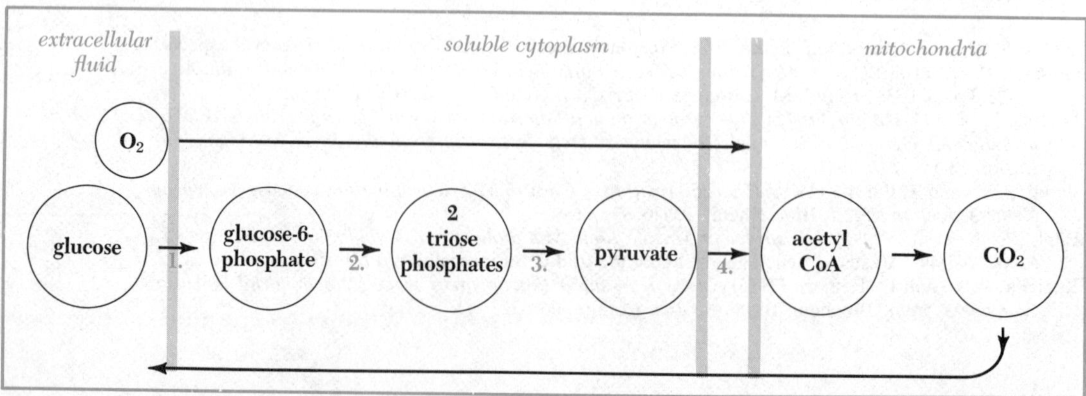

FIGURE 25–1 **General outline of the oxidation of glucose. The numbered segments of the scheme are discussed in the text.**

in this pathway, as well as by oxidative phosphorylation during the subsequent mitochondrial oxidations.

CELLULAR UPTAKE OF GLUCOSE

Transport

Most tissues obtain glucose by facilitated diffusion from the blood. This is possible because the concentration in blood is maintained at a relatively high level, between 4 and 6 mM in most humans after an overnight fast (M.W. of glucose is 180). It may go as low as 3 mM or as high as 9 mM at other times in normal individuals, depending upon diet and activity, but the range is usually small, considering the amount of the compound that is metabolized each day and the intervals between meals.

We have seen that the small bowel and the kidney take up glucose by Na^+-linked active transport (p. 369), but the driving force for the passage of glucose into all other cells, with the possible exception of those in the nervous system, is the higher concentration in the blood compared to the intracellular concentration. The polar glucose molecules cannot freely cross the plasma membrane, and the rate of their passage down the concentration gradient is facilitated by specific carrier proteins in the membrane.

The carrier in the liver functions without any known external controls. Glucose rapidly equilibrates between the liver cytosol and the extracellular fluid, with the rate and direction of equilibration dependent only upon the concentration gradient across the plasma membrane. The liver is an organ that both uses and produces glucose, although it uses glucose mainly to convert an excess supply to fats. The red blood cells, which have lost most of their organelles, and tissues such as the lens of the eye and bone, which have only slight demands for fuel, also have an unregulated equilibration of internal and external glucose concentrations.

Tissues with potentially high demands for fuel do not take up glucose so readily. The brain has an excess capacity for transport under most circumstances, with a V_{max} of 0.7 to 2.8 μmoles per minute per gram of tissue compared to a fuel requirement of 0.26 μmoles min^{-1} g^{-1}. However, the transport system is normally not saturated, with an apparent K_M of 3.5 to 8.7 mM. According to best estimates, the result is that the transport capacity is three times the demand when blood glucose concentrations are within the usual range, but transport is calculated to become limiting at concentrations below 1.44 mM. Symptoms of deprivation, such as confusion or hallucinations may appear in some at even higher concentrations (2.5 mM, or more), whereas others may tolerate concentrations approaching 1 mM for brief periods without effect. Sustained deprivation of the brain leads to profound coma.

Transport into muscles, including the heart, skeletal muscles, and smooth muscles, is tightly regulated by the insulin concentration. Little glucose moves into resting muscle fibers unless insulin is present. The pancreas increases its output of insulin in response to a higher blood glucose concentration, enabling muscles to obtain more glucose from the blood. At lower glucose concentrations, much of this fuel is reserved for the nervous system, which is not dependent upon insulin for glucose uptake. In other words, glucose utilization in muscles becomes limited by the rate of transport from the blood when the insulin level is low. How the interaction of insulin with the plasma membrane aids transport is unknown.

The uptake of glucose in muscle fibers is also increased by some unknown mechanism when they are excited to contract, a mechanism perhaps involving the Ca^{2+} that is released to initiate contraction (p. 356). In any event, the effect serves to make more glucose available to the muscles when the demand for fuel is great.

Phosphorylation of Glucose

The metabolism of glucose begins by transferring a phosphate group to it from ATP in a reaction catalyzed by a hexokinase:

$$\alpha\text{- or }\beta\text{-D-glucose} \xrightarrow[\textit{hexokinase}]{\text{ATP} \;\; Mg^{2+} \;\; \text{ADP}} \alpha\text{- or }\beta\text{-D-glucose 6-phosphate}$$

HO—CH₂ … (H,OH) ; $^{2-}O_3P—O—CH_2$ … (H,OH)

A route for generating high-energy phosphate begins by spending a molecule of high-energy phosphate! However, this is not an accidental quirk of evolution. ATP is being expended to trap glucose as glucose 6-phosphate, which does not pass freely through the membranes. The reaction is essentially irreversible at physiological concentrations of reactants and products because the pyrophosphate bond of ATP that is being cleaved has a much higher standard free energy of formation than does the phosphate ester bond in glucose 6-phosphate. It is this irreversibility that enables the intracellular concentration of glucose to be maintained at a low level so that glucose will flow into the cell without further expenditure of energy. Indeed, it is theoretically possible to convert all of the glucose in the body almost quantitatively into glucose 6-phosphate, or at least to exhaust the phosphate supply. This does not happen in part because transport is regulated, but regulation of hexokinase activity, to be discussed later in the chapter, also helps.

THE FORMATION OF TRIOSE PHOSPHATES

The conversion of glucose 6-phosphate to two molecules of triose phosphates involves three reactions (Fig. 25–2). Glucose 6-phosphate (an aldose phosphate) is converted to fructose 6-phosphate (a ketose phosphate), which can then be phosphorylated on C-1. The resultant fructose 1,6-bisphosphate is then cleaved in the middle to create two molecules of triose phosphates. One is glyceraldehyde 3-phosphate (an aldose phosphate) and the other is dihydroxyacetone phosphate (a ketose phosphate). These two triose phosphates are equilibrated by another reaction. Each of these reactions is an example of a type that occurs repeatedly in metabolism and it is worth looking at them in some detail.

Glucose 6-Phosphate → Fructose 6-Phosphate. This freely reversible interconversion involves shifting the potential carbonyl group of the glucose residue from C-1 to C-2. It is an isomerization of the aldohexose into a ketohexose in the

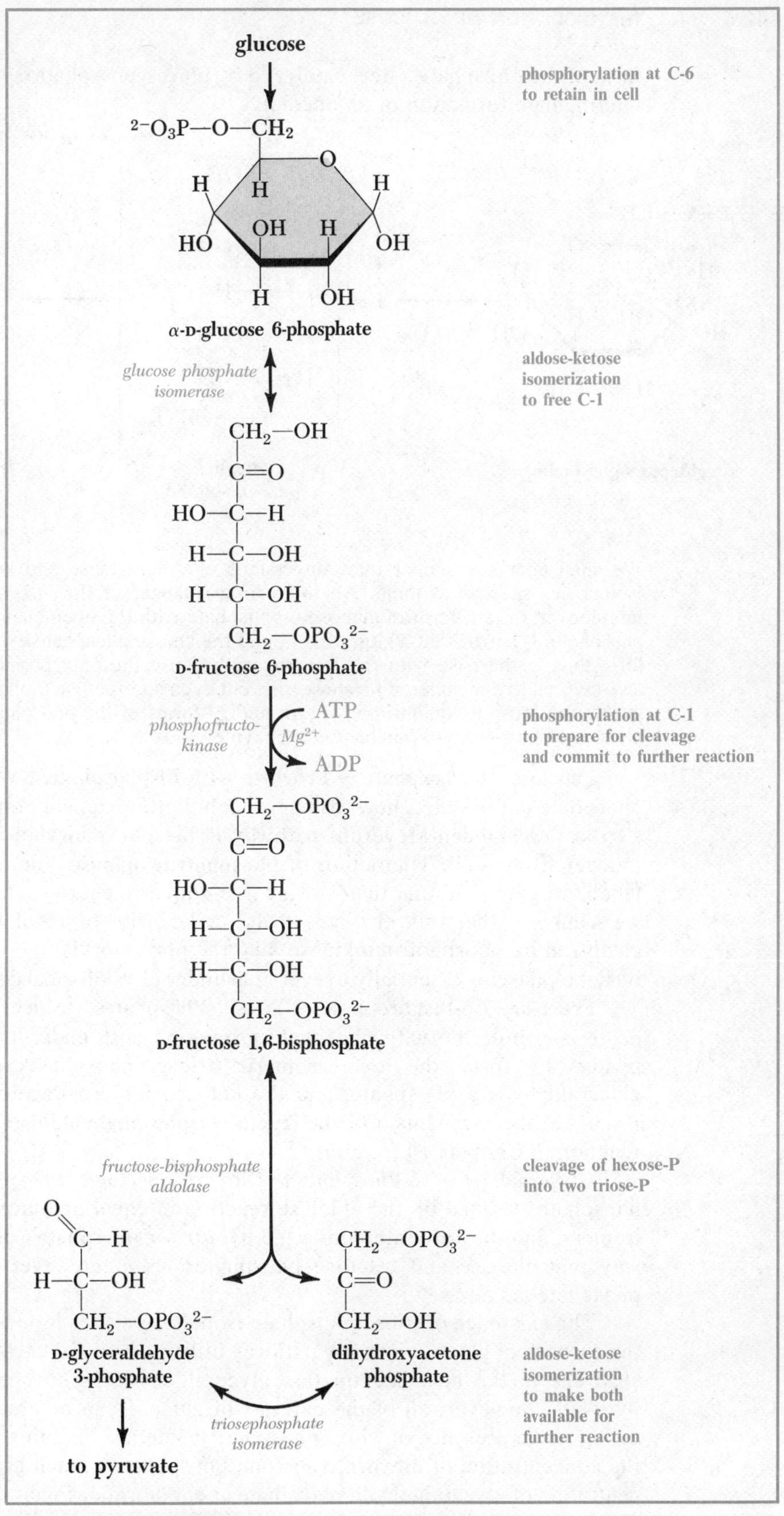

FIGURE 25–2 **The conversion of glucose 6-phosphate to two molecules of triose phosphate.**

form of the phosphate ester, catalyzed by **glucosephosphate isomerase** through the intermediate formation of an enediol:

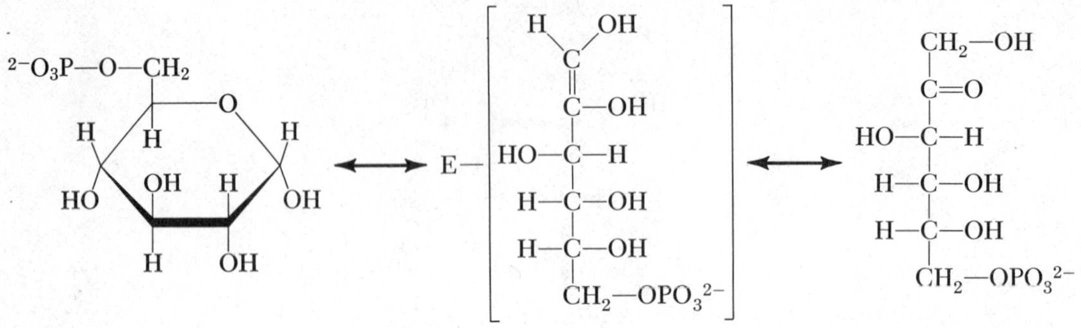

$^{2-}O_3P—O—CH_2$

α-D-glucose 6-phosphate ⟷ E— [H, OH >C=C—OH; HO—C—H; H—C—OH; H—C—OH; $CH_2—OPO_3^{2-}$] ⟷ $CH_2—OH$; C=O; HO—C—H; H—C—OH; H—C—OH; $CH_2—OPO_3^{2-}$

α-D-glucose 6-phosphate — enediol (enzyme-bound) — D-fructose 6-phosphate

We shall later see similar interconversions of other aldose and ketose phosphates by isomerases specific to them. An interesting sidelight of the enzyme at hand is that it interconverts the α-form of glucose 6-phosphate with the open-chain form of fructose 6-phosphate. It is true that a larger fraction of the ketose phosphate exists in the open chain form than is the case with the free sugars because the only possible ring closure is a less-favored five-membered **furanose** ring, but even so, fructose 6-phosphate exists mainly in the ring form at equilibrium. The α- and β- forms of the phosphate equilibrate, so all glucose 6-phosphate is available to the enzyme.

Fructose 6-Phosphate → Fructose 1,6-Bisphosphate. When fructose 6-phosphate is formed, a new primary alcohol group appears on C-1, and the next step in the Embden-Meyerhof pathway is the phosphorylation of this group by transfer from ATP. The action of **phosphofructokinase** commits the cell to metabolizing glucose rather than storing it or converting it to some other hexose, and we shall see that this enzyme is also a key site of regulation. The reaction catalyzed by phosphofructokinase, like the phosphorylation of glucose catalyzed by hexokinase, is essentially irreversible under physiological conditions.

Fructose 1,6-Bisphosphate → Triose Phosphates. Since fructose bisphosphate is a molecule with phosphate groups on both ends, it can be split in the middle to form the two isomeric triose phosphates — the aldose, D-glyceraldehyde 3-phosphate, and the ketose, dihydroxyacetone phosphate. We discussed the mechanism of the **fructosebisphosphate aldolase** that catalyzes this reaction in Chapter 14 (p. 266).

Glyceraldehyde 3-Phosphate ↔ Dihydroxyacetone Phosphate. The triose-phosphates formed by the aldolase reaction in equal amounts are aldose-ketose isomers, and they are interconverted by a **triose phosphate isomerase** in the same way that glucose and fructose phosphates are interconverted by the glucose phosphate isomerase.

The existence of triose phosphate isomerase makes it possible to divert all of the carbons of glucose into any pathway utilizing one of the triose phosphates. We shall see in the next section that glyceraldehyde 3-phosphate is converted to pyruvate; however, all of the carbons of glucose can be converted to pyruvate owing to the presence of triose phosphate isomerase, and this is true even though the concentration of dihydroxyacetone phosphate is much greater than the concentration of glyceraldehyde phosphate at equilibrium. Figure 25–3 represents the pools of compounds in a tissue as tanks, connected by reactions in the form of pipes. Fructose bisphosphate and the triose phosphates are equilibrated freely.

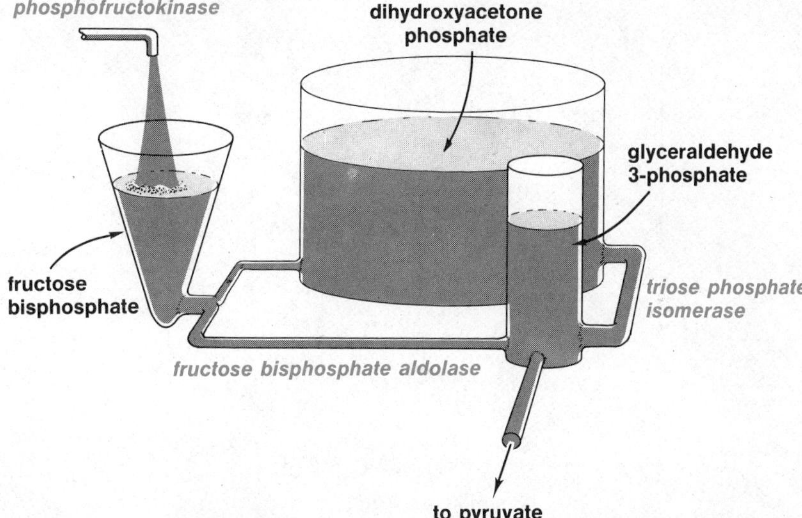

FIGURE 25–3 **Fructose bisphosphate and the triose phosphates are maintained near equilibrium concentrations, with dihydroxyacetone phosphate present in the highest concentration. The pool of these three compounds is fed by the nearly irreversible formation of fructose bisphosphate. The removal of glyceraldehyde 3-phosphate as the direct precursor of pyruvate causes part of it to be replaced from dihydroxyacetone phosphate in order to maintain equilibrium. The result is that all of the hexose carbons flow toward pyruvate to an equal extent. (The diagram shows the proportion of fructose bisphosphate increasing when the total concentration of the compounds rises, since it is one molecule in equilibrium with two.)**

We can see from the diagram that conversion to pyruvate has the immediate effect of lowering the concentration of glyceraldehyde 3-phosphate, but this will cause replacement of glyceraldehyde 3-phosphate from the dihydroxyacetone phosphate equilibrated with it, which is therefore also being consumed. Both triose phosphates are converted to pyruvate in equal amounts.

THE FORMATION OF PYRUVATE FROM TRIOSE PHOSPHATES

The conversion of the triose phosphates to pyruvate in the soluble cytoplasm of cells initiates the actual recovery of energy from the metabolism of glucose. Since both triose phosphates are involved, two molecules of pyruvate are formed from each molecule of glucose 6-phosphate consumed. Essentially, the sequence involves an oxidation of glyceraldehyde 3-phosphate with capture of the energy as high-energy phosphate, followed by rearrangements that result in the formation of still further high-energy phosphate (Fig. 25–4). The high-energy phosphate is used to form ATP from ADP.

Glyceraldehyde 3-Phosphate ↔ 3-Phosphoglycerate. The aldehyde group on glyceraldehyde 3-phosphate is oxidized by NAD to form a high-energy phosphate bond in 1,3-bisphosphoglycerate. The high-energy phosphate is then transferred to ADP, releasing 3-phosphoglycerate. Oxidation of an aldehyde normally forms a carboxylic acid, 3-phosphoglyceric acid in this case. When we discussed

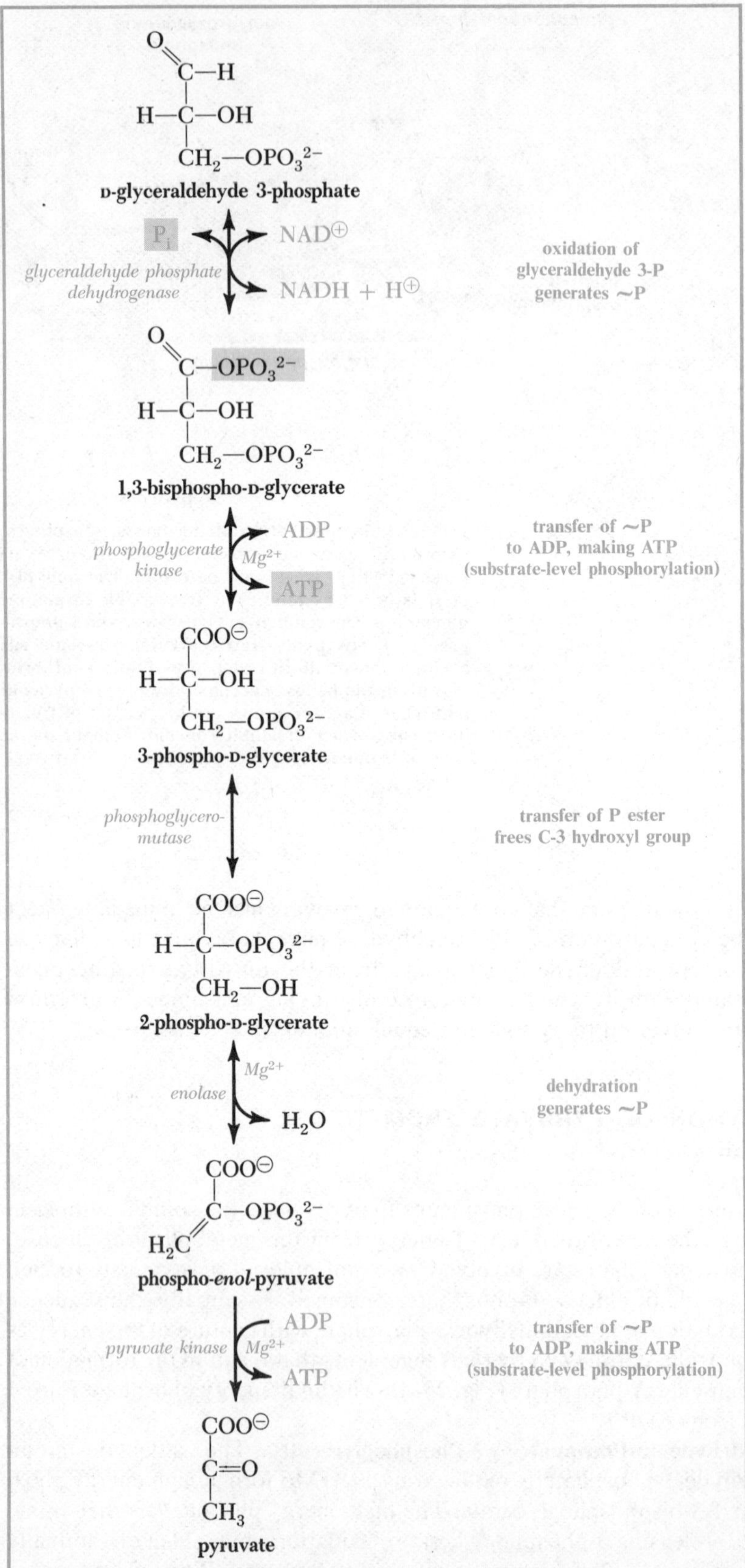

FIGURE 25–4 *See legend on opposite page.*

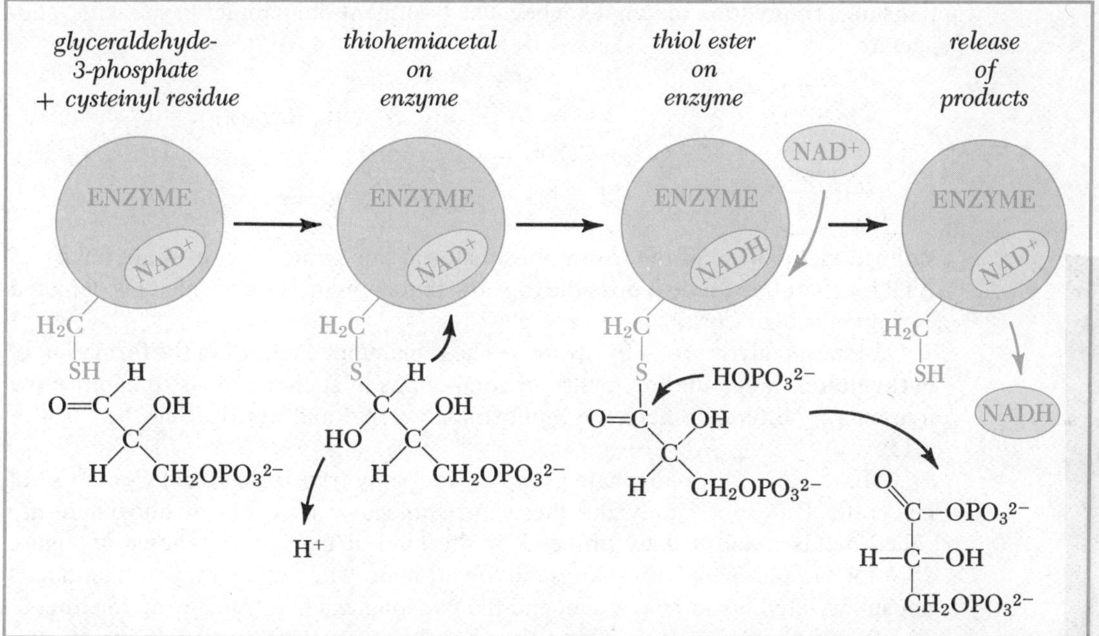

FIGURE 25–5 **The oxidation of glyceraldehyde 3-phosphate involves its combination with a sulfhydryl group on the dehydrogenase, forming a thiohemiacetal. This structure is oxidized by NAD, also on the enzyme, to form a thiol ester, which is cleaved by inorganic phosphate to produce 1,3-bisphosphoglycerate. The NADH produced on the enzyme is less tightly bound and is displaced by a molecule of NAD from the solution.**

the oxidation of α-ketoglutarate to succinyl coenzyme A in the citric acid cycle (p. 408), we noted that such oxidations have a standard free energy change sufficiently negative to support the formation of a high-energy bond at physiological concentrations. In the example at hand, the energy of oxidation of glyceraldehyde 3-phosphate is captured by causing the simultaneous formation of an anhydride bond between the carboxyl group and phosphate. The products of the reaction catalyzed by **glyceraldehyde-3-phosphate dehydrogenase** are NADH and 1,3-bisphosphoglycerate, and the 1-phosphate anhydride is a high-energy bond.

Glyceraldehyde-3-phosphate dehydrogenase is like α-ketoglutarate dehydrogenase in creating a thiol ester as the initial high-energy bond, but the participating sulfhydryl group is provided by a residue of cysteine in the enzyme peptide rather than by coenzyme A or dihydrolipoate (Fig. 25–5). The aldehyde group of the substrate reversibly combines with the cysteinyl residue as a thiohemiacetal, which is converted to a high-energy thiol ester upon oxidation with NAD. Phosphorolysis of the ester creates the high-energy phosphate bond in 1,3-bisphosphoglycerate and restores the original sulfhydryl group on the enzyme.

Glyceraldehyde-3-phosphate dehydrogenase is susceptible to inhibition by reagents reacting with thiols. This provided a powerful tool for early investiga-

---

FIGURE 25–4 **The conversion of glyceraldehyde 3-phosphate to pyruvate. For each molecule of pyruvate produced, two molecules of ATP are formed by substrate-level phosphorylations.**

tions in carbohydrate metabolism because treatment of an intact tissue with iodoacetate:

$$\underset{\boxed{\text{peptide}}}{\overset{\text{SH}}{|}} + \text{I—CH}_2\text{—COO}^{\ominus} \longrightarrow \underset{\boxed{\text{peptide}}}{\overset{\text{S—CH}_2\text{—COO}^{\ominus}}{|}} + \text{H}^{\oplus} + \text{I}^{-}$$

stopped metabolism of the triose phosphates at this point, The resultant failure of ATP regeneration made it possible to show that high-energy phosphate is depleted during muscular work.

3-Phosphoglycerate → Pyruvate. The remaining reactions in the formation of pyruvate rearrange the three-carbon compounds in such a way as to capture the free energy difference between 3-phosphoglycerate and pyruvate in the form of ATP.

MUTASES. The phosphate group is reversibly transferred from C-3 to C-2 of glycerate. Enzymes catalyzing these intramolecular transfers of phosphate are called mutases and usually proceed by the kind of mechanism shown in Figure 25–6 for the **phosphoglyceromutase** involved here. The active enzyme contains a phosphorylated histidine residue, and the phosphate is first transferred to either 3- or 2-phosphoglycerate. This forms the transient intermediate, 2,3-bisphosphoglycerate, on the enzyme surface, which in turn reacts to transfer one of its phosphate groups back onto the histidine residue. However, either of the

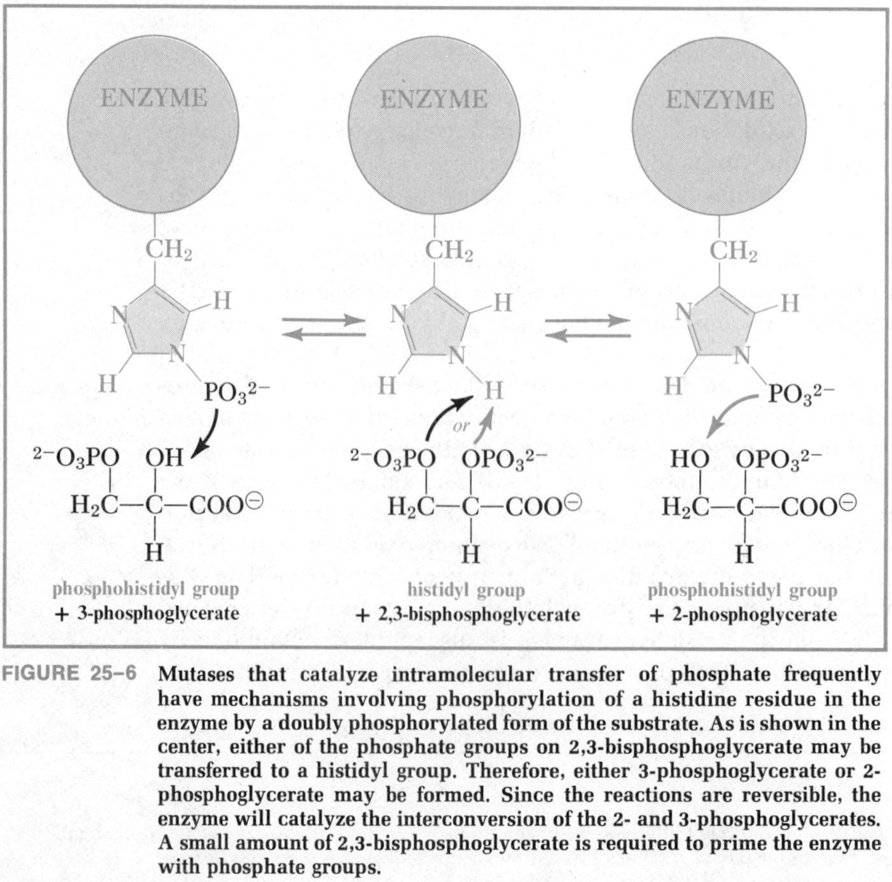

FIGURE 25–6 **Mutases that catalyze intramolecular transfer of phosphate frequently have mechanisms involving phosphorylation of a histidine residue in the enzyme by a doubly phosphorylated form of the substrate. As is shown in the center, either of the phosphate groups on 2,3-bisphosphoglycerate may be transferred to a histidyl group. Therefore, either 3-phosphoglycerate or 2-phosphoglycerate may be formed. Since the reactions are reversible, the enzyme will catalyze the interconversion of the 2- and 3-phosphoglycerates. A small amount of 2,3-bisphosphoglycerate is required to prime the enzyme with phosphate groups.**

phosphate groups may be transferred, and the remaining phosphate on the glycerate may be on either C-2 or C-3, so the effect of a continuation of the process is to bring 2-phosphoglycerate and 3-phosphoglycerate to equilibrium.

The small concentrations of the intermediate 2,3-bisphosphoglycerate necessary in the solution to prime the reaction are made by a bisphosphoglycerate mutase in the cytoplasm, which catalyzes a transfer of the phosphate from the first to the second carbon of 1,3-bisphosphoglycerate:

$$\underset{\text{1,3-bisphospho-D-glycerate}}{\begin{array}{c} O{=}C{-}OPO_3^{2-} \\ | \\ H{-}C{-}OH \\ | \\ CH_2OPO_3^{2-} \end{array}} \xrightarrow[\text{bisphosphoglycerate mutase}]{Mg^{2+}} \underset{\text{2,3-bisphospho-D-glycerate}}{\begin{array}{c} COO^{\ominus} \\ | \\ H{-}C{-}OPO_3^{2-} \\ | \\ CH_2OPO_3^{2-} \end{array}}$$

This priming reaction ought not to be confused with the major pathway we are discussing, in which most of the 1,3-bisphosphoglycerate is used to form ATP by the phosphoglycerate kinase reaction discussed above.

ENOLASE. The 2-phospho-D-glycerate created by the mutase reaction is next dehydrated by the action of an enzyme, enolase, to form phospho-*enol*-pyruvate. Phospho-*enol*-pyruvate is a high-energy phosphate compound because release of the enol by cleavage of the phosphate allows it to revert spontaneously to the keto form:

$$\underset{\text{phospho-}enol\text{-pyruvate}}{\begin{array}{c} COO^{\ominus} \\ | \\ H_2C{=}C{-}OPO_3^{2-} \end{array}} \underset{\text{pyruvate kinase}}{\overset{ADP \quad ATP}{\xrightarrow{Mg^{2+}}}} \left[\underset{\text{pyruvate (enol form)}}{\begin{array}{c} COO^{\ominus} \\ | \\ H_2C{=}C{-}OH \end{array}}\right] \xrightarrow{\text{spontaneous}} \underset{\text{pyruvate (keto form)}}{\begin{array}{c} COO^{\ominus} \\ | \\ C{=}O \\ | \\ CH_3 \end{array}}$$

The position of equilibrium is far in the direction of the ketone form of pyruvate, and the large amount of free energy released by this equilibration is added to the free energy of phosphate ester hydrolysis.

PYRUVATE KINASE. The final reaction of the Embden-Meyerhof pathway to pyruvate is the transfer of the high energy phosphate from phospho-*enol*-pyruvate to ADP, thereby forming ATP. The equilibrium position of the reaction is so far toward ATP formation that it is not reversible to a significant extent under physiological conditions ($K'_{eq} = 6{,}500$ at pH 7.4, 30°C).

THE COMPLETE EMBDEN-MEYERHOF PATHWAY

If we begin with a molecule of glucose and add all of the reactions involved in its conversion to two molecules of pyruvate according to the Embden-Meyerhof scheme, the result is:

$$\text{glucose} + 2\ NAD^{\oplus} + 2\ (ADP + P_i) \longrightarrow 2\ \text{pyruvate}^{\ominus} + 2\ NADH + 2\ ATP.$$

(You should convince yourself that this is so, but note that this, and the equations that follow, are not balanced for H^+ and H_2O.) Three things happen simultaneously: glucose is oxidized to pyruvate, NAD is reduced to NADH, and ADP is phosphorylated to form ATP. There can be no Embden-Meyerhof pathway without all three events, which means that NAD, ADP, and P_i, as well as glucose, must be present.

Oxidation of Cytoplasmic NADH. Since the Embden-Meyerhof pathway consumes NAD and produces NADH, it cannot proceed without some means of regenerating the oxidized nucleotide. The NADH of eukaryotic cells cannot enter mitochondria to be oxidized by molecular oxygen through the apparatus of oxidative phosphorylation, and it therefore must be handled in the cytosol. The only significant means of doing this is through displacement of the equilibrium of alcohol dehydrogenases by the increasing concentration of NADH; this isn't difficult because most of the alcohol dehydrogenases have equilibria favoring formation of the alcohol:

$$\mathrm{H{-}\overset{|}{\underset{|}{C}}{-}OH + NAD^{\oplus} \rightleftharpoons \overset{|}{\underset{|}{C}}{=}O + NADH + H^{\oplus}}.$$

But what to do with the accumulating alcohols? They could leave cells as end products in metabolism. We shall discuss the way some cells use this method of handling the problem in the next two chapters; but many cells, particularly those of the skeletal muscles and nervous tissue, remove the resulting alcohol by transporting it into mitochondria where it can be oxidized by molecular oxygen. This effectively transports electrons from NADH in the cytosol to oxygen in mitochondria. Let us look at two important examples.

THE GLYCEROL PHOSPHATE SHUTTLE. The cytosol contains a dehydrogenase that equilibrates glycerol 3-phosphate and dihydroxyacetone phosphate with NAD and NADH (Fig. 25–7). This is a straightforward equilibration of secondary alcohol and the corresponding ketone. (We saw that dihydroxyacetone phosphate is an intermediate in the conversion of glucose to pyruvate and is also in equilibrium with glyceraldehyde 3-phosphate, but we are here talking about an entirely

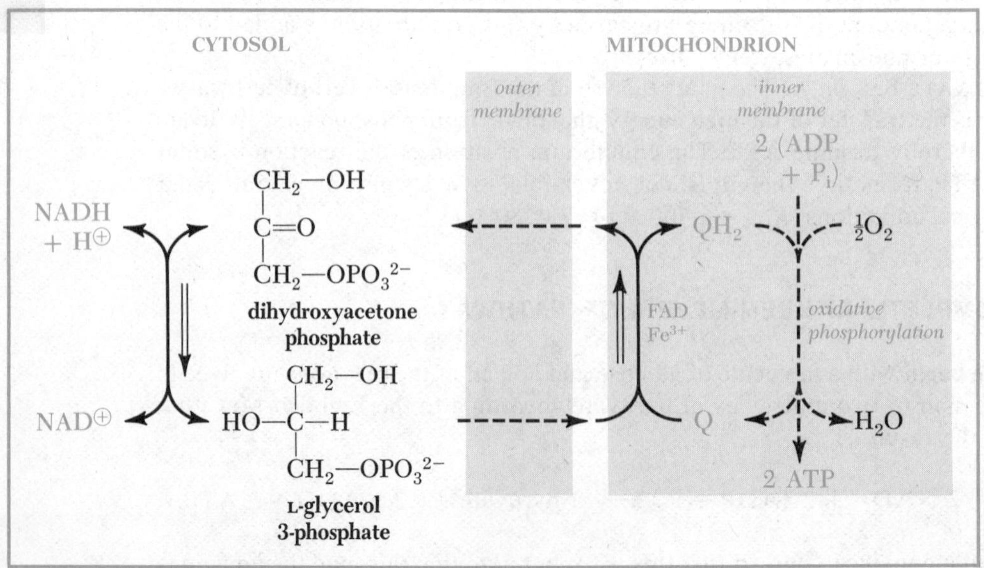

FIGURE 25–7 The glycerol phosphate shuttle of electrons.

different function of the compound.) When the concentration of NADH tends to rise owing to the action of the Embden-Meyerhof pathway, there will be a shift from dihydroxyacetone phosphate to glycerol 3-phosphate that oxidizes part of the increased NADH to NAD.

Now, glycerol 3-phosphate can cross the outer mitochondrial membranes and become exposed to the action of a dehydrogenase present on the outer surface of the inner membrane. This dehydrogenase also catalyzes the interconversion of glycerol 3-phosphate and dihydroxyacetone phosphate; however, it is an iron-containing flavoprotein similar to succinate dehydrogenase, and the use of this stronger oxidizing agent makes the equilibrium strongly favor the formation of the ketone. In addition, the electrons removed are directly transferred to ubiquinone, and the re-oxidation of reduced ubiquinone by O_2 results in the generation of two high-energy phosphate bonds per glycerol 3-phosphate oxidized.

Here we have it. Dihydroxyacetone phosphate acts as a catalytic carrier for electrons in the glycerol phosphate shuttle, much in the way that oxaloacetate acts as a catalytic carrier for acetyl groups in the citric acid cycle. The shuttle occurs because the appearance of NADH tends to raise the ratio [glycerol 3-P] /[dihydroxyacetone P] in the cytosol and the mitochondrial oxidation lowers the same ratio in the mitochondria. Therefore, there is a concentration gradient for glycerol 3-phosphate into mitochondria and for dihydroxyacetone phosphate out of mitochondria.

Since there is no net consumption or production of dihydroxyacetone phosphate in the shuttle, this function does not detract from utilization of the triose phosphate in the Embden-Meyerhof pathway. The glycerol phosphate shuttle is especially important in white muscle fibers of the higher animals and the flight muscles of insects.

The overall reaction of the glycerol phosphate shuttle is nothing more than:

$$NADH_{cytosol} + \tfrac{1}{2}\, O_2 + 2\,(ADP + P_i) \longrightarrow NAD + 2\, ATP.$$

When the two molecules of NADH produced in the Embden-Meyerhof pathway are oxidized by the shuttle, the effective result is the oxidation of glucose to pyruvate by O_2:

$$\text{glucose} + O_2 + 6\,(ADP + P_i) \longrightarrow 2\ \text{pyruvate}^{\ominus} + 6\, ATP.$$

Four of the six ATP are produced by oxidative phosphorylation as a result of the shuttle, and two are produced by substrate-level phosphorylations in the Embden-Meyerhof pathway itself.

THE MALATE-ASPARTATE ELECTRON SHUTTLE. Red skeletal muscles and the heart muscle transport electrons into mitochondria by a mechanism that involves amino acids and the malate dehydrogenase reaction. We encountered malate dehydrogenase in our discussion of the mitochondrial citric acid cycle, but the same reaction also occurs in the cytosol:

$$\text{malate}^{2-} + NAD^+ \longleftrightarrow \text{oxaloacetate}^{2-} + NADH + H^+.$$

The complete system is sketched in Figure 25–8, but let us see if we can reason how it works, numbering the steps as shown in the figure.

(1) Accumulation of NADH in the cytosol will shift the malate dehydrogenase reaction to the left, regenerating NAD by reducing oxaloacetate (the ketone) to malate (the alcohol).

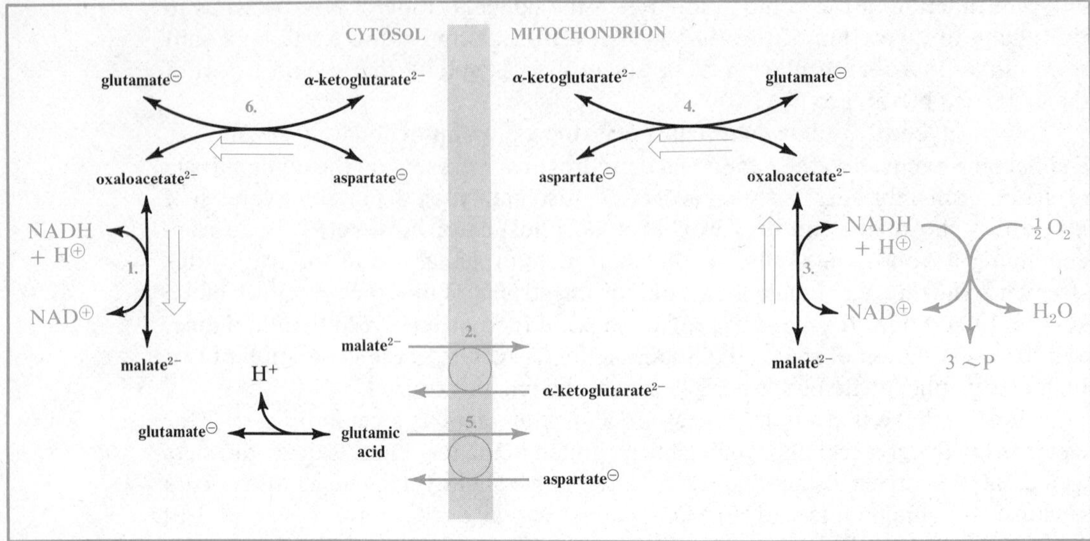

FIGURE 25–8 **The malate aspartate shuttle of electrons. The reactions are numbered in the sequence discussed in the text.**

(2) The malate can move into the mitochondrial matrix by exchange for α-ketoglutarate; this exchange is catalyzed by a specific antiport protein and requires no energy.

(3) When malate appears within the mitochondrial matrix, it is again equilibrated with oxaloacetate by oxidation with NAD, but the resultant NADH within mitochondria is attacked by electron transport to oxygen, generating 3 ATP.

(4) In order for the exchange of malate to occur, α-ketoglutarate must be supplied within the mitochondrial matrix. This is done by transamination of glutamate with the oxaloacetate. We discussed this transamination reaction as an example of enzyme mechanism (p. 267), and it causes the formation of α-ketoglutarate and aspartate.

(5) The transamination reaction requires glutamate to be supplied and aspartate to be removed within the mitochondria. This is done by an antiport exchange. However, this antiport is specific for the neutral ionic form of glutamate and the anionic form of aspartate. It is, therefore, moving negative charges out of the mitochondria and dissipating the electrochemical gradient created by electron transport in the inner membrane. This is the same as saying that the antiport is driven by part of the energy of oxidation in the mitochondrial inner membrane. To the extent that the high-energy state is dissipated in this way, oxidative phosphorylation will be diminished.

(6) Aspartate aminotransferase, like malate dehydrogenase, also occurs in the cytosol. Therefore, the aspartate pumped out of mitochondria will react with the α-ketoglutarate that moves out of mitochondria to regenerate oxaloacetate and glutamate in the cytosol. This completes the sequence. The overall result is an oxidation of NADH to NAD in the cytosol and a reduction of NAD to NADH in the mitochondrial matrix. The transfer of electrons is made possible even though the [NADH]/[NAD] ratio is much lower in the cytosol than in the mitochondrial matrix, because energy is expended to move the intermediates across the inner membrane against this concentration gradient. The amount of energy used cannot

be stated exactly, but a minimal value is equivalent to the synthesis of 1/4 ATP.

THE OXIDATION OF PYRUVATE

The final steps in the total combustion of glucose by plants and animals involve the transport of pyruvate from the cytosol into the mitochondria. Electron transport creates an OH^- concentration gradient across the inner membrane, with $[H^+]$ high and $[OH^-]$ low on the cytosol side. OH^- moves down this gradient by an antiport exchange with pyruvate. Once inside the mitochondrion, pyruvate is oxidized to acetyl coenzyme A by a pyruvate dehydrogenase complex, and then to CO_2 and H_2O by the citric acid cycle:

$$\underset{\text{pyruvate}}{COO^{\ominus}\text{—}C{=}O\text{—}CH_3} \xrightarrow[\textit{pyruvate dehydrogenase complex}]{NAD^{\oplus}\ \ CoA{-}SH \ \ NADH \ \ CO_2} \underset{\text{acetyl coenzyme A}}{O{=}C(CH_3)\text{—}S\text{—}CoA} \xrightarrow[\textit{citric acid cycle}]{} CO_2, H_2O$$

These mitochondrial oxidations account for over 80 per cent of the total ATP obtained from the complete oxidation of glucose.

The Pyruvate Dehydrogenase Complex

Pyruvate is an α-ketocarboxylate, which is oxidized by the same sort of mechanism as is α-ketoglutarate (p. 408). A decarboxylase polypeptide in the pryuvate dehydrogenase complex catalyzes the removal of CO_2 through combination of the remaining two carbons with an attached molecule of thiamine pyrophosphate as a hydroxyethyl group (Fig. 25–9). A dehydrogenase polypeptide then catalyzes the transfer of the hydroxyethyl group, which is equivalent to acetaldehyde in oxidation state, to a **lipoyllysyl** side chain on still another pep-

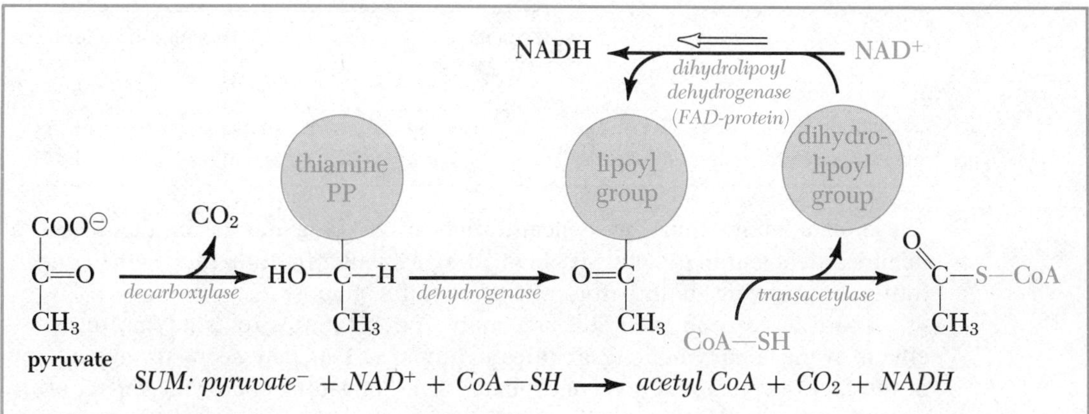

FIGURE 25–9 **The oxidative decarboxylation of pyruvate uses the same kind of mechanism seen with α-ketoglutarate dehydrogenase, except that the enzyme has separate decarboxylase and dehydrogenase polypeptide chains.**

tide. The transferred group is thereby oxidized to an **acetyl** group, combined with what is now a dihydrolipoyllysyl side chain. Finally, the acetyl group is transferred to coenzyme A, and the remaining dihydrolipoyllysyl group is reoxidized to its disulfide form by transferring electrons through **FAD** to NAD.

The pyruvate dehydrogenase differs from α-ketoglutarate dehydrogenase in animal tissues in having the decarboxylase and dehydrogenase activities on separate polypeptide chains.

Disturbances in Pyruvate Oxidation

Thiamine Deficiency. Thiamine must be supplied in the diet as a precursor for thiamine pyrophosphate. (If you guess that the coenzyme is formed by a transfer of the pyrophosphate group from ATP, leaving AMP, you are absolutely right. There is a small amount of a thiamine pyrophosphokinase in animal tissues to catalyze this reaction.) If the supply of thiamine is restricted, then one or more enzymes requiring thiamine pyrophosphate will also be deficient.

Thiamine pyrophosphate is required for the oxidation of pyruvate into acetyl coenzyme A and for the oxidation of α-ketoglutarate in the citric acid cycle, as well as for other reactions that we shall encounter later. A defect in the citric acid cycle will impair the metabolism of fatty acids and amino acids, as well as glucose; yet, all of the evidence we have at hand links the deficiency primarily to disturbances of carbohydrate metabolism, especially in the brain. For example, the thiamine requirement of a human is dependent upon carbohydrate intake. On typical mixed diets in this country, an adult will need about 1.5 micrograms per gram of carbohydrate ingested (0.8 micromole per mole). Perhaps this is an indication of a greater loss of coenzyme from the intermediate states of actively working enzymes. In any event, this would appear to indicate that the thiamine deficiency is primarily being manifested in carbohydrate metabolism, rather than in other routes, and this in turn indicates that pyruvate dehydrogenase loses thiamine pyrophosphate more readily than does α-ketoglutarate dehydrogenase of the citric acid cycle.

This interpretation is supported by studies on the concentration of pyruvate and α-ketoglutarate in the blood of normal and of thiamine-deficient individuals during fasting, and following ingestion of 100 grams of glucose (data from Metabolism, *14:* 141 [1965]):

	Normal		**Thiamine-deficient**	
	Fasting	*After glucose*	*Fasting*	*After glucose*
[pyruvate]	31 μM	42 μM	49 μM	115 μM
[α-ketoglutarate]	6 μM	7 μM	11 μM	14 μM

These data show that the concentration of α-ketoglutarate in the blood of thiamine-deficient individuals is elevated by a smaller fraction than is the concentration of pyruvate, both before and after eating glucose.

There have been, and still are, ample opportunities for studying the gross effects of these enzymatic disabilities in humans. This may seem strange in view of the fact that thiamine pyrophosphate is an obligatory coenzyme for all of the organisms from which natural human foods are derived, and therefore ought to be a constant constituent of the diet. Indeed, a primary thiamine deficiency, as opposed to a deficiency secondary to a starvation diet, would be a rare thing were it not for two human traits. The first of these is a dislike of coarse food. Hence, humans developed the technique of removing the hard outer layers from seeds, leaving the soft, starchy interior for cooking directly or making into flour. Unfortunately, most of the thiamine pyrophosphate is in the cells of the outer layers. Therefore, those people heavily dependent upon seeds for food are liable to a deficiency of thiamine, and this is most acute in the rice-consuming areas of the Orient. The deprivation is not complete, so that the deficiency develops relatively slowly in many cases, and the resultant illness is called **beri-beri.**

The second human trait resulting in thiamine deficiency is the desire to diminish unpleasant stimulations through the use of ethanol as a depressant. In plain English, a drunk may live only with his bottle and eat so little food that an acute deficiency of thiamine rapidly develops. The manifestations differ from those of beri-beri, and the illness is known as **Wernicke's encephalopathy.** In addition to its occurrence in alcoholics, it was well known in Japanese prisoner camps during the Second World War and has developed quite rapidly in elderly patients maintained on intravenous glucose solutions without vitamin supplementation.

The first symptoms of beri-beri are usually abnormal sensations in the limbs. The early disturbance in nervous function may grow worse, finally resulting in paralysis and wasting of the limbs (dry beri-beri). This is consistent with the major dependence of the nervous system on glucose metabolism as a source of energy. However, the nerves of other individuals may successfully compete for the limited thiamine supply, and the cardiovascular system may be caught short. This may be manifested by congestive heart failure with an **edema** — a seepage of liquid into the tissues so that they become puffy (wet beri-beri). There may be a combination of neural and cardiac symptoms, in which the neurological symptoms can be a life-saver by keeping the patient bed-ridden so that he doesn't overload his damaged heart.

Infantile beri-beri is a leading cause of death in some areas, causing vomiting and a peculiar aphonia — a soundless crying.

The more sudden deficiency of Wernicke's disease primarily appears as defects of the central nervous system, with acute mental disturbances and failures of motor control, but even in these cases, the mental impairment sometimes masks effects on the circulation that can result in sudden cardiac failure.

Here, then, are different clinical entities with deprivation of thiamine as a common fundamental cause, but with the tissue predominantly affected depending upon the rate of deprivation. We have dwelt upon this deficiency at some length to illustrate how difficult it is to recognize all of the factors that influence the rate of a single enzymatic reaction in a real animal.

Arsenite and Mercuric Ion Poisoning. During the action of pyruvate dehydrogenase, dihydrolipoyl groups are formed, which have a closely spaced pair of sulfhydryl groups. Metallic ions with a high affinity for sulfhydryl groups, such as mercuric or arsenite ions, are bound much more tightly to a pair on one molecule than they are to two sulfhydryl groups on separate molecues. For example, trivalent arsenic was formerly in common use by murderers and suicides, essentially because the formation of a dihydrolipoyl–arsenite chelate prevents the re-oxidation of the dihydrolipoyl group necessary for continued activity of the α-keto acid dehydrogenase complexes. Indeed, arsenite poisoning mimics some effects of thiamine deficiency on the nervous system and causes an accumulation of pyruvate in the blood. An American ambassadress to Rome had her pyruvate dehydrogenase inhibited in this way by her bedroom wallpaper, which had a patrician design partially created with the mellow green of cupric arsenite.

The high affinity of arsenite for compounds containing neighboring thiol groups means that it accumulates in the protein keratin. This constituent of hair, skin, horns, and the like, contains a high concentration of disulfide bonds. Some of the parent sulfhydryl groups tightly bind arsenite during maturation of the protein. Enough of the hair and fingernails survived dissolution in the otherwise well-aged corpse of Charles Francis Hall, an explorer who died in Northern Greenland in 1871 after a two-week illness, to enable proof in 1968 that he had been murdered by his associates, because the hair and nails grown during his last days of life contained high levels of arsenic and older portions did not.*

Treatment of acute poisoning by either arsenite or mercuric ions uses 2,3-dimercaptopropanol, also known as British Anti-Lewisite, BAL, because it was originally developed as an antidote for that arsenical war gas. BAL, with its adjacent sulfhydryl groups, can compete with dihydrolipoyl residues for binding with the metallic ions, forming a soluble chelate that is excreted in the urine:

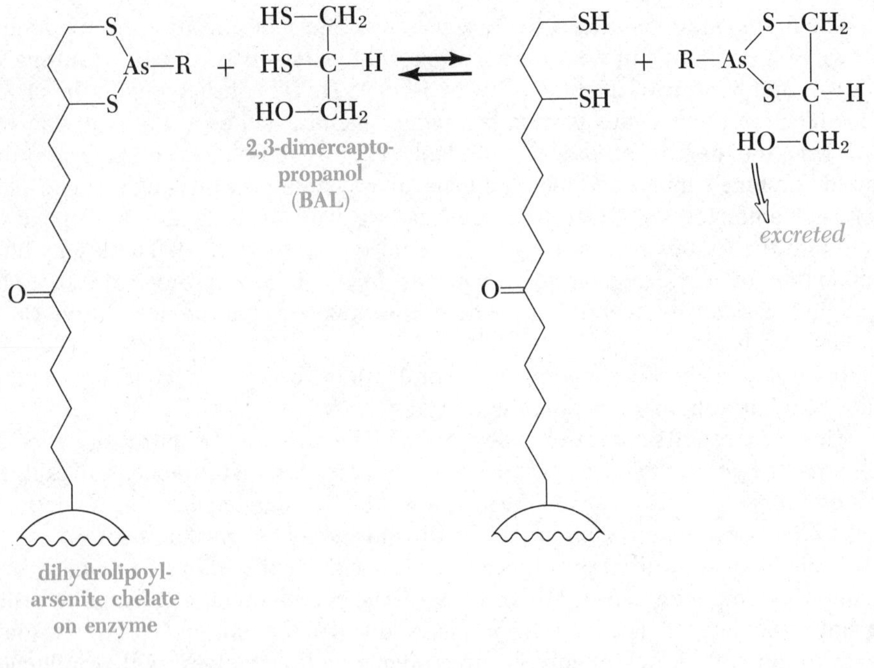

THE COMPLETE PROCESS

We have now seen all of the reactions necessary for the total oxidation of glucose: the oxidation of glucose to pyruvate in the cytosol, the transfer of the resultant electrons from the cytosol to mitochondria, the oxidation of the pyruvate to acetyl coenzyme A in mitochondria, and finally the total oxidation of acetyl coenzyme A by the citric acid cycle in mitochondria. What is the result?

If we add the individual equations when the glycerol phosphate shuttle is used, we arrive at an overall stoichiometry:

---

*The wife of a patient hospitalized at the University of Virginia with suspected arsenic poisoning was seen briskly cutting his hair after being advised that samples were needed for analysis.

$$\text{glucose} + 6\ O_2 + 35.5\ (\text{ADP} + P_i) \rightarrow 6\ CO_2 + 35.5\ \text{ATP}.$$

Nearly 36 molecules of high-energy phosphate are generated per molecule of glucose burned, and the P:O ratio is 3.0. Slightly less oxygen is consumed to generate a given quantity of ATP when glucose is burned than is consumed when fatty acids are the fuel (P:O ratio ~ 2.8).

Stoichiometry of High-Energy Phosphate

Let us examine exactly how the figure of 35.5 high-energy phosphates produced per glucose consumed was obtained. It is important to remember that two moles of triose phosphates are produced from each mole of glucose so that the stoichiometry for all of the reactions beginning with the oxidation of glyceraldehyde 3-phosphate must be multiplied by two.

1. One high-energy phosphate is consumed to phosphorylate glucose:	−1
2. One high-energy phosphate is consumed to phosphorylate fructose 6-phosphate:	−1
3. Two high-energy phosphates are produced by oxidative phosphorylation in mitochondria for each of the two pairs of electrons transported from glyceraldehyde 3-phosphate via NADH and the glycerol phosphate shuttle:	+4*
4. One high-energy phosphate is gained by transfer to ADP from each of two molecules of 1,3-bisphosphoglycerate:	+2
5. One high-energy phosphate is gained by transfer to ADP from each of two molecules of phospho-*enol*-pyruvate:	+2
6. One H^+ is discharged across the inner mitochondrial membrane during the transport of each pyruvate from the cytosol to the mitochondrial matrix. This is equivalent to the loss of ¼~P per H^+ discharged (perhaps more):	−0.5
7. Three high-energy phosphates are produced by oxidative phosphorylation in mitochondria from reoxidation of each of the two molecules of NADH formed by oxidation of pyruvate:	+6
8. Twelve high-energy phosphates are produced by the complete oxidation of each of the two molecules of acetyl coenzyme A in the citric acid cycle:	+24
TOTAL	+35.5

*This value may be higher in tissues utilizing the malate-aspartate shuttle for electron transfer. Oxidation of each NADH created in the mitochondria will produce three ATP, for a total of six; however, some potential ATP is dissipated in the shuttle itself, amounting to at least 0.5 mole per mole of glucose oxidized. The total yield from oxidation of glucose in these tissues will, therefore, be no more than 37 ATP. This yield is close to the maximum that can be obtained when the [ATP]/[ADP] ratio is near 100, while still permitting the $[P_i]$ to fall below 1mM, as it does in some organisms. (This is true when $[CO_2]/[O_2] = 100$ and the [glucose] = 1mM.)

Regulation of Glucose Oxidation

The metabolism of glucose is so central to the life of the organism that several controls are built into it so as to balance consumption and demand. When the use of ATP increases in tissues such as muscle, the combustion of glucose also

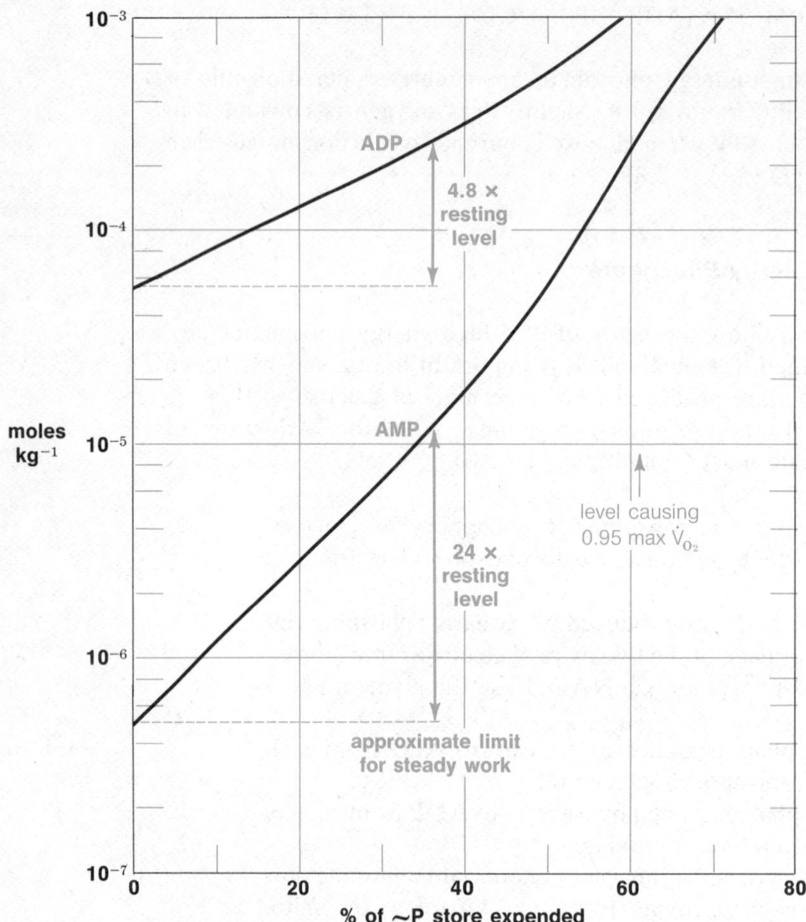

FIGURE 25–10 **The changes in concentrations of ADP and AMP in skeletal muscles as the total high-energy phosphate stores are discharged. The extent of discharge is a measure of the power output of a muscle. Although the absolute concentrations are low with light activity, the relative changes are very sensitive to changes in power output, especially in the case of AMP.**

increases, if it is available. The primary controls over the rate of combustion are the concentrations of ADP and P_i. An increased use of ATP by muscular contraction or movement of ions raises the ADP concentration and accelerates those reactions using ADP to regenerate ATP.

Regulation by Changes in ADP Concentration. Perhaps we can most easily understand the interplay of these various regulatory devices by following the sequence of events when a muscle begins working harder. The increased work load accelerates the hydrolysis of ATP to ADP + P_i, and the rising ADP concentration is the initial signal. The way in which this concentration changes is shown in Figure 25–10. Although the absolute magnitude of the changes in ADP concentration is kept low in muscles by the presence of phosphocreatine (p. 360), the relative changes are large as work begins. Indeed, the creatine kinase equilibrium may be used to accelerate mitochondrial response to changing ADP concentrations near the myofibrils because that enzyme also occurs between the outer and inner mitochondrial membrane. Diffusion of creatine to the intermembrane space therefore would generate ADP in the space and augment the diffusion of ADP.

The rising ADP concentration will accelerate the rate of all of the reactions that utilize ADP to generate ATP, which can be summarized as follows:

CYTOSOL *(triose P ⟶ pyruvate, Fig. 25–11*D):

$$\text{1,3-bisphosphoglycerate} + \text{ADP} \longleftrightarrow \text{ATP} + \text{3-phosphoglycerate}$$
$$\text{phospho-}\textit{enol}\text{-pyruvate} + \text{ADP} \longrightarrow \text{ATP} + \text{pyruvate}$$

CYTOSOL-MITOCHONDRIAL ANTIPORT *(not shown in Fig. 25–11)*:

$$\text{ADP}_{\text{cytosol}} + \text{ATP}_{\text{mitochondria}} \longrightarrow \text{ADP}_{\text{mitochondria}} + \text{ATP}_{\text{cytosol}}$$

MITOCHONDRIA *(pyruvate ⟶ CO_2, H_2O, Fig. 25–11*F,G):

$$\text{NADH} + \text{Q} + \text{ADP} \longleftrightarrow \text{ATP} + \text{QH}_2 + \text{NAD}^+$$
$$\text{QH}_2 + 2(\text{cyt c}) + \text{ADP} \longrightarrow \text{ATP} + 2(\text{reduced cyt c}) + \text{Q}$$
$$2(\text{reduced cyt c}) + \tfrac{1}{2}\text{O}_2 + \text{ADP} \longrightarrow \text{ATP} + \text{H}_2\text{O} + 2(\text{cyt c})$$

The ADP concentration will continue to rise, causing greater and greater acceleration of the oxidation of triose phosphates until the rate of ATP production matches the rate of ATP utilization, at which point the concentration will stop rising. The effects of changes in ADP concentration are huge. The oxygen consumption of human thigh muscles increases some 200-fold upon going from rest to maximum effort, and this is mainly a result of the effect of increased concentrations on successive reactions in the cytosol and the mitochondria. When two reactions utilizing the same substrate in a sequence are near equilibrium, such as they are in resting muscle, the net reaction velocity will vary as the square of the substrate concentration; when three such reactions are in sequence, as in oxidative phosphorylation, the velocity will vary as the cube of the substrate concentration. Hence, the great effect of changes in ADP concentration on rate of oxidation.

Regulation of Phosphofructokinase. Other controls are necessary to insure that the ATP-generating steps are not limited by a lack of fuel. They regulate some of the early steps in glucose metabolism by changing the activity of appropriate enzymes with allosteric effectors. We have seen how the combustion of triose phosphate is accelerated by rising ADP concentrations; now let us examine how the supply of triose phosphates is increased to meet the demand. The phosphofructokinase reaction (Fig. 25–11*C*) is a critical control point. The general strategy of control is to keep the enzyme inhibited until demands for its product develop. This strategy is necessary because the reaction catalyzes an essentially irreversible reaction that commits hexose units to the Embden-Meyerhof pathway. Its potential activity must be high if it is to supply sufficient fuel for times of rapid energy expenditure, but if the activity remained high at rest, the enzyme would continue to use ATP and hexose phosphates until one or the other was exhausted, while accumulating fructose bisphosphate and triose phosphates.

To prevent inappropriate activity, the enzyme is almost totally inhibited by physiological concentrations of ATP. Now, ATP is a substrate for the enzyme and absolutely necessary for the reaction, but ATP also combines with allosteric sites on the enzyme and makes it inactive.

The binding of ATP to the enyzme can be blocked by ADP or AMP, and AMP is the most effective in the physiological range of concentrations. The result is an activation of phosphofructokinase by AMP, which we used as an example in

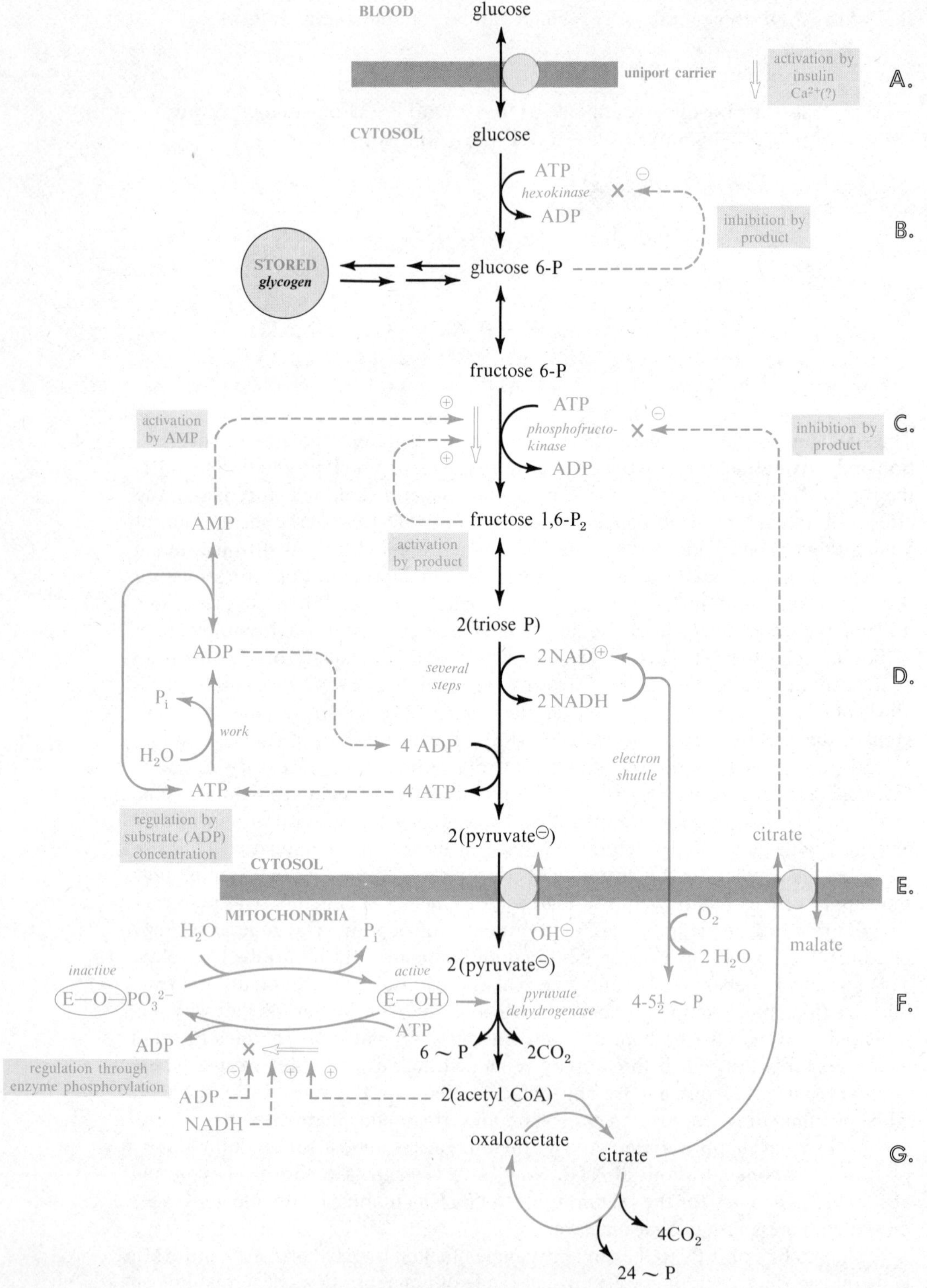

FIGURE 25–11 *See legend on opposite page.*

Chapter 16 (p. 301), but it is more accurately described as a relief of the inhibition by ATP. Let us now examine in more detail how the concentration of AMP changes during work.

AMP as an Effector. The concentration of AMP becomes a very sensitive indicator of the demand for high-energy as a result of the AMP kinase reaction:

$$2\ \text{ADP} \longleftrightarrow \text{ATP} + \text{AMP}; \qquad K'_{eq} = \frac{[\text{ATP}][\text{AMP}]}{[\text{ADP}]^2} \cong 1.$$

Dividing both numerator and denominator by $[\text{ATP}]^2$ and rearranging, we see that:

$$\frac{[\text{AMP}]}{[\text{ATP}]} \cong \left(\frac{[\text{ADP}]}{[\text{ATP}]}\right)^2.$$

Since the concentration of ATP changes very little until muscles are nearly exhausted, the AMP concentration will be varying nearly as the square of the ADP concentration, and AMP is therefore a magnified signal of demands for high-energy phosphate, as is shown in Figure 25–10. This is the reason why some enzymes, such as phosphofructokinase, are constructed to respond best to AMP as an effector. As a muscle works and its phosphocreatine store falls, the ADP concentration will rise, but there will be much larger increases in AMP concentration. This will accelerate the action of phosphofructokinase proportionally more, so as to match the accelerated removal of triose phosphates with an accelerated production of triose phosphate.

Phosphofructokinase is also regulated by other effectors. Citrate inhibits, and this is reasonable, since an accumulation of citrate means that the citric acid cycle in the muscle is being supplied with excess substrate, and the use of glucose residues as a source of acetyl coenzyme A ought to be slowed.

The enzyme is also activated by its own product, fructose 1,6-bisphosphate, when AMP is present. Here we have a rare example of positive feedback with an acceleration of the reaction tending to accelerate it still further. This is evidently a device for magnifying the activating effects. (Remember that the activity of this enzyme must be regulated over a several hundred-fold range in order to meet the fuel demand as muscles using glucose go from rest to maximum work.)

FIGURE 25–11 **Regulation of glucose oxidation. *A.* Activation of glucose transport into muscle by insulin or Ca^{2+}.**

***B.* Inhibition of hexokinase by its product, glucose 6-phosphate.**

***C.* Phosphofructoskinase is inhibited by ATP (not shown); the inhibition is relieved by AMP and relief is augmented by the product of the reaction, fructose 1,6-bisphosphate. The enzyme is inhibited by citrate, which enters the cytosol from mitochondria by an antiport mechanism.**

***D.* The conversion of triose phosphates to pyruvate depends upon availability of ADP, which is also a substrate. Unless ATP is being utilized, the ADP concentration will fall and triose phosphate oxidation will slow.**

***E.* The extent of regulation of pyruvate exchange across the inner mitochondrial membrane is not known.**

***F.* The pyruvate decarboxylase component of pyruvate dehydrogenase is inactivated by phosphorylation. Phosphorylation is accelerated by NADH or acetyl coenzyme A, and is inhibited by ADP.**

***G.* The citric acid cycle, as well as the electron transport shuttles, hinges upon the availability of ADP to maintain oxidative phosphorylation (not shown).**

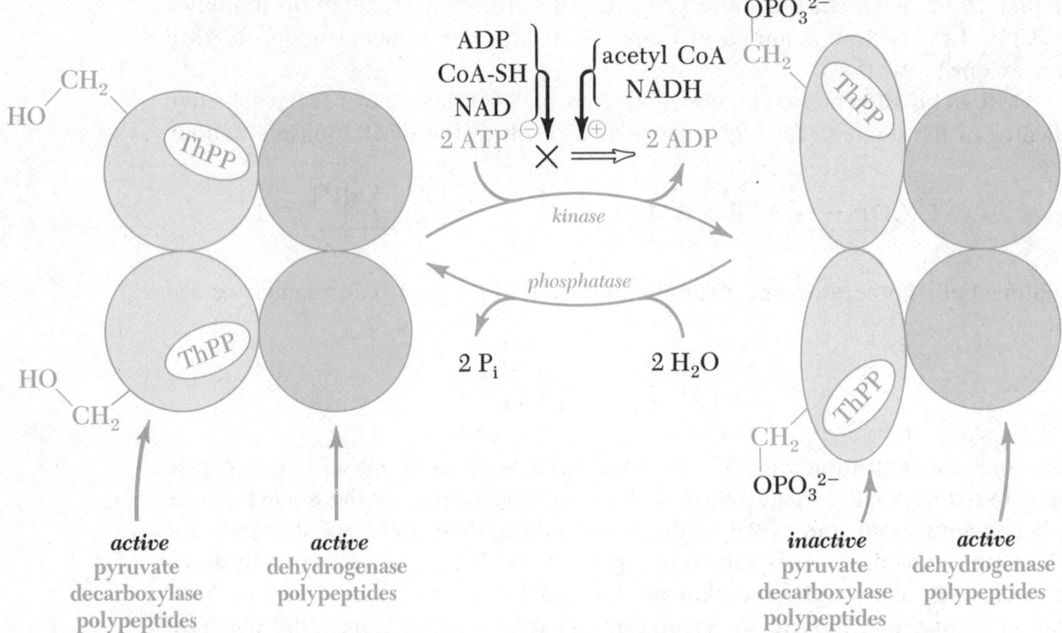

FIGURE 25–12 **The pyruvate dehydrogenase complex contains a regulatory kinase that catalyzes the phosphorylation of a serine residue in the decarboxylase component making it inactive, and a phosphatase that removes the inactivating phosphate group.**

Regulation of Pyruvate Dehydrogenase by Phosphorylation. The pyruvate dehydrogenase reaction (Fig. 25–11*F*) is a bridge between the Embden-Meyerhof pathway in the cytosol and the citric acid cycle in the mitochondria. It produces acetyl coenzyme A from pyruvate and, therefore, from glucose. However, acetyl coenzyme A also can be made from fatty acids, and there is no point in utilizing glucose for this purpose if there is already a good supply of acetyl groups. Part of the necessary adjustments are made by regulating pyruvate dehydrogenase.

Pyruvate dehydrogenase is the first specific example we have encountered of regulation through phosphorylation of the enzyme (Fig. 25–12). This covalent modification causes the enzyme to lose activity. The polypeptide that is modified is the pyruvate decarboxylase component of the complex, which catalyzes the initial and rate-determining step*.

The phosphorylation of the pyruvate decarboxylase component is catalyzed by another enzyme, a **protein kinase,** that is present in the dehydrogenase complex. There is approximately one molecule of the pyruvate decarboxylase kinase present for each 10 pyruvate decarboxylase polypeptides.

The kinase phosphorylates the enzyme and causes it to lose activity; activity can be restored by hydrolysis of the phosphate group, catalyzed by a **phosphoprotein phosphatase** that is also present in the complex. The overall activity of pyruvate dehydrogenase therefore depends upon the relative activities of the kinase and phosphatase within it.

The best evidence at hand is that it is the activity of the kinase that is controlled, while the phosphatase probably has a nearly constant and a substan-

*α-Ketoglutarate dehydrogenase is not regulated in this way, and the decarboxylating activity is not on a separate polypeptide in that enzyme complex.

tially smaller V_{max}. The kinase appears to be activated by an increase in the concentration ratios [acetyl CoA]/[CoA] or [NADH]/[NAD^+], which signal, respectively, an overload on the citric acid cycle or the mitochondrial electron transfer system. These activations may be overridden by an increase in the ratio [ADP]/[ATP], which signals a demand for more high-energy phosphate and which inhibits the kinase.

To recapitulate: a rise in either acetyl coenzyme A or NADH concentrations will lower the activity of pyruvate dehydrogenase and diminish the rate of oxidation of pyruvate to acetyl coenzyme A. They do this by increasing the activity of a protein kinase that phosphorylates the dehydrogenase and makes it inactive. A rise in ADP concentration will increase the activity of pyruvate dehydrogenase by blocking the protein kinase; this diminishes phosphorylation of the enzyme, and more of the enzyme will be converted to its active dephospho form by the action of the phosphoprotein phosphatase.

Regulation of the Glucose 6-Phosphate Supply. We are not in a position to discuss the regulation of the early steps in glucose metabolism in complete detail, because the glucose residues in glucose 6-phosphate may be stored as glycogen, or obtained from glycogen, as we shall discuss in Chapter 27.

The utilization of glucose from the blood, either for storage or for combustion, is affected by the regulation of transport across the plasma membrane, with insulin and perhaps Ca^{2+} as activators (Fig. 25–11*A*), and by the regulation of hexokinase. The phosphorylation of glucose is a nearly irreversible reaction (Fig. 25–11*B*).

The hexokinases of all cells except liver parenchymal cells are inhibited by glucose 6-phosphate. Unless this compound is used to form glycogen or fructose 1,6-bisphosphate, its concentration will rise and stop its own formation. The result will be a slowing of glucose uptake even with high insulin and glucose concentrations in the blood, because the intracellular concentration of glucose will also rise and decrease the concentration gradient across the plasma membrane.

FURTHER READING

Dickens, F., P. J. Randle, and W. J. Whelan: (1968) *Carbohydrate Metabolism and Its Disorders*. Vol. 1, Chaps. 1, 2, 3. Academic Press. Contains detailed discussions of processes in animal tissues.

Clausen, T.: (1975) *The Effect of Insulin on Glucose Transport in Muscle Cells*. Curr. Top. Memb. Res., *6*: 169.

Elbrink, J., and I. Bihler: (1975) *Membrane Transport: Its Relation to Cellular Metabolic Rate*. Science, *188*: 1177.

Betz, A. L., D. D. Gilboe, and L. R. Drewes: (1976) *The Characteristics of Glucose Transport Across the Blood-brain Barrier and Its Relation to Cerebral Glucose Metabolism*. Adv. Exp. Biol. Med., *69*: 133.

Purich, D. L., H. J. Fromm, and F. B. Rudolph: (1973) *The Hexokinases: Kinetics, Physical and Regulatory Properties*. Adv. Enzymol., *39*: 249.

Tornheim, K., and J. M. Lowenstein: (1976) *Control of Phosphofructokinase from Rat Skeletal Muscle*. J. Biol. Chem., *251*: 7322.

Mansour, T. E.: (1972) *Phosphofructokinase*. Curr. Top. Cell. Reg., *5*: 1.

Ottoway, J. H., and J. Mowbray: (1977) *The Role of Compartmentation in the Control of Glycolysis*. Curr. Top. Cell. Reg., *12*: 107.

Reed, L. J., et al.: (1972) *Pyruvate Dehydrogenase Complex: Structure, Function and Regulation*. p. 253 *in* M. A. Mehlman, and R. W. Hanson, eds.: *Energy Metabolism and the Regulation of Metabolic Processes in Mitochondria*. Academic Press.

Hansford, R. G.: (1976) *Studies on the Effects of Coenzyme A-SH:Acetyl Coenzyme A . . . on the Interconversion of Active and Inactive Pyruvate Dehydrogenase. . .* J. Biol. Chem., *251*: 5483.

Kerbey, A. L., et al.: (1976) *Regulation of Pyruvate Dehydrogenase in Rat Heart*. Biochem. J., *154*: 327.

Nadel, A. M., and P. G. Burger: (1976) *Wernicke Encephalopathy Following Prolonged Intravenous Therapy*. J.A.M.A., *235*: 2403.

McGilvery, R. W.: (1975) *Use of Fuels for Muscular Work*. p. 12 *in* H. Howald and J. Poortmans, eds.: *Metabolic Adaptations to Prolonged Physical Exercise*. Birkhauser Verlag Basel. Discussion of regulation by ADP and AMP.

Paddock, F. K., C. C. Loomis, and A. K. Perkons: (1970) *An Inquest on the Death of Charles Francis Hall*. N. Engl. J. Med., *282*: 784.

26 GLYCOLYSIS AND GLUCONEOGENESIS

Glycolysis literally means the splitting of glucose. Glucose can be cleaved to form two molecules of lactic acid according to the formal stoichiometry:

$$\underset{\text{D-glucose}}{C_6H_{12}O_6} \longrightarrow 2\ \underset{\text{L-lactate}}{HO-\overset{\displaystyle COO^{\ominus}}{\overset{|}{C}}\underset{\underset{\displaystyle CH_3}{|}}{}-H} + 2\,H^+$$

This conversion of glucose to lactate and H^+ is used by some tissues in place of the complete combustion of glucose to CO_2 and H_2O; the replacement may be partial or nearly complete. The formation of lactate provides a device for generating ATP without using O_2.

Lactate is formed by the action of lactate dehydrogenase:

$$\underset{\text{pyruvate}}{\overset{\displaystyle COO^{\ominus}}{\overset{|}{C}}{=}O \atop {|\atop CH_3}} + NADH + H^{\oplus} \xrightleftharpoons{\text{lactate dehydrogenase}} \underset{\text{L-lactate}}{HO-\overset{\displaystyle COO^{\ominus}}{\overset{|}{C}}-H \atop {|\atop CH_3}} + NAD^{\oplus}$$

The cytosol of most, if not all, cells of the mammalian body contains enough of this enzyme to maintain the reaction near equilibrium. The direction in which the reaction goes depends upon the relative concentrations of the reactants. Some cells have a rapid Embden-Meyerhof pathway and relatively few mitochondria. For example, rapidly contracting white fibers in skeletal muscles convert glucose residues and NAD to pyruvate and NADH faster than the mitochondrial oxidations can handle them. The high pyruvate concentration and the increased ratio of [NADH]/[NAD] then cause a rapid formation of lactate, with the lactate pouring out into the blood (Fig. 26–1).

Glycolysis is therefore nothing more than the Embden-Meyerhof pathway for the conversion of glucose residues to pyruvate with the added action of lactate dehydrogenase. It differs from combustion of glucose in that the NADH generated in the cytosol is converted back to NAD by reducing pyruvate to lactate, rather than by use of the glycerol phosphate or malate-aspartate electron shuttles (Fig. 26–2).

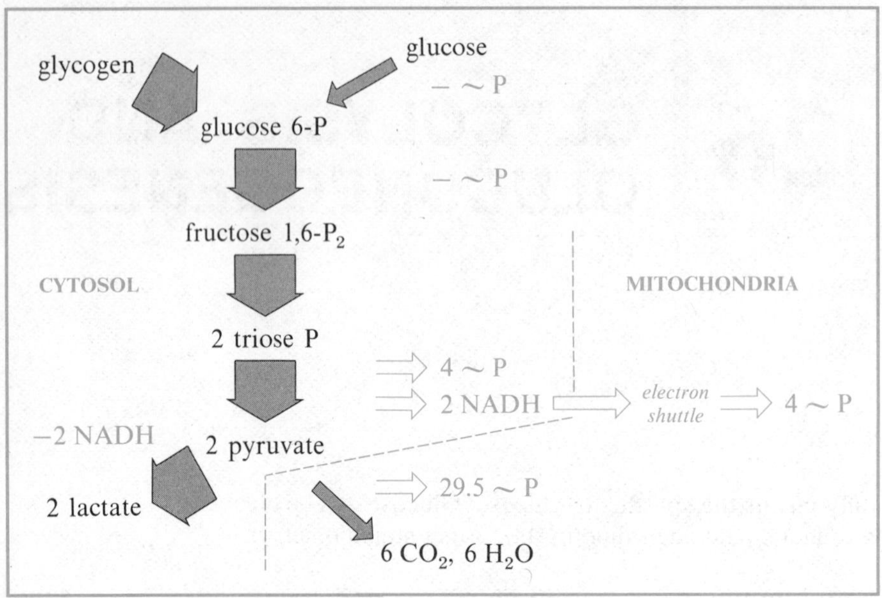

FIGURE 26–1 **Carbohydrate metabolism in white skeletal muscle fibers. The width of the arrows indicates relative flow during exercise.**

Yield of ATP. When glucose residues are converted to lactate as the end product, ATP is formed only by the reactions of the Embden-Meyerhof pathway. The disposal of the NADH by reduction of pyruvate to lactate does not affect the high-energy phosphate balance:

$$\text{glucose} + 2(\text{ADP} + \text{P}_i) + 2\,\text{NAD}^{\oplus} \longrightarrow 2\,\text{pyruvate}^{\ominus} + 2\,\text{ATP} + 2\,\text{NADH} + 4\,\text{H}^{\oplus}$$
$$2\,\text{pyruvate}^{\ominus} + 2\,\text{NADH} + 2\,\text{H}^{\oplus} \longrightarrow 2\,\text{lactate}^{\ominus} + 2\,\text{NAD}^{\oplus}$$
$$\text{SUM:}\quad \text{glucose} + 2(\text{ADP} + \text{P}_i) \longrightarrow 2\,\text{lactate}^{\ominus} + 2\,\text{H}^{\oplus} + 2\,\text{ATP}^*$$

When lactate is being produced at a high rate, the glucose residues are usually supplied from stored glycogen, which can be converted to glucose 6-phosphate without an initial expenditure of ATP in the hexokinase reaction. Conversion of glucose residues in glycogen to lactate therefore results in the formation of one more ATP than when free glucose is used, a total of three per glucose residue. To summarize:

$$\left.\begin{array}{r}\text{free glucose} \longrightarrow 2\text{ lactate} + 2\text{ ATP}\\ \text{glucose in glycogen} \longrightarrow 2\text{ lactate} + 3\text{ ATP}\end{array}\right\}\text{glycolysis}$$
$$\left.\begin{array}{r}\text{free glucose} + 6\,O_2 \longrightarrow 6\,CO_2 + 35.5\text{–}37\text{ ATP}\\ \text{glucose in glycogen} + 6\,O_2 \longrightarrow 6\,CO_2 + 36.5\text{–}38\text{ ATP}\end{array}\right\}\text{combustion}$$

Despite its low yield of ATP, glycolysis is advantageous to white muscle fibers because it can go fast. Rates for individual fibers are not known in humans, but a typical muscle with a mixed fiber population can make pyruvate some 25 times faster than it can oxidize it. These muscles therefore can make twice as much ATP per second by converting glycogen to lactate as they can by oxidizing it completely ($25 \times 3 = 75$ ATP compared to ~ 38 ATP).

*These equations do not include the fractional stoichiometry for H^+ resulting from ATP generation, which is balanced by H^+ given off during ATP utilization.

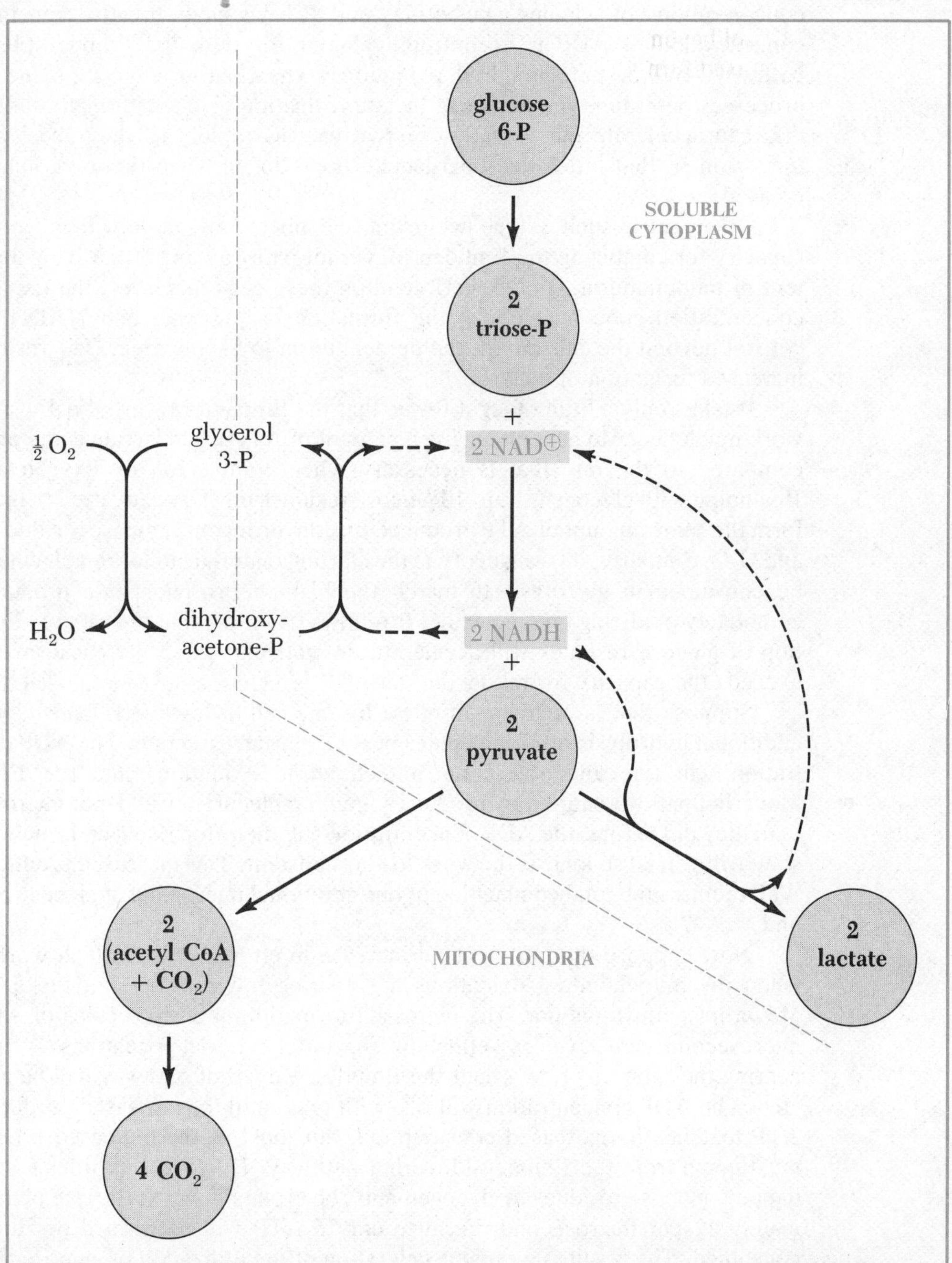

FIGURE 26–2 NADH generated in the cytosol may be converted back to NAD by an electron shuttle, such as the glycerol phosphate shuttle shown, or by reducing pyruvate to lactate.

Regulation of Lactate Formation

Regulation by ADP Concentration. The only known signals that directly interrelate muscular contraction and the Embden-Meyerhof pathway are the concentrations of adenine nucleotides and P_i. We have already seen that the concentration of ADP is a controlling factor for both the Embden-Meyerhof pathway and for mitochondrial oxidations. The relative effects on these two processes determine the extent of lactate formation. Cells with many mitochondria can accelerate the oxidation of pyruvate as rapidly as they accelerate its formation so that little additional lactate is produced when the work load is increased.

Other cells, such as the white muscle fibers, are built with a very high capacity for catalyzing the Embden-Meyerhof pathway and a relatively low content of mitochondria. As the work load on these cells increases, the rising ADP concentration soon accelerates the formation of pyruvate and NADH in the cytosol beyond the capacity of the mitochondria to handle them. The result is an increased formation of lactate.

It takes only a little thought to see that the Embden-Meyerhof pathway must work much faster to generate a given amount of ATP when lactate is the product, compared to the rate that is necessary when combustion by oxygen occurs. Beginning with glycogen, 12 to 13 glucose residues must be converted to lactate to form the same amount of ATP produced by converting one glucose residue to CO_2 and H_2O. Similarly, if free glucose is the starting material, 18 to 19 molecules must be consumed in glycolysis to match the high-energy phosphate produced by completely oxidizing one molecule. It follows from this that the rate of consumption of glucose residues will accelerate dramatically as an increasing workload exceeds the capacity for mitochondrial ATP production in white muscle fibers.

Suppose that an increase in effort by one cell at low work loads causes an additional hydrolysis of 37 molecules of ATP per microsecond. The ADP concentration will rise and accelerate mitochondrial oxidations and the Embden-Meyerhof pathway until they produce 37 more molecules of ATP per microsecond than they did before; the ADP concentration will then stop rising and a new steady state will persist so long as the work load is constant. This increased production of ATP requires the total combustion of one additional micromole of glucose per second.

Now suppose that an identical increase in effort occurs at high work loads when the mitochondrial oxidations are nearing full capacity but the Embden-Meyerhof pathway is not. The increased formation of 37 molecules of ADP per microsecond can have less effect on the mitochondrial oxidations — they are nearing their limiting rate — but the Embden-Meyerhof pathway will be accelerated. The ADP concentration will rise with time until the increased production of ATP matches its increased consumption, but most of the increased production must come from the Embden-Meyerhof pathway. However, in order for that to happen, glucose residues in glycogen must be broken down at the rate of approximately 12 per microsecond, because only 3 ATP will be formed per molecule consumed. The result is a rapid acceleration of the utilization of glucose residues at higher workloads (Fig. 26–3). At its maximum, the rate of the Embden-Meyerhof pathway in working muscles can be 1,000 times that of resting muscles. Notice that lactate production and oxygen consumption both increase in working muscle; the appearance of lactate is not indicating loss of the oxygen supply.

Regulation by AMP Concentration; Role of the Fructose Phosphate Cycle. A rising concentration of ADP accelerates the Embden-Meyerhof pathway by its

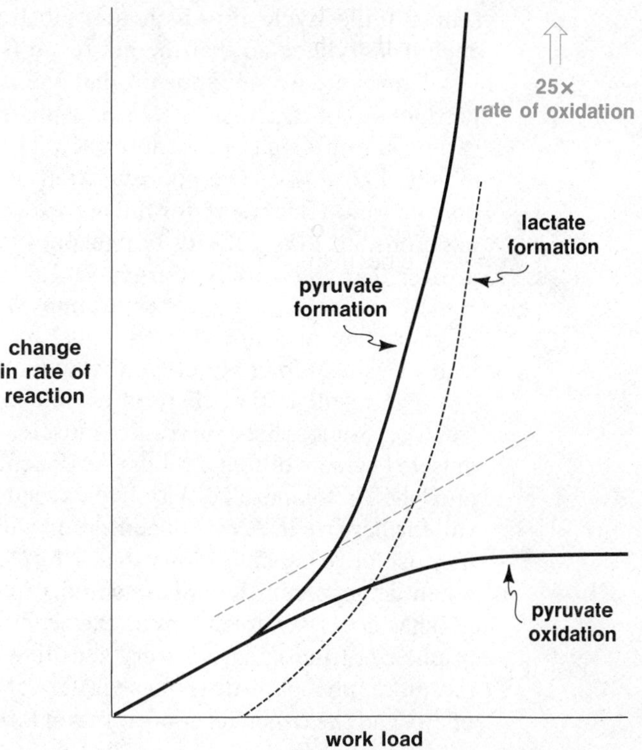

FIGURE 26–3

Changes in lactate production as the power output of muscles increases. At low power outputs, increased rates of pyruvate formation can be matched by increased rates of pyruvate oxidation. At power outputs bringing the rate of pyruvate oxidation near its V_{max}, any increase in power output is fueled by steep rises in the rate of pyruvate formation and its conversion to lactate. The steep rise is necessary to compensate for the low yield of ATP.

effects on the phosphoglycerate kinase and pyruvate kinase reaction (p. 457); that is, it accelerates the conversion of triose phosphates to pyruvate. This increased consumption of triose phosphates as fuel cannot continue unless there is a concomitant acceleration of the formation of triose phosphates from glucose residues.

We noted earlier (p. 459) that phosphofructokinase catalyzes the rate-determining step in the conversion of hexose phosphates to triose phosphates, and that it is controlled by the AMP concentration, which is a sensitive indicator of the demand for high-energy phosphate. The enzyme is also responsive to other effectors, such as citrate. These regulatory devices are sufficient in most tissues, but the extreme range of control (1,000-fold) necessary in white skeletal muscle fibers requires a mechanism for amplifying the response to low concentrations of AMP.

White, fast-twitch, muscle fibers contain fructose bisphosphatase, an enzyme that catalyzes the hydrolysis of the phosphate on C-1:

$^{2-}O_3PO—CH_2$ H OH $CH_2—OPO_3^{2-}$ / H OH / HO H — H_2O → P_i (*fructose bisphosphatase*) → $^{2-}O_3PO—CH_2$ H OH $CH_2—OH$ / H OH / HO H

D-fructose 1,6-bisphosphate → D-fructose 6-phosphate

At first glance, this reaction appears to defeat the purpose of the tissue, since the actions of fructose bisphosphatase and phosphofructokinase taken together represent a futile cycle in which the substrate is continually phosphorylated and dephosphorylated so that the net result is nothing but the hydrolysis of ATP.

However, it now appears that the cycle is a device for insuring a prompt production of fructose 1,6-bisphosphate, and therefore triose phosphates, in response to a demand for more fuel. The fructose bisphosphatase of muscles is inhibited by AMP. The concentration of AMP required is low compared to the concentration necessary for full activation of phosphofructokinase (Fig. 26–4). In addition, the total activity of phosphofructokinase is many-fold (15 to 40 times) greater than the activity of fructose-1,6-bisphosphatase, so the balancing concentration of AMP — the concentration at which the kinase forms fructose 1,6-bisphosphate at a rate exactly equal to the rate of hydrolysis by the phosphatase — is also low, although it is above the resting level.

The result is that there is no net formation of fructose bisphosphate from fructose 6-phosphate in resting muscle, and there will be little change when a muscle begins working until the AMP concentration rises to the point at which the phosphofructokinase activity will exceed the fructose bisphosphatase activity. A still further rise in AMP concentration will then cause a very steep acceleration of fructose bisphosphate formation. Why? Because there is a high concentration of potential phosphofructokinase activity in these muscles, and the AMP concentration has now risen to a point on the sigmoidal response curve at which the activity of phosphofructokinase is very sensitive to changes in AMP concentration. Furthermore, phosphofructokinase is activated by its own product, fructose bisphosphate, and the rise in its concentration further accelerates its own production.

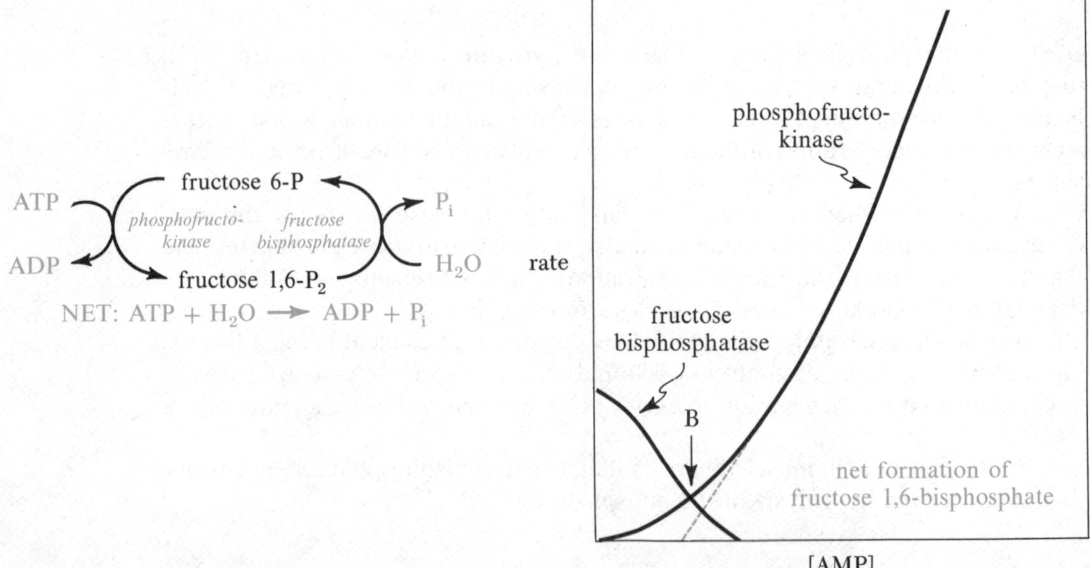

FIGURE 26–4 **The presence of fructose 1,6-bisphosphatase in white skeletal muscle fibers creates the potential for a futile cycle (*left*) in which the net result is the hydrolysis of ATP. However, phosphofructokinase is activated by AMP and fructose bisphosphatase is inhibited. The concentration of AMP required to inhibit phosphatase activity is much lower than that required to activate the kinase (*right*), and this serves to limit the futile cycle. When the concentration of AMP is very low, as in resting muscle, the activity of the phosphatase equals or exceeds the activity of the kinase, and there is no net formation of fructose 1,6-bisphosphate. When the muscles begin to work, the AMP concentration sharply rises beyond point B, where the effect of changing concentrations of phosphofructokinase activity is steep. The result is a large increase in the rate of formation of fructose 1,6-bisphosphate.**

Here we have it; the fructose phosphate cycle is a device permitting phosphofructokinase to be sensitive to changes in AMP concentration at levels not far above the resting level without having fructose bisphosphate piling up in the cell. The cycle wastes ATP but the amount involved is small potatoes compared to the expenditure in a fully working muscle. Red muscles contain little fructose bisphosphatase because they do not have the extreme changes in rate of glucose metabolism from rest to activity, and ordinary allosteric controls are sufficient.

Lactate Formation and the Oxygen Supply

Anything that impairs the supply or utilization of oxygen will tend to increase the formation of lactate. Remember that a rising lactate concentration is a signal that a rising ADP concentration has increased the formation of pyruvate and NADH more than it has the oxidation of pyruvate and NADH. If the ability to oxidize pyruvate and NADH is impaired in any way, even light work loads will cause excessive formation of pyruvate.

A common way to diminish the utilization of oxygen is to slow its supply. This happens, for example, in isometric contraction of muscles. Muscles that develop high tension without shortening squeeze the blood supply out of contact with many fibers so that they promptly produce lactate.

The oxygen supply also can be diminished by simply moving to a higher altitude. The mechanism of delivering oxygen to the tissues is ordinarily adjusted so that its maximum capacity is near the maximum demand by the tissues. They are so nearly balanced that it has been difficult to determine if the limiting factor for oxidative phosphorylation in the whole body during maximum work is the rate of oxygen supply or the capacity of the tissues to use oxygen. If the oxygen tension in the atmosphere is diminished, there is no longer any doubt; the oxygen supply becomes limiting and the production of lactate is accelerated by more modest work loads than those necessary at sea level.

Acute interruptions of the oxygen supply will cause lactate production in organs that normally consume the compound rather than producing it, such as the heart, brain, or liver. The interruption may be localized as in obstruction or rupture of an artery, or it may be more general, as in pulmonary edema (accumulation of fluid in the lungs), obstructions of the airway, exposure to carbon monoxide, and so on. The formation of lactate may extend the life of the affected organs under these conditions, but the quantitative magnitude of its contribution is not clear. On the one hand, a lot of lactate must be produced to generate the same energy as does the normal oxidative metabolism, but on the other, a little extra ATP may make the difference between potential recovery and irreversible damage.

Glycolysis in Other Animals

Not all air-breathing animals live on the land; those that lack gills and dive under water close their access to oxygen, and the small store they can carry with them is not adequate to sustain their activity through oxidative phosphorylation for more than a few minutes in most cases; the diver does not differ from the surface-dweller in this respect.

Since the diver is deprived of oxygen, his muscles produce lactate from glycogen. Diving reptiles are extreme examples; they produce large quantities of

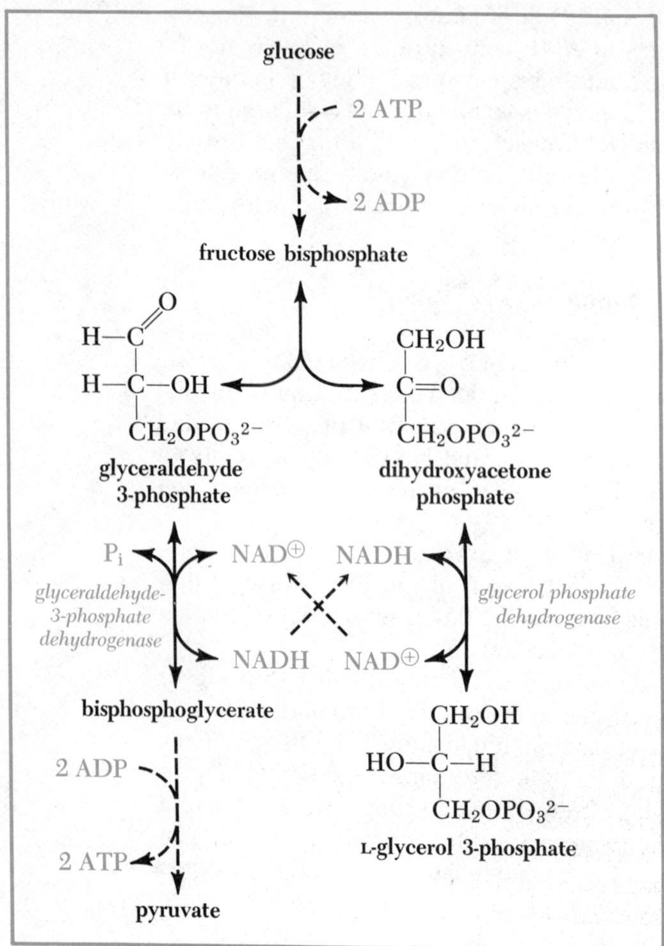

FIGURE 26–5

The dismutation of glucose to pyruvate and glycerol 3-phosphate. The NADH formed during the production of pyruvate is used to reduce dihydroxyacetone phosphate to glycerol 3-phosphate. There is no net gain of high-energy phosphate. (A dismutation is a simultaneous oxidation and reduction of one compound.)

lactate and are built to tolerate the associated massive acidosis. The blood pH of a diving turtle may fall to 6.7 from its normal 7.4 as a result of lactate accumulation, but the ATP generated in this way is sufficient to permit survival for as long as 12 hours at 22° without oxygen. Even the lizards and snakes can survive as long as 45 minutes under these conditions.

Whether it be mammal, reptile, or bird, the diving animal has a brain that is as dependent upon oxidative phosphorylation as is ours, and it can stay below only as long as there is oxygen in the blood circulating through its brain. The animals are able to achieve long dives because they conserve the small amount of oxygen carried with them for use by the vital tissues through a reflex response to immersion. The circulation to the skeletal muscles is partially or completely closed, and the heart rate drastically slows. Adjustments of this sort occur to a lesser extent in trained human divers. Indeed, nearly all of us were divers when we entered the world; the temporary deprivation of oxygen during delivery produces a similar reflex in the fetus.

Ordinarily, loss of the oxygen supply for more than four minutes causes permanent damage to the human brain. It has recently been noted that immersion in cold water may slow brain metabolism sufficiently so that it will survive much longer. Resuscitation has been successful on apparently dead victims of drowning in cold water.

Glycerol Phosphate in Insects. The flight muscles of insects lack lactate dehydrogenase and therefore do not carry out glycolysis. However, incubation of the muscles in the absence of oxygen causes an accumulation of glycerol 3-phosphate and pyruvate (Fig. 26–5), and it is sometimes said that these are normal end-products of glucose metabolism in insects. However, this is wrong. The flight muscles have a very active oxidative metabolism, including an efficient glycerol phosphate shuttle. Insects couldn't afford to lift the large quantities of fuel that would be required to sustain a long flight by glycolysis and therefore dropped lactate dehydrogenase. The appearance of glycerol 3-phosphate and pyruvate is an experimental artifact caused by the lack of oxygen. (The same compounds would appear in mammalian muscles in the absence of oxygen if the muscles lacked lactate dehydrogenase: they don't appear because the equilibria are such as to favor the formation of lactate from glycerol 3-phosphate and pyruvate.)

Fermentations

Mechanisms similar to glycolysis are used by those microorganisms that have evolved to fill ecological niches in which there is a supply of carbon compounds and little, or no, oxygen. These organisms are properly named **anaerobes.** They may be **obligatory** and use only anaerobic pathways, or they may be **facultative,** able to switch from aerobic to anaerobic metabolism when needed. The overall anaerobic fuel metabolism of microorganisms is known as fermentation.

Some bacteria produce lactate as the primary product of glucose metabolism by the same route used in mammalian muscles, but the best known fermentation is the cleavage of glucose by some yeasts to form ethanol and CO_2. These yeasts also use the reactions of the Embden-Meyerhof pathway for the formation of pyruvate, but they lack lactate dehydrogenase. Instead, they contain a pyruvate decarboxylase that converts pyruvate to acetaldehyde and CO_2. (The enzyme utilizes thiamine pyrophosphate as a coenzyme, and its mechanism is the same as that of the pyruvate decarboxylase component of pyruvate dehydrogenase in animals.) Acetaldehyde is then reduced by NADH to ethanol, through an equilibration catalyzed by alcohol dehydrogenase (Fig. 26–6).

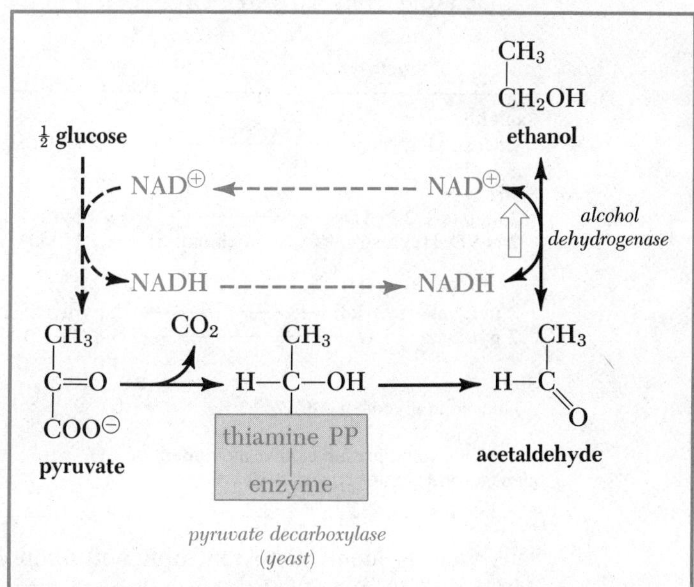

FIGURE 26–6

Yeasts form ethanol from glucose by first making pyruvate and NADH through the Embden-Meyerhof pathway. The pyruvate is decarboxylated to acetaldehyde, and the NADH is used to reduce acetaldehyde to ethanol.

There are many other and more complicated fermentations known in microorganisms, producing mixtures of products such as butanol, butyrate, glycerol, acetone, and acetate. Hydrogen gas is also a frequent product.

THE UTILIZATION OF LACTATE: OXIDATION OR GLUCONEOGENESIS

The lactate and lesser amounts of pyruvate diffusing into blood from rapidly contracting muscles represent incompletely oxidized glucose residues and are potential fuels in themselves. Animals use the compounds in other tissues, thereby improving the overall efficiency of the fuel metabolism.

Combustion of Lactate

One way of disposing of lactate is to oxidize it to pyruvate, and this occurs in red muscle fibers, the heart, and the brain. We have been emphasizing the formation of lactate from pyruvate by the lactate dehydrogenase in rapidly contracting muscles, but the actual direction of the pyruvate-lactate interconversion depends upon the concentrations of the compounds involved. Tissues with a high oxidative capacity, such as the red muscle fibers, heart, and brain, can maintain a ratio of [NAD]/[NADH] high enough to favor the oxidation of lactate to pyruvate, which is then further oxidized in mitochondria in the usual way. Lactate is, therefore, a substitute for glucose as a fuel for combustion in those tissues. Lactate may pass directly from white to red fibers within a muscle.

We noted before (p. 283) that tissues with high oxidative capacity contain an isozyme of lactate dehydrogenase that is fully effective only when concentrations favor the oxidation of lactate, rather than its formation.

Efficiency. Glycolysis is often said to be very inefficient compared to direct combustion because the yield of ATP is low. But is it? Let us calculate the total yield when a glucose residue is converted to lactate in a whole muscle fiber, and the lactate is converted to CO_2 and H_2O in a cardiac muscle fiber:

Muscle			~P Yield
Skeletal			
glucose (in glycogen)	⟶	2 lactate	+3
Heart			
2 lactate + 2 NAD^+	⟶	2 pyruvate + 2 NADH	0
2 NADH (cytosol) + O_2 (mitochondria)	⟶	2 NAD^+ (cytosol) + 2 H_2O	+5.5* (using malate-aspartate shuttle)
2 pyruvate (cytosol)	⟶	2 pyruvate (mitochondria)	−0.5*
2 pyruvate + 5 O_2	⟶	6 CO_2 + 6 H_2O	+30
Sum			
glucose in glycogen + 6 O_2	⟶	6 CO_2 + 6 H_2O	38

*This value presumes that movement of 4 H^+ across the inner mitochondrial membrane is equivalent to one high-energy phosphate.

This yield is identical to that obtained upon complete combustion of glycogen within a single cell. To the extent that other cells can oxidize the lactate produced by skeletal muscle fibers, the only price paid for the rapid supply of energy from glycolysis is a temporary acidification.

Conversion of Lactate to Glucose

Only part of the lactate can be burned. The remainder is converted back to glucose. The reactions used are part of the route for formation of glucose from non-carbohydrate precursors. This formation is known as gluconeogenesis, and it occurs mainly in the liver and kidneys. The entire process, in which glucose is converted to lactate in skeletal muscles by glycolysis and the lactate is converted back to glucose in the liver and kidneys by the gluconeogenic pathway, is known as the Cori cycle*.

Glycolysis, in which glucose is converted to lactate, is an irreversible process; otherwise it wouldn't have much value as a means of generating ATP. Something different must be done to reverse this process and convert lactate to glucose, and this something different must include new reactions creating equilibria favoring gluconeogenesis rather than glycolysis. The general differences between gluconeogenesis and glycolysis are shown in Figure 26–7, and these involve two sets of reactions:

(1) The hexokinase and phosphofructokinase used in glycolysis to push glucose toward the triose phosphates are replaced in gluconeogenesis by glucose-6-phosphatase and fructose 1,6-bisphosphatase, which catalyze the irreversible hydrolysis of the phosphates. The push is then from triose phosphates to glucose.

(2) Gluconeogenesis also involves two new reactions inserted between phospho-*enol*-pyruvate and pyruvate. These reactions cause the transient formation of oxaloacetate with the expenditure of an additional molecule of high-energy phosphate per phospho-*enol*-pyruvate formed. The pyruvate kinase reaction of the Embden-Meyerhof pathway, in which phosphate is transferred from phospho-*enol*-pyruvate to ADP, has an equilibrium favoring the formation of pyruvate, but the change in chemical potential is not great enough to support the formation of any more ATP. When the reaction is changed so as to include an additional transfer of high-energy phosphate, the position of equilibrium is changed to favor the reverse reaction in which pyruvate is converted to phospho-*enol*-pyruvate, and gluconeogenesis occurs.

Let us now look at the distinctive reactions of gluconeogenesis more closely.

Mechanism of Gluconeogenesis

Pyruvate → Oxaloacetate. The formation of phospho-*enol*-pyruvate begins with carboxylation of pyruvate at the expense of ATP to form oxaloacetate:

$$\underset{\textbf{pyruvate}}{\begin{matrix}CH_3\\ |\\ C{=}O\\ |\\ COO^{\ominus}\end{matrix}} + CO_2 + ATP + H_2O \xrightarrow[Mg^{2+},\ Mn^{2+},\ K^{+}]{\textit{pyruvate carboxylase}} \underset{\textbf{oxaloacetate}}{\begin{matrix}COO^{\ominus}\\ |\\ CH_2\\ |\\ C{=}O\\ |\\ COO^{\ominus}\end{matrix}} + ADP + P_i$$

*Carl F. Cori (1896–) and Gerty T. Cori (1908–1955): Austrian-born American biochemists and Nobel laureates.

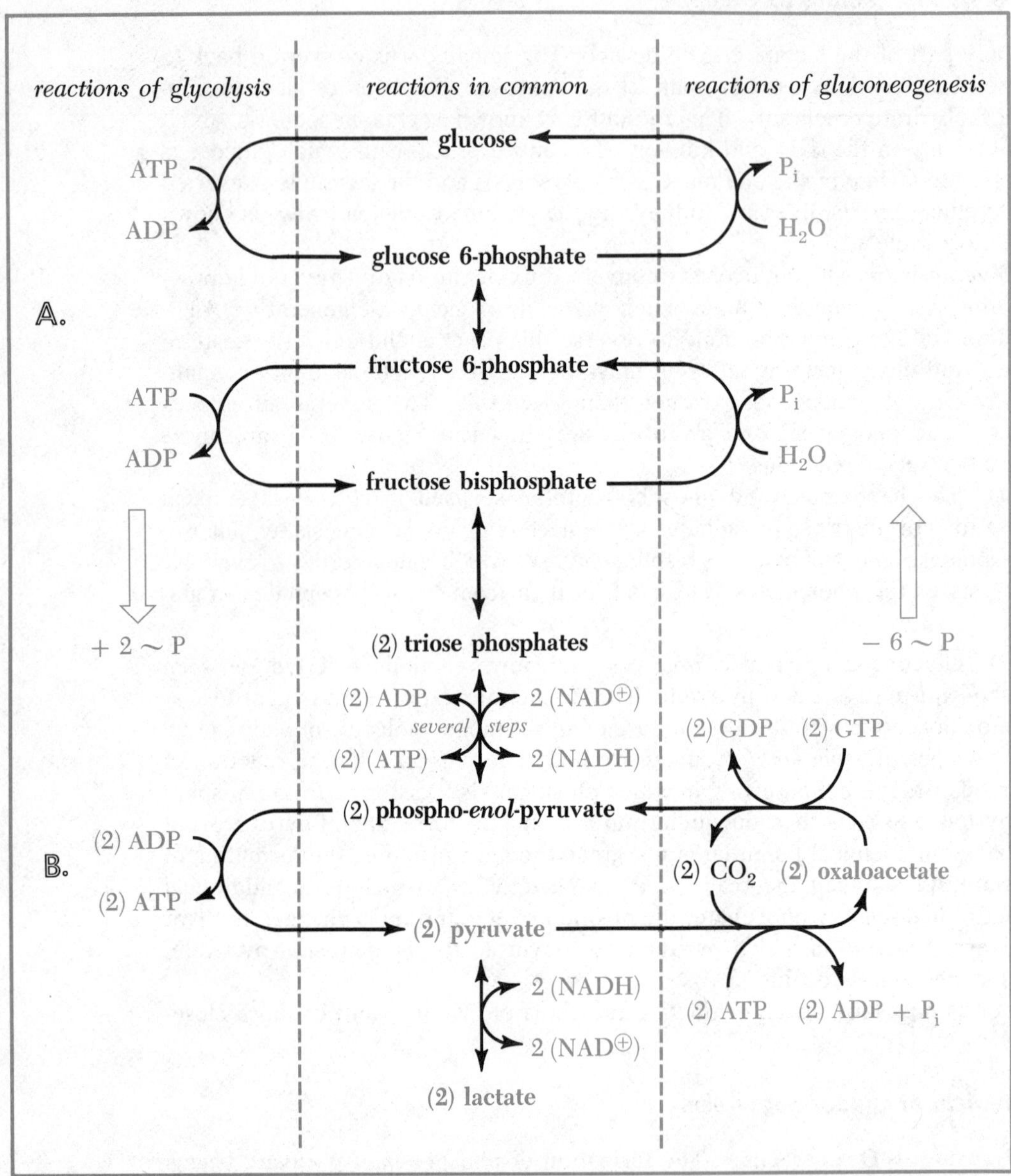

FIGURE 26–7 **Since the conversion of glucose to lactate by glycolysis is not reversible, different reactions must be used to make the formation of glucose from lactate exergonic. *A.* The kinases used to phosphorylate glucose and fructose 6-phosphate during glycolysis (*left*) are replaced by phosphatases that catalyze hydrolyses to form these compounds (*right*). *B.* The use of phospho-*enol*-pyruvate to generate ATP by the pyruvate kinase reaction *(left)* is replaced by a pair of reactions that convert pyruvate back to phospho-*enol*-pyruvate (*right*). This is done by an ATP-driven carboxylation to form oxaloacetate, which is then phosphorylated by GTP and decarboxylated to phospho-*enol*-pyruvate. (The expenditure of 0.5 ~ P to transport pyruvate is not shown.)**

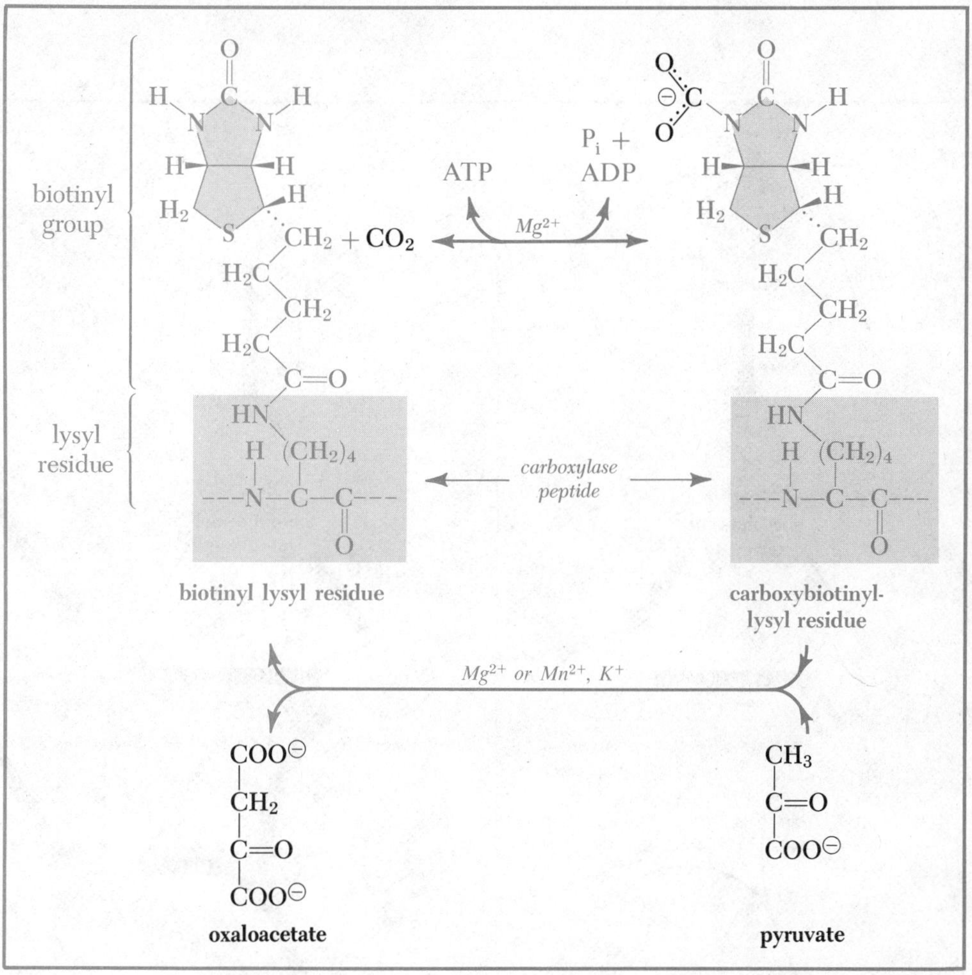

FIGURE 26–8 **The mechanism of pyruvate carboxylase involves a biotinyl group attached to a lysine residue. The biotinyl group is carboxylated at the expense of high-energy phosphate.**

Dissolved CO_2 is in equilibrium with carbonic acid and the bicarbonate ion; the bicarbonate ion is the form used by pyruvate carboxylase.

Carboxylation reactions of this type involve the participation of biotin, a coenzyme attached to lysine residues of the specific enzymes. Biotin serves as a carrier for the "active" CO_2 that is transferred as a substrate (Fig. 26–8). Biotin cannot be synthesized by animals and is required in the diet.

Pyruvate carboxylase has an unusual metal requirement; in addition to a requirement for a monovalent cation, which is satisfied by K^+ within the cell, it also requires a divalent cation, and this requirement appears to be partially met by either Mg^{2+} or Mn^{2+} *in vivo*. The enzyme isolated from calf liver, as well as from tissues of lower animals, contains both of these ions. However, the enzyme from chickens fed a manganese-deficient diet has mainly Mg^{2+}, and it works equally well.

Pyruvate carboxylase also has a nearly absolute requirement for acetyl coenzyme A as an activator. This compound is not directly involved in the reaction, but the rate of conversion of pyruvate to oxaloacetate is governed by the extent to which acetyl coenzyme A has accumulated. Why would acetyl coenzyme A

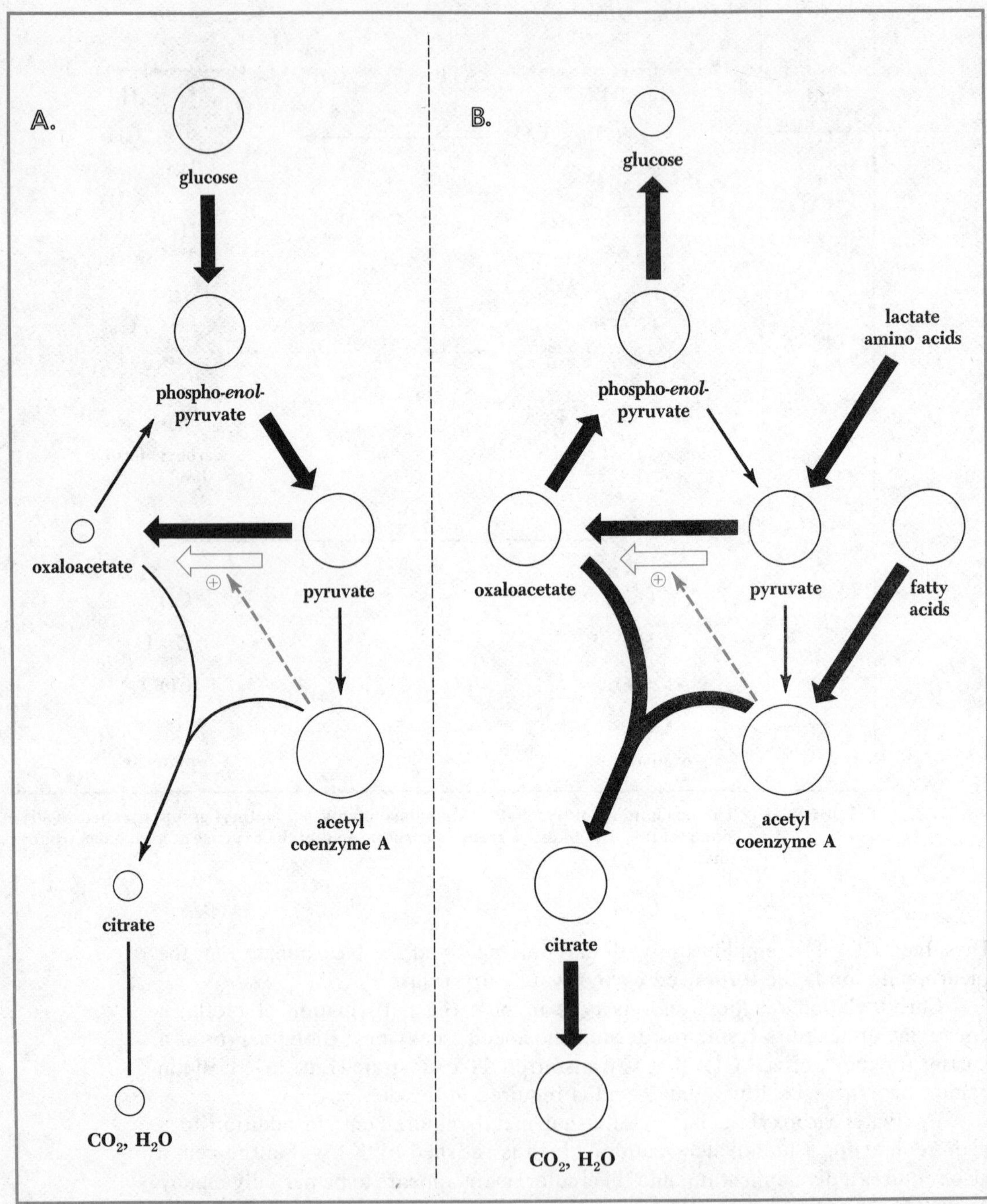

FIGURE 26–9 **Two of the circumstances that might cause an accumulation of acetyl coenzyme A, with a resultant activation of pyruvate carboxylase to accelerate oxaloacetate formation are: *A.* a deficit of oxaloacetate necessary to remove acetyl coenzyme A as citrate for total combustion, and *B.* an accumulation of acetyl coenzyme A from fatty acids. In the latter case, pyruvate derived from lactate and amino acids is not being oxidized, and is available for gluconeogenesis. Relative concentrations and rates are sketched by the size of the circles and arrows.**

accumulate? We can at this point visualize two possible circumstances (Fig. 26–9). There might be a shortage of oxaloacetate to combine with the acetyl coenzyme A and form citrate. Increased activity of the pyruvate carboxylase, thereby causing the formation of more oxaloacetate, primes the citric acid cycle in this case.

Acetyl coenzyme A may accumulate because enough is being formed from fatty acids to supply the energy requirements of the cell. Increasing the conversion of pyruvate to oxaloacetate then diverts the carbons of pyruvate to the formation of glucose.

Oxaloacetate → Phospho-*enol*-pyruvate. Oxaloacetate is converted to phospho-*enol*-pyruvate by phosphorylation with GTP, accompanied by a simultaneous decarboxylation:

$$\begin{array}{l} COO^{\ominus} \\ | \\ C{=}O \\ | \\ CH_2 \\ | \\ COO^{\ominus} \end{array} + GTP \xrightleftharpoons[Mn^{2+},\ K^{+}]{\textit{phosphopyruvate carboxykinase}} \begin{array}{l} COO^{\ominus} \\ | \\ C{-}OPO_3^{2-} \\ \| \\ CH_2 \end{array} + GDP + CO_2$$

oxaloacetate | **phospho-*enol*-pyruvate**

This reaction effectively discharges the chemical potential gained at the expense of ATP in the preceding carboxylation of pyruvate. This discharge drives the energetically unfavorable transfer of phosphate from GTP to the pyruvate moiety.

The **phosphopyruvate carboxykinase** catalyzing this reaction differs from pyruvate carboxylase in mechanism. The reaction does not involve biotin, and CO_2 is not "activated." The enzyme is present in the mitochondria of some species and in the cytosol of others; it is in both cellular compartments of human liver. The result in either case is the appearance of phospho-*enol*-pyruvate in the cytosol. (The details of intracellular transport need not concern us at this time; no expenditure of energy is required.)

Phospho-*enol*-pyruvate → Fructose 1,6-bisphosphate. All of the reactions between these two compounds in the Embden-Meyerhof pathway are freely reversible. The preceding irreversible formation of phospho-*enol*-pyruvate pushes these reactions toward the formation of fructose bisphosphate, with a concomitant consumption of ATP (to form 1, 3-bisphosphoglycerate) and NADH (to form glyceraldehyde 3-phosphate).

Fructose 1,6-bisphosphate → Glucose. The liver and other gluconeogenic organs contain an isozyme of **fructose bisphosphatase,** differing from the one in muscle. The liver contains a much higher activity of the enzyme than do the skeletal muscles because its function is to cause a net formation of fructose 6-phosphate, not to participate in a weak futile cycle for regulatory purposes:

$$\text{D-fructose 1,6-bisphosphate } ({}^{2-}O_3PO{-}CH_2 \ldots CH_2{-}OPO_3^{2-}) + H_2O \xrightarrow{\textit{fructose bisphosphatase}} \text{D-fructose 6-phosphate } ({}^{2-}O_3PO{-}CH_2 \ldots CH_2{-}OH) + P_i$$

D-fructose 1,6-bisphosphate | **D-fructose 6-phosphate**

These gluconeogenic organs also contain another hydrolase, **glucose-6-phosphatase,** which catalyzes the release of glucose from glucose 6-phosphate:

D-glucose 6-phosphate + H_2O —(*glucose-6-phosphatase*)→ D-glucose + P_i

Glucose-6-phosphatase occurs in the endoplasmic reticulum; it is often used as an identifying marker for this organelle. It is missing in all vertebrate muscles, as is pyruvate carboxylase, and this absence of the initial and terminal enzymes necessary for gluconeogenesis from pyruvate explains why the participation of the liver and similar organs is necessary for the completion of the Cori cycle. (Some invertebrate muscles have complete complements of gluconeogenic enzymes and may be able to re-utilize their own lactate.)

Energy Cost of Gluconeogenesis and the Cori Cycle

In the Cori cycle, glucose residues in the muscles are converted to lactate, and the lactate is converted back to glucose in the liver. If all of the glucose reappeared during the cycle, we would be getting energy in the muscle for nothing, which obviously doesn't happen.

Energy is supplied to this system during gluconeogenesis; the reactions are tabulated in Table 26–1. A total of 6.5 moles of high-energy phosphate is consumed in the liver for each mole of glucose produced from 2 moles of lactate. How is the liver to provide this high-energy phosphate? It does so by oxidizing some fuel, which is equivalent to the loss of part of the glucose. The calculations will be summarized in the next chapter for those with especial interest, but it is possible to arrive at the approximation that for each 100 millimoles of glucose residues in glycogen converted to lactate by glycolysis in the muscles, only 80 millimoles will come back and be stored again as glycogen. However, the muscle originally obtained 300 millimoles of ATP from glycolysis, and only 20 millimoles of glucose have been lost in the end, so the net yield is 15 moles of ATP gained to drive the muscles for each mole of glucose consumed. This is a lot better than the

TABLE 26–1 STOICHIOMETRY OF GLUCONEOGENESIS FROM LACTATE (H^+ IS OMITTED)

Reactants		Products
2 lactate$^-$ + 2 NAD^+	→	2 pyruvate$^-$ + 2 NADH(*cytosol*)
2 pyruvate$^-_{(cytosol)}$ + 0.5 ~ P*	→	2 pyruvate$^-_{(mitochondria)}$
2 pyruvate + 2 CO_2 + 2 ~ P	→	2 oxaloacetate(*mitochondria*)
2 oxaloacetate + 2 ~ P	→	2 phospho-*enol*-pyruvate(*mitochondria* or *cytosol*)
2 phospho-*enol*-pyruvate + 2 NADH + 2 ~ P	→	fructose bis-P + 2 NAD^+(*cytosol*)
fructose bis-P + 2 H_2O	→	glucose + 2 P_i
SUM: 2 lactate$^-$ + 6.5 ~ P	→	glucose

*Assuming discharge of 4 H^+ across inner membrane is equivalent to loss of one ATP.

2 moles obtained when lactate is the final product, as it is in fermenting bacteria. To summarize:

1. Energy-yielding glycolysis in muscle:

$$100 \text{ glucose (as glycogen)} \longrightarrow 200(\text{lactate}^{\ominus} + \text{H}^{\oplus}) + 300 \sim\text{P}$$

2. Energy-consuming resynthesis of glucose in liver and storage in muscle, using part of lactate as fuel:

$$200 \text{ lactate} + 120 \text{ O}_2 \longrightarrow 80 \text{ glucose (as glycogen)} + 120 \text{ CO}_2$$

$$\text{SUM } 20 \text{ glucose (as glycogen)} + 120 \text{ O}_2 \longrightarrow 120 \text{ CO}_2 + 300 \sim\text{P}$$

$$\sim\text{P/glucose} = 15$$

The final conclusion is that lactate production in animals is approximately 41 per cent as efficient as direct complete combustion in producing ATP. Put another way, 59 per cent of the potential ATP is sacrificed in order to have a rapid, emergency production by glycolysis. In total amounts, this loss is really not as great as it seems. Maximum lactate production is used only for brief periods, of the order of a minute. Even in our civilized times, being able to go all-out for a minute is sometimes life-saving, and the waste of what might have been another two minutes' worth of ATP isn't a great price.

LACTIC ACIDOSIS

Transitory increases in lactate concentration, and an associated increase in H^+ concentration, are a normal consequence of exercise. Values up to 17 mM in blood are common, and may be higher in some individuals after extreme exertion. Even excitement will cause more lactate to be formed, although at much lower levels.

However, some conditions cause a more persistent elevation in lactate concentration; the underlying metabolic disturbance, as well as the change in acid-base balance (Chapter 39), is frequently a cause for concern. Any concentration greater than 2mM after resting is considered abnormal.

The lactate concentration may rise because of increased production from glucose. We have already noted that any impairment of the oxygen supply will lead to a general increase in lactate production. This may occur during anesthesia. A similar increase is seen with any inhibitor of the oxidative processes, so long as glucose is available. For example, cyanide poisoning can cause a massive lactic acidosis.

The lactate concentration also may rise because of decreased gluconeogenesis in the liver. A congenital absence of any of the necessary isozymes that are peculiar to the liver (or other gluconeogenic tissues) would cause lactate accumulation. Infants have been found with persistent lactate levels as high as 20mM who lacked the hepatic fructose bisphosphatase, but not the enzyme in muscle.

Gluconeogenesis can also be impaired by drinking ethanol. The metabolism of ethanol begins by oxidation with NAD in the liver. The resultant increase in the [NADH]/[NAD] ratio makes the equilibrium for lactate dehydrogenase shift toward lactate (Fig. 26–10).

Lactic acidosis also accompanies a variety of physiological and pathological states (Table 26–2). Phenformin, a drug formerly used for the oral treatment of diabetes mellitus, sometimes provoked lactic acidosis in an unknown way.

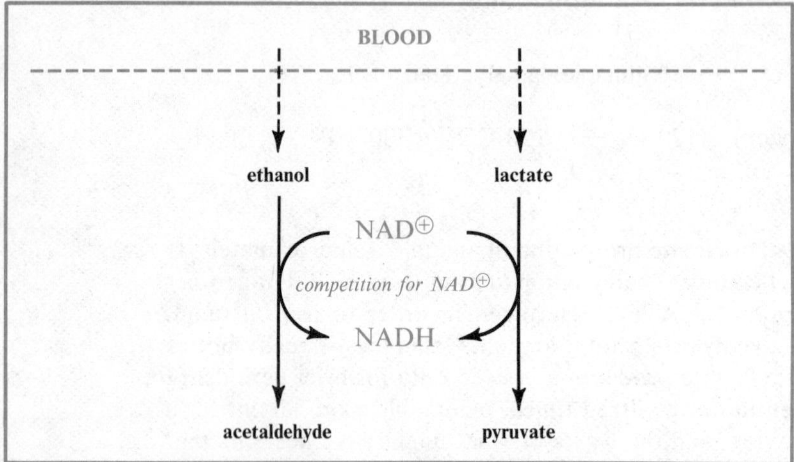

FIGURE 26–10 **Ingestion of ethanol tends to prevent gluconeogenesis from lactate in the liver because the oxidation of ethanol competes for the NAD necessary for the oxidation of lactate.**

TABLE 26–2 LACTIC ACIDOSIS

Characteristics

blood lactate concentration >2 mM without exercise
blood pH low to variable extent
[lactate]/[pyruvate] ratio often increased over resting

Causes and Resultant Lactate Concentrations*

1. low blood [O_2] (3–4 mM)
2. severe anemia (<2 mM)
3. congenital absence of enzymes necessary for gluconeogenesis (e.g., hepatic fructose bisphosphatase 3–23 mM)
4. leukemia (12–25 mM)
5. cardiovascular shock (2–35 mM)
6. surgical bypass of the heart (2–8 mM)
7. anesthesia
8. diabetes mellitus (10–31 mM)
9. ethanol ingestion (2–8 mM)
10. phenformin intoxication (10–31 mM)

*Based on Oliva (1970), Am. J. Med., *48*: 209.

FURTHER READING

Margaria, R.: (1976) *Biomechanics and Energetics of Muscular Exercise.* Clarendon Press. Includes discussion of lactate production.

McGilvery, R. W.: (1975) *Use of Fuels for Muscular Work.* p. 12 *in* H. Howald, and J. Poortmans, eds.: *Metabolic Adaptation to Prolonged Physical Exercise.* Birkhauser Verlag Basel. An extension and amplification of the treatment given here. (Neglects propulsion of mitochondrial ADP-ATP antiport by membrane potential, however.)

Keul, J., E. Doll, and D. Keppler: (1972) *Energy Metabolism of Human Muscle.* Vol. 7 of E. Jokl, ed.: *Medicine and Sport.* University Park Press. Contains good and quite readable reviews.

Andersen, H. T.: (1966) *Physiological Adaptations in Diving Vertebrates.* Physiol. Rev. *46*: 212.

Siebke, H., et al.: (1975) *Survival After 40 Minutes' Submersion Without Cerebral Sequelae.* Lancet, *1*: 1275.

Newsholme, E. A.: *The Regulation of Phosphofructokinase in Muscle.* Cardiology, *56*: 22. Evaluation of the futile cycle as a regulatory device.

Utter, M. F., R. E. Barden, and B. L. Taylor: (1975) *Pyruvate Carboxylase: An Evaluation of the Relationships Between Structure and Mechanism and Between Structure and Catalytic Activity.* Adv. Enzymol., *42*: 1.

McGilvery, R. W.: (1975) *Biochemical Concepts.* W. B. Saunders. Chapter 16 is a more extensive discussion of fermentations. Not bad — not bad at all.

Horecker, B. L., E. Melloni, and S. Pontremoli: (1975) *Fructose-1, 6-bisphosphatase: Properties of the Neutral Enzyme and Its Modification by Proteolytic Enzymes.* Adv. Enzymol., *42*: 193.

Everse, J., and N. O. Kaplan: (1973) *Lactate Dehydrogenases: Structure and Function.* Adv. Enzymol., *37*: 61.

Hanson, R. W., and M. A. Mehlman, eds.: (1976) *Gluconeogenesis.* Wiley.

Hermansen, L., and I. Stensvold: (1972) *Production and Removal of Lactate during Exercise in Man.* Acta Physiol. Scand., *86*: 191.

Oliva, P. B.: (1970) *Lactic Acidosis.* Am. J. Med., *48*: 209.

Fulop, M., et al.: (1976) *Lactic Acidosis in Diabetic Patients.* Arch. Intern. Med., *136*: 987.

Conlay, L. A., and J. E. Loewenstein: (1976) *Phenformin and Lactic Acidosis.* J.A.M.A., *235*: 1575.

Gram, D. L., et al.: (1977) *Acute Cyanide Poisoning Complicated by Lactic Acidosis and Pulmonary Edema.* Arch. Intern. Med., *137*: 1051.

Pagliara, A. S., et al.: (1972) *Hepatic Fructose-1, 6-diphosphatase Deficiency.* J. Clin. Invest., *51*: 2115.

Ramberg, C. F., and M. Vranic, eds.: (1977) *Glucose Recycling and Gluconeogenesis.* Feder. Proc., *36*: 225.

27 STORAGE OF GLUCOSE AS GLYCOGEN

The ability of an animal to interrupt its food intake — to eat meals rather than repeated small snacks and to sleep for long intervals — depends upon a capacity to store temporary excesses of foods for later use. The demand for high-energy phosphate is not transient, and carbon compounds supplying the processes by which high-energy phosphate is generated must always be available from internal stores. Among these stores are the polymers of glucose: starch in plants and glycogen in animals.

We considered the molecular nature of these glucans in Chapter 10. Glycogen appears in cells as discrete particles (Fig. 27–1), which are evidently aggregates of molecules with molecular weights ranging at least to 2×10^7. An estimated 0.7 of the mass is water; the many hydroxyl groups make the molecule quite polar. The particles also contain the enzymes concerned with the formation and breakdown of glycogen.

The amount of glycogen varies widely, not only among different tissues but also within a tissue, depending upon the supply of glucose and the metabolic demand for energy. Liver and muscle contain the largest stores, and we shall mainly be concerned with these two tissues. Glycogen is important to the economy of other tissues, where it is usually in lower concentration and is a small fraction of the total in the body.

In general, the content of glycogen in the liver will vary greatly with the diet, while that in the muscles will be less affected. In contrast, exercise rapidly depletes the glycogen of the muscles, while that in the liver is less affected.

Much of our knowledge about the glycogen concentration of human tissues comes from analysis by Swedish investigators of biopsy specimens from normal volunteers. (We are all indebted to university students and members of the fire department in Stockholm for contributing pieces of themselves to satisfy our curiosity.) The data have some limitations, because only readily accessible regions of the organs can be explored in living subjects, but the values are consistent with other quantitative information on human fuel metabolism. Results under various conditions from several laboratories are summarized in Figure 27–2.

A typical value for the liver glycogen content in a well-fed human is near 400 millimoles of glucosyl residues (65 grams dry weight) per kilogram of tissue; there is less after fasting and more upon eating a high-carbohydrate meal.

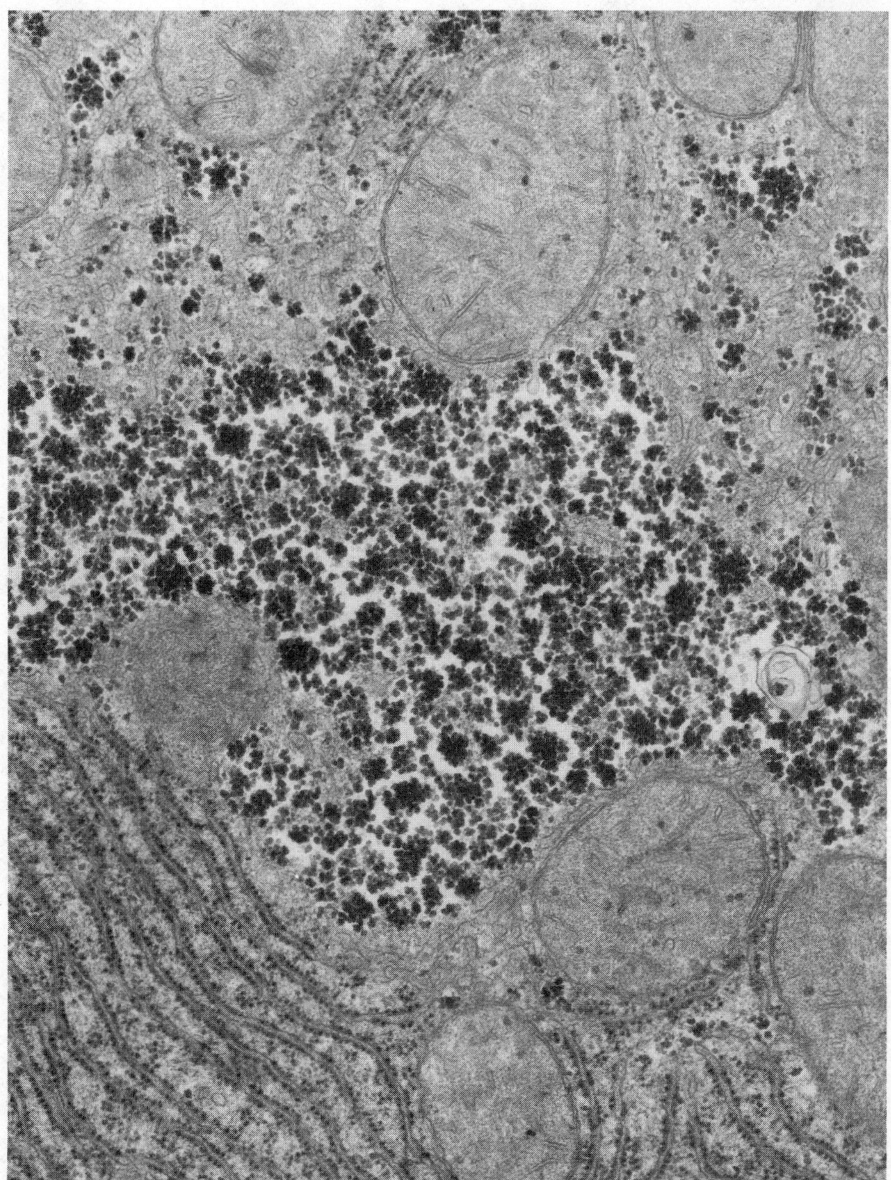

FIGURE 27–1 Dense clumps of glycogen fill the *center* of this view of a liver cell in a rat fed a high-carbohydrate diet. The large organelles are mitochondria. Rough endoplasmic reticulum and associated ribosomes are at the *lower left*; smooth endoplasmic reticulum is at the *upper right*. Magnification 31,000×. (Courtesy of Dr. Robert R. Cardell, Jr.)

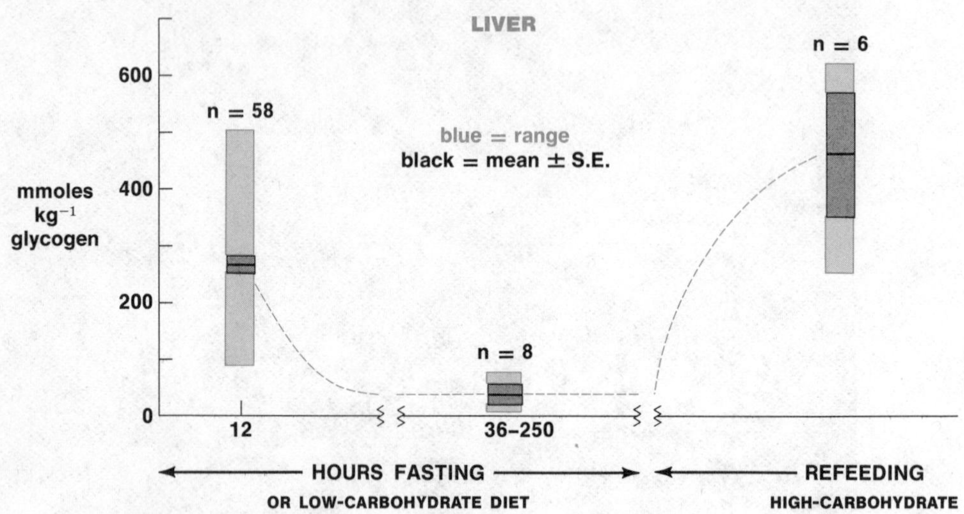

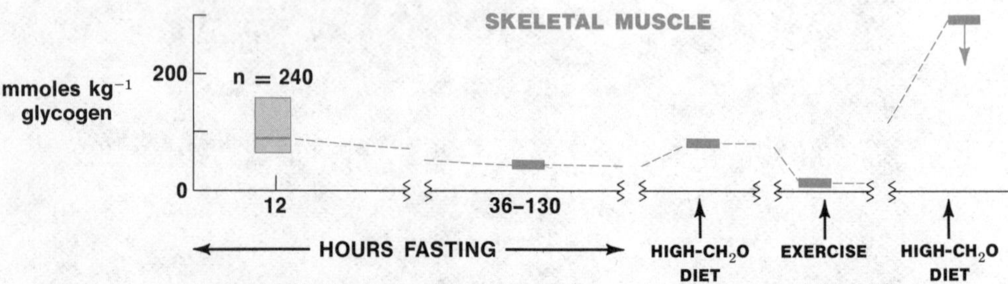

FIGURE 27–2 **Changes in the glycogen content of human liver and skeletal muscles with fasting and high-carbohydrate diets. Changes are also shown for skeletal muscles with exercise. This is a composite of data from several sources. Solid bars indicate values obtained from a small number of subjects. Note that the glycogen content of muscle responds to glucose feeding more sharply after exercise.**

Values based on data from the following:

Nilsson, L. H:S.: (1973) *Liver Glycogen Content in Man in the Postabsorptive State.* Scand. J. Clin. Lab. Invest., *32*: 317.

Nilsson, L. H:S., and E. Hultman: (1973) *Liver Glycogen in Man—the Effect of Total Starvation or a Carbohydrate-poor Diet Followed by Carbohydrate Refeeding.* Scand. J. Clin. Lab. Invest., *32*: 325.

Hultman, E.: (1967) *Studies on Muscle Metabolism of Glycogen and Active Phosphate in Man with Special Reference to Exercise and Diet.* Scand. J. Clin. Lab. Invest., *19* (suppl. 94).

Hultman, E., J. Bergstrom, and A. E. Roch-Norlund: (1971) *Glycogen Storage in Human Skeletal Muscle.* Adv. Expt. Biol. Med., *11*: 273.

Saltin, B., and J. Karlsson: (1971) *Muscle Glycogen Utilization during Work of Different Intensities.* Adv. Exp. Biol. Med., *11*: 289.

The skeletal muscles typically have 85 millimoles of glucosyl residues (14 grams) per kilogram of tissue, which does not change much with overnight fasting or with a high-carbohydrate diet. However, the content falls to 1 millimole per kilogram, or lower, after an hour or two of heavy work. After depletion in this way, high-carbohydrate diets on subsequent days can cause the level to rise as high as 300 millimoles per kilogram.

Even though the liver usually has a higher concentration of glycogen than do the skeletal muscles, the muscles contain the bulk of the total glycogen store, owing to their large mass. A man weighing 70 kilograms will have about 28 kilograms of skeletal muscle, but only about 1.6 kilograms of liver. Given typical contents per kilogram of tissue, 400 millimoles in the liver and 85 millimoles in the muscles, the total store is 0.6 moles in the liver and 2.4 moles in the muscles. The total in the body, considering all organs, will, therefore, range somewhat over 3 moles in a well-fed individual and near 3 moles after an overnight fast (the postabsorptive state).

MECHANISM OF GLYCOGEN STORAGE

Glycogen is made by adding one glucosyl residue at a time to an existing glycogen molecule, creating amylose chains that are then rearranged to make branches. However, it isn't necessary to pass precursor molecules of glycogen from parent to daughter cells; new molecules can slowly arise through the action of the same enzymes we are about to describe, but simple chance favors extension of existing chains when they are abundant.

The glucosyl residues that are transferred come from a nucleotide, **uridine diphosphate glucose (UDP-glucose),** which in turn is derived from glucose 6-phosphate. The overall process is conveniently thought of in three stages: (1) the conversion of glucose 6-phosphate to UDP-glucose; (2) the transfer of glucosyl units from UDP-glucose onto residues in glycogen chains, so as to make growing amylose chains composed of α-1,4-linkages; and (3) the creation of branches by shifting portions of the chains onto the C-6 hydroxyl groups of adjacent chains.

The Formation of UDP-Glucose. Since it is the C-1 of glucose that will later be attached to glycogen, the sequence begins by transferring the phosphate group of glucose 6-phosphate to form glucose 1-phosphate (Fig. 27–3). This freely reversible reaction is catalyzed by **phosphoglucomutase,** which utilizes the corresponding bisphosphate compound in small concentrations as an intermediate (compare phosphoglycerate mutase, p. 446).

Glucose 1-phosphate then reacts with UTP to form UDP-glucose and inorganic pyrophosphate. As is the case with other reversible reactions involving pyrophosphate, hydrolysis by inorganic pyrophosphatase keeps the concentration low, and pulls the reaction in the direction of UDP-glucose formation. The UTP utilized in this reaction is generated by a **nucleoside diphosphokinase** reaction:

$$\text{ATP} + \text{UDP} \longleftrightarrow \text{ADP} + \text{UTP}$$

so that the ultimate source of the high-energy phosphate is ATP.

Formation of Amylose Chains. UDP-glucose donates glucose residues used for extending the terminal branches of glycogen in a reaction catalyzed by glycogen synthase. The reaction is specific for the hydroxyl group on C-4 of the

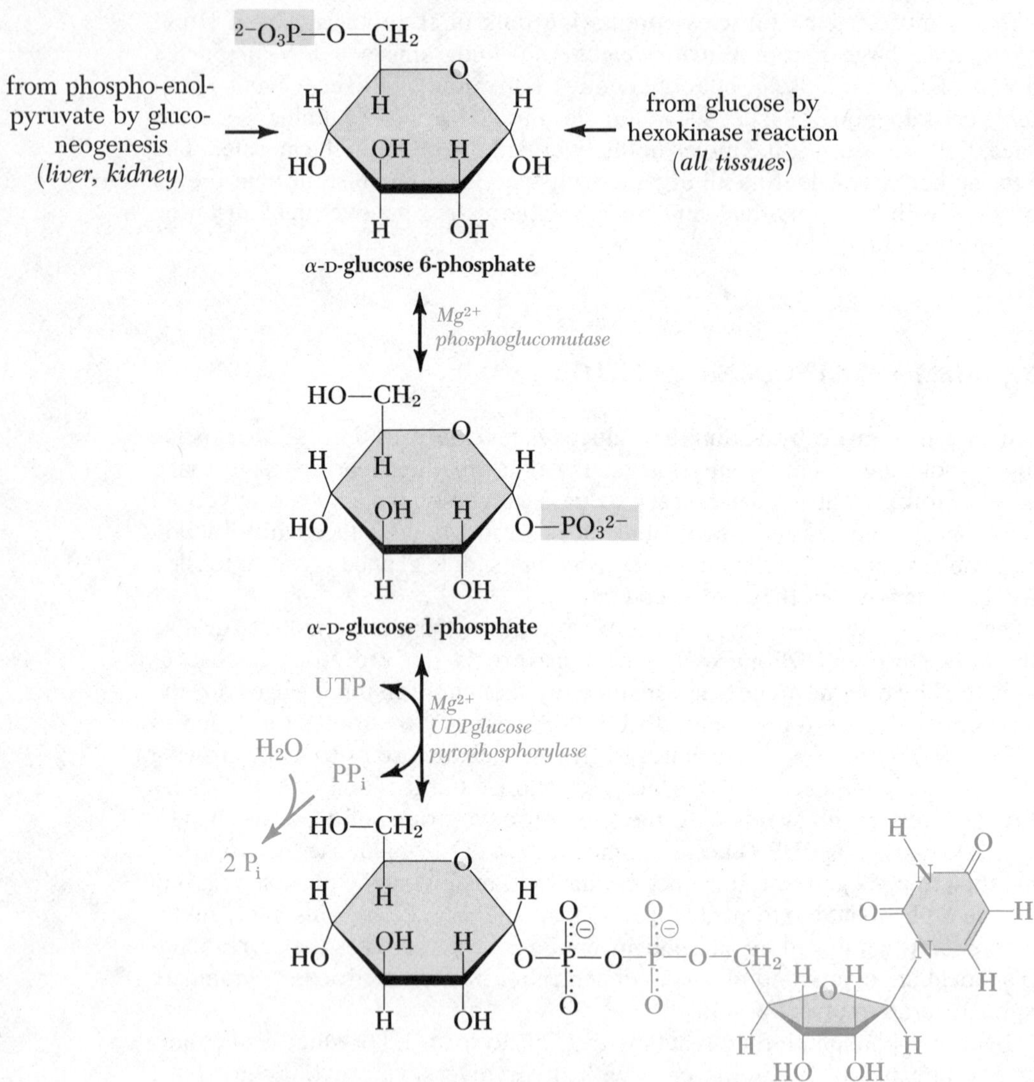

FIGURE 27–3 **Storage of excess glucose residues as glycogen involves the formation of uridine diphosphate glucose (UDP-glucose).**

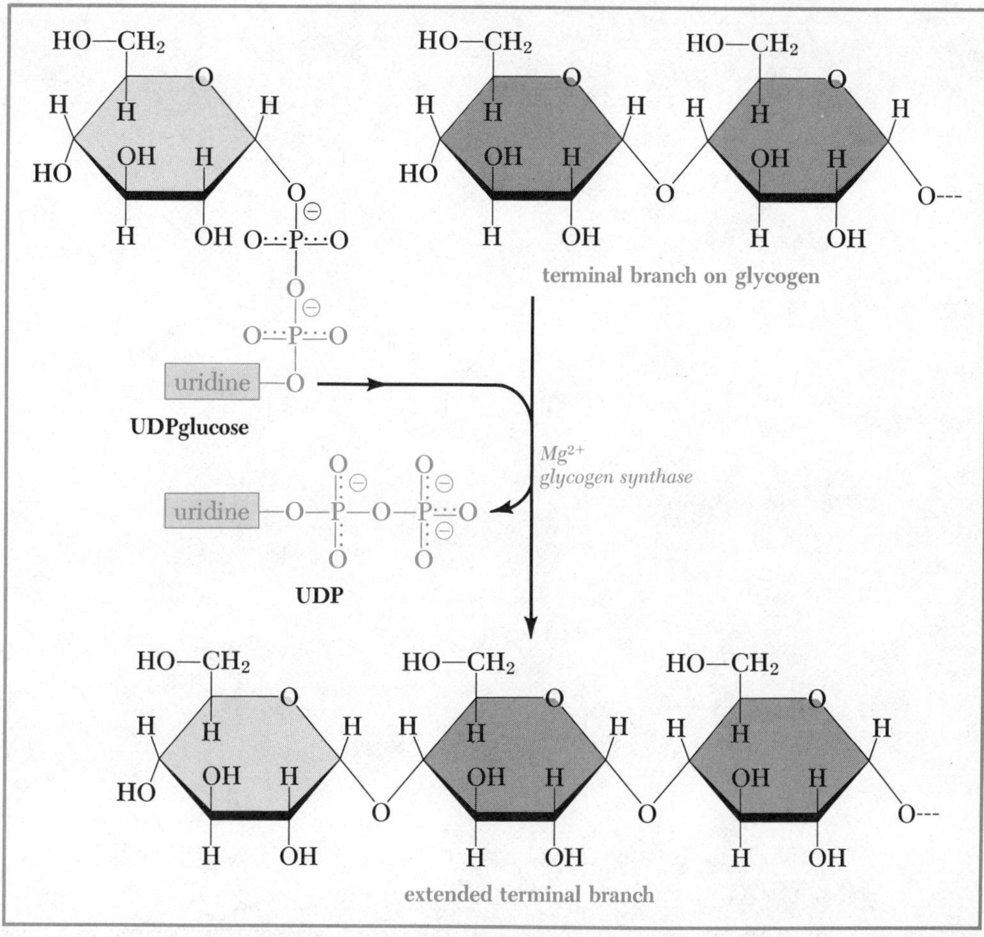

FIGURE 27–4 **Terminal amylose chains on glycogen are extended by transfer of glucose residues from UDP-glucose to the C-4 hydroxyl groups of the chains.**

glycogen residues, so the new glucosyl residue simply extends the 1,4-chain (Fig. 27–4). Since the nature of the chain is not changed by extension, the **glycogen synthase** reaction can be repeated indefinitely. A molecule of UDP is released for each glucosyl residue added.

Formation of Branches. If nothing else happened, the result of the glycogen synthase reaction would be long amylose chains composed only of 1,4-bonds. However, cells storing glycogen also have a branching enzyme, a **glycosyl-4:6-transferase,** that catalyzes the transfer of a segment of amylose chain onto the C-6 hydroxyl of a neighboring chain (Fig. 27–5). This enzyme moves a block of seven 1,4-residues from a chain at least 11 residues in length and transfers it onto another segment of amylose chain at a point four residues removed from the nearest branch. (Transfer of seven residues is favored, but specificity is not absolute, and this number is emphasized for clarity.) The new branch is therefore seven residues in length, and the remaining stub from which it was removed is at least four residues in length, but is more commonly six to nine residues long. 1,6-Glucosides have about 4,800 joules per mole less standard free energy content than do 1,4-glucosides, so the equilibrium of reaction favors branching.

Further Growth. The new branch and the remaining stub from which it was obtained can grow in length by addition of more 1,4-glucosyl residues from

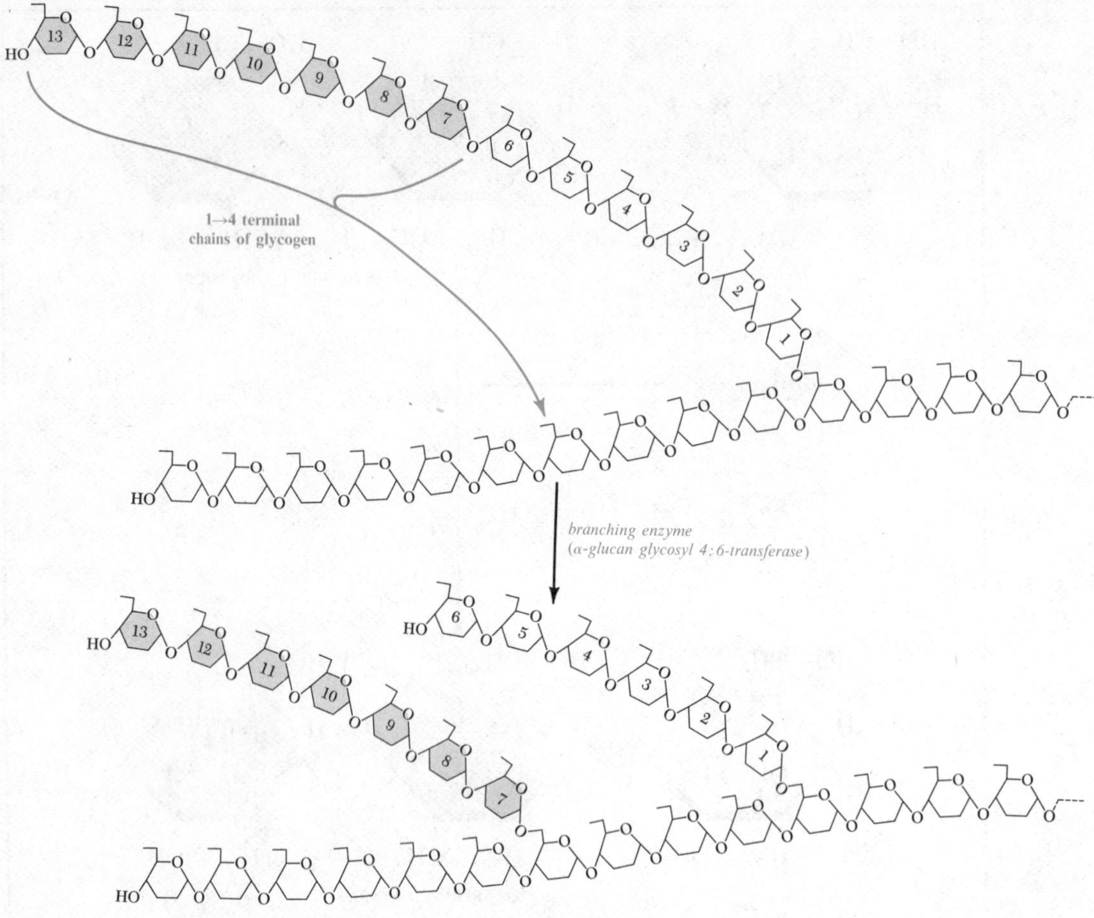

FIGURE 27–5 **Branches are created in glycogen by transferring seven-residue segments of amylose terminal chains to hydroxyl groups on carbon 6 of glucose residues that are four residues removed from existing branches. A terminal branch must be at least 11 residues long before a segment is transferred from it.**

UDP-glucose through the action of glycogen synthase. When enough have been added to extend them to at least 11 residues beyond the branch, a further transfer by the branching enzyme may create still another branch, which also can grow in length. What limits the size of the molecule and the number of particles is not known. There is little information to justify speculation, although large glycogen particles appear to be made of subunits of approximately the theoretical limit for tightly packed spheres. What is certain is that skeletal muscle does not accumulate glycogen indefinitely, even with a high concentration of blood glucose. The content of glycogen in the liver can rise markedly under heavy dietary loading, but even in this tissue glucose is absorbed slowly enough with more usual conditions so that an increased fraction of the excess supply is used for making fat (next chapter) after the glycogen content rises above 50 to 60 grams per kilogram of tissue. The mashed livers of force-fed Strasbourg geese, from which the pâte de foie gras favored by epicures is made, owe their special properties to high contents of both glycogen and fat.

UTILIZATION OF GLYCOGEN

Phosphorolytic Cleavage. Glycogen is mainly degraded by a simple phosphorolysis of the 1,4-glucosidic bonds to form glucose 1-phosphate, accompanied by reactions to remove the 1,6-branches. The primary reaction, catalyzed by **glycogen phosphorylase**, cleaves the terminal 1,4-glucosidic bond with inorganic phosphate so as to shorten the chain by one residue (Fig. 27–6). The reaction is freely reversible, with the equilibrium ratio of P_i to glucose 1-phosphate concentrations near 3.5. However, the concentration of P_i in the liver is some 30-fold greater than the concentration of glucose 1-phosphate, so the reaction always proceeds in the direction of formation of glucose 1-phosphate.

When a glucose residue is removed by phosphorolysis, the shortened 1,4-branch on the glycogen molecule can be attacked again to remove another residue, followed by further attack on the remainder. However, phosphorylase by itself will not catalyze the removal of all of the terminal residues. In anthropomorphic terms, it won't go near the branches. Expressed more mechanistically, the specificity of the enzyme is such that the 1,4-bond attacked must be at least four residues removed from a 1,6-branch. Therefore, phosphorylase alone will not make the outer branches any shorter than four glucosyl residues. The function of this specificity is not clear. Glycogen completely sheared by phosphorylase to the four-residue stubs is called the phosphorylase limit dextrin, or ϕ-dextrin. (Dextrin is a term borrowed from starch chemistry, describing partially hydrolyzed glucose homopolymers, and it is widely used for describing intermediates obtained during chemical investigation of structure, but this does not imply the occurrence of these discrete types of molecules in cells.)

Removal of Branches. Further degradation of glycogen requires a second enzyme, **amylo-1,6-glucosidase.** This enzyme catalyzes two successive reactions. In the first reaction, it acts as a **glucosyl transferase** akin to the branching enzyme

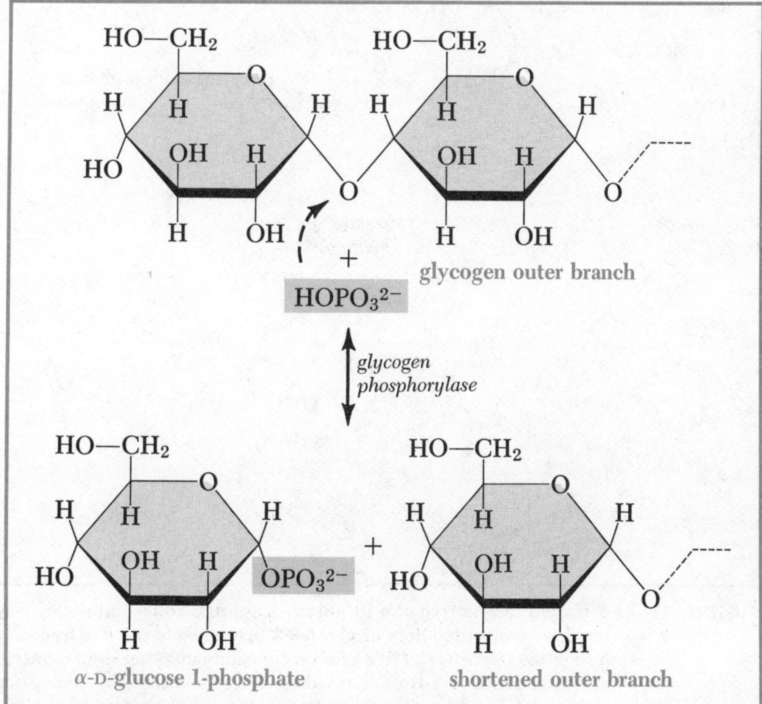

FIGURE 27–6

The recovery of the glucose residues stored as glycogen begins with phosphorolysis of terminal glucosidic bonds to form glucose 1-phosphate.

functioning during glycogen synthesis, but differing in that it transfers three glucosyl residues from a branch onto a chain terminus so as to lengthen it (Fig. 27–7). Suppose that adjacent chains are shortened as much as possible at a branch by phosphorylase. Each will be four glucosyl residues in length. The glucosyl-4:4-transferase activity in amylo-6-glucosidase catalyzes the removal of three of the residues of the stub of the branch, leaving a single residue attached to C-6. The enzyme transfers the three-residue package to the end of the other stub, which now has seven residues in 1,4-linkage and can therefore be attacked again by phosphorylase.

The amylo-6-glucosidase then catalyzes its second reaction: hydrolysis of the single residue remaining on C-6 at the branch to yield free glucose. The hydrolytic activity is absolutely specific for a 1,6-linkage to a single residue, thereby protecting longer branches on glycogen from internally disruptive attack by this enzyme.

Removal of the single branched residue exposes a straight chain of 1,4-linked residues down to the next branch, so phosphorylase can catalyze the removal of more residues as glucose-1-phosphate before reaching its limit of four residues

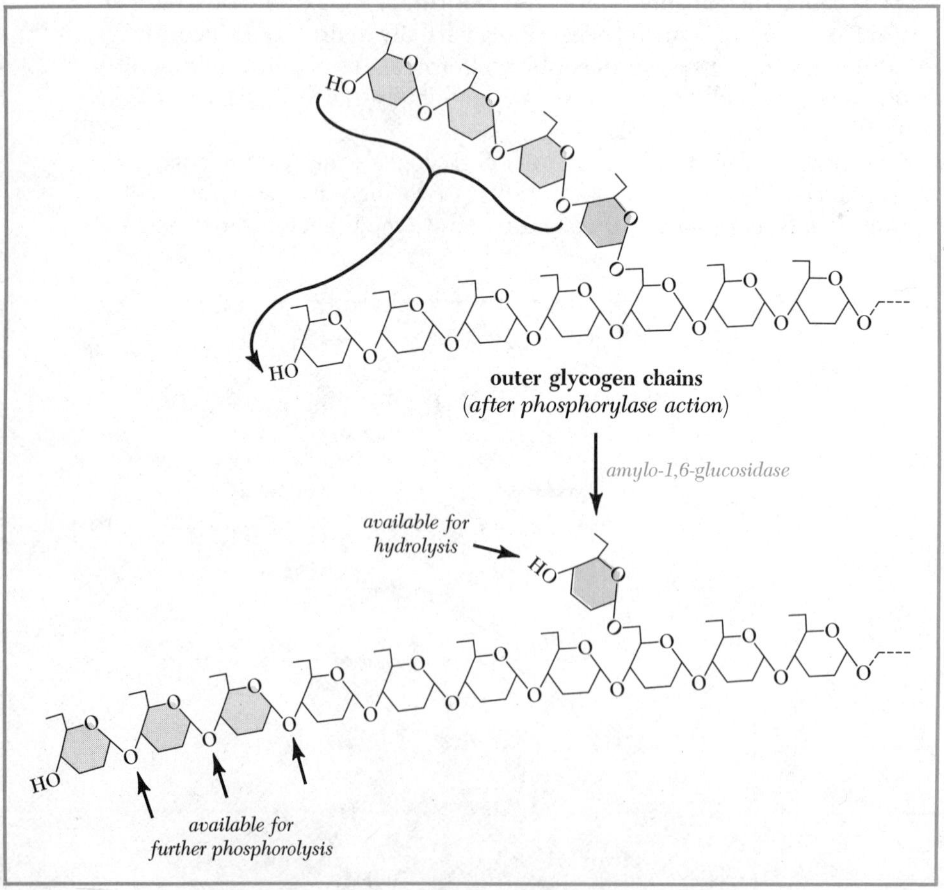

FIGURE 27–7 **A transferase active site in amylo-1,6-glucosidase catalyzes the transfer of three residues from the stubs of glycogen branches to the C-4 hydroxyl groups of amylose stubs. Another active site on the same enzyme then catalyzes the hydrolysis of the exposed 1,6-linked residue. Further attack by phosphorylase on the extended 1,4-amylose chain can then remove the three added glucose residues.**

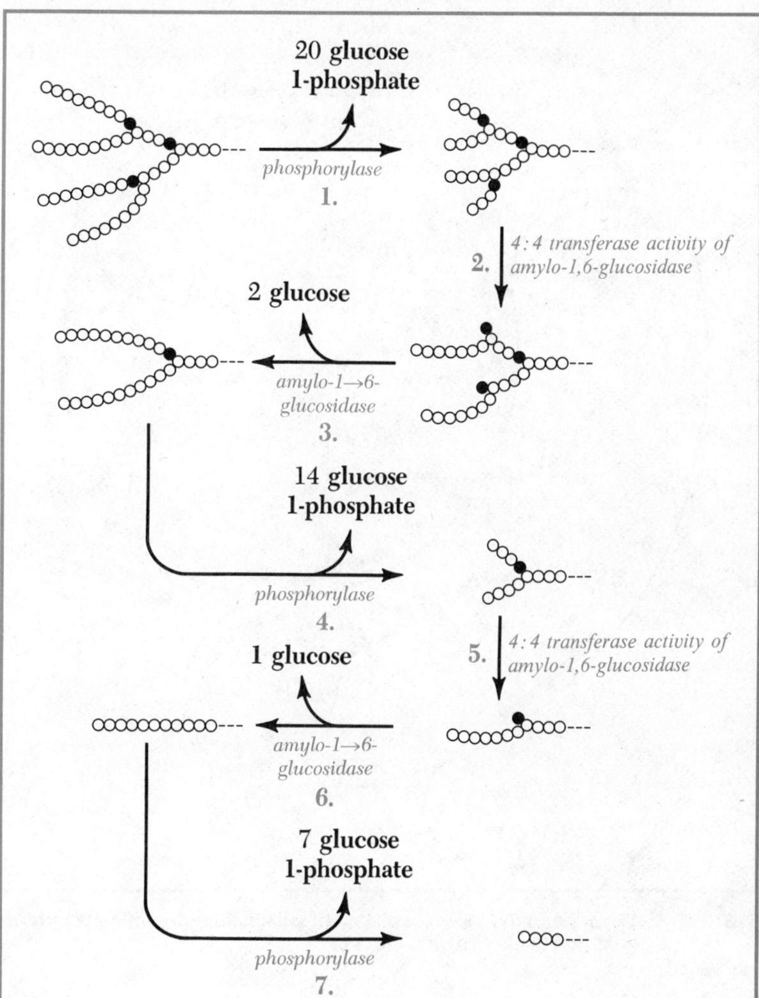

FIGURE 27–8

Summary of the sequence of glycogen breakdown. The diagram shows the fate of four outer chains and the next two branches from which they arise. Each chain is shown with nine residues beyond the branch. The residues linked by 1,6 bonds and released as free glucose are indicated by *solid circles*.

***SUM:* Out of the 44 glucosyl residues removed, three have appeared as free glucose and 41 as glucose 1-phosphate.**

next to a branch. What is the result? The whole sequence (Fig. 27–8) causes a continual stripping of glucosyl residues as glucose 1-phosphate, except for the single residues at branches that are removed as free glucose. The degree of branching is such that 11 to 14 molecules of glucose 1-phosphate are formed for each molecule of free glucose released.

REGULATION OF GLYCOGEN STORAGE

Since any normal cell in which glycogen is stored contains the enzymes necessary to catalyze both its formation and its degradation, it has the potential for a major futile cycle in which glucose residues are rapidly being added to and stripped from glycogen. The ultimate result of such a cycle is nothing more than the hydrolysis of ATP, and the possibility of this major wastage of chemical potential may have been sufficient in itself to have required the evolution of regulatory mechanisms preventing simultaneous full-velocity synthesis and degradation.

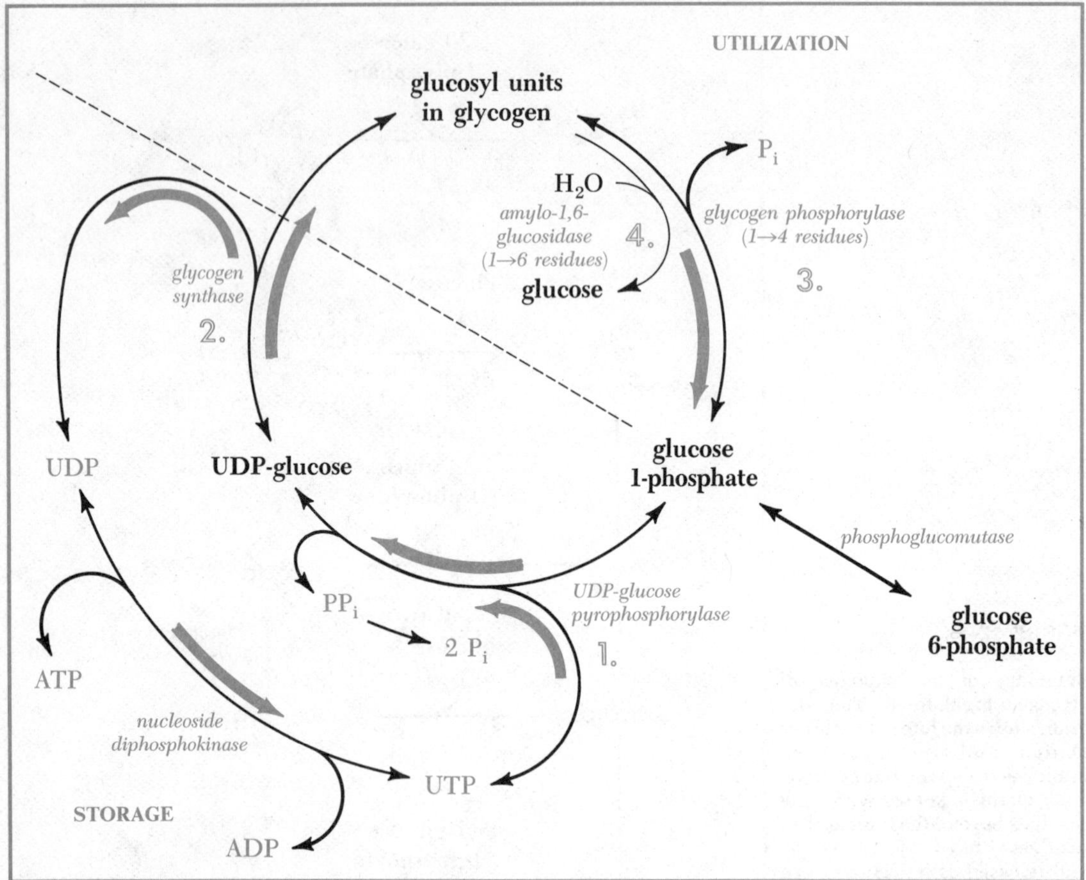

FIGURE 27–9 **A summary of glycogen metabolism. Simultaneous occurrence of all reactions would create a futile cycle hydrolyzing ATP.**

Beyond all of this, the different functions of various organs in the higher animals demand different controls over the deposition and mobilization of their respective glycogen stores. For example, most of the blood circulating through the small bowel passes into the liver by way of the portal veins, and only then is it returned to the heart and lungs for general circulation. The glucose absorbed from a meal in the small bowel therefore first traverses the liver where some of it is taken up and stored as glycogen. The balance leaves the liver in the venous return to the heart and is then distributed to the peripheral tissues either for immediate use as a fuel, as in the brain, or for further storage as glycogen, as in the skeletal muscles. We see that under these conditions of plentiful glucose supply both the liver and muscles must store glucose; something must occur in both tissues to promote the action of UDP-glucose pyrophosphorylase and glycogen synthase (reactions 1 and 2 in Fig. 27–9) while slowing the reactions catalyzed by glycogen phosphorylase and amylo-1,6-glucosidase (reactions 3 and 4 in Fig. 27–9).

Between meals, the situation is changed. Liver glycogen is mobilized to maintain the concentration of glucose in the blood going to the peripheral tissues, so the enzymes degrading glycogen in the liver must be activated while the enzymes catalyzing synthesis are inhibited. Muscles may at the same time be responding differently. When circumstances call for strong contractions, glycogen is degraded to provide fuel, especially in the white fibers, with the glycogen rebuilt during times of rest.

CONTROL BY PHOSPHORYLATION OF ENZYMES

The primary control over glycogen metabolism involves regulation of synthesis by changes in the activity of glycogen synthase, and regulation of degradation by changes in the activity of glycogen phosphorylase. These changes result from the existence of the respective enzymes in active and inactive conformations. The conformations may be made to change by direct interaction of the enzymes with some metabolites, but the principal cause of a shift in conformation is a phosphorylation of the respective enzymes. Indeed, we discussed the phosphorylation of glycogen phosphorylase as an example of this mode of control (p. 307). The phosphate groups are added by transfer from ATP in reactions catalyzed by kinases, and they are removed by hydrolysis in reactions catalyzed by phosphatases.

Phosphorylation May Inhibit or Activate. Now comes the important point. Efficient metabolism requires that synthesis of glycogen diminishes when degradation accelerates, and *vice versa*. The activities of glycogen synthase and glycogen phosphorylase must therefore change in opposite directions in response to the same environmental stimulus. Translating this into molecular terms, if control is exerted because some stimulus causes phosphorylation of these enzymes, the phosphorylation must make one enzyme more active and the other enzyme less active than the corresponding non-phosphorylated forms, and this is exactly how it happens.

Let us designate the forms of these enzymes with non-phosphorylated seryl side chains as glycogen synthase-H and glycogen phosphorylase-H, and designate the corresponding phosphate ester forms as glycogen synthase-P and glycogen phosphorylase-P, analogous to the nomenclature of other ester phosphates. (H or P is used to indicate a proton or a phosphate group on the serine side chain O atom. This is a departure from approved nomenclature, and it is used to clarify an intricate series of events.)

Under most conditions:

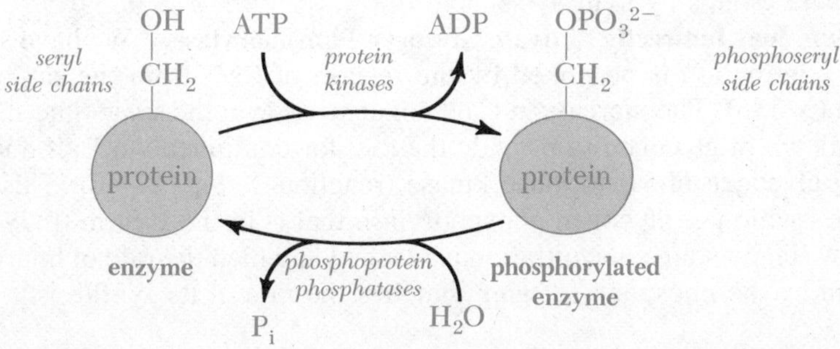

Glycogen synthase-H is more active than glycogen synthase-P.
Glycogen phosphorylase-H is less active than glycogen phosphorylase-P.

(It is common to designate the form usually most active as **a,** and the other form as **b,** ignoring the state of phosphorylation.) Glycogen metabolism is regulated by controlling the addition or removal of phosphate groups to these two enzymes.

Addition promotes degradation of glycogen; removal promotes synthesis of glycogen.

The mechanisms for regulating these interconversions of active and inactive conformations of enzymes involved in glycogen metabolism are considerably more complex than anything we have encountered before. They differ somewhat from one tissue to another; let us first dissect the circumstances in mammalian skeletal muscles.

REGULATION OF GLYCOGEN PHOSPHORYLASE IN MUSCLE

Effects of Metabolites

The proportion of active and inactive conformations of glycogen phosphorylase may be changed by three fundamentally different mechanisms. A compound may bind to the phosphorylase so as to favor one conformation. A compound may affect the action of the kinase that catalyzes the conversion of phosphorylase-H to phosphorylase-P. Finally, a compound may affect the action of a phosphatase that catalyzes the hydrolysis of the phosphate groups from phosphorylase-P.*

AMP Activates Glycogen Phosphorylase. This occurs with the nonphosphorylated form (reaction 1, Fig. 27–10). It does so because the inactive conformation of the enzyme has a low affinity for AMP, whereas the active conformation, although present in low concentrations in the free form, has a high nity for AMP. Rising concentrations of AMP therefore shift the equilibrium ay from the inactive conformation that usually predominates for the free yme toward the active form of the AMP complex, even though the enzyme is not phosphorylated. Since a rising AMP concentration is a sensitive indicator of a fall in high-energy phosphate concentration (p. 459), its effect on glycogen phosphorylase is an important means of adjusting the mobilization of glycogen as a fuel to match the demand for high-energy phosphate.

AMP is also bound by glycogen phosphorylase-P, which is already nearly fully active. Although the binding has little additional effect on enzymatic activity, it does make the phosphoprotein a poor substrate for phosphorylase phosphatase (reaction 2, Fig. 27–10). AMP therefore stabilizes the active P-form against loss of its phosphate groups by hydrolysis.

Calcium Ions Indirectly Activate Glycogen Phosphorylase. We have seen that muscular contraction is provoked by the release of Ca^{2+} from the sarcoplasmic reticulum (p. 356). The increase in Ca^{2+} concentration at the same time stimulates the breakdown of glycogen to provide the fuel for continued work; it does so by activating glycogen phosphorylase kinase (reaction 3, Fig. 27–10). This will increase the fraction of glycogen phosphorylase that is in the P-form (fully active), until a new steady state concentration is reached at which the rate of hydrolysis of the P-form by the phosphatase again matches the rate of its synthesis.

Adrenaline and Cyclic AMP

Now we must consider still another level of control, of a kind we have not seen before. Efficient use of glycogen as a fuel involves a coordination of events in

*More than one protein phosphatase is present; their separate functions are not known. We shall speak of phosphorylase phosphatase as a single separate entity, but there may be more than one enzyme not necessarily specific for phosphorylase.

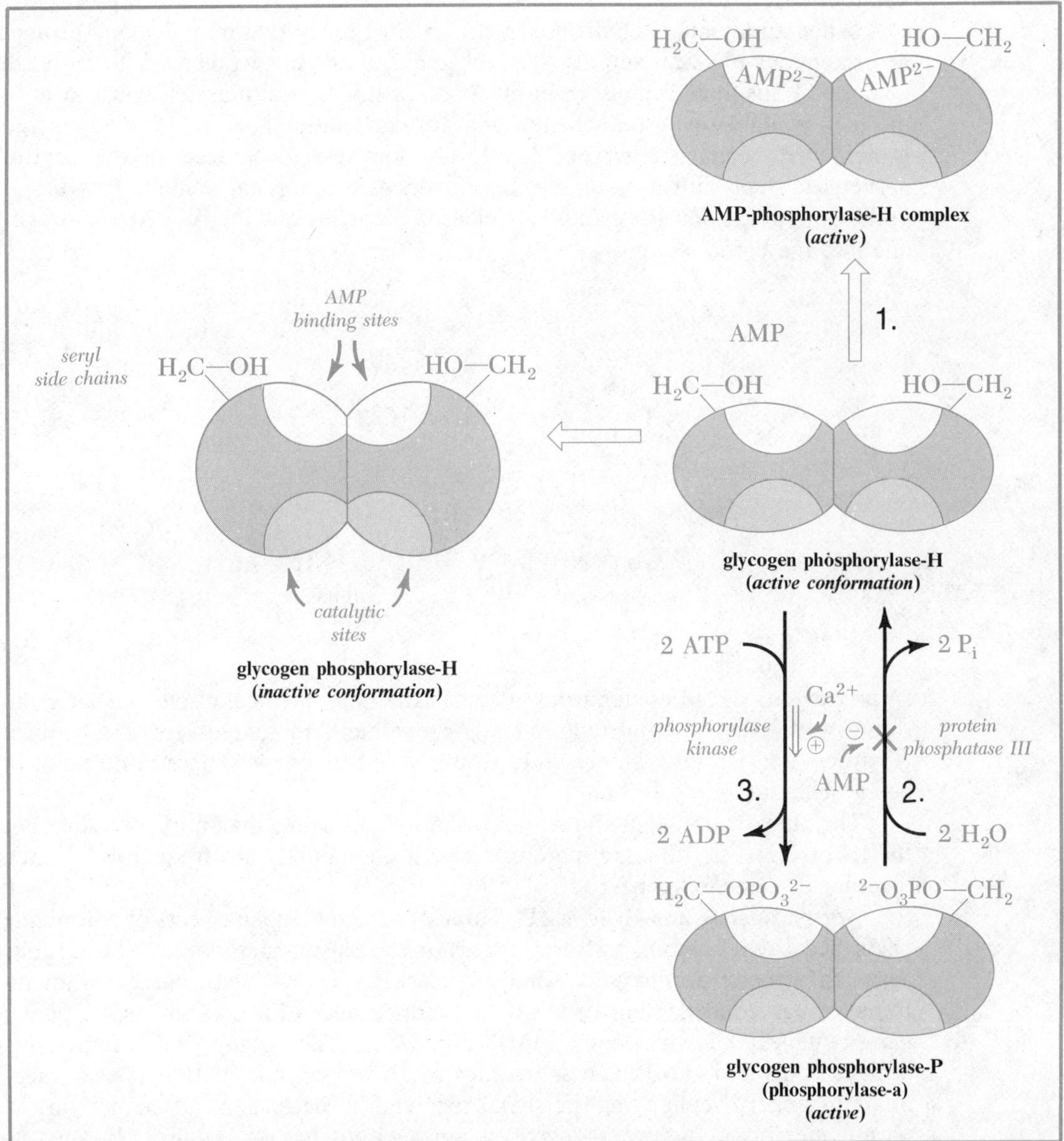

FIGURE 27–10 The glycogen phosphorylase-H of skeletal muscles is a dimer existing mainly in an inactive conformation. The active conformation may be stabilized by combination with AMP (*top right*) or through phosphorylation of a seryl group on each monomeric unit to form phosphorylase-P (*bottom right*). The kinase catalyzing the phosphorylation is activated by Ca^{2+}, and the phosphatase catalyzing the dephosphorylation is inhibited by AMP. Both Ca^{2+} and AMP therefore promote the breakdown of glycogen.

different tissues, especially the liver and skeletal muscles. The coordination is obtained by the use of hormones as signals to the different tissues, and part of the regulation of glycogen metabolism consists of responses to the appearance of particular hormones in the circulation. Adrenaline is such a hormone, and one of its important effects is to increase the formation of glycogen phosphorylase-P in striated muscles.

When an animal is confronted with an emergency that may demand prompt and strenuous physical activity–flee or fight, it is advantageous for it to mobilize glucose-1-phosphate rapidly from glycogen in its skeletal muscles so as to have the fuel available even before the signal for contraction is given. The advantage is achieved because emergencies of this sort are recognized in the central nervous system with a resultant stimulation of the adrenal medulla (the tissue in the core of the adrenal glands) to release adrenaline and the related noradrenaline into the blood stream:

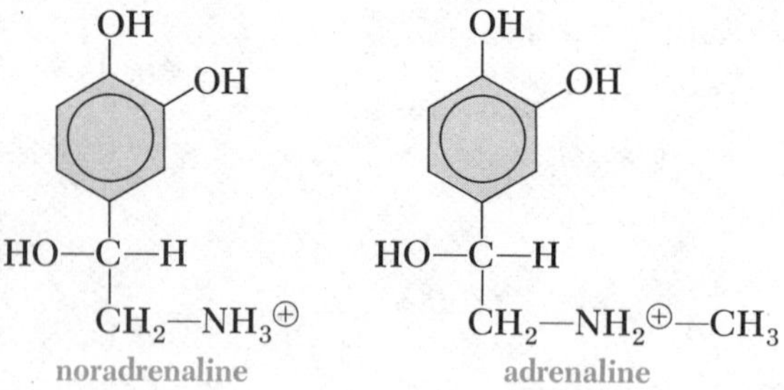

OH OH HO—C—H $CH_2—NH_3^{\oplus}$
noradrenaline

OH OH HO—C—H $CH_2—NH_2^{\oplus}—CH_3$
adrenaline

(Adrenaline is called epinephrine in some American medical circles because the former term used to be a trade name. Adrenaline is the name used in Chemical Abstracts and is more widely understood, so let us use it. There is no point in being deliberately mysterious.)

The human adrenal medulla releases more adrenaline than noradrenaline, but the reverse is true in some mammals. Both compounds are frequently lumped together as catecholamines.

Adenyl Cyclase and Cylic AMP. Many of the metabolic effects of adrenaline occur because it combines with receptors on the plasma membrane of target cells so as to activate an enzyme, adenyl cyclase. Adenyl cyclase catalyzes an intramolecular condensation of ATP to produce **adenosine 3′,5′-monophosphate,** more familiarly known as cyclic AMP (Fig. 27–11). (This compound is sometimes designated as 3′,5′-AMP, and sometimes as cAMP.) Cyclic AMP acts as a "second messenger," conveying the signal received in the form of adrenaline on the plasma membrane to the responsive enzymes within the cell. It does so by activating a particular protein kinase in muscle fibers, and this is perhaps its only effect in other eukaryotic cells in which it serves as a mediator of endocrine responses.

Activation of cAMP-Sensitive Protein Kinase. The important changes in the activity of the cyclic AMP-sensitive protein kinase occur by changing the association of its two pairs of polypeptide chains. One kind of chain has catalytic activity in monomeric form (C-peptide), and the other pair serves as a regulator dimer (R-peptides). The combination of the two pairs in a tetramer has no enzymatic activity, so the regulatory polypeptide acts as an inhibitor. However, the R-peptides will also bind cyclic AMP; when they do, the conformation changes so that they no longer will bind with the catalytic polypeptide. In other words, cyclic

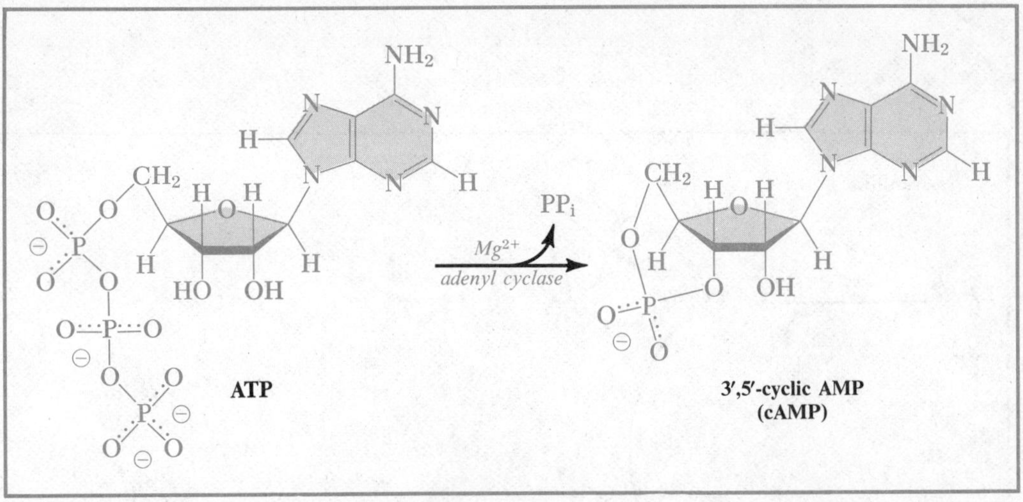

FIGURE 27–11 **The formation of cyclic AMP from ATP.**

AMP and the C-peptide are antagonistic for binding with the R-peptide (Fig. 27–12). When cyclic AMP concentration rises, it forms more of the complex with R-peptide, thereby leaving more active C-peptide free in the solution. When the cyclic AMP concentration falls, more R-peptide will be available to inhibit the C-peptide by combining with it. The proteins kinase acts on many different enzymes, but we are now concerned with its direct effects on glycogen metabolism.

Effect of Protein Kinase on Glycogen Phosphorylase Kinase. What we are dealing with here is a cascade sequence in which an enzyme activated by cyclic AMP acts on enzyme that acts on an enzyme, and the cyclic AMP is the product of still another enzyme activated by adrenaline. Let us summarize the activating events here and in Figure 27–12.

1. A small amount of adrenaline stimulates adenyl cyclase to produce larger amounts of cyclic AMP.

2. The cyclic AMP removes an inhibiting polypeptide from a protein kinase, liberating the active C-polypeptide.

3. The C-polypeptide acts on glycogen phosphorylase kinase, which also exists in free (H-) and phosphorylated (P-) forms. The phosphorylated form is more active than the free enzyme.

4. The newly activated glycogen phosphorylase kinase causes the phosphorylation of glycogen phosphorylase-H, which is active only in the presence of AMP, to glycogen phosphorylase-P, which has greater activity, even in the absence of AMP.

5. The more active glycogen phosphorylase catalyzes the formation of orders of magnitude more glucose 1-phosphate from glycogen.

Since there is a magnification by successive enzyme action, a minute amount of adrenaline has a very large effect, so large that 2 micromoles (364 μg) is a potent dose for humans. The normal blood concentration of adrenaline is in the nanomolar range. It has been estimated that a rise in the concentration of cyclic AMP to 1 micromole per kg of muscle causes the formation of 25,000 times as much glucose 1-phosphate per minute.

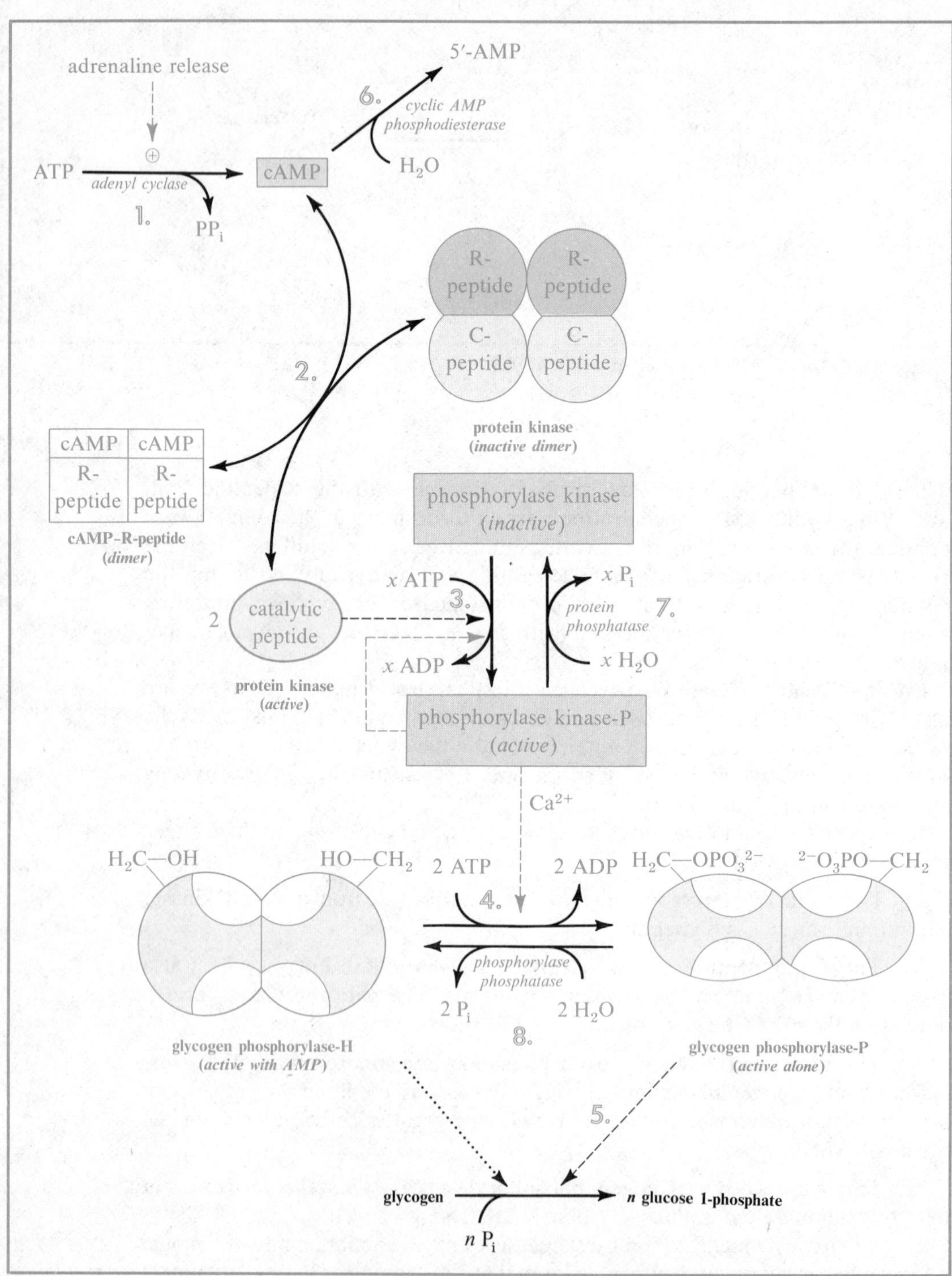

FIGURE 27–12 *See legend on opposite page.*

FIGURE 27–12 **The triggering of glycogen breakdown by adrenaline. Reactions are numbered as discussed in the text. A protein kinase dimer has regulatory subunits with a high affinity for cyclic AMP. In the absence of cyclic AMP, they combine with the catalytic subunits, making them inactive (*top center*). When cyclic AMP is formed in response to an alarm signal (*top left*), the regulatory subunit preferentially combines with it (*center left*) and changes conformation to a form that releases the two catalytic subunits as active monomers.**

The active catalytic subunit catalyzes the phosphorylation of inactive phosphorylase kinase (*center*), making it active. The fully phosphorylated form of this enzyme also will catalyze its own phosphorylation. Active phosphorylase kinase then catalyzes the phosphorylation of glycogen phosphorylase-H, converting it to the active phosphorylase-P (*lower right*). Phosphorylase-H will catalyze the breakdown of glycogen in the presence of AMP, but phosphorylase-P will do so without AMP (*bottom*).

The activating phosphate groups are removed by the action of protein phosphatases.

Reversal of Adrenaline Action. When the stimulus disappears and adenyl cyclase is no longer activated, the effects disappear as a result of the action of phosphatases (steps 6,7, and 8 in Fig. 27–12). A cyclic phosphodiesterase catalyzes the hydrolysis of cyclic AMP to ordinary 5′-AMP. (This enzyme is inhibited by **caffeine** and related stimulatory purines, resulting in an increased concentration of cyclic AMP. Coffee appeals to us for this reason.)

Removal of cyclic AMP releases the regulatory peptide for inhibition of protein kinase. The phosphorylated forms of glycogen phosphorylase and glycogen phosphorylase kinase are then formed at a slower rate than they are attacked by phosphatases, and they are converted to their free forms.

Specificity of Adenyl Cyclase Activation. It is important to realize that the formation of cyclic AMP is not a specific indication of the presence of adrenaline. A cell may have receptors for other hormones that activate adenyl cyclase, and the effects of a change in cyclic AMP concentration will be the same, regardless of the hormone that caused the change.

The specificity of the effect of a hormone acting through cyclic AMP arises from the specificity of the receptor sites in the target cells. In the case at hand, striated muscle fibers have receptors sites with a high affinity for catecholamines, and binding at these sites causes an activation of adenyl cyclase with the production of an increased concentration of cyclic AMP within the fibers. Adrenaline therefore acts as a hormone stimulating cycle AMP formation in skeletal muscles, while it will not do so in other cells lacking receptors for the compound, even though such cells may have adenyl cyclase in their plasma membranes (Fig. 27–13).

Nature of Phosphorylase Kinase. The regulation of phosphorylase kinase hinges on its construction as a tetramer of three different polypeptides $(\alpha\beta\gamma)_4$. The β chains contain the catalytic sites, while the α chains bind the β chains in an

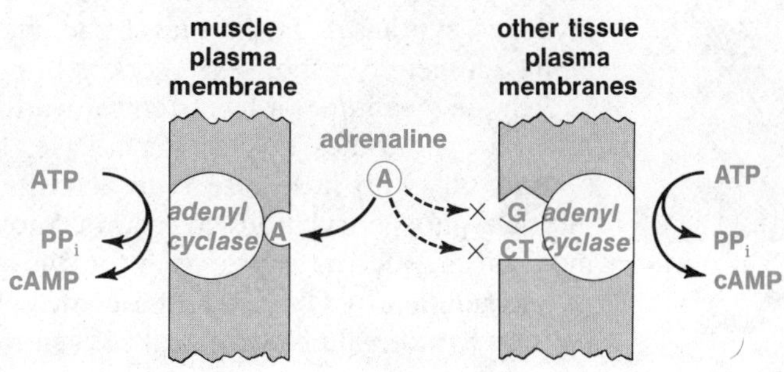

FIGURE 27–13

The plasma membrane of muscles has a site that specifically binds adrenaline (A) to activate adenyl cyclase. The plasma membranes of other tissues have sites that fit other hormone, such as glucagon (G) or corticotropin (CT), and some have no sites that fit adrenaline.

inactive configuration. However, phosphorylation of the β chains causes a shift in conformation that enables them to become active. (The binding of Ca^{2+} will also cause the β chains to assume an active conformation. Two Ca^{2+} are bound per $\alpha\beta\gamma$, perhaps one on each β and each γ subunit.)

According to one laboratory, after the β chains are phosphorylated, the α chains also become phosphorylated, but at a slower rate, and other kinases may participate. In this view, the extra phosphorylation is a device for regulating inactivation of the enzyme. The presence of phosphate groups on the α chains makes the β chain phosphate groups susceptible to hydrolysis by a protein phosphatase, which makes the enzyme inactive! (This phosphatase is believed to be identical to phosphorylase phosphatase and is designated protein phosphatase III by this laboratory. The α chain phosphates are then removed by another phosphatase.)

The point of this back-and-forth sequence is to provide means for restoring the status quo, but not until the rapid phosphorylation of β chains generates enough active phosphorylase kinase to make glycogen phosphorylase fully active; and the slower phosphorylation of the α chains then causes loss of kinase activity to enable the system to relax.

The function of the γ chains in phosphorylase kinase is unknown. They resemble actin and tend to polymerize, which has led to the suggestion that they may be necessary for proper association with other proteins in the glycogen particle.

Other Effects of Adrenaline. Now that we have firmly in mind the effects of adrenaline mediated via cyclic AMP, perhaps it is safe to admit that the hormone also activates phosphorylase in other ways. There is more than one kind of receptor for adrenaline. These receptors are characterized as alpha or beta, depending upon the nature of drugs that will mimic the effects of adrenaline at the receptor site (p. 710). In muscles, adenyl cyclase is activated by combination with beta receptors. The effects of combination with alpha receptors are not characterized completely, but they include a release of Ca^{2+}, which in itself will activate phosphorylase kinase.

REGULATION OF GLYCOGEN SYNTHASE IN MUSCLE

Glycogen synthase also exists with active and inactive conformations in muscles, and it is the inactive form that is stabilized in this case by phosphorylation. However, the story is not so simple because each of the identical subunits in glycogen synthase can be phosphorylated at two, and perhaps several, sites *in vivo,* and only one of these phosphorylations is very effective in making the enzyme inactive. A protein kinase is present in the muscle that phosphorylates this regulatory site in glycogen synthase, but the kinase is not dependent on cyclic AMP for activity, and it is not activated by Ca^{2+} (its only known substrate is glycogen synthase). Is the kinase the site of insulin action or of some effect of alpha adrenergic responses? Work is in progress to test these possibilities.

In the meantime, it has been shown that the cAMP-dependent protein kinase also phosphorylates glycogen synthase, but rapid phosphorylation occurs at a second site with little effect on activity *in vitro*. What is the purpose of the multiple phosphorylation? We don't know, but it may be some sort of a latch mechanism, such as is proposed for phosphorylase kinase.

Regulation by Glucose 6-Phosphate. Glycogen synthase-P is sensitive to inhibition by normal physiological concentrations of ATP, but glucose 6-phosphate

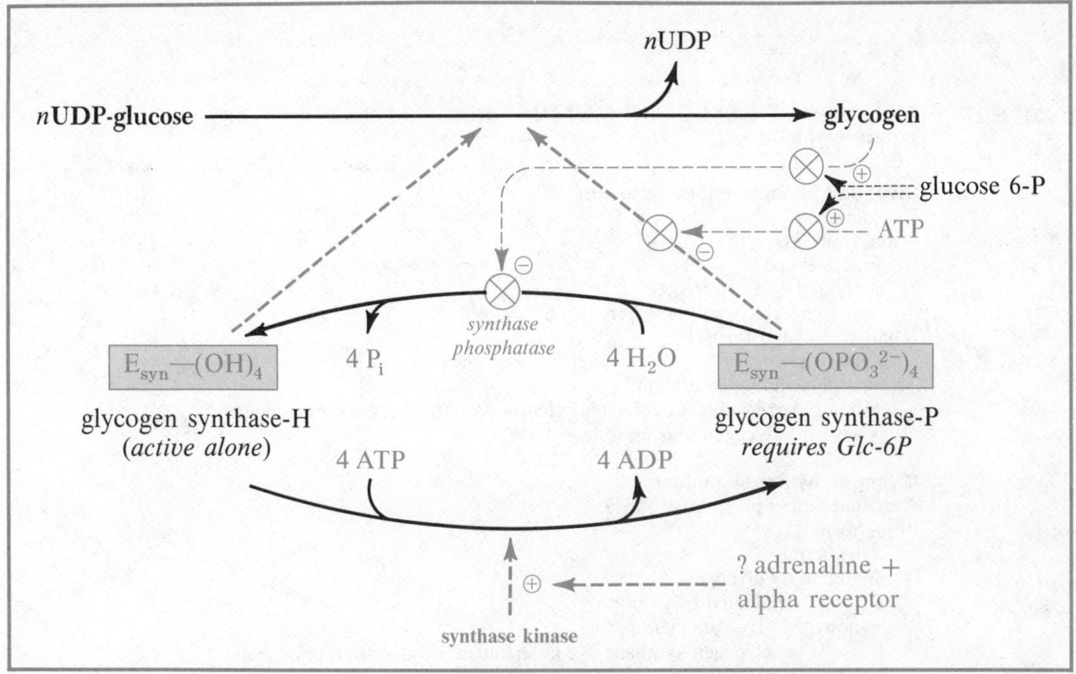

FIGURE 27–14 **Glycogen synthase exists in a non-phosphorylated H-form (*left*) and a phosphorylated P-form (*right*). The H-form easily catalyzes the transfer of glucose residues from UDP-glucose to build glycogen. The P-form is inhibited by ATP; however, this inhibition is blocked by glucose 6-phosphate. Glucose 6-phosphate also prevents the inhibition by glycogen of the protein phosphatase that removes the phosphate groups from glycogen synthase-P. The drawing indicates one molecule with four groups phosphorylated. There is disagreement over the degree of further polymerization of these molecules in the active enzyme. The molecule itself contains two polypeptide subunits (not indicated).**

competes for the binding site and prevents the inhibition (Fig. 27–14). Glycogen synthase-H is active without the presence of glucose 6-phosphate, whereas glycogen synthase-P is active only in the presence of glucose 6-phosphate. It seems likely that this effect would be useful for gathering up glucose 6-phosphate and storing the glucosyl group as glycogen at any time an accumulation of the compound isn't being utilized for energy production. Whether this is true for mammalian muscle is disputed by many, but it so happens that frogs have a glycogen synthase with an absolute requirement for glucose 6-phosphate for activity, even in the free form, and it is, therefore, difficult to doubt the physiological relevance of the effect to the croakers.

Regulation by Glycogen Concentration. Glycogen inhibits the removal of phosphate from glycogen synthase-P by a protein phosphatase. As glycogen accumulates in the muscle, it therefore shuts off its own production by decreasing the conversion of glycogen synthase-P to its more active free form. This effect is the most likely explanation of why animals on high carbohydrate diets don't store massive quantities of glycogen in their muscles.

TROPONIN

Phosphorylation and dephosphorylation may be used to control the activity of the contractile apparatus itself in skeletal muscles. Phosphorylase kinase also

TABLE 27–1 ENZYMES OF GLYCOGEN METABOLISM IN MUSCLE

Glycogen phosphorylase (muscle)
- α_2 dimer
- free form (H)
  - inactive alone
  - activated by bound AMP
  - synonym: glycogen phosphorylase b
- phosphorylated form (P)
  - fully active
  - one phosphate per subunit
  - binding AMP inhibits dephosphorylation by phosphorylase phosphatase*
  - synonym: glycogen phosphorylase a

Glycogen synthase (muscle)
- α_2 dimer (may polymerize further)
- free form (H)
  - fully active
  - inhibited by glycogen
  - activity augmented by insulin
  - synonyms: glycogen synthase a
    glycogen synthase (I); independent of glucose 6-phosphate
- phosphorylated form (P)
  - has 2 phosphate groups on each subunit of dimer
  - inhibited by ATP
  - activated by glucose 6-phosphate
  - synonyms: glycogen synthase b
    glycogen synthase (D); dependent upon glucose 6-phosphate

Glycogen phosphorylase kinase (muscle)
- tetramer of 3 subunits $(\alpha\beta\gamma)_4$
- free form (H)
  - requires Ca^{2+}
  - synonym: phosphorylase kinase b
- β subunit phosphorylated
  - $(\alpha,\beta\text{-P},\gamma)_4$
  - requires Ca^{2+}
  - activity 30–50 times that of H-form
  - synonym: phosphorylase kinase a(β)
- α and β subunits phosphorylated
  - $(\alpha\text{-P},\beta\text{-P},\gamma)_4$
  - target for attack by protein phosphatase*
  - synonym: phosphorylase kinase a($\alpha\beta$)
- α subunit phosphorylated
  - $(\alpha\text{-P},\beta,\gamma)_4$
  - low activity
  - synonym: phosphorylase kinase b (α)

cAMP-dependent protein kinase
- contains catalytic (C) and regulatory (R) subunits
- R subunit has high affinity for cAMP

Glycogen synthase kinase (muscle)
- phosphorylates specific site conveying Glc 6-P dependency
- not activated by cAMP

*According to P. Cohen, the same phosphatase attacks the phosphorylated forms of glycogen phosphorylase, glycogen synthase, and the β subunit of phosphorylase kinase. He labels this enzyme protein phosphatase III.

causes a phosphorylation of the inhibitory component of troponin (p. 354). Since the kinase is activated by Ca^{2+} directly and by adrenaline indirectly, its action may be a physiological device for promoting increased contraction of the actomyosin complex upon stimulation. It has also been known that the phosphate groups are removed from troponin-P by a phosphorylase phosphatase.

GLYCOGEN METABOLISM IN THE LIVER

The turnover of glycogen in the liver is regulated through the same kind of mechanism seen in striated muscles. However, the enzymes involved are isozymes of those seen in the muscle and are regulated by metabolites in different ways. The protein phosphatases acting on glycogen phosphorylase-P and glycogen synthase-P are clearly different enzymes. In addition, the activating receptors for adenyl cyclase on the hepatic cell plasma membrane respond to different hormones. Let us first consider the endocrine effects.

GLUCAGON AND INSULIN

Secretion by Pancreas. The pancreas of higher animals is both a digestive organ that elaborates enzymes to digest food and an endocrine organ that secretes hormones into the blood. The pancreas contains nests of cells known as the **islets of Langerhans,** which play a major role in regulating metabolism by secreting three hormones: insulin, glucagon, and somatostatin. We have already encountered insulin, which consists of two polypeptide chains linked by disulfide bonds, as an example of modification of a polypeptide after its synthesis (p. 142); the cells forming insulin are known as **beta cells.** Glucagon is also a polypeptide but has only some 29 residues in the known examples, and it is secreted by different cells in the islets, the **alpha cells.** (We shall discuss somatostatin, another polypeptide, in Chapter 38.)

These hormones have opposing effects on glucose metabolism. The alpha cells are stimulated to release glucagon when the blood glucose concentration falls, and glucagon activates the adenyl cyclase in liver to cause an increased rate of degradation of glycogen. Most glucose 6-phosphate is produced from liver glycogen to be hydrolyzed by glucose 6-phosphatase so that free glucose is released for transport to the peripheral tissues. Glucagon has no effect on the adenyl cyclase of muscles; it does not control the utilization of glucose, but rather its production by the liver.

Insulin, on the other hand, promotes the storage of glucose in both liver and the skeletal muscles. It does so in the muscles by permitting more rapid transport of glucose, but it also augments the maintenance of glycogen synthase in its active free form in both the muscles and the liver. The mechanism of these effects is unknown. Even though it is more than 50 years since the announcement of the discovery of insulin, we still don't know how the hormone works. However, we can say positively that insulin promotes the storage of glycogen in both the muscles and the liver. The beta cells are stimulated to secrete insulin when the blood glucose concentration rises, and the insulin causes the excess glucose to be

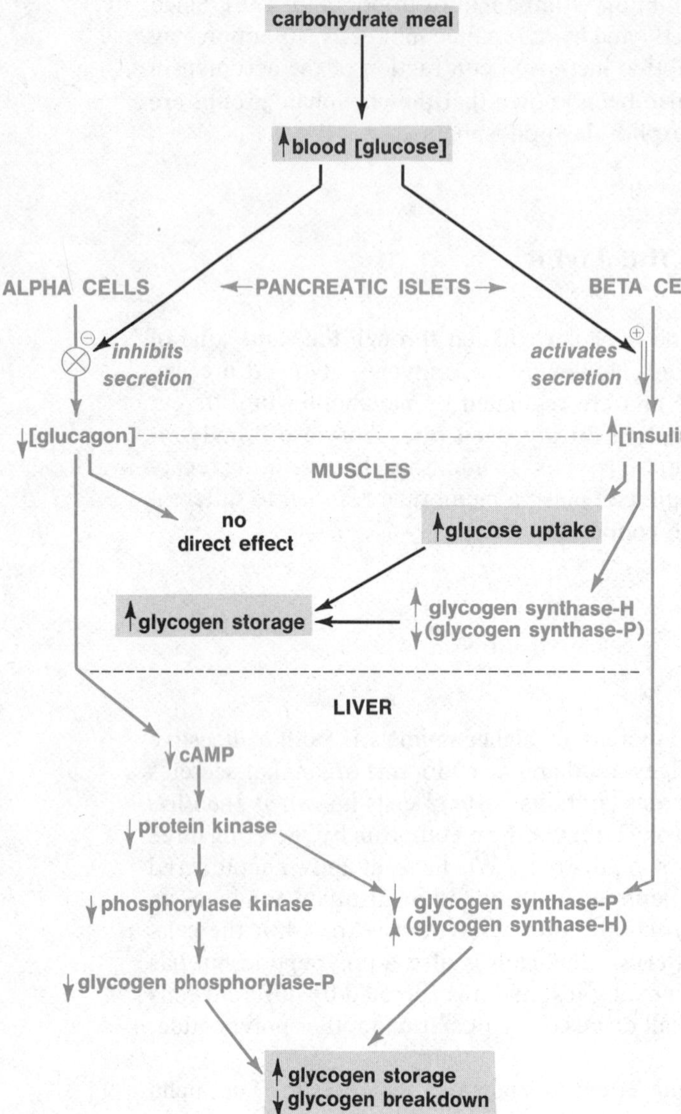

FIGURE 27–15

Changing blood glucose concentrations have opposing effects on the secretion of glucagon by alpha cells (*left*) and insulin by beta cells (*right*) in pancreatic islets. An elevated glucose concentration following a carbohydrate meal inhibits glucagon secretion and stimulates insulin secretion. The lowered glucagon concentration has no direct effect on muscles, but it diminishes the formation of cyclic AMP in the liver with a consequent diminution of glycogen breakdown and increased glycogen storage. The increased insulin concentration promotes uptake and storage of glucose in the muscles, and also promotes glycogen storage in the liver.

laid down as glycogen. In general, insulin secretion rises when the secretion of glucagon falls and vice versa (Fig. 27–15).

Metabolite Regulation

Perhaps the most important modulation of enzyme activity in the liver is due to a combination of glycogen phosphorylase-P with glucose (reaction 1, Fig. 27–16). The combination is enzymatically inactive, which means that elevated glucose concentration will prevent the breakdown of liver glycogen to form more glucose. In addition, the glucose phosphorylase-P complex is a better substrate for glycogen phosphorylase phosphatase than is phosphorylase-P alone, so the presence of glucose accelerates the conversion of glycogen phosphorylase to its less active free form (reaction 2).

Isn't this getting to be a bag of well-regulated worms! The active form of glycogen phosphorylase, the P-form, inhibits the conversion of the inactive form of glycogen synthase, which is also the P-form, to its active H-form state. Glucose enables the formation of the active H-form of glycogen synthase by accelerating the removal of the active P-form of glycogen phosphorylase.

SUMMARY OF GLUCOSE HOMEOSTASIS

Let us pull all of this together by recapitulating the events that occur when the blood glucose concentration rises and falls.

Excess Glucose

We eat and the blood glucose concentration rises. There are three important direct results:

1. The pancreatic islets put out less glucagon and more insulin. Less glucagon means less cyclic AMP formed in the liver, and less active protein kinase. With less active protein kinase, there is less formation of active glycogen phosphorylase (reaction 4, Fig. 27–16) and of inactive glycogen synthase (reaction 5). More insulin also results in more active glycogen synthase in both the liver and the muscles and more rapid transport of glucose into the muscles.

2. Glucose combines with glycogen phosphorylase in the liver. This causes increased loss of active glycogen phosphorylase and also promotes the formation of active glycogen synthase (reaction 3).

3. Increased blood glucose concentration results in the net formation of glucose 6-phosphate in the liver, since the liver glucokinase is not saturated with substrate and therefore can respond to changes in blood glucose concentration.

All of these things taken together mean an elevation in blood glucose concentration is itself a sufficient signal to cause further storage of glucose as glycogen in both the liver and muscles. We mentioned only the increased glycogen synthase activity in the muscles, but don't forget that this will cause increased removal of glucose 6-phosphate: glucose 6-P $\leftrightarrow$ glucose 1-P $\rightarrow$ UDP-glucose $\rightarrow$ glycogen. Since glucose 6-phosphate is an inhibitor of its own formation by the hexokinase reaction (p. 461), its increased removal to form glycogen will also tend to promote increased uptake of glucose in the muscle.

The deposition of glycogen in the muscles will proceed only until a characteristic level is reached even if elevated concentrations of glucose persist. However, the liver will continue to store more glycogen, although at an increasingly slower rate. We shall see in the following chapters that more and more glucose is converted to fat for storage as the glycogen reserves begin to be filled.

Deprivation of Glucose

The blood concentration falls as more time elapses from the preceding meal. Here we have the opposite effects:

1. The pancreatic islets put out more glucagon and less insulin. More glucagon means more cyclic AMP formed in the liver and a more active protein kinase. The protein kinase causes the conversion of glycogen synthase to its inactive phosphorylated form and the conversion of glycogen phosphorylase to its active phosphorylated form.

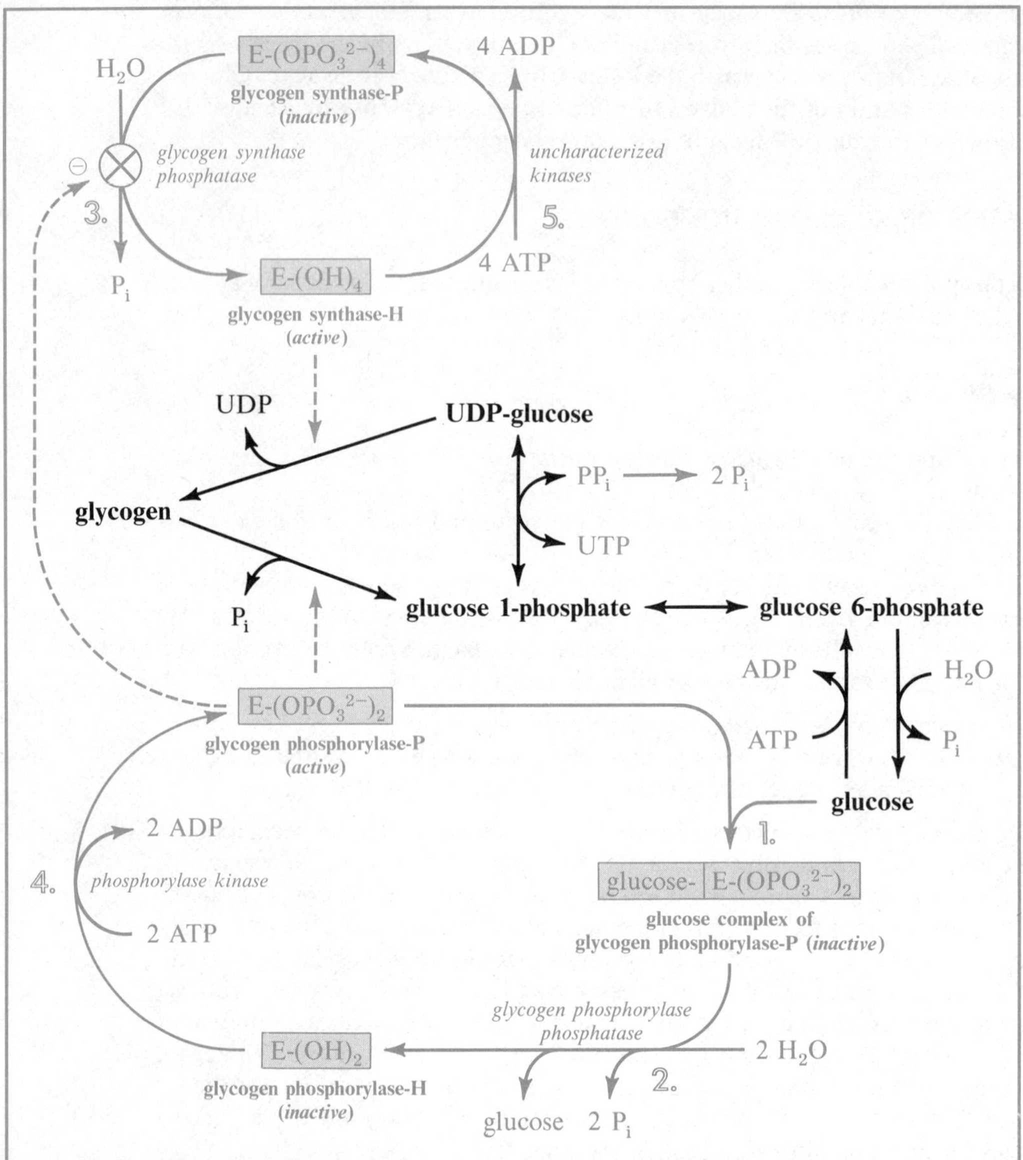

FIGURE 27–16 **Important features of the regulation of glycogen metabolism in the liver hinge on a dual function of glycogen phosphorylase-P. Numbered reactions are discussed in the text.**

2. The fall in the amount of glucose slows the hydrolysis of glycogen phosphorylase-P in the liver because phosphorylase-P without glucose is a poorer substrate for phosphorylase-P phosphatase than is the glucose complex of the phosphoprotein. The resultant higher concentration of glycogen phosphorylase-P also inhibits the conversion of inactive glycogen synthase-P to its active free form.

3. The low concentration of glucose results in a low rate of the liver glucokinase reaction so that the liver adds glucose to the blood, rather than removing it.

What are the results? The changes in enzyme activity in the liver stop the formation of glycogen and accelerate its degradation. The resultant rise in the concentration of glucose 6-phosphate will accelerate the hydrolysis of this compound to form free glucose, which will diffuse into the blood to prevent a further drop in concentration. The maintenance of the blood glucose concentration will enable the brain to continue using this fuel at its normal rate, and the supply for this purpose is conserved by the simultaneous slowing of glucose uptake and of glycogen formation in the muscles, owing to the declining insulin levels.

THE EFFICIENCY OF GLYCOGEN STORAGE

What is the price the organism pays for storage of excess glucose as glycogen? We can show it is only 3 per cent of the potential for generating ATP. Let us put the question in this way: How much ATP will be generated per glucosyl group that disappears during combustion if the glucose has been stored as glycogen, compared to the amount generated when glucose is used directly without storage?

It requires two high-energy phosphates to store one glucose residue as glycogen, according to the balance:

$$\begin{aligned}
\text{glucose} + \text{ATP} &\longrightarrow \text{glucose 6-phosphate} + \text{ADP} \\
\text{glucose 6-phosphate} &\longrightarrow \text{glucose 1-phosphate} \\
\text{glucose 1-phosphate} + \text{UTP} &\longrightarrow \text{UDP-glucose} + \text{PP}_i \\
\text{UDP-glucose} + \text{glycogen} &\longrightarrow \text{glucosyl-glycogen}^* + \text{UDP} \\
\text{PP}_i + \text{H}_2\text{O} &\longrightarrow 2\ \text{P}_i \\
\text{UDP} + \text{ATP} &\longrightarrow \text{UTP} + \text{ADP}
\end{aligned}$$

$$\text{SUM: glucose} + 2\ \text{ATP} + \text{glycogen} + \text{H}_2\text{O} \longrightarrow \text{glucosyl-glycogen} + 2\ (\text{ADP} + \text{P}_i)$$

Where is the high-energy phosphate to be obtained? Assuming that glucose is the only available fuel, part of the supply must be burned immediately in order to store the remainder, and the complete combustion of one glucose molecule will generate 35.5 ATP in white fast-twitch fibers. From this, we can estimate that 5.3 per cent of the glucose will be burned to store the remaining 94.7 per cent as glycogen:

$$.947 \times 2\ \text{ATP consumed} \cong .053 \times 35.5\ \text{ATP produced}$$

When the stored glycogen is later used, approximately 95 per cent of it will be converted to glucose 1-phosphate; the remaining 5 per cent will be released as free glucose from hydrolysis of the 1,6 branches.†

Suppose that there were 100 millimoles of glucose originally available. 94.7 millimoles were stored as glycogen, while 5.3 millimoles were burned to provide the energy for storage. When the 94.7 stored millimoles are recovered, 5 per cent, or 4.7 millimoles, is released as free glucose, and 90 millimoles are obtained as glucose 1-phosphate. Oxidation of glucose 1-phosphate generates 36.5 ATP; therefore $90 \times 36.5 = 3{,}285$ millimoles of ATP will be formed.

---

*An extended glycogen chain is designated glucosyl glycogen to distinguish the added residues (s) from the primer.

†These figures are based on the assumption that the proportion of 1,4 bonds is higher in the portion of the glycogen molecule degraded than it is in the entire molecule.

Out of the original 100 millimoles of glucose, 4.7 millimoles reappeared as glucose during the breakdown of glycogen; therefore, only 95.3 have disappeared to generate 3,285 millimoles of ATP, for a yield of 34.5 ATP per glucose molecule disappearing. This compares with 35.5 that could be obtained by direct oxidation of glucose. The cost of glycogen storage is therefore only 3 per cent of the potential ATP — a small price for an immediately available supply of rapidly combustible fuel. (This analysis neglects the cost of carrying the extra weight of the stored fuel.)

Efficiency of the Cori Cycle

A similar analysis can be made for the Cori cycle, in which glycogen is broken down in the muscles to lactate, and the lactate is converted to glucose in the liver for return to the muscles, and re-storage as glycogen. It goes like this:

Skeletal muscles: 100 millimoles of glucose residues in glycogen yield 200 millimoles of lactate plus 300 millimoles of ATP.

Liver: 200 millimoles of lactate arrive in the blood from the muscles.

1. 31.3 millimoles are oxidized to CO_2 to generate 548 millimoles of ATP (17.5 millimoles per lactate, utilizing the malate-aspartate electron shuttle).
2. The remaining 168.7 millimoles of lactate are converted to 84.4 millimoles of glucose, utilizing the 548 millimoles of ATP generated in step 1.

Skeletal muscles: 84.4 millimoles of glucose arrive in the blood from the liver.

1. 94.7 per cent of the glucose of 79.9 millimoles can be stored as glycogen by burning the other 5.3 per cent to provide the energy.
2. This replaces all but 20.1 millimoles of the original glucose residues used in the muscle; therefore, the muscle gained the 300 millimoles of ATP used initially for contraction at the expense of 20.1 millimoles of glucose, for a yield of 14.9 ATP per glucose residue consumed.

This value is 40.9 per cent of the 36.5 ATP generated by total oxidation in the muscle of a glucose residue in glycogen.

GENETIC DEFECTS IN GLYCOGEN METABOLISM

A number of humans have been discovered who are deficient in one of the enzyme activities necessary for glycogen metabolism. The combined incidence of these defects is about 1 in 40,000 births. One case might arise per year in Los Angeles. Although the diseases are not of quantitative importance in the catalogue of human ills, their effects have promoted our understanding of normal carbohydrate metabolism.

A defect may occur in the formation of any of the proteins necessary for either the breakdown or the synthesis of glycogen, so that the result could be either a high or a low content of glycogen with a normal structure, or the presence of glycogen with abnormalities in the degree of branching.

Increased accumulation of glycogen (**glycogen storage disease,** or **glycogenosis)** will occur when there is a deficiency of glycogen phosphorylase, or of any component in the phosphorylase activating system, such as adenyl cyclase, protein kinase, or phosphorylase kinase. A classification of the resultant clinical conditions is given in Table 27–2. However, it ought to be remembered that

GLYCOGEN STORAGE DISEASES TABLE 27–2

Deficient Enzyme	Name	Type	Site of Deficiency	Incidence
glucose 6-phosphatase	von Gierke's disease	I	liver, kidneys	1:200,000
lysosomal α-glucosidase	Pompe's disease	II	all tissues	1:200,000
amylo-1,6-glucosidase	limit dextrinosis	III	all tissues	1:200,000
branching enzyme	amylopectinosis	IV	all tissues	very low
glycogen phosphorylase	McArdle's disease	V	skeletal muscle	low
glycogen phosphorylase	Hers' disease	VI	liver	?
phosphofructokinase		VII	skeletal muscles, erythrocytes	very low
adenyl cyclase(?)		VIII	brain, liver	very low
phosphorylase kinase		IX	liver, other tissues (not muscle)	1:100,000
cAMP-sensitive protein kinase		X	liver, muscles	very low

genetic errors can affect the formation of a protein in various ways; the rate of synthesis of normal protein may be changed, or mutations of coding may affect amino acid composition at any point in the molecule. The result is that there may be many types of deficiency of a given enzyme, some more severe in their consequences than others, so the rigid classification of the table does not adequately convey the actual spectrum of diseases.

A genetic defect may affect all tissues with an active glycogen metabolism. However, in some cases, a defective gene will disturb glycogen metabolism in only a few tissues because normal genes for different isozymes are still being expressed in other organs. Thus, defects in glycogen phosphorylase may be peculiar to the liver or to the skeletal muscles, because different isozymes are made in these organs.

Two decades ago it was believed that glycogen is synthesized by glycogen phosphorylase, because its reaction is readily reversed in the test tube. A young man was discovered who was weak and who developed severe pain in his muscles after modest exercise, and he was shown to have high concentrations of glycogen and an absence of glycogen phosphorylase in his muscles. This proved that he was not synthesizing glycogen by the phosphorylase reaction and provided strong supporting evidence for the then newly-discovered UDP-glucose pathway. (The condition is known as **McArdle's disease,** after its discoverer. A psychiatrist engagingly confesses [Ann. Inter. Med., *62*: 412 (1965)] that he thought McArdle's patient was displaying classic hysteria owing to an unhappy childhood. We are not told how much the phosphorylase deficiency may have contributed to the unhappy childhood.)

McArdle's disease tells us something else. Impairment of glycogen utilization in skeletal muscles is not fatal. The impairment is very real; there isn't any unsuspected route for utilizing the polysaccharide, and this is easily demonstrated by shutting off the blood supply to an arm with a tourniquet. Clenching of the fist causes a sharp rise in lactate concentration in a normal individual ending in painful tetany, but there is no rise in a patient with McArdle's disease. However, the importance of carbohydrate metabolism for full efficiency is shown by the weakness of the patient and is supported by the uncommon occurrence of the genetic lesion in the population — this is a mutation that is eliminated rapidly.

This view of glycogen metabolism as a convenience for full efficiency of muscles rather than as an absolute necessity is reinforced by the discovery of individuals with a deficiency of phosphofructokinase in skeletal muscles, who have the same clinical picture found in McArdle's disease. They not only accumulate glycogen because of an inability to use it; they can't even use glucose taken up from the blood stream directly, and they still survive.

Accumulation of normal glycogen in the liver causes massive enlargement of the organ, so much so that it may occupy a large fraction of the abdomen in affected children. This in itself causes surprisingly little difficulty; infants without glycogen phosphorylase in the liver grow to be adults. However, if the failure is due to an absence of glucose-6-phosphatase, which prevents the use of stored glycogen to maintain the blood glucose concentration between meals, the consequences without treatment are grave. The blood glucose concentration falls, while the lactate concentration rises. The brain is usually damaged, perhaps owing in part to the absence of glucose as a fuel and in part to the lactic acidosis. The treatment is repeated feeding of carbohydrate at two- to three-hour intervals night and day. If the infant can be brought through the first four years without intellectual impairment, his chances for gradual adjustment to a less heroic feeding schedule are good.

Defects in phosphorylase kinase are interesting because two patterns of inheritance are known. One is an autosomal recessive, and the other is an X-linked recessive. If we had only this information, we could infer that the kinase contains at least two polypeptide chains coded by genes in different chromosomes, and we have already seen that the kinase contains some three different chains.

One of the most damaging storage diseases results from accumulation of glycogen by lysosomes. The enzyme affected is an α-glucosidase that attacks either 1,4 or 1,6 linkages and is not involved in the major routes of glycogen metabolism. It probably is a device for removing glycogen trapped during the normal scavenging function of lysosomes, which become filled to the bursting point when the enzyme is missing. All tissues are affected, but damage to the heart is usually the immediate cause of death, which occurs in infancy.

FURTHER READING

Fischer, E. H., et al.: (1976) *Concerted Regulation of Glycogen Metabolism and Muscle Contraction.* p. 137 *in* L. M. G. Heilmeyer, Jr., J. C. Ruegg, and T. Wieland: *Molecular Basis of Motility.* Springer-Verlag.

Hers, H. G.: *The Control of Glycogen Metabolism in the Liver.* Annu. Rev. Biochem., *45*: 167.

Cohen, P., D. B. Rylatt, and G. A. Nimmo: (1977) *The Hormonal Control of Glycogen Metabolism: The Amino Acid Sequence at the Phosphorylation Site of Protein Phosphatase Inhibitor-1.* FEBS Letters, *76*: 182. The introduction to this paper summarizes events in the activation and inactivation of phosphorylase kinase.

Huijing, F.: (1975) *Glycogen Metabolism and Glycogen Storage Diseases.* Physiol. Rev., *55*: 609.

Larner, J.: (1976) *Mechanisms of Regulation of Glycogen Synthesis and Degradation.* Circ. Res., *38*(Suppl. 1): 2.

Villar-Palasi, C., J. Larner, and L. C. Shen: (1971) *Glycogen Metabolism and the Mechanism of Action of Cyclic AMP.* Ann. N. Y. Acad. Sci., *185*: 74.

Stalmans, W.: (1976) *The Role of the Liver in Homeostasis of Blood Glucose.* Curr. Top. Cell Regul., *11*: 51. Includes discussion of regulation of glycogen metabolism.

Mahler, R. F.: (1976) *Disorders of Glycogen Metabolism.* Clin. Endocrinol. Metabol., *5*: 579.

Greene, H. C., et al.: (1976) *Continuous Nocturnal Intragastric Feeding for Management of Type I Glycogen Storage Disease.* N. Engl. J. Med., *294*: 423.

(The action of cyclic AMP is discussed more fully in Chapter 38, and further references are cited).

28 STORAGE OF FATS

NATURE AND DISTRIBUTION OF FAT STORES

Fat is a fuel that is laid down for long-term storage. Glycogen and starch are fuels for short-term storage or for the maintenance of organisms in the presence of limited amounts of oxygen. The human epitomizes this dual storage; the ordinary adult has only enough glycogen to maintain activity for one day or less, but he can live from his fat for nearly a month. A human being is built for a daily routine in which he oxidizes glucose residues for energy immediately after meals, while rebuilding glycogen reserves and converting any excess glucose to fatty acids. As the time of the last meal recedes, and the glycogen supply again becomes depleted, more and more of his energy is obtained by oxidizing fatty acids previously stored as triglycerides. Even the overnight fast is sufficient to cause the amount of oxygen used for the oxidation of fatty acids from fats to be twice that used for the oxidation of glucose from glycogen.

The great advantage of fat as a stored fuel is that it is light in weight, and the appearance of organisms with large fat stores evidently coincided with the development of the ability to move over relatively long distances without an intake of food. Salmon and ducks are alike in building up large stores of fat before they begin their long migrations, but vertebrates of more fixed domicile, along with many insects, also can store fat for less dramatic exertion.

Location of Fat. Most people have a general idea of the character of animal fat and give it an equally diffuse anatomical role. Let us take a moment to discuss the sites of deposition so that we can more fully appreciate the biochemistry of these tissues. Fat deposition began to be important with the evolution of vertebrates, and the liver was the initial site of deposition. Modern sharks frequently have massive livers containing cells loaded with triglycerides. With the appearance of bony fish, fat began to be deposited to a greater extent in and around the muscle fibers, creating the oily flesh we see in salmon and sardines. Insects went on another pathway and created a multi-purpose organ with many of the functions of the vertebrate liver, but which contains so much fat that it is known as the **fat body**. The advanced vertebrates, beginning with some fish, developed a discrete **adipose tissue** by modifying the same kind of cells from which the blood cells are derived. These **adipocytes** contain globules of triglyceride that may constitute 90 per cent or more of the mass of the cell. Adipose tissue is especially prevalent in subcutaneous tissues, around deep blood vessels, and in the abdominal cavity.

Adipose tissue appears relatively formless and is difficult to handle experimentally because of its high content of fat, but is well-organized with an active metabolism appropriate for its important function as an internal larder. Adipose tissue can become the largest in the body, comprising half or more of the total mass of some individuals. Humans can become tubs of lard. Such people are objects of humor, disdain, or concern in our society, but in societies subject to famine they may be happily living on their own fat while burying the last of their formerly trim companions.

Plants also store fat, especially in the tissues surrounding the embryos in seeds. The light weight of this kind of a fuel no doubt aids in the dispersal of small seeds, but fat is also the predominant stored fuel in large seeds, where its hydrophobic character may be of primary importance in protecting the embryo until time for development.

THE ORIGIN OF FATTY ACIDS

The stored fatty acids originate mainly from the diet or by synthesis from glucose, the proportion depending upon the relative amount of fats or glucose ingested. Worldwide, carbohydrates represent the largest source of chemical potential in the human diet, but those with a high fat intake, such as a typical American, may convert little of their glucose to fatty acids for long-term storage. We do most of our conversion of glucose to fatty acids in the liver, and then store the products in our adipose tissue, but the rats and mice commonly used for laboratory studies synthesize fatty acids in both liver and adipose tissue.

Overall Mechanism

The first step in making fatty acids from glucose is to oxidize the glucose to acetyl coenzyme A. This is done in exactly the same way as it is when the acetyl groups are destined to be burned in the citric acid cycle (Chapter 25). That is, glucose is oxidized to pyruvate along the Embden-Meyerhof pathway in the cytosol, and the pyruvate is then oxidized to acetyl coenzyme A by the pyruvate dehydrogenase complex in mitochondria. The differences between glucose oxidation and conversion of glucose to fat come after the formation of acetyl coenzyme A. A surfeit of glucose is accompanied by regulatory events that cause acetyl groups to be diverted to fat synthesis in the cytosol, rather than being used for combustion or for the formation of 3-oxybutyrates in the mitochondria.

Fatty acids are synthesized from acetyl coenzyme A by a process that is similar to a reversal of fatty acid oxidation. That is, the chain is made longer by successive additions of acetyl groups, each addition being followed by reduction to an increasingly longer fatty acid residue. We shall now consider the nature of these two processes — chain lengthening and reduction to fatty acids.

The Enzymes

Before discussing the mechanism of fatty acid synthesis, let us examine the enzymes involved.

Acetyl-CoA Carboxylase. Rebuilding a fatty acid chain from acetyl groups requires expenditure of energy. The energy for carbon-to-carbon condensation is supplied by a process of carboxylation and decarboxylation of acetyl groups in the cytosol. (This process resembles the carboxylation and decarboxylation of pyruvate that is used to drive the synthesis of glucose from lactate (p. 473). Each two-carbon unit that is added to the growing fatty acid chain is supplied as malonyl coenzyme A, which is obtained by carboxylating acetyl coenzyme A:

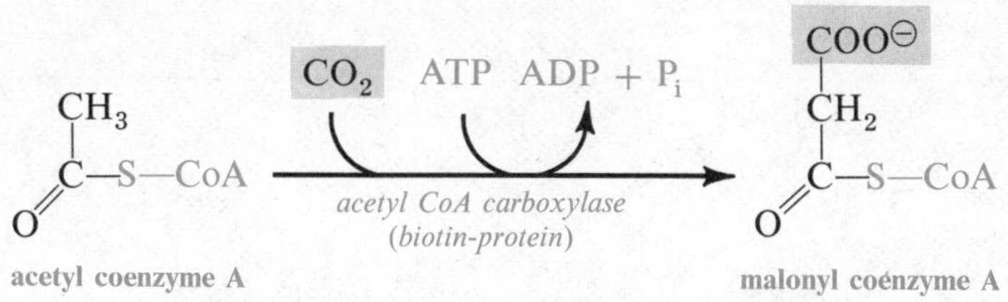

$$CH_3-C(=O)-S-CoA \xrightarrow[\text{acetyl CoA carboxylase (biotin-protein)}]{CO_2 \quad ATP \quad ADP + P_i} {}^{\ominus}OOC-CH_2-C(=O)-S-CoA$$

acetyl coenzyme A → malonyl coenzyme A

Acetyl-CoA carboxylase contains biotin to act as a CO_2 carrier in the same way as it does in pyruvate carboxylase. Malonyl coenzyme A is used to make only fatty acids; therefore, its formation has been made the rate-controlling step in fatty acid synthesis.

Fatty acid synthase, also occurring in the cytosol, is an especially interesting enzyme.* At least seven different catalytic activities are required to convert malonyl groups into fatty acids and all are contained in one molecule (M.W. = 410,000 in human liver). It is difficult to separate these activities as distinct proteins. Some believe that the entire complex is composed of only two differing polypeptide chains, each containing more than one kind of catalytic site.

Others believe that the chains are partially hydrolyzed to separate at least some of the active sites into distinct subunits within the enzyme complex. Bacteria do contain separate proteins to catalyze the individual reactions of fatty acid synthesis, including the two stages of malonyl coenzyme A formation. This lucky circumstance made it much easier to determine the nature of the entire sequence, since each reaction could be studied separately. Demonstration that the eukaryotic enzyme complex operated in a similar fashion then followed.

The fatty acid synthase complex contains two all-important sulfhydryl groups. One of these is relatively fixed in position because it is on a cysteine residue. The other can apparently swing across the seven different catalytic sites because it is located in a residue of **phosphopantetheine** (Fig. 28–1). Phosphopantetheine is the same structure found in coenzyme A, except that it is bound to a

*The enzyme is frequently named fatty acid synthetase, but its action does not fit the definition of a synthetase.

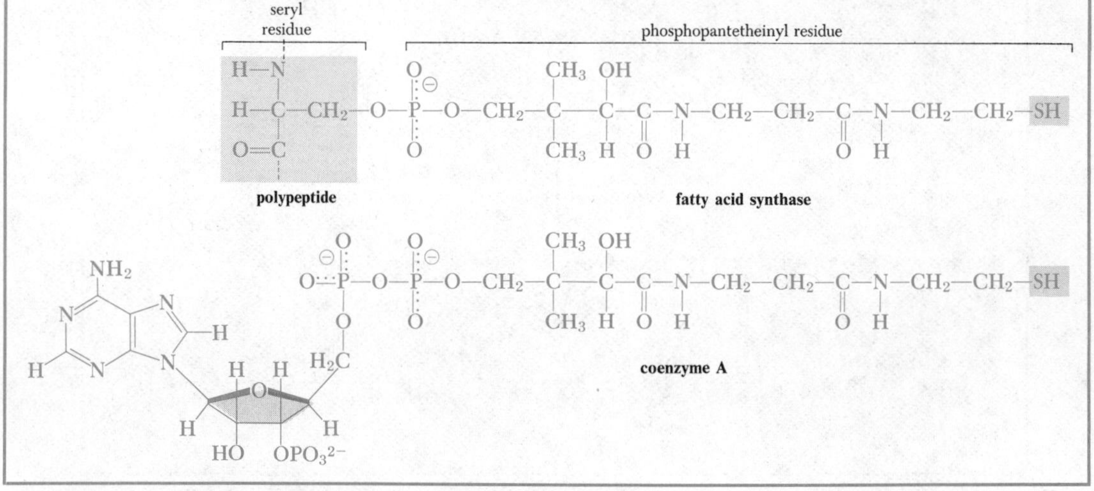

seryl residue — phosphopantetheinyl residue

polypeptide — fatty acid synthase

coenzyme A

FIGURE 28–1 **A residue of phosphopantetheine is found in both fatty acid synthase and in coenzyme A. It is bound to the synthase by formation of a phosphate ester with a serine residue.**

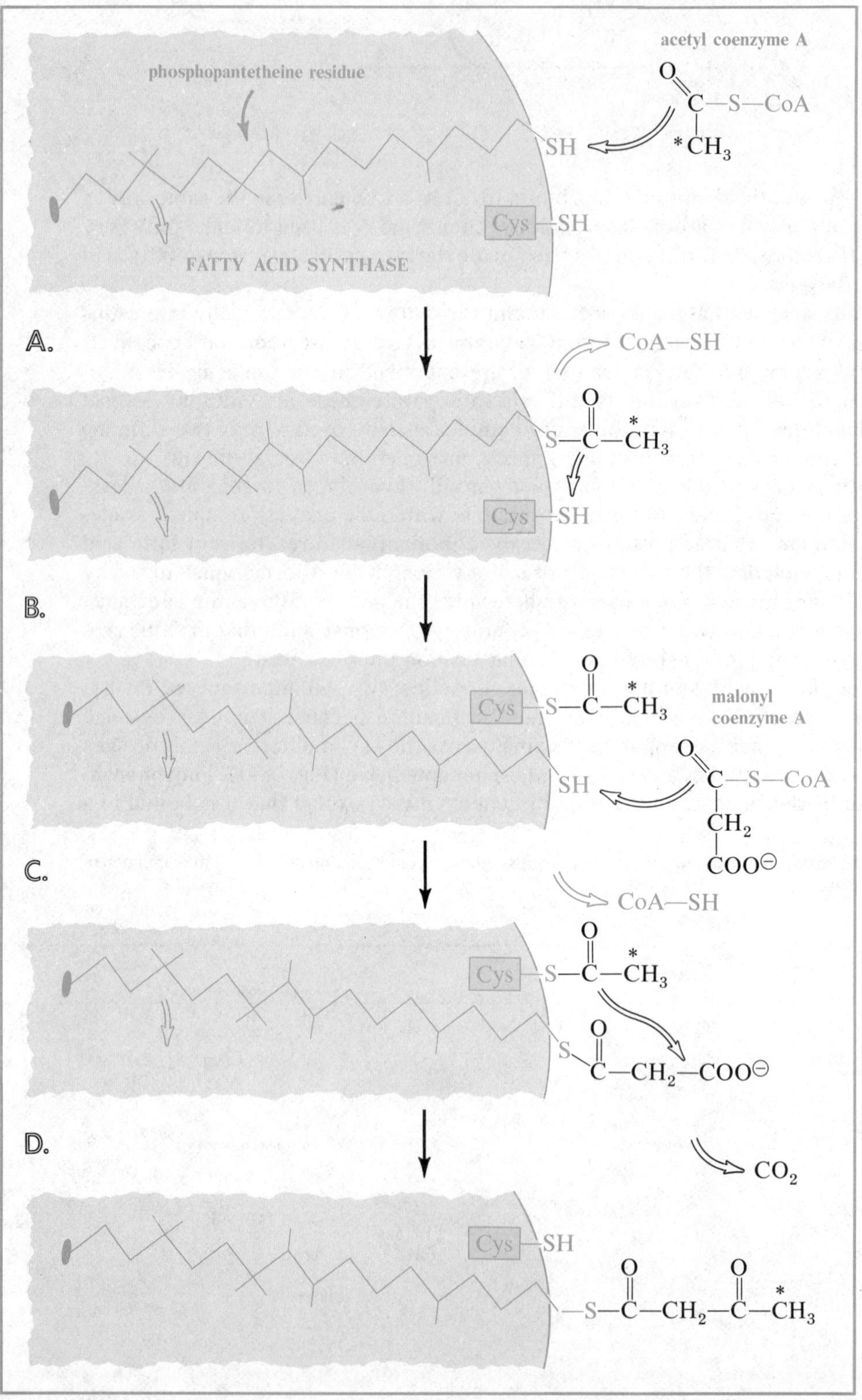

FIGURE 28–2 *See legend on opposite page.*

serine residue in one of the synthase polypeptide chains. The phosphopantetheinyl group in the synthase has the same acyl-carrying function during synthesis that it does in coenzyme A during fatty acid oxidation. The important difference is that the group is covalently bound to the fatty acid synthase, whereas coenzyme A is in solution and is free to migrate from one enzyme to another.

Assembly of the Carbon Chain

The synthesis of fatty acids begins with the binding of an acetyl or butyryl group to fatty acid synthase (Fig. 28–2*A*). This short-chain acyl group is the tail upon which a long-chain fatty acid will be constructed toward the head by addition of two-carbon units. It is obtained by transfer from acetyl or butyryl coenzyme A to the phosphopantetheine residue, catalyzed by one of the active sites on the fatty acid synthase. (The transfer involves intermediate bonding as an ester with a serine residue, not shown.) The acyl group is then further transferred to a cysteine residue in the synthase, which acts as a temporary parking place (Fig. 28–2*B*). (Acetyl groups are used as a starter by mammalian adipose tissue, bacteria, and yeasts, whereas butyryl groups are used by mammalian liver and mammary gland. The origin of butyryl coenzyme A is described below.)

The next step is a transfer of a malonyl group from coenzyme A to the now vacant phosphopantetheine residue (Fig. 28–2*C*); this transfer is catalyzed by a second active site on the synthase. The synthase is thus loaded with an acyl group and a malonyl group. The malonyl group is then moved into proximity with the acyl group (initially an acetyl or butyryl group) at another site, which catalyzes condensation of the two groups (Fig. 28–2*D*). The condensation involves transfer of the acyl group (the tail) to the malonyl group (the new head), which is simultaneously decarboxylated. The loss of free energy upon decarboxylation drives the reaction toward completion. The 3-ketoacyl product is two carbon atoms longer than the starting acyl group.

Reduction to an Acyl Group

The next steps in fatty acid synthesis convert the 3-ketoacyl group to the corresponding saturated acyl group by a pair of reductions. The first reduction forms a D-3-hydroxyacyl group (Fig. 28–3*A*). (It is not clear why synthesis involves the D-isomer, when oxidation of fatty acids proceeds through the L-hydroxyacyl coenzyme A compounds. Perhaps metabolic confusion was avoided in primordial organisms lacking intracellular compartmentation.) Water is removed to form the unsaturated enoyl group (Fig. 28–3*B*), which is then reduced to form the saturated acyl group two carbon atoms longer than the original starter (Fig. 28–3*C*). The example shown is the sequence occurring after use of an acetyl group as a starter, with the product being a butyryl group on the enzyme. If a

---

FIGURE 28–2 **Formation of the carbon chains in fatty acids involves the use of sulfhydryl groups on a phosphopantetheine residue and on a cysteine residue. *A.* An acetyl group is transferred from acetyl coenzyme A to the phosphopantetheine residue. *B.* The acetyl group is transferred to the cysteine residue for temporary storage. *C.* A malonyl group is then transferred to the phosphopantetheine from malonyl coenzyme A. *D.* Condensation occurs by displacement of the carboxyl group with the stored acetyl group, forming an acetoacetyl group on the enzyme.**

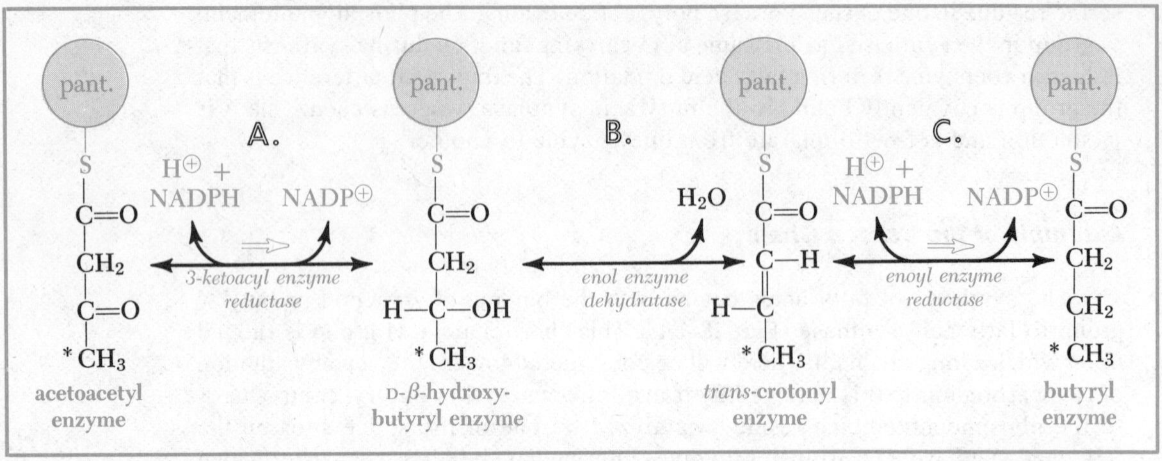

A. $H^{\oplus}$ + NADPH → $NADP^{\oplus}$ (3-ketoacyl enzyme reductase)

B. H_2O (enol enzyme dehydratase)

C. $H^{\oplus}$ + NADPH → $NADP^{\oplus}$ (enoyl enzyme reductase)

acetoacetyl enzyme — D-β-hydroxybutyryl enzyme — *trans*-crotonyl enzyme — butyryl enzyme

FIGURE 28–3 **The reduction of 3-ketoacyl groups, in this case the acetoacetyl group, occurs during fatty acid synthesis while the groups are attached to the phosphopantetheine residue in the enzyme. This residue shifts position so as to expose the substrate group to different catalytic sites in the enzyme complex.**

butyryl group, obtained from butyryl coenzyme A, had been used as a starter, the same sequence would form caproyl coenzyme A (hexanoyl coenzyme A).

Each of the steps in the sequence is catalyzed by a distinct site of fatty acid synthase; the substrate group is moved from one site to the next while bound to the phosphopantetheine residue of the enzyme.

NADPH as an Electron Carrier. The reduction of ketoacyl groups to acyl groups on fatty acid synthase uses electrons from the reduced form of a coenzyme we have not encountered previously:

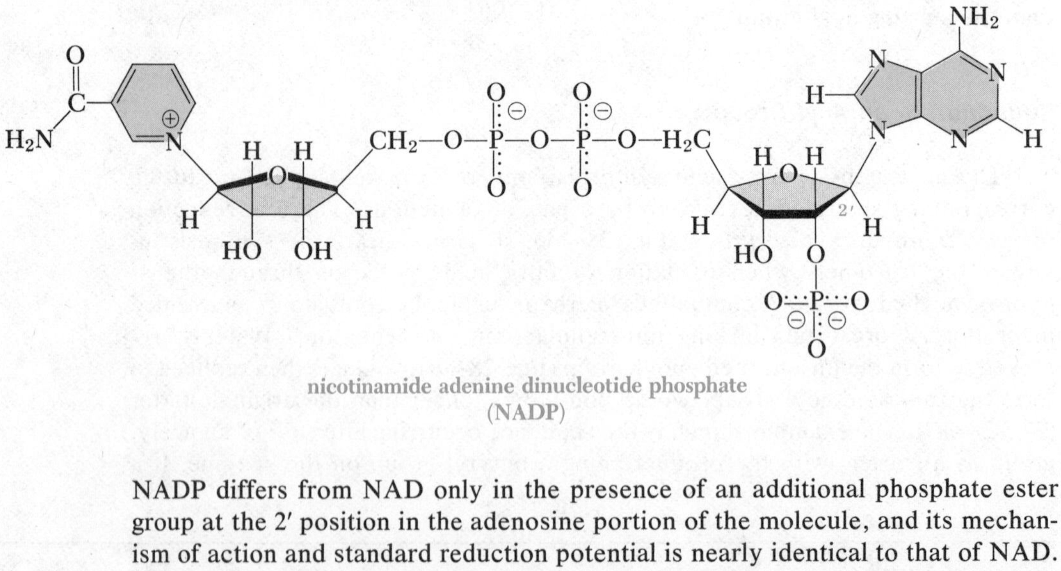

nicotinamide adenine dinucleotide phosphate
(NADP)

NADP differs from NAD only in the presence of an additional phosphate ester group at the 2′ position in the adenosine portion of the molecule, and its mechanism of action and standard reduction potential is nearly identical to that of NAD. It accepts two electrons on the nicotinamide ring to become NADPH. (The older name for NADP is triphosphopyridine nucleotide, abbreviated TPN.)

Why does this coenzyme exist? The reason is that the extra phosphate group enables the construction of enzymes to discriminate between NAD and NADP. NADP is maintained in a highly reduced state compared to NAD. We shall soon see that NADP is involved in only a few energetically favorable oxidations, with equilibrium positions far in favor of formation of NADPH. NAD, on the other hand, is used for reactions in which the equilibrium is less favorable for formation of NADH. We can see the result in the calculated ratios of oxidized and reduced forms in liver cytosol from well-fed rats*:

$$\frac{[\text{NADPH}]}{[\text{NADP}]} = 77 \qquad \frac{[\text{NADH}]}{[\text{NAD}]} = 0.00083$$

The proportion of the reduced form of NADP is 10^5 that seen with NAD! Given these ratios, reduction of other compounds by NADPH will liberate 30,000 more joules of free energy per pair of electrons, the energetic equivalent of driving each reduction by the hydrolysis of ATP. The high concentration of NADPH is used to push synthetic reactions, of which the reduction of acetyl groups to fatty acids is an example.

Chain Growth

The product of the initial sequence of condensation and reduction is an acyl group — butyryl or caproyl — on the phosphopantetheine residue. This acyl group, like the original acyl starter group, is then transferred to the cysteine residue for temporary storage. The phosphopantetheine residue is again loaded with a malonyl group obtained from malonyl coenzyme A, and the entire sequence of condensation and reductions is repeated, except that each intermediate is two carbon atoms longer than it was in the initial sequence. All of this happens again and again, making a growing fatty acid residue in which all but the terminal two or four carbon atoms have passed through malonyl coenzyme A.

Chain Termination. The chain grows on the fatty acid synthase until a 16-carbon palmitoyl residue is produced. The final step is a hydrolysis to release the chain as free palmitate. It is not known how the chain length is limited to 16 carbon atoms; the fatty acid synthase is evidently constructed to expel the fatty acid once it reaches this size. The mammary glands of some animals, such as the cow, synthesize shorter chain fatty acids, but the synthase isolated from these tissues also makes palmitate in the laboratory. The apparent paradox was resolved when those mammaries were discovered to contain a protein that combines with fatty acid synthetase and causes it to hydrolyze shorter chain acyl derivatives. It is as if the additional protein occupies part of a notch for the growing chain in such a way as to activate early hydrolysis.

In any event, the stoichiometry for the formation of palmitate from acetyl coenzyme A, allowing for the intermediate formation of seven moles of malonyl coenzyme A, becomes (neglecting balance of H^+ and H_2O):

(1) 7 acetyl-S-CoA + 7 CO_2 + 7 ATP ⟶
7 malonyl-S-CoA + 7 ADP + 7 P_i

(2) acetyl-S-CoA + 7 malonyl-S-CoA + 14 NADPH ⟶
palmitate + 7 CO_2 + 14 $NADP^{\oplus}$ + 8 CoA-SH

SUM: 8 acetyl-S-CoA + 7 ATP + 14 NADPH ⟶
palmitate + 7 ADP + 7 P_i + 8 CoA-SH + 14 $NADP^{\oplus}$

*Data from H. A. Krebs, and R. W. Veech, (1969) Adv. Enzym. Regul., *7*: 397.

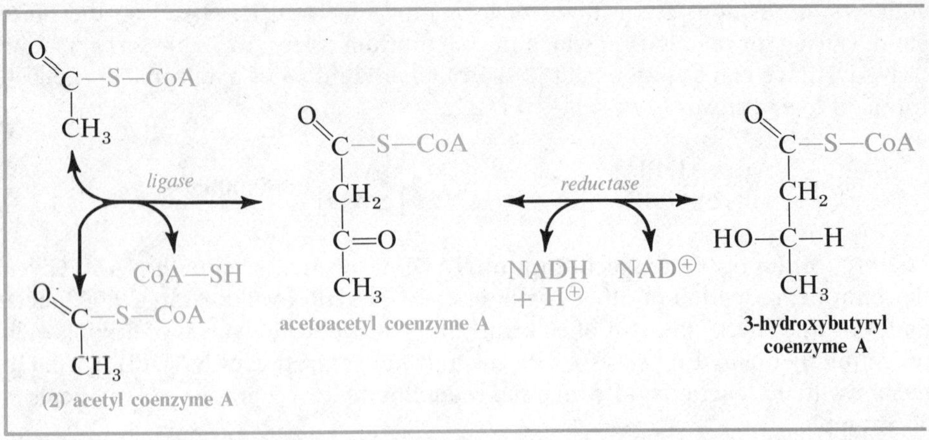

CH_3—C(=O)—S—CoA

ligase

CoA—SH

CH_3—C(=O)—S—CoA

(2) acetyl coenzyme A

CH_3—C(=O)—CH_2—C(=O)—S—CoA

acetoacetyl coenzyme A

reductase

NADH + $H^{\oplus}$ NAD$^{\oplus}$

CH_3—CH(OH)—CH_2—C(=O)—S—CoA

3-hydroxybutyryl coenzyme A

FIGURE 28–4 **Tissues using butyryl coenzyme A as a primer for fatty acid synthesis generate it in the cytosol from acetyl coenzyme A, which is equilibrated with crotonyl coenzyme A by reactions similar to those occurring during fatty acid oxidation. The energetically less favorable reduction of crotonyl coenzyme A differs by using NADPH as the electron donor.**

Illustration continued on opposite page.

Butyryl Primers. The liver and mammary glands can convert acetyl coenzyme A to butyryl coenzyme A in the soluble cytoplasm (Fig. 28–4). Three of the steps are like the final reactions of fatty acid oxidation in mitochondria; the final step is a reduction of crotonyl coenzyme A to butyryl coenzyme A by NADPH, which fatty acid synthase in some way can catalyze.

When butyryl groups are used as primer by fatty acid synthase, only six malonyl groups need be consumed to make the final palmitate, and the overall reaction in this case is:

$$\text{8 acetyl-S-CoA} + \text{6 ATP} + \text{NADH} + \text{13 NADPH} \longrightarrow \text{palmitate} + \text{6 ADP} + 6\ P_i + \text{8 CoA-SH} + \text{NAD}^{\oplus} + \text{13 NADP}^{\oplus}$$

MODIFICATION OF THE FATTY ACID CHAIN

Both the triglycerides and the membrane lipids of cells contain a variety of fatty acids. Most are created by elongating the palmitoyl carbon chain, and by introducing double bonds. **Oleic acid**, 18:1(9), is the most abundant fatty acid in mammalian fat, and it is formed from palmitate by the addition of two carbons and the creation of one double bond. Animals also contain smaller proportions of some polyunsaturated fatty acids that they cannot synthesize from glucose. These are obtained from the diet, and are in part further modified after absorption. Let us see how changes, both of newly synthesized palmitate and of dietary fatty acids, are made.

CHAIN ELONGATION

Elongation on Endoplasmic Reticulum. Palmitate is converted to palmitoyl coenzyme A, and the chain is then elongated by the addition of one or more acetyl

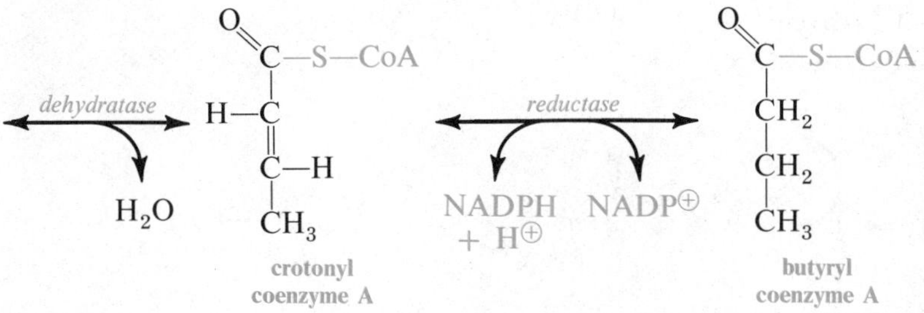

dehydratase
H_2O
crotonyl coenzyme A
reductase
NADPH + $H^{\oplus}$ NADP$^{\oplus}$
butyryl coenzyme A

FIGURE 28–4 *Continued.*

groups from malonyl coenzyme A. The sequence of reactions is exactly the same as we outlined for the action of fatty acid synthase, except that the elongation enzymes are located on the endoplasmic reticulum. Two carbon units can be added to various saturated or unsaturated fatty acids by this system until the total length of the chain reaches 24 carbon atoms. However, the specificity is such that saturated chains are most easily elongated if they contain 16, or fewer, carbon atoms, so the principal saturated product is stearoyl coenzyme A (18:0). Increasing the number of double bonds makes it easier to lengthen the chain, so the residues with 24 carbon atoms are mainly polyunsaturated.

Elongation in Mitochondria. There is also a chain elongation system in mitochondria, which presumably produces the fatty acid residues necessary for the formation of structural lipids in that organelle. The lower [NAD^+]/[NADH] ratio in the mitochondria makes it possible to use some of the enzymes of fatty acid oxidation for chain elongation. One additional enzyme is required, an NADPH-coupled enoyl coenzyme A reductase. This sequence appears to be sufficient to provide the relatively small amounts of acyl residues required.

Chain Desaturation

Plants have a special propensity for introducing *cis*-double bonds at C-9 and then at three-carbon intervals toward the tail (C-12 and C-15), although they can also introduce double bonds toward the head (C-6). Animals, on the other hand, can introduce double bonds at C-9 and toward the head (C-6, C-3), but not toward the tail. Animals therefore cannot make linoleoyl coenzyme A ($\Delta^{9,12}$-octadecadienoyl coenzyme A), but they have no problem in making oleoyl coenzyme A from stearoyl coenzyme A (Fig. 28–5), or in putting an additional double bond toward the carbonyl group to make $\Delta^{6,9}$-octadecadienoyl coenzyme A.

Mixed-Function Oxidases. Double bonds are made by the action of acyl coenzyme A desaturases on the endoplasmic reticulum. These enzymes belong to the class of mixed-function oxidases, which are enzymes that simultaneously oxidize two substrates, one of which frequently is a reduced coenzyme that serves to "activate" molecular oxygen for attack on the second substrate. We shall see

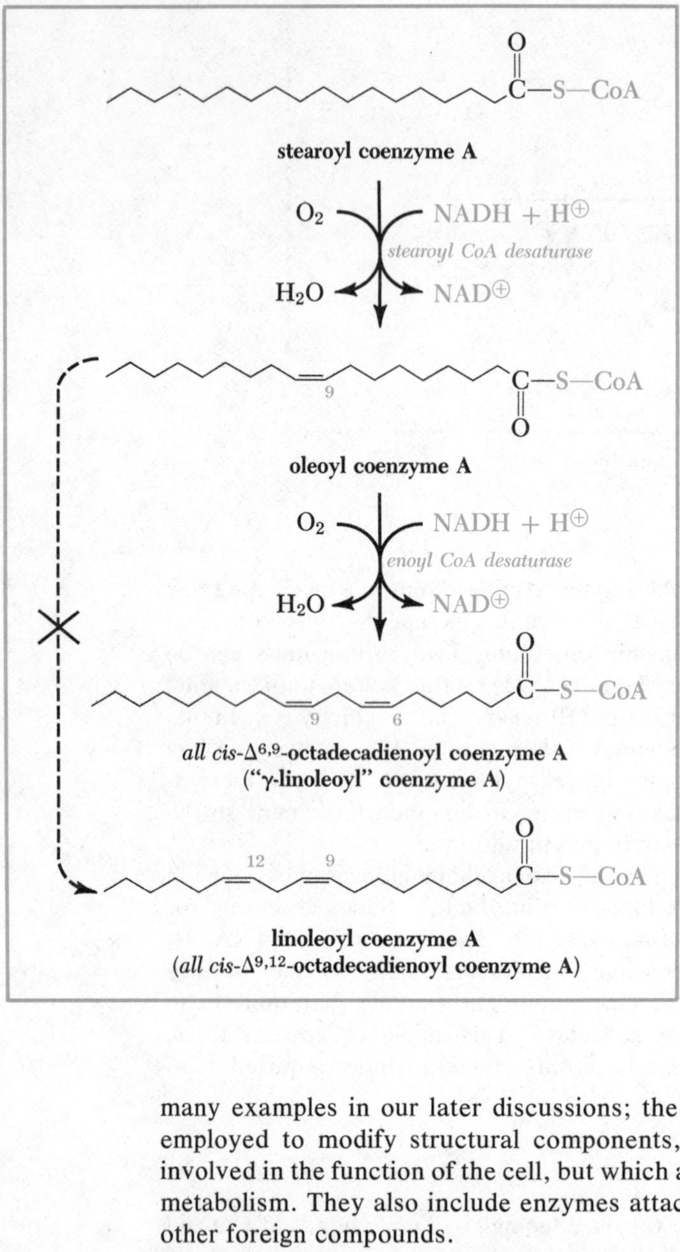

FIGURE 28–5

Animals can introduce double bonds into the coenzyme A esters of palmitic or stearic acids by oxidation at C-9, with subsequent oxidations at 3-carbon intervals toward the carboxyl end. The mammalian enzymes cannot introduce double bonds beyond C-9, and therefore cannot form linoleic acid residues.

many examples in our later discussions; the mixed-function oxidases are often employed to modify structural components, hormones, and other compounds involved in the function of the cell, but which are outside of the mainstream of fuel metabolism. They also include enzymes attacking some carcinogens, drugs, and other foreign compounds.

In general, the endoplasmic reticulum activates oxygen by transfering electrons from NADH or NADPH through a special cytochrome, either **cytochrome b_5** or **cytochrome P-450**. The acyl coenzyme A desaturases employ NADH and cytochrome b_5 in this way (Fig. 28–6). (Cytochrome b_5 is known to have a small hydrophobic domain, which fixes the molecule to the membrane core, and a larger polar domain containing the reacting hemin.) As the figure shows, the oxidase complex contains an uncharacterized component that is also involved in these transfers. Electron transfer in the endoplasmic reticulum is used only for modification of substrates; there is no associated phosphorylation or other means of capturing the released energy.

Families of Fatty Acids. Since animals cannot introduce double bonds beyond the 9-10 position of palmitoyl CoA, forming 16:1(9), any fatty acid with fewer than seven carbon atoms beyond the double bond must have originated by

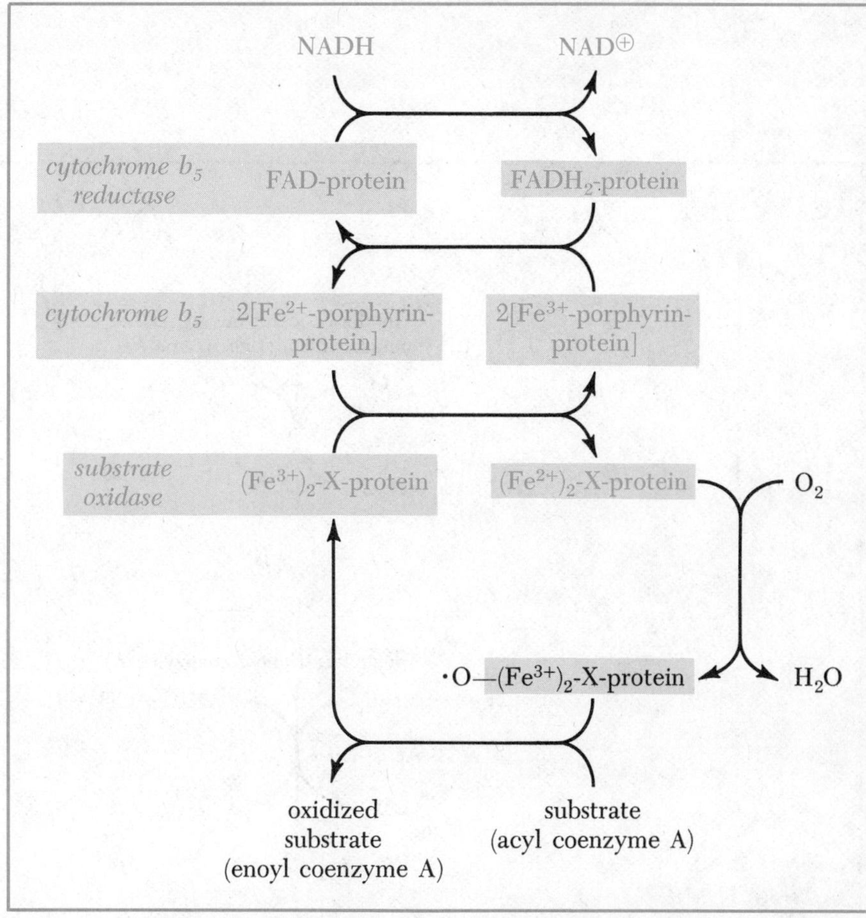

FIGURE 28–6 **The electron transport system of the endoplasmic reticulum used in desaturation of acyl coenzyme A. Molecular oxygen must be partially reduced in order to activate it for attack on the acyl substrate. The electrons for the partial reduction are supplied by NADH and pass through a flavoprotein and a cytochrome before appearing in the substrate oxidase, which is an uncharacterized iron-containing protein. Enzymes that use a reduced substrate or coenzyme to activate oxygen so that it will oxidize a second substrate are said to be mixed function oxidases.**

modifying a dietary fatty acid of plant origin. Thus, dietary linoleate, 18:2(9,12), can be converted to arachidonate, 20:4(5,8,11,14), the precursor of the prostaglandin hormones, but palmitate cannot (Fig. 28–7). The closest the animals can come is to begin with palmitoleoyl coenzyme A, 16:1(9), and successively add carbons and desaturate to create 20:4(4,7,10,13) or to begin with oleoyl coenzyme A and create 20:4(2,5,8,11) (Fig. 28–8). (The pathway from palmitoleoyl CoA also creates residues of vaccenic acid, 18:1(11), which is a common constituent of animal fats, although not as abundant as oleoyl residues, 18:1(9).)

A family of precursor and its products generated in animals through elongation and desaturation can be recognized by subtracting the number designating the last double bond from the total number of carbon atoms. The result is the same within a family. For example, with linoleate, 18:2(9,12), and arachidonate, 20:4(5,8,11,14): $18 - 12 = 6 = 20 - 14$.

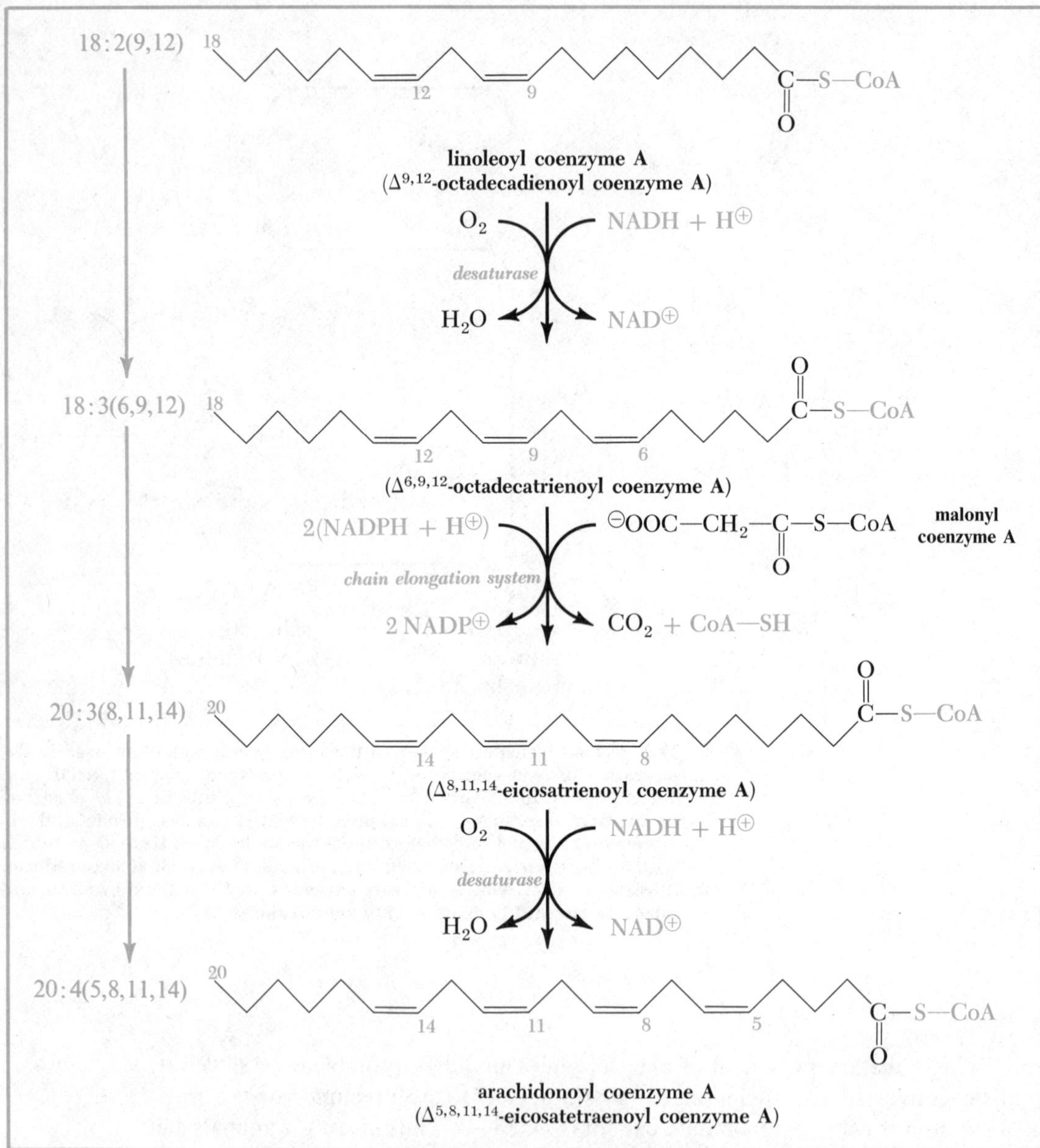

FIGURE 28–7 **The synthesis of arachidonic acid residues from linoleic acid residues involves a combination of chain lengthening and further desaturation toward C-1 in the coenzyme A derivatives. All of the illustrated compounds have a 6-carbon tail beyond the final double bond, a feature that identifies long-chain fatty acids formed from dietary linoleic acid because animals cannot introduce double bonds in that portion of the molecule.**

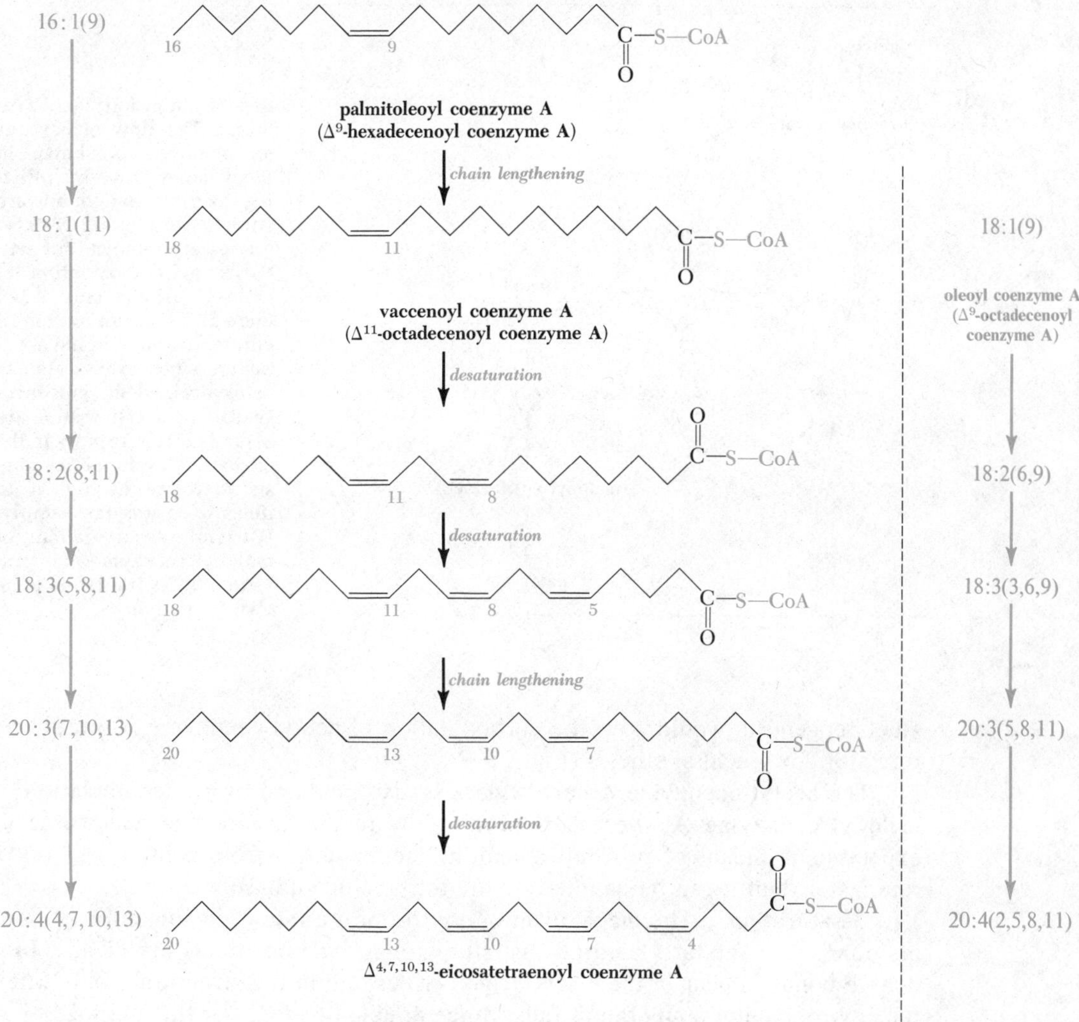

FIGURE 28–8 **When the diet lacks linoleic acid as a precursor for fatty acids, substitutes with 7-carbon tails beyond the last double bond may be synthesized in increased amounts for incorporation into membrane lipids. The scheme for synthesizing fatty acids with 9-carbon tails is also indicated.**

REGULATION OF FATTY ACID SYNTHESIS

Chain Synthesis. The principal regulation over fatty acid synthesis involves acetyl CoA carboxylase, because the conversion of acetyl coenzyme A to malonyl coenzyme A commits the carbon atoms to the formation of fatty acids. This enzyme has a nearly absolute requirement for citrate (or isocitrate) as an activator. The binding of citrate appears to change the conformation so as to bring the biotinyl prosthetic group into effective proximity to the substrate.

The point is that there must be a sufficient supply of both acetyl coenzyme A and of oxaloacetate from which citrate can be formed, if fatty acid synthesis is to proceed, and an accumulation of both of these compounds occurs when there is a surfeit of glucose. A very active citric acid cycle, a lack of ample supply of pyruvate as a precursor of acetyl coenzyme A, or a depletion of oxaloacetate for

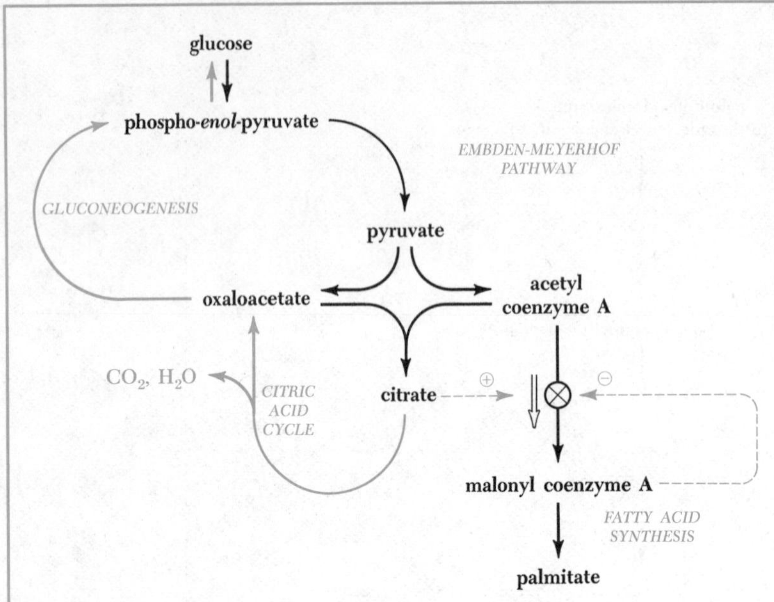

FIGURE 28–9

Regulation of fatty acid synthesis. The flow of carbons for synthesis is shown in *black*; other processes utilizing the same compounds are shown in *blue*. Acetyl groups are removed for synthesis by carboxylation to malonyl groups only when there is an accumulation of citrate. Citrate will not accumulate when oxaloacetate is being drained for gluconeogenesis or acetyl groups are being oxidized rapidly in the citric acid cycle. Fat synthesis therefore is limited to times of excess fuel supply. Wasteful accumulation of malonyl coenzyme A is prevented by its inhibition of its own formation.

gluconeogenesis would stop the carboxylation of acetyl coenzyme A and therefore stop fatty acid synthesis (Fig. 28–9).

The acetyl coenzyme A carboxylase is also inhibited by an accumulation of malonyl coenzyme A, the end-product of its reaction, and it is reasonable to expect some means of preventing malonyl coenzyme A from being formed at a rate faster than it can be handled by the fatty acid synthase.

Desaturation. The mechanism of control of the rate of introduction of double bonds into the fatty acid chains is unknown, but the effects are clear. More double bonds appear in the stored triglycerides and in the structural lipids when the environmental temperature falls. Since double bonds lower the melting point, this provides a mechanism to prevent crystallization of the lipids in the tissue.

Furthermore, polyunsaturated compounds that can be completely synthesized from acetyl coenzyme A — those with a seven- or nine-carbon tail — are not formed if there is an adequate dietary supply of linoleate and linolenate, or some other precursors with only three- or six-carbon saturated tails. These essential fatty acids are preferentially used for the construction of structural lipids when they are available; when they are not, more of the synthetic substitutes will be formed.

ELECTRON SUPPLY IN FATTY ACID SYNTHESIS

The reduction of acetyl groups to fatty acids is accomplished by oxidizing NADPH to NADP. The question at hand is: what supplies the NADPH?

Knowing that NADP is structurally and mechanistically very similar to NAD, and knowing that the [NADPH]/[NADP] is high, we might predict the actual circumstance. NADP is converted to NADPH by being used in specific dehydrogenase reactions that oxidize hydroxyl groups; the reactions differ, however, from many of those involving NAD in that their equilibrium lies far in the direction of reduction of NADP to NADPH.

Malate Dehydrogenases and Acetyl Transfer

A major source of electrons for fatty acid synthesis comes from an oxidation of L-malate in the cytosol by an NADP-specific malate dehydrogenase, which catalyzes the oxidation of L-malate with a simultaneous decarboxylation to pyruvate:

$$\underset{\text{L-malate}}{\mathrm{^{-}OOC{-}CH(OH){-}CH_2{-}COO^{-}}} \xrightarrow[\text{NADP—malate dehydrogenase}]{\mathrm{NADP^{\oplus}}\ \to\ \mathrm{NADPH}} \underset{\text{pyruvate}}{\mathrm{^{-}OOC{-}C(=O){-}CH_3}} + CO_2$$

The cytosol also contains an NAD-coupled malate dehydrogenase, which, like the mitochondrial enzyme, catalyzes an equilibration of malate and oxaloacetate without decarboxylation. The existence of both malate dehydrogenases provides a neat device for transferring electrons from NADH to form NADPH in the cytosol, with the transfer driven by the decarboxylation:

$$\underset{\text{oxaloacetate}}{\mathrm{^{-}OOC{-}C(=O){-}CH_2{-}COO^{-}}} \underset{\mathrm{NADH}\ \rightleftharpoons\ \mathrm{NAD^{\oplus}}}{\longleftrightarrow} \underset{\text{L-malate}}{\mathrm{^{-}OOC{-}CH(OH){-}CH_2{-}COO^{-}}} \xrightarrow[\mathrm{NADP^{\oplus}}\ \to\ \mathrm{NADPH} + CO_2]{} \underset{\text{pyruvate}}{\mathrm{^{-}OOC{-}C(=O){-}CH_3}}$$

SUM: NADH + NADP$^{\oplus}$ + oxaloacetate ⟶ NAD$^{\oplus}$ + NADPH + pyruvate + CO_2

Malate acts only as a transient intermediate in this process. Since the ratio [NADH]/[NAD] is low, energy is required to drive the electron transfer so as to create a high [NADPH]/[NADP] ratio. The energy is provided in this sequence by the decarboxylation, much in the way that a simultaneous decarboxylation pulls the reaction from isocitrate to α-ketoglutarate in the citric acid cycle.

The two malate dehydrogenases reinforce the principle that oxidation of 3-hydroxycarboxylic acids may or may not involve a simultaneous decarboxylation. We pointed out (p. 407) that the oxidative decarboxylation of isocitrate may involve transitory formation of the unstable 3-keto acid, but the oxidation of malate in the citric acid cycle produces a 3-keto acid, oxaloacetate, as the product.

This rather tricky transfer of electrons fits very smoothly into the overall mechanism of fatty acid synthesis, because the conversion of glucose to fatty acids begins with the formation of pyruvate and NADH by the Embden-Meyerhof pathway in the cytosol. Since the malate dehydrogenases provide a means of utilizing NADH in the cytosol to form NADPH, glucose is, therefore, supplying some of the electrons as well as the carbon atoms for fatty acid synthesis.

Transfer of Acetyl Groups. The malate dehydrogenases fit into the synthetic scheme in still another way. We must keep an eye on the requirement for NADPH, but synthesis also requires a transfer of acetyl groups from the mi-

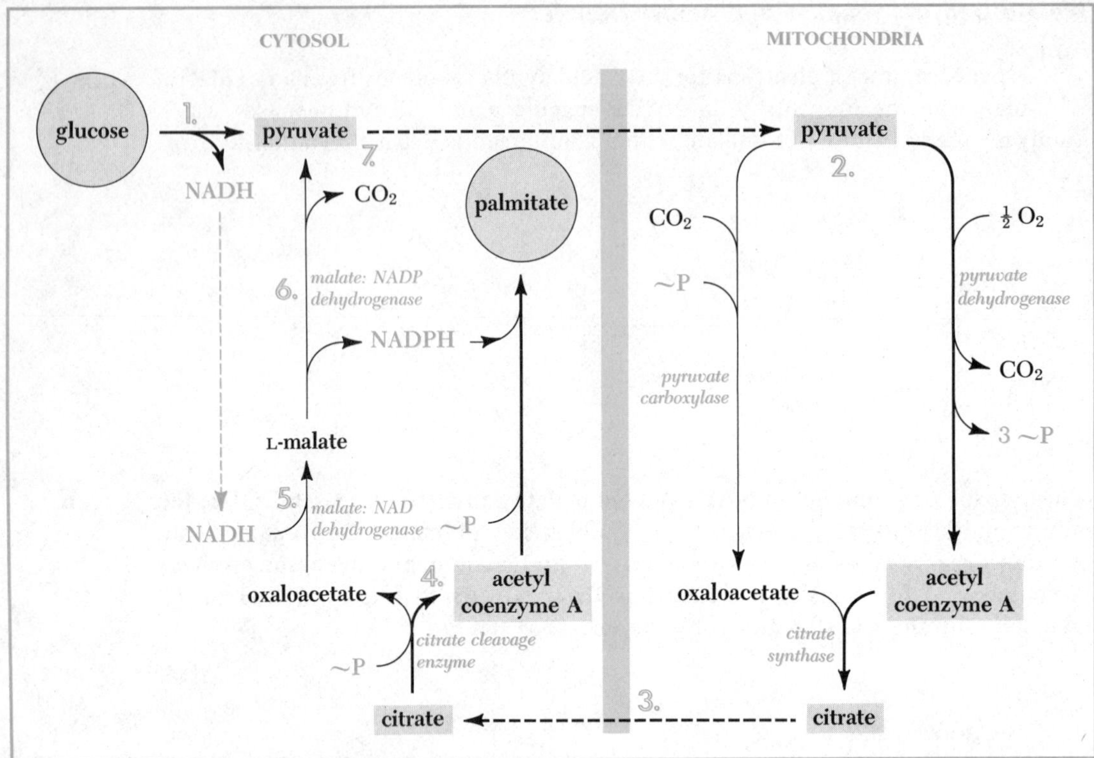

FIGURE 28–10 **Fatty acid synthesis from glucose involves a cyclical flow of pyruvate and citrate across the inner mitochondrial membrane. Pyruvate is generated in the cytosol and oxidized within the mitochondria. The resultant acetyl group is combined with oxaloacetate for transport as citrate into the cytosol, where citrate is cleaved into acetyl coenzyme A used for fatty acid synthesis, and into oxaloacetate. The oxaloacetate is converted to extra pyruvate during a transfer of hydride ions from NADH to NADP. This pyruvate replaces that used to make oxaloacetate in the mitochondria. (Citrate and pyruvate cross the inner mitochondrial membrane by antiport exchanges, which are not shown.)**

tochondria, where they are formed by the oxidation of pyruvate, to the cytosol. This is done by an ingenious scheme in which citrate is formed in the mitochondria and transported to the cytosol. Citrate is then cleaved in the cytosol, supplying acetyl groups for the fatty acid chain, and also providing the oxaloacetate for driving electron transfer via malate. In this scheme, citrate is formed from oxaloacetate and acetyl coenzyme A by the familiar citrate synthase reaction of the citric acid cycle in mitochondria. Instead of being oxidized, it is transported to the cytosol by exchange for P_i or other anions. It is cleaved to form acetyl coenzyme A and oxaloacetate in the cytosol by a separate enzyme, and the cleavage occurs at the expense of ATP.

The Citrate — Pyruvate Cycle (Fig. 28–10). Now comes the clever part of the mechanism. The acetyl coenzyme A released in the cytosol by the cleavage of citrate is used to create fatty acid chains. The oxaloacetate that is also released is used by the malate dehydrogenase route for transfer of electrons from NADH to NADP, and it is converted to pyruvate and CO_2 as a result. The original source of the oxaloacetate was in the mitochondria, where it was used to make citrate. We have seen in our discussion of gluconeogenesis from lactate (p. 473) that oxaloacetate is made in the mitochondria by carboxylation of pyruvate:

$$\underset{\text{pyruvate}}{\begin{array}{c}CH_3\\|\\C{=}O\\|\\COO^{\ominus}\end{array}} + CO_2 + ATP + H_2O \xrightarrow[Mg^{2+},\ Mn^{2+}]{\text{biotin, pyruvate carboxylase}} \underset{\text{oxaloacetate}}{\begin{array}{c}COO^{\ominus}\\|\\CH_2\\|\\C{=}O\\|\\COO^{\ominus}\end{array}} + ADP + P_i$$

Therefore, the pyruvate formed in the cytosol by the sequence citrate→oxaloacetate→pyruvate is transported into mitochondria to replace the pyruvate used in the sequence pyruvate→oxaloacetate→citrate!

The citrate — pyruvate cycle simultaneously moves an acetyl group from mitochondria to the cytosol and shifts a pair of electrons from NADH to NADP. Two molecules of ATP are expended per turn of the cycle, one in carboxylating pyruvate and the other in cleaving citrate.

Integration of Citrate — Pyruvate Cycle and Fatty Acid Synthesis. This is a fascinatingly interwoven set of reactions; let us go through it again, this time in a step-by-step analysis: (The corresponding reactions are illustrated in Fig. 28–10.)

1. An excess supply of glucose to the liver or adipose tissue causes an accelerated formation of pyruvate and NADH in the cytosol by the Embden-Meyerhof pathway.

2. Pyruvate is transported into mitochondria. Part is oxidized to acetyl coenzyme A, and part is carboxylated to form oxaloacetate. Acetyl coenzyme A and oxaloacetate condense to form citrate.

3. Tissues synthesizing fatty acids from glucose have a relatively slow citric acid cycle, and most of the citrate formed is available for exchange into the cytosol. Here we see the first of many examples of the use for other purposes of individual reactions also employed in the citric acid cycle.

4. Citrate is cleaved in the cytosol to form oxaloacetate and acetyl coenzyme A at the expense of ATP. The acetyl coenzyme A is utilized for fatty acid synthesis.

5. The oxaloacetate released from citrate in the cytosol is reduced to malate by the NADH formed in the Embden-Meyerhof pathway.

6. The malate is oxidatively decarboxylated to form NADPH, pyruvate, and CO_2. The NADPH is utilized in forming fatty acids.

7. The pyruvate produced from malate mixes with the pyruvate directly formed by the Embden-Meyerhof pathway. When it moves into mitochondria, it is effectively replacing the pyruvate formerly utilized to make oxaloacetate, so the only pyruvate now missing is that which was oxidized to form acetyl coenzyme A.

The sum of all of the reactions involved shows that the conversion of 8 molecules of triose phosphate to palmitate via 8 molecules of acetyl coenzyme A will also form 8 molecules of NADPH. We have seen that 14 molecules of NADPH are required. Let us now consider how the additional 6 molecules of NADPH are obtained.

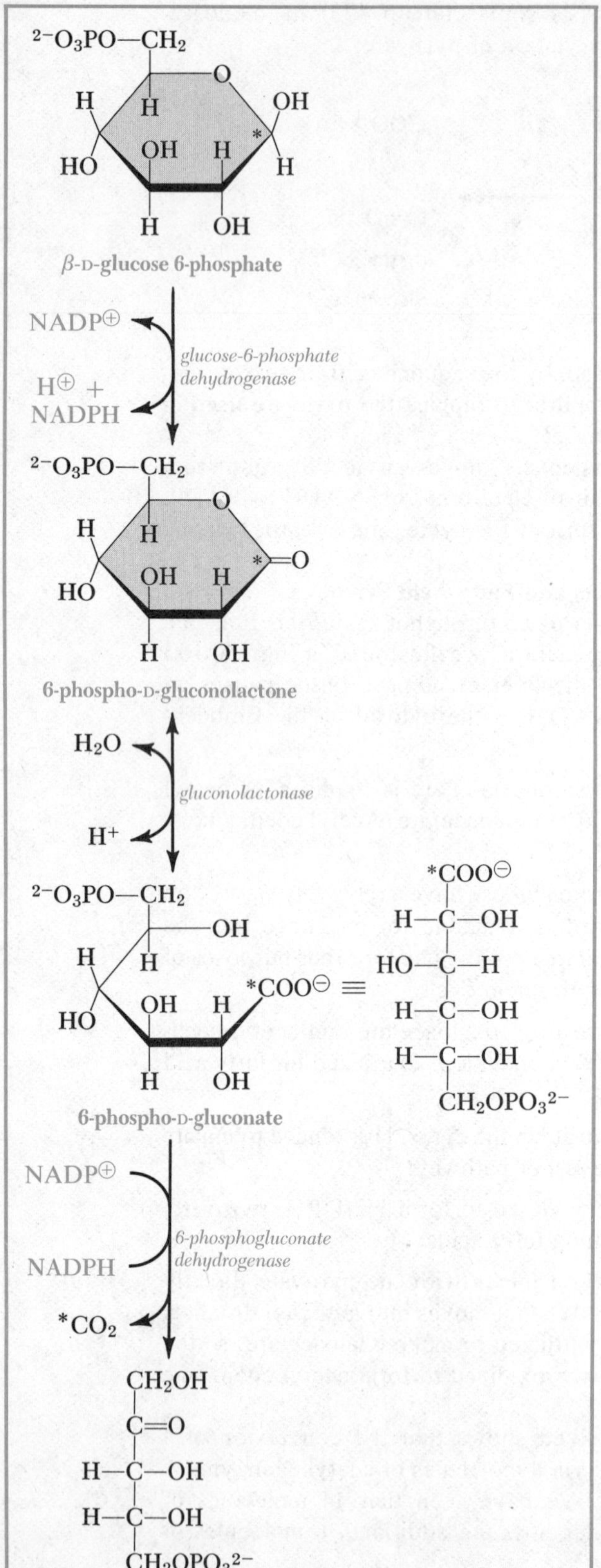

$^{2-}O_3PO{-}CH_2$

H H O OH

HO OH H * H

H OH

β-D-glucose 6-phosphate

$NADP^{\oplus}$

glucose-6-phosphate dehydrogenase

$H^{\oplus}$ + NADPH

$^{2-}O_3PO{-}CH_2$

H H O

HO OH H *=O

H OH

6-phospho-D-gluconolactone

H_2O

gluconolactonase

H^+

$^{2-}O_3PO{-}CH_2$

H H OH

HO OH H $^*COO^{\ominus}$ ≡

H OH

6-phospho-D-gluconate

$^*COO^{\ominus}$

H—C—OH

HO—C—H

H—C—OH

H—C—OH

$CH_2OPO_3^{2-}$

$NADP^{\oplus}$

6-phosphogluconate dehydrogenase

NADPH

$^*CO_2$

CH_2OH

C=O

H—C—OH

H—C—OH

$CH_2OPO_3^{2-}$

D-ribulose 5-phosphate

FIGURE 28–11

The pentose phosphate pathway is a device for generating NADPH by the oxidation of glucose 6-phosphate to ribulose 5-phosphate + CO_2. It is used to supply NADPH for many reactions in addition to fatty acid synthesis.

THE PENTOSE PHOSPHATE PATHWAY

NADPH is also generated in the cytosol of many cells by oxidizing glucose 6-phosphate. The sequence of reactions is quite different from the Embden-Meyerhof pathway we have hitherto emphasized. The route is variously known as the **pentose shunt**, the **hexosemonophosphate pathway**, and other permutations of these words. Only a small fraction of the carbons of glucose pass through this route, even in the liver, but it is a major source of NADPH and also provides pentose phosphates for the synthesis of nucleotides. Some of the reactions are involved in plant photosynthesis. The general idea of the pentose phosphate pathway is very simple — glucose 6-phosphate undergoes two oxidations by NADP, the final one being an oxidative decarboxylation to form a pentose phosphate. These oxidations are sufficiently exergonic to create a high [NADPH]/[NADP] ratio. The remainder of the reactions are concerned with a transformation of the pentose phosphate into triose and hexose phosphates that can be re-used.

Glucose 6-phosphate contains a number of hydroxyl groups and could in principle be the substrate for a variety of dehydrogenases. In fact, only one is of importance in mammalian tissues, and this *glucose-6-phosphate dehydrogenase* catalyzes the oxidation of the β-anomer by NADP (Fig. 28–11). The oxidation may be regarded as a conversion of an alcohol group to a carbonyl group, but the substrate is really a potential aldehyde in the form of its hemiacetal and the product is a potential acid in the form of its lactone, or inner ester. The ester bond of this 6-phosphogluconolactone is hydrolyzed by a gluconolactonase to form **6-phospho-D-gluconate**. (Gluconic acid is the 1-carboxyl analogue of glucose.)

6-Phosphogluconate is a substrate for another dehydrogenase, which uses NADP to oxidize C-3, with a simultaneous decarboxylation. The reaction is a typical oxidative decarboxylation of a 3-hydroxycarboxylate without release of the intermediate 3-keto compound and is therefore of the same type as the reactions catalyzed by isocitrate dehydrogenase and malate:NADP dehydrogenase (decarboxylating). The product D-**ribulose 5-phosphate** is the ketose isomer of D-ribose 5-phosphate.

This sequence of reactions produces two moles of NADPH for each mole of CO_2. This is the crucial part of the overall economy of the pentose phosphate pathway: one carbon of glucose appears as CO_2 for each pair of NADPH generated. The pentose phosphate, representing five sixths of the original glucose, is recovered for further use by the remaining reactions of the pathway, which we shall now consider.

Disposition of the Pentose Phosphate

The problem solved by evolution in handling the pentose phosphates is essentially one of reshuffling the carbon atoms so as to wind up with hexose and triose phosphates. The redistribution is accomplished mainly through the action of a **transketolase**, which catalyzes the transfer of two-carbon units from one sugar to another, and a **transaldolase**, which catalyzes the transfer of three-carbon units.

5 + 5 = 3 + 7. Transketolase uses aldose phosphates and ketose phosphates as substrates, and it contains thiamine pyrophosphate as a coenzyme.

The first two carbons of the ketose phosphate are transferred to the aldose

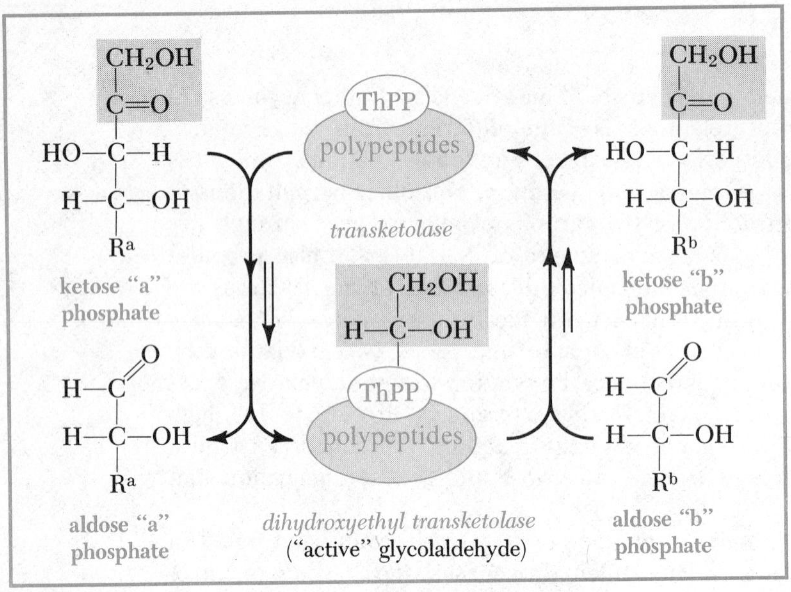

FIGURE 28–12

The action of transketolase. A two carbon group is transferred from ketose phosphate "a" to thiamine pyrophosphate on the enzyme, leaving the remaining carbons as aldose phosphate "a." The two-carbon group is then transferred to another aldose phosphate ("b") converting it to a new ketose phosphate. The configuration of the asymmetric carbons in the ketose phosphate is said to be D-*threo* because this is the configuration found in the four-carbon aldose, D-threose (p. 169).

phosphate with intermediate carriage on the thiamine pyrophosphate (Fig. 28–12). Almost any ketose and aldose phosphate are substrates, so long as the ketose has a D-*threo* configuration on the carbons adjacent to the carbonyl group, and the aldose has a D-configuration on the carbon adjacent to its carbonyl group.

Both aldose and ketose phosphate substrates for transketolase are generated from ribulose 5-phosphate (Fig. 28–13). The aldose, **D-ribose 5-phosphate,** is formed by a straightforward isomerase equilibrating ketose and aldose phosphates, analogous to the isomerase that equilibrates glucose and fructose phosphates. The ketose, **D-xylulose 5-phosphate**, is formed by an **epimerase** that equilibrates the D- and L-configurations on C-3. Transketolase then catalyzes the transfer of a two-carbon unit from xylulose phosphate to ribose phosphate. The result is the conversion of two pentose phosphates into a triose phosphate and a heptose phosphate.

3 + 7 = 6 + 4. The **sedoheptulose 7-phosphate** and **glyceraldehyde 3-phosphate** formed by transketolase are in turn substrates for **transaldolase,** which is an enzyme that catalyzes the transfer of a three-carbon unit from the ketose phosphate. The products are **fructose 6-phosphate** and **erythrose 4-phosphate.**

What do we have now? Two molecules of the original ribulose 5-phosphate have been consumed, and their 10 carbon atoms have been converted to a molecule each of fructose 6-phosphate and erythrose 4-phosphate. Fructose phosphate is a compound of the Embden-Meyerhof pathway and represents full rehabilitation, as it were, of six carbons derived from pentose phosphates. However, four carbons still remain in the form of erythrose 4-phosphate. What do we do with these?

FIGURE 28–13 **Ribulose 5-phosphate formed during the pentose phosphate pathway is recovered by converting it to fructose and glyceraldehyde phosphates. This is accomplished by successive transfers of two- and three-carbon groups, catalyzed by transketolase and transaldolase. The result is the conversion of 15 carbon atoms in the form of three molecules of ribulose 5-phosphate to two molecules of fructose 6-phosphate and one molecule of glyceraldehyde 3-phosphate.**

See illustration on opposite page.

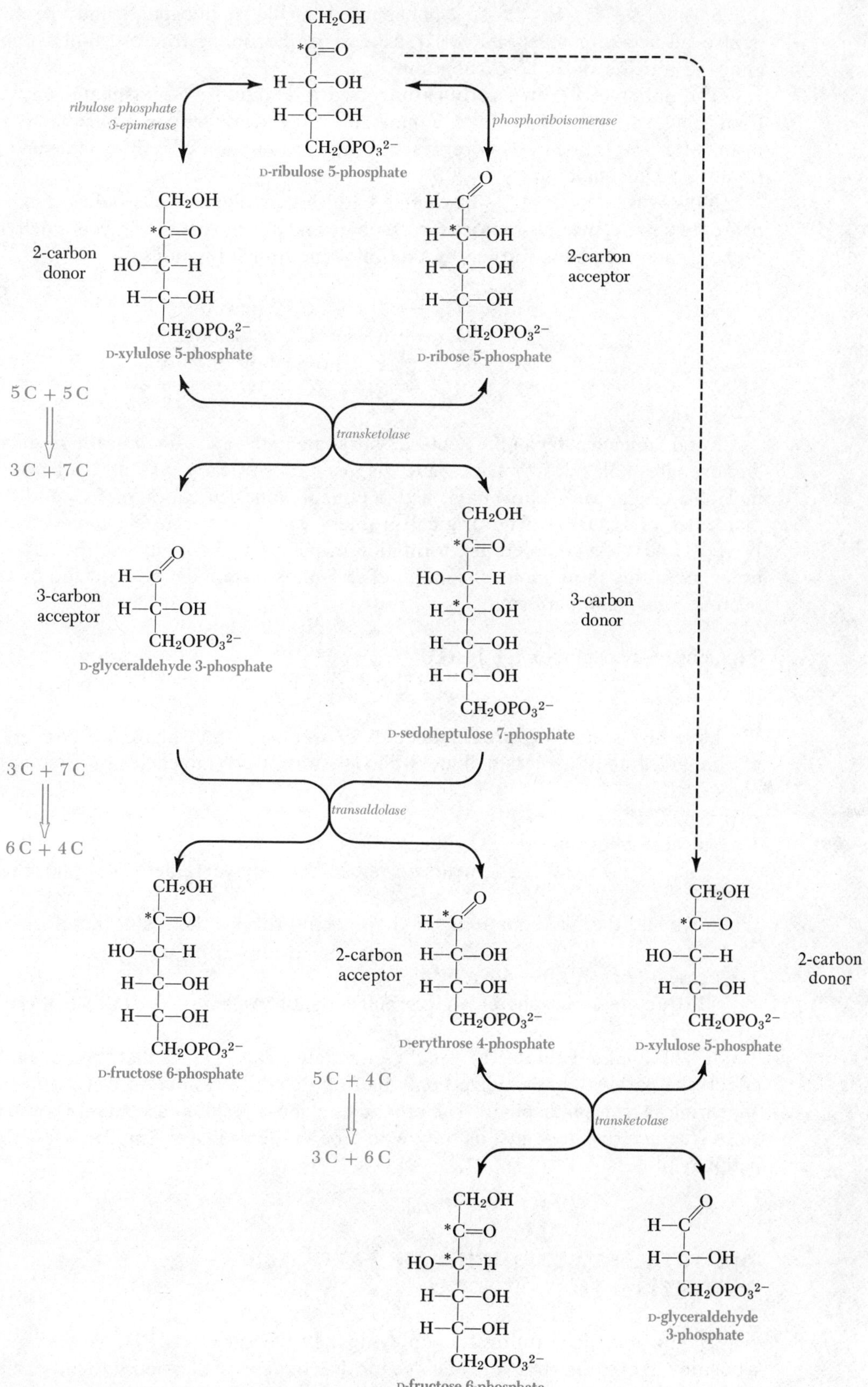

FIGURE 28–13 *See legend on opposite page.*

5 + 4 = 3 + 6. Erythrose 4-phosphate is an aldose phosphate and therefore is also an acceptor substrate for transketolase, becoming fructose 6-phosphate after the transfer of a two-carbon unit.

The donor of the two carbons once more is xylulose 5-phosphate, derived from ribulose 5-phosphate, and a molecule of glyceraldehyde 3-phosphate remains after the transfer. This represents the consumption of a third molecule of ribulose 5-phosphate.

Now what do we have? Beginning with three molecules of ribulose 5-phosphate, two molecules of fructose 6-phosphate and one molecule of glyceraldehyde 3-phosphate have been formed by the following transformations.

$$5C + 5C \longrightarrow 3C + 7C \text{ (transketolase)}$$
$$3C + 7C \longrightarrow 6C + 4C \text{ (transaldolase)}$$
$$5C + 4C \longrightarrow 3C + 6C \text{ (transketolase)}$$
$$\text{SUM:} \quad 5C + 5C + 5C \longrightarrow 6C + 6C + 3C$$

Total Stoichiometry of the Pentose Phosphate Pathway. Each of the ribulose-5-phosphate molecules that we have disposed of so neatly was derived from a molecule of glucose 6-phosphate, with a concomitant generation of CO_2 and two molecules of NADPH. The original problem was to generate six molecules of NADPH so as to complete the formation of palmitate from glucose (p. 527). In order to do this three molecules of glucose 6-phosphate must be oxidized by the pentose phosphate pathway:

$$3 \text{ (glucose 6-phosphate)} + 6 \text{ NADP}^+ \longrightarrow 3 \text{ (ribulose 5-phosphate)} + CO_2 + 6 \text{ NADPH}$$

We have just seen that three molecules of ribulose 5-phosphate are converted to one molecule of glyceraldehyde-3-phosphate and two molecules of fructose 6-phosphate:

$$3 \text{ (ribulose 5-phosphate)} \longrightarrow 2 \text{ (fructose 6-phosphate)} + \text{glyceraldehyde 3-phosphate}$$

When we add the two equations for the overall process, the result therefore is:

$$\mathbf{3 \text{ (glucose 6-phosphate)} + 6 \text{ NADP}^+ \longrightarrow 2 \text{ (fructose 6-phosphate)} + \text{glyceraldehyde 3-phosphate} + 3\, CO_2 + 6 \text{ NADPH}}$$

This is the final stoichiometry for the complete pentose phosphate pathway. In effect, the pathway has oxidized only half of a glucose residue to CO_2, with half appearing as triose phosphate. The other two glucose residues are merely converted to fructose residues, just as they would be in the ordinary Embden-Meyerhof pathway.

THE TOTAL STOICHIOMETRY OF FATTY ACID SYNTHESIS

Having examined routes for supplying the necessary NADPH, we are now in a position to cast the complete balance for the conversion of glucose to fatty acids.

The result is a beautiful demonstration that metabolism is an array of reactions, ordered in character of the intermediates and in anatomical location of the enzymes, and susceptible to rational, albeit not easy, analysis. Let us go through this one step at a time, not because the process is something to be memorized in detail, but rather so we can see the magnificent harmony it displays.

To begin, one molecule of palmitate is made from eight molecules of acetyl coenzyme A (neglecting ionic charges, water, and H^+) (p. 517).

$$\text{8 (acetyl-S-CoA)} + \text{7 ATP} + \text{14 NADPH} \longrightarrow \text{palmitate} + \text{7 (ADP} + \text{P}_i) + \text{14 NADP}^+ + \text{8 (CoA-SH)}$$

The eight molecules of acetyl coenzyme A are provided in the cytosol along with a transfer of hydrogen from NADH to NADP by the pyruvate-citrate cycle (p. 526):

$$\text{8 (acetyl-S-CoA}_{(mitoch.)}) + \text{8 NADH} + \text{8 NADP}^+ + \text{20 ATP} \longrightarrow \text{8 (acetyl-S-CoA}_{(cytosol)}) + \text{8 NAD}^+ + \text{8 NADPH} + \text{20 (ADP} + \text{P}_i)$$

(The high-energy phosphate balance here includes an assessment of 4 ~ P for the transport of eight pyruvate and eight citrate molecules that were omitted to simplify an already intricate discussion. This assumes a loss of one H^+ from the mitochondrial gradient for each molecule transported, with the loss of four H^+ equivalent to a loss of one ATP. Eight ~ P were used to carboxylate pyruvate, and eight more to cleave citrate.)

These acetyl coenzyme A molecules in the mitochondria are formed by oxidizing pyruvate with a concomitant oxidative phosphorylation (p. 451):

$$\text{8 pyruvate} + \text{4 O}_2 + \text{24 (ADP} + \text{P}_i) + \text{8 CoA-SH} \longrightarrow \text{8 (acetyl-S-CoA)} + \text{8 CO}_2 + \text{24 ATP}$$

The pyruvate in turn arises from glyceraldehyde 3-phosphate by the Embden-Meyerhof pathway (p. 444):

$$\text{8 (glyceraldehyde 3-phosphate)} + \text{8 NAD}^+ + \text{16 ADP} + \text{8 P}_i \longrightarrow \text{8 pyruvate} + \text{8 NADH} + \text{16 ATP}$$

One molecule of glyceraldehyde phosphate, alone with the remaining six molecules of NADPH that are required, is formed by the pentose phosphate pathway:

$$\text{3 (glucose 6-phosphate)} + \text{6 NADP}^+ \longrightarrow \text{glyceraldehyde 3-phosphate} + \text{2 (fructose 6-phosphate)} + \text{3 CO}_2 + \text{6 NADPH}$$

The other seven molecules of glyceraldehyde phosphate are formed by the usual steps of the Embden-Meyerhof pathway:

$$3^1/_2 \text{ (fructose 6-phosphate)} + 3^1/_2 \text{ ATP} \longrightarrow \text{7 (glyceraldehyde 3-phosphate)} + 3^1/_2 \text{ ADP}$$

Of the $3^1/_2$ fructose 6-phosphate required, only 2 are made by the pentose phosphate pathway; the other $1^1/_2$ are made by the usual isomerase reaction:

$$1^1/_2 \text{ (glucose 6-phosphate)} \longrightarrow 1^1/_2 \text{ (fructose 6-phosphate)}$$

Finally, the total of 4½ molecules of glucose-6-phosphate are made from glucose:

$$4^1/_2 \text{ glucose} + 4^1/_2 \text{ ATP} \longrightarrow 4^1/_2 \text{ (glucose 6-phosphate)} + 4^1/_2 \text{ ADP}$$

Adding all of these steps together we get a grand total:

$$4^1/_2 \text{ glucose} + 4\ O_2 + 5\ (\text{ADP} + P_i) \longrightarrow \text{palmitate} + 11\ CO_2 + 5\ \text{ATP}$$

How about that! Everything works out, and there is even a small production of high-energy phosphate. Essentially, the balance states that adipose tissue consumes 27 carbon atoms as glucose to store 16 carbon atoms as palmitate, with the remaining 11 appearing as CO_2. It does so in such a way that the eight moles of acetyl coenzyme A that must be transported from mitochondria to cytosol exactly

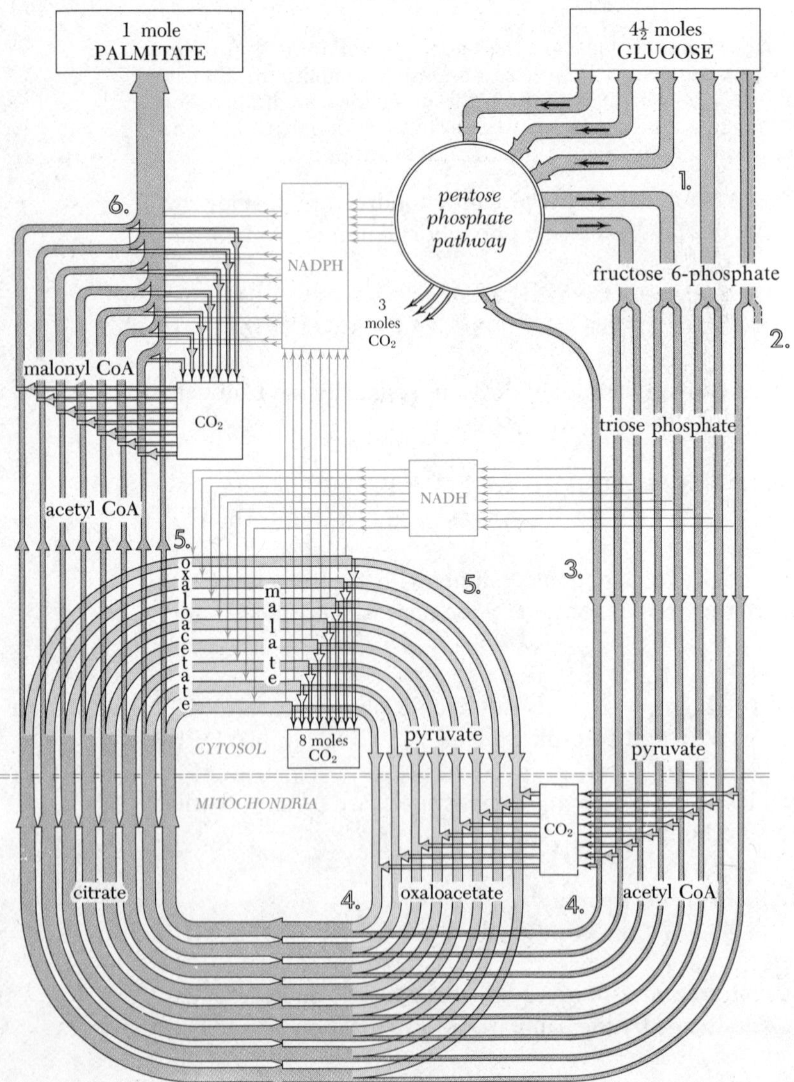

FIGURE 28–14 *See legend on opposite page.*

equals the number of pairs of electrons that must be transferred from NADH to NADP in the cytosol. The citrate moving out of mitochondria supplies both needs in the cytosol, and pyruvate returning replaces the oxaloacetate used in making citrate, so there is no need for a diffusion of oxaloacetate across the mitochondrial membrane. The entire process is shown as a flow sheet in Figure 28–14.

FORMATION OF TRIGLYCERIDES

Fatty acids, whether obtained by hydrolysis of dietary fat or by synthesis from glucose, are converted to triglycerides for transport to, and for deposition in, adipose tissue. The fatty acids are first converted to their coenzyme A thioesters (p. 422). The acyl groups are then transferred to the hydroxyl groups of glycerol, which is initially provided in the form of dihydroxyacetone phosphate. Let us look at these processes in detail (Fig. 28–15).

The Initial Acylation. There are two routes for placing an initial acyl group on what will be the glyceryl backbone of a triglyceride. In one, the acyl group is transferred to dihydroxyacetone phosphate before reduction to create a glycerol residue. In the other route, reduction of dihydroxyacetone phosphate to glycerol phosphate occurs first, with the acyl group then added. The ultimate source of the glycerol residue is glucose, or some gluconeogenic precursor in either case. The monoacyl glycerol phosphate product is trivially named as a **lysophosphatidate.** (The lyso- prefix is given by analogy with lysolecithin, a monoacyl glyceryl phosphate ester of choline, which has the property of causing lysis of cells by its detergent action.)

The initial acylation preferentially involves a saturated fatty acid, and this is true with either dihydroxyacetone phosphate or glycerol 3-phosphate as the acceptor for the acyl group. Figure 28–15 shows a palmitoyl group, the most common at position 1, being transferred.

FIGURE 28–14 **Flow sheet for the synthesis of palmitate from glucose. The flow of carbons appearing in palmitate is shown by *heavy shading. Lighter shading* indicates the recycling of oxaloacetate via pyruvate from the cytosol to the mitochondria. Transfers of electrons are shown in *blue.***

1. Four and a half moles of glucose are required to make one mole of palmitate; three of these pass through the pentose phosphate pathway to generate six moles of NADPH. The remainder are converted directly to fructose 6-phosphates by the Embden-Meyerhof pathway.

2. Two additional moles of fructose 6-phosphate are formed by the pentose phosphate pathway, making a total of 3½ moles, which are converted to seven moles of triose phosphate. Another mole of triose phosphate is formed by the pentose phosphate pathway.

3. The eight moles of triose phosphate created in step 2 are converted to pyruvate by the Embden-Meyerhof pathway, with the formation of eight moles of NADH.

4. Eight moles of pyruvate are oxidized to acetyl coenzyme A in the mitochondria, and the acetyl coenzyme A combines with oxaloacetate to form citrate (8 moles).

5. The citrate passes from the mitochondria into the cytosol, where it is cleaved to acetyl coenzyme A and oxaloacetate. The oxaloacetate is reduced to malate, using the eight moles of NADH formed in step 2. The malate is then oxidized to pyruvate by the NADP-coupled dehydrogenase to form an additional eight moles of NADPH. The pyruvate moves back into the mitochondria, where it is carboxylated to regenerate oxaloacetate consumed in step 3.

6. The eight moles of acetyl coenzyme A transported into the cytosol as citrate are carboxylated (7 moles) and condensed to form palmitate. The necessary reductions consume the NADPH generated in steps 1 and 4.

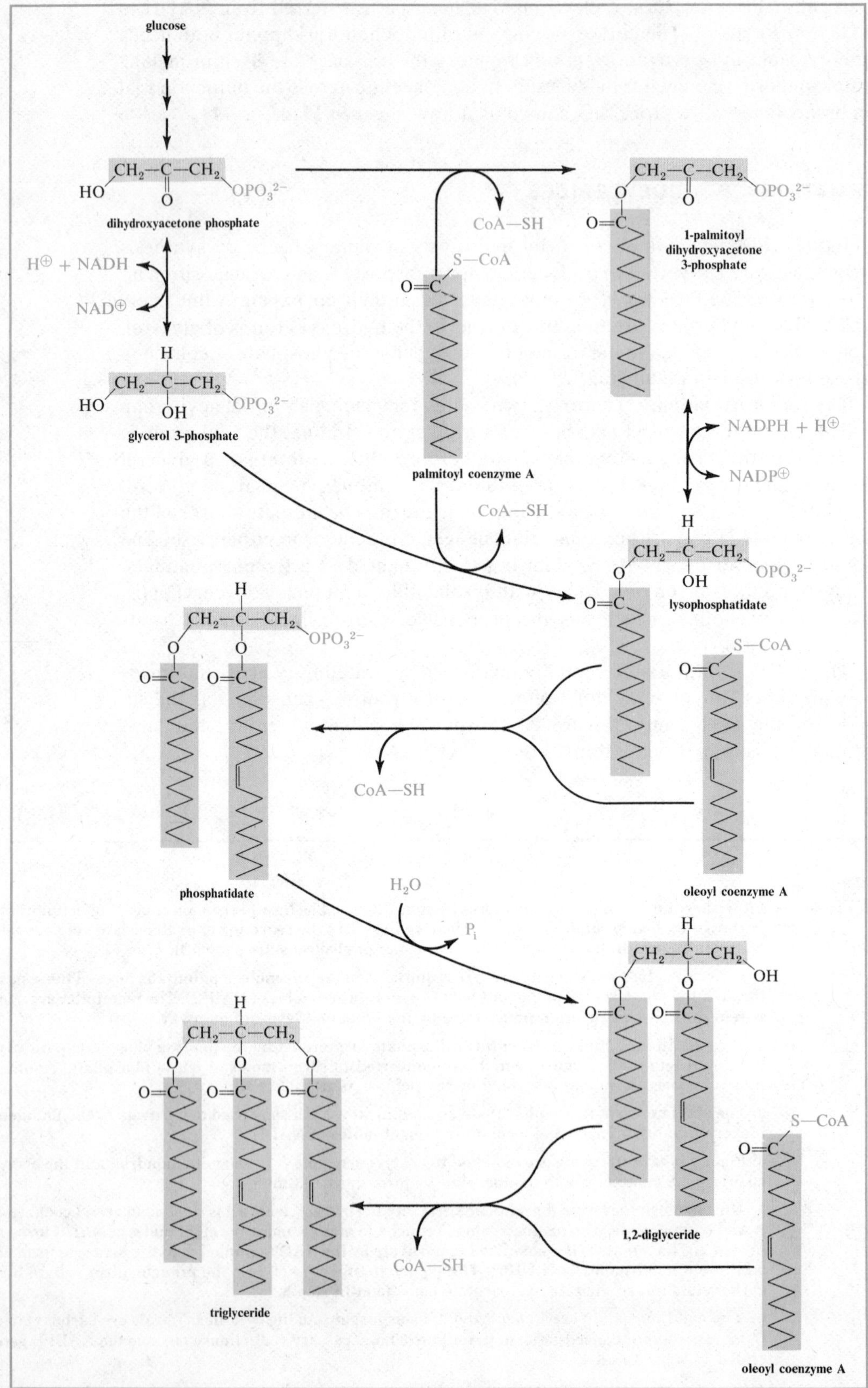

FIGURE 28–15 *See legend on opposite page.*

Since the ultimate precursors and products appear to be the same in the two routes, except for the electron carriers, it is not certain why both routes exist. There is some evidence that the direct acylation of dihydroxyacetone phosphate is preferred at times of glucose deprivation.

The Second and Third Acylations. An unsaturated fatty acid residue is transferred from its coenzyme A derivative to the free hydroxyl group at position 2 of lysophosphatidates, forming **phosphatidates,** the 1,2-diacyl esters of glycerol 3-phosphate (Fig. 28–15, *center*). This is true in nearly all tissues, but there are interesting exceptions, including human mammary glands, in which a palmitate residue is transferred. A protein may be present in these tissues to combine with the acyl transferase and modify its specificity during triglyceride synthesis.

The phosphatidate formed by the second transfer is then attacked by a specific phosphatase, exposing the final hydroxyl group of the glyceryl backbone. Still another acyl transferase, this one specific for a diglyceride acceptor, then adds the final fatty acid residue, which may be either saturated or unsaturated.

The result in a typical human will be a mixture of triglycerides with the following acyl groups comprising most of the total:

20% 16:0 (palmitoyl)
7% 16:1 (palmitoleoyl)
50% 18:1 (oleoyl and vaccenoyl)
10% 18:2 (linoleoyl)

This is not completely a reflection of the relative synthesis of the various types of fatty acid. It should not be forgotten that dietary fatty acids may be incorporated into triglycerides without change. In fact, the composition of the diet partly determines the character of the stored fat. This is shown strikingly by experiments in which individuals were fed corn oil, linseed oil, or coconut oil for periods of a year or more, and the fatty acid composition of their subcutaneous adipose tissue was compared with that of individuals eating the usual random American diet. The fatty acids of these oils and the resultant changes in the adipose tissue are shown in Table 28–1.

Several things emerge from these data. There is little chain lengthening beyond 18 carbon atoms. The data with a random diet show that the liver isn't very fancy in its synthesis of fatty acids. The palmitoyl and oleoyl residues make up nearly two thirds of the total and linoleate from the diet makes up the next most abundant residue. There is little introduction of additional double bonds beyond the first needed to supply the oleoyl and palmitoleoyl residues. The fact that a diet high in 18:2 or 18:3 fatty acids leads to a storage of these residues in adipose tissue suggests that there is no inherent screen against polyunsaturated fatty acids. Eating coconut oil, with its abundance of short chain residues, does lead to an increased storage of 12:0 and 14:0 residues, but most of the 8:0 and 10:0 residues found in the food have evidently been modified or disposed of elsewhere.

FIGURE 28–15 **Triglycerides are synthesized with acyl groups from coenzyme A derivatives of fatty acids. The first acyl group may be added to C-1 of dihydroxyacetone phosphate, followed by reduction with NADPH to form a glycerol residue (*top center and right*), or it may be added to C-1 of glycerol phosphate already obtained by reduction with NADH (*top left and middle center*). The second and third acyl groups are then added, with an intervening hydrolysis of the phosphate ester.**

TABLE 28–1 COMPOSITION OF TRIGLYCERIDES IN PLANT OILS AND IN ADIPOSE TISSUE OF HUMANS EATING THE OILS*

	Percentage of Total Fatty Acid Residues						
Fatty Acid	*Humans on Random American Diet*	*Corn Oil*	*Humans with 40% of Energy Source as Corn Oil for 3 Years*	*Linseed Oil*	*Humans Eating 83 g of Linseed Oil Daily for 1 Year*	*Coconut Oil*	*Humans Eating 60 g of Coconut Oil for 1.5 Years*
8:0						8	
10:0						10	
12:0	0.7		0.1		0.1	47	14.5
14:0	3.3		0.7		1.2	16	13.9
16:0	19.5	13	15.3	6	14.7	8	17.5
16:1	6.9		2.2		5.8	1	7.6
18:0	4.2	4	2.2	4	5.4	2	2.7
18:1	46.3	29	32.1	22	35.5	6	30.4
18:2	11.4	54	45.2	16	20.5	2	9.3
18:3	0.4			52	13.7		0.1

*Data for adipose tissue from J. Hirsch: (1965) *Handbook of Physiology*. p. 148, section 5. American Physiological Society. Data for oil composition from H. E. Longenecker: (1959) J. Biol. Chem., *130*: 167, and E. Fedeli, and G. Jacini: (1971) Adv. Lipid Res., *9*: 335.

THE MOBILIZATION AND TRANSPORT OF FATS

Triglycerides are transported in the blood stream as small droplets coated with a mixture of specific proteins, phospholipids, and cholesterol. The droplets may be of molecular size (**lipoproteins**), or so large as to give a milky appearance to the blood plasma (**chylomicrons**).

The phospholipids in the surface shells of fat droplets are mainly phosphatidylcholines and sphingomyelins, which, along with cholesterol, are also typical membrane components. We can reason that the assembly of fat droplets for transport must depend upon the proteins with which they are coated. Membranes contain proteins that differ from those used to transport fat, and the difference must be the decisive factor in determining whether the associated phospholipids

TABLE 28-2 PLASMA LIPOPROTEINS

Property	Chylomicrons	Very Low-density Lipoproteins	Low-density Lipoproteins	High-density Lipoproteins
Typical Per Cent Composition				
fat	87	55	8	0.6
phospholipid	8	20	24	21
free cholesterol	1.5	10	8	5
cholesteryl esters	2	5	35	15
protein	1.7	9	25	55
Density	0.92–0.96	0.95–1.006	1.019–1.063	1.063–1.21
Molecular Weight	5×10^8 (typical)	$5–100 \times 10^6$	$2.2–3.5 \times 10^6$	3.2×10^5 (HDL_2) 1.75×10^5 (HDL_3)
Diameter (nm)	75–600	28–75	17–26	10.8 (HDL_2)
Function	triglyceride transport from small bowel	triglyceride transport from liver	cholesterol transport	cholesterol transport

will form a bilayer sheet or vesicle, or a monolayer coating over a hydrophobic core.

The blood lipoproteins differ in the quantity and nature of associated lipids; they range from the **high-density lipoproteins (HDL)**, which are only about one-half lipid, to the **very low-density lipoproteins (VLDL),** which are mostly fat. The nature of these vehicles for moving lipids is summarized in Table 28–2, and we shall have more to say about them when we discuss cholesterol metabolism (p. 671). Note that triglycerides are transported almost exclusively as chylomicrons or very low-density lipoproteins.

Apolipoproteins. Five families of apolipoproteins, A, B, C, D, and E, are known to occur in the circulating lipoproteins. All of these are made in the liver, but at least one, the B apolipoprotein, is also made in the intestinal mucosa. Specific functions are known for only a few.

The association of apolipoproteins and lipids involves the same sort of non-covalent interactions seen in membrane formation, and the components are quite free to move about, not only within a droplet, but also between different lipoprotein molecules or droplets that meet within the circulation. The apoplipoproteins fix the general character of the aggregates, but they do not create exactly reproducible assemblies. The result is that the lipoproteins are made as heterogeneous mixtures, which become even more heterogeneous through changes in composition during transit.

The amino acid sequences are known for some apolipoproteins, but not the detailed structures. The sequences are consistent with formation of segments of α-helix with both hydrophobic and hydrophilic faces, suggesting a role in bridging two liquid phases.

Chylomicrons

Digested fats are absorbed as a mixture of free fatty acids and monoglycerides (p. 315), which the intestinal mucosal cells rebuild into triglycerides for transport, using the coenzyme A derivatives and triose phosphate in the usual way. However, these cells are also able to use the 2-monoglycerides directly. They contain an enzyme that catalyzes direct formation of 1,2-diglycerides:

$$\text{acyl coenzyme A} + \text{2-monoglyceride} \longrightarrow \text{CoA-SH} + \text{1,2-diglyceride.}$$

The smooth endoplasmic reticulum of the mucosal cells then coats droplets of triglycerides with apolipoproteins, phospholipids, and cholesterol to generate chylomicrons. Chylomicrons emerge from the cell after passage through the Golgi apparatus, perhaps by reverse pinocytosis — that is, enclosure by membrane vesicles, which fuse with the plasma membrane and discharge their contents outside the cell into the central lacteals (lymph channels).

Chylomicrons may contain apolipoprotein B as the major protein component at the time of synthesis; it is difficult to say because they shortly acquire representatives of the other apolipoprotein families, even during passage through the lymphatic drainage to the thoracic duct and on to the venous circulation. At least one of these, CII, which is made by the liver, is of critical importance for utilization of the chylomicrons.

The triglycerides in chylomicrons are sources of fatty acids for the liver, the muscles, and the adipose tissue. Each of these tissues contains a **lipoprotein lipase**

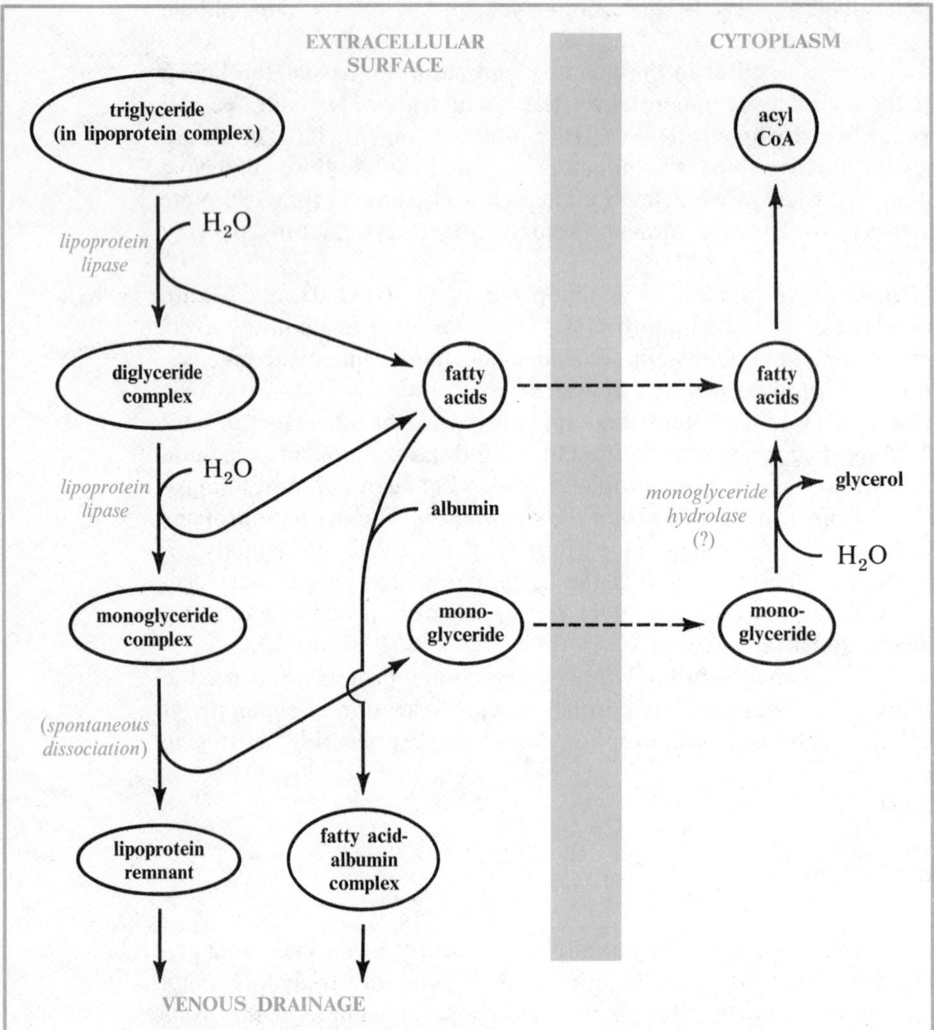

FIGURE 28–16 **Lipoprotein lipase on the surface of endothelial cells hydrolyzes either the 1- or the 3-acyl group from triglycerides in chylomicrons or very low-density lipoproteins. Much of the resultant free fatty acid escapes in the blood as an albumin complex, but the remaining monoglyceride passes into neighboring parenchymal cells for further hydrolysis and utilization.**

on the surface of the endothelial cells in the capillary bed, and perhaps on the outer surface of adjoining cells. This enzyme is an acyl glycerol hydrolase that will attack either the 1 or the 3 ester bond with release of a fatty acid (as the anion) and first diglycerides, then monoglycerides (Fig. 28–16). Triglycerides and diglycerides only are attacked by this enzyme if they are in a droplet that has the CII apolipoprotein in its shell. Many of the free fatty acids escape into the circulation, but the monoglycerides pass into the parenchymal cells, where they are further hydrolyzed to free fatty acids and glycerol. Although some triglycerides are taken up intact through pinocytotic engulfment, hydrolysis by the lipase is the major route of terminal transport in the peripheral tissues. The chylomicrons shrink as their triglycerides are hydrolyzed and removed. The remnants are believed to be taken up by the liver.

Very Low-density Lipoprotein (VLDL)

The liver is another source of triglycerides for export to the adipose tissue and muscles. Any intracellular elevation of fatty acid concentration will tend to increase the synthesis of triglycerides for this purpose; it makes no difference if the fatty acids have come to the liver or have been made within it. The triglycerides enter the blood as droplets resembling chylomicrons, except that they are much smaller and are made with a more complex mixture of apolipoproteins. These droplets are the very low-density lipoproteins, the low density arising from their high content of fat.

The very low-density lipoproteins are attacked by the lipoprotein lipase in the same way as are chylomicrons, liberating monoglycerides and fatty acids for absorption and utilization by the peripheral tissues. The lipid-poor remnants appear in the circulation, first as intermediate-density (IDL), and then as low-density lipoproteins (LDL). The fate of these proteins is involved with cholesterol metabolism (p. 671).

Free Fatty Acids

Release from Stored Triglycerides. The stored fat in adipose tissue is hydrolyzed to mobilize it for use as a fuel. Hydrolysis is catalyzed by a battery of lipases, including enzymes specific for triglycerides, diglycerides, and monoglycerides (Fig. 28–17). (The intracellular triglycerides are not accessible for attack by lipoprotein lipase; that enzyme is concerned with delivery of fatty acids to the tissue, not with their release from the tissue.) The breakdown of fats, like the breakdown of glycogen, is under strict control, and there is further similarity; the triglyceride lipase catalyzing the initial step is phosphorylated by the same sort of cyclic AMP-sensitive protein kinase regulating glycogen metabolism in other tissues. Phosphorylation converts the inactive lipase **b** to an active lipase **a**. Once

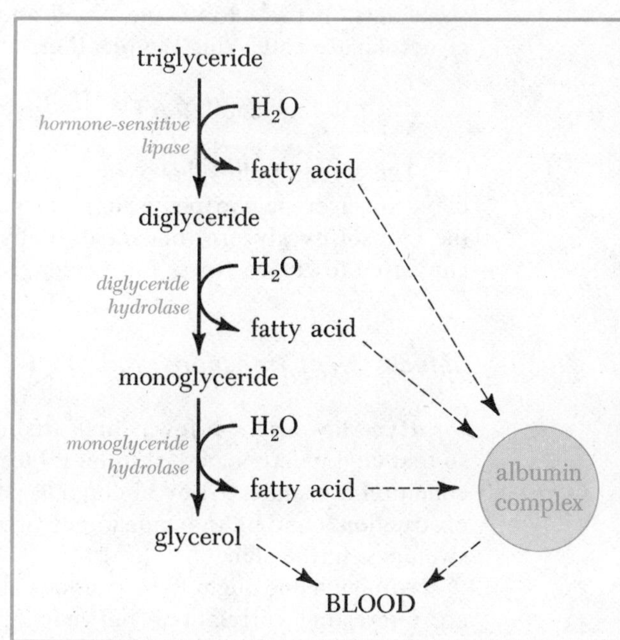

FIGURE 28–17

Hydrolysis of stored triglyceride in adipose tissue is triggered by a hormone-sensitive lipase. The resultant diglyceride is rapidly hydrolyzed by successive action of diglyceride and monoglyceride lipases. The liberated fatty acids go into the blood for transport as an albumin complex; glycerol is transported as such in solution.

hydrolysis of triglyceride begins, the diglyceride and monoglyceride hydrolases rapidly complete the job.

The formation of cyclic AMP in adipose tissue is stimulated by adrenaline or noradrenaline as it also is in muscles. Therefore, an alarm signal or neural stimulation causes the release of free fatty acids from the adipose tissue into the circulation for use by the muscles. However, this signal can be overridden by insulin. Since the pancreas secretes insulin in response to a high concentration of glucose in the blood, its effect on the adipose tissue presumably serves to delay the release of fatty acids in those individuals who have an ample store of glycogen available in the skeletal muscles for emergency use.

The formation of cyclic AMP in adipose tissue is also stimulated by glucagon and by corticotropin (sometimes called adrenocorticotrophin, ACTH), as well as by other hormones, but there is some question about the physiological significance of these effects in humans.

Transport of Fatty Acids. Free fatty acids are transported in combination with **serum albumin** — the most abundant protein in blood plasma, constituting about 40 of the 70 grams of protein per liter. Serum albumin has two sites with high affinity for fatty acids, and five more sites of moderate affinity. The molecular weight is 69,000, so more than one millimole of fatty acids can be strongly bound by albumin per liter of blood plasma, but the actual loading ranges from 0.5 to 0.8 millimole per liter in an individual at rest after an overnight fast, and half or less of these values after eating.

These relatively low concentrations do not reflect the rate of fatty acid transport through the blood; turnover is rapid, and the concentration is determined by the balance between rate of release from adipose tissue and consumption by other tissues, especially working muscles.

The fate of the fatty acids depends upon circumstances. During exercise, most are used by the muscles. At rest after an overnight fast, approximately one third of the circulating fatty acids are removed by the liver, which then sends about 60 per cent of the carbon atoms back into the blood — 20 per cent as triglycerides in very low-density lipoproteins and 40 per cent in 3-oxybutyrates.

Fate of Glycerol. Adipose tissue cannot use the glycerol released by hydrolysis of fats. It passes into the blood and is taken up by the liver, which has a glycerokinase catalyzing the reaction:

$$\text{glycerol} + \text{ATP} \longrightarrow \text{glycerol 3-phosphate} + \text{ADP}$$

The glycerol phosphate may be used to form triglycerides or to add to the dihydroxyacetone phosphate supply, since the two compounds are equilibrated in the cytosol by glycerol phosphate dehydrogenase. Glycerol can, therefore, be converted to glucose or to fatty acids, or it can be used as a fuel.

Defects in Fat Transport

Hyperlipemias. A few people are born with a deficiency of lipoprotein lipase so that chylomicrons persist in their blood long after ingestion of a fat meal. The condition is manifested by abdominal pain from an inflamed pancreas (capillary obstruction?) and by the appearance of xanthomas, which are yellow, lipid-filled swellings, in the skin.

Although one ought to be cautious about adopting too simplistic a view, there are interesting correlations between blood triglyceride concentration and the occurrence of myocardial infarction. A large number of American adults have

LIPOPROTEIN LIPASE DEFICIENCY

Synonyms:	familial exogenous hyperlipemia familial hyperchylomicronemia type I lipoprotein phenotype
Blood:	normal to elevated cholesterol levels elevated triglycerides with creamy layer on top of clear serum upon standing
Incidence:	rare, autosomal recessive
Clinical findings:	xanthoma abdominal pain pancreatitis no demonstrated premature atherosclerosis
Treatment:	low fat diet (10–15 g/day) add medium chain triglycerides

high blood concentrations of very low-density lipoproteins, and the condition is especially common in relatively young survivors of a myocardial infarction. For example, of 500 consecutive patients who survived infarction for three months, 31 per cent had blood triglyceride concentrations greater than the 95th percentile level of a control population. A similarly high concentration was found in 60 per cent of the male survivors of age 40 or less, and female survivors of age 50 or less.

Many appear to have elevated blood triglyceride concentrations as an inherited trait transmitted as an autosomal dominant, but there may be several distinct conditions involving many genes. It has been estimated that 1 per cent of the general population and 5 per cent of those surviving myocardial infarction will have a hypertriglyceridemia that occurs without an accompanying elevation in cholesterol concentration, while 1.5 per cent of the general population and 11 to 20 per cent of the survivors will have an elevation in both triglycerides and cholesterol. The long known familial susceptibility to myocardial infarction may be explained at least in part by these data.

Analbuminemia. Given the important role of serum albumin in transport of fatty acids and other important functions we shall encounter later, it is disconcerting to discover that some people have a congenital lack of the protein and yet survive without any distinctive set of physical or metabolic abnormalities assignable to the condition, except for a high blood triglyceride concentration, resulting in xanthomas. Other abnormalities are known in the patients, but not consistently enough to assign a causal relationship in the small number of cases at hand. The rarity of the condition in itself suggests that it is ordinarily highly disadvantageous, and those surviving it may be unusual in other respects.

BROWN ADIPOSE TISSUE

Animals have another type of adipose tissue that has a high content of mitochondria, giving it a distinctive brown hue. Adult humans have small amounts of the tissue, but it is prominent in the newborn, especially in their upper torso and neck. It is also common in the adults of animals that hibernate. The tissue acts as a means of producing heat, which is quite different from the function of white adipose tissue as a reservoir of potential energy.

Larger animals appear to be limited in the rate of metabolism by the difficulty in disposing of heat. The late George Gamow estimated that an elephant would be literally red-hot if its metabolic rate equalled that of a mouse. The point here is that the limitations on the elephant's metabolism come not from some inherent theoretical limit of the metabolic scheme, but from a built-in regulation adjusting its energy metabolism to its capacity for heat disposal. However, infants of a given species, including humans, have a high ratio of surface area to tissue mass as compared to adults, and are more vulnerable to cold. They are equipped with brown adipose tissue in which metabolism is made deliberately inefficient. The device for doing this appears to be a futile cycle in which stored triglycerides are constantly hydrolyzed within the cells to the constituent fatty acids, which are then converted to the coenzyme A derivatives and back to triglycerides, with a concomitant expenditure of high-energy phosphate. ATP is continually being hydrolyzed to AMP and 2 P_i by the futile cycle, with a concomitant continual regeneration of ATP by oxidative phosphorylation. The net result is a high rate of fuel combustion that accomplishes nothing but the generation of heat.

FURTHER READING

Bloch, K., and D. Vance: (1977) *Control Mechanisms in the Synthesis of Saturated Fatty Acids.* Annu. Rev. Biochem., *46*: 263.

Volpe, J. J., and R. Vagelos: (1976) *Mechanism and Regulation of Biosynthesis of Saturated Fatty Acids.* Physiol. Rev., *56*: 339.

Bazan, N. G., R. R. Bremer, and N. M. Guisto: (1977) *Function and Biosynthesis of Lipids.* Adv. Exp. Biol. Med., vol. 83. Symposium includes section on biosynthesis with emphasis on desaturation.

Qureshi, A. A., et al.: (1976) *Subunits of Fatty Acid Synthetase Complex.* Arch Biochem. Biophys., *177*: 364.

Carey, E. M.: (1977) *The Interaction of Fatty Acid Synthetase with Cytoplasmic Protein in the Control of the Chain Length of Fatty Acids Synthesized by the Lactating Mammary Gland.* Biochem. Biophys. Acta, *486*: 91.

Osborne, J. C., Jr., and H. B. Brewer, Jr.: (1977) *The Plasma Lipoproteins.* Adv. Prot. Chem., *31*: 253.

Eisenberg, S., and R. E. Levy: (1975) *Lipoprotein Metabolism.* Adv. Lipid Res. *13*: 1.

Morrisett, J. D., R. L. Jackson, and A. M. Gotto, Jr.: (1975) *Lipoproteins: Structure and Function.* Annu. Rev. Biochem., *44*: 183.

Boman, H., et al.: (1976) *Analbuminemia in an American Indian Girl.* Clin. Genet., *9*: 513.

Goldstein, J. L., et al.: (1973) *Hyperlipidemia in Coronary Artery Disease.* J. Clin. Invest., *52*: 1533 ff.

Motulsky, A. G.: (1976) *The Genetic Hyperlipidemias.* N. Engl. J. Med., *294*: 823.

Seemayer, T. A., et al.: (1975) *On the Ultrastructure of Hibernoma.* Cancer, *36*: 1785. Includes discussion of brown adipose tissue.

Snyder, F., ed.: (1977) *Lipid Metabolism in Mammals.* Plenum.

29 AMINO ACIDS: DISPOSAL OF NITROGEN

We opened our discussion of biochemistry by examining the amino acids as constituents of proteins. The use of amino acids as precursors of proteins must constantly be kept in mind, even though amino acid metabolism is used for other purposes.

It is easy to show that just maintaining a supply of amino acids from which proteins can be made isn't a simple process. We eat proteins, which are hydrolyzed to the constituent amino acids in the gut. At the same time, tissue proteins are also being hydrolyzed to form amino acids, which mix with those derived from food as an **amino acid pool** in the extracellular fluid upon which all tissues may draw for their requirements.

Part of the amino acid pool is used to rebuild tissue proteins, but the total amount of protein remains relatively constant in most adults from day to day, which means that the quantity of amino acids used to make tissue proteins is no greater than the quantity obtained by the breakdown of tissue proteins. Therefore, the usual adult will have a surplus of amino acids equivalent to the amount he has ingested. Amino acids aren't excreted in significant quantities, and the surplus must be disposed of in other ways. In short, there is an active metabolism of amino acids that constantly must be reconciled with maintenance of a supply for protein synthesis. Let us summarize the nature of this metabolism (Fig. 29–1).

The amino acid pool is constantly drained for the synthesis of other nitrogenous constituents. One of these, urea, is deliberately made as a device for disposal of excess nitrogen in the urine. Others are vital tissue components: purines, pyrimidines, porphyrins, amines, creatine, and so on. These compounds are constantly being broken down and re-made, as are the proteins themselves. Part of their nitrogen can be recovered as ammonia, which is a precursor of several amino acids, but part is lost through excretion of other products.

The bulk of the excess amino acid intake is degraded through oxidative pathways associated with the formation of high-energy phosphate: Amino acids are used as fuel. How important a fuel are they? They are usually less significant than fat or carbohydrate but are sometimes of major importance. The extent of the contribution to the energy supply obviously depends upon the composition of the diet. The rib roast eaten by an affluent human contains protein and fat to the extent of one quarter each of its cooked weight, with virtually no carbohydrate. The fish or fried chicken more likely to be on most tables has three to six times as much protein as fat in the original tissues, but the fat will be increased in the final food to some extent through the cooking oil used in its preparation.

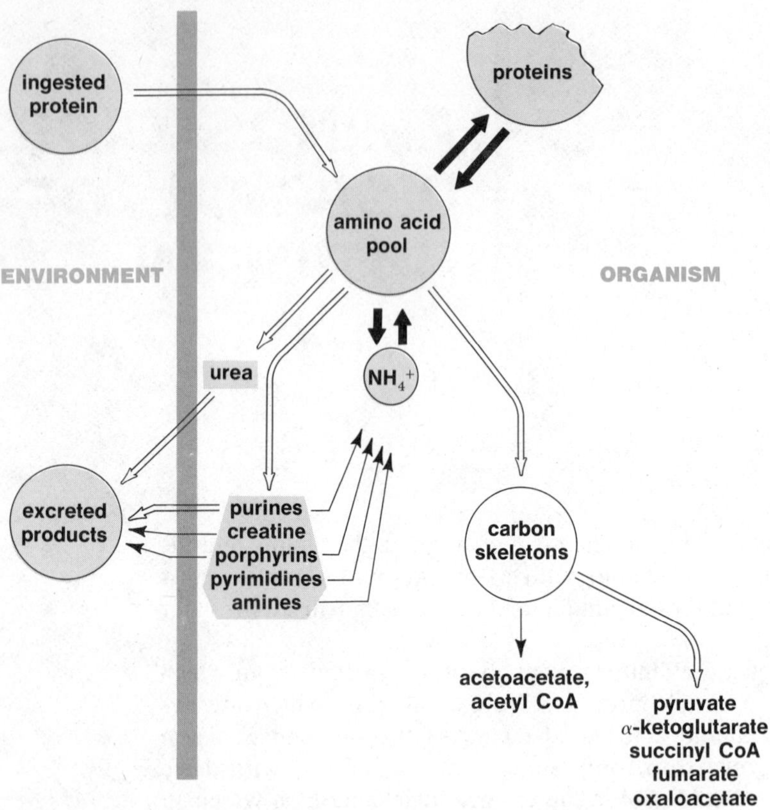

FIGURE 29–1 **Schematic outline of the flow of nitrogen through the body.**

Considering the total intake, most humans obtain only one tenth, rarely more than one fifth, of their high-energy phosphate by oxidizing amino acids, but purely carnivorous animals generate nearly one half of their supply in this way.

Since there are some 20 amino acids commonly metabolized during the utilization of proteins, the description of protein metabolism inevitably becomes more complicated than the description of the major routes of carbohydrate and fat metabolism, and it is tempting to skip over the subject lightly with a few superficial generalities, but the temptation ought to be, and will be, resisted. The metabolism of amino acids is complicated, and it is important. Furthermore, aberrations of the nitrogen economy accompany many of the most pressing of current medical problems.

We shall deal in this chapter with some of the general processes of amino acid metabolism, and discuss the fate of the individual amino acids in the next chapters.

TRANSPORT OF NITROGEN

Uptake of Amino Acids

Many cells are capable of concentrating amino acids from the extra-cellular environment, but the processes of transport have not been explored in detail in most cases. Separate carriers are known to exist for some neutral amino acids, cationic amino acids, and anionic amino acids. For example, patients have been recognized with a hereditary defect causing them to excrete abnormal quantities of cystine, arginine, lysine, and ornithine. The formation of stones from the slightly-soluble cystine is the usual presenting complaint, hence the condition is known as **cystinuria,** but the mutation affects a single carrier for all amino acids with two positively charged groups so as to prevent their reabsorption from the glomerular filtrate in the kidney, permitting their escape in the urine. Such mutant genes have a frequency greater than 1:200, at least in England, with 8 homozygotes found upon screening 142,000 newborn.

A Na^+-coupled symport mechanism for the absorption of neutral amino acids is also present in the brush borders of the intestinal mucosa and the kidney tubules. The passage of Na^+ down its concentration gradient causes simultaneous transport of amino acids against their concentration gradient in the same way that glucose is moved in these cells (p. 369). (Some amino acids are transported into striated muscles by a Na^+-dependent process that is stimulated by insulin.)

There is strong evidence that a known set of reactions involving 5-glutamyl peptides (Fig. 29–2) is used for amino acid transport into many tissues, including the brush borders of the kidney and the small bowel, the brain, and red blood cells. These peptides, commonly known as γ-glutamyl peptides because they involve the carboxyl group attached to the γ-carbon atom, include glutathione (5-glutamylcysteinylglycine). The tissue content of glutathione ranges as high as eight millimoles per kilogram. (Glutathione is also used as a reducing agent, but this function is not involved in amino acid transport.)

The amino acid to be transported is converted to a γ-glutamyl derivative by moving the γ-glutamyl group of glutathione or another γ-glutamyl peptide onto it. The γ-glutamyl transferase is believed to be exposed on the external side of the plasma membrane, with the products released on the internal side. In any event, the γ-glutamyl amino acid is attacked by a cyclotransferase that releases the amino acid inside the cell and forms pyroglutamate (5-oxoproline) from the glutamyl group.

As Figure 29–2 shows, the starting materials can be synthesized from the other product of the amino acid transfer, creating a cycle. When glutathione is used as a glutamyl group donor, the other product is cysteinylglycine, which can be hydrolyzed to its constituent amino acids. The ring in pyroglutamate is opened to form glutamate by a reaction requiring ATP. Glutathione can then be rebuilt from glutamate, cysteine, and glycine by two ATP-requiring synthetases.

According to the mechanism given, the transport of one molecule of amino acid requires the expenditure of three high-energy phosphate bonds, but there are alternative possibilities that are less expensive involving γ-glutamylcysteine itself as the glutamyl group donor. The significance of these pathways remains to be seen.

Transport Between Tissues

The skeletal muscles, the intestines, and the liver are particularly important in disposing of excess amino acids, but much of the nitrogen is channeled into only

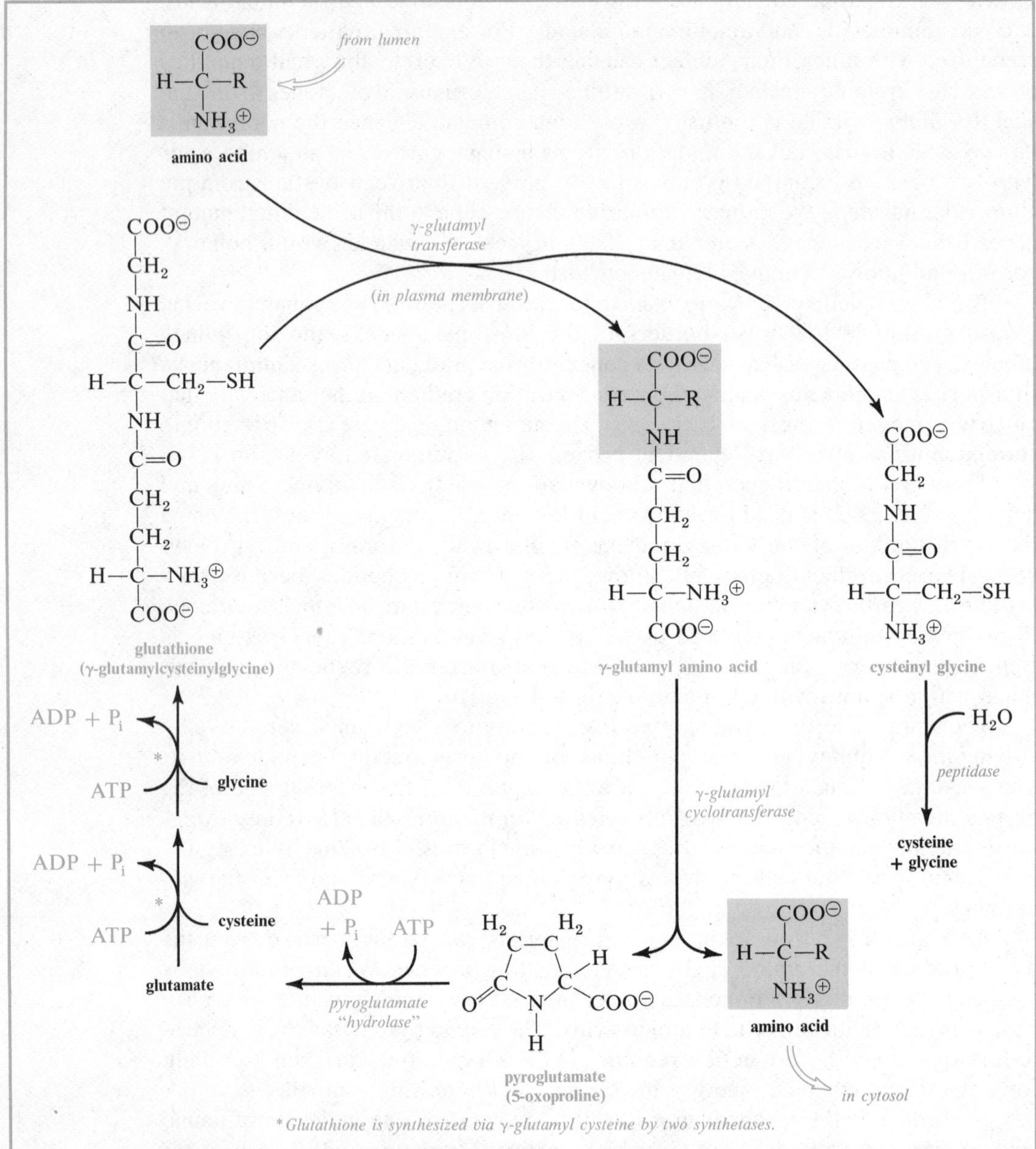

FIGURE 29–2 **Transport of amino acids by the γ-glutamyl transfer cycle of A. Meister.**

a few compounds for transport between these tissues (Fig. 29–3). For example, the skeletal muscles export nitrogen mainly in the form of **glutamine** and **alanine.**

Glutamine is neutral, non-toxic, and crosses plasma membranes readily. It is synthesized as a device for storing and transporting ammonium ions, as well as for incorporation into proteins, and this function is especially important in the brain and the striated muscles. Muscles, with their great mass, are more significant quantitatively.

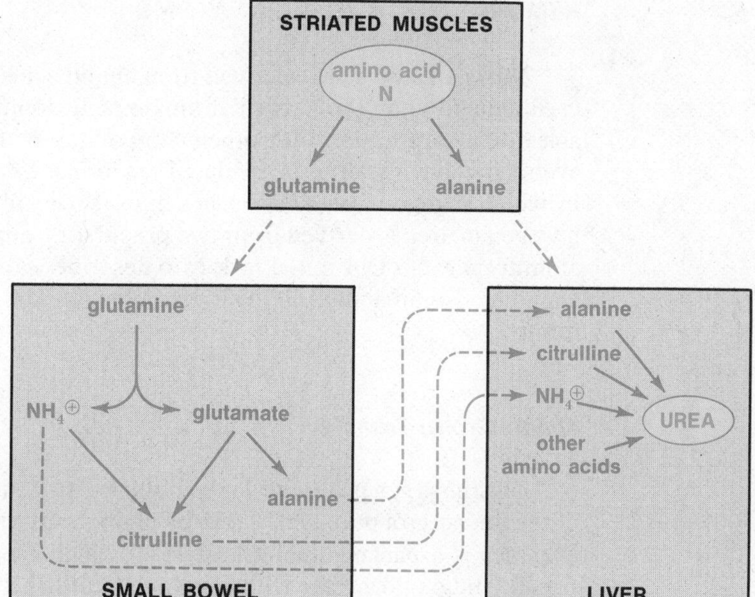

FIGURE 29–3

Transfer of nitrogen through the blood. The nitrogen of amino acids degraded in striated muscles leaves mainly as glutamine and alanine. Much of the glutamine is removed from the blood in the small bowel, reappearing in the portal blood as ammonium ions, alanine, and citrulline. The liver converts the nitrogen of these various compounds to urea for excretion.

Glutamine synthetase catalyzes the combination of ammonium ions and glutamate at the expense of ATP:

$$\underset{\text{L-glutamate}}{^{\ominus}OOC{-}CH_2{-}CH_2{-}CH(NH_3^{\oplus}){-}COO^{\ominus}} + NH_4^{\oplus} + ATP \xrightarrow{\textit{glutamine synthetase}} \underset{\text{L-glutamine}}{H_2N{-}C({=}O){-}CH_2{-}CH_2{-}CH(NH_3^{\oplus}){-}COO^{\ominus}} + P_i + ADP$$

We will see how the nitrogen from many amino acids can appear in glutamate and in ammonium ions, which can then be combined in glutamine for transport from the muscles.

Most of the glutamine is extracted from the blood by the small intestine, which contains a **glutaminase** catalyzing the hydrolysis to glutamate and NH_4^+. Some of the ammonium ions leave the small bowel as such, but some are converted to citrulline (p. 556) for export. Both the ammonium ions and the citrulline go directly from the bowel to the liver through the portal blood and are used to make urea.

The nitrogen from amino acids in the muscles or the small bowel appears first in glutamate, and then in alanine by transamination. Alanine is the amino acid with the highest concentration in blood. The liver takes up alanine and removes the nitrogen for urea synthesis through the same chain of transamination.

REMOVAL OF NITROGEN ATOMS

Nitrogen atoms are removed from amino acids in several ways; the pathways used depend upon the particular amino acid. Removal makes the nitrogen available for incorporation into other compounds or for excretion, and subsequent events involve a web of interrelated reactions. Let us focus for now upon routes by which nitrogen from amino acids finds its way into urea for excretion. All of the nitrogen in urea is derived from two precursors, **ammonium ion** and **aspartate,** one atom from each. Our initial task is to describe ways in which nitrogen from other amino acids can appear in these two compounds; we shall then see how urea is formed.

Amino Group Transfer

The most general route for removing nitrogen from amino acids is a transfer of the amino group to a keto acid by aminotransferases. The most active of these enzymes is **aspartate aminotransferase,** which catalyzes the equilibration of glutamate and oxaloacetate with α-ketoglutarate and aspartate:

$$\underset{\text{L-glutamate}}{\begin{array}{c} COO^{\ominus} \\ | \\ {}^{\oplus}H_3N-C-H \\ | \\ CH_2 \\ | \\ CH_2 \\ | \\ COO^{\ominus} \end{array}} + \underset{\text{oxaloacetate}}{\begin{array}{c} COO^{\ominus} \\ | \\ C=O \\ | \\ CH_2 \\ | \\ COO^{\ominus} \end{array}} \xrightleftharpoons[\text{(pyridoxal } PO_4)]{\text{aspartate aminotransferase}} \underset{\alpha\text{-ketoglutarate}}{\begin{array}{c} COO^{\ominus} \\ | \\ C=O \\ | \\ CH_2 \\ | \\ CH_2 \\ | \\ COO^{\ominus} \end{array}} + \underset{\text{L-aspartate}}{\begin{array}{c} COO^{\ominus} \\ | \\ {}^{\oplus}H_3N-C-H \\ | \\ CH_2 \\ | \\ COO^{\ominus} \end{array}}$$

We considered the mechanism of this readily reversible reaction earlier (p. 267).

Since aspartate aminotransferase is present in high concentrations in most cells, the reaction remains near equilibrium, and any removal of aspartate for urea synthesis will be withdrawing nitrogen from glutamate. Now, there are many other aminotransferases that catalyze transfer of amino groups from different amino acids to α-ketoglutarate, forming glutamate. These reactions are also readily reversible, so nitrogen from these amino acids will also appear, first in glutamate and then in aspartate, as aspartate is used in urea synthesis. This is the route through which at least half of the excess nitrogen is converted to urea. Figure 29–4 illustrates this sequence of events with alanine (the alanine enzyme is the second most active aminotransferase), but the same sort of thing occurs with many amino acids.

Clinical literature commonly refers to these aminotransferases as glutamate-oxaloacetate transaminase (SGOT; S for serum), and glutamate-pyruvate transaminase (SGPT).

Formation of Ammonium Ion

Oxidative Deamination. Much of the nitrogen of the amino acids sooner or later appears in the form of glutamate because of the action of aminotransferases,

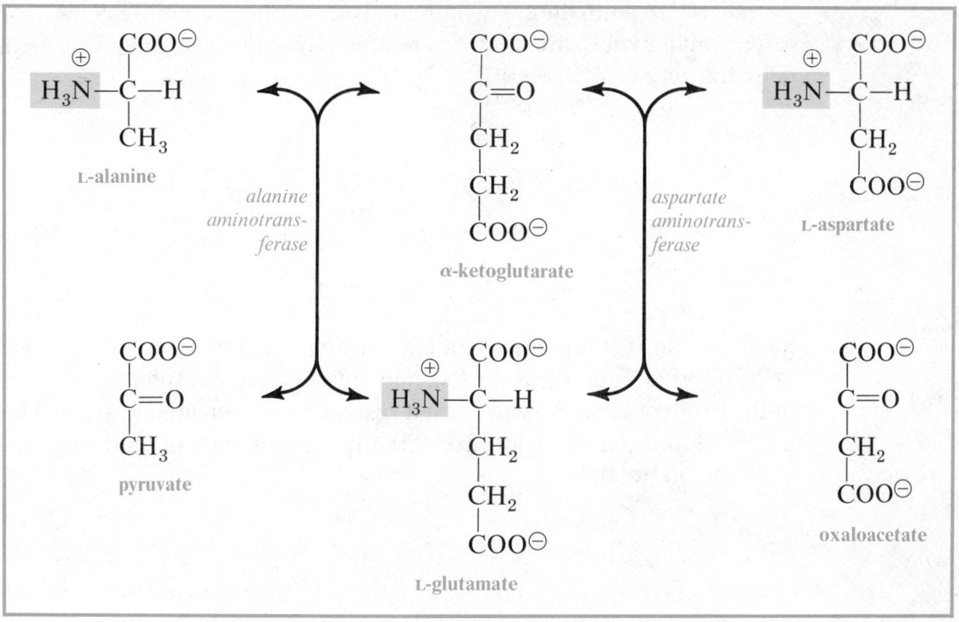

$$\text{L-alanine } (\overset{\oplus}{H_3N}-CH(CH_3)-COO^{\ominus}) + \alpha\text{-ketoglutarate } (COO^{\ominus}-C(=O)-CH_2-CH_2-COO^{\ominus}) \underset{\text{alanine aminotransferase}}{\rightleftharpoons} \text{pyruvate } (COO^{\ominus}-C(=O)-CH_3) + \text{L-glutamate } (\overset{\oplus}{H_3N}-CH(COO^{\ominus})-CH_2-CH_2-COO^{\ominus})$$

$$\text{L-glutamate} + \text{oxaloacetate } (COO^{\ominus}-C(=O)-CH_2-COO^{\ominus}) \underset{\text{aspartate aminotransferase}}{\rightleftharpoons} \alpha\text{-ketoglutarate} + \text{L-aspartate } (\overset{\oplus}{H_3N}-CH(COO^{\ominus})-CH_2-COO^{\ominus})$$

FIGURE 29–4 **The action of alanine aminotransferase makes it possible for the nitrogen of alanine to appear first in glutamate and then in aspartate.**

but all of glutamate nitrogen need not be used to form aspartate. Glutamate is also subject to oxidative deamination in a reaction catalyzed by glutamate dehydrogenase:

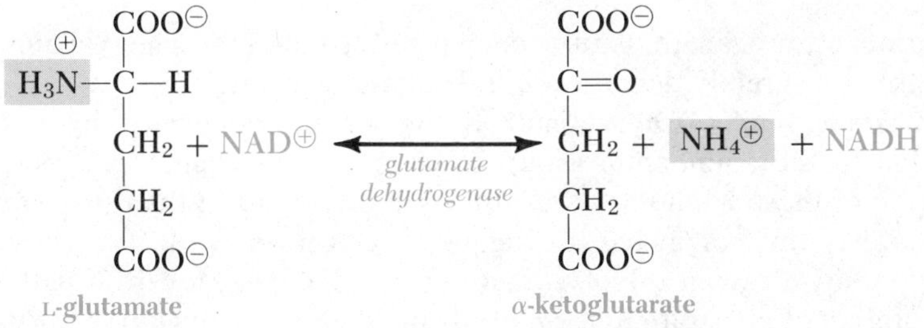

$$\underset{\text{L-glutamate}}{\overset{\oplus}{H_3N}-CH(COO^{\ominus})-CH_2-CH_2-COO^{\ominus}} + NAD^{\oplus} \underset{\text{glutamate dehydrogenase}}{\rightleftharpoons} \underset{\alpha\text{-ketoglutarate}}{COO^{\ominus}-C(=O)-CH_2-CH_2-COO^{\ominus}} + NH_4^{\oplus} + NADH$$

Glutamate dehydrogenase is present in the mitochondria of most, if not all, tissues. The enzyme will catalyze the reaction with either NAD or NADP, and it is readily reversible. We are now concerned primarily with the oxidation of glutamate and the accompanying release of ammonium ion, but should note that the reaction provides a capability for creating glutamate from ammonia by a reversal of the reaction.

The combined action of glutamate dehydrogenase and the various aminotransferases provides a route for making ammonium ion from the same amino acids that can be used to make aspartate. Here we get another glimpse of the versatility of metabolic routes: the nitrogen from many amino acids appears as glutamate; the glutamate may be used to generate aspartate by an aminotransferase reaction, or it may be used to generate ammonium ions by the dehydrogenase reaction. Which way the reactions go depends upon the relative balance of the components.

Direct Deamination. Compounds containing amino groups, such as amino acids, could hypothetically lose ammonia by a direct deamination analogous to the dehydration of an alcohol:

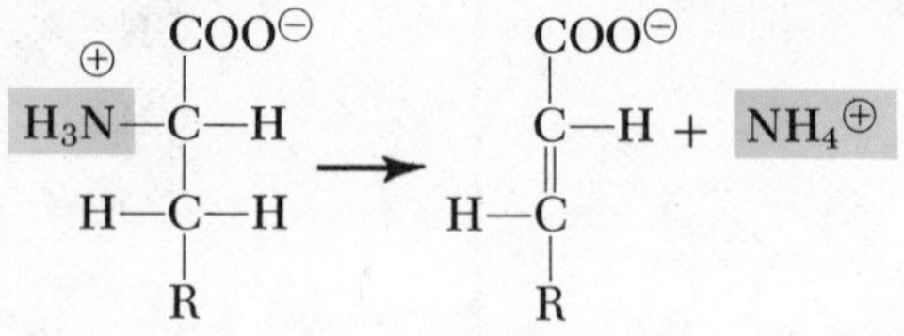

$$\overset{\oplus}{H_3N}-\underset{\underset{R}{|}}{\overset{\overset{COO^{\ominus}}{|}}{C}}(-H)-\underset{}{C}(H)(H) \longrightarrow H-C(R)=C(H)(COO^{\ominus}) + NH_4^{\oplus}$$

Such reactions with simple amines or amino acids have an unfavorable position of equilibrium, and they are feasible only in aqueous solution when the character of the product is such as to form a resonating conjugated system. Therefore, only a few of the amino acids are directly deaminated in this way, and only one, histidine, in animals:

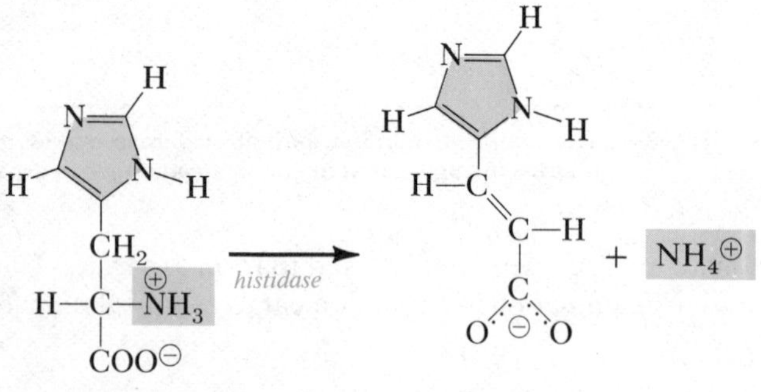

$$\text{L-histidine} \xrightarrow{\text{histidase}} \text{urocanate} + NH_4^{\oplus}$$

L-histidine urocanate

The product, urocanate, is fully conjugated. (Aspartate, phenylalanine, and tyrosine are also directly deaminated in bacteria or plants.)

Deamination by Dehydration. Serine and threonine have hydroxyl groups in addition to the ammonium group on their carbon chain. In effect, the carbon skeleton of these amino acids is already more oxidized than that of many amino acids. They are somewhat analogous to glycerate, with its adjacent hydroxyl group, and we saw in the discussion of the Embden-Meyerhof pathway that the dehydration of glycerate to form pyruvate liberates enough free energy to sustain the formation of a high-energy phosphate. Similarly, the deamination of serine and threonine to form the corresponding 2-keto compounds liberates enough free energy to be essentially irreversible:

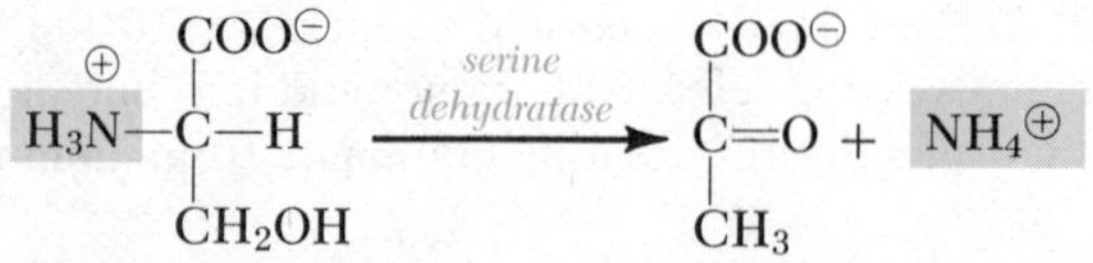

$$\overset{\oplus}{H_3N}-CH(COO^{\ominus})-CH_2OH \xrightarrow{\text{serine dehydratase}} COO^{\ominus}-C(=O)-CH_3 + NH_4^{\oplus}$$

L-serine pyruvate

$$\overset{\oplus}{H_3N}-CH(COO^{\ominus})-CH(OH)-CH_3 \xrightarrow{\text{serine dehydratase}} COO^{\ominus}-C(=O)-CH_2-CH_3 + NH_4^{\oplus}$$

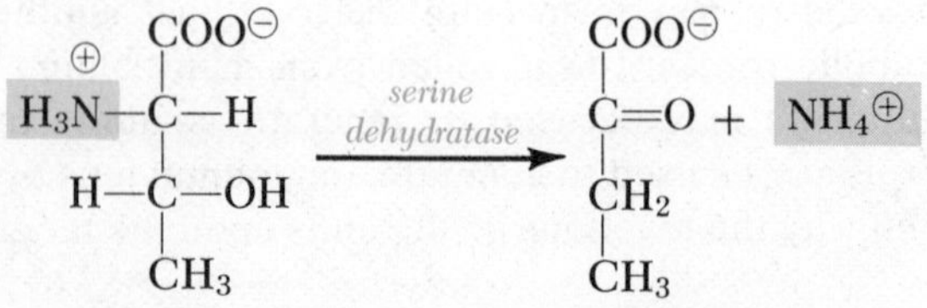

L-threonine 2-ketobutyrate

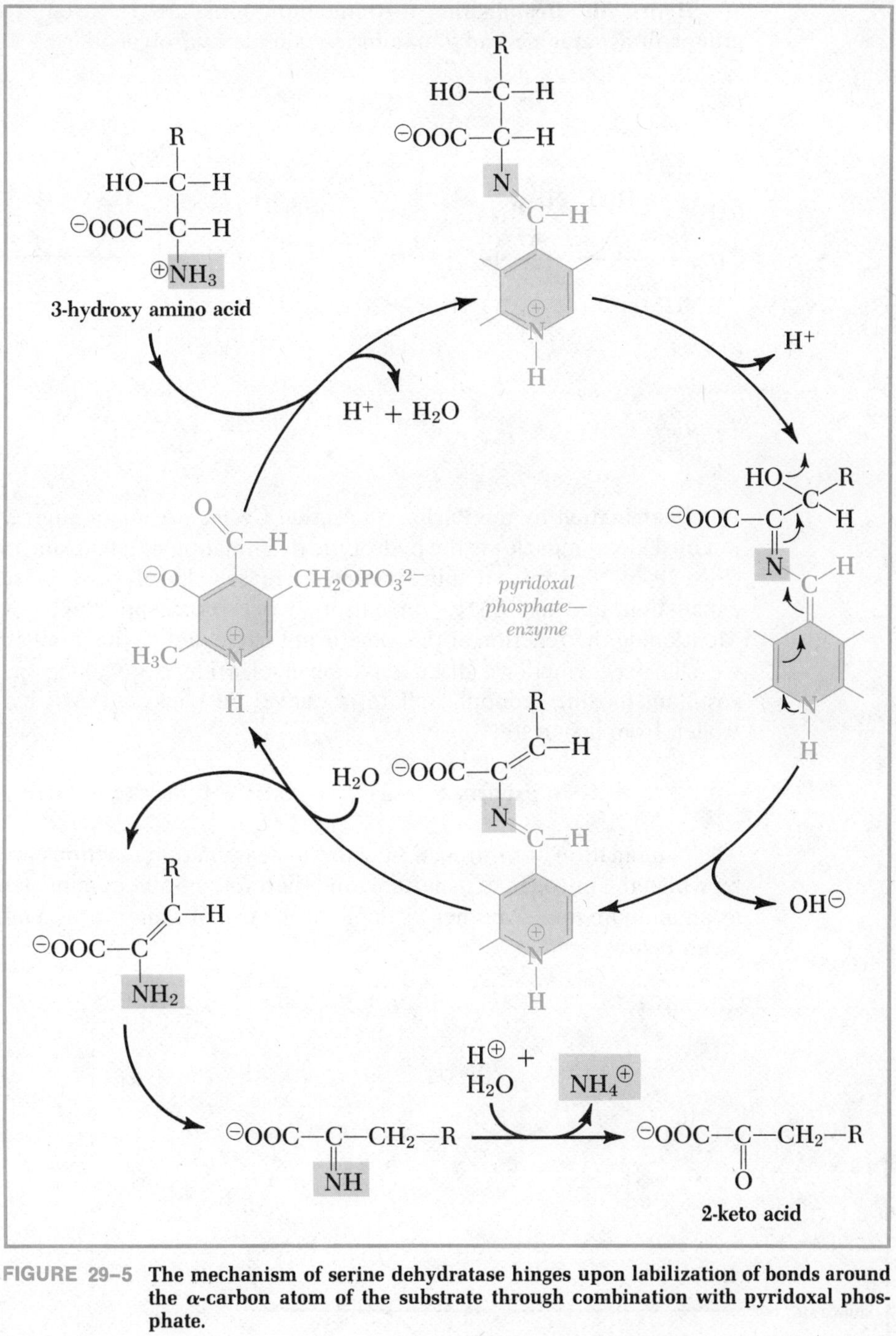

FIGURE 29–5 **The mechanism of serine dehydratase hinges upon labilization of bonds around the α-carbon atom of the substrate through combination with pyridoxal phosphate.**

The enzyme is called a dehydratase rather than a deaminase, because the reaction proceeds by the initial loss of the elements of water (Fig. 29–5). Serine dehydratase, like the aminotransferases and some other kinds of enzymes involved in amino acid metabolism, contains **pyridoxal phosphate** as a coenzyme. This coenzyme, through the formation of a Schiff's base, labilizes all of the bonds around the α-carbon of the amino acid, and which bond is broken depends upon the surrounding catalytic groups contributed by the enzyme protein.

Hydrolytic Deamination. Ammonium ions are released from the amide groups of asparagine and glutamine by simple hydrolysis:

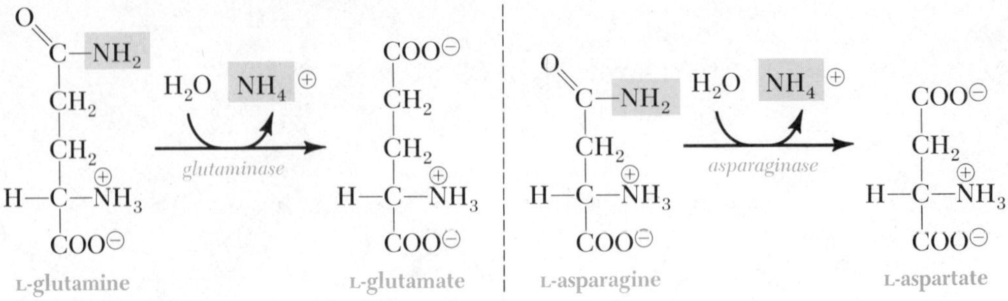

L-glutamine + H_2O → (glutaminase) → L-glutamate + NH_4^+

L-asparagine + H_2O → (asparaginase) → L-aspartate + NH_4^+

Deamination by the Purine Nucleotide Cycle. A major source of ammonium ion in skeletal muscles is the hydrolytic deamination of adenosine monophosphate (Fig. 29–6). The rate of this reaction increases sharply as a muscle approaches exhaustion, and the AMP concentration rises correspondingly sharply (p. 459). Mentioning the reaction at this time is not irrelevant to the point at hand, because we shall see when we discuss purine nucleotide metabolism (p. 654) that the resultant inosine monophosphate is converted back to AMP by transfer of nitrogen from aspartate:

$$\text{IMP} + \text{aspartate} + \text{GTP} \rightarrow \text{AMP} + \text{fumarate} + \text{GDP} + \text{P}_i$$

The combination of this reaction with the deamination reactions provides a device by which the nitrogen of aspartate, and therefore of other amino acids, can appear as ammonium ion. (Another example of the use of aspartate as a nitrogen donor is given below.)

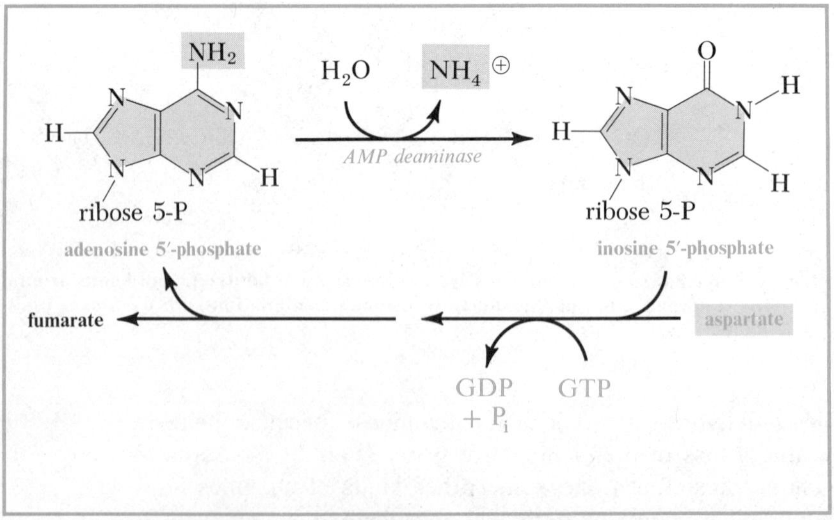

FIGURE 29–6 **The purine nucleotide cycle of J. M. Lowenstein for deamination of aspartate. AMP is deaminated by a hydrolysis, and the resultant IMP is converted back to AMP by transfer of nitrogen from aspartate, with the energy for the transfer supplied by cleavage of GTP.**

DISPOSAL OF NITROGEN — UREA SYNTHESIS

One of the fundamental facts about animals is that they are intolerant of even modest concentrations of ammonium ion in the cellular environment for reasons that are not entirely clear. Simple organisms living in water have no problem with nitrogen disposal. Ammonia diffuses out freely and is thereby diluted to a very low concentration. The first organisms probably developed ionic gradients and the basic metabolic processes of amino acid metabolism to fit this circumstance. Adaptation for more efficient disposal of nitrogen only became necessary with the development of larger size and an enclosed circulation with impermeable skin,which were necessary preludes for movement from marine to terrestrial environments.

In any event, modern animals of the higher phyla have efficient means of maintaining the ammonium ion concentration at very low levels — 70 μM is the upper limit for the normal range in human blood. Many animals, including the mammals, accomplish this by converting ammonia to urea. (It is common to speak of ammonia and ammonium ion somewhat interchangeably — they are in equilibrium. At the pH of blood, only 1 per cent of the total is present as the base, so it is better to express concentrations as ammonium ion concentrations. Ammonia is the form that passes through cell membranes. The particular ionic form used in many enzymatic reactions has not been determined.)

The liver is the principal site of urea synthesis. Now, urea is made by a simple hydrolysis of the amino acid, arginine, catalyzed by the enzyme, **arginase:**

$$H_2N{\cdots}\overset{\oplus}{C}(NH_2){\cdots}NH{-}CH_2{-}CH_2{-}CH_2{-}CH(NH_3^{\oplus}){-}COO^{\ominus} + H_2O \xrightarrow[Mn^{2+}]{arginase} H_2N{-}C(=O){-}NH_2 + {}^{\oplus}NH_3{-}CH_2{-}CH_2{-}CH_2{-}CH(NH_3^{\oplus}){-}COO^{\ominus}$$

L-arginine — urea — L-ornithine

The other product of the hydrolysis is the dibasic amino acid, ornithine. (Ornithine is the next lower homologue of lysine, but it is not used for protein formation.)

Urea synthesis therefore requires some mechanism for producing arginine. This is done by a cyclical process in which the ornithine portion of the molecule is used over and over again, acting only as a carrier for the carbon atom and two nitrogen atoms released as urea. This is analogous to the continual reuse of the oxaloacetate portion of citrate in the citric acid cycle.

We have considered the citric acid cycle earlier because of its central role in metabolism. Historically, the urea cycle became known earlier, being described in the late 1930's by H. A. Krebs (who later discovered the citric acid cycle), and this description of a metabolic process in terms of a cycle was one of the major milestones in modern biochemistry.

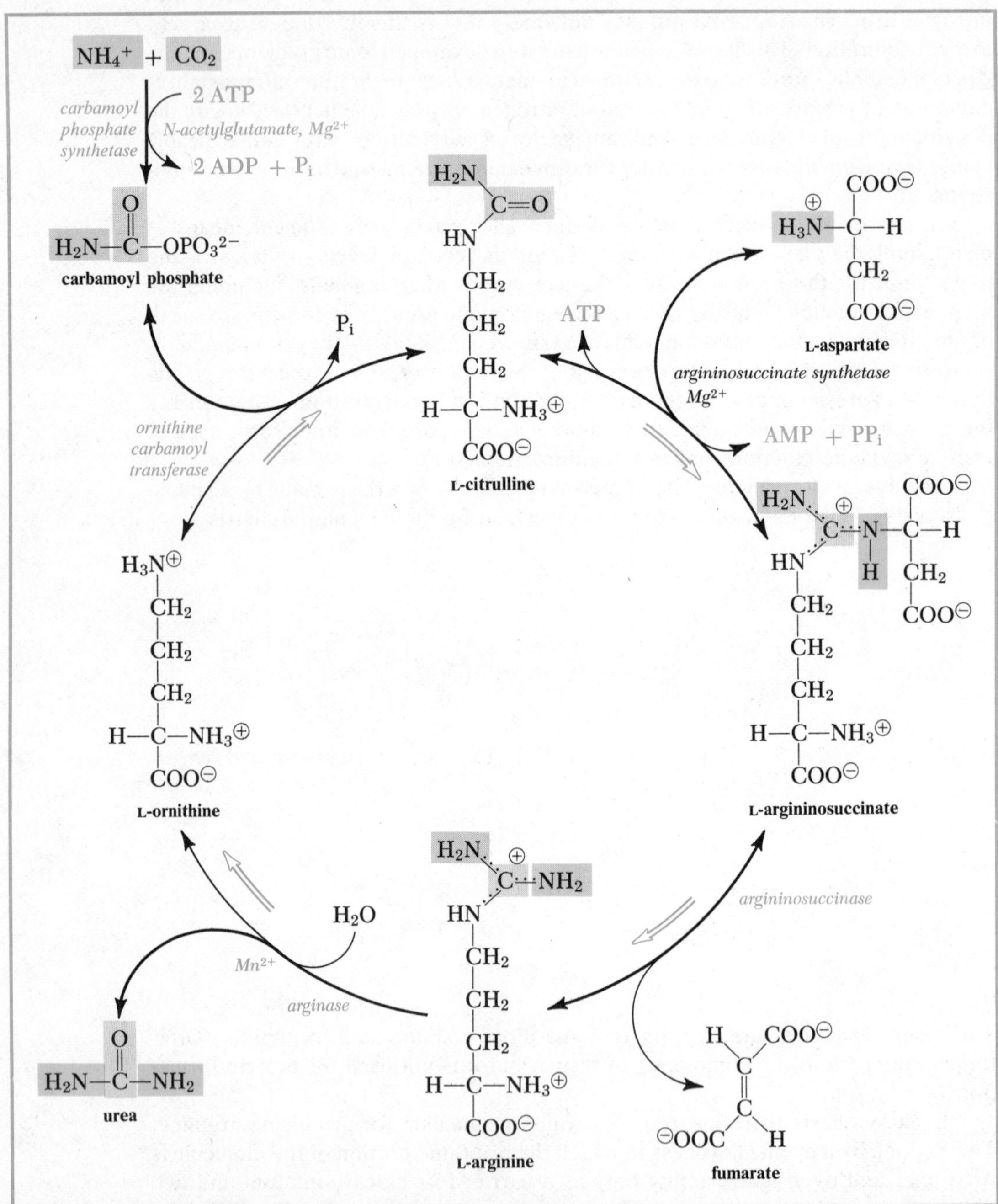

FIGURE 29–7 **The synthesis of urea by the Krebs-Henseleit cycle. Kurt Henseleit was a student who worked for his M.D. thesis with Krebs.**

As we mentioned earlier, one of the nitrogen atoms in urea is derived from ammonium ion and the other from aspartate. The carbon is obtained from CO_2. The complete process of urea synthesis is shown diagrammatically in Figure 29–7.

Synthesis of Carbamoyl Phosphate. The first step in rebuilding arginine from ornithine involves the combination of NH_4^+ and CO_2 (as bicarbonate ion) to make **carbamoyl phosphate***. This is the critical step for lowering the ammonium ion concentration, which is assured in two ways: *First,* the reaction is made irreversible by cleaving two molecules of ATP, even though the product, carbamoyl phosphate, contains a high-energy phosphate bond. *Second,* the enzyme, carbamoyl phosphate synthetase, is present in more than adequate amounts for coping with the ammonia load. Its K_M is 250 μM for NH_4^+, but the high activity of the enzyme enables it to handle the load at $[NH_4^+]$ well below this value. Since the physiological concentration of NH_4^+ (*ca.* 70 μM) *is* well below the K_M, any elevation in its concentration will greatly accelerate its removal.

The synthetase has an absolute requirement for an activator, **N-acetylglutamate,** which is not otherwise involved in urea synthesis. This activator is synthesized from acetyl coenzyme A and glutamate as a signal of elevated amino acid concentrations. (Arginine is especially effective.) Eating a high-protein diet causes more acetylglutamate to be made, and it in turn increases the activity of carbamoyl phosphate synthetase, anticipating the need for increased disposal of excess nitrogen.

Formation of Arginine via Citrulline. The carbamoyl group of carbamoyl phosphate is transferred to ornithine, forming citrulline. (Citrulline is primarily of importance as an intermediate in arginine synthesis; some residues of the amino acid occur in proteins, for example, in hair, but they are formed by modification of arginine residues, not by insertion during translation.) The reaction is driven by the loss of the high-energy phosphate in carbamoyl phosphate.

Citrulline then gains a nitrogen atom from aspartate to form arginine. The condensation of aspartate and citrulline is driven by the effective cleavage of two high-energy phosphate bonds. (PP_i is an initial product.) The resultant **argininosuccinate** is then cleaved so as to liberate arginine as one product and fumarate as the other. The effect of the sequence is to form a —C=N—C— bridge on one side of the central nitrogen atom and cleave it on the other; the condensation and cleavage effectively transfer the nitrogen of aspartate to citrulline.

Complete Cycle. In sum, the formation of one molecule of urea requires the consumption of one molecule each of NH_4^+, aspartate, and CO_2, with a cleavage of four high-energy phosphate bonds:

1. $NH_4^+ + CO_2 + 2\,ATP \rightarrow$ carbamoyl phosphate + 2 ADP + P_i
2. carbamoyl phosphate + ornithine → citrulline + P_i
3. citrulline + aspartate + ATP → argininosuccinate + AMP + PP_i
4. AMP + ATP → 2 ADP
5. $PP_i + H_2O \rightarrow 2\,P_i$
6. argininosuccinate → arginine + fumarate
7. arginine + H_2O → urea + ornithine

SUM: $NH_4 + CO_2 + 4\,ATP$ + aspartate → urea + fumarate + 4 ADP + 4 P_i

*Carbamoyl is official nomenclature for the —CO—NH_2 group, but carbamyl is sometimes used.

Compartmentation of Urea Synthesis

Nitrogen metabolism also involves transport between cellular compartments. The reactions of urea synthesis are divided between mitochondria and the cytosol in such a way as to facilitate utilization of the carbon skeletons (Fig. 29–8). We now encounter an architectural arrangement that is in many ways even more beautiful and sophisticated than the design of fatty acid synthesis, so much so that the consideration of the nuances still occupies the attention of many subtle minds. The major features are the assignment of citrulline synthesis to the mitochondrial matrix, and arginine synthesis and hydrolysis to the cystosol. Citrulline is made from ammonium ions *via* carbamoyl phosphate in the matrix, and the sources of the ammonium ions are the glutamate dehydrogenase and glutaminase reactions, both of which also occur in the matrix.

Citrulline leaves the mitochondria before it is converted, first to argininosuccinate, and then to arginine. Why? The reason is that the transfer of nitrogen from aspartate forms fumarate as the other product, and the reactions for the disposal of fumarate occur in the cytosol. Fumarate equilibrates with malate, and we have already recited how malate may be directly converted to pyruvate for use in fatty acid synthesis during times of glucose plenitude, or converted to phospho-*enol*-pyruvate *via* oxaloacetate as a source of glucose during times of glucose deprivation. After cleavage of arginine to urea and ornithine, transport of the ornithine cation back to the mitochondrial matrix is driven by the membrane potential, so

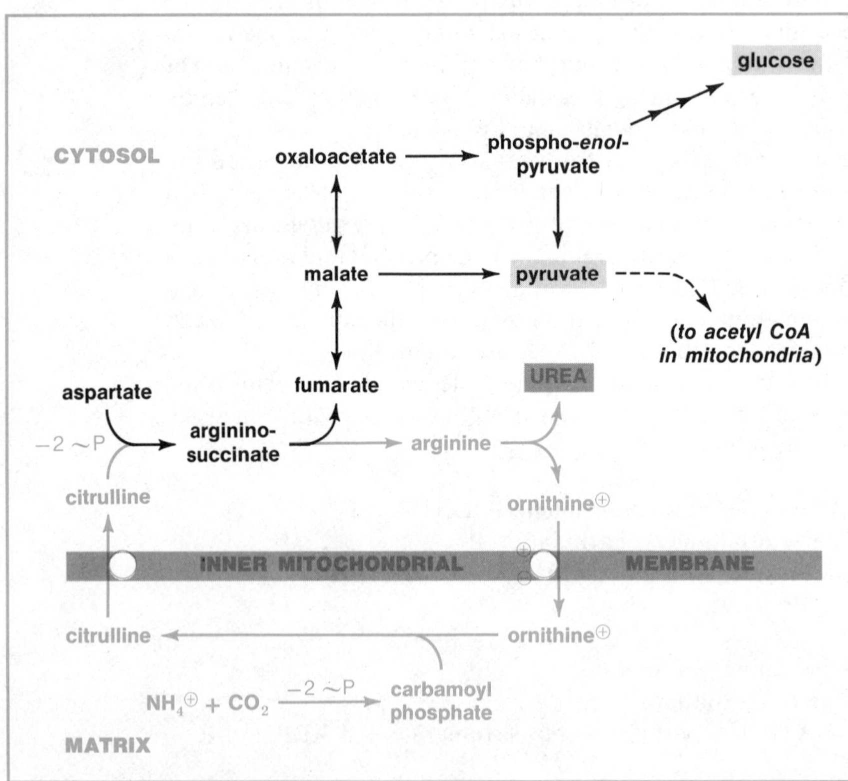

FIGURE 29–8 **The partition of urea synthesis between the mitochondrial matrix and the cytosol. Citrulline synthesis occurs in the mitochondria, and arginine synthesis and hydrolysis occur in the cytosol. After transfer of nitrogen from aspartate, the remaining fumarate can be used as a precursor of pyruvate or of glucose.**

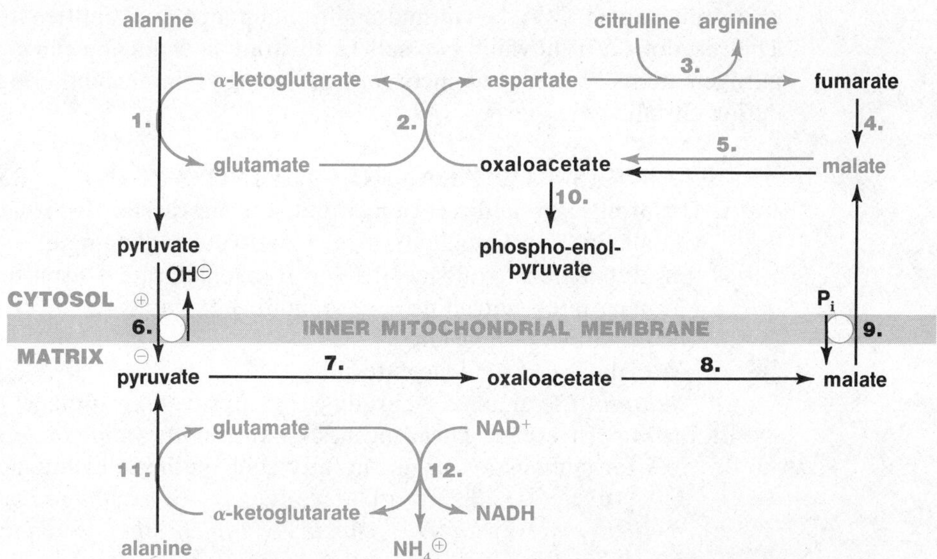

FIGURE 29–9 **Coordination of alanine metabolism in the liver with other metabolic processes. Nitrogen may be removed by alanine aminotransferase in either the mitochondria or the cytosol to appear as NH_4^+ or aspartate, respectively. In either case, the pyruvate produced can be converted to malate and then to phospho-*enol*-pyruvate as a precursor of glucose. Reactions are discussed by number in the text.**

there is an additional expenditure of *ca* 0.25~P per molecule of urea synthesized.

The message here is that urea synthesis is organized within a liver cell in such a way as to coordinate nitrogen disposal with the other reactions required for generating precursors and disposing of other products. We can see this more clearly by examining the fate of specific amino acids; let us use alanine and glutamate as examples — alanine because it comes to the liver in the highest concentration, and glutamate because it is the most active intermediate.

Disposal of Excess Alanine (Fig. 29–9). Alanine aminotransferase is contained within both the cytosol and mitochondrial matrix of human liver cells. Therefore, the nitrogen atoms will appear on glutamate, and the carbon skeleton will appear as pyruvate in both compartments. If sufficient **ammonium ions** are not available from other sources for urea synthesis, these ions can be generated from glutamate, and therefore from alanine, by the glutamate dehydrogenase reaction within mitochondria. If sufficient **aspartate** is not available from other sources, it can be generated from glutamate, and therefore from alanine, within the cytosol. (The oxaloacetate necessary for aspartate synthesis can be obtained by using the fumarate produced during urea synthesis.)

The excess **pyruvate** generated from excess alanine can be removed by the routes of fuel metabolism we discussed in earlier chapters. For example, it can be converted to acetyl coenzyme A for combustion or fatty acid synthesis (not shown in Fig. 29–9), or it may be converted to malate *via* oxaloacetate for transport out of the mitochondria into the cytosol, followed by conversion to glucose *via* phospho-*enol*-pyruvate.

The balance of the overall processes can be shown by tabulating a likely sequence of reactions. The numbers correspond to those in Figure 29–9. (High-energy phosphate, CO_2, NAD, and similar components are omitted for simplicity.*) The reactions are shown in two sets to illustrate how alanine can contribute both nitrogen atoms of urea, one incorporated during arginine synthesis and the other during citrulline synthesis.

I. METABOLISM OF ALANINE DURING ARGININE SYNTHESIS

A. **Use of nitrogen and recycling of four-carbon carrier** (*cytosol*).

1. alanine + α-ketoglutarate ⟶ pyruvate + glutamate
2. glutamate + oxaloacetate ⟶ α-ketoglutarate + aspartate
3. aspartate + citrulline ⟶ arginine + fumarate
4. fumarate ⟶ malate
5. malate ⟶ oxaloacetate

Subtotal IA: alanine + citrulline ⟶ pyruvate + arginine

B. **Use of pyruvate for gluconeogenesis.** (An equally valid conversion to acetyl CoA for combustion or use in fatty acid synthesis is not shown.)

6. $\text{pyruvate}_{cytosol}$ ⟶ $\text{pyruvate}_{mitochondria}$ (*An exchange for OH^-, which then reacts with H^+, effectively utilizes the proton concentration gradient to pump pyruvate.)*
7. pyruvate ⟶ oxaloacetate
8. oxaloacetate ⟶ malate
9. $\text{malate}_{mitochondria}$ ⟶ $\text{malate}_{cytosol}$ (*An exchange for P_i. Since P_i and malate are equally charged for this exchange, it is not actively pumped.*)

Cytosol:

5. malate ⟶ oxaloacetate
10. oxaloacetate ⟶ phospho-*enol*-pyruvate

Subtotal IB: pyruvate ⟶ phospho-*enol*-pyruvate

TOTAL I: alanine + citrulline ⟶ arginine + phospho-*enol*-pyruvate

II. METABOLISM OF ALANINE DURING CITRULLINE SYNTHESIS

A. **Use of nitrogen** (*mitochondria*).

11. alanine + α-ketoglutarate ⟶ pyruvate + glutamate
12. glutamate ⟶ α-ketoglutarate + $NH_4^{\oplus}$

not shown: $NH_4^{\oplus}$ + CO_2 ⟶ carbamoyl phosphate
carbamoyl phosphate + ornithine ⟶ citrulline

Subtotal IIA: alanine + ornithine ⟶ pyruvate + citrulline

B. **Use of pyruvate for gluconeogenesis**—same as IB except for transfer of pyruvate across inner membrane not necessary.

TOTAL II: alanine + ornithine ⟶ phospho-*enol*-pyruvate + citrulline

Disposal of Excess Glutamate. (Fig. 29–10). Excess glutamate can be handled in two ways within mitochondria. It may be oxidized by glutamate dehydrogenase to release NH_4^+, or its nitrogen may be transferred to oxaloacetate to form aspartate. In either case, the carbon skeleton of the excess glutamate appears as α-ketoglutarate, and an excess of this compound within mitochondria is oxidized to malate by the reactions of the citric acid cycle. (Here we have another illustration of the extent to which these reactions are used for other

*An even more interesting picture emerges when the reactions are tabulated in detail, and the use of all components including NAD and NADH in the two cellular compartments is accounted for. The results of such lengthy arithmetic are used in the next chapter.

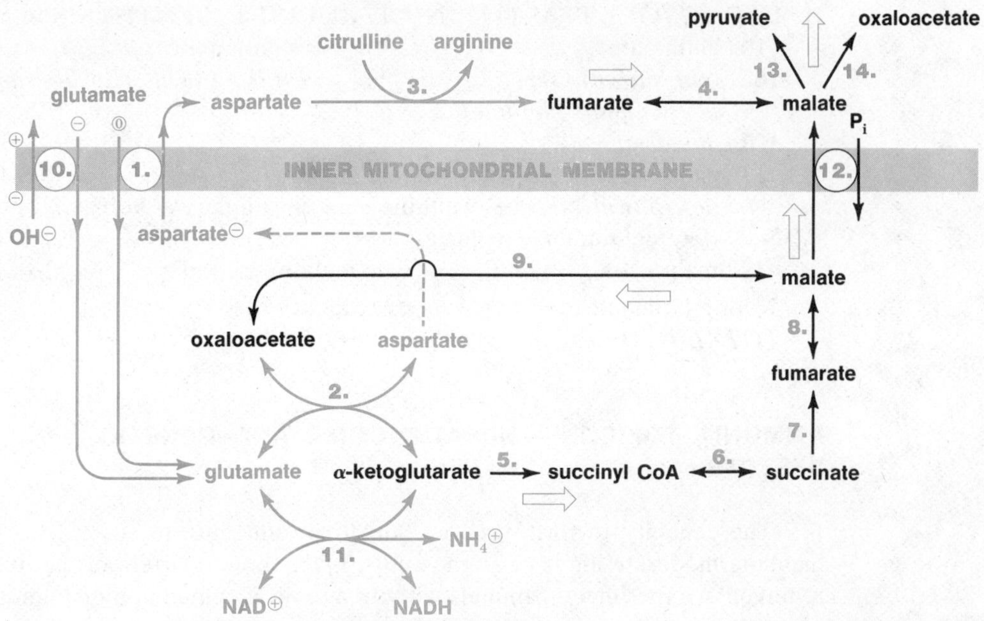

FIGURE 29–10 **Coordination of glutamate metabolism in the liver with other metabolic processes. Reactions are discussed by number in the text.**

purposes within the liver.) The fate of malate depends upon what else is happening. When aspartate is being withdrawn for use in the cytosol, the depletion of oxaloacetate to make aspartate will accelerate the oxidation of malate to oxaloacetate. The carbon skeletons will then appear as malate *via* fumarate within the cytosol, owing to urea synthesis. When glutamate is removed by oxidation, the accumulating malate will pass directly into the cytosol. By either route, malate in the cytosol is available for synthesis of fatty acids or of glucose.

A likely balance can also be cast for glutamate metabolism in the liver; again omitting high-energy phosphate, CO_2, NAD, and NADH:

I. USE OF GLUTAMATE IN ARGININE SYNTHESIS

1. $\text{glutamate}_{cytosol} + \text{aspartate}^{\ominus}_{mitochondria} \longrightarrow \text{glutamate}_{mitochondria} + \text{aspartate}^{\ominus}_{cytosol}$

(This electrogenic carrier utilizes the difference in charge across the membrane to pump unequally charged forms of the amino acids against their concentration gradients.)

Mitochondria:

2. glutamate + oxaloacetate ⟶ α-ketoglutarate + aspartate

5–9. α-ketoglutarate ⟶ oxaloacetate (reactions of citric acid cycle*)

Cytosol:

3. aspartate + citrulline ⟶ fumarate + arginine

4. fumarate ⟶ malate

13 or 14. malate ⟶ pyruvate (*fat synthesis*) or
malate ⟶ oxaloacetate (*gluconeogenesis via phospho-enol-pyruvate*)

TOTAL I: glutamate + citrulline ⟶ arginine + pyruvate (or oxaloacetate)

---

*The sum of reactions 2 and 5–9 is a mitochondrial oxidation of glutamate to aspartate:
$\text{glutamate} + 2\ O_2 + 12\ (\text{ADP} + P_i) \longrightarrow \text{aspartate} + CO_2 + 12\ \text{ATP}$

II. USE OF GLUTAMATE IN CITRULLINE SYNTHESIS

10. glutamate$^{\ominus}_{cytosol}$ + OH$^{\ominus}_{mitochondria}$ ⟶ glutamate$^{\ominus}_{mitochondria}$ + OH$^{\ominus}_{cytosol}$

10a. (*not shown*) OH$^{\ominus}_{cytosol}$ + H$^{\oplus}_{cytosol}$ ⟶ H_2O (*This pair of reactions actively pumps glutamate.*)

Mitochondria:

11. glutamate ⟶ α-ketoglutarate + $NH_4^{\oplus}$
not shown: $NH_4^{\oplus}$ + ornithine ⟶ citrulline (via carbamoyl phosphate)

5–8. α-ketoglutarate ⟶ malate

12. malate$_{mitochondria}$ + $P_{i_{cytosol}}$ ⟶ malate$_{cytosol}$ + $P_{i_{mitochondria}}$

13 or 14. malate ⟶ pyruvate or oxaloacetate

TOTAL II: glutamate + ornithine ⟶ citrulline + pyruvate (or oxaloacetate)

AMMONIA TOXICITY — HEPATIC COMA AND GENETIC DEFECTS

The capacity to form urea is more than adequate to sustain the organism handling moderate loads of amino acids. Over three quarters of the liver can be removed in experimental animals without a toxic accumulation of ammonium ions because the animals eat little and the increase in breakdown of tissue proteins is not too great.

However, extensive loss of functional liver causes a dangerous loss of ability to handle proteins in even normal quantities. Degeneration of the liver arises from a number of causes, but perhaps the most common is too much alcohol coupled with too little of other dietary constituents, leading to chronic hepatic cirrhosis*. More acute damage is caused by toxic agents, such as halogenated hydrocarbons (e.g., halothane or carbon tetrachloride), or by viral infections. A high concentration of ammonium ions results, even though there is a concomitant accumulation of glutamine to store the excess, and this may be a major factor in causing the coma frequently seen with both chronic and acute damage. (The glial cells in the brain — the cells surrounding neurons — are particularly affected, and assume the same abnormal forms that can be created experimentally by toxic doses of ammonium ions. These cells are apparently responsible for much of the ammonium ion turnover in the brain; they are the sites of glutamine synthesis.)

There have been various reports of improvement in hepatic coma through the administration of glutamate, ornithine, or citrulline to bolster the inadequate rate of ammonium ion removal, but the consequences of hepatic degeneration are usually so manifold that little can be done other than to diminish the sources. The easiest way to do this is to cut down on the protein intake, but this is not without its own hazards, since many of the individuals with chronic hepatic failure have already been on a diet deficient in protein.

Some help is obtained by diminishing the quantity of ammonium ion entering the portal circulation from the bowel. This can be done by giving antibiotics to decrease the intestinal bacteria, because they produce large bursts of ammonium ion after a protein meal and also form a urease that hydrolyzes urea back to ammonium and bicarbonate ions. (A substantial amount of the excess nitrogen in the body normally recirculates through urea in this way.)

Absorption of ammonium ions is also diminished by feeding a non-absorbable synthetic sugar, lactulose, that is fermented by intestinal bacteria so as to make

*In older days when the University of Virginia had a certain reputation for bibulosity, an in-house pathologist said that no member of the faculty came to autopsy with a normal liver. Now, not even the Big Ten can boast a more soberly dedicated collection of savants.

the contents more acidic. Only neutral ammonia will go across cell membranes, and increasing the $[H^+]$ increases the $[NH_4^+]/[NH_3]$ ratio, thereby diminishing the rate of absorption even though the total ammonia-ammonium ion pool is unchanged.

None of these measures solves the overall problem completely, because half of the nitrogen atoms must pass through ammonium ion on their way to urea, and elimination of all ammonium ion formation would also eliminate all urea formation.

Genetic Defects in Urea Synthesis. Although not common, such defects provide evidence for the deleterious effects of ammonium ion accumulation. First, it must be emphasized that infants with a complete absence of an enzymatic activity necessary for urea synthesis die promptly after the first feeding of a protein meal. (These total impairments are described as "neonatal" for this reason.)

Partial defects permitting continued, albeit often uncertain, survival are known for each of the enzymes involved in urea synthesis. The most damaging are those affecting the first two enzymes, carbamoyl phosphate synthetase or ornithine carbamoyl transferase, which result in a hyperammonemia. Surprisingly, those who survive a partial impairment usually have a normal, or nearly normal, excretion of urea. That seems strange. What, then, is the difficulty? The problem is that the rate of urea synthesis is normal only because the ammonium ion or carbamoyl phosphate concentrations have risen to a higher fraction of the K_M value so that fewer enzyme molecules can work faster. It is the higher ammonium ion concentration (as high as one millimolar), not a failure in urea synthesis, that is responsible for the symptoms, which include lethargy, stupor, vomiting, convulsions, and other indications of central nervous system impairment. Fewer than 50 patients with these two deficiencies have been described.

Partial impairments of the enzymes converting citrulline to arginine sometimes have less severe consequences, probably because the ammonium ion concentrations do not rise to such high levels. Symptoms may appear gradually over weeks or months, but some degree of mental retardation is common. A substantial fraction of the nitrogen in the urine appears as citrulline (argininosuccinate synthetase deficiency), or as argininosuccinate (argininosuccinase deficiency), although urea is still the major product. In these, as with other point mutations, a spectrum of abnormalities is created, depending upon the part of the enzyme molecule affected. It was demonstrated through examination of the enzyme in cultured fibroblasts that the mutation of argininosuccinate synthetase in one patient resulted in an altered K_M for citrulline, from 0.4 to over 10 millimolar, which would in itself lead to an increased accumulation of citrulline before its rate of removal balanced its rate of formation.

FURTHER READING

Meister, A.: (1965) *Biochemistry of the Amino Acids,* 2nd ed. Academic Press. Authoritative two-volume source for details of reaction. No more recent publication supplants it.

Campbell, J. W., and L. Goldstein, eds.: (1972) *Nitrogen Metabolism and the Environment.* Academic Press. Discusses comparative biochemistry of nitrogen excretion.

Grisolia, S., R. Báguena, and F. Mayor: (1976) *The Urea Cycle.* Wiley. Many general reviews of pertinent biochemistry and clinical applications. The best recent source.

Prusiner, S., and E. Stadtman: (1973) *The Enzymes of Glutamine Metabolism.* Academic Press. Those with the time and interest ought to read the accounts of the regulation of bacterial glutamine synthetase.

Felig, P.: (1975) *Amino Acid Metabolism in Man*. Annu. Rev. Biochem., *44*:933. Describes movement of amino acids through the circulation.

Rognatad, R.: (1977) *Sources of Ammonia for Urea Synthesis in Isolated Rat Liver Cells*. Biochem. Biophys. Acta, *496*: 249.

Lowenstein, J. M.: (1972) *Ammonia Production in Muscle and Other Tissues: The Purine Nucleotide Cycle*. Physiol. Rev., *52*: 382. Review by discoverer of the cycle.

McGivan, J. D., and J. B. Chappell: (1975) *On the Metabolic Function of Glutamate Dehydrogenase in Rat Liver*. FEBS Lett., *52*: 1.

Martinez-Hernandez, A., K. P. Bell, and M. D. Norenberg: (1977) *Glutamine Synthetase: Glial Localization in Brain*. Science, *195*: 1356.

DeRosa, G., and R. W. Swick: (1975) *Metabolic Implications of the Distribution of Alanine Aminotransferase Isoenzymes*. J. Biol. Chem., *250*: 7961. Includes discussion of mitochondrial transport processes related to urea synthesis.

Krebs, H. A., P. Lund, and M. Stubbs: (1976) *Interrelations Between Gluconeogenesis and Urea Synthesis*. Pages 269–292 *in* R. W. Hanson, and M. A. Mehlman. eds.: *Gluconeogenesis: Its Regulation in Mammalian Species*. Wiley.

Hochachka, P. W., and G. N. Somero: (1973) *Strategies of Biochemical Adaptation*. W. B. Saunders.

Fischer, J. E., and R. J. Baldessarini: (1976) *Pathogenesis and Therapy of Hepatic Coma*. Progr. Liver Dis., *4*: 363. Biased, but useful, review.

Meister, A., and S. S. Tate: (1976) *Glutathione and Related γ-Glutamyl Compounds: Biosynthesis and Utilization*. Annu. Rev. Biochem., *45*: 559. Includes discussion of amino acid transport.

Marstein, S. V., et al.: (1976) *Biochemical Studies of Erythrocytes in a Patient with Pyroglutamic Acidemia*. N. Engl. J. Med., *295*: 406. Also see editorial on p. 441. More discussion of amino acid transport.

Oxender, D. L., et al.: (1977) *Neutral Amino Acid Transport Systems of Tissue Culture Cells*. J. Biol. Chem., *252*: 2675 ff.

30 AMINO ACIDS: DISPOSAL OF THE CARBON SKELETONS

When amino acids are directly degraded without being used to make other cellular constituents, all of the carbon atoms appear as CO_2 or as eight familiar intermediates of fuel metabolism (Fig. 30–1). These are acetoacetate, acetyl coenzyme A, crotonyl coenzyme A, pyruvate, α-ketoglutarate, succinyl coenzyme A, oxaloacetate, or fumarate. How has this come about? It seems reasonable to assume that the 20 amino acids found in proteins were selected from many possible structures during early evolution, not only because of the particular properties they contribute to polypeptides, but also because their metabolism can be integrated easily with the metabolism of glucose and fatty acids.

The ultimate disposition of the carbon atoms from amino acids therefore depends upon what is happening to fatty acids and carbohydrates, with these exceptions: acetyl groups formed within muscles, and acetoacetate from any source, are oxidized to CO_2 and H_2O at all times. Within the liver, where most carbon skeletons are handled, they may at some times be used to form glucose and 3-oxybutyrates, and at other times to form fatty acids, as was noted for alanine and glutamate in the preceding chapter. Let us review the choices.

During fasting or times of low carbohydrate intake when fatty acids are the principal fuel, pyruvate and intermediates in the citric acid cycle are converted to glucose:

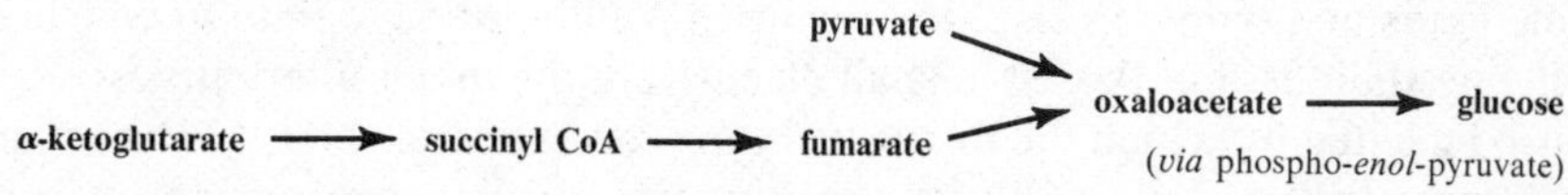

At the same time, intermediates of fatty acid oxidation (crotonyl coenzyme A and acetyl coenzyme A) are converted to the 3-oxybutyrates:

crotonyl CoA ⟶ **acetyl CoA** ⟶ **acetoacetate** ⟶ **D-3-hydroxybutyrate**

Glucose and the 3-oxybutyrates are then used by the other tissues. Amino acids acting as precursors of glucose under these conditions are said to be glucogenic; those acting as precursors of the 3-oxybutyrate are said to be ketogenic.

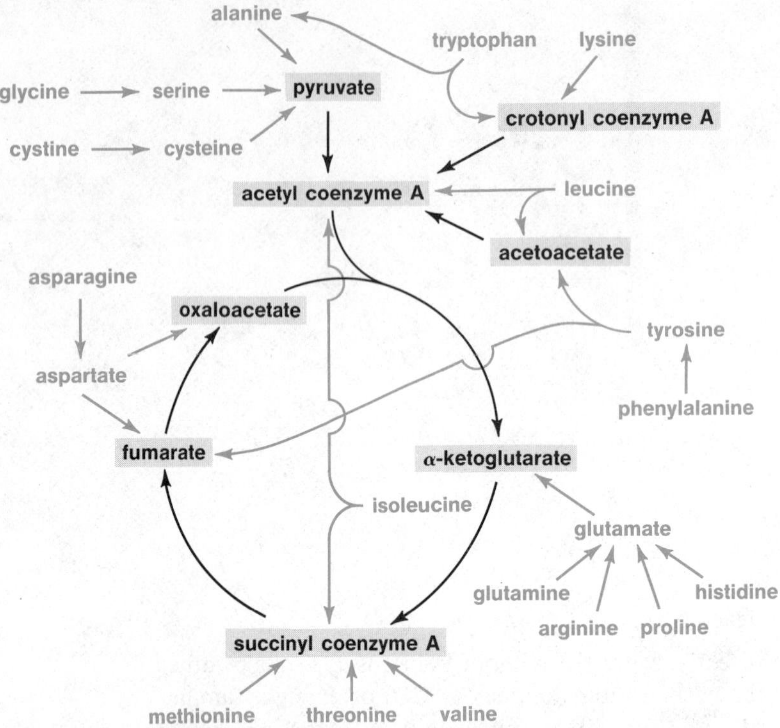

FIGURE 30–1 **Outline of the fate of the carbon skeletons of amino acids when used as fuels.**

After a carbohydrate meal when glucose is already present in ample supply, part of the intermediates may be converted first to acetyl coenzyme A and then to fatty acids for storage:

$$\text{α-ketoglutarate} \longrightarrow \text{succinyl CoA} \longrightarrow \text{fumarate} \longrightarrow \text{malate} \longrightarrow \text{pyruvate} \longrightarrow$$
$$\text{acetyl CoA} \longrightarrow \text{palmitate} \longrightarrow \text{oleoyl CoA}$$

With this background, let us consider in this and the following chapter some routes by which each amino acid's carbon skeleton is converted to these common intermediates and then correlate the results. (We do not have time to consider all known possibilities, but those we shall discuss are the major alternatives.) Routes will also be listed by which the nitrogen atoms can appear as NH_4^+ or as aspartate for incorporation into urea. All of the amino acids are glucogenic during starvation unless otherwise specified.

Aspartate

Nitrogen: Used directly in arginine synthesis, or appears as NH_4^+ in purine nucleotide cycle (pp. 554 and 556).

Carbons: Appear as fumarate in arginine synthesis or in purine nucleotide cycle. Also may appear as oxaloacetate as a result of transamination in peripheral tissues.

GLUTAMATE AND ITS PRECURSORS

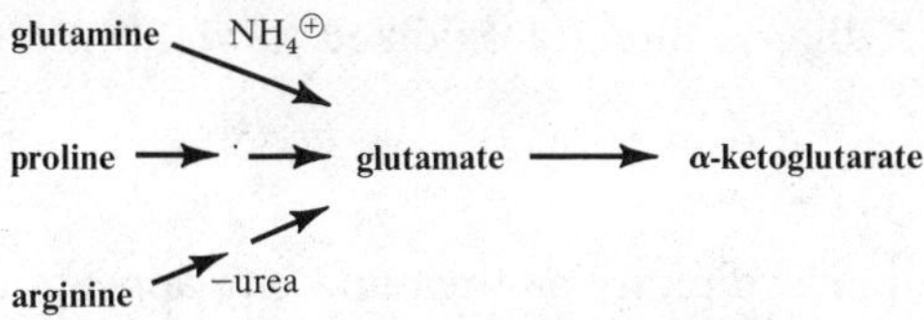

Glutamate

Nitrogen: Appears directly as NH_4^+ (glutamate dehydrogenase, p. 551) or as aspartate (aspartate aminotransferase, p. 550).

Carbons: Appear directly as α-ketoglutarate. Therefore, three of the five carbon atoms may appear as glucose, or two as fatty acids, with the remainder appearing as CO_2.

Glutamine

Nitrogen: Amide nitrogen appears as NH_4^+ (glutaminase, p. 549).

Carbons: Appear as glutamate, which is then metabolized as given above.

Proline

Nitrogen and carbon skeleton: Appear as glutamate.

The transformation of L-proline to L-glutamate is simple. Proline is oxidized by a cytochrome-linked enzyme in mitochondria, presumably a flavoprotein,

L-proline

proline dehydrogenase (mitochondrial)

Q

FAD-protein(?)

QH_2

H_2O

Δ^1-pyrroline-5-carboxylate

glutamate semialdehyde

Δ^1-pyrroline-5-carboxylate dehydrogenase

$NAD^{\oplus} + 2\,H_2O$

$NADH + H^{\oplus}$

L-glutamate

FIGURE 30–2 **Oxidation of proline to glutamate.**

forming an unsaturated ring. Although simple hydrolysis of the ring at the double bond spontaneously yields the semialdehyde of glutamate, it is the ring form, pyrroline carboxylate, that is directly oxidized to yield free glutamate (Fig. 30–2).

Arginine

Nitrogen: Two appear directly as urea and one appears in glutamate.
Carbons: One appears directly as urea and five appear in glutamate.

As we have seen, arginine is hydrolyzed to urea and ornithine as a part of the process of urea synthesis. Ornithine is re-used in the urea cycle, but excess arginine derived from the diet or tissue proteins causes the formation of excess ornithine beyond the amount needed to sustain the urea cycle.

The terminal amino group of ornithine is transferred to α-ketoglutarate:

arginase, Mn^{2+}; H_2O; $H_2N-C(=O)-NH_2$ urea

L-arginine → L-ornithine (ornithine aminotransferase; α-Kg* → Glu*) → glutamate semialdehyde ($H_2O + NAD^{\oplus}$ → $2H^{\oplus} + NADH$; ?) → L-glutamate

*α-Kg = α-ketoglutarate
Glu = glutamate

The product is the semialdehyde of glutamate, which is oxidized to glutamate in an essentially irreversible reaction. (Recall that the equilibrium position for the oxidation of aldehydes is far in the direction of the acid (p. 408). A hereditary deficiency of the aminotransferase has been shown to cause atrophy of the retina, resulting in blindness.

PRECURSORS OF PYRUVATE

Alanine

Nitrogen: Appears in glutamate by transamination (p. 551).
Carbons: Appear as pyruvate.

Serine

Nitrogen: Directly deaminated to NH_4^+ by serine dehydratase (p. 552).
Carbons: Appear as pyruvate.

Cysteine and Cystine

Nitrogen: Appears in glutamate by transamination.
Carbons: Appear as pyruvate.
Sulfur: Appears as sulfate.

When we discussed the formation of proteins, we noted that cysteine residues in proteins (sulfhydryl groups) are interconverted with cystine residue (disulfide groups) by exchange reactions (p. 142). This is a more general type of reaction within cells so that the cystine present in proteins undergoing degradation is interconvertible with cysteine, which is the form that enters the major pathways of metabolism.

It is clear that the metabolism of cysteine produces pyruvate and inorganic sulfate as the major products, but there is still uncertainty over the routes. In what may be the major pathway, cysteine is oxidized in the cytosol to cysteine sulfinate (Fig. 30–3), which undergoes transamination. The resultant 2-keto compound releases SO_2 as sulfite; sulfite is oxidized to inorganic sulfate which may be excreted or incorporated into other cellular constituents.

Sulfite oxidase is a complex protein, containing molybdenum and a hemoprotein resembling cytochrome b_5. It is located in the intermembrane space of mitochondria, where its electrons can be transferred to cytochrome c on the outer surface of the inner membrane. As a result, one high-energy phosphate bond is generated per sulfite oxidized. This is a trivial part of the metabolic economy, but the oxidation of sulfite is evidently not at all trivial to the function of the brain. Infants born with a congenital deficiency of sulfite oxidase have such a severe neurological impairment that they shortly become functionally decerebrate.

There are also uncharacterized routes by which cysteine directly loses H_2S, which is oxidized to sulfite and then to sulfate. (Plants and bacteria have an enzyme analogous to serine dehydratase, which catalyzes loss of H_2S from cysteine with formation of NH_4^+ and pyruvate, but this enzyme has not been demonstrated clearly in animals.)

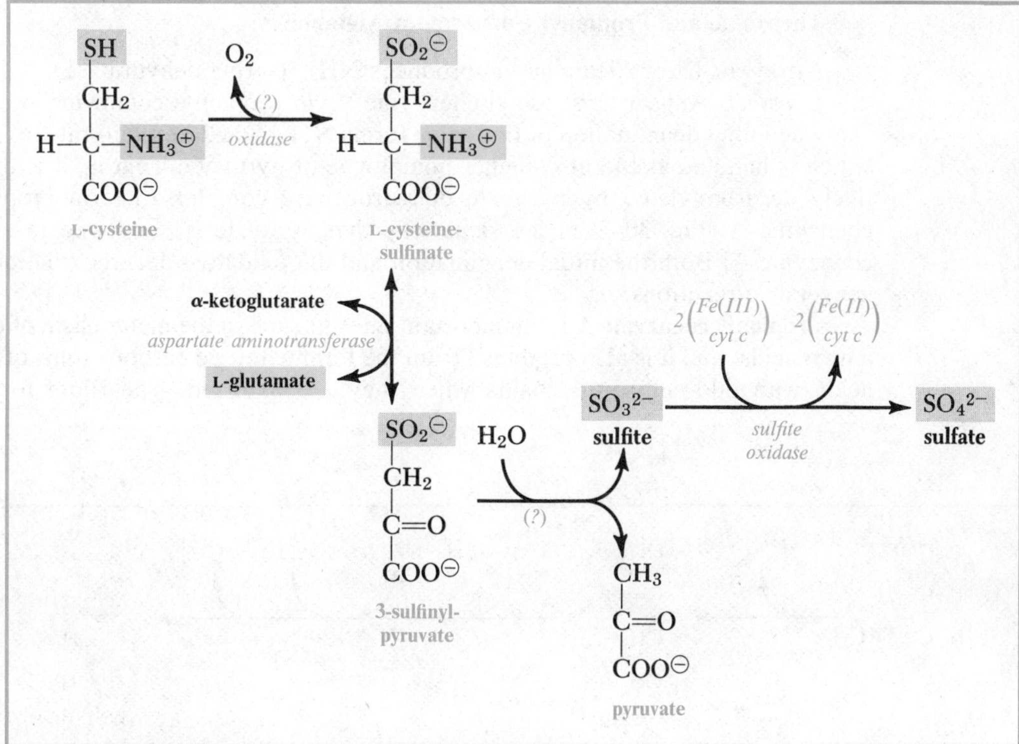

FIGURE 30–3 **The oxidation of cysteine to pyruvate and sulfate.**

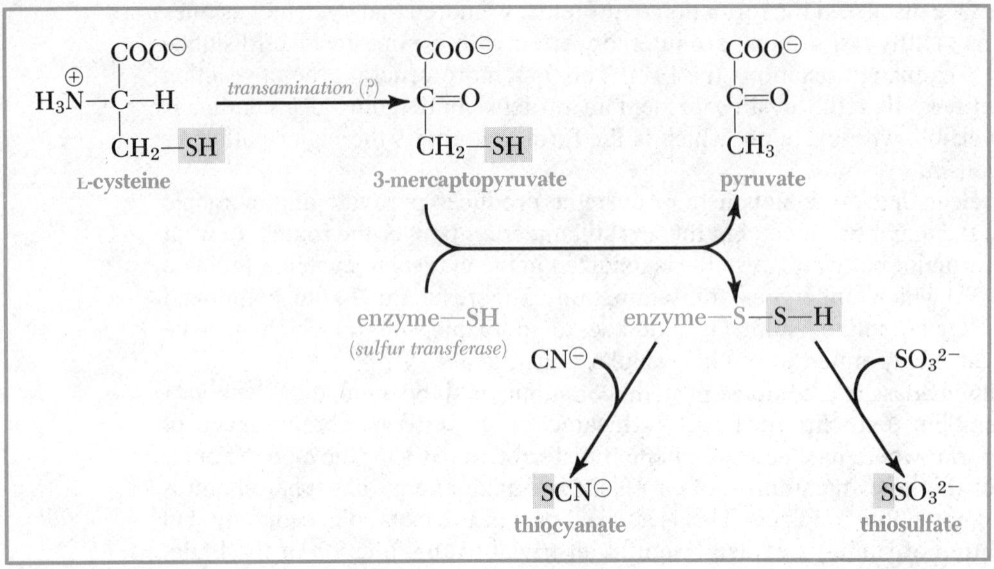

FIGURE 30–4 **Minor routes of metabolism of cysteine include the formation of thiocyanate and thiosulfate.**

A variety of other possible routes of cysteine metabolism have been reported, many quantitatively minor and of uncertain purpose. For example, a small fraction of the sulfur of the body is excreted in the urine as **thiosulfate** and **thiocyanate**, which apparently arise by the route shown in Figure 30–4. The carbon skeleton appears as pyruvate in most of these routes,

Threonine and Propionyl Coenzyme A Metabolism

Nitrogen: Direct deamination produces NH_4^+ (serine dehydratase).

Carbons: Appear as succinyl coenzyme A *via* propionyl coenzyme A.

The initial deamination of threonine forms NH_4^+ and 2-ketobutyrate (p. 552), which is handled as the next higher homologue of pyruvate. That is, it is oxidatively decarboxylated by pyruvate dehydrogenase complex to form propionyl coenzyme A (Fig. 30–5) in the same way that pyruvate is converted to acetyl coenzyme A. Both the initial deamination and the oxidative decarboxylation are irreversible reactions.

Propionyl coenzyme A is an important intermediate in the metabolism of other amino acids, and it is also produced from the terminal three carbon atoms of fatty acids with odd-numbered chains when they are oxidized. The route for me-

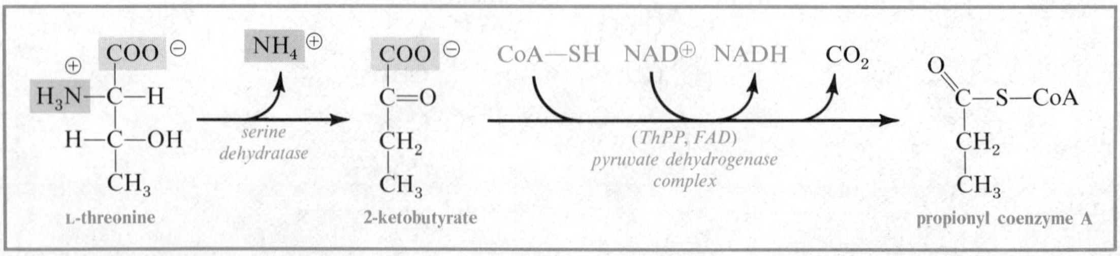

FIGURE 30–5 **2-Ketobutyrate arising from threonine is oxidatively decarboxylated to propionyl coenzyme A.**

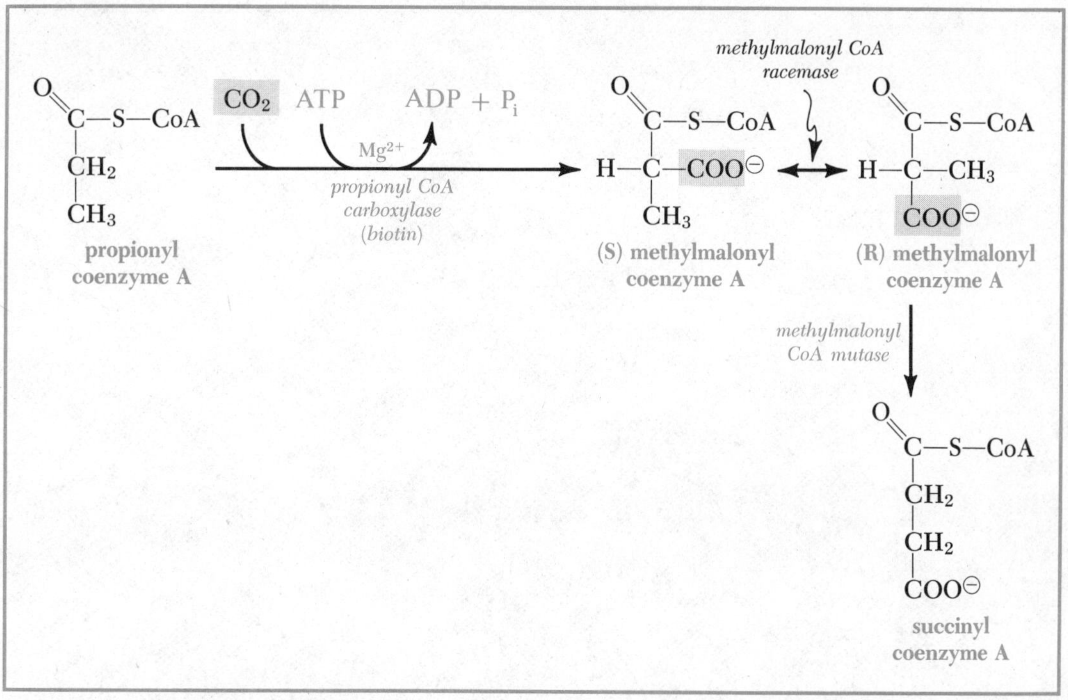

FIGURE 30–6 **Propionyl coenzyme A is metabolized by a carboxylation followed by rearrangements to form succinyl coenzyme A.**

tabolism of this compound is summarized in Figure 30–6. The first reaction is catalyzed by propionyl CoA carboxylase, which contains biotin and acts by the same sort of mechanism seen with acetyl CoA carboxylase and pyruvate carboxylase. The product is **methylmalonyl coenzyme A,** which is an optically active compound. A racemase exists to catalyze the equilibration of the two isomers; the (R) isomer is the substrate for the next reaction. (It isn't at all clear why different isomers are used in this sequence.)

The final reaction is the rearrangement of (R)-methylmalonyl coenzyme A to succinyl coenzyme A, catalyzed by a mutase. Since succinyl coenzyme A is converted to oxaloacetate, it follows that threonine and all other compounds giving rise to propionyl coenzyme A or methylmalonyl coenzyme A are glucogenic.

We have not seen anything comparable to the mutase reaction heretofore. It involves a coenzyme derived from vitamin B_{12}, and let us consider this important compound.

Cobalamin Coenzyme and Vitamin B_{12}

The coenzyme for methylmalonyl CoA mutase is **adenosylcobalamin** (Fig. 30–7) — a very complicated molecule containing **cobalt** bound to nitrogen or carbon on all six coordination positions. The salient part of the molecule is the large tetrapyrrole ring surrounding the cobalt, which resembles a porphyrin ring superficially, but differs in being more saturated and in lacking one methylene bridge. This is known as the **corrin** ring. The cobalamins are corrinoid compounds.

The cobalamin coenzyme has two unusual chemical features. First, it contains a metal-carbon bond, the only known biological example of this linkage.

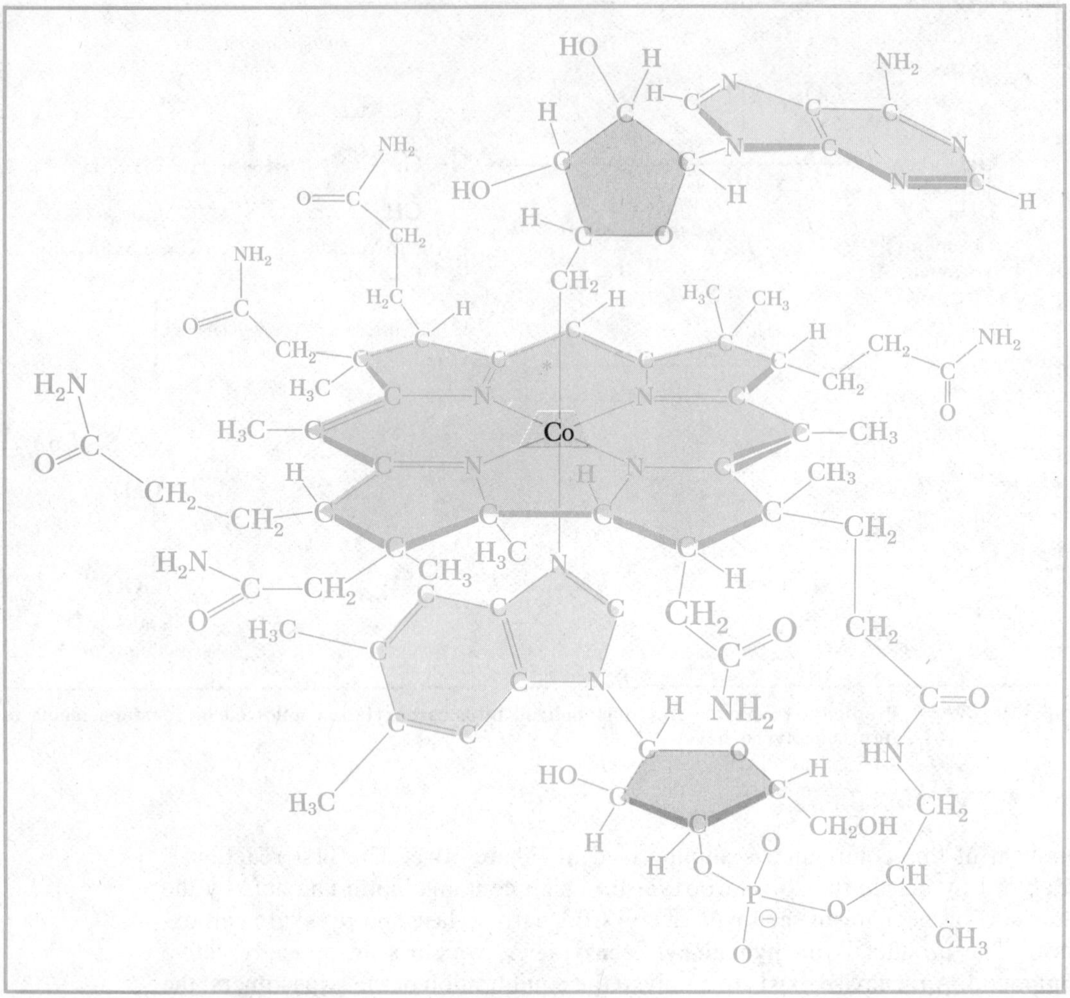

FIGURE 30–7 **Adenosylcobalamin. The adenosyl moiety *(top)* is attached to cobalamin through its cobalt atom *(center)*. Cobalamin includes a corrin ring *(center blue shading)* and a benzimidazole group *(bottom blue shading)*.**

Secondly, the cobalt is present at a univalent oxidation state and cob(I)alamin is the most powerful nucleophile known. The detailed mechanism of action of the coenzyme is still unknown, but it involves rupture of one of the cobalt coordination bonds by the substrate, with formation of a free radical and migration of the substituent groups (Fig. 30–8).

What is commonly referred to as vitamin B_{12} is the same as adenosylcobalamin except that the adenosyl group is replaced by a cyanide ion, and the valence of the cobalt is +3 so that there is a net charge of +1 remaining on the metal. This is cyanocobalamin, which is the form of the compound that was originally isolated during the search for the vitamin. Although we ordinarily think of cyanide as a dangerous poison, it is produced by many microorganisms in small amounts, and the affinity of the cobalamins for cyanide is so great that cyanocobalamin may be created during the handling of the microbial cultures from which it is isolated. Indeed, large doses of hydroxycobalamin have been used in treating cases of cyanide poisoning. Most of the cobalamin in animal tissues is present as the adenosyl or methyl derivatives, which are therefore the natural dietary forms of the vitamin.

The search for the vitamin was originally spurred because it is deficient in the tissues of humans with **pernicious anemia**. The name implies that the untreated anemia is ultimately fatal — pernicious. The effects include serious disturbances of the central nervous system that result in abnormal sensation, motion, behavior, or thought.

Curiously, pernicious anemia turned out to be a defect of the stomach, rather than a dietary deficiency disease. The absorption of dietary cobalamins depends upon the formation by the gastric mucosa of a carbohydrate-rich protein known as the **intrinsic factor**. People with pernicious anemia do not make this protein. Before purified cobalamins were available, treatment consisted of eating hog stomach preparations to supply the missing intrinsic factor and liver to supply the vitamin (the extrinsic factor).

The amount of cobalamin required in the diet is very small — about one nanomole a day for a normal individual. Even those with pernicious anemia can absorb enough if the oral dose is raised to the level of one micromole or so. The

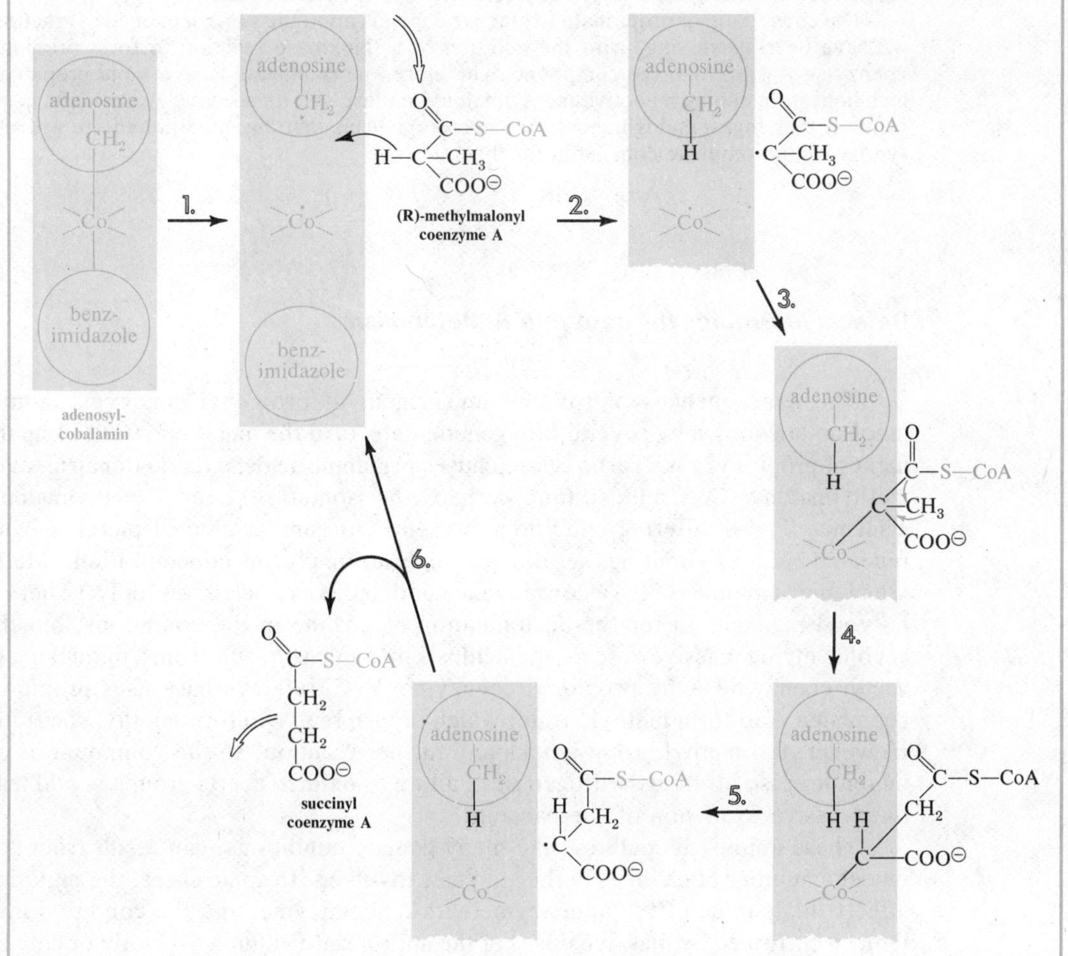

FIGURE 30–8 **A postulated mechanism of action for methylmalonyl CoA mutase.**

entire body contains approximately two micromoles. The cobalamins therefore represent some of the most potent biological agents known. The formation of these compounds is the only known biological function of cobalt, but it is a critical one for animals, and this brings us to a striking paradox. It is almost a truism that the metabolic processes of plants and animals proceed with the same kinds of reactions, with the important exception of photosynthesis. It is therefore startling to discover that the yeasts, the green algae, and all of the higher plants have no need for cobalt and contain no cobalamins, despite the fact that these organisms deal with the same kinds of compounds metabolized by cobalamin-dependent reactions in animals. This drastic evolutionary parting of the metabolic way has few direct consequences over most of the world. However, there are some areas, particularly in Australia, in which the soils have a very low cobalt content. Plants grow, despite their low cobalt content, but some animals that eat the plants don't.

The need for cobalt is especially pronounced in ruminants, and it is the proclivity of the Australians to herd sheep that exposed the unsuspected deficiency in their land. Cud-chewing animals depend upon microbial fermentations in their rumen to break down cellulose in the diet. (Cellulose is a glucose polysaccharide, but unlike starch and glycogen, it is made of β-glucosyl residues and is not attacked by the usual battery of enzymes in mammalian intestines.) The fermentations produce acetate and butyrate, which can serve as sources of fatty acids, but they also convert a large part of the carbohydrate to propionate, and this is the major source of glucose for the animals.

The conversion of propionate to glucose depends upon the same sequence of reactions we have been discussing, with the addition of a thiokinase reaction to form propionyl coenzyme A from the free compound. Therefore, the ruminants have a rapid propionate metabolism, including an active methylmalonyl mutase and its necessary cobalamin coenzyme. If they ingest enough cobalt, the microorganisms that digest carbohydrate will also synthesize the requisite cobalamin for them.

Defects in Propionyl Coenzyme A Metabolism

The consequences of isolated impairment of propionyl coenzyme A metabolism are shown by several rare genetic defects in the metabolic route. Impairment of propionyl CoA carboxylase causes **propionic acidemia**, and impairment of methylmalonyl CoA mutase (and perhaps the isomerase) causes **methylmalonic acidemia**. These differing conditions present with similar clinical pictures, with repeated bouts of vomiting, ketosis, and high serum glycine concentration. (Methylmalonic acidemia is also a consequence of defective cobalamin supply.) There is no good explanation for the accumulation of glycine in the conditions, but the accompanying massive ketosis and acidosis may in part result from a mimicking of acetyl coenzyme A by propionyl coenzyme A. Citrate synthase uses propionyl coenzyme A to form methylcitrate, which in turn reacts to form methylisocitrate. However, the methyl group blocks any further reaction, so the compound is an inhibitor of isocitrate dehydrogenase. Failure to oxidize acetyl groups would lead to excessive formation of acetoacetate.

These **aminoacidopathies,** like many genetic conditions, can result from any one of a number of changes in the enzymes involved. In some cases, the mutation affects the loading of the apoenzyme with its coenzyme, and the condition may then be corrected by massive doses of the appropriate vitamin — biotin or cobalamin in these cases, pyridoxine in others. Such therapy should always be tried in aminoacidopathies.

Branched-Chain Amino Acids — Leucine, Isoleucine, and Valine

Nitrogen: Transferred from all to α-ketoglutarate, forming glutamate.
Carbons: Leucine: Converted to 3-hydroxy-3-methylglutaryl coenzyme A, the precursor of acetoacetate (p. 433).
Isoleucine: Converted to equal amounts of succinyl coenzyme A (*via* propionyl coenzyme A) and acetyl coenzyme A.
Valine: Converted to succinyl coenzyme A *via* propionyl coenzyme A

The branched-chain amino acids, unlike most of the others, are taken up by the striated muscles after a protein meal and are at least partially oxidized in those tissues. The brain also may use significant amounts. Still higher concentrations are taken up by the liver. Whatever the partition may be between the tissues, the general features of the routes seem clear.

All of the branched-chain amino acids transaminate with α-ketoglutarate. Each of the resultant 2-ketocarboxylates then undergoes an oxidative decarboxylation to form the coenzyme A esters (Fig. 30–9).

Some human genetic defects in the enzymes catalyzing these first two steps are illuminating. One rare defect causes **hypervalinemia,** in which there is an accumulation of valine without an accompanying accumulation of leucine or isoleucine. Another rare defect causes an accumulation of both leucine and isoleucine. Evidently there is one aminotransferase for valine and another for leucine and isoleucine in humans. (This is contrary to experimental findings with other animals.)

Approximately one out of 250,000 infants is born with maple syrup urine disease, in which the 2-keto acids accumulate. The name of the disease is derived from the odor of the urine, which connoisseurs say is similar to that of certain vintages of maple syrup. Since no other condition is known to create a urine of even remotely similar aroma, physicians are not faced with the necessity for such acute olfactory discrimination. The odor is actually the result of polymerization of 2-hydroxybutyrate derived as a by-product of 2-ketobutyrate.

Although the name has a certain puckishness, the consequences of the classic form of the condition are grim. Affected infants develop severe neurological damage within a few days after birth, and die within several weeks unless treated. The damage may be due to toxic effects of the accumulating 2-keto compounds, or of the amino acids themselves. Treatment of these, like many other aminoacidopathies, involves rigid dietary restrictions to reduce the intake of the affected amino acids to the minimal level consistent with growth. Continued rigid control occupies much of the family's attention and is expensive.

Let us now go on to consider the further metabolism of the branched-chain skeletons just to see how variations on familiar types of reactions have been developed to cope with these compounds. The coenzyme A esters produced by oxidative decarboxylation are typical fatty acid derivatives, except for the branches, and they undergo the same kind of oxidation by flavoproteins seen with other acyl coenzyme A compounds. (The existence of a few infants with an apparently normal fatty acid metabolism who excrete isovalerate, but not isobutyrate or 2-methylbutyrate, shows that a separate enzyme must exist for at least one of the branched-chain derivatives. Some observers say the infants smell like cheese, others say dirty feet.)

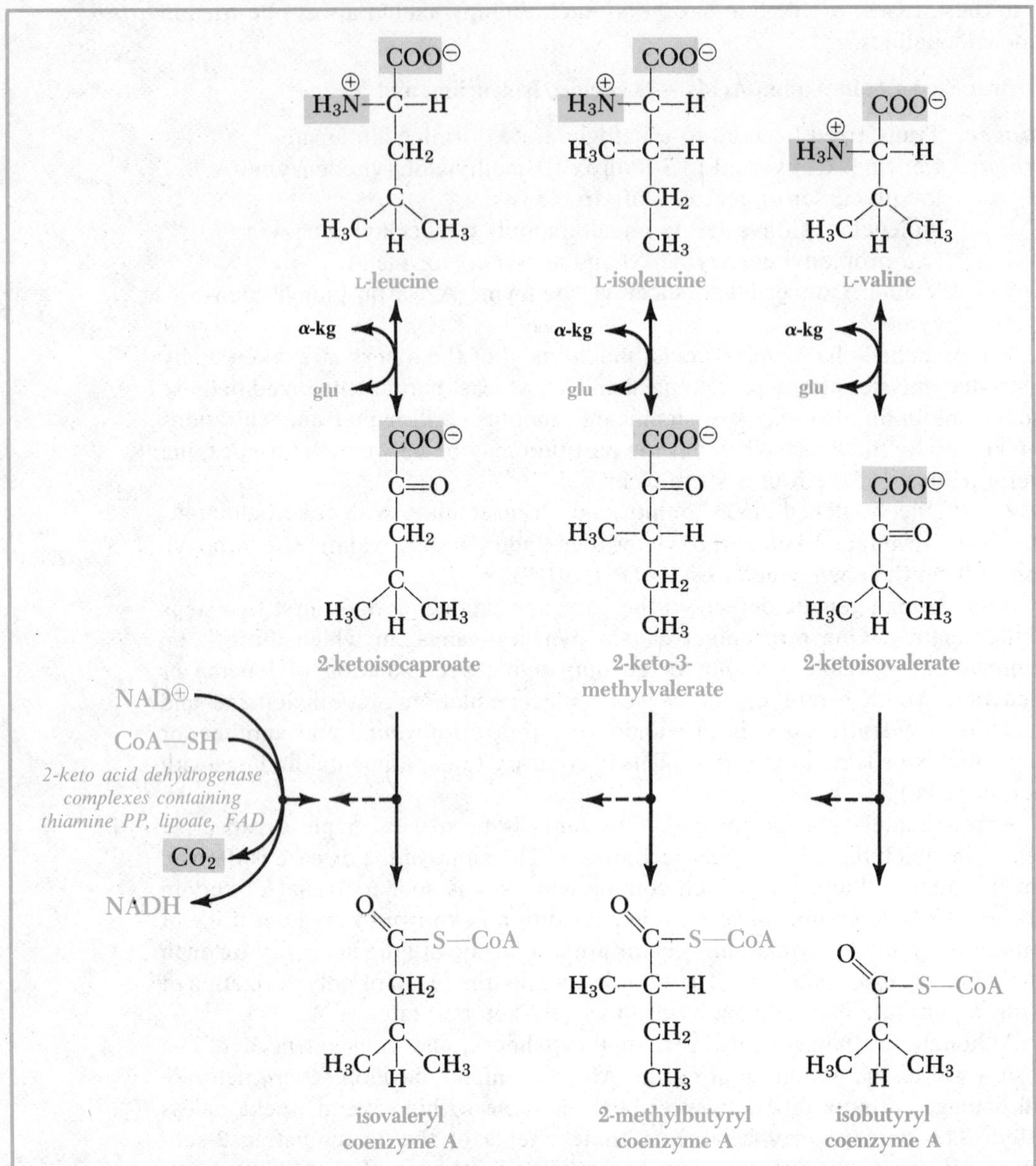

FIGURE 30–9 **The branched-chain amino acids are metabolized by similar initial steps, beginning with a transamination. The resultant 2-ketocarboxylates are oxidatively decarboxylated by enzyme complexes resembling pyruvate dehydrogenase to form branched-chain fatty acids (as the coenzyme A esters), containing one less carbon atom than the original amino acid. The succeeding steps are detailed in the next three figures.**

The predicted initial oxidation of the isovaleryl coenzyme A derived from **leucine** results in the formation of the corresponding enoyl coenzyme A ester (Fig. 30–10). However, the hydration followed by oxidation to yield a 3-ketoacyl coenzyme A compound that would be expected with straight-chain fatty acid derivatives cannot occur because of the branch on the third carbon.

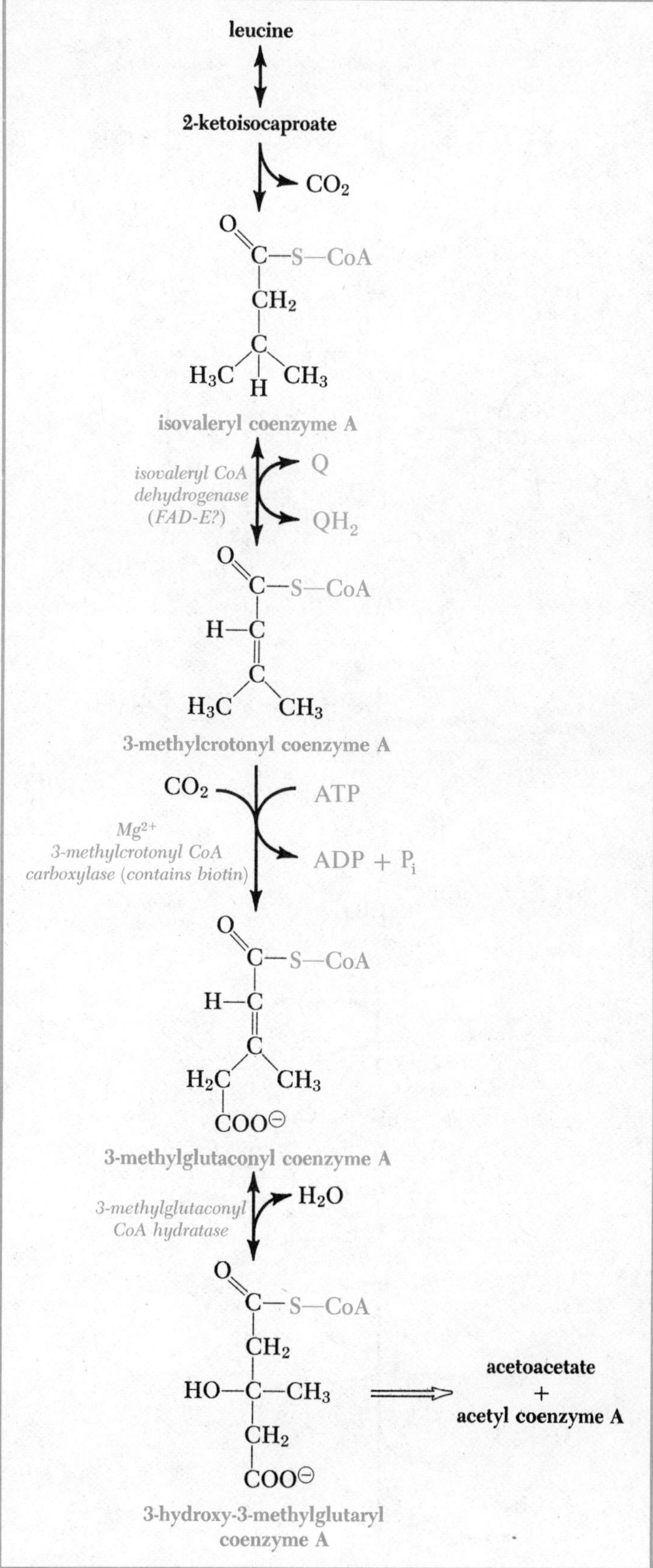

FIGURE 30–10

Isovaleryl coenzyme A obtained from the metabolism of leucine is converted to acetyl coenzyme A and acetoacetate, with the intermediate formation of 3-hydroxy-3-methylglutaryl coenzyme A.

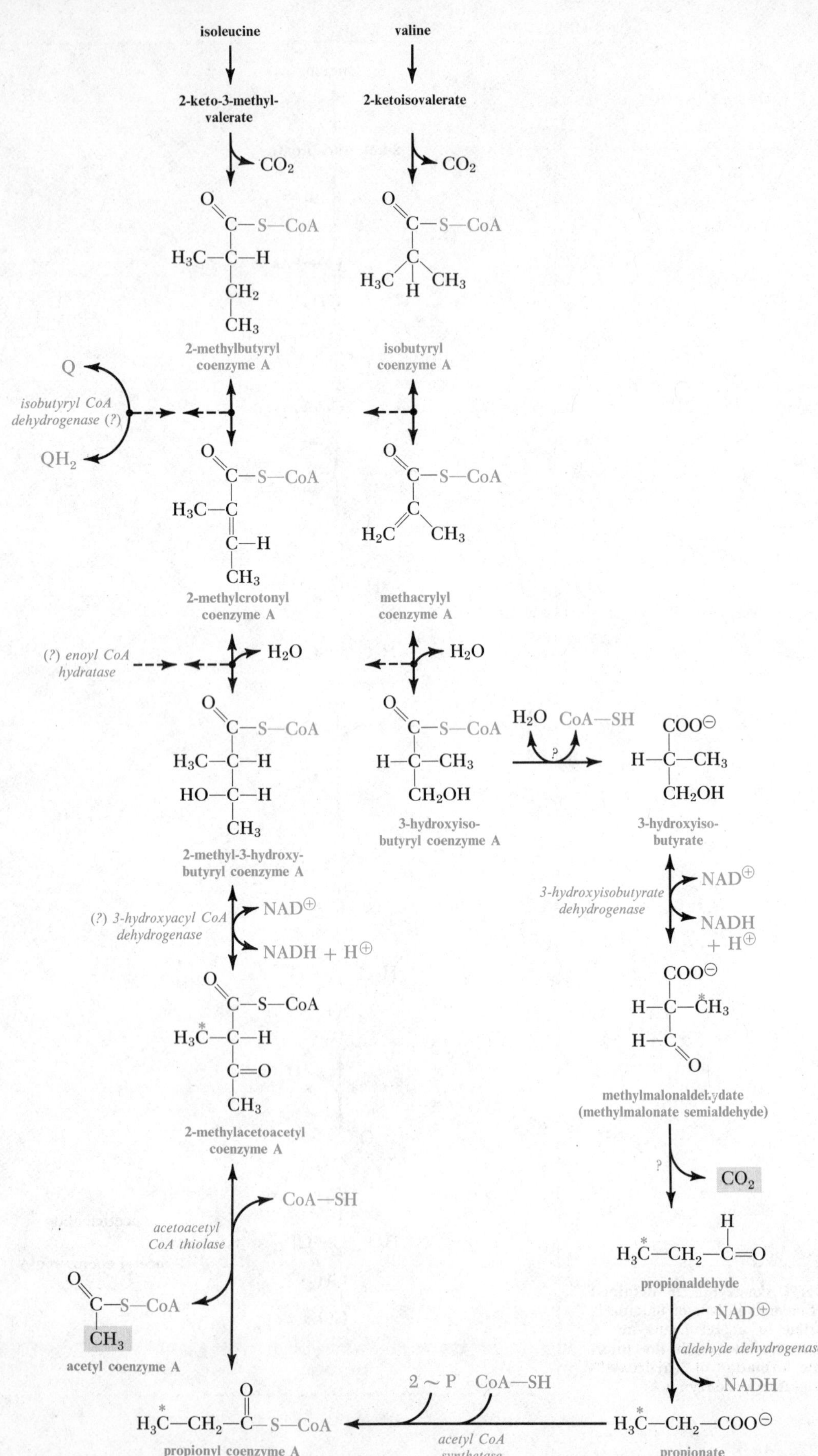

FIGURE 30–11 *See legend on opposite page.*

A different route has evolved, in which the compound is carboxylated by a typical biotin-containing carboxylase at the expense of a high-energy phosphate. The result is the coenzyme A ester of an unsaturated dicarboxylic acid. This is then hydrated to form 3-hydroxy-3-methylglutaryl coenzyme A., which is also the intermediate used in the production of acetoacetate. Therefore, the compound will be cleaved in either the liver or muscles to yield acetyl coenzyme A and acetoacetate by routes we have already considered. Leucine, therefore, is unlike the amino acids considered earlier. It is a ketogenic, rather than a glucogenic amino acid, and all six carbons may appear as acetoacetate.

The branched-chain acyl coenzyme A compounds derived from **valine** and **isoleucine** can, and do, undergo the standard sequence of oxidation, hydration and oxidation seen with ordinary fatty acid derivatives, with the qualification that the valine derivative loses its coenzyme A along the route (Fig. 30–11). The methylacetoacetyl coenzyme A derived from isoleucine is cleaved just as acetoacetyl coenzyme A is, except that one of the products of the cleavage is propionyl coenzyme A rather than acetyl coenzyme A. The semialdehyde of methylmalonate is derived from valine. It also is converted to propionyl coenzyme A.

Since both isoleucine and valine give rise to the formation of one mole of succinyl coenzyme A *via* propionyl coenzyme A, three of the carbons of these amino acids may appear as glucose. Isoleucine also forms a mole of acetyl coenzyme A, and two of its carbons may appear as acetoacetate *via* that compound. There is also a net production of one mole of CO_2 from isoleucine and two from valine.

AROMATIC AMINO ACIDS

Phenylalanine and Tyrosine

Nitrogen: Transferred to α-ketoglutarate, producing glutamate.

Carbons: Appear as acetoacetate and fumarate.

Phenylalanine is oxidized to form tyrosine, and tyrosine is further metabolized with the ultimate formation of fumarate and acetoacetate. The sequence of oxidations by which the aromatic ring is broken is catalyzed by a number of enzymes localized in the soluble cytoplasm and the endoplasmic reticulum. These reactions therefore do not lead to the generation of high-energy phosphate. Of course, the fumarate and the acetoacetate eventually produced are intermediates of fuel metabolism, and since a mole of each is produced, the potential production of high-energy phosphate in the whole body is considerable, with three carbons convertible to glucose, four appearing as acetoacetate, and two appearing as CO_2.

The conversion of phenylalanine to tyrosine involves an irreversible oxidation by **phenylalanine hydroxylase**. The reaction consumes molecular oxygen and

---

FIGURE 30–11 **The metabolism of the branched-chain coenzyme A esters obtained from isoleucine and valine begins with the same main-line sequence of reactions seen with ordinary fatty acids. An oxidation by a flavoprotein is followed by hydration to an alcohol, which is oxidized to a 3-carbonyl derivative. However, the metabolism of the valine skeleton deviates in that the coenzyme A is lost before oxidation of the alcohol group. Metabolism of the isoleucine skeleton continues on the main-line sequence with a thiolytic cleavage to form acetyl coenzyme A and propionyl coenzyme A. The methylmalonic semialdehyde derived from valine is decarboxylated and oxidized to propionate, which also appears as propionyl coenzyme A.**

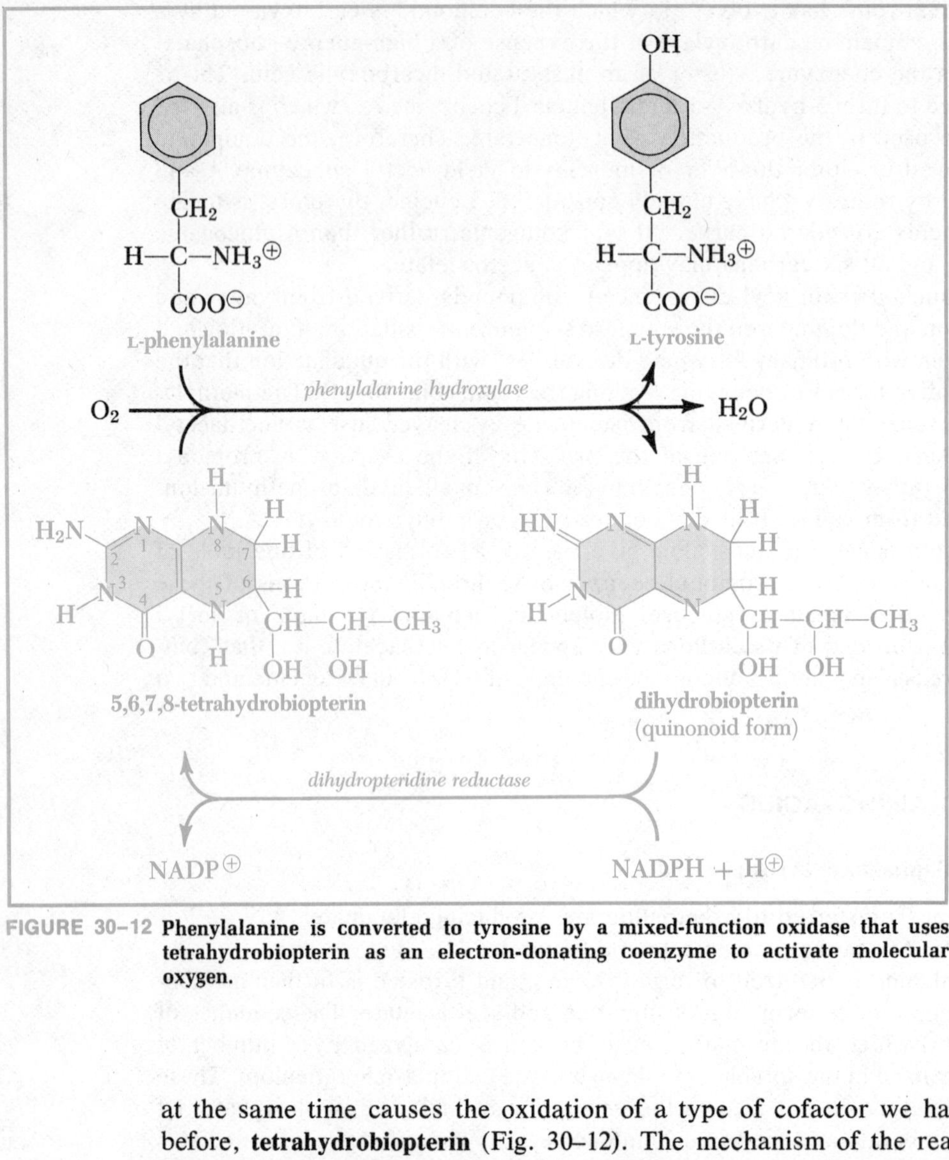

FIGURE 30–12 **Phenylalanine is converted to tyrosine by a mixed-function oxidase that uses tetrahydrobiopterin as an electron-donating coenzyme to activate molecular oxygen.**

at the same time causes the oxidation of a type of cofactor we have not seen before, **tetrahydrobiopterin** (Fig. 30–12). The mechanism of the reaction is unknown, but the cofactor presumably "activates" oxygen in some way typical of mixed-function oxidases. In any event, the reduced pterin must be regenerated by a second enzyme at the expense of NADPH.

We shall consider other pteridine coenzymes, which are **folic acid** derivatives, in some detail in the next chapter. The biologcial significance of the compounds to animals was first noted in an odd way. The pterins were known to occur in animal pigments, for example on the wings of butterflies. The basis for their chemistry was laid back in the late 1800's, but the subject remained stagnant until the 1930's when one of the compounds, xanthopterin, was synthesized. This compound was shown to alleviate an anemia produced in rats fed on goat milk — a curious observation. More significantly, salmon in Washington hatcheries fed on diets in which yeast was used as a vitamin supplement also became anemic, but this anemia was corrected by feeding small amounts of xanthopterin. Obviously, the fish needed the insect pigment. With this background, it was much easier to connect the pteridines with the vitamin folic acid, which was isolated in the 1940's, and to show that it was chemically related. So much attention was concentrated on folate derivatives that the simpler pterins were neglected, and it was only in the last few years that the use of these compounds in oxidation-reduction reaction, such as the oxidation of phenylalanine, was demonstrated.

The major pathway for the metabolism of **tyrosine** begins with a transamination with α-ketoglutarate, catalyzed by a specific transaminase (Fig. 30–13). The resultant *p*-hydroxyphenylpyruvate is oxidized by a **dioxygenase**, a kind of enzyme that causes the addition of a full molecule of O_2, and which usually contains iron (II).

This particular dioxygenase is typical of a subclass with an unusual mechanism. Enzymes of its sort use a second substrate to generate an intermediate peroxide as the active oxidant:

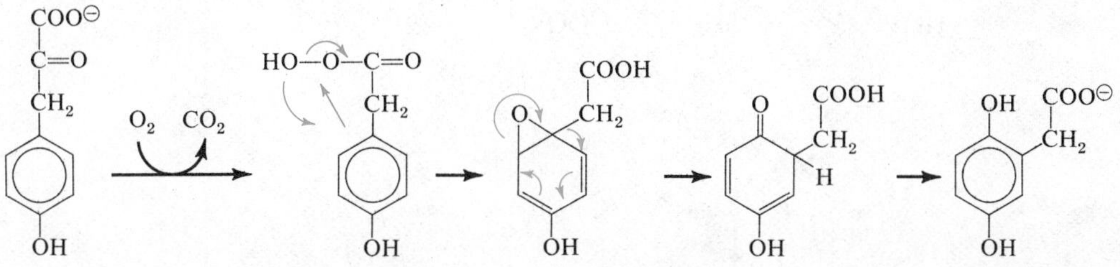

In the case at hand, the second substrate is the 2-keto side chain on the aromatic ring, but with other enzymes, such as the prolyl hydroxylase that converts proline residues in collagen to hydroxyproline residues, α-ketoglutarate is used as the second substrate.

The dioxygenase causes a hydroxylation of the ring, an oxidative decarboxylation of the side chain, and a shift of the acetyl group on the ring. The resultant 2,5-dihydroxyphenylacetate has the trivial name of **homogentisate**. Homogentisate is attacked by another dioxygenase containing iron (II), which does not require a second substrate. It cleaves the ring with dioxygen to form the *cis*-unsaturated compound C-maleoylacetoacetate. An isomerase converts this to the *trans*-compound, C-fumaroylacetoacetate, and this is hydrolyzed to yield fumarate and acetoacetate.

This is a remarkably complex series of oxidations. Rupture of an aromatic ring so as later to yield the same metabolites encountered in fat and carbohydrate metabolism is in itself somewhat of a stunt. Does the fact that this can take place explain why proteins have phenylalanine rather than phenylglycine or phenylaminobutyrate? Who knows? Furthermore, we shall see that similar kinds of reactions are used to make several other kinds of aromatic compounds in small quantities, so the type of reaction has been evolved for more general purposes than the particular sequence we are considering.

A number of genetic defects in the metabolism of aromatic amino acids are known in humans. The most common one is the absence of phenylalanine hydroxylase. Phenylalanine accumulates in the blood and tissues. Part of it is transaminated, forming phenylpyruvate. This compound, as well as phenylalanine, is excreted in the urine. Hence the condition is called **phenylketonuria**. It is also known as phenylpyruvic oligophrenia, because the condition results in early neurological damage preventing normal intellectual development. Treatment consists of early avoidance of high concentrations of phenylalanine in the diet, but the problem is the same as in maple syrup urine disease. The body cannot synthesize phenylalanine, so some must be present to build normal proteins; it is difficult to gain a proper balance, and an expensive diet is required.

Phenylketonuria is more common than most genetic defects. We earlier discussed it as an example (p. 124). The full blown condition is only seen in homozygotes, yet it occurs once in each 20,000 births, which means that the

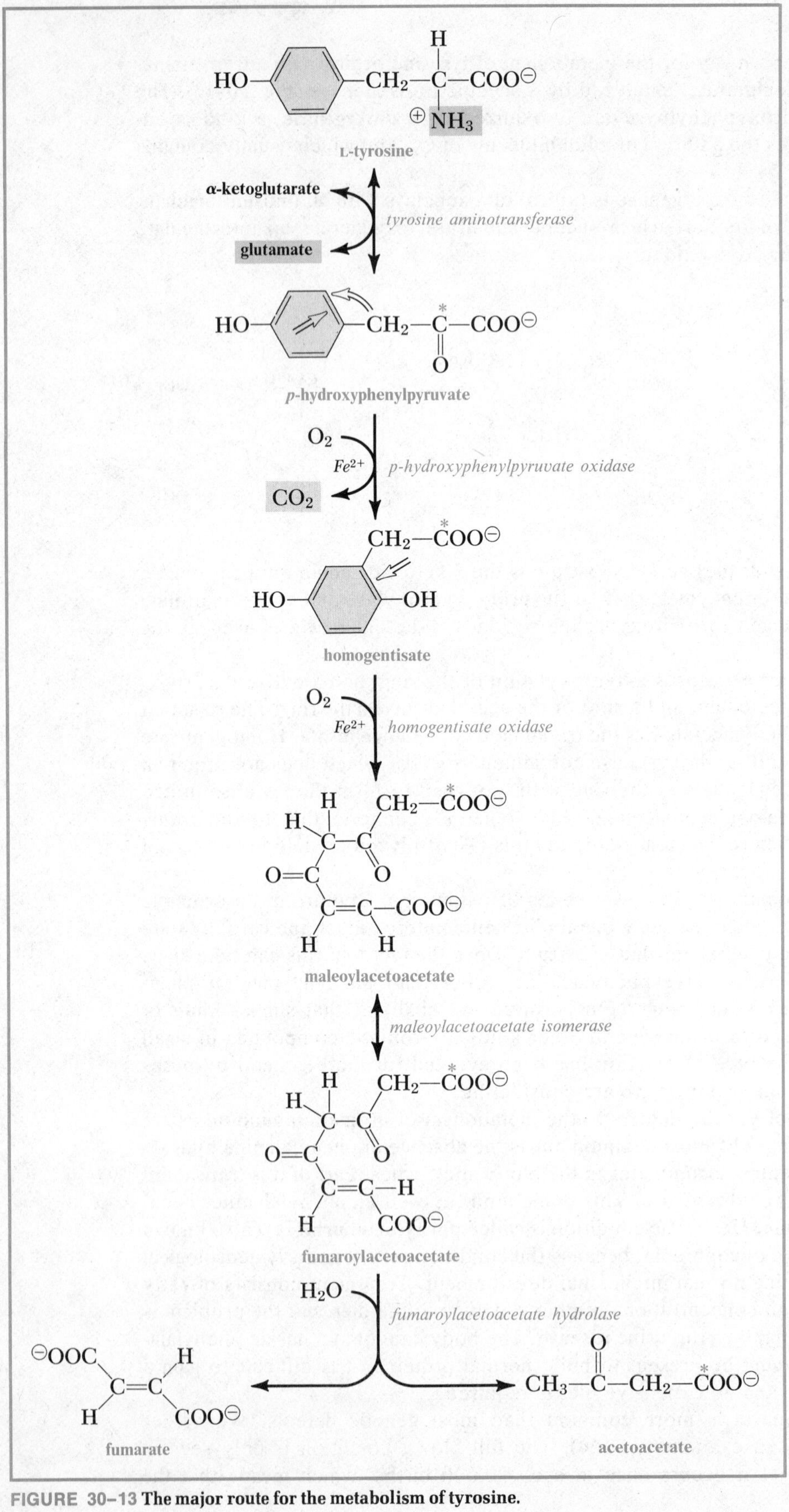

L-tyrosine

α-ketoglutarate

tyrosine aminotransferase

glutamate

p-hydroxyphenylpyruvate

O_2

Fe^{2+} *p-hydroxyphenylpyruvate oxidase*

CO_2

homogentisate

O_2

Fe^{2+} *homogentisate oxidase*

maleoylacetoacetate

maleoylacetoacetate isomerase

fumaroylacetoacetate

H_2O

fumaroylacetoacetate hydrolase

fumarate

acetoacetate

FIGURE 30–13 The major route for the metabolism of tyrosine.

heterozygotes, detectable only by deliberately loading them with phenylalanine and testing for blood levels of phenylalanine and phenylpyruvate, must be about one in 70 of the population. Such a prevalence makes one immediately suspicious that the prevalence is due to a balanced polymorphism similar to some hemoglobinopathies. But under what circumstances could this condition possibly be advantageous? We don't know. Perhaps the conservation of phenylalanine had an advantage on some peculiar diet*. Perhaps it is still advantageous in some regions of the world where food is frequently in short supply.

People with phenylketonuria also excrete some other aromatic compounds (Fig. 30–14), representing aberrations of the normal process of metabolism due to

*The incidence among Celts is especially high — one in 5,000 births. Perhaps the condition was a defense against haggis.

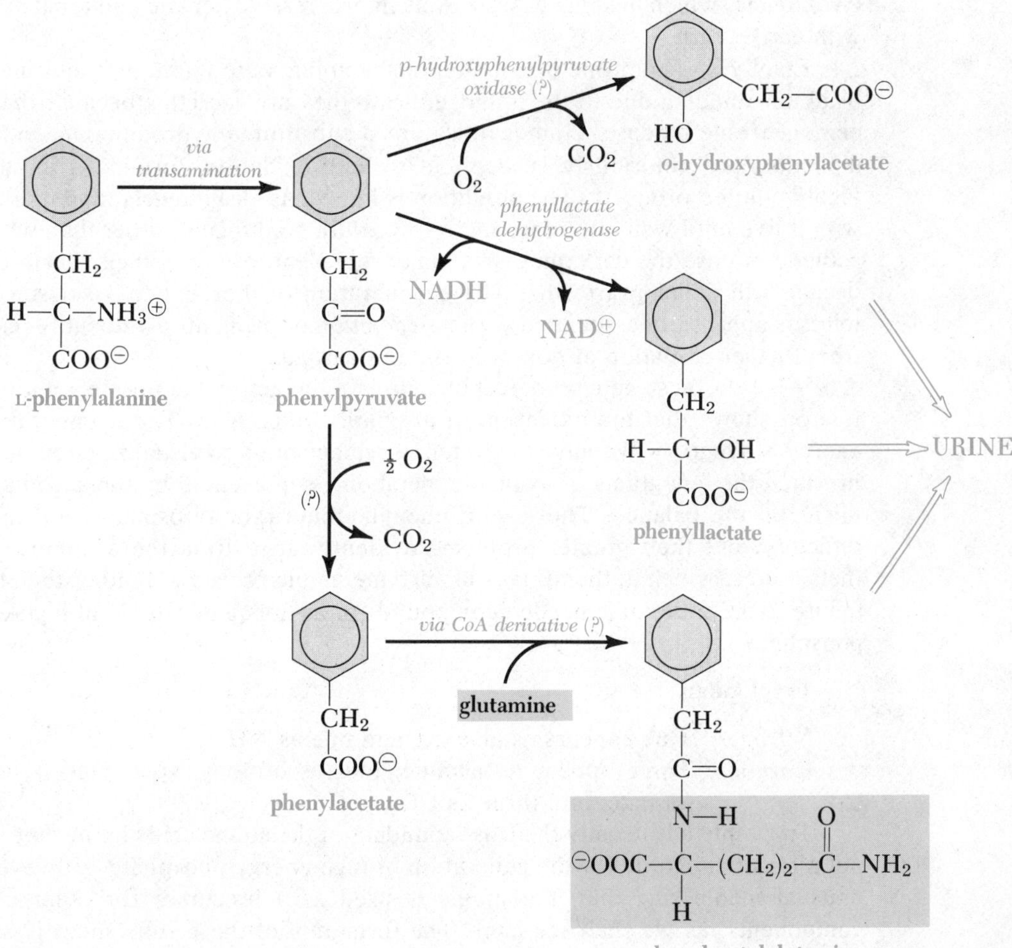

FIGURE 30–14 **Ordinarily trivial routes of phenylalanine metabolism become more important in patients lacking phenylalanine hydroxylase, and a variety of products appear in the urine in increased amounts.**

the high concentration of phenylpyruvate. The compound is reduced by a phenyllactate dehydrogenase in the liver, forming phenyllactate. Phenylpyruvate is also oxidized in the same way as is the normal *p*-hydroxyphenylpyruvate, except that the product is *o*-hydroxyphenylacetate, rather than the 2,5-dihydroxyphenylacetate that is homogentisate. The monohydroxy compound is not further metabolized and appears in the urine.

The suspicion that the prevalence of phenylketonuria may reflect some occasional advantage to the heterozygotes is deepened by the fact that the other known genetic abnormalities of tyrosine metabolism are much less common, although their consequences are sometimes less grave.

A single clear-cut case of **tyrosinosis** was reported in an adult by Grace Medes in 1927. The facts are consistent with the absence of *p*-hydroxyphenylpyruvate oxidase in the patient who excreted both tyrosine and the keto acid. The excretion of both increased with phenylalanine feeding, but homogentisate was metabolized at the normal rate. The case is of historical interest in that the metabolism of the aromatic acids was not well understood until painstaking studies were made of the metabolism of this one human. The compounds that did and didn't appear in the urine now appear obvious in light of our knowledge, but much of that knowledge is a result of just those measurements. This condition ought not be confused with **tyrosinemia**, which in some cases results in progressive hepatic and renal damage, with early death.

Finally, in about one in 200,000 births, infants are found whose urine turns black on standing due to the homogentisate they are secreting because they lack homogentisate oxidase. Homogentisate is a substituted hydroquinone, and these diphenols are notoriously susceptible to auto-oxidation, forming a mixture of highly-colored products. The condition is known as **alcaptonuria**, and individuals with it live until well into reproductive age with no difficulty other than whatever esthetic offense the darkening urine may represent. Many in their fourth or fifth decade will develop arthritis. The degeneration of the connective tissue in the joints is apparently associated with a deposition of pigment, presumably resulting from further oxidation of homogentisate in cartilage.

What do these genetic defects teach us? The active life possible with alcaptonuria shows that the oxidation of aromatic amino acids is not imperative for energy production. We have to stretch our imagination to visualize circumstances in which the few grams of available metabolite represented by these compounds might tip the balance. Those with phenylketonuria or tyrosinemia are in more difficulty, but their greater problems evidently arise from the accumulation of metabolites, which in themselves disturb metabolic processes, rather than from a failure to produce some needed compounds or an adequate supply of high-energy phosphate.

Tryptophan

Nitrogen: One appears as alanine, and one as NH_4^+.

Carbons: Three appear as alanine, four as crotonyl coenzyme A, one as formate, and three as CO_2.

Tryptophan is usually the least abundant of the amino acids in the diet, and is not a major substrate for the generation of high-energy phosphate. However, the unusual indole ring that it contains is used as a precursor for other cellular components, as we shall see later. The formation of these substances is usually satisfied by a small fraction of the dietary consumption. The balance of the carbon skeleton is metabolized to CO_2 by way of alanine made from the sidechain and crotonyl CoA formed from the ring, which can be converted to glucose and

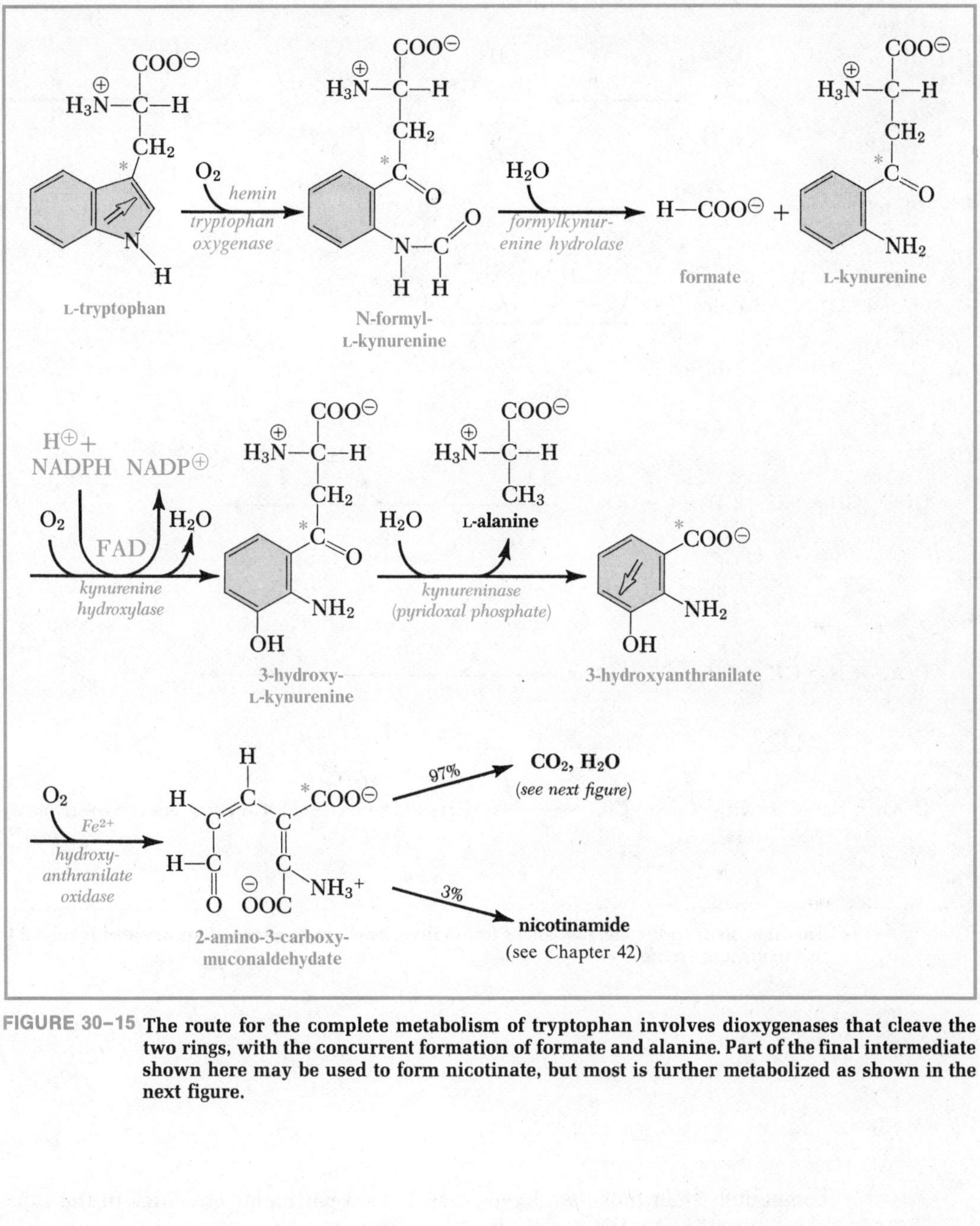

FIGURE 30–15 **The route for the complete metabolism of tryptophan involves dioxygenases that cleave the two rings, with the concurrent formation of formate and alanine. Part of the final intermediate shown here may be used to form nicotinate, but most is further metabolized as shown in the next figure.**

acetoacetate, respectively. Metabolism begins with an irreversible oxidation by O_2, catalyzed by **tryptophan oxygenase**, an enzyme containing heme. This reaction opens the indole ring, and succeeding reactions oxidize the phenyl ring and open it with the intermediate removal of carbons as formate and alanine (Fig. 30–15). The oxidations are of the same sort encountered in the metabolism of phenylalanine and tyrosine. We have already considered the disposal of alanine. We shall consider the fate of formate more completely in the next chapter. Suffice it to say now that it can be oxidized to CO_2 and H_2O.

The remaining reactions by which the carbons are converted to CO_2 and crotonyl CoA, which is an intermediate of fatty acid metabolism, involve different

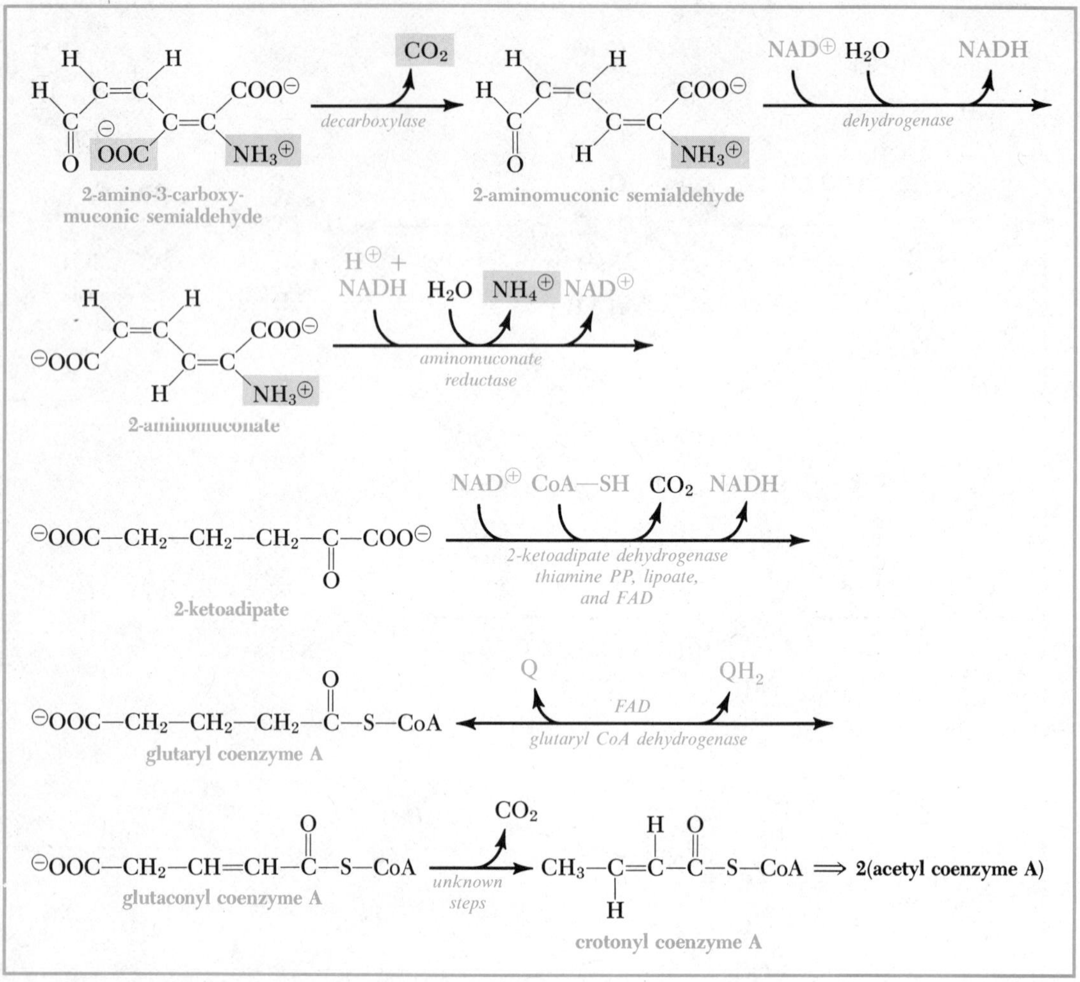

FIGURE 30–16 **The final steps in the metabolism of tryptophan. The last three reactions are also involved in the metabolism of lysine.**

compounds than those we have seen, but no particular novelties in the types of reactions (Fig. 30–16).

LYSINE

Nitrogen: Both transferred to α-ketoglutarate forming glutamate.

Carbons: Two appear as CO_2 and four as crotonyl coenzyme A.

Lysine, like tryptophan, is metabolized *via* 2-ketoadipate, eventually forming crotonyl CoA. However, lysine is a relatively abundant amino acid. The amount consumed on most diets is comparable to the intake of the branched-chain amino acids. It is metabolized in mitochondria with the generation of high-energy phosphate.

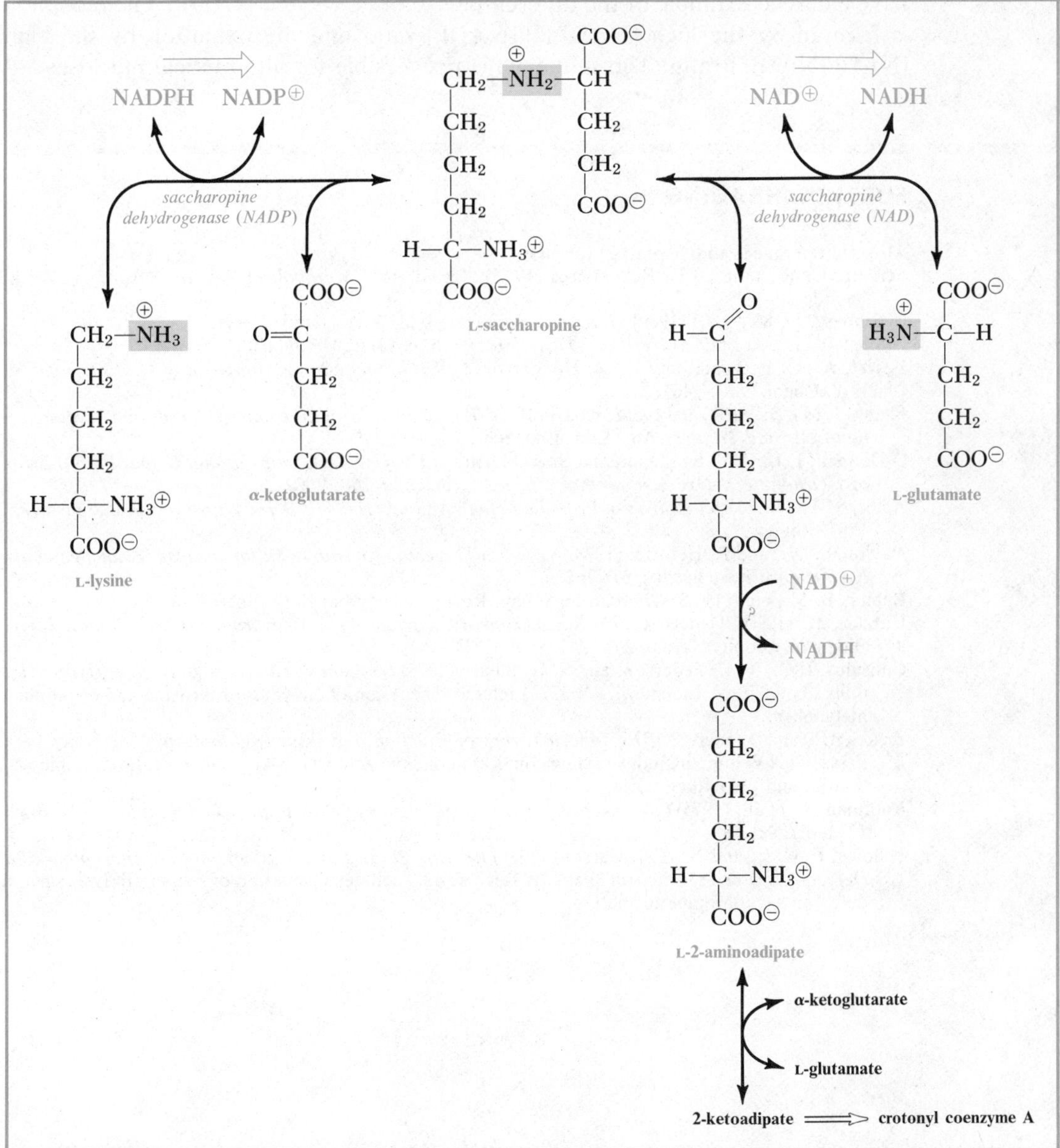

FIGURE 30–17 **The major route for the metabolism of lysine.**

The route involves a condensation with α-ketoglutarate and a reduction to form the compound, saccharopine, which is then cleaved with an oxidation on the opposite side of the nitrogen bridge so as to release glutamate and the semialdehyde of 2-aminoadipate (Fig. 30–17). (The semialdehyde spontaneously cyclizes much like sugars.) The semialdehyde is then oxidized to form 2-aminoadipate. This is the next higher homologue in the dicarboxylic amino acid series; it does not occur in proteins, but it will transaminate, forming 2-ketoadipate. Here we have a classic example of the differential use of NAD and NADP. The reduction is favored by the high [NADPH]/[NADP] ratio and the oxidation by the high [NAD]/[NADH] ratio. The sequence is irreversible for all practical purposes.

FURTHER READING

Also see references to Chapter 29 (p. 563).

Scriver, C. R., and L. E. Rosenberg: (1973) *Amino Acid Metabolism and its Disorders.* W. B. Saunders.

Greenberg, D. M., ed.: (1969) *Metabolic Pathways,* vol. 3. Academic Press.

Nyhan, W. L., ed.: (1974) *Heritable Disorders of Amino Acid Metabolism.* Wiley.

Besrat, A., C. E. Polan, and L. M. Henderson: (1969) *Mammalian Metabolism of Glutaric Acid.* J. Biol. Chem., *244*: 1461.

Phang, J. M., D. Valle, and E. M. Kouriloff: (1975) *Proline Biosynthesis and Degradation in Mammalian Cells and Tissues.* Am. Clin. Lab. Sci., *5*: 298.

O'Donnell, J. J., R. P. Sandman, and S. R. Martin: (1978) *Gyrate Atrophy of the Retina: Inborn Error of* L-*Ornithine: 2-Oxo-acid-aminotransferase.* Science, *200*: 200.

Adibi, S. A.: (1976) *Metabolism of Branched-chain Amino Acids in Altered Nutrition.* Metabolism, *25*: 1287. Review.

Naylor, E. W., and R. Guthrie: (1978) *Newborn Screening for Maple Syrup Disease (Branched-chain Ketoaciduria).* Pediatrics, *61*: 262.

Babior, B. M., ed.: (1975) *Cobalamin.* Wiley. Review of vitamin B-12 functions.

Hutzler, J., and J. Dancis: (1970) *Saccharopine Cleavage by a Dehydrogenase of Human Liver.* Biochem. Biophys. Acta, *206*: 205.

Gunsalus, I. C., T. C. Pederson, and S. G. Sligar: (1975) *Oxygenase-catalyzed Biological Hydroxylations.* Annu. Rev. Biochem., *44*: 377. Includes discussion of enzymes of tyrosine and tryptophan metabolism.

Bickel, H., and J. Stern: (1976) *Inborn Errors of Calcium and Bone Metabolism.* University Park Press. This volume includes reviews on α-aminoadipic aciduria, α-ketoadipic aciduria, homocystinuria, and phenylketonuria.

Kaufman, S., et al.: (1975) *Phenylketonuria due to Deficiency of Dihydropteridine Reductase.* N. Engl. J. Med., *293*: 785.

Fellows, F. C. I., and N. A. J. Carson: (1974) *Enzyme Studies on a Patient with Saccharopinuria: a Defect of Lysine Metabolism.* Pediatr. Res., *8*: 42. Includes discussion of relation to lysinemia, a defect in saccharopine formation.

31 AMINO ACIDS: ONE-CARBON POOL AND TOTAL BALANCE

The metabolism of some amino acids involves an associated metabolism of one carbon units. Groups containing single carbon atoms are transferred for many different purposes, including the introduction of carbon atoms into purine rings and the formation of methyl groups on compounds such as choline, creatine, and the bases of nucleic acids. Of course, we have already seen many examples of the utilization or formation of CO_2 through carboxylations or decarboxylations, but we are now concerned with the transfer of carbon atoms in a less fully oxidized state. It is convenient to lump together the sources of these more reduced carbon groups as the one-carbon pool, but it will become apparent as the story unfolds that the pool is more like a series of interconnected basins.

NATURE OF ONE-CARBON GROUPS

What precisely are we talking about? Single carbon atoms can exist in various oxidation states, which are represented by the series of simplest organic compounds: methane, methanol, formaldehyde, formate, and carbon dioxide. It is possible to incorporate carbon units at each of these oxidation states, except that of methane, into other compounds by the formal elimination of water. The results are given in Figure 31–1. All of the possible structures of one-carbon groups shown in the figure are found in biological compounds. The figure uses the general R-designation for groups bearing one-carbon units, but the actual nature of the groups is limited by their chemical potential in most cases to those containing N, O, or S atoms. Furthermore, the figure includes carbonic acid, the hydrated form of CO_2, for completeness even though compounds at this oxidation state are handled by carboxylation and decarboxylation, not by the reactions of the one-carbon pool.

Figure 31–1 is organized so that each row differs from its neighbors by the equivalent of two electrons; it follows that the rows are interconvertible only by oxidations or reductions. Those groups on the same row are of equivalent oxidation-reduction state and are interconvertible by removal or addition of the elements of water (p. 271).

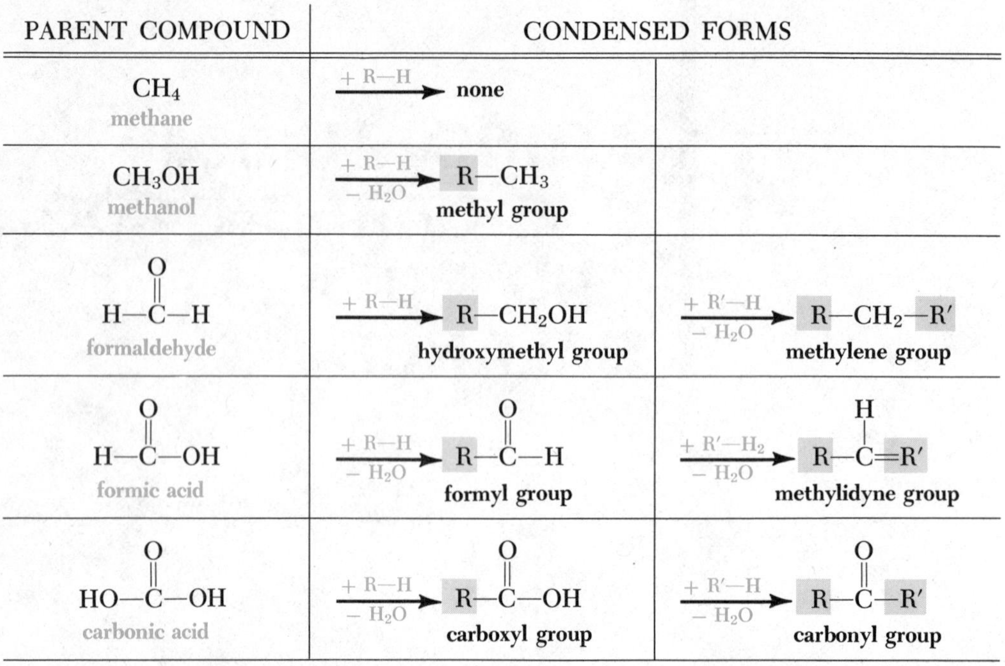

PARENT COMPOUND	CONDENSED FORMS	
CH_4 methane	$\xrightarrow{+R-H}$ none	
CH_3OH methanol	$\xrightarrow[-H_2O]{+R-H}$ $R-CH_3$ **methyl group**	
$H-C(=O)-H$ formaldehyde	$\xrightarrow{+R-H}$ $R-CH_2OH$ **hydroxymethyl group**	$\xrightarrow[-H_2O]{+R'-H}$ $R-CH_2-R'$ **methylene group**
$H-C(=O)-OH$ formic acid	$\xrightarrow[-H_2O]{+R-H}$ $R-C(=O)-H$ **formyl group**	$\xrightarrow[-H_2O]{+R'-H_2}$ $R-C(H)=R'$ **methylidyne group**
$HO-C(=O)-OH$ carbonic acid	$\xrightarrow[-H_2O]{+R-H}$ $R-C(=O)-OH$ **carboxyl group**	$\xrightarrow[-H_2O]{+R'-H}$ $R-C(=O)-R'$ **carbonyl group**

FIGURE 31–1 **One-carbon compounds of various oxidation states and their formal relationships to groups of the same oxidation states. The reverse reactions involving hydrations or hydrolyses are not shown.**

The division of the one-carbon groups according to oxidation state also divides them according to their typical types of reactions. For example, we have already seen that CO_2 is carried by biotin for the formation of carboxylate groups. Now we shall see a pteridine compound, tetrahydrofolate, used for the transfer of single carbon groups equivalent in oxidation state to formate, formaldehyde, and sometimes methanol. That is, tetrahydrofolate carries methylidyne groups, methylene groups, and sometimes methyl groups.

Finally, we shall see how methyl groups, the most reduced of the transferrable single-carbon groups, are usually derived from an activated form of the amino acid methionine.

TETRAHYDROFOLATE

Tetrahydrofolate is a reduced pteridine like tetrahydrobiopterin (p. 580), but it has the specific function of carrying one-carbon units and does not participate in mixed-function oxidase reactions. Plants and many protists can combine the pteridine with p-aminobenzoic acid to form pteroic acid (Fig. 31–2), followed by conjugation of the tetrahydro derivative with a molecule of glutamate to make **tetrahydropteroylglutamate,** which is a more systematic name for tetrahydrofolate. Still more molecules of glutamate may be attached through the 5-carboxyl group to make tetrahydropteroyl polyglutamates. Animals, including us, usually depend upon plants for the principal supply, although the protists are sometimes of importance. (Rats, for example, disgust us by eating their own feces, but this enables them to use the folate and other vitamins synthesized by their intestinal bacteria. We aren't built to take advantage of this practice, and it is not recom-

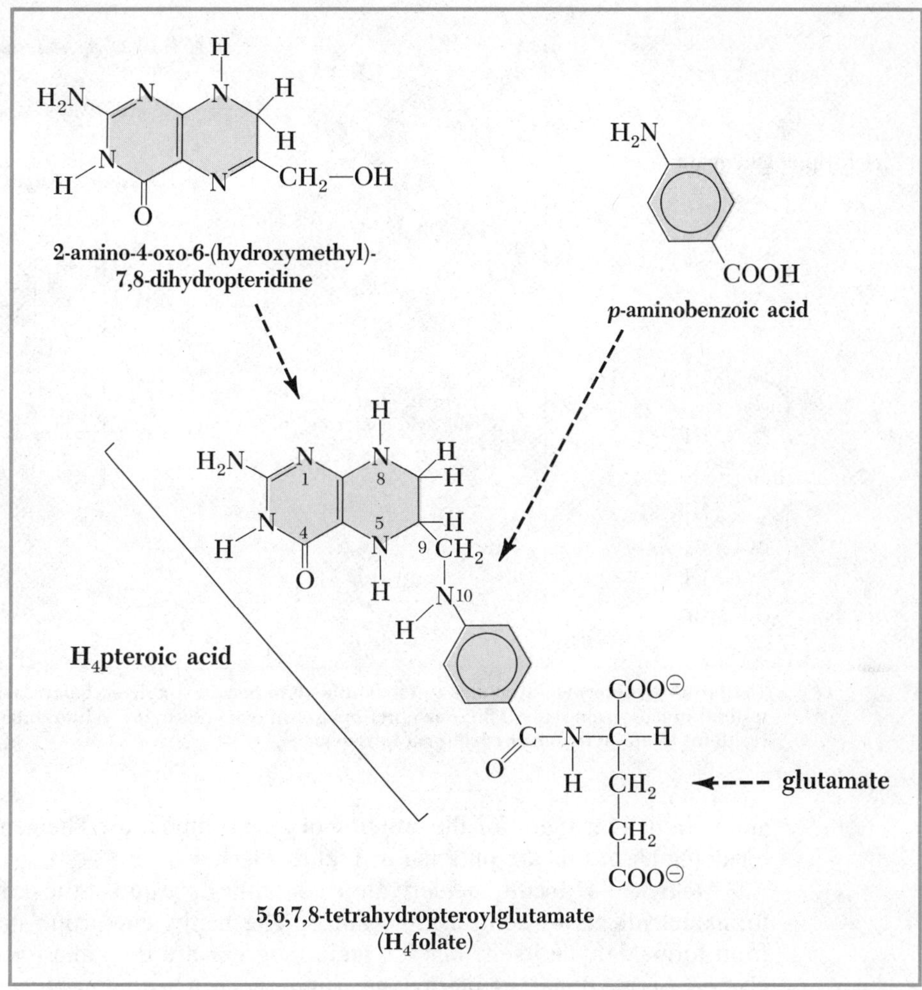

FIGURE 31–2 Tetrahydrofolate (H_4 folate) and its precursors. It is more formally named as tetrahydropteroylglutamate, abbreviated as H_4PteGlu.

mended, even for the most avid devotee of health foods.) We shall say more about the origin of folate and its turnover when we discuss nutrition (p. 809).

The tetrahydro form of folate and its derivatives are readily oxidized to folate and its derivatives by atmospheric oxygen during preparation and digestion of foods, and it is mostly these forms that reach the intestinal mucosa. The intestines have enzymes that remove all but one glutamyl group from polyglutamyl forms, leaving only folate, which is reduced by NADPH after absorption, first to dihydrofolate and then to tetrahydrofolate, the active form of the coenzyme (Fig. 31–3). For convenience in writing reactions or in specifying derivatives, the reduced forms are designated as H_2folate and H_4folate.

ONE-CARBON GROUPS ON TETRAHYDROFOLATE

General Outline

The function of tetrahydrofolate is to receive one-carbon groups from various sources, retain them while they are oxidized or reduced from one form to another,

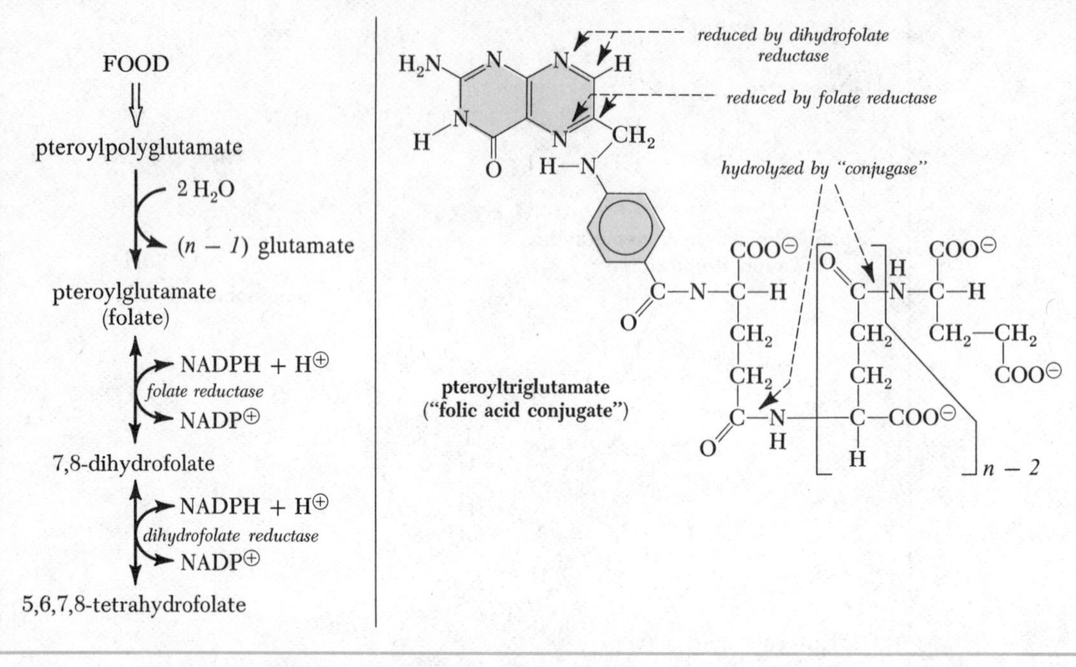

FIGURE 31–3 **The tetrahydropteroylglutamates in foodstuffs often become oxidized before absorption. The intestinal mucosa removes all but one glutamyl group from pteroylpolyglutamates and reduces the resultant folate to its tetrahydro form in two steps.**

and then deliver them for the creation of new compounds. The general pathways for doing all of this are outlined in Figure 31–4.

Methylene-H_4folate, the derivative containing a group at the oxidation level of formaldehyde, is of central importance. The methylene group may be derived from formaldehyde itself, but the major sources are the amino acids serine and glycine. Serine donates a methylene group, thereby becoming glycine, and glycine is a source of an additional methylene group.

The methylene groups are used to form the side chains of thymine incorporated into DNA (Chapter 34), and are sources of methyl groups (by reduction) and methylidyne groups (by oxidation).

Methylidyne-H_4folate and the equivalent formyl-H_4folate are made by oxidation of methylene-H_4folate and by combination of H_4folate with formate. Formate is not only ingested by ant-fanciers, but it also is a normal intermediate of metabolism. Histidine, too, is a source of a methylidyne group.

Methylidyne and formyl H_4folate are obligatory precursors of the purines.

Methyl-H_4folate is made by reduction of methylene-H_4folate and is used to regenerate methyl groups on methionine, which is the precursor of most methylated metabolites of the sort we shall consider later in this chapter. The degradation of such compounds results in the oxidation of the methyl groups to formaldehyde, which is recovered by combination with H_4folate.

Let us now look at some of the details of transfer of one-carbon units *via* H_4folate.

PRIMARY SOURCES FOR THE ONE-CARBON POOL

Methylene Groups from Serine and Glycine. Serine combines with pyridoxal phosphate on a transferase that transfers the hydroxymethyl group to nitrogen

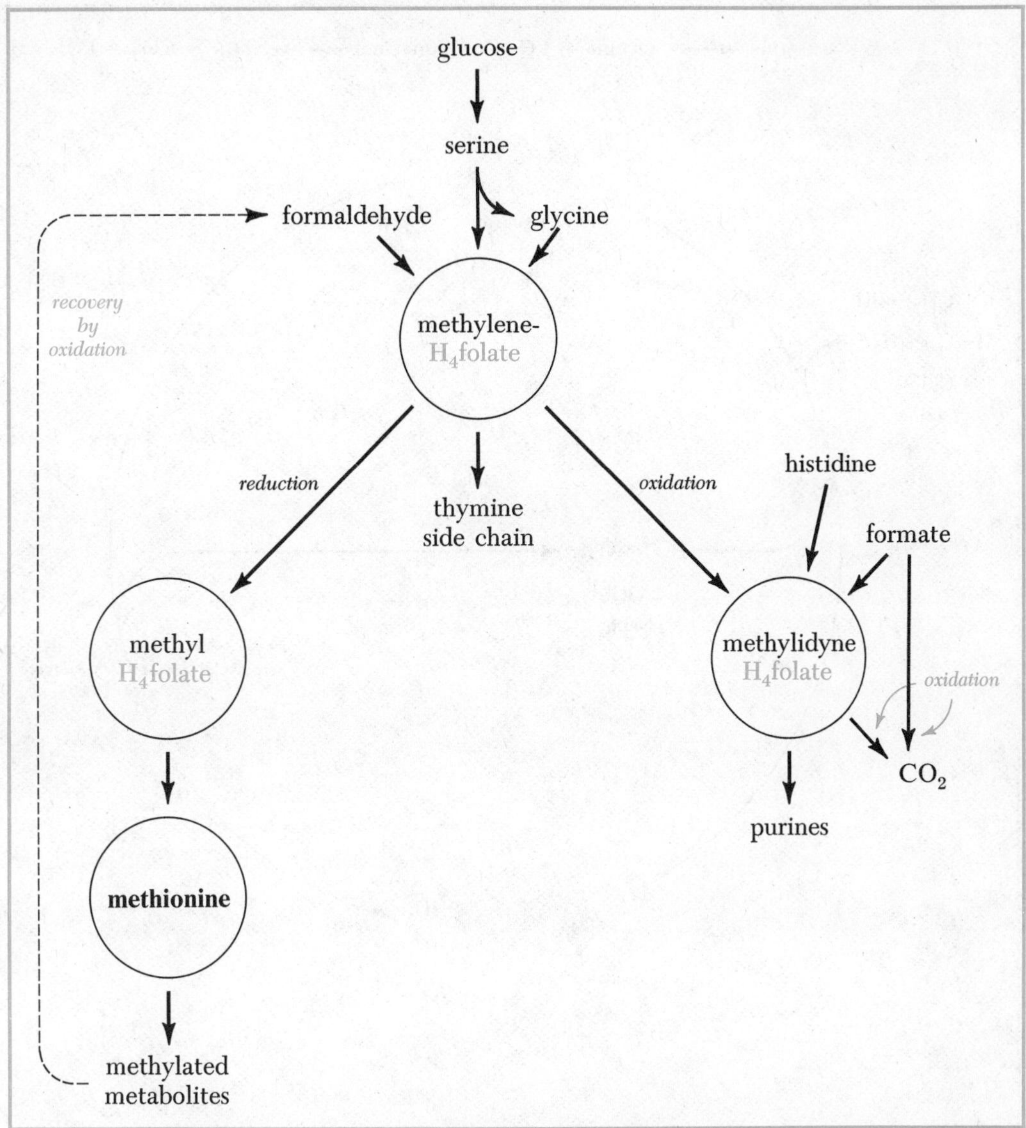

FIGURE 31–4 **Overall view of one-carbon metabolism. Glucose is the ultimate endogenous source of one-carbon groups, which are formed *via* serine and glycine. The principal uses of the groups are to form methylated compounds, the side chain of thymine, and carbons in the purine ring. One-carbon groups can be recovered by oxidation to formaldehyde or formate and by direct transfer from histidine. Excess one-carbon groups are removed by oxidation to CO_2.**

atoms at positions 5 and 10 of H_4folate (Fig. 31–5). The other product is glycine, and the equilibration of serine and glycine in this way within mitochondria is one of the most important reactions in nitrogen metabolism:

$$\text{serine} + \text{H}_4\text{folate} \longleftrightarrow \text{glycine} + \text{5,10-methylene-H}_4\text{folate}$$

This reaction is rapid enough in most animals to provide their needs for glycine, which is therefore not an essential amino acid for them.

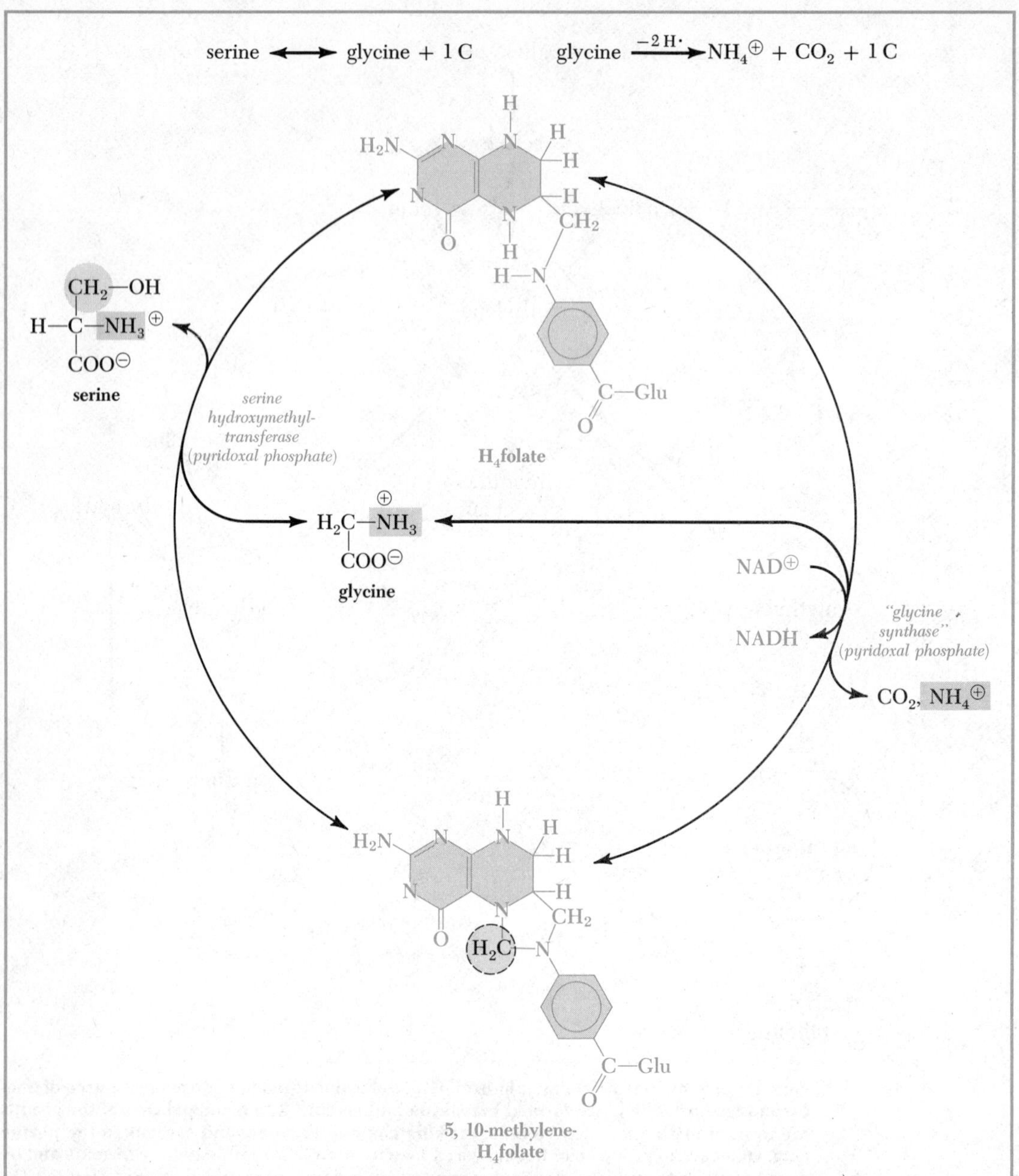

FIGURE 31–5 **A methylene group can be transferred reversibly from serine to H_4folate, forming glycine. Glycine can be oxidized to form another methylene group, along with CO_2 and NH_4^+.**

Glycine is degraded within mitochondria by a reaction sequence that is equivalent to an oxidative deamination with an associated decarboxylation:

$$\text{glycine} + \text{NAD}^{\oplus} + \text{H}_4\text{folate} \longrightarrow \text{NH}_4^{\oplus} + \text{CO}_2 + \text{5,10-methylene-H}_4\text{folate} + \text{NADH}.$$

The resultant NADH supplies electrons for oxidative phosphorylation.

A high demand for one-carbon units may be met by converting serine to glycine, by oxidizing glycine, or by both processes. The reactions also provide a means of converting excess glycine to serine in the absence of a demand for one-carbon groups. The resultant serine can then be catabolized via pyruvate:

1. glycine + $1/2\ O_2$ + H_4folate $\longrightarrow$ $NH_4^{\oplus}$ + CO_2 + 5,10-methylene-H_4folate + 3 ~ P
2. 5,10-methylene-H_4folate + glycine $\longrightarrow$ H_4folate + serine
3. serine $\longrightarrow$ pyruvate + $NH_4^{\oplus}$

SUM: 2 glycine + $1/2\ O_2$ $\longrightarrow$ 2 $NH_4^{\oplus}$ + CO_2 + pyruvate + 3 ~ P

We can therefore state:

Glycine

Nitrogen: appears as NH_4^+.

Carbons: one fourth (one from every two glycine) appear directly as CO_2, three fourths appear in pyruvate.

Glycine is clearly a glucogenic amino acid during fasting.

Source of Serine. Serine is a metabolically active amino acid that is the precursor of several cellular constituents in addition to being a major component of proteins. Animals have, therefore, preserved the ability to synthesize the amino acid from glucose, using routes that perforce differ from the irreversible deamination used in catabolism and discussed in the preceding chapter.

The immediate precursor of serine is hydroxypyruvate, or phosphohydroxypyruvate:

amino acid keto acid

$COO^{\ominus}$–C=O–CH_2—O(H, PO_3^{2-})

hydroxypyruvate *or* 3-phosphohydroxypyruvate

$\longleftrightarrow$ *serine-pyruvate* or *phosphoserine aminotransferase*

$COO^{\ominus}$–$H_3N^{\oplus}$—C—H–CH_2—O(H, PO_3^{2-})

serine *or* phosphoserine

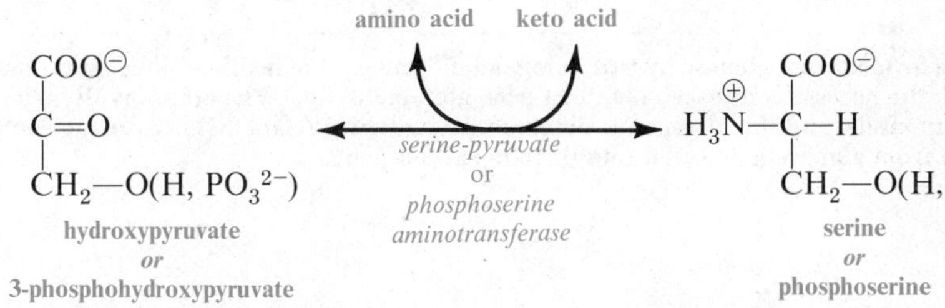

Amino groups are transferred from alanine to hydroxypyruvate, or from glutamate to the phosphorylated form.

Phosphohydroxypyruvate can be made directly from glucose, hydroxy pyruvate indirectly *via* glycerol (Fig. 31–6), so methylene groups on H_4folate can also rise from glucose by these routes. The reason for the existence of two routes for synthesis is not clear.

Methylidyne Groups

Tracer studies with ^{14}C-labeled serine and glycine show that carbon atoms from these compounds are rapidly incorporated at the oxidation state of formate into purine nucleotides and the like. The route by which this is accomplished most likely is an oxidation of 5,10-methylene-H_4folate to 5,10-methylidyne-H_4folate by an NAD-coupled dehydrogenase (Fig. 31–7). (There is a dehydrogenase catalyzing the same reaction with NADP as a coenzyme, but the equilibrium is such that the high [NADPH]/[NADP] ratio would make the reaction almost irreversible in favor of the methylene derivative.)

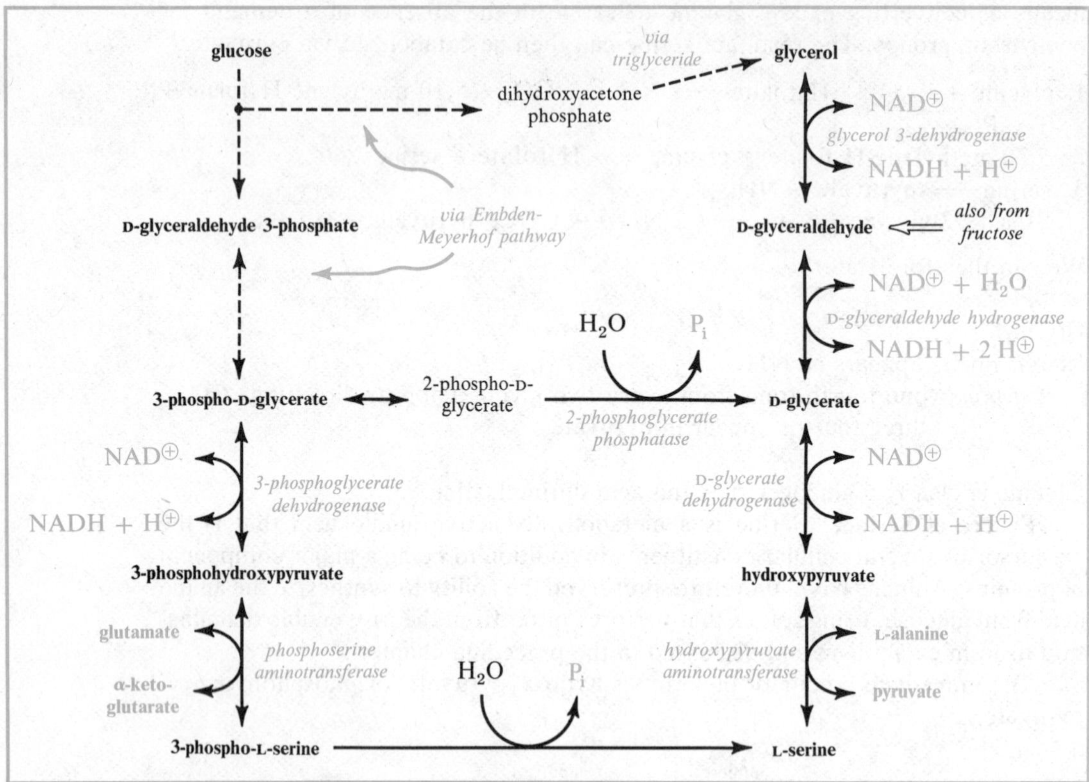

FIGURE 31–6 **Serine may be formed from glucose by two independent routes. One involves phosphorylated derivatives, with the necessary nitrogen obtained from glutamate *(left)*. The other involves the nonesterified compounds, and the nitrogen is obtained from alanine *(right)*. These are the routes by which carbon from glucose is injected into the one-carbon pool.**

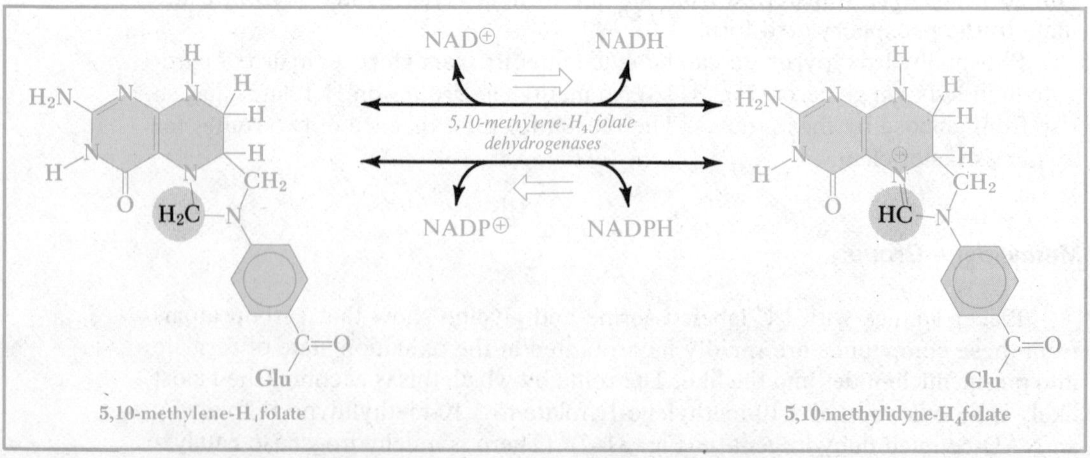

NAD⊕ NADH

5,10-methylene-H_4folate dehydrogenases

NADP⊕ NADPH

5,10-methylene-H_4folate 5,10-methylidyne-H_4folate

FIGURE 31–7 **Interconversion of 5,10-methylene-, and 5,10-methylidyne-H_4folate.**

Formation from Histidine. We noted in Chapter 30 that the catabolism of histidine begins with a deamination and ends with the formation of glutamate. Intermediate reactions involve opening the imidazole ring, and doing this creates **N-formiminoglutamate.** The formimino group is the nitrogen analogue of a formyl group, and it is transferred intact to H_4folate (Fig. 31–8), followed by loss of NH_4^+ and cyclization to form 5,10-methylidyne-H_4folate. To summarize:

Histidine

Nitrogens and carbons: One nitrogen and five carbons appear as glutamate, two nitrogen as NH_4^+, and one carbon as methylidyne-H_4folate.

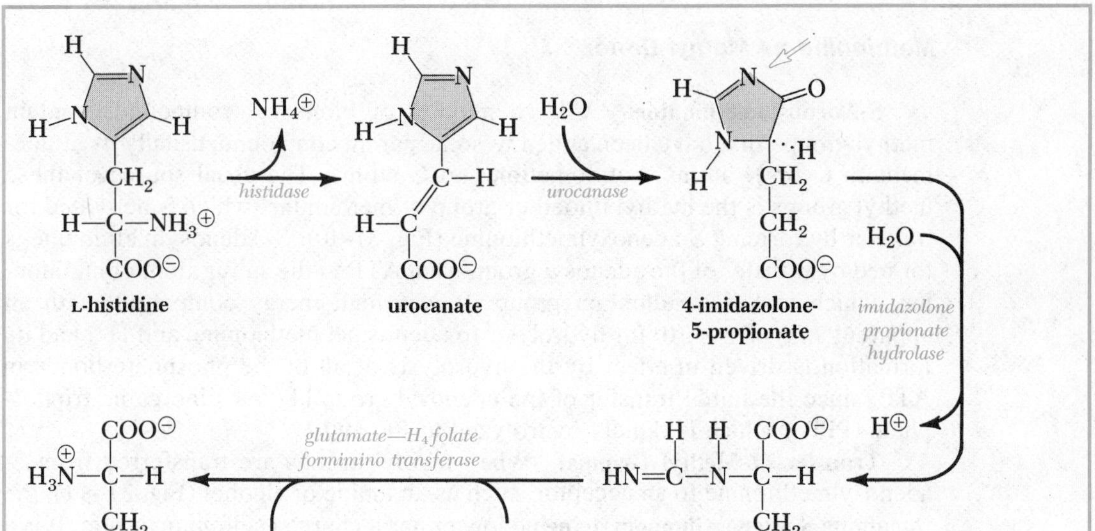

FIGURE 31–8 **The degradation of histidine.**

Since histidine is one of the less abundant amino acids, this route is not of great quantitative importance, but it provides a useful guide to the state of the H_4 folate pool in this way: The formimino transferase reaction is slowed by a deficiency of H_4folate, which therefore causes an accumulation of **formimino glutamate (FIGLU)** and its excretion in the urine, especially after ingestion of extra histidine.

Formation from Formate. Free formate is incorporated into the one-carbon pool *via* 10-formyl-H_4folate by the action of a synthetase (Fig. 31–9). A major source of formate is the oxidation of formaldehyde, which we shall shortly see is obtained by oxidation of methyl groups.

METABOLISM OF METHYL GROUPS

Methionine as Methyl Donor

S-Adenosylmethionine. A large number of biological compounds contain methyl groups that have been added to some parent compound, usually by attachment to O or N atoms, but sometimes to C atoms. The usual source of these methyl groups is the methyl thioether group of methionine, which is activated for transfer by forming S-adenosylmethionine (Fig. 31–10). S-Adenosylmethionine is formed by transfer of the adenosyl group from ATP to the sulfur atom of methionine, which makes a **sulfonium** group. It is a high-energy compound, with an apparent K'_{eq} of 2×10^4 for hydrolysis to adenosine, methionine, and H^+, and its formation is driven in effect by the hydrolysis of all of the phosphate bonds in ATP, since the initial transfer of the adenosyl group liberates inorganic triphosphate (PPP_i), which is rapidly hydrolyzed to PP_i and P_i.

Transfer of Methyl Groups. When methyl groups are transferred from S-adenosylmethionine to an acceptor, such as an amine or alcohol (Fig. 31–11), the remaining **S-adenosylhomocysteine** no longer has a charged sulfonium group. It is a simple thioether, analogous to methionine, and the resultant loss of free energy makes the methyl transfers essentially irreversible. Let us now consider important examples of these transfers.

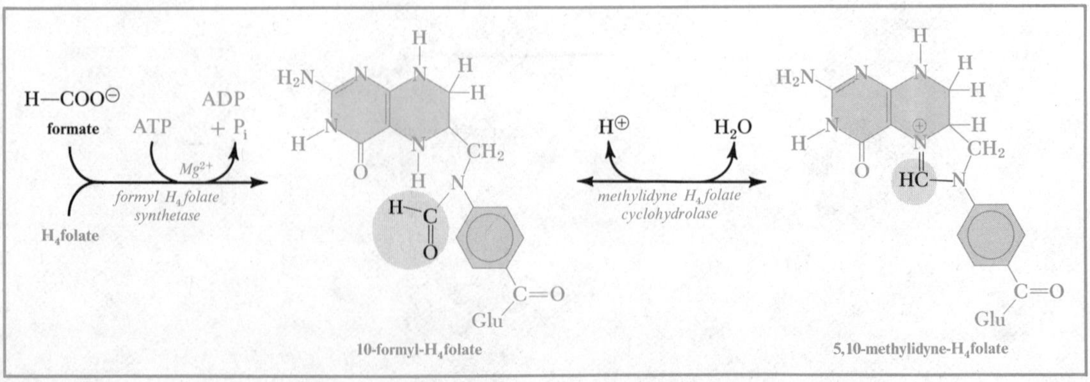

FIGURE 31–9 **The interconversion of carbon groups at the oxidation state of formic acid.**

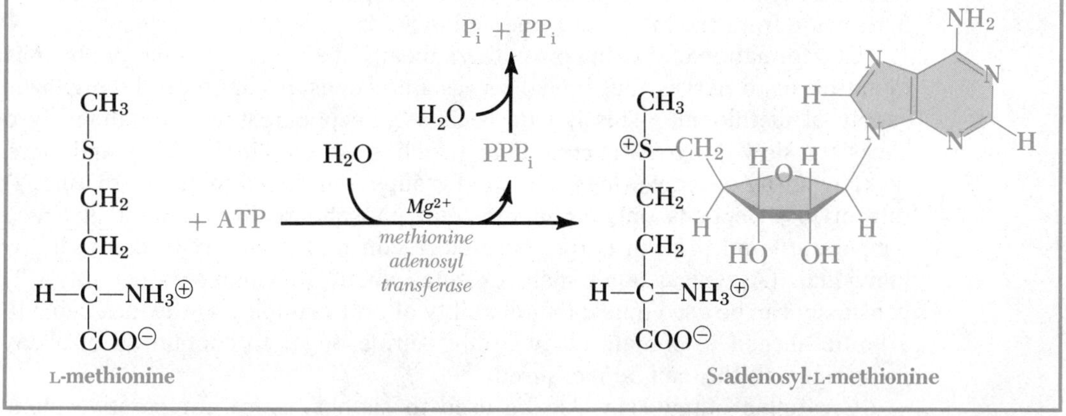

FIGURE 31–10 **The formation of S-adenosylmethionine is, in effect, driven by the hydrolysis of two high-energy and one low-energy phosphate bonds, since the concentrations of PPP_i and PP_i are kept at low levels by hydrolases.**

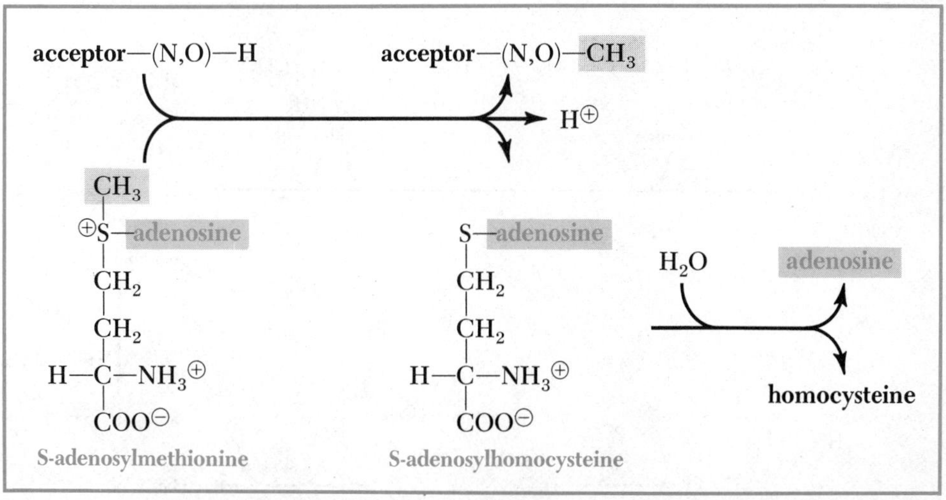

FIGURE 31–11 **Methyl groups are transferred from S-adenosylmethionine to various acceptors, usually groups containing N or O atoms. The S-adenosylhomocysteine that remains is hydrolyzed to release adenosine and homocysteine.**

Formation of Creatine

Creatine is made by transfer of a methyl group from S-adenosyl-methionine to guanidinioacetate (Fig. 31–12). The guanidinioacetate is formed by transfer of an amidinio group from arginine to glycine, leaving ornithine. Creatine is, therefore, made from fragments of three amino acids.

The formation of creatine provides a means of storing high-energy phosphate in muscles and nerves, but it also represents a constant drain on the organism's supply of methionine. This is true because phosphocreatine spontaneously cyclizes at a slow rate to form **creatinine,** which is excreted in the urine, and there is no known way of recovering the methyl groups transferred to make creatine. The rate of loss depends only upon the total phosphocreatine content at a given temperature and pH and is therefore quite constant from day to day in a given individual. (A typical adult male excretes about 15 millimoles per day.) The constancy can be used to test the reliability of urine samples — the idea being that a bottle alleged to contain all of a day's urine ought to contain all of a day's creatinine and this can be measured.

Creatinine output can also be used to simplify some comparative clinical measurements by making it unnecessary to collect timed samples of urine. Since the quantity of creatinine in a conveniently collected urine sample is in itself an index of the volume of blood filtered to excrete that sample, one compares the ratio of the urinary output of some other metabolite and the output of creatinine. The ratio will change in proportion to the relative rate of excretion of the metabolite.

Since creatinine excretion depends upon phosphocreatine content, it can also be used to assess muscle mass. When muscle degenerates for any reason — from paralysis, or from muscular dystrophy — the creatinine content of the urine falls.

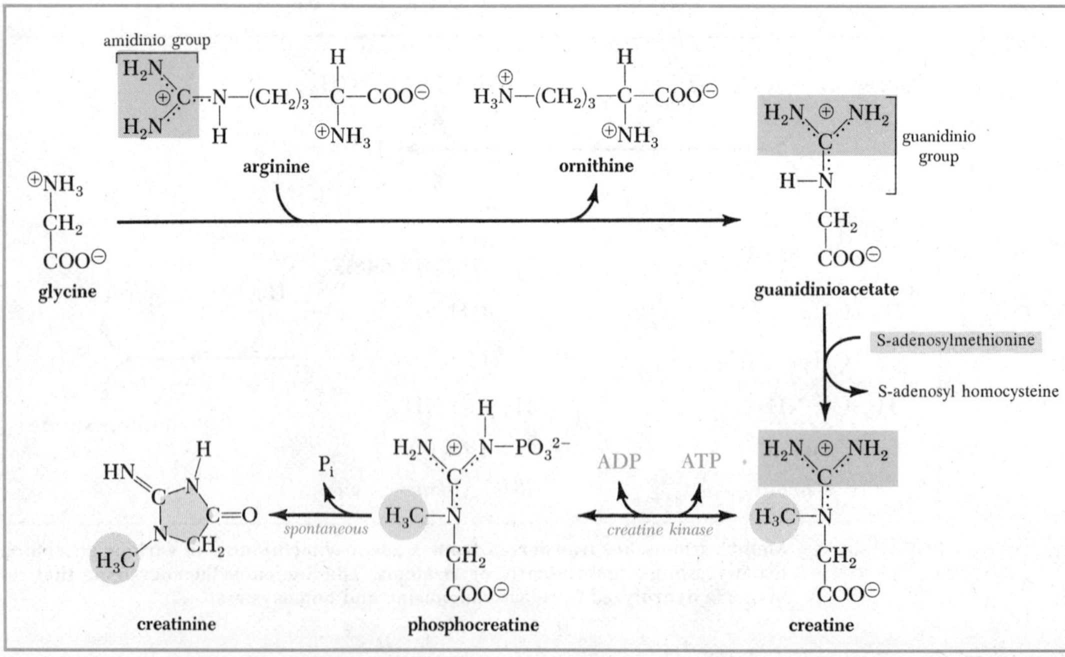

FIGURE 31–12 Creatine is constructed from glycine, an amidinio group from arginine, and a methyl group from S-adenosylmethionine. There is a continual loss of these groups as creatinine in the urine, which is formed by spontaneous cyclization of phosphocreatine.

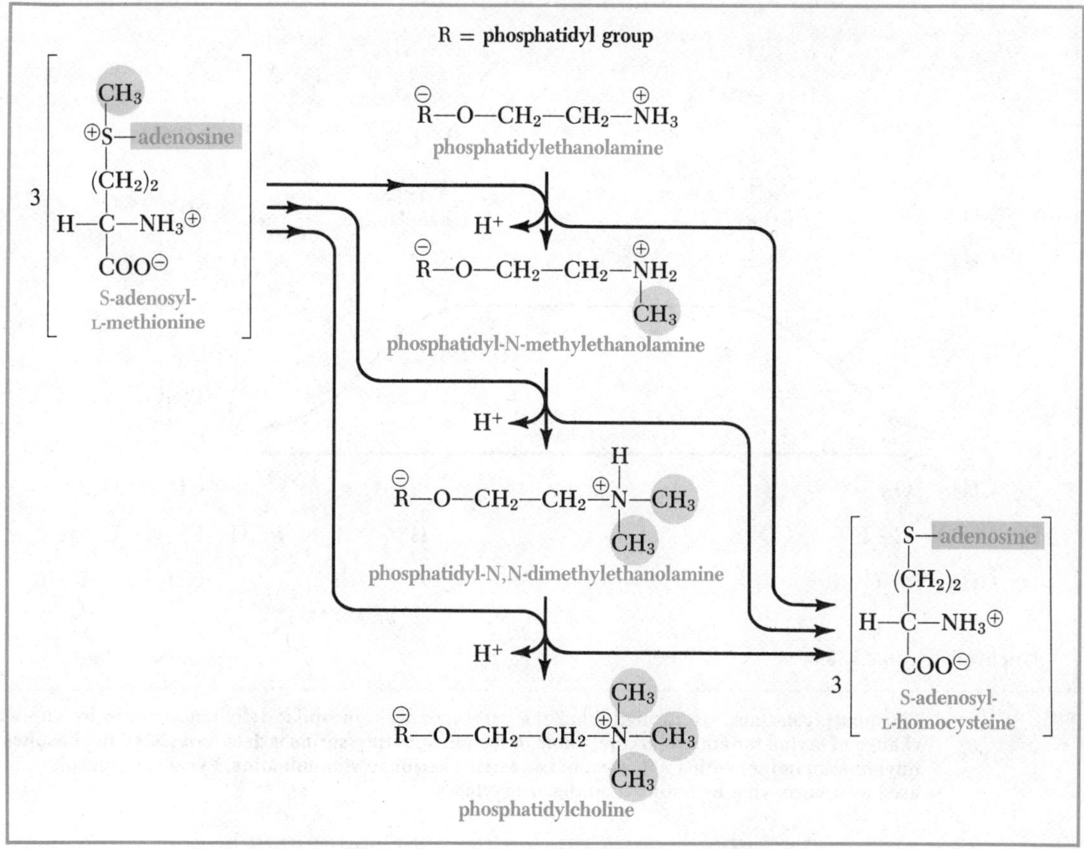

FIGURE 31–13 **Phosphatidylcholines are formed from the corresponding phosphatidylethanolamines by three successive transfers of methyl groups from S-adenosylmethionine.**

In addition, any rise in blood creatinine concentration is a sensitive indicator of kidney malfunction, since it is ordinarily eliminated rapidly.

Choline and Phospholipids

Choline is the most abundant N-methyl compound in the body. Although it is qualitatively important as the precursor of acetylcholine, which is a transmitter in some nerves, choline occurs mainly in phospholipids — phosphatidylcholines, sphingomyelins, and the like. Indeed, choline is synthesized in the form of a phosphatidylcholine (Fig. 31–13).

Synthesis of choline involves a transfer of three methyl groups from S-adenosylmethionine to phosphatidylethanolamine. The phospholipids are constantly turning over through hydrolysis, so choline generated in this way is later released.

The ethanolamine moiety is created from serine by an ingenious arrangement (Fig. 31–14). The phosphatidyl group in phosphatidylethanolamine is transferred to serine, creating phosphatidyl serine and releasing free ethanolamine. A second enzyme catalyzes the decarboxylation of phosphatidylserine, leaving phosphatidylethanolamine as a product. Here we have an example of another class of enzymes, the **amino acid decarboxylases,** that utilize the labilizing effect of

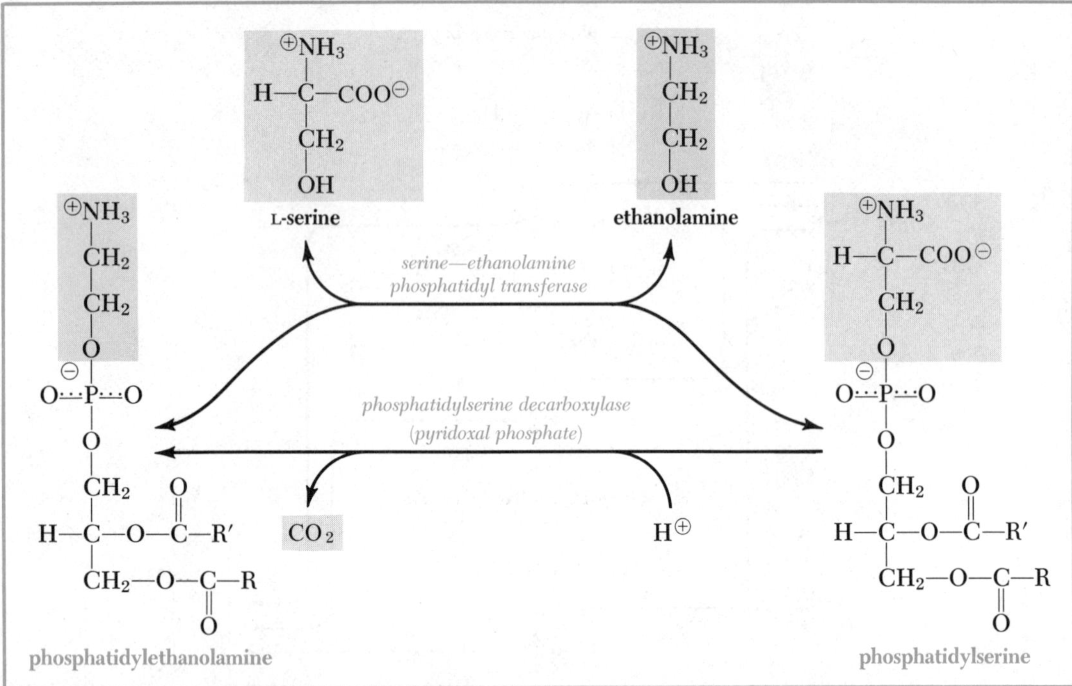

FIGURE 31–14 **Phosphatidylserines are formed from the corresponding phosphatidylethanolamines by an exchange of serine for ethanolamine. Some of the phosphatidylserine is decarboxylated to phosphatidylethanolamine, with a net effect of converting serine to ethanolamine. Pyridoxal phosphate is used as a coenzyme by amino acid decarboxylases.**

condensation with pyridoxal phosphate to promote cleavage of one of the C-2 bonds.

The sum of these two enzymatic reactions amounts to a net formation of ethanolamine from serine:

1. serine + phosphatidylethanolamine ⟶ phosphatidylserine + ethanolamine
2. phosphatidylserine + H^+ ⟶ phosphatidylethanolamine + CO_2

SUM: serine + H^+ ⟶ ethanolamine + CO_2

Since serine can be made from glucose and ammonia, it follows that ethanolamine can also be completely synthesized from these compounds, and a dietary supply is not necessary. Now let us see how ethanolamine is converted to phosphatidylethanolamine.

Use of the Free Base. In addition to the ethanolamine and choline obtained from the syntheses we outlined, the pool of the free compounds is also supplemented by the diet. (Indeed, the dietary supply can be sufficient to supply most of the needs.) The free compounds are used to make phospholipids through an intermediate conversion to cytidine diphosphate derivatives (Fig. 31–15), at a total cost of 3 ~P per phosphatidyl base made.

Summarizing the picture at this point, methyl groups from methionine can be used to make choline in the form of phosphatidylcholines, provided that phosphatidylethanolamines are available. However, the demand for synthesis can be alleviated to the extent that choline and ethanolamine are obtained from the diet.

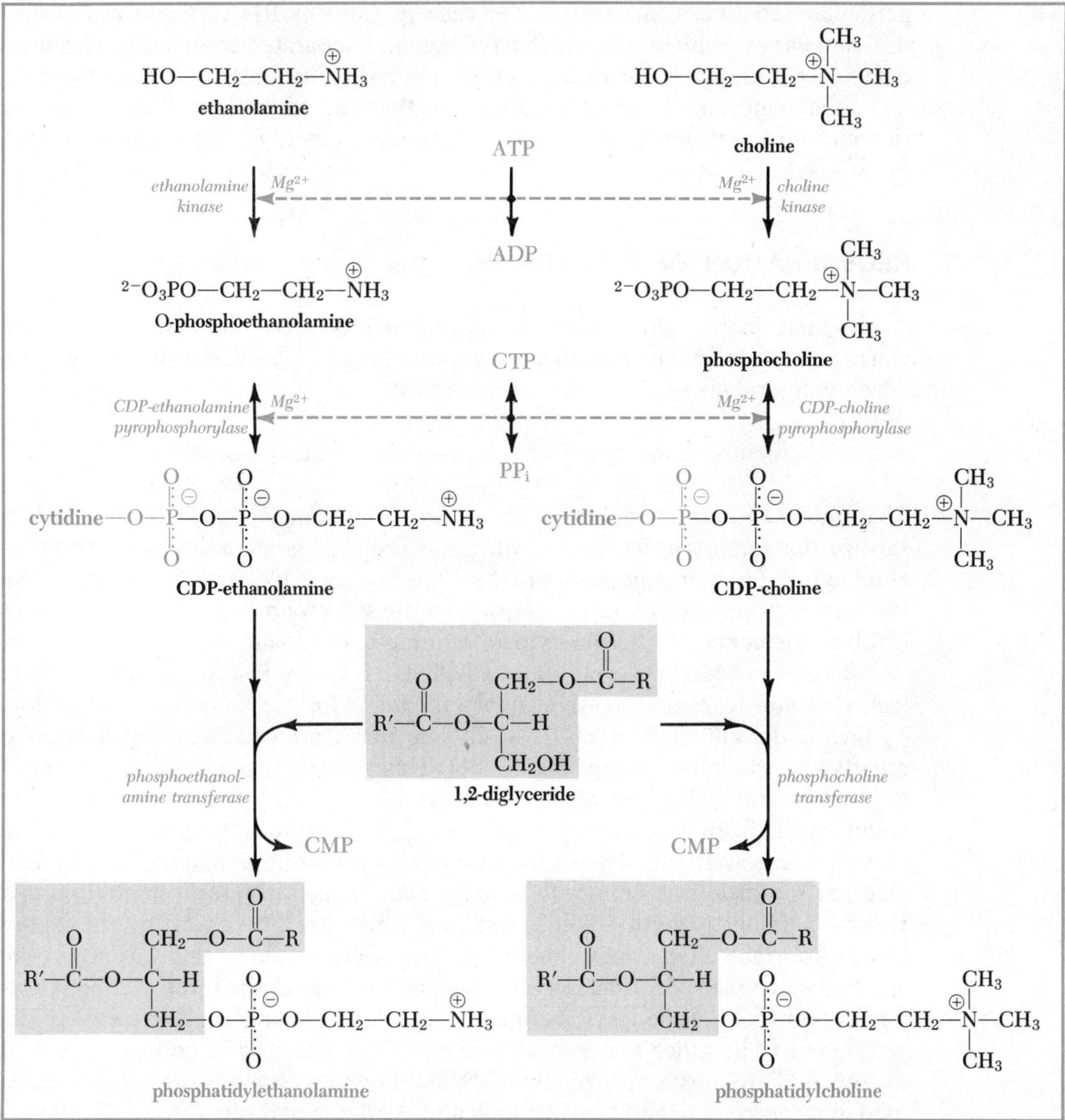

FIGURE 31–15 **Phosphatidylcholines and phosphatidylethanolamines can be made from the free choline and free ethanolamine by converting them to the CDP-derivatives, from which the phospho-base moieties are transferred to 1,2-diglycerides.**

Methylation of RNA

We noted in our discussion of protein synthesis that transfer RNA and ribosomal RNA contain bases other than the usual guanine, adenine, cytosine, and uracil. The distinctive bases frequently are methylated derivates of the more common compounds (p. 79). All of these methyl derivatives are created by enzymatic transfer of methyl groups from S-adenosylmethionine to the ribonucleates, which originally contain only the non-methylated bases at the time of assembly on DNA templates. Even the thymine found in RNA is made by methylation of uracil, not by direct incorporation of thymidine at the time of nucleic acid formation. The bases to be methylated appear to be located in particular sequences, at least in the case of transfer RNA. There are several different enzymes involved, which have not been separated completely. (Separate enzymes are known to occur for methylating individual bases.) For our purposes now, the important thing to recognize is that the formation of these specific ribonucleates represents a drain upon the supply of methyl groups represented by S-adenosylmethionine.

REGENERATION OF METHIONINE

In these methylation reactions, only the methyl group in methionine is being consumed; the remainder of the molecule is intact as S-adenosylhomocysteine, which is hydrolyzed to release free homocysteine:

$$\text{S-adenosylhomocysteine} + H_2O \rightarrow \text{adenosine} + \text{homocysteine}.$$

There is no route for making homocysteine in mammals, and it is important to salvage this important molecule. All that is required is methylation of the sulfur atom to regenerate methionine, and there are two ways by which this is done. One involves the recovery in mitochondria of a methyl group from choline; the other involves the generation in the cytosol of a methyl-H_4folate.

Recovery of Methyl Groups from Choline. It is possible to salvage one of the methyl groups from an excessive supply of choline for use elsewhere. This is done by raising the chemical potential of choline through an oxidation of its alcohol group to a carboxylate group (Fig. 31–16). This creates a **betaine** (''bay-tah-een''), an N,N,N-trimethylamino acid.* The loss of free energy upon removing one methyl group from the quaternary nitrogen is sufficient to support a transfer of the group to homocysteine. The tertiary amino group in the remaining dimethylglycine has too little free energy to support additional transfers to methyl groups. However, many N-methyl compounds are catabolized by oxidizing the methyl group to formaldehyde, and dimethylglycine is converted in this way to glycine and two molecules of formaldehyde. In short, one out of the three methyl groups transferred from S-adenosyl methionine to make choline can be recovered as methionine. The other two can also re-enter the one-carbon pool, but as 5,10-methylene-H_4folate *via* formaldehyde. Note, however, the dietary choline can be used as a source of a methyl group in lieu of synthetic pathways. Note also that an average yield of 3 ~P per methyl group is obtained during the degradation of choline.

Formaldehyde in excess can be oxidized to CO_2 and H_2O. It is first oxidized to formate, which is then converted to the H_4folate derivative before the final oxidation.

Although H_4folate and its derivatives cross the inner mitochondrial membrane only very slowly, there is interaction between the one-carbon pool inside and outside the mitochondria through transfer of amino acids bearing the groups. The hydroxymethyl transferase equilibrating glycine and serine occurs both in the mitochondria and in the cytosol, so passage of these two amino acids is in effect carrying methylene groups.

***De Novo* Synthesis from Methylene-H_4folate.** The supply of methyl groups obtained from the diet in the form of methionine and choline is supplemented by generating them from the one-carbon pool. 5,10-Methylene-H_4folate is irrevers-

*Named for its occurrence in *Beta vulgaris,* the common beet. More specifically, it is glycine betaine, distinguishing it from other trimethylamino acids.

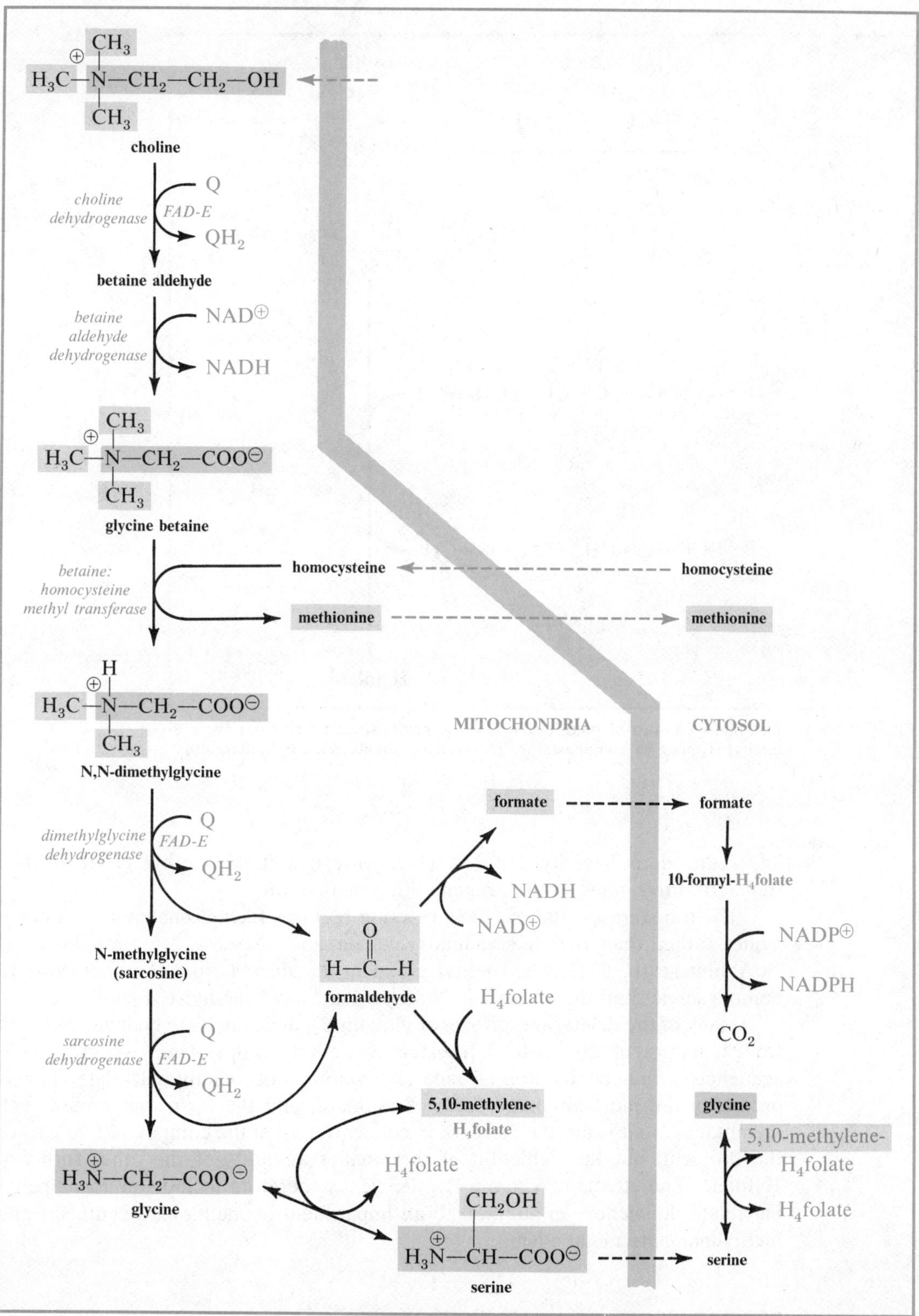

FIGURE 31–16 Choline is constantly degraded in mitochondria by oxidation of the alcohol group to a carboxylic acid, followed by removal of the methyl groups. One methyl group is transferred to homocysteine *(center)*, and the other two are oxidized to formaldehyde. The one-carbon groups are believed to move from mitochondria to the cytosol in the form of methionine and serine.

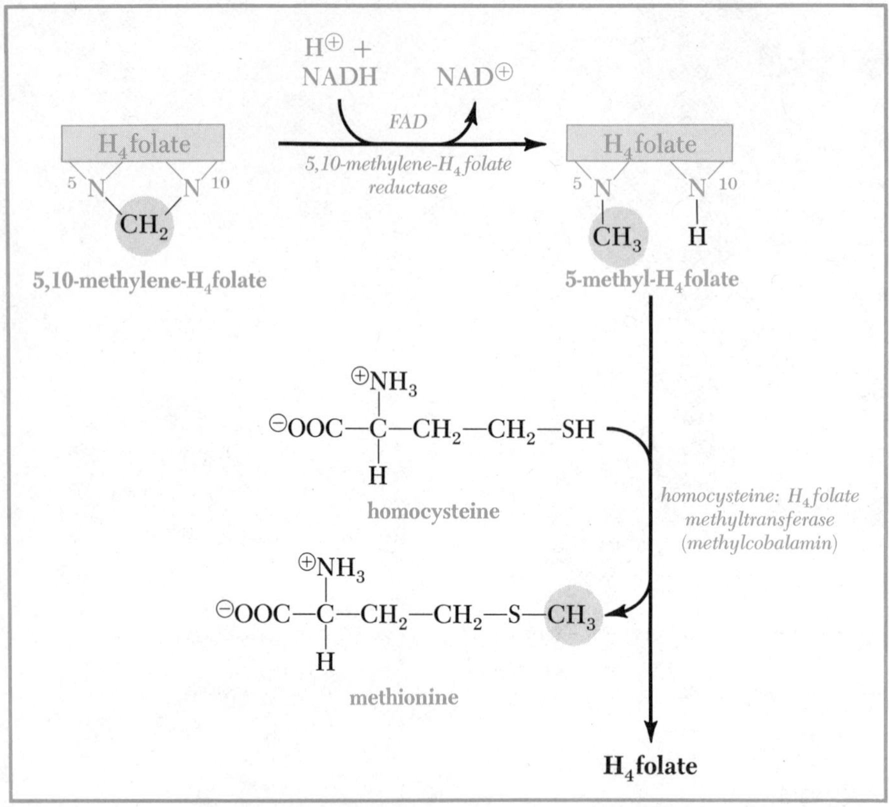

FIGURE 31–17 **The methyl group of methionine may be synthesized by transfer from 5-methyl-H_4folate to homocysteine. The transfer involves methylcobalamin.**

ibly reduced to 5-methyl-H_4folate (Fig. 31–17), and the methyl group is then transferred to homocysteine, regenerating methionine.

The transferase catalyzing the reaction requires **methylcobalamin** as a coenzyme, rather than the deoxyadenosylcobalamin necessary for methylmalonyl CoA mutase (p. 571). The methyl group is transferred from the coenzyme to homocysteine, and the coenzyme is recharged from 5-methyl-H_4folate.

Many of the deleterious effects of vitamin B_{12} deficiency are believed to result from a failure of this critical transfer of methyl groups. One important consequence is an accumulation of excessive amounts of 5-methyl-H_4folate. This is ordinarily the most abundant form of H_4folate, and the form that crosses cell membranes, so it is not the increase in concentration of the compound that causes the problem, but the depletion of the smaller amounts of the other forms of H_4folate. The effects of vitamin B_{12} deficiency therefore include an accompanying folate deficiency, in addition to an impairment of methionine synthesis and methylmalonate metabolism.

DEGRADATION OF HOMOCYSTEINE

Nitrogen: Released as NH_4^+
Carbons: Converted to succinyl coenzyme A.
Sulfur: Appears as cysteine.

Although some homocysteine can be recovered as methionine by methylation reactions, part is constantly lost through degradation by an irreversible route in which the sulfur appears as cysteine and the carbon skeleton as 2-ketobutyrate, which we have already seen is metabolized *via* propionyl coenzyme A and CO_2 to succinyl coenzyme A. Therefore, methionine is a source of cysteine and methyl groups and is a glucogenic amino acid.

The actual degradation of homocysteine is a simple two-step process involving the formation of **cystathionine** and its cleavage to form cysteine, 2-ketobutyrate, and ammonium ion (Fig. 31–18). Both of the enzymes involved have pyridoxal phosphate as a coenzyme.

Defects in both these enzymatic steps occur in humans. **Homocystinuria,** the excretion of the oxidized form of homocysteine, occurs in a hereditary absence of cystathionine synthase, which also results in mental retardation. **Cystathioninuria,** the excretion of cystathionine, also is associated with mental defects. When one considers the unusually high content of cystathionine in the brains of normal individuals (0.2 to 0.5 mg per g of tissue, as contrasted with 7 to 8 μg per g in other tissues) and the disturbances in the nervous system apparently created by an inability to make or destroy this amino acid, one suspects it may have some presently unknown function in nerve physiology. Primates have a higher concentration of cystathionine in the brain than do other mammals, whereas birds have much lower levels than do mammals. This is not to imply that the term "hypocerebrocystathioninic" could rationally be substituted for bird-brained as an epithet.

Low levels of cystathionine synthase activity have another consequence of great interest: damages to the endothelial cells in arteries. Indeed, administration of homocysteine has been used as an experimental tool to demonstrate that endothelial cell damage is probably an essential preliminary to the development of atherosclerotic plaques. The mechanism for damage is not clear. Accumulation of homocysteine causes the formation of a mixed disulfide with cysteine; this is the principal form in which homocysteine appears in the blood, and the formation of the compound further diminishes an already low supply of cysteine resulting from the failure to metabolize homocysteine.

The question arises: Is there any connection between coronary artery disease and a disturbance in homocysteine metabolism? Very preliminary results suggest there may be in at least some cases, but this is by no means conclusive. A very,

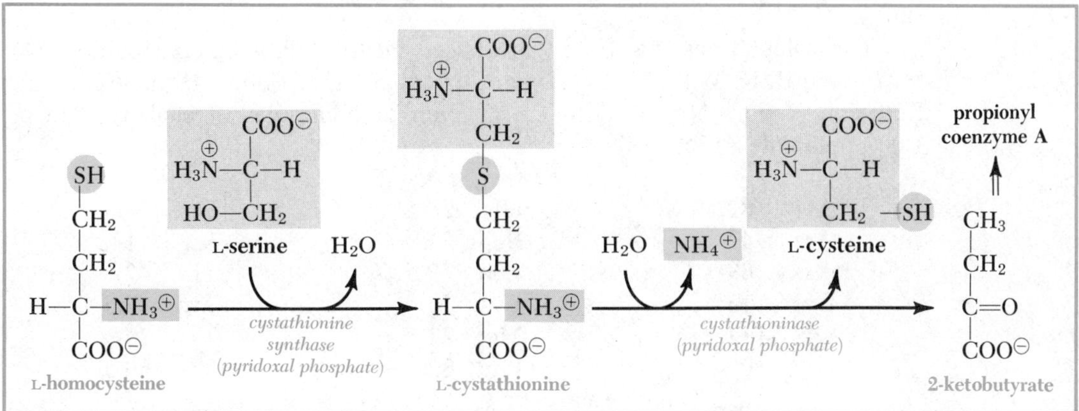

FIGURE 31–18 **The degradation of homocysteine. The effective result is the transfer of the sulfur to serine, forming cysteine, and a deamination with the carbon skeleton appearing as 2-ketobutyrate.**

very small sample (n = 25) of coronary patients had significantly higher concentrations of the homocysteine-cysteine mixed disulfide than did control patients. In the general population, the incidence of homocystinuria may be as high as 1:50,000, implying that 1 in 100 people are heterozygotes, so it is possible that additional people have less severe impairments in homocysteine metabolism, but sufficient to cause an elevated concentration as the preliminary data suggest.

Before leaving this subject, let us re-emphasize that methionine is an important dietary constituent for three purposes. Most importantly, it is **required as an amino acid for the formation of proteins,** and no other dietary component will substitute for this purpose, since the homocysteine chain cannot be made. As a **source of methyl groups** it is important, but not imperative, since choline in the diet can provide methyl groups, and a further supply is available through the one-carbon pool by the formation of methyl-H_4folate. Finally, it is a **source of sulfur** available for the formation of cysteine. This relieves the necessity of an absolute requirement for cysteine in the diet, but it means that sufficient methionine must be present to provide an excess sulfur supply, if the content of cysteine is low.

TOTAL BALANCE OF AMINO ACID DEGRADATION

Now that we have completed a survey of amino acid degradation, let us take stock. What is the net result of the use of amino acids as fuels? It is not possible to give a single answer to that question because several tissues are involved in ways that are influenced by other metabolic events. However, we can make an approximation by defining conditions. Let us develop a model for the fate of the amino acids in a piece of beefsteak eaten by an individual who has been fasting for one day. Assume that this hungry person has eaten 1,000 milliomoles of amino acids. The average formula weight of the amino acid residues in beef muscle is near 110, so this quantity of amino acids will be obtained from 110 grams of protein, which is the amount in about 530 grams of raw lean meat. Since some of the amino acids contain more than one nitrogen atom, the total nitrogen content of the protein is near 1,390 milliatoms. To make our calculations, we shall specify exact amounts, even though they are not known with this precision for real foods. As a frame of reference it is helpful to know that the quantity we are dealing with is close to the mean total daily intake of protein for young white male adults in the United States.

The calculations that follow are based on the following composition (Orr, M. L., and B. K. Watt: [1957] *Amino Acid Content of Foods.* Home Economics Research Report No. 4, U.S.D.A.), given in millimoles of each residue per 1,000 millimoles of all residues:

Ala	82.5	His	28.5	Pro	54.5
Arg	47	Ile	51	Ser	51
*Asx	89	Leu	79.5	Thr	47
Cys	13	Lys	76	Trp	7.5
*Glx	131	Met	21	Tyr	24
Gly	105	Phe	32	Val	60.5

amide nitrogen (=Asn + Gln) = 110

*Asx = Asn + Asp; Glx = Gln + Glu.

Events in Skeletal Muscles

The branched chain amino acids from proteins eaten by a fasting individual are mainly taken up by the skeletal muscles. The muscles put out nitrogen as glutamine and alanine, for the most part. Assuming that the muscles metabolize the carbon skeletons, and supplying a small amount of glutamate from the liver, the following stoichiometry can be developed:

Consumed By Muscles

51 isoleucine	21 glutamate
$79\frac{1}{2}$ leucine	$919\frac{1}{2}$ O_2
$60\frac{1}{2}$ valine	

Produced By Muscles

53 alanine	634 CO_2
$79\frac{1}{2}$ glutamine	4,855 ~ P

Events in the Small Bowel

Most of the aspartate, asparagine, glutamate, and glutamine from the diet is processed in the intestinal mucosa, and the mucosa also is the tissue mainly responsible for clearing glutamine from the blood for use as a fuel. Much of the ammonium ion released by hydrolysis of asparagine and glutamine appears as such in the portal blood. Alanine is the other predominant nitrogenous product. (Citrulline and other compounds are also produced, but we can simplify the stoichiometry by considering only alanine without serious distortion of the final result.)

Consumed By Small Bowel

from the diet	*from the blood*
89 (aspartate + asparagine*)	$79\frac{1}{2}$ glutamine
131 (glutamate + glutamine*)	$315\frac{3}{4}$ O_2

Produced By Small Bowel

$299\frac{1}{2}$ alanine	510 CO_2
$189\frac{1}{2}$ NH_4^+	1,838 ~ P

Events in Liver

The alanine poured out by the muscles and the small bowel and the balance of the dietary amino acids are mainly handled in the liver. The nitrogen appears as urea, the sulfur as sulfate, and the carbon skeletons as CO_2 and glucose in a fasting individual, except for a small amount of acetoacetate. (We shall neglect D-3-hydroxybutyrate formation, since the amount involved is minor.)

*dietary asparagine + glutamine = 110.

Consumed By Liver

435 alanine	$54\frac{1}{2}$ proline
47 arginine	51 serine
13 cysteine	47 threonine
105 glycine	$7\frac{1}{2}$ tryptophan
$28\frac{1}{2}$ histidine	24 tyrosine
76 lysine	$189\frac{1}{2}$ NH_4^+
21 methionine	$1,166\frac{1}{2}$ O_2
32 phenylalanine	

Produced By Liver

$695\frac{3}{4}$ urea	$367\frac{1}{2}$ CO_2
21 glutamate	56 acetoacetate
396 glucose	150 ~ P
34 SO_4^{2-}	

The striking thing here is the neglible yield of high-energy phosphate from a large consumption of oxygen. Amino acid metabolism by itself will not sustain the energy-requiring processes of the liver under these conditions. Note also that the stoichiometry allows a small output of glutamate to replace the amount used in skeletal muscles to make glutamine. This is consistent with the observed events *in situ*.

Events in Brain

To complete the picture, we must account for the glucose and acetoacetate produced by the liver. A reasonable assumption in a fasting man is the use of the fuels by the brain, for which the stoichiometry is:

$$396 \text{ glucose} + 56 \text{ acetoacetate} + 2,600\ O_2 \rightarrow 2,600\ CO_2 \times 15,346 \sim P.$$

Total Stoichiometry

If we add the balances within the individual tissues, the complete equation for the metabolism of the amino acids in beefsteak becomes:

$$1,000 \text{ amino acids} + 5,001^3/_4\ O_2 \rightarrow 695^3/_4 \text{ urea} + 4,111^1/_4\ CO_2 + 22,188 \sim P + 34\ SO_4^{2-}.$$

The overall P:O ratio in the whole body is 2.22.

The inevitable question is whether this is an academic exercise without relevance to real events. We have a clue from old observations: Animals can be caused to excrete glucose in the urine by removing the pancreas so as to create diabetes, or by treating them with phlorhizin, which prevents the reabsorption of glucose from the filtered blood plasma in the kidney. (Phlorhizin is a polyphenolic glycoside found in the bark of apples, cherries, and other rosaceous plants.) When properly performed, the experiments resulted in nearly quantitative recovery of any glucose formed in the animal. The amount of glucose formed by the metabolism of amino acids could then be measured by analyzing the excretion of nitrogen and of glucose. The results were expressed as a D:N ratio, the weight of glucose (dextrose) formed per weight of nitrogen excreted.

Now, the theoretical D:N ratio, according to our stoichiometry, would be 396 × 180 grams of glucose per 696¾ × 2 × 14 grams of excreted N, which calculates out as 3.66. The actual observed values on animals fed meat were 3.63 in dogs, 3.68 in humans treated with phlorhizin, and 3.63 to 3.73 in humans with diabetes!

Another test of theory is through the observed relative value of proteins as fuels, compared to fats and carbohydrates. This will be discussed more extensively in Chapter 37, when we shall use the theoretical values in reasoning about the total metabolic economy.

Origin of the Stoichiometries

A detailed discussion of the calculations made in developing the preceding equations is beyond our scope. Those with special interest might try to reconstruct the reasoning. Simplified versions of the events in skeletal muscle and the small bowel are presented in Figures 31–19 and 31–20. (The model for skeletal muscle metabolism was developed to

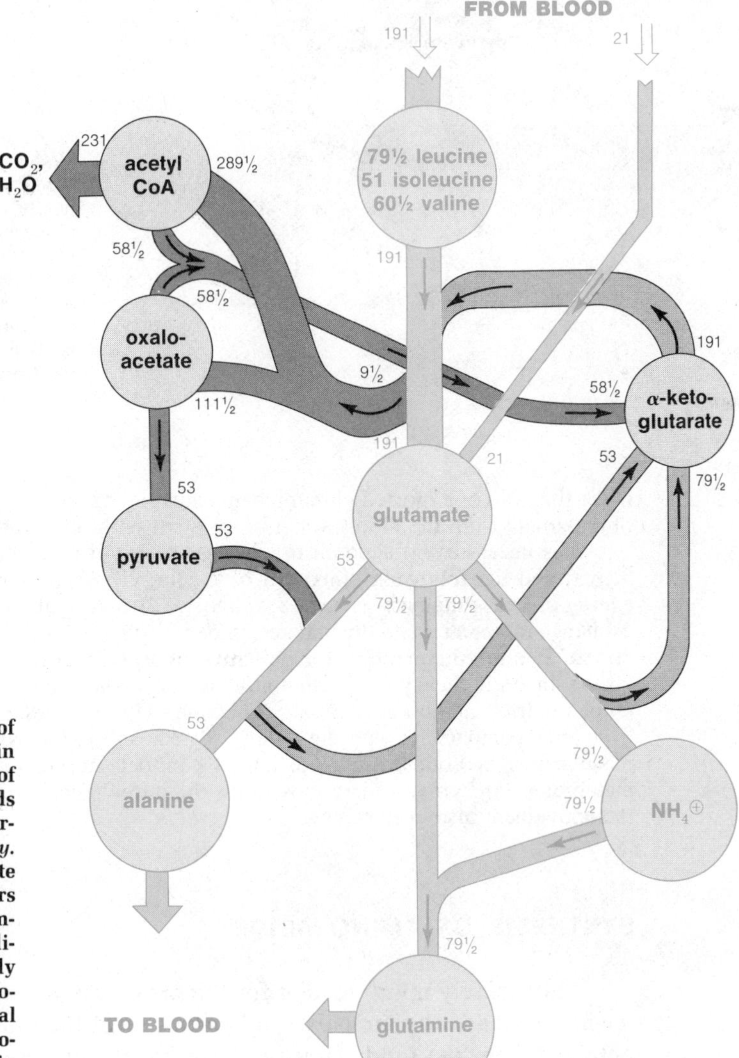

FIGURE 31–19

Model for the metabolism of branched-chain amino acids in skeletal muscles. The path of nitrogen-containing compounds is in *blue*, and that for the carbon skeletons is in *dark gray*. The recycling of α-ketoglutarate is shown in *light gray*. Numbers indicate the millimoles of compound reacting per 1,000 millimoles of amino acids originally in the diet. Passage of components across the mitochondrial membranes is not shown. Evolution of CO_2 is not shown in many reactions.

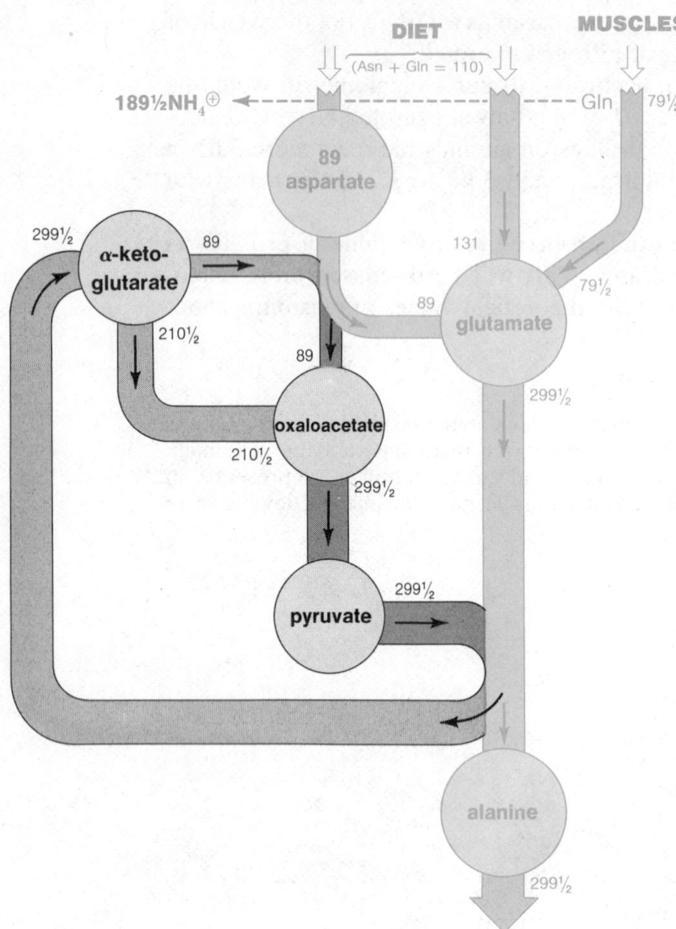

FIGURE 31–20

Model for the metabolism of dicarboxylic amino acids and their amides in the intestinal mucosa. The path of nitrogen-containing compounds is in *blue*, and the route for carbon skeletons from aspartate and asparagine is in *dark gray*. The route for carbon skeletons from glutamate and glutamine, which includes a recycling during transamination, is in *lighter gray*.

make the ratio of exported glutamine and alanine 3:2 and minimize the requirement for use of glutamate from the blood; this is consistent with observations *in situ*.)

It is necessary to account for the movement of metabolites within cellular compartments, and the following hints will be useful: (1) Aspartate enters mitochondria by the energy-driven glutamate-aspartate antiport. (2) α-Ketoglutarate leaves mitochondria by exchange for malate, and the malate can be returned by exchange for P_i. There is a constant flow of P_i into mitochondria for oxidative phosphorylation. (3) Alanine aminotransferase occurs in both the cytosol and mitochondria, and alanine can move across the inner mitochondrial membrane without exchange. This enables movement of nitrogen atoms from one compartment to the other. (4) Phospho-*enol*-pyruvate can be formed from oxaloacetate in both the cytosol and the mitochondria, and it moves across the inner membrane. (5) Pyruvate may move into the mitochondria by exchange for OH^-, making this movement also energy-driven.

SYNTHESIS OF AMINO ACIDS

Animals rely upon the diet for the provision of some amino acids. Indeed, it might be asked why animals synthesize any of the amino acids, since all 20 are obtained from the food. However, we have seen that some of the amino acids have important metabolic roles, which in themselves involve a constant synthesis and degradation of the compounds in quantities much beyond those needed for protein synthesis. Animals have maintained an ability to synthesize these amino acids *de novo,* that is, from glucose as a carbon source and ammonium ion as a nitrogen source.

The premier example is **glutamate,** which can be formed by a reversal of the glutamate dehydrogenase reaction:

$$\underset{\alpha\text{-ketoglutarate}}{COO^{\ominus}-C{=}O-CH_2-CH_2-COO^{\ominus}} + NH_4^{\oplus} + \text{NADPH or NADH} \underset{\text{dehydrogenase}}{\overset{\text{glutamate}}{\rightleftharpoons}} \underset{\text{L-glutamate}}{COO^{\ominus}-\overset{\oplus}{H_3N}{-}C{-}H-CH_2-CH_2-COO^{\ominus}} + \text{NADP}^{\oplus} \text{ or NAD}^{\oplus}$$

This is the point of entry for ammonia nitrogen into other nitrogenous compounds in animal tissues. The α-ketoglutarate required for the reaction can be made from glucose *via* citrate. (Both the oxaloacetate and the acetyl coenzyme A used for citrate formation can be made from pyruvate.) Since glutamate dehydrogenase can function with either NAD or NADP, some presume that it uses NAD for oxidation and NADPH for the synthetic reduction, but a rational explanation for this discrimination and a demonstration of its occurrence has yet to be made.

Any increase in the concentration of glutamate will automatically cause an increase in the concentration of **aspartate** and **alanine,** because the aminotransferases constantly equilibrate these amino acids and the available supply of glutamate, oxaloacetate, and pyruvate. Therefore, aspartate and alanine need not be supplied in the diet (Fig. 31–21).

Arginine is continually being synthesized by the reactions of the urea cycle, but if arginine is withdrawn from the cycle, ornithine must be provided from some other source to replace it. It is known that ornithine can be made from glutamate by animals, but it is not known how. The reaction of ornithine aminotransferase (p. 568) is reversible, so ornithine can be made from glutamate semialdehyde, but how is glutamate semialdehyde made? The oxidation of glutamate semialdehyde

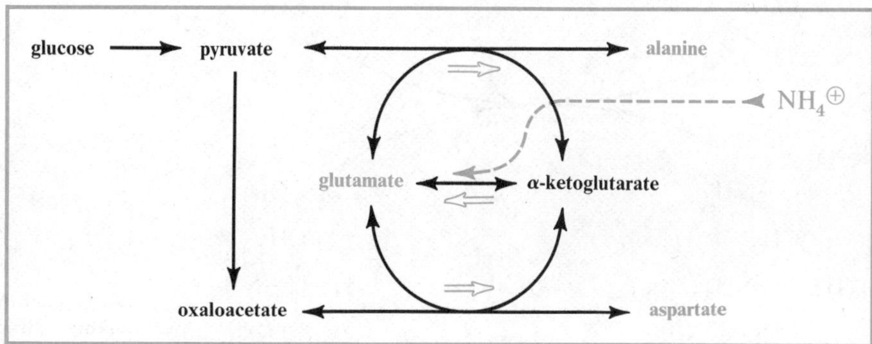

FIGURE 31–21 **Carbon from glucose and nitrogen from ammonium ions can be used for total synthesis of alanine, aspartate, and glutamate in animals.**

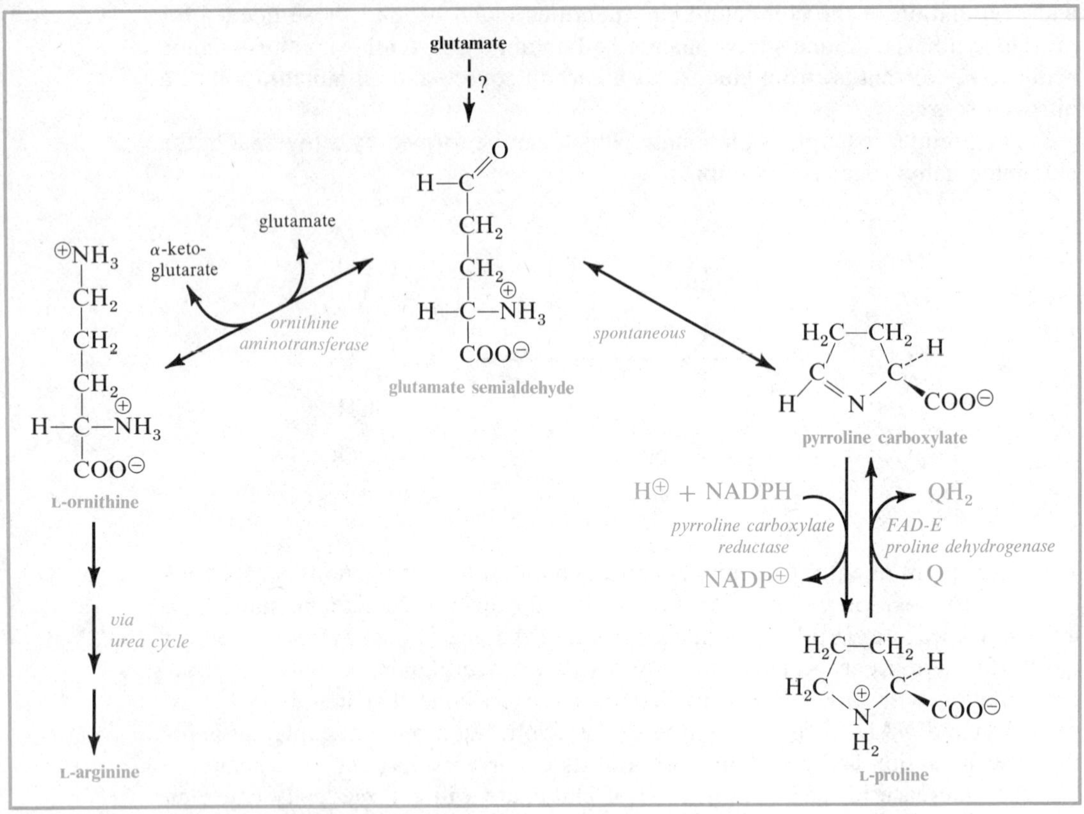

glutamate

?

glutamate

α-keto-glutarate

ornithine aminotransferase

spontaneous

glutamate semialdehyde

L-ornithine

pyrroline carboxylate

$H^{\oplus}$ + NADPH

QH_2

pyrroline carboxylate reductase

FAD-E proline dehydrogenase

$NADP^{\oplus}$

Q

via urea cycle

L-arginine

L-proline

FIGURE 31–22 **Arginine and proline can be synthesized from glutamate, but the mechanism of formation of the intermediate glutamate semialdehyde is not known.**

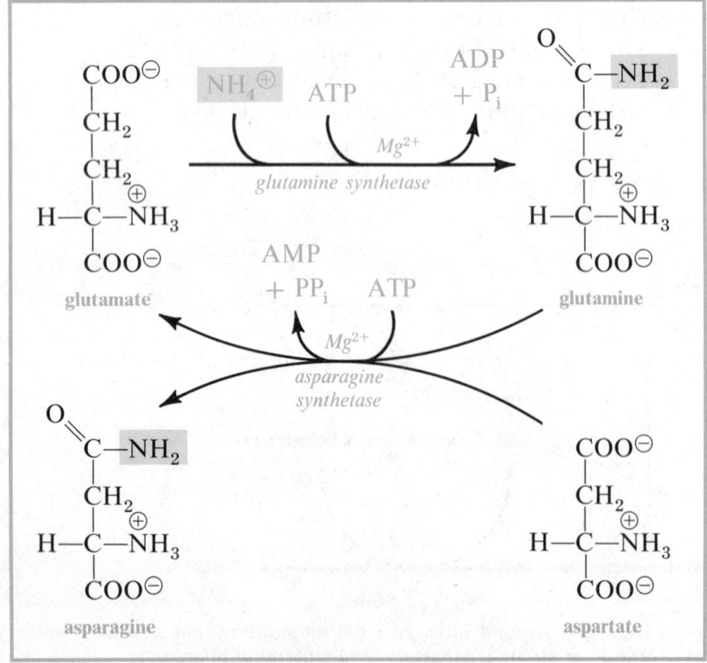

FIGURE 31–23

The amide nitrogen of glutamine is obtained directly from ammonium ion, and once formed, it may be transferred to aspartate to generate asparagine.

to glutamate is irreversible, like most aldehyde oxidations, and there is no known alternative route from glutamate to the aldehyde in animals (Fig. 31–22).

Proline is also known to be synthesized from glutamate by a process involving the reduction of the cyclic form of glutamate semialdehyde with NADPH, but the same uncertainty exists over the source of the semialdehyde. Here we also see another example of the selective use of electron carriers. Proline is degraded through the use of the flavoprotein to favor oxidation; it is synthesized through the use of NADPH to favor reduction.

The retention of arginine and proline synthesis by animals reflects their important metabolic roles. Arginine is involved in the synthesis of urea, creatine, and other compounds. Proline is an important fuel for the muscles of insects and other animals; it is also turned over more rapidly than many amino acids in human tissues, although a specific function of its metabolism has not been described.

Serine, and therefore **glycine,** can be made from glucose and glutamate through routes described earlier in the chapter. They have a rapid turnover in animals reflecting their participation in the one-carbon pool and in the formation of other nitrogenous compounds. (We shall see later that glycine is required for purine synthesis.)

Glutamine is made from glutamate and ammonium ion at the expense of high-energy phosphate. **Asparagine** is made from aspartate by the transfer of the glutamine amide nitrogen, also at the expense of high-energy phosphate (Fig. 31–23). Here we see the first example of a general function of glutamine: It often acts as a donor of an amino group during the synthesis of other compounds.

None of the other 11 amino acids can be made *de novo* by mammals. However, **tyrosine** can be made from phenylalanine, and **cysteine** can be made from methionine, so these two amino acids are not essential constituents of the diet, so long as the supply of their precursor amino acids is adequate. The practical implications of this are discussed in Chapter 41.

Finally, we come to a hard-core list of nine amino acids that must be present in the human diet:

Essential Amino Acids in Human Diets

histidine	lysine	threonine
isoleucine	methionine	tryptophan
leucine	phenylalanine	valine

Inspection of the routes of degradation of these amino acids will show that they include effectively irreversible steps; animals have no other reactions to bypass these steps. We depend upon plants and autotrophic bacteria to make our supply of these compounds.

FURTHER READING

Also see the references at the end of Chapters 29 and 30.

Brown, G. M.: (1971) *The Biosynthesis of Pteridines.* Adv. Enzymol., *35*: 35; (1970) *Biogenesis and Metabolism of Folic Acid. In* D. M. Greenberg, ed.: *Metabolic Pathways.* Vol. 4, p. 383, Academic Press.

Pfleiderer, W., ed.: (1976) *Chemistry and Biology of Pteridines.* W. de Gruyter.

Turner, A. J.: (1977) *The Roles of Folate and Pteridine Derivatives in Neurotransmitter Metabolism.* Biochem. Pharmacol., *26*:1009. Includes more general discussion.

Clinics Haematology, *5*: 471 ff. Several reviews on folate and cobalamin function.

Cybolski, R. L., and R. R. Fisher: (1976) *Intramitochondrial Localization and Proposed Metabolic Significance of Serine Transhydroxymethylase.* Biochemistry, *15*: 3183.

Todhunter, E. N., ed.: (1971) *Evolution of Present Concepts Concerning the Action of Lipotropic Agents.* Fed. Proc., *30*: 130. A symposium.

32 AMINES

The amino acids are precursors of a wide variety of small molecules containing nitrogen that are much more important than their relatively low concentrations would indicate, because they have potent physiological actions. Many have been studied intensively because of their function in the nervous system; others are hormones or constituents of coenzymes; still others are known mainly from the effects of excessive concentration with their physiological role still unknown. With the major types of reactions of amino acids fresh in mind, this is a good time to consider the formation of some of these compounds.

NEUROTRANSMITTERS

Transmission of a nervous impulse between cells occurs by diffusion of specific chemical compounds. The junctions at which communication occurs are **synapses** (neuron to neuron) or **neuromuscular junctions,** also known as motor end-plates. The end of a nerve fibril at the junction contains vesicles loaded with the transmitting substance. When the wave of depolarization constituting a nerve impulse reaches the region of the vesicles, the vesicles fuse with the cell membrane (pre-synaptic membrane) at the junction and discharge their contents into the gap (synaptic cleft) between the cells.

The transmitter diffuses across the synaptic cleft and combines with receptor sites on the membrane of the affected cell so as to increase the selective permeability of the membrane for particular ions. At some synapses, it is a flow of Na^+ that is increased, making it easier for a traveling wave of depolarization to be triggered; such synapses stimulate the receptor cell to fire. At other synapses, it is the flow of K^+ or Cl^- that is increased by the transmitter; these synapses are inhibitory, and their action makes a larger stimulation at other sites necessary before the nerve will fire.

After passage of an impulse, status quo is restored by removal of the released transmitter and restocking of the vesicles. In most cases, the bulk of the transmitter is pumped back into the presynaptic cell through the action of an Na^+-coupled symport, but we shall see one case in which the transmitter is destroyed to remove it.

Glycine or glutamate, and perhaps aspartate, are used by some nerves as transmitters. Their effectiveness depends upon storage of high concentrations in the vesicles — much higher than is usually present for protein synthesis and other metabolic functions of the amino acids.

Acetylcholine

Acetylcholine is a general transmitting agent between effector neurons that synapse outside the central nervous system. It is also the transmitting agent at the motor end-plates of skeletal muscle fibers, and in terminals of the parasympathetic nervous system.

Acetylcholine is formed by a simple reaction between choline and acetyl coenzyme A in the cytoplasm. The mechanism of concentration in the vesicles is not known. (In general, synaptic vesicles contain a mixture of proteins and other compounds of uncharacterized function.)

When acetylcholine is released into the synaptic cleft, it is destroyed by **acetylcholinesterase.** We noted earlier (p. 264) that this enzyme uses a seryl group in its mechanism and that combination with this seryl group is the basis of action of phosphate insecticides and war gases. Prevention of the hydrolysis of acetylcholine causes constant stimulation.

Black widow spider venom also causes stimulation, but in a different way. The fusion of synaptic vesicles with the pre-synaptic membrane is probably caused by an influx of Ca^{2+} from the extracellular fluid, this influx is in turn caused by the depolarizing effect of the nerve stimulus arriving at the synaptic region. The spider venom appears to increase the cationic permeability of the membrane bilayer, thereby initiating discharge of acetylcholine.

Curare blocks the action of acetylcholine in motor end-plates, and it is used by South American Indians and by anesthesiologists of various ancestries to relax muscles. Its effect is counteracted by neostigimine, which inhibits acetylcholinesterase and therefore raises the concentration of acetylcholine within the synaptic cleft.

Neurons cannot synthesize choline. They concentrate the compound from the blood and recover it from the synaptic cleft by a Na^+-coupled symport.

4-Aminobutyrate

Glutamate is decarboxylated at certain inhibitory synapses by a specific enzyme, which contains pyridoxal phosphate as a coenzyme:

$$\underset{\text{L-glutamate}}{{}^{\oplus}H_3N\text{—}CH(COO^{\ominus})\text{—}CH_2\text{—}CH_2\text{—}COO^{\ominus}} + H^+ \xrightarrow[\text{(pyridoxal phosphate)}]{\text{glutamate decarboxylase}} CO_2 + \underset{\text{4-aminobutyrate}}{{}^{\oplus}H_3N\text{—}CH_2\text{—}CH_2\text{—}CH_2\text{—}COO^{\ominus}}$$

Here we have another example of the action of the general class of amino acid decarboxylases (p. 601). The resultant 4-aminobutyrate (commonly referred to in neurological literature as GABA, for gamma-aminobutyrate) is concentrated in the synaptic vesicles as a transmitter. Its release increases passage of Cl^- through the post-synaptic membrane.

The compound is recovered from the synaptic cleft by active transport.

However, it is also metabolized within the neurons by a route involving an initial transamination followed by an NAD-linked dehydrogenation to succinate:

$$\underset{\text{4-aminobutyrate}}{\overset{NH_3^{\oplus}}{|}\atop CH_2-CH_2-CH_2-COO^{\ominus}} \xrightarrow[\substack{\text{aminobutyrate}\\ \text{aminotransferase}\\ \text{(pyridoxal phosphate)}}]{\alpha\text{-keto-glutarate} \;\rightarrow\; \text{L-glutamate}} \underset{\text{succinic semialdehyde}}{H-C(=O)-CH_2-CH_2-COO^{\ominus}} \xrightarrow[\substack{\text{aldehyde}\\ \text{dehydrogenase}}]{NAD^{\oplus} + H_2O \;\rightarrow\; NADH + 2\,H^{\oplus}} \underset{\text{succinate}}{COO^{\ominus}-CH_2-CH_2-COO^{\ominus}}$$

On the face of it, the route represents a bypass of the α-ketoglutarate dehydrogenease reaction in the citric acid cycle:

1. glutamate → 4-aminobutyrate + CO_2
2. 4-aminobutyrate + α-ketoglutarate → succinaldehydate + glutamate
3. succinaldehydate + NAD^+ → succinate + CO_2 + NADH

SUM: α-ketoglutarate + NAD^+ → succinate + CO_2 + NADH

Much has been made of this as a major route of glucose metabolism in brain, but optimistic estimates allow no more than 0.1 of the α-ketoglutarate to be handled by this route. Beyond that, it seems purposeless as a bypass in the citric acid cycle, because the generation of GTP ordinarily occurring from succinyl coenzyme A would be lost.

Adrenaline, Noradrenaline, and Melanin

Most of the tyrosine in the body that is not used for protein synthesis is metabolized to CO_2 and urea by the route beginning with transamination (p. 582). A small part of the tyrosine is diverted to make other components in some tissues by pathways beginning with an oxidation. These tissues are derived from the ectodermal layer of an embryo and include some nerves, the skin, and the adrenal medulla. While the routes are trivial in a quantitative sense, they are of major qualitative importance for proper function.

The oxidation of tyrosine in nerves and the adrenal medulla is catalyzed by a **tyrosine hydroxylase,** which resembles phenylalanine hydroxylase (p. 580). It utilizes molecular oxygen, which is activated by tetrahydrobiopterin so that one atom of oxygen goes onto the aromatic ring of the substrate and the other oxidizes the coenzyme (Fig. 32–1). The product is **3,4-dihydroxyphenylalanine (dopa).** This amino acid is then decarboxylated to form **3,4-dihydroxyphenylethylamine (dopamine),** which is used as a transmitter in some neurons of the brain, for example, in the basal ganglia.

In other nerves and the adrenal medulla, dihydroxyphenylethylamine is again oxidized, this time on the β-carbon of the side chain, producing **noradrenaline.** The hydroxylase is a copper-protein and is an example of a class of mixed-function oxidases that apparently use ascorbate (vitamin C) as a second reducing

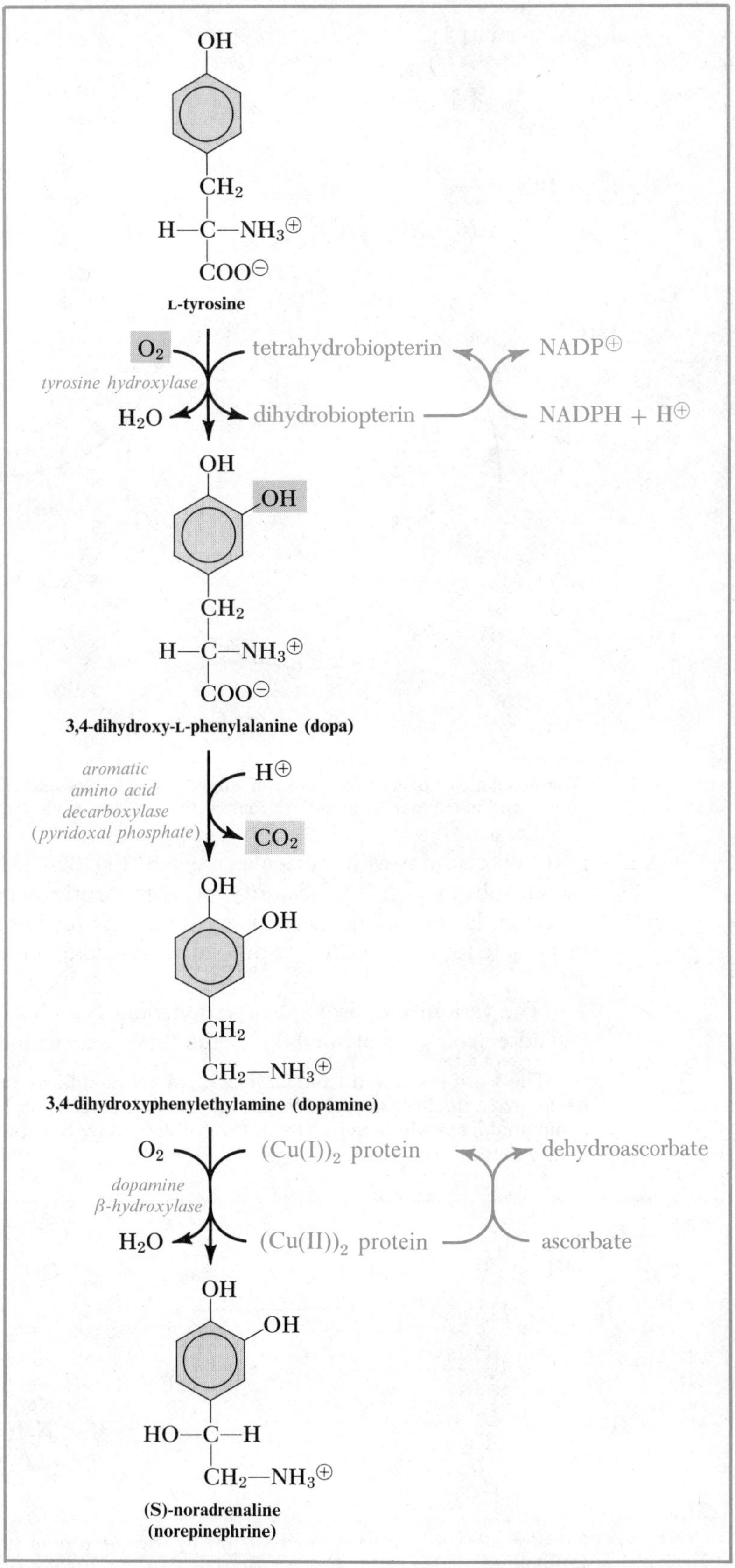

FIGURE 32–1

The formation of noradrenaline from tyrosine.

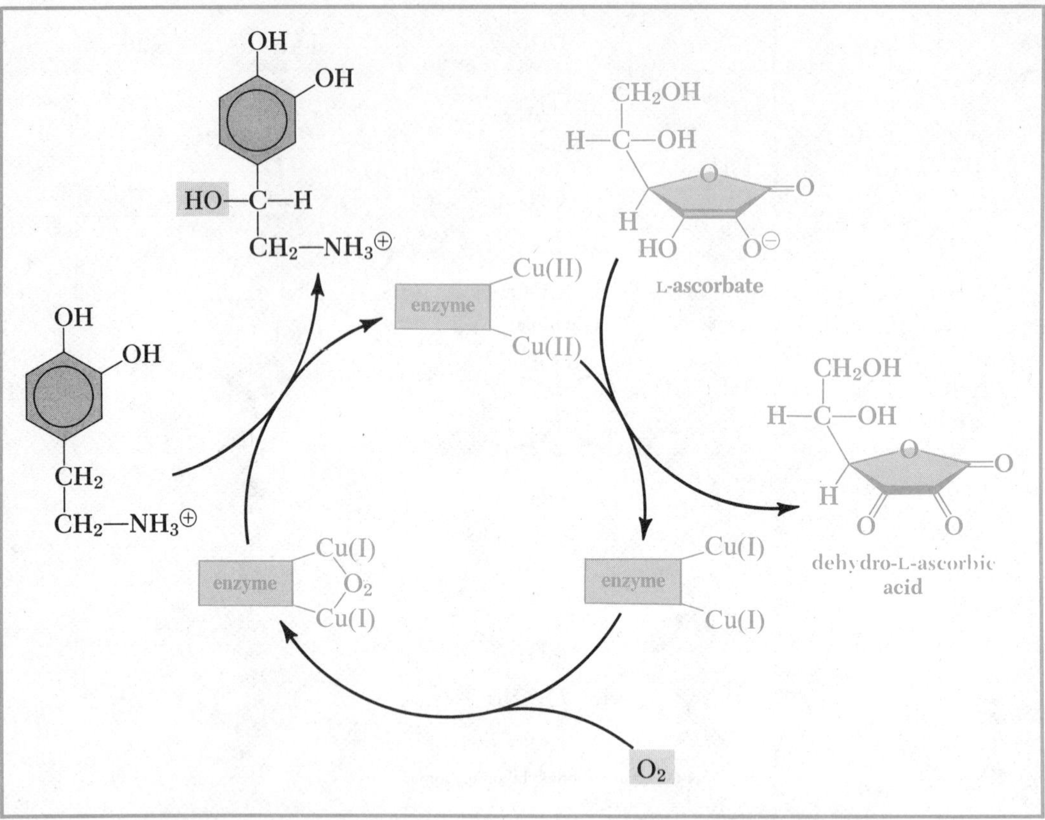

FIGURE 32–2 **The mechanism of hydroxylation of dopamine involves ascorbate as the second electron donor and participation of copper ions in the enzyme as electron carriers.**

agent in reactions with molecular oxygen (Fig. 32–2). Noradrenaline is the transmitter substance in the majority of sympathetic nerve terminals and at some synapses in the central nervous system. It is further methylated in the adrenal medulla to form **adrenaline,** with S-adenosyl methionine as methyl donor (Fig. 32–3).

The various dihydroxyphenylethylamines, such as dopamine, noradrenaline, and adrenaline, are lumped under the term **catecholamine.**

The word is derived from cathechol, which is a fairly recent shortening of the trivial name pyrocatechol, or 1,2-dihydroxybenzene (catechol was a more complex aromatic compound from which pyrocatechol was obtained by heating). The nomenclature is unfortunate, but it is in common usage.

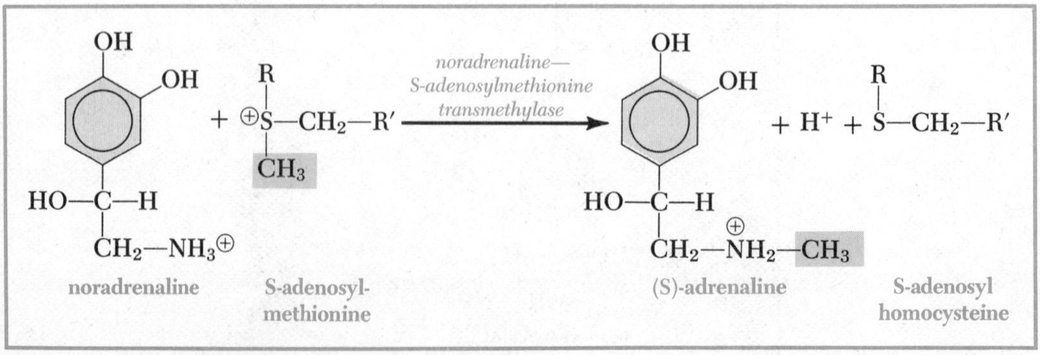

noradrenaline + S-adenosylmethionine ($^{\oplus}$S(R)(CH$_3$)—CH$_2$—R′) —[noradrenaline—S-adenosylmethionine transmethylase]→ (S)-adrenaline (HO—C—H, CH$_2$—$\overset{\oplus}{N}H_2$—CH$_3$) + H$^+$ + S-adenosyl homocysteine (S(R)—CH$_2$—R′)

FIGURE 32–3 **Noradrenaline is converted to adrenaline by transfer of a methyl group from S-adenosylmethionine.**

The vesicles containing catecholamines are also known to have high concentrations of ATP. The chromaffin granules* seen in the adrenal medulla and related tissues may contain 0.5 M adrenaline and 0.11 M ATP. The compounds strongly interact in some undefined way because the osmotic pressure in the granules is the same as that of blood.

After discharge at synapses as transmitters, the catecholamines are recovered into the cytosol by an Na^+-coupled symport and then concentrated into the vesicles. Catecholamines in the cytosol are subject to destruction by a **monoamine oxidase** in the outer membrane of mitochondria. This enzyme is an important protective device; its activity is high enough to rapidly remove amines, many of which have potent physiological effects. It is a flavoprotein that uses dioxygen and produces hydrogen peroxide as a product (Fig. 32–4). We shall encounter other examples of oxidases that form H_2O_2. As with other enzymes utilizing O_2 outside of mitochondria, no high-energy phosphate is generated as a result of the electron transfers: the flow of electrons is relatively small compared to that in mitochondria. The hydrogen peroxide produced by such enzymes is used as an oxidizing agent in **peroxidase** reactions, which we shall encounter later, or it is destroyed by the action of **catalase,** a nearly ubiquitous hemoprotein catalyzing the reaction:

$$2\ H_2O_2 \rightarrow 2\ H_2O + O_2.$$

The aldehyde produced by the action of monoamine oxidase may be either reduced to the alcohol or oxidized to the acid by NAD-coupled dehydrogenases. Any of the dihydroxyphenylethyl compounds — the amines, alcohols, or acids — are subject to methylation when they leave the vicinity of the synapse.

The same monoamine oxidase, dehydrogenases, and methyl transferase occur in other tissues, especially the liver, so any of the intermediates escaping into the blood will undergo the same degradation in those tissues.

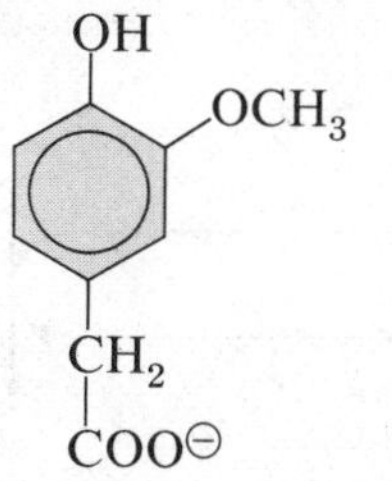

OH
OCH$_3$
CH$_2$
COO$^{\ominus}$

homovanillate

NEUROBLASTOMA

Malignant neoplasm arising from cells derived from neural crest and sympathetic nervous system
Frequently originates in adrenal medulla
Comprises 15 to 50 per cent of neonatal malignancies and 5 per cent of all childhood cancer deaths
Some produce dopamine; some dopamine and noradrenaline
Serum level of dopamine β-hydroxylase correlates well with level of urinary vanillyl mandelic acid
Dopamine excreted as homovanillic acid:

The determination of 3-methoxy-4-hydroxymandelate excretion is a useful guide to the clinical management of tumors such as neuroblastoma and pheochromocytoma that form adrenaline, noradrenaline, and dopamine. The compound is known clinically as vanillmandelate or vanillylmandelate, an excruciating exam-

*Chromaffin granules stain brown upon fixation in dichromate solutions, hence the name.

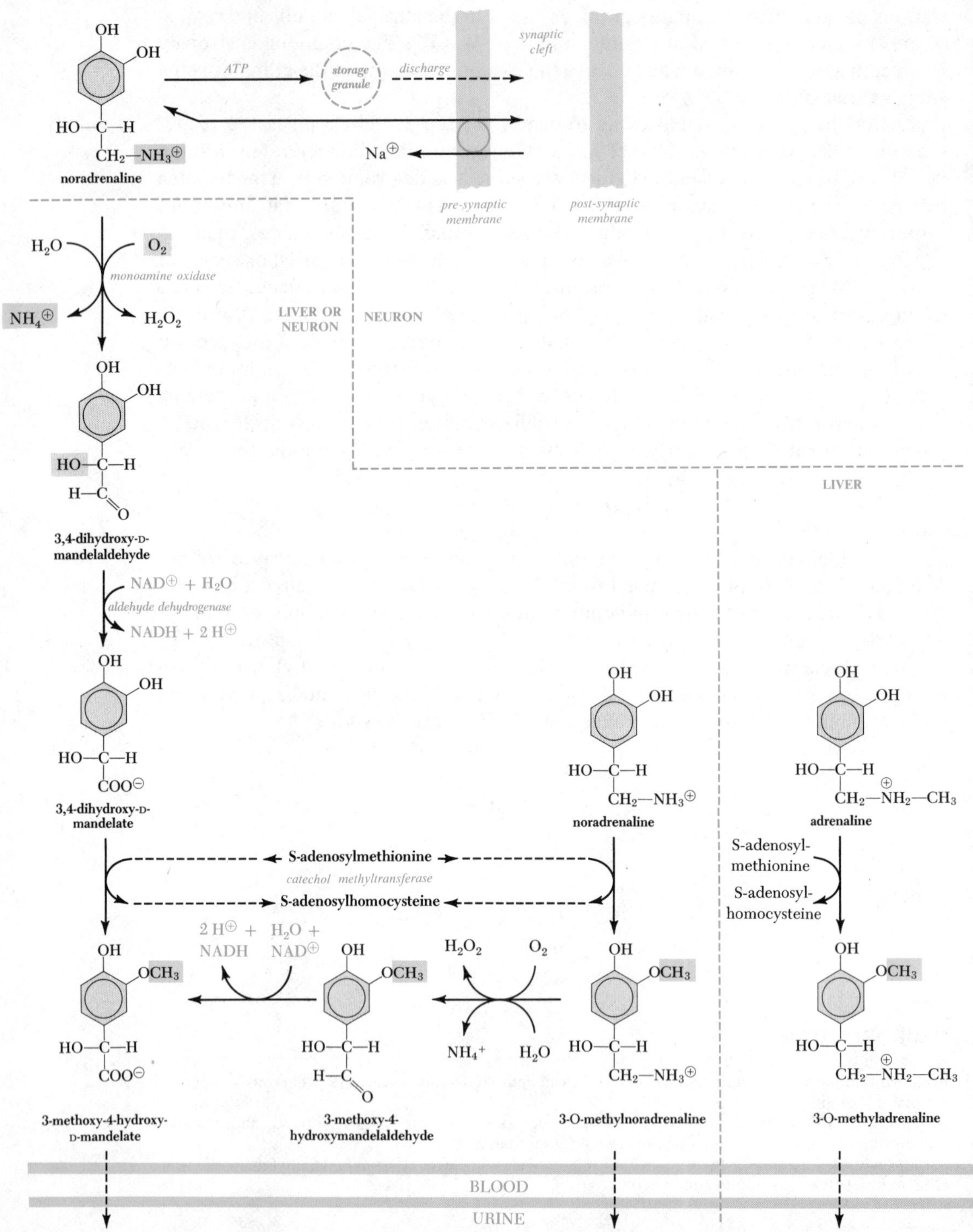

FIGURE 32–4 **The degradation of noradrenaline involves oxidation of the side chain by the successive action of monoamine oxidase and an aldehyde dehydrogenase and methylation of one of the phenolic hydroxyl groups. Degradation can begin with the oxidation *(upper left)* or the methylation *(center)* in either the neurons or the liver. Adrenaline is first methylated *(lower right)*. The methoxy derivative may be excreted as such, or attacked by monoamine oxidase *(not shown)* to form the same mandelate derivatives derived from noradrenaline.**

ple of bastard nomenclature whose origin lies in the widespread occurrence of methylated 3,4-dihydroxyphenyl derivatives in plants. These frequently have a pleasant odor; among them is vanillin in vanilla and eugenol in cloves, which are responsible for most of the flavor in these spices:

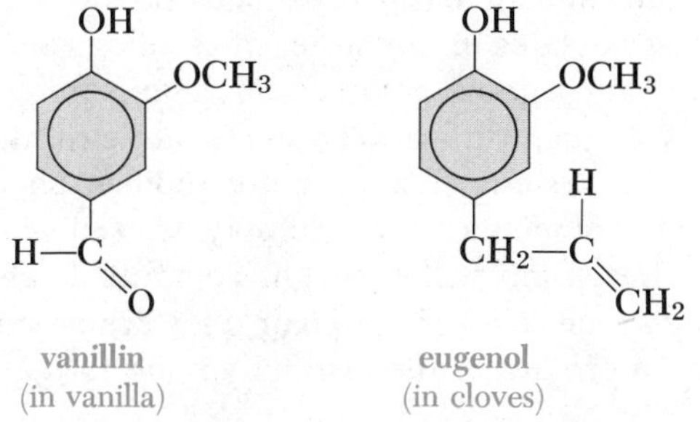

vanillin
(in vanilla)

eugenol
(in cloves)

The melanins, the pigments of skin and hair, are complex polymers in which a major constituent is formed from tyrosine *via* dihydroxyphenylalanine (Fig. 32–5). Synthesis occurs in specialized vesicles, the melanosomes of melanocytes, and the initial reactions involve oxidations by **tryosinase,** a phenol monooxygenase peculiar to those structures. Tyrosinase is a copper-containing mixed function oxidase that carries out a tricky sequence. The product of the initial hydroxylation, dihydroxyphenylalanine, is also the oxygen-activating electron donor! It

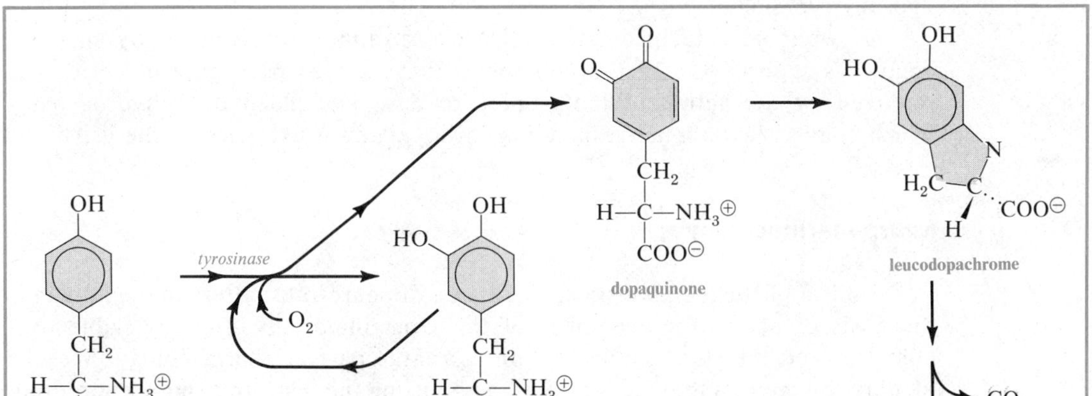

FIGURE 32–5 **Formation of melanin from tyrosine. Tyrosinase is a mixed function oxidase utilizing tyrosine and dopa as the two electron donors. Since dopa is also the product of tyrosine oxidation, the result of this enzyme's action is the successive oxidation of tyrosine to dopaquinone. Dopaquinone goes through a series of spontaneous reactions to yield indole quinones that polymerize to form melanin.**

serves the same function during its own formation that tetrahydrobiopterin serves in the tyrosine hydroxylase reaction of nerves, and in the process it becomes further oxidized to a quinone, in the same way that tetrahydrobiopterin becomes oxidized to dihydrobiopterin by the other monooxygenase.

The dopaquinone formed by the oxidations undergoes a series of fast spontaneous reactions in which an indole ring is formed and the carboxyl group is lost as CO_2. The various intermediates in this series of reactions participate in condensations to form a three-dimensional polymer. Melanins are complex structures of varying and unknown composition. There are at least three different types, including the yellow-brown eumelanins, more reddish pheomelanins, in which some sulfur-containing compounds are also polymerized, and brilliant red trichochromes, which are lower molecular weight compounds also containing sulfur. (Although the trichochromes occur in red hair, they are probably not as important as the pheomelanins in producing the distinctive hue.)

5-Hydroxytryptamine (Serotonin)

5-Hydroxytryptamine is a transmitter in some neurons of the central nervous system. However, about 80 per cent of the body content is outside the nervous system. The route to the amine is like the route to noradrenaline in that it begins by the direct hydroxylation of an aromatic ring, in this case of tryptophan (Fig. 32–6). The enzyme utilizes tetrahydrobiopterin as a coenzyme and hydroxylates the indole ring of tryptophan at C-5. The 5-hydroxytryptophan formed can be acted on by dihydroxyphenylalanine decarboxylase in the neurons to form 5-hydroxytryptamine.

As with the catecholamines, released serotonin is recovered by an Na^+-coupled symport. Any of the compound that escapes packaging in vesicles is removed by the action of monoamine oxidase and aldehyde dehydrogenase, which convert it to the corresponding carboyxlate for excretion in the urine.

Neuropharmacology

Control of the metabolism and function of neurotransmitters in the human is the basis of action for a number of the most important drugs available to a clinician, and the study of this action is a large part of pharmacology. We can identify the major site of action of drugs by using the neurotransmitter noradrenaline as an example; drugs are known that do the following:

(1) Interfere with noradrenaline synthesis by inhibiting tyrosine hydroxylase: alpha-methyl-*p*-tyrosine;

(2) Act as a false transmitter: alpha-methyl-DOPA is a precursor of such a compound. It is converted to a α-methyldopamine, and transported into granules, where it is converted to α-methyl-noradrenaline, which displaces noradrenaline and is not a neurotransmitter;

(3) Interfere with transport of the transmitter: reserpine blocks uptake of noradrenaline into vesicles; imipramine and chlorpromazine interfere with reutilization of noradrenaline released outside of the cell;

(4) Inhibit breakdown of the neurotransmitter: pargyline inhibits monoamine oxidase and potentiates the effects of noradrenaline (and other amines) by blocking their degradation;

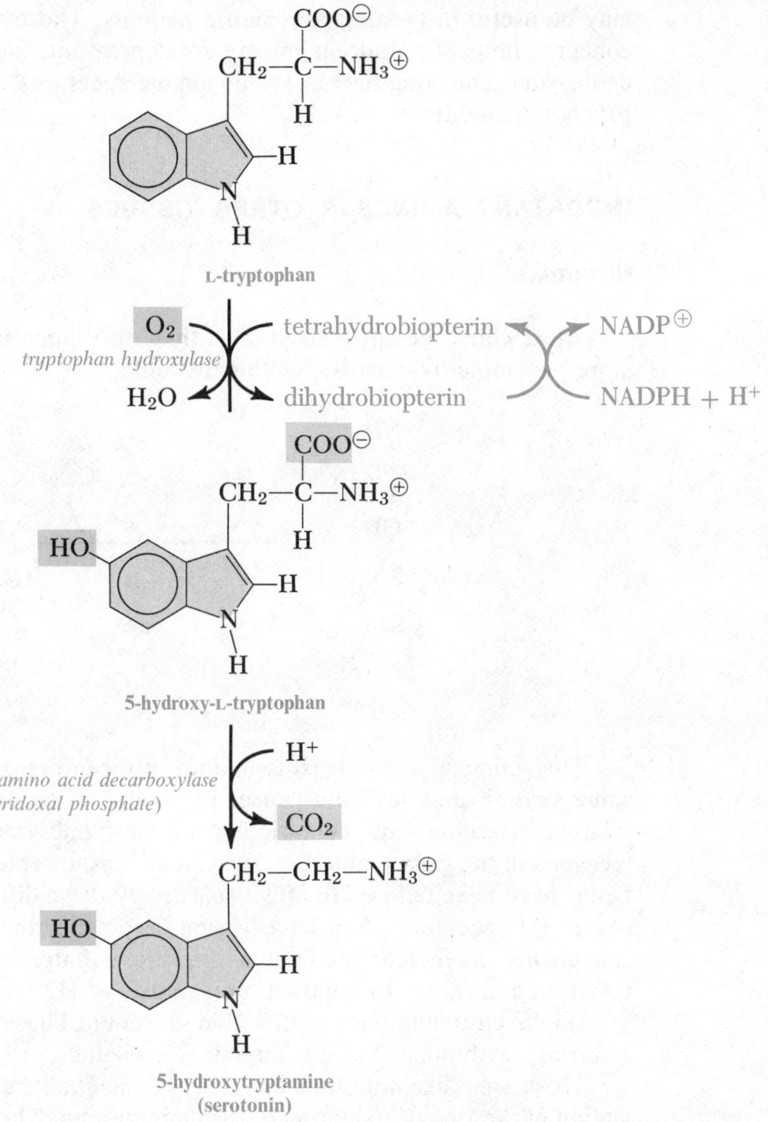

$$\text{L-tryptophan} \xrightarrow[\text{tryptophan hydroxylase}]{O_2,\ \text{tetrahydrobiopterin} \to H_2O,\ \text{dihydrobiopterin}} \text{5-hydroxy-L-tryptophan} \xrightarrow[\text{(pyridoxal phosphate)}]{\text{aromatic amino acid decarboxylase};\ H^+ \to CO_2} \text{5-hydroxytryptamine (serotonin)}$$

Dihydrobiopterin + NADPH + H^+ → tetrahydrobiopterin + $NADP^{\oplus}$

FIGURE 32–6 The formation of serotonin from tryptophan.

(5) Stimulate release of the neurotransmitter: tyramine, the product of decarboxylation of tyrosine, does this and can precipitate a hypertensive crisis. Aged cheese and Chianti wine, which are rich in tyramine, must be avoided by patients taking pargyline for this reason;

(6) Deplete the supply of the neurotransmitter by causing a slow release — too slow to stimulate the receptors, but too rapid for replenishment: guanethidine;

(7) Inhibit normal release in response to the action potential: bretylium;

(8) Interfere with the receptor site: phenoxybenzamine;

(9) Mimic the normal transmitter: phenylephrine.

In considering this battery of drugs, it is of interest to note that many of them may be useful in treating psychiatric patients. Those that increase the synaptic concentrations of noradrenaline are antidepressant; those that lower it aggravate depression. Blocking agents for dopamine receptors are useful in more severe psychotic conditions.

IMPORTANT AMINES IN OTHER TISSUES

Histamine

Mast cells, the specialized cells that store heparin (p. 185), also make and store histamine by decarboxylating histidine:

$$\text{L-histidine} + H^+ \xrightarrow[\text{(pyridoxal phosphate)}]{\text{histidine decarboxylase}} \text{histamine} + CO_2$$

L-histidine histamine

Histamine affects at least two kinds of receptors. **H1 receptors** are present in some smooth muscles, and cause, for example, an expansion of capillaries by dilating arterioles and constricting venules, and constriction of bronchi. **H2 receptors** in the gastric mucosa cause an increased secretion of hydrochloric acid. Drugs have been tailored to affect selectively these different receptors. Antagonists of H1 receptors often have various aromatic rings with positively charged side-chains and include the familiar over-the-counter antihistaminics that counter the stuffed-up nose. In contrast, antagonists of H2 receptors such as cimetidine are employed to minimize gastric acid secretion. These specifically contain imidazole rings with polar, but uncharged, side chains.

Histamine, like noradrenaline, may be metabolized by routes involving methylation of the ring or oxidation of the amino group. The methyl derivatives or the imidazoleacetates appear in the urine. The oxidation of the amino group of the parent compound is catalyzed by a relatively non-specific enzyme, **diamine oxidase,** that also oxidizes a number of aliphatic compounds containing two amino groups, whereas the methylated derivative is a substrate for monoamine oxidase.

Putrescine, Spermidine, and Spermine

Putrescine received its ugly name because it was discovered as a product of the action of bacteria in decaying meat, being formed by decarboxylation of ornithine, which in turn arises from bacterial hydrolysis of arginine in the meat. (The product of decarboxylation of lysine was named **cadaverine** for the same reason.)

The polyamino compounds, spermidine and spermine, were named for their discovery in human semen. Indeed, crystals of spermine phosphate in semen were

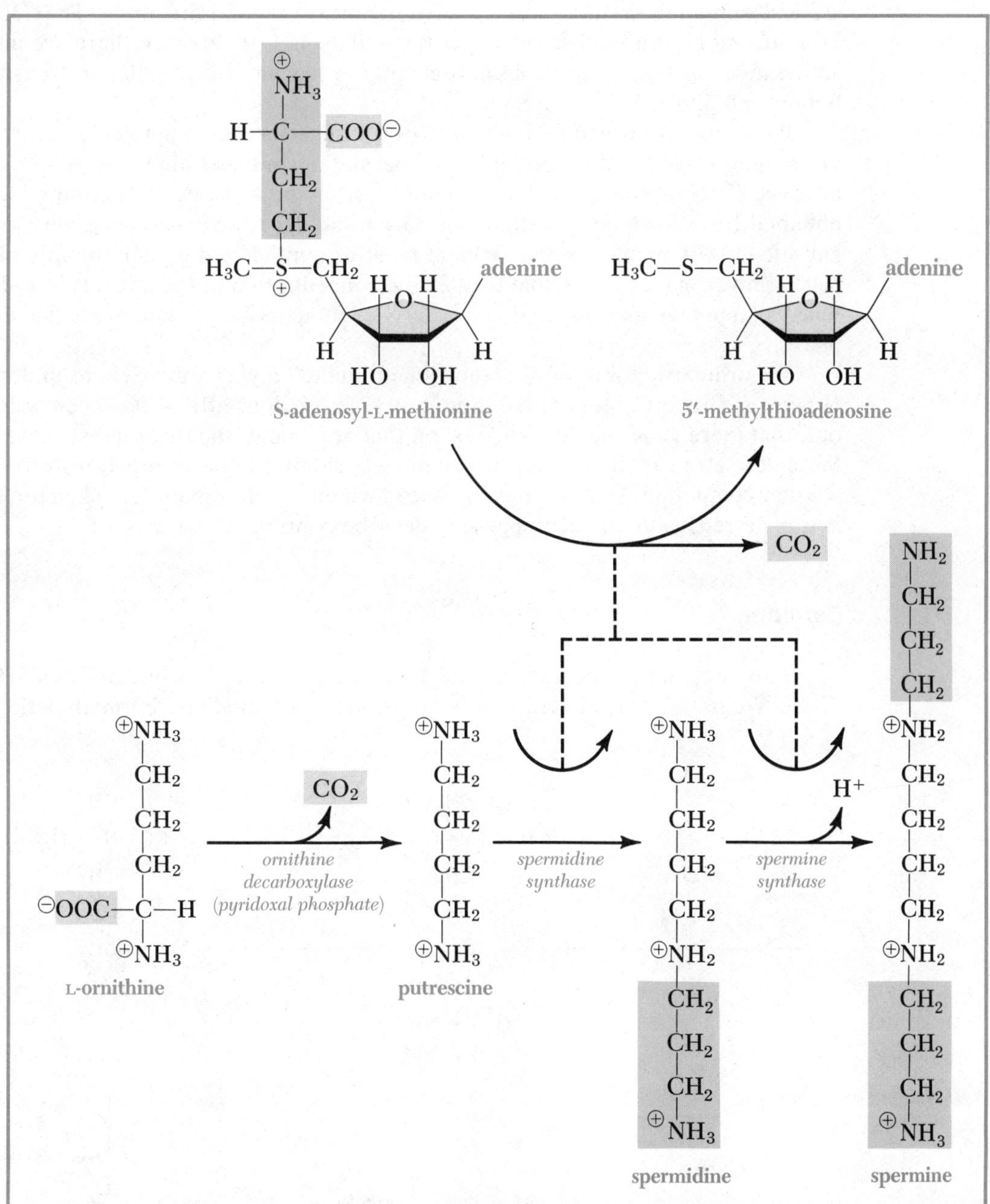

FIGURE 32–7 **The synthesis of putrescine, spermidine, and spermine. The figure indicates the greater dissociation of one of the ammonium groups in spermine.**

one of the things noted by van Leeuwenhock with his newly invented microscope, and their presence is still used as a part of the legal identification of suspect stains, but the structure of the compound was only worked out in 1926. It is only in the last decade that it has been realized that these compounds, along with putrescine, are widespread in tissues and may have important general functions. They occur in association with nucleic acids, as we noted (p. 110), and are therefore under active investigation as possible participants in mechanisms regulating transcription or cell division.

Putrescine is formed by the action of a straightforward amino acid decarboxylase on ornithine and is converted to spermidine and spermine (Fig. 32–7). The necessary aminopropyl groups — one in spermidine and two in spermine — are obtained from S-adenosylmethionine. Ornithine decarboxylase has a very short half-life (10–20 minutes in the rat) and must be synthesized equally rapidly. This fast turnover makes it possible to alter the concentration of the enzyme in a short time. A large increase in activity is an early event in tissue growth, preceding the S phase.

We ordinarily think of S-adenosylmethionine only as a methyl group donor. However, Guilio Cantoni noted, and it appears obvious after it has been pointed out, that there is no mechanistic reason that any one of the three substituents on the sulfur atom of S-adenosylmethionine couldn't be the group transferred to another compound. Here we have a case in which it is the amino acid skeleton that is transferred, with an accompanying decarboxylation.

Carnitine

The immediate precursor of this acyl carrier is N,N,N-trimethyllysine (Fig. 32–8). We have earlier noted that some proteins are modified by methylation of

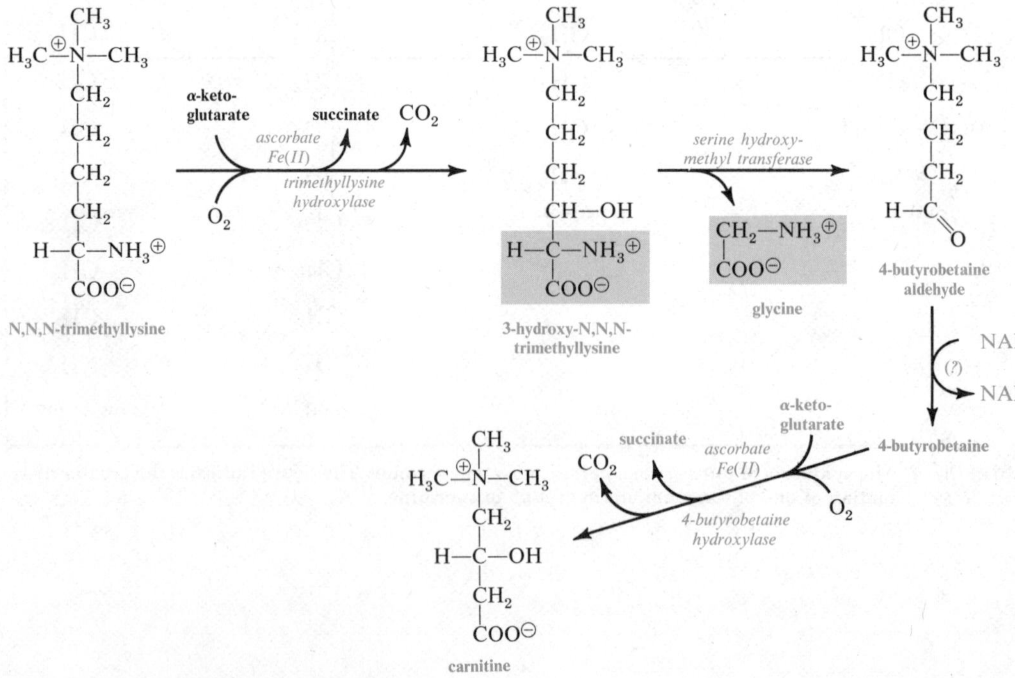

N,N,N-trimethyllysine: $H_3C-N^{\oplus}(CH_3)_2-CH_2-CH_2-CH_2-CH_2-CH(NH_3^{\oplus})-COO^{\ominus}$

α-ketoglutarate + O_2 → succinate + CO_2 (*trimethyllysine hydroxylase*; *ascorbate, Fe(II)*)

3-hydroxy-N,N,N-trimethyllysine: $H_3C-N^{\oplus}(CH_3)_2-CH_2-CH_2-CH_2-CH(OH)-CH(NH_3^{\oplus})-COO^{\ominus}$

serine hydroxymethyl transferase → glycine ($CH_2-NH_3^{\oplus}$, $COO^{\ominus}$)

4-butyrobetaine aldehyde: $H_3C-N^{\oplus}(CH_3)_2-CH_2-CH_2-CH_2-CHO$

$NAD^{\oplus}$ → NADH (?)

4-butyrobetaine

α-ketoglutarate + O_2 → succinate + CO_2 (*4-butyrobetaine hydroxylase*; *ascorbate, Fe(II)*)

carnitine: $H_3C-N^{\oplus}(CH_3)_2-CH_2-CH(OH)-CH_2-COO^{\ominus}$

FIGURE 32–8 **The synthesis of carnitine.**

the terminal ammonium group in lysine residues; S-adenosyl methionine is the methyl group donor. The trimethyllysine used for carnitine synthesis is probably obtained by the normal breakdown of these proteins. The conversion to carnitine begins with an oxidation to form 3-hydroxy-trimethyllysine by an example of a special class of mixed function oxidases that use α-ketoglutarate as a second electron donor. The mechanism of action is similar to that of *p*-hydroxyphenylpyruvate oxidase (p. 581), with the formation of an intermediate peroxide through the addition of O_2. One of the atoms of oxygen combines with trimethyllysine and the other is used to oxidatively decarboxylate α-ketoglutarate to succinate. (The approved name for this enzyme, which was only recently reported, would be trimethyllysine, oxoglutarate dioxygenase.)

The 3-hydroxy compound can be regarded as a derivative of serine, and it is attacked by serine hydroxymethyl transferase to form glycine. The other product is an aldehyde, commonly known as 4-butyrobetaine aldehyde. (This is the analogue of the potential formaldehyde — the hydroxymethyl group — transferred from serine to H_4folate by this enzyme.) The aldehyde is then oxidized to the corresponding acid, which is hydroxylated by still another representative of the dioxygenases utilizing α-ketoglutarate as an electron acceptor. The product is carnitine.

Before going on, we ought to note that these oxygenases, like others we have mentioned, require Fe(II) for activity. Their activity is also increased by ascorbate, although it is not absolutely required. (Other reducing agents can replace ascorbate in the test tube, but it is probably the compound used for maintaining the enzyme in active form within cells.)

Other Examples

Some compounds are embarrassing because they occur at too high concentrations in specialized tissues to ignore them completely, and yet their function is unknown. For example, skeletal muscles contain up to 30 millimoles of **carnosine** per kilogram, which is a peptide of β-alanine and histidine:

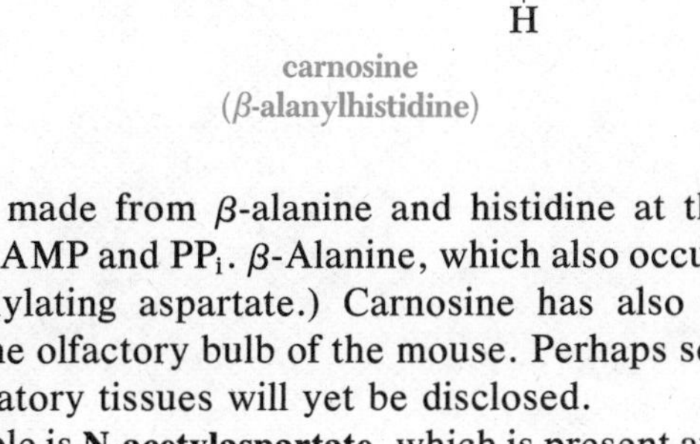

$$^{\oplus}H_3N{-}CH_2{-}CH_2{-}\overset{\overset{\displaystyle O}{\|}}{C}{-}\overset{\overset{\displaystyle H}{|}}{N}{-}\underset{\underset{\displaystyle H}{|}}{\overset{\overset{\displaystyle CH_2\text{(imidazole ring: N=CH, N–H, C–H)}}{|}}{C}}{-}COO^{\ominus}$$

carnosine
(β-alanylhistidine)

(The compound is made from β-alanine and histidine at the expense of ATP, which is cleaved to AMP and PP_i. β-Alanine, which also occurs in coenzyme A, is made by decarboxylating aspartate.) Carnosine has also been found in high concentrations in the olfactory bulb of the mouse. Perhaps some general function in specialized excitatory tissues will yet be disclosed.

Another example is **N-acetylaspartate,** which is present at a concentration of 5 millimoles per kilogram of human brain, a level exceeded only by glutamine and glutamate among the amino acid derivatives, but no physiological role has been demonstrated.

FURTHER READING

Siegel, G. J., et al.: (1976) *Basic Neurochemistry,* 2nd ed. Little, Brown. An excellent introductory survey of all aspects, including the metabolism of neurotransmitters.

Casey, R. P., et al.: (1977) *The Biochemistry of the Uptake, Storage, and Release of Catecholamines.* Horiz. Biochem., *3*: 224.

Moskowitz, M. A., and R. J. Wurtman: (1975) *Catecholamines and Neurologic Diseases.* N. Engl. J. Med., *291*: 707.

Beaven, M. A.: (1976) *Histamine.* N. Engl. J. Med., *294*: 30, 320.

Bright, H. J., and D. J. T. Porter: (1975). *Monoamine Oxidase. In* P. D. Boyer, ed.: *The Enzymes,* 3rd ed. Vol. 12, p. 446. Academic Press.

Wolstenholme, G. E. W., and J. Knight, eds.: (1976) *Monoamine Oxidase and Its Inhibitors.* Ciba Found. Symp., *39*. Elsevier.

Hearing, V. J., and T. M. Ekel: (1976) *Mammalian Tyrosinase.* Biochem. J., *157*: 549.

Jimbow, K., et al.: (1976) *Some Aspects of Melanin Biology: 1950–1975.* J. Invest. Dermatol., *67*: 72.

Prota, G., and R. H. Thomson: (1976) *Melanin Pigmentation in Mammals.* Endeavour, *35*: 32.

Hulse, J. D., S. R. Ellis, and L. M. Henderson: (1978) *Carnitine Biosynthesis.* J. Biol. Chem., *253*: 1654.

Lindstedt, G., and S. Lindstedt: (1969) *Cofactor Requirmenets of γ-Butyrobetaine Hydroxylase from Rat Liver.* J. Biol. Chem., *245*: 4187. Papers by discoverers of α-ketoglutarate-linked hydroxylases.

Barchas, J. D., et al.: (1978) *Behavioral Neurochemistry: Neuroregulators and Behavioral States.* Science, *200*: 964.

Berger, P. E.: (1978) *Medical Treatment of Mental Illness.* Science, *200*: 974.

33 TURNOVER OF PORPHYRINS AND IRON

Hemoproteins and other iron-containing proteins are constantly synthesized and degraded with a concomitant synthesis and degradation of the associated porphyrins and a reuse of the ligated iron atoms. While more is known about the turnover of the abundant hemoglobin, the other hemoproteins are important because of the central role of the iron-containing electron carriers in metabolism. The synthesis and degradation of protoporphyrin IX in hemoglobin is a significant part of the nitrogen economy and concomitantly involves a major part of the economy of iron in the body.

Nomenclature of the Porphyrins

Before going further, it is necessary to say something about the classification of porphyrins. The basic unit of porphyrins is the pyrrole ring, with four of these linked to form the large porphyrin ring. Each of the four pyrrole rings may have two side chains attached, and these side chains may differ among the four pyrrole groups. The name of the porphyrin indicates the kinds of side chains; it does not indicate the arrangement of the particular pyrrole groups. Some porphyrins contain the kinds of pyrrole groups shown in Figure 33–1, and the composition of physiologically important examples may be summarized as follows:

uroporphyrins — each pyrrole group has an **acetate** and a **propionate** side chain.

coproporphyrins — each pyrrole group has a **methyl** and a **propionate** side chain.

protoporphyrins — each of two pyrrole groups has a **methyl** and a **propionate** side chain; each of the other two has a **methyl** and a **vinyl** side chain.

The porphyrins of a given name can vary among themselves in the order in which the pyrrole rings are put together. Suppose we designate the acetate and propionate groups in the uroporphyrins as A and P. There are four possible ways of combining the pyrroles bearing these groups so as to make a porphyrin:

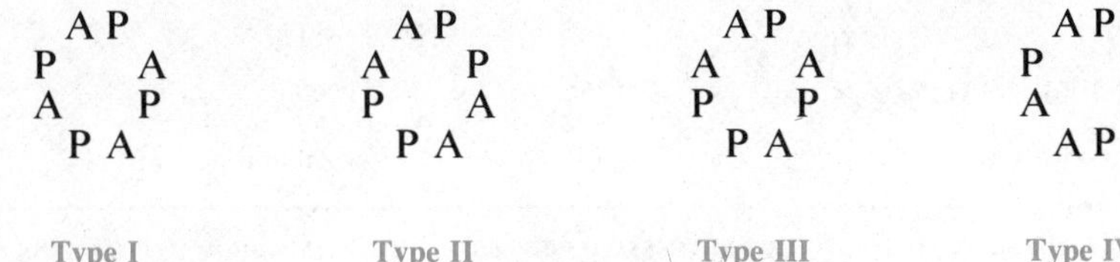

Type I (-AP-AP-AP-AP-) Type II (-AP-PA-AP-PA-) Type III (-AP-AP-AP-PA-) Type IV (-AP-PA-PA-AP-)

	$^{\ominus}OOC-H_2C$ / $CH_2-CH_2-COO^{\ominus}$ (pyrrole)	H_3C / $CH_2-CH_2-COO^{\ominus}$ (pyrrole)	H_3C / $CH=CH_2$ (pyrrole)	H_3C / CH_2-CH_3 (pyrrole)	H_3C / $HO-CH-CH_3$ (pyrrole)	H / CH_3 (pyrrole)
uroporphyrin	4	—	—	—	—	—
coproporphyrin	—	4	—	—	—	—
protoporphyrin	—	2	2	—	—	—
etioporphyrin	—	—	—	4	—	—
hematoporphyrin	—	2	—	—	2	—
mesoporphyrin	—	2	—	2	—	—
deuteroporphyrin	—	2	—	—	—	2

FIGURE 33–1 **Kinds of porphyrins. The nature of the constituent pyrroles as listed at the top defines the porphyrin, and the number of the various pyrroles in different kinds of porphyrins is given. It is not hard to visualize sequences of decarboxylations, oxidations, hydrations, and reductions by which all of these could be formed from the parent uroporphyrins listed first.**

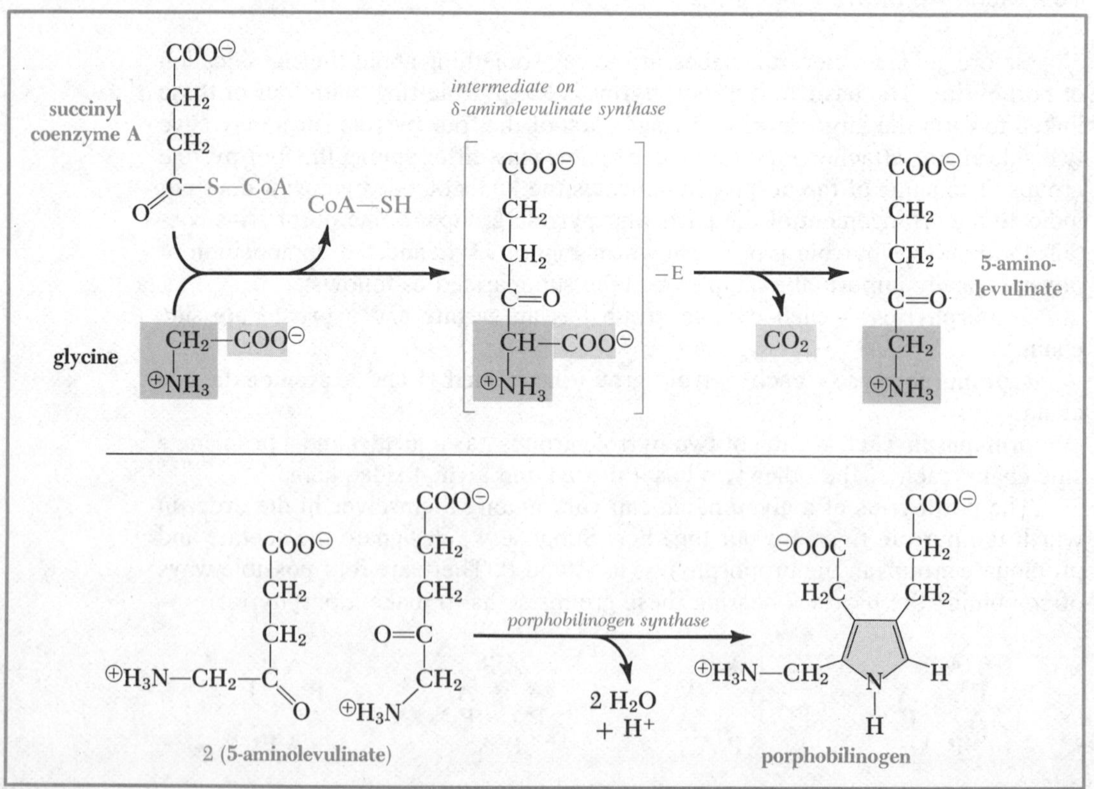

FIGURE 33–2 **Porphyrin synthesis begins with two successive condensations by which a pyrrole ring is generated from two molecules each of succinyl coenzyme A and glycine.**

Each of these four uroporphyrins is designated by a Roman numeral. (There are only four uroporphyrins because any other reversal of pyrrole groups beyond those shown is superimposable on one of the four by turning the ring over.)

Now, if two of the groups in uroporphyrin are changed into a third kind of group, which is the circumstance seen in protoporphyrins, then there are 15 possible combinations. Hans Fischer* wrote down the 15 possibilities, and showed that the porphyrin in hemoglobin had the same arrangement as the ninth he had tabulated. Hence, the porphyrin in heme is designated as protoporphyrin IX.

All natural porphyrins are derived from uroporphyrin I, in which there is a regular alternating sequence of groups, as might be expected if the pyrroles are combined head-to-tail, and from uroporphyrin III, which represents a reversal — an isomerization — of one of the pyrrole groups.

PORPHYRIN SYNTHESIS

The complex porphyrin molecule is made from two simple precursors, **succinyl coenzyme A** and **glycine** (Fig. 33–2). The initial reaction is a condensation of these compounds within mitochondria, where they are readily available, to form **5-aminolevulinate.** This is the rate-controlling step in porphyrin biosynthesis. The 5-aminolevulinate passes into the cytosol for the next step.

The reaction involves an intermediate condensation of glycine with pyridoxal phosphate. The mechanism is not shown; one of the H atoms on C-2 of glycine leaves after condensation; the resultant carbanion then unites with the electropositive carbonyl carbon of succinyl coenzyme A.

Two molecules of 5-aminolevulinate condense to form **porphobilinogen.** This is the parent pyrrole compound, and four molecules of it are combined to make **uroporphyrinogen III** (Fig. 33–3).

*Hans Fischer (1881–1945): German biochemist and Nobel Laureate. Not to be confused with Emil Fischer (1852–1919), also a German biochemist and Nobel Laureate, discoverer of much of the fundamental knowledge of the chemistry of proteins, carbohydrates, and nucleic acids; nor with Emil Fischer's late son, H. O. L. Fischer, a carbohydrate chemist of distinction at Toronto and Berkeley; nor with E. H. Fischer, very much alive at Seattle, and not bad as a biochemist, either. (This list is by no means exhaustive.)

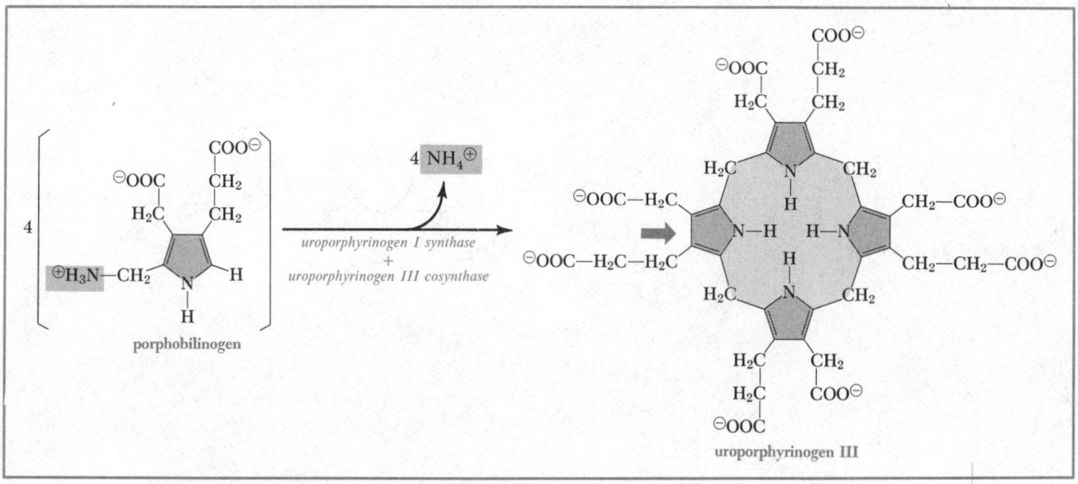

FIGURE 33–3 **Four molecules of porphobilinogen are condensed to form uroporphyrinogen III. One of the molecules *(arrow)* condenses head-to-head in the presence of a cosynthase; the others condense head-to-tail.**

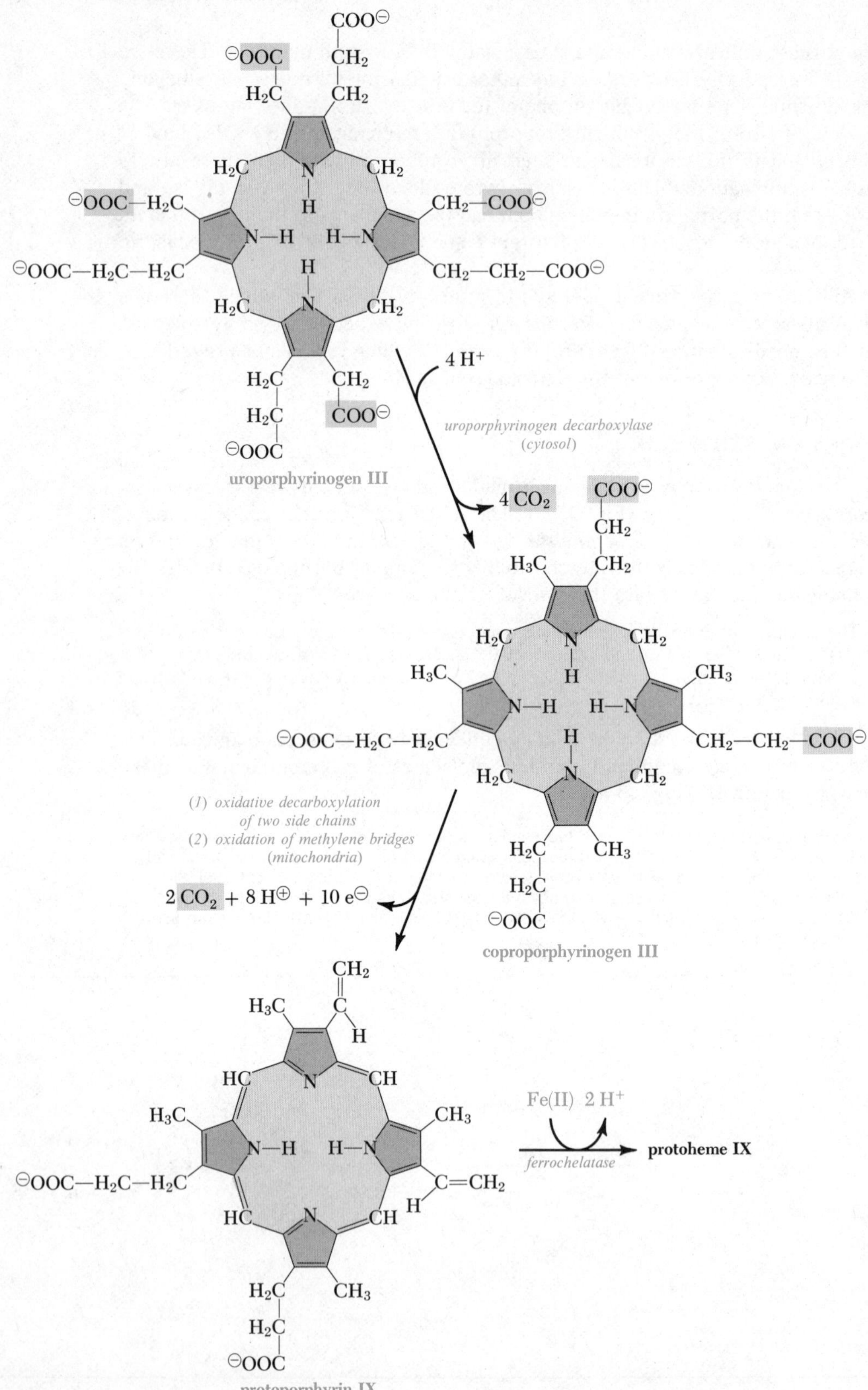

FIGURE 33–4 The conversion of uroporphyrinogen III to protoporphyrin IX and then to protoheme IX.

(The **porphyrinogens** are porphyrins in which the bridge atoms between pyrrole rings are in the reduced, or methylene, state, whereas these atoms are in the methylidyne state in porphyrins.) Two proteins are involved in this condensation. Uroporphyrinogen I synthase by itself would catalyze a simple head-to-tail condensation of porphobilinogen units, forming uroporphyrinogen I. The second protein, a uroporphyrinogen III cosynthase, has no apparent catalytic activity, but it somehow combines with the synthase so as to alter its specificity, causing one of the porphobilinogen molecules to condense head-to-head, creating a type III porphyrin. The mechanism is unknown.

The remaining steps (Figs. 33–4) involve decarboxylation of the aceto side chains to form methyl groups (coproporphyrinogen III), oxidative decarboxylation of two of the propiono side chains to form vinyl groups (protoporphyrinogen IX), and the oxidation of the methylene bridges to methylidyne bridges (protoporphyrin IX). The latter two steps are catalyzed by mitochondria, but it is not known where the enzymes are localized within the organelle. However, the final step of

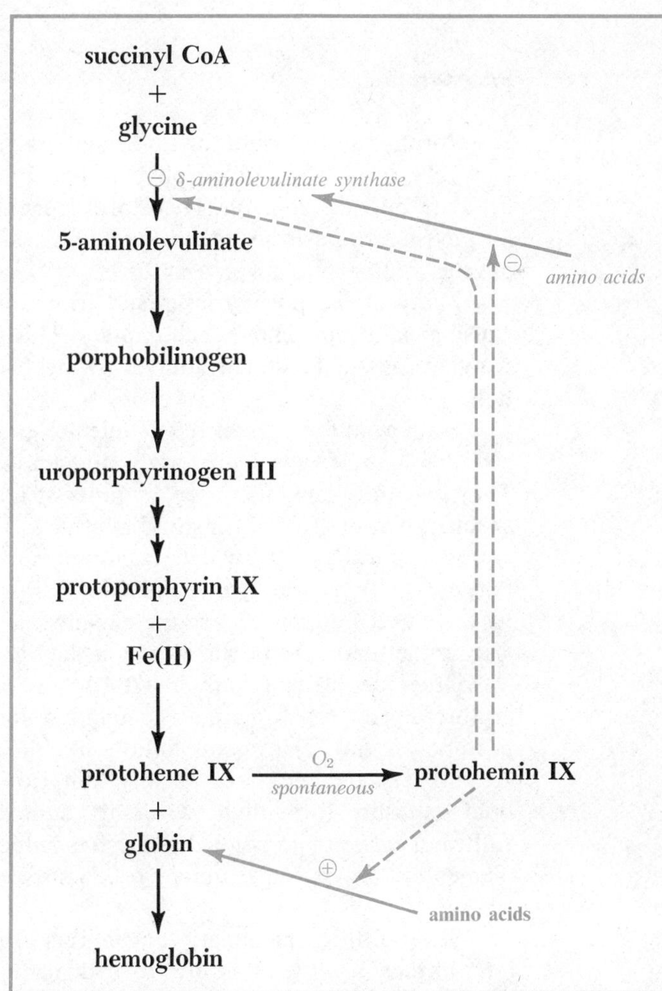

FIGURE 33–5

Regulation of hemoglobin synthesis by protohemin IX. Hemin forms when the supply of heme exceeds the supply of globin. The hemin suppresses formation of additional protoporphyrin, probably by direct inhibition of aminolevulinate synthase and also by repression of the enzyme's synthesis. The hemin also promotes synthesis of globin polypeptides.

heme synthesis, the addition of Fe(II) to protoporphyrin IX, is catalyzed by an enzyme localized on the inside of the inner membrane, at least in liver.

Regulation of hemoglobin synthesis involves control of both porphyrin and polypeptide synthesis. The regulating factor is an accumulation of heme, which in the free form is spontaneously oxidized to hemin (Fe(III)-porphyrin). An accumulation of hemin diminishes the activity of 5-aminolevulinate synthase, probably by repressing synthesis of this enzyme protein, as well as directly inhibiting the existing enzyme (Fig. 33–5), thereby shutting off porphyrin synthesis. Hemin has this effect both in the primitive red blood cells synthesizing hemoglobin and in other cells in which the cytochromes and other hemoproteins are being made.

Hemin activates the synthesis of globin polypeptide chains in this way: The erythropoietic cells contain an inhibitory protein that prevents translation of mRNA specifying the globin polypeptide chains. This protein will combine with hemin, and the combination no longer blocks formation of the globin chains. (There is no similar regulatory effect on the synthesis of the apoproteins of the cytochromes and other hemoproteins.)

The result of these regulatory mechanisms is to strike a balance between the synthesis of the heme and protein components of hemoglobin. When heme production gets ahead of globin production, hemin will slow its own formation and accelerate globin formation. The converse is true when globin production gets ahead of heme production.

Porphyrias

Porphyrias are conditions in which, as the name implies, porphyrins accumulate. The production of porphyrins by erythroblasts and by the liver is under separate genetic control. Therefore, porphyrias may be caused by defects in porphyrin synthesis in either kind of tissue, and there are **erythropoietic porphyrias** and **hepatic porphyrias.**

Many of the porphyrias result from some defect in the synthetic pathway causing an intermediate to accumulate. This situation is aggravated by the accompanying loss of feedback control owing to diminution of the heme concentration.

Somewhat fewer than 100 people are known whose erythroblasts form excessive amounts of type I uroporphyrinogen (hereditary erythropoietic porphyria). They evidently lack adequate amounts of normal cosynthase protein to alter the specificity of uroporphyrinogen I synthase. The product is oxidized spontaneously to uroporphyrin I, which accumulates. Part is converted to other type I porphyrins by metabolism of the side chains. These porphyrins are red to brown; they are excreted mainly as zinc chelates, and give a pink color to the urine, and also to the teeth and bones in which they accumulate.

Other people are known with a hereditary lack of ferrochelatase in their erythroid cells, resulting in an accumulation of protoporphyrin IX (erythropoietic protoporphyria). This accumulation, like the accumulation of type I porphyrins in cosynthase deficiency, causes sensitivity to light, with blistering of the skin upon brief exposure to sunlight. Ordinary suntan lotions are ineffective, but giving sufficient carotene to color the skin has helped. (The carotenes are common plant pigments; they are also dietary precursors of the visual pigments in animals (p. 814).)

Several different hepatic porphyrias are known. The most prevalent, acute intermittent porphyria, is due to a deficiency in uroporphyrinogen I synthase.

(The condition is an autosomal dominant; therefore, those afflicted are heterozygous to the altered gene.) The condition is more prevalent in Scandinavia and the British Isles. One in a thousand people in Lapland is afflicted, and one in 13,000 in Sweden, perhaps reflecting a higher degree of consanguinity in that area, whereas the incidence is about 15 in a million people elsewhere. They make excessive amounts of porphobilinogen and 5-aminolevulinate. The condition causes acute abdominal pain and other neurological symptoms; the ''madness'' of George III was an expression of this condition.

The symptoms of porphyrias have successfully been alleviated in recent trials by injecting hemin intravenously (as hematin, the hydroxide of hemin) to diminish the activity of 5-aminolevulinate synthase.

RED-CELL SURVIVAL AND METHEMOGLOBIN

The mean life-time of a red blood cell in adult humans is 120 days, and most of the constituent hemoglobin molecules survive unscathed for that time. The hemoproteins of other tissues turn over much more rapidly, with half-lives of a few days. Even so, the greater mass of hemoglobin present in the body makes it the greater source of degraded porphyrins, contributing about 70 per cent of the total.

Although hemoglobin is remarkably stable, it is slowly, but continually, oxidized to methemoglobin, which contains Fe(III). The processes involved are very slow compared to the enzymatically-catalyzed metabolic events we have been considering, but their cumulative effects would be disabling without some means of counteraction. The mechanisms of methemoglobin formation are not completely understood; they may include a feeble action of hemoglobin as a mixed function oxidase, in which it and some of the many compounds appearing in the blood accept electrons from O_2. Other mechanisms may involve the generation of H_2O_2 from separate oxidation.

One way in which the erythrocyte is protected from an accumulation of methemoglobin is by minimizing its rate of formation. Hydrogen peroxide is removed within the erythrocyte by the action of catalase, but a more effective route involves a reaction with glutathione:

$$2\ \text{G-SH} + H_2O_2 \xrightarrow{\textit{glutathione peroxidase}} \text{G-S-S-G} + 2\ H_2O$$

The oxidized glutathione is then reduced by reaction with NADPH, which is formed in the erythrocyte by the pentose phosphate pathway:

$$\text{G-S-S-G} + \text{NADPH} + H^{\oplus} \xrightarrow{\textit{glutathione reductase}} \text{2-G-SH} + \text{NADP}^{\oplus}$$

Another way in which accumulation of methemoglobin is minimized is by reducing it to hemoglobin with NADH:

$$\underset{\text{methemoglobin}}{2\ (\text{Fe(III)-porphyrin-protein})} + \text{NADH} + H^{\oplus} \longrightarrow \underset{\text{hemoglobin}}{2\ (\text{Fe(II)-porphyrin-protein})} + \text{NAD}^{\oplus}$$

Curiously, the enzyme catalyzing this sequence is the same one that transfers electrons from NADH to cytochrome b_5 on the endoplasmic reticulum — the enzyme involved in desaturation of stearoyl coenzyme A and other saturated fatty acids. During maturation of the erythrocyte, it apparently becomes a soluble enzyme upon loss of the endoplasmic reticulum. There is also a reduction of methemoglobin, although slower, by NADPH catalyzed by a separate enzyme.

Defects in red-cell survival may involve any part of these protective devices. If methemoglobin accumulates, it causes the formation of granules (Heinz bodies), which cause the cell to be destroyed during its passage through the spleen. We already noted that some genetic variations in amino acid sequence cause such an accumulation. The accumulation could also be caused by excessive oxidative activity of hemoglobin, by excessive peroxide formation, or by a hereditary deficiency in one of the enzymes involved, or by a combination of these effects.

Any chemical compound that readily accepts electrons from O_2 under the conditions prevailing within the erythrocyte will accelerate methemoglobin formation. The effects of the acceleration are exaggerated in those with deficiency of glucose-6-phosphate dehydrogenase. This deficiency was discovered when a large number of American blacks developed hemolytic crises after taking primaquine, an antimalarial drug. The enzyme deficiency in their red cells prevented them from generating NADPH as rapidly as normal cells. It was shortly found that there are many different variants in this enzyme, much as there are in hemoglobin, with some quite rare and some quite common; some are innocuous, while others are seemingly deleterious. The common mutants were found to be present in the same populations as the common hemoglobin mutants: those that were exposed to falciparum malaria. It is believed that the increased rate of methemoglobin formation somehow protects against malaria. Serious problems occur only with the development of illness from other causes, or upon ingestion of compounds promoting methemoglobin formation. Such compounds occur in nature, for example, in a certain vetch known as the fava ''bean'' (*Vicia favia*), but they are also found among many commonly used drugs, such as the sulfonamides, and even aspirin, in addition to the anti-malarials.

The gene for glucose-6-phosphate dehydrogenase occurs on the X-chromosome. All of the red blood cells in an affected male will be deficient, but only part in the heterozygous female. (The condition was used to demonstrate the Lyon hypothesis that only one of the X chromosomes is functional in female somatic cells.) An estimated 13 per cent of American black males and 20 per cent of the females carry a gene for glucose-6-phosphate dehydrogenase deficiency. The incidence is 50 per cent or greater among Kurdish Jews, 14 per cent in Sardinia, and 2.7 per cent in Malta. (It is only 0.4 per cent in mainland Italy.)

Rare deficiencies also causing hemolytic anemia include defects in various enzymes of the Embden-Meyerhof pathway, with decreased generation of NADH, and defects in the enzymes of glutathione metabolism — the peroxidase, reductase, and synthase.

PORPHYRIN DEGRADATION

The principal sites of heme catabolism are the spleen and the liver. Which site is used to handle hemoglobin depends upon the circumstances of red cell destruction. Usually, the aged red blood cells are destroyed in the spleen, and heme degradation also begins there.

Degradation of hemin begins by reduction to heme with NADPH, followed by rupture of one of the methylidyne bridges between pyrrole rings, specifically between the two rings carrying vinyl groups (Fig. 33–6). A sequence of oxidations

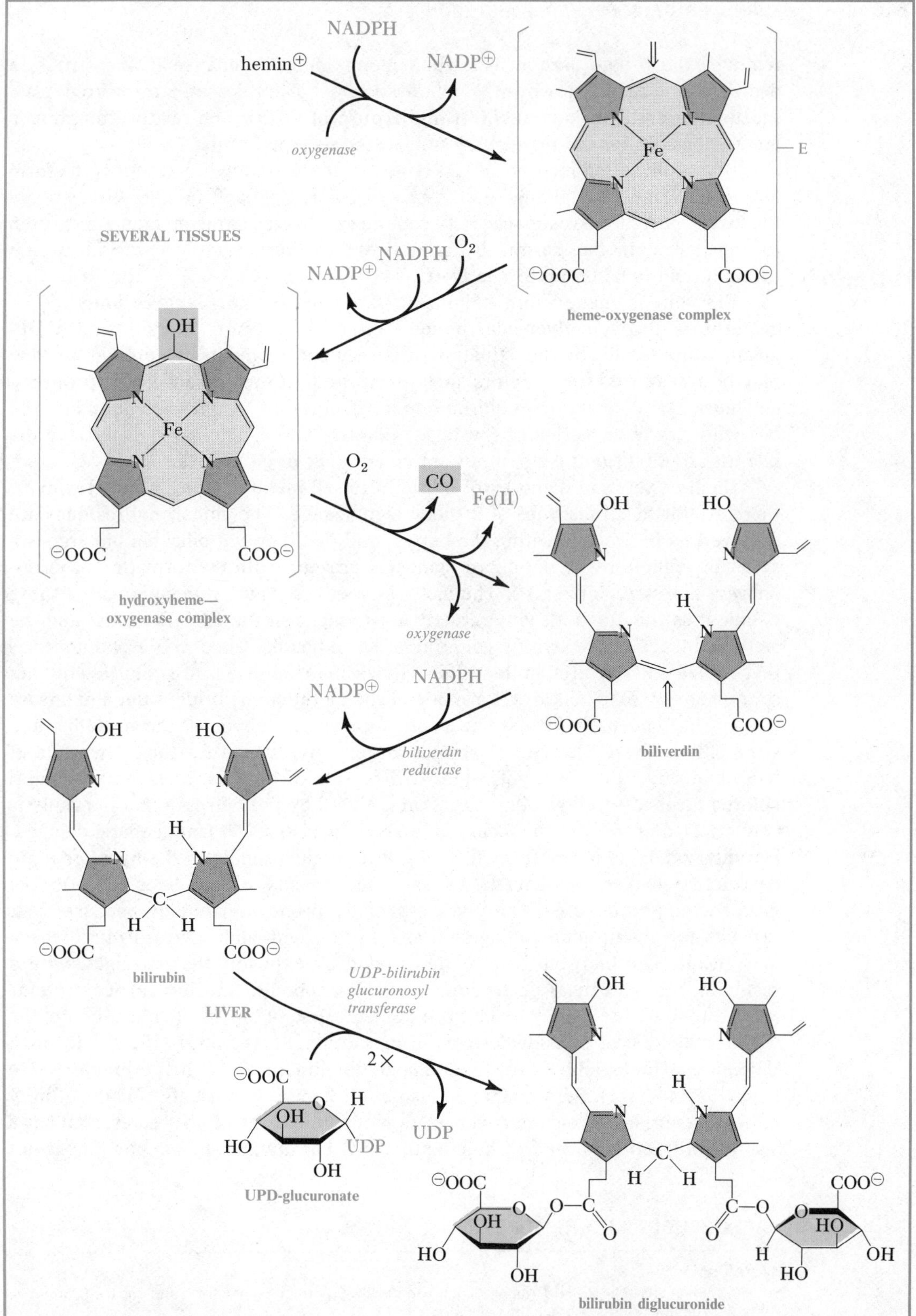

FIGURE 33–6 **Degradation of hemin begins by reduction to heme and simultaneous formation of a complex with an oxygenase system in the endoplasmic reticulum. A mixed-function oxidase reaction hydroxylates one of the methylidyne bridges, followed by a dioxygenase reaction that cleaves the bridge, liberating carbon monoxide and the iron atom. The biliverdin formed is reduced to bilirubin, which travels through the blood to the liver for conjugation with glucuronosyl groups and passage into the bile.**

removes the bridge carbon as carbon monoxide. The change in the structure decreases the affinity for iron, which dissociates. The released tetrapyrrole has a green color and was named **biliverdin** because of this. It and related compounds are responsible for the pigmentation of well-developed bruises.

Biliverdin is reduced by NADPH at the center methylidyne bridge to form **bilirubin,** which is named for its red-brown color. The shift in color comes from the loss of part of the resonance between the two halves of the molecule. Bilirubin is transported in the plasma by attachment to that versatile carry-all, serum albumin, and is taken up by the liver.

Bilirubin is made more soluble in the liver by attaching residues of D-**glucuronate** through glycosidic bonds. Suffice it for now to note that UDP-glucuronate is made by the oxidation of UDP-glucose, and its glucuronate residue may be transferred to acceptors such as bilirubin in the presence of appropriate enzymes. The conversion of bilirubin to its glucuronide enables secretion into the bile without the formation of crystals. Even so, there is enough hydrolysis in the bile that free bilirubin is an important constituent of gallstones.

Biliary excretion disposes of bilirubin for all metabolic purposes. However, there are further transitions of esthetic significance. The glucuronyl residues are removed by hydrolysis within the bowel, and the liberated bilirubin undergoes a series of reductions by the microorganisms present, with the formation of **urobilinogens** and **stercobilinogens**. The methylidyne bridges are reduced in all of these compounds; one or both vinyl groups are reduced in the urobilinogens, and the stercobilinogens have two pyrrole rings also reduced. Part of the urobilinogen is reabsorbed and excreted in the urine. The various bilinogens are colorless but are spontaneously oxidized to re-form one of the methylidyne bridges upon exposure to oxygen. The partially oxidized compounds are known as the **urobilins** and **stercobilins.** These are the compounds largely responsible for the brown hue of feces and the yellow color of urine. Obstruction of the bile ducts leads to light-colored feces and yellow skin. The skin is tinted by the bilirubin that normally is converted to urobilin in the feces. Jaundice may also result from hepatic damage, hereditary defects in the formation of bilirubin glucuronide in the liver, or rapid destruction of red blood cells beyond the capacity of the liver to form the glucuronide. Premature infants are especially prone to jaundice because their capacity for glucuronide conjugation has not reached adult levels. Brain damage may result. The bilirubin level is diminished by exposing them to light, with a resultant photochemical destruction of the compound. Specific carrier proteins are present in the blood to carry oxidized hemoglobin or hemin, should the erythrocyte be broken elsewhere in the circulation, and minimize the loss through the kidney. **Haptoglobins** carry methemoglobin dimers, and **hemopexins** carry free hemin (Table 33–1). Serum albumin also binds hemin, but at only 1/60 the affinity of hemopexin; however, the concentration of albumin is 50-fold greater, so it has a significant fraction of the free hemin attached. The albumin-hemin complex is not

TABLE 33–1 HAPTOGLOBIN AND HEMOPEXIN

Haptoglobin

Two binding sites for $\alpha\beta$ Hb^{+} dimers; each site itself a pair of sites, one for α and one for β subunits
M. W. 100,000; 0.4–1.8 g per liter of blood

Hemopexin

One binding site for hemin
M. W. 57,000; 0.6–0.8 g per liter of blood
Hemin complex taken up by liver

taken up by the liver; it may act as a reservoir, with the hemin transferred to hemopexin for removal.

Turnover of Iron

Iron atoms are not allowed to move about the body unescorted. Specific carriers are known to be employed for extracellular transport and intracellular storage of iron. Even if there were no such carriers, few iron ions could exist in the free form in the body. This is true because these ions have a high affinity for complexing groups containing O, N, or S, and the free ions therefore react readily with a variety of cellular constituents. Indeed, if enough inorganic iron salt is consumed to swamp the capacity of the approved carriers, the excess is acutely toxic. (Children have died after eating their mothers' attractively coated iron tablets; it is claimed to be the second most common cause of accidental poisoning in childhood — aspirin is still number one.)

The known forms in which the available iron pool — as opposed to the specific heme proteins and other iron-containing electron carriers — exists are summarized in the following and in Table 33–2:

Iron is transported in the blood between tissues by combination with specific proteins, **transferrins,** which have a high affinity for two Fe(III) ions. This affinity is so high that these proteins are half-saturated at a [Fe(III)] of only 25 atoms of free Fe(III) per liter of blood.

Within cells, Fe(III) is stored by combination with other proteins, apoferritins, which can ligate many atoms of the metal per molecule, so that the iron content of **ferritins** approaches one quarter of the total mass. This is possible because the apoprotein is composed of 24 subunits in a spherical shell that is penetrated by six channels. Iron enters the channels and is deposited in the core as a hydroxy phosphate. These complex ferritin molecules associate to form small granules in the cytoplasm.

As the iron content of a cell represented by ferritin grows, an increased amount of another type of granule, **hemosiderin,** appears. Hemosiderin is an ill-defined complex of ferric ions with hydroxide ions, various polysaccharides, and proteins, with a third of its weight as iron.

TABLE 33–2

Transferrins

Many variants in one individual
Single polypeptide chain; M. W. 76,000–80,000, 6 per cent carbohydrate
Two binding sites for Fe(III); no cooperative interaction
Anion binding required for iron binding; bicarbonate or carbonate is usual anion

$$K' = \frac{[\text{Fe chelate}][\text{H}^+]^3}{[\text{apoprotein}][\text{Fe(III)}][\text{HCO}_3^-]} = 10^3\ \text{M}$$

$$\text{Effective } K' \text{ in blood} = \frac{[\text{chelate}]}{[\text{apoprotein}][\text{Fe(III)}]} = 5 \times 10^{23}\ \text{M}^{-1}$$

Ferritins

May be composed of two similar polypeptide chains in varying proportions
24 subunit hollow shell; subunit M. W. 18,500–19,000; 12.4–13.0 nm external dia., 7–8 nm internal dia.
Core with up to 4,000 Fe atoms, mainly $(FeO(OH))_8(FeO \cdot PO_4H_2)$
Small amount of apoferritin in blood from normal tissue turnover; level may be guide to iron store, with 1 μg apoferritin per liter of blood representing *ca.* 8 mg stored iron

The nature of the intracellular carriers of iron is not known, although the presence of compounds with a high affinity, including some relatively small molecules, has been shown. It is likely that the iron must be present as Fe(II) before it can pass through any membranes. Indeed, only Fe(II) enters the channels in apoferritin, which acts as a ferro-oxidase, using O_2 to oxidize the trapped iron to Fe(III) for deposition. The way in which iron passes out of the liver and spleen is not known.

Once in the blood, iron is oxidized to Fe(III) by O_2, the reaction here being catalyzed by a **ferro-oxidase** known as **ceruloplasmin.** (It is bright blue owing to the presence of copper ions.) Fe(III) is held by transferrin. Cells receiving iron have receptor sites in their plasma membranes for transferrin, which exchanges between the blood and intercellular fluid. The receptor-transferrin complex may be engulfed by pinocytosis, with the transferrin apoprotein expelled after release of the iron.

Iron losses from the body are usually low, and most of the iron released from degraded hemoproteins is retained within the body. There is no specific secretory mechanism for iron within the kidney, and the amount in the urine is less than 10 μmoles per day. In all, men lose about 20 μmoles per day with moderate activity, of which about two thirds is lost from the bowel. Women lose more through menstruation, the loss ranging from 50 to 15,000 μmoles per period, with an average of 250 μmoles, which is equivalent to about 9 additional μmoles per day on a long-term basis.

It is obvious that if the absorption of iron exceeds these various small losses, the total iron content of the body will rise with age, and this in fact is a frequent occurrence in males. The only regulation of the quantity of iron in the body is a regulation of absorption from the intestine, and this is the site of control, although its mechanism is not known.

Iron is most readily absorbed when it occurs in hemin; the absorbed hemin is attacked within the intestinal mucosa by heme oxygenase to liberate iron and biliverdin in the same way that it is attacked by the spleen or liver. Inorganic iron is much less readily absorbed, although the fraction taken in is increased by simultaneously eating meat, for unknown reasons. Much probably depends upon the form of the iron. It may be that only certain chelates can readily enter mucosal cells; this may explain why ingestion of ascorbate with iron salts improves absorption, since the ascorbate would reduce iron to Fe(II), and form a soluble chelate. (Certain chelating agents diminish iron absorption, for example, phytate, an inositol polyphosphate occurring in plants such as spinach.)

Whatever the mechanism of absorption, the fraction of the iron passing through the mucosa into the blood depends upon the amount of iron already in the body. The intestinal mucosal cells form ferritin, and this deposit is lost when the cells slough into the lumen. This may be part of the control mechanism, but there also may be more direct regulation of transit through the cells.

Iron Overload. A chronic accumulation of iron causes excessive deposits of ferritin and hemosiderin to appear within cells, especially within the liver. This is believed to be damaging because there is a high correlation of cirrhosis of the liver with these deposits. (Other nutritional problems may sometimes contribute to the damage.) Three kinds of circumstances are likely to cause a dangerous accumulation. One is a hereditary deficiency in control over iron absorption known as **hemochromatosis.** The deficiency may be in regulation by the intestinal mucosa, or in the regulation of any aspect of hemoprotein synthesis or degradation. (Anemia causes increased iron absorption.) These people absorb a larger quantity of iron than they need when the dietary intake is normal; any increase in the intake magnifies their problem. It takes many years to build these abnormal iron stores; consequently, hemochromatosis is typically a disease of mature men. It takes longer for it to appear in women owing to their larger iron losses.

Excessive iron may also accumulate from repeated transfusions given to compensate for hemolytic conditions, such as thalassemia. This is perhaps the most common source of excessive iron deposits in children.

Finally, excessive iron may accumulate from excessive intake. The intestinal control over absorption is not perfect. It limits the fraction of dietary iron that is absorbed, but any increase in the absorbable iron content of the diet will cause some increase in the absolute amount taken up. This is classically illustrated by the Bantus of South Africa, who prepare meals and make beer in large iron pots so that the average of 50 μmoles originally in the foodstuffs ingested per day is increased to 2,000 μmoles or more. Many of these people are ill, and the condition is designated as the iron pot syndrome. We need not travel so far afield, because wines sold in the United States have contained as much as 500 μmoles per liter. Finally, a few areas of the world have extraordinary amounts of iron in water used for drinking, and some people in those areas accumulate excessive stores*.

*I once drank beautifully clear and cold water from a Forest Service well in Colorado and then had the disquieting experience of seeing a heated pot of the water become absolutely opaque from a massive precipitate of iron hydroxide. Much lower concentrations than this would be dangerous upon lengthy exposure.

FURTHER READING

Dolphin, D., ed.: (1978) *The Porphyrins.* Academic Press. Multi-volume treatise.

Muller-Eberhard, U., ed.: (1977) *Iron Excess. Aberrations of Iron and Porphyrin Metabolism.* Sem. Hematol., vol. 14, nos. 1,2. Includes surveys of porphyrin and iron metabolism.

DeMatteis, F., and W. N. Aldridge, eds.: (1978) *Heme and Hemoproteins.* vol. 44 of *Handbuch der Experimentellen Pharmakologie.* Springer-Verlag.

Bergsma, D., et al., eds.: (1976) *Iron Metabolism and Thalassemia.* Liss.

London, I. M., et al.: (1976) *The Role of Hemin in the Regulation of Protein Synthesis in Erythroid Cells.* Fed. Proc., *35*: 2218.

deHaro, C., A. Datla, and S. Ochoa: (1978) *Mode of Action of the Hemin-Controlled Inhibitor of Protein Synthesis.* Proc. Natl. Acad. Sci. U.S.A., *75*: 243

Yoshida, T., and G. K. Kuchi: (1974) *Sequence of the Reaction of Heme Catabolism Catalyzed by the Microsomal Heme Oxygenase System.* FEBS Lett., *48*: 256.

Dhar, G. J., et al.: (1975) *Effects of Hematin on Hepatic Porphyria. Further Studies.* Ann. Intern. Med., *83*: 20.

Kaufman, L., and H. S. Marver: (1970) *Biochemical Defects in Two Types of Human Hepatic Porphyrias.* N. Engl. J. Med., *283*: 954.

Meyer, U. A., and R. Schmid: (1973) *Hereditary Hepatic Porphyria.* Fed. Proc., *32*: 1649.

Beutler, E., et al.: (1974) *Prevalence of G-6-PD Deficiency in Sickle-cell Disease.* N. Engl. J. Med., *290*: 826.

Crichton, R. R., ed.: (1975) *Proteins of Iron Storage and Transport in Biochemistry and Medicine.* Elsevier.

Cook, J. D., and D. A. Upschitz: (1977) *Clinical Measurements of Iron Absorption.* Clin. Haematol., *6*(3): 567.

VanCampen, D.: (1974) *Regulation of Iron Absorption.* Fed. Proc., *33*: 100.

Jacobs, A.: (1977) *Serum Ferritin and Iron Stores.* Fed. Proc., *36*: 2024.

Zapolski, E. J., and J. V. Princiotto: (1977) *Preferential Utilization in vitro of Iron Bound to Diferric Transferrin by Rabbit Reticulocytes.* Biochem. J., *166*: 175.

Wochner, R. D., et al.: (1974) *Hemopexin Metabolism in Sickle-cell Disease Porphyrias and Controls.* N. Engl. J. Med., *290*: 822.

Macara, I. G., and T. G. Hoy, and P. M. Harrison: (1972) *The Formation of Ferritin from Apoferritin.* Biochem. J., *126*: 151.

Frieden, E., and H. S. Hsieh: (1976) *Ceruloplasmin: The Copper Transport Protein with Essential Oxidase Activity.* Adv. Enzymol., *44*: 187.

Anon.: (1977) *Chelation May Reduce Iron Excesses due to Transfusion.* Hosp. Pract., *12*(9): 49.

Crosby, W. H.: (1978) *Prescribing Iron? Think Safety.* Arch. Intern. Med., *138*: 766.

34 TURNOVER OF NUCLEOTIDES

We have seen two general functions of nucleotides: to participate in metabolic reactions, and to act as precursors of the nucleic acids. Both the metabolic pool of nucleotides and the ribonucleic acids are constantly being synthesized and degraded. The attrition of these compounds and the associated requirements for synthesis of replacements is most acute in those tissues in which normal function involves loss of entire cells, such as the skin, the intestinal mucosa, and the blood. These tissues also require the constant formation of new DNA.

The turnover of RNA is also rapid in cells that secrete proteins as a major part of their function. Prominent examples are cells of various parts of the gastrointestinal tract that pour out large volumes of digestive juices, rich in enzymes and mucoproteins. In addition to secreting many proteins, the liver also rapidly changes its intracellular protein composition, which imposes an additional load on the formation of RNA. Even the stable cell population in adult muscles and nerves has an internal turnover of RNA although at a slower rate than that of most tissues.

An Overall View

Let us summarize the general feature of nucleotide turnover:

(1) Both the pyrimidine and the purine nucleotides are constantly degraded to the free bases and ribose phosphates.

(2) The free bases and ribose phosphates can be used in a salvage pathway to rebuild nucleotides.

(3) The base components also can be synthesized *de novo* on ribose phosphate from common metabolic intermediates so as to increase the size of the nucleotide pool or replace any degraded components.

(4) The ribose phosphates can be synthesized from glucose or metabolized by the usual reactions of the pentose phosphate pathway.

(5) There is a constant attrition of bases through degradative reactions:

(a) The pyrimidines are degraded to common metabolic intermediates.

(b) The purines are converted to uric acid, a purine that is excreted in the urine.

The low solubility of uric acid and its monosodium salt will cause much of our attention to be focused on purine nucleotide turnover; precipitation of these compounds causes trouble. Formation of sodium urate crystals damages the joints, kidneys, and other tissues, causing gout. Uric acid is a common constituent of stone in the urinary tract. A variety of hereditary conditions and disease states causes dangerous elevations in uric acid production.

DEGRADATION AND SALVAGE

Nucleic acids are hydrolyzed by **endonucleases,** which attack bonds in the middle of the polynucleotide chains, and **exonucleases,** which hydrolyze successive terminal bonds. Many such enzymes have been identified in mammalian cells, some specific for DNA, some for RNA, and some attacking both. Several of the enzymes in the nucleus appear to be concerned with repair of DNA or maturation of RNA after initial transcription. Other nucleases occur in the lysosomes, presumably to handle foreign nucleic acids and damaged cellular components. These include a 3′-exonuclease, which forms oligonucleotides with terminal 2′,3′-cyclic phosphodiester bonds, much like the intermediate in the action of bovine pancreatic ribonuclease (p. 258). A cyclic phosphodiester hydrolase then cleaves these bonds. However, there is an important gap in our knowledge, and this concerns the enzymes responsible for the constant turnover of RNA in the cytoplasm, which results in the release of the constituent 5′-nucleotides. A 5′-endonuclease occurs in the intermembrane space of mitochondria, and it will attack either ribo- or deoxyribonucleotides, but this appears to be an unlikely location for an enzyme attacking mRNA, tRNA, and rRNA, in the cytosol.

In any event, the ribonucleic acids are constantly being converted to 5′-nucleotides. (The repair of DNA in surviving cells is a small part of the quantitative balance and will not be considered further.) Many of the modified bases in tRNA and rRNA are excreted as such, but it is mRNA that turns over most rapidly, and this larger source of unmodified nucleotides contributes to the general nucleotide pool of the cell, which is also constantly degraded and rebuilt.

The general pathway for degradation of 5′-nucleotides begins with hydrolysis of the phosphate ester group (Fig. 34–1), releasing the nucleosides. The nucleosides are then cleaved with inorganic phosphate, releasing the free bases and forming ribose 1-phosphate. Ribose 1-phosphate is interconvertible with ribose 5-phosphate through the action of phosphoglucomutase, the same enzyme acting on glucose phosphates. The bases are used to rebuild nucleotides, or are destroyed.

Provision of Ribose Phosphate Groups. When nucleotides are synthesized, either by salvage of pre-formed bases or by *de novo* synthesis, the ribose phosphate moiety is supplied in the form of **5-phospho-α-D-ribose diphosphate,** commonly referred to as **phosphoribosyl pyrophosphate (PRPP).** This compound is made from ribose 5-phosphate, which can be generated from glucose 6-phosphate by the pentose phosphate pathway (p. 529) or from ribose 1-phosphate (above):

$^{2-}O_3PO$–H_2C … (ring: H, O, H, H; HO, OH; HO, OH) — ATP → AMP, Mg^{2+}, *ATP phosphoribosyl transferase* (⊕ P_i; ⊖ NTP) → $^{2-}O_3PO$–H_2C … (ring: H, O, H, H; H, $OP_2O_6^{3-}$; HO, OH)

α-D-ribose 5-phosphate → **5-phospho-α-D-ribosyl pyrophosphate**

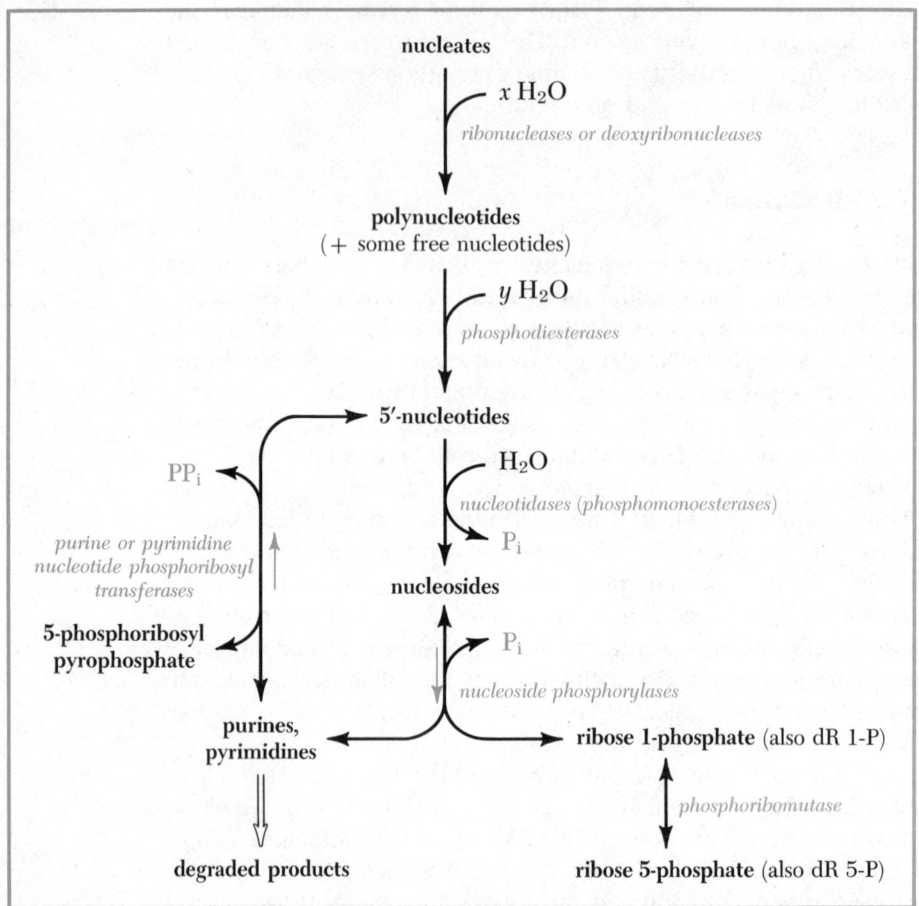

FIGURE 34–1 **Outline of the general processes for degradation of nucleic acids and other nucleotides. The purines and pyrimidines arising in this way or obtained from the diet may be re-incorporated into nucleotides by reactions with phosphoribosyl pyrophosphate *(lower left)*, or they may be further degraded. The ribose portion appears as ribose 5-phosphate.**

The ATP phosphoribosyl transferase catalyzing this reaction is an important site of regulation of nucleotide synthesis. The enzyme has an absolute requirement for inorganic phosphate; the effect is to increase the synthesis of nucleotides for the metabolic pools whenever their degradation becomes excessive, as manifested by accumulation of P_i. The enzyme is inhibited by nucleotides, thereby preventing their accumulation through either the salvage pathway or *de novo* synthesis.

The salvage pathway involves a condensation of free bases with phosphoribosyl pyrophosphate to form a nucleotide:

$^{2-}O_3PO$, H_2C, H, H, O, H, H, $OP_2O_6^{3-}$, HO, OH + Pu *or* Py, N, H $\xrightarrow[\text{phosphoribosyl transferase}]{Mg^{2+}}$ PP_i + $^{2-}O_3PO$, H_2C, H, H, O, Pu *or* Py, N, H, H, HO, OH

5-phospho-α-D-ribosyl pyrophosphate **base** **nucleotide (nucleoside 5′-phosphate)**

PURINE METABOLISM

Degradation (Fig. 34–2)

The concentration of adenine nucleotides is several-fold greater than the concentration of other nucleotides, and their turnover is correspondingly greater. We already noted (p. 554) that AMP is deaminated by a simple hydrolysis to produce inosine monophosphate as part of the purine nucleotide cycle. This is also the first step in degradation of the nucleotide. Inosine monophosphate and guanosine monophosphate are then hydrolyzed to form the nucleosides and cleaved with P_i to liberate hypoxanthine and guanine.

Hypoxanthine is oxidized to xanthine and then to uric acid by molecular oxygen. The enzyme catalyzing these successive reactions, **xanthine oxidase,** occurs in the soluble cytoplasm of the cell, and yet it is a molybdenum-containing iron-sulfide protein. The presence of **molybdenum** in this, as well as the related aldehyde oxidase also present in the cytoplasm, makes it necessary to have traces of the inorganic element in the diet. The molybdenum, ordinarily hexavalent, is reduced to the pentavalent compound during the passage of electrons to oxygen. Oxygen is reduced to hydrogen peroxide as the other product.

Guanine is at the same oxidation level as xanthine, so simple hydrolytic deamination of the free base liberated by degradation of its nucleotides will form xanthine. Xanthine therefore is an intermediate in the degradation of both major purines, and uric acid is the end product.

Since the first ionization of uric acid has a pK′ near 5.4, it exists mainly as the urate anion in the body. It is cleared efficiently from the blood by the kidney. (Remember that the side group nitrogen of both adenine and guanine appears as NH_4^+, so the purines also make a small contribution to urea synthesis.) In an acidic urine, much of the compound will be present as uric acid.

Most mammals other than the primates degrade urate further, using uricase, a copper-containing enzyme, to generate allantoin as an end-product:

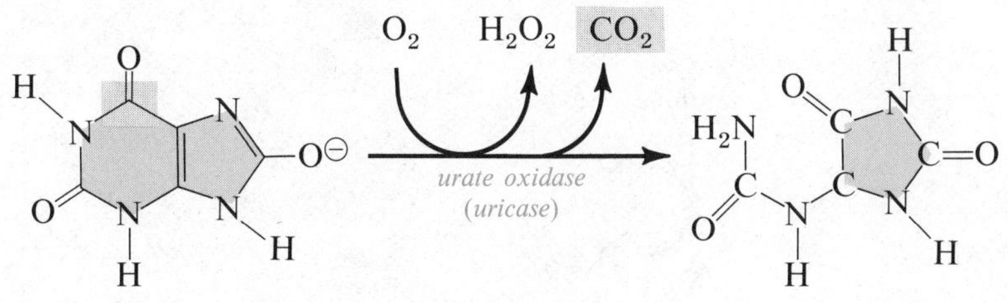

This compound has a solubility in excess of 5,000 mg per liter. Compare this with the solubilities of the purines (mg per liter):

Compound	Blood Serum	Water	Urine at pH 5
sodium urate	68	1,200	—
uric acid	—	65	150
xanthine	100	?	50
hypoxanthine	1,150	?	1,400
guanine	?	39	?

We must ask why the primates, including humans, stop purine metabolism with the formation of uric acid, since the solubility product of the sodium salt is so

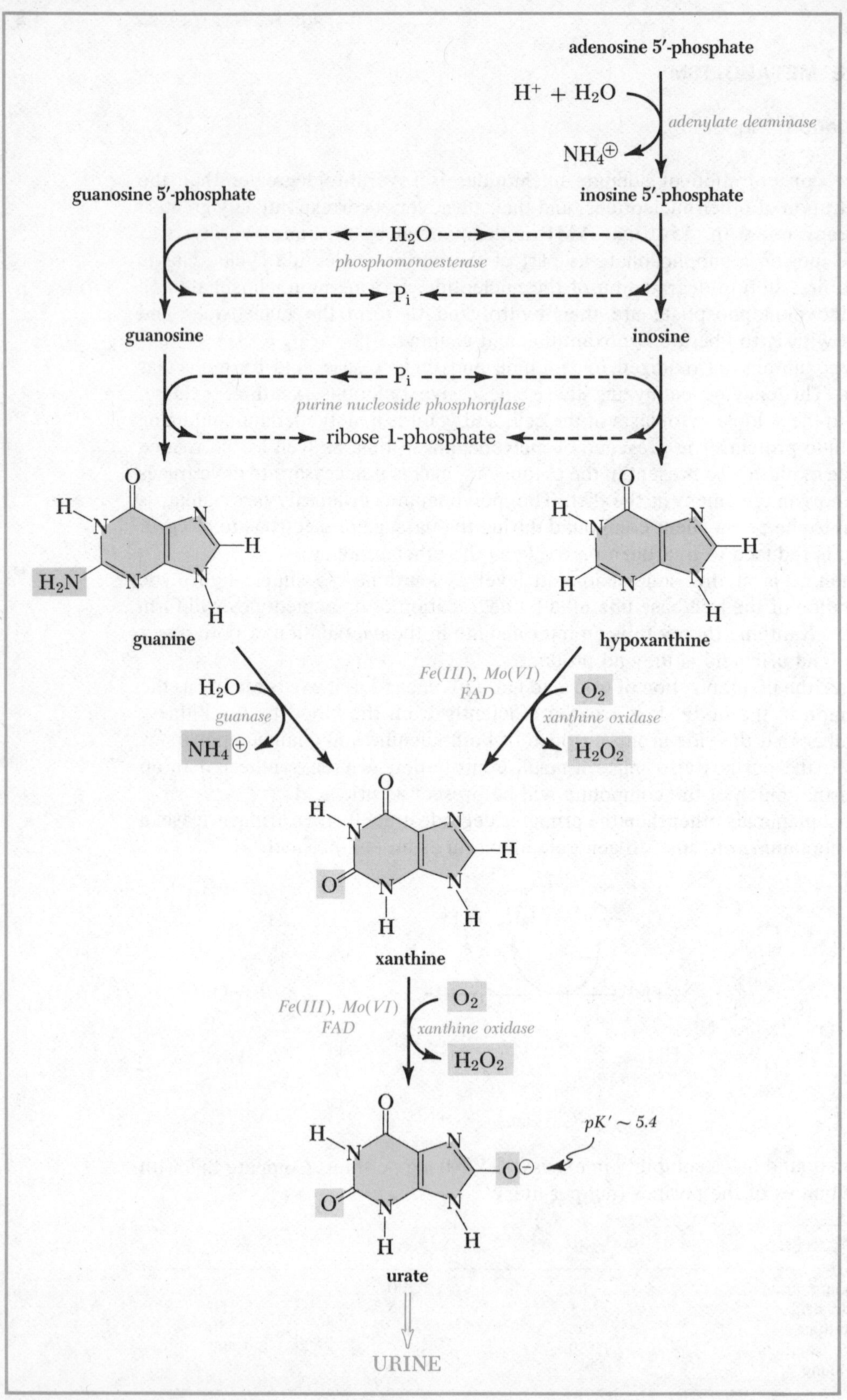

FIGURE 34–2 **The degradation of purine residues to urate. The product exists in the urine as a mixture of urate and uric acid *(not shown).***

low that concentrations over 70 mg per liter in the sodium-rich extracellular fluids are supersaturated, and the free acid is little more soluble in urine. If some intact purine must be excreted, why not choose hypoxanthine, which is much more soluble?

The answer is clear-cut with birds and reptiles, which conserve water in their closed eggs and during flight by converting nearly all of the waste nitrogen to solid masses of uric acid. That is, they synthesize purines instead of urea to dispose of amino acid nitrogen. Many invertebrates make either uric acid or guanine for the same purpose.

The primates, however, make urea as the principal excretory product of nitrogen metabolism, and dilute it with a constant supply of water. Even so, their purine turnover is rapid enough to make troublesome amounts of uric acid, and we shall continue this discussion later.

Purine Salvage

Salvage of pre-formed purines is catalyzed by two enzymes (Fig. 34–3). By far the most active enzyme is the **guanine phosphoribosyl transferase** specific for either guanine or hypoxanthine. The **adenine phosphoribosyl transferase** is specific for adenine; it is less active.

The use of hypoxanthine is especially important in those cells that cannot synthesize purines *de novo*. The source of their purines is the nucleotides and nucleosides arriving through the blood. However, the nucleotides cannot be used directly, and 5′-nucleotidase is exposed on the outer surface of the plasma membrane to form nucleosides. An adenosine deaminase converts the resultant adenosine to inosine, and nucleoside phosphorylases liberate hypoxanthine. (There is also an adenosine kinase to convert adenosine to AMP.)

The importance of pre-formed purines in the economy of some cells is demonstrated by hereditary deficiencies of nucleoside phosphorylase and of adenosine deaminase in the blood-forming cell lines. Either condition results in deficiencies in the immune response. On the other hand, an excessive formation of adenosine deaminase activity is associated with a hemolytic anemia.

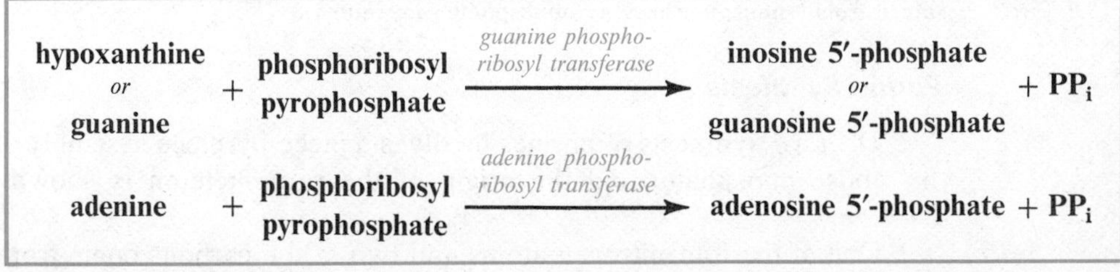

FIGURE 34–3 **Salvage of free purines. The enzyme recovering guanine or hypoxanthine is the more active.**

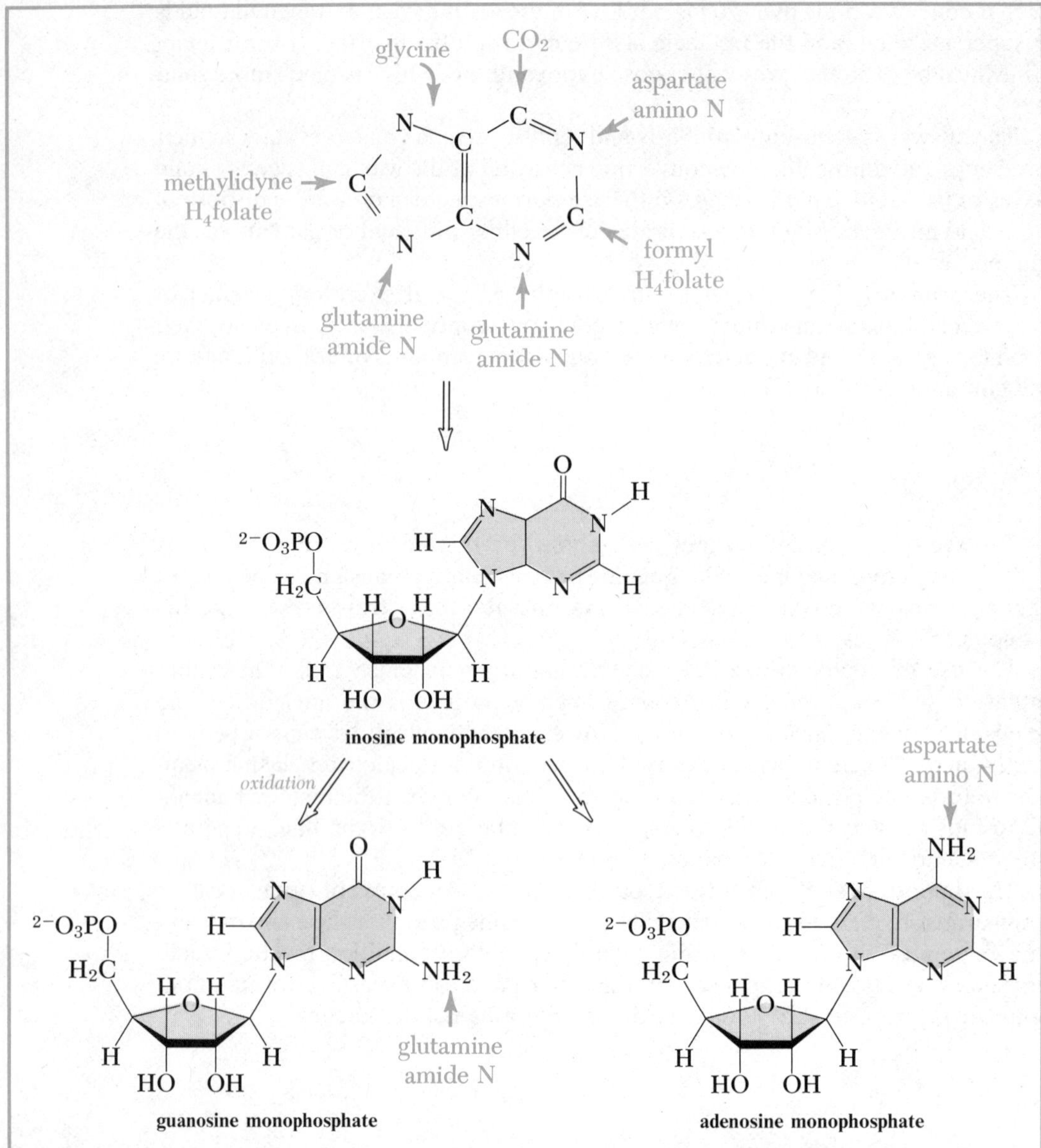

FIGURE 34–4 **Sources of the atoms of purines. The rings are assembled on a ribose 5-phosphate residue obtained from 5-phosphoribosylpyrophosphate** *(not shown).*

Purine Synthesis

De novo synthesis of purines involves a piece-by-piece assembly of the rings on ribose phosphate, and the origin of the ring skeleton is shown in Figure 34–4.

One of the four nitrogen atoms and two of the carbons come from an intact **glycine** molecule. Another nitrogen is contributed by **aspartate,** and two more by **glutamine.** Two of the carbons come from the **one-carbon pool** as tetrahydrofolate derivatives, and the remaining carbon is added as $\mathbf{CO_2}$. The initial product when the ring is assembled is inosine monophosphate.

Inosine monophosphate (IMP) is the precursor of both guanosine and adenosine monophosphates. Simple replacement of the oxygen by an amino group derived from aspartate makes the adenine nucleotide. Oxidation of the ring and replacement of the added oxygen by the amide group of glutamine make the guanine nucleotide.

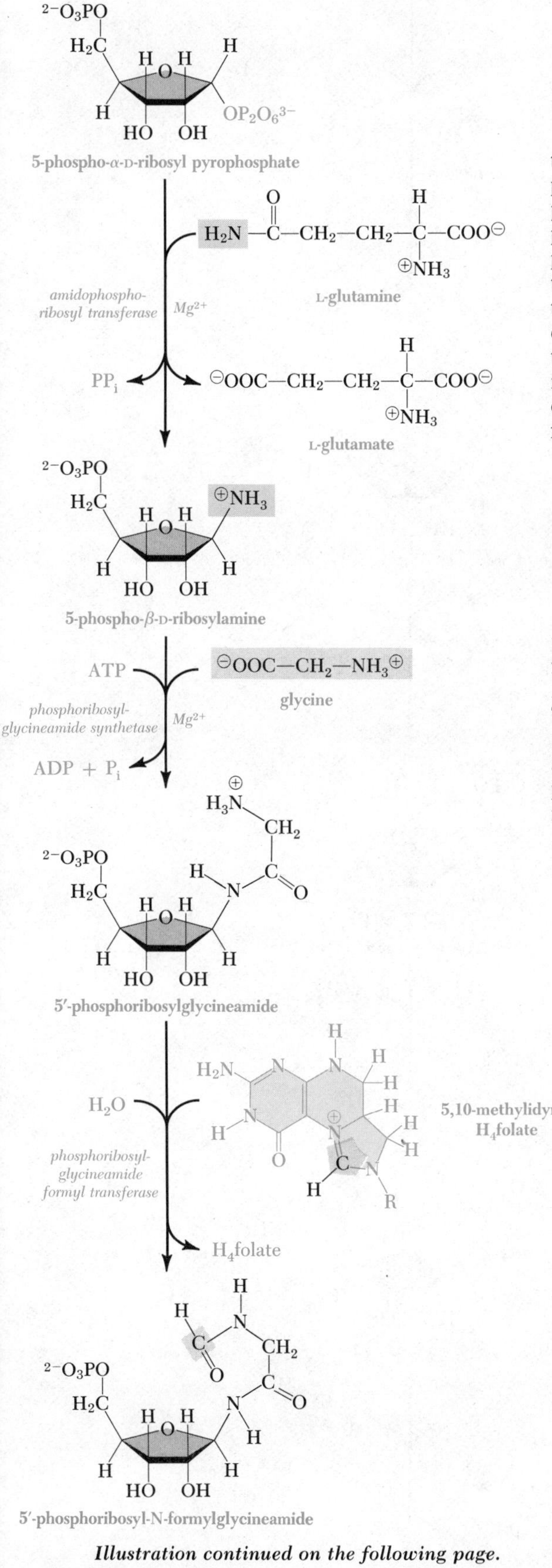

$^{2-}O_3PO$
H_2C H
H H
O
H $OP_2O_6^{3-}$
HO OH
5-phospho-α-D-ribosyl pyrophosphate

O H
H_2N—C—CH_2—CH_2—C—COO⊖
⊕NH_3
amidophosphoribosyl transferase Mg^{2+} L-glutamine

H
PP_i ⊖OOC—CH_2—CH_2—C—COO⊖
⊕NH_3
L-glutamate

The initial reaction is a transfer of the amide group of glutamine to phosphoribosylpyrophosphate forming **phosphoribosylamine** and liberating inorganic pyrophosphate. Since the ribosylpyrophosphate has the α-configuration and the ribosylamine the β-configuration, there is an inversion of configuration during the reaction. The remainder of the purine ring is built around this added nitrogen, so the products will have the characteristic β-ribosyl configuration found in the nucleotides.

$^{2-}O_3PO$
H_2C ⊕NH_3
H H
O
H H
HO OH
5-phospho-β-D-ribosylamine

ATP ⊖OOC—CH_2—NH_3⊕
glycine
phosphoribosylglycineamide synthetase Mg^{2+}
ADP + P_i

A glycyl group is attached intact to the amino group on ribose by a straightforward synthetase reaction in which ATP is cleaved to drive the reaction. The reaction is reversible because of the similar free energies of hydrolysis of the aminoacyl amide and ATP, but the product is constantly removed by the next reaction.

$H_3\overset{\oplus}{N}$
CH_2
$^{2-}O_3PO$ H C
H_2C N O
H H
O
H H
HO OH
5′-phosphoribosylglycineamide

H_2O
phosphoribosylglycineamide formyl transferase
H_2N N H N H H H H H H O C N H R
5,10-methylidyne-H_4folate
H_4folate

A formyl group is then added to the free amino group of the glycyl residue. This completes the atoms necessary for the five-membered ring of the purines. The formyl donor is 5,10-methylidyne-H_4folate in the one-carbon pool.

H
H N
C CH_2
$^{2-}O_3PO$ O C
H_2C N O
H H H
O
H H
HO OH
5′-phosphoribosyl-N-formylglycineamide

Illustration continued on the following page.

A further nitrogen is next transferred from the amide group of glutamine, and the reaction is driven by the hydrolysis of ATP. Since the free energy of hydrolysis of ordinary amides, in contrast to aminoacyl amides, is less than the free energy of hydrolysis of ATP, the reaction is essentially irreversible.

The five-membered imidazole ring of the purine is then formed in a reaction driven by ATP. The amino group added in the preceding reaction now sticks out.

In the next step, CO_2 adds to the imidazole ring. The carboxylase catalyzing this reaction has no biotin, pyridoxal phosphate, or thiamine pyrophosphate attached, and therefore differs in mechanism from most of the enzymes catalyzing addition or removal of carboxylate groups.

In the next pair of reactions, the amino group of aspartate is transferred to form an amide of the carboxyl group created by the preceding reaction. The first step is the formation of the amide between the aminoimidazole carboxylate and the amino group of aspartate, using ATP as an energy donor. Again, the reaction is reversible because an aminoacyl amide is the product.

Illustration continued on the opposite page

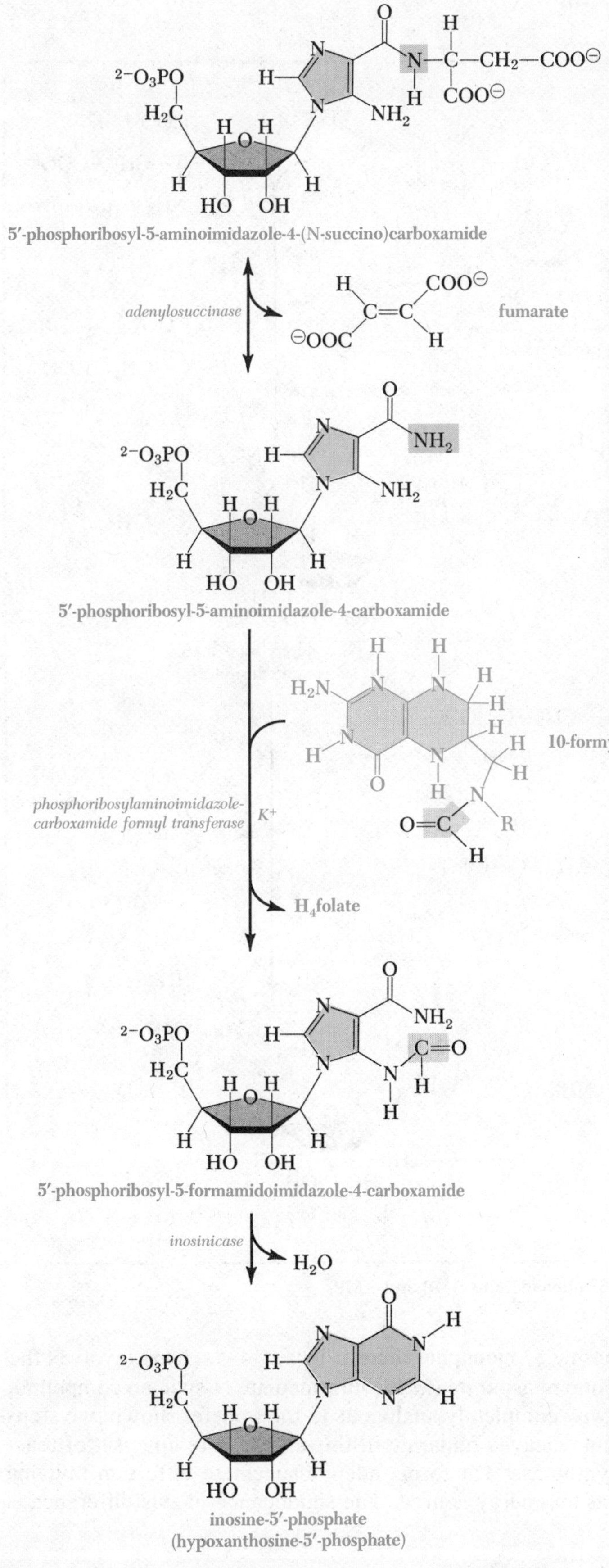

$^{2-}O_3PO$ H N N H NH CH_2 COO⊖ COO⊖ NH_2 H_2C H O H H H HO OH

5′-phosphoribosyl-5-aminoimidazole-4-(N-succino)carboxamide

adenylosuccinase

fumarate

The product then is cleaved on the other side of the connecting nitrogen atom to liberate fumarate and leave the aminoimidazole carboxamide as a product. The pair of reactions is analogous to the mechanism in urea synthesis by which the nitrogen of aspartate is transferred to citrulline, forming arginine (p. 556), with an intermediate N-succino compound formed in both cases. (However, ATP is cleaved to form pyrophosphate in arginine synthesis, with the constant removal of pyrophosphate insuring that the reaction will be drawn toward arginine.)

5′-phosphoribosyl-5-aminoimidazole-4-carboxamide

phosphoribosylaminoimidazole-carboxamide formyl transferase K^+

10-formyl-H_4folate

H_4folate

The final atom necessary for the purine ring is transferred from 10-formyl-tetrahydrofolate by an irreversible reaction.

5′-phosphoribosyl-5-formamidoimidazole-4-carboxamide

inosinicase

H_2O

inosine-5′-phosphate
(hypoxanthosine-5′-phosphate)

With the final atom for the purine in place, the ring is formed by an enzyme catalyzing the simple removal of the elements of water to produce the purine nucleotide, inosine 5′-phosphate.

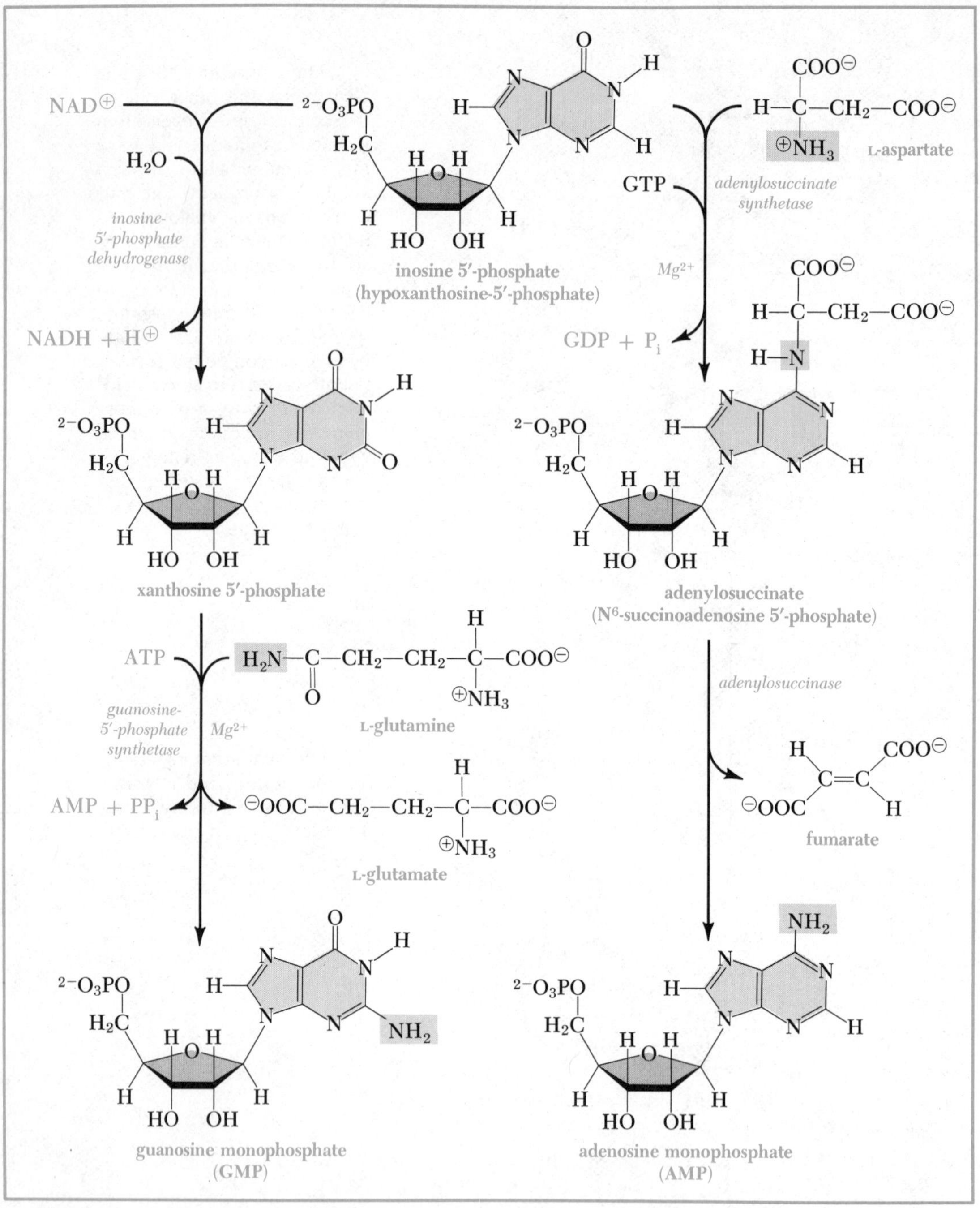

FIGURE 34–5 **The conversion of inosine 5′-phosphate to GMP and AMP.**

The route to **adenosine 5′-monophosphate** (Figure 34–5, *right*) involves the transfer of the amino group of asparate *via* the intermediate N-succino compound, adenylosuccinate, in a way completely analogous to the transfer shown two steps earlier. The same enzyme cleaves fumarate in this and the previous set of reactions. However, the synthetase that forms adenylosuccinate differs in utilizing GTP rather than ATP as an energy source. The significance of this difference is not clear.

Guanine is a more oxidized purine than is adenine. The first step in the formation of **guanosine monophosphate** is therefore the oxidation of inosine 5′-phosphate by NAD to form **xanthosine 5′-phosphate,** the nucleotide of xanthine (2,6-dioxopurine). Guanosine monophosphate is then formed by a transfer of the amide group of glutamine, with the reaction driven by ATP. This reaction differs from the earlier ones in which glutamine is an amide donor in that ATP is cleaved to AMP and inorganic pyrophosphate, at least in bacterial systems. The reaction is irreversible, even without removal of inorganic pyrophosphate, so the functional value of the loss of the extra high-energy phosphate is not clear.

Regulation and Hyperuricemia

Of each million otherwise-normal adults in the United States, approximately 2,750 will have frank **gout.** This condition is caused by deposition of sodium urate crystals initially in the joints, with the subsequent ingestion of the crystals by leukocytes causing an exquisitely painful inflammation. (The joint at the base of one great toe is frequently affected in initial attacks). If untreated, the accumulating deposits can cause disabling damage to the kidneys, as well as to the joints; nodules (tophi) of sodium urate crystals appear under the skin. Chronic gout is a life-shortening condition.

In the absence of other disease, gout is a condition of the male, with only 3 to 7 per cent of the primary cases occuring in women. It often first appears in the fourth decade of life. The normal range of blood serum urate concentrations in adult males runs up to 70 mg per liter — the limit of solubility, whereas it only ranges to 60 mg l^{-1} in adult pre-menopausal females. (The upper limit is defined here as 2 S.D. above the mean value.) Many males have even higher values without ever developing gout.

The defects in primary gout are not well-characterized, and it may be a condition of diverse causes. The known points of control of purine nucleotide formation are the generation of phosphoribosyl pyrophosphate and its conversion to phosphoribosylamine (Fig. 34–6). A failure in control at either point, or an excessively rapid rate of breakdown of the purine nucleotides, or a failure of the kidney to clear urate from the blood could contribute.

In addition, gout is a secondary consequence of many conditions that disturb purine metabolism, for example, any condition that causes an increased rate of cell destruction. These include chronic hemolytic anemia, pernicious anemia, starvation, the toxemia of pregnancy, glucose-6-phosphatase deficiency, endocrine disturbances, and so on. Malignancies are in themselves likely to contribute to hyperuricemia, and the use of cytotoxic drugs to control malignancy exacerbates the condition. Loss of kidney function causes hyperuricemia, as will the commonly used thiazide diuretics. Additionally, hyperuricemia can result from acidosis, as in diabetic ketoacidosis or lactic acidosis.

A dramatic, although rare, cause of hyperuricemia is a hereditary deficiency of the salvage enzyme, guanine phosphoryibosyl transferase. This causes a loss of control over the synthesis of phosphoribosyl pyrophosphate. It is not clear why, but the result is a rapid *de novo* synthesis of IMP, with a consequent formation of large amounts of urate, more per unit of body mass than is seen in any other condition. Children with this condition, known as the **Lesch-Nyhan syndrome,** have grossly distorted behavior. While said to be very likeable and open, quick to laugh and capable of warm affection, they lapse into excessive aggression, directed at themselves as well as others. They bite; characteristically they bite off the

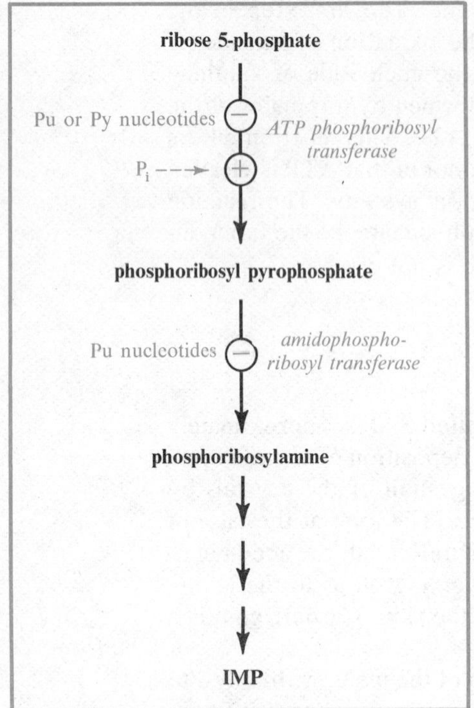

FIGURE 34-6

De novo **synthesis of purine nucleotides is regulated at the formation of phosphoribosyl pyrophosphate and its conversion to phosphoribosylamine. Either purine or pyrimidine nucleotides inhibit phosphoribosyl pyrophosphate formation.**

tips of their own fingers, and their lips if unprotected. In later years, they curse, use obscene gestures, fling feces, and so on. Why does this simple enzyme defect have these bizarre effects? No one knows.

Although it may be unrelated, the Lesch-Nyhan syndrome returns us to a more fundamental question. Why have the primates lost their ability to make the more soluble allantoin? Folklore has persistently associated gout with high living. More recent understanding of the regulation of purine metabolism has discredited an intake of a diet rich in purines as an important cause of primary gout, but there is still a residuum of a belief that people with gout are likely to be superior folk. (Chauvinists take note.)

This notion was given only a faint breath of life by comparison of the blood urate concentrations and intelligence test scores in a large sample of World War I draftees; there was a significant, but very small, correlation. However, a more recent study of male high-school students supplied a new twist. The results indicated a correlation between urate concentration and drive and ambition. People with high urate tended to be over-achievers who got ahead, not by greater abilities, but by using what they had to the utmost. (It is interesting to compare the emotional responses people have when they learn of this finding, and it will be even more interesting to see the results of more comprehensive studies if they can be done and interpreted in an unbiased way.)

Treatment of Hyperuricemia. Gout was treated in ancient times by eating corms of the autumn crocus, *Colchicum autumnale,* and more recently with the active principle of the corms, **colchicine.** This yellow compound combines with tubulin in cells, preventing motility. It arrests mitosis, but it also is very effective in diminishing gouty inflammation. (The patient is effectively titrated with 500 μg doses at hourly intervals until nausea or diarrhea demonstrate a dangerous cessation of cell division in the gastrointestinal tract.) Colchicine is still a useful tool for differential diagnosis, but it has been replaced for chronic therapy by drugs that promote excretion of urice acid, or by allopurinol:

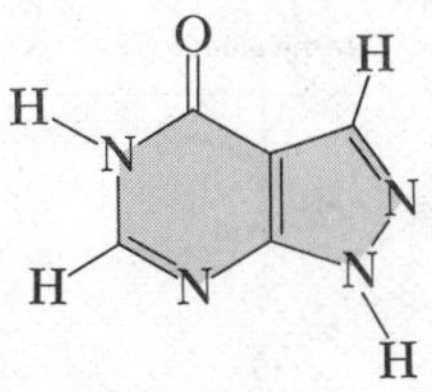

allopurinol

This compound is an analogue of hypoxanthine, differing in the distribution of ring nitrogens, and it was developed as an inhibitor of xanthine oxidase. The rationale for its use is that hypoxanthine will not and xanthine may not precipitate in the joints. Hypoxanthine is more soluble than sodium urate. Xanthine is not, and people with hereditary deficiencies of xanthine oxidase, as well as those treated with allopurinol, sometimes have trouble with xanthine stones and kidney damage. (The incidence of difficulty is much less than it is with untreated gout.)

PYRIMIDINE METABOLISM

Degradation (Fig. 34–7)

The pyrimidine nucleotides, like the purine compounds, are converted to nucleosides. The amino group is then removed from cytidine by hydrolysis to form uridine. Uracil is released by phosphorolysis. Thymine nucleotides are handled in the same way to release thymine; both thymine and uracil then undergo similar reactions, in which the ring is reduced, and the amide groups are cleaved by successive hydrolysis. The 3-amino acids β-alanine and β-aminoisobutyrate, along with NH_4^+ and CO_2, are the products. The amino acids are known to undergo transamination, and the carbon skeletons are oxidized to CO_2, but the exact route is not known.

β-Aminoisobutyrate is only slowly metabolized, and a surprisingly large proportion of the human race excrete significant quantities of the compound in the urine. Five to 10 per cent of Caucasians are hyper-excretors, and up to 90 per cent of Orientals. Hyperexcretion has no known deleterious effects.

The metabolic pool of pyrimidine nucleotides is much smaller than that of the purine nucleotides; although their turnover is rapid and of great qualitative significance, they make little dent in the total fuel or nitrogen economy.

Synthesis of Pyrimidine Nucleotides

Synthesis of the pyrimidines, unlike that of the purines, involves several steps before attachment of the ribose phosphate moiety. The elements of the ring are contributed completely by **glutamine, CO_2**, and **aspartate.**

The process begins by a reaction between carbamoyl phosphate and aspartate to form carbamoylaspartate (Fig. 34–8). The formation of carbamoyl phosphate is also the initial step in urea synthesis (p. 556). We saw that urea formation largely occurs in the liver, but many tissues are able to make the pyrimidines for use in

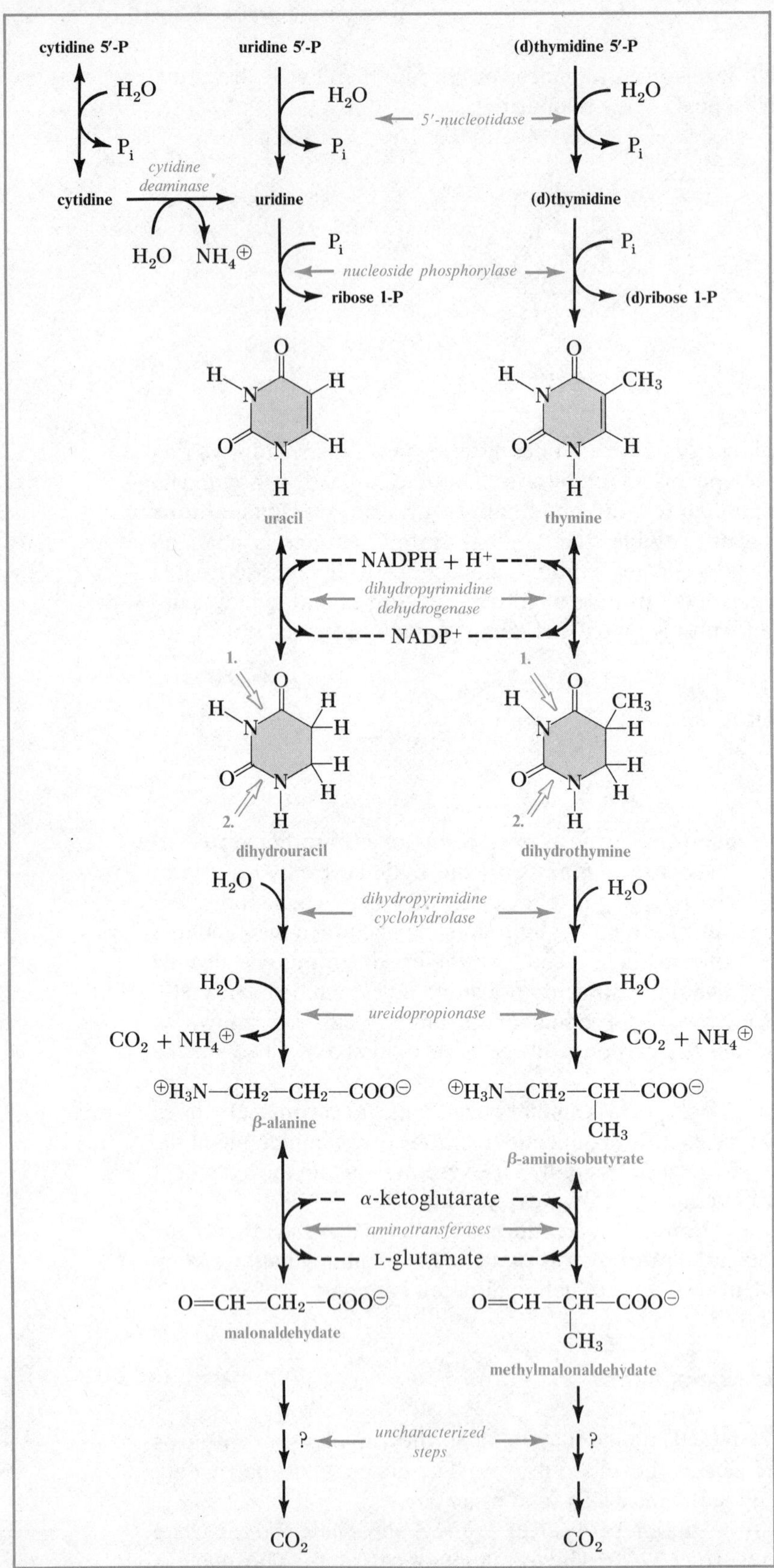

FIGURE 34–7 **Degradation of pyrimidine nucleotides. Thymine occurs in both ribo- and deoxyribonucleotides.**

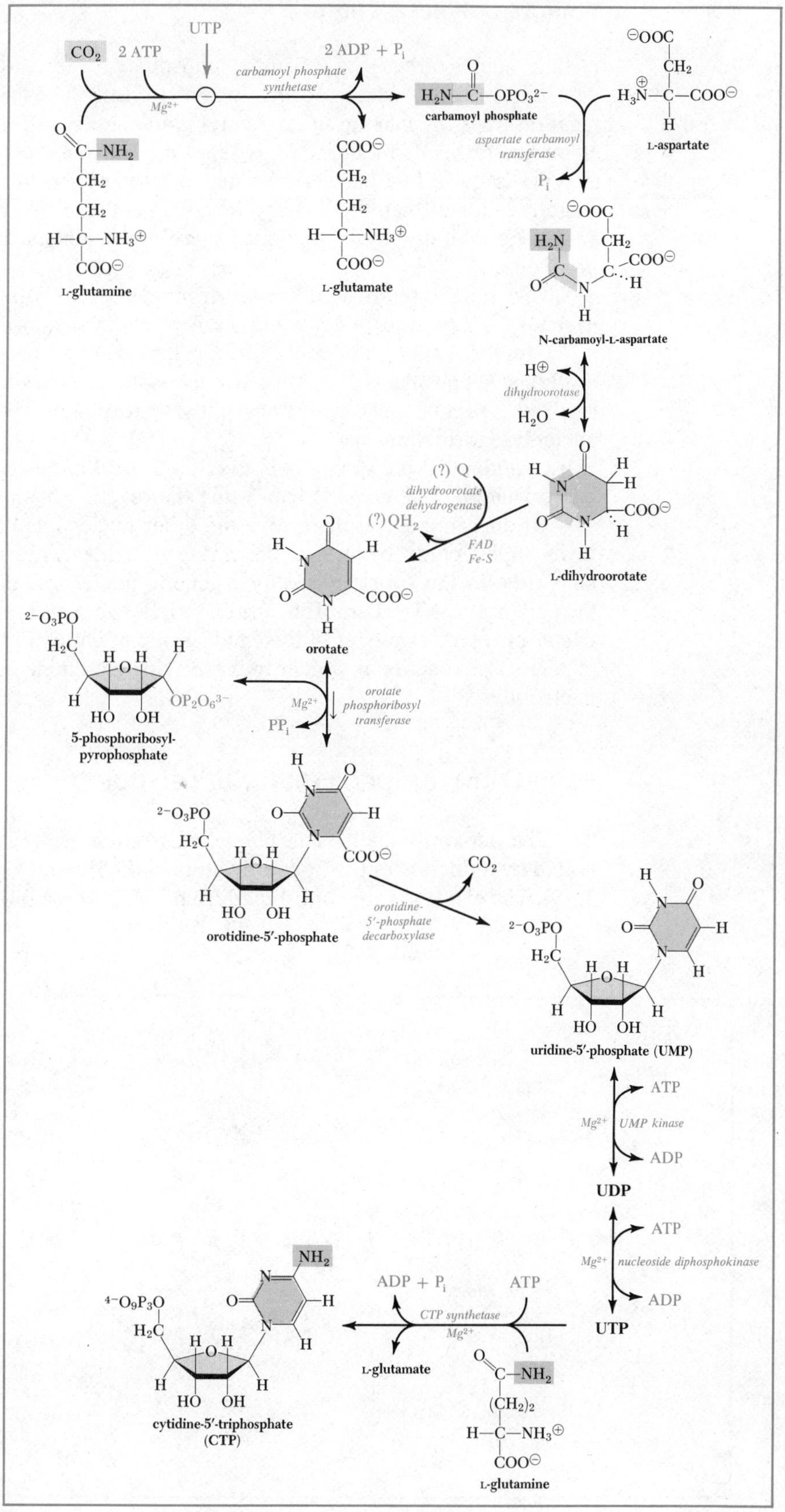

UTP

CO_2 2 ATP

2 ADP + P_i

carbamoyl phosphate synthetase

Mg^{2+}

L-glutamine

L-glutamate

carbamoyl phosphate

aspartate carbamoyl transferase

P_i

L-aspartate

N-carbamoyl-L-aspartate

$H^{\oplus}$

dihydroorotase

H_2O

(?) Q

dihydroorotate dehydrogenase

(?) QH_2

FAD Fe-S

L-dihydroorotate

orotate

5-phosphoribosyl-pyrophosphate

PP_i

orotate phosphoribosyl transferase

orotidine-5′-phosphate

CO_2

orotidine-5′-phosphate decarboxylase

uridine-5′-phosphate (UMP)

ATP

UMP kinase

ADP

UDP

ATP

nucleoside diphosphokinase

ADP

UTP

ADP + P_i

ATP

CTP synthetase

L-glutamate

L-glutamine

cytidine-5′-triphosphate (CTP)

FIGURE 34–8

Synthesis of pyrimidine nucleotides. The carbamoyl phosphate synthetase catalyzing the first step is in the cytosol and differs from the mitochondrial enzyme involved in urea synthesis. Dihydroorotate dehydrogenase is a mitochondrial enzyme, perhaps occurring on the outer surface of the inner membrane.

nucleic acid synthesis. Carbamoyl phosphate is formed for this purpose by a different enzyme, which occurs in the cytosol. The process differs from the reaction used for making urea. Acetyl glutamate is not a required cofactor, and glutamine rather than ammonia is the direct nitrogen donor. The content of the enzyme is quite low because the demand for pyrimidine synthesis is orders of magnitude lower than the demand for urea synthesis. Most importantly of all, the enzyme is inhibited by UTP, which therefore regulates the initial step in its own production.

The simple removal of water from carbamoylaspartate forms a reduced pyrimidine, dihydroorotate, which is then oxidized to **orotate.**

Orotate reacts with phosphoribosylpyrophosphate to form the nucleotide, **orotidine 5′-phosphate,** liberating pyrophosphate. As with other pyrophosphorylases, the removal of the inorganic pyrophosphate drives the reaction toward nucleotide formation.

Orotidine 5′-phosphate is converted to **uridine 5′-phosphate** by a simple decarboxylation. The enzyme catalyzing the reaction has no known cofactors.

Uridine monophosphate, like the other nucleoside phosphates, is converted to its diphosphate by a specific kinase utilizing ATP, and the diphosphate is converted to the triphosphate by a general nucleoside diphosphokinase, also at the expense of ATP. Part of the uridine triphosphate is consumed to make **cytidine triphosphate** by a transfer of the amide group of glutamine, driven by the cleavage of ATP. This reaction is similar to those using glutamine in the formation of purine nucleotides.

FORMATION OF DEOXYRIBONUCLEOTIDES

The deoxyribonucleoside phosphates required for DNA synthesis are generated by reduction of the 2′-carbon atom in the ribonucleoside diphosphates (Fig. 34–9). The electrons are contributed from NADPH *via* **thioredoxin,** a small protein

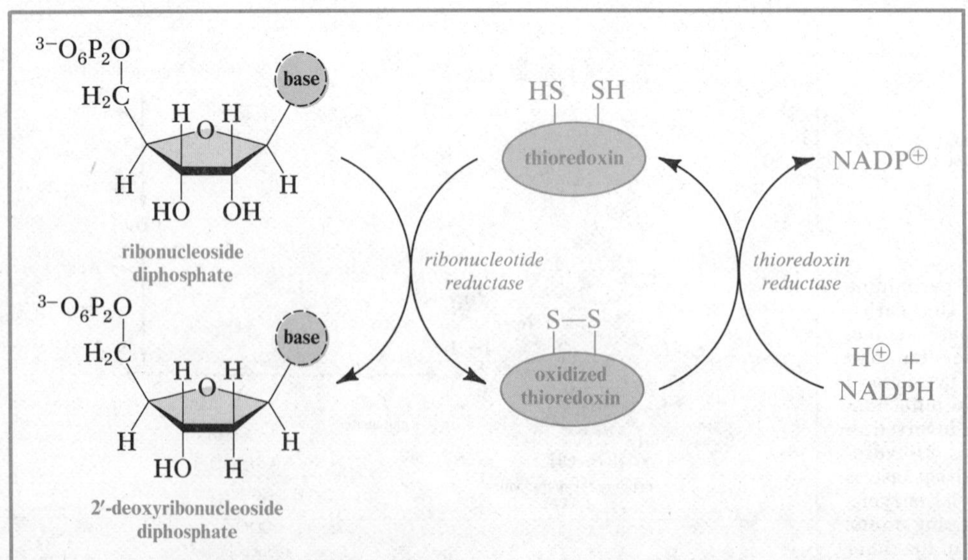

FIGURE 34–9 **Formation of deoxyribonucleoside diphosphates by reduction of ribonucleoside diphosphates.**

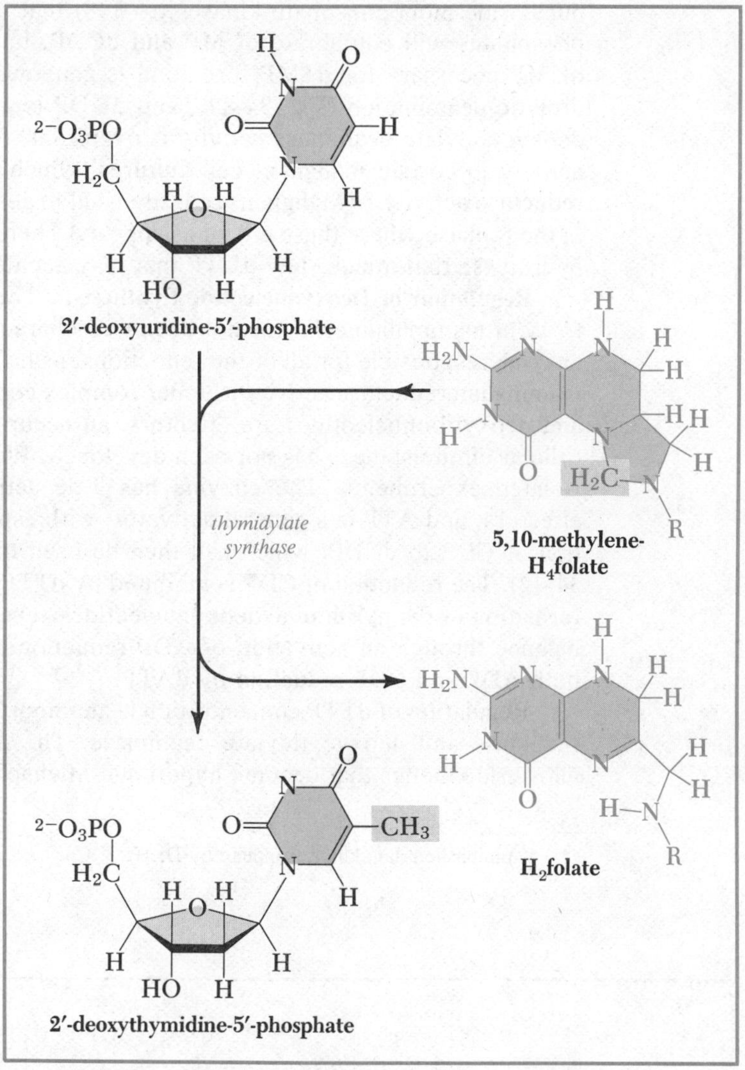

2′-deoxyuridine-5′-phosphate

thymidylate synthase

5,10-methylene-H_4folate

H_2folate

2′-deoxythymidine-5′-phosphate

FIGURE 34–10

dUMP is converted to dTMP by transfer of a methylene group and two electrons from 5,10-methylene-H_4folate.

(103 residues) containing a protruding disulfide group. dADP, dGDP, and dCDP are made directly in this way and are converted to the triphosphates for use in DNA synthesis by the same nucleoside diphosphate kinase that equilibrates ribonucleoside diphosphates and triphosphates.

The formation of dTTP as the fourth precursor of DNA involves a special route. The thymine moiety is created by an unusual reaction catalyzed by an enzyme named **thymidylate synthase.** In this reaction (Fig. 34–10), the methylene group of methylene-H_4folate is transferred to dUMP. If nothing else happened, the result would be the appearance of a hydroxymethyl group. In order to make a methyl group, there must be a simultaneous reduction, and the reducing agent in the reaction is the tetrahydrofolate that also serves as a one-carbon carrier. Tetrahydrofolate contributes two electrons and is converted to dihydrofolate. Cells producing DNA must therefore use dihydrofolate reductase (p. 592) to regenerate H_4folate.

The other reactions of the route for producing dTTP are designed to minimize formation of dUTP, which would otherwise be incorporated into DNA. The nucleoside monophosphate kinases (p. 348) that interconvert the mono- and diphosphates will equilibrate dCMP and dCDP, but not dUMP and dUDP. The dUMP necessary for dTMP formation is generated instead from dCMP by hydrolytic deamination (Fig. 34–11), and dUDP is made from it only slowly. The deoxycytidylate deaminase activity is over 1,000-fold that of the CDP reductase activity in certain malignant cell cultures, which is in turn 5-fold that of UDP reductase activity.* (Malignant cells are used to get the necessary high proportion in the S phase where these enzymes appear.) There is also evidence for a specific hydrolyase that attacks any dUTP that may accidentally appear.

Regulation of Deoxynucleotide Synthesis. The ribonucleotide reductase activity of mammalian cells has not been well-characterized. There may be a single enzyme responsible for all of the reductions, as has been shown for bacteria. The mammalian reductase activity is under complex control in which many of the ribo- and deoxyribonucleotides are effectors; an accurate picture applicable to intracellular circumstances has not been developed. We can get the general idea from isolated experiments. The enzyme has little activity without the presence of effectors, and ATP is a general activator, with especial stimulation of the reduction of CDP to dCDP, which can then be used to form dCTP and dUMP (Fig. 34–12). The reduction of CDP is inhibited by dTTP, thereby preventing excessive formation of the pyrimidine deoxynucleotides. The purine compounds are kept in balance through an activation of ADP reduction by dGTP, and an inhibition of both ADP and CDP reduction by dATP.

Regulation of dTTP concentration is an important device that affects both the reductase and deoxycytidylate deaminase. The deaminase has typical sigmoid allosteric kinetics that become hyperbolic Michaelis-Menten kinetics in the pres-

*Unpublished data kindly supplied by Dr. J. G. Cory.

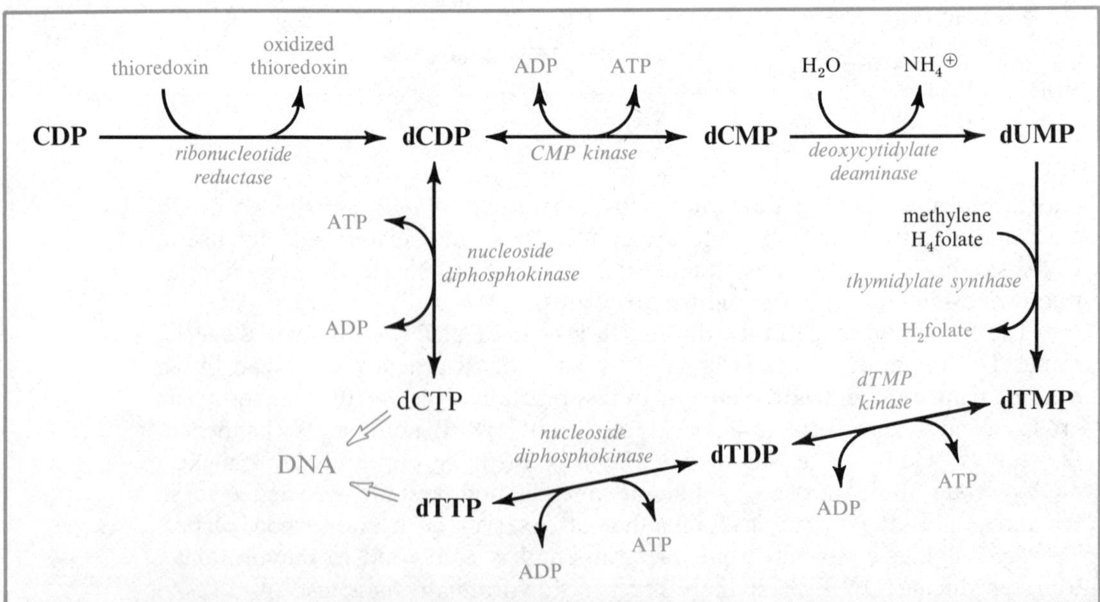

FIGURE 34–11 The formation of deoxythymidine triphosphate (dTTP) for DNA synthesis. The irreversible conversion of dUMP to dTMP and the inability of dTMP kinase to interconvert dUMP and dUDP insure that there is little formation of dUTP for incorporation in place of dTTP.

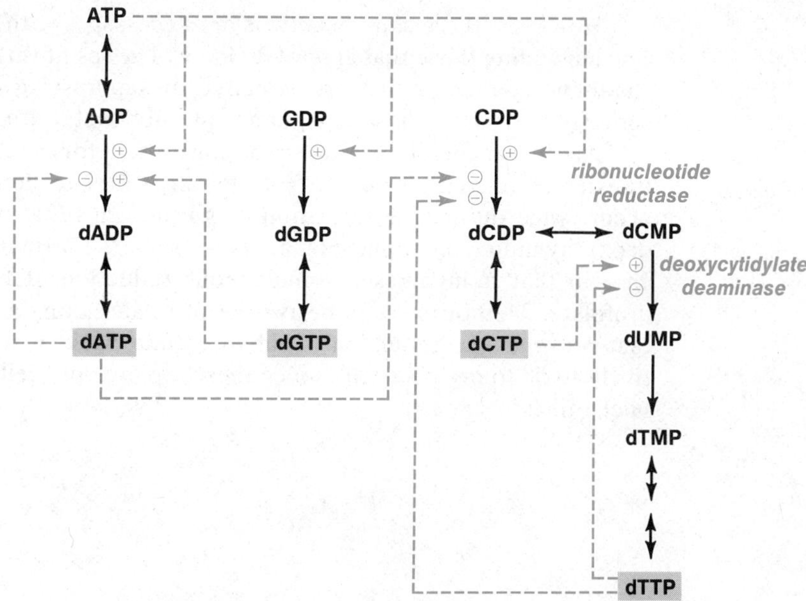

FIGURE 34–12 **The precursors of DNA *(blue boxes)* are allosteric effectors of ribonucleotide reductase and deoxycytidylate deaminase.**

ence of dCTP as a positive effector. This activation is not only overcome by dTTP as a negative effector, but micromolar concentrations of this nucleotide make the kinetics even more exaggeratedly sigmoid. The result of all this is to achieve a balanced mixture of the four ingredients of DNA — not too much or too little of any component.

CHEMOTHERAPY AND NUCLEOTIDE SYNTHESIS

The requirements of rapidly dividing cells for synthesis of nucleotides as precursors of nucleic acids has led to extensive exploration of selective inhibitors of the synthetic routes as potential chemotherapeutic agents (see p. 113).

The least successful attempts involve the inhibition of the general types of reactions used in the formation of the purine and pyrimidine rings. For example, nitrogen is transferred twice from glutamine in the formation of the purine ring, again in forming GMP from XMP, and again in forming CTP from UTP. Certain diazo compounds, such as azaserine and 6-diazo-5-oxo-norleucine, are potent inhibitors of most of these reactions:

$$\begin{array}{c} N{\equiv}N^{\oplus} \\ | \\ CH_2 \\ | \\ C{=}O \\ | \\ CH_2 \\ | \\ CH_2 \\ | \\ H{-}C{-}NH_3^{\oplus} \\ | \\ COO^{\ominus} \end{array}$$

6-diazo-5-oxo-L-norleucine

$$\begin{array}{c} N{\equiv}N^{\oplus} \\ | \\ CH_2 \\ | \\ C{=}O \\ | \\ O \\ | \\ CH_2 \\ | \\ H{-}C{-}NH_3^{\oplus} \\ | \\ COO^{\ominus} \end{array}$$

O-(2-diazoacetyl)-L-serine (azaserine)

However, these same reactions are necessary for the maintenance of normal cells, including those that are not dividing. The result is that concentrations of the glutamine analogues that are effective in suppressing cancer growth are dangerously toxic, and these compounds are not useful drugs.

Another general class of reactions used for nucleotide synthesis is the transfer of one-carbon units from tetrahydrofolate derivatives. Such transfers occur twice during the formation of purine rings and once in the formation of deoxythymidine monophosphate. As we saw, the formation of dTMP is a special case in that it involves a simultaneous reduction of tetrahydrofolate to dihydrofolate. Methotrexate, a derivative of folate acting as a competitive inhibitor that prevents the reduction of dihydrofolate back to tetrahydrofolate, is selectively toxic to dividing cells, since these are the only cells rapidly forming deoxynucleotides:

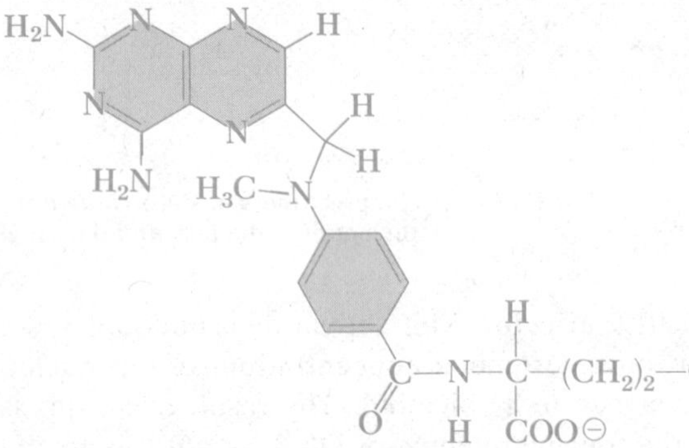

methotrexate
(4-deoxy-4-amino-10-methylfolate)

Methotrexate also inhibits the reduction of folate to dihydrofolate. The drug has a general toxicity because it affects all rapidly dividing cells, such as those in the intestinal mucosa, and it creates a general deficiency of H_4folate, but it is a valuable chemotherapeutic agent. A strategy sometimes used is to give massive doses — many times the lethal dose — followed by rescue of the normal cells through administration of 5-formyl H_4folate (''folinic acid'').

Other successful drugs are analogues of the purine and pyrimidine bases themselves. These include, for example, 6-mercaptopurine and 5-fluorouracil:

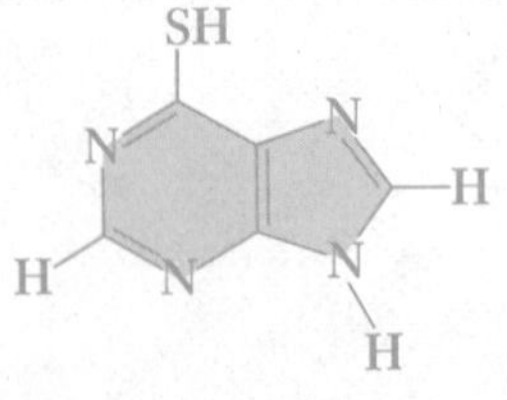

6-mercaptopurine

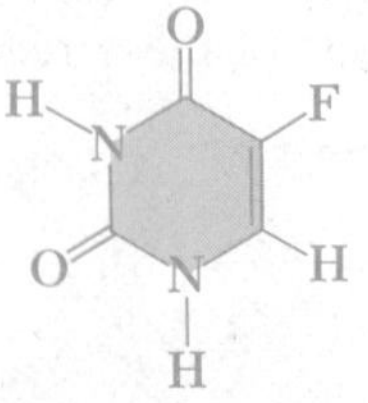

5-fluorouracil

6-Mercaptopurine is sometimes used in adult leukemias, and it has its effect because it is incorporated into nucleic acids (as thioguanosine), thereby preventing normal protein synthesis, and also because its nucleotides inhibit the formation of phosphoribosylamine and the conversion of inosine monophosphate to adenosine monophosphate. 5-Fluorouracil is also incorporated into RNA, but its major effect comes from the formation of deoxyfluorouridine monophosphate, which is a potent inhibitor of thymidylate synthase, and therefore prevents normal DNA synthesis.

FURTHER READINGS

Kit, S.: (1970) *Nucleotides and Nucleic Acids. In* Greenberg, D. M., ed.: *Metabolic Pathways.* Vol. 4, p. 69. Academic Press. Useful survey of older literature on nucleotide metabolism.

Müller, M. M., E. Kaiser, and J. E. Seegmiller, ed.: (1977) *Purine Metabolism in Man.* Adv. Exptl. Med. Biol., Vols, 76A and B. Extensive symposium articles with emphasis on hyperuricemic conditions.

Newcombe, D. S.: (1975) *Inherited Biochemical Disorders and Uric Acid Metabolism.* University Park Press.

Kelley, W. N., and J. B. Wyngaarden: (1974) *Enzymology of Gout.* Adv. Enzymol., *41*: 1.

Klinenberg, J. R.: (1977) *Hyperuricemia and Gout.* Med. Clin. N. Am., *61*: 299, Readable survey.

Kasl, S. V., G. W. Brooks, and W. L. Rodgers: (1970) *Serum Uric Acid and Cholesterol in Achievement Behavior and Motivation.* J.A.M.A., *213*: 1158, 1291.

Hirschhorn, R.: (1977) *Defects of Purine Metabolism in Immunodeficiency Diseases.* Prog. Clin. Immunol., *3*: 67.

Stoop, J. W., et al.: (1977) *Purine Nucleoside Phosphorylase Deficiency Associated with Selective Cellular Immunodeficiency.* N. Engl. J. Med., *296*: 651.

Sierakowska, H., and D. Shugar: (1977) *Mammalian Nucleolytic Enzymes.* Prog. Nucl. Acid Res., *20*: 60.

Chabner, B. A., et al.: (1975) *Clinical Pharmacology of Antineoplastic Agents.* N. Engl. J. Med., *292*: 1107, 1159. Includes discussion of methotrexate and 5-fluorouracil.

Smith, L.: (1973) *Pyrimidine Metabolism in Man.* N. Engl. J. Med., *288*: 764. Review.

Cory, J. G.: (1976) *Control of Ribonucleotide Reductase in Mammalian Cells.* Adv. Enzy. Regul., *14*: 45.

Thelander, L.: (1977) *Mechanism and Control of Ribonucleoside Diphosphate Reductase.* Biochem. Soc. Trans., *5*: 606. Discusses enzyme in *E coli.*

Anderson, E. P.: (1973) *Nucleoside and Nucleotide Kinases. In* P. Boyer, ed.: *The Enzymes,* 3rd ed. Vol. 9, p. 49. Academic Press.

35 METABOLISM OF CHOLESTEROL AND PHOSPHOLIPIDS

CHOLESTEROL

The layman has been conditioned to regard cholesterol as a nasty word. We know better, having already seen its importance in membrane structure (p. 206). In addition, a large part of the flow of cholesterol is used for the production of bile salts. A smaller but exceedingly important use is as a precursor of several hormones (Chapter 38).

Cholesterol is synthesized in many tissues, especially the liver and the small bowel, and it is also obtained from the diet. Cholesterol in the diet induces extreme disquiet in many owing to an apparent association between serum cholesterol concentration and the incidence of coronary artery disease in some populations. Others regard the emphasis on dietary cholesterol as a misguided, even dangerous, diversion of attention from more manageable contributory factors to coronary artery disease.

Synthesis

The steroid skeleton is made by condensing six prenyl groups, otherwise known as isoprene groups:

$$\begin{array}{c} \quad\;\; CH_3 \\ \quad\;\; | \\ -CH_2-C{=}CH-CH_2- \end{array}$$

Many other polyprenyl compounds occur in nature. Animals, for example, make the side chain of ubiquinone (p. 382), and plants make a wide variety of compounds, including the carotenes, rubber, and the cyclic terpenoid compounds (camphor, pinene, limonene, and many others).

The precursor of all these compounds is 3-hydroxy-3-methylglutaryl coenzyme A (Fig. 35–1), which is formed from acetyl coenzyme A in both the cytosol and the mitochondria. It is used for acetoacetate synthesis in mitochondria (p. 433), where it also appears as an intermediate in leucine metabolism. In the cytosol, it is used for steroid synthesis. The rate-controlling step in cholesterol synthesis is the initial reduction of 3-hydroxy-3-methylglutaryl coenzyme A to

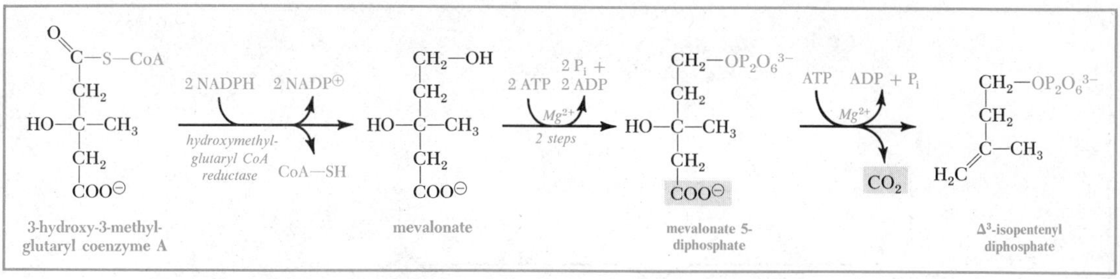

FIGURE 35–1 **Formation of isopentenyl diphosphate from 3-hydroxy-3-methylglutaryl coenzyme A.**

mevalonate on the endoplasmic reticulum. Mevalonate is then phosphorylated in three stages, with the third stage creating an intermediate that loses phosphate and is decarboxylated to create an isoprenol (as a pyrophosphate ester).

This **Δ^3-isopentenyl diphosphate** is the general donor of prenyl groups for the synthesis of all polyprenyl compounds. The acceptor molecules are Δ^2-isopentenyl compounds (Fig. 35–2). The initial acceptor, **dimethylallyl diphosphate,** is made by simple conversion of the Δ^3 compound to its Δ^2 isomer. An enzyme then catalyzes transfer of prenyl groups to it, first making the diprenyl (C_{10}) compound **geranyl diphosphate,** and then the triprenyl (C_{15}) compound **farnesyl diphosphate.** Farnesyl diphosphate is used for cholesterol synthesis; another transferase is present to add an additional prenyl group to make a C_{20} precursor of longer chain polyprenyl compounds. (This reaction is especially important to plants, which use the C_{20} compound to make C_{40} carotenes.)

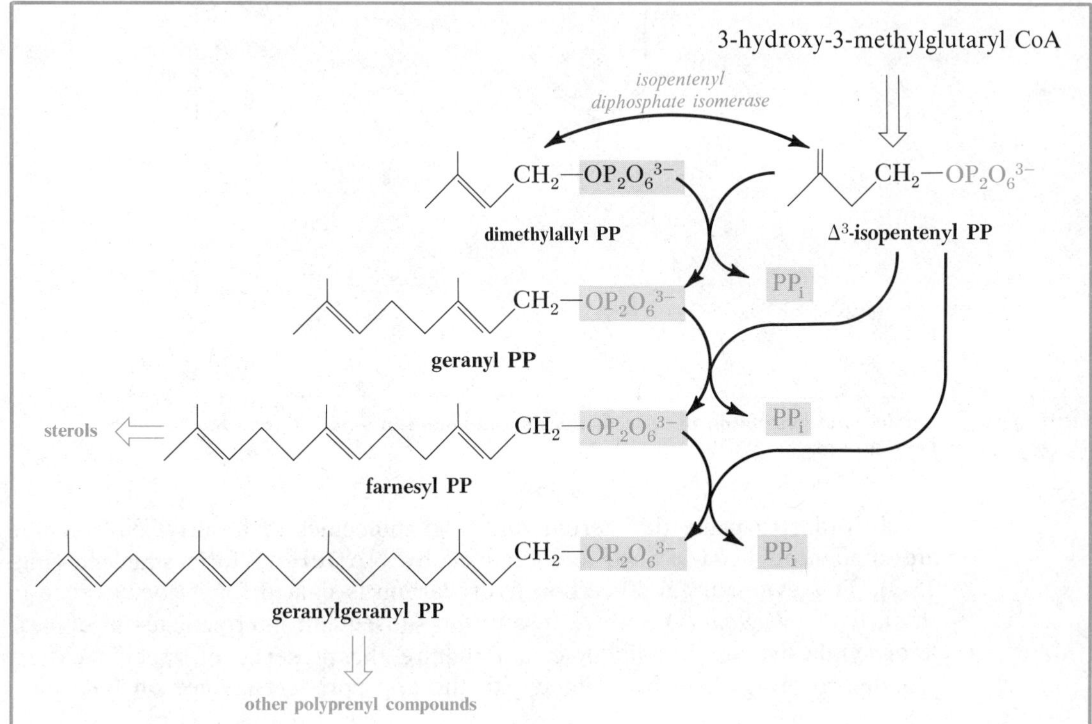

FIGURE 35–2 **Polyprenyl alcohols are formed as the pyrophosphate esters by successive addition of isopentenyl groups.**

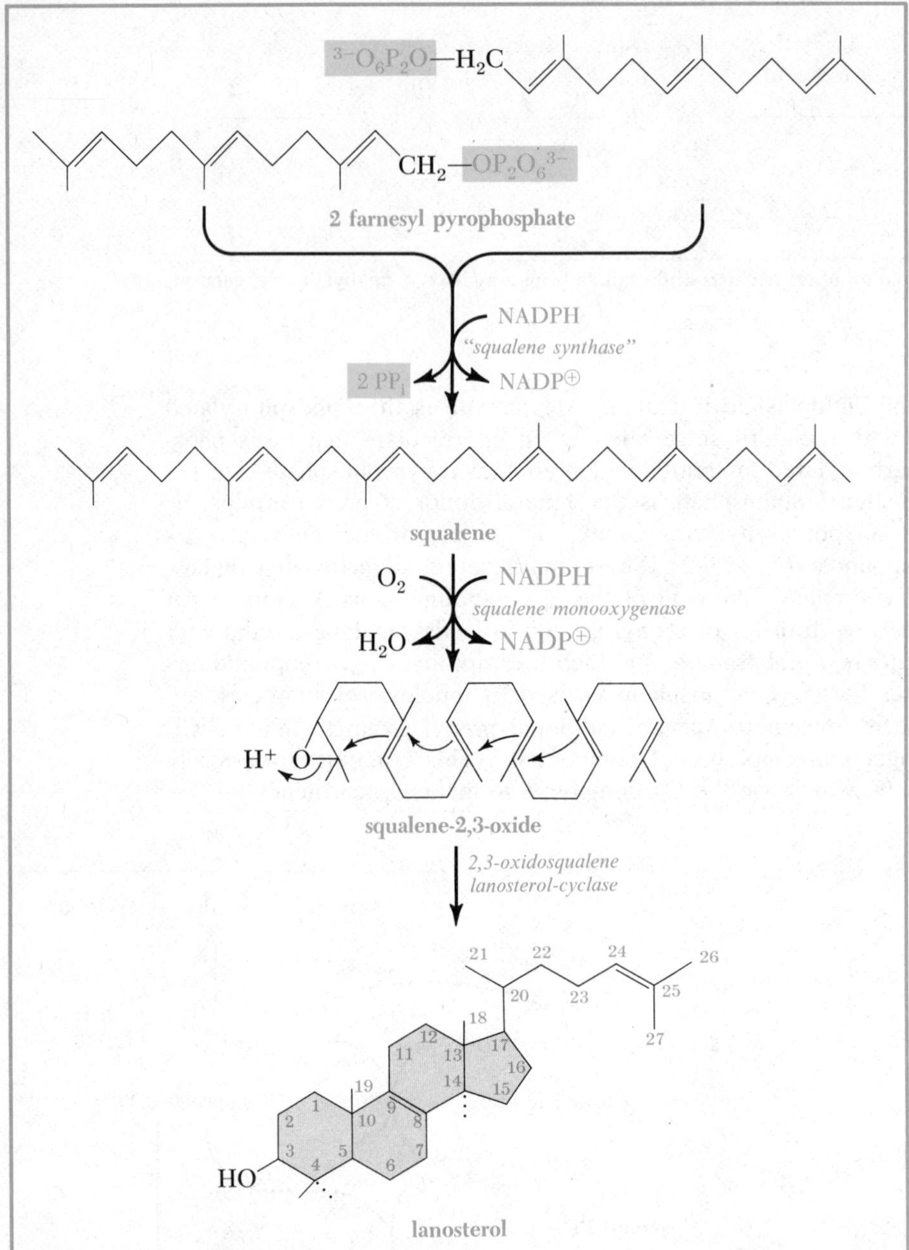

FIGURE 35–3 **Synthesis of lanosterol, the parent steroid compound in animals, from farnesyl groups.**

In order to make the steroid ring, two molecules of farnesyl diphosphate are condensed head-to-head and reduced by NADPH to form **squalene** (Fig. 35–3). This symmetrical 30-carbon hydrocarbon is named for its occurrence in shark liver oil (*Squalus* spp.)*. It, and the subsequent intermediates in cholesterol synthesis, are hydrophobic and require the presence of specific carrier proteins to bring them in contact with the appropriate enzymes on the endo-

*Konrad Bloch, who received the Nobel Prize for elucidation of the pathway of cholesterol synthesis, combined work and pleasure by traveling to Bimini for collection of shark livers.

plasmic reticulum. One of these is mixed-function oxidase that converts squalene to a cyclic oxide, which then undergoes a concerted internal condensation to form **lanosterol,** the parent steroid in animals.

Lanosterol differs from cholesterol in having three additional methyl groups, and in the location of double bonds. The methyl groups are removed by a succession of oxidations (Fig. 35–4) interspersed with relocation of the double bonds by appropriate reductions and oxidations (Fig. 35–5) to form cholesterol.

Regulation of Cholesterol Synthesis. The rate of cholesterol synthesis is controlled by altering the activity of the rate-limiting hydroxymethylglutaryl CoA reductase, and not all of the factors involved have been identified. The cholesterol concentration has a somewhat slow effect on the enzyme concentration, probably through repression of synthesis. There are also more rapid responses of unknown mechanism. Work currently being reported indicates the existence of an enzyme that catalyzes inactivation of the hydroxymethylglutaryl CoA reductase, and suggests control of both the inactivating enzyme and the reductase itself through phosphorylation and dephosphorylation. Whatever the mechanism, the synthesis of cholesterol is adjusted under normal circumstances so as to maintain concentrations near a constant level day-to-day, and

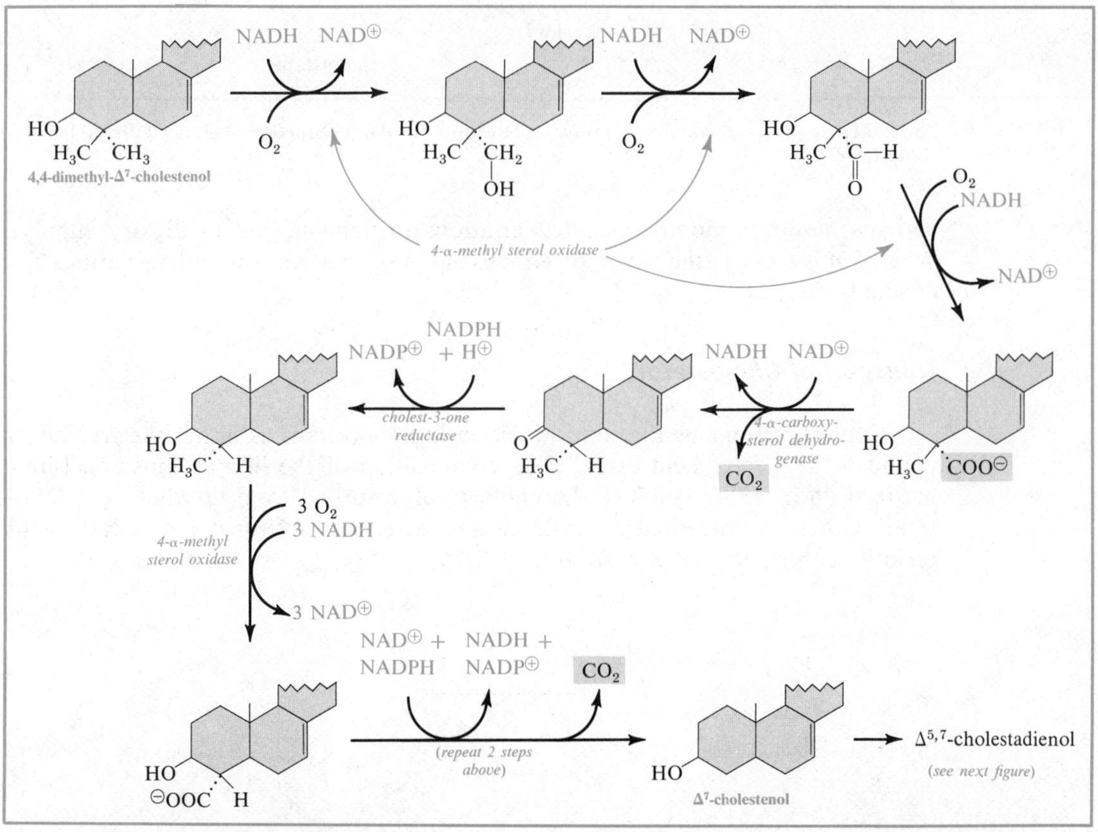

FIGURE 35–4 **The conversion of lanosterol to cholesterol involves removal of the two 4-methyl groups. A 4-α-methyl sterol oxidase catalyzes three successive oxidations of the α-methyl group (the group behind the plane of the ring as shown by the dashed bond) by O_2 (activated by NADH). The resultant carboxyl group is removed by an oxidative decarboxylation, forming a ketone. The remaining methyl group is placed in the α-configuration during the decarboxylation and can also be attacked by the sterol oxidase, after the ketone is reduced back to an alcohol.**

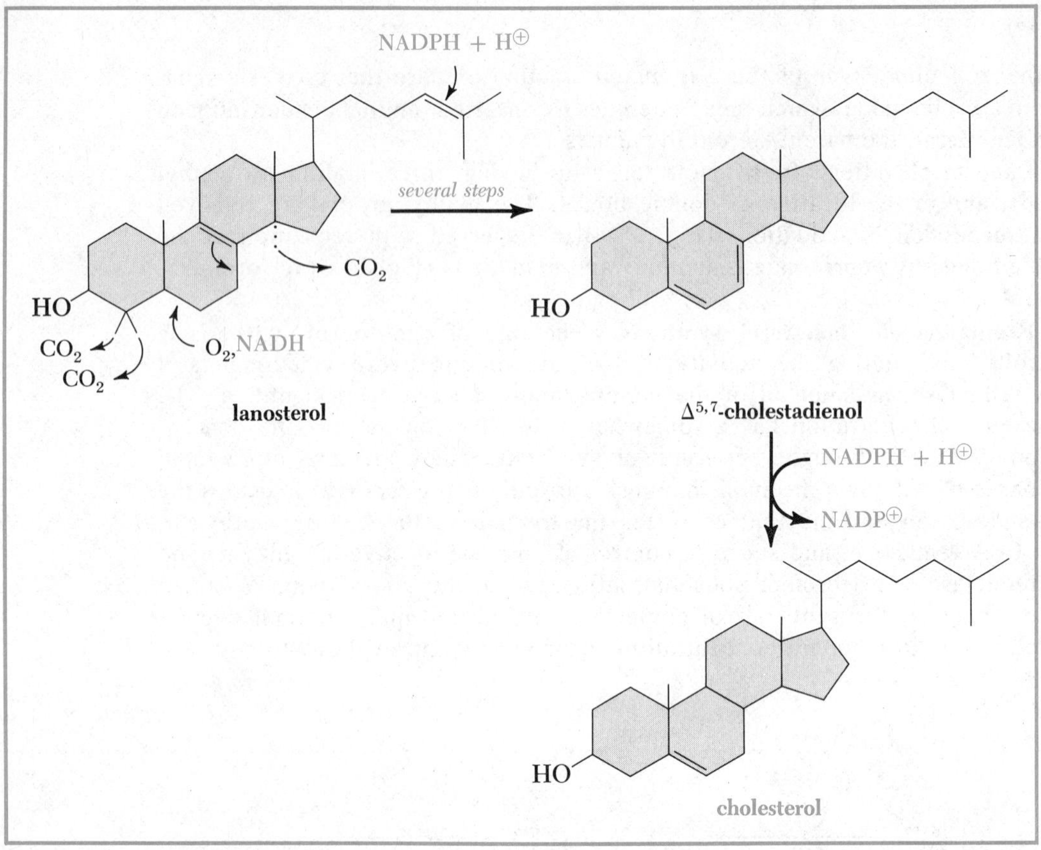

NADPH + $H^{\oplus}$

several steps

CO_2

HO

CO_2

CO_2

O_2,NADH

lanosterol

HO

$\Delta^{5,7}$-cholestadienol

NADPH + $H^{\oplus}$

$NADP^{\oplus}$

HO

cholesterol

FIGURE 35–5 **The final step in the conversion of lanosterol to cholesterol is a reduction of cholestadienol to cholesterol.**

perhaps month-to-month, despite variations in demand and in dietary supply. What is not clear is the range of effectiveness of this control among various individuals.

Transport of Cholesterol

Cholesterol moves through the blood in lipoproteins as either the free compound or as a fatty acid ester. (The composition of the lipoproteins was summarized on p. 538.) Much of the cholesterol absorbed from the diet or synthesized within the intestinal mucosal cells is esterified by transfer of a fatty acid residue, either 16:0 or 18:0, from acyl CoA:

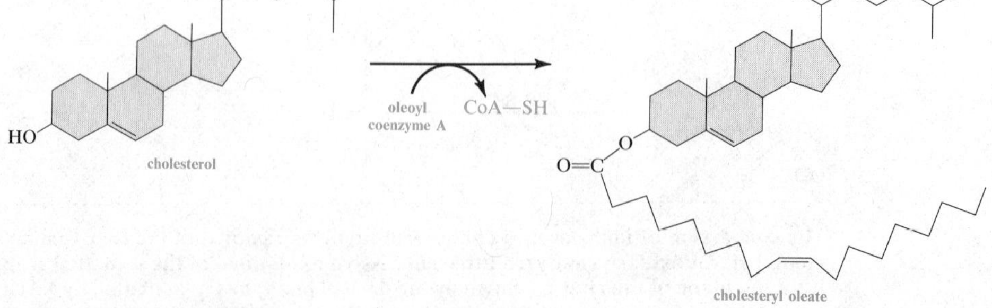

HO

cholesterol

oleoyl coenzyme A

CoA—SH

O

O=C

cholesteryl oleate

The ester is then secreted within chylomicrons.

Similarly, the liver produces very low-density lipoproteins that contain cholesteryl esters and free cholesterol. When the triglycerides of chylomicrons and very low-density lipoproteins are removed by the action of lipoprotein lipase in the peripheral tissues, the cholesteryl esters and cholesterol remain behind, along with many of the apolipoproteins, first in the form of intermediate-density lipoproteins, and then as low-density lipoproteins. (Many of the C apolipoproteins are lost when LDL is formed.) These low-density lipoproteins are the principal sources of cholesterol for peripheral tissues, since most of the cholesterol in the body is either obtained from the diet or made in the liver.

Specific receptor sites for the low-density lipoproteins are present in the plasma membranes of most cells. Upon attachment of the lipoprotein to the site, it is engulfed by pinocytosis and transferred to a lysosome within the cell, where its constituent proteins and lipids are hydrolyzed. After discharge of cholesterol from the lysosome, it may be transported by a carrier protein for incorporation within membrane structures, or any excess may be converted to an ester for storage. The appearance of cholesteryl esters has two regulatory effects — suppression of hydroxymethylglutaryl CoA reductase, as we mentioned before, and suppression of the formation of the receptor sites on the plasma membrane. The proteins constituting these sites have a half-life of *ca.* 20 hours, so the ability of a cell to use cholesterol from the blood will decline within a day after an internal rise in its concentration.

The high-density lipoproteins of the blood are in some way necessary for the prevention of an abnormal accumulation of cholesteryl esters in some peripheral tissues, although the mechanism of this effect is not clear. They are mainly composed of A-I and A-II apolipoproteins made by the liver. The A-I component is an activator of a **lecithin-cholesterol acyl transferase** in the blood. This enzyme catalyzes a transfer of an acyl group, usually 18:2, from phosphatidyl choline to cholesterol:

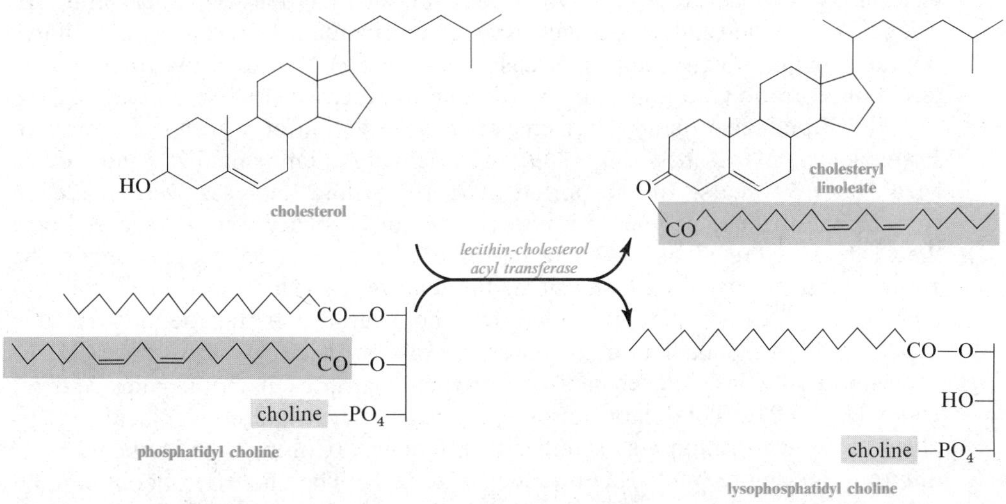

HO
cholesterol
CO—O
CO—O
choline —PO_4
phosphatidyl choline
lecithin-cholesterol acyl transferase
O
CO
cholesteryl linoleate
CO—O
HO
choline —PO_4
lysophosphatidyl choline

Cholesterol is esterified by this enzyme only when it is present in the high-density lipoprotein. It may be that the cholesteryl esters in VLDL are obtained only by transfer from HDL after the VLDL reaches the blood. (HDL also acts as a reservoir of the C-apolipoproteins, binding them as they are released during conversion of chylomicrons and VLDL to LDL, and in turn releasing them for incorporation into newly-formed chylomicrons and VLDL.)

Tangier* disease provides evidence for the importance of the high-density lipoproteins. This is a rare hereditary defect in the formation of the A apolipoproteins with a resultant absence of HDL, which causes accumulation of cholesteryl esters in smooth muscle outside the blood vessels, in reticuloendothelial cells, especially in the tonsils, and in the Schwann's cells around peripheral neurons, among others. (Orange-colored tonsils are among the pathognomonic signs.) The total cholesterol content of the patients' blood is abnormally low. All of this suggests that the high-density lipoproteins are either necessary for removal of excess cholesterol from the tissues, or in preventing its storage.

Cholesterol and Atherosclerosis. Evidence is accumulating for an association between changes in cholesterol metabolism and the incidence of atherosclerosis with resultant myocardial infarction. A principal feature of the atherosclerotic plaque on an arterial wall is the mass of cholesteryl esters within it. There may be two factors necessary for plaque formation — damage to the vessel wall and some disturbance in transport of cholesterol.

It has been known for some time that conditions causing elevated concentrations of low-density lipoproteins (LDL) are associated with atherosclerosis; a more recent development is even stronger evidence for an association of *decreased* concentrations of high-density lipoproteins with atherosclerosis, as manifested by myocardial infarction.

Hypercholesterolemias are hereditary conditions causing elevated LDL concentrations. One type results from the absence or deficiency of the LDL-receptor site protein in plasma membranes; another is caused by a failure in the mechanism for endocytosis of the attached LDL from the receptor site. A total deficiency of receptor sites causes atherosclerosis in infants, with myocardial infarction observed as early as 18 months after birth. The condition is an autosomal dominant, and the heterozygous parents, although only mildly affected, are prone to coronary artery disease in the middle years. On the other hand, individuals with a rare excessive production of high-density lipoproteins are claimed to be unusually long-lived with a low incidence of myocardial infarction, and patients with Tangier disease, who lack HDL, have low total cholesterol concentrations, but appear to be prone to atherosclerosis.

The Framingham study is a long-term observation of a group of people in Framingham, Mass., begun in 1949, and originally comprising 5,209 individuals aged 30 to 59 years. In the period 1968–1971, lipid analyses were made on some 2,500 of these people with no evidence of coronary artery disease. Since the analyses were made, 142 new cases of the disease have appeared in the group. A recent statistical analysis of the data reveals that the cholesterol content from high-density lipoproteins (HDL-cholesterol) is by far the most reliable indicator of the likelihood of later coronary disease — the more HDL-cholesterol, the less the chance of obstructed arteries in both males and females ($P < 0.01$). Total cholesterol as a single variant had no significance. Triglyceride concentration was significant in females. (Contrast the implications of different, and more limited data, cited on p. 607.) The cholesterol content contributed by low-density lipoprotein (LDL-cholesterol) was marginally significant as a risk factor in both sexes ($P < 0.05$). The sum of HDL-cholesterol, total cholesterol, and triglyceride concentrations was somewhat more reliable than HDL-cholesterol concentration alone, even though the latter two factors were

---

*Tangier Island, Virginia.

not separately reliable. These data would indeed raise questions about the desirability of emphasis on cholesterol intake to the exclusion of other factors. In other studies the HDL content appears to be significantly lower in the obese and diabetics. Some believe obesity and lack of exercise are the risk factors that deserve most attention. It must be emphasized that all of these studies are demonstrating association of conditions, not a causal mechanism.

The Bile Acids and Salts

Synthesis of the bile acids, **cholate** and **chenodeoxycholate,** begins with a hydroxylation at C-7 (Fig. 35–6), and this is the rate-limiting step. The concentration of chenodeoxycholate regulates this step as well as the synthesis of cholesterol, itself. (Remember that cholesterol synthesis is also subject to regulation by cholesterol concentration and other uncharacterized factors.) A series of additional oxygenase reactions then results in the introduction of another hydroxyl group, and the partial cleavage of the side chain. These are inter-

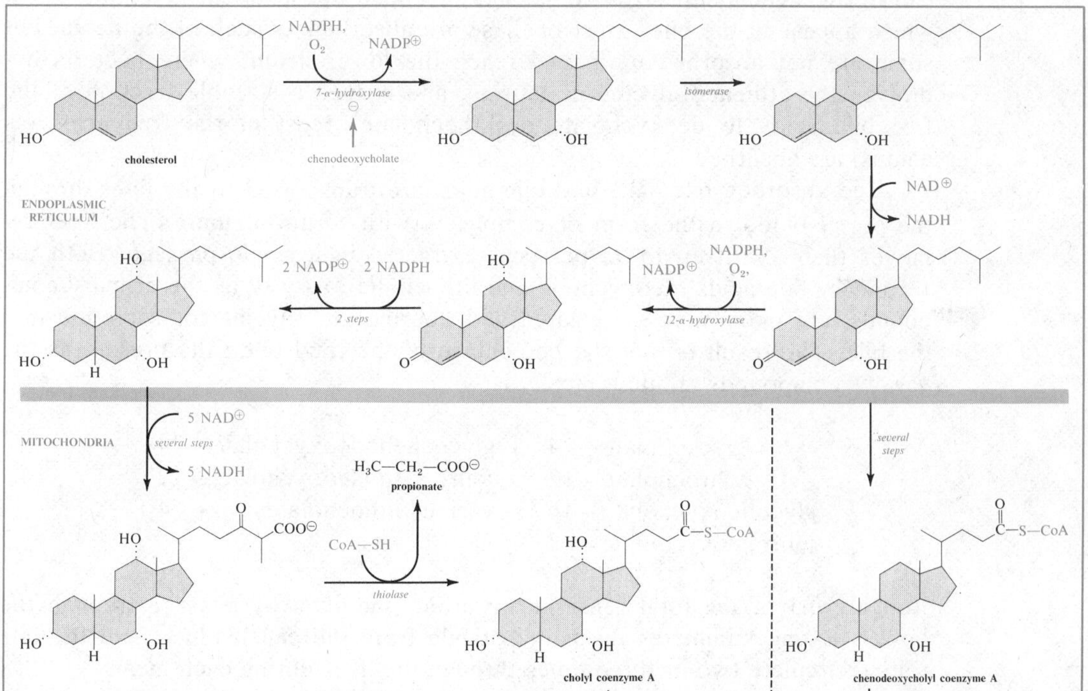

FIGURE 35–6 **The synthesis of bile acids from cholesterol involves hydroxylations by mixed function oxidases in the endoplasmic reticulum and oxidative cleavage of part of the side chain in mitochondria. Note the inversion of configuration of the 3-hydroxyl group by its oxidation by NAD to a ketone *(top right)*, which is later reduced by NADPH to the alcohol of opposite configuration *(center left)*. The formation of chenodeoxycholate (as the coenzyme A ester) is not shown in detail.**

spersed with an oxidation of the already existing β-hydroxyl group at C-3 to a carbonyl group, and its reduction to form an α-hydroxyl group. (This change brings all of the polar groups to the same face of the molecule, making it amphipathic.)

Finally, the steroid acids are converted to the coenzyme A derivatives, and the acyl group is transferred to either taurine or glycine making the bile salts (p. 313). Taurine is formed by decarboxylation of cysteic acid — which in turn is made by oxidation of cysteine.

Most of the cholesterol synthesized or absorbed from the diet each day is used to replace the bile acids and cholesterol lost in the feces. Bile contains a micelle of phosphatidylcholines, free cholesterol, and bile salts. The concentration of the bile salts in the bile is 8 to 12 times the concentration of cholesterol, but only 0.2 to 0.4 of the cholesterol in the intestinal lumen is absorbed, whereas nearly all of the bile salts, or their constituent bile acids, are absorbed. The result is that the steroid fecal losses are roughly equally partitioned between cholesterol and the bile salts (and their bacterial degradation products). Typically, an adult must replace 2 millimoles of total steroids per day, of which 1.8 millimoles will be synthesized as cholesterol, and 0.2 millimoles obtained from the diet (assuming the diet is not rich in cholesterol).

This enterohepatic circulation of the bile salts (Fig. 35–7) is strikingly efficient. The liver synthesizes conjugates of cholic and chenodeoxycholic acids, which appear in the bile. Most of these are absorbed as such in the ileum, but some are not absorbed until they reach the lower ileum, where bacteria hydrolyze the attached glycine or taurine, and reduce a sizeable fraction of the free bile acids to deoxycholate or lithocholate. Most of the free acids (as anions) are absorbed.

The absorbed bile salts and bile acids are transported to the liver through the portal blood in the form of complexes with serum albumin. The liver removes them, with up to 95 per cent extracted in a single passage. Both the secondary bile acids (deoxycholic and lithocholic), as well as the primary compounds, are once more conjugated with taurine and glycine for secretion into the bile. The result is that the percentage of the total bile salts present as the various compounds in bile typically is:

glycocholate – 24	glycochenodeoxycholate – 24
taurocholate – 12	taurochenodeoxycholate – 12
glycodeoxycholate – 16	various lithocholates – 4
taurodeoxycholate – 8	

Roughly 0.3 of the total represents steroids modified by passage through the bowel. Eating stimulates discharge of bile from the gall bladder, and the bile salts recirculate two or three times through the liver during each meal.

Chenodeoxycholate and Gallstones. The formation of cholesterol stones is an indication that the bile remains supersaturated to cholesterol for extended periods of time. (Supersaturation after an overnight fast is common.) Administration of chenodeoxycholic acid, but not cholic acid, over long periods (several months) may cause cholesterol stones to disappear. The reason is that chenodeoxycholic acid specifically diminishes the activity of hydroxymethylglutaryl CoA reductase, and it is therefore an important source of feedback control over cholesterol synthesis. With less being synthesized, there is less export in the bile.

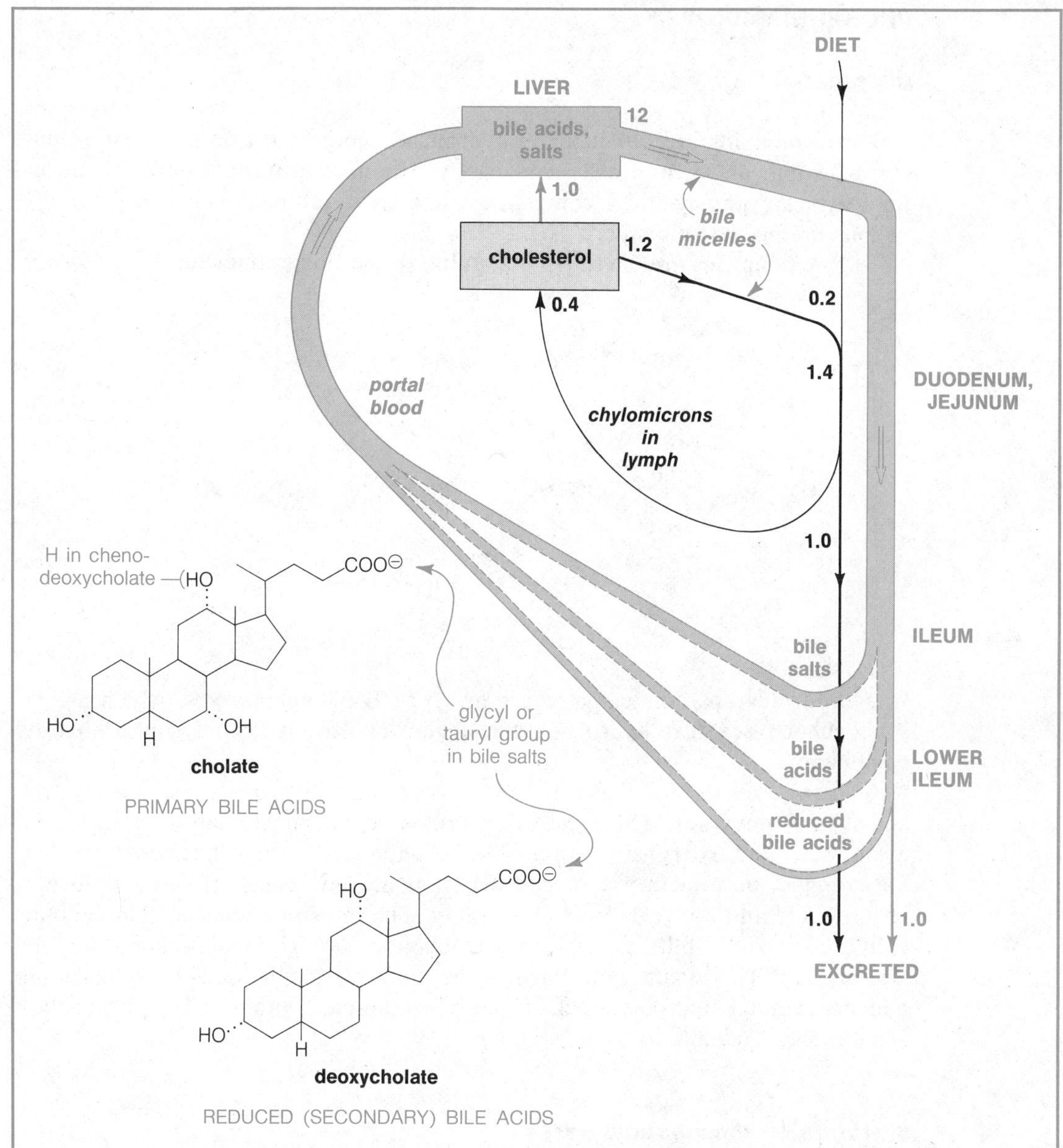

FIGURE 35–7 **Recirculation of bile salts from the small bowel to the liver is much greater than the recirculation of cholesterol. The numbers indicate approximate millimoles per day. Cholesterol circulation is shown in *black*, bile salts in *blue*.**

The importance of this regulatory mechanism has been demonstrated in another way. The bile flow of two patients with intractable hereditary hypercholesterolemia was diverted outside the body. The objective was to cause a large steroid loss and diminish the body pool of cholesterol. The post-operative loss indeed amounted to 100 millimoles per month, a six-fold increase over their previous value. Despite this increase, the operation was a failure because the *de novo* synthesis of cholesterol promptly increased in each patient, owing to the loss of feedback inhibition, so as to maintain the abnormally high concentrations of cholesterol and its esters within the blood.

PHOSPHOLIPIDS

Degradation

Phospholipids are attacked by hydrolases, both in the small intestine during digestion, and within cells as a part of the normal turnover. We shall deal here with the phosphatidyl compounds, and treat sphingomyelins and the glycolipids in the next chapter.

Phospholipases are classified according to the bond attacked:

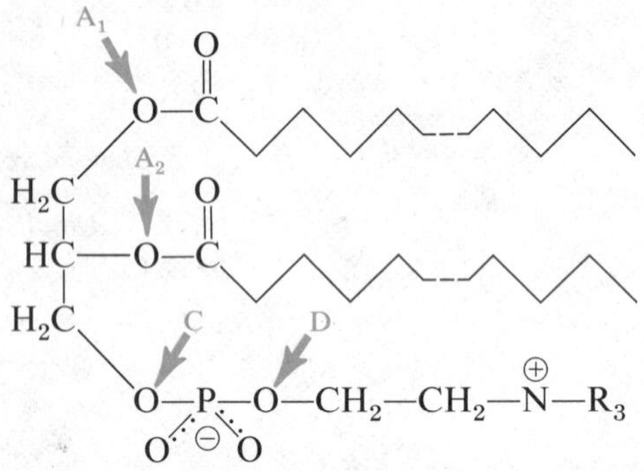

phosphatidyl base

In addition to those indicated, there are type B phospholipases, which are lysophospholipases that hydrolyze the single acyl group from lysophosphatidyl compounds.

The best known of these hydrolases in mammalian tissues are the type A and type B enzymes. The pancreas secretes an A_2 phospholipase as a proenzyme, activated by trypsin, and a lysophospholipase. When these enzymes act, for example, on a dietary phosphatidyl choline, they release the two molecules of fatty acid and a glycerylphosphocholine. The intestinal mucosa also contains a hydrolase that splits glycerylphosphocholine into glycerol-3-phosphate and free choline. The result is cleavage of the dietary phospholipids into their components. Similar hydrolases act on the phospholipids within cells, making their constituents available for re-use.

Synthesis of Phosphatidyl Bases

We already discussed the complete synthesis of the phosphatidyl bases, beginning with free choline or ethanolamine (p. 602), which are converted to the cytidine diphosphate derivatives, which in turn act as phospho-base donors for phospholipid synthesis in the same way that UDP-glucose acts as a glucosyl donor for polysaccharide synthesis. That is, the phosphorylated base is transferred to a 1,2-diglyceride, creating a phosphatidyl base and leaving CMP as the other product.

However, this transfer does not complete the synthesis for all purposes. For example, membrane phosphatidylcholines are rich in arachidonoyl groups (20:4) on C-2. We have already seen that the major diglycerides have mainly 18:1 on C-2 (p. 537). The same diglycerides are incorporated into phospholipids, and something else must happen to change the phospholipid fatty acid composition before they are incorporated into membranes.

The intracellular phospholipase A_2 removes the 18:1 acyl group from newly synthesized phosphatidyl cholines. Another enzyme is present that selectively transfers a 20:4 group from arachidonoyl coenzyme A, creating a phosphatidyl choline of proper composition:

arachidonoyl coenzyme A

palmitoyl
oleoyl
P-choline
H_2O
oleate
phospholipase A_2
palmitoyl
HC—OH
P-choline
CoA—SH
arachidonoyl transferase
palmitoyl
arachidonoyl
P-choline

phosphatidylcholine (*newly synthesized*)
lysophosphatidylcholine
phosphatidylcholine (*modified for membranes*)

Cardiolipins, the diphosphatidylglycerols, are made in mitochondria by a sequence involving the formation of CDP-diglycerides (Fig. 35–8). This compound is also a phosphatidyl donor for the synthesis of **phosphatidyl inositol.**

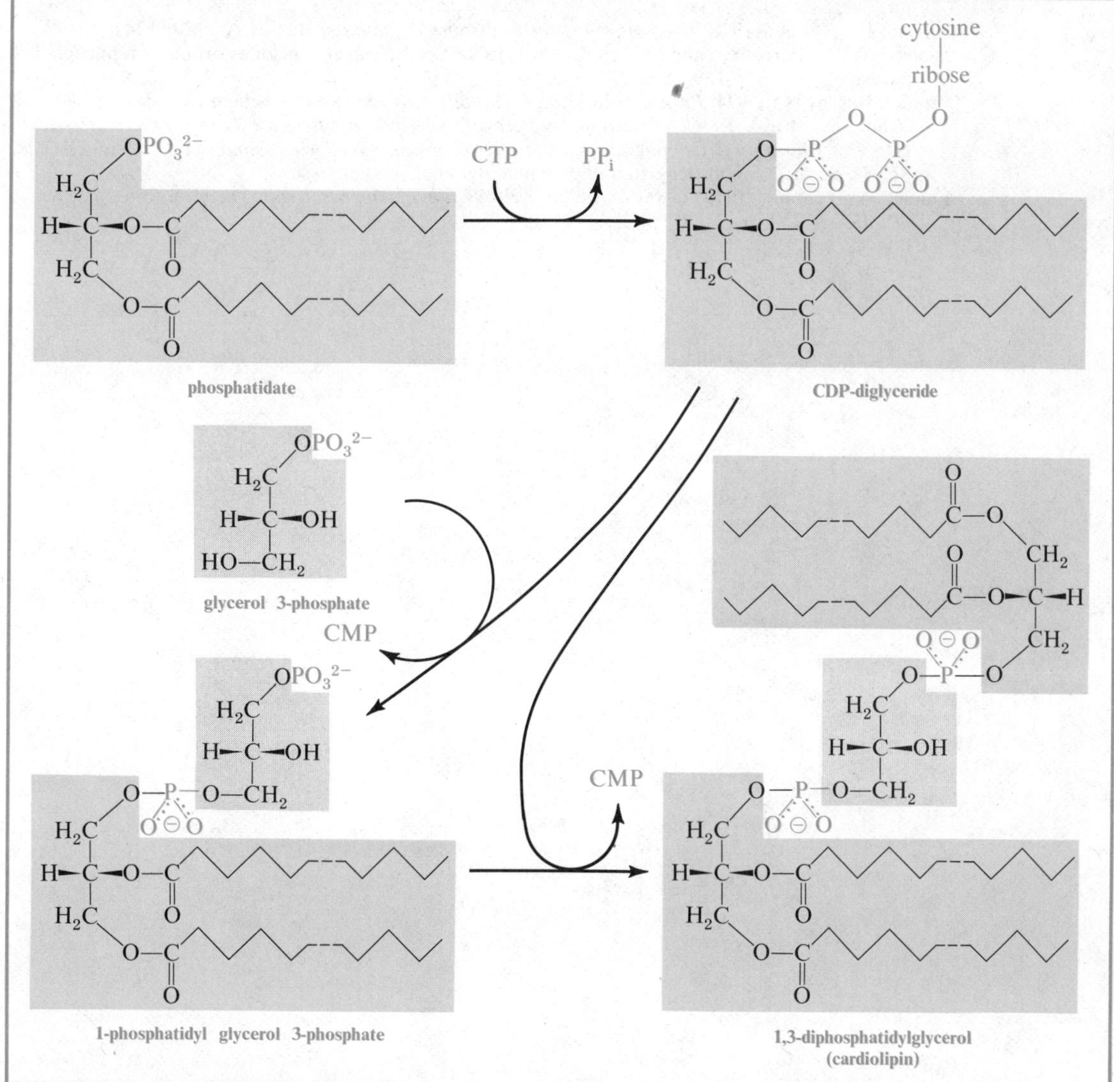

FIGURE 35–8 **Phosphatidate groups are converted to the CDP-diglyceride for transfer to a glycerol backbone, generating mono- or di-phosphatidyl glycerol derivatives.**

FURTHER READING

Beytia, E. D., and J. W. Porter: (1976) *Biochemistry of Polyisoprenoid Biosynthesis.* Annu. Rev. Biochem., *45*: 113.

Gaylor, J. L., Y. Miyaki, and T. Yamano: (1975) *Stoichiometry of 4-Methyl Sterol Oxidase of Rat Liver Microsomes.* J. Biol. Chem., *250*: 7159.

Jackson, R. L., J. D. Morrisett, and A. M. Gotto, Jr.: (1976) *Lipoprotein Structure and Metabolism.* Physiol. Rev., *56*: 259.

Goldstein, J. L., and M. S. Brown: (1977) *The Low-density Lipoprotein Pathway and Its Relation to Atherosclerosis.* Annu. Rev. Biochem., *46*: 897.

Manning, G. W., and M. D. Haust, eds.: (1977) *Atherosclerosis.* Adv. Exptl. Biol. Med., Vol. 82. Lengthy symposium includes articles on individual serum lipoproteins.

Gordon, T., et al.: (1977) *High Density Lipoprotein as a Protective Factor Against Coronary Heart Disease. The Framingham Study.* Am. J. Med., *62*: 707.

Havel, R. J.: (1977) *Classification of the Hyperlipidemias.* Annu. Rev. Med., *28*: 195.

Kritchevsky, D., ed.: (1977) *Symposium on Disorders of Cholesterol Metabolism.* Am. J. Clin. Nutr., *30*: 965.

Assmann, G., et al.: (1977) *The Lipoprotein Abnormality in Tangier Disease.* J. Clin. Invest., *59*: 565.

Blum, C. H., et al. (1977) *High-density Lipoprotein Metabolism in Man.* J. Clin. Invest. *60*: 795.

Sabine, J. R.: (1977) *Cholesterol.* Dekker.

Paumgartner, G., ed.: (1977) *Bile Acids.* Clin. Gastroenterol., *6*: 1. Valuable summaries of sterol turnover.

Thistle, J. L., et al.: (1978) *Chenotherapy for Gallstone Dissolution.* J.A.M.A., *239*: 1041.

Paoletti, R., G. Porcellati, and G. Jacini, eds.: (1976) *Lipids.* Raven. Includes articles on phospholipases.

van den Bosch, H.: (1974) *Phosphoglyceride Metabolism.* Annu. Rev. Biochem., *43*: 243.

Yamashita, S., et al.: (1975) *Separation of 1-Acyl Glycerophosphate Acyl Transferase and 1-Acyl Glycerol Phosphoryl Choline Acyl Transferase of Rat Liver Microsomes.* Proc. Natl. Acad. Sci. U. S. A., *72*: 600. Key step in phosphatidyl choline synthesis.

Kannel, W. B., et al.: (1979) *Cholesterol in the Prediction of Atherosclerotic Disease.* Ann. Int. Med., *90*: 85.

36 GLYCOPROTEIN AND GLYCOLIPID METABOLISM

PROTEOGLYCANS

The proteoglycans consist of heteropolysaccharide chains laid down on a specific polypeptide (p. 181).

Carbohydrate Precursors

The carbohydrate groups found in the structural components are all obtained by transfer from nucleotide derivatives such as UDP-glucose, which we have already seen as the precursor of glycogen. The nature of the nucleotide carrier to which the carbohydrate is attached varies with the kind of carbohydrate involved.

Hexosamines. The amino group of the hexosamines is obtained by transfer of the amide group of glutamine to fructose 6-phosphate (Fig. 36–1). The ketose simultaneously isomerizes to the aldose, creating **glucosamine 6-phosphate, which is the progenitor of all other hexosamine residues.** The amino group is acetylated, using acetyl coenzyme A, and the phosphate is moved from C-6 to C-1 by a specific mutase. Transfer of a uridyl group from UTP then forms UDP-N-acetyl-D-glucosamine, from which acetylglucosaminyl groups can be transferred to form polysaccharide chains. The final part of the synthetic process is analogous to steps in glycogen synthesis:

glycogen: glucose 6-P → glucose 1-P → UDP-glucose → glycogen
heteropolysaccharide: GlcNAc 6-P → GlcNAc 1-P → UDP-GlcNAc → polysacch.

Other hexosamine groups can be created by epimerization. For example, a galactosamine residue (as the UDP-N-acetyl derivative) is made from a glucosamine residue by inverting the configuration at C-4. The epimerase requires NAD as a cofactor, and it is really an oxidoreductase. (Contrast ribulose-5-phosphate epimerase, p. 531.) That is, it uses NAD to oxidize the hydroxyl group at C-4 of the glucosamine residue to a carbonyl group, but this intermediate keto compound and the NADH produced are not released. Instead the

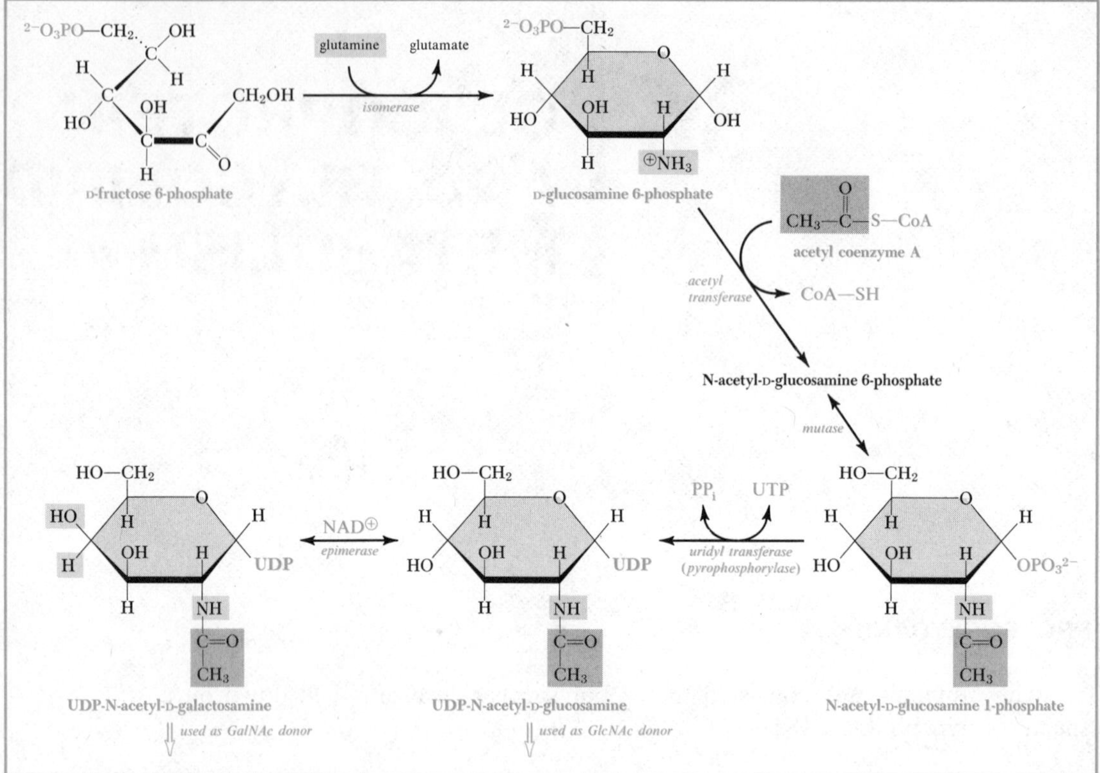

FIGURE 36–1 **The synthesis of hexosamines as the UDP N-acetyl derivatives. Enzyme types, not full names, are indicated.**

electrons are transferred back to re-form NAD and a hydroxyl group, which may be in either configuration:

$$\text{H-C(OH)} + \text{NAD}^{\oplus} \longrightarrow \text{—C(=O)—} + \text{NADH} \rightarrow \begin{cases} \text{H-C-OH} + \text{NAD}^{\oplus} \\ \text{HO-C-H} + \text{NAD}^{\oplus} \end{cases}$$

Uronates and Other Precursors. Another epimerase with a similar mechanism acts directly on UDP-glucose to form UDP-galactose as a donor of galactosyl groups (Fig. 36–2).

UDP-glucuronate is formed from UDP-glucose by an enzyme catalyzing two successive oxidations of C-6 without releasing the intermediate 6-aldehyde compound. Still another NAD-coupled epimerase inverts the configuration of C-5 in UDP-glucuronate to form UDP-L-iduronate.

A small amount of UDP-glucuronate is decarboxylated to form UDP-xylose.

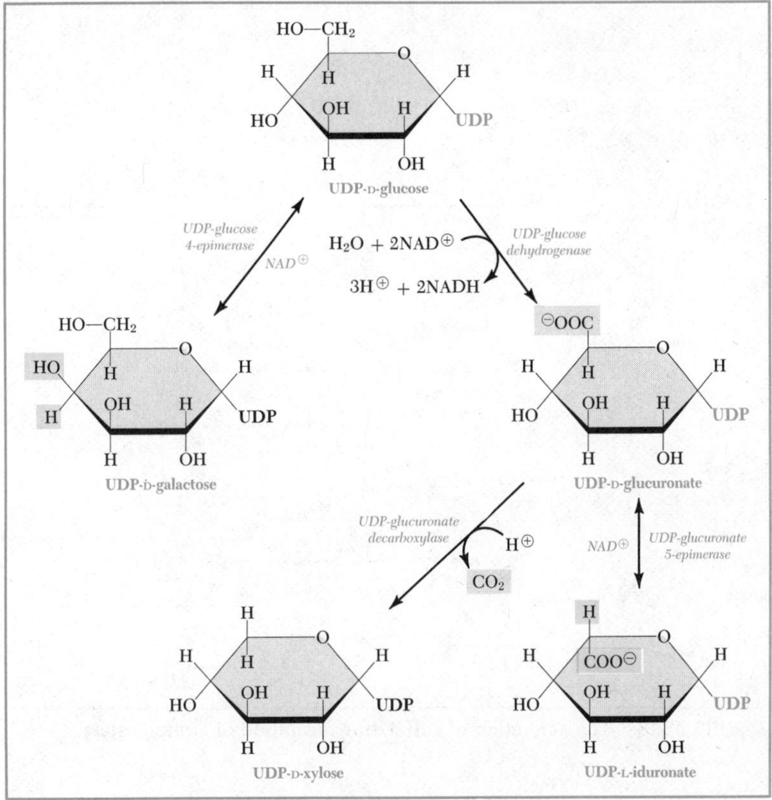

FIGURE 36–2 **The formation of glucuronic acid, iduronic acid, galactose, and xylose residues as UDP derivatives.**

Sulfate Groups

The sulfate ester groups found in some heteropolysaccharides (pp. 181, 185) and other compounds are obtained by transfer from **3′-phosphoadenosine 5′-phosphosulfate** (Fig. 36–3). This compound is a high-energy sulfate in the same sense that pyrophosphates are high-energy phosphates; it is a mixed anhydride of sulfuric and phosphoric acids, esterified with a phosphoadenosine group. The source of the sulfate group is inorganic sulfate, which is obtained mostly by metabolism of the sulfur in cysteine and methionine residues, but also from the diet as such. An enzyme cleaves pyrophosphate from ATP with sulfate ions, forming adenosine 5′-phosphosulfate. The free energy of hydrolysis of the phosphosulfate anhydride bond is more negative than the free energy of hydrolysis of a pyrophosphate bond, so that equilibrium is against this reaction. However, the product is removed by a transfer of phosphate from a second molecule of ATP to the 3′-hydroxyl group of the adenosine moiety. Two high-energy phosphate bonds therefore are cleaved to make one molecule of the sulfate donor.

Various transferases catalyze transfer of the sulfate group to specific acceptors. Adenosine 3′,5′-bisphosphate is the other product, and a phosphatase removes the 3′-group, leaving 5′-AMP to be re-utilized in the adenine nucleotide pool.

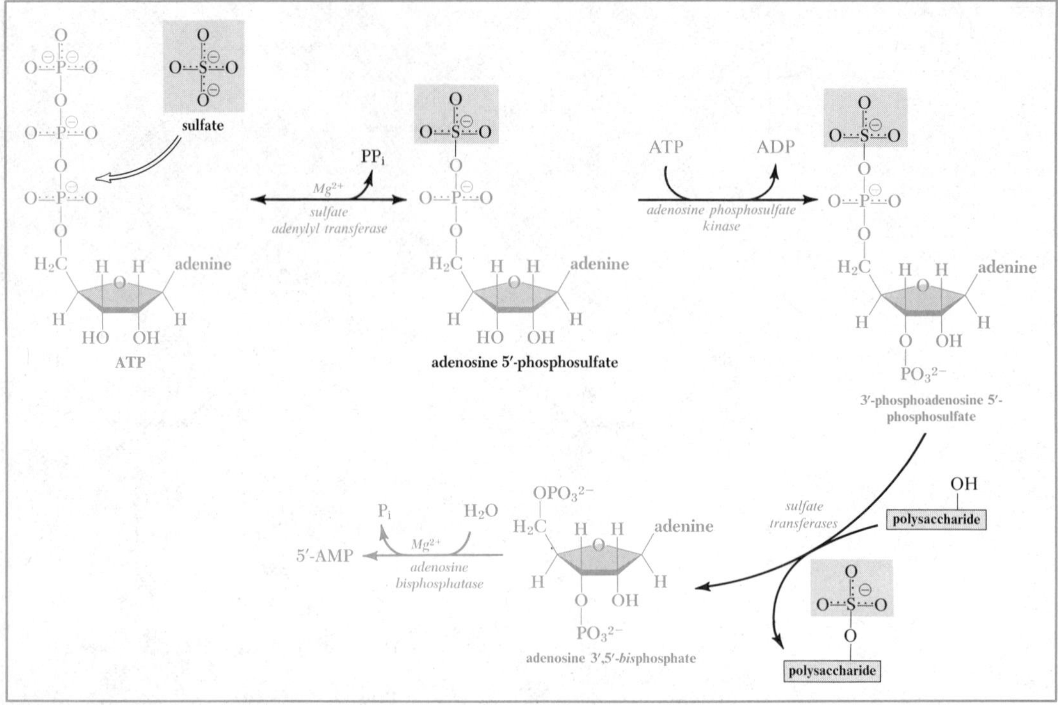

FIGURE 36–3 The activation of sulfate for formation of sulfate esters.

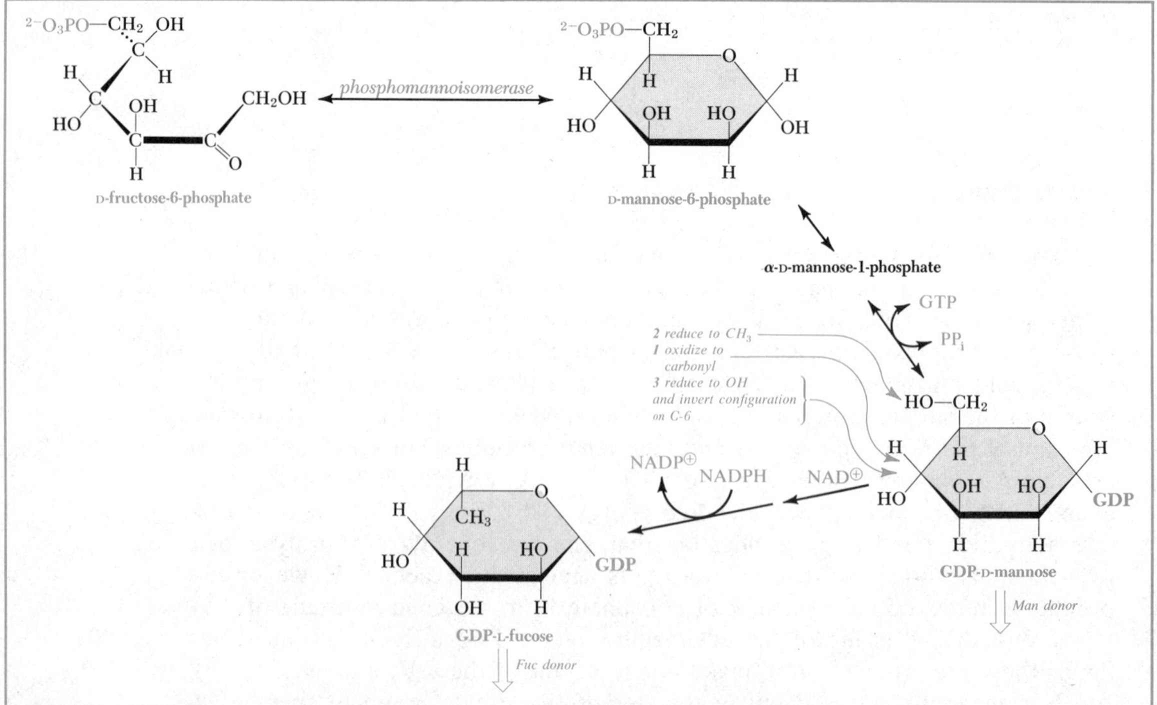

FIGURE 36–4 The formation of mannose and L-fucose residues as GDP derivatives. Only a summary of the steps for conversion of GDP-mannose to GDP-fucose is given.

Synthesis of Proteoglycans

We have seen that proteoglycans contain various repeating units, but attachment to the polypeptide chain is frequently made by the sequence -Gal-Gal-Xyl-Ser (p. 182). The factors are not known that determine which serine residues are the acceptors. The xylose and galactose residues are successively transferred by specific enzymes from UDP-xylose and UDP-galactose. In these transfers, as in all of the transfers mentioned in this chapter, there is an inversion of configuration, so the α-xylosyl group in UDP-xylose and the α-galactosyl group in UDP-galactose become β-xylosyl and β-galactosyl groups in the polysaccharide. We know that such inversions of configuration are not obligatory for transfer, because they do not occur in the synthesis of glycogen from UDP-glucose. The desired properties are obviously obtained in the heteropolysaccharides by incorporating β-hexosyl units, so the enzymes are built to create this configuration when transfer occurs.

Once the Gal-Gal-Xyl sequence is in place, construction of the long heteropolysaccharides begins by alternate transfer of uronic acid (glucuronic or iduronic) and N-acetylgalactosamine residues in the case of chondroitin and dermatan chains, or galactose and N-acetylglucosamine residues in the case of keratan chains, each from the corresponding UDP-derivative. Sulfate groups are added to the growing heteropolysaccharide by the action of specific transferases.

GLYCOPROTEINS

Additional Precursors

The diverse carbohydrate side chains of glycoproteins include residues other than those found in the proteoglycans, and this is especially true of those side chains attached to asparagine residues (p. 187). The **mannose** residues originate from fructose 6-phosphate much in the same way as the residues of glucose in UDP glucose do (Fig. 36–4), except that the isomerization of fructose 6-phosphate creates a different configuration on C-2, and GTP is used as a nucleotidyl donor instead of UTP:

fructose 6-P → glucose 6-P → glucose 1-P → UDP-glucose
fructose 6-P → mannose 6-P → mannose 1-P → GDP-mannose.

The **fucose** residues sometimes used to cap the polysaccharide branches are made from GDP mannose by a series of reactions involving oxidations and reductions at C-5 and C-6, also summarized in Figure 36–4.

Some oligosaccharide side chains have one or more **acetylneuraminate** residues attached, often as the terminal group. This group is made by combination of the pyruvate moiety of phospho-*enol*-pyruvate with N-acetylmannosamine 6-phosphate (Fig. 36–5). The N-acetylmannosamine is made by an unusual reaction in which UDP is cleaved from UDP-N-acetylglucosamine with a simultaneous inversion of configuration on C-2. The reaction is made irreversible by the loss of UDP. (It is an interesting exercise to speculate what might go wrong if the mannosamine residue were made by a reversible epimerization of a glucosamine residue.)

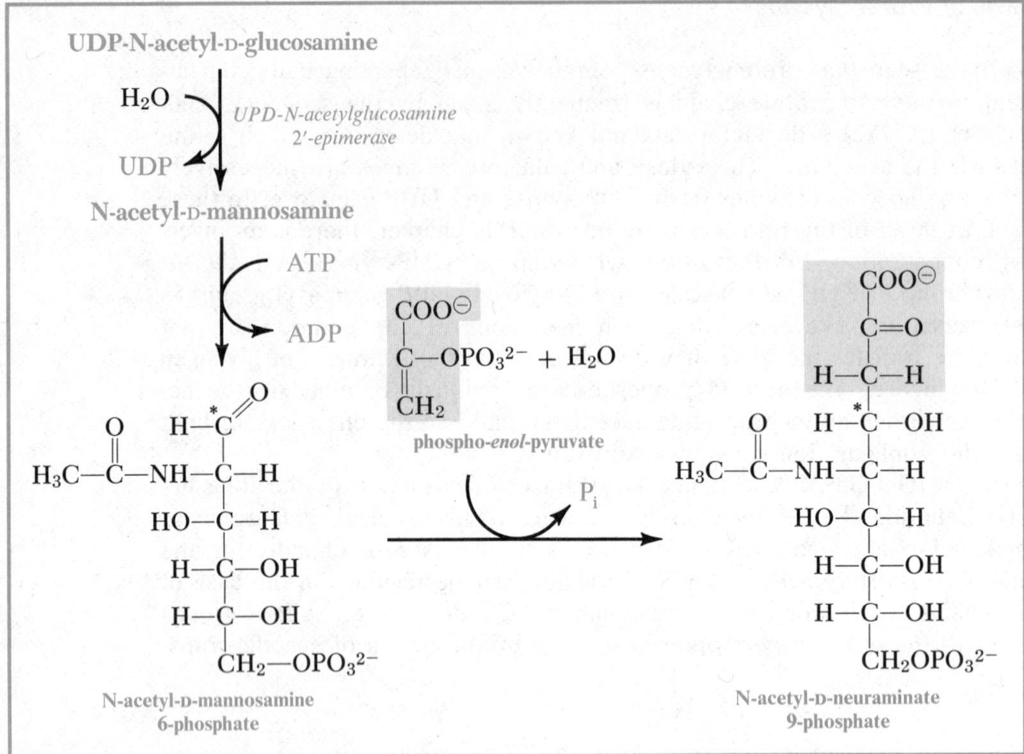

FIGURE 36–5 **Formation of neuraminic acid residues as the N-acetyl phosphate ester.**

Once the N-acetylneuraminate residue is made, it is activated for transfer, first by hydrolysis of the phosphate ester group, and then by attachment of a CMP residue:

H_2O P_i

N-acetyl-D-neuraminate 9-phosphate ⟶ N-acetyl-D-neuraminate

CTP

PP_i

cytosine

CMP-N-acetyl-D-neuraminate

Here we see a variation on the usual theme: the nucleotide residue is placed directly on the sugar, not on a phosphate, creating a nucleosidemonophosphate sugar, rather than the usual nucleosidediphosphate sugar.

Synthesis of Glycoproteins

Those glycoproteins containing saccharide groups attached to serine or threonine residues are made by direct transfer of the appropriate carbohydrate groups from the UDP-derivatives, frequently UDP-glucose or UDP-galactose. Those that have the saccharide groups attached to asparagine residues are made by another route, in which the core saccharide is first put together in combination with dolichol. Dolichol is a long-chain polyprenyl alcohol, with 80 to 100 carbon atoms:

CH_2—OH

a dolichol (C_{80}-C_{100})

Dolichol is converted to its phosphate ester for use as a sugar carrier by transfer of phosphate from CTP (Fig. 36–6). As we saw, many of the mannose-

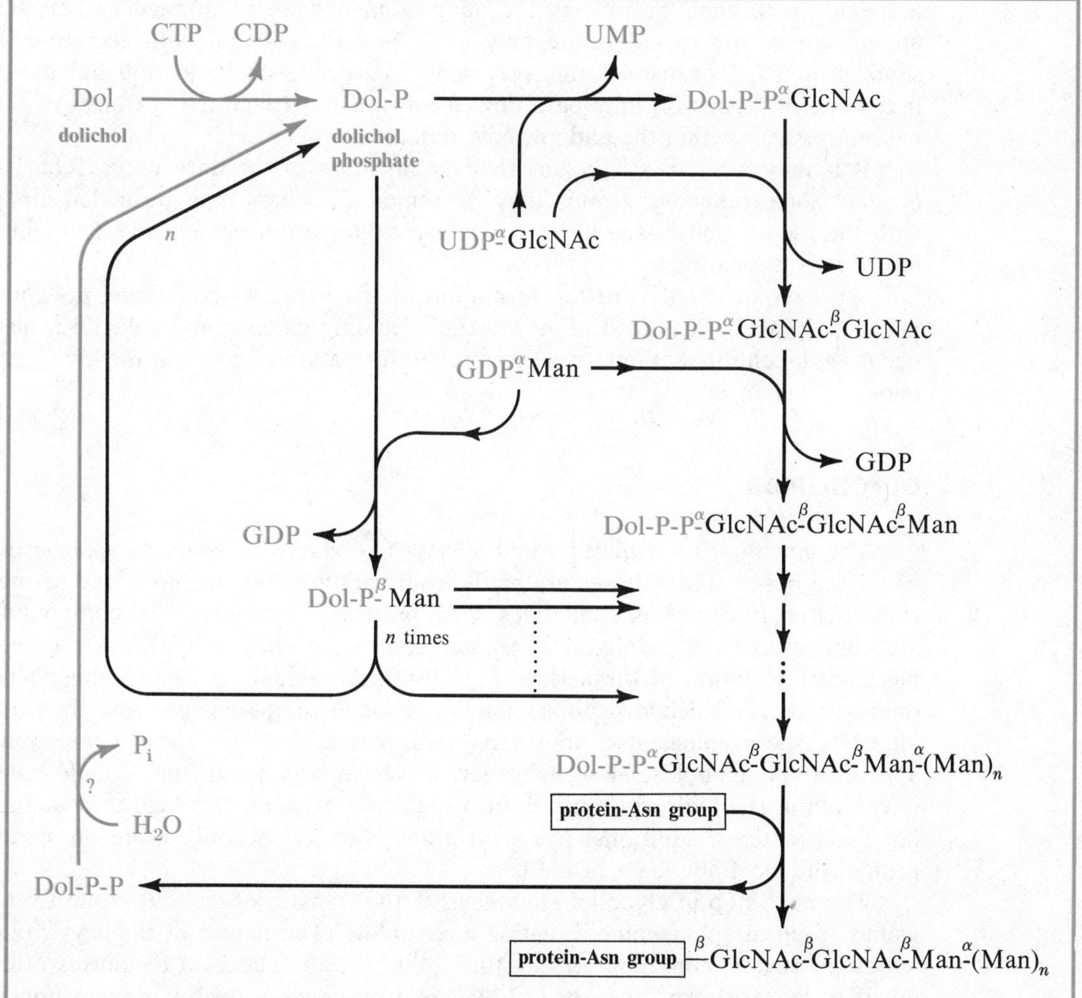

FIGURE 36–6 **Construction of heterooligosaccharide chains on dolichol for transfer to asparagine residues in glycoproteins. Note the different transfers from UDP-GlcNAc at the top. In one, UMP is a product; in the other UDP is a product.**

rich glycoprotein side chains are attached to proteins through Man-GlcNAc-GlcNAc-Asn, and it is this sequence that is constructed on dolichol phosphate. The first step is a transfer of a GlcNAc-phosphate group — not the GlcNAc alone — from UDP-GlcNAc. This unusual transfer leaves UMP as a product, and more importantly, it does not disturb the α-configuration of the attached hexosamine.

The initial sequence is then completed by transfer of another GlcNAc group from UDP-GlcNAc, and a mannose residue from GDP-mannose. These typical transfers both involve inversion of configuration to the β-isomers. Then begins a series of transfers of mannose residues to complete the core, but the donor for these transfers is not GDP-mannose itself, but mannosyl esters of dolichol phosphate. These esters are made by transfer of mannose from GDP-mannose, with inversion of configuration; when the mannose residues are again transferred from Dol-P-Man, there is another inversion of configuration. The result of this molecular game of musical chairs is that the final mannose residues in the polysaccharide have the α-configuration, which presumably conveys some benefit yet to be rationalized.

After the chain is put together, it is transferred from dolichol phosphate to a specific asparagine residue in the apoprotein. As we mentioned earlier, the attachment of the carbohydrate may have something to do with secretion of some proteins. The use of the very long hydrocarbon tail in dolichol as an initial site of formation may be a device for somehow facilitating transit of the nascent protein within the endoplasmic reticulum.

It is only after transit begins that modification of the core polysaccharide occurs. Some mannose groups may be removed; others may be added along with the fucose and N-acetylneuraminate residues often present in asparagine-linked polysaccharides.

We have stressed here the formation of the mannose-containing dolichol-linked polysaccharides, but other sugars, including glucose and galactose, also occur as dolichol-linked intermediates in the formation of some of the glycoproteins.

GLYCOLIPIDS

The glycolipids are built by adding acyl and glycosyl groups to **sphingosine** or related bases. These bases are made from palmitoyl coenzyme A and serine. (Stearoyl coenzyme A is sometimes used, but less commonly.) The compounds are condensed by a pyridoxal phosphate-containing enzyme with a simultaneous decarboxylation of the serine (Fig. 36–7). The resultant 3-ketosphinganine (more properly, 3-dehydrosphinganine) is reduced to **sphinganine** and then oxidized to **sphingenine,** also known as sphingosine. Both of these compounds appear in sphingolipids, along with derivatives in which a further double bond is created in the hydrocarbon tail, or a molecule of water is added to a double bond to create an additional hydroxyl group, but let us concentrate on sphingenine, the most abundant of the bases.

The next step in glycolipid synthesis is the transfer of an acyl group to the amino group of sphingenine, creating a **ceramide.** (The nature of the acyl group varies in tissues; commonly it is either 24:0 or 18:0. The 24:0 residue is often modified by oxidation to form a 2-hydroxyl group or a central double bond.) The ceramide is in turn a precursor of the **sphingomyelins,** which are made by

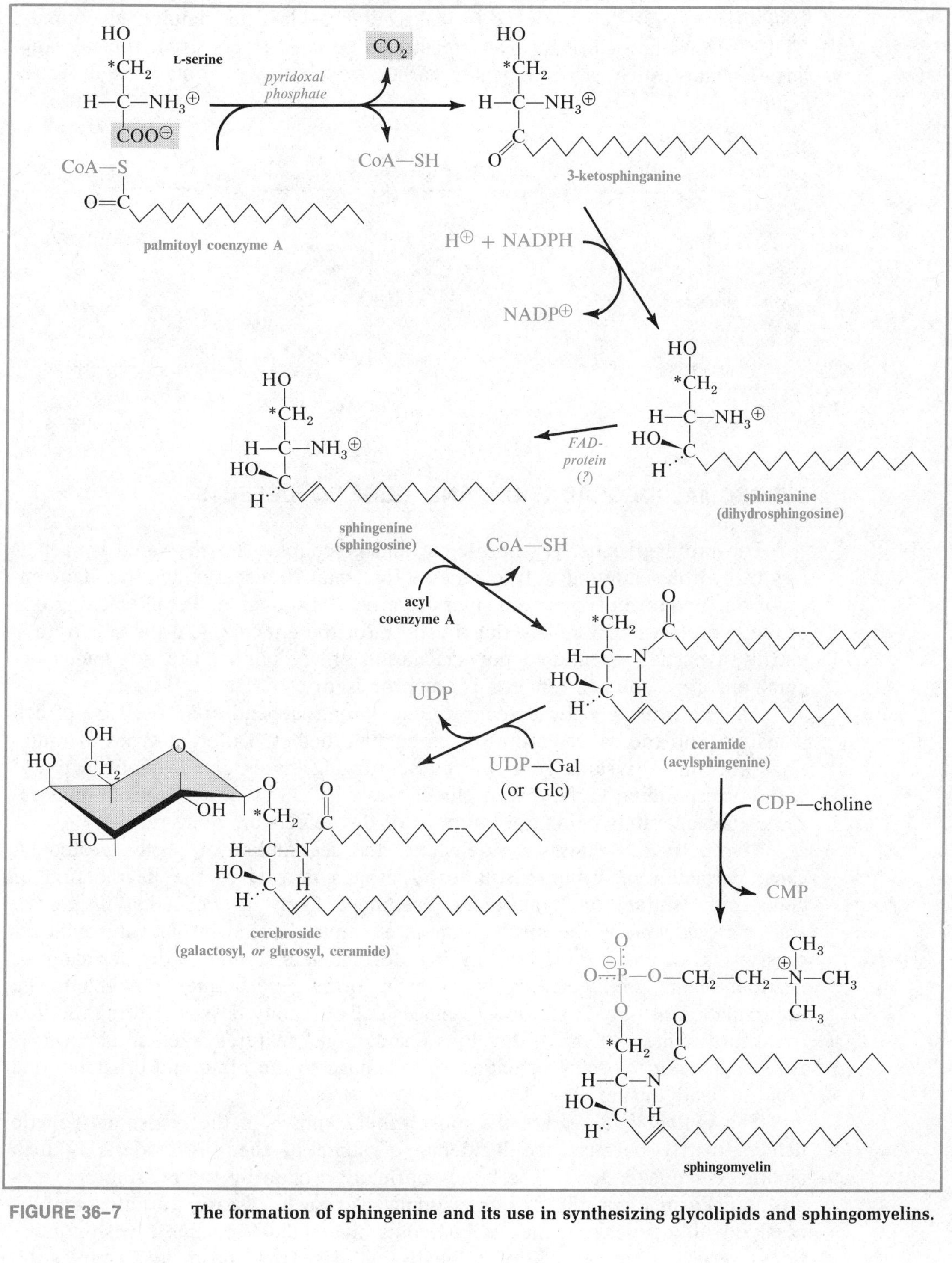

FIGURE 36–7 The formation of sphingenine and its use in synthesizing glycolipids and sphingomyelins.

transferring phosphorycholine from CDP-choline, and of the glycolipids, which are made by transferring carbohydrate residues from nucleotide sugars.

The same nucleotide sugars that are precursors of proteoglycans and glycoproteins are also the donors of carbohydrate residues in glycolipids. Thus, UDP-Glc or UDP-Gal are used to make cerebrosides, the simplest glycolipids. UDP-GalNAc and CMP-NeuNAc must also be used to construct typical ganglioside chains, such as we indicated earlier, which always contain NeuNAc residues:

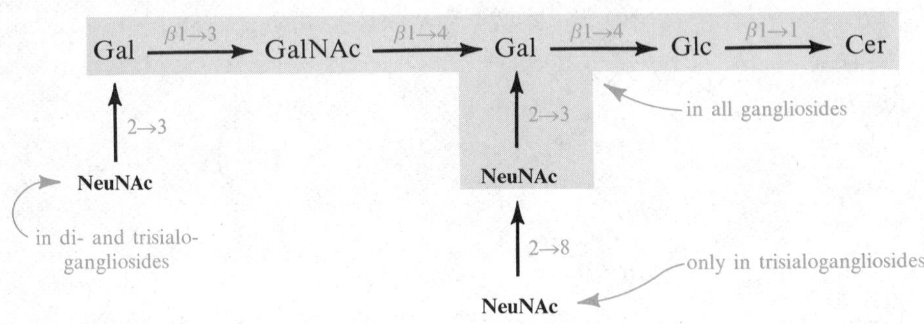

LYSOSOMAL DEGRADATION AND GENETIC DEFECTS

The proteoglycans, glycoproteins, and glycolipids are degraded by a battery of hydrolases found in lysosomes. More than 20 diseases due to a deficiency of one or more of these enzymes have been discovered. People lacking one of the hydrolases accumulate the substrate for the enzyme, and the substrate is usually a partially degraded polysaccharide or glycolipid. The lysosomes become engorged and cell function is compromised.

As with other genetic defects, the symptoms depend upon the type of cell damaged, and the exact nature of the enzyme defect. Different types of mutations affecting the same gene will cause varying losses of enzymatic activity, with corresponding variations in clinical severity. Loss of activity of some enzymes is moderately common, but most of the defects are quite rare.

Mucopolysaccharidoses are defects in the degradation of certain proteoglycans: dermatan or heparan sulfates. (Perhaps defects in the degradation of chondroitin sulfate or hyaluronate are more likely to be lethal *in utero*.) **Hurler's syndrome** is the most common example, caused by an autosomal recessive loss of the enzyme hydrolysing iduronic acid residues from dermatan or heparan chains. The incidence is 1:40,000, so the gene frequency is 1:200. The accumulation of polysaccharides damages cells in many tissues. During the first two years of life, the infant develops coarse facial features, skeletal abnormalities, ridged skin, corneal opacities, and damage to the brain and other internal organs. Death ensues.

The **sphingolipidoses** are the most widely known of the lysosomal genetic defects, partly because the incidence of some of them is moderately high among Ashkenazic Jews. The bonds normally broken by the missing enzymes are outlined in Figure 36–8. For example, **Tay-Sachs disease,** an autosomal recessive with a gene frequence estimated as high as 0.03 in some Jewish populations, results from a lack of a hexosaminidase that normally removes N-acetylgalactosamine residues from gangliosides. This causes intracellular accumulation of a modified (GM_2) ganglioside. The affected lipids are especially

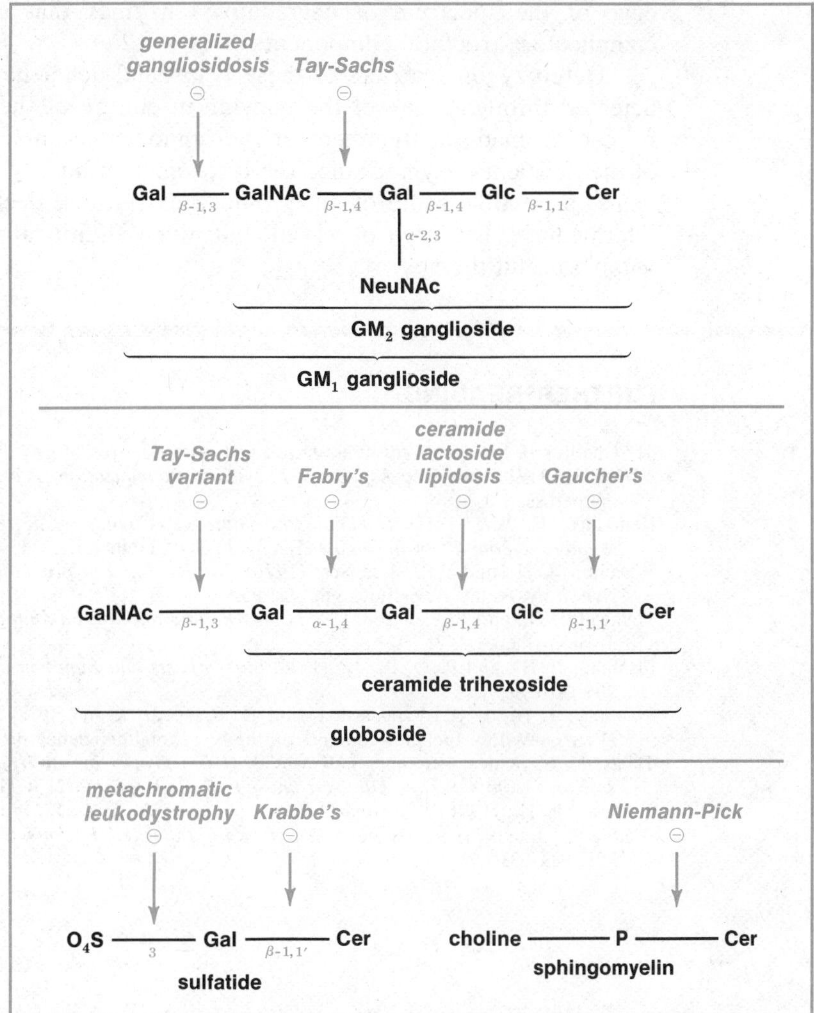

FIGURE 36–8 **Sites of action of enzymes deficient in some genetic defects of lysosomal function. The names of the compounds accumulating in the deficiencies are shown below the structures.**

abundant in the nervous system, and Tay-Sachs disease causes progressive blindness and degeneration of the central nervous system.

Gaucher's disease, an autosomal recessive with a gene frequency approaching 0.01 in some Jewish populations, is a defect in the hydrolysis of the G1c-ceramide bond that occurs in both gangliosides and globosides. Gaucher's disease affects many tissues, but the accumulation of glucosylceramide is especially conspicuous in the spleen and liver, which are enlarged owing to the turnover of the globoside-rich erythrocytes and leukocytes.

Other, rarer, sphingolipidoses affect the removal of sulfate groups from sulfatides, and the hydrolysis of the ceramide phosphate ester bond in sphingomyelin.

There are also rare defects in the handling of carbohydrate fragments derived from both glycoproteins and glycolipids. Failure to remove fucose residues and mannose residues result in fucosidosis and mannosidosis, respectively. Still other conditions will no doubt be uncovered to enlarge our knowledge of the spectrum of degradative enzymes that exists to remove these complicated structural components.

Heterozygous carriers of these lysosomal deficiency diseases can often be detected through assay of the enzyme in culture of their fibroblasts. Attempts have been made at treatment of the homozygous infants through replacement of the deficient enzyme. Since the lysosomes ordinarily take up cellular constituents, a possible route of entry indeed exists, and decline in the accumulated intermediates has been observed, but much additional work is required to develop a useful therapy.

FURTHER READING

See Chapter 10 for carbohydrate structure and synthesis (p. 166).

Adams, E.: (1976) *Catalytic Aspects of Enzymatic Racemization.* Adv. Enzymol., *44*: 69. Discusses epimerases.

Hemming, F. W.: (1977) *Dolichol Phosphate, a Coenzyme in the Glycosylation of Animal Membrane-bound Glycoproteins.* Biochem. Soc. Trans., *5*: 1223. A review.

Waechter, C. J., and W. J. Lennarz; (1976) *The Role of Polyprenol-linked Sugars in Glycoprotein Synthesis.* Annu. Rev. Biochem., *45*: 95.

Heinegard, D., and I. Axelsson: (1977) *Distribution of Keratan Sulfate in Cartilage Proteoglycans.* J. Biol. Chem., *252*: 1971.

Fishman, P. H., and R. O. Brady: (1976) *Biosynthesis and Function of Gangliosides.* Science, *194*: 906. A review.

Stanbury, J. B., J. B. Wyngaarden, and D. S. Fredrickson: (1978) *Metabolic Basis of Inherited Disease.* Wiley. Includes chapters on the lysosomal deficiency diseases.

Tabas, I., S. Schlesinger, and S. Kornfeld: (1978) *Processing of High Mannose Oligosaccharides to Form Complex Type Oligosaccharides.* J. Biol. Chem., *253*: 716.

Kolodny, E. H.: (1976) *Lysosomal Storage Diseases.* N. Engl. J. Med., *294*: 1217–1220.

Brady, R. O.: (1974) *Replacement Therapy for Inherited Enzyme Deficiency.* N. Engl. J. Med., *291*: 989–993.

37 ENERGY BALANCE

MEASUREMENT OF ENERGY DEMAND AND SUPPLY

One of the simplest and, at the same time, most fundamental approaches to the metabolic economy of an entire organism is to assess its energy exchange with the environment. The evaluation of this basic parameter was begun during the nineteenth century, and it is a tribute to the wide-ranging imagination of the pioneers that the approach they developed has persisted to this date. Knowledge of the biochemical processes was very fragmentary in that century. Metabolic sequences, the idea of high-energy phosphate, and even the amino acid composition of proteins were yet to come. What was known was that organisms oxidized ingested foods and evolved heat, and this knowledge was used to develop the basic concept that the liberated heat could be used as an index of the energy requirements of the organism and the heat of combustion of foods as an index of their relative ability to meet those requirements.

This was in many ways a bold step. Classic thermodynamics developed as a study of heat engines, and it was clear at an early date that heat could not be used for work in organisms because of their relatively constant temperature. Therefore, heat of combustion would be valuable as a guide only if it had a constant relationship to free energy with all types of nutrients.

Proof that heat exchange was useful as an approximation took much experimentation. It was necessary to build large calorimeters for measuring the heat exchange of animals, thereby providing proof that the heat output for the combustion of foods within an organism was the same as the output obtained by burning the same materials in a bomb calorimeter. Humans, because of their cooperativeness, proved to be the most valuable subjects for this exacting type of experimentation.

The agreement between heat evolved during biological combustion and during bomb calorimetry proved to be excellent for readily absorbed fats and carbohydrates. However, there was an important and unsatisfactory correction necessary in the case of proteins because nitrogen is converted to nitrogen gas and oxides of nitrogen in the bomb, whereas it is converted to various other products, mainly urea, in the mammal.

The next step was establishing that heats of combustion did bear some relationship to the relative yield of utilizable energy from the major fuels in the biological system. Perhaps the most reassuring finding was the result of comparing fats and carbohydrates for the maintenance of constant weight. If the heat of combustion of the dietary intake was held constant, it was found that there was little change in weight if fat was substituted for a substantial amount of carbohydrate in the diet. Variation of protein content gave less consistent results. We shall also see that heat of combustion, even after correction for urea output, is an inherently less satisfactory guide to the energy available from utilization of protein.

Despite its limitations, heat of combustion is still used for equating food intake with the energy requirements of the organism. Tabulations of caloric values pervade nutrition and are the basis for innumerable dietary fads. These tables are calculated after use of additional fudge factors intended to allow for differences in digestibility of various foods, and the normal loss of organic compounds in the feces. They are therefore empirical comparisons rather than true heats of combustion. Even so, they are useful, as we shall see.

An important point must be made clear. The standard unit of heat energy in scientific circles formerly was the calorie, then defined as the amount of heat necessary to raise the temperature of one gram of water by 1° at 15° C. (The calorie is now defined in terms of the joule: 1 cal = 4.184 J.) The heat of combustion of the common foodstuffs lies in the range of several thousand calories per gram. Because of a reluctance to write large numbers, it became the practice to specify the values in **kilocalories.** Someone made the mistake of designating this as a large calorie, spelled with a capital letter (Calories or Cal.). This trivial typographical distinction was soon neglected (what is a thousand-fold among friends?), and we must face the fact that laymen and dieticians mean kilocalories when they say calories. There is no excuse for such license in professional circles. When we say calories, we mean it, and 1,000 calories is a kilocalorie, not a Calorie. It will take time to think in terms of megajoules (MJ) rather than kilocalories, but we shall make a beginning on this.

The Respiratory Quotient and Indirect Calorimetry

The relative amounts of fats, carbohydrates, and proteins being used for the production of energy can be estimated from three simple measurements: (1) the consumption of oxygen; (2) the production of carbon dioxide; and (3) the excretion of nitrogen. Since nothing more is involved than the collection of urine and feces and the small discomfort of periodically breathing through a system in which the disappearance of oxygen and appearance of carbon dioxide can be measured, the monitoring of energy balance in this way has largely replaced the use of cumbersome calorimeters. Let us now analyze how these values can be used.

The first step is to determine how much of the gas exchange — the consumption of oxygen and production of carbon dioxide — results from the metabolism of proteins. If we know the nitrogen excretion and the kinds of proteins being metabolized, the corresponding gas exchange can be calculated from simple stoichiometry. For example, the stoichiometry for the metabolism of some proteins, if all of the nitrogen appears as urea, is as follows:

STOICHIOMETRY OF PROTEIN METABOLISM

	Moles per Kilogram of Food				
	CO_2	$-O_2$	N	CO_2/N	$-O_2/N$
Beef muscle proteins	37	45	12.6	3.0	3.6
Casein*	37	45	11.2	3.3	4.0
Zein*	39	47	11.4	3.4	4.1

*Casein is the principal protein of milk and zein is the principal protein of corn.

If we assume that an individual is metabolizing a mixture of amino acids approximating the composition of muscle proteins, we can multiply the measured nitrogen excretion by 3.0 and 3.6 to obtain the respective values for carbon dioxide output and oxygen uptake due to the combustion of amino acids. Of course, the values may be in error by 0.1 or more if the mixture of amino acids consumed more closely approximates the composition of casein or zein.

The second step in the calculation is to subtract the values for gas exchange due to protein metabolism, estimated as above, from the total gas exchange. This leaves the values of oxygen consumption and carbon dioxide production due to the metabolism of fats and carbohydrates. Now, the stoichiometry for the metabolism of typical examples of these fuels is as follows:

STOICHIOMETRY OF FAT AND CARBOHYDRATE METABOLISM

	Moles per Kilogram of Food		
	CO_2	$-O_2$	$CO_2/-O_2$
Starch	37	37	1.00
Sucrose	35	35	1.00
Lactose	35	35	1.00
Corn oil	65	90	0.72
Pig fat	64	91	0.71

We see right away that the ratio of carbon dioxide production and oxygen consumption is greatly different from the carbohydrates and for fats. This ratio is the **respiratory quotient,** and it is a valuable guide to the relative amounts of oxygen being used for the combustion of fats and of carbohydrates. It may be used both with intact animals and with isolated tissue preparations for this purpose.

Since the respiratory quotient (RQ) for the oxidation of carbohydrates is 1.00, and the value for the oxidation of most natural fats is near 0.71, we can estimate the fraction of the oxygen consumption used for carbohydrate metabolism by the following relationship (after correction for amino acid metabolism):

$$\frac{-O_{2\,(\text{carbohydrate})}}{-O_{2\,(\text{fat + carbohydrate})}} = \frac{RQ_{(\text{observed})} - RQ_{(\text{fats})}}{RQ_{(\text{carbohydrates})} - RQ_{(\text{fats})}} = \frac{RQ_{(\text{obs.})} - 0.71}{0.29}$$

Knowing the stoichiometry for high-energy phosphate production from fats, carbohydrates, and proteins, we can then use the oxygen consumption for the metabolism of these three kinds of fuel, as calculated by the above methods, and estimate the total energy production in the entire organism (or tissue) from the gas exchange and nitrogen excretion.

The pioneers in the field knew nothing about high-energy phosphate and the mechanisms of utilization of the energy of combustion. Indeed, they employed the calculated partition of oxygen consumption among the fuels to estimate the total caloric yield. Hence, the method is termed **indirect calorimetry,** because heat production is calculated from gas exchange rather than by direct observation in a calorimeter. This name gives the implication of lesser reliability, but in fact the measure of oxygen consumption is in many ways a more useful guide to the metabolic economy than is heat evolution.

The factors commonly used in indirect calorimetry are shown in the following table. (The data are given in terms of liters of gas as well as moles, because these units are still in common use in many laboratories.)

ENERGY PRODUCED

	RQ	Per Gram of Food		Per Mole O_2		Per Liter O_2	
		kJ	*kcal*	*kJ*	*kcal*	*kJ*	*kcal*
Proteins	0.80	18.1	4.32	418	100	18.7	4.46
Fat	0.71	39.6	9.46	439	105	19.6	4.69
Starch	1.00	17.5	4.18	473	113	21.1	5.05

1 mole of urinary N = 1.56 MJ, 372 kcal; 3.00 moles CO_2 produced and 3.70 moles O_2 consumed.
1 gram of urinary N = 0.11 MJ, 26.5 kcal; 4.8 liters CO_2 produced and 5.9 liters O_2 consumed.

The Basal Metabolic Rate (BMR)

The basal metabolic rate (BMR) is the rate of oxygen consumption, or the calculated equivalent heat production, of an awake individual lying at rest, who has had no food for at least 12 hours. This measurement is an important physiological tool because it assesses the energy requirement for maintenance of tissues and any dissipation for heat production through oxidations not coupled to phosphorylations.

The usual determination of BMR is really only a determination of the rate of oxygen consumption. Much of the success of this approach results from the specified conditions, which minimize many potential sources of error. A previously well-fed individual will depend mainly on his fat stores for energy after an overnight fast, with a respiratory quotient near 0.82 if he has been eating the usual American diet. Only 15 per cent of the oxygen consumption will be used for the metabolism of amino acids, thereby minimizing error due to lack of knowledge of the exact composition of the amino acids being metabolized.

It is therefore reasonably safe to assume that most people in the basal condition will be metabolizing a proportion of fuels such that **1 mole of O_2 is equivalent to 452 kJ, 108 kcal; one liter of O_2 is equivalent to 20.2 kJ, 4.82 kcal.** (The latter value is frequently quoted as 4.8205 kcal per liter, which is a beautiful example of confusing accuracy in arithmetic with experimental precision. One part in 100 is an optimistic expectation for precision in this field. When measurements of BMR were commonly used in assessing thyroid function, any value within 10 per cent of the accepted mean was taken as normal.)

The Use of Surface Area

It is common practice to compare the metabolism of individuals on the basis of the area of the body surface. This practice is even extended to such things as the calculation of drug dosages. The advantages and defects of the custom deserve examination.

It ought to be apparent from what we have learned to this point that metabolic processes do not occur predominantly at the body surface; therefore, metabolism and body surface do not have a direct anatomical relationship. However, it was realized over a century ago that mammals have a proportionately slower basal metabolism as their size increases. A large mammal has a greater total metabolism than does the smaller mammal, but the metabolism per unit of body weight decreases with size. Many later studies have shown that the basal metabolism is more nearly proportional to the surface area of the adult animals in various species than it is to the weight of the animals.

We really ought to expect this to be so. As we mentioned in the earlier discussion of brown adipose tissue, there is no inherent reason that the cells of an elephant could not be conducting oxidations at the same rate as do the cells of a mouse. The limiting factor is the ability to dispose of the heat generated by metabolism. The primary mechanisms for heat loss for land animals are direct radiation to the environment and evaporation of water on the surface, both of which can be expected to be proportional to the surface area. (There are other mechanisms, such as evaporation through the lungs, so heat loss is not directly proportional to surface area among all animals.)

In short, we are dealing with an evolutionary development in which the metabolic processes of many kinds of animals have been adjusted to conform, at least roughly, to the surface area of the animal rather than to its tissue mass. This is the basis for using the surface area in comparing metabolic rates of species.

The justification for the use of surface area in comparing individuals of one species is considerably more shaky. One would have to argue that the metabolism of individuals somehow adapts to keep their heat output proportional to area, and there is no real evidence that this is so. The evidence at hand suggests that metabolic rate is more nearly proportional to the total amount of protein in the body, and if we could estimate the number of mitochondria, this would probably be the best guide of all. It is well known that, given two men of equal height and weight, the more muscled specimen has the higher metabolic rate. It is known that metabolic rate per unit surface area differs among various groups of people, and differs with age and sex in the same group of people. This is shown in Figure 37–1.

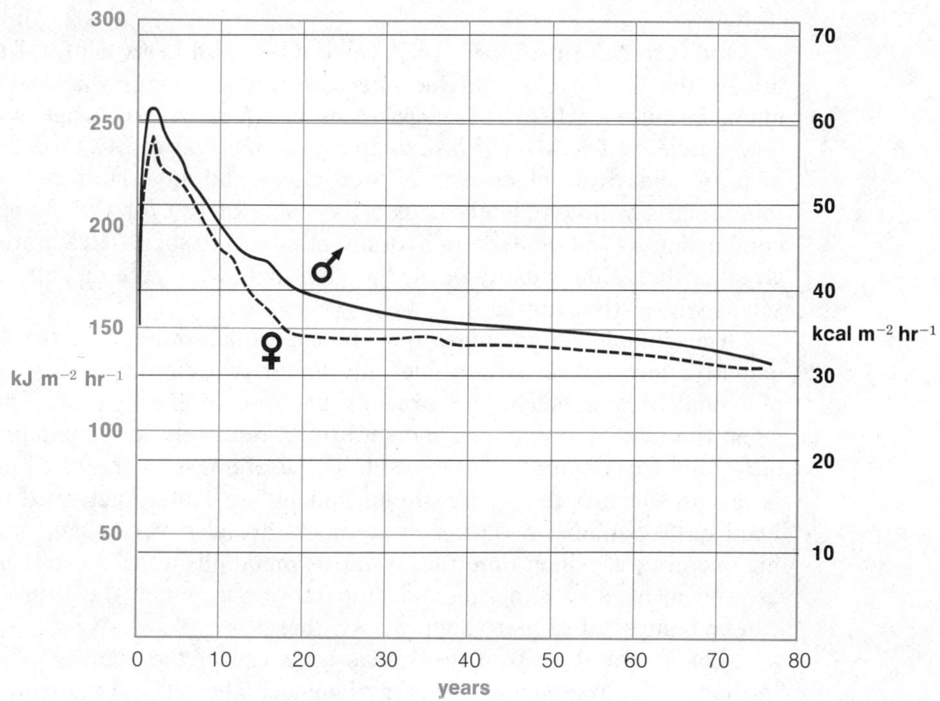

FIGURE 37–1 **The basal metabolic rate of males and females of varying ages in the United States, given per square meter of calculated body surface area.**

Why, then, aren't metabolic measurements made by simply measuring oxygen consumption per unit time and comparing the results with the average range of values obtained for people of similar age, sex, height, and weight? Perhaps the major reason is the striving for logical unity in science. Since we obviously can't compare the oxygen uptake of a lean person with that of an obese person on the basis of weight — the droplets of triglyceride have no oxygen uptake — we seize on the surface area as the best available means of minimizing individual variations. This evidently comforts many even though the practice has no biochemical basis, and the calculated values still must be compared with empirically determined standards for similar people.

It is exceedingly difficult to measure surface area accurately. Hence, there have been many attempts to estimate this value from more easily measured parameters, such as the overall length of the body and the body weight. Metabolic laboratories have charts for this purpose, which have been constructed from the relationship:

$$\log A = 0.425 \log W + 0.725 \log H - 2.144$$

in which A = area in square meters; W = weight in kilograms; H = height in centimeters.

One has to be alert for some curious examples of circular logic in the literature on surface area measurements. The object is to determine how closely metabolism is related to surface area. Some proceed to contend that their formulas for determining surface area are better because they result in more constant values of metabolic rate per unit of surface.

The Validity of Calorimetry

Indirect calorimetry is only valid when synthesis of fat from carbohydrates, the Cori cycle, and the interconversion of amino acids are at a minimum, because each of these processes involves gas exchange with a respiratory quotient of its own. For example, we saw that the RQ for the production of fatty acids from glucose is 2.75 (p. 534), and any significant occurrence of this process will completely negate the value of RQ for estimating the relative combustion of fat and carbohydrate in other tissues. (Respiratory quotients greather than 1.00 actually occur in individuals who have recently stuffed themselves with carbohydrates.)

These qualifications mean that the use of calorimetry for the assessment of oxidative metabolism is of value only under two circumstances: when the rate of oxidative metabolism far exceeds the rate of the synthetic processes and when the rate of amino acid metabolism is relatively low compared to the rate of fat and carbohydrate metabolism. The usefulness of the basal metabolic rate is due to the fact that these circumstances are largely achieved under the defined basal conditions. The measurements are also very useful in animals during exercise, at which time the oxidative metabolism far exceeds other processes, and in isolated experimental preparations in which the supply of nutrients can be controlled so as to suppress syntheses.

Let us now turn to a theoretical appraisal of the validity of heats of combustion in the assessment of energy balance. The real energy balance is a result of the formation and utilization of high-energy phosphate, which we cannot measure with present techniques in a living animal. However, we do have estimates of the theoretical stoichiometry for the complete oxidation of the various fuels, and we have noted in the preceding chapters various lines of reasoning that suggest that the actual yield of high-energy phosphate is not far from the theoretical value.

The following table compares various kinds of foods, both in terms of high-energy phosphate and the classic values for heat derived from complete oxidation. (The yields from glycogen and fat stored within the individual are included for later use. The yield from *ingested* glycogen is the same as that from starch.) The high-energy phosphate yields were computed on the following assumptions:

(1) The bulk of the ingested glucose is directly stored as glycogen in skeletal muscles, and completely oxidized in that tissue, so the Cori cycle can be neglected.

(2) Ingested triglycerides are hydrolyzed and stored in adipose tissue without modification by the liver.

(3) Amino acids are metabolized according to the stoichiometry given in Chapter 31 (p. 610). All of the nitrogen is converted to urea. The carbon skeletons are converted to glucose and acetoacetate, which are metabolized in other tissues without intermediate storage.

(4) The amount of high-energy phosphate necessary for transport across cell membranes is negligible compared to the total yield from oxidation. This is the most shaky premise, and we can hope only that the relative loss on transport is roughly the same with all types of compounds.

The last column of the table gives the ratio of heat production to high-energy phosphate formation. This ratio ought to be the same for all types of foods if heat production is a valid guide for utilizable energy from oxidations. The values for fats and carbohydrates are indeed nearly the same for both the ingested foods and the stored fuels, although they differ for the same food when ingested and when utilized from the body stores.

The value for heat production from proteins obviously gives too high an assessment for the utilizable energy. This is true even though the caloric yield of proteins has been corrected in a way that gives inherently low values, as we discussed earlier. Early experiments on feeding for maintenance of weight showed this discrepancy, which gave rise to an extraordinary rationalization that will be discussed in the next section.

ENERGY FROM COMMON FUELS **TABLE 37–1**

		kJ Liberated Heat			**Moles ~P**		**Ratio**
Fuel	**RQ**	*per gram of food*	*per mole of O_2*	*per liter of O_2*	*per gram of food*	*per mole of O_2*	**kJ/mole ~P**
Ingested starch	1.00	17.5	473	21.1	0.222	6.0	78.8
Ingested fat	0.71	39.6	439	19.6	0.502	5.5	79.9
*Ingested protein	0.82				0.202	4.44	94.1
†Ingested protein	0.80	18.1	418	18.7			
††Stored glycogen	1.00	17.5	473	21.1	0.235	6.3	74.7
Stored fat	0.71	39.6	439	19.6	0.510	5.6	78.7

*Calculated from amino acid composition of beef muscle proteins; grams protein = 5.65 × grams N.
†Accepted values for mixed proteins, corrected for fecal losses; grams protein = 6.25 × grams N.
††Muscle glycogen.

What is the practical result of our re-examination of the caloric values? The usefulness of caloric values for comparing fats and carbohydrates has long been established through experimentation without the aid of our theoretical justification. Indeed, the tables are turned, and this large body of observations generates confidence in the stoichiometric calculation of high-energy phosphate yields for assessing the potential energy of foods. This is important both for immediate use and for its future potentialities. The immediate applications are in adjusting the accepted values for the energy equivalent of protein as a fuel and in making a distinction between the ingested and stored fuels, which have identical heats of combustion but different capacities for supporting work. The future potentialities lie in assessing energy balance under conditions in which processes other than oxidation are proceeding at significant rates and in being able to take advantage of the more detailed analyses of foods now becoming available for computing their nutritional value.

We come, then, to the following conclusions at this point in the discussion:

1. The multitude of tables showing caloric equivalents of foods are useful and sound, provided that the protein content of the food is not unusually high (>0.2 or so of the calculated caloric equivalent). The major objection lies in the implication of using true thermochemical values and of being able to use heat for work.

2. The energy yield from stored glycogen is significantly higher than that of ingested starch, and the "caloric" value of oxygen consumed for the metabolism of glycogen ought to be raised by 5.5 per cent.

3. The accepted values for the energy yield from meat proteins are too high. If the caloric basis is to be used, and if the caloric yield from starch is taken as 17.5 kJ per gram, the corresponding yield from meat proteins would be near 1.18 MJ (282 kcal) per mole of N, or 84.5 kJ (20.2 kcal) per g of N. (The current accepted value is 26.5 kcal per g of N. Meat has a relatively high content of glycine, owing to the presence of collagen, and the energy yield per mole of N from glycine is low. It so happens that the fudge factors applied to the original calculation for meat proteins make the value of 26.5 kcal per g of N about right for proteins such as casein and zein, with their larger content of high molecular weight amino acids. This occurred by accident, not design, and illustrates the need for re-evaluation based on amino acid content of proteins as fuels.)

Specific Dynamic Action

The oxygen consumption and heat output of an individual rise upon eating, even if he remains at rest. This elevation in metabolic rate over the basal rate was named the specific dynamic action of foods. It is well established that the specific dynamic action is something above and beyond the result of the exertion of mastication, swallowing, and the increased motility of the gastrointestinal tract. The increased metabolism follows the absorption of the digested foods.

There have been many differing estimates of the magnitude of the specific dynamic action of various foods. These range from near zero for the fats, through 5 to 10 per cent of the total caloric equivalent of ingested carbohydrate, to 20 to 30 per cent of the total caloric equivalent of ingested proteins, when these foodstuffs are tested individually.

The phenomenon is real, and much effort has been expended in attempting to explain it. The problem has been complicated by the inconsistency of the results under varying conditions, and by a general increase in metabolic rate caused by eating any meal. For example, the specific dynamic action is greater for foods given to a starving animal than for foods given to an animal that is fed repeatedly. The specific dynamic action of a mixed meal is less than the sum of the effects seen with the constituents given individually.

A compromise in interpretation has been reached by which a specific dynamic action of 10 per cent of the total caloric value is assigned by many to the usual mixed meal. They then rationalize that the immediate output of heat represents something extra, not useful, and additional fuels must be supplied to compensate. Therefore, in computing the necessary food intake for a given individual, it is common to estimate his caloric requirements from his basal metabolic rate, and then add an extra 10 per cent to the intake to compensate for the specific dynamic action.

When we examine the various metabolic pathways previously discussed, it is somewhat difficult to see what all the fuss has been about, because we now have the advantage of being able to balance metabolic processes in terms of high-energy phosphate. It is self-evident that the handling of nutrients places variable demands on the oxidative metabolism. Synthesis of the constituents of the digestive juices is in itself an energy-consuming process. Beyond that, there are inherent expenditures of energy in the processing and storage of nutrients. In the case of fats, we can see that storage in adipose tissue may involve nothing more than re-forming triglycerides, involving the expenditure of approximately 1.6 per cent of the potential ~ P yield. Glucose will be converted to glycogen, which involves the immediate expenditure of ATP amounting to 5 per cent of its potential ~ P yield. These extra loads on the ATP supply will cause corresponding increases in oxidative phosphorylation, which account for the increased O_2 consumption.

Amino acids, if given to an individual who has fasted, will be metabolized relatively rapidly. According to the stoichiometry we cast in Chapter 31, the liver will use over 10 millimoles of O_2 per gram of protein with little net formation of ATP. This "extra" consumption of one mole of O_2 per mole of mixed amino acids amounts to 23 per cent of the total oxygen consumption for their complete metabolism, and it will be even greater if conditions are such that the amino acid load increases the total oxygen consumption of the liver. It will be less on mixed diets, due to the temporary storage of amino acids in the tissue pools. It is not surprising that the specific dynamic action varies with conditions — the routes and rates of disposal of nutrients depend upon the particular state of the metabolic economy.

Should the specific dynamic action be allowed for in planning diets? Essentially, the phenomenon is nothing more than the initial stages of metabolism of the particular nutrients, which are included in the total stoichiometry for ~ P balance, and the increased energy production associated with eating and digestion. What is the point, then, of adding an extra 10 per cent to the dietary intake to compensate for specific dynamic action? There wouldn't be any point if we calculated fuel value on the basis of ~ P, or used caloric values adjusted to that basis, and included eating as an energy-expending activity. However, adding an extra 10 per cent to the dietary requirement calculated from the customary values for the caloric equivalent of foods has the same effect as lowering these values so that they are more nearly correct in relation to stored fuels, which we have seen is particularly necessary for ingested protein and carbohydrate.

In summary, the adjustment of dietary intakes for specific dynamic action is one of those lucky accidents. No one has been able to give a sound explanation of the logic for the practice because it isn't logically sound, but making this adjustment fortuitously compensates for discrepancies between true useable energy and heats of combustion. This is especially true in the case of proteins; good correction of the accepted caloric yield can be achieved by assignment of a specific dynamic action of 25 per cent, which is in the range of frequently quoted values.

CHANGES IN ENERGY DEMAND

Measurements of oxygen consumption show in a general way the partition of energy demand among the various tissues and the changes in demand with work. The following relative values were calculated from a tabulation by Dr. Richard Havel (quoted by permission):

TABLE 37–2 RELATIVE OXYGEN CONSUMPTION: (whole body at rest = 1.00; actual value near 0.17 mmoles min^{-1} kg^{-1})

	At Rest	Light Work	Heavy Work
Skeletal muscles	0.30	2.05	6.95
Abdominal organs	0.25	0.24	0.24
Kidneys	0.07	0.06	0.07
Brain	0.20	0.20	0.20
Skin	0.02	0.06	0.08
Heart	0.11	0.23	0.40
Other	0.05	0.06	0.06
SUM	1.00	3.00	8.00

These data reinforce a point already made — the oxygen consumption of working muscles far outstrips all other demands. The demands of the viscera account for about one third of the total oxygen consumption at rest. The brain utilizes one fifth. These demands remain relatively constant during work, but the utilization by the skeletal muscle and heart increases dramatically.

Essentially, we are seeing in these data the requirements for fuel represented by the basal metabolism and the additional fuel required for daily activities. The additional amounts required have been measured through calorimetry for a wide range of human activities, and such measurements provide the basis for estimates of the food consumption required for maintenance of body weight. As a first approximation, activities can be sorted into a few categories (energy expenditures given as fractions of basal):

lying at rest	1.0
sedentary work (reading, writing, talking, waiting, either sitting or standing)	1.6
light movement (dressing, sorting, strolling)	3
moderate work (gardening, shoveling, tennis, coitus (male))	5
heavy work (hand sawing, climbing, stacking 50 kg loads, swimming, skiing on level)	9

The time actually spent on activities in each category can then be estimated. (It is important that this be done carefully. Eight hours at the manufacturing plant is not necessarily eight hours of continuous activity, and one hour at the club is certainly not one hour of continuous swimming.) The errors will be large, so large that finicky nitpicking over specific dynamic action and the like is a useless refinement. Even so, the results are a valuable guide.

To illustrate the principles, assume two 22-year-old males, with a median 177 cm height and 73 kg weight, and a basal metabolic rate of 310 kJ hr^{-1} (74 kcal hr^{-1}). The hours each devotes to particular grades of activity and the correspond hourly equivalents of basal energy expenditure are given in the following table*:

Activity	Fraction of Basal Energy	Student hrs	Student hrs × fraction	Warehouse Laborer hrs	Warehouse Laborer hrs × fraction
Asleep	1.0	7	7.0	8	8.0
Sedentary	1.6	14.5	23.2	8	12.8
Light movement	3	2	6.0	2	6.0
Heavy work		0.5	4.5	6	54.0
SUM		24	40.7	24	80.8

The total daily expenditure of a student is 40.7 times his hourly basal expenditure, or 12.6 MJ (3,010 kcal), whereas the total for the laborer is 25.0 MJ (5,980 kcal), nearly twice as much. The lesson here is that a half-hour of heavy workout per day, whatever its other advantages, will not require the fuel intake of a working man. (The recommended daily allowance for caloric intake of young men in the United States is 2,800 kcal per day. This reflects the sedentary character of our population before the jogging epidemic, and is calculated for a man smaller than the current median. We shall discuss this, and other recommendations, in the last section of the book.)

Another factor that influences the energy requirement of a given person is the environmental temperature. This is really self-evident. Since the body temperature must be maintained within narrow limits, and exposure to cold increases the heat loss, there must be a corresponding increase in metabolism. Part of the increase comes from brown adipose tissue, part from shivering, and part from some increase in metabolism of muscle through mechanisms that have not been defined. Acute exposure to cold, such as immersion in ice-water, will increase metabolic rate as much as two-fold.

Utilization During Work

Let us now translate the energy demand into biochemical events. The sequence of utilization of potential substrates for the generation of ATP during heavy activity is simply described: Glycogen is used preferentially, the fats are used next, and the proteins contribute little.

Let us get a clear picture of the changing fuel economy by studying what happens in a single individual working over an extended period of time. The table on the next page shows what happened in a young man accustomed to physical work who ran intermittently over a period of six hours. His respiratory exchange and nitrogen excretion were measured during this period, and these data enable us to calculate the amount and kind of fuels being used.

*X-rated work is quantitatively insignificant.

TABLE 37–3 THE MARATHON RUNNER

Rate of Running	Body Weight	Blood Sugar	$-O_2$	RQ	~P Production				Weight of Fuels	
					Total	From Glycogen	From Fats	Fraction From Fat	Glycogen	Fats
km hr^{-1}	*kg*	*mM*	*moles*		*moles*	*moles*	*moles*		*g*	*g*
0	59.61	5.6								
11.3	59.32	4.6	2.75	0.97	17.1	15.5	1.6	0.09	67	3
9.3	59.03	4.9	2.34	0.96	14.5	12.7	1.8	0.12	55	4
11.3	58.82	4.8	2.71	0.94	16.6	13.5	3.1	0.19	58	6
9.3	58.65	4.4	2.31	0.88	13.8	8.5	5.3	0.38	37	11
11.3	58.25	4.6	2.76	0.86	16.5	9.0	7.5	0.45	39	15
9.3	58.15	4.5	2.40	0.82	14.1	5.7	8.4	0.60	25	16
11.3	57.93	4.2	2.85	0.82	16.7	6.8	9.9	0.59	29	19
9.3	57.78	4.3	2.44	0.79	14.1	4.2	9.9	0.70	18	20
11.3	57.48	3.8	2.85	0.82	16.7	6.8	9.9	0.59	29	19
9.3	57.30	4.1	2.49	0.79	14.3	4.3	10.0	0.70	19	20
11.3	57.55	3.7	2.88	0.81	16.9	6.3	10.6	0.63	27	21
9.3	57.35	3.2	2.48	0.77	14.2	3.2	11.0	0.77	14	22
					185.5	96.5	89.0		415	175

1. Blood sugar concentrations were measured by methods now known to give high values owing to the detection of compounds other than glucose in the assay. The values are still useful as a relative guide.

2. The listed totals differ somewhat from the sum of the figures given, owing to the rounding off of the latter figures.

3. This is an experiment reported by Edwards, Margaria, and Dill, Am. J. Physiol. *108*:203 (1934), and their data have been recalculated to determine high-energy phosphate production. Protein metabolism accounted for only 0.02 of the total energy production and has been neglected.

First let us describe the circumstances. The man was previously well-fed on a mixed diet, which means that we can expect his glycogen reserves were reasonably typical. He was somewhat on the lean side, and therefore did not have gross rolls of fat. He worked quite hard, running at 3.1 meters per second for 25 minutes out of each hour and at 2.6 meters per second for another 25 minutes. However, this is not the maximum possible effort for six hours of work, because the comparable world record for continuous running is 5.4 meters per second.

Now, let us look at the data. They are simple, but they have a wealth of information concealed within them that is applicable to the general question of fuel economy.

(1) Despite his leanness, the man used his stored triglycerides and glycogen for 98 per cent of his energy production, with only 2 per cent supplied by protein. This is the usual circumstance in the well-fed.

(2) His body weight (column 2) fell by 2.3 kilograms, even though he was supplied with water. This is a guide to the magnitude of the effort.

(3) His blood sugar concentration (column 3) fell by nearly one half, and the drop was precipitate during the last hour.

(4) His oxygen consumption (column 4) went up and down with the varying rates of running, as might be expected, but increased during periods of equivalent effort as time went on, despite the fall in body weight. This would indicate that more oxygen was needed to maintain the same supply of ATP.

(5) His respiratory quotient (column 5) declined throughout, indicating a shift from carbohydrate oxidation to fat oxidation. The RQ was generally higher during the periods of greater exertion, indicating increased mobilization of glycogen.

(6) The total yield of high-energy phosphate (column 6) calculated from

the gas exchange was remarkably constant throughout for periods of equivalent effort, even though the nature of the fuel was changing.

(7) During the first period, 90 per cent of the ATP was supplied by oxidation of glycogen, but the fraction declined steadily, and the oxidation of fat increased until it was supplying nearly 80 per cent of the total ATP during the final period (columns 7, 8, and 9). This shift accounts for the increased oxygen consumption, because the ATP yield per mole of O_2 is less for fats than it is for carbohydrates. Using the values of 6.3 and 5.6 moles of ATP produced per mole of O_2 consumed in oxidizing glycogen and stored triglycerides, respectively, the calculated yield per mole of O_2 consumed for the first and final periods of running at 9.3 km hr^{-1} are 6.20 and 5.73 moles ~P/mole O_2. To maintain constant ATP supply, the O_2 consumption of 2.34 moles during the first period at that rate would have had to rise to 2.53 moles during the final period. The observed consumption during the final period was 2.48 moles. This is remarkable agreement. (There would be a slight drop in the requirement for ATP owing to the decline in body weight, but the variation with body weight during running is small — not at all proportional to the weight.)

(8) The final two columns show the calculated weights of the glycogen and fat consumed. In discussing the storage of glycogen (Chapter 27), we noted that the total in the skeletal muscles and liver of a male adult of this size would be in the neighborhood of 450 grams. The total consumption shown in the table is over 90 per cent of this value. This agrees with the conclusion that might be drawn from the sharp fall in blood sugar concentration: The man had almost exhausted his carbohydrate reserves, and any further effort would have had to be sustained almost completely by the oxidation of triglycerides.

(The weight of the glycogen and fat consumed, together with the loss of the water associated with glycogen, accounts for approximately 60 per cent of the observed weight loss. However, it ought not to be assumed that loss in body weight can be predicted accurately from such data for a few hours of exertion. Changes in water balance have a large effect, and are not directly dependent on the loss of fuels.)

The conclusions we can come to with this experiment are reinforced by much more evidence than we have shown. Glycogen reserves are preferentially used for physical activity, but this activity is sustained with equal efficiency by either glycogen or triglycerides. That is, the metabolism of carbohydrate produces the same fraction of the predicted high-energy phosphate as does the metabolism of triglycerides. *There is no basis for believing that either fat or carbohydrate is an inherently inefficient fuel.* This in itself cuts the ground from under most dietary fads, which reverse the usual ambition and try to get nothing for something.

There is an important qualification that must be added. The availability of a glycogen reserve enables more intense short-term activity. We pointed this out in Chapter 26, showing that lactate production reflects a more rapid formation of pyruvate and associated production of ATP than can be accommodated by oxidative phosphorylation. In addition, the production of ATP is greater per mole of O_2 consumed during the oxidation of carbohydrates. It follows that mitochondria in which electron transport is proceeding as rapidly as possible will generate more ATP from the oxidation of pyruvate, which is derived from glycogen, than they will from the oxidation of fatty acids. The availability of glycogen therefore enables a greater maximum effort up to the time the glycogen is exhausted. This accounts in part, but not completely, for the inability of individuals with glycogen storage diseases of the muscle to do heavy work.

Normal humans can get along reasonably well on a diet containing little or no carbohydrate. Their RQ falls to levels consistent with almost total dependence on fatty acids as fuels. They cannot be expected to win short races in competition with people on a mixed diet, but they can carry on normal physical activities without difficulty.

High Power Exertion

Even in these civilized times, many humans are concerned with the rate at which they can do work. Delivering maximum power for a set period of time may gain fame and fortune in the arena, earn a livelihood in more prosaic ways, or be life-saving, especially in the metropolitan jungles. The level of maximum power depends upon the time it is to be sustained. We can picture this by plotting the rate of running necessary to establish world records as a function of the logarithm of time required for different distances (Fig. 37–2). The rate is used as an index of power output. Everyone knows that a person must pace himself to complete a given distance. An ordinary person can go all out for perhaps 100 meters, but must begin at a slower pace if he is to complete a kilometer, and still slower if he is to travel 20 kilometers. Even so, with world-class athletes there are two sharp breaks in the decline of sustained

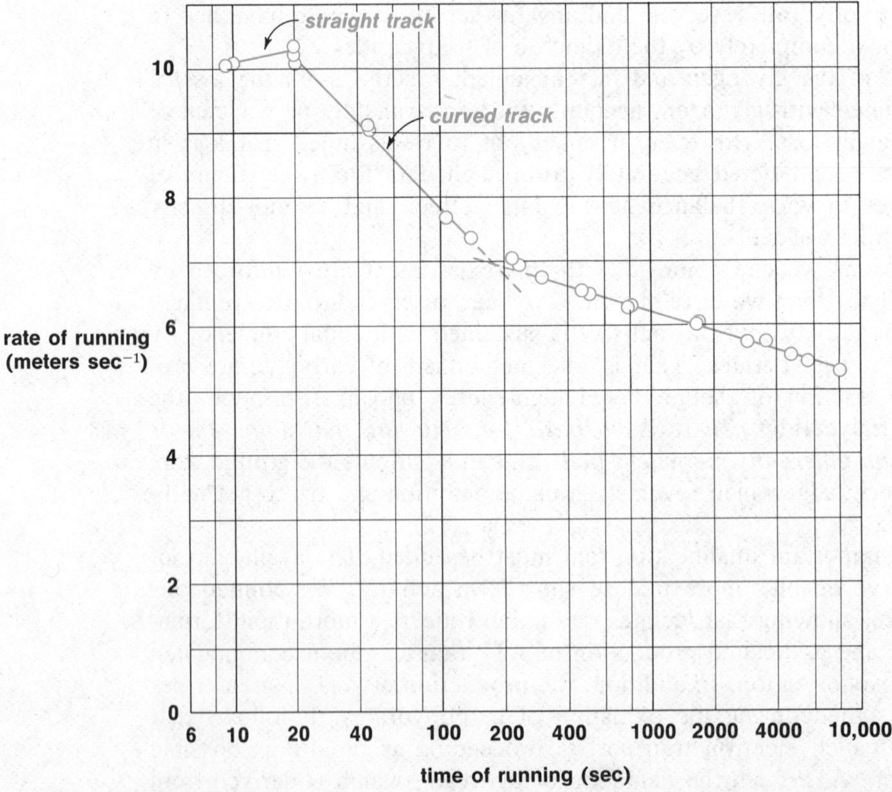

FIGURE 37–2 **World record track performances, plotted to show the relationship between the rate of running as an index of power output and the time for which a given power output can be sustained. The data suggest that humans exhaust some source of energy when running at a given rate for 20 seconds, and that there is some additional source that will suffice for 180 seconds at the most.**

power output with time it is to be sustained, one near 20 seconds, and the other near 180 seconds.

Two factors are involved. One is the power output of single muscle fibers, and the other is the type of fiber recruited for the effort. We have noted before that the white fibers — fibers of low oxidative capacity and fast twitch — will produce more lactate than do the red fibers of high oxidative capacity, some of which are slow twitch for sustained effort and some fast twitch. The power output in each case sets the drop in phosphate potential necessary before metabolism is stimulated enough to make the production of high-energy phosphate balance its consumption.

For very fast efforts, the high-energy phosphate store suffices. Leaping from a crouch discharges high-energy phosphate at a rate of approximately 6 millimoles per kg of muscle per second, but the efforts lasts for only 0.5 second, or so, and the resultant decline in phosphate potential (p. 394) is barely sufficient to cause a noticeable acceleration of oxygen consumption, with little lactate formation.

The maximum power output over several seconds is one that will bring the phosphate potential to the level of exhaustion, with essentially total discharge of phosphocreatine. It is not a simple matter of dividing the high-energy phosphate store by the rate of discharge, because the falling phosphate potential activates the generation of high-energy phosphate, finally reaching a level at which there is massive lactate formation. The break near 20 seconds represents the power output which will in that time exhaust the muscle despite maximum activation of lactate formation.

The next limitation on the power output is the availability of the glycogen supply for lactate formation. If a muscle were all white fibers, lactate formation would theoretically sustain contractions for 240 seconds with total dissipation of the glycogen. It is not all white fibers, and there is a transition from dependence on lactate formation to dependence on oxidative phosphorylation as power output is diminished. When the run is at a rate that can last longer than 180 seconds, steady state is being maintained completely by oxidative phosphorylation after the first few seconds of effort (the phosphate potential must fall before the rate of oxidation reaches its maximum). Beyond that point, the metabolic factor governing maximum sustainable power is the necessary shift from glycogen to fat oxidation as the supply of glycogen becomes exhausted.

Persistent Oxygen Consumption. Increased oxygen consumption will persist so long as the phosphate potential is low. The mitochondria do not "know" that muscles have ceased contracting. There is an immediate demand for continued high-energy phosphate production to restore the phosphocreatine concentration and ionic balances, but there are longer-lived demands for rebuilding of glycogen and triglycerides, and the conversion of lactate to glucose. It is unfortunate that this continued oxygen consumption was associated with lactate formation, which in turn was regarded as something occurring primarily in the absence of oxygen, because it led to the term "oxygen debt" for the state of the tissue at the end of exercise. One could equally well talk about an "oxygen credit," since the use of phosphocreatine and the formation of lactate diminish the necessity for supplying oxygen to meet metabolic demands. The phrase ought to be abandoned.

Utilization During Starvation

Starvation differs from heavy exertion primarily in the protracted demand on the stored fuels. The glycogen supply is largely depleted within a few days during total starvation, with nearly all of the glycogen disappearing from the liver, and the content in the muscle falling to the level that can be sustained by glucose derived from the metabolism of amino acids in the liver.

After the loss of the original glycogen stores, the starved person becomes totally dependent upon the metabolism of triglycerides from the adipose tissue and amino acids derived from his tissue proteins. The length of time he can survive is a function of the amount of triglycerides he carries, and the ability to exist on stored fat is demonstrated nicely by the use of starvation in the treatment of the obese. Patients have been deprived of all food for periods as long as eight months, taking only water and vitamin supplements. One such patient lost 33.7 kilograms of body weight. This proved to be a risky procedure with occasional sudden deaths, but in the main, most of the obese people who were subjected to such a regimen got along well. Within 30 days, the daily loss of nitrogen fell to less than 0.3 mole, so there was only rarely any difficulty from failure to maintain protein synthesis. (However, the adjustment of the kidney to the decreased demands on excretion did lead to an increase in the uric acid concentration, occasionally causing gout.)

Many of the starved have blood glucose concentrations as low as 2 mM without ill effect. This leads us to a consideration of a major change in the metabolic economy occurring with starvation. The production of acetoacetate and 3-hydroxybutyrate by the liver sharply increases, and these compounds become a major fuel for the brain.

Glucose is the major fuel for the brain in a well-fed individual on a mixed diet. Since the RQ of the brain remains near 1.0 under all circumstances, it was thought until very recently that the brain always depended upon an available glucose supply, believed to be derived by gluconeogenesis from amino acids during starvation. However, simple calculations with long-available data showed that the total supply of glycogen at the beginning of a fast and the amount of degradation of amino acids during a fast could account for only a small fraction of the total metabolism of the brain, which is responsible for 20 per cent of the total oxygen uptake at rest. It was then shown by placing catheters in the neck vessels of volunteers and measuring the arterial-venous differences in metabolite concentrations, that the brains of starved humans use the 3-oxybutyrates as fuels. The RQ values for the oxidation of acetoacetate and 3-hydroxybutyrate are 1.00 and 0.89 respectively, which partially explains why the utilization of these compounds was not observed earlier.

We saw earlier that acetoacetate may be formed from several compounds. It is a direct product of the metabolism of tyrosine and phenylalanine. It is also formed from 3-hydroxy-3-methylglutaryl coenzyme A, which in turn is created by condensation of acetyl coenzyme A and acetoacetyl coenzyme A, in addition to being a product of the metabolism of leucine. We previously emphasized the utilization of these routes in disposing of the carbons of amino acids. However, acetyl coenzyme A and acetoacetyl coenzyme A are also formed by the oxidation of fatty acids, and acetoacetate and 3-hydroxybutyrate are potential end products of this process. Elevation of the free fatty acid concentration causes increased production of 3-oxybutyrate by the liver.

Under ordinary circumstances, the output of the 3-oxybutyrates by the liver is relatively small, rising in the morning before breakfast, and subsiding after carbohydrates and proteins are eaten. The heart can effectively use these compounds at low concentrations, as will the skeletal muscles at somewhat higher concentrations. Together, these tissues will remove the compounds from the blood fast enough to keep the concentrations low. This explains why little is found in the blood during exercise strenuous enough to cause rapid mobilization of fatty acids from adipose tissue; the level sharply rises shortly after the contractions are stopped because synthesis continues while consumption diminishes. Current estimates indicate that this process accounts for only a few per cent of the total energy yield on mixed diets.

Starvation causes a sustained massive mobilization of fatty acids from the adipose tissue. At the same time, owing to the declining supply of glucose, an increased fraction of any citric acid cycle intermediates available in the liver is diverted to the formation of glucose. This combination of circumstances results in an increased oxidation of fatty acids by the liver with a resultant increased elaboration of the 3-oxybutyrates. Although the exact mechanism of these events is still in doubt, there are likely contributing factors. One is a decline in the concentration of oxaloacetate, owing to its removal for gluconeogenesis, which would slow the oxidation of acetyl coenzyme A. Another is increased oxidation of fatty acids to supply ATP within the liver for protein synthesis and other normal demands, thereby compensating for the decreased utilization of lactate and amino acids. Still another is an increase in the glucagon/insulin ratio (next chapter).

Estimates indicate that the metabolism of the 3-oxybutyrates may account for as much as one-quarter of the total energy production during starvation. The formation of these compounds provides a mechanism for feeding the brain and conserving the meager supply of glucose generated from amino acids.

Whether obese or not, total deprivation of food must eventually result in the exhaustion of the triglyceride supply. As this occurs, the only remaining source of energy is in the metabolism of the amino acids derived from tissue proteins. This is not adequate to sustain anything near normal activity, which would require the loss of something like 400 grams of protein (2 kg of tissue) per day, and death occurs rapidly after the triglyceride supply is gone.

In less severe starvation, with some food available, a person with ordinary amounts of adipose tissue gradually becomes more apathetic as the triglycerides disappear, thereby diminishing his energy requirements. He may lose so much tissue and still be alive that most of the external anatomical features of his skeleton are plainly molded by his skin.

FURTHER READING

Lusk, G.: (1928) *The Science of Nutrition,* 4th ed. W. B. Saunders. The best source for the foundations of calorimetry.

Swift, R. W., and K. H. Fisher: (1946) *Energy Metabolism. In* G. H. Beaton and E. W. McHenry, eds.: *Nutrition, A Comprehensive Treatise.* Vol. 1, p. 181. Academic Press.

Kleiber, M.: (1947) *Body Size and Metabolic Rate.* Physiol. Rev., *27*: 511.

Merrill, A. L., and B. K. Watt: (1973) *Energy Value of Foods.* U.S.D.A. Handbook No. 74.

Southgate, D. A. T., and J. V. G. A. Durnin: (1970) *Calorie Conversion Factors.* An experimental reassessment. Br. J. Nutr., *24*: 517.

Garrow, J. S.: (1976) *Energy Balance and Obesity in Man.* North-Holland. An informative and splendidly contentious paperback.

McGilvery, R. W.: (1975) *Use of Fuels for Muscular Work. In* H. Howald and J. R. Poortsmans, eds.: *Metabolic Adaptation to Prolonged Physical Exercise.* p. 12. Birkhauser Verlag Basel.

Keul, J., E. Doll, and D. Keppler: (1972) *Energy Metabolism of Human Muscle.* Med. Sport, *5*: 7.

McClellan, W. S., et al.: (1930–31) *Clinical Calorimetry: XLV. Prolonged Meat Diets with a Study of Kidney Function and Ketosis.* J. Biol. Chem., *87*: 651.

XLVI. Prolonged Meat Diets with a Study of the Metabolism of Nitrogen, Calcium and Phosphorus. J. Biol. Chem., *87*: 669.

XLVII. Prolonged Meat Diets with a Study of Respiratory Metabolism. J. Biol. Chem., *93*: 419.

Kinsell, L. W.: (1964) *Calories Do Count.* Metabolism, *13*: 195.

Keys, A., et al.: (1950) *Human Starvation.* Vols. 1 and 2. University of Minnesota Press.

Owen, O. E., et al.: (1967) *Brain Metabolism During Fasting.* J. Clin. Invest., *46*: 1589.

Altman, R. L., and D. S. Dittmer: (1968) *Metabolism.* Fed. Am. Soc. Exp. Biol. This handbook contains many useful tables.

38 HORMONES

It ought to be evident at this point in our study that survival of a multicellular, multiorgan animal frequently hinges upon some organization of responses by its constituent parts. The cells in the engine room and the cells on the quarterdeck must communicate. Signals may pass through the nerves as impulses or through the blood as changes in concentration. The substance changing in concentration may be a metabolite or it may be a specific message in the form of a hormone produced and secreted by an endocrine gland.

Our purpose in this chapter is to develop the general principles of the action of hormones and the regulation of their formation through a survey of specific examples. This will be accomplished in part by drawing together some information already introduced during our discussion of metabolism, but we also shall be considering some additional compounds. We want to lay a basis for the more detailed consideration of their physiology and clinical applications that comes later in your studies.

We shall see considerable diversity, ingenuity, dependence, independence, and organ cooperativity as we examine features of the biochemistry of the hormones. Some of the endocrine organs are arrayed in a pecking order in which the chain of command runs from the hypothalamus to the anterior pituitary gland and from there to individual endocrine glands such as thyroid, adrenal cortex, gonads, and others.*

GENERAL MECHANISM OF ACTION

Most, if not all, hormones bind with receptors specific for the particular hormone in target cells. The density of the receptors, as well as the concentration of the hormone in the interstitial fluid, is an important factor in determining the extent to which a particular cell will be influenced by a hormone. Hormones may be divided into two general classes according to the kind of event that follows binding:

(1) Some hormones primarily affect the properties of the plasma membrane itself. These include all the peptide hormones, such as glucagon, insulin, and the

---

*While some glands (parathyroid, juxtaglomerular cells in the kidney) appear independent of the action of other glands, their responses are influenced through the nervous system.

hormones of the pituitary gland. They also include the catecholamines and prostaglandins.

(2) Some hormones are taken into the cell and transported to the nucleus, where they influence the nature and rate of gene expression. These include triiodothyronine from the thyroid gland and all of the steroid hormones.

Action on the Plasma Membrane

Activation of adenyl cyclase, as we have seen, is a major mechanism for endocrine action, with a primary effect being the activation of specific protein kinases. Many of the effects of adrenaline and glucagon result from this process; we shall see that some of the effects of other peptide hormones and the prostaglandins may also appear in this way.

Activation of a guanyl cyclase may also be important in endocrine action. This enzyme generates 3′,5′-cyclic GMP in the same way that adenyl cyclase generates cyclic AMP. However, it has proved difficult to pin down exactly what is the function of cyclic GMP and the effects of hormones on its formation. It is premature to make definitive statements.

Alterations in membrane permeability are also important consequences of hormone action. We have already mentioned the possibility that some of the effects of catecholamines are due to an increased flow of Ca^{2+} into the cell, perhaps related to activation of a distinct protein kinase. Calcium, rather than cyclic AMP, may act as a second messenger in triggering some metabolic events. Insulin also increases the permeability of the plasma membrane in some cells to glucose and amino acids; this effect may or may not be related to an effect on calcium ion flow or on a protein kinase.

Action on Gene Expression

The number of different kinds of mRNA being made at a given moment by a mammalian cell is only a small fraction, sometimes less than 3 per cent, of the genetic information stored within the cell. Although differentiated cells have the complete complement of genes necessary to construct all of the diverse cells in the body, they differ from the other cells in both kind and quantity of protein being made.

The fertilized ovum divides into identical, or nearly identical, cells, which change their character with successive divisions. The cells begin to differentiate, first into precursors of various classes of mature cells, and then through intermediate forms into the individual types present in the completed organism. Each stage in this process, from fertilized ovum to fully differentiated cell, involves the transcription of a different set of genes, all of which must be present in the original fertilized ovum. The presence of some hormones is a necessary part of this selective transcription, both during differentiation and the subsequent life of the differentiated cells.

It is not known how the expression of eukaryotic genes is regulated. In order for a gene to be transcribed, it must be accessible to RNA polymerase. Exposure of DNA involves unfolding of the chromatin and dissociation of histones, and perhaps other proteins. The histones cannot in themselves be the primary regulating proteins because they occur at regular intervals along the DNA fiber and are not associated with specific genes.

Current speculations revolve around an interaction of an effector molecule, that is, a signal, with a regulating protein in such a way as to cause the protein to be bound at a specific site on a chromatin fiber, akin to the mechanisms of operon regulation in prokaryotes (p. 62), but not identical to them. The binding may occur on DNA itself, or to still another protein that is one of the non-histone components of chromatin. The critical feature is that the binding in some way affects the transcription of a neighboring segment of DNA by RNA polymerase. The effector that precipitates the sequence may be any molecule, such as a metabolite, but in many cases it is a hormone.

Those hormones directly affecting gene expression combine with receptor proteins after entering the cell. (These are not to be confused with receptor sites on the cell surface.) The hormone-protein complex combines with the chromatin so as to change the transcription of selected DNA sequences into mRNA.

ADRENALINE

Summarizing our earlier observations, the chromaffin cells of the adrenal medulla form adrenaline and lesser amounts of noradrenaline. (The proportions are reversed in some animals.) These catecholamines combine with receptor sites on the target cells. Phenomenological evidence indicates the existence of different kinds of receptors. The first clue was a differential response toward a synthetic agonist, isoproterenol, and the natural agonists, adrenaline and noradrenaline. A given response was said to be due to combination with alpha receptors if adrenaline was most effective, and with beta receptors if isoproterenol was most effective. Later subdivisions were made on the basis of the action of inhibitors (blocking agents):

Receptor	Agonists	Antagonists
α	adrenaline $\geqq$ noradrenaline > isoproterenol	phentolamine
β_1	isoprotenenol > noradrenaline $\geqq$ adrenaline	practolol
β_2	isoproterenol > adrenaline > noradrenaline	propranolol

Most of the actions ascribed to beta receptors now appear to be due to activation of adenyl cyclase. The actions ascribed to alpha receptors are perhaps in part due to internal release of Ca^{2+} and in part to inhibition of adenyl cyclase, depending upon the cells involved.

Distinct structures corresponding to alpha and beta receptors have not been isolated, and the separate effects may in some instances be due to two or more groupings of proteins in the membrane adjacent to the site of catecholamine binding. It is not difficult to visualize arrangements by which the presence of one molecule of catecholamine can have more than one effect within the neighboring areas of plasma membrane.

The prominent metabolic effects of the catecholamines are the mobilization of stored glycogen and triglycerides, but catecholamines also affect the actions of other hormones in their target cells. We do not have time to discuss the interactions of the various hormones, but we can sense the possible permutations by realizing that some cells may have both beta adrenergic receptors and receptors for other hormones that activate adenyl cyclase. Release of adrenaline may in those instances augment the action of the other hormones.

TRIIODOTHYRONINE AND THYROXINE (T3 and T4)

The action of these hormones and the regulation of their production by the thyroid gland provide examples of many general principles, in addition to being important in their own right. The thyroid regulates metabolic activity and promotes growth and development through the synthesis and release of thyroxine (tetraiodothyronine, T4) and triiodothyronine (T3):

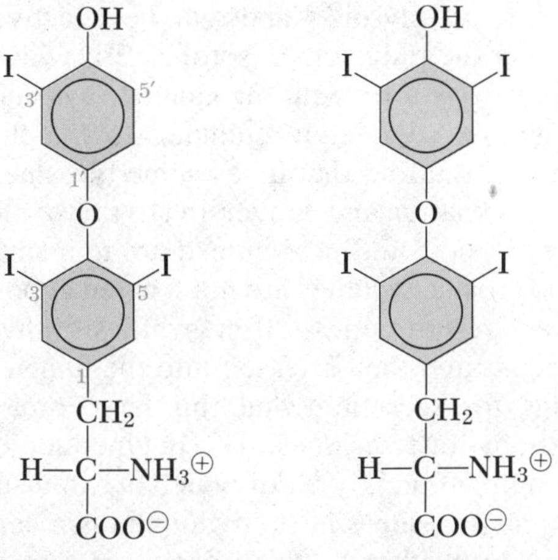

3,3′,5-triiodothyronine (T3)

thyroxine (T4, 3,3′,5,5′-tetraiodothyronine)

Thyroxine is the major product of the gland, but triiodothyronine is more active and may be the only form bound to receptor proteins in the nucleus, thereby altering gene expression. We shall see that these hormones are made by iodinating specific tyrosine residues in a protein, **thyroglobulin**.

Specific effects of thyroid hormone include striking changes in gross appearance and activity. Lack of hormone can produce a listless, constipated, coarse-haired slow-pulsed individual who complains of being cold. Oversecretion of thyroxine or triiodothyronine leads to a garrulous, hyperkinetic individual with rapid heart rate, diarrhea, and huge appetite who complains of being hot. While these "classical" findings are easily discerned, the manner in which thyroid hormone produces these effects is not clear. The increased basal metabolic rate led to a search for many years for uncoupling of oxidation from phosphorylation in mitochondria. Early claims of positive results were quietly abandoned, and we have no valid substitute. However, the absolute necessity of the hormone for normal fetal development, together with its specific binding by a nuclear protein, makes alterations in genetic expression the more likely possibility.

The thyroid gland is a bilobed organ in the anterior portion of the neck. It is really a collection of individual glands in the form of follicles, which are circular in cross section with a central lumen in which the newly synthesized thyroid hormone still present in thyroglobulin is stored. Apical portions of the cuboidal cells comprising the follicles contain numerous microvilli and secretory granules on the luminal side.

Synthesis of Thyroxine

Transport of iodide. The thyroid consumes about 70 to 100 μg of iodide per day for hormone synthesis, which it obtains by re-utilizing the iodide released upon degradation of the hormones, making up any deficit from the dietary intake. The daily intake of iodine in the United States typically ranges from 200 to 500 μg a day. Dietary iodine is reduced to iodide and almost completely absorbed into the blood stream from the intestinal tract. Iodide is actively transported into thyroidal cells through linkage to a (Na^+ + K^+)-ATPase system. A thyroid in a normal adult contains *ca*. 6,000 μg of iodide, whereas all of the rest of the body has only about 75 μg of inorganic iodide and 500 μg of organic iodide. Follicular cells are avid collectors of iodide from the blood stream, much more so than any other cells in the body. Indeed, hyperactive thyroid glands can be selectively and therapeutically destroyed by drinking the radioactive isotope, ^{131}I, which is concentrated in the thyroid gland and destroys it by emitting gamma rays and electrons.

Thyroglobulin is almost completely confined to the thyroid gland, which synthesizes it to serve as a scaffold that holds some tyrosine residues in specific configurations for ready iodination and conversion to thyroxine and triiodothyronine. Other proteins can be iodinated, and indeed are to a small extent within the thyroid gland, but their tyrosine residues are not located at positions favorable for combination into the active hormones. After synthesis, thyroglobulin is transferred to secretory vesicles and then released into the lumen. Carbohydrates are added during packaging for secretion, and this large protein (670,000 M.W.) contains some 280 carbohydrate residues. The polypeptide chain is rich in cysteine residues, with about 200, nearly all of which are in disulfide linkage.

Iodination of the tyrosyl residues in thyroglobulin is a complex process (Fig. 38–1), not yet completely elucidated, which occurs in the apical portion of the cells — the part next to the lumen. The iodinating enzyme is a heme-containing peroxidase, which also travels through the cell as if it is to be secreted. However, it is probably retained in the plasma membranes or other structures at the lumen-cell interface. (This interface has a complexly interdigitated morphology.) The sources of the necessary H_2O_2 also have not been defined; likely possibilities are the transfer of electrons from NADPH through cytochrome c to oxygen, or from NADH through cytochrome b_5 to oxygen by extramitochondrial enzymes. The mechanism of iodination may involve free radical forms of both iodine and the phenolate portion of tyrosine residues, which combine to form monoiodotyrosine residues. Further reaction forms diiodotyrosine residues.

The coupling of two molecules of diiodotyrosine to form thyroxine (tetraiodothyronine) may follow the scheme shown in Figure 38–2. Thyroglobulin is ob-

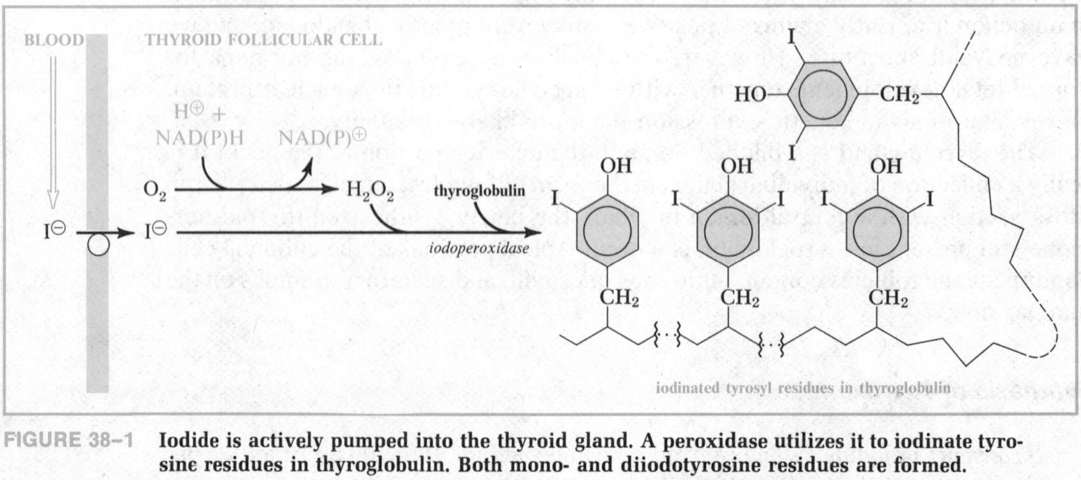

FIGURE 38–1 **Iodide is actively pumped into the thyroid gland. A peroxidase utilizes it to iodinate tyrosine residues in thyroglobulin. Both mono- and diiodotyrosine residues are formed.**

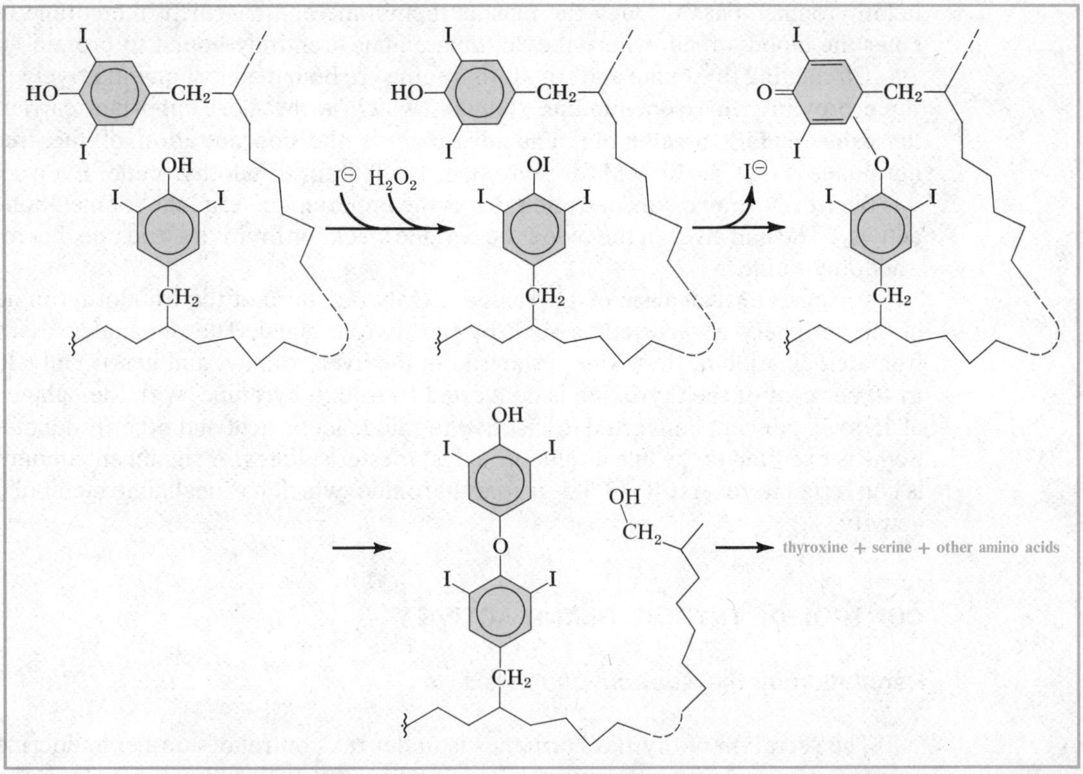

FIGURE 38–2 **Possible mechanism for coupling of two adjacent iodinated tyrosine residues in thyroglobulin. The example shows the formation of a thyroxine residue; triiodothyronine residues are created by a similar condensation of a monoiodotyrosine residue with a diiodotyrosine residue. According to this postulated mechanism, iodine also acts as the necessary oxidant for the condensation.**

viously constructed in a fashion that will facilitate iodination of residues that are in favorable positions for coupling to make the iodothyronines. Triiodothyronine is generated by coupling monoiodo- and diiodotyrosine in a similar fashion. Many proteins can be iodinated in vitro by the thyroid peroxidase system, but little thyroxine is formed. Human thyroglobulins from normal glands, on the other hand, had on the average only 15 tyrosine residues iodinated out of the 118 present per molecule, as analyzed by one laboratory. Of these, roughly five residues were still present as monoiodotyrosine, and three as diiodotyrosine, but six had been converted to three residues of thyroxine and one had been converted to triiodothyronine (one residue in two molecules of thyroglobulin). The efficiency of formation of the iodothyronines increases as the iodine content increases *in vivo,* indicating that those residues are preferentially iodinated that are in appropriate positions for conjunction to form the iodothyronines, of which more than 80 per cent will be the tetraiodo compound (thyroxine) in individuals with an adequate iodine supply.

Secretion of the iodothyronines. Secretion is initiated by the return of iodinated thyroglobulin to the cell through fusion of droplets of the lumen contents with lysososomes to form phagosomes, in which the protein is hydrolyzed to its constituent amino acids. The released iodinated residues include both mono- and diiodotyrosine, as well as the coupled tri- and tetraiodothyronines. Iodine is removed from the iodotyrosines and becomes available for re-utilization. The iodothyronines pass through the plasma membrane and basement membrane to enter the blood stream where they circulate almost entirely bound to protein.

Circulating thyroxine and triiodothyronine are bound almost quantitatively to three proteins: **thyroxine-binding globulin,** which is the most important carrier, **thyroxine-binding prealbumin**, and **albumin**, so the concentration of the free hormones is only 4×10^{-11} M for thyroxine, 1×10^{-11} for triiodothyronine. Even so, it is the free hormone concentration that is the important determinant of metabolic activity. The half-lives in the blood are *ca.* one week for thyroxine and one day for triiodothyronine.

Peripheral Metabolism of Thyroxine. Only one third of the triiodothyronine in the periphery is secreted as such by the thyroid gland. The remainder arises from deiodination of thyroxine, primarily in the liver, kidney, and heart. Only 30 to 40 per cent of the thyroxine is converted to triiodothyronine, with the balance of 15 to 20 per cent converted to inactive tetraiodoacetic acid and other products. Some is excreted in the bile as glucuronides or ester sulfates. A significant amount is converted to reverse T_3 (3,3′5′ triiodothyronine) which has negligible metabolic activity.

CONTROL OF THYROID GLAND ACTIVITY

Regulation by the Adenohyphophysis

The secretion of thyroid hormones is under the control of another endocrine gland, the **adenohypophysis** or **anterior pituitary gland**. Some of the cells in the adenohypophysis secrete a polypeptide hormone, **thyrotropin (thyroid stimulating hormone, TSH)**, which reaches the thyroid gland through the blood and stimulates it to release thyroxine and triiodothyronine. The thyrotropin-forming cells of the anterior pituitary gland are in turn stimulated by another hormone, the **thyrotropin releasing hormone**, which is an oligopeptide formed in the **hypothalamus** and is transported to the anterior pituitary gland through a portal circulation in the pituitary stalk. This sequence of cascade activations, hypothalamus to anterior pituitary gland to thyroid gland, is typical of a sequence affecting other endocrine glands, and therefore deserves more detailed attention. Like other cascade mechanisms, it greatly amplifies signals, with one nanogram of hypothalamic hormone causing the release of many times as much thyrotropin, which in turn stimulates the release of much more thyroxine from the thyroid gland.

The pituitary gland is a collection of differentiated cells that act as a message center. Signals reach it from the hypothalamus, cerebrospinal fluid, blood plasma, and nerve terminals. In response to these signals, the involved cells transmit their messages in the form of peptide hormones. Anatomically, the pituitary gland is enclosed in a bony box, the sella turcica, with a stalk connecting the gland to the hypothalamus. It is really two distinct glands. The **posterior pituitary gland**, or **neurohypophysis**, secretes the hormones **vasopressin** and **oxytocin**, which reach the gland for storage in secretory vesicles through the axons of specialized nerves arising in the hypothalamus, where these hormones are made in the cell bodies. We shall say more about the neurohypophyseal hormones in the next chapter.

The anterior pituitary gland synthesizes, as well as secretes, several polypeptide hormones. The controlling messages in the form of hypothalamic hormones reach it through a portal system of capillary vessels draining the median eminence of the hypothalamus and passing the blood through the anterior pituitary gland before it is returned to the heart. All of the hormones of the anterior pituitary

HORMONES OF THE ANTERIOR PITUITARY GLAND

TABLE 38–1

Name	Abbreviation	Target Organ	Subunits
thyrotropin	TSH	thyroid gland	2
luteinizing hormone	LH	gonads	2
follicle stimulating hormone	FSH	gonads	2
corticotropin (adrenocorticotropic hormone)	ACTH	adrenal cortex	1
prolactin		mammary glands	1
somatotropin (growth hormone)	GH	many	1

gland are polypeptides (Table 38–1), and their secretion is under the control of other factors in addition to the hypothalamic hormones. Now let us consider the steps in this general scheme that directly affect the formation of thyroid hormones.

Thyrotropin releasing hormone is a tripeptide; it is almost certainly made by cleaving a larger precursor because it contains a pyroglutamyl group:

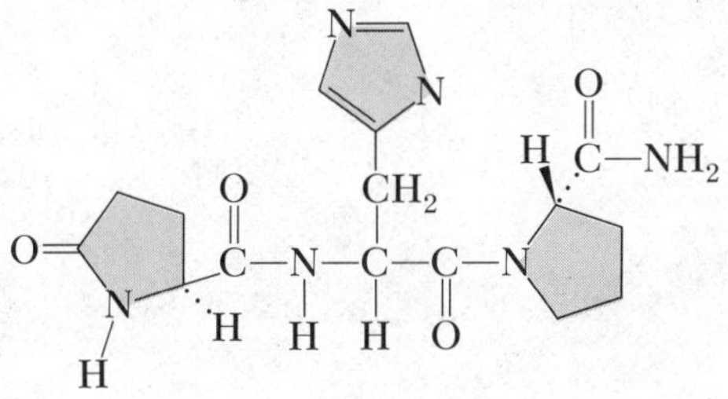

thyrotropin releasing factor
(pyroglutamyl histidyl proline amide)

The cells synthesizing this hormone in the hypothalamus release it upon stimulation of alpha adrenergic receptors by noradrenaline; this, therefore, is an important locus of control over thyroid action through the nervous system. Other hypothalamic hormones are listed in Table 38–2. These include somatostatin, which has an inhibitory action on secretion of hormones by the anterior pituitary gland. Somotostatin is also made in the pancreatic islet cells and specialized cells of the upper gastrointestinal tract, and will be discussed further in connection with the pancreas.

Thyrotropin is made by particular basophilic cells in the anterior pituitary gland known as thyrotrophs. (There is a curious admixture in the literature of the stems **tropic**, meaning turning, and **trophic**, meaning feeding, in connection with these hormones.) These cells are stimulated to release thyrotropin upon the binding of thyrotropin releasing hormone to their plasma membranes. The mechanism is not clear-cut; as was mentioned in the general discussion, these peptide

HYPOTHALAMIC HORMONES OR FACTORS*

TABLE 38–2

thyrotropin releasing hormone
luteinizing hormone and follicle stimulating hormone releasing hormone
somatostatin
prolactin inhibitory factor
corticotropin releasing factor
prolactin releasing factor

*The term *factor* is used when the structure is not known. The list is not complete.

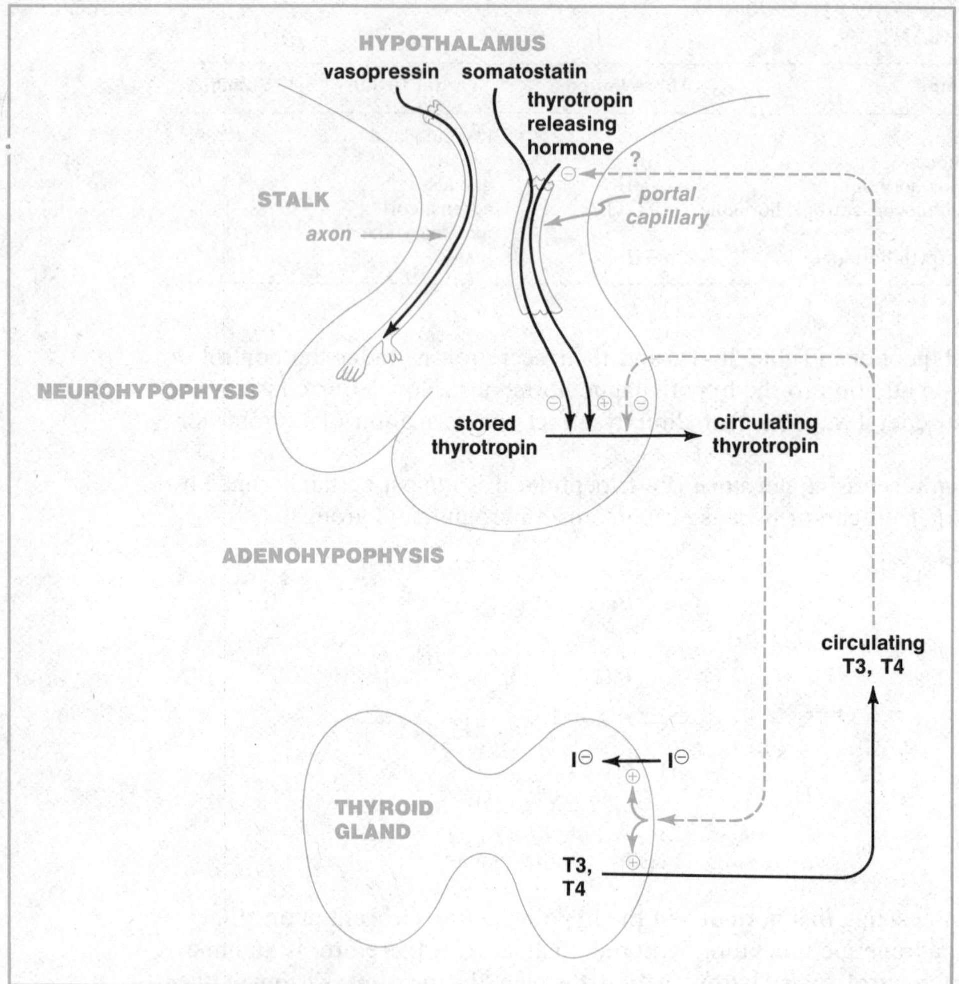

FIGURE 38–3 **Regulation of the release of thyroid hormones through the hypothalamus and anterior pituitary gland.**

hormones may act by initiating several changes, including activation of adenyl cyclase, activation of protein kinases by other routes, and altering the permeability and release of Ca^{2+}.

A major control of thyrotropin secretion is an inhibition by triiodothyronine or thyroxine. As the circulating iodothyronines increase in concentration, they shut off the release of thyrotropin, the signal for their own formation (Fig. 38–3). This very sensitive feedback loop is the device by which the blood hormonal concentration is kept relatively constant; similar devices are used for regulating the concentrations of hormones produced by other glands under control of the anterior pituitary. The action of the hypothalamic thyroid releasing hormone and other regulating factors may be regarded as devices for overriding the primary control through feedback inhibition. The secretion of thyrotropin is also inhibited by somatostatin from the hypothalamus.

Thyrotropin consists of an α and a β subunit. The same polypeptide chain is used to make the α subunit of other hormones from the anterior pituitary (luteinizing hormone and follicle stimulating hormone), which resemble thyrotropin in being glycopeptides. A variable number of residues are removed from the end of the α chain in these hormones. Thyrotropin and the other hormones get their distinctive characters from their β subunits.

Within minutes of administering thyrotropin to experimental animals, the cells of the thyroid gland begin synthesis of mRNA, active transport of iodide into the cells, and reabsorption of thyroglobulin from the lumen. Again, these responses may be mediated in part by activation of adenyl cyclase and in part by other effects on the plasma membrane of the thyroid cells.

Thyrotropin has less well-defined functions in other tissues. Perhaps the clearest demonstration came from the discovery that the hormone could be partially hydrolyzed by pepsin to produce a large fragment containing most of the β chain, but only part of the α chain. This fragment was devoid of activity on the thyroid gland, but it stimulated development of the retro-retinal tissues in the guinea pig to produce the exophthalmos (protruding eyeball) sometimes associated with hyperthyroidism. These tissues were being stimulated by thyrotropin, not by the iodothyronines.

Regulation by Iodide Concentration

Changes in the concentration of circulating iodide cause opposite changes in the release of the iodothyronines. Part of the effect comes from a direct inhibition of the follicle cells by iodide; part can be indirect. Iodide has an inhibitory effect on the thyrotrophs in the anterior pituitary; as its concentration rises, less thyrotropin is released.

Clinical Disruption of Thyroid Hormone Production. Administering radioactive ^{131}I or plying cold steel are widely used and effective treatments for hyperactive thyroid glands. More sophisticated attacks upon the biochemical pathways involve blocking of specific sites with drugs. Monovalent anions (thiocyanates, perchlorates, and nitrates) inhibit active transport of iodide. Perchlorate can be used in humans. Propylthiouracil and methimazole are clinically useful drugs that interfere with iodination of tyrosyl residues:

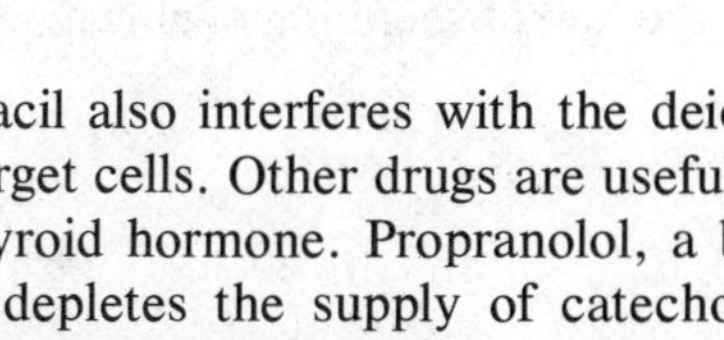

propylthiouracil

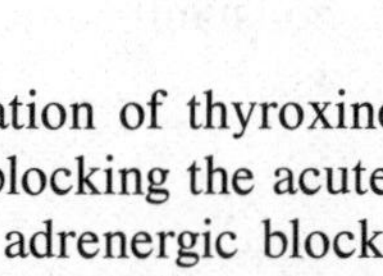

methimazole

Propylthiouracil also interferes with the deiodination of thyroxine to triiodothyronine in target cells. Other drugs are useful in blocking the acute symptoms of excessive thyroid hormone. Propranolol, a beta adrenergic blocker, and reserpine, which depletes the supply of catecholamines, will relieve nervousness, fever, and hyperkinetic activity.

PANCREATIC ISLET HORMONES

Hormones of the islets of Langerhans play a vital role in the regulation of fuel metabolism. One of the commonest serious disorders of man, diabetes mellitus, affecting 5 per cent of the United States population, represents a disturbance of

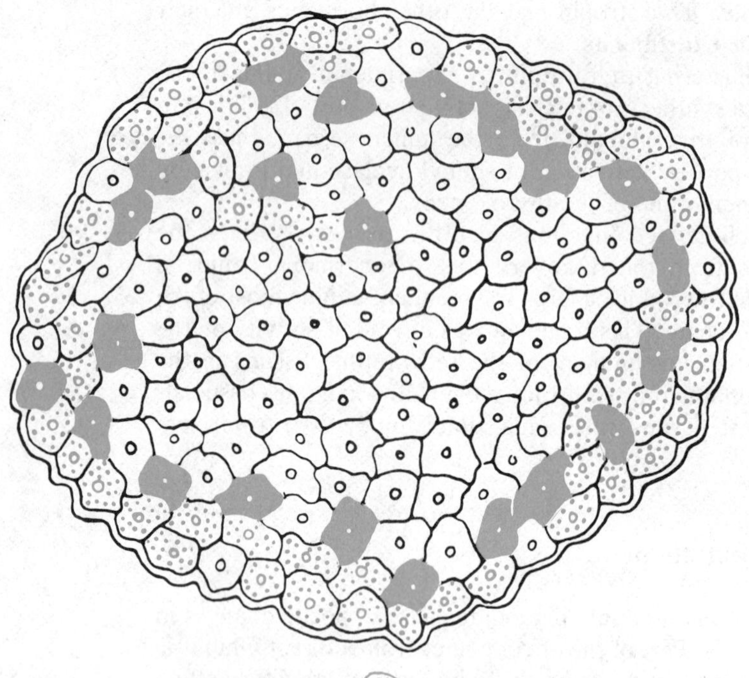

A CELLS Glucagon
D CELLS Somatostatin
B CELLS Insulin

FIGURE 38–4

Arrangement of cells in the pancreatic islets of Langerhans. Reproduced by permission from R. H. Unger and L. Orci: (1977) *Role of Glucagon in Diabetes.* Arch. Intern. Med., *137*: 484. Copyright 1977, American Medical Association.

the ability of the islet cells to properly regulate the glucose concentration. The distribution of the various cell types in the islet is shown in Figure 38–4. Glucagon is synthesized by the A cells, insulin by the B cells, and somatostatin by the D cells. About 60 per cent of the cells, occupying the central zone of the gland, are B cells. Cell to cell contact is largely B to B cell. At the margin of the organ, a thick rim of A cells, one to two cells thick, makes up about 30 per cent of the total. Interspersed between A and B cells, or occasionally between A cells, are the somatostatin-secreting D cells. There are numerous gap and tight junctions between cells. The areas where the three cell types meet is invested with a rich blood and nerve supply. The islets act as a sensor of the glucose concentration and its rate of change, constantly adjusting the rate of secretion of glucagon and insulin to match conditions.

Glucagon

Release of glucagon is stimulated by decreasing glucose concentrations. The level of glucagon in plasma after fasting ranges from 30 to 200 ng/l and is lowered after a carbohydrate meal. Increases in amino acid concentration or stimulation of the sympathetic nervous system also stimulate glucagon secretion. Glucagon is usually measured by a radioimmunoassay technique (Chapter 12). Four immunoreactive forms have been reported in human plasma. A form with a molecular weight of 3,500 appears to have the greatest hormonal activity. The others include a smaller product (M.W. = 2,000), a possible proglucagon (M.W. = 9,000), and a very large form (M.W. = 180,000).

We have already seen that glucagon stimulates adenyl cyclase in the liver and adipose tissue, thereby causing mobilization of glucose from the liver glycogen stores and of fatty acids from the adipose tissue triglyceride stores. The cAMP-dependent protein kinase in the liver also phosphorylates pyruvate kinase, making it inactive, and thereby hangs a tale.

The conversion of phospho-*enol*-pyruvate to pyruvate by the pyruvate kinase reaction and the conversion of pyruvate back to phospho-*enol*-pyruvate via oxaloacetate both occur in hepatocytes and constitute a futile cycle (Fig. 38–5). The cycle is evidently controlled by alterations in the pyruvate kinase activity, as well as by the effect of acetyl coenzyme A in activating pyruvate carboxylase (p. 475). At times of carbohydrate surplus, fructose bisphosphate activates the enzyme by relieving an inhibition by ATP, and this promotes the utilization of triose phosphates to make pyruvate, and then acetyl coenzyme A. Glucagon, on the other hand, is a signal of glucose deprivation, and it shuts off the flow of triose phosphates toward acetyl coenzyme A by tripping an inhibitory phosphorylation of pyruvate kinase.

Glucagon therefore stimulates gluconeogenesis from lactate or amino acids. In sum, glucagon stimulates the delivery of glucose to the extracellular fluid by both glycogenolysis and gluconeogenesis in the liver. By raising the glucose

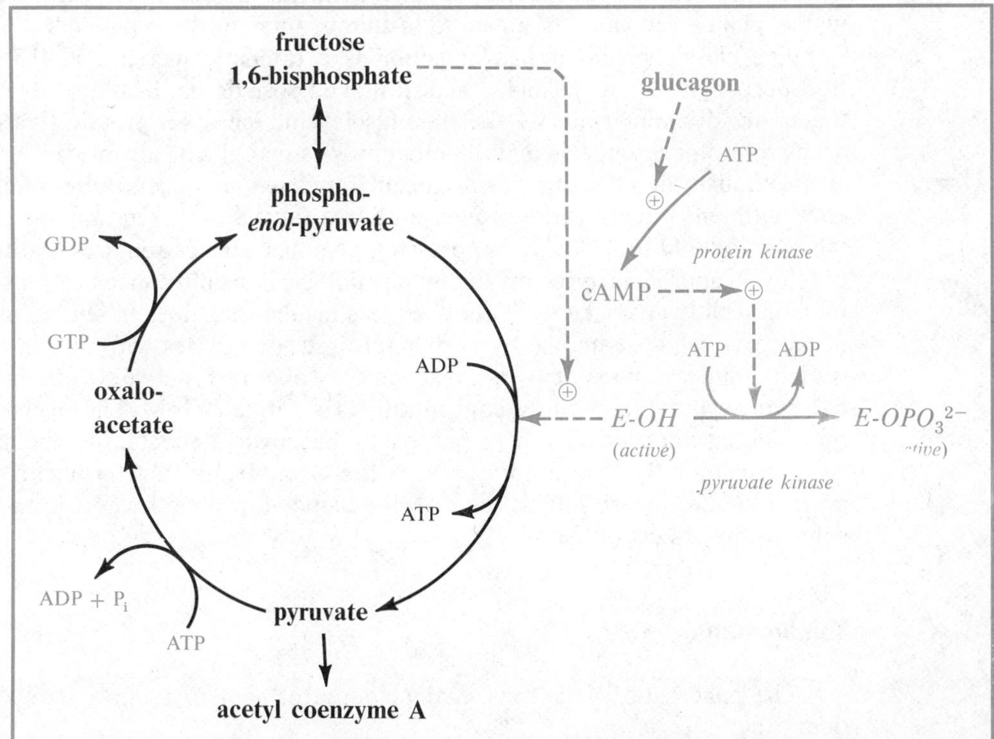

FIGURE 38–5 **The flow of pyruvate carbons toward fructose 1, 6-bisphosphate in gluconeogenesis or toward acetyl coenzyme A in lipogenesis is regulated by the predominating reactions in a futile cycle. Glucagon triggers a conversion of pyruvate kinase to an inactive phosphorylated form, leaving the conversion of pyruvate to phospho-*enol*-pyruvate *(left)* as the predominant route, and thereby promoting gluconeogenesis.**

concentration, glucagon can indirectly stimulate insulin release to enable use of the glucose.

The diversion of oxaloacetate to gluconeogenesis by the action of glucagon probably augments the ketosis seen with starvation, low-carbohydrate diets, or diabetes, although the presence of glucagon is not obligatory for development of ketosis.

Insulin

Active insulin is made by progressive modification of preproinsulin (Chapter 8). A rise in the concentration of blood glucose is a primary signal for secretion of insulin, and the release is prompt, beginning within one minute. Insulin is also secreted in response to a rise in blood amino acid concentration; arginine is the most effective signal, and arginine-loading tests are used to test beta cell function. The sympathetic nervous system inhibits insulin secretion through alpha adrenergic receptors.

Receptors for insulin are present in a variety of tissues, although the dependence of the tissues on insulin varies. Even the brain has some receptors, but it can use glucose very well in the absence of insulin. The number of receptors in cells of a given tissue varies between individuals, and an inverse relationship has been demonstrated between receptor concentration and the ambient insulin level. Where non-obese normal individuals had a basal insulin level of 35 to 145 pM*, obese non-diabetic individuals had the high basal insulin levels of 180 to 440 pM in their blood, but they had a decrease in the concentration of insulin receptors. This may explain a seeming resistance to insulin in some of these patients.

The classic result of insulin action is a dramatic increase in the rate of transport of glucose into skeletal muscle and adipose tissue. Insulin also promotes the uptake of amino acids by skeletal muscles and increases protein synthesis. It accelerates lipid synthesis and inhibits lipolysis and gluconeogenesis.

Although the effects of insulin occur rapidly, the concentration required to elicit different effects varies. Release of free fatty acids from adipose tissue is inhibited by 200 to 350 pM. Suppressing gluconeogenesis requires 700 to 1,400 pM, levels approaching the maximum physiological insulin concentrations. Inhibition of hepatic glycogenolysis requires less insulin than does inhibition of gluconeogenesis. Glucose uptake by peripheral tissues increases with insulin concentration until a maximum is reached at about 1,400 pM. Disposal of D-3-hydroxybutyrate requires concentrations of 350 to 700 pM. The mechanism(s) by which insulin produces these purposeful but diverse effects, like the mechanisms of many polypeptide hormones, is not clear. It may act through a distinct protein kinase; it may inhibit the cAMP-dependent protein kinase; it may block some of the effects of Ca^{2+}.

Somatostatin

The pancreatic islets are a major source of somatostatin, a tetradecapeptide:

---

*Insulin quantities are frequently expressed in units (U), with a unit defined as the amount giving the same biological activity as 42 micrograms of pure insulin (7.2 nanomoles). Blood concentrations are often given in microunits per ml; 1 μU ml^{-1} = 7.2 picomolar.

$$\begin{array}{l} \overset{\oplus}{(H_3N)}\overset{1}{\text{Ala}}\text{—Gly—Cys—Lys—Asn—Phe—Phe} \\ \qquad\qquad\qquad\quad | \qquad\qquad\qquad\qquad\quad | \\ \qquad\qquad\qquad\quad \text{S} \qquad\qquad\qquad\quad\;\; \text{Trp} \\ \qquad\qquad\qquad\quad | \qquad\qquad\qquad\qquad\quad | \\ \qquad\qquad\qquad\quad \text{S} \qquad\qquad\qquad\quad\;\; \text{Lys} \\ \qquad\qquad\qquad\; {}^{14}| \qquad\qquad\qquad\qquad\quad | \\ (^{\ominus}\text{OOC})\text{Cys—Ser—Thr—Phe—Thr} \end{array}$$

somatostatin (sheep, pig)

Here we have an example of a hormone synthesized and secreted in different regions of the body, including the hypothalamus. Somatostatin inhibits release of the following hormones, some of which we shall not discuss: thyrotropin, corticotropin, and somatotropin (growth hormone) by the adenohypophysis; insulin and glucagon by the pancreas; gastrin by the gastric mucosa; secretin by the intestinal mucosa; and renin by the kidney (next chapter). Somatostatin also inhibits the emptying of the stomach and secretion of both gastric acid and pancreatic enzymes.

Diabetes Mellitus

Diabetes* is a group of diseases having in common an insulin effect that is inadequate for the uptake of glucose from the blood. The resultant **hyperglycemia** frequently causes glucose to appear in the glomerular filtrate at rates exceeding the capacity of the kidney to reabsorb it (next chapter). This results in **glucosuria** and the voiding of large volumes of urine (**polyuria**), frequently at night (**nocturia**). Large volumes of water are drunk to replace the losses (**polydipsia**). Polyuria and polydipsia are also symptoms of diabetes insipidus, but they demand further attention in any event. Even in the absence of such overt symptoms, diagnosis is easily made when blood glucose concentrations are measured after an overnight fast. Single fasting values in excess of 7mM (120 mg/dl), or in excess of 9 mM (160 mg/dl) one hour after a breakfast containing 100 grams of carbohydrate, are suggestive, and repeated excessive values are diagnostic. (The reported concentrations vary among clinical laboratories, owing to differing responses of the methodology to other compounds in the samples and to calibration errors. Each laboratory must establish its own normal upper limit of blood glucose concentrations by experience. Furthermore, concentrations of glucose are higher in serum or plasma samples than they are in the whole blood.)

Diabetes is also accompanied by abnormalities in fat and protein metabolism. Not only is there an increased mobilization of fatty acids, but the liver converts a larger fraction to the oxybutyrates rather than to the triglycerides and phospholipids:

$$\text{palmitoyl residue} + 7\ O_2 \longrightarrow 4\ \text{acetoacetate}^{\ominus} + 4\ H^{\oplus}$$

Up to one mole of oxybutyrates may be excreted per day, and the concomitant production of H^+ results in acidosis, or ketoacidosis as it is commonly called in view of the simultaneous production of both ketone bodies and hydrogen ions.

---

*The word *diabetes* means an excessive volume of urine. Diabetes mellitus is sweet urine. Diabetes insipidus is dilute urine, and so on. Used alone, diabetes is taken to mean diabetes mellitus, by far the most common of the conditions.

Diabetes usually occurs in two forms, commonly designated as juvenile and adult-onset. Perhaps a better terminology would be to speak of a rapid-onset severe form and a slow-onset mild form, since there is age overlap in the appearance of the forms. Even so, many of the cases appearing in adolescence are severe and difficult to control with insulin, whereas most of the much more common cases appearing in middle and late life develop stealthily and can be controlled by dietary management for years without insulin or other drugs. Although the genetic factors involved in diabetes are complex and not resolved, it appears that the severe and mild forms do have different genetics. The severe form in many cases is provoked by other events, such as a viral infection.

A patient with diabetes may present with **diabetic ketoacidosis** — comatose, dehydrated, and with acidosis, glucosuria, and ketonuria. This frequent medical emergency may be the first indication of the juvenile form of the disease, or it may have occurred in an insulin-dependent patient who failed to take his insulin, or who developed an infection. In any event, it is a lethal condition, with a mortality remaining between 5 and 10 per cent in major medical centers.

A diabetic patient may become unconscious with **hypoglycemia** caused by taking too much insulin. This insulin shock represents starvation of the brain, and patients on insulin may carry supplies of readily absorbed sugar to counteract its first indications.

Complications. While the mortality from diabetic ketoacidosis has been dropping with better management and patient education, there has been a disappointing persistence of morbidity and mortality from other complications of the disease. Even with the best control, vascular disease involving any and all organs and regions of the body is likely to appear, causing blindness, renal failure, coronary artery disease, gangrene, and so on. We mentioned earlier (p. 164) the considerable thickening of the basement membrane in blood vessels likely to be seen upon microscopic examination in patients with diabetes.

It is not even known if tight regulation of the glucose level aids in avoiding these complications, since it would require an almost heroic effort to monitor closely the hourly fluctuations of glucose concentration, even if a large number of unbelievably cooperative subjects could be corralled for this purpose. Sensors permitting continuous measurement are being developed, but since we presently lack this technology, the use of concentrations of hemoglobin Alc or Alb (p. 243) has been proposed as a device for assessing average glucose concentrations over long intervals. The level of Hb Alc was observed to decrease within a few weeks in patients under strict control in a hospital.

Therapeutic Measures. We must remember that the basic disturbance in diabetes is a failure to achieve adequate results of insulin action within the receptor tissues that are appropriate to the metabolic state. It is not a low insulin concentration *per se*; the circulating insulin concentration may be high in a definitely diabetic person who requires supplemental insulin for control. Many obese diabetic patients regain effective tissue responses to the level of insulin that they can produce simply by losing weight. Exercise also helps through adaptive increases in the capacity to use fuels effectively.

Patients with severe diabetes lose weight; they are losing fuel in the urine and have a diminished effectiveness for utilizing amino acids to make proteins. Not only is amino acid transport into the tissues impaired, but increased amounts of the amino acids are being degraded to make glucose, which is spilled into the urine. In contrast to the usual maturity-onset diabetic, these patients require insulin to avoid loss of weight.

Giving exogenous insulin one or more times during the day does not provide the adjustment in insulin level that normal pancreatic islet cells do as they constantly adjust the delivery of insulin (and glucagon) in response to the changing glucose level and other signals. Although insulin has been modified to provide different times of action (immediate, 6–8 hours, etc.) and although intake of food can be altered to try to match insulin effect and glucose level, proper control of diabetes with avoidance of both hyperglycemia and hypoglycemia can be extremely difficult in some people. An artificial pancreas is now in the stage of human experimental application. In this system venous blood is continuously analyzed for glucose level. A computer is programmed to respond to the glucose concentration and rate of change of concentration by releasing insulin or glucagon into the blood stream.

STEROID HORMONES

Common to all steroids is a cyclopentenoperhydrophenanthrene ring system. We have previously encountered cholesterol (p. 203) and the bile acids (p. 314). Just as cholesterol is the parent of the bile acids in the liver, it is also the parent of a variety of steroid hormones in some endocrine glands. They are classified on the basis of activity:

glucocorticoids made by the adrenal cortex modify certain metabolic reactions and have an anti-inflammatory effect;

mineralocorticoids made by different cells in the adrenal cortex modify excretion of salt and water by the kidney (next chapter);

androgens, made principally by the testis, affect male sexual characteristics;

estrogens, made principally by the ovary, affect female sexual characteristics;

progesterone, made by the ovary and the placenta, affect uterine development.

hydroxylated cholecalciferols, made from an intermediate of cholesterol synthesis by the combined action of several tissues, affect calcium metabolism. This is a somewhat simplistic list, but it will serve the purpose.

Those human steroid hormones made from cholesterol are synthesized by sequences involving reactions already familiar to you. Debranching enzymes, hydratases, dehydrogenases, monooxygenases (mixed-function oxidases), and isomerases are used in the steroid synthetic pathways to modify specific sites. Frequently more than one pathway may be used to manufacture a given hormone, as the reactions can occur in different permutations of sequence.

There is a confusing array of natural steroids with endocrine activity, but we can sort out the principal ones into three main groups based only on the carbon skeleton:

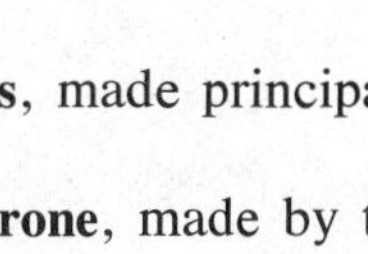

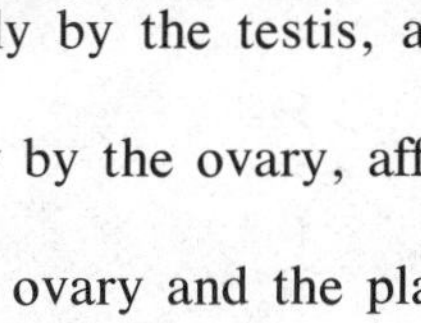

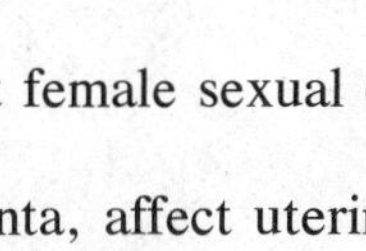

androgens

glucocorticoids
mineralocorticoids
progesterone

estrogens

It is important, however, to recognize that not all steroids with these unadorned skeletons have endocrine activity. The skeleton is a device for placing groups capable of polar interactions or H-bond formation in particular arrangements on a non-polar background that has its own specific geometry. The nature of these groups is critical for activity — the black non-polar sky is arranged so as to make these polar stars seem especially bright.

Adrenal Steroids

The adrenal cortex is made of different types of cells. Cells in a thin outer zone, the **zona glomerulosa,** make the mineralocorticoids; those in the wide middle zone, the **zona fasciculata,** and a narrower inner zone, the **zona reticularis,** make glucocorticoids.

The Biosynthetic Pathways. Synthesis of all steroid hormones made from cholesterol begins with the conversion of cholesterol to pregnenolone in mitochondria. This conversion is catalyzed by a complex containing cytochrome P-450, and it involves three successive monooxygenase reactions, utilizing NADPH as the second electron donor:

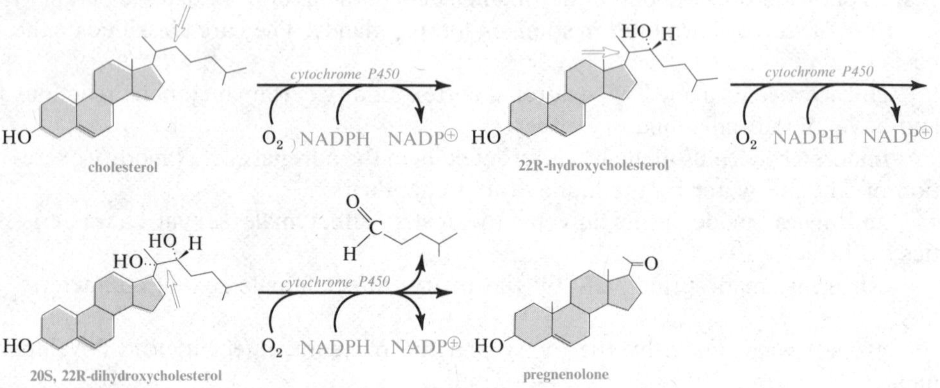

cholesterol → (cytochrome P450; O_2, NADPH → NADP⊕) → 22R-hydroxycholesterol → (cytochrome P450; O_2, NADPH → NADP⊕) → 20S, 22R-dihydroxycholesterol → (cytochrome P450; O_2, NADPH → NADP⊕; released aldehyde) → pregnenolone

The first, and an important, choice of alternative pathways occurs in the metabolic disposition of pregnenolone (Fig. 38–6). One begins with an oxidation to a ketone at C-3 and a shift of the double bond from the 5- to the 4-position, producing progesterone. These reactions occur on the endoplasmic reticulum of cells in all regions of the adrenal cortex and lead to both mineralo- and glucocorticoids. The other alternative is to hydroxylate C-17, a reaction that leads to the glucocorticoids and therefore occurs primarily in the zona fasciculata and reticularis. This, like the other hydroxylases used in adrenal steroid synthesis, is a monooxygenase (mixed-function oxidase) that uses NADPH as a second electron donor and transfers electrons to oxygen through **adrenodoxin,** an iron-sulfide protein, and cytochrome P-450. The 17-hydroxylation may precede or follow the oxidation and isomerization by which progesterone is created. In either case, 17-hydroxypregnenolone is formed.

The route to the glucocorticoids in the inner zones continues with the successive action of a 21-hydroxylase and an 11β-hydroxylase to form **cortisol:**

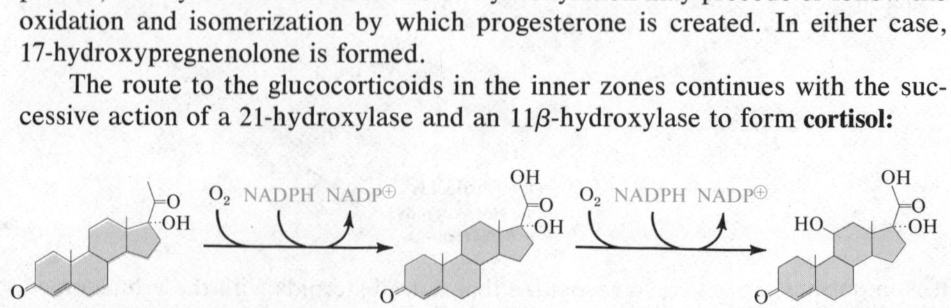

17α-hydroxyprogesterone → (O_2, NADPH → NADP⊕) → 11-deoxycortisol → (O_2, NADPH → NADP⊕) → cortisol

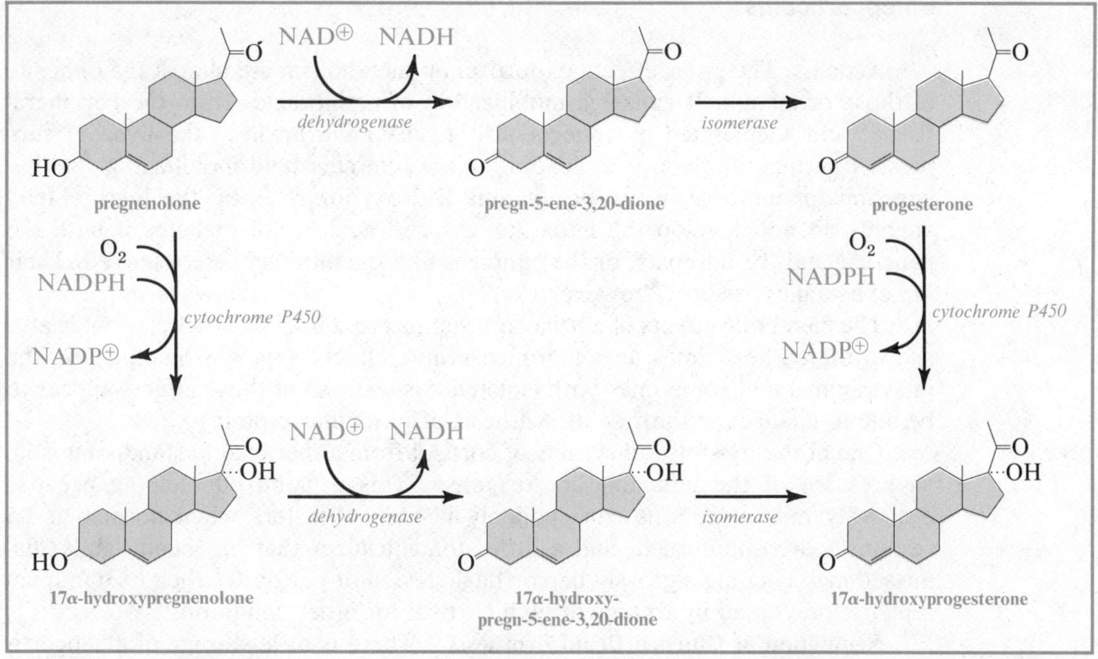

FIGURE 38–6 **Pregnenolone can be converted to 17α-hydroxyprogesterone by either of two routes in the adrenal cortex.**

The route to the mineralocorticoids in the zona glomerulosa also involves hydroxylations at C-21 and C-11, but without the hydroxylation at C-17.

The 18-methyl group side chain is then attacked by another mitochondrial complex containing cytochrome P-450 to produce an aldehyde group linked with the 11-hydroxyl group as a hemiacetal. The result is aldosterone:

O_2 NADPH NADP⊕ — progesterone → 11-deoxycorticosterone → corticosterone → 18-hydroxycorticosterone → (?) aldosterone

The 18-hydroxylase also occurs in the inner zones of the cortex, resulting in the formation of other 18-hydroxy derivatives. The estimated 24-hour production of the major compounds by the human adrenal gland is :

cortisol	8–24 mg
corticosterone	1.5–4 mg
11-deoxycortisol	0.5 mg
11-deoxycorticosterone	0.2 mg
aldosterone	0.04–0.20 mg
18-hydroxycorticosterone	0.15–0.45 mg
18-hydroxy, 11-deoxycorticosterone	*ca.* 0.1 mg

Glucocorticoids

Action. The gross effects of cortisol on metabolism are almost the opposite of those of insulin. It causes a mobilization of amino acids from the peripheral tissues and accelerated gluconeogenesis from amino acids in the liver. It suppresses peripheral glucose utilization and accelerates lipid mobilization, with a concomitant increase in the production of 3-oxybutyrates by the liver. (Many animals do not develop the lethal ketosis and acidosis of diabetes if both the pancreas and the adrenals, or the pancreas and the pituitary, are removed. Their life expectancy is short, however.)

The metabolic effects of cortisol are summarized in Table 38–3. The table also shows the earliest times at which measurable effects appear (sometimes in the intact animal and sometimes with isolated tissues). All of these effects appear to be due to changes in the rate of synthesis of particular proteins.

One of the most useful actions of cortisol from a therapeutic standpoint is its suppression of the inflammatory response. This is a mixed blessing because ordinarily mild infections can suddenly become disasters when normal tissue responses are suppressed, and a little stomach ulcer that has been lightly dismissed may become a grossly hemorrhagic lesion or perforate when its continual repair is prevented by treatment with cortisol for other conditions.

Regulation of Glucocorticoid Synthesis. There is little storage of glucocorticoids within the adrenal gland so that the blood level of glucocorticoids closely reflects their rate of synthesis. In the blood, glucocorticoids are mostly bound to plasma proteins, particularly **transcortin,** a corticosteroid-binding globulin. Less than 10 per cent is free. The blood half-life of cortisol, the major glucocorticoid in humans, is 80 minutes. (The blood half-life is considerably shorter than the tissue half-life.)

TABLE 38–3 EFFECTS OF CORTISOL

Liver	Hours
Increased glycogen	4–6 (*in vivo*)
Increased glucose production	2–6 (*in vitro*)
	2–6 (*in vivo*)
Increased oxybutyrate production	3–24 (*in vivo*)
Increased urea production	4–8 (*in vivo*)
Increased amino acid uptake	1.5–2 (*in vitro*)
	2–4 (*in vivo*)
Increased RNA synthesis	2–4 (*in vivo*)
Increased protein synthesis	8–20 (*in vivo*)
Adipose Tissue	
Increased fatty acid release	1–2 (*in vitro*)
Decreased glucose utilization	2–4 (*in vitro*)
Muscle	
Decreased glucose utilization	2–4 (*in vitro*)
Lymphatic Tissue	
Decreased glucose utilization	2–4 (*in vitro*)
Decreased nucleic acid synthesis	2–4 (*in vitro*)

Claims have been made elsewhere for suppression of glucose uptake by thymus cells in 15 to 20 minutes.

Adapted from J. Ashmore (1967) *in* A. D. Eisenstein, ed.: *The Adrenal Cortex.* Little Brown. By permission.

The blood cortisol level varies frequently during a 24-hour period, because it is discharged in pulses rather than in a continuous flow, but a pattern of diurnal variation is clearly seen. Around midnight cortisol levels are at their lowest (~ 0.15 μM, ~ 5 μg/dl) and begin to rise at about 0200 reaching their peak (~0.3 to 0.7 μM; 10 to 25 μg/dl) at 0600–0800 hours.

Stress or other unpleasant stimuli can rapidly alter the rate of cortisol synthesis. Within minutes after the start of a surgical procedure, the patient's cortisol blood level rises.

The Pituitary-Adrenal Axis. These changes in cortisol secretion are brought about through a system of control resembling the system described for the thyroid gland. The hypothalamus under appropriate stimulation releases an as yet uncharacterized **corticotropin-releasing hormone,** which stimulates the anterior pituitary to synthesize and release the hormone **corticotropin (adrenocorticotropic hormone, ACTH).** Stress stimulates liberation of corticotropin releasing hormone, whereas release may be inhibited by increasing cortisol concentrations. (A neurohypophyseal hormone, vasopressin, also stimulates synthesis and secretion of corticotropin. This hormone primarily serves as a regulator of water balance, as is discussed in the next chapter.)

Corticotropin (ACTH) is a 39-residue polypeptide that stimulates the adrenal cortex to synthesize and secrete glucocorticoids. It acts on the plasma membrane of the affected cells so as to increase the rate of conversion of cholesterol to pregnenolone. There is the same uncertainty about mechanism seen with the other polypeptide hormones. (It also causes an increased accumulation of cholesterol under some circumstances, but this is not the primary site of action, because the adrenal cells are usually loaded with cholesteryl esters and free cholesterol. They accumulate cholesterol from the blood and also can synthesize their own *de novo*.)

The half-life of corticotropin in the blood is less than 10 minutes. Normal adult concentrations range between 2 and 18 picomolar (10 and 80 ng/1) in the morning and are less than 2 picomolar (10 ng/l) by midnight. Corticotropin synthesis and secretion are directly controlled through negative feedback by free cortisol. High levels of cortisol inhibit and low levels stimulate corticotropin synthesis. External stimulation is absolutely necessary for adrenal function; in its absence, the fasciculata and reticularis zones of the cortex atrophy. This may occur upon prolonged therapeutic use of glucocorticoids in high doses because the medication suppresses synthesis of corticotropin in the pituitary. Subsequent discontinuation of the medication may then result in acute adrenal insufficiency.

This untoward event is avoided by using glucocorticoids that have a short duration of action and by giving them in the morning. The glucocorticoid level again becomes low at around midnight and remains so during the small hours, without interference in the normal diurnal appearance of early morning corticotropin secretion, and the resultant stimulation of the adrenal cortex.

The structure and synthesis of corticotropin is of considerable interest in its own right. It may be derived from "big ACTH," a prohormone. (The possible existence of a preprohormone will not be examined here.) Corticotropin bears similarities in structure to a 91-residue polypeptide, β-**lipotropin,** also made by the adenohypophysis. Lipotropins have the ability to mobilize fat and also to darken skin by stimulating pigment cells called melanocytes.

Residues 4 to 10 of corticotropin, Met-Glu-His-Phe-Arg-Trp-Gly, occur as residues 47 to 53 of β-lipotropin. β-Lipotropin also contains a sequence with potent analgesic-pain relieving property. This sequence (61 to 91) is called β-endorphin. Residues 61 to 65, Tyr-Gly-Gly-Phe-Met, had previously been isolated and characterized as a morphine or opiate-like agonist, **methionine enkephalin.** Leucine enkephalin is similar except for the substitution of leucine for methionine. The enkephalins seem to be neurotransmitters that integrate pain information.

Cushing's syndrome is a condition in which effects of excessive glucocorticoid secretion dominate the clinical picture. It may result from the growth of secreting tumors of the anterior pituitary or the adrenal cortex, from nodular hyperplasia of the adrenal cortex, from carcinomas, frequently carcinoma of lung, that produce an ectopic corticotropin (ectopic = out of place). The most common cause of the symptoms of Cushing's disease is the prolonged administration of glucocorticoids for medical treatment. There is a general depletion of proteins to supply amino acids for gluconeogenesis with resultant loss of muscle mass and thinning of skin and bones. As a result, there is general weakness, easy bruisability, bone fractures, and a redistribution of fat, creating a characteristic moon-face and buffalo hump. Frequently serious mental changes are present. The excessive output of glucose causes hyperglycemia, and the picture of full-blown diabetes. Excessive secretion of aldosterone also occurs.

Failure of the adrenal cortex causes **adrenal insufficiency (Addison's disease** of the adrenal). Weakness, fatigue, weight loss, increased skin pigmentation, nausea, vomiting, diarrhea, muscle pain, and salt craving are common symptoms. The patients are hypotensive, hypothermic, and usually have a low glucose and sodium level. Their grip on life is tenuous and any stress, infection, cold, or even noise can precipitate a crisis leading to death.

Inborn errors of steroid metabolism result from absence or impaired function of one of the many synthetic enzymes. The low production of cortisol leads to high secretion of corticotropin by the anterior pituitary gland, which in turn causes bilateral hyperplasia of the adrenal gland. The enzymes commonly involved are 21-hydroxylase, 11β-hydroxylase, 17α-hydroxylase, 18-hydroxylase, and 3β-hydroxysteroid dehydrogenase.

Drugs are available that can inhibit some of these enzymes:

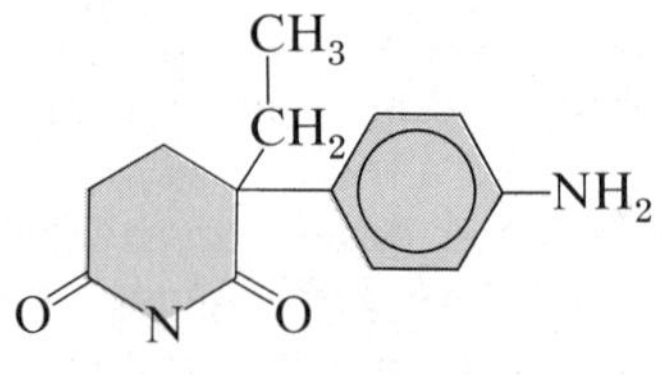

aminoglutethimide

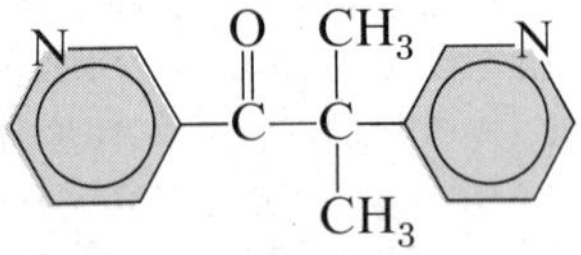

metyrapone

Aminoglutethimide inhibits hydroxylation at C-20 and blocks all adrenal steroid hormone synthesis. Metyrapone inhibits 11β-hydroxylase activity, thereby interfering with synthesis of aldosterone and the glucocorticoids. (Sufficient 11-deoxy steroids with mineralocorticoid activity are made to preserve normal salt metabolism.)

The treatment of advanced breast cancer can include the removal of the adrenal glands. Some patients will demonstrate a remarkable destruction of the cancer following adrenalectomy. Recently a "medical" adrenalectomy has been performed by giving metyrapone and aminoglutethimide, and a synthetic glucocorticoid dexamethasone. The dexamethasone exerts a negative feedback upon the pituitary so that ACTH, which could override the blocks created by the drugs, is not released.

Gonadal Steroids

Both the androgens and the estrogens are made to some extent by the adrenal cortex, although in much lesser amounts than are the other steroid hormones. The major sources of these steroids are the gonads, which carry pregnenolone through the same initial steps involved in adrenal steroid synthesis, but control the rates of the reactions by varying enzyme concentrations so as to arrive at a different spectrum of products. The differences in function of the glands producing steroid hormones is, therefore, more of a matter of quantitative variation of gene expression than it is of presence or complete absence of particular enzymes.

Testosterone, the male androgen, is mainly made by the interstitial cells of the testis. A typical pathway of synthesis is:

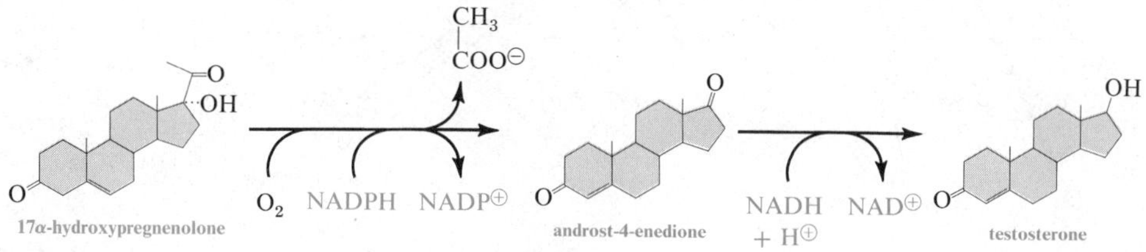

$CH_3COO^{\ominus}$

O_2 NADPH NADP$^{\oplus}$ NADH + H$^{\oplus}$ NAD$^{\oplus}$

17α-hydroxypregnenolone androst-4-enedione testosterone

Many cells responsive to testosterone contain a 5α-reductase to convert testosterone to a more potent androgen **dihydrotestosterone,** which combines with a receptor protein in the target cells. As with other steroid hormones, the receptor-steroid complex is translocated to the nucleus, where it affects gene transcription. In this case, phosphorylation of non-histone proteins in the nucleus has been observed within 30 minutes of the administration of testosterone to castrated animals.

Not only is testosterone necessary for the development of adult male sexual characteristics, its presence during fetal development is necessary for normal development of both male physique and psyche. More remarkably, administration of testosterone results in almost complete elimination of attacks of hereditary angioedema, a life-threatening disease characterized by acute localized episodes of soft tissue swelling, such as in the upper respiratory passage. (It is caused by a deficiency of an inhibitor of the activation of Cl in the complement system (Chapter 18). Testosterone restores the inhibitor concentrations to normal.)

Estrogens are mainly made by thecal cells in the ovarian follicles, through modification or extension of the routes by which testosterone is made:

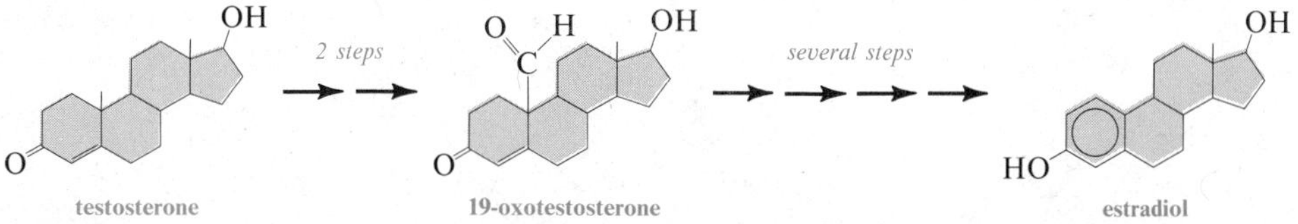

2 steps *several steps*

testosterone 19-oxotestosterone estradiol

A branched methyl group (C_{19}) is removed and ring A is made aromatic.

Progesterone is a precursor of the other steroids we have mentioned, but it is also an important hormone in its own right, being supplied primarily by the corpus luteum that develops in ovarian follicles after release of the ova. Its synthesis and secretion is therefore cyclical, increasing about two days before ovulation, and surging after ovulation to a peak about eight days later, then returning to basal levels by 12 to 14 days if fertilization and implantation have not occurred.

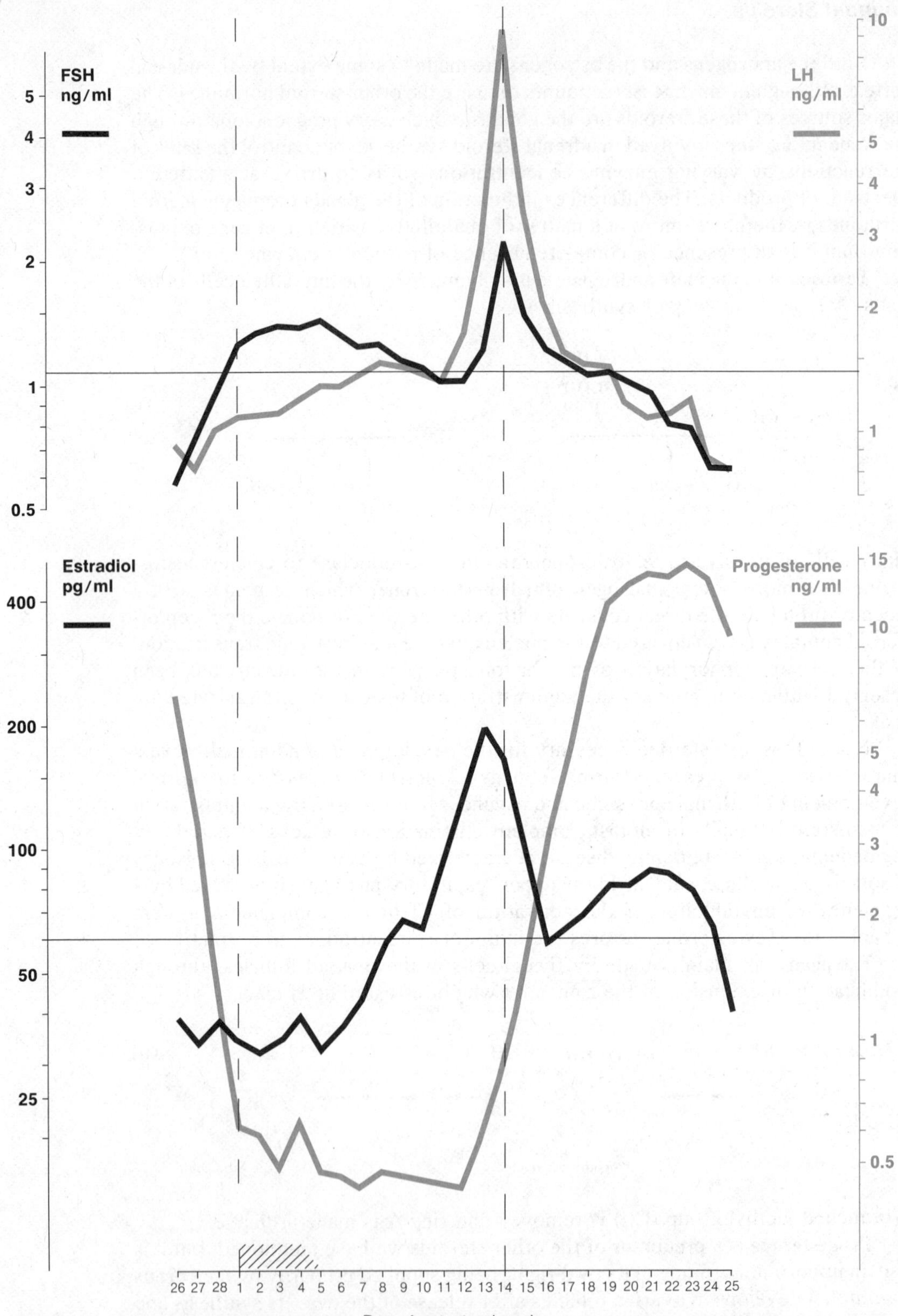

FIGURE 38-7 **Concentrations of circulating hormones during the menstrual cycle. Reproduced by permission from L. Wide (1976), p. 87, *in* J. A. Loraine and E. T. Bell, eds.: *Hormone Assays and Their Clinical Applications*, 4th ed. Churchill.**

Detection of cytosol receptors for estrogen and progesterone has become an important part of the management of patients with carcinoma of the breast. The presence of **estrophilin,** an estrogen receptor protein, in the malignant tissue is an indicator that the cancer will be adversely affected by an alteration of the steroid hormone milieu. The correlation between presence of receptor and a good clinical response to hormonal manipulation is far from perfect. The higher the level of the receptor, the more likely a favorable response will be elicited. There is also evidence to indicate that one action of the estrogen translocated to the nucleus by the receptor is activation of chromatin leading to the synthesis of progesterone receptors. High response rates occur when the breast cancer has both estrogen and progesterone receptors. Scientific mysteries lurk here. It is hard to explain why such diverse and even contradictory procedures as giving estrogens or antiestrogens, removing the ovaries, or removing the adrenals while giving glucocorticoids and mineralocorticoids can produce regression in breast cancer that contains estrophilin.

Hypothalamic-Pituitary-Gonadal Axis. The hypothalamus forms a single releasing hormone that stimulates the synthesis and secretion of two pituitary hormones, **luteinizing hormone (luteotropin, LH)** and **follicle-stimulating hormone (follitropin, FSH).** Both of the pituitary hormones, which have different effects, are important in the male and female. The plasma levels vary independently of one another, and the mechanism of control is not clear.

Variations of these hormones, estradiol and progesterone, during a normal menstrual cycle are shown in Figure 38–7. Near mid-cycle prior to ovulation, estradiol has a positive feedback effect upon secretion of luteinizing and follicle-stimulating hormones. At other times in the cycle estradiol has a negative feedback effect. Estradiol appears to act upon both the hypothalamus and the anterior pituitary gland.

In the male, luteinizing hormone causes testosterone secretion, while follicle-stimulating hormone causes spermatogenesis.

HORMONAL CONTROL OF CALCIUM METABOLISM

The concentration of calcium in blood is controlled closely: normal values lie between 2.1 and 2.6 millimoles per liter of serum. Part of the ion is chelated with proteins, so the concentration of the free ion is near 1.2 millimoles per liter. It is the free ion that determines the balance with the tissues.

Close control appears to be necessary because of the importance of calcium ion in the function of tissues generally. A role for calcium ion has been proposed for many of the hormone target cell effects described in this chapter. We have mentioned its importance in muscular contraction, and a fall in calcium ion concentration to 50 per cent of its normal value will cause tetany. Calcium appears to be required for normal function of most membranes.

The turnover of bone tissue enables adjustment of form to the kind of loads an individual routinely imposes upon his skeleton. Bone also serves as a reservoir of calcium ions; there are around 30 moles of calcium in the bones of a 73 kg man available for maintaining the 20 millimoles (0.8 gram) found in the body fluids.

The turnover of calcium is regulated by three hormones. We have already mentioned **1,25-dihydroxycholecalciferol,** a steroid derivative that is made by the combined action of the skin, liver, and kidneys. **Parathormone,** a polypeptide secreted by the parathyroid glands is another. The third is **calcitonin,** a polypeptide secreted by cells located in the connective tissue between follicles in the thyroid gland.

1,25-Dihydroxycholecalciferol

It was independently shown in 1919 that the childhood disease rickets, characterized by failure of proper development of the bones, could be prevented by a factor in the diet or by irradiation with ultraviolet light. The emphasis on the dietary factor led to the term **vitamin D** for any compound with curative action on rickets. It was later demonstrated that compounds with vitamin D activity were generated by the action of ultraviolet light on steroids containing conjugated double bonds, and that one of these, then designated vitamin D_3, was normally formed in the skin by the action of light on 5,7-cholestadienol, a normal intermediate in cholesterol synthesis (Fig. 38–8). The recognition that rickets was not a dietary deficiency disease, but a sunlight deficiency disease led to the introduction of the name cholecalciferol to replace vitamin D_3.*

Cholecalciferol is transported in the blood in combination with a specific transport globulin, and it is taken up and stored in the liver. A 25-hydroxylase in the endoplasmic reticulum forms **25-hydroxycholecalciferol.** This enzyme is a typical monooxygenase, utilizing NADPH as the second electron donor. It is regulated by feedback inhibition, so the product is maintained at a relatively low concentration.

The 25-hydroxy derivative is carried by the same transport globulin to the kidney. The kidney contains both a 1α- and a 24-hydroxylase that act on 25-hydroxycholecalciferol. The resultant **1,25-dihydroxycholecalciferol** is physiologically active, whereas the 24,25-dihydroxy compound is inert, and the 1,24,25-trihydroxy compound is less active.† The 1-hydroxylase is regulated by parathormone and by $[P_i]$ (next section); like the adrenal mitochondrial steroid hydroxylases, it acts through cytochrome P-450, and contains an iron-sulfide protein (renal ferredoxin) for transport of electrons to oxygen from NADPH. The mechanism of the 24-hydroxylase is not as well known.

Action. 1,25-Dihydroxycholecalciferol is carried through the blood to target cells by the same cholecalciferol-transporting globulin; it is carried within the cells to the nucleus by attachment to an intracellular transport protein, and the complex is bound to chromatin. The apparent action of the compound is to promote the transcription of genes that facilitate transport of calcium and phosphate ions through the plasma membranes. In any event, this hormone has effects upon the intestine and bone that are clearly established. It increases absorption of calcium and phosphate ions from the digested food in the small intestine. It causes resorption of bone, with release of calcium and phosphate into the blood (along with citrate and other ions present in the bone). It appears that the prime role of this hormone is to maintain the calcium ion concentrations above hypocalcemic levels, thereby preventing tetany, with concomitant effects on phosphate metabolism. (The action on other tissues has been difficult to ascertain clearly.)

Parathormone

The parathyroid glands respond to a fall in extracellular calcium ion concentration by secreting an 84-residue polypeptide hormone, parathormone, which

---

*There has been an unfortunate effort in recent years to resurrect the misleading vitamin D_3 terminology. It is sometimes useful to speak of vitamin D activity when talking about the several compounds that may be included in the diet to replace cholecalciferol, but there is no reason to use this name for a compound made by the body in adequate amounts under normal conditions.

†The trivial name calcifediol has been proposed for 25-hydroxycholecalciferol, which is a dihydroxy compound. By analogy, 1,25-dihydroxycholecalciferol would be calcifetriol.

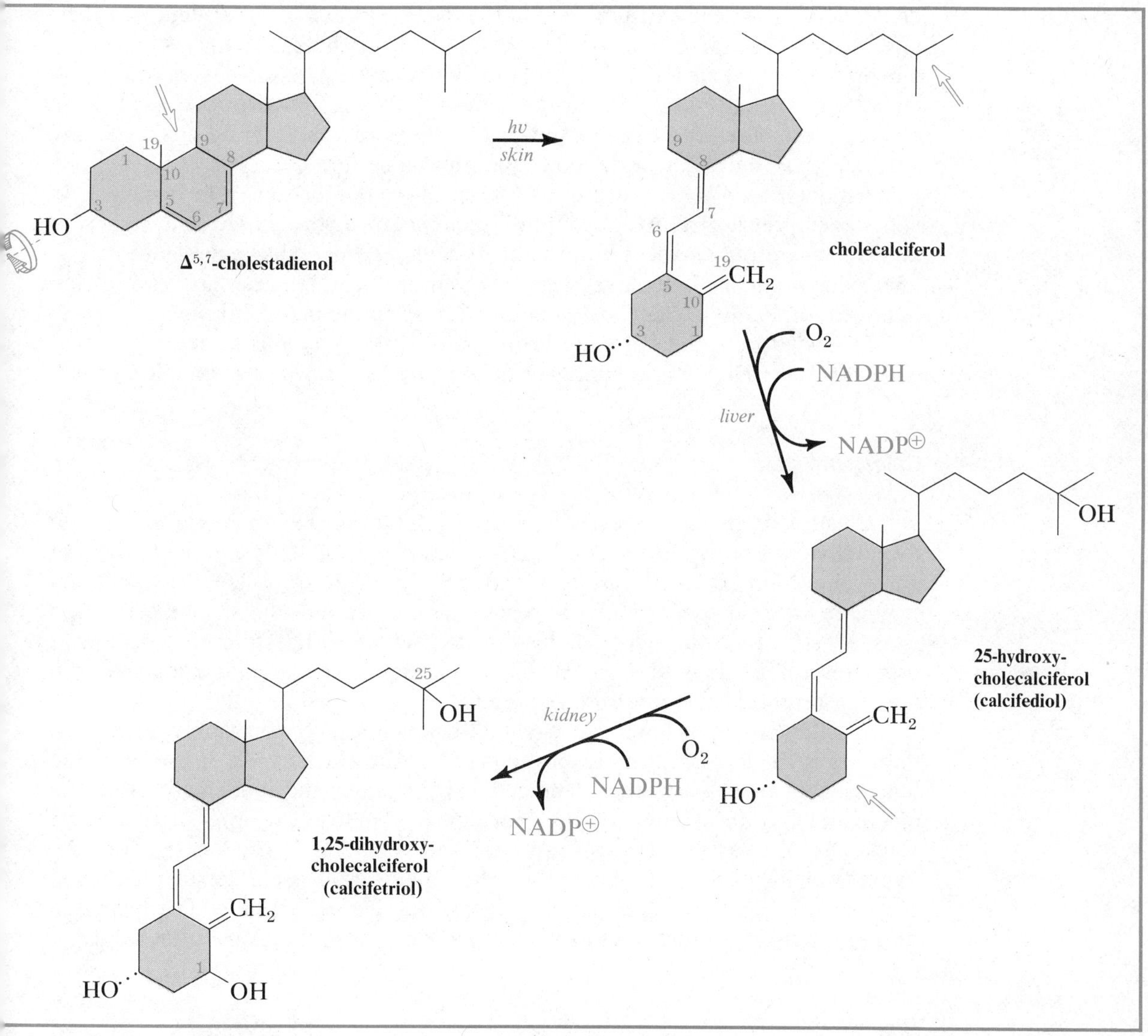

HO

$\Delta^{5,7}$-cholestadienol

$h\nu$
skin

HO

cholecalciferol

CH_2

O_2
NADPH
liver
NADP⊕

OH

25-hydroxy-
cholecalciferol
(calcifediol)

CH_2

HO

kidney
O_2
NADPH
NADP⊕

OH

1,25-dihydroxy-
cholecalciferol
(calcifetriol)

CH_2

HO

OH

FIGURE 38–8 **The synthesis of the hormone 1,25-dihydroxycholecalciferol involves the successive action of three tissues.**

acts upon the bone and kidney. Parathormone acts in concert with 1,25-dihydroxycholecalciferol to stimulate bone resorption. It also has important effects on the kidney: It causes increases in the activity of the 1α-hydroxylase that produces 1,25-dihydroxycholecalciferol, and it increases the absorption of calcium from the glomerular filtrate and decreases the absorption of phosphate. A low [P_i] is an independent stimulus for increasing the 1-hydroxylase activity, and it may be that the parathormone effect is an indirect result of its effect on phosphate absorption. Parathormone and 1,25-dihydroxycholecalciferol therefore independently act to raise the calcium level in the blood and extracellular fluid, but their effects are also interrelated. Parathormone stimulates synthesis of 1,25-dihydroxycholecalciferol, thereby indirectly assuring increased absorption of calcium by the small intestine. On the other hand, 1,25-dihydroxycholecalciferol evidently promotes the transport of calcium into the parathyroid cells, thereby enhancing the signal for cessation of parathormone formation.

Parathormone is synthesized as a preprohormone of some 115 residues. As with other preproproteins, the "pre" sequence is hydrophobic and serves to attach the protein to the endoplasmic reticulum. Following emergence, a *ca.* 25-residue sequence is removed leaving the proprotein. Another six residues are removed, probably in the Golgi apparatus, to form the active hormone.

Degradation of the hormone in the parathyroid glands is affected by intracellular [Ca^{2+}]. High levels stimulate breakdown, and low levels inhibit.

Calcitonin

Clusters of calcitonin-secreting cells are found in the connective tissue between thyroid follicles; these cells, unlike the follicular cells, are derived from the embryonic neural crest and are related to cells in the adrenal medulla. Calcitonin is a 32-residue polypeptide that is secreted when the calcium ion concentration rises. Gastrin (p. 140), secreted by the stomach after ingestion of a meal, also promotes calcitonin secretion. Perhaps this aids in preventing hypercalcemia upon absorption of the ion from the food.

Calcitonin acts to lower extracellular calcium ion concentrations in at least three ways: It decreases the resorption of bone by elastic cells, it increases the formation of osteoblasts that deposit bone minerals, and it increases the loss (decreases the absorption) of both calcium and phosphate ions in the urine. Calcitonin decreases the activity of the 1α-hydroxylase acting on 25-hydroxycholecalciferol. In general, calcitonin appears to be a device for protection against hypercalcemia, causing increased deposition of bone, decreased absorption from the diet, and decreased recovery from the glomerular filtrate.

Rickets

Infants and young children who are not exposed to ultraviolet light and who also lack cholecalciferol or related compounds in the diet develop rickets, a condition characterized by deficient calcification of the bones. The relationship between lack of sunlight and the incidence of rickets is quite clear. It is said that in the days before dietary supplementation all of the children in a New York hospital during the winter months had some signs of rickets, whereas few had the signs in the summer. The condition was mainly confined to the very young, who had the least exposure to daylight, especially in the tenement districts.

Now, 1,25-dihydroxycholecalciferol is necessary for the mobilization of calcium from the bone, as well as for absorption from the gut, and one might think that the deficient children would be depositing too much calcium. Two factors are operating: The young have relatively high requirements for calcium because they are forming new bones, and the calcium deficiency resulting from defective intestinal absorption will be correspondingly more critical. Secondly, the parathyroid gland will respond to the fall in calcium supply with an increased output of parathormone. While this will tend to conserve calcium in the kidneys, it increases the urinary loss of phosphate, which in turn causes even more mobilization of the bone mineral. Rickets is therefore to a considerable extent a **hypophosphatemia,** and its consequences can be mimicked by dietary deprivation of phosphate.

Because of the prevalence of rickets, manufacturers were encouraged to fortify common foods such as milk, bread, and margarine. In addition, mothers were urged to give routine daily doses of vitamin D, principally as modified fish liver oils such as oleum percomorpheum (8,500 International units, equivalent to 212 micrograms of cholecalciferol, per gram of oil). Infants require something in the neighborhood of 5 micrograms of cholecalciferol per day if they synthesize none, and 10 micrograms (400 I.U.) is usually specified to be safe.

The result of the program was a rapid decline in the incidence of rickets. This decline was greatly augmented by changes in the care of infants, who now receive more exposure to sunlight. However, cases of rickets still appeared and the pressure for more widespread use of fortified foods continued. The apparent consequences of this effort to supply vitamin D to those remaining deficient by increasing the intake of the whole population have tarnished the luster obtained from the nearly complete victory over rickets.

It has long been known and repeatedly stated in textbooks that large doses of vitamin D are toxic. The few who stopped to think saw that the intake of some children could be approaching the toxic level, since they had a high content of the compound in the diet, plus the amount synthesized internally during the summer, plus the large doses easily given from preparations commonly found in most households. Unfortunately, the few who saw were not heeded until frank cases of poisoning were finally recognized in a number of children. Most of the severe cases turned out to be children unusually prone to hypercalcemia, but further inquiry suggested the existence of marginal cases in children of normal response. The result has been a quiet de-emphasis on fortification of foods with cholecalciferol-like compounds, a drop in the amount of those compounds in preparations sold without prescription, and a plea to physicians to make some effort to find out how much an individual is already ingesting before prescribing more.

PROSTAGLANDINS AND THROMBOXANES

Swedish investigators exploring the effects of semen on the uterus discovered that it indeed contained compounds affecting uterine motility. Believing them to be formed by the prostate gland, the investigators called them prostaglandins. (They are actually elaborated in quantity by the seminal vesicle.) This small beginning led to what is now an explosion of effort, with a journal devoted solely to work on these compounds. It is now known that the prostaglandins and their relatives the thromboxanes are probably made by all cells and affect all cells.

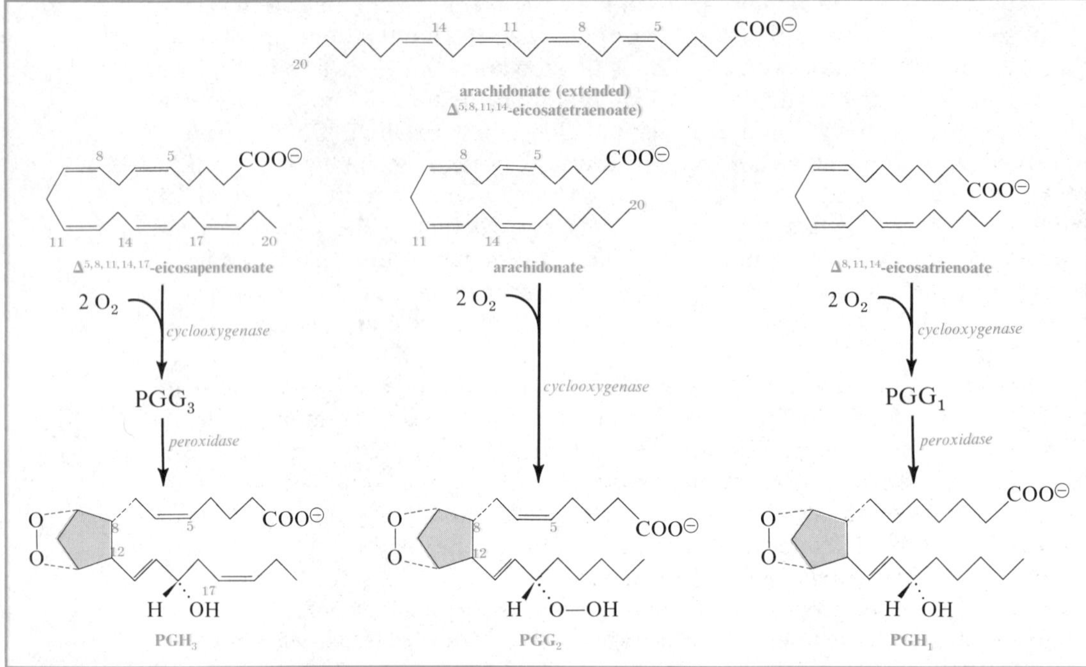

FIGURE 38–9 **The prostaglandins may be synthesized from 20-carbon polyunsaturated fatty acids with varying numbers of double bonds. The synthesis first forms a 15-hydroperoxy derivative of a cyclic endoperoxide, designated as a PGG** ***(illustrated at bottom center)***, **which is converted to a 15-hydroxy derivative by a peroxidase** ***(illustrated at bottom left and right).***

The prostaglandins and thromboxanes are families of cyclic fatty acid derivatives derived from polyunsaturated fatty acids, especially arachidonic acid. It is perhaps misleading to speak of them as hormones; although they are effective at low concentrations, their main effects may be upon the cells that make them, perhaps upon neighboring cells. They are at best local hormones, but in terms of mode of action, they fit in the group. These compounds modify plasma membrane responses, and in bewilderingly diverse ways. Their locus of action may lie between the primary endocrine effects on membranes and the modification of cyclic AMP formation, Ca^{2+} flux, or other intracellular changes caused by these effects. There are many different prostaglandins and relatives, often with quite different physiological actions. The net effect on a given tissue is the resultant of these actions rather than a definitive change we can describe in simple terms.

Our goal is to introduce these structures and point out the features in which they differ. First a note on nomenclature. The prostaglandins are designated by symbols such as PGE_1. The PG stands for prostaglandin, and the third capital letter indicates the type of substituents found on the hydrocarbon chain. The subscript number indicates the number of double bonds in the molecule.

The first step in the synthesis of all of these compounds is the addition of oxygen catalyzed by a cyclizing dioxygenase (Fig. 38–9), which produces a hydroperoxy prostaglandin (PGG) that is converted to a hydroxy prostaglandin (PGH) by a peroxidase. (We indicate here the formation of the 1, 2, or 3 series, but let us concentrate on the more abundant 2-series derived from arachidonate.) The anti-inflammatory action of agents such as aspirin and indomethacin is due to their inhibition of this initial step in prostaglandin formation. (Some even go as far as to

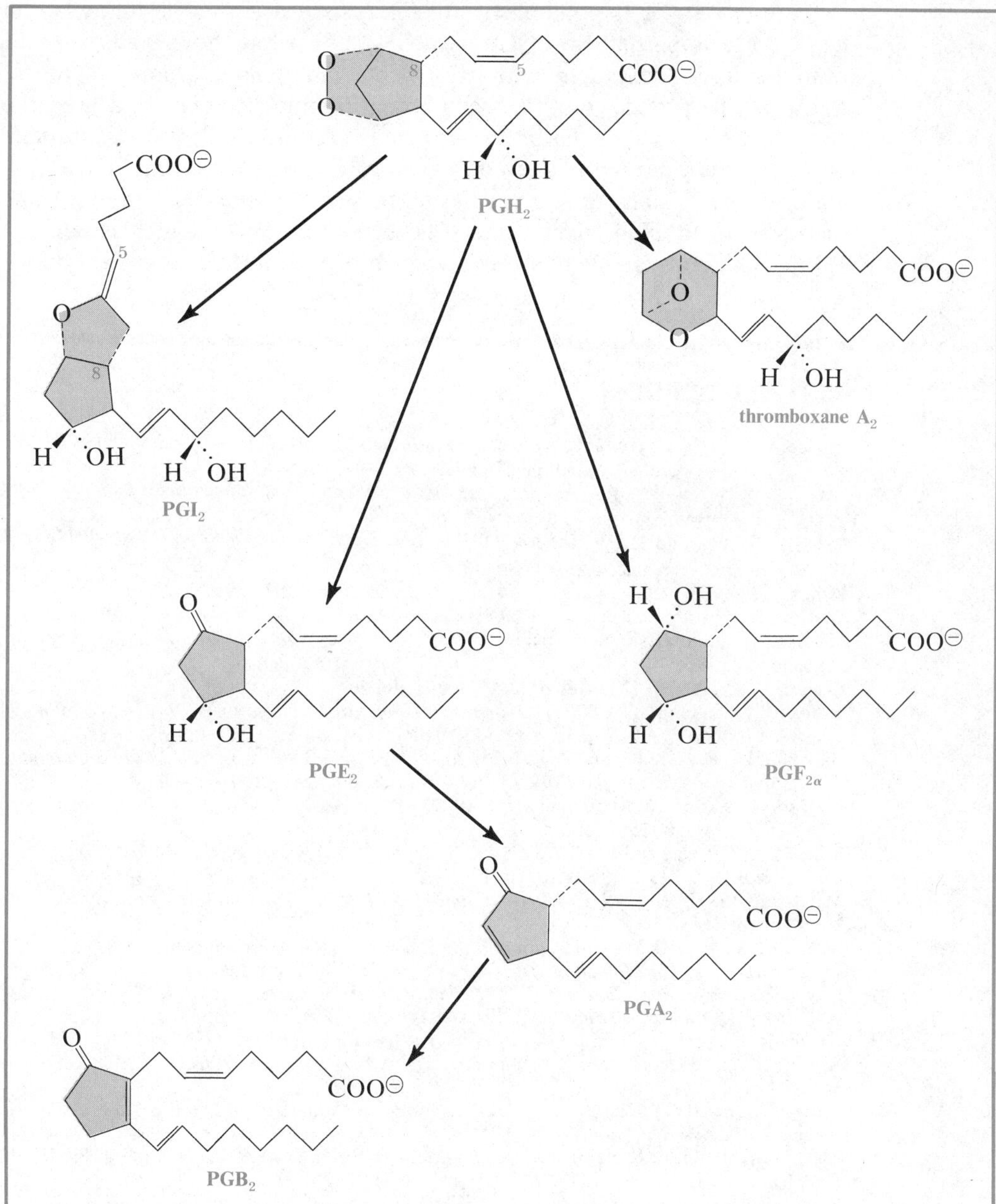

FIGURE 38–10 **Conversion of prostaglandins G to thromboxanes (*upper right*) and various other types of prostaglandins.**

say that this also accounts for their pain-relieving effects. PGE injection causes pain.)

The endoperoxide produced by the cyclizing dioxygenase is the precursor of the thromboxanes and other prostaglandins (Fig. 38–10). The regulation of these diverse pathways is not understood. Study of the compounds is difficult owing to their low concentrations, their very short half-life (the thromboxanes have half-lives on the order of seconds), and to the presence of so many different compounds.

We can savor the complexity of the responses from the following observations: the endoperoxides, PGG_2 and PGH_2, as well as the thromboxanes derived from them, induce rapid and irreversible clumping of blood platelets. Contrariwise, the prostacyclin PGI_2 is a potent inhibitor of platelet aggregation. The opportunities for speculation on how it is that normal platelets are kept from aggregating and promptly caused to aggregate upon exposure to damaged vessels are obvious. Turning to another tissue, the smooth muscle of the bronchus or of the uterus in non-pregnant animals is relaxed by PGE_2, but it is contracted by PGF_2. Both compounds cause contraction of the uterus in pregnant animals.

FURTHER READING

Williams, R. H., ed.: (1974) *Textbook of Endocrinology*. 4th ed. W. B. Saunders. *The* source on endocrine physiology and clinical applications.

O'Malley, B. W., and L. Birnhaumer, eds.: (1978) *Receptors and Hormone Action*. Academic Press. A new multi-volume work.

Goldberg, N. D., and M. K. Haddox: (1977) *Cyclic GMP Metabolism and Involvement in Biological Regulation*. Annu. Rev. Biochem., *46*: 823.

Kunos, G.: (1978) *Adrenoreceptors*. Annu. Rev. Pharmacol., *18*: 291.

Sterling, K., and J. H. Lazarus: (1977) *The Thyroid and Its Control*. Annu. Rev. Physiol., *39*: 349.

Morrison, M., and G. R. Schonbaum: (1976) *Peroxidase-catalyzed Halogenation*. Annu. Rev. Biochem., *45*: 861.

Evered, D. C.: (1976) *Diseases of the Thyroid*. Wiley.

Dillman, W. H., et al.: (1978) *Triiodothyronine-stimulated Formation of Poly (A)-containing Nuclear RNA and mRNA in Rat Liver*. Endocrinology, *102*: 568.

Bjokman, U., R. Ekholm, and L. E. Ericson: (1978) *Effects of Thyrotropin on Thyroglobulin Exocytosis and Iodination in the Rat Thyroid Gland*. Endocrinology, *102*: 460.

Degroot, L. J., and H. Niepomniszcze: (1977) *Biosynthesis of Thyroid Hormone: Basic and Clinical Aspects*. Metab. *26*: 665.

Vale, W., C. River, and M. Brown: (1977) *Regulatory Peptides of the Hypothalamus*. Annu. Rev. Physiol., *39*: 473.

McGowan, G. K., ed.: (1977) *Hypothalamic and Pituitary Hormones*. J. Clin. Pathol., 30(suppl. 7). Short reviews.

Sachs, B. A., ed.: (1978) *The Brain and the Endocrine System*. Med. Clin. N. Am., *62*: 227.

Czech, M. P.: (1977) *Molecular Basis for Insulin Action*. Annu. Rev. Biochem., *46*: 359.

Unger, R. H., and R. E. Dobbs: (1978) *Insulin, Glucagon, and Somatostatin Secretion in the Regulation of Metabolism*. Annu. Rev. Physiol., *40*: 307.

Unger, R. H., and L. Orci.: (1977) *Role of Glucagon in Diabetes*. Arch. Int. Med., *137*: 482.

Bar, R. S., and J. Roth: (1977) *Insulin Receptor Status in Disease States of Man*. Arch. Int. Med., *137*: 474.

Carey, R. M., et al. 1978. *Diabetes Mellitus Updated: Standards of Quality Care in Office and Hospital Practice*. Va. Med. Mon., *105*: 195.

Zonana, J., and D. L. Rimoin: (1976) *Inheritance of Diabetes Mellitus*. N. Engl. J. Med., *295*: 603.

Symposium on Diabetes. (1977) Arch. Intern. Med., *137* (Oct.)

Bunn, H. F., K. H. Gabbay, and P. M. Gallop: (1978) *The Glycosylation of Hemoglobin: Relevance to Diabetes Mellitus*. Science, *200*: 21.

Ganda, O. P., et al.: (1977) *Somatostatinoma: A Somatostatin-containing Tumor of the Endocrine Pancreas*. N. Engl. J. Med., *296*: 963.

Albisser, A. M., et al.: (1977) *Studies with an Artificial Endocrine Pancreas*. Arch. Int. Med., *137*: 639.

Chan, L., and B. W. O'Malley: (1978) *Steroid Hormone Action: Recent Advances*. Ann. Int. Med., *89*: 694.

Jones, C. I., and I. W. Henderson, eds.: (1976) *General, Comparative, and Clinical Endocrinology of the Adrenal Cortex*. Academic Press.

Ganong, W. F., L. C. Alpert, and T. C. Lee: (1974) *ACTH and the Regulation of Adrenocortical Secretion*. N. Engl. J. Med., *290*: 1006.

Jefcoate, C. R.: (1977) *Cytochrome P-450 of Adrenal Mitochondria*. J. Biol. Chem., *252*: 87, 88.

Chan, L., and B. W. O'Malley: (1976) *Mechanism of Action of the Sex Steroid Hormones*. N. Engl. J. Med., *294*: 1322, 1372, 1430.

Horwitz, K. B., and W. L. McGuire: (1978) *Estrogen Control of Progesterone Receptor in Human Breast Cancer*. J. Biol. Chem., *253*: 2223.

Jensen, E. V.: (1977) *Estrogen Receptors in Human Cancers*. J.A.M.A., *238*: 59.

McCann, S. M.: (1977) *Luteinizing Hormone-releasing Hormone*. N. Engl. J. Med., *296*: 797.

Valenta, L. J., and J. C. Zolman: (1977) *Gonadotropin-releasing Hormone.* N. Engl. J. Med., *297*: 725.

Frantz, A. G.: (1978) *Prolactin.* N. Engl. J. Med., *298*: 201.

Avioli, L. V., ed.: (1978) *Vitamin D Metabolites: Their Clinical Importance.* Arch. Intern. Med., *138*: 835. Several articles, some not as recent as the date would indicate.

DeLuca, H. F., and H. K. Schnoes: (1976) *Metabolism and Mechanism of Action of Vitamin D.* Ann. Rev. Biochem., *45*: 631. The hydroxylation of cholecalciferol was discovered in the senior author's laboratory. He is also responsible for revival of the vitamin D_3 nomenclature.

Haussler, M. R., and T. A. McCain: (1977) *Basic and Clinical Concepts Related to Vitamin D Metabolism and Action.* N. Engl. J. Med., *297*: 974.

Report of a committee of the American Academy of Pediatrics: (1963) *The Prophylactic Requirement and the Toxicity of Vitamin D.* Pediatrics, *31*: 512.

Arnaud, C. D., ed.: (1978) *Parathyroid Hormone, Calcitonin, and Vitamin D.* Fed. Proc., *37*: 2557. Symposium.

Lakdawala, D. R., and E. M. Widdowson: (1977) *Vitamin D in Human Milk.* Lancet, *1*: 167.

Habener, J. B., et al.: (1977) *Biosynthesis of Parathyroid Hormone.* Recent Progr. Horm. Res., *33*: 249.

Kadowitz, P. D., P. D. Joiner, and A. L. Hyman: (1975) *Physiological and Pharmacological Roles of Prostaglandins.* Annu. Rev. Pharmacol. *15*: 285.

Needleman, P., and G. Kaley: (1978) *Cardiac and Coronary Prostaglandin Synthesis and Function.* N. Engl. J. Med., *298*: 1122.

Martini, L., and G. M. Besser, eds.: (1977) *Clinical Neuroendocrinology.* Academic Press. Includes discussion of hypothalamic hormones.

Tager, H. S., and D. F. Steiner: (1974) *Peptide Hormones.* Annu. Rev. Biochem., *43*: 509.

39 CONTROL OF WATER AND ION BALANCE

Maintenance of a constant concentration of most mineral ions hinges upon balancing intake and excretion, because they are not chemically modified during their sojourn in the body. (Sulfate and phosphate compounds are sometimes exceptions.) Control is exerted at both ends, but the physiological mechanisms concerned with thirst and salt hunger are beyond our scope. Let us concentrate on the biochemical devices for regulating water and ionic excretion by the kidney.

THE KIDNEYS

A kidney is essentially a device for filtering a large volume of blood and passing the filtrate through a long tubule, which is lined with cells that selectively transport substances into and out of the filtrate. Most of the selective transport involves uptake of water and solutes from the filtrate for re-use within the organism. Some of the transport is by active secretion from the cells into the filtrate. The product of all this processing is the urine, which, if all goes well, contains any surplus of ingested water and electrolytes, along with the daily production of urea, uric acid, creatinine, and other waste products not eliminated elsewhere. The result is shown in Figure 39–1.

Anatomy

The anatomical features most important for our discussion are summarized in Figure 39–2. We are concerned with the flow of two fluids — the blood and the filtrate. A veritable torrent of blood, one fifth of the cardiac output at rest, enters the kidneys, where it passes into tufts of capillaries within glomeruli. A glomerulus is a filtration chamber; the space around the capillary tuft is at relatively low pressure, but the blood within the tuft is kept at relatively high pressure because the efferent arteriole carrying blood from the tuft is constricted more than the afferent arteriole. This high pressure gradient causes a rapid weeping of fluid through the capillary walls. Small solutes pass with little constraint; only a very

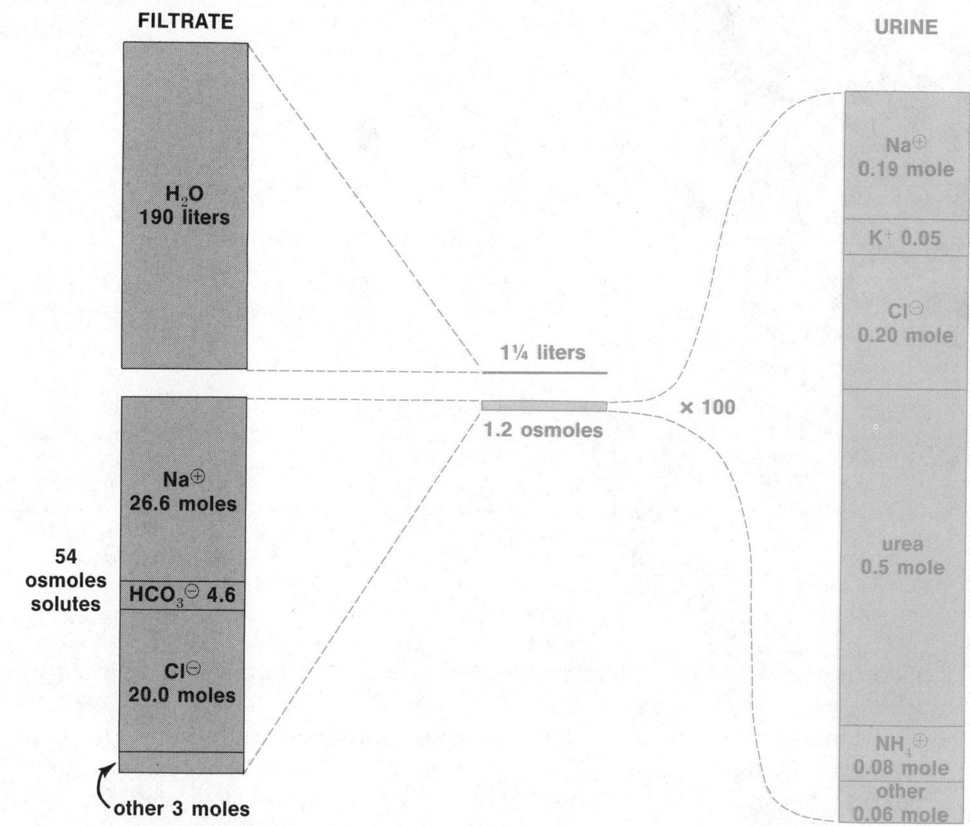

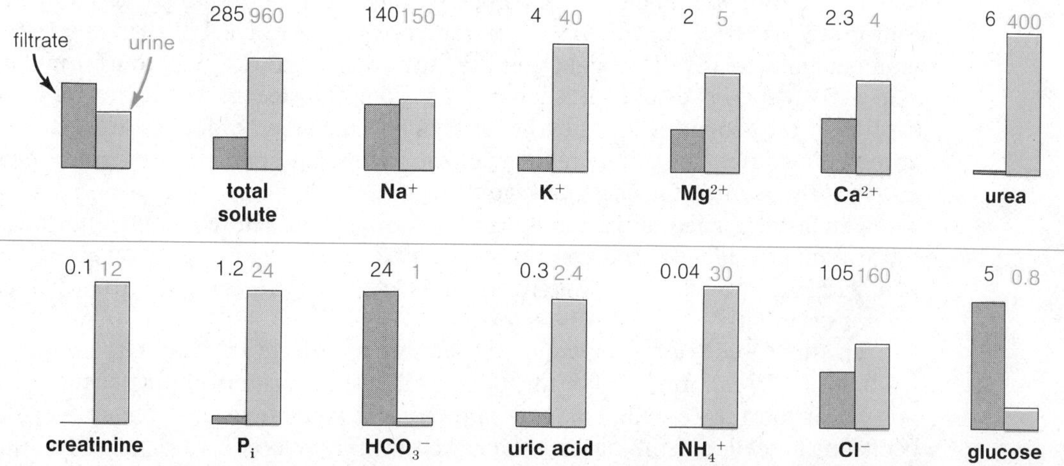

FIGURE 39–1 *A.* **Changes in the total composition of the daily output of glomerular filtrate and urine. The scale of solute composition of the urine is expanded on the right. Typical values are given, but the quantity of water and solutes in the urine of normal individuals depends upon their dietary intake and varies widely from that shown.**

B. **Relative concentrations of individual components in the filtrate and urine are shown by bars. Actual millimolarities are also listed: filtrate is in black, urine in blue.**

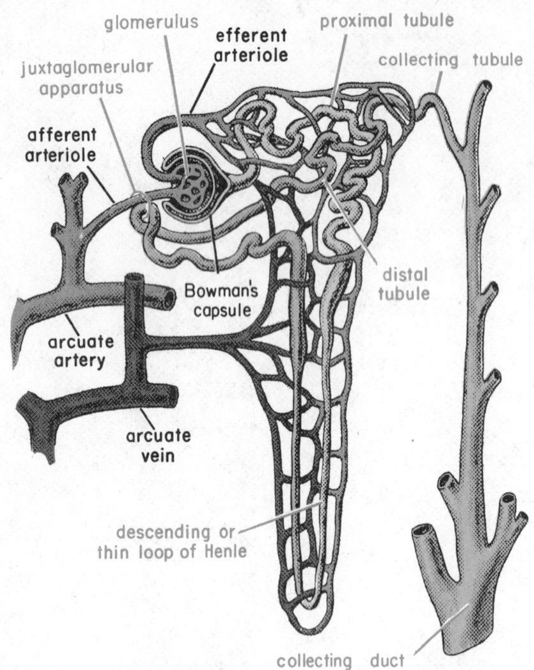

FIGURE 39-2

Schematic outline of the anatomy of a nephron in the kidney. Modified from A. C. Guyton: (1971) *Textbook of Medical Physiology*, 4th ed. W. B. Saunders.

minor portion of serum albumin and other proteins get through, and virtually no cells.

The volume of the resultant **glomerular filtrate** is roughly one fifth of the volume of the blood plasma passing through the kidney. The cardiac output of a median adult male at rest is about 6 liters per minute, of which one fifth, or 1.2 liters, passes through the kidneys. In that volume are 0.65 liter of plasma, of which one fifth, or 0.13 liter, appears as glomerular filtrate.

The filtrate is drained from each glomerulus by a **tubule,** which has a highly **convoluted proximal portion** located in the cortex (outer layer) of the kidney that then becomes relatively straight and dips toward the interior, where it forms the **loop of Henle.** Those tubules issuing from glomeruli deeper in the cortex have especially long loops, going nearly through the medulla before returning to the cortex. The descending limb and part of the ascending limb of the loop of Henle are very thin-walled for easy passage of water.

The **distal portion** of the tubule is again convoluted before turning once more toward the medulla as a **collecting tubule** of different structure. Collecting tubules join within the medulla as **collecting ducts** of still different properties, which deliver urine into the renal pelvis.

The tubule cells are surrounded throughout their length by capillary networks fed by the efferent arterioles of the glomeruli. The plasma membrane of the basal portion of the tubule cells contains numerous invaginations that create a space between the cell and the basement membrane. The blood is at a relatively low pressure in these networks, which facilitates absorption. To summarize, there is a high pressure gradient from capillaries to lumen in the glomerulus to promote filtration, and a low adverse pressure gradient from lumen to capillaries in the tubules to promote absorption. Most of the blood passes through networks around the proximal and distal tubules; a smaller portion, a few per cent, passes through the **vasa recta** ("straight vessels"), which carry the blood deep into the medulla and back again much like the flow of filtrate in the neighboring loops of Henle.

Absorption in the Proximal Tubule

Let us first deal with the management of Na^+, Cl^-, and water balance, and consider those events affecting H^+ concentration later in the chapter. **The proximal convoluted tubule removes about 80 per cent of the water and salt from the filtrate passing through it,** so that the 130 ml min^{-1} entering the tubules diminishes to *ca.* 26 ml min^{-1} at the beginning of the loops of Henle. The absorption requires energy, part of which is provided by the contraction of the heart, creating the pressure that produces an almost protein-free filtrate. The total concentration of solutes is less in the filtrate than it is in the blood, owing to this lack of protein; the filtrate in the lumen therefore has a lower osmotic pressure than is present in the surrounding tubule cells and interstitial fluid, which is in equilibrium with the blood. The second source of energy is the (Na^+ + K^+)-ATPase in the basal and lateral membranes of the tubule cells, which is used to create concentration gradients between the interior of the cells and both the filtrate and the interstitial fluid.

Let us summarize events more or less in causal sequence (Fig. 39–3):

(1) Water flows from the region of low osmotic pressure in the lumen toward the capillaries. Some of the flow occurs through the tight junctions into the spaces between tubule cells and some through the cells. The loss of water causes the

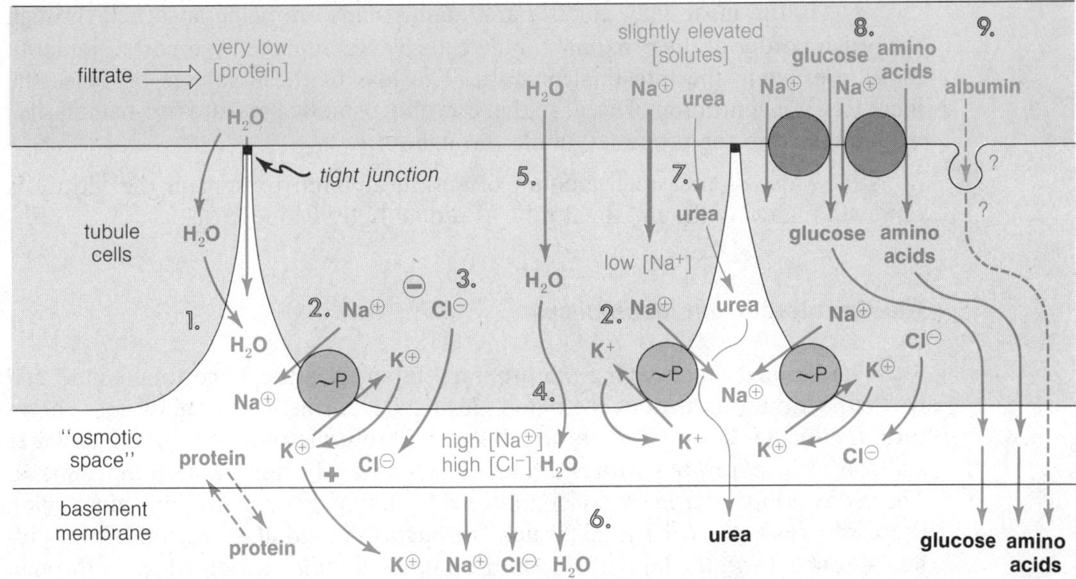

FIGURE 39–3 **Absorption of filtrate in the proximal convoluted tubule. Numbers correspond to description in the text. (1) Passage of water down osmotic pressure gradient created by differences in protein concentration. (2) ($Na^+ + K^+$)-ATPase action, and subsequent leakage of K^+ back out of cell. (3) Movement of Cl^- down potential gradient created by step 2. (4) Flow of water down solute concentration gradient created by steps 2 and 3. (5) Influx of water down concentration gradient created by step 4. (6) Solutes and water pushed into capillaries by the hydraulic pressure created in the confined pericellular "osmotic" spaces by steps 2, 3, and 4. (7) Movement of Na^+ into cell from filtrate down concentration gradient created by step 2. Lesser amounts of urea also move down the gradient. (8) Absorption of glucose and amino acids is driven by simultaneous transport of Na^+ down the concentration gradient created by preceding steps. (9) The small amount of serum albumin passing into the filtrate is absorbed by unknown mechanisms.**

filtrate rapidly to reach the same total solute concentration (same osmolarity) as the blood plasma. Some Na^+ and Cl^- move with the water (not shown).

(2) Independently, the ATPase pumps Na^+ out of the cells into the confined spaces (osmotic spaces) provided by the invaginations of the basal plasma membrane. K^+ is simultaneously pumped into the cells, but in lesser amounts. The K^+ leaks back out, preventing its internal accumulation.

(3) Cl^- passively moves with the Na^+ and K^+ out of the cells into the surrounding spaces. The driving force for this movement is the electrical potential created by cation movement.

(4) The increased concentration of Na^+ and Cl^- outside the cells causes a movement of water into the confined osmotic spaces, creating hydraulic pressure between them and the blood.

(5) Water moves from the tubular lumen into the cells to replace water moving into the osmotic spaces.

(6) The combination of increased salt concentration and hydraulic pressure in the osmotic spaces causes both salt and water to move across the basal membrane into the capillaries. (The basal and the lateral membranes of the tubular cells, unlike the luminal membrane, are impermeable to Na^+, so the salt cannot move back into the cells.)

(7) The loss of water from the tubular lumen raises the concentration of all solutes in it. As a result, Na^+ diffuses so rapidly that its concentration is held nearly constant throughout the proximal tubule. Urea passes less freely across the luminal membrane, so its concentration becomes higher.

(8) In the meantime, glucose and amino acids are being absorbed through the brush border of the proximal tubule cells by Na^+-coupled symports, similar to those present in the intestinal mucosa. The loss of these solutes balances the increased concentration of urea, so that constant osmotic pressure (iso-osmolarity) is maintained in the proximal tubule throughout.

(9) Most of the small amount of serum albumin present in the filtrate is removed intact by the proximal tubule, probably by endocytosis.

The Countercurrent Mechanism

The luminal fluid leaving the proximal tubule has the same total solute and Na^+ concentrations as does the blood plasma. The urine usually does not. Therefore, the further conversion of the tubular fluid to urine requires additional water and salt absorption, but with the relative amounts adjusted to fit circumstances. These deviations of urinary concentrations from blood concentrations hinge upon the triple passage of the luminal fluid through the medullary region of the kidney — once down the loop of Henle, back up, and then back down again through the collecting ducts. During each of these passages, the fluid encounters luminal membranes with different transport properties (Fig. 39–4). Let us list these properties and then discuss how they are used to create urine.

The thin descending limb of the loop is freely permeable to water, but only slightly permeable to Na^+, Cl^-, or urea. The thin ascending limb is almost the opposite; it is nearly impermeable to water, and very permeable to Na^+ and Cl^-, although much less permeable to urea. The thick-walled section of the ascending limb, nearer the cortex, is different. The cells are loaded with mitochondria to

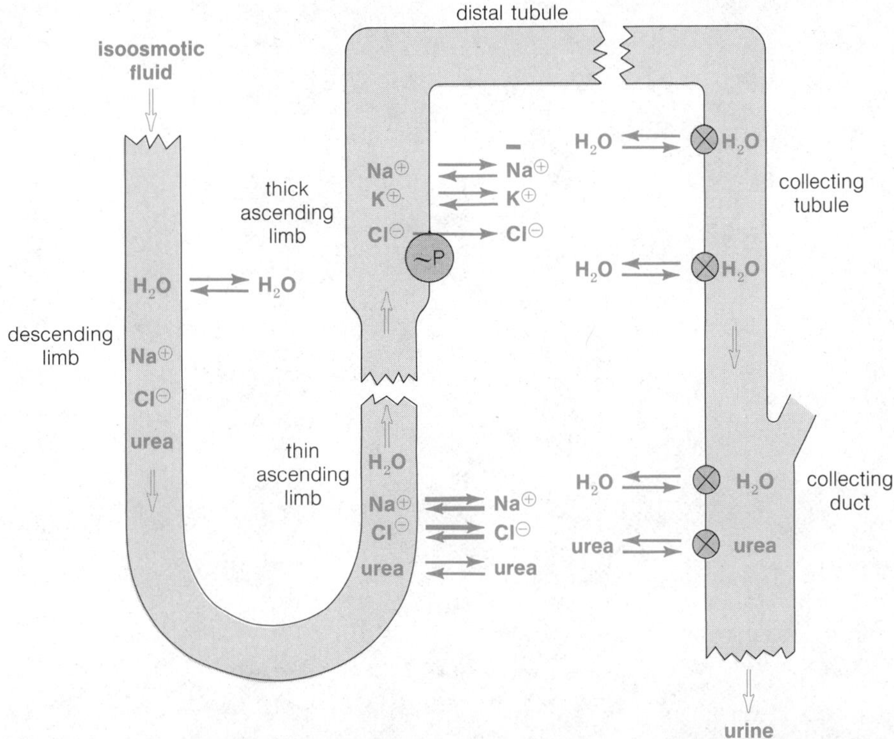

FIGURE 39–4 **Diagram of permeabilities of different sections of the nephron after the proximal convoluted tubule, which delivers an iso-osmotic fluid to the loop of Henle. This schematic diagram does not show the tubule cells and indicates only the luminal membrane. However, the greater diameter of the thick ascending limb is due to larger cells, not a thicker membrane. ⊗ indicates vasopressin-regulated channels for water passage.**

drive an active pump for Cl^-. (The mechanism is unknown.) They may also have some (Na^+ + K^+)-ATPase activity, but most of the Na^+ that leaves the cells does so because of the electrical potential created by the Cl^- pump. This section of the tubule is also nearly impermeable to water and urea.

Most of the distal tubule has little water permeability. However, the collecting tubules, and perhaps the adjacent segments of the distal tubules, have a regulated permeability to water, which depends upon the presence of an antidiuretic hormone vasopressin discussed below. The collecting tubule has almost no permeability to urea; however, the collecting ducts in the inner medulla do permit some urea to leave.

The Concentration Gradient in the Medulla. There is an all-important result of the different transport properties of tubular segments. Let us simply state it as a fact for the moment:

> **Whenever the kidney is producing a concentrated urine, the interstitial fluid surrounding the loop of Henle and the collecting ducts becomes more and more concentrated with respect to NaCl and urea from the cortex through the medulla to the pelvis.**

The result is that the fluid within the tubules is also more and more concentrated as it moves down the descending limb of the loop, less and less concentrated as it moves up the ascending limb, and again more and more concentrated as it moves

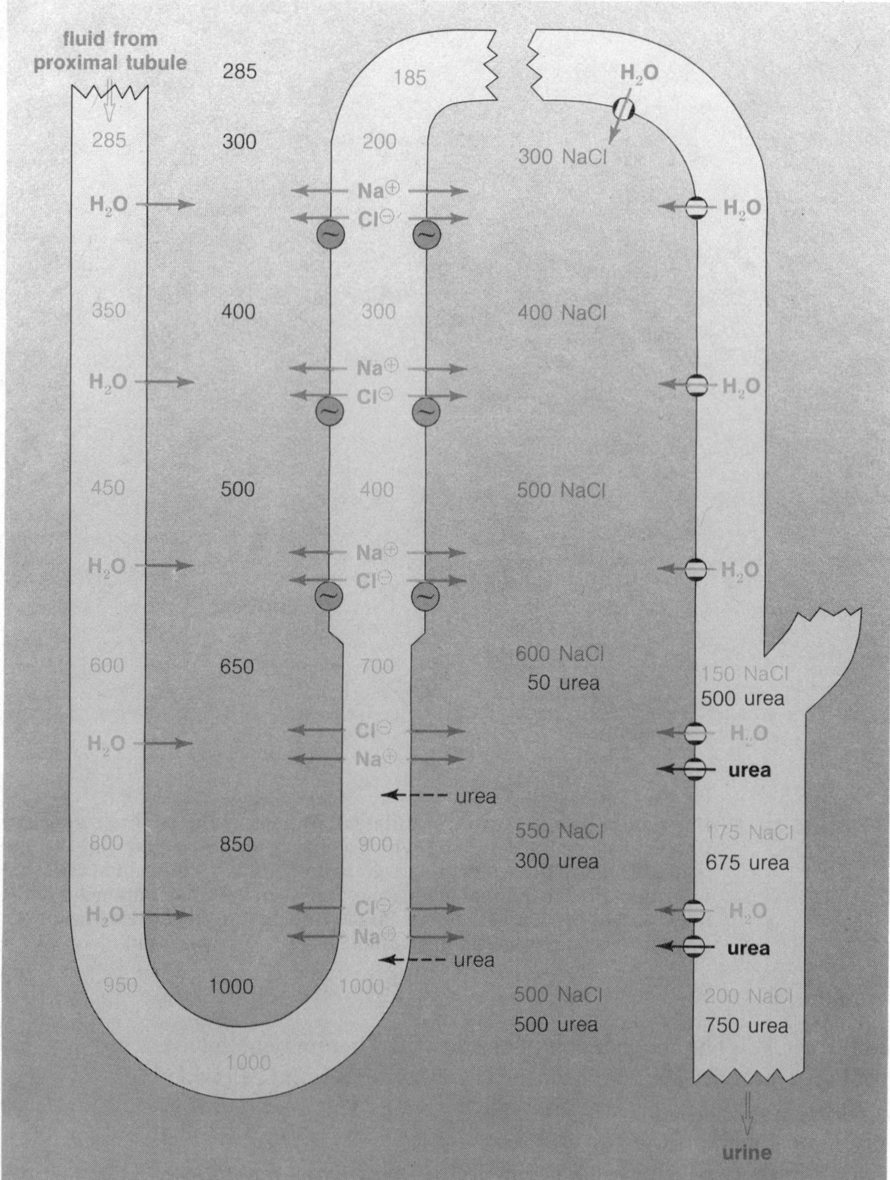

FIGURE 39–5 **The countercurrent concentration mechanism. The numbers are hypothetical concentrations selected to illustrate the principles when urine is being concentrated; the values chosen for the fluid at the top and the bottom of the diagram are realistic under conditions of low water intake. The volume of the interstitial space is grossly exaggerated in diagrams of this type.**

Concentrations are given in milliosmoles per liter (total millimolarity of all solutes, including individual ions). Concentrations within the tubular fluid are given in blue; the total milliosmolarity of the interstitial fluid is shown by the left center column in black. The contributions of NaCl and urea to that total are shown in blue and black, respectively, in the right center column in the interstitial space. The important features are: *(left)* water passes from the tubule lumen into the interstitial space, with the solute concentration constantly increasing to a maximum of 1,000 mOsm. *(Bottom center)* The fluid in the lumen of the thin ascending tubule is now more concentrated than the surrounding interstitial fluid, and it loses NaCl, but does not gain water. It does gain some urea. *(Top center)* Chloride ions are actively pumped out in the thick-walled ascending tubule, accompanied by sodium ions. This pump is the driving force by which the concentration from cortex to pelvis is established.

Legend continued on the opposite page

down the collecting tubules and collecting ducts. However, these fluctuations in total concentrations within the tubules involve *changing proportions of salt and urea,* and this is also an important part of the story that follows (Fig. 39–5).

(1) As fluid moves through the lumen of the **descending limb,** it continuously loses water to the increasingly concentrated interstitial fluid. It becomes more concentrated to both NaCl and urea, but it is always a little more dilute than the interstitial fluid around the tubule until it reaches the turn of the loop and starts back toward the cortex. However, the NaCl concentration is higher than that in the interstitial fluid, because the interstitial fluid is rich in urea; the luminal fluid becomes correspondingly rich in NaCl as it loses water.

(2) As the fluid passes up the **ascending limb,** it shortly reaches regions where the interstitial fluid has a lower osmolarity. However, this limb has a low permeability to water and a high permeability to salt. The result is that Na^+ and Cl^- move out, along with a lesser amount of urea, which does not pass so readily through the luminal plasma membrane. The total solute concentration therefore progressively diminishes in the luminal fluid as it passes up the limb, being only a little greater than that of the interstitial fluid at any point, but most of the loss in solute is in NaCl, not urea.

(3) This sequence culminates upon reaching the thick portion of the ascending limb, where Cl^- is actively pumped out and accompanied by Na^+. **This is the source of energy for creation of the concentration gradient in the medulla.** As a result of its action, the osmotic pressure in the luminal fluid falls below that of blood plasma — it becomes hypoosmolar, although its urea concentration is high.

(4) An important feature of the countercurrent mechanism is the small solute gradient between the ascending and descending limbs of the loop of Henle at any level, even though there is a large gradient along the length of each limb. The interstitial fluid at any point is a little more concentrated than the fluid within the descending limb and a little more dilute than the fluid within the ascending limb. The pumps in the ascending limb do not have to work against impossible concentration gradients, even though the total gradient they create is very large.

(5) When the hypoosmolar fluid reaches the region of controlled water permeability in the **distal and collecting tubule,** water may be permitted to pass out of the lumen into the surrounding interstitial fluid, which has a higher solute concentration. Water loss will continue, if permitted, upon passage through the increasingly concentrated regions of the medulla, and the result will be recovery of most of the water that entered the loop of Henle and excretion of a concentrated urine. If water loss is not permitted, the urine will remain at the hypoosmolar concentration of the fluid leaving the loop of Henle, and a large volume of dilute urine will be excreted.

(6) Free passage of water through the medullary collecting duct membrane is associated with passage of urea, and this is an obligatory step in the establish-

---

FIGURE 59–5 Continued **(*Right top*) Water passes out of the fluid in the distal and collecting tubules through channels that are open in the presence of vasopressin. (*Right bottom*) Vasopressin also opens channels for passage of urea out of the collecting ducts. Urea leaves because its concentration in the urine is greater than that in the interstitial fluid, even though the total solute concentration is lower in the urine.**

See illustration on the opposite page

ment of the high concentration gradient in the medulla necessary for water absorption. Although the total solute concentration of urine in the collecting duct is lower throughout its course than it is in the surrounding interstitial fluid, the urea concentration is higher. Therefore, urea will move from the urine into the region of lower concentration in the interstitial space, if its passage is permitted. NaCl cannot flow back into the collecting duct to restore osmotic balance, so the movement of urea creates a progressively higher osmotic pressure in the surrounding interstitial fluid.

The counter-current concentration gradient that can be established is much lower if an adequate supply of urea is not available. Persons on a low-protein diet excrete a higher volume of dilute urine as a result.

(7) The water and salts entering the interstitial fluid in the medulla are removed by blood flowing through the vasa recta loops. Blood flowing toward the region of high concentration accumulates solutes; on the return trip toward the cortex, it loses most, but not all, of the solutes and accumulates water. The concentrations in the blood parallel those in the loop of Henle as it flows through the vasa recta. Transport is efficient because the actual volume of the interstitial fluid between the tubule cells and the blood is small, and it need not contain high quantities of solute to establish large concentration gradients.

Sodium and Potassium Balance

So far, we have been dealing with the massive movements of Na^+, and the much lesser movement of K^+, associated with recovery of the glomerular filtrate and concentration of the urine. There are additional mechanisms in the distal tubule and collecting tubule for regulation of the Na^+ and K^+ concentrations of the blood. These mechanisms need have only the capacity for handling ions in the amounts likely to accumulate from dietary intake or dehydration.

The major mechanism appears to be a regulated (Na^+ + K^+)-ATPase on the interstitial side of the tubular cells. It raises $[K^+]$ and lowers $[Na^+]$ within the cells. K^+ leaks out, into both the lumen and the interstitial fluid. The loss into the lumen is a major source of excreted K^+. The leakage also makes the interior of the cell more negative causing more Na^+ to move from the lumen into the cell (Fig. 39–6). In addition to this passive loss, there appears to be secretion of K^+ into the lumen by an active pump. The result of these mechanisms is that there is a constant loss

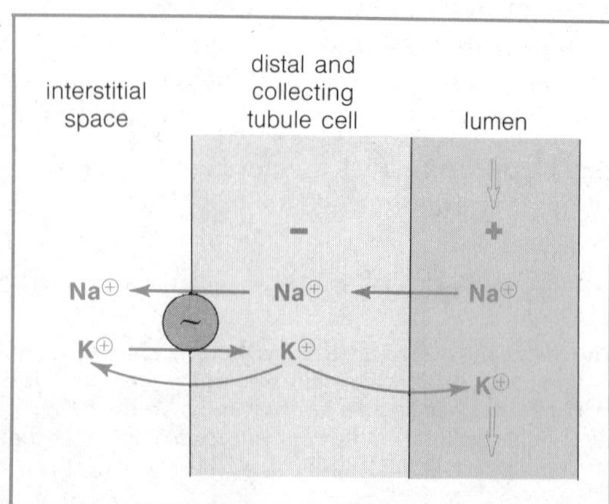

FIGURE 39–6

Recovery of Na^+ and loss of K^+ in the distal and collecting tubules. The large + and − signs indicate the potential gradient.

of K^+ regardless of intake, whereas the loss of Na^+ can be reduced to very low levels if there is little intake.

REGULATION OF SALT AND WATER BALANCE

Vasopressin, the Antidiuretic Hormone

A specialized group of neurons located above the pituitary gland and behind the optic chiasm (the supraopticoneurohypophyseal tract) synthesizes two small polypeptide hormones, vasopressin and oxytocin, which differ in only two amino acid residues:

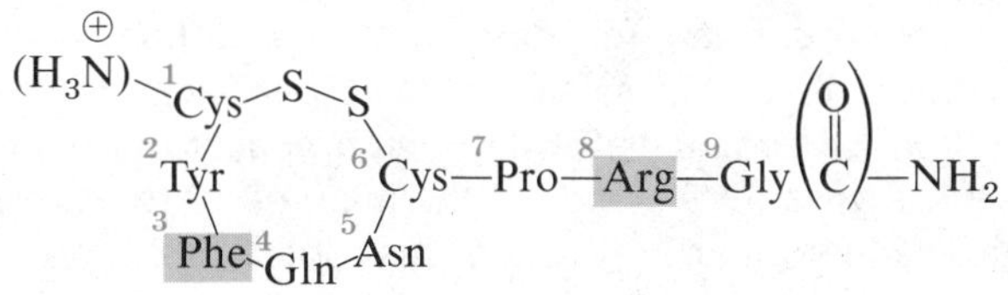

$\oplus$
(H_3N)—^{1}Cys—S—S
^{2}Tyr ^{6}Cys—^{7}Pro—^{8}**Arg**—^{9}Gly(C=O)—NH_2
^{3}**Phe**—^{4}Gln—^{5}Asn

vasopressin

$\oplus$
(H_3N)—Cys—S—S
Tyr Cys—Pro—**Leu**—Gly(C=O)—NH_2
Ile—Gln—Asn

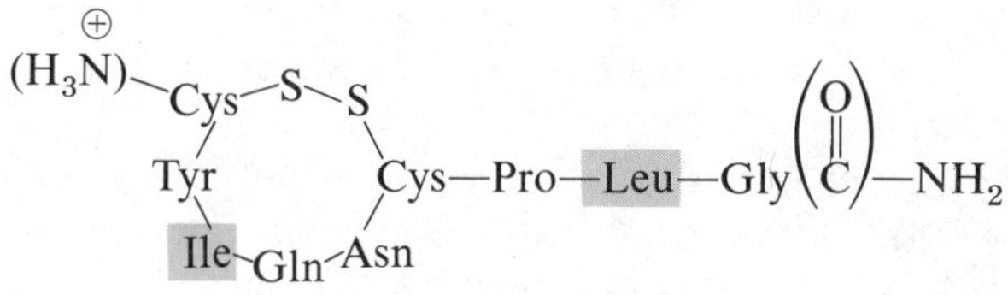

oxytocin

This difference is sufficient to change the physiological activity. The physiological function of oxytocin is not known. Therapeutic doses cause contraction of the uterus and expulsion of milk from the mammary gland.

Vasopressin causes water to be absorbed from the distal and collecting tubules of the kidney, and this is its physiological function. Administration of higher doses causes constriction of arteries, resulting in higher blood pressures in anesthetized animals; this is the origin of the name. The hormone from humans and most mammals is sometimes named Arg-vasopressin to distinguish it from the pig hormone, which contains Lys in place of Arg.

Vasopressin is synthesized in the cell bodies of the specialized neurons, along with a larger polypeptide, a **neurophysin** (M.W. = 10K), which binds the vasopressin and forms granules. The granules pass down the axons to their terminals, some of which occur in the median eminence, and others in the body of the neurohyphophysis (posterior pituitary).

The granules are discharged from the neurohypophysis in response either to hypertonicity (increased solute concentration in the interstitial fluid) or to lowered blood volume. The released vasopressin travels through the blood, where its concentration is only 10^{-11} to 10^{-12} M, and binds to receptors on the distal nephrons. (Nephron is used as a collective term for the entire glomerulus-tubule apparatus.) These specific receptors activate adenyl cyclase in the target cells, and the resultant increase in 3′,5′-cyclic AMP causes the luminal membranes to become water permeable. This is the permitted passage of water that we referred to earlier. The mechanism is not known, but it seems likely that a cyclic AMP-sensitive protein kinase causes a modification of protein conformation in the luminal membrane.

Aldosterone, Renin, and Angiotensin

Vasopressin provides a prompt response to transient alteration in salt and water balance. A slower regulation of basal conditions is obtained by the effect of aldosterone (p. 725) on cation exchange in the distal nephron. This steroid hormone affects gene expression in the distal nephron so as to increase the recovery of Na^+ and increase the loss of K^+. The protein(s) whose synthesis is being affected is not known. The result could be partially rationalized by effects on luminal permeability or on the (Na^+ + K^+)-ATPase, or on both.

Salt balance is therefore partially determined by the rate of secretion of aldosterone by the zona glomerulosa of the adrenal cortex. Corticotropin must be present for secretion to occur, but it is not the primary regulator. An increase in [K^+] increases aldosterone synthesis, and a large decrease in [Na^+] also does; these responses are appropriate for maintaining cation homeostasis.

One of the most effective signals for aldosterone secretion is the formation of an oligopeptide known as **angiotensin II,** which appears in response to changes within the kidney. Figure 39–2 sketched the location of the specialized structure known as the juxtaglomerular apparatus; this structure forms a proenzyme that upon stimulation is secreted into the blood as an active protease, **renin.** The signals causing release are (1) a loss of stretch of receptors in the afferent arteriole of the glomerulus indicating diminished blood volume, (2) a drop in Na^+ concentration in the adjacent distal tubular cells, (3) catecholamines in the circulation, and (4) stimulation of renal sympathetic nerves.

Renin attacks a particular protein, angiotensinogen, in the blood (Fig. 39–7) so as to release only the decapeptide initial segment from this *ca.* 100,000 M. W. glycoprotein. The decapeptide, angiotensin I, has only modest physiological activity, but when it passes through the lungs, it is attacked by a relatively nonspecific enzyme that removes the two amino acid residues at the carboxyl terminus as a dipeptide. (This is the same enzyme that inactivates bradykinin, p. 334.) The remaining octapeptide, angiotensin II, has profound physiological effects, including an increase in cardiac output and constriction of the vascular smooth muscles, thereby raising the blood pressure but diminishing fluid loss in the kidney by restricting flow. The effect of most interest for our purposes is a stimulation of the conversion of cholesterol to pregnenolone in the zona glomerulosa, thereby increasing aldosterone production and release.

Angiotensin II is also hydrolyzed by aminopeptidases in the blood and various tissues to create angiotensin III, a heptapeptide that is more active on the adrenal cortex, but less active on the vascular smooth muscle. It is not clear if this is a physiologically important transition. Further hydrolysis of angiotensin III by tissue peptidases destroys the activity.

The renin-angiotensin system therefore conserves fluid by constricting flow through the kidney, increases blood pressure so as to counteract the effects of low blood volume, and conserves salt by increasing the recovery of Na^+ in the distal tubule. (The accompanying loss of K^+ is quantitatively smaller, especially when the urine flow is low.)

Diuresis

Disturbances in almost any part of the concentrating mechanism can cause an increased loss of water and salts from the kidney. Diuresis is a consequence of uncontrolled diabetes mellitus because the amount of glucose in the glomerular

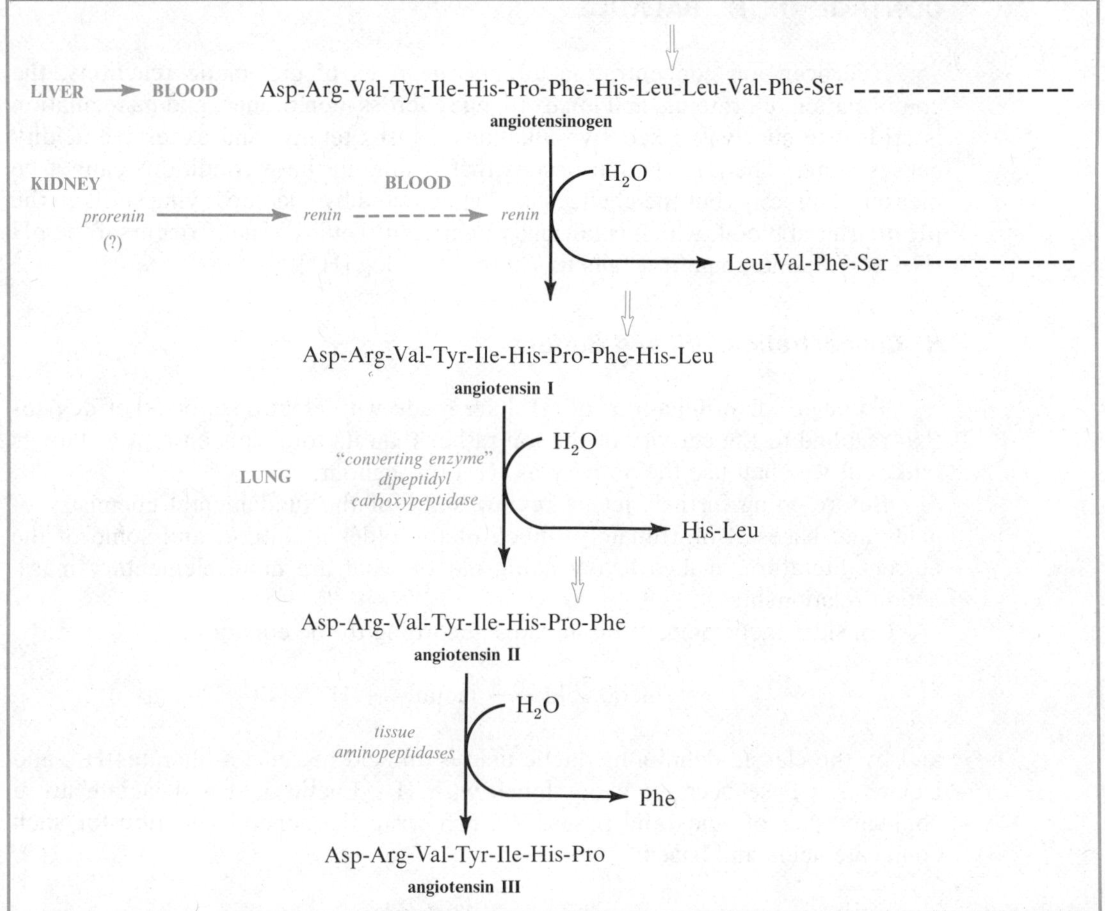

FIGURE 39–7 **Formation of angiotensins.**

filtrate exceeds the capacity of the proximal tubule to absorb it. This extra solute in the tubular fluid raises its osmotic pressure to the point that the distal nephron cannot reabsorb as much water.

Damage to the supraopticoneurohypophyseal stalk, especially above the median eminence, prevents the normal output of vasopressin. Without vasopressin, the distal nephron is nearly impermeable to water, so there is a massive flow of very dilute urine. The flow is so fast that glucose cannot be completely reabsorbed at even normal blood concentrations, and the condition is named diabetes insipidus. (A rare congenital lack of vasopressin receptors in the distal nephron has the same result.) An opposite effect is seen in patients with some neoplasms, which secrete peptides with antidiuretic activity. These people have a high urine osmolality and a low blood osmolality.

Drugs are frequently used to decrease the body salt content, and therefore tissue fluids, as a means of lowering the blood pressure. These diuretic agents, such as the thiazides, furosemide, and ethacrynic acid, inhibit the chloride pump in the thick ascending limb of the loop of Henle. This not only inhibits absorption of NaCl, it diminishes the medullary concentration gradient necessary for water reabsorption, so both salt and water excretion are enhanced.

CONTROL OF H^+ BALANCE

Hydrogen ion concentration affects the rates of enzymatic reactions, the conformation of proteins, and ion exchanges across membranes, and its regulation is critical to survival. Excessive alkalinity causes tetany, and excessive acidity causes coma. The $[H^+]$ within tissues that results in these conditions cannot be measured directly, but the changes can be detected by accompanying shifts in the pH of arterial blood, which is normally near 7.40. Tetany usually occurs if the pH rises to 7.8, and coma if it falls to 7.0 ($pH = -\log [H^+]$).

H^+ Concentration, pH, and Buffers

To begin, determinations of $[H^+]$ are made with electrodes or other devices that respond to the activity of the ion rather than its total concentration. Purists balk, but we shall use the activity as $[H^+]$ throughout.

Before going further, let us review some of the fundamental chemistry of acids and bases. Unfortunately, much of the older literature, and some of the current literature, makes heavy going out of what are quite elementary mass-action relationships.*

Consider lactic acid. It dissociates according to the equation:

$$\text{lactic acid} \longleftrightarrow \text{lactate}^- + H^+,$$

and by the classic definitions, lactic acid is an acid because it liberates H^+, and lactate is a base because it combines with H^+. Lactic acid and lactate are a conjugate pair of acids and bases. We can write the general equation for such conjugate acids and bases:

$$\text{acid} \longleftrightarrow \text{base} + H^+$$

and state a general mass-action equilibrium expression for the relationship between them:

$$K' = \frac{[\text{base}][H^+]}{[\text{acid}]} \quad (1)$$

or, by dividing by $[H^+]$:

$$\frac{K'}{[H^+]} = \frac{[\text{base}]}{[\text{acid}]} \quad (2)$$

This simple equation is the basis for all of our discussion. If any three of the terms are known, the fourth can be determined. Some people use it as is; others prefer to take the negative logarithm so as to cast the expression in terms of pK and pH, which are $-\log K'$ and $-\log [H^+]$:

$$pH - pK' = \log \frac{[\text{base}]}{[\text{acid}]} \quad (3)$$

$$\text{or:} \quad pH = pK' + \log \frac{[\text{base}]}{[\text{acid}]} \quad (4)$$

*I still remember the feeling of total frustration as a young biochemist trying to make sense out of the textbook descriptions of acid-base balance, even though I had no trouble dealing with buffer relationships in the laboratory.

The equilibrium expression in this form is known as the Henderson-Hasselbalch equation. The advent of pocket calculators that readily handle negative fractional exponents has largely destroyed whatever advantage this equation possessed. Since pH and p*K* are still with us, some find it handy to use the logarithmic form.

Here are some examples of the value of the equilibrium expression, and the essential simplicity of its use:

Problem 1. The K' for ammonium ion at 38° C and 0.3 M salt is 3.98×10^{-10}. Estimate the proportion present as ammonia in capillaries at pH 7.40 and in a urine at pH 5.62.
The equilibrium in question is: $NH_4^+ \longleftrightarrow NH_3 + H^+$.

Therefore, $$\frac{K'}{[H^+]} = \frac{[\text{base}]}{[\text{acid}]} = \frac{[NH_3]}{[NH_4{}^+]}$$

So, at pH 7.40, $$\frac{[NH_3]}{[NH_4^+]} = \frac{K'}{[H^+]} = \frac{3.98 \times 10^{-10}}{10^{-7.40}} = \frac{3.98 \times 10^{-10}}{3.98 \times 10^{-8}} = 0.010.$$

(Many pocket calculators will solve the ratio as first given without recasting $10^{-7.4}$ as 3.98×10^{-8}.) Therefore, there is 100 times as much ammonium ion as ammonia in the arterial blood, no matter what the total concentration of the two. At pH 5.62,

$$\frac{K'}{[H^+]} = \frac{3.98 \times 10^{-10}}{10^{-5.62}} = \frac{3.98 \times 10^{-10}}{2.40 \times 10^{-6}} = 1.66 \times 10^{-4} = \frac{1}{6{,}027},$$

and there is only one ammonia molecule for each 6,027 ammonium ions.

Here is the same problem solved by the logarithmic expression:
3.98×10^{-10} is the same as $10^{-9.40}$. so pK = 9.40

$$\log\frac{[\text{base}]}{[\text{acid}]} = \text{pH} - \text{p}K' = 7.40 - 9.40 = -2; \quad \frac{[\text{base}]}{[\text{acid}]} = 10^{-2} \quad \text{(blood)}$$

$$= 5.62 - 9.40 = -3.78; \quad \frac{[\text{base}]}{[\text{acid}]} = 10^{-3.78} \quad \text{(urine)}$$

Problem 2. How much 1 M NaOH must one add to one liter of 0.1 M NH_4Cl to bring the pH to 8.9?

At pH 8.9 $$\frac{[\text{base}]}{[\text{acid}]} = \frac{K'}{[H^+]} = \frac{10^{-9.40}}{10^{-8.90}} = 10^{-0.50} = 0.316 = \frac{[NH_3]}{[NH_4^+]}.$$

The result says that for every 316 molecules of ammonia formed in the final solution, there must be 1,000 molecules of ammonium ion remaining. That is, of 1,316 original molecules of $NH_4{}^+$, only 316, or 24 per cent, are to be converted to ammonia. Therefore, using the relationship $V_1M_1 = V_2M_2$, we find one wants to add 0.24×0.1 or 0.024 liter of 1 M NaOH.

Buffers are solutions that minimize a change in $[H^+]$ upon addition of acid or base. They consist of roughly comparable and moderately high concentrations of both a conjugate acid and its conjugate base. That is, a solution must contain both lactic acid and lactate in concentrations within an order of magnitude of each other to be a good buffer. Another good buffer solution at a much higher pH would contain both ammonium ions and ammonia at concentrations within the same order of magnitude. It is necessary that both the acid and the base be present in order to resist a change in $[H^+]$ upon addition or removal of H^+ by some other source. The more nearly equal their concentrations, the better the buffer action.

Example: Consider two solutions of lactic acid ($K' = 10^{-3.74}$) and sodium lactate:

	A	B
[lactic acid]	0.05 M	0.015 M
[sodium lactate]	0.05 M	0.085 M
pH	3.74	4.49

(It is a useful exercise to verify the pH values.)

Now suppose that NaOH is added to each solution in an amount equivalent to 0.005 M. The hydroxide ion will remove H^+ and more lactic acid will dissociate to replace it. As a first approximation, the new situation will be:

	A	B
[lactic acid]	0.045 M	0.010 M
[sodium lactate]	0.055 M	0.090 M
pH	3.83	4.69

The $[H^+]$ in solution A has changed by 23 per cent ($10^{0.09}$), whereas it has changed by 58 per cent in solution B ($10^{0.2}$). (To do a better approximation, we would have to allow for part of the OH^- being used to react with H^+ in solution, but the error is not great at these H^+ concentrations.)

A truism emerges: Conjugate acids and bases are effective buffers only in solutions in which the $[H^+]$ is within an order of magnitude of the K' of the acid, if the total concentration of acid and base combined is held constant. That is, the pH must be within 1.0 of the pK'. The lactic acid-lactate pair buffers between pH 2.74 and 4.74. Ammonium ion-ammonia buffers between pH 8.40 and 10.40.

The internal pH of the organism is usually near 7. What then are compounds with pK' values near 7 that resist a change in $[H^+]$ within cells or in the extracellular fluids? By far the most important are groups in proteins, especially the imidazolium group in histidine residues. Its pK' value varies with location in a protein, but it frequently is not far from 7.0.

Inorganic phosphate and phosphate monoesters such as glucose 6-phosphate and glycerol phosphate are effective as buffers. The pK' values are near 6.8 for the dissociations:

$$H_2PO_4^- \rightarrow HPO_4^{2-} + H^+$$
$$R-O-PO_3H^- \rightarrow R-O-PO_3^{2-} + H^+.$$

Notice it is the second dissociations in each case that are in the appropriate range. Phosphodiesters with only one dissociation, such as phosphatidyl compounds and the phosphate groups in nucleic acids, are too strong acids for effectiveness as a buffer. Nucleoside polyphosphates, such as ATP, contribute little buffering power because of their high affinity for Mg^{2+}, which displaces H^+.

The Carbonic Acid-Bicarbonate Equilibrium

Another buffer system, the carbonic acid-bicarbonate pair, requires special discussion. The effective pK value for carbonic acid in blood plasma is only 6.10, but it still is very important for [H+] homeostasis because the concentration of the conjugate acid is held relatively constant through regulation of respiration. Additions of acid and base therefore tend to affect only the bicarbonate concentration. The equilibria of interest are:

$$CO_2(gas) \longleftrightarrow CO_2(solution) \overset{\pm H_2O}{\longleftrightarrow} H_2CO_3 \longleftrightarrow H^+ + HCO_3^-.$$

The usual equilibrium constant defines the relationship between the concentrations of carbonic acid, H^+, and HCO_3^-:

$$K' = \frac{[H^+][HCO_3^-]}{[H_2CO_3]}$$

However, it is difficult to determine just how much carbonic acid is present in a solution compared to the amount of dissolved CO_2 that isn't hydrated. It is therefore customary to lump together the concentrations of dissolved gas and of carbonic acid and treat them as a single entity, and this sum is what we shall mean when we speak of [dissolved CO_2].

The second practical problem is that we are dealing with blood, which is a complex tissue, and it is not possible to sort out all of the possible equilibria involving CO_2, such as the combination with amino groups of proteins that occurs to a small extent spontaneously:

$$R{-}NH_2 + CO_2 \rightleftharpoons R{-}NH{-}COO^- + H^+.$$

These unknown reactions prevent us from applying equilibrium constants determined with more pure compounds to the relationship between dissolved CO_2, H^+, and HCO_3^- in blood. The solution is to be pragmatic and not worry about the side reactions, and this is what is done. An apparent equilibrium constant is determined by measuring the actual relationship between the three components of the carbonic acid equilibrium in the plasma compartment of whole blood, and it is designated K'' to indicate its empirical nature:

$$K'' = \frac{[H^+][HCO_3^-]}{[\text{dissolved } CO_2]} = 10^{-6.10} \text{ (blood at 38°C)}$$

in which all of the various forms of CO_2 are lumped together as an entity. The concentration of these forms can be calculated from the partial pressure of the gas in equilibrium with the blood:

$$[\text{dissolved } CO_2] / p_{CO_2} = 2.25 \times 10^{-7} \text{ M pascal}^{-1} = 3 \times 10^{-5} \text{ M torr}^{-1}.$$

We are primarily interested in the way in which the concentration of H^+ changes with changes in the concentration of dissolved CO_2, so let us rearrange the ionization equation by dividing by [H^+] so as to obtain:

$$\frac{K''}{[H^+]} = \frac{[HCO_3^-]}{[\text{dissolved } CO_2]}$$

Now, let us apply the equation by examining normal arterial blood. The pH is 7.40, so $[H^+] = 10^{-7.40}$. Since we know that $K'' = 10^{-6.10}$, we can now determine the ratio of concentrations of bicarbonate ions and of dissolved CO_2 in the plasma of that blood:

$$\frac{10^{-6.10}}{10^{-7.40}} = 10^{1.3} = 20 = \frac{[HCO_3^-]}{[\text{dissolved } CO_2]}.$$

We find that there is 20 times as much bicarbonate as dissolved CO_2 in arterial blood. (We round off $10^{1.3}$ = 19.953 as 20.)

The arterial blood is equilibrated with CO_2 at the partial pressure existing in the alveoli of the lungs, which is normally near 40 torr (5.3 kPa). Therefore,

$$[\text{dissolved } CO_2] = 3 \times 10^{-5}\, p_{CO_2} = 3 \times 10^{-5} \times 40 = 1.2 \text{ mM}.$$

The concentration of bicarbonate ions is 20 times greater than this, or 24 mM.

The carbonic acid-bicarbonate pair is a useful indicator of acid-base status because of the accessibility of blood. All buffers equilibrate within the same cellular compartment according to the relationship:

$$[H^+] = \frac{K'[\text{acid}]}{[\text{base}]}.$$

The ratios of acidic and basic forms at a given pH of each buffer is inversely proportional to their dissociation constants, but all of these ratios will shift by the same factor when there is a change in $[H^+]$, so a shift in one is an index of a shift in all. Hydrogen ions do not fully equilibrate in all cellular compartments; even so,

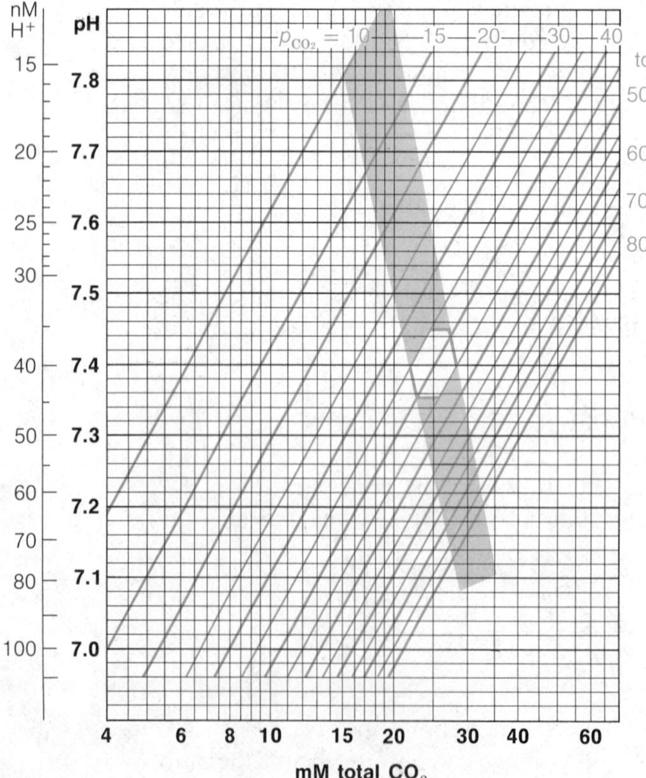

FIGURE 39–8

The carbonic acid-bicarbonate equilibrium in blood. The blue shading is the normal buffer band; rapid rises or falls in the carbon dioxide tension will create pH values lying within this band. Concentrations in normal individuals usually lie within the clear area in the center of the buffer band.

changes in the readily accessible blood plasma are a guide to changes within the tissues, and the blood is important in itself as a carrier of H^+.

Three measurements of clinical interest are commonly made: the pH and p_{CO_2} of arterial blood, and the total CO_2 released from venous blood by acidification, which includes the dissolved CO_2 and the bicarbonate. Any two of these parameters in the same blood sample defines the third, and the relationship is shown in Figure 39–8. Therefore, the total CO_2 in arterial blood can be calculated from the pH and p_{CO_2} values. As a rough check on the validity of the laboratory measurements:

$$[H^+] \text{ in nM} = \frac{24 \times p_{CO_2} \text{ in torr}}{[HCO_3^-] \text{ in mM}}$$

ORIGIN OF CHANGES IN pH

Respiratory Loads

The high influx of CO_2 from the combustion of fuels and its discharge in the lungs inevitably causes fluctuation of the $[H^+]$ in the blood. Figure 39–9 shows the changes that occur in exercising men. The components of blood, especially hemoglobin, are designed to minimize these changes. In addition to these cyclical changes from arterial to venous circulation, changes also occur in the level of CO_2 that returns from the lungs, that is, in the basal level around which the arterial-venous fluctuations occur. These shifts in basal level occur where there is an alteration in the effectiveness of removal of CO_2 by the lungs. Increasing the rate of breathing when there is no corresponding increase in delivery of CO_2 to the lungs will diminish the CO_2 tension in the alveoli, and therefore in the blood returned to the peripheral tissues. The $[H^+]$ falls (pH rises), and a **respiratory alkalosis** has occurred. Holding the breath or impairment of gas exchange by partial obstruction of the bronchi or by drug overdose will diminish the discharge of CO_2 and elevate the CO_2 tension in the blood. The resultant rise in $[H^+]$ (fall in pH) is a **respiratory acidosis.**

The Bohr effect provides part of the compensation for changes in CO_2 concentration from arterial to venous blood. When oxygen is released, the affinity of hemoglobin for H^+ increases (p. 236). Approximately 0.4 H^+ is taken up for each O_2 released.* Approximately 10 per cent of the extra CO_2 put into the blood is transported to the lungs in this form in humans. In a resting person liberating 0.8 CO_2 per O_2 consumed, over 50 per cent of the CO_2 can be transported without any change in other buffer systems:

$$\begin{array}{lrcl} \text{Tissue:} & O_2 + \text{fuel} & \longrightarrow & 0.8\ CO_2 \\ \text{Blood:} & HbO_2 + 0.4\ H^+ & \longrightarrow & Hb + O_2 \\ & 0.4\ CO_2\ (+H_2O) & \longrightarrow & 0.4\ H^+ + 0.4\ HCO_3^- \\ \text{SUM:} & HbO_2 + \text{fuel} & \longrightarrow & Hb + 0.4\ HCO_3^- + 0.4\ CO_2 \end{array}$$

*This value is an approximation determined in the presence of CO_2 and bisphosphoglycerate. It therefore includes an adjustment for the change in carbamoylation of hemoglobin by CO_2, which releases H^+:

$$^+H_3N\text{—}Hb + CO_2 \rightarrow {}^-OOC\text{—}HN\text{—}Hb + 2\ H^+$$

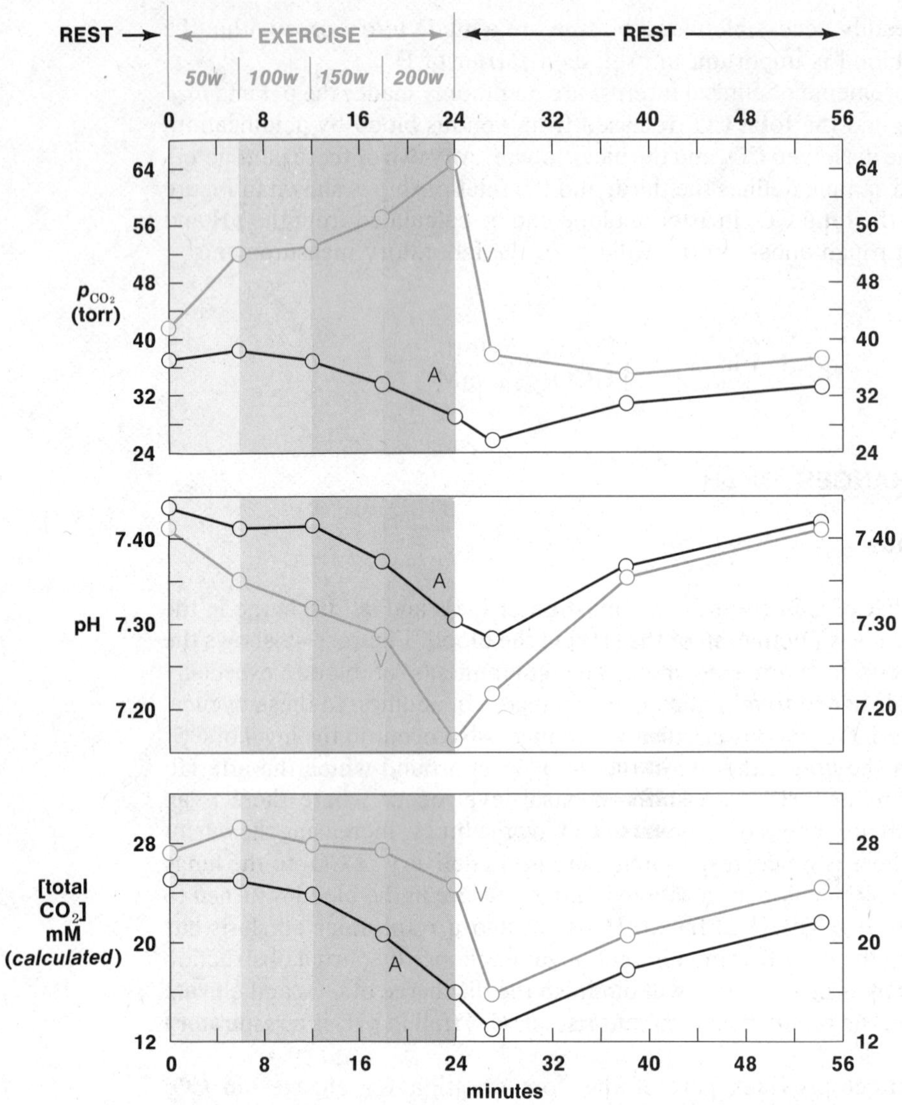

FIGURE 39–9 **Changes in the carbonic acid–bicarbonate equilibrium in blood circulating through exercising muscles. Individuals were exercised on a bicycle ergometer for six-minute periods at increasing power outputs, ranging from 50 to 200 watts *(top left)*. Analyses were made of arterial (A) and femoral vein (V) blood at the beginning and at the end of each exercise period, and during the resting period thereafter. Data from J. Keul, E. Doll, and D. Keppler: (1972) *Energy Metabolism of Human Muscle.* Med. Sport, 7: 72–74.**

Since the ratio of dissolved CO_2 to bicarbonate is 1:20 at pH 7.4, the 0.4 HCO_3^- will balance 0.4/20 = 0.02 of the remaining CO_2 without changing pH. The total **isohydric carriage** (carriage without change in [H^+]) is therefore 0.4 + 0.02 = 0.42 CO_2 out of the 0.8 formed.

The balance of the CO_2 reacts with the buffer systems until they are all equilibrated at constant pH:

$$CO_2 + H_2O + \text{buffer base} \rightarrow HCO_3^- + \text{buffer acid.}$$

As we noted, hemoglobin itself is the most powerful buffer in the blood other than the bicarbonate system. (We are adding carbonic acid to the carbonic acid-bicarbonate system; it cannot buffer itself.) The shaded band in Figure 39–8 shows the changes in pH that occur in normally buffered individuals with changes in p_{CO_2}. That is, the pH of blood lies within this band with respiratory acidosis or alkalosis until compensatory mechanisms occur. There is little shift in the pH upon going from arterial to venous circulation and back again in resting individuals. Here are typical values for normal young adult males at rest:

	Arterial Blood	Femoral Venous Blood
pH	7.43	7.41
p_{CO_2} (torr)	37	42
[total CO_2] (mM)	25	27

The bicarbonate concentration increases in venous blood owing to the reaction of CO_2 with the buffer systems, mainly hemoglobin. This involves a series of events. CO_2 moves into the blood from the tissues as such. It will spontaneously react with water to form carbonic acid, but this is a relatively slow reaction in comparison to the rate of blood circulation, and there is an enzyme, **carbonic anhydrase,** within the red blood cells to accelerate the reaction. This is convenient, because the carbonic acid is generated in proximity to hemoglobin, which acts as the principal buffer system:

$$\begin{aligned} \text{Erythrocyte:}\quad CO_2 + H_2O &\longrightarrow H_2CO_3 \\ H_2CO_3 &\longrightarrow H^+ + HCO_3^- \\ H^+ + Hb &\longrightarrow HHb^+ \\ \text{SUM:}\quad CO_2 + H_2O + Hb &\longrightarrow HCO_3^- + HHb^+ \end{aligned}$$

(We are distinguishing the buffer action of hemoglobin from the Bohr effect shift by indicating the acidic forms as HHb^+.) Additional events occur to restore osmotic balance. The formation of bicarbonate within the erythrocyte not only raises the total solute concentration, but it also increases the ratio $[HCO_3^-]_{in}/[HCO_3^-]_{out}$ over the corresponding ratio for chloride ions. Bicarbonate diffuses out of the erythrocyte and chloride in until these ratios match. Osmotic balance is restored by movement of water into the erythrocyte. This sequence of events, sometimes called the **chloride shift,** is detailed as a matter of general interest and to call attention to the importance of carbonic anhydrase in equilibrating CO_2 and bicarbonate in the presence of buffer systems.

Dietary and Metabolic Loads

Hydrogen ions may be added or removed as a result of eating particular foods, or by metabolic changes. An important point must be emphasized: **There is no substitute for writing a complete stoichiometry for any physiological event in analyzing the effects of the event on [H⁺].** This is neglected with surprising frequency, even by some with no small reputation in the field, with consequent mistakes in logic. It is critical in establishing the stoichiometry that the ionic state of the reactants and products be stated at least approximately, taking note of the fact that most meals are near neutrality, or slightly acidic. Some examples will make the idea clear, and show some important principles:

(1) What is the effect on H^+ balance of ingesting starch and oxidizing it to CO_2 and H_2O? The overall reaction is:

$$(C_6H_{10}O_5)_n + 6n\ O_2 \rightarrow 6n\ CO_2 + 5n\ H_2O.$$

Note that all of the reactants come from outside the body and all of the products are excreted from the body; note also that nothing is said about high-energy phosphate, the citric acid cycle and so on. Including these reactions would be a mistake, because all of the intermediate compounds are recycled in relatively steady-state concentrations that do not affect the overall result. There is no net production or consumption of H^+ so long as the CO_2 passes out in the lungs as fast as it is formed in the tissues.

(2) What is the effect of ingesting acetic acid (vinegar) and oxidizing it to CO_2 and H_2O? The ingestion of an acid increases the total body $[H^+]$ until it is removed. After it has been oxidized, the overall stoichiometry becomes:

$$C_2H_4O_2 + 2\ O_2 \rightarrow 2\ CO_2 + 2\ H_2O.$$

Again, all of the components are outside of the body, and there is no net *long-term* change in H^+ concentration from the ingestion of a metabolizable acid. (This includes citric acid, malic acid, succinic acid, and so on.)

(3) What is the effect of ingesting disodium hydrogen citrate ($Na_2C_6H_6O_7$) and oxidizing the citrate to CO_2 and H_2O? If we write a stoichiometry for the combustion as such, we must formally balance the charges in some way:

$$2\ Na^+ + C_6H_6O_7^{2-} + 2\ H^+ + 4\tfrac{1}{2}\ O_2 \longrightarrow 2\ Na^+ + 6\ CO_2 + 4\ H_2O$$

or: $$2\ Na^+ + C_6H_6O_7^{2-} + 4\tfrac{1}{2}\ O_2 \longrightarrow 2\ Na^+ + 6\ CO_2 + 2\ H_2O + 2\ OH^-$$

or; $$2\ Na^+ + C_6H_6O_7^{2-} + 4\tfrac{1}{2}\ O_2 \longrightarrow 2\ Na^+ + 4\ CO_2 + 2\ H_2O + 2\ HCO_3^-$$

Any way we do it, we come to the conclusion that Na^+ is accumulating and the body will become more alkaline — H^+ is removed directly, or by combination with OH^-, or by a shift in the buffer pairs to match an increased bicarbonate concentration. The only difference between ingesting sodium bicarbonate and the sodium salts of metabolizable carboxylic acids is that the latter increase the bicarbonate concentration more slowly. Sodium lactate has been used therapeutically to diminish $[H^+]$ in this way. Citric acid cycle intermediates are common constituents of fruits and vegetables, where they are partially or completely ionized. Their solutions are acid, but complete combustion takes up H^+.

(4) What is the effect of ingesting alanine and oxidizing it to CO_2, H_2O, and urea, which are excreted?

$$\underset{\text{alanine}}{C_3H_7O_2N} + 3\ O_2 \longrightarrow \tfrac{1}{2}\ \underset{\text{urea}}{CH_4ON_2} + 2\tfrac{1}{2}\ CO_2 + 2\tfrac{1}{2}\ H_2O$$

In general, metabolism of amino acids or polypeptides containing only C, H, O, and N that are ingested near their isoelectric points have no effect on H^+ balance if urea is excreted.

(5) What is the effect of ingesting methionine and oxidizing it to CO_2, H_2O, urea, and sulfate, which are excreted?

$$C_5H_{11}O_2NS + 7\tfrac{1}{2}\ O_2 \rightarrow \tfrac{1}{2}\ CH_4ON_2 + 4\tfrac{1}{2}\ CO_2 + 3\tfrac{1}{2}\ H_2O + SO_4^{2-} + 2\ H^+$$

Oxidation of cysteine and methionine-containing proteins causes a production of H^+. (This is one of the major sources of H^+.)

(6) What is the effect of ingesting phosphate esters and oxidizing the carbon skeletons to CO_2 and H_2O? The phosphate esters may be monoesters like glycerol 3-phosphate, or diesters as in the nucleic acids and phospholipids. (Remember that pK_2' for the monoesters and P_i is approximately 6.8, whereas the single pK' of the diesters is much lower.) The approximate changes in neutral foods (pH 6.8) and the resultant stoichiometries to excrete P_i in a urine at pH 6.8 are:

$$R{-}PO_4H_{0.5}^{1.5-} + xO_2 \rightarrow yCO_2 + 2\ H_2O + H_{1.5}PO_4^{1.5-}$$
$$R,R'{-}PO_4^- + xO_2 \rightarrow yCO_2 + 2\ H_2O + H_{1.5}PO_4^{1.5-} + 0.5\ H^+.$$

(These equations were obtained by calculating the ratios of acidic to basic forms of the compounds at pH 6.8.) If food and urine are at the same $[H^+]$, then the monoesters will cause no H^+ production. (This point is missed by many.) However, the nucleic acids and phospholipids are the most abundant phosphate compounds in foods, and their metabolism is another major source of H^+ production. (Question: Compare the effects of the two compounds if the urine is at pH 5.8.)

In addition to the continuing loads of the sort just outlined, alterations in H^+ balance also occur from changes in the metabolic processes, either as normal transients or as pathological events. For example, the accumulation of the anion of any carboxylic acid with a pK' value well below 7 is accompanied by a concomitant production of H^+. Here we must distinguish between acute and chronic loads.

The H^+ produced by the formation of lactate during strenuous exercise is an example of an acute load. (Part of the changes during heavy exercise shown in Figure 39–9 were due to lactate formation.) The constant formation of the 3-oxybutyrates in uncontrolled diabetes mellitus, or of propionate or methylmalonate in the aminoacidopathies is an example of a chronic load in which there is constant spillage of the anion in the urine, and there must be an equally constant discharge of H^+ if it is not to accumulate to lethal levels.

The base forms of the protein and phosphate buffers bind H^+ liberated by metabolic events, just as they do H^+ formed by respiratory acidosis, but the metabolic pH changes are also minimized by alterations in the bicarbonate level. The increased H^+ is partially dissipated by conversion of bicarbonate to CO_2, which is blown off by the lungs:

$$H^+ + HCO_3^- \rightarrow H_2CO_3 \rightarrow H_2O + CO_2.$$

If nothing else happens, a **metabolic acidosis** will be characterized by a drop in bicarbonate, or total CO_2 concentration, whereas an uncompensated respiratory acidosis will be characterized by an increase in bicarbonate, or total CO_2 concentration.

Conditions that cause a loss of H^+, thereby creating a **metabolic alkalosis,** include direct losses through vomiting of the dilute HCl normally present in the stomach. Hereditary defects in urea formation that result in excessive excretion of ammonium ions also cause metabolic alkalosis. For example, the total stoichiometry of alanine combustion then becomes:

$$C_3H_7O_2N + 3\ O_2 + H^+ \rightarrow NH_4^+ + 3\ CO_2 + 2\ H_2O.$$

There is a loss of one H^+ for each NH_4^+ excreted that does not occur when urea is synthesized and excreted.

COMPENSATORY MECHANISMS

Buffer action can only *minimize* the change in $[H^+]$ that occurs with a given gain or loss of H^+. Other compensatory mechanisms exist for diminishing the burden on the buffers by instituting processes that have the contrary effect on the turnover of H^+. In crude terms, acidosis is corrected by processes that would in themselves create an alkalosis and *vice versa*. The compensatory mechanisms involve changes in the rate of three processes: CO_2 exchange through the lungs, elimination of H^+ through the kidneys, and metabolic reactions that create or consume H^+.

Respiratory Compensation

The rate of pulmonary ventilation is controlled by receptors that are separately sensitive to concentrations of CO_2, H^+, and O_2, listed in the order of decreasing sensitivity to deviations from physiological levels. That is, the O_2 concentration must change relatively more than the CO_2 concentration to cause equal changes in the rate of breathing. The effect of increasing $[H^+]$ is to diminish the $[CO_2]$ necessary to achieve a given rate. This provides a compensatory device for changes in $[H^+]$; any increase makes the respiratory centers more sensitive to CO_2. Therefore, the breathing rate will increase until the $[CO_2]$ has fallen. The drop in carbonic acid concentration will cause a corresponding shift of all buffers from their acidic to basic forms at the expense of bicarbonate, thereby eliminating part of the H^+ load.

Picture the sequence this way: A metabolic acidosis creates high $[HHb^+]$ and low [Hb]. Compensation by blowing off additional CO_2 creates low $[CO_2]$ leaving $[HCO_3^-]$ at its normal high level. This cannot persist because the two buffer systems must be at equilibrium; H^+ will exchange between them with the net effect being a reaction of the components in high concentrations:

$$HCO_3^- + HHb^+ \rightarrow H_2O + CO_2 + Hb.$$

The extra HHb^+ created by the acidosis has now been partially removed. The contrary events occur, although to a lesser extent, in compensation for metabolic alkalosis. The rate of ventilation is decreased in order to elevate the $[CO_2]$, which contributes H^+.

Renal Compensation

The normal course of metabolism in most people produces an acidic urine, depending upon their diet. The more protein, the more H^+ formed; the more fruits

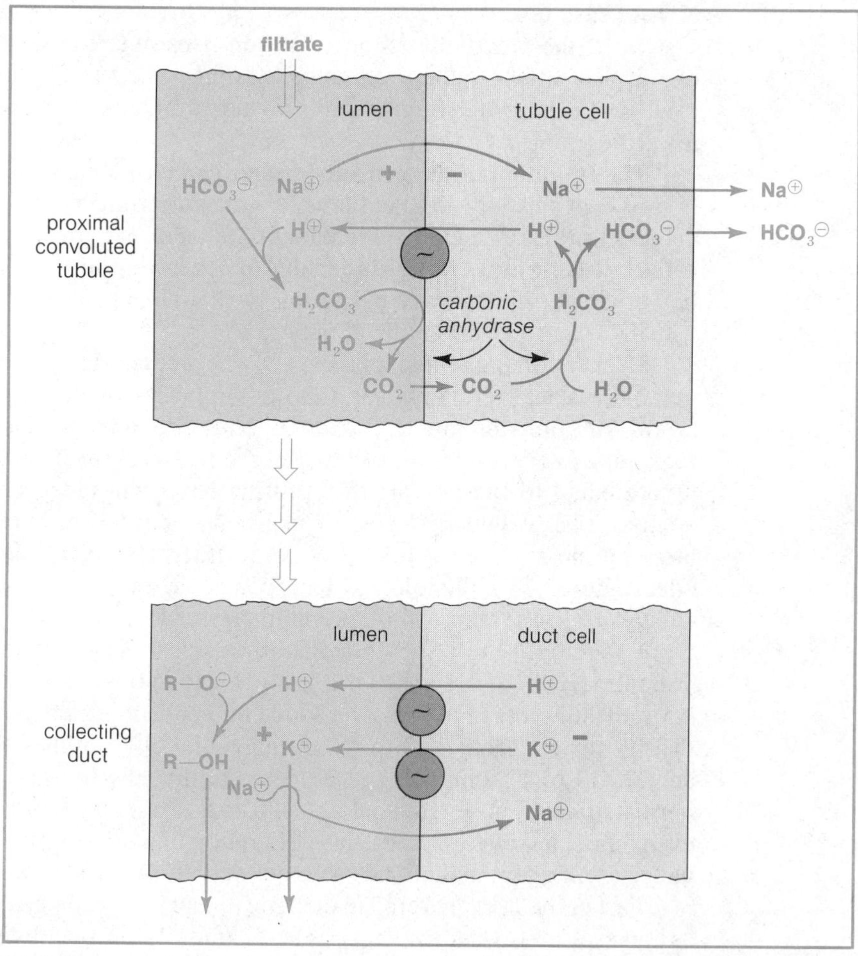

FIGURE 39–10 **Action of H^+ pumps in the proximal convoluted tubule and in the collecting ducts. H^+ is recycled in the proximal tubule as a device for recovery of bicarbonate and associated Na^+. The pump can be used without recycling to excrete H^+. H^+ is also excreted in the collecting duct, where it combines with anions to form undissociated forms that are excreted. The collecting duct also pumps out K^+, and there is some competition between H^+ and K^+ excretion, since each creates an adverse potential gradient.**

and vegetables, with their partially ionized carboxylic acids, the more H^+ removed. The kidney is constructed to eliminate H^+, or conserve it, in order to maintain constant $[H^+]$ in the face of this metabolic load. Two mechanisms involve the proximal convoluted tubule and the collecting ducts.

The proximal tubule has an active pump that will secrete H^+ into the lumen (Fig. 39–10) as a part of a cycle for recovering bicarbonate. The secreted H^+ reacts with filtered bicarbonate to produce carbonic acid. A carbonic anhydrase in the brush border catalyzes equilibration with CO_2, which passes into the cells where another carbonic anhydrase again equilibrates it with carbonic acid, which dissociates to replace the H^+ originally pumped out and at the same time forms bicarbonate inside the cell. **The net result is a recovery of bicarbonate from the filtrate.** The electrical potential created by these ionic movements causes Na^+ and some K^+ to enter the tubule cells. (This augments the recovery of Na^+ by action of the (Na^+ +

K^+)-ATPase discussed earlier.) However, there is a limit to the capacity of the system. If the bicarbonate concentration rises over the range of 26 to 28 mM in blood plasma, the excess continues through the tubule into the urine. Excess bicarbonate is created by metabolic alkalosis, and its elimination restores the [H^+] toward normal.

The H^+ pump of the proximal tubule can also be used to eliminate excess H^+. However, it can only do so until the H^+ concentration in the lumen is four times that in the cells (pH 0.6 unit lower in lumen). **Its action is supplemented by another pump in the collecting ducts,** which is capable of generating almost a 1,000-fold concentration gradient, with urinary pH as low as 4.5. (Increases in H^+ excretion interfere with K^+ excretion in the distal nephron.)

Even at the minimal pH of 4.5, the actual [H^+] in the urine is only 32 micromoles per liter. However, this does not mean that excretion of an acidic urine is removing only this much H^+ from the body. For one thing, there is no bicarbonate remaining in the urine; all of the bicarbonate present in a volume of glomerular filtrate equal to the volume of the urine has been added back to the remaining volume of body fluid. If the blood bicarbonate concentration is 10 millimolar, each liter of urine excreted at pH 5.2 or below (the pH of a carbonic acid solution) is in effect adding 10 millimoles of bicarbonate to the remaining body fluid, almost equivalent to the removal of 10 millimoles of H^+.

In addition, the urine contains other acids and bases, especially phosphate. The ratio $[HPO_4^{2-}]/[H_2PO_4^-]$ is 2.0 at pH 7.10 in the blood, but it is almost 0 at pH 4.5, and 0.67 mole of H^+ must be added for each mole of phosphate excreted before the pH can fall from 7.10 to 4.5 during metabolic acidosis. This can account for another 13 to 17 millimoles of H^+ excretion at the levels of phosphate found in normal urines, but severe acidosis also causes a mobilization of bone that leads to even more phosphate in the urine. Excretion of a urine at pH 5.5 to 6.0 is adequate under normal metabolic loads to maintain blood pH at 7.40.

Part of the H^+ generated in diabetic ketoacidosis can be eliminated in the urine as undissociated 3-hydroxybutyric acid. Since its dissociation constant is $10^{-4.39}$, 44 per cent of the compound is present as the undissociated acid at pH 4.5, which means that an additional 0.44 mole of H^+ must be added to the urine for each mole of 3-hydroxybutyrate excreted before the pH can fall to 4.5. (The dissociation constant of acetoacetic acid is too low for it to be a significant factor in this way. However, the spontaneous decarboxylation of acetoacetate may contribute a small loss of H^+:

$$\text{acetoacetate}^- + H^+ \rightarrow CO_2 + \text{acetone.})$$

Metabolic Compensation

Alterations in body composition and in metabolic processes can also be used to consume or produce additional H^+. **The dissolution of bone** during metabolic acidosis is an example. Much of the bone consists of hydroxyapatite, which has a lattice with a repeating unit of $Ca_{10}(PO_4)_6(OH)_2$, but it also contains substantial amounts of carbonate ions — of the order of one carbonate per seven phosphates in a typical long bone. This mass of crystalline salt represents a large reservoir of buffer capacity. Within a few hours of the appearance of metabolic acidosis, there is substantial loss of carbonate from the bones although with little loss of calcium or phosphate. If the acidosis persists over several days, more substantial alterations in composition occur. Part of the hydroxyapatite may then be converted to calcium phosphate, with loss of the hydroxyl ions, and part is dissolved. During metabolic alkalosis, there is increased deposition of carbonate in the bone, in both the short- and long-term conditions.

Diminution in the formation of urea and increase in the excretion of ammonium ions is an important adaptation to severe metabolic acidosis. Half, or more, of the nitrogen in the urine may be present as NH_4^+. In our sample stoichiometries, we saw that the conversion of nitrogen in amino acids to urea results in no change in H^+ balance, whereas the conversion to NH_4^+ results in a consumption of one H^+ per NH_4^+ formed.

It makes no difference whether the shift from urea excretion to ammonium ion excretion occurs because of changes in the rate of urea formation in the liver or from a depletion of ammonia nitrogen in the kidney; the overall result is the same: **there is one less H^+ to handle when nitrogen is excreted as NH_4^+**. However, a persistent misconception in the literature of renal physiology that the effect hinges on the use of glutamine in the mechanism is worth noting, if for no other reason than to avoid similar errors. Here are the actual events (Fig. 39–11):

(1) Cells in the tubule of the kidney take in glutamine from the blood, and use their mitochondrial glutaminase to hydrolyze it to glutamate and NH_4^+. (Normally, most of the glutaminase occurs in the distal portion of the nephron. Some animals adapt to continued acidosis by synthesizing an equal activity in the proximal tubule.)

(2) Only the small fraction of NH_3 in equilibrium with NH_4^+ will diffuse into the lumen, and it does so, leaving H^+ within the cells.

(3) Upon entry into the lumen, the NH_3 immediately reacts with H^+ to regenerate NH_4^+. The net effect so far is to **gain** one H^+ within the cell and to lose one in the lumen per NH_4^+ formed.

(4) H^+ is pumped out of the cell to restore the original status quo.

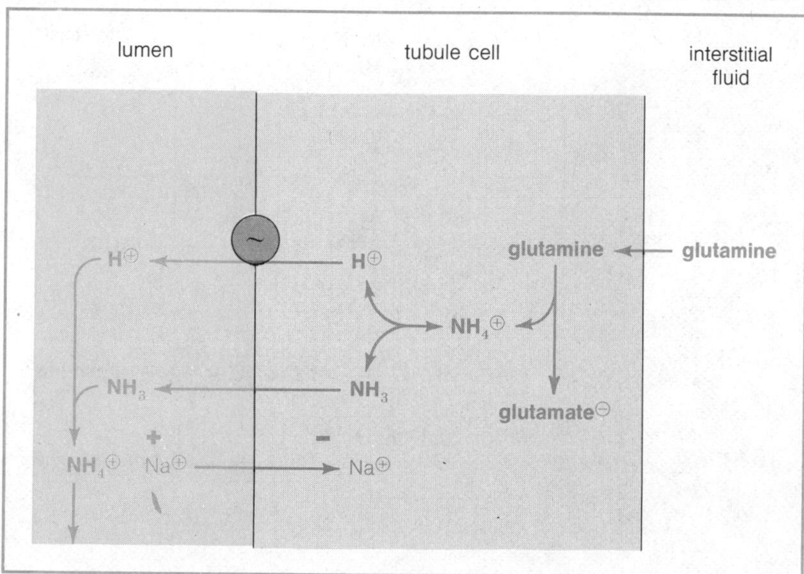

FIGURE 39–11 **The exchange of ammonium ion for sodium ion in the renal tubule. Glutamine is the source of the ammonium ion in this example, but a similar stoichiometry would result when other amino acids are the source. There is no net change in H^+ balance within the kidney from this process by itself.**

(5) Na^+ diffuses into the cell down the potential gradient created. **There is no net change in H^+ balance as a result of the mechanism.** The overall reaction is:

$$\text{glutamine (blood)} + Na^+ \text{ (filtrate)} \rightarrow \text{glutamate}^- \text{ (cell)} + Na^+ \text{ (cell)} + NH_4^+ \text{ (filtrate)}.$$

Further metabolism of glutamate has no effect other than what it would have in the absence of this exchange mechanism. The picture is not altered in substance by hypothesizing other sources of NH_4^+ within the tubule cells.

Here is the common explanation seen in the literature and in the textbooks (Fig. 39–12):

(1) Glutaminase in the tubule cells hydrolyzes glutamine to glutamic acid and ammonia:

$$\begin{array}{c} O{=}C{-}NH_2 \\ | \\ (CH_2)_2 \\ | \\ H{-}C{-}NH_2 \\ | \\ COOH \end{array} \xrightarrow{+H_2O} \begin{array}{c} COOH \\ | \\ (CH_2)_2 \\ | \\ H{-}C{-}NH_2 \\ | \\ COOH \end{array} + NH_3$$

glutamine → glutamic acid

(2) The ammonia diffuses into the lumen of the tubules, where it reacts with H^+ to form NH_4^+. The neutralization of H^+ in the lumen diminishes the gradient for

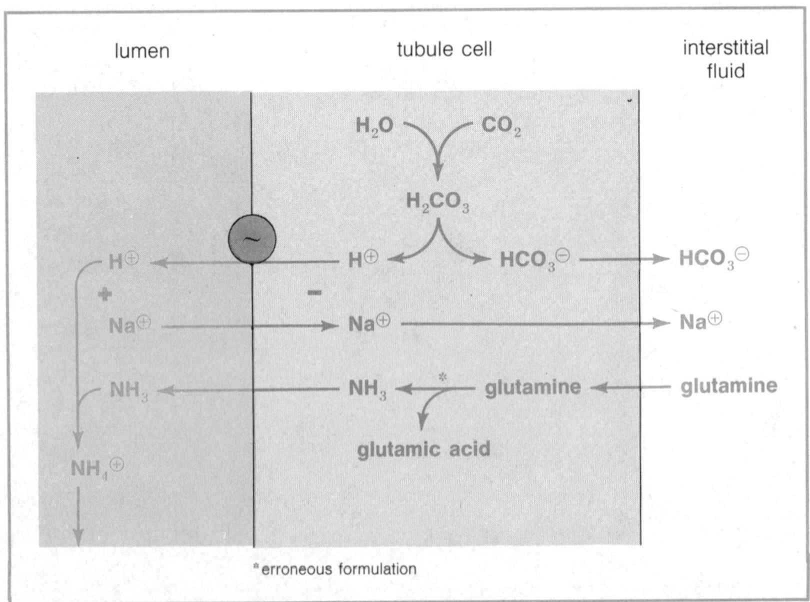

FIGURE 39–12 **A common erroneous formulation of ammonium ion excretion in the kidney, purporting to show a net loss of H^+ and gain of bicarbonate ion as a result. The fallacy lies in improper formulation of the hydrolysis of glutamine (compare p. 765).**

H^+ across the wall of the lumen and permits more H^+ to be pumped into the filtrate in exchange for Na^+. The overall stoichiometry becomes:

$$(\text{glutamine} + H_2O + CO_2)_{\text{blood}} + Na^+_{\text{filtrate}} \longrightarrow$$
$$(\text{glutamic acid } (sic) + Na^+ + HCO_3^-)_{\text{blood}} + NH_4^+{}_{\text{lumen}}.$$

Two things ought to have prevented the persistence of this explanation: writing the proper stoichiometry for the postulated events and considering the fact that some animals adapt to acidosis by excreting NH_4^+ but have little change in the glutaminase activity in their kidneys.

Accumulation of carboxylates provides partial compensation for alkalosis. For example, respiratory alkalosis is often accompanied by an increased accumulation of lactate, and metabolic alkalosis causes increased accumulation of the 3-oxybutyrates and citrate. It is not known what causes these accumulations.

Uncontrolled formation of lactate (lactic acidosis, p. 479) often accompanies a variety of conditions and is sometimes caused by a hereditary defect of unknown mechanism. A severe acidosis would tend to suppress lactate utilization through an adverse shift of the lactate dehydrogenase equilibrium:

$$\text{lactate}^- + NAD^+ \longleftrightarrow \text{pyruvate}^- + NADH + H^+$$

which forms H^+ when lactate is converted to pyruvate, and this may account in part for the lactic acidosis that sometimes aggravates severe diabetic ketoacidosis. (The concomitant shift of the equilibrium from acetoacetate to D-3-hydroxybutyrate sometimes causes trouble because measurement of acetoacetate is used as an index of the degree of ketosis.)

Assessment of Disturbances of H^+ Balance from Laboratory Data

The blood analyses of interest in interpreting acid-base changes are those correlated in Figure 39–6: pH, p_{CO_2}, and total CO_2, along with additional measurements of K^+, Na^+, and Cl^- concentrations. The pH value, obtained directly or by calculation, tells whether a significant acidosis or alkalosis exists. The effects of acute uncompensated changes are shown in Figure 39–13. Sudden changes in ventilation rate shift the pH and p_{CO_2} values within a characteristic band on the chart, as we noted earlier. The position of this band is determined by the total buffer capacity and the relative proportions of acidic and basic forms present at a given p_{CO_2}.

If a metabolic acidosis or alkalosis appears without respiratory compensation, the proportion of acidic and basic forms of the buffers is being changed without altering the p_{CO_2}. The effect is then a shift of the buffer band up or down the line of constant p_{CO_2}.

Compensated Changes. The compensation for alkalosis and acidosis in effect superimposes respiratory changes and the opposite renal metabolic changes. That is, the kidney creates a metabolic acidosis by conserving H^+ in order to compensate for a respiratory alkalosis caused by the hyperventilation in an overly anxious person. Contrariwise, the ventilation is sharply increased in the lungs to create a respiratory alkalosis to compensate for the increased production of H^+ associated with 3-oxybutyrate formation in an uncontrolled diabetic. If there is a clear-cut

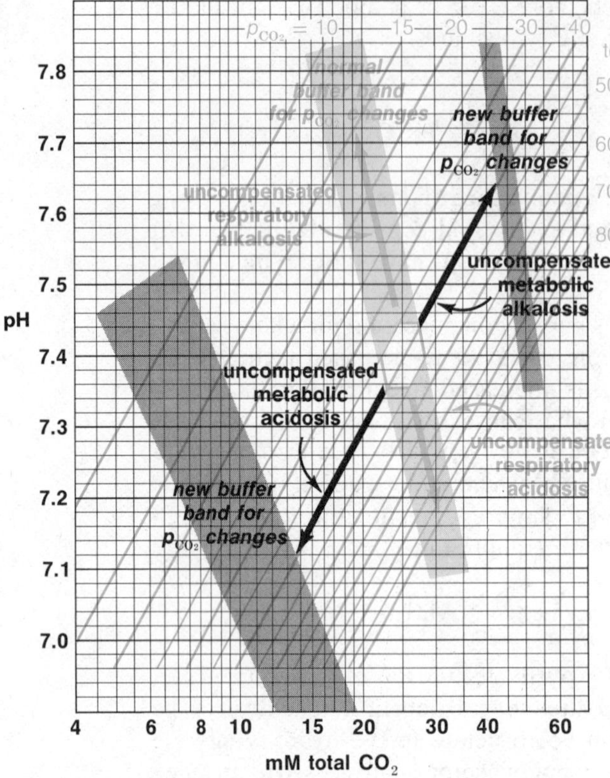

FIGURE 39–13

The effects of uncompensated changes in acid-base balance. Faster or slower exchange of CO_2 produces a respiratory alkalosis or acidosis within the normal buffer band, as shown by the *heavy blue line.* A rapid addition of acid or alkali produces a metabolic acidosis or alkalosis along the line of normal P_{CO_2}, as shown by the *heavy black line.* Compensatory mechanisms promptly begin to shift these values from the lines shown.

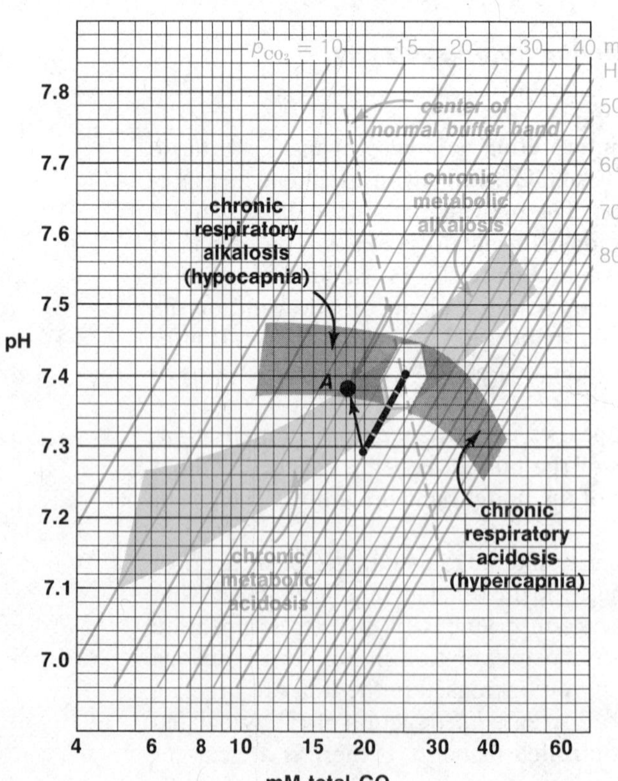

FIGURE 39–14

Typical ranges of values for chronic acidosis or alkalosis after compensation has proceeded. Decisions as to the primary events are frequently difficult. Even with values within the ranges shown, as at point A, similar concentrations could be produced by two routes—a respiratory alkalosis followed by renal compensatory acidosis, as shown by the *blue band,* or a metabolic acidosis, followed by respiratory compensatory alkalosis, as shown by the *black band.*

deviation from normal pH, the primary cause and the compensating event can be distinguished by making use of data on p_{CO_2} or [total CO_2] (Fig. 39–14):

$[H^+]$	pH	[total CO_2]*	p_{CO_2}	Primary Event	Compensation†
(1) high	low	low	low	metabolic acidosis	respiratory alkalosis
(2) high	low	high	high	respiratory acidosis	metabolic alkalosis
(3) low	high	high	high	metabolic alkalosis	respiratory acidosis
(4) low	high	low	low	respiratory alkalosis	metabolic acidosis

*The concentration of total CO_2 is regarded as high or low when it lies outside the normal buffer band for the measured pH or p_{CO_2}.

†Renal compensations are regarded as included in metabolic compensations.

Problems in interpretation arise when the $[H^+]$ is nearly normal, and the total CO_2 and p_{CO_2} are not. Which is the compensating acidosis or alkalosis, and which is the primary event? The concentrations of the other ions, in addition to giving important information on the water balance and the possible existence of deficits or surpluses of K^+ in particular, provide important clues to the existence of ions in the blood that are ordinarily not measured. That is, the sum of the measured anions (bicarbonate and chloride) is normally less than the sum of the measured cations (sodium and potassium), because there are more unmeasured anions, such as lactate, phosphate, and so on, than there are unmeasured cations like calcium and magnesium. This **anion gap** becomes much larger in metabolic acidosis caused by accumulations of 3-oxybutyrates, lactate, and so on, and this is an important diagnostic clue in some conditions. If there are no previous measurements to give a clue as to the direction things are going, measurement of the anion gap can be helpful. In practice, the normal range of the excess of $[Na^+]$ + $[K^+]$ over $[Cl^-]$ + [total CO_2] is taken as 11 to 17 mM. (This range will vary from laboratory to laboratory.) A low total CO_2 concentration and a high anion gap is in itself an indication of primary metabolic acidosis.

FURTHER READING

The following small books give informative summaries; they differ in their coverage of various aspects:

Masoro, E. J., and P. D. Siegel: (1977) *Acid-Base Regulation*. 2nd ed. W. B. Saunders.

Maude, D. L.: (1977) *Kidney Physiology and Kidney Disease*. Lippincott.

Rose, B. D.: (1977) *Clinical Physiology of Acid-base and Electrolyte Disorders*. McGraw-Hill.

Schwartz, A. B., and H. Lyons, eds.: (1977) *Acid-base Balance and Electrolyte Balance*. Grune & Stratton.

The Annual Review of Physiology for 1978 (vol. 40) *contains the following pertinent articles:*

Grantham, J. J., J. M. Irish, and D. A. Hall: *Studies of Isolated Renal Tubules* in vitro, p. 249.

Seif, S. M., and A. G. Robinson: *Localization and Release of Neurophysins*, p. 345.

Reid, I. A., B. J. Morris, and W. F. Ganong: *The Renin-angiotensin System*, p. 377.

Additional References:

Barger, A. C., and J. A. Herd: (1971) *The Renal Circulation*. N. Engl. J. Med., *284*: 482, Photographs of beautiful casts.

Jamison, R. L., and R. H. Maffly: (1976) *The Urinary Concentrating Mechanism*. N. Engl. J. Med., *295*: 1059. Unusually lucid for this subject.

Peach, M. J.: (1977) *Renin-angiotensin System. Biochemistry and Mechanism of Action*. Physiol. Rev., *57*: 313.

Ott, M. S., and H. J. Carroll: (1977) *The Anion Gap*. N. Engl. J. Med., *297*: 814.

Siggaard-Andersen, O.: (1971) *An Acid-base Chart for Arterial Blood with Normal and Pathophysiological Reference Areas*. Scand. J. Clin. Lab. Invest., *27*: 239. An excellent chart with many references to the literature upon which it is based. It is useful to practice appraisal of clinical data with charts that differ in the variables placed on the various axes so as to become familiar with the concepts rather than the position of points on a particular piece of paper. That can come later!

Burnell, J. M.: (1971) *Changes in Bone Sodium and Carbonate in Metabolic Acidosis and Alkalosis in the Dog*. J. Clin. Invest., *50*: 327.

Soffer, R. L.: (1978) *Converting Enzyme, Angiotensin II and Hypertensive Disease*. Am. J. Med., *64*: 147.

40 NUTRITION: FUELS

Nutrition is the study of the effect on the organism of changes in the diet. It is fitting that we close our discussion of biochemistry with this aspect of the subject because the search for food is one of the things life is all about, and proper analysis of what happens when we consume the results of the search requires the broadest sort of background. We are all aware of the growing problems of feeding the human population, and we are also aware of the many admonitions to the affluent against overeating. The very importance of nutrition creates some ancillary problems that we may as well face head-on.

Until recently, many biological scientists and clinicians avoided discussing or even thinking about nutrition, despite its obvious importance as a practical matter and as an ultimate test of abstract rationalization. Nutrition acquired the taint of a pseudo-science in the eyes of many, who therefore denied themselves one of the more useful bodies of information in understanding the living organism. We now frequently encounter the converse problem. Nutrition has become chic to the public and an attractive source of funds for the research administrator. The desire to be a true believer in the new faith has led even professionals to occasional uncritical acceptance of dogmatic pronouncements on nutrition from an astonishing spectrum of people. To avoid falling into the same pits, old or new, let us take the time to discuss how these problems came about.

Part of the difficulty comes from confusing the science of nutrition with the art of dietetics, a confusion that is augmented by the little maxims about good food that are first delivered as the revealed word to nearly everyone in the primary schools and repeated at frequent intervals thereafter without any examination of their basis. To put it bluntly, most people talking about hygiene, health, or menu-planning know little about nutrition.

The next part of the problem paradoxically comes from the eminent practicality of much nutritional knowledge. It provides a basis for value judgments on the character of the food supply. The provision of the supply and its delivery is a major part of the economy of all nations; changes in any aspect can bring financial ruin to some people and great wealth to others, and can make or break a political career. Even a President of the United States publicly drank a glass of milk and proclaimed its necessity for good health when a decline in the market threatened the income of the producers. Spectacles in the form of White House conferences and Senate hearings have been convened with intensive press coverage announcing that a large fraction of the U.S. population is ''suffering'' from nutritional inadequacy. The result of all this is that the hard facts are sometimes misstated, ignored, or presented out of context to promote a particular product, or justify a particular action, and for some reason these misuses are rarely challenged by those who know better. (It doesn't help objective appraisal to have some of the best-staffed university centers of nutritional research totally dependent upon their own fund-raising activities.) Our task here is not to assess economic or political decisions, but to uncover some of the scientific bases upon which they can be made.

The task is made more difficult by our own tendency to confuse objectives with proven facts. We talk about malnutrition and frequently forget that it means bad nutrition, with its implication of failure to reach some desired goal. The science of nutrition does not define goals; it only gives information on effects. When we talk about a *minimum requirement* for a nutrient we are stating an estimate based on scientific information to achieve some purpose that is not defined by the science, and there will inevitably be discrepancies in the definition of the requirement because of discrepancies in the objectives. An adviser in a rich nation may strive to keep all of his people saturated with the nutrient as defined by laboratory measurements while one in a poor nation might be content with avoiding obvious functional impairment in 99 per cent of his people. The major defined objective for children frequently is maximum growth — somewhat as if people were Angus cattle. This is an easily measured criterion; however, there is also some evidence from experimental animals that maximum intelligence and maximum longevity are achieved at something less than the maximum possible growth rate. Here is a case in which the science of nutrition has yet to give us all the necessary information we need to make a considered judgement in light of our personal aims.

Perhaps the most difficult misconception for a layman to shake is the idea that there is some essential value in a given type of food — that without fruit, or leafy vegetables, or meat, or milk, and so on, a diet is inadequate, if not immoral. **There is no nutritional basis for an absolute requirement of any particular food.** Humans have been around for some time, and we have evidence from archeological explorations, historical records, and current observations to show that cultures can thrive and their adherents perpetuate their kind on a wide variety of dietary intakes — in some cases almost completely deficient in one or more of the "basic foods" touted in some quarters.

Having now shaken our preconceptions, let us go on to consider some of the truly usable information. Nearly all of the various components of foods have been encountered in the earlier chapters, and we have seen how they are used within the body. Our remaining problem is one of organizing the information and gaining a more quantitative concept. Let us begin by considering the dietary fuel supply.

THE FUEL SUPPLY IN FOODS

Most animals fit in ecological niches by confining themselves to foods of restricted types, and their survival hinges upon the maintenance of an appetite for tissues from particular classes of organisms, plant or animal. We don't think of cows eating meat or cats eating grass as a significant part of their dietary intake. This is in part due to biochemical differences. Cats don't have the intestinal flora necessary for degradation of cellulose in grass, and probably don't have the enzymatic capacity to handle propionate in the concentrations absorbed by the intestines of cows. Cows probably can't handle amino acids in the concentrations absorbed by cats. Much of the difference, however, is a result of adaptation and habits to fit the ecology, with the qualitative nature of the biochemistry not being so altered as might appear. Cows can utilize meat and cats can utilize grains if they are presented in forms that the animals will eat.

The problem of the relative value of energy sources in food comes into its own with the omnivores such as man, who may be gulping down the still-warm hindquarters of a wild pig (1.5 per cent carbohydrate, 25 per cent fat, and 16 per cent protein) brought down by a spear to provide the week's ration, while aspiring to a life of self-contemplation in the land of milk (4.9 per cent carbohydrate, 3.9 per cent fat, and 3.5 per cent protein) and honey (79.5 per cent carbohydrate, 0.3 per cent fat, 0 per cent protein).

Since man is capable, both by anatomy and by taste, of eating a wide variety of foods, we might expect the proportion of carbohydrate, fat, and protein available from the diet for utilization as fuels to vary quite widely.*

Actual Consumption. The estimated production of major foodstuffs for the human race is plotted in Figure 40–1 in terms of the number of people that could be supported by the crop at an energy expenditure of 8 megajoules per day, a maintenance level only a little above basal. The figure is misleading in that it assumes total ingestion of the crops without waste or diversion for animal feed or other purposes. Even so, it is informative.

The majority of the world's population lives on diets in which cereals are the primary nutrient, and this has probably been true for a long time. The maintenance of the ever-normal granary was a desideratum in most early civilizations; cities and grain grew together. Most cereals are rich in starch and low in fat, so

*A recent lead article in *Science* stated:

"The traditional precompetition meal of athletes continues to be a protein-rich one, and the favorite main dish is a large steak. The fact that protein is not used as an energy source by the body except in starvation or when the diet is grossly deficient in fat and carbohydrate seems to have little impact on the practice."

Whatever the merits of the traditional meal, the grounds given here for attacking it are wrong. An adult maintaining constant weight uses all of his dietary amino acids as fuel, except for the small fraction excreted in the form of uric acid, creatinine, and the like.

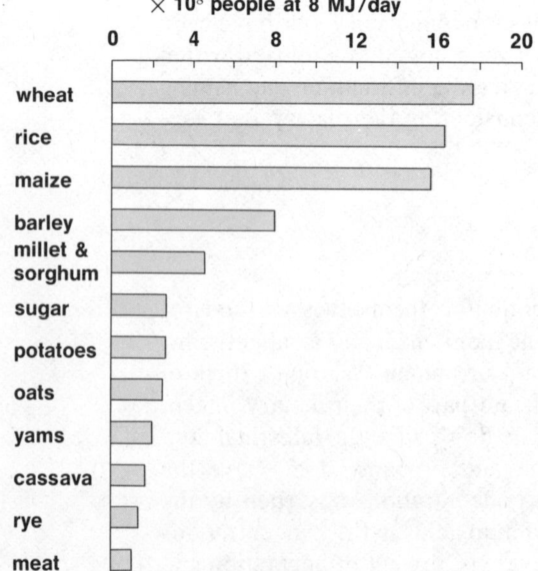

FIGURE 40–1

World production of major food energy sources, shown as the number of people who could be given 8 megajoules of energy per day from the crop if it were possible to use it with perfect efficiency. Calculated from rough estimates by the Food and Agriculture Organization of the United Nations. The estimated world population will reach 4.5×10^9 shortly after this book appears.

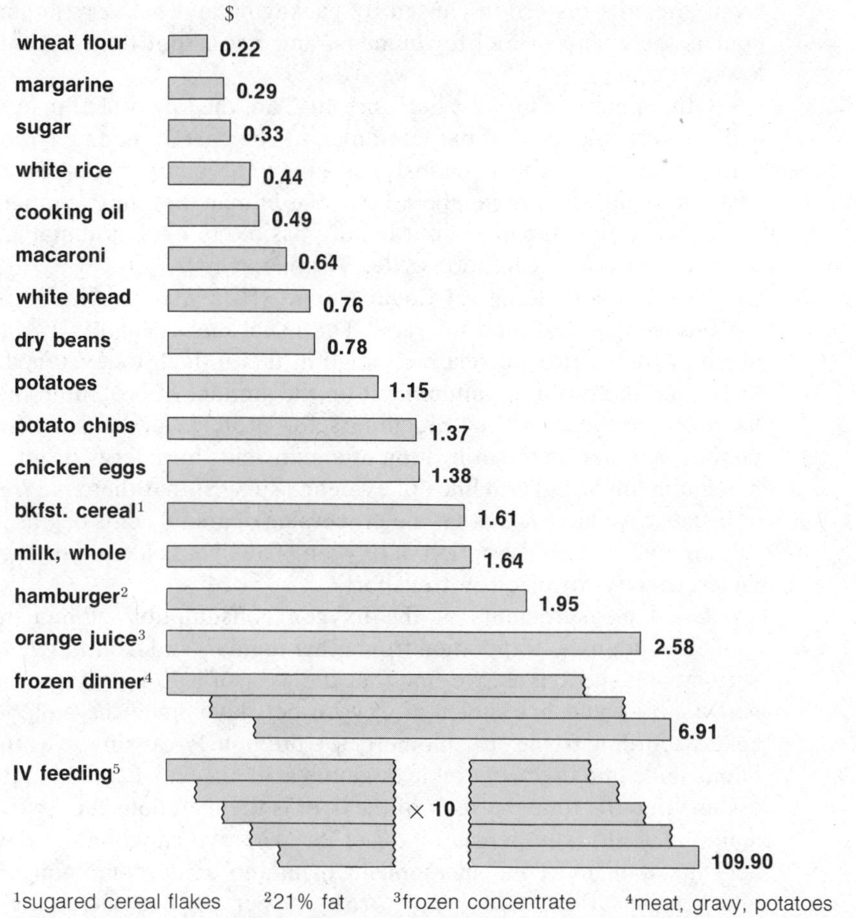

[1]sugared cereal flakes [2]21% fat [3]frozen concentrate [4]meat, gravy, potatoes
[5]glucose, amino acids, fat emulsion, minerals, and vitamin mixture; exclusive of consultation fees

FIGURE 40–2 **The cost of supplying 10 megajoules of food energy in the form of various foods available at retail stores. Calculated from lowest price items available in Charlottesville, Virginia, in June, 1978.**

oxidation of carbohydrate generates more than half of the total energy for most humans. Even children, for whom the dietary requirements are most demanding, will thrive on diets in which the content of carbohydrate is 13 times greater than the content of fat. If we define a normal human diet as the diet of most adults alive today, then the normal partition of potential energy is approximately two thirds from carbohydrate, one quarter from fat, and one tenth from protein. For comparison, over 40 per cent of the energy yield from the average United States adult diet is derived from fats, and approximately 14 per cent from protein.

Regional preferences in foods differ widely. In parts of Africa, the small grains (millet, sorghum, and teff) and the tubers (yams and cassava) constitute the major part of the fuel supply. Much of the maize* supply is used as animal feed, causing large losses of the potential energy upon combustion before reaching the human mouth.

The Cost of Fuel. The economic factors making the cereals the major human foods can be appreciated from Figure 40–2, which plots the retail cost in Virginia supermarkets of 10 megajoules of potential fuel energy in individual foods as of June, 1978, considering economy rather than quality as the criterion of choice. (Ten megajoules is a day's supply for a lightly active adult woman of median size.)

*The word corn is used for the principal grain in English-speaking countries. In the United States it means maize, in England wheat, and in Scotland oats.

Even after the distortion caused by packaging and delivery in small lots, wheat flour is the cheapest fuel for humans, and rice is not too far behind, despite its lower demand.

Other points of interest in Figure 40–2 are the low cost of margarine and sugar and the very high cost of prepared meals. The cost of feeding patients a complete diet parenterally* (intravenously) is given for comparison. In considering the costs, it ought to be remembered that the human diet must contain other components than a fuel supply, and it is not possible to keep a human alive on $0.22 a day, at least not in Charlottesville, Virginia.

The Tolerable Range of Composition. How much variation in the nature of the fuel supply can man tolerate? The usual diet is high in carbohydrates over much of the world and relatively high in fats in highly developed countries. We know that there is a minimum limit on the amount of protein in the diet, which is fixed by the necessity of precursors for proteins and other nitrogenous compounds, and not by the utilization of amino acids for energy production. Is there a maximum limit? Several lines of evidence suggest that there is. We should expect it, because we have seen that the processing of amino acids begins in the liver, and this involves the consumption of oxygen. The liver is like other organs in having a finite capacity for electron transport.

Good measurements of the oxygen consumption of human liver are not available, but by extrapolation from other animals and from *in vitro* measurements with human specimens, we find that the liver of a 73 kg man probably consumes between two and five moles of oxygen per day. Since the metabolism of amino acids according to the stoichiometry we previously cast involves the consumption of one mole of oxygen per mole of amino acid nitrogen handled in the liver, we can assume that the total capacity of the liver is the metabolism of two to five moles of amino acid nitrogen per day if *all* of its oxidative capability is devoted to amino acid metabolism. Total metabolism of amino acids containing this quantity of nitrogen is sufficient to supply from 40 per cent to all of the basal requirements.

Since the metabolism of the liver cannot be completely devoted to amino acids, we might expect that the actual ability of the body to handle amino acids as fuels is limited to somewhere near one-half of the total energy requirements. (Of course, any degeneration of liver tissue would reduce this value even further.)

Human experience reinforces the idea of a limited capacity to handle protein. William Clark and Meriwether Lewis, two boys from Albemarle County, Virginia, led a group from the Mississippi River to the Pacific Ocean and back in 1803–1805. They lacked scientific credentials, but proved themselves to be reasonably accurate and astute observers. They were dependent upon animals as the sole source of food during much of the journey, and fat dogs became not only a staple, but the preferred diet. However, they were able to obtain only lean deer at some stages during the early spring. They found that, although the meat was available in sufficient quantity, they lost weight and developed gastrointestinal distress along with other symptoms, justifying the conclusion they made: fat is a necessary component of a meat diet. (The word protein hadn't been invented.)

The conclusion is supported in a somewhat better study in more recent times in which an Arctic explorer, Vilhjamur Stefannson, and friends ate an animal tissue diet for more than a year to counter skepticism that such diet could be consistent with health. The results of the study are published in the papers by McClellan *et al.* cited at the end of the chapter. In McClellan's words, "At our

---

*Parenteral is any route outside the gastrointestinal tract, including subcutaneous, intramuscular, intraperitoneal, and intravenous injection or infusion. In practice, parenteral feeding is done intravenously.

request, he began eating lean meat only, although he had previously noted, in the North, that very lean meat sometimes produced digestive disturbances. On the 3rd day nausea and diarrhea developed. When fat meat was added to the diet, a full recovery was made in 2 days. This disturbance was followed by a period of persistent constipation lasting 10 days. The subject had a craving for calf brain, of which he ate freely. [Calf brain has 9 grams of lipid and 11 grams of protein per 100 grams.] On March 12, poor appetite, nausea, and abdominal discomfort were present and a second but milder attack of diarrhea occurred which responded quickly to a proper proportionment of lean and fat meat."

As is always true of these kinds of observations, there is room for conjecture. Was the discomfort with lean meat a result of mental bias or of biochemical aberration caused by the diet? In light of other observations, and the general proclivity of those in pure hunting cultures to gulp fat, it seems likely that protein can not be used as the primary source of energy by humans.

The discussion is somewhat academic as it pertains to most people in the world. We shall see that the principal problem is supplying enough, rather than too much, protein. Beyond this, the amount of protein consumed by choice is amazingly constant with humans from a variety of cultures, accounting for 10 to 15 per cent of the total energy yield. This is not always true, because many humans still live almost entirely on meat. The Eskimos are prime examples. However, the protein content in such cases rarely reaches one-third of the potential energy yield, with the remainder being mainly fat.

The hunter's diet thus represents another of the potential extremes, in which there is little carbohydrate and fat represents the bulk of the fuel. The "proper proportionment" of protein and fat referred to by McClellan in the quotation above was such that protein represented approximately one-quarter of the energy yield, and fat represented three-quarters. What happens in these circumstances? The evidence at hand suggests that adult humans can get along reasonably well for extended periods on such diets. Beyond the McClellan experiments, little has been done in careful observation, and there are no reliable data on relative morbidity and mortality of people living on meat compared to others in similar circumstances. Northern American Indians relied heavily on pemmican, which is ground dried meat in melted buffalo fat, as the primary food on long journeys, and the production of pemmican later became a quite substantial enterprise because the trappers of the Canadian trading companies relied on it exclusively during the winter. (There was a little-known "Pemmican War" between two of these companies for control of the supply during the first half of the nineteenth century. Interestingly enough, settlers on Hudson Bay supplied with conventional food from England developed serious signs of nutritional inadequacy, whereas no indication of this sort of difficulty was reported by the trappers living on pemmican.)

However, the meat-fat diet does cause the appearance of an asymptomatic ketosis, with the excretion of some 50 millimoles of 3-oxybutyrates per day in the urine. (There are great individual and daily variations in that amount.) This is akin to the ketosis seen in the starved obese, except that the obese frequently adapt so as to have little ketosis. The excretion of 3-oxybutyrates during ketosis represents the loss of potential fuel, but the quantity involved is only a few per cent of the total.

Adaptations to Fuel Changes

Whenever there is a shift in the quantity or the nature of the fuel supply, the enzymatic constitution changes so as to make most efficient use of the materials at hand. This is especially true of the liver, which plays a major role in balancing the flow of different metabolites. Most of the changes are a result of altered rates of gene transcription, sometimes caused by changes in hormone secretion, and sometimes by changes in metabolite concentration.

Adaptations to Starvation. The study of dietary adaptations is particularly satisfying because the observed changes are nearly always what we think they ought to be. Table 40–1 shows some examples from starved rats (everything we know about the omnivore, man, leads us to believe he has similar changes). When an animal doesn't eat, many of the enzymes responsible for handling excesses of foodstuffs become excess baggage, and we might expect them to disappear so that their constituent amino acids can be used for more pressing purposes. Starvation sharply diminishes the need for digestion, for conversion of glucose to fatty acids, for storage of fatty acids in adipose tissue, and for adjustments in fatty acid composition. At the same time, there is an increased mobilization of fatty acids. Even though the total utilization of amino acids as a fuel may fall with starvation, it is important that the liver divert the carbon skeletons as much as possible toward glucose formation when gluconeogenesis provides the only source of carbohydrate for emergency effort or for the construction of cellular components.

It is especially impressive to note the selective cut-off in synthesis of enzymes peculiar to fat storage during starvation, even including the liver desaturase that introduces more double bonds into fatty acids and the lipoprotein lipase that serves to clear transported fats from the blood for storage. The pyruvate carbox-

TABLE 40–1 ENZYME ADAPTATIONS TO STARVATION*

Increased	Decreased
Enzymes Secreted By The Pancreas	
None	All hydrolytic enzymes
Enzymes of Fatty Acid Synthesis (Adipose Tissue and Liver)	
None	Acetyl CoA carboxylase Fatty acid synthase Acyl CoA desaturase (liver) Lipoprotein lipase (adipose tissue) NADP-malate dehydrogenase Citrate cleavage enzyme
Enzymes of Fatty Acid Utilization	
Carnitine palmitoyl transferase (liver)	None
Enzymes of Glucose Utilization	
None	Glucokinase (liver)
Enzymes of Gluconeogenesis From Amino Acids	
Serine dehydratase Alanine aminotransferase (liver cytosol) Pyruvate carboxylase Phospho-*enol*-pyruvate carboxykinase Glucose-6-phosphatase	None

*Only some of the more significant changes are listed. This is also true of later tabulations.

ENZYME ADAPTATIONS TO GLUCOSE OR FAT DIETS **TABLE 40–2**

Increased with High Glucose, Decreased with High Fat	Increased with High Fat, Decreased with High Glucose
Amylase (pancreas)	Lipase (pancreas)
Glucokinase (liver)	Glucose-6-phosphatase (liver)
Glucose-6-phosphate dehydrogenase (liver and adipose tissue)	Fructose bisphosphatase (liver)
6-Phosphogluconate dehydrogenase (liver and adipose tissue)	Carnitine palmitoyl transferase (liver)
Acetyl CoA carboxylase (liver and adipose tissue)	Serine dehydratase (liver)
Fatty acid synthase (liver and adipose tissue)	Tyrosine aminotransferase (liver)
Citrate cleavage enzyme (liver and adipose tissue)	Ornithine aminotransferase (liver)
NADP-malate dehydrogenase (liver and adipose tissue)	

ylase of liver does not decrease even though it is involved in fatty acid synthesis from glucose because it is also an essential component of the route of gluconeogenesis. Those enzymes peculiar to gluconeogenesis, such as phosphopyruvate carboxykinase, are made in increased amounts during starvation. Glucose-6-phosphatase, responsible for releasing glucose from the liver, increases during starvation, but the glucokinase that takes up glucose in the liver when the concentration is high declines. This is as it ought to be, because a starving animal does not have a high concentration of glucose.

When we think on it, we see that most animals go through a period of food deprivation every day while they sleep. The levels of many of the enzymes in the preceding tabulation have been shown to rise and fall every day in response to the diurnal variation in habits. It is a little unsettling to know that our metabolic machinery is being taken apart and rebuilt to new specifications while we sleep, only to have the alterations rescinded when we eat, but if it weren't so we should really be stuck in a rut.

Adaptations to Changes in Dietary Glucose and Fat. There are only two degrees of freedom for variation in the proportion of major fuels in a diet. A diet can't be simultaneously rich in carbohydrates, fats, and proteins. If the carbohydrate content is high, the fat content frequently will be low, and the enzymes necessary for handling carbohydrates will increase, while those peculiar to routes utilizing *dietary* fat and to routes of gluconeogenesis from amino acids decrease. The reverse effects occur if there are little glucose and abundant fat.

The responses shown in Table 40–2 are very much along the lines that would be predicted, with especially dramatic responses in the enzymes necessary for synthesizing fatty acids from glucose and for synthesizing glucose from amino acids. As a minor point, the selective response of the pancreas to the two types of diet, so that only the appropriate hydrolytic enzyme is elaborated in increased amounts, indicates that the decline in synthesis of all hydrolytic enzymes during total starvation mentioned earlier is a real adaptation and not merely the result of some general debilitation of the secretory cells.

Adaptations to Diets Rich in Protein. Eating diets in which proteins supply a major fraction of the oxidizable substrates mainly causes increases in the activity of enzymes involved in nitrogen metabolism. Many of the changes are exaggerated if the diet also is low in carbohydrates. All of the following changes occur in the liver unless otherwise noted. In some cases a specific amino acid is required for the change, rather than a general increase in nitrogen intake, and these are also noted parenthetically:

Increased

peptidases (pancreas)
ornithine carbamoyl transferase (Arg)
argininosuccinate synthetase
argininosuccinase
arginase
serine dehydratase (repressed by glucose)
tyrosine aminotransferase (separate inductions by Tyr and by mixed amino acids)
alanine aminotransferase (cytosol)
aspartate aminotransferase (cytosol)
cystathionase (Met)
glutaminase (kidney)
tryptophan dioxygenase (Trp)
fructose-bisphosphatase
ornithine aminotransferase

Decreased

methionine adenosyl transferase (Cys)

FAMINE

There is an inherent dichotomy of approach to nutritional problems. Shall we concentrate on feeding the multitude, or note each sparrow's fall? Let us first sketch the demographic problems and then examine individual problems of clinical interest.

Lack of sufficient food is so common that it may be considered a part of the normal human condition, now and at all past times for which we have evidence. More people die from starvation than are killed by war injuries, and even more people experience starvation and then recover. Demand for food and its supply run so close together that a wide variety of causes can cause a mismatch. Too little rain, too much rain, a plague of locusts — we in the United States grasp that such disasters befall people in the sub-Sahara, or Bangladesh, or similarly remote regions. We forget that famine nearly always accompanies war — Norwegians, Dutch, Russians, Germans, all have experienced varying periods of deprivation in this century and as a consequence have provided some of the best studies of the physiological effects of starvation. Nearly one quarter of the eight million Irish died between 1845 and 1851 when a blight wiped out the potato crop and relief from the English was too little and too late. On a much lesser scale, but closer to home, the United States had a "year of no summer" in the 19th century that resulted in widespread hunger, but with poorly documented consequences.

The metabolic changes occurring during starvation were summarized in Chapter 37: An initial shift from carbohydrate to fat oxidation, followed by increased utilization of protein as the fat stores are depleted. However, there is a rapid loss of protein for gluconeogenesis up to a maximum of about 3 per cent of the total store of 80 moles of N during the initial days of acute starvation, before adaptation has occurred. The loss mainly comes from the liver in experimental animals, and the amount depends upon the previous protein intake, with those having ingested high protein diets losing more upon sudden deprivation. The proteins lost quickly without measurable impairment of function are designated "labile" proteins. After the initial high rate of loss, nitrogen excretion tapers to a level ranging around 0.14 mole per day at the close of a month of total starvation. The tissues losing protein most rapidly during this period are those with high rates of protein synthesis at times of normal nutrition: the liver, pancreas, and intestinal mucosa. The skeletal muscles lose a smaller fraction but represent a major supply during the initial weeks because of their large mass. The duration of starvation that can be withstood obviously depends upon the quantity of stored fat. An extremely lean person faces a quick death because he will begin to lose as much as 6 per cent of his protein per day when the fat is exhausted. Young children are also in a precarious position because of a more limited fat store and a greater metabolic rate.

People adjust to less than total starvation. Part of the adjustment comes from adaptations of the biochemical economy that we noted earlier. Part comes from a slowing of processes utilizing high-energy phosphate at rest, as is reflected by a drop in the basal metabolic rate. Another part comes from changes in habit and temperament. Starving people don't like to move and become self-centered and withdrawn. The often-invoked specter of starving hordes ravaging their well-fed neighbors probably never occurs; the starved aren't good candidates for group efforts. (A fear of starvation is something else again.) Charles Dickens was quite correct in suggesting that rambunctious boys might be more tractable if they were fed less meat and more gruel. The increased apathy explains some of the isolated cases of death from starvation in the United States, many of which occur among elderly recluses.

Perhaps the most surprising thing about starvation is the relative lack of specific deficiencies in many instances. Emaciation may be severe without symptoms ascribable to the lack of a particular coenzyme or amino acid, although this is not always true.

Consequences of Famine. We have statistics that are informative about the effect of natural famines on the very young. The Dutch experience has been especially valuable in this regard, because detailed data are available for those born or conceived during the famine of World War II at the time of their later compulsory military service. The conclusions, which are borne out by data on other starved populations, are that starvation of a pregnant woman leads to the birth of a shorter, lighter baby with a smaller head, who is somewhat more likely to die in the early years. However, given adequate food in the subsequent years, babies who survive grow to normal stature and weight without any impairment of intelligence.

Similarly, a study of Korean orphans adopted in the United States within three years after birth suggested that severe postnatal deprivation might result in slight residual impairment of physique or intelligence, but not of the magnitude implied by experiments with laboratory animals. The Dutch data show a drastic drop in fertility during famine; it is possible that humans have evolved so as to ride out famine by conceiving few children, but with preferential utilization of the limited nutrient supply for development of the brain in those who are conceived.

Problems can appear from dependence upon particular foods to stave off hunger. For example, cassava roots, a major source of fuel in Africa, are rich in nitrile-containing glycosides that yield thiocyanate upon degradation. The thiocyanate inhibits iodine uptake, so inhabitants of regions in which foods are low in iodine, such as parts of Zaire, have a high incidence of hypothyroidism with a consequent high incidence of mental retardation (cretinism).

INDIVIDUAL DIETARY PROBLEMS

Obesity. An obese person by definition is one who has too large a fat store. What is too large depends upon who makes the definition. There is a continuum from what is the cultural ideal to an obviously disabling accumulation of masses of triglyceride. There is an equal continuum in the responses of clinician and patient. Eating, like cigarette smoking and sexual intercourse, is a difficult habit to change, once a pattern is established. Gross obesity is most common among women from lower income families in the United States. The usual protective defenses erected against feelings of guilt are hardly alleviated by those physicians who approach even the moderately fat with all the zeal of a Puritan divine confronting a candidate for the scarlet letter. This zeal has led to such barbarisms as wiring the jaws closed to prevent surreptitious snacks. As might be expected, those with motivation so poor as to require such measures promptly begin to gain weight upon removal of the barriers. Given a life-threatening condition and a willing patient, surgical bypass of long segments of the small bowel can prevent effective absorption of food, but this is not a procedure for casual application.

Experience seems to indicate that the prospects for becoming sufficiently motivated to diminish food intake permanently are inversely proportional to the degree of obesity. Assuming that no reason for heroic measure exists, the necessary diminution may be more acceptable if achieved in this fashion: A lean body weight is estimated from height and build, and the caloric requirement for a person of that weight with similar activities is estimated. (The estimates ought not be deliberately sloppy, but they cannot be expected to be precise.) A reasonable schedule of fat loss is then projected. In most instances, a loss of 0.5 kilogram of body weight per week is acceptable. This will require a deficit of approximately 2.4 megajoules of potential fuel energy per day (570 kcal). Food tables are then used to adjust the amount of the usual diet to that range. For many, it is helpful to take four small meals per day of approximately equal size, except that the customary main meal is made somewhat larger than the others. Unless the usual diet is bizarre, it is not necessary to pay attention to vitamin and mineral content at this point.

If the patient realizes that weight loss at this modest pace goes in irregular spurts, depending upon water intake, activity, bowel motility, and so on, he will not expect to see much happening in less than two weeks. If there isn't a clear loss of at least 1.5 kilograms after a month, re-examination of the eating habits is encouraged. The problem is nearly always one of social dining, or of between-meal snacks, especially in the case of those doing the cooking, but the estimates of energy needs also may be far from the mark. Unusually large losses may signal a potentially dangerous self-starvation.

A critical test of motivation is made upon reaching a desired weight. The additions to the food intake necessary for maintenance of the weight are small, and a self-congratulatory splurge can undo all of the control over appetite achieved with such difficulty. It is probably wise to have in mind well in advance exactly what those additions are to be, so they can constitute the reward.

Food Intolerances. Some people cannot tolerate particular foods because of allergies. Others have partial or complete deficiencies of an enzyme necessary for utilization of a particular fuel. We have already mentioned the problems created by lack of a digestive hydrolase attacking disaccharides such as lactose or sucrose, with microbial fermentation of the undigested sugar causing flatulence, diarrhea, and cramps. A more serious problem is **celiac disease**, which appears to be caused by a defect in the hydrolysis of some polypeptide sequences. It is alleviated by elimination from the diet of gluten, a mixture of gliadin and glutenin, the principal endosperm proteins in wheat. The alcohol-soluble gliadin appears to be the offender. The condition causes atrophy of the villi of the intestinal mucosa, sometimes severe enough to be life-threatening. Even those with mild cases are plagued by passage of watery, foul-smelling stools.

Other defects affect the intracellular metabolism of sugars. The problems fall in two classes, those involving loss of a kinase, and those involving loss of an enzyme attacking a phosphorylated sugar derivative. Consider the metabolism of **galactose**, derived from the lactose in every nursing infant's diet. Galactose is utilized by first phosphorylating it and then exchanging the galactose phosphate for a glucose phosphate moiety in uridine diphosphate glucose (Fig. 40–3). An epimerase then regenerates UDP-glucose from the UDP-galactose, with the net result being a conversion of galactose to glucose 1-phosphate.

The effects of a deficiency of the specific galactokinase are akin to some of the effects of diabetes. The accumulating galactose is reduced to **galactitol** by a non-specific **aldose reductase** using NADPH. There is no route for disposal of the galactitol other than loss in the urine; its accumulation in the lens of the eye, which has an active aldose reductase, results in the formation of cataracts.

Loss of the uridyl transferase activity that utilizes galactose 1-phosphate is even more serious. Not only is there an accumulation of galactose and galactitol, owing to feedback inhibition of the kinase, there is also a loss of inorganic phosphate, and perhaps some deleterious effects from galactose 1-phosphate itself. In any event, **galactosemia**, as the condition is called, results in liver failure and mental retardation, in addition to cataracts. The defective gene has an estimated incidence near 1 per cent in the United States; 1 in 35,000 infants in New York were found to have a full-blown galactosemia. There is no known benefit from the heterozygous condition to explain the high gene incidence. Treatment consists of scrupulous avoidance of all foods containing galactose, either free or in oligosaccharides.

Similar defects occur in the metabolism of fructose (Fig. 40–4). Deficiency of a specific kinase leads to an accumulation of the sugar after ingestion of sucrose or other fructose-containing foods. However, this ketose is not affected by aldose reductase. Indeed, one of the functions of the aldose reductase is to convert glucose to sorbitol from which fructose can then be generated by a specific dehydrogenase:

H—C(=O)
H—C—OH
HO—C—H
H—C—OH
H—C—OH
CH₂—OH
D-glucose

$\xrightarrow[\text{"aldose reductase" (several tissues)}]{H^+ + \text{NADPH} \rightarrow \text{NADP}^+}$ (reversible)

CH₂—OH
H—C—OH
HO—C—H
H—C—OH
H—C—OH
CH₂—OH
D-sorbitol
(D-glucitol)

$\xrightarrow[\text{NAD—sorbitol 2-dehydrogenase (seminal vesicle, placenta)}]{\text{NAD}^+ \rightarrow H^+ + \text{NADH}}$ (reversible)

CH₂—OH
C=O
HO—C—H
H—C—OH
H—C—OH
CH₂OH
D-fructose

(Fructose is the main carbohydrate fuel in semen and is also important for fetal nourishment in many mammals.) A deficiency of fructokinase therefore leads only to fructose accumulation, which causes the compound to appear in the urine. This **essential fructosuria** is a benign condition.

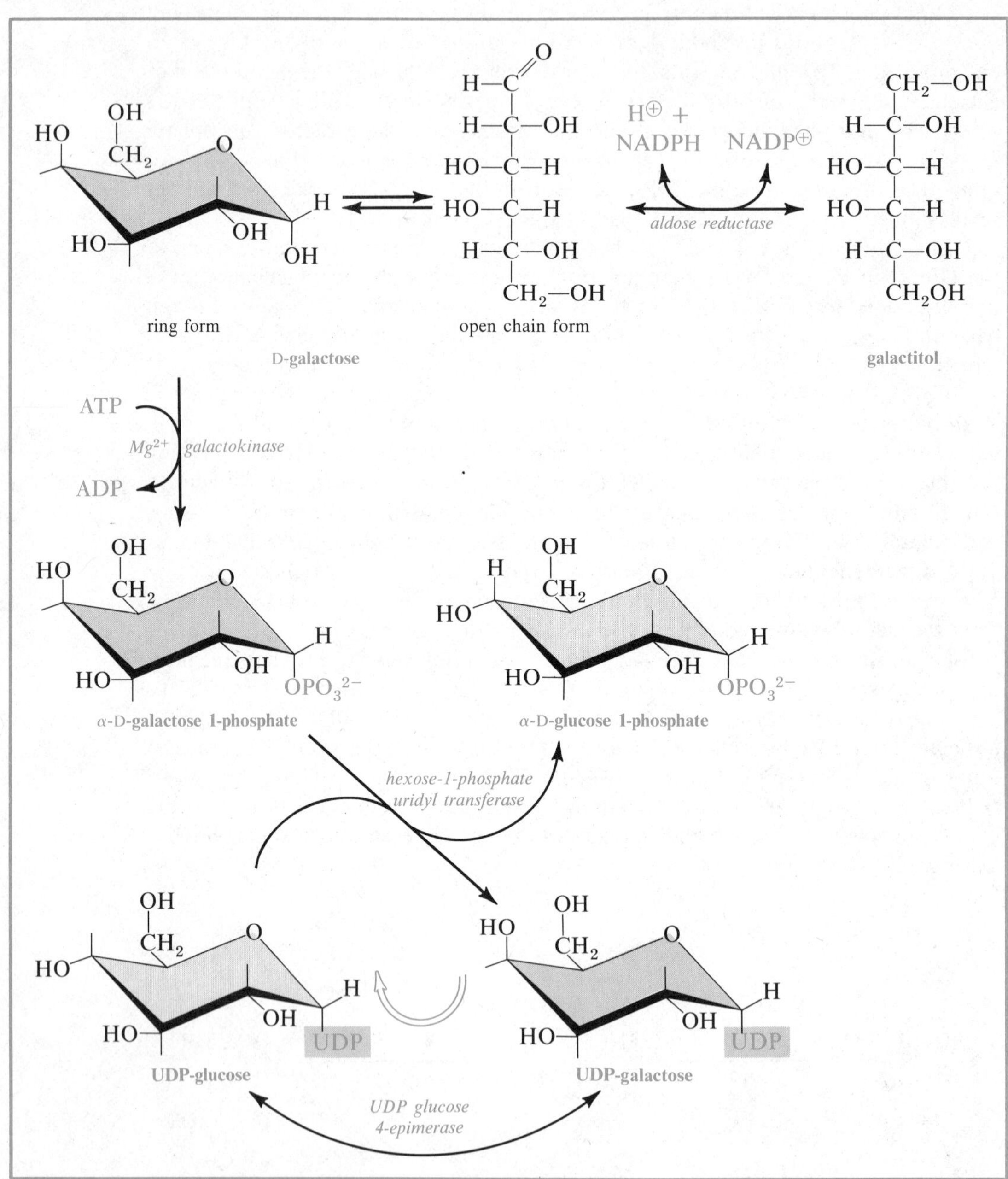

FIGURE 40–3 **The routes of galactose catabolism in humans.**

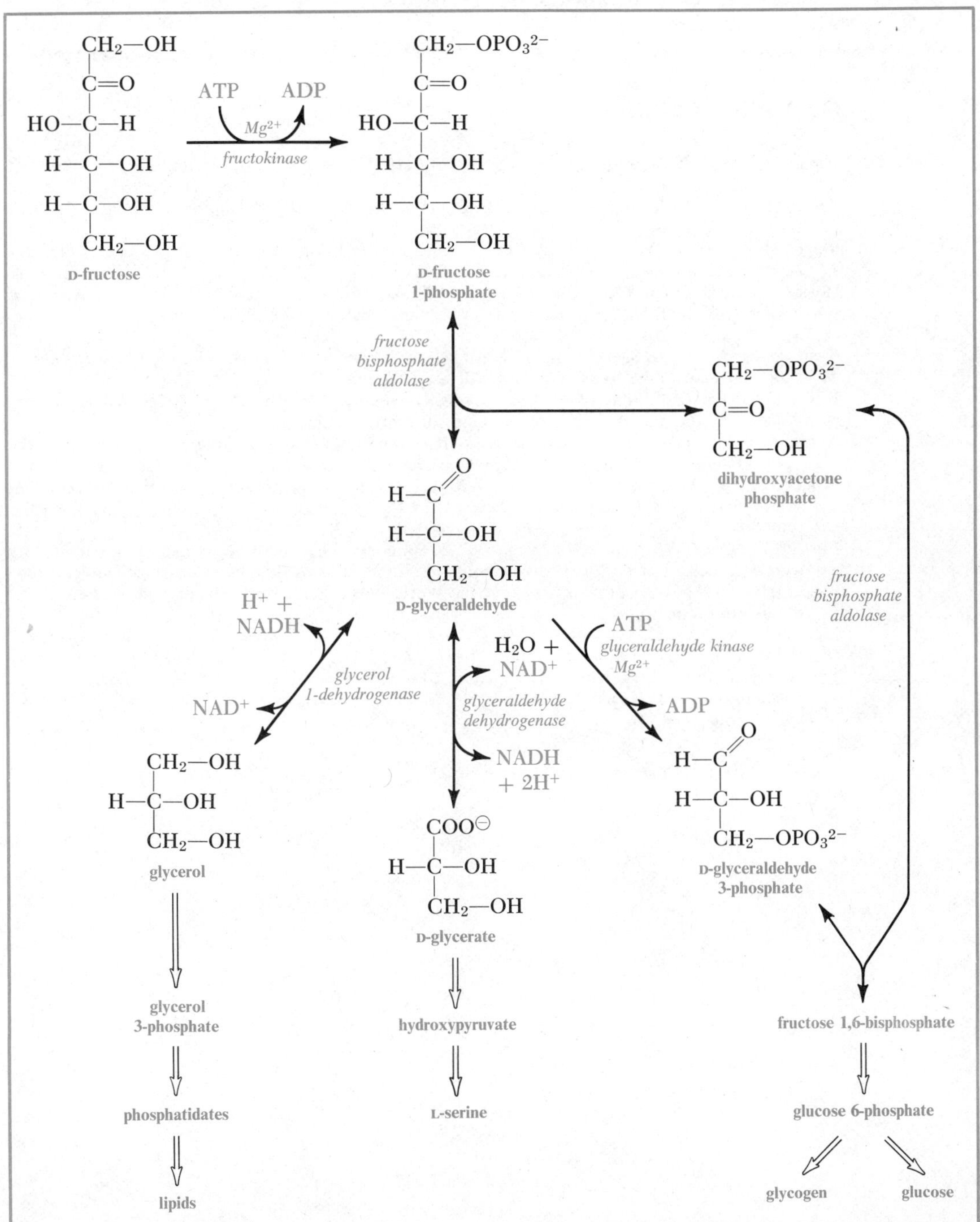

FIGURE 40–4 **The routes of fructose catabolism in humans.**

In contrast, deficiency of the aldolase that removes fructose 1-phosphate is a highly deleterious condition. Inorganic phosphate is removed to make the accumulating phosphate ester, and again the ester itself may have deleterious effects. The resultant **fructose intolerance** causes liver failure.

FURTHER READING

Davidson, S. S., R. Passmore, and J. F. Brock: (1975) *Human Nutrition and Dietetics.* 6th ed. Williams & Wilkins.

Schneider, H. A., C. E. Anderson, and D. B. Coursin: (1977) *Nutritional Support of Medical Practice.* Harper & Row.

Sipple, H. L., and K. W. McNutt, eds.: (1974) *Sugars in Nutrition.* Academic Press. The introductory historical survey of sucrose is a particular delight.

Hegsted, D. M.: (1976) *Current Knowledge of Energy, Fat, Protein, and Amino Acid Needs of Adolescents. In* J. I. McKigney and H. N. Munro, eds.: *Nutrient Requirements in Adolescence.* MIT Press, p. 107.

Stein, Z., M. Susser, G. Saenger, and F. Marolla: (1975) *Famine and Human Development. The Dutch Hunger Winter of 1944–1945.* Oxford. A classic study.

Keys, A., et al.: (1950) *Human Starvation.* 2 vols. University of Minnesota Press. Detailed description of lengthy experiments with conscientious objectors as subjects.

Aykroyd, W. R.: (1974) *The Conquest of Famine.* Dutton. Includes an illuminating account of the Bengal famine of 1943, in which 1.5 million lives were lost according to an underestimate.

Winick, M., K. Katchadurian Meyer, and R. C. Harris: (1975) *Malnutrition and Environmental Enrichment by Early Adoption.* Science, *190*: 1173. Another good study, but of a much smaller sample.

Lusk, G.: (1931) *The Science of Nutrition.* W. B. Saunders. This highly recommended guide to early work includes a discussion of several bouts of prolonged starvation by professional subjects, and of the efforts of the Germans to relieve the World War I famine through addition of sawdust and the like to food.

McClellan, W. S., et al.: (1930–1) *Clinical Calorimetry:* XLV. *Prolonged Meat Diets with a Study of Kidney Function and Ketosis.* J. Biol. Chem., *87*:651. XLVI. *Prolonged Meat Diets with a Study of the Metabolism of Nitrogen, Calcium, and Phosphorus.* J. Biol. Chem., *87*:669. XLVII. *Prolonged Meat Diets with a Study of Respiratory Metabolism.* J. Biol. Chem., *93*:419.

41 NUTRITION: THE NITROGEN ECONOMY

Proteins in the food are the major precursors of most of the nitrogenous compounds in the body; the only exceptions are the nitrogen-containing vitamins. While it is true that variable amounts of purines, pyrimidines, creatine, choline, etc. are obtained as such from the diet, they represent a relatively small portion of the total flow of nitrogen through the body. The error is not large if the food proteins are considered to be the only sources of nitrogen.

NITROGEN BALANCE

We are concerned with the requirements of the organism for nitrogen compared to its supply, that is, with its state of nitrogen balance. **Nitrogen equilibrium** is the characteristic condition in the adult, with the losses just balanced by the intake so that the body composition remains relatively constant, although there will be moment to moment and day to day variations. The need for dietary proteins to maintain nitrogen equilibrium is a result of metabolic losses. (The hibernating bear cuts its losses to nearly zero; the higher plants have no devices for excreting nitrogen. Loss is not an obligatory feature of metabolism.) Some nitrogen is lost in the urine, in feces, and from the skin even if none is present in the diet. Milk production is a major loss of nitrogen in women. Losses from bleeding, ejaculation of semen, and exhalation of ammonia are less important.

Negative nitrogen balance, or losses exceeding the supply, is an obvious result of an inadequate intake. However, it is also characteristic of the ill and the injured, in whom cellular damage causes more nitrogen to be lost than is taken in. There is also a small negative nitrogen balance associated with aging, so small as to be imperceptible by direct comparisons of intake and excretion.

Positive nitrogen balance occurs when tissues are growing. It is the characteristic state of the pregnant woman and the convalescent adult, as well as the young. Maintenance of positive nitrogen balance requires sufficient dietary protein from which to construct the additional tissues being formed in addition to the amount required to replace metabolic losses.

DIETARY NITROGEN REQUIREMENT

As with other dietary requirements, a definition of the nitrogen requirement hinges upon the desired objectives. Most would agree that a basal value must at least maintain adults of a desired size in nitrogen equilibrium, or maintain a desired rate of tissue growth in the young or the convalescent. However, it must be emphasized that there can be no rigorous specification of the amount of dietary protein that will fulfill the defined objectives for one person at all times — there is no number that can be memorized with its basis safely forgotten. The reason for this is that the amount of nitrogen necessary depends upon the form in which it is supplied and upon the other components of the diet that come with it.

Three factors are involved: the content of essential amino acids, the total amount of nitrogen, and the extent to which protein must be utilized as a fuel. Each influences the nitrogen balance.

Essential Amino Acids. The mixed population of proteins comprising tissues contain all of the 20 amino acids. When tissue proteins are being synthesized, either as replacements of existing molecules or as additional components, all of the 20 must be present. If the supply of any one runs low, then the concentration of the corresponding aminoacyl tRNA will fall below the level necessary for prompt incorporation into newly synthesized polypeptide chains. The rate of protein synthesis will decline, and the functions of the tissue will suffer. In that important sense, all of the 20 amino acids are essential. However, we have seen that most cells are capable of making some of the amino acids from glucose and almost any source of amino nitrogen. Some of those that cannot be made by a particular tissue can be supplied through the blood by the liver, which can make them. A total of eleven amino acids can be made in this way. All of the other nine must be obtained from the diet and are therefore essential dietary components. They are:

histidine	**phenylalanine**
isoleucine	**threonine**
leucine	**tryptophan**
lysine	**valine**
methionine	

The critical experiments on the amino acid requirements of human adults were performed by W. C. Rose at the University of Illinois with volunteer graduate students maintained on artificial diets. One amino acid at a time was omitted, and if negative nitrogen balance resulted, the compound was restored until the requirement, as indicated by a slightly positive balance, was satisfied. In most cases, deprivation of one or more of these essential amino acids has immediate deleterious effects. However, this is not necessarily true; Rose could find no effects from omission of histidine, and this amino acid was regarded as non-essential for humans. This was contrary to the observations with every other mammalian species tested, and it seemed highly unlikely that young Midwestern men had suddenly remade the genes that were gone so long. Later experiments with human infants (much criticized on ethical grounds) established that histidine is indeed essential in the human diet. The failure to discover the inability of adults to make histidine may be due to its relatively low abundance in proteins, a slow catabolism, and the large reservoir present in the form of carnosine in the muscles (p. 629).

Tissues contain different amounts of the various amino acids, and they are catabolized at different rates. It follows that there is a pattern of dietary amino acid composition that will enable the most efficient use of nitrogen for repair and growth, a pattern in which no one of the essential amino acids is in great excess or deficit of the demand for it when compared to the other amino acids. If the amino acid composition of food follows that pattern, a total nitrogen intake including enough of one of the essential amino acids to meet its requirement will also include enough of all of the others. The balance of the necessary total nitrogen intake can then be supplied in any convenient form. (Indeed, it has been shown experimentally that it can be supplied as ammonium acetate.)

If one of the essential amino acids is in short supply compared to the others, protein synthesis will continue only until the supply is used, and other amino acids then remaining will be used as fuel. Remedying the deficit of that one will require the ingestion of more of the unbalanced mixture of amino acids with most of the additional intake except for the needed amino acid burned as fuel.

Contrariwise, increasing the total nitrogen intake diminishes the requirements for the essential amino acids. Less of these compounds will be used for making other nitrogenous compounds when the total supply of nitrogen is high. Similarly, the requirements for total nitrogen and for the essential amino acids depend upon the total supply of fuel ingested. If the fuel supply is low, more of the amino acids will be burned; it if is high, more amino acids will be left intact for protein synthesis. Indeed, the amino acid requirement is lowest in someone who is becoming fat.

Definition of the Requirement

Quantitative Expression. It is clumsy to express the nitrogen requirement in terms of the 20 amino acids, even if analyses for them were always available. The pioneers in the field had no such analyses, and developed analyses for total nitrogen content instead. Since the proteins and free amino acids contain over 95 per cent of the total nitrogen in most foods, this is a useful guide to the content of amino acids. Unfortunately, these early workers decided to translate the nitrogen analysis into weights of protein, and after comparing the samples available to them, they decided that an "average" protein contains 16 per cent nitrogen. To this day, we speak about a content of protein when we really mean a content of nitrogen multiplied by 6.25 ($100 \div 16$). Modern tables of food composition include columns of protein contents that have been laboriously calculated by multiplying the analysis for total nitrogen by factors estimated for the particular type of food. Although these numbers are sometimes instructive in making assessments of the proportion of the mass occupied by protein, they create extra work for most purposes. It is the nitrogen content itself that must be used for assessing quantitative metabolism.

Direct comparisons of the relative supplies of different amino acids are best facilitated by speaking of moles of amino acids in the individual cases, and of moles of protein nitrogen when discussing total balance. (Conventional calculated "weights" of protein will sometimes be given parenthetically for comparison.) The relationships between the common units are:

Moles N	Grams N	Grams "Protein"
0.071	1	6.25
1	14	87.5
1.14	16	100

(moles N) × 87.5 = (grams "protein")
(grams "protein") × 11.4 = (millimoles N)
(grams "protein") × 0.16 = (grams N)

To get an appreciation of scale, one mole of animal protein N is a generous daily intake for most adults.

Calculation of the Requirement. The most desirable determination of the nitrogen requirement would be a measurement of the minimum amount that must be ingested in the form of particular foods in order to achieve the defined objective. As a practical matter, this is very difficult to do over the extended periods that would be necessary to minimize transient changes, including those caused by the novelty of the experiment itself. However, it must be done in the future if reliable information is to be obtained. In the meantime, most estimates hinge on the postulate that a person placed on a protein-free diet will lose the same amount of nitrogen that he will on ordinary diets through normal wear-and-tear. That is, the urea, uric acid, etc. in the urine and the nitrogen lost in the feces, along with losses from the skin when a person is not eating any source of nitrogen are assumed to represent a minimum turnover of body constituents, a minimum that must be replaced by nitrogen absorbed from the diet. (The measurements are made after a week or more, after the rapid loss of labile protein.)

The first task, then, is to estimate the nitrogen losses. This varies, but values for a 73 kg man are something like these, given in millimoles of N per day:

urine	175
feces	65
skin	20
other	10
SUM	270

The fraction of nitrogen lost from the skin depends upon circumstances. It is increased by frequent bathing and by heavy sweating. Approximately 8.5 millimoles more are lost for each megajoule of increased heat production during exercise. The value for miscellaneous losses is approximately 50 per cent higher in women, owing to menstruation.

The next task is to add to this minimum value any additional requirement created by special circumstances. It is especially difficult to define the required intake for growing children. Hegsted approached this question in a purely pragmatic way by defining the way things actually seem to be, rather than the way they ought to be, with a sample of American children presumably in reasonable health and growing the way a random sample of the population in Boston does. He used the expected basal loss of nitrogen and the increment of the total nitrogen body content of the body because of growth to estimate the required dietary supply. A plot of the estimates is given in Figure 41–1; two curves are shown, one for girls in the tenth percentile of weight (only 10 per cent of the girls are smaller) and one for boys in the 90th percentile (only 10 per cent of the boys are larger), so these two curves encompass a range including the requirements for all but 10 per cent of the children. The calculated minimum amount of nitrogen absorbed is given per kilogram body weight and as a total. The body weights are also plotted. The relatively high requirement for protein during the early years is especially striking; small girls need half as much nitrogen absorbed when they are 3-kg infants as they do when they are 47-kg adults. It is also apparent that girls and boys have nearly then same protein requirement per unit body weight despite their different sizes, and the value declines throughout life.

All of the estimates hinge upon the amount of nitrogen absorbed. Real foods are not totally digested and absorbed. Furthermore, there is a variation of requirements within the population, so it is customary to add two standard deviations to the estimated mean requirement in order to satisfy the needs of nearly all people.

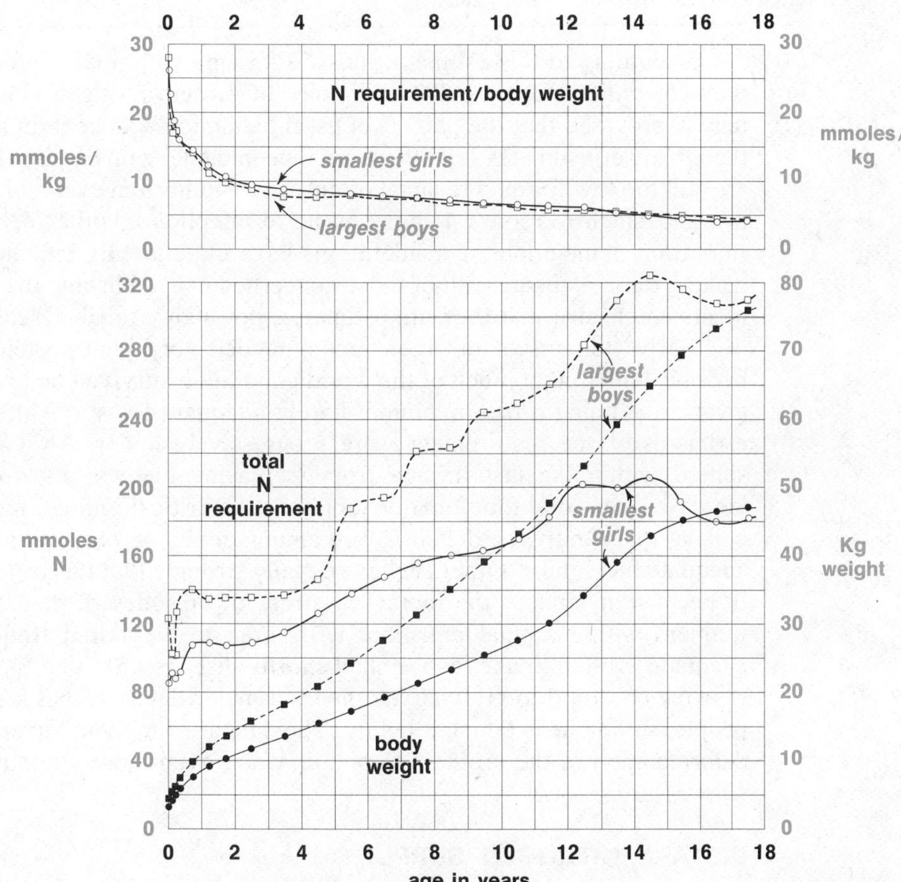

FIGURE 41–1 **Theoretical estimates of the protein requirement of children as a function of age. The amount of protein nitrogen that must be absorbed in order to balance the utilization for tissue formation and the losses in the urine and feces is shown for the smallest girls (tenth percentile of weight) and for the largest boys (ninetieth percentile of weight) in terms both of the amount per kg of body weight and of the total amount. The body weights are plotted with solid points *(bottom graphs)*. (Data from Hegsted: (1957) J. Am. Diet. Assn., *33*: 225).**

Having made these adjustments, a recent estimate of the minimum amount of food nitrogen that must be eaten each day (Williams, et al. 1974) is:

Age	N Intake (millimoles kg^{-1} d^{-1})
0–2 mo	25
2–4 mo	21
2 yr	10
12 yr	8
adult	5.5

According to this estimate, most 73 kg men will obtain sufficient nitrogen if they eat a diet including 400 millimoles of protein nitrogen (35 grams of "protein"), provided that the pattern of essential amino acid content is optimal. Even though lower estimates than these have been made, some authorities believe they are still too low for maintenance of nitrogen equilibrium even in healthy individuals, and much too low for those prone to infection or other debilitating illness. Important demographic considerations have undoubtedly influenced attitudes in making these estimates; all of those concerned in establishing the protein requirements emphasize that they are estimates, not highly reliable determinations.

Early students of nutrition recommended very high protein intakes. As it became evident that much of the world population survived on less, attention was given to defining a true minimum barely adequate for survival, and succeeding estimates of the requirement were progressively lower. As a result, attention shifted within the last decade from increasing the world's protein supply to increasing the total food energy supply as the critical limiting factor. Some now believe that a nitrogen deficit is rare among adults on natural diets who have an adequate fuel intake. Others believe equally strongly that this is not so, particularly because it ignores the increased stress of episodes of illness that are more frequent with marginal nutrition. (All agree that children frequently have inadequate protein intake, as we shall shortly discuss.) Shifting the emphasis from growing one food to growing another is a major decision that affects millions of people, so no one is taking it lightly. The data are simply not available for precise determination of the problem — solutions are even more tenuous.

DIETARY NITROGEN SUPPLY

The ability of a natural food to satisfy the nitrogen requirement hinges upon its total protein content, its amino acid composition, and its digestibility. Two important methods are used to gauge this ability in the absence of controlled long-term feeding experiments. One involves feeding test subjects a protein-free diet to deplete their labile protein and to enable measurement of the basal nitrogen loss. The subject is then fed measured amounts of the food in question, and the nitrogen loss is again measured. The difference between the amount of nitrogen fed and the *increase* in nitrogen loss after feeding is taken as the amount of nitrogen retained and used to rebuild tissue proteins. The fraction of the dietary protein nitrogen retained is the **net protein utilization.** (This is sometimes given as a fraction of the dietary nitrogen absorbed — the nitrogen fed after correction for increased fecal loss due to the feeding — and is then called the **biological value** of the protein.)

Another way of assessing the value of a food protein is to compare its amino acid composition to the composition of some protein known to be used effective-

ESSENTIAL AMINO ACIDS

TABLE 41–1

Amino Acid	Reference Pattern millimoles/1,000 millimoles of total N	Requirements (millimoles $kg^{-1} d^{-1}$) *Infants*	*Adults*
His	10	0.21	0.02 (?)
Ile	28	0.63	0.073
Leu	47	1.03	0.095
Lys	31	0.68	0.064
Met*	9	0.20	0.010
Met + Cys	20	0.36	0.098
Phe†	16	0.36	0.019
Phe + Tyr	35	0.81	0.069
Thr	26	0.57	0.055
Trp	4.7	0.10	0.014
Val	36	0.79	0.091

*With cysteine also present.
†With tyrosine also present.

ly. The proteins of hen's eggs were formerly used as a standard, but more recently a pattern based upon the composition of human milk has been developed as a reference because nitrogen deficiency is most likely to occur in the very young, and the infant utilizes human milk more efficiently than any other known protein source. The composition is adjusted so that the amount of protein satisfying the minimum requirement for total nitrogen will at the same time just satisfy the minimum requirement for the essential amino acids. The reference values are given in Table 41–1, together with the estimated minimum requirements of the essential amino acids in both infants and adults. Other foods can be given a **chemical score** based on a comparison of their amino acid composition with that of the reference pattern. The chemical score is the relative quantity of the most limiting amino acid. Table 41–2 compares the chemical score and net protein utilization index for important foodstuffs.

An example will illustrate how the chemical score is determined. Whole wheat flour contains 16 millimoles of lysine residues per mole (1,000 millimoles) of total nitrogen. The reference pattern contains 31 millimoles. Therefore, wheat

VALUE OF SOME FOODS AS NITROGEN SOURCES*

TABLE 41–2

	Chemical Score	Net Protein Utilization†
human milk	100	95
whole hen egg	100	87
cow's milk	95	81
soya (bean)	74	–
(flour)	–	54
peanuts	65	57
maize	49	36
polished rice	67	63
whole wheat	53	49

*Taken from p. 67 in WHO Technical Report No. 522.

†Not all measurements were made under comparable conditions. The last three were made with children aged 8 to 12 years, the others with children aged 3 to 7 years. The percent of the food energy derived from protein also varied, although no marked effect on the values were obtained when one food was studied at different levels of intake.

flour has a chemical score of 16/31 × 100 = 53, with respect to lysine. The score for no other amino acid is this low; therefore, lysine is the limiting amino acid in wheat flour, and 53 is taken as the chemical score for the entire food. By this index, one ought to feed enough wheat flour so that 100/53 = 1.89 times as much nitrogen is absorbed as is absorbed from human milk in order to meet the protein requirement. The measured net protein utilization of wheat flour in young children is 49 per cent, which is in close agreement with the chemical score.

However, chemical score and net protein utilization are not always close, even if the score is corrected for digestibility of the protein. A protein completely lacking an essential amino acid would have a chemical score of zero. This is fitting if it is the only nitrogen source, because it cannot sustain life. However, such a protein has a measured net protein utilization of as much as 30 per cent. The reason for this is that the test subjects adapt to a limited supply of any essential amino acid except threonine by diminishing the activity of the catabolic enzymes for that amino acid. (The content of threonine has never been shown to be limiting in natural foods, which probably explains why no protection against its deficiency has ever been evolved.) As a result, they conserve the amino acid from their own proteins and re-utilize it along with the dietary amino acids to make complete proteins. (Continued lack is eventually lethal, however.)

We see that neither index of protein quality is completely satisfactory. The chemical score is a more reliable indicator of defects in a single source of protein, whereas the net protein utilization index assigns some value to any source of nitrogen, which is appropriate when one is considering a food as an addition to other components in a diet.

Tissue and Seed Proteins. Since most living cells carry out much the same general classes of reactions, we might guess that most of their proteins, taken as a whole, have similar amino acid compositions. Actively metabolizing cells, plant or animal, ought to have an amino acid mixture not too far different from the mixture necessary to construct human cells. This is indeed the case. Most animal tissue and leaf proteins have high chemical scores. The exceptions among the animal and plant proteins are those present in relatively high concentration as structural or storage proteins, some of which have unusual amino acid compositions. For example, collagen and the gelatin derived from it contain very little tryptophan, but have high contents of proline and glycine. The most important examples are the seed proteins, which include maize, the cereal grains, and the pulses. Most of them have a low chemical score, owing to a deficiency of lysine, methionine, methionine + cysteine, or tryptophan, compared to the reference pattern. This is a matter of great practical importance. The protein content of the dried seeds is often not very large to begin with, ranging from 1.1 to 1.4 millimoles of N per gram of dried seeds. This compares to the 2 millimoles per gram *wet weight* of beef. In addition, more of the protein must be eaten to satisfy the protein requirement because of the low net protein utilization index. People forced to use cereal as the sole source of protein frequently will not eat enough to maintain nitrogen balance, even if the whole grain is consumed, and the problem is aggravated by refinement to remove the active cells of the plant embryo in the seeds. The appetite of even the undernourished might falter before three liters of corn mash. This problem has led to massive and encouraging efforts to alter the amino acid composition of the principal grains through genetic changes that alter the proportion of the proteins. Another solution to the problem is the use of mixtures of seed proteins, such as corn and beans, which have differing patterns of deficiency in the essential amino acids. Such mixtures are still deficient in some amino acids compared to the reference pattern for children, but experience indicates that adults may be efficient enough in using them to supply their nitrogen requirments.

KWASHIORKOR: PROTEIN STARVATION

Simple starvation is due to a generalized restriction of food supply. In some areas of the world, especially in or near the tropics, the total quantity of food may not be so obviously limiting, but the material available may consist mostly of starchy plant substances without an adequate content of protein. The diet is usually one of necessity rather than choice, and in an undernourished family it is frequently the children who are most affected, not only because lack of food tends to destroy any feeling of selflessness by the adults, but also because the very young have not built up a reserve of protein and fat upon which to live.

A long-standing inadequacy of protein intake in children leads to a condition known as kwashiorkor, from the Bantu word meaning displaced child; the symptoms appear in infants after they are no longer suckled by their mother (owing to the appearance of still another baby). Lack of protein is a central part of the condition — if the intake of carbohydrates and fats is also deficient, the child is simply being starved and wastes away. Such infant starvation has been termed *marasmus*, and there obviously is a spectrum of conditions ranging from total starvation with "pure" marasmus to a "pure" kwashiorkor in which the total supply of ingested fuels would be ample for maintenance of high-energy phosphate under ordinary conditions, but in which there is not sufficient protein to maintain cellular constituents. The latter results when the mother has access to quantities of starchy vegetables or sugar but not to foods containing enough protein, and the starchy part of the original food is frequently fed to the infant as a thin gruel. This circumstance is common in parts of Latin America and Africa.

Unlike children who are simply starved, those fed a protein-deficient diet may live for a considerable period, perhaps surviving into adulthood even though irreversibly impaired by the consequences of inadequate cellular development. Since kwashiorkor essentially represents a failure to synthesize normal amounts of protein, its consequences could theoretically appear in every metabolic process of the body, and in many ways do. However, some of the more striking phenomena can be given speculative interpretations in terms of isolated segments of the metabolic economy. The metabolic load in kwashiorkor is quite different than that in marasmus. Carbohydrates are still being supplied and the metabolic machinery for handling these compounds will tend to remain intact. At the same time, it is not possible to maintain all of the proteins in the carcass because of the lack of dietary amino acids. The result is an uneven depletion affecting some processes more than others, rather than the relatively smooth, general decline seen in marasmus.

A striking finding in kwashiorkor is the deceptively plump appearance of the youngsters; they are called "sugar babies" in the Caribbean area because they are fed on sugar and starches, but the name also evokes an image of round cheeks and bellies. The plumpness is not an expression of overfeeding and storage of fat, but is due to edema. The youngsters have a general accumulation of water to such a degree that their weight actually falls when they are put on restorative diets.

What can cause edema? One suspects congestive failure of the heart, but that is usually not the case in kwashiorkor; a mild failure is more likely to occur during initial recovery than it is during the active condition. Another possibiliity is a fall in the protein concentration of the blood, especially of serum albumin, causing a drop in the osmotic pressure. The synthesis of serum albumin is indeed impaired in kwashiorkor, sometimes falling from its normal level of 40 mg per ml to less than 10 mg per ml, and this kind of fall will invariably cause loss of fluid from the blood into the tissues. However, edema may appear without a precipitous fall in the albumin concentration (this is usually the case in adult starvation), and additional factors must be sought. Another possibility would be disturbances in the electrolyte and water balance, and these events also occur, both at the cellular level and in the kidney. There is an especially marked loss of potassium from the cells. It may well be that some of the proteins involved in ion transport are being destroyed and utilized for the formation of other proteins elsewhere, or to put it more accurately, there may be less of the transport proteins formed from the diminishing amino acid pool than there is of some other proteins.

Diarrhea is almost invariably a result of kwashiorkor and contributes to the potassium loss. Runny bowels are so frequent an accompaniment of disease in general that we give little thought to cause. Many nutritional deficiencies are accompanied by diarrhea. This indicates a failure of intestinal function. The intestinal mucosa has the highest known rate of cell turnover in the whole body and, therefore, is especially vulnerable to any failure in the supply of nutrients needed for constructing cells or to an interference in protein synthesis.

The stools in kwashiorkor may also be fatty. Fatty stools may result from the loss of emulsifying agents secreted in the bile, which are steroid derivatives formed by the liver, or from a failure in secretion of lipase by the pancreas. Since the pancreas requires a high rate of protein synthesis for activity, one might expect a decrease in function when there is protein deficiency, and this is the case. The pancreas, along with the intestinal mucosa, atrophies in kwashiorkor.

The skin, which is constantly being shed and replaced in a normal individual, develops gray and scaly or ulcerating patches. This is probably due to a mixture of protein deficiency and nicotinamide deficiency (see the next chapter). Similarly, the growing hair becomes fine, dry, brittle, and abnormally light in color. Most of the people in areas where kwashiorkor occurs have naturally dark hair, so that the reddish or even blonde sparse hair of kwashiorkor is striking. In some cases the hair will be banded with light color along its length; the bands indicate times of inadequate protein supply in the way that tree rings indicate the passage of the seasons.

Failure of hemoglobin synthesis causes anemia, and the iron accumulating from continued erythrocyte destruction appears as deposits of hemosiderin. The accumulation of fat in the liver that often occurs is probably due to a failure in apolipoprotein synthesis necessary for fat transport. (A low supply of methyl groups accompanying a methionine deficiency may also contribute by failure of phosphatidylcholine synthesis.)

Finally, kwashiorkor is frequently accompanied by the symptoms of deficiency of one or more other nutrients, including many of the mineral and vitamins, as described in the next chapter. (Cholecalciferol is rarely deficient.) While the dietary supply of these nutrients may be low, a general failure of gastrointestinal function is probably the most important factor, preventing absorption of proper amounts.

SPECIAL REQUIREMENTS

Pregnancy. Women need additional protein with which to construct a placenta and fetus. From calculation of the average increment of tissue for which nitrogen must be supplied by foods with a net protein utilization index of perhaps only 70 per cent, followed by a further 30 per cent allowance for two standard deviations from the mean, one arrives at the following values:

Quarter of Pregnancy	Extra Nitrogen Requirement* total millimoles per day	g "protein" per day
1	10	1
2	50	4
3	90	8
4	100	9

*Values rounded off to nearest gram or 10 millimoles

Lactation. Lactating women need even more protein than pregnant women to sustain the outpouring of milk proteins. Provision of an extra 190 millimoles of N (17 grams of "protein") per day is suggested as a safe allowance.

Injury or Surgery. Massive intravenous feeding of nitrogen in the form of amino acids, up to 20 millimoles $kg^{-1}d^{-1}$ has been used to minimize negative nitrogen balance. We shall discuss the full composition of solutions for parenteral nutrition in Chapter 43. Suffice it now to note that less than adequate thought has been used in devising many of these solutions. Some are hydrolysates of proteins. Not only do they contain some unhydrolyzed oligopeptides of unknown physiological effects, but they also contain most of the amino acids originally present in the proteins. Remember that the intestinal mucosa normally processes dietary amino acid mixtures, removing much of the aspartate and glutamate before the compounds reach other tissues. The liver further modifies the mixture.

Other solutions used for intravenous feeding contain mixtures of the crystalline amino acids. These have their faults. One contains DL-methionine, causing accumulation of D-methionine in the blood. They frequently contain large amounts of glycine, which is a cheap source of nitrogen. The deleterious effects that go along with accumulation of glycine from hereditary conditions give one pause before deliberately creating a hyperglycinemia, even for short periods.

Liquid Protein Diets. A current fad involves the use of protein hydrolysates as supplements to partial starvation in the correction of obesity. Many of these are hydrolysates of collagen, and therefore have grossly unbalanced amino acid compositions, including a very high concentration of glycine. The contribution of the imbalance to the sudden deaths that sometimes accompany this regimen is unknown.

Renal Disease. The accumulation of urea and other products of nitrogen metabolism is an especially serious consequence of renal failure. The accumulation can be relieved by dialysis. Considerable attention has been given to elimination of the need for dialysis in patients retaining a small fraction of normal kidney function, or to prolongation of the interval between dialyses in others, by restricting the protein intake to the minimal level consistent with survival.

Early regimens† involved restriction of amino acid intake to as little as 200 millimoles of N per day (18 grams of "protein"), sometimes as essential amino acids, sometimes in the form of two eggs, but cooperation of the patients was greatly improved by increasing the allowance to 450 millimoles of N (40 grams of "protein"), and cooperation is critical for any treatment that must be continued indefinitely.

†These regimens are sometimes known as G-G diets, after C. Giordano and S. Giovannetti, who (together with Q. Maggiore) originally developed them.

A promising new approach to treatment has been the administration of the 2-keto analogues of the essential amino acids, which cause part of the nitrogen to be utilized through transamination to re-form the amino acids. This is both experimental and expensive at present and probably would not be useful for extended periods.

FURTHER READING

The following reports enable one to follow the development of the current rationale on amino acid requirements:

Protein Requirements. (1957) FAO Nutritional Study No. 16.

Protein Requirements. (1965) WHO Technical Report No. 301.

Energy and Protein Requirements. (1973) WHO Technical Report No. 522.

Waterlow, J. C., and P. R. Payne: (1975) *The Protein Gap.* Nature, *258*: 113. An assertion that protein starvation is not a general problem.

Scrimshaw, N. S.: (1976) *Strengths and Weaknesses of the Committee Approach.* N. Engl. J. Med., *294*: 136, 198. An important student of kwashiorkor defends the necessity for higher intakes.

Scrimshaw, N. S.: (1977) *Through a Glass Darkly: Discerning the Practical Implications of Human Dietary Protein-energy Interrelationships.* Nutr. Rev., *35*: 321. A reiteration of Scrimshaw's position.

Whitaker, J. R., and S. R. Tannenbaum, eds.: (1977) *Food Proteins.* AVI. A generally valuable source.

Orr, M. L., and B. K. Watt: (1957) *Amino Acid Content of Foods.* Home Economics Research Report No. 4. U.S. Government Printing Office. The important source of older data.

Schneider, H. A., C. E. Anderson, and D. B. Coursin, eds.: (1977) *Nutritional Support of Medical Practice.* Harper and Row.

Rose, W. C., et al.: (1955) *The Amino Acid Requirements of Man.* J. Biol. Chem., *217*: 987.

Hegsted, D. M.: (1957) *Theoretical Estimates of the Protein Requirements of Children.* J. Am. Diet. Assn., *33*: 225.

Munro, H. N., and J. B. Allison, eds.: (1964) *Mammalian Protein Metabolism.* 4 vols. Academic Press.

Olson, R. E., ed.: (1975) *Protein-calorie Malnutrition.* Academic Press.

Munro, H. N., ed.: (1978) *Nutrition and Muscle Protein Metabolism.* Fed. Proc., *37*: 2281.

42 NUTRITION: MINERALS AND VITAMINS

MINERAL ELEMENTS

Humans, like other animals, have evolved so as to retain a sufficient supply of the needed mineral components from their customary diets. Even so, circumstances and individual genetic variations sometimes prevent maintenance of adequate concentrations of one or more of the minerals, necessitating additional supplementation in order to restore normal function.

What are the essential mineral nutrients? We have already encountered most of them in connection with a discussion of their functions. There are others with no known function, but which appear to be necessary in that experimental animals do not thrive when fed a diet deficient in the element. The confidence with which these findings are extrapolated to humans varies with the degree of observed impairment and the ease of reproducibility of the experiments.

The essential mineral elements may be divided into classes on the basis of either function or the magnitude of the daily turnover. **The bulk minerals** include the major electrolytes and the constituents of bone and teeth. Moles are convenient units with which to describe their estimated content in a 73 kg man:

Ca – 34 moles	Na – 4.8 moles	Cl – 2.5 moles
P – 24 moles	K – 3.6 moles	Mg – 1.2 moles

The major prosthetic minerals, those most commonly occurring in proteins, are present in millimole quantities:

Fe – 70 mmoles	Cu – 2 mmoles
Zn – 40 mmoles	Mn – 0.2 mmoles

The more specialized prosthetic minerals, those occurring as constituents of only a few proteins, are probably present in micromole quantities. Good analyses are not available for most:

I – 50 micromoles	Mo – ?
Se – ?	Cr – ?

Other elements that may be essential include **tin** and **vanadium**. In addition, there is evidence that normal bone and tooth formation is aided by the presence of **silicon** and **fluoride.**

TABLE 42–1 TYPICAL MINERAL REQUIREMENTS OF YOUNG ANIMALS*

Given in millimoles of element per kilogram dry weight of diet per day.

P – 160	Fe – 1	Se – 0.001
Ca – 150	Zn – 0.75	Cr – ?
K – 80	Mn – 0.6	Sn – ?
Na – 45	Cu – 0.08	Ni – ?
Cl – 30	Mo – 0.002	V – ?
Mg – 25	I – 0.0015	Si – ?

*Calculated from a compilation by W. G. Hoekstra, (1972) Ann. N.Y. Acad. Sci., *199*: 182.

The minimal requirements of most mineral elements for normal human development have not been clearly defined. Table 42–1 gives typical amounts required for normal growth of young domestic animals in terms of the amount of element per kilogram of diet (dry weight). If human requirements are comparable, a four-year-old child requires roughly the amount in one-half kilogram of diet per day. Adults would require less in most cases owing to previous accretion of the necessary total content, even though the dry weight of their total daily food intake approximates one kilogram.

Continued ingestion of high levels of many of the mineral elements causes toxicity, even death, in domestic animals. The minimal toxic intake has been shown to be 40 to 50 times the minimal level for normal growth in several cases. It is to be expected that evolution will have adapted mineral metabolism so as to cope with dietary variations. It is only when animals are living in areas where the soil content of an element is especially high or low that toxicity or deficiency becomes common.

Phosphorus

Since phosphorus in the form of phosphate compounds is a universal constituent of living cells, it is present in all natural diets of animals. As is the case with other minerals and vitamins considered individually, the question of interest is whether or not isolated deficiencies of the substance occur. That is, are people likely to become depleted in phosphate because it is specifically lacking in their diets? The answer is no. Although eating too little food or a menu of drastically limited variety obviously causes a depletion of a variety of components, such dietary practices do not cause an isolated phosphate deficiency. We have seen that the absorption of phosphate is stimulated in the intestine by 1,25-dihydroxycholecalciferol and is inhibited in the renal tubule by parathormone. The close endocrine regulation of phosphate concentration and the univeral dietary occurrence prevent the occurrence of an isolated deficiency, even though the total amount of phosphate required for development is quite large.

Phosphate deficiency does occur in infants with genetic defects in the mechanism of absorption of phosphate in the renal tubule. They develop a **vitamin D-resistant rickets.** Repeated daily feedings of phosphate appear to be a more successful treatment for this condition than the massive doses of vitamin D previously used, the excess vitamin D being both toxic and without important effect on the renal loss.

Hypophosphatemia in general is a result of some pathological disturbance of the metabolism rather than of grossly limited dietary intake. It occurs when phosphate is transiently taken up by cells for phosphorylation of carbohydrates or deposition as bone and when there is chronic excessive loss in the urine, which will occur with continued metabolic acidosis. Correction of a diabetic ketoacidosis may, therefore, give a double whammy; the acidosis itself has caused loss of phosphate, and the action of insulin results in uptake of phosphate for phosphorylation of glucose. Attention to phosphate and also K^+ balance is therefore a necessary part of the correction. Refeeding after any dietary deprivation is likely to have a similar effect. Other causes include chronic alcoholism and the chronic use of magnesium or aluminum-containing antacids, which form unabsorbable complexes with dietary phosphate.

Calcium

The necessity for calcium in the diet is so heavily emphasized to the layman that one would suspect a primary dietary deficiency to be common. It is not. This is rather surprising, since a human infant contains only some 0.7 mole of calcium at birth, and must have an average daily increment of 5 millimoles over a period of 18 years in order to achieve an adult content of 34 moles. This increment is supplied by the diet. The nursing infant obtains its supply from its mother; human milk is 8 mM calcium. After weaning, the supply comes from other foods.

Hypocalcemia does occur within the first two weeks of life, when it is the most common cause of convulsions. Why it happens in the first day or two is not clear, but a contributing cause to a second flurry near the end of the first week is the use of cow milk in formulas. This milk, and some commercial formulas based on it, has a relatively high content of phosphate compared to calcium. The molar calcium: phosphate ratio is 1.7 in cow milk and 2.7 in human milk. The increase in phosphate lowers the solubility of calcium both in the bowel and the blood. Phosphate and calcium concentrations tend to vary inversely because a rise in either exceeds the solubility product of the calcium phosphates, causing increased formation of insoluble complexes, which pass into feces in the case of the bowel.* Less of the calcium from cow milk may be available even though its total concentration is greater than it is in human milk. This may well be a transient failure of normal regulatory mechanisms during the first days of life, since experimental changes in the calcium:phosphate ratio did not affect calcium balance in small experimental samples of older infants or adults.

Signs of calcium deficiency in people with normal cholecalciferol content and parathyroid function are rare because the regulation of balance is so good. People adapt to the dietary intake of calcium. If it is small, little passes into the feces; if it is large, only a small fraction is absorbed. This adaptation has confused efforts to define the calcium requirement through balance studies. An adult ingesting 8 millimoles per day in his customary diet will "require" 8 millimoles in short-term studies to maintain balance, whereas one ingesting 15 millimoles will "require" 15 millimoles to maintain balance. Over a long period, intakes of 5 millimoles per day may be sufficient for an adult, although some believe the average intake of this amount by the Japanese before World War II may account in part for their short stature. Intakes of 10 to 12 millimoles (0.4 to 0.5 gram) appear to offer sufficient margin for losses and incomplete absorption, except that pregnant or lactating

*Adults can get into trouble from severe hypocalcemia caused by using phosphate-containing cathartics. At least one death has been ascribed to repeated use of such a preparation to remove stubborn feces prior to study of a patient. The reason for the use is that feces in the bowel make the radiologist howl.

women will require more. Approximately 6 millimoles per day will be incorporated into the fetus during the last trimester of pregnancy and secreted into the milk during lactation. The total drain on the calcium supply is approximately 1.5 moles due to pregnancy and subsequent lactation. This is of the order of 5 per cent of the supply in the skeletal reservoir, which is not an excessive loss in the absence of any additional source of calcium, but repeated pregnancies on a low calcium diet can lead to detectable losses of bone mineral **(osteomalacia).** The loss of bone mineral that occurs in the aged, especially in post-menopausal women, is of a different nature **(osteoporosis).** It is not due to a deficient intake of calcium, and it is not corrected by increasing the intake.

Sodium, Potassium, and Chloride

The urinary loss of sodium can be diminished to trivial levels when the supply is low. Sweat contains greater amounts, up to 80 mM, but this can be diminished to near 30 mM upon acclimatization in a hot environment. The dietary requirement for sodium chloride is therefore very low in cold climates, as little as 10 mmoles, depending upon the potassium intake, and it is easily met by natural foods for all except those who sweat large volumes. Man's craving for salt is a matter of taste rather than a reflection of some desperate nutritional need. The use of salt as such is a relatively recent event in human history. It began as a luxury, with sitting below the salt equivalent to social inconsequence. While the demand for salt became an important political and economic force in post-Roman cultures, its major use was as a preservative. Perhaps it was an acquired taste for salt pork and the like that set the stage for the current habit of heavily salting food. Since large quantities of salt are frequently added to foods during cooking or preparation, the major practical problem is now one of preparing low-sodium diets for hypertensive patients and patients with heart failure who tend to retain extracellular fluid with sodium rather than one of anticipating a deficiency state. Those who lose salt and water through excessive sweating can easily compensate by eating salted foods and drinking water. Young people in generally good health who exercise vigorously on hot days may lose enough salt to develop painful cramps, which are relieved by NaC1 tablets. (Cramping of abdominal muscles has on occasion been so severe as to cause surgical exploration for a non-existent acute abdomen.) Severe dehydration and salt loss causes coma, and infusion of isotonic sodium chloride is then indicated.

Potassium is continually lost in the urine of normal people and must be replaced. However, it is a constituent of all cells and therefore present in all normal diets. The amount is adequate except in the presence of abnormal losses, such as occur with continued vomiting, or diarrhea and excessive aldosterone secretion, or treatment with some diuretics. Mild supplementation of potassium can be provided by feeding orange juice (5mM K^+). More intensive supplementation with potassium salts requires care, not only because of local irritation of the gastrointestinal tract, but also because the amount of potassium necessary to restore intracellular balance can create toxic extracellular levels as it passes through the blood if it is absorbed too fast.

Magnesium

Magnesium ions are also present in all cells; this wide distribution and close regulation of its concentration make primary dietary deficiencies very rare. However, excessive loss or failure in absorption is known to occur. Conditions in which these may occur include alcoholism, congenital heart failure, use of diuretics, intoxication by digitalis, diabetic acidosis, malabsorption syndrome, and aldosteronism. The tissue content of magnesium is sometimes low without a corresponding fall in blood concentration. The effects of magnesium deficiency in humans are rather diffuse. The nervous system bears the brunt, as shown by weakness, nausea, tremor, stupor, coma, and cardiac arrhythmia. This is in sharp contrast to the effects of magnesium deficiency in some animals. Experimental deprivation in rats causes them to develop strikingly red ears and an excessive response to stimulus manifested first by biting the hand that feeds them and then by running fits in response to loud noises. Veterinarians in general practice are well aware of the magnesium-deficient cow, who changes from placid Bossie to a horn-waving candidate for the corrida. The nearest counterpart in humans is an apparent exacerbation of delerium tremens caused by concomitant magnesium deficiency.

Iron

The body content of iron both as a prosthetic component and in storage forms increases from birth to maturity; indeed, the amount of stored iron increases in many males throughout life. As we noted in Chapter 33, the requirement for iron is fixed by the amount necessary to build new tissue from the time of conception to adulthood, and by replacement of losses, mostly due to bleeding. Typical values of iron content and the requirement for absorbed iron are given in Table 42–2. These data could be used as a guide for the dietary requirement of iron were it not for the fact that the efficiency of absorption and the magnitude of the losses are highly variable, not only from person to person, but also from day to day, depending upon the nature of the diet and the physiological state of the gastrointestinal tract, as well as the variations in the reproductive cycle in women.

If the absorption of iron fails to match losses in those without a large store of iron or fails to provide for growth in the young, the replacement of the total hemoproteins and iron-sulfide proteins will lag behind their destruction. This will lead to a diminished content of these proteins, usually hemoglobin in the first instance, followed by impairment of function.

Some consequences of **negative iron balance** are clearly related to loss of function when the hemoglobin concentration falls: breathing is more labored (dyspnea), especially upon exertion, the heart rate increases (tachycardia), palpi-

BODY CONTENT OF IRON **TABLE 42–2**

	Body Content *millimoles*				**Physiological Requirement** *millimoles/year*	
Age Years	*Storage* male	female	*Total* male	female	male	female
birth	1.0	1.0	5.1	5.3	2.6	2.0
4	2.7	2.7	13.9	13.3	2.5	2.6
19*	13.0	8.3	62	40	5.0	8.3†
60–70	12.3	9.0	60	44	1.2	1.4

*Values for virgin females.
†World Health Organization estimate is 18 mmoles/yr for menstruating adults.

tations may occur, and there is general fatigue. The effects of deprivation of other iron-containing proteins are less clearly ascribed to their function, but they include the loss of papillae and inflammation of the tongue (glossitis), causing it to be smooth and bright red, appearance of fissures at the angles of the mouth, and changes in the growth of the nails.

How common is chronic negative iron balance? Perhaps no aspect of nutrition currently represents a more serious challenge to objective appraisal than this question. It will be an important guide to our thinking about nutrition in general if we can effectively meet this challenge. The problem essentially is one of defining terms and eliminating some of the connotations. Direct assessment of iron balance is not practical. The indicators that are routinely available include measurements of the blood hemoglobin concentration, the proportion of the blood volume contributed by the cells (the hematocrit or packed cell volume, given in per cent), the iron-binding capacity of the serum (mainly transferrin), and the percentage saturation of that capacity with iron. (The serum ferritin level may also be low in untreated iron-deficiency anemia.)

Normal ranges have been established for these measurements. Table 42–3 lists the minimal values given by a committee of the American Medical Association and those published by Massachusetts General Hospital. Now come the all-important definitions. If the quoted values are taken to be the lower limit of the normal range, then a person with a lower packed cell volume or hemoglobin content has an **anemia** by definition. Similarly, by definition, a person with a low saturation of iron-binding capacity (and usually with an elevated total capacity) is **iron deficient.** (Still another definition has a person deficient in iron who has not ingested the recommended daily allowance.) These definitions are impeccable. They give the words clear meaning in terms that are susceptible to measurement, although it is apparent that iron deficiency may mean different things to different people. (Iron ingestion can only be approximated by questioning the subject.)

TABLE 42–3 MINIMUM NORMAL HEMATOLOGICAL VALUES

Committee on Iron Deficiency*

Age *years*	[Hemoglobin] *g/dl*	[Hemoglobin] *mM*	Per cent Packed Cell Volume
0.6–4	11	1.7	33
5–9	11.5	1.8	34.5
10–14	12	1.9	36
adult male	14	2.2	42
adult female			
not pregnant	12	1.9	36
pregnant	11	1.7	33

Massachusetts General Hospital†

	[Hemoglobin] *g/dl*	[Hemoglobin] *mM*	Per cent Packed Cell Volume	Iron-binding Capacity *μg/dl*	Iron-binding Capacity *μM*	[Iron] *μg/dl*	[Iron] *μM*
adult male	13	2.0	42	250	45	50	9
adult female	12	1.9	40				

*J.A.M.A., *203*: 407 (1968).
†N. Engl. J. Med., *298*: 34 (1978).

The critical point that must not be missed is that none of these definitions is equivalent to a chronic negative iron balance or to a deficit in iron needed for tissue growth, and they do not in themselves imply morbidity. In other words, the existence of an anemia or of an iron deficiency as defined by the measurements does not establish the presence of a functional impairment. According to their definitions, they are conditions like pulse rate or blood glucose concentrations and are not in themselves diseases. Once this is comprehended, the question then becomes: To what extent are these conditions indicative of disease? The present answer is that we do not know. The data at hand are conflicting and often reflect the pre-existing bias of the investigator.

All of this came to a head with the publication of preliminary results of a useful and estimable project of the Public Health Service, known as the Health and Nutrition Survey, and their misuse by some, both within and without the Health Service, as justification for a crash program of increased fortification of the American diet with iron. (Hematologists sensitized to the problem of managing patients with already existing iron overloads forced a more balanced appraisal. See Further Reading for summary reports.) The flavor of the misuses can be appreciated from this excerpt of a report in *Science,* the official organ of the American Association for the Advancement of Science: "The preliminary report [of the survey] details the results from the first two tests; dietary intake and biochemical findings from a sample of 10,126 people. The most striking finding, which confirms those of earlier, smaller surveys, is that the population suffers from widespread iron deficiency. According to the report, about 95 per cent of all preschool children and women of childbearing age have iron intakes below the standards set by the Food and Nutrition Board of the National Academy of Sciences." This quotation is misleading in two important respects. It equates iron deficiency defined by less than a prescribed level of dietary intake with a disease state ("suffer"), and it equates the recommended daily allowances of the Food and Nutrition Board, which we shall discuss in the next chapter, with a requirement, an identification the Board specifically warned ought not be made.

In practical terms, clear indications for action do not exist in many cases. No one disputes the requirement for prompt attention, including iron supplementation, in a woman with 1.2 mM of hemoglobin (8 g/dl), or less. In the Health Service survey, 16 people, all women, out of 3,444 aged 18 to 44 examined (0.24 per cent) had hemoglobin concentrations below 1.4 mM (9 g/dl). This is a small percentage, but it represents a large number of women in the country, and some with higher levels will also have a functional impairment. A cautious physician would like to rule out unsuspected loss of blood when there is any indication of anemia or iron deficiency. (Anemia is frequently the first indication of intestinal malignancy in either male or female.) At the same time, he might well avoid vigorous supplementation with iron salts in the absence of organic defect or overt symptoms when the anemia or deficiency is only mild according to the usual tests. (It is clear that many of those defined as anemic or deficient by these tests show no significant changes upon iron supplementation and continue to live their normal lives with persistently deficient concentrations of hemoglobin or iron.) The anemia of pregnancy is a physiological event that is not altered by iron supplementation. In general, questions have been raised about the potential advantages of a mild anemia or iron deficiency as defined by current standards. Too little iron makes one susceptible to infection owing to failure of normal cell function, but too much iron may make one more susceptible to infection by providing a better culture medium for the invading organisms.

Zinc

Zinc is a component of a number of enzymes, including RNA and DNA polymerases, carbonic anhydrase, and carboxypeptidase. Blood serum zinc is ordinarily 14 to 19 micromolar, but the concentration falls sharply with several pathological conditions, including malignancies, myocardial infarction, and infections. Frank deficiencies of zinc are rare. Several dwarf males in Iran and Egypt were found to have retarded sexual development, anemia, enlargement of the liver and spleen, and mental lethargy. The anemia could be corrected by iron supplementation, but improvement of the arrested development was found to require added zinc. These people had been subsisting on a diet mainly composed of bread and beans, and like many people on a deficient diet, they ate large amounts of earth (as much as 400 grams of clay per day).

Zinc deficiency also is caused by hereditary defects in its absorption, resulting in **acrodermatitis enteropathica,** a disease marked by severe chronic diarrhea, loss of skin around anus and mouth, and rash on the extremities. This condition is alleviated by human milk, but not by cow milk, and the apparent reason is that the zinc in human milk is bound to polypeptides of lower molecular weight, making the zinc more accessible for absorption. A similar condition has been observed upon parenteral feeding with formulas deficient in zinc.

The effects and incidence of more moderate zinc deficiencies are not known. The blood and hair concentrations in some American infants appear low compared to those observed in other countries, and some American formulas have been low in zinc concentration, as well as being based on cow milk with its higher molecular weight binding protein. Supplementation of the diet with zinc has in some cases aided growth.

Deficiencies in adults cause a loss of normal taste. The taste buds have no circulation and the saliva is evidently an important source of nutrients for them. The saliva includes **gustin,** a 27K molecular weight polypeptide that is high in histidine (8 per cent) and contains two zinc atoms. (Histidine supplementation can be used to create an experimental zinc deficiency.) Gustin is a close relative of, if not identical to, a nerve growth factor. The presence of this protein, perhaps as a source of zinc, appears to be necessary for normal development of taste buds. Some individuals have a hereditary defect in taste function that is correctable by large doses of zinc. (This would be consistent with a mutation causing a lowered affinity of some protein for zinc.)

Copper

The occurrence of copper in several oxidative enzymes, including cytochromes a + a_3, lysyl oxidase, ferroxidase, and dopamine hydroxylase makes its presence in the diet essential. However, a dietary copper deficiency is exceedingly rare except in infants and in patients given parenteral feeding with deficient formulas. Anemia is an important early symptom. Copper moves through the body as Cu(II), and the body content is regulated through excretion of an excess in the bile. Absorbed copper is mainly transported by combination with serum albumin (one tight binding site per molecule). It is probably stored for use by combination with one or more intracellular proteins that have been shown to have a high affinity for copper (II). (Some also bind Zn(II).)

Two genetic disturbances of copper metabolism are known. **Wilson's disease,** or **hepatolenticular degeneration,** is an autosomal recessive condition with a gene frequency that may be as high as 0.02 (equivalent to a carrier frequency of 0.04

and a disease frequency of 0.0004). This condition involves some uncharacterized disturbance of the movement of copper within the liver, so that biliary excretion and the formation of ferroxidase (ceruloplasmin), which are the principal means of export from the liver, are below normal. The failure in excretion causes an accumulation of copper in various tissues. A pathognomonic sign is the occurrence of greenish-brown deposits in a ring around the outside edge of the cornea (Kayser-Fleischer ring), which occurs in no other condition. The disease usually is diagnosed in the first or second decade, and early diagnosis is important to minimize the cirrhosis of the liver and neurological degeneration it causes. Treatment involves the administration of a chelating agent, usually penicillamine, to remove the excess copper.

Menke's disease, the kinky-hair or steely-hair syndrome, we have previously mentioned as one or more defects in the utilization of dietary copper, which affects the formation of normal connective tissue because of the resultant loss of lysyl oxidase activity (p. 165). The loss of other enzymatic activities has more widespread effects, and death within the first three years is expected. This X-linked recessive may have an incidence as high as 1 in 35,000 births. Diagnosis has usually been made too late to determine if intravenous copper supplements would be an effective treatment.

Other Minerals

The only known function of **selenium** is as a constituent of **glutathione peroxidase.** This enzyme is found in most tissues as an agent for removing hydrogen peroxide and organic peroxides:

$$2\ GSH + ROOH \rightarrow G\text{-}S\text{-}S\text{-}G + H_2O + ROH$$

Human deficiencies are unknown, but animals reared in areas where the selenium content of soils is low have a muscular dystrophy (white muscle disease). Selenium is unusual among the essential minerals in that it commonly occurs in toxic levels in several kinds of plants growing in regions where the soil level is high. Animals grazing in these areas lost hoof and hair from chronic exposure and had severe damage to many organs in acute cases. Although people in those areas are alleged to be affected on occasion with the loss of hair, brittle nails, and a garlic odor on the breath characteristic of human selenium poisoning, clear confirmation is lacking.

Deficiencies of **manganese, molybdenum,** or **chromium** are not known in humans, except possibly in some patients fed intravenously for prolonged periods. Manganese is a cofactor for several enzymes; molybdenum is a constituent of some oxidases; the only known function of chromium is its occurrence in a complex that facilitates the effect of insulin on glucose transport in experimental animals. This **glucose tolerance factor** contains glutamate, glycine, and cysteine along with two moles of some nicotinate derivative, and it is evidently volatile enough to be removed from samples dried at 100° C or less. It has not been identified in humans.

WATER-SOLUBLE VITAMINS

The functions of the vitamins were discussed as we encountered them, and page references are noted in the following. As with other nutritional components, the vitamins are subject to bouts of over-attention, sometimes followed by a period of rebound neglect. We first see assertions that a deficiency within the United States is a prime cause of a variety of frank illnesses and less well-defined general malaise. Then, after much discussion and an occasional experiment, the claims prove to be ill-founded. However, occasional deficiencies of some vitamins do occur even in our well-fed nation, and with sufficient frequency to warrant awareness on the part of the physician, but not so often as to warrant political interference.

Ascorbate (Vitamin C—pp. 164, 620)

Ascorbate is a newcomer to the ranks of vitamins in terms of geological time. Invertebrates do not make the compound. The vertebrates developed a synthesis of the compound from glucose (Fig. 42–1), but this synthetic route was discarded in both the primate and the higher birds. The evolutionary loss probably resulted from a combination of unfailing supply and diminished demand. Unlike the other water-soluble vitamins, ascorbate is not an established component of any enzyme, and its exact function is not known, except that it is involved in some way with the action of prolyl and lysyl hydroxylases, and p-hydroxyphenylpyruvate oxidase, and in noradrenaline formation. It has an unusually varied distribution for a vitamin, both within an organism and among species. There are large concentrations in the adrenal gland and the aqueous humor of the eye. The high adrenal content suggests some relationship with the many hydroxylases involved in hormone production, a notion reinforced by the large discharge of ascorbate caused by corticotropin. However, cultured adrenal cells make the steroid hormones very well in the apparent absence of ascorbate. This does not completely kill the idea of some role as an electron donor for the hydroxylases, since the other ascorbate-sensitive oxygenases also function *in vitro* without it, but it chills it.

The best known sources of ascorbate are fruits (especially citrus fruits, although berries and green peppers are better), but it also occurs in broccoli in high concentrations. The potato is a good source. Pure carnivores gain an ample supply by eating liver, brain, or kidney; the muscles provide only marginal amounts.

Either ascorbate or dehydroascorbate, the oxidized form, can be utilized within the body, but formation of dehydroascorbate in foods results in a subsequently slower, but irreversible, isomerization to 2,3-diketo-L-gulonate (Fig. 42–2). The formation of dehydroascorbate by reaction with oxygen, which is accelerated by trace metals, therefore results in loss of biological activity. (Moral: Don't store large opened containers of orange juice, and keep it cold.)

Laymen are frequently exhorted to have a daily supply of ascorbate because it is not stored. This is false. The total content in adults varies from a high of approximately 23 millimoles (4 grams) to as little as 1.7 millimoles (0.3 gram) without signs of deficiency. This level is determined by the intake, with the metabolic losses being approximately 3 per cent. The pool size adjusts until intake balances loss. According to this estimate, an intake of 0.7 millimole (140 mg) per day would maintain maximum content, whereas an intake of 0.05 millimole (10 mg) per day would prevent deficiency. A person saturated with the vitamin would require 90 days before developing signs of scurvy on an absolute, deficient diet. (Trials in England during World War II arrived at estimates of 0.036 millimole (6.5 mg) per day as a minimum requirement. Other studies showed a period of six months elapsing in saturated individuals before scurvy appeared, but in both of these

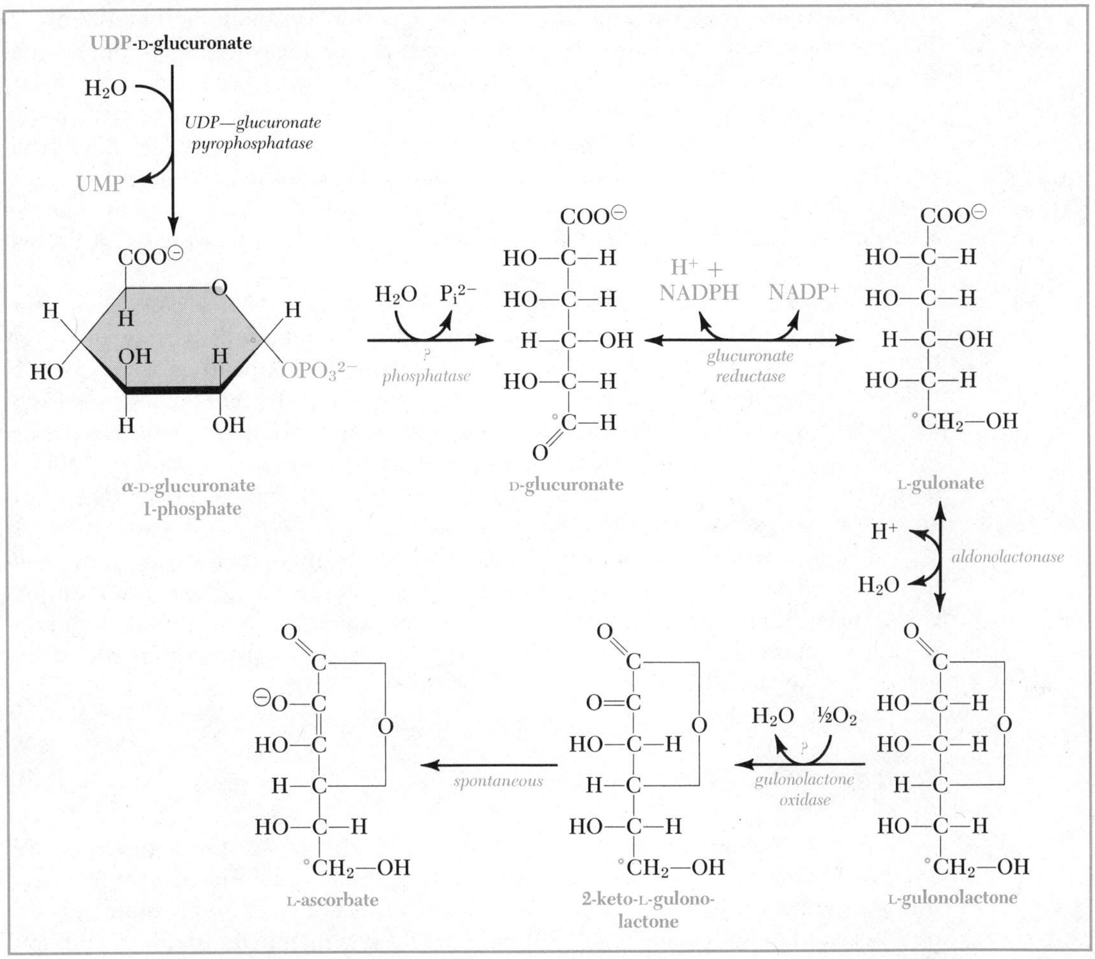

UDP-D-glucuronate

H_2O → UDP—glucuronate pyrophosphatase → UMP

α-D-glucuronate 1-phosphate → (H_2O, P_i^{2-}; ? phosphatase) → D-glucuronate → (H^+ + NADPH, $NADP^+$; glucuronate reductase) → L-gulonate → (H^+, H_2O; aldonolactonase) → L-gulonolactone → (H_2O, ½O_2; ? gulonolactone oxidase) → 2-keto-L-gulonolactone → (spontaneous) → L-ascorbate

FIGURE 42–1 The synthesis of ascorbate utilizes UDP-glucuronate derived from UDP-glucose as a precursor. Primates lack the gulonolactone oxidase that catalyzes the final enzymatic reaction in the sequence. (Note that the gulonate configuration is the reverse of the glucose configuration; that is, C-1 of glucuronate becomes C-6 of gulonate, and vice versa.)

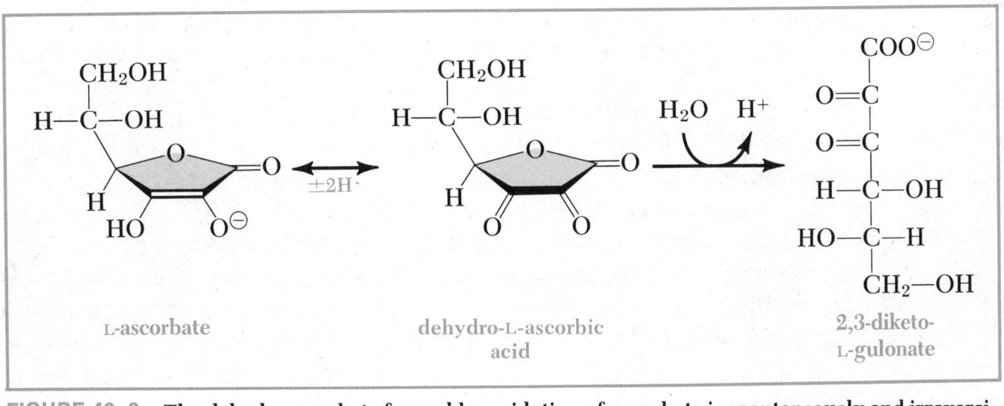

L-ascorbate ⇄ (±2H·) dehydro-L-ascorbic acid → (H_2O, H^+) → 2,3-diketo-L-gulonate

FIGURE 42–2 The dehydroascorbate formed by oxidation of ascorbate is spontaneously and irreversibly hydrolyzed to diketogulonate.

cases there may have been small intakes of ascorbate in the experimental diet.) These observations explain why people can survive the winter without eating fresh fruits, vegetables, or meat in any quantity.

Excess ascorbate is excreted or metabolized. No particular benefit or harm has been detected from the very large doses that are currently the fad. (The benefits may be marginal, but the damage may also be difficult to detect, except that those prone to oxalate stones have their problem exacerbated by the increased formation of oxalate, which is one of the metabolic products of ascorbate.)

In experiments with convict volunteers fed an ascorbate-free diet, the extent of depletion of the body content necessary for development of particular symptoms varied widely among the individuals, sometimes appearing when over 2 millimoles was still present, and in other cases not until the level had dropped to nearly 0.6 millimole. In general, a whole blood concentration of 17 micromolar or less (3 mg l^{-1}) is a signal for attention. The early symptoms of scurvy include petechiae, hyperkeratosis, congested hair follicles with coiled hairs, joint effusions, swollen gums, and arthralgia.

Scurvy is now mainly seen in the elderly, especially in men living alone, and in infants. There is some indication that large intakes by the mother during gestation cause an adaptive high rate of ascorbate turnover in the infant, so that it must have more ascorbate after birth to avoid deficiency. It is not known to what extent the adult adapts.

Biotin (p. 475)

Only minute amounts of this cofactor of some carboxylase proteins are necessary to balance losses, and part of this small requirement may be met through synthesis by intestinal bacteria; deficiencies are rare. Some infants develop a dermatitis that is corrected by biotin, and a few men had a similar condition caused by eating large amounts of raw eggs, which contain **avidin,** a protein binding biotin in a tight complex that escapes absorption.

Cobalamins (vitamin B_{12}—p. 572)

In the earlier discussion, we noted that these large molecules require the presence of a glycoprotein, the **intrinsic factor,** to be absorbed and that the daily requirement is of the order of one nanomole or less. (Some estimates place it as low as 200 picomoles per day for adults.) Since plants do not use cobalt and do not make the cobalamins, the ultimate sources of the compound in the food chain are those microorganisms that have evolved routes that require it. The compound enters animals through ingestion of the microorganisms or of other animals. The liver can accumulate a milligram or so, a generous supply that will stave off appearance of symptoms for years in people who have lost their capacity to absorb the compound (for example, by removal of the stomach).

Deficiencies of cobalamins cause a combined system disease with lack of acid secretion by the stomach (and an increased risk of developing carcinoma of the stomach), neurologic degeneration primarily in the posterior columns of the spinal cord, and anemia in which large red cells appear in the circulation and the bone marrow contains large megaloblasts instead of the normal precursor of erythrocytes, the erythroblast. The megaloblastic anemia is at least in part due to a concomitant deficiency of folate, owing to accumulation of 5-methyltetrahydrofolate (p. 606). The anemia, but not the neurological defects, is relieved in many cases by administration of folate. (It is for this reason that supplementation with folate is discouraged in this country; it may mask the presence of pernicious anemia or other causes of cobalamin deficiency until irreversible neural defects have appeared.) Methionine also diminishes the folate deficiency by inhibiting the reduction of methylene H_4folate to methyl H_4folate.

Cobalamin deficiency is a possible adjunct of any chronic disturbance of gastrointestinal function. Among normal people, those on vegetarian diets are at risk, especially if they are purists who shun even milk as an animal tissue. The many people in the world who subsist on a mainly cereal diet frequently have multiple dietary deficiencies rather than an isolated cobalamin deficiency. How those in the developed countries who are vegetarians by choice rather than necessity escape deficiency is not clear. Perhaps the lack of deficiency is an indication of backsliding, or sufficient time has not elapsed since they were more carnivorous. Scandinavians sometimes become deficient owing to infestations of fish tapeworms that compete successfully for the incoming supply.

Folate (p. 590)

Folate deficiency, rather than cobalamin deficiency, is the usual cause of megaloblastic anemia. It is common in those areas of the world in which the diet is limited and mainly composed of cereals, and it also occurs in both the chips-and-cola set and the vegetarians in more affluent societies.

Megaloblastic change in the bone marrow is common among pregnant women. The problem here is similar to that of iron deficiency: It is difficult to draw the line between natural changes without deleterious consequences and sufficient deficiency to cause functional impairment. There is no doubt that frank impairment does occur in the United States and may result in death of the mother. This is true even though the requirements for the compound are low (200 nanomoles or less per day) and it is widespread in both animal and plant tissues. Even if the diet includes a substantial amount of foods other than refined cereals, prolonged cooking destroys the compound.

Folate mainly occurs in plants as the heptaglutamyl conjugate and in animals as the pentaglutamyl, or smaller, conjugates. In order to be absorbed, the excess glutamyl groups must be removed by an enzyme in the intestinal mucosa. Any of the compound that has been oxidized is reduced to the tetrahydro form in the intestinal mucosa, and most of it is methylated for transport through the blood. Glutamyl groups are added back within the cells, probably to aid in retaining the compound. It may be that the tetrahydropolyglutamyl derivatives are the active forms for some purposes, since they function very well as cofactors in vitro. It is only the free tetrahydro form, and not the methyl derivative, that can be conjugated with glutamate within the cells, and this will cause even more depletion of the total folate pool during cobalamin deficiency.

Nicotinamide (p. 273)

Nicotinamide, the active component in NAD and NADP, can be formed from

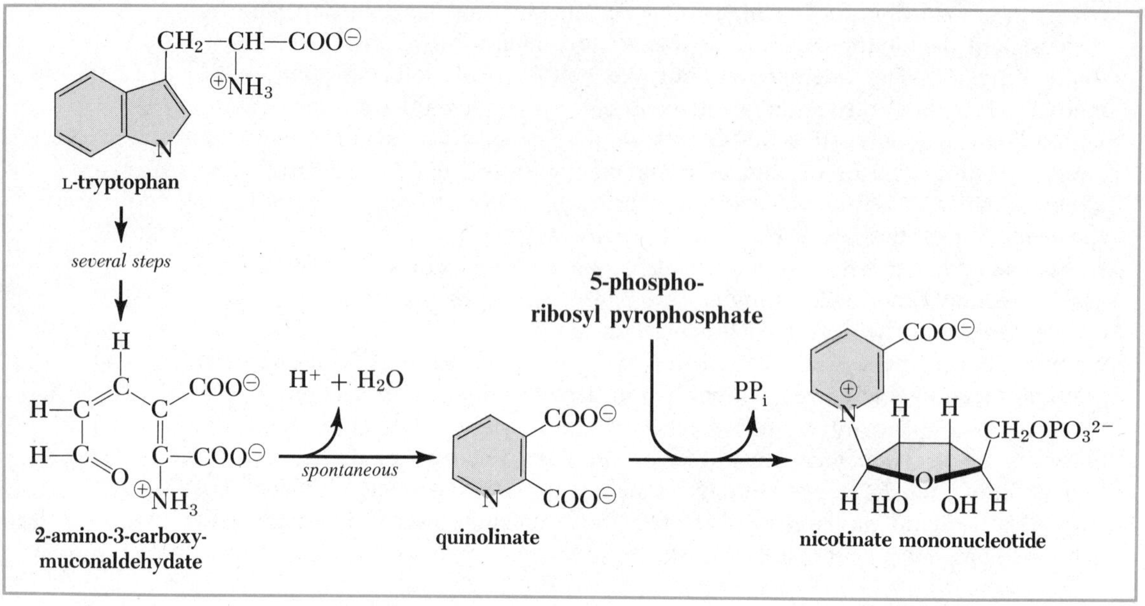

L-tryptophan → several steps → 2-amino-3-carboxy-muconaldehydate → (spontaneous; $H^+ + H_2O$) → quinolinate → (5-phosphoribosyl pyrophosphate; PP_i) → nicotinate mononucleotide

FIGURE 42–3 **The nicotinate moiety of NAD and NADP is formed from tryptophan in animals.**

tryptophan in the liver (Fig. 42–3). The reactions by which this is done diverge from the main route of catabolism at 2-amino-3-carboxymuconic semialdehyde (p. 585). In the main route, this compound is decarboxylated. However, it also spontaneously cyclizes to form quinolinate, which condenses with 5-phosphoribosyl pyrophosphate and loses CO_2 to form nicotinate mononucleotide. This nucleotide is converted to NAD by acquisition of an adenylyl group from ATP and an amide nitrogen from glutamine.

Dietary nicotinate and nicotinamide are also used as precursors of NAD and NADP through direct reaction with phosphoribosyl pyrophosphate. It is obvious that the diet need not include these compounds if it contains a supply of tryptophan sufficient to meet the needs for protein synthesis and for nicotinate formation, and if the rate of conversion of tryptophan to nicotinate matches the rate of loss of the nicotinate moiety. Man is an example of an animal that has no dietary requirement for nicotinate in the presence of an adequate supply of tryptophan. Cats, on the other hand, have a high level of the enzymes in the main route of tryptophan catabolism and are able to form little nicotinate.

In human adults, the molar conversion of extra tryptophan to NAD is roughly 3 per cent of the total degraded, so that 60 milligrams of tryptophan yield 1 milligram of nicotinate as NAD. The daily turnover of NAD is equivalent to about 1 mg of nicotinate per megajoule equivalent of dietary intake, so an adult with a 10 megajoule daily requirement would require roughly 600 milligrams of tryptophan to supply his nicotinate requirement if that was the only source. This estimate is probably high, because there is some reason to believe that the efficiency of nicotinate formation becomes greater when a deficiency threatens. This amount of tryptophan would be supplied by approximately 250 grams of beefsteak, 450 grams of whole oats, or 1,330 grams of whole maize.

These figures tell the tale. People subsisting mainly on maize (corn in the United States) are subject to tryptophan deficiency. Corn contains somewhat over 10 milligrams of pre-formed nicotinate per kilogram, but for some reason much of this is not available. However, treatment of the corn with mild alkali not only releases the nicotinate in a useable form, but it also may improve the amino acid composition of the product. (Corn has an excess of leucine compared to isoleucine, which further impairs tryptophan utilization. The alkali treatment appears to remove part of the leucine.) In any event, most of the Amerindians who depended upon maize as an important foodstuff had developed procedures for its preparation involving exposure to lime. Farmers in the southern United States, who also became dependent upon maize at times of deprivation, either ate it without alkali treatment or steeped it in sodium hydroxide with discard of the liquor. As a result, pellagra was widespread. This condition is characterized by the three D's: diarrhea, dermatitis, and dementia. However, the poor diet leading to pellagra is lacking in components other than nicotinate and tryptophan so that the condition is usually caused by a number of concurrent deficiencies.

The malignancy known as carcinoid can also provoke the appearance of a pellagric dermatitis because the tumor cells have a high capacity for converting tryptophan to 5-hydroxytryptamine (serotonin). The major end product, 5-hydroxyindolacetate, may be excreted in amounts up to 1.5 millimoles per day compared to the normal maximum of 0.05 millimole.

Large doses of nicotinate are toxic. The compound causes vasodilation by some unknown mechanism and blocks the mobilization of fatty acids from triglycerides. (There is a concomitant fall in cholesterol production, leading to tests of the compounds as a therapeutic agent in atherosclerosis, but it has not proved useful.) Nicotinamide does not have this effect. Increased doses of nicotinamide do cause increased excretion of N-methyl-nicotinamide, a normal product of NAD turnover, which will cause methyl group deficiencies in the absence of excess methionine and choline supplies. (This probably accounts for the liver damage seen in a schizophrenic who was ingesting nine grams per day.)

Pantothenate (p. 403)

Deficiencies of this widespread constituent of coenzyme A and fatty acid synthase are not known in humans except through experimental administration of a synthetic antagonist.

Pyridoxal (p. 270)

Natural deficiencies of pyridoxal are also unknown in adults owing to the widespread occurrence of it and its precursors, pyridoxine and pyridoxamine. Isolated cases occurred in infants fed deficient formulas, which led to convulsions. Induced deficiencies are created in adults either by administration of an antagonist, or through the use of drugs reacting with aldehydes, such as isonicotinic acid hydrazide used in tuberculosis or hydralazine used for hypertension. Very high doses of pyridoxine are toxic and also cause convulsions.

Riboflavin (p. 379)

Isolated deficiencies of riboflavin are rare, although deficiencies do occur in conjunction with other deficiencies in impoverished areas or in patients with bizarre eating habits. The symptoms are nondescript, including inflamed tongue, lesions at the corners of the lip, dermatitis, and anemia. In general, suspicion of riboflavin deficiency ought to be equated with suspicion of general malnutrition and treated accordingly. The estimated requirement for adults is approximately 0.2 micromole per megajoule energy equivalent. It is lost only slowly from the body owing to the tight binding of the flavin coenzymes to their apoenzymes.

Thiamine (p. 409)

We discussed thiamine deficiency in connection with glucose catabolism. It is the only deficiency of a single dietary component known to cause heart disease (wet beri-beri). Thiamine deficiency is still common in parts of the world, especially those in which polished rice is the major food. It is seen in the developed countries mainly in association with alcoholism or a chronic disease of the gastrointestinal tract. Thiamine occurs in the body only as the pyrophosphate and the triphosphate in association with proteins. The function of the triphosphate is unknown. The total thiamine content of an adult is approximately 80 micromoles, and it is not increased by ingestion of large doses. (As Davidson, Passmore, and Brock put it, the body is an efficient machine for dissolving thiamine pills and transferring them to the urinal.)

LIPID-SOLUBLE VITAMINS

The hydrophobic vitamins as a class will not be absorbed unless fat digestion and absorption are proceeding normally, and deficiencies ought to be excluded in anyone with malabsorptive disease. Chronic use of mineral oil as a laxative, formerly more common than it is today, may also provoke deficiencies.

Retinol, Vitamin A, and Vision

The image that we "see" in our brain is constructed by the organization of impulses that are generated by the absorption of quanta of light in the rods and cones of the retina. In order to translate incoming light into impulses, the eye must contain compounds that will absorb quanta of particular wavelengths, in this case 400 to 700 nanometers. This is a useful, indeed a necessary, range of wavelengths for good vision — useful not only because the peak energy of sunlight at the earth's surface lies in the middle of the range, but also because it is a range enabling great distinction of objects. Light of much shorter or longer wavelength is absorbed by a variety of chemical structures, and we distinguish objects because of selective absorption by their less common constituents. In the biological realm, peptides, carbohydrates, and lipids are colorless. In the surroundings, water and the common minerals, except those containing iron, are colorless. We distinguish rubies and sapphires not by the colorless aluminum oxide crystal lattice, but by contaminants of the lattice, and we detect the jaundiced person through the selective absorption of quanta by the relatively small concentrations of bilirubin.

It is apparent from this that the light-sensitive compound in the eye must also have a distinctive structure; it must exist in a number of electronic states of relatively small energy difference so that quanta of light in the visible range can cause excitation, and at the same time we would expect this important compound to differ sufficiently in structure from other components of the retinal cells so that it is not easily removed by the usual metabolic reactions.

We have noted that animals frequently do not synthesize unusual structures when compounds containing the structure are available from the diet, and this is the case with the visual pigment. The necessary highly resonant structure is an isoprenoid hydrocarbon chain containing conjugated double bonds that is contributed by compounds in the diet having vitamin A activity. Let us consider the nature of these compounds and then go on to consider how they contribute to vision.

The Nature of Vitamin A. Compounds having vitamin A activity in the diet of mammals are precursors of a 20-carbon polyprenol, **retinol** (Fig. 42–4). Some of the precursors are fatty acid esters of the *all-trans* isomer of retinol, the most abundant being **retinyl palmitate.** These esters are hydrolyzed in the small bowel to form free retinol, which passes into the intestinal mucosa.

The other dietary components with vitamin A activity are **carotenes,** the original polyprenyl precursors of retinol. The carotenes are synthesized by plants as light-harvesting pigments for photosynthesis. (The synthesis resembles that of squalene in that two polyprenyl pyrophosphates are condensed head-to-head. However, they are 20-carbon compounds rather than the 15-carbon precursors of squalene.) The most abundant carotene, β-carotene, is a symmetrical compound that is oxidized in the intestinal mucosa to form two molecules of retinal, the aldehyde analogue of retinol. Retinal and retinol are equilibrated by an NAD-coupled alcohol dehydrogenase present in most tissues. Some other carotenes have the retinal configuration in only one half of the molecule, while others lack it completely.

Part of the carotenes in the diet is absorbed as such and dissolved in the lipid phase of lipoproteins. The normal plasma concentration is between 2 and 4 micromolar, but it becomes so high in some leporine people from their proclivity for eating carrots and the like in wholesale quantities that they turn yellow. This carotenemia is harmless, as is the lycopenemia rarely seen in some tomato enthusiasts. The latter become a bright red-orange from absorbed lycopene, a polyprenyl compound not useable by humans. In any case, the absorbed carotene can later be oxidized as a precursor of retinal.

The retinol generated in the intestinal mucosa from either retinyl esters or from carotenes is converted to the palmitoyl ester and passes into the circulation by solution in the triglyceride phase of chylomicrons. The liver removes it for storage in Kupffer cells, and there may be as much as 2 micromoles of retinyl esters per gram of tissue (0.25 micromole is more common). The ester is constantly broken down and rebuilt by the liver, and the parenchymal cells export free retinol in combination with a specific transport protein, which in turn combines with a protein that also transports thyroxine. There are only 40 mg of the retinol-binding protein per liter of blood plasma, and the normal total plasma retinol concentration is between 1 and 2.5 micromolar.

The Visual Pigment. The transport of retinol into the retina appears to involve the removal of the compound from its plasma protein carrier and the formation of a fatty acid ester within the retinal cells. The mechanism of concentration is quite effective, because the level of retinol must drop to half its normal range before any loss of the store in the eye becomes apparent. The retinyl esters of the eye are subject to hydrolysis, and the retinol liberated with the cells is oxidized by NAD in a reaction catalyzed by a specific dehydrogenase.

The equilibrium position of retinol dehydrogenase does not favor retinal production. However, the retinal that is formed is rapidly removed to form complexes with proteins, the *opsins,* of the rods and cones, and oxidation of retinol continues until no more of these complexes, which are the light-sensitive

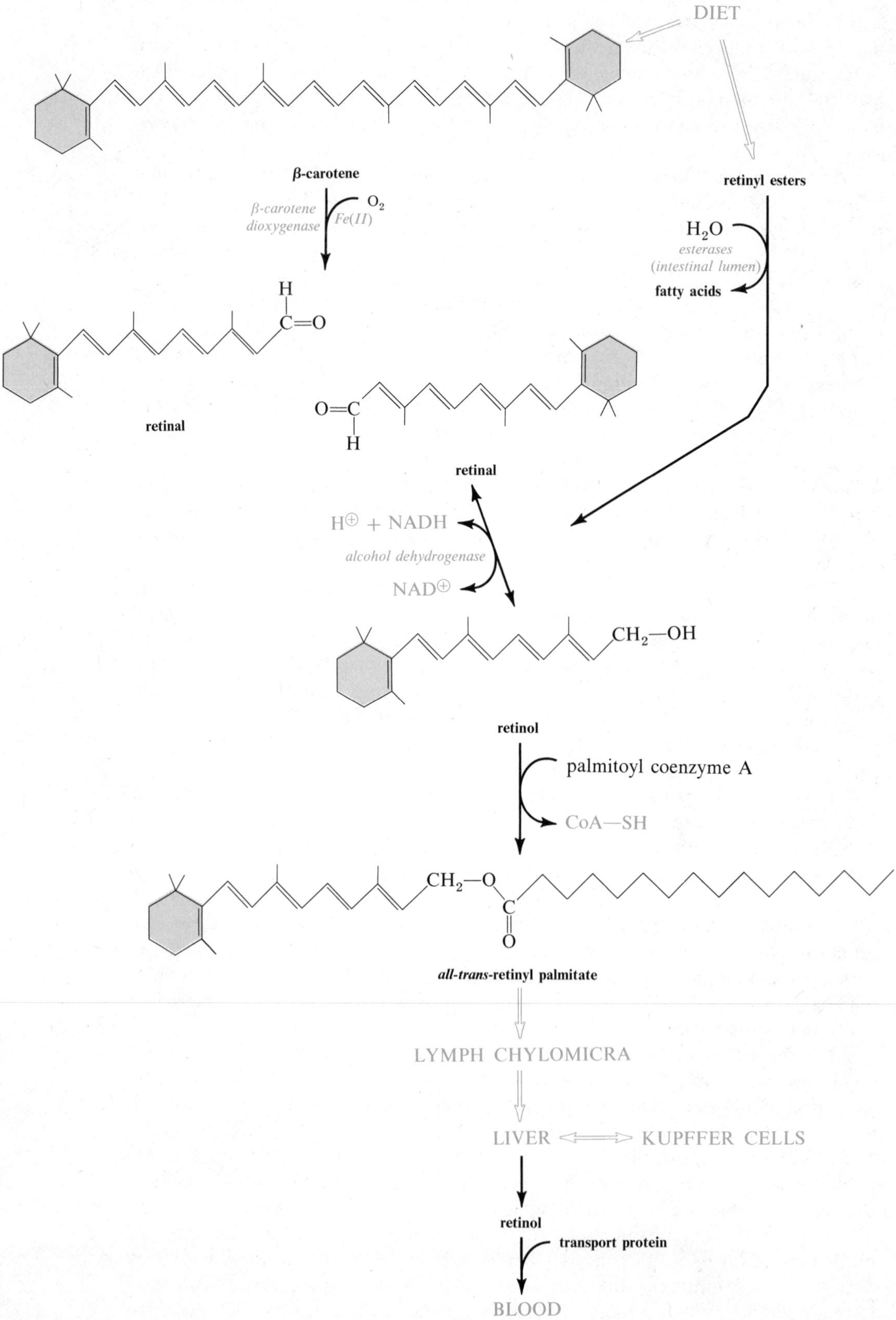

DIET

β-carotene

β-carotene dioxygenase

O_2

Fe(II)

retinal

retinyl esters

H_2O

esterases (intestinal lumen)

fatty acids

retinal

$H^{\oplus}$ + NADH

alcohol dehydrogenase

$NAD^{\oplus}$

CH_2—OH

retinol

palmitoyl coenzyme A

CoA—SH

all-trans-retinyl palmitate

LYMPH CHYLOMICRA

LIVER ⟺ KUPFFER CELLS

retinol

transport protein

BLOOD

FIGURE 42–4 *See legend on the opposite page.*

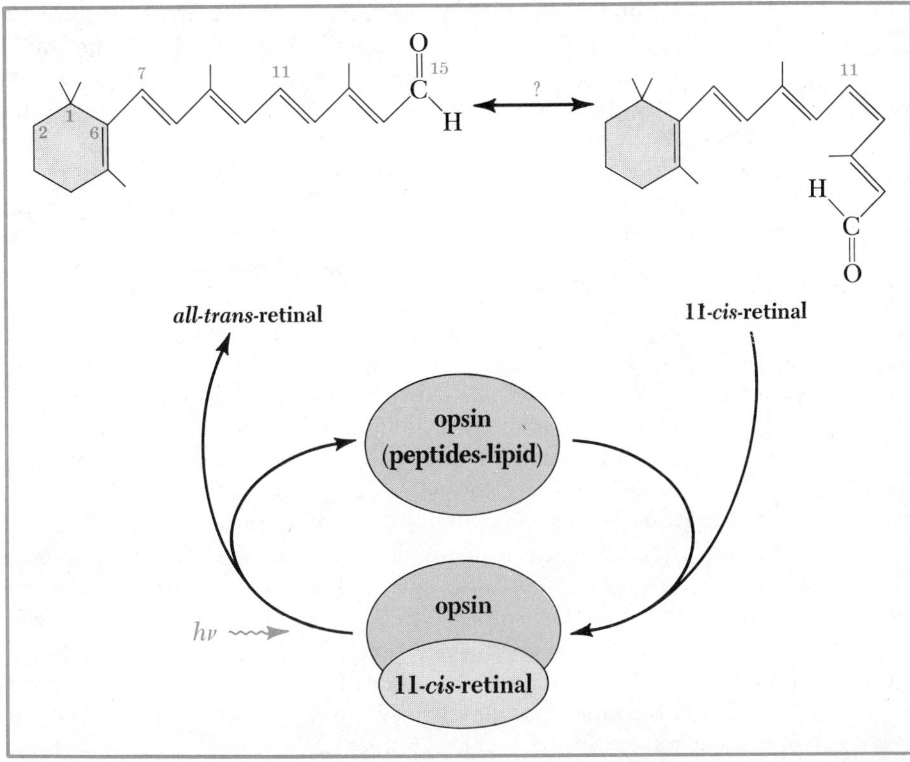

FIGURE 42–5 **The visual pigments are complexes of opsin, a lipophilic polypeptide in retinal membranes, with 11-*cis* retinal. Excitation of the pigment causes a conformational shift that results in release of the retinal in an *all-trans* configuration. Isomerization of 11-*cis* and *all-trans* retinal may be spontaneous or catalyzed by an enzyme.**

pigments, can be formed — until the opsins are saturated. The combination of retinal and opsin may involve Schiff's base formation with a lysine residue. The rod pigment, which has been studied most, is named **rhodopsin.**

One additional step is necessary before the visual pigment is made. The opsins are constructed so that they combine with the 11-*cis* isomer of retinal. There may be an isomerase in the retina catalyzing the equilibration of the *all-trans* isomer with this form, or the equilibration may be spontaneous. Although the position of equilibrium also favors the original *trans* isomer, the strong affinity of the 11-*cis* form for the opsins pulls the whole sequence toward pigment formation (Fig. 42–5).

The mechanism by which light absorption stimulates nerve transmission is another membrane-related phenomenon involving changing affinity for protons. Rhodopsin occurs in the membranes of disk-like structures, which are stacked as cylinders at the interior end of rods. (The stacked array increases the likelihood

---

FIGURE 42–4 **Dietary sources of retinol include β-carotene *(upper left)*, which is oxidized by the intestinal mucosa to retinal. Retinal and retinol equilibrate through the action of an alcohol dehydrogenase. The retinyl esters in the diet *(upper right)* are hydrolyzed in the intestinal lumen, and the resultant retinol is absorbed. The retinol is converted to a palmitate ester for transport and storage.**

See illustration on the opposite page.

that a photon will be absorbed as it moves the length of the rod.) The absorption of light triggers a series of conformational changes in the rhodopsin that cause proton uptake, owing to decreases in the ionization constants of some constituent groups. The process is driven by a simultaneous change in configuration of the attached retinal to the *all-trans* form, which has only a weak affinity for opsin.

The decreased negative charge on the opsin is believed by some to facilitate in some way a movement of calcium into the cytoplasmic space, which in turn decreases the conductance of sodium across the plasma membrane of the rods and generates an action potential. The calcium ion also promotes a slower phosphorylation of rhodopsin, as it regenerates from *cis*-retinal and opsin, making it more negative and causing proton release, thereby ending the stimulation. Later removal of the phosphate again sensitizes the system to a light-induced conformational change. Others believe the intermediate transfer of the signal involves changes in the formation of cyclic GMP, including light-activation of a phosphodiesterase.

Other Functions of Vitamin A. Animals deprived of vitamin A not only go blind; they die. Before they die, they develop abnormal deposition of keratin in the mucous membranes, a failure of bone remodelling leading to thick, solid long bones like the bones in the skull, lesions of the nerves, an increased pressure of the cerebrospinal fluid causing hydrocephalus, testicular degeneration in the males, and abortion or malformed offspring in the females. Obviously, vitamin A is involved in some critical function in most tissues. We don't know what it is, but we do know that it probably doesn't involve the retinol-retinal interconversion, because the corresponding acid, *retinoate,* will prevent malfunction in all of the tissues except the reproductive tract and the eye. Since it won't save vision, it is apparent that retinoate is not readily reduced to retinal. We may be dealing here with a considerably more fundamental and primitive function of the carotenes, with their use in the development of vision being a fortuitous result of the accompanying absorption spectrum.

Retinol itself has been shown to participate as a phosphate ester in the transfer of glycosyl groups, much like dolichol (p. 685), but this is not believed to be a major function of the compound.

Supply of Vitamin A. It is difficult for two reasons to make any one with normal gastrointestinal function, except the very young, deficient in vitamin A. One is the common occurrence of the vitamin or its carotene precursors in plants, in most fish tissues, in eggs, and in mammalian and chicken liver. Most weaned humans eat one or more of these food classes. The other reason is the large capacity for storing the retinyl esters in the liver. The estimated requirement for humans past the weaning age is on the order of 6 micrograms per kilogram body weight per day, perhaps 1.5 micromoles per day for an adult. At this rate, the usual adult has a nine-month supply in his liver, and some may have a four-year supply.

Potential vitamin A deficiency is something to consider in any chronic disease of the pancreas, intestine, or liver. A failure of rod vision (night-blindness) is an early symptom, but it is also a result of other conditions, and vitamin A deficiency must be established by careful measurement, not by subjective report. The really characteristic sign is *xerophthalmia,* a dry keratinization of the cornea and conjunctiva; this is a late symptom and an indication for prompt therapy.

Retinol, like cholecalciferol, is toxic in excess. Acute poisoning has long been known from the fatal result of eating polar bear livers, which contain as much as 30 micromoles of retinyl esters per gram — a five-year supply for a human in each 100 grams. Severe headache, vomiting, and prostration result within a few hours. Doses of 300 micromoles in infants produced a transient hydrocephalus.

The more usual cases are those in which the intake has been relatively high for a long period of time, and chronic toxicity is frequently manifested by fatigue, loss of appetite, an enlarged liver, diffuse pains in the muscles, coarsening and loss of hair, scaly skin eruptions, and attenuation of the long bones that sometimes results in fractures. Diagnoses of pyschoneurosis or even schizophrenia seem to be made quite frequently. Most of the severe cases have been ingesting on the order of 500 micromoles of retinol per day for a few years — sometimes from overzealous treatment for other conditions, and sometimes from self-dosage with easily available commerical preparations (or from maternal over-dosage, in the case of infants). The minimum quantity necessary for the appearance of chronic toxicity has been suggested as 50 micromoles per day for 18 months (about 15 times the current recommended daily allowance).

Vitamin E – Tocopherols

Polyunsaturated fatty acids are subject to spontaneous attack by molecular oxygen through an autocatalytic mechanism involving free radicals (Fig. 42–6). The mechanism may be initiated through exposure to light or by complexes of the transition metal ions. The reason that the polyunsaturated compounds are vulnerable is a stabilization of the intermediate free radical through resonance. The tocopherols are compounds that have the ability to interrupt the free radical cycle by donating electrons, although the mechanism is not clear. They therefore supplement the action of glutathione peroxidase, which removes the peroxides formed by the cycle. (This explains why selenium and the tocopherols diminish, but do not eliminate, each other's dietary requirement.

The tocopherols occur widely, and there is no evidence for a deficiency in adults. However, this may change, because the requirement for tocopherols depends upon the rate of oxidation of polyunsaturated fatty acids, which in turn depends upon the concentration of these acids and their exposure to initiators of free radical formation. Infants who are fed diets rich in polyunsaturated acids and supplemented with high levels of iron have been shown to develop frank tocopherol deficiencies, as manifested by hemolytic anemias. (The high peroxide levels make the erythrocytes and other membranes more fragile.) Similar circumstances are not hard to visualize in some adults.

Vitamin K (p. 330)

Dietary deficiencies of vitamin K have not been demonstrated in normal adults. The compound is widely distributed in plants and is also synthesized by intestinal bacteria. Although a major portion of the compound made by the bacteria is probably inaccessible for absorption, this source is probably imperative for people on a meat diet. In any event, experimental animals treated with antibiotics to suppress the intestinal flora become deficient. Newborn infants, especially the premature, are sometimes deficient.

The natural compounds have long polyprenyl side chains and a correspondingly limited solubility in water. They are relatively non-toxic in excess, whereas the more water-soluble synthetic analogues are oxidized to reactive intermediates that cause hemolysis. Since these compounds are not available without prescription and are usually administered for brief intervals, this has not been a practical problem.

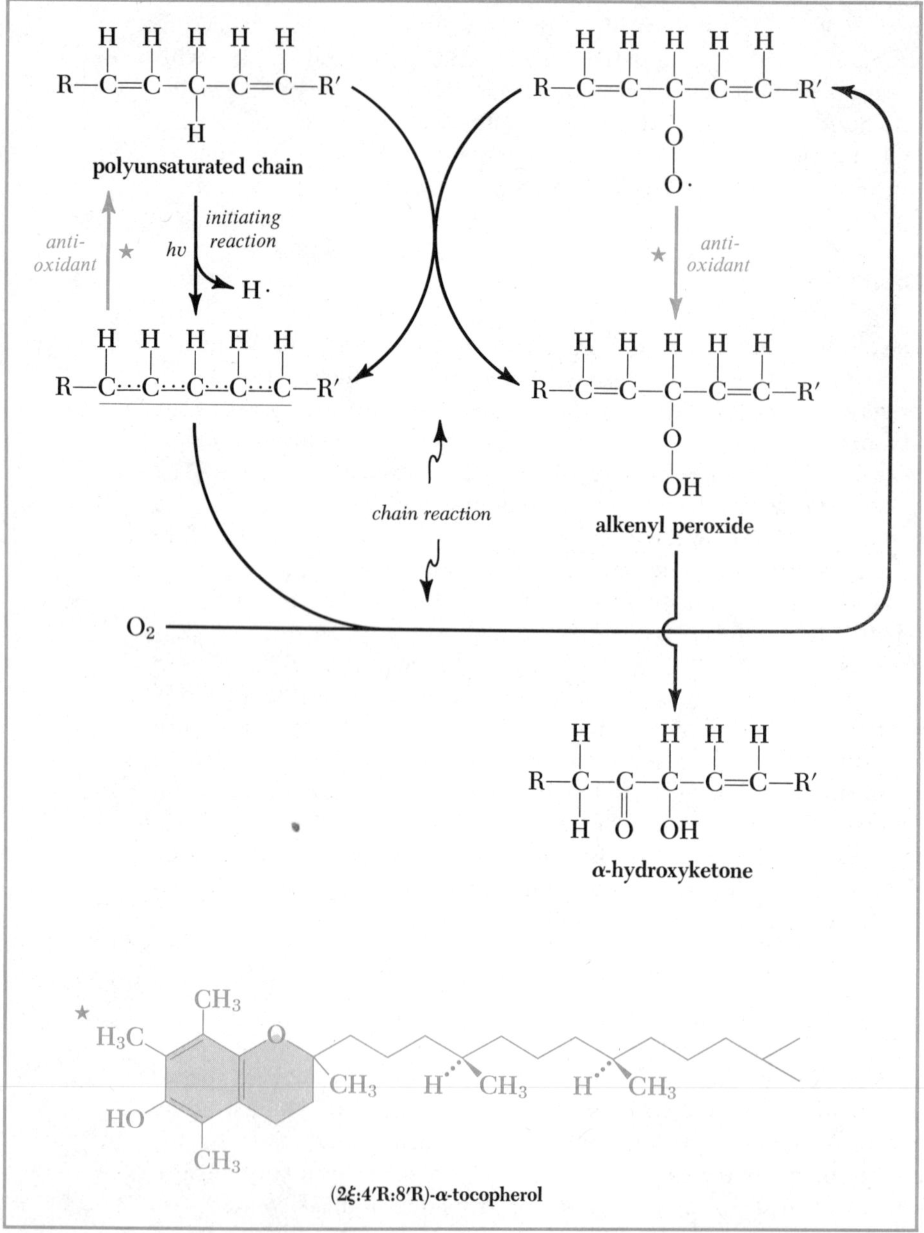

FIGURE 42–6 **Polyunsaturated fatty acids are subject to oxidation through a free radical mechanism. This cyclical process can be interrupted by antioxidants that act as scavengers of the free radicals. The tocopherols *(bottom)* serve this function.**

FURTHER READING

General Sources

Davidson, S., R. Passmore, and J. F. Brock: (1975) *Human Nutrition and Dietetics.* 6th ed. Williams and Wilkins.

Latner, A. L.: (1975) *Clinical Biochemistry.* W. B. Saunders. Includes chapters on minerals and vitamins.

Public Health Service: (1974) *Preliminary Findings of the First Health and Nutrition Survey, United States, 1971–1972.* DHEW publication No. (HRA) 74:1219–1. Valuable for its data, but not for its interpretations.

Bone Minerals and Electrolytes

Nordin, B. E. C., ed.: (1976) *Calcium, Phosphate, and Magnesium Metabolism.* Churchill Livingston.

Paterson, C. R.: (1974) *Metabolic Disorders of Bone.* Blackwell. Includes discussion of calcium metabolism.

Avioli, L. V., and S. M. Krane, ed.: (1977) *Metabolic Bone Disease.* Academic Press. Includes discussion of calcium metabolism.

Knochel, J. P.: (1977) *The Pathophysiology and Clinical Characteristics of Severe Hypophosphatemia.* Arch. Int. Med., *137*: 203.

Iseri, L. T., J. Freed, and A. R. Bures: (1975) *Magnesium Deficiency and Cardiac Disorders.* Am. J. Med., *58*: 837.

Meneely, G. R., and H. D. Battarbee: (1976) *Sodium and Potassium.* Nutr. Rev., *34*: 1.

Wiberg, J. J., G. G. Turner, and F. Q. Nuttall: (1978) *Effect of Phosphate or Magnesium Cathartics on Serum Calcium.* Arch. Inter. Med., *138*: 1114.

Iron

Leavell, B. S., and O. A. Thorup, Jr.: (1976) *Fundamentals of Clinical Hematology.* 4th ed. W. B. Saunders. Includes discussion of iron deficiency anemia.

Wood, M. M., and P. C. Elmwood: (1966) *Symptoms of Iron Deficiency Anemia.* A community survey. Br. J. Prev. Soc. Med., *20*: 117.

Committee on iron deficiency of the American Medical Association: (1968) *Iron Deficiency in the United States.* J.A.M.A., *203*: 407. Includes definition of normal concentrations.

Nutrition Today: (1972) (Mar/Apr) Vol 7, no. 2 is devoted to an instructive round-table discussion of the dietary iron requirement and the occurrence of negative iron balance.

Elwood, P. C.: (1977) (July/Aug) *The Enrichment Debate.* Nutr. Today, *12*(4): 18. Logical dissection of the questions involved in discussions of iron deficiency.

Anonymous: (1978) (Jan/Feb) *Anatomy of a Decision.* Nutr. Today, *13*(1): 6. A description of the scientific politics involved in the debate over iron enrichment.

Zinc and Copper

Eckhert, C. D., et al.: (1977) *Zinc Binding: A Difference Between Human and Bovine Milk.* Science, *195*: 789.

Prasadi, A. S., ed.: (1976) *Trace Elements in Human Health and Disease.* Vol. 1. *Zinc and Copper.* Academic Press.

Hamidge, K. M., and B. L. Nichols, Jr., eds.: (1978) *Zinc and Copper in Clinical Medicine.* Spectrum.

Roueché, B.: (1977) *Annals of Medicine. All I Could Do Was Stand in the Woods.* New Yorker (Sept. 12), p. 97. Interesting account of apparent zinc deficiencies.

Other Minerals

Stadtman, T.: (1977) *Biological Function of Selenium.* Nutr. Rev., *35*: 161.

Leach, R. M., Jr., ed.: (1977) *Biochemical Functions of Selenium and Its Interrelationships with Other Trace Elements and Vitamin E.* Fed. Proc., *34*: 2082. A symposium.

Hamidge, K. M.: (1977) *Trace Elements in Pediatric Nutrition.* Adv. Pediatr., *24*: 191.

Tuman, R. W., J. T. Bilbo, and R. J. Doisy: (1978) *Comparison and Effects of Natural and Synthetic Glucose Tolerance Factor in Normal and Genetically Diabetic Mice.* Diabetes, *27*: 49.

Miller, W. J., ed.: (1974) *Newer Candidates for Essential Trace Elements.* Fed. Proc., *33*: 1747. A symposium including discussions of tin, fluorine, silicon, nickel, and vanadium.

Vitamins, General

Sebrell, W. H., Jr., and R. S. Harris, ed.: (1967 ff.) *The Vitamins*. Academic Press. Multi-volume treatise.

Dipalma, J. R., and D. M. Ritchie: (1977) *Vitamin Toxicity*. Annu. Rev. Pharmacol., *17*: 133.

World Health Organization Technical Report No. 452: (1970) *Requirements for Ascorbic acid, Vitamin D, Vitamin B_{12}, Folate, and Iron*.

World Health Organization Technical Report No. 362: (1967) *Requirements for Vitamin A, Thiamine, Riboflavin, and Niacin*.

Ascorbate

King, C. G., and J. J. Burns, eds.: (1975) *Second Conference on Vitamin C*. Ann. N.Y. Acad. Sci., vol. 258. Includes several useful reviews.

Hodges, R. E., et al.: (1971) *Clinical Manifestations of Ascorbic Acid Deficiency in Man*. Am. J. Clin. Nutr., *24*: 432.

Passmore, R.: (1977) *How Vitamin C Deficiency Injures the Body*. Nutr. Today, *12(2)*: 6, Review by one of the more lucid authorities.

Coulehan, J. L., et al.: (1976) *Vitamin C and Acute Illness in Navajo Schoolchildren*. N. Engl. J. Med., *295*: 973.

Enloe, C. F., Jr., ed.: (1978) *To Dose or Megadose. A Debate About Vitamin C*. Nutr. Today, *13(2)*. Revealing exposition of opinion.

Cobalamin and Folate

Babior, B. M., ed.: (1975) *Cobalamin. Biochemistry and Pathophysiology*. Wiley.

Scott, J. M., and D. G. Weir: (1976) *Folate Composition, Synthesis, and Function in Natural Materials*. Clin. Haematol., *5*: 347.

Rosenberg, I. H.: (1976) *Absorption and Malabsorption of Folates*. Clin. Haematol, *5*: 589.

Reynolds, E. H.: (1976) *The Neurology of Vitamin B_{12} Deficiency*. Lancet, *2(7990)*: 832.

Hoffbrand, A. V.: (1975) *Synthesis and Breakdown of Natural Folate*. Prog. Hematol., *9*: 85.

Herbert, V., and K. C. Das: (1976) *The Role of Vitamin B_{12} and Folic Acid in Hemato- and Other Cell-poiesis*. Vitam. Horm., *34*: 1.

Erbe, R. W.: (1975) *Inborn Errors of Folate Metabolism*. N. Engl. J. Med., *293*: 753, 807.

Other Vitamins

Katz, S. H., M. L. Hediger, and L. A. Valleroy: (1974) *Traditional Maize Processing Techniques in the New World*. Science, *184*: 765.

Rivlin, R.: (1970) *Riboflavin Metabolism*. N. Engl. J. Med., *284*: 463.

Marks, J., ed.: (1974) *Symposium on Fat-soluble Vitamins*. Vitam. Horm., *32*: 131.

Ostroy, S. E.: (1977) *Rhodopsin and the Visual Process*. Biochem. Biophys. Acta, *463*: 91.

Tappel, A. L.: (1973) *Vitamin E*. Nutr. Today, *8(4)*: 4.

Bieri, J. G., and B. M. Farrell: (1976) *Vitamin E*. Vitam. Horm., *34*: 31.

Williams, M. L., et al.: (1975) *Role of Dietary Iron and Fat on Vitamin E Deficiency Anemia of Infants*. N. Engl. J. Med., *292*: 887.

DeLuca, L. M. (1977) *The Direct Involvement of Vitamin A in Glycosyl Transfer Reactions in Mammalian Membranes*. Vitam. Horm., *35*: 1.

43 THE COMPLETE DIET

Given an understanding of the requirements for the individual components of the diet, the formidable problem remains of translating this information into usable form. Even the most dedicated professional cannot calculate the contribution of every morsel a person might consume. Both the weight and the composition of foods often must be estimated. Every crop of wheat and every patty of hamburger is not identical, and even if they were, people forget or conceal what they do.

However, the person who desires to do so can achieve satisfactory control over his diet by the use of relatively simple guides. These prepared lists have food portions broken into categories such as meat, bread, or fat in a way that permits exchange of one food for another within the category. The Mayo Clinic Diet Manual is a good example of a collection of exchange lists for various purposes that permits considerable latitude in satisfying individual tastes while avoiding detailed calculations that are beyond the reach of most people.

However, such lists can only be prepared by calculation, and detailed calculation is sometimes necessary for other purposes. The basic information is provided by tables of food composition, especially the Department of Agriculture Handbooks number 8 and 456 or the World Health Organization tables, supplemented by information provided by some manufacturers. (The Campbell Soup Company and the H. J. Heinz Company have been helpful in publishing detailed tabulations of the composition of their products.)

Recommended Daily Allowances (Table 43–1)

Individuals vary in their requirements for each nutrient. If all consumed only the average minimum daily requirement for prevention of functional impairment, signs of deficiency might be expected in a large part of the population. Where data permit, the requirement is usually estimated as the mean plus two standard deviations, a value that would be expected to avoid deficiency in 97.5 per cent of the population. Even this number of deficient individuals would be regarded as intolerable in a developed country. The Food and Nutrition Board of the National Academy of Sciences (U.S.) has, therefore, applied expert judgment to gauge the amount of each nutrient that would diminish the occurrence of deficiency symptoms to an acceptable level. These are the Recommended Daily Allowances, which are reviewed periodically, with the last assessment made in 1973.

It is apparent that these are not minimum daily requirements. Many individuals consume less than the recommended amount without developing a deficiency. The tables are intended as a guide for developing standards for large groups and for minimizing the risk of deficiency to an individual. In general, the expert

TABLE 43–1

	Age	Weight *kg*	Height *cm*	Energy Yield *MJ*	Protein N *mmoles*	Retinol Equiv. *μmoles*	Cal-ciferol Equiv. *nmoles*	Toco-pherol Equiv. *μmoles*	Ascor-bate *μmoles*
Infants	0–6 mo	6	60	0.49/kg	23/kg	1.5	25	9	200
	6–12 mo	9	71	0.45/kg	23/kg	1.4	25	12	200
Children	1–3 yr	13	86	5.5	260	1.4	25	16	230
	4–6	20	110	7.5	340	1.8	25	16	230
	7–10	30	135	10.0	400	2.4	25	21	230
Men	11–14	44	158	11.7	500	3.5	25	28	260
	15–18	61	172	12.5	620	3.5	25	28	260
	19–22	67	172	12.5	620	3.5	25	35	260
	23–50	70	172	11.3	640	3.5		35	260
	51+	70	172	10.0	640	3.5		35	260
Women	11–14	44	155	10.0	500	2.8	25	23	260
	15–18	54	162	8.8	550	2.8	25	26	260
	19–22	58	162	8.8	530	2.8	25	28	260
	23–50	58	162	8.4	530	2.8		28	260
	51+	58	162	7.5	530	2.8		28	260
	Pregnant			+1.2	+340	3.5	25	35	340
	Lactating			+2.1	+230	4.2	25	35	450

Notes: In converting the Board allowances to a molar basis, the increments between ages used by the Board have been translated to the closest molar increment. That is, an increase of 10 mg of a nutrient might be converted to the nearest 0.05 millimole increase. The Board's estimate for the protein allowance is on the basis of proteins with a 75 per cent efficiency of utilization. The fat-soluble vitamin allowances have been translated to equivalent moles of retinol, calciferol, and α-tocopherol. The Board regards β-carotenes to be 1/6 as effective as retinol in providing vitamin A activity on a weight basis. Those carotenes in which only one half of the molecule yield retinol are 1/12 as effective per unit weight. The nicotinate allowance is calculated on the basis of 1 micromole of nicotinate being equivalent to 36 micromoles of tryptophan. The folate allowance is determined on the basis of a microbiological assay of the foods, and less than one quarter of the stated amounts given as pure folate will be equally effective.

committee has been generous in its margin of safety. However, it ought to be noted that the estimates for required fuel intake are adjusted to a sedentary life style, and were made before the current emphasis on physical exercise was so widespread. We already know that the fuel intake always ought to be appraised individually.

It also should be noted that some individuals are still likely to become deficient to a nutrient even with conscientious consumption of the recommended daily allowance. There are many possibilities for impairment of utilization of a nutrient — during absorption, transport, metabolism, excretion — and mutations thus can easily magnify the amount of the nutrient required to maintain function. It is probable that there is no level of supplementation that would completely eliminate deficiencies from the population. Indeed, extraordinary increases in the dietary content of one component are likely to cause trouble for another group with different mutations — those who are especially sensitive to the toxic effects of the component.

When to Calculate

Full scale calculation of the composition of a diet is justified in at least three circumstances other than the preparation of guides. One is in the design of

RECOMMENDED DAILY ALLOWANCES

Folate μmoles	Nicotinate Equiv. μmoles	Riboflavin μmoles	Thiamine μmoles	Pyridoxine μmoles	Cobalamin nmoles	Calcium mmoles	Phosphate mmoles	Iodine μmoles	Iron mmoles	Magnesium mmoles	Zinc mmoles
0.1	40	1.1	1.0	2.1	0.2	9	7.7	0.3	0.18	2.5	0.05
0.1	65	1.6	1.7	2.5	0.2	13.5	13	0.35	0.27	3.0	0.08
0.2	75	2.1	2.3	3.5	0.7	20	26	0.5	0.27	6.0	0.15
0.5	100	2.9	3.0	5.5	1.1	20	26	0.65	0.18	8.5	0.15
0.7	130	3.0	4.0	7	1.5	20	26	0.85	0.18	10.5	0.15
0.7	145	4.0	4.7	9.5	2.2	30	39	1.0	0.32	14.5	0.25
0.7	165	4.8	5.0	12	2.2	30	39	1.2	0.32	17.0	0.25
0.7	165	4.8	5.0	12	2.2	20	26	1.1	0.18	14.5	0.25
0.7	145	4.3	4.7	12	2.2	20	26	1.0	0.18	14.5	0.25
0.7	130	4.0	4.0	12	2.2	20	26	0.9	0.18	14.5	0.25
0.7	130	3.5	4.0	9.5	2.2	30	39	0.9	0.32	12.5	0.25
0.7	115	3.7	3.7	12	2.2	30	39	0.9	0.32	12.5	0.25
0.7	115	3.7	3.7	12	2.2	20	26	0.8	0.32	12.5	0.25
0.7	105	3.2	3.3	12	2.2	20	26	0.8	0.32	12.5	0.25
0.7	100	2.9	3.3	12	2.2	20	26	0.65	0.18	12.5	0.25
1.8	+20	+0.8	+1.0	15	3.0	30	39	1.0	0.32+	19.0	0.30
1.4	+35	+1.3	+1.0	15	3.0	30	39	1.2	0.29	19.0	0.40

The stated protein nitrogen allowance may be converted to grams of "protein" by multiplying by 0.0875. The fat-soluble vitamin allowances may be converted to I.U. by multiplying by the following conversion factors:

I.U. vitamin A = μmoles retinol × 950
I.U. vitamin D = nmoles calciferol × 16
I.U. vitamin E = μmoles tocopherol × 1.6

The molar allowances for the other vitamins and the minerals can be converted to a weight basis by multiplying by the molecular weights: ascorbate = 176, folate = 441, riboflavin = 376, thiamine = 301, pyridoxine = 170, cobalamin = 1,355, calcium = 40, phosphate = 31 (as P), iodine = 127, iron = 55.6, magnesium = 24, zinc = 65.4.

solutions for intravenous feeding, discussed further below. Another is the assessment of a diet that is highly restricted in composition or quantity. The third is for assessment of population food supplies. Anyone who anticipates offering advice on nutrition ought to make such a full-scale calculation using commonly available foods in order to gain an appreciation of the actual circumstances.

Additional Consideration

Assessment of the content of the nutrients in the Recommended Daily Allowances is not enough. Important technical and personal factors remain to be solved, and available information is inadequate. For example, we have said little about the requirement for polyunsaturated fatty acids in the diet, such as the linoleic acid that is necessary for prostaglandin synthesis and which facilitates the formation of membrane lipids with proper functional characteristics. Some state that the polyunsaturated acids should constitute 5 per cent of the total energy intake; others say that one can get by nicely with only one gram per day, an amount that is difficult to avoid consuming. This discussion is complicated by evidence that polyunsaturated fats diminish plasma cholesterol and triglyceride levels. It is best to keep in mind how little we know. The advantages of polyunsaturated fats seemed very clear, but wait! Most of the oils and spreads on the market so rich in the polyunsaturated compounds are prepared with partially hydrogenated oils in order to have the desired consistency, flash point, and resistance to rancidity. The hydrogenation process causes isomerization of some of the double bonds from the natural *cis* configuration to the unnatural *trans*. Some people say these are harmful. Not only that, but the high intake of polyunsaturated compounds increases the formation of free radicals and peroxides, which increases the requirement for tocopherols and also accelerates what other people allege are the principal processes responsible for aging and senility. And so it goes. Low iron, anemia, and inanition; high iron, more infections, and still more generation of free radicals and peroxides. It's enough to make one hate to swallow.

Here we have an illustration of another point; We are not concerned with factories that need be supplied only with proper raw material in order to hum along their appointed ways. The quality of life can be destroyed for many people by overemphasis on their diet, as well as by underemphasis. This is especially true for the obese. To take an extreme example, the zealous resident who places an elderly obese patient with limited life expectancy on a severely restricted diet is punishing a sinner, not treating a human. Milder examples are commonplace; flourishes and alarums about some speculative dietary danger, perhaps affecting one in 10,000 people at the most, abound to the point that one suspects a mass neurosis.

This is not to disparage a constant examination of the components of our foods and inquiry into their effects. It is only to suggest mildly that the attention of the well-fed American middle- and upper-class might be diverted to more demonstrably valid problems, and all efforts to change the diet of the entire population ought to be examined with great suspicion. On the other hand, most practicing physicians might well learn how to extract a detailed dietary history quickly and how to make a rough, but quick appraisal of the quality of the diet.

Another factor of potential importance is the quantity and nature of indigestible fiber in the diet. The formation of feces nearly ceases with completely soluble synthetic diets. Insoluble materials, especially cellulose and other nonhydrolyzable polysaccharides, promote the formation and movement of fecal material. This is believed by some to account for the lesser incidence of lower bowel malignancies in primitive populations that use cereals as their major foods. The merits and demerits have not been fully assessed. Ever since the Germans experimented with the addition of sawdust to food as a device for alleviating the pangs of famine during World War I, it has been known that the presence of fiber diminishes the absorption of nutrients. This obviously increases the problem of sustaining a population with marginal resources. However, most populations with precariously low intakes already have an ample supply of fiber because of their dependence on cereals, and the question of fiber supplementation is moot for them.

Some Difficult Problems

Nutritional advice for the very poor, the people who could perhaps benefit most, is frequently deficient or faulty in the United States. Most advisers are trained to think in more affluent terms. (An excellent test is to see how often lettuce appears in the posters and homilies. Lettuce is a poor buy in nutritional terms, even when its price is low.) The basic problem is first an adequate caloric intake, and second an adequate protein intake: both need to be at low price and in forms that the people concerned will make the effort to prepare and will eat after preparation. For those who can cook, enriched flour and the oils and margarine are the first line of defense, supplemented with an egg and some milk to bolster the quality of the protein. A presently more expensive, but almost complete alternative is the potato.* Eaten in quantity, this food alone will supply the energy requirement, almost enough protein, and most of the necessary vitamin and mineral intake, including ascorbic acid. Local knowledge is required to adjust these kinds of fare to the prevailing culture and the season, but assurance that one can survive and even thrive on much less than the prevailing norm may do wonders.

The food faddists of various kinds represent another serious problem. An incredible array of grossly deficient and bizarre diets are enthusiastically adopted and quietly discarded by the American public. Although some deaths frequently ensue, each of these fads is transient, and the major damage is to confidence in the population's intelligence and education. A recent example is the liquid protein fad adopted by many obese women. Among the 40 people using this diet who died during the latter part of 1977 were 15 with no other conditions that might have contributed. Twelve of them were under medical supervision and 14 were taking vitamin and mineral supplements. The use of gelatin, collagen, or similar proteins with seriously unbalanced amino acid composition as the sole source of nitrogen shows defective judgment, as we touched on in Chapter 41. To make such a protein the major fuel in a restricted diet over a period of weeks or months is rank stupidity.

A more serious problem is presented by the growing number of vegetarians. Their practices run the gamut from the ignorant and lethal (the Zen cults) to the considered and conscientious. A purely vegetarian diet, eschewing any animal products, can be adequate in all respects except in its content of cobalamins. The only known way to provide the cobalamins without eating animal products or the isolated compound is to consume some microorganisms. Eating some soil rich in organic matter will do it; perhaps more esthetic ways of resolving conscience and hereditary requirements will be found. The provision of adequate protein intake is not difficult if a mixture of plant materials, especially the pulses, is eaten. Reliance on the cereals or on maize is to be avoided.

Intravenous nutrition is an area in which much is yet to be learned. It has become more sophisticated with the shift from the 5 per cent glucose solution commonly employed in the past to mixtures containing electrolytes, other minerals, vitamins, fat emulsions, and amino acids, as well as glucose. As we noted in discussing the amino acid composition of such solutions, intravenous feeding has more potential dangers than oral feeding because it bypasses the processing mechanism of the intestine and the liver and directly exposes the brain and other organs to the components of the solution. The circumstances in which this can be a danger, and the most desirable formulations, are yet to be developed.

Nutritional Experimentation

Space and time prevent the discussion in this already lengthy book of the experimental methods used to discover the chemistry of living systems. However, a word on nutritional experiments is appropriate because the results are useless

*The quality of potatoes available in the retail market has deteriorated owing to diverson of the better crop to fast-food outlets — an unfortunate development.

without adequate procedural controls that are often missing. The problem is one of limiting the experiment to a single variable. Perhaps the best way of accomplishing this is by feeding solutions through tubes, thereby controlling the quantity administered and time of feeding, and avoiding any change in gustatory stimuli that might affect the result. This is rarely done because of the inconvenience and expense. The next best way is to present two animals with identical quantities of diets differing in only a single component, the quantities being such that both animals completely consume the quantity offered. This technique of paired feeding is often neglected, so that the experimental and control animals differ both in fuel intake and in intake of the varied component. (This is an especial problem when one of the major components is changed. Given a free choice, the animals almost always will eat more of one diet than another.)

A recent conspicuous example of defective experimentation can be found in Consumer Reports, in which different samples of breakfast cereals were offered rats ad libitum. The cereal was adjudged best for human consumption that resulted in greatest weight gains in the rats consuming it. Several readers pointed out that few foods, especially cereals, are expected to be a complete diet in themselves, and that the experiment also tested the taste appeal of the cereals for rats. (Consumer Reports later stated that there was little difference in the consumption of the choices, but offered no data.) Despite this guidance in seeing the error of their ways, the organization repeated the same mistakes in assessing the nutritional quality of breads.

Let these be the closing words: The knowledge gained by your study of biochemistry enables you to improve many aspects of the human condition and to bolster the human spirit, if you will but use it. I hope the book has helped.

FURTHER READING

National Academy of Sciences: (1974) *Recommended Dietary Allowances,* 8th ed.

Harper, A. E.: (1974) *Those Pesky Recommended Dietary Allowances.* Nutr. Today, *9(2)*: 17. An explanation of the values by the chairman of the committee developing them.

Rynearson, E. H.: (1974) *Americans Love Hogwash.* Nutr. Rev., *32 (suppl. 1).* A valuable collection of articles on food fads and nutrition quacks, despite occasional inclusion of a little hogwash of their own in the form of outdated dogma.

Higginbottom, M. C., L.. Sweetman, and W. L. Nyhan: (1978) *B_{12} Deficiency Syndrome in a Breast-fed Infant of a Vegan.* N. Engl. J. Med., *299*: 317. Also see editorial on p. 355.

Anonymous: (1976) *Bread. You Can't Judge a Loaf by Its Color.* Consumer Reports. May, p. 256. How not to guide the consumer on nutritional matters.

Michiel, R. R., et al.: (1978) *Sudden Death in a Patient on a Liquid Protein Diet.* N. Engl. J. Med., *298*: 1005.

Anonymous: (1978) *Protein Diets.* FDA Drug Bulletin, *8(1)*: 2.

ABBREVIATIONS

A adenosine
ACTH adrenocorticotropic hormone (corticotropin)
ADP adenosine diphosphate
Ala alanyl
AMP adenosine monophosphate
Arg argininyl
Asn asparaginyl
Asp aspartyl
ATP adenosine triphosphate
ATPase adenosine triphosphatase

C cytidine
cAMP 3′,5′-cyclic AMP
CDP cytidine diphosphate
Cer ceramide
CMP cytidine monophosphate
CoA coenzyme A (in names of compounds)
CoA-SH coenzyme A (in reactions)
CTP cytidine triphosphate
Cys cysteinyl
cyt cytochrome

dAMP 2′-deoxyadenosine monophosphate
dATP 2′-deoxyadenosine triphosphate
dCMP 2′-deoxycytidine monophosphate
dCTP 2′-deoxycytidine triphosphate
dGMP 2′-deoxyguanosine monophosphate
dGTP 2′-deoxyguanosine triphosphate
DNA deoxyribonucleic acid(s)
Dol dolichyl
DOPA dihydroxyphenylalanine
dTMP 2′-deoxythymidine monophosphate
dTTP 2′-deoxythymidine triphosphate
dUMP 2′-deoxyuridine monophosphate

E enzyme
EDTA ethylenedinitrilotetraacetate

FAD(H_2) flavin adenine dinucleotide (reduced)
FDP fructose 1,6-diphosphate
Fru fructosyl
FMN(H_2) riboflavin 5′-phosphate (reduced)
Fuc fucosyl

G guanosine
Gal galactosyl
GalNAc N-acetylgalactosaminyl
GDP guanosine diphosphate
Glc glucosyl
Glc 6-P glucose 6-phosphate
GlcNAc N-acetylglucosaminyl
GlcUA glucuronyl
Gln glutaminyl
Glu glutamyl
Gly glycyl
GMP guanosine monophosphate
GSH, GSSG glutathione (reduced and oxidized)
GTP guanosine triphosphate

Hb Hemoglobin
HbO_2 oxyhemoglobin
HDL high-density lipoprotein
His histidyl
hnRNA heterogenous nuclear RNA
Hyl hydroxylysyl
Hyp hydroxyprolyl

I inosine
IdUA iduronyl
Ig immunoglobulin
Ile isoleucyl
IMP inosine monophosphate

K_{eq} equilibrium constant
K_M Michaelis constant

LDL low-density lipoprotein
Leu leucyl
Lys lysyl

Man mannosyl
Met methionyl
MgATP magnesium ATP
mRNA messenger RNA
M.W. molecular weight

NAD (NADH) nicotinamide adenine dinucleotide (reduced)
NADP (NADPH) nicotinamide adenine dinucleotide phosphate (reduced)
NeuNAc N-acetylneuraminyl

~P high-energy phosphate
$\mathbf{P_i}$ inorganic phosphate
PAPS phosphoadenosine phosphosulfate
PCr phosphocreatine
pI isoelectric pH
Phe phenylalanyl
P : O high-energy phosphate/atom of oxygen consumed
$\mathbf{PP_i}$ inorganic pyrophosphate
$\mathbf{PP\cdots P_i}$ inorganic polyphosphates
Pro prolyl

RNA ribonucleic acids
R.Q. respiratory quotient
rRNA ribosomal RNA

S Svedberg unit
S substrate
SDS sodium dodecyl sulfate
Ser seryl

SGOT aspartate aminotransferase (serum glutamic-oxaloacetic transaminase)
SGPT alanine aminotransferase (serum glutamic-pyruvic transaminase)

T thymidine (ribosylthymine)
ThPP thiamine pyrophosphate
Thr threonyl
tRNA transfer RNA
Trp tryptophanyl
TSH thyroid-stimulating hormone (thyrotropin)
Tyr tyrosyl

U uridine
UDP uridine diphosphate
UMP uridine monophosphate
UTP uridine triphosphate

$\mathbf{V}_{max}$ maximum velocity
Val valyl
VLDL very-low-density lipoprotein

XMP xanthosine monophosphate
Xyl xylosyl
Ψ pseudouridine

INDEX

Page numbers in **boldface** indicate figures illustrating structures of listed compounds, reactions catalyzed by listed enzymes, or other listed phenomena. Isomeric designations are neglected in the primary alphabetization. Thus, *cis*-Aconitate, S-Adenosyl-L-methionine, L-Alanine, and *p*-Aminobenzoate are listed as beginning with A.

RANGE OF CONCENTRATIONS IN NORMAL HUMAN BLOOD

Clinical analyses may be made on whole blood, separated plasma, or the serum remaining after clotting, as indicated by (B), (P), or (S). Venous blood is used unless otherwise specified. Standard values differ from one laboratory to another according to the conditions of analysis; those cited here are from the University of Virginia Hospital, except those designated by an asterisk are from an extensive tabulation in the N. Engl. J. Med. (1978) *298:* 34. The more useful SI units are calculated where appropriate, but the commonly used clinical units are also shown. The values are for adults. U = International Units of enzyme activity (one micromole of substrate reacting per minute).

Acetoacetate + acetone (S)*	0.05–0.35 mM	0.3–2.0 mg/dl
Alanine aminotransferase, SGPT (S)		4–30 mU/ml
Aldolase (S)		1.2–7.6 mU/ml
Ammonia (P)	11–35 μM	20–60 μg/dl
Amylase (S)		15–90 mU/ml
Aspartate aminotransferase, SGOT (S)		4–30 mU/ml
Bilirubin (S):		
direct	0–4 μM	0–0.25 mg/dl
total	3–20 μM	0.2–1.2 mg/dl
Blood volume (B)*	8.5–9.0 liters/100 kg body weight	
Calcium (S)	2.1–2.6 mM	8.5–10.5 mg/dl
CO_2, total (S)	24–30 mM	24–30 meq/liter
Chloride (S)	96–106 mM	96–106 meq/liter
Cholesterol (S):		
total, age 20–39 yrs	3.6–7.0 mM	140–270 mg/dl
esterified*	60–75% of total	
Cobalamins (S)*	0.1–0.3 nM	150–450 pg/ml
Copper (S)	11–26 μM	70–165 μg/dl
Corticotropin, ACTH (P)*	3–15 pM	15–70 pg/ml
Creatine kinase, CPK (S)		0–110 mU/ml

RANGE OF CONCENTRATIONS

Creatinine (S)	0.06–0.13 mM	0.7–1.5 mg/dl
Fat, triglycerides (S), age 20–39	0.1–1.8 mM	10-150 mg/dl
Fatty acids (S)*	7.0–15.5 mM	190–420 mg/dl
Ferritin (S)		10–300 ng/ml
Folate (S)*	14–34 nM	6–15 ng/ml
Glucose, fasting (P)	3.6–6.1 mM	65–110 mg/dl
Haptoglobin (S)	10–30 μM	100–300 mg/dl
Hematocrit (B):		
males	40–52%	
females	37–47%	
Hemoglobin (B):		
males	2.2–2.8 mM	14–18 g/dl
females	1.8–2.5 mM	12–16 g/dl
Hydrogen ion, arterial (B)	35–45 nM	
Immunoglobulins (S):		
Ig A	6–28 μM	90–450 mg/dl
Ig D	0–1.7 μM	0–30 mg/dl
Ig E	0.1–5 nM	20–1,000 ng/ml
Ig G	55–125 μM	800–1,800 mg/dl
Ig M	0.7–3 μM	60–275 mg/dl
Insulin (S or P)*	40–190 pM	6–26 microunits/ml
Iron (S)	11–29 μM	60–160 μg/dl
Iron-binding capacity (S)	52–74 μM	290–410 μg/dl
Lactate (P)	0.5–2.2 mM	0.5–2.2 meq/liter
Lactate dehydrogenase, LDH (S)		100–350 mU/ml
Magnesium (S)	0.7–1.2 mM	1.8–2.8 mg/dl
Osmolality	295–315 mOsm/kg water	
P_{CO_2}, arterial (B)	4.7–6.0 kPa	35–45 torr
pH, arterial (B)		7.35–7.45